Book of Abstracts of the 71st Annual Meeting of the European Federation of Animal Science

T0269415

EAAP
European Federation of Animal Science

The European Federation of Animal Science wishes to express its appreciation to the
Ministero delle Politiche Agricole Alimentari e Forestali (Italy) and the
Associazione Italiana Allevatori (Italy)
for their valuable support of its activities.

Book of Abstracts of the 71st Annual Meeting of the European Federation of Animal Science

Virtual Meeting, 1st–4th December, 2020

EAAP Scientific Committee:

E. Strandberg
L. Pinotti
S. Messori
H. Sauerwein
M.R.F. Lee
J.F. Hocquette
J. Conington
S. Millet
A.S. Santos
T. Veldkamp
I. Halachmi
G. Pollott

OASES
Online Academic Submission and Evaluation System

EAN: 9789086863495
e-EAN: 9789086869008
ISBN: 978-90-8686-349-5
e-ISBN: 978-90-8686-900-8
DOI: 10.3920/978-90-8686-900-8

ISSN 1382-6077

First published, 2020

© Wageningen Academic Publishers
The Netherlands, 2020

Wageningen Academic
P u b l i s h e r s

Welcome

On behalf of EAAP – the European Federation of Animal Science – it is my pleasure to welcome you to the 71st EAAP Annual Meeting, the first to be held as in virtual form.

EAAP was founded in 1949 and entering in its eighth decade EAAP remains the organiser of one of the world première conferences on animal science. In its previous 70 editions, it has been attended by thousands of animal scientists and presented an incredible amount of papers, many of which reported world-first results that have been setting the standards for livestock industry and animal science.

This 71st edition will be no exception with its plenary session, 72 scientific sessions, and close to 1,100 technical contributions.

This year's conference is being run under exceptional circumstances. We started planning for the 2020 Annual Meeting nearly 2 years ago, the Portuguese colleagues even years before. The expectation was to run this in Porto with an expected record attendance close to 1,700 participants and prepared for a conference that maintained EAAP excellent technical standards and fun social events. Following much debate, we had to take the decision to proceed with the conference in the now realised alternative format. This was not without its own challenges.

With the support of the Scientific Commissions and, naturally, of the scientists sending abstracts submitted as presentations or posters, we have created both an exciting programme and conference structure which was designed to stimulate discussion and participation of attendees. While all papers are required to have a pre-recorded video presentation, the conference sessions will have life in them, presenters will be there to answer questions and technical discussions will be encouraged.

We certainly encourage all participants and attendees to attend and engage with sessions as-though you were here in person. In addition to attending scientific sessions conferences are always good for networking, to meet old friends and make new connections and friendships. Something which normally make EAAP meeting so unique. However, the Virtual Platform we are using has many features to allow you to contact and network with other attendees. Please do engage with our Annual Meeting as if you were in the traditional on-site meeting. Please look at the website and try the technology.

I hope that the use of this year virtual conference technology will offer us the experience for different types of future conferences combining physical and virtual therefore giving wider opportunities to animal scientists to participate.

I do hope that better times will come and that in 2021 we will meet in person in Davos (Switzerland). The Swiss organisers are already perfectly prepared to welcome us!

Matthias Gauly
President EAAP

European Federation of Animal Science (EAAP)

President:	M. Gauly
Secretary General:	A. Rosati
Address:	Via G. Tomassetti 3, A/I
	I-00161 Rome, Italy
Phone/Fax:	+39 06 4420 2639
E-mail:	eaap@eaap.org
Web:	www.eaap.org

Council Members

President	Matthias Gauly (Italy)
Secretary General	Andrea Rosati
Vice-Presidents	Johannes Sölkner (Austria)
	John Carty (Ireland)
Members	Isabel Casasus Peyo (Spain)
	Ilan Halachmi (Israel)
	Stephane Ingrand (France)
	Denis Kučevič (Serbia)
	Martin Lidauer (Finland)
	Veronika Maurer (Switzerland)
	Bruno Ronchi (Italy)
	Mogens Vestergaards (Denmark)
Auditor	Zdravko Barac (Croatia)
	Gerry Greally (Ireland)
Alternate Auditor	Andreas Hofer (Switzerland)
FAO Representative	Badi Besbes

European Federation of Animal Science has close established links with the sister organisations American Dairy Science Association, American Society of Animal Science, Canadian Society of Animal Science and Asociación Latinoamericana de Producción Animal.

Friends of EAAP

By creating the 'Friends of EAAP', EAAP offers the opportunity to industries to receive services from EAAP in change of a fixed sponsoring amount of support every year.
- The group of supporting industries are layered in three categories: 'silver', 'gold' and 'diamond' level.
- It is offered an important discount (one year free of charge) if the sponsoring industry will agree for a four years period.
- EAAP will offer the service to create a scientific network (with Research Institutes and Scientists) around Europe.
- Creation of a permanent Board of Industries within EAAP with the objective to inform, influence the scientific and organizational actions of EAAP, like proposing choices of the scientific sessions and invited speakers and to propose industry representatives for the Study Commissions.
- Organization of targeted workshops, proposed by industries.
- EAAP can represent and facilitate activities of the supporting industries toward international legislative and regulatory organizations.
- EAAP can facilitate the supporting industries to enter in consortia dealing with internationally supported research projects.

Furthermore EAAP offers, depending to the level of support (details on our website: www.eaap.org):
- Free entrances to the EAAP annual meeting and Gala dinner invitation.
- Free registration to journal *animal*.
- Inclusion of industry advertisement in the EAAP Newsletter, in the banner of the EAAP website, in the Book of Abstract and in the Programme Booklet of the EAAP annual meeting.
- Inclusion of industry leaflets in the annual meeting package.
- Presence of industry advertisements on the slides between presentations at selected standard sessions.
- Presence of industry logos and advertisements on the slides between presentations at the Plenary Sessions.
- Public Recognition by the EAAP President at the Plenary Opening Session of the annual meeting.
- Discounted stands at the EAAP annual meeting.
- Invitation to meetings (at every annual meeting) to discuss joint strategy EAAP/Industries with the EAAP President, Vice-President for Scientific affair, Secretary General and other selected members of the Council and of the Scientific Committee.

Contact and further information

If the industry you represent is interested to become 'Friend of EAAP' or want to have further information please contact jean-marc.perez0000@orange.fr or EAAP secretariat (eaap@eaap.org, phone : +39 06 44202639).

The Association

EAAP (The European Federation of Animal Science) organises every year an international meeting which attracts between 900 and 1,500 people. The main aims of EAAP are to promote, by means of active co-operation between its members and other relevant international and national organisations, the advancement of scientific research, sustainable development and systems of production; experimentation, application and extension; to improve the technical and economic conditions of the livestock sector; to promote the welfare of farm animals and the conservation of the rural environment; to control and optimise the use of natural resources in general and animal genetic resources in particular; to encourage the involvement of young scientists and technicians. More information on the organisation and its activities can be found at www.eaap.org

What is the YoungEAAP?

YoungEAAP is a group of young scientists organized under the EAAP umbrella. It aims to create a platform where scientists during their early career get the opportunity to meet and share their experiences, expectations and aspirations. This is done through activities at the Annual EAAP Meetings and social media. The large constituency and diversity of the EAAP member countries, commissions and delegates create a very important platform to stay up-to-date, close the gap between our training and the future employer expectations, while fine-tuning our skills and providing young scientists applied and industry-relevant research ideas.

Committee Members at a glace

- Dr Christian Lambertz (President)
- MSc Torun Wallgren (Vice President)
- Ines Adriaens (Secretary)

YoungEAAP promotes Young and Early Career Scientists to:
- Stay up-to-date (i.e. EAAP activities, social media);
- Close the gap between our training and the future employer expectations;
- Fine-tune our skills through EAAP meetings, expand the special young scientists' sessions, and/or start online webinars/trainings with industry and academic leaders;
- Meet to network and share our graduate school or early employment experiences;
- Develop research ideas, projects and proposals.

Who can be a Member of YoungEAAP?

All individual members of EAAP can join the YoungEAAP if they meet one of the following criteria: Researchers under 35 years of age OR within 10 years after PhD-graduation

Just request your membership form (christian.lambertz@fibl.org) and become member of this network!!!

72nd Annual Meeting of the European Federation of Animal Science

Davos, Switzerland,
August 30th to September 3rd 2021

Swiss Organizing Committee

The 72nd EAAP annual meeting is organized by the Swiss Association of Animal Sciences (SAAS) in partnership with the Federal Office for Agriculture (FOAG) and Global Agenda for Sustainable Livestock.

It is organized in close collaboration with other Swiss institutions active in animal sciences, including: ETH Zürich | VetSuisse Bern | BFH-HAFL Zollikofen | FiBL Frick | Agroscope Posieux | Qualitas AG Zug | Breeding Organizations

Members:

Name	Function	Affiliation
Lucas Casanova	President, Finance, Sponsorships	Braunvieh Schweiz, Zug
Andreas Hofer	Infrastructure, Congress venue	SUISAG, Sempach
Beat Bapst	Secretariat and coordination with EAAP	Qualitas AG, Zug
Adrian Aebi	Relations with Federal Office for Agriculture (FOAG)	FOAG, Bern
Veronika Maurer	Scientific program and contact with the Scientific Committee of EAAP	FiBL, Frick
Stefan Probst	Social Events, Technical Tours	BFH-HAFL, Zollikofen
Léonie von Tavel	Communication	Swissgenetics, Zollikofen

Conference website: www.eaap2021.org

Commission on Animal Genetics

Erling Strandberg	President	Swedish University of Agricultural Sciences
	Sweden	erling.strandberg@slu.se
Eileen Wall	Vice-President	Scotland's Rural College
	United Kingdom	eileen.wall@sruc.ac.uk
Filippo Miglior	Vice-President	Ontario Genomics
	Canada	fmiglior@ontariogenomics.ca
Han Mulder	Vice-President	Wageningen University
	The Netherlands	han.mulder@wur.nl
Alessio Cecchinato	Secretary	Padova University
	Italy	alessio.cecchinato@unipd.it
Tessa Brinker	Industry rep.	EFFAB
	The Netherlands	tessa.brinker@effab.info
Marcin Pszczola	Young Club	Poznan University of Life Sciences
	Poland	marcin.pszczola@gmail.com

Commission on Animal Nutrition

Luciano Pinotti	President	University of Milan
	Italy	luciano.pinotti@unimi.it
Sam de Campaneere	Vice-President	ILVO
	Belgium	sam.decampeneere@ilvo.vlaanderen.be
Geert Bruggeman	Secretary	Nuscience Group
	Belgium	geert.bruggeman@nusciencegroup.com
Roselinde Goselink	Secretary	Wageningen Livestock Research
	The Netherlands	roselinde.goselink@wur.nl
Latifa Abdenneby-Najar	Secretary	IDELE
	France	latifa.najar@idele.fr
Daniele Bonvicini	Industry Rep	Prosol S.p.a
	Italy	d.bonvicini@prosol-spa.it
Susanne Kreuzer-Redmer	Young Club	Vetmed Vienna
	Austria	susanne.kreuzer-redmer@vetmeduni.ac.at

Commission on Health and Welfare

Stefano Messori	President	STAR-IDAZ International Research Consortium
	France	stefano.messori@yahoo.it
Gürbüz Das	Vice-President	Leibniz Institute for Farm Animal Biology
	Germany	gdas@fbn-dummerstorf.de
Evangelia N. Sossidou	Vice-President	Hellenic Agricultural Organization
	Greece	sossidou.arig@nagref.gr
Stefanie Ammer	Secretary	Goettingen University
	Germany	stefanie.ammer@uni-goettingen.de
Giulietta Minozzi	Secretary	University of Milan
	Italy	giulietta.minozzi@unimi.it
Olivier Espeisse	Industry Rep	Ceva Santé Animale
	France	olivier.espeisse@ceva.com
Delphine Gardan	Industry Rep	Groupe CCPA
	France	dgardan-salmon@ccpa.com
Mariana Dantas de Brito Almeida	Young Club	University of Tras-os-Montes and Alto Douro
	Portugal	mdantas@utad.pt

Commission on Animal Physiology

Helga Sauerwein	President	University of Bonn
	Germany	sauerwein@uni-bonn.de
Isabelle Louveau	Vice president	INRA
	France	isabelle.louveau@inra.fr
Chris Knight	Secretary	University of Copenhagen
	Denmark	chkn@sund.ku.dk
Yuri Montanholi	Secretary	North Dakota State University
	USA	yuri.montanholi@ndsu.edu
Arnulf Troescher	Industry rep.	BASF
	Germany	arnulf.troescher@basf.com
Akos Kenez	Young Club	City University of Hong Kong
	Germany	akos.kenez@cityu.edu.hk

Commission on Livestock Farming Systems

Michael Lee	President	Harper Adams University
	United Kingdom	mrflee@harper-adams.ac.uk
Monika Zehetmeier	Vice President	Institute Agricultural Economics and Farm Management
	Germany	monika.zehetmeier@lfl.bayern.de
Tommy Boland	Secretary	University College Dublin
	Ireland	tommy.boland@ucd.ie
Raimon Ripoll-Bosch	Secretary	Wageningen University
	The Netherlands	raimon.ripollbosch@wur.nl
Vincent Thenard	Secretary	INRA
	France	vincent.thenard@inra.fr
Alfredo J. Escribano	Industry rep.	Independent researcher and consultant
	Spain	alfredojescribano@gmail.com
Tiago T. da Silva Siqueira	Young Club	INRA
	France	tiago.squeira@purpan.fr
Maria-Anastasia Karatzia	Young Club	Hellenic Agricultural Organization
	Greece	karatzia@rias.gr

Commission on Cattle Production

Jean François Hoquette	President	INRA
	France	jean-francois.hocquette@inra.fr
Sven König	Vice-President	University of Kassel
	Germany	sven.koenig@uni-kassel.de
Andreas Foskolos	Vice-President	Aberystwyth University
	United Kingdom	anf20@aber.ac.uk
Massimo De Marchi	Secretary	Padova University
	Italy	massimo.demarchi@unipd.it
Ray Keatinge	Secretary	Agriculture & Horticulture Development Board
	United Kingdom	ray.keatinge@ahdb.org.uk
Karsten Maier	Industry rep.	UECBV
	Belgium	info@uecbv.eu
Cătălin Necula	Industry rep.	Alltech
	Ireland	cnecula@alltech.com
Akke Kok	Young Club	Wageningen Livestock Research
	The Netherlands	akke.kok@wur.nl

Commission on Sheep and Goat Production

Joanne Conington	President United Kingdom	SRUC joanne.conington@sac.ac.uk
Nóirín McHugh	Vice-President Ireland	Teagasc noirin.mchugh@teagasc.ie
Ouranios Tzamaloukas	Secretary Cyprus	Cyprus University of Technology ouranios.tzamaloukas@cut.ac.cy
Vasco Augusto Pilão Cadavez	Secretary Portugal	CIMO - Mountain Research Centre vcadavez@ipb.pt
Neil Keane	Industry rep. Ireland	Alltech nkeane@alltech.com

Commission on Pig Production

Sam Millet	President Belgium	ILVO sam.millet@ilvo.vlaanderen.be
Paolo Trevisi	Vice president Italy	Bologna University paolo.trevisi@unibo.it
Giuseppe Bee	Vice president Switzerland	Agroscope Liebefeld-Posieux ALP giuseppe.bee@alp.admin.ch
Katja Nilsson	Secretary Sweden	Swedish University of Agricultural Science katja.nilsson@slu.se
Egbert Knol	Industry rep. The Netherlands	TOPIGS egbert.knol@topigs.com
Stafford Vigors	Young Club Ireland	University College Dublin staffordvigors1@ucd.ie

Commission on Horse Production

Ana Sofia Santos	President Portugal	CITAB-UTAD assantos@utad.pt
Rhys Evans	Vice president	Norwegian University College of Agriculture and Rural Development
	Norway	rhys@hlb.no
Katharina Stock	Vice president Germany	IT Solutions for Animal Production (vit) friederike.katharina.stock@vit.de
Klemen Potočnik	Vice president Slovenija	University of Ljubljana klemen.potocnik@bf.uni-lj.si
Isabel Cervantes Navarro	Secretary Spain	Complutense University of Madrid icervantes@vet.ucm.es
Pasquale De Palo	Secretary Italy	University of Bari pasquale.depalo@uniba.it
Melissa Cox	Industry rep. Germany	CAG GmbH – Center for Animal Genetics melissa.cox@centerforanimalgenetics.de
Juliette Auclair- Ranzaud	Young Club France	Institut français du cheval et de l'équitation juliette.auclair-ronzaud@ifce.fr

Commission on Insects

Teun Veldkamp	President	Wageningen Livestock Research
	The Netherlands	teun.veldkamp@wur.nl
Laura Gasco	Vice president	University of Turin
	Italy	laura.gasco@unito.it
Jorgen Eilenberg	Secretary	University of Copenhagen
	Denmark	jei@plen.ku.dk
Christoph Sandrock	Secretary	Research Institute of Organic Agriculture FiBL
	Switzerland	christoph.sandrock@fibl.org
Marian Peters	Industry rep.	IIC (International Insect Centre)
	The Netherlands	marianpeters@ngn.co.nl
Matteo Ottoboni	Young Club	UNIMI
	Italy	matteo.ottoboni@unimi.it
Marwa Shumo	Young Club	University of Bonn
	Germany	mshummo@hotmail.com

Commission on Precision Livestock Farming

Ilan Halachmi	President	Agriculture research organization (ARO)
	Israel	halachmi@volcani.agri.gov.il
Jarissa Maselyne	Vice president	ILVO
	Belgium	jarissa.maselyne@ilvo.vlaanderen.be
Matti Pastell	Vice president	Natural Resources Institute Finland (Luke)
	Finland	matti.pastell@luke.fi
Claire Morgan-Davies	Vice president	Scotland's Rural College (SRUC)
	United Kingdom	claire.morgan-davies@sruc.ac.uk
Shelly Druyan	Secretary	ARO, The Volcani Center
	Israel	shelly.druyan@mail.huji.ac.il
Radovan Kasarda	Secretary	Slovak University of Agriculture in Nitra
	Slovakia	radovan.kasarda@uniag.sk
Stephane Ingrand	Secretary	INRA
	France	stephane.ingrand@inra.fr
Malcolm Mitchell	Industry rep.	SRUC
	United Kingdom	malcolm.mitchell@sruc.ac.uk
Ines Adriaens	Young Club	KU Leuven
	Belgium	ines.adriaens@kuleuven.be

Friends of EAAP

Feedase

FROM
INDIGESTIBLE TO > **DIGESTIBLE**

EVERYTHING YOU NEED TO KNOW
ABOUT ADISSEO'S FEEDASE APPROACH
IS ON THE WEBSITE WWW.FEEDASE.COM

Thank you
to the 71st EAAP Annual Congress Sponsors and Friends

━━━━━━━━ *Platinum sponsor* ━━━━━━━━

━━━━━━━━ *Silver sponsors* ━━━━━━━━

━━━━━━━━ *Session sponsor* ━━━━━━━━

sessions 37 and 47

Acknowledgements

Scientific programme EAAP 2020

Tuesday 1 December 10.00 – 12.30	Tuesday 1 December 13.45 – 17.30	
	Session 02 Joint EAAP/ADSA/Interbull session: Breeding and genetics – ten years of genomic selection Chair: Miglior	
	Session 03 Genetic diversity Chair: Fernández Martín	
	Session 04 PLF for saving feed Chair: Maselyne / Halachmi	
	Session 05 Climate change impact on livestock health and welfare Chair: Messori / Sossidou	
	Session 06 Energy and protein efficiency in pigs, from conception to slaughter Chair: Bee / Millet	
Session 01 **Plenary session** Responsibility of animal production on climate change Chair: Gauly	**Session 07** Responsiveness of the dairy cattle industry to social needs Chair: Foskolos	
	Session 08 Physiological basis of PLF technologies to infer on animal health, welfare and production efficiency in cattle Chair: Montanholi	
	Session 09 Insects as feed and health effects for livestock Chair: Gasco / Ottoboni	
	Session 10 Sustainable land use for healthy humans and a healthy planet Chair: Zehetmeier / Ripoll Bosch	
	Session 11 Dairy sheep and goat systems: new research in genes, nutrition and management Chair: Tzamaloukas / Keane	
	Session 12 What livestock has to offer to biodiversity & healthy soils (ATF) Chair: Peyraud	

Wednesday 2 December 9.00 – 12.45	Wednesday 2 December 13.45 – 16.30
Session 13 Governance of the genetic resources Chair: Leroy / Baumung	**Session 22** Can you have your cake and eat it too – tools to get the most out of animal resilience and efficiency Chair: Friggens
Session 14 Genetics free communications Chair: Formoso-Rafferty	**Session 23** Breeding for improved animal health and welfare Chair: Minozzi / Ammer / Cecchinato
Session 15 Fitting PLF to species and animal size, possibilities for sheep, goats, poultry, horses, pigs and 'exotic' animals Chair: Morgan-Davies / Druyan / Grisot	**Session 24** Data-based PLF solutions for the cattle sector: opportunities, challenges and pitfalls Chair: Adriaens / Pastell / Fürst-Waltl
Session 16 'Net zero-carbon' - how is it possible in livestock farming systems? Chair: M. Lee	**Session 25** Feed production facing climate change (new feeds, alternative fibre sources) Chair: De Campeneere
Session 17 Early career competition: stressing innovation in pig production Chair: Nilsson / Vigors	**Session 26** Early career competition: stressing innovation in pig production Chair: Vigors / Nilsson
Session 18 Young Train: Innovative research and extension in cattle Chair: Lambertz	**Session 27** Integration of beef and dairy production Chair: Keatinge
Session 19 SMART NUTRITION – the role of nutrition in the circular economy: improving the use of feed not competitive with human nutrition Chair: Pinotti	**Session 28** Aquaculture Working Group Chair: Zupan
Session 20 Insect production: techniques, substrates and health issues Chair: Eilenberg / Peters	**Session 29** Understanding and combatting antimicrobials and anthelmintics resistance Chair: Das / Espeisse
Session 21 Omics in animal nutrition and physiology (The European Joint Doctorate Program Molecular Animal Nutrition 'MANNA') Chair: Eckersall / Sauerwein	**Session 30** Genetic evaluations and breeding programs in horses and their development Chair: Cervantes / Vicente
	16.30 – 17.30 **Poster session 1** (session 1 – 30)

Thursday 3 December 9.00 – 12.45	Thursday 3 December 13.45 – 16.30
Session 31 Recent advances in livestock physiology: cellular, organoid and whole-organism studies Chair: Louveau / Stadnicka	**Session 41** Breeding programmes Chair: Y.L. Lee
Session 32 Breeding for improved animal health and welfare Chair: Minozzi / Wall	**Session 42** New technological challenges (including nanotechnology) facing future feeding industry – how to value? Chair: Bruggeman
Session 33 Opportunities and risks in animal management applying AI, big data, data science tools, API economy and digital agriculture Chair: Pastell / Adriaens	**Session 43** Impact of PLF and sensors on phenotyping, animal breeding, animal selection and animal genetics Chair: Kasarda / Druyan
Session 34 Genetics poster session Chair: Strandberg	**Session 44** Dairy4Future project Chair: Le Gall

ONE-DAY SYMPOSIUM **Modulating gut function and microbiome for resilient livestock production**	
Session 35 Chair: Trevisi / Ammer	**Session 45** Chair: Ammer / Vigors

Session 36 Housing of cattle Chair: Kuipers	

ONE-DAY SYMPOSIUM **Combining the diversity of resources and farming practices to ensure resilience at different scales**	
Session 37 Chair: Thenard	**Session 47** Chair: Hadjipavlou

Session 38 Standardization of methods, parameters and terminology in insect research and socio-economic impact Chair: Deruytter / Shumo	**Session 48** Insect genetics: opportunities and challenges Chair: Lefebvre / Bouwman
Session 39 New perspectives and approaches in equidae production Chair: Evans / Auclair	**Session 49** New perspectives and approaches in equidae production Chair: Vial / Fradinho
Session 40 Dairy sheep and goat systems: new research in genes, nutrition and management Chair: Tzamaloukas / Keane	**Session 50** Sheep and goat free communications Chair: Cadavez / Keane
	Session 51 Young Scientist session: jobs of the future Chair: Lambertz

11.45 – 12.45 **Commission meeting** • Physiology • Insects	16.30 – 17.30 **Commission meeting** • Sheep and Goat **Poster session 2** (session 31 – 74 minus 34)

Friday 4 December 9.00 – 11.45	Friday 4 December 13.45 – 17.30
Session 52 Long-term selection and limits Chair: Wientjes	**Session 63** Breeding, climate and sustainability Chair: Pszczola
Session 53 Free communications - cattle Chair: Hocquette	**Session 64** Genomic evaluation Chair: Cecchinato
Session 54 PLF as a tailor-made way to care for individuals within groups Chair: Maselyne	**Session 65** PLF and sensing animal health, welfare, physiological measures and behaviour of animals Chair: Ingrand
Session 55 PLF and sensing animal health, welfare, physiological measures and behaviour of animals Chair: Ingrand	**Session 67** Mineral and vitamin nutrition in pigs and poultry Chair: Millet / Bee
Session 56 Pig, poultry and rabbit husbandry for improved product quality Chair: Stadnicka / Millet	**Session 68** Collaborative international research related to beef quality Chair: Hocquette
Session 57 Lameness and metabolic aspects in cattle Chair: Fürst-Waltl	**Session 69** Dietary methyl donor supplementation in dairy ruminants Chair: Pinotti
Session 58 Neonatal nutrition/ early life programming Chair: Kreuzer-Redmer	**Session 70** Applied animal behaviour Chair: Boyle / Dantas de Brito Almeida
Session 59 Impact of different farming practices on animal health and welfare Chair: Sossidou / Messori	**Session 71** Combating the increasingly negative consumer opinion on livestock's contribution to climate change Chair: Pulina
Session 60 Horse registration, legal status of horses and its consequences Chair: Potocnik	**Session 72** Resilient sheep and goats: breeding & management strategies to overcome disease and environmental challenges Chair: Conington / McHugh
Session 61 SMARTER: small ruminants breeding for efficiency and resilience Chair: Conington / Moreno	**Session 73** Innovative farm systems to meet societal demand Chair: Siqueira
Session 62 Livestock farming systems free communications Chair: M. Lee	

11.45 – 12.45
Commission meetings

- Genetics
- Cattle
- PLF
- Pig

- Nutrition
- Health and Welfare
- Horse
- LFS

Scientific programme

Session 01. Plenary session – responsibility of animal production on climate change

Date: Tuesday 1 December 2020; 10.00 – 12.30
Chair: Gauly

Theatre Session 01

Session 02. Joint EAAP/ADSA/Interbull session: Breeding and genetics – ten years of genomic selection

Date: Tuesday 1 December 2020; 13.45 – 17.30
Chair: Miglior

Theatre Session 02

Session 03. Genetic diversity

Date: Tuesday 1 December 2020; 13.45 – 17.30
Chair: Fernández Martín

Theatre Session 03

Poster Session 03

Session 04. PLF for saving feed

Date: Tuesday 1 December 2020; 13.45 – 17.30
Chair: Maselyne / Halachmi

Theatre Session 04

Poster Session 04

Session 05. Climate change impact on livestock health and welfare

Date: Tuesday 1 December 2020; 13.45 – 17.30
Chair: Messori / Sossidou

Theatre Session 05

Poster Session 05

Session 06. Energy and protein efficiency in pigs, from conception to slaughter

Date: Tuesday 1 December 2020; 13.45 – 17.30
Chair: Bee / Millet

Theatre Session 06

Session 07. Responsiveness of the dairy cattle industry to social needs

Date: Tuesday 1 December 2020; 13.45 – 17.30
Chair: Foskolos

Theatre Session 07

Poster Session 07

Session 08. Physiological basis of PLF technologies to infer on animal health, welfare and production efficiency in cattle

Date: Tuesday 1 December 2020; 13.45 – 17.30
Chair: Montanholi

Theatre Session 08

Poster Session 8

Session 09. Insects as feed and health effects for livestock

Date: Tuesday 1 December 2020; 13.45 – 17.30
Chair: Gasco / Ottoboni

Theatre Session 09

Poster Session 09

Session 10. Sustainable land use for healthy humans and a healthy planet

Date: Tuesday 1 December 2020; 13.45 – 17.30
Chair: Zehetmeier / Ripoll Bosch

Theatre Session 10

Poster Session 10

Session 11. Dairy sheep and goat systems: new research in genes, nutrition and management

Date: Tuesday 1 December 2020; 13.45 – 17.30
Chair: Tzamaloukas / Keane

Theatre Session 11

Poster Session 11

Session 12. What livestock has to offer to biodiversity & healthy soils (ATF)

Date: Tuesday 1 December 2020; 13.45 – 17.30
Chair: Peyraud

Theatre Session 12

Session 13. Governance of the genetic resources

Date: Wednesday 2 December 2020; 9.00 – 12.45
Chair: Leroy / Baumung

Theatre Session 13

Session 14. Genetics free communications

Date: Wednesday 2 December 2020; 9.00 – 12.45
Chair: Formoso-Rafferty

Theatre Session 14

Poster Session 14

Session 15. Fitting PLF to species and animal size, possibilities for sheep, goats, poultry, horses, pigs and 'exotic' animals

Date: Wednesday 2 December 2020; 9.00 – 12.45
Chair: Morgan-Davies / Druyan / Grisot

Theatre Session 15

Poster Session 15

Session 16. 'Net zero-carbon' - how is it possible in livestock farming systems?

Date: Wednesday 2 December 2020; 9.00 – 12.45
Chair: M. Lee

Theatre Session 16

Poster Session 16

Session 17. Early career competition: stressing innovation in pig production

Date: Wednesday 2 December 2020; 9.00 – 12.45
Chair: Nilsson / Vigors

Theatre Session 17

Poster Session 17

Session 18. Young Train: Innovative research and extension in cattle

Date: Wednesday 2 December 2020; 9.00 – 12.45
Chair: Lambertz

Theatre Session 18

Poster Session 18

Session 19. SMART NUTRITION – the role of nutrition in the circular economy: improving the use of feed not competitive with human nutrition

Date: Wednesday 2 December 2020; 9.00 – 12.45
Chair: Pinotti

Theatre Session 19

Effect of replacing soybean meal by *Lupinus luteus* on meat fatty acids composition of growing lambs   230
M. Almeida, S. Garcia-Santos, A. Nunes, S. Rito, C.M. Guedes, L.M. Ferreira, V. Santos, R. Dominguez, J.M. Lorenzo, S. Silva and J. Azevedo

Effect of *Camelina sativa* cake and *Cynara cardunculus* supplements on goat milk functional quality   231
C. Giromini, F. Omodei Zorini, S. Sandrini, R. Rebucci, G. Savoini, A. Baldi, A. Buccioni, A.A.K. Salama and G. Invernizzi

Effects of dietary dried food waste addition to broiler diets on growth performance   231
E. Giamouri, A.C. Pappas, G. Papadomichelakis, E. Tsiplakou, K. Sotirakoglou, K. Feggeros, T. Manios and G. Zervas

Carbohydrase enzyme complex (CEC) and agro-industrial byproducts in broiler nutrition   232
W. Al-Izzi and Y. Beckers

High-moisture grain meals for swine: proximal analyses, *in vitro* digestibility and estimated energy   232
C.F. Martins, M. Vasconcelos, D.M. Ribeiro, A.M. Almeida and J.P.B. Freire

Effects of adding Sake lees in milk replacer on growth, faecal traits and blood metabolites in calves   233
S. Katsumata, Y. Hayashi, S. Yamanaka, A. Obata, H. Tsutsumi, H. Aoki, T. Tsukahara, K. Oishi, H. Hirooka and H. Kumagai

Invited   Salty or sweet former food based diets for piglets: effect on growth performance and digestibility   233
L. Pinotti, A. Luciano, M. Comi, F. Fumagalli, N. Rovere, M.C. Rulli and M. Ottoboni

Plasma amino acid profile after feeding a balanced or unbalanced sustainable protein source in pigs   234
F.A. Eugenio, J. Van Milgen, J. Duperray, R. Sergheraert and N. Le Floc'h

Energy value of ensiled lucerne and red clover leaves and their impacts on performance in pigs   234
S.R. Habit, L. Stødkilde-Jørgensen, S. Krogh Jensen and D. Renaudeau

Can rumen micro-organisms transformed dietary selenite into elemental selenium?   235
M.A. Hachemi, E. Pinloche, M. De Marco and M. Briens

Hermetia illucens reared on seaweed or selenium enriched substrates for designing a feed specialty   235
M. Ottoboni, V. Sele, M. Silva, A. Grisendi, M. Della Porta, R. Cazzola, F. Selmin, R. Ørnsrud and L. Pinotti

Poster Session 19

Turning fruit and vegetable waste from the wholesale market into feed   236
D. Tedesco, S. Scario, E. Biazzi, A. Tava and L. Baschenis Begnis

Performance of 4 handheld NIR spectrometers to predict the dry matter, starch and NDF of corn silage   236
N. Chamberland, P.H. Vermeulen, B. Lecler, O. Minet, V. Decruyenaere, E. Froidmont and V. Baeten

Ruminal carbohydrase enzyme complex (CEC) vs a commercial cellulolytic enzyme in broiler nutrition   237
W. Al-Izzi and Y. Beckers

Session 20. Insect production: techniques, substrates and health issues

Date: Wednesday 2 December 2020; 9.00 – 12.45
Chair: Eilenberg / Peters

Theatre Session 20

Poster Session 20

Session 21. Omics in animal nutrition and physiology (The European Joint Doctorate Program Molecular Animal Nutrition "MANNA")

Date: Wednesday 2 December 2020; 9.00 – 12.45
Chair: Eckersall / Sauerwein

Theatre Session 21

Poster Session 21

Session 22. Can you have your cake and eat it too – tools to get the most out of animal resilience and efficiency

Date: Wednesday 2 December 2020; 13.45 – 16.30
Chair: Friggens

Theatre Session 22

Session 23. Breeding for improved animal health and welfare

Date: Wednesday 2 December 2020; 13.45 – 16.30
Chair: Minozzi / Ammer / Cecchinato

Theatre Session 23

Session 24. Data-based PLF solutions for the cattle sector: opportunities, challenges and pitfalls

Date: Wednesday 2 December 2020; 13.45 – 16.30
Chair: Adriaens / Pastell / Fürst-Waltl

Theatre Session 24

Poster Session 24

Session 25. Feed production facing climate change (new feeds, alternative fibre sources)

Date: Wednesday 2 December 2020; 13.45 – 16.30
Chair: De Campeneere

Theatre Session 25

Poster Session 25

Session 26. Early career competition: stressing innovation in pig production

Date: Wednesday 2 December 2020; 13.45 – 16.30
Chair: Vigors / Nilsson

Theatre Session 26

Session 27. Integration of beef and dairy production

Date: Wednesday 2 December 2020; 13.45 – 16.30
Chair: Keatinge

Theatre Session 27

Poster Session 27

Session 28. Aquaculture Working Group

Date: Wednesday 2 December 2020; 13.45 – 16.30
Chair: Zupan

Theatre Session 28

Poster Session 28

Session 29. Understanding and combatting antimicrobials and anthelmintics resistance

Date: Wednesday 2 December 2020; 13.45 – 16.30
Chair: Das / Espeisse

Theatre Session 29

Poster Session 29

Session 30. Genetic evaluations and breeding programs in horses and their development

Date: Wednesday 2 December 2020; 13.45 – 16.30
Chair: Cervantes / Vicente

Theatre Session 30

Poster Session 30

Session 31. Recent advances in livestock physiology: cellular, organoid and whole-organism studies

Date: Thursday 3 December 2020; 9.00 – 11.45
Chair: Louveau / Stadnicka

Theatre Session 31

Poster Session 31

Adiponectin in the mandibular glands of cows fed a diet supplemented with essential fatty acids 354
C. Dall'Aglio, F. Mercati, P. Scocco, E. De Felice, A. Tröscher, H.M. Hammon and M. Mielenz

Autophagy is required for in vitro myogenesis in piglets 354
A. Vincent, M.H. Perruchot, I. Louveau and F. Dessauge

Session 32. Breeding for improved animal health and welfare

Date: Thursday 3 December 2020; 9.00 – 12.45
Chair: Minozzi / Wall

Theatre Session 32

Poster Session 32

Session 33. Opportunities and risks in animal management applying AI, big data, data science tools, API economy and digital agriculture

Date: Thursday 3 December 2020; 9.00 – 12.45
Chair: Pastell / Adriaens

Theatre Session 33

Poster Session 33

Session 34. Genetics poster session

Date: Thursday 3 December 2020; 9.00 – 12.45
Chair: Strandberg

Poster Session 34

Session 35. Modulating gut function and microbiome for resilient livestock production

Date: Thursday 3 December 2020; 9.00 – 12.45
Chair: Trevisi / Ammer

Theatre Session 35

Poster Session 35

Session 36. Housing of cattle

Date: Thursday 3 December 2020; 9.00 – 12.45
Chair: Kuipers

Theatre Session 36

Poster Session 36

Session 37. Combining the diversity of resources and farming practices to ensure resilience at different scales

Date: Thursday 3 December 2020; 9.00 – 12.45
Chair: Thenard

Theatre Session 37

Poster Session 37

Session 38. Standardization of methods, parameters and terminology in insect research and socio-economic impact

Date: Thursday 3 December 2020; 9.00 – 11.45
Chair: Deruytter / Shumo

Theatre Session 38

Session 39. New perspectives and approaches in equidae production

Date: Thursday 3 December 2020; 9.00 – 12.45
Chair: Evans / Auclair

Theatre Session 39

Poster Session 39

Session 40. Dairy sheep and goat systems: new research in genes, nutrition and management

Date: Thursday 3 December 2020; 9.00 – 12.15
Chair: Tzamaloukas / Keane

Theatre Session 40

Session 41. Breeding programmes

Date: Thursday 3 December 2020; 13.45 – 16.30
Chair: Y.L. Lee

Theatre Session 41

Poster Session 41

Session 42. New technological challenges (including nanotechnology) facing future feeding industry – how to value?

Date: Thursday 3 December 2020; 13.45 – 16.30
Chair: Bruggeman

Theatre Session 42

Poster Session 42

Session 43. Impact of PLF and sensors on phenotyping, animal breeding, animal selection and animal genetics

Date: Thursday 3 December 2020; 13.45 – 16.30
Chair: Kasarda / Druyan

Theatre Session 43

Poster Session 43

Session 44. Dairy4Future project

Date: Thursday 3 December 2020; 13.45 – 16.30
Chair: Le Gall

Theatre Session 44

Poster Session 44

Session 45. Modulating gut function and microbiome for resilient livestock production

Date: Thursday 3 December 2020; 13.45 – 16.30
Chair: Ammer / Vigors

Theatre Session 45

Poster Session 45

Session 47. Combining the diversity of resources and farming practices to ensure resilience at different scales

Date: Thursday 3 December 2020; 13.45 – 16.30
Chair: Hadjipavlou

Theatre Session 47

Poster Session 47

Session 48. Insect genetics: opportunities and challenges

Date: Thursday 3 December 2020; 13.45 – 16.30
Chair: Lefebvre / Bouwman

Theatre Session 48

Session 49. New perspectives and approaches in equidae production

Date: Thursday 3 December 2020; 13.45 – 16.30
Chair: Vial / Fradinho

Theatre Session 49

Poster Session 49

Session 50. Sheep and goat free communications

Date: Thursday 3 December 2020; 13.45 – 16.30
Chair: Cadavez / Keane

Theatre Session 50

Poster Session 50

Session 51. Young Scientist session: jobs of the future

Date: Thursday 3 December 2020; 13.45 – 16.30
Chair: Lambertz

Theatre Session 51

Session 52. Long-term selection and limits

Date: Friday 4 December 2020; 9.00 – 11.45
Chair: Wientjes

Theatre Session 52

Session 53. Free communications - cattle

Date: Friday 4 December 2020; 9.00 – 11.45
Chair: Hocquette

Theatre Session 53

Poster Session 53

Session 54. PLF as a tailor-made way to care for individuals within groups

Date: Friday 4 December 2020; 9.00 – 11.45
Chair: Maselyne

Theatre Session 54

Blood parameters predictions based on milk MIR spectral data 506
L.M. Dale, K. Drössler, F. Gollé-Leidreiter, H. Spiekers, P. Hertel-Böhnke, E. Stamer, F. Onken and A. Werner

Methane, energy balance and feed efficiency MIR predictions evaluation on dairy cows population 507
L.M. Dale, A. Werner, K. Drössler, F. Gollé-Leidreiter, H. Spiekers, P. Hertel-Böhnke, E. Stamer, B. Khula, F. Onken and A. Vanlierde

Poster Session 54

Use of precision livestock farming tools to evaluate the performance of growing-finishing pigs 507
V.F.C. Cruz, R. Charneca, F. Baptista, T. Morgado and J. Rico

Faecal-NIRS for predicting animal-to-animal variation in feed organic matter digestibility in cattle 508
D. Andueza, F. Picard, J. Pourrat, A. De La Torre, M. Devant, C.K. Reynolds, E. Froidmont, L. Bernard, C. Martin, P. Nozière and G. Cantalapiedra-Hijar

Session 55. PLF and sensing animal health, welfare, physiological measures and behaviour of animals

Date: Friday 4 December 2020; 9.00 – 11.45
Chair: Ingrand

Theatre Session 55

Better control of the sow health using a novel sensor embedded in an ear tag 508
A. Herlin, S. Dasen, P. Liechti, A. Peñaranda, I. Cuevas Martinez, A. Jara, A. Bugueiro, I. Gomez, C. Ollagnier, P. Renevey and C. Verjus

Thermal imaging cameras vs mobility scoring for early detection of digital dermatitis in dairy cows 509
T. Gerrard, J. Dodgson and A. Westland

Using a pecking detection system to test for correlations between pecking and injuries in turkeys 509
J.J. Gonzalez, A. Nasirahmadi and U. Knierim

PLF to better understand the behaviour of young beef cattle at fattening 510
A. Philibert, B. Mounaix, E. Vanbergue and A. Aupiais

Invited Long-term BCS monitoring by 3D camera assistant system 510
V. Bloch, M. Lidauer, P. Mäntysaari, T. Mehtiö, A. Kemppainen and M. Pastell

Mask R-CNN based detection of group housed cows and automated definition of social contacts 511
J. Salau and J. Krieter

Cattle activity recognition using Recurrent Neural Networks and TensorFlow on Android Smartphones 511
J.F. Ramirez-Agudelo, S. Bedoya-Mazo, R. Muñoz-Tamayo, S.L. Posada-Ochoa and J.R. Rosero-Noguera

Poster Session 55

Air sample as a welfare friendly method to detect swine influenza A virus in pigs? 512
C. Fablet, S. Herve, V. Dorenlor, F. Eono, E. Eveno, N. Barbier, S. Gorin, A. Keita, G. Simon and N. Rose

Automatic body position and behaviour detection of sows based on video images 512
S. Küster, P. Nolte, B. Stock and I. Traulsen

Session 56. Pig, poultry and rabbit husbandry for improved product quality

Date: Friday 4 December 2020; 9.00 – 11.45
Chair: Stadnicka / Millet

Theatre Session 56

Poster Session 56

Session 57. Lameness and metabolic aspects in cattle

Date: Friday 4 December 2020; 9.00 – 11.45
Chair: Fürst-Waltl

Theatre Session 57

Poster Session 57

Session 58. Neonatal nutrition/ early life programming

Date: Friday 4 December 2020; 9.00 – 11.45
Chair: Kreuzer-Redmer

Theatre Session 58

Poster Session 58

Session 59. Impact of different farming practices on animal health and welfare

Date: Friday 4 December 2020; 9.00 – 11.45
Chair: Sossidou / Messori

Theatre Session 59

Poster Session 59

Session 60. Horse registration, legal status of horses and its consequences

Date: Friday 4 December 2020; 9.00 – 11.45
Chair: Potocnik

Theatre Session 60

Session 61. SMARTER: small ruminants breeding for efficiency and resilience

Date: Friday 4 December 2020; 9.00 – 11.45
Chair: Conington / Moreno

Theatre Session 61

Poster Session 61

Session 62. Livestock farming systems free communications

Date: Friday 4 December 2020; 9.00 – 11.45
Chair: M. Lee

Theatre Session 62

Poster Session 62

Session 63. Breeding, climate and sustainability

Date: Friday 4 December 2020; 13.45 – 17.30
Chair: Pszczola

Theatre Session 63

Session 64. Genomic evaluation

Date: Friday 4 December 2020; 13.45 – 17.30
Chair: Cecchinato

Theatre Session 64

Session 65. PLF and sensing animal health, welfare, physiological measures and behaviour of animals

Date: Friday 4 December 2020; 13.45 – 17.30
Chair: Ingrand

Theatre Session 65

Poster Session 65

Session 67. Mineral and vitamin nutrition in pigs and poultry

Date: Friday 4 December 2020; 13.45 – 17.30
Chair: Millet / Bee

Theatre Session 67

Poster Session 67

Session 68. Collaborative international research related to beef quality

Date: Friday 4 December 2020; 13.45 – 17.30
Chair: Hocquette

Theatre Session 68

The relationship between plasma biomarkers and dark cutting in beef   596
*K.M.W.L. Loudon, G.E. Gardner, D.W. Pethick, P. McGilchrist, F.R. Dunshea, R. Polkinghorne, G. Tarr
and I.J. Lean*

DEXA prediction of beef retail cut weights   596
H.B. Calnan and G.E. Gardner

Invited Beef eating quality categorisation in Poland based on Meat Standards Australia   597
G. Pogorzelski, J. Wierzbicki, A. Onopiuk and A. Wierzbicka

Automated grading of eye muscle area and intramuscular fat in Australian beef   597
S.M. Stewart, T. Lauridsen, H. Toft, D.W. Pethick, G.E. Gardner, P. McGilchrist and M. Christensen

Invited Global collaboration to encourage a consumer focused beef industry: the Meat Research 3G Foundation 598
R. Polkinghorne, L. Farmer, D. Troy, M. Miller, J.F. Hocquette, J. Wierzbicki and A. Wierzbicka

BeeQ – building capacity for beef eating quality assessment in Wales   598
P.K. Nicholas-Davies, N.D. Scollan, T. Rowe, E. Thomas, E. Williams and R. Polkinghorne

Welsh beef carcase survey – exploring proxies for predicting eating quality   599
S. Siphambili, P.K. Nicholas-Davies and R. Polkinghorne

Contributions of tenderness, juiciness and flavour liking to overall liking of beef   599
*J.J. Liu, M.P. Ellies-Oury, S. Chriki, I. Legrand, G. Pogorzelski, J. Wierzbicki, L. Farmer, D. Troy,
R. Polkinghorne and J.F. Hocquette*

The French Meat Dictionary (collective work of the French Meat Academy)   600
F. Landrieu, M. Kombolo-Ngah and J.F. Hocquette

European conformation and fat scores of bovine carcasses are not good indicators of marbling   600
*J.J. Liu, S. Chriki, M.P. Ellies-Oury, I. Legrand, G. Pogorzelski, J. Wierzbicki, L. Farmer, D. Troy,
R. Polkinghorne and J.F. Hocquette*

Factors influencing conformation and fatness of bovine carcasses in Italy and France   601
N. Rampado, E. Hocquette, D. Roux, C. Montanari, K. De Roest and J.F. Hocquette

Poster Session 68

Feasibility of visible-near infrared spectroscopy to discriminate between steer and bull beef   601
L. Moran, L.J.R. Barron, N. Aldai and A.P. Moloney

Using real-time ultrasonography to identify the optimum subcutaneous fat depth in Arouquesa breed   602
C. Venâncio, D. Moreira, J.C. Almeida, R. Vieira, J.A. Silva and S.R. Silva

Rumen protected fat in early stage altered adipogenic gene expression and carcass quality in cattle   602
*K.H. Kim, K. Thirugnanasambantham, R. Bharanidharan, J.Y. Kim, G.D. Kim, X. Panyavong and
S.H. Beak*

Genetic trends in the most important beef cattle breeds in Poland   603
B. Szymik and G. Skrzyński

Session 69. Dietary methyl donor supplementation in dairy ruminants

Session 70. Applied animal behaviour

Date: Friday 4 December 2020; 13.45 – 17.30
Chair: Boyle / Dantas de Brito Almeida

Theatre Session 70

Poster Session 70

Session 71. Combating the increasingly negative consumer opinion on livestock's contribution to climate change

Date: Friday 4 December 2020; 13.45 – 17.30
Chair: Pulina

Theatre Session 71

Poster Session 71

Session 72. Resilient sheep and goats: breeding & management strategies to overcome disease and environmental challenges

Date: Friday 4 December 2020; 13.45 – 17.30
Chair: Conington / McHugh

Theatre Session 72

Poster Session 72

Session 73. Innovative farm systems to meet societal demand

Date: Friday 4 December 2020; 13.45 – 17.30
Chair: Siqueira

Theatre Session 73

Poster Session 73

+1.5 ° – global climate and climate politics

G. Kaser
University of Innsbruck, Faculty of Geo- and Atmospheric Sciences, Innrain 52f, 6020 Innsbruck, Austria;
georg.kaser@uibk.ac.at

In their Paris agreement, global national governments have consented to 'Holding the increase in the global average temperature to well below 2 °C above pre-industrial levels and to pursue efforts to limit the temperature increase to 1.5 °C above pre-industrial levels, recognising that this would significantly reduce the risks and impacts of climate change'. In the same agreement they have 'invited the Intergovernmental Panel on Climate Change (IPCC) to prepare a special report in 2018 on the impacts of global warming of 1.5 °C above pre-industrial levels and related global greenhouse gas emission pathways'. The IPCC SR1.5 was prepared and presented and, meanwhile, the World has further warmed. Based on this report and on recent observations, the presentation summarises the most recent changes in climate, emphasises the risks increasing towards a +1.5 °C, a +2.0 °C, and an even warmer world, and outlines the pathways to possibly keep control over the warming climate system.

Rethinking methane – livestock's path to climate neutrality

F. Mitloehner
University of California, Davis, Department of Animal Science, One Shields Ave, Davis, CA 95616, USA;
fmmitloehner@ucdavis.edu

As the global community actively works to keep temperatures from rising beyond 1.5 °C, predicting greenhouse gases (GHGs) by how they warm the planet – and not their CO_2-equivalence – provides information critical to developing short- and long-term climate solutions. Livestock, and in particular cattle, have been broadly branded as major emitters of methane (CH_4) and significant drivers of climate change. Livestock production has been growing to meet the global food demand; however, increasing demand for production does not necessarily result in the proportional increase of CH_4 production. The present paper intends to evaluate the actual effects of the CH_4 emission from U.S. dairy and beef production on temperature and initiate a rethinking of CH_4 associated with animal agriculture to clarify long-standing misunderstandings and uncover the potential role of animal agriculture in fighting climate change. Two climate metrics, the standard 100-year Global Warming Potential (GWP_{100}) and the recently proposed GWP*, were applied to the CH_4 emission from the U.S. cattle industry to assess and compare its climate contribution. Using GWP*, calculations show that since 1986, CH_4 emissions from the U.S. cattle industry have not contributed to warming but have been actively removing atmospheric carbon through the biogenic carbon cycle. The projected climate impacts show that the California dairy industry will approach climate neutrality in the next ten years if the current cow inventory holds constant, with the possibility to be CH_4 negative if there are further reductions of emissions. GWP* should be used in combination with GWP to provide informative strategical suggestions on fighting SLCPs-induced climate change. The potential cooling effects from the biogenic carbon cycle are a short-term solution to fight climate warming that the global community can leverage while developing long-term solutions for fossil fuel carbon emissions.

Adaptation actions to reduce environmental impact (global warming) on livestock farming

U. Bernabucci

Department of Agriculture and Forests Sciences, University of Tuscia, Via San Camillo De Lellis, s.n.c., 01100 Viterbo, Italy; bernab@unitus.it

Environmentally induced periods of heat stress decrease productivity with devastating economic consequences to global animal agriculture. The increasing concern with the thermal comfort of agricultural animals is justifiable not only for countries occupying tropical zones, but also for nations in temperate zones where high ambient temperatures are becoming an issue. Although, the effects of global warming will not be adverse everywhere, a relevant increase of drought is expected across the world affecting forage and crop production. Moreover, it is well known, hot environment impairs production (growth, meat and milk yield and quality, egg yield, weight, and quality) and reproductive performance, metabolic and health status, and immune response. At a World level, animal production has to increase in the next decades to satisfy the growing demand. We have to expect that the livestock systems (based on grazing, the mixed farming systems or industrialised system) will be more and more negatively affected by climate changes (namely global warming). Other than the production (growth, meat and milk, egg), also quality of animal products is strongly and negatively affected by hot environment. Regarding livestock systems, it will be strategic to optimise productivity of crops and forage (mainly improving water and soil management), and to improve the ability of animals to cope with environmental stress by management and genetic selection. To guide the evolution of livestock production systems under the increase of temperature and extreme events, better information is needed regarding biophysical and social vulnerability, and this must be integrated with agriculture and livestock components.

Session 02

Theatre 1

How to implement genomic selection

P.M. Vanraden and J.B. Cole

Agricultural Research Service, USDA, Animal Genomics and Improvement Laboratory, 10300 Baltimore Avenue, Beltsville, MD 20705-2350, USA; john.cole@usda.gov

Key features of genomic selection remain essential, and new features continue to be added 10 years after first implementation. Breeding values with higher reliability earlier in life are estimated by combining DNA genotypes for many thousands of loci using existing identification, pedigree, and phenotype databases for millions of animals. Quality control for both new and previous data greatly improves by comparing genomic and pedigree relationships to fix parent-progeny conflicts and discover many additional ancestors. Many quantitative trait loci and gene tests have been added to previous assays that used only evenly spaced, highly polymorphic markers. Imputation now combines genotypes from many assays of differing marker density. Prediction models have gradually advanced from normal or Bayesian distributions within breed and trait to single-step, multi-trait, multibreed, or other more complex models. Genomic selection was initially applied to males to predict progeny performance but now is widely applied to females or even embryos to predict their own later performance. The initial focus on additive merit has expanded to include mating programs, genomic inbreeding, and recessive alleles. Many producers now use DNA testing to decide which heifers should be inseminated with elite dairy, beef, or sex-sorted semen; be embryo donors or recipients; or be sold or kept for breeding. Because some of these decisions are expensive to delay, predictions are now provided weekly instead of every few months. International genomic databases and predictions are often more accurate than within-country genetic evaluations previously designed for progeny testing. Obtaining predictions from an international database is usually more effective than computing predictions from only local data unless local breeds, conditions, or traits differ greatly from the large database. Selection indexes include many new traits, often with lower heritability or requiring large initial investments to obtain phenotypes, which provide further incentive to cooperate internationally. The genomic prediction methods developed for dairy cattle are now applied widely to many other animal, human, and plant populations.

Role of Interbull in the successful implementation of genomic selection in breeds other than Holstein

E. Santus
Anarb, Loc. Ferlina, 204, 37012 Bussolengo, Italy; enrico.santus@anarb.it

The use of DNA information in dairy cattle is not a complete novelty as specific markers for interesting genes have been used in selection schemes for specific needs since the late 1980s. As we all know, the breakthrough occurred at the turn of the millennium. First with the envision and later with the practical application of the massive analysis of thousands of markers on cattle genome. Since then, the approach has changed completely passing from the Mendelian identification of desirable alleles to the studies on the associations between markers and phenotypes. At that time, numbers started to matter and the risk of facing the 'new opportunities, old inequalities' paradox became a real scenario as the new mechanisms of exploitation of the new technology called for a minimum size population and set a limit for entry into this new market. The Brown Swiss breed is a small breed compared with the Holstein breed, but all breeds are small if you use the Holstein breed as the term of reference. Thus, it is of some relevance to use the Brown Swiss example as a possible model. The greatest advantage of this breed in Europe was the existence of a very active international community. Another advantage was the fact that no single country had alone the size of the population needed for a reliable implementation of genomic selection. Political aspects were not, as always, of secondary importance when the discussion about sharing valuable information started. Since the very beginning, Interbull – recognised by all as an independent, reliable and competent third party – was identified as a natural plat- form that could solve most of the political issues. The development of an agreement among all parties, called 'Intergenomics', was the first necessary step, which was followed by specific agreements between each party and Interbull. Thanks to that approach, genomic selection has become a reality for the Brown Swiss breed and has revolutionised the way the genetic management of the population has been carried out since then. After only a few years, in 2014, Intergenomics evolved into Intergenomics 2.0 and now it has become a routine service provided by Interbull to the member countries that joined the agreement.

Exploiting homozygosity in the era of genomics: selection, inbreeding and genomic mating programs

C. Maltecca[1], F. Tiezzi[1], J.B. Cole[2] and C. Baes[3]
[1]North Carolina State University, Animal Science, Raleigh, NC 27695, USA, [2]Agricultural Research Service, USDA, Animal Genomics and Improvement Laboratory, Beltsville, MD 20705, USA, [3]University of Guelph, Centre for Genomic Improvement of Livestock, Department of Animal Biosciences, Guelph, Ontario, Canada; cmaltec@ncsu.edu

The advent of genomic selection paved the way for an unprecedented acceleration in genetic progress. The increased ability to select superior individuals has been coupled with a drastic reduction in the generation interval for most dairy populations. This represents both an opportunity and a challenge. Homozygosity is now accumulating at a faster rate in dairy populations. Currently, inbreeding depression is managed mostly through culling at the farm level and by controlling the overall accumulation of homozygosity at the population level. A better understanding of how homozygosity and genomic load are related will guarantee continued genetic improvement while curtailing the accumulation of harmful recessives and maintain enough genetic variability to ensure the possibility of selection in the face of changed environmental conditions. Here we present a snapshot of the current dairy selection structure as it relates to response to selection and accumulation of homozygosity. We briefly outline the main approaches currently used to manage inbreeding and overall variability and finally, we present a few possible approaches that could be used in the short term to control the accumulation of harmful recessives while maintaining a sustained selection pressure.

Single-step genomic evaluations in dairy cattle

E.A. Mäntysaari, M. Koivula and I. Strandén
Natural Resources Institute Finland (Luke), Myllytie 1, 31600 Jokioinen, Finland; ismo.stranden@luke.fi

Single-step approach enables genetic evaluations in populations where some animals, but not all, have genotypes available. The single-step models are either based on genomic relationships or direct modelling of the marker effects. Large part of single step research has concentrated on reducing computing costs, where the key methodologies are alternative presentations of the inverse genomic relationship matrix (APY-G and GTBLUP), and the algorithms combining the pedigree relationships and the marker effects (hybrid models). In addition to computing approaches, we review single-step validation results and discuss possible reasons for biases in single step evaluations.

What's next for dairy cattle breeding?

G. Gorjanc
The Roslin Institute, University of Edinburgh, Easter Bush, Midlothian, EH259RG, Edinburgh, United Kingdom; gregor.gorjanc@roslin.ed.ac.uk

I will highlight some recent trends that will impact future dairy cattle breeding. Genomic selection improves the rate of genetic improvement via shortened generation interval and decoupling of selection from phenotyping. However, further improvements are possible via the other three factors of the breeder's equation. Accuracy will improve with ever larger training populations. Genotyped animals are now counted in millions and soon all breeding, if not all production, animals may be genotyped and maybe also phenotyped. Genome re-sequencing is becoming cheaper and many AI bulls are sequenced routinely. New genome assemblies, strategies and imputation will deliver sequence for millions of animals. While SNP arrays are sufficient for genomic selection within breeds, sequence will enable multi-breed analysis. While this huge data is challenging in dimension and structure, techniques such as graph genomes and succinct trees show a way forward. Intensity of selection will increase via the same avenues as the accuracy. This will impact developing countries, albeit their structural issues of low connectedness, large environmental variability and small herds still exist. I will show our attempts to use genomic and spatial modelling to overcome these issues. Genetic diversity is an issue in some breeding programmes. I will show how genomic data allows us to quantify the change of genetic variance over time, which chromosomes and regions harbour variation for traits and pairs of traits. New scalable methods are needed to harness this information in breeding. There is a host of other developments. Agriculture is increasingly data-driven with the increasing volume and depth of phenotypic data. Combination of this phenome data with genomic, but also gene expression, proteome, metabolome and microbiome data will provide a rich resource for machine learning such as (organismal) deep-neural networks. There is a shift in dairy breeding from semi-open to closed programmes. This opens role for evaluating and exploiting non-additive genetic variation and sustained profits for programmes. Distributed Ledger Technologies could maintain semi-open programmes for the benefit of all. Lastly, new reproductive tech, genome editing, and drives promise even brighter future.

Session 03

Theatre 1

Genetic variability in four genetic lines of guinea pigs in Peru

J.I. Cedano-Castro[1], R. Jiménez[2], A. Huaman[2], B. Fuerst-Waltl[3], M. Wurzinger[3] and G. Gutiérrez[1]
[1]*National Agrarian University La Molina, Av. La Molina S/N, 15024 Lima, Peru,* [2]*National University of San Marcos, Instituto Veterinario de Investigaciones Tropicales y de Altura, Carretera Central, 12150 Junin, Peru,* [3]*University of Natural Resources and Life Sciences, Sustainable Agricultural Systems, Division of Livestock Sciences, Gregor-Mendel-Straße 33, 1180 Vienna, Austria; joiscecas@gmail.com*

This research was carried out in order to assess the genetic variability within the population of four genetic lines of guinea pigs, two paternal (P1 and P2) and two maternal lines (M1 and M2). The pedigree dataset for P1, P2, M1 and M2 lines consisted of 9,047, 6,228, 5,661 and 5,650 animals, respectively. Pedigree data were available for the years 2011 to 2018 for paternal and 2009 to 2018 for maternal lines. Each genetic line was managed as a closed population and under selection. For all four populations the number of founders (f), effective number of founders (f_e), effective number of ancestors (f_a), average inbreeding coefficient considering all generations (F), average complete generation equivalent, increase of inbreeding (ΔF) and effective population size (N_e) were estimated using the software ENDOG v4.8. The f was 274, 973, 378, 426 in the P1, P2, M1 and M2 lines, respectively. The f_e was 74 and 343 in P1 and P2 line, 58 and 58 in M1 and M2 line. The f_a was 72 and 271 in P1 and P2 line, 52 and 49 in M1 and M2 line. The F was 4.30, 0.81, 4.36 and 4.22% in P1, P2, M1 and M2 line, respectively. The average of complete generation equivalent was 5.83 and 2.71 in P1 and P2 line, 4.51 and 4.42 in M1 and M2 line. The ΔF and N_e was 0.71%, 70.16 in P1 line, 0.66%, 75.42 in P2 line, 1.16%, 42.93 in M1 line and 1.44%, 34.74 in M2 line, respectively. The lower values of number of founders, f_e, f_a and N_e and the higher values of the F, ΔF for P1, M1 and M2 lines indicate the loss of genetic diversity due to the closed population management and selection. The lower average complete generation equivalent and F for P2 line is due to incomplete pedigree information. Thus, it is not possible to get accurate results for the parameters mentioned above for this line. However, the similar management of all four lines suggests also the same trend of loss of genetic diversity for P2 line as shown for P1, M1 and M2 lines.

Session 03

Theatre 2

Effect of genomic selection on inbreeding and effective population size of Holsteins

B. Makanjuola[1], F. Miglior[1], E.A. Abdalla[1], C. Maltecca[2], F.S. Schenkel[1] and C.F. Baes[1,3]
[1]*University of Guelph, Animal Biosciences, 50 Stone Road East, N1G 2W1, Canada,* [2]*North Carolina State University, Animal Science, Campus Box 7621, 27695-7621, USA,* [3]*University of Bern, Institute of Genetics Vetsuisse Faculty, Bremgartenstrasse 109a, 3012, Switzerland; cbaes@uoguelph.ca*

Genetic diversity in livestock populations contributes to selection. In this study, we investigated the rate of inbreeding (ΔF), rate of coancesry (Δf) and effective population size (N_e) and evaluated the effect of genomic selection on genetic diversity in North American Holstein and Jersey dairy cattle. Inbreeding and coancestry coefficients were calculated using the traditional pedigree method and genomic methods estimated from segment and marker based approaches. Effective population size was estimated from the rate of inbreeding and coancestry. A total of 205,755 and 89,238 pedigreed and genotyped animals born between 1990 and 2018 were available for Holsteins and Jerseys, respectively. Estimated average pedigree inbreeding coefficients were 7.74 and 7.20% for Holsteins and Jerseys, respectively, with values for segment and marker-by-marker genomic inbreeding coefficients were 13.61, 15.64 and 31.40% for Holsteins and 21.16, 22.54 and 42.62% for Jerseys, respectively. Generation interval averaged approximately 5 years for all selection pathways combined. The ΔF per generation based on pedigree, segment and marker-by-marker genomic measures was estimated to be 0.75, 1.10, 1.16 and 1.02% for Holstein animals and 0.67, 0.62, 0.63 and 0.59% for Jersey animals, respectively; the Δf was estimated to be 0.98 and 0.98% for Holsteins and 0.73 and 0.78% for Jerseys with pedigree and genomic measures, respectively. These ΔF and Δf estimated resulted in an N_e that ranged from 46 to 66 animals for Holsteins and 75 to 85 animals for Jerseys. The effective population size based on linkage disequilibrium was 58 and 120 for Holsteins and Jerseys, respectively. The 10-year period that involved the application of genomic selection, resulted in an increased ΔF per generation with ranges from 1.19 to 2.06% for pedigree and genomic measures. Given the rate at which inbreeding increases after implementation of genomic selection, there is a need to implement measures and means for controlling the rate of inbreeding.

The genomic inbreeding landscape of Italian heavy pig breeds over the last decades

G. Schiavo[1], S. Bovo[1], A. Ribani[1], S. Tinarelli[1,2], V.J. Utzeri[1], M. Cappelloni[2], M. Gallo[2] and L. Fontanesi[1]
[1]University of Bologna, Department of Agricultural and Food Sciences, Viale G. Fanin 46, 40127 Bologna, Italy,
[2]Associazione Nazionale Allevatori Suini (ANAS), Via Nizza 53, 00198 Roma, Italy; giuseppina.schiavo2@unibo.it

The level of inbreeding in a population should be limited to avoid the inbreeding depression and maintain variability needed to assure long term genetic progress. Inbreeding coefficient (F_{PED}) is usually calculated using pedigree information. Single nucleotide polymorphism (SNP) genotyping tools give the possibility to calculate genomic inbreeding parameters. The proportion of the autosomal genome covered by runs of homozygosity (ROH) can be used to estimate the level of genomic inbreeding (F_{ROH}). The breeding program of Italian pig breeds started about 25 years ago. In this work, we retrospectively analysed F_{ROH} over this period in Italian Large White (ILW) and Italian Duroc (ID) heavy pig breeds and compared this parameter with F_{PED}. A total of 1,420 ILW and 572 ID pigs born over the last 25 years have been genotyped with the Illumina PorcineSNP60 BeadChip. ROH were identified with PLINK version 1.9, considering a minimum of 15 SNPs, and a minimum length of 1 Mbp. Then F_{ROH} was computed for all animals and averaged over all animals born by year. Averaged F_{ROH} over all considered years was higher in ID than in ILW and was constant over all analysed time window. F_{ROH} and F_{PED} were parallel for both breeds over the same period. The results indicated that both F_{ROH} and F_{PED} can be used to manage inbreeding levels in Italian heavy pig breeds.

Whole genome re-sequencing reveals adaptation prior to the divergence of buffalo subspecies

M. Rafiepour[1,2,3], E. Ebrahimie[3], M.F. Vahidi[2], G. Hosseini Salekdeh[2], A. Niazi[3], M. Dadpasand[3], D. Liang[1], J. Si[1], X. Ding[1], J. Jianlin Han[4], Y. Zhang[1] and S. Qanbari[2,5]
[1]China Agricultural University, College of Animal Science and Technology, 100193 Beijing, China, P.R., [2]Agricultural Biotechnology Research Institute of Iran (ABRII), Seed & Plant Improvement Campus, 31535-1897 Karaj, Iran, [3]Shiraz University, Ghasro Dasht St, 71946 Shiraz, Iran, [4]Chinese Academy of Agricultural Sciences (CAAS), Institute of Animal Science, 100193 Beijing, China, P.R., [5]Leibniz Institute for Farm Animal Biology (FBN), Wilhelm-Stahl-Allee 2, 18196 Dummerstorf, Germany; qanbari@fbn-dummerstorf.de

The application of high throughput genotyping or sequencing data helps us to understand the genomic response to natural and artificial selection. In this study, we scanned the genome of five indigenous buffalo breeds, adapted to different geographical and agro-ecological zones in Iran, to unravel the extent of genomic diversity and to localise genomic regions and genes underwent past selection. A total of 46 river buffalo whole genomes, from West and East Azarbaijan, Gilan, Mazandaran and Khuzestan provinces, were re-sequenced. We identified 20.55 million SNPs, including >63,097 missense, 707 stop-gain and 159 stop-loss mutations that might have functional consequences. Genomic diversity analyses showed modest structuring among Iranian buffalo populations following frequent gene flow or admixture among these populations in the recent past. Evidence of positive selection was investigated using both differentiation (F_{st}) and fixation (Pi) metrics. Analysis of fixation revealed three genomic regions with aberrant polymorphism contents on BBU2, 20 and 21. Fixation signal on BBU2 overlapped with the *OCA2-HERC2* genes, suggestive of adaptation to UV exposure through pigmentation mechanism. Further validation using re-sequencing data from other bovine species as well as the Axiom® Buffalo Genotyping Array 90K data of river and swamp buffaloes indicated that these fixation signals persisted across river and swamp buffaloes and extended to cattle, implying an ancient evolutionary event occurred before the speciation of buffalo and cattle. These results contributed to our understanding of major genetic switches that took place during the evolution of modern buffaloes.

Inferring ancestral alleles of cattle from outgroup species

M.M. Naji[1], Y.T. Utsunomiya[2], B.D. Rosen[3], J. Sölkner[1] and G. Mészáros[1]
[1]University of Natural Resources and Life Sciences(BOKU), Vienna, Sustainable Agricultural Systems, Gregor Mendel Strasse 33, 1180 Vienna, Austria, [2]Sao Paulo State University(UNESP), School of Veterinary Medicine, Aracatuba-Jaboticabal, Sao Paulo, Brazil, [3]United States Department of Agriculture, Animal Genomics and Improvement Laboratory, Baltimore Ave, Beltsville, MD 10300, USA; agis.maulana12@gmail.com

In evolutionary theory, divergence and speciation can arise from long periods of reproductive isolation, genetic mutation, selection and environmental adaptation. After divergence, alleles can either persist in their initial state (ancestral allele; AA) or co-exist or be replaced by a mutated state (derived alleles; DA). Defining AA and DA at polymorphic sites is useful to test hypotheses regarding molecular evolutionary processes, including estimation of allelic age, formation of linkage disequilibrium patterns and selection signatures. In this study, we mapped whole genome sequences of Bovinae subfamily (gaur, yak, bison, wisent, gayal, and cattle) to the cattle reference, ARS.UCD-1.2, using BWA-Mem. We then used the GATK best practices pipeline to produce variants in vcf format. The indep algorithm in PLINK was used with default parameters to select a set of independent variants for principal components analysis. Phylogenetic trees were built by considering genetic distance information between individuals of different species for each chromosome, we then defined yak as the closest relative of cattle. We defined AA as fixed alleles in biallelic sites from yak population. We defined AA in ~6.3 million variants based on ~2.7 Gb cattle genome. Using non-overlapping scanning windows of 10 Kb, we counted the AA observed within a sample of eleven Holstein cattle. We observed several peaks across the genome. We focused on the top 0.001% with the highest concentration of AA and found 14 regions with functional genes. These include the LOC789121 gene the protein of which acts as motor for organelles motility in microtubules, LOC100337390, associated with multidrug resistance protein, LOC513969, encoding protein for diverse protein-protein interactions, LOC100336984 encoding for zinc finger protein. CD1A, LOC100297263 and LOC783151 genes encoding for protein in immunology domain. OR6C3, LOC107131130, and LOC618675 genes are responsible for olfactory receptors. Our study suggests that despite the long period of divergence, a portion of genomic regions were kept in the ancestral form, harbouring genes related to sensory and fitness traits.

Copy number variation analysis of the East Adriatic sheep breeds

M. Ferenčaković[1], M. Spehar[2], V. Brajkovic[1], I. Drzaic[1], I. Curik[1] and V. Cubric-Curik[1]
[1]Faculty of Agriculture, University of Zagreb, Department of Animal Science, Svetošimunska cesta 25, 10000 Zagreb, Croatia, [2]Ministry of Agriculture, Vukovarska 78, 10000 Zagreb, Croatia; mferencakovic@agr.hr

Copy number variations (CNV) are structural variations in the genome of an individual in the form of losses or gains of DNA fragments greater than 1 kb in size, or in recent times, greater than 50 bp. CNVs are a significant source of genetic and phenotypic variation. More overlapping CNVs detected in two different samples are copy number variation regions (CNVRs). The existence of CNV in the protein coding region alters the protein function, whereas in the regulatory region alters the gene expression level. So far, studies of diversity of CNV distribution in worldwide sheep populations revealed differentiation in CNV between geographical area as well as between diverse groups. Gene ontology (GO) analysis of CNVRs genes gave association with genes related to environmental response and biological functions. Along the East Adriatic, we have analysed 200 individuals belonging to eight sheep breeds (Istria sheep, Krk Island sheep, Cres Island sheep, Rab Island sheep, Lika sheep, Pag Island sheep, Dalmatian Pramenka sheep, Dubrovnik Ruda sheep, 25 individuals from each breed). Animals were genotyped with Ovine Infinium® HD SNP BeadChip (606,006 SNPs). Quality control of SNP data was done using Golden Helix SVS v8.8.3 software with the following parameters: call rate higher than 0.9, a departure from Hardy-Weinberg equilibrium at the 0.001 level, missing genotype rate less than 0.05 and missing data rate less than 0.1. Only markers with GC scores higher than 0.7 were used. Wave detection and correction on autosomes were based on the Oar_v4.0 with a minimum training marker distance of 1 Mb and a recommended wave factor threshold of 0.05. Detected CNVs and CNVRs were used to explore lineage-specific CNVs as well as for GO enrichment analysis. To our knowledge, CNV analysis of any of breeds from Balkan was never performed and our study is a first attempt to perform a comprehensive study of the CNV population genetic properties of the East Adriatic sheep breeds. Our findings contribute to published sheep CNVs and will be helpful for future studies of the CNVRs associated with traits, history, migrations and genetic diversity in sheep.

Genomic analysis of indigenous Croatian sheep breeds based on a high-density SNP chip

I. Drzaic[1], V. Brajkovic[1], M. Gianni[1], B. Lukić[2], V. Orehovacki[3], M. Cacic[3], M. Ferenčaković[1], I. Curik[1] and V. Cubric-Curik[1]
[1]Faculty of Agriculture, University of Zagreb, Department of Animal Science, Svetosimunska 25, 10000, Croatia, [2] Faculty of Agrobiotechnical Sciences Osijek, J.J, Department for Animal Production and Biotechnology, Vladimira Preloga 1, 31000, Croatia, [3]Ministry of Agriculture, Directorate of Livestock and Food Quality, Ulica grada Vukovara 78 (Ilica 101), 10000, Croatia; vcubric@agr.hr

In an extensive genomic analysis, performed recently, based on 50K SNP profiles it has been shown that the regional Balkan sheep populations have a considerable genetic overlap, but are clearly different from surrounding breeds. All eight Croatian sheep breeds were represented by a small number of individuals per breed. Here, we have genotyped 200 individuals representing indigenous Croatian breeds (Istria sheep, Krk Island sheep, Cres Island sheep, Rab Island sheep, Lika sheep, Pag Island sheep, Dalmatian Pramenka, Dubrovnik Ruda sheep, 25 individuals per breed) by Ovine Infinium® HD SNP BeadChip (606,006 SNPs). Our analyses revelled fine population structure of Croatian sheep breeds and their dependence on the geographical origin and barriers (island versus mainland). Estimated effective population size varied from small to relatively high, thus pointing to the need for improved conservation management toward preservation of genetic diversity in some breeds. We used graphical approach to illustrate relationship of analysed breeds with respect to their effective population size, inbreeding and population structure. In addition, we have derived migration pattern among all eight breeds with respect to their geographical positions. Our results will contribute to the breeding and conservation management strategies of indigenous Croatian sheep breeds.

Genetic diversity study in the Indian Attappady Black and Malabari goats using SNP data

T. Marykutty[1], W. Gorssen[2], G. Radhika[1], C.B. Bimal[2], R.T. Venkatachalapathy[1], T.V. Aravindakshan[1], R. Meyermans[2], N. Buys[2] and S. Janssens[2]
[1]KVASU, Department of ABG, CoVAS, Thrissur, Kerala, India, 680651, India, [2]KU Leuven, Livestock Genetics, Department of Biosystems, Kasteelpark Arenberg 30 – box 2472, 3001 Leuven, Belgium; bimal.chakkingalbhaskaran@kuleuven.be

Local goat breeds form a strong basis for livelihood and nutritional security in developing regions. In these regions goat rearing is preferred for its high production with minimal input. Attappady Black (A) and Malabari (M), two of the native goat breeds from the South Indian State of Kerala are well adapted to humid tropical stressors. Moreover, they are vital for the small holder goat production system in this socio-economic setting. However, a little is known about the genetic diversity status of these breeds and their relationship with other goat breeds worldwide. We conducted a genetic diversity analysis in these breeds (n=24 each), using the Caprine SNP50 BeadChip. Although both breeds were not represented in the chip design, ascertainment bias was found to be low. A high proportion of polymorphic loci ranging from 0.87 (A) to 0.89 (M) was found, implying that majority of these SNP loci were segregating in the native goats. Mean heterozygosity (H_O and H_E) was 0.38±0.03 and 0.39±0.04, respectively. The mean genomic inbreeding coefficient (F_{IS}), computed based on excess of heterozygosity, was 0.032±0.081. The F_{IS} of A was comparatively lower (0.030±0.038) than that of M (0.035±0.011). The mean Fixation Index, F_{ST}, among A and M was 0.018±0.057, denoting a low level of genetic differentiation across these two populations. The average genomic inbreeding based on runs of homozygosity (F_{ROH}) indicates M (0.037±0.063) had higher inbreeding compared to A (0.019±0.034). Malabari showed higher recent inbreeding based on $F_{ROH>16Mb}$. Principal component analysis (PCA) showed distinct clustering of the two breeds, however clustering was tighter in case of A. Further, a PCA including genotype data of 133 goat breeds from the International Goat Genome Consortium showed that these Indian goats clearly clustered together with goats from the Middle East and Pakistan. Interestingly, some Malabari goats appear more related to commercial breeds, indicating introgression during the development of the breed.

Runs of homozygosity and inbreeding in Swiss Large White pigs

H. Signer-Hasler[1], A. Hofer[2] and C. Flury[1]
[1]Bern University of Applied Sciences, School of Agricultural, Forest and Food Sciences HAFL, Länggasse 85, 3052 Zollikofen, Switzerland, [2]SUISAG, Allmend 8, 6204 Sempach, Switzerland; heidi.signer@bfh.ch

In 2002, the Swiss Large White pig population was split into the Large White maternal line (LWM) and the Large White paternal line (LWP). The breeding objectives of the two populations differ significantly. In LWM the focus is on reproductive characteristics and in LWP the major interest is on fattening performance, carcass composition and meat quality. Since their separation the populations developed in different directions. This differentiation could be confirmed based on imputed SNP genotypes of 5,400 LWM and 4,932 LWP animals born between 1999 and 2019. Today the genotypes of the two populations cluster into clearly separated groups. Runs of homozygosity (ROH) were derived using software PLINK 1.9 based on 57,627 SNP. Except of – homozyg-density 120 and – homozyg-snp 60 the default parameters were used. As genotypes were imputed, call rate filtering was not necessary. Filtering for minor allele frequencies was not applied according to suggestions from other studies. The increasing trend in genomic inbreeding coefficient (FROH) from 1999 to 2019 was almost parallel to the pedigree-based inbreeding coefficient (FPED). The average FPED for animals born in 2019 was 6.3% (for 181 LWM animals) and 8.9% (for 1,109 LWP animals), whereas the average FROH was 22.6% (LWM) and 25.4% (LWP). Around 80% of LWM animals had an ROH on SSC6 at ~109 Mb and on SSC13 at ~152 Mb. About 80% of LWP animals had an ROH on SSC1 at ~207 Mb and on SSC13 at ~148 Mb.

Genetic characterisation by genealogical analysis of the Marinhoa breed of cattle

A. Vitorino, E. Guicho, M. Silveira and N. Carolino
INIAV, Fonte Boa, Vale Santarém, Portugal; andreia.vitorino93@gmail.com

The bovine breed Marinhoa has its natural range in the Baixo Vouga region. It is used mainly in the Aveiro district, although it is spread all over the country. There are currently less than 1,100 females and 63 males in the Herdbook, in around 179 farms, from which 146 keep only pure-bred females. The objective of this study is to evaluate the demography and genetic variability existent in the breed and its evolution through generations. All the information in the Marinhoa Herdbook database was analysed using diverse informatics applications, having the following demographic parameters being studied: Evolution of the records, Generations interval, Level of fulfilment of the genealogies, Number of generations known, Individual inbreeding (ΔF), Degree of relatedness, ΔF/year and ΔF/generation, Effective size of the population, Genetic conservation index, Average relatedness, Effective number of founders and ascendants. Data included 13,446 males and 17,880 females, resulting in a population in study of 31,326 individuals. There has been a significant decrease in the number of births since 1995, where there occurred more than 1,300. In 2018 there were born 653 animals able to be signed up in the Herdbook. Animals recently born have, in average, almost 5 generations known. The average generation interval is 6.52 years. The distribution of the number of females according to the number of descendents indicates that the majority of the cows (51%) have less than 4 calves throughout their life and that only about 10% of the females have more than 10 calves. Approximately 72% of the bulls are, in total, parents of less than 14% of the total stock of animals. Some farms show an average inbreeding coefficient above 10% in the last years (2015-2019). More than 85% of the current births are inbred. The average relatedness is around 2.4%. For the 2010-2019 period, the annual increase of inbreeding per year was 0.259% and per generation 1.749%, leading to an effective size of the population of 28.59 (way below the FAO recommended value of 50). There is a positive evolution in the monitoring of the Marinhoa breed, mainly in the management of the information from the genealogic records. The annual increase in inbreeding, especially in the last years, deserves some attention.

The breed-wise structure and genomic inbreeding in Estonian dairy cattle

S. Värv, E. Sild, K. Rooni, T. Kaart and H. Viinalass
Estonian University of Life Sciences, Fr. R. Kreutzwaldi 1, 51006, Estonia; sirje.varv@emu.ee

The aim of the study was to assess the genetic structure of two dairy breeds, the Estonian Red and the Estonian Holstein, and to estimate genomic inbreeding using SNP data. Analyses of two datasets were performed. Samples of 96 cows related to each other as distantly as possible from 60 herds were selected from the Estonian Red (n=42) and the Estonian Holstein (n=54) breed. These cows were genotyped using the Bovine50K v2 SNP chip (Illumina) and were grouped according to their paternal origin (EE, CA, DE, DK, SE, NL, US) in order to examine the expectedly heterogeneous population structure due to crossbreeding. In addition, one single herd of the Estonian Holstein (n=42) was genotyped using BovineHD SNP chip to quantify the genomic inbreeding both, on individual and on herd level. The principal component analysis showed that despite the cross-usage of sires and holsteinisation of the red breed still the strongest pattern (PC1 6.7%) distinguishes the two breeds. While among the Estonian Holstein cows no clear structure was revealed, the Estonian Red cows formed two groups based on paternal origin: an EE-DK cluster and another, the DE-NL cluster, closer to the Holstein area. The inbreeding calculations based on runs of homozygosity (ROH) revealed differences between the two breeds. Of the determined ROHs exceeding 2 MB in length, there was almost three-fold difference in the average length of ROH segments per animal (9.2 MB in Estonian Holstein and 3.4 MB in Estonian Red). The average number of ROHs per animal was in the Estonian Holstein twice as much as in the Estonian Red – 2.7 and 1.3, respectively. The analysis based on the high-density genotype data of a Holstein herd showed relatively high levels of inbreeding estimates. 21% of the individuals showed genomic inbreeding (F_{ROH}) exceeding 0.10 and being on average 0.078 (SD 0.02). The majority of ROHs amongst the five length categories (2-4 MB, 4-8 MB, 8-16 MB, 16-32 MB and >32 MB) were found in the shortest category (48%). The least ROHs were identified in the longest category (0.8%). This work was supported by Estonian Research Council (PRG554).

Trends of inbreeding in a population of Polish HF young bulls before and after genomic selection

P. Topolski and W. Jagusiak
National Research Institute of Animal Production, Balice, Poland, Genetic and Breeding of Cattle, ul. Krakowska 1, 32-083 Balice k. Krakowa, Poland; piotr.topolski@izoo.krakow.pl

Inbreeding was analysed in a population of 14,144 Polish Black-and-White Holstein-Friesian (PBWHF) young bulls born between 1994 and 2017 and bred under both conventional and genomic breeding programmes. The inbreeding coefficients were computed using a model with genetic groups, according to the algorithm given by VanRaden. It was found that in the analysed population all bulls are inbred (100% of the population), with the mean coefficient of inbreeding ranging from 0.09 to 26.95%. Pedigree analysis also showed a relationship between the changing number of bulls over the years and the dynamics of population inbreeding. These trends are connected with changes in the breeding scheme, related to the implementation of genomic selection in the breeding programme for PBWHF cattle in 2014. The increasing number of weaned young bulls in Poland was paralleled by a fairly consistent increase in the mean inbreeding, but the inbreeding dynamics were relatively small. A reverse trend was observed in the group of young bulls born after 2013. As the number of bulls very rapidly decreased in successive birth years, the mean inbreeding for successive birth-year groups very rapidly increased. As a result, the estimated linear trend was equal to 0.02% inbreeding per year of birth in the group of bulls raised before genomic selection (~20 birth-year) whereas in the group bulls raised after genomic selection (~4 birth-year) the trend was much higher and amounted to 0.56% inbreeding per year of birth. The high mean inbreeding found in the group of the genomically selected young bulls may translate into higher inbreeding in the whole population of PBWHF cattle, because these bulls are now intensively used as sires. The results of our study also show that the implementation of genomic selection in the breeding programme caused a very rapid increase in the inbreeding rate per birth-year in young bulls.

The CLARITY project: a combined physical-genetic map for a dairy cattle breed in Germany

D. Wittenburg and S. Qanbari

Leibniz Institute for Farm Animal Biology, Institute of Genetics and Biometry, Wilhelm-Stahl-Allee 2, 18196 Dummerstorf, Germany; wittenburg@fbn-dummerstorf.de

Genetic maps are built based on the relative positioning of genes or genetic markers and are species-specific. The distance between loci/markers can be measured on two scales: physical and genetic. The latter is of more interest to breeders as it is related to the probability that variants at two markers jointly transmit from parent to offspring. This probability is determined by the so-called recombination rate. The CLARITY project aims at constructing a combined physical-genetic map for a main German dairy breed by bringing together the physical and genetic measures of the proximity of molecular markers. The software toolkit developed will take into account the family structure for estimating recombination rates and other population genetic parameters. The project furthermore will facilitate data integration from different sources/breeds in order to compare genome features. This workflow can later be adapted to other species where molecular markers are investigated at a large scale, e.g. honey bee or fish, among others. An open access data base will be developed that comprises all results and will be capable of integrating end-users' expectations, e.g. what parameters are relevant and in which way should they be available best. Eventually, this data base will be open for extensions to other cattle breeds and allows comparison with other species.

Assessment of the risk status of Polish local breeds under conservation programmes

G.M. Polak, J. Krupinski, E. Martyniuk, J. Calik, A. Kawęcka, J. Krawczyk, A. Majewska, J. Sikora, E. Sosin-Bzducha, M. Szyndler-Nędza and I. Tomczyk-Wrona

National Research Institute of Animal Production, Department of Horse Breeding, ul. Krakowska 1, 31-047 Balice, Poland; grazyna.polak@izoo.krakow.pl

The importance of conservation of farm animal genetic resources has been accepted widely, under the Convention on Biological Diversity (CBD). Farm animal genetic resources are the primary biological capital for food security and sustainable rural development. Their conservation and sustainable use are essential to conserve biodiversity and to keep rural ecosystems in balance. A risk status assessment methods are the key components of early warning systems in every country. The aim of the projects was to develop the model for risk status assessment in Poland, using available information on 64 endangered breeds of cattle, horse, gouts, sheep, swine and poultry. A proposed model based on two main factors: number of females (L) and effective population size (N_e), and an additional factor (D) composed of 6 elements (sub-factors): 1. geographical concentration in a country; 2. demographic trend; 3. cultural value; 4. *ex situ* conservation; 5. anthropogenic factors (breeders organisations, financial support, activity and age of breeders); 6. epidemiological risk. The results showed that the current state of the conserved populations is worrying: 10.8% of analysed populations were designated critical status, 81.5% endangered, and 7.7% are in need of monitoring.

Serra-da-Estrela sheep breed: a Portuguese genetic resource

J. Oliveira[1,2], N. Carolino[3,4,5], M. Silveira[6], M. Miranda[7] and R. Dinis[7]
[1]*Centre for the Research and Technology of Agro-Environmental and Biological Sciences (CITAB), Universidade de Trás-os-Montes e Alto Douro, Vila Real, 5000-801 Vila Real, Portugal,* [2]*Escola Superior Agrária de Viseu, Quinta da Alagoa, 3500-606, Portugal,* [3]*Centre for Interdisciplinary Research in Animal Health, Faculdade de Medicina Veterinária (CIISA), Faculdade de Medicina Veterinária, Avenida da Universidade Técnica, 1300-477 Lisboa, Portugal,* [4]*Escola Universitária Vasco da Gama, Estrada da Conraria, 3040-714 Coimbra, Portugal,* [5]*Instituto Nacional de Investigação Agrária e Veterinária, I.P., Fonte Boa, 2005-048 Vale de Santarém, Portugal,* [6]*Ruralbit, Lda, Av. Dr. Domingos Gonçalves Sá, 132, Ent 1, 5º Esq, 4435-213 Rio Tinto, Portugal,* [7]*Associação Nacional de Criadores de Ovinos Serra da Estrela (ANCOSE), Quinta da Tapada, 3405-006 Bobadela OHP, Portugal; joliveira@esav.ipv.pt*

The Serra-da-Estrela sheep breed is native from the central region of Portugal, explored in extensive system, grazing on pastures around the Serra-da-Estrela mountain. This breed is considered the main Portuguese dairy sheep breed, being the second autochthonous sheep breed more explored in this country. Currently, the main productive aptitude is milk production (for PDO Serra-da-Estrela Cheese and PDO Serra-da-Estrela curd cheese) but it's also explored for meat (PDO Serra-da-Estrela lamb). This breed has two varieties: white and black. The main goal is to characterise this animal genetic resource through demographic analysis, providing useful information about the breed current status and there perspectives for genetic improvement. The database from the Herdbook provided by ANCOSE (National Association of the Serra-da-Estrela Sheep Breeders), with a total of 709,855 animals was analysed in order to characterise the sheep population registered between 1990 and 2019. In recent years, an effective reduction of this breed has been detected to the detriment of exotic breeds. Although the number of animals is decreasing, following the decrease in the number of farms till 200, in the last decade the number of active females remain stable in about 15,000, promoting a slight increase in the sheep/farm. Paradoxically, this can provide a better logistic way to promote this breed and consolidate the strategy to its genetic breeding.

What are the values and beliefs that drive cattle breeding in south-western Burkina Faso?

B. Zoma-Traoré[1,2], S. Ouédraogo-Koné[1], A. Soudré[3], D. Ouédraogo[1,2], B. Yougbaré[2,4], A. Traoré[4], A.M. Okeyo[5], P.A. Burger[6], G. Mészáros[2], J. Sölkner[2], M. Wurzinger[2] and L. Probst[2]
[1]*Université Nazi Boni, Bobo-Dioulasso, Burkina Faso,* [2]*University of Natural Resources and Life Sciences, Gregor-Mendel-Straße 33, 1180, Austria,* [3]*Université Norbert Zongo, Koudougou, Burkina Faso,* [4]*Institut de l'Environnement et de Recherches Agricoles, Ouagadougou, Burkina Faso,* [5]*International Livestock Research Institute, Nairobi, Kenya,* [6]*University of Veterinary Medicine, Savoyenstraße 1, 1160, Austria; bienvenue.zoma@students.boku.ac.at*

Lobi taurine cattle, an endangered breed of south-western Burkina Faso, are known to be tolerant to trypanosomosis. Community-based breeding programs (CBBPs) currently running in the area could be a viable option to preserve the breed while improving productivity. Research has shown that an adaptation of production strategies is more likely to succeed when resonating with producers' beliefs and values. Accordingly, we investigated the beliefs and values of 32 purposively sampled cattle keepers in south-western Burkina Faso using in-depth interviews. The data were analysed following established conceptual frameworks (values-beliefs-norms theory and rural livelihood transitions) in a combination of deductive and inductive coding strategies. The respondents characterised their livelihoods as being under pressure by the scarcity of resources, the dependence on a volatile climate, and cattle diseases. These pressures are only partly mitigated by a weak institutional framework and the market for cattle products. Producers react by diversifying their livelihood strategies, but cattle and crop production remain central to food security. Cattle have a critical role as savings and insurance mechanism. Against this background, the data established *security* as the dominant value of farmers, closely linked to *achievement* in terms of harvest and animal quantity. *Conformity* with accepted social roles is valued, and *power* can be established through visible success. We conclude that a successful CBBP will need to convincingly reduce risks for participants (e.g. in bull sharing), and make benefits of participation immediately visible. Appealing to the sense of being a professional breeder may help to overcome the strong status quo orientation in breeding strategies. Finally, a trusted institutional set-up of the CBBP will be crucial to reach the necessary buy-in by the cattle keeping communities.

The need for hybrid neural network models in precision livestock farming systems

L.O. Tedeschi
Texas A&M University, Department of Animal Science, 230 Kleberg Center – TAMU 2471, College Station, TX 77843-2471, USA; luis.tedeschi@tamu.edu

The adequate nutrition of livestock requires an accurate determination of the animal's requirements for energy and nutrients and the feedstuff's bromatological attributes to supply the energy and nutrients needed by the animal. The advancements in the biosensor technology in the last decade have enabled the livestock industry to gather a multitude of measurements, including electronic feeder, rumination (acoustics), in-cow sensors to measure rumen temperature and pH, methane emission, automated-scale technologies to measure milk yield and body weight, off-cow sensors to measure skin-surface temperature, and body condition score (3D image analysis), infrared cameras (heat production), in-line sensors to measure milk components, on-cow sensors to measure physical activity (pedometers, accelerometers), infrared spectroscopy (chemical composition of feeds), and 'Internet of Things.' The assembly of big data is innocuous by itself; it is complex and multiscale and requires proper data interpretation to provide meaningful and timely information. The interpretation of big data in many science fields by artificial intelligence (AI: machine learning and deep learning) allows for its integration with mechanistic nutrition modelling (MNM) with big data to more precisely formulate and balance diets. The combination of AI with MNM facilitates the adoption of precision livestock farming (PLF), also referred to as smart farming technology, at many different levels within the farm. We propose a novel approach that uses statistical relationships (i.e. AI) combined with conceptual MNM to interpret, process, and provide inputs to different layers (e.g. submodels). This novel approach (hybrid neural network, HNN) is more advanced and possibly more accurate because it uses accumulated scientific knowledge that has been interpreted and systematically combined to improve the predictability of MNM in animal agriculture for many decades.

Monitoring and evaluation of cattle individual feed efficiency in commercial farms

I. Halachmi
Agricultural Research Organization, the Volcani Centre, Israel; halachmi@volcani.agri.gov.il

Low-cost eating behaviour sensors will soon be available for commercial use in dairy farms. The aim of this study was to develop a feed intake model for the individual dairy cow that includes eating behaviour. In a research farm, the individual cows' voluntary feed intake and feeding behaviour were monitored at every meal. A feed intake model was developed based on data that exist in commercial modern farms such as 'BW,' 'milk yield' and 'days in milking' parameters. Over and above, at the individual cow level, eating behaviour seemed to be correlated with feed intake (R^2=0.93 to 0.94). The eating behaviour coefficient varied among individuals, ranging from 150 to 230 g/min per cow. The contribution of eating behaviour (0.28) to the dry matter intake (DMI) prediction model was higher than the contribution of BW (0.20), similar to the contribution of fat-corrected milk (FCM)/BW (0.29) and not as large as the contribution of FCM (0.49). Incorporating eating behaviour into the DMI model improved its accuracy by 1.3 (38%) kg/cow per day. The model was implemented in commercial farm. The proposed invited review presentation in the EAAP will explore and compare also other monitoring cattle individual feed efficiency methods such as machine vision currently running in the Israeli PLF lab.

Gestating sows' feeding behaviour with precision feeding

C. Gaillard, R. Gauthier and J.Y. Dourmad
PEGASE, INRAE, Agrocampus Ouest, Le Clos, 35590 Saint-Gilles, France; charlotte.gaillard@inrae.fr

Precision feeding (PF) of gestating sows, with the individual and daily mixing of two diets with different nutrient content was previously reported to reduce protein intake, feed costs and environmental losses compared to a conventional single diet feeding (CF). The effect of PF on sows' feeding behaviour has not been reported yet as this strategy has mainly been studied via modelling. The objective of this study, realised within the Horizon 2020 EU Feed-a-Gene program (grant agreement no. 633531), was to compare feeding behaviour of sows in gestation fed with a PF or a CF strategy. The experimental trial included 5 successive batches of 12-18 sows, housed in groups in a room (8×8 m) equipped with two automatic feeders able to daily mix two diets for each animal. The total number of visits to the feeder per day per sow (4.9 visits/d) and the total daily time spend in the feeder (64 min/d) was not affected by parity. After 70 days of gestation, PF sows visited the feeder more often than CF sows (5.1 vs 4.1 visits/d, P<0.01), and PF sows stayed longer in the feeder than CF sows (79.8 vs 55.6 min/d). The number of feeding visits to the feeder was on average of 1.1 per day per sow, and did not vary with feeding strategy. This number was slightly higher for multiparous sows compared to primiparous (1.08 vs 1.02 ± 0.01 (SE), P<0.01). The average daily eating time was not affected by the feeding strategy or parity, and was on average of 36 min per day per sow. The higher frequency of feeding visits to the feeder occurred after 00:30 at the start of the new feeding day in this farm. Then the frequency of visits decreased exponentially until 10:00 when all sows had already eaten their daily ration. The afternoon (from 13:00 to 21:00) was quiet with rare visits to the feeder. The frequency of non-feeding visits increased before the start of the new feeding day at 00:30 and in the morning (peaks reached at 23:30 and 9:00). In conclusion, independently of their feeding strategy, most of the sows only need one visit to eat their entire ration, usually in the first hours of the new feeding day. After 70 days of gestation, PF sows visited the feeder more often than CF sows.

Individual precision feeding and lowering crude protein might help to decrease environmental burden

A. Remus[1], L.S. Santos[2], C. Pomar[1] and L. Hauschild[3]
[1]Agriculture and Agri-Food Canada, Sherbrooke, QC, J1M 0C8, Canada, [2]UFRJ, Seropédica,RJ, 23890070, Brazil, [3]UNESP, Jaboticabal, SP, 14884900, Brazil; aline.remus@canada.ca

This study aimed to evaluate the effect of individual precision feeding (IPF) compared to conventional 2 phase-feeding (2P) system in growing-finishing pigs receiving diets varying in crude protein (CP). Sixty crossbreed barrows were randomly distributed in a 2×2 factorial arrangement (IPF vs 2P and 2 CP levels: reduced CP (RCP) and high CP (HCP)). Each treatment had 15 replicates. An individual electronic chip allowed all pigs to be housed in the same pen and to be fed individually using computerised feeding stations. The trial comprised a growing 40-65 kg BW and a finishing 65-100 kg BW period of 28 d each. Body composition was measured by dual-energy X-ray (DXA) densitometry on d 1, 28, and 56 of the trial. Nitrogen (N) and phosphorus (P) excretion and efficiency were estimated by accounting for the ingested and retained nutrients. Data was analysed using the Tukey-test in the MIXED procedure of SAS. Pigs in 2P with RCP received 16.5%, and 14.4%, while HCP pigs received 22.2%, and 18.6% CP during the growing and finishing periods, respectively. Pigs in IPF with RCP received 13.7%, and 11.9% CP, and HCP pigs received 18.5%, and 16.7% CP during the growing and finishing periods, respectively. During the growing period IPF pigs receiving RCP diets presented the lowest (feeding system and CP interaction; P=0.03) N excretion (17 g/d) compared to other treatments. Additionally, IPF pigs retained 20% more N (P<0.001) and excreted 46% less P (P<0.001) than 2P pigs, independent of the CP level. During the finishing phase, IPF pigs were 28 and 18% more efficient retaining N and P, respectively, than 2P pigs. Overall both periods, pigs receiving RCP were 27%, and 4% more efficient (P<0.001) retaining N and P, respectively, than HCP pigs. Disregarding CP level, IPF pigs were 22 and 11% more efficient (P<0.001) retaining N and P, respectively, than 2P pigs. In conclusion, even when decreasing CP in diets, pigs in conventional 2P systems are less efficient in nutrient utilisation than IPF pigs. Additionally, the use of low CP diets within IPF might decrease further the environmental burden originated from N and P excretion in pig production systems.

A Bayesian comparison of individual growth response models for precision-feeding of growing pigs

M.M. Misiura[1], J.A.N. Filipe[1,2], L. Brossard[3], E.F. Knol[4], M.R. Bedford[5] and I. Kyriazakis[6]

[1]Newcastle University, Agriculture Building, NE1 7RU, Newcastle upon Tyne, United Kingdom, [2]University of Aberdeen, Ashgrove Rd W, AB25 2ZD, Aberdeen, United Kingdom, [3]Inra Agrocampus Ouest, 16, le clos, 35590 Saint-Gilles, France, [4]Topigs Norsvin, Schoenaker 6, 6641 SZ Beuningen, the Netherlands, [5]AB Vista, Woodstock Ct, SN8 4AB, Marlborough, United Kingdom, [6]Queen's University, Belfast, Biological Sciences Building, BT9 5DL, Belfast, United Kingdom; m.m.misiura@ncl.ac.uk

Precision feeding of growing-finishing pigs require a mathematical model to forecast individual responses to different nutrient supplies. The two most common models assume linear growth trends: (1) double exponential smoothing, a type of moving average of past observations; and (2) dynamic linear regression, where the regression parameters can vary in time. The statistical approaches that are currently used to fit these models (i.e. maximum likelihood) give limited information on parameter uncertainty and correlations, which could limit the usefulness of the forecasts. Here, we developed and evaluated alternative models of growth response to nutrient supply that are non-linear (allometric, monomolecular and rational). We utilised a Bayesian inference approach to account for the uncertainty and correlations in parameter estimates, which outputs distributions rather than point estimates. We fitted the current linear and the alternative non-linear models to individual pig data from two distinct populations for different estimation scenarios: (1) in-sample (using all available individual data); and (2) out-of-sample (using the training subset to infer parameters and a testing subset to validate the forecasts). We found that: (1) all models gave similar in-sample goodness-of-fit; (2) forecast of future growth differed between the models, with the allometric model generating the most reliable forecasts across individual pigs, especially when forecasts were made over more than 1-2 days ahead. The results of this study could help to forecast individual nutrient requirements through a robust estimation of their average requirements and a suitable estimation of their uncertainty given the known past performance of the animal. The approach could also be utilised to inform management strategies, such as pen allocation and slaughter weight prediction.

The limitations of actual mathematical models to predict individual real-time nutrient requirements

C. Pomar and A. Remus

Agriculture and Agri-Food Canada, Sherbrooke Research and Development Center, 2000, rue Collège, J1G 5E4, Sherbrooke, Quebec, Canada; candido.pomar@canada.ca

Energy, amino acids (AA), minerals, vitamins and water must be provided to animals in adequate amount to live, grow, and produce. Providing the required amount of nutrients to animals reduces energy expenditure, increases nutrient utilisation efficiency, reduces nutrient excretion and feeding costs. Mathematical models are proposed to estimate nutrient requirements and to represent the animal responses in different nutritional, genetic and environmental contexts. However, these models use comprehensible nutritional principles to represent the biological phenomena of a single average animal. The success of feeding individual growing animals with diets tailored in real-time to their requirements is limited by the inaccuracy of the principles used to estimate AA and other nutrient requirements. For example, one of these principles states that for growth, AA efficiency (e.g. 72% for Lys) and body protein AA concentration (e.g. 7% for Lys) is constant. These parameters are, however, affected by nutrient restriction while pigs may respond differently to nutrient provisions. Also, pigs are raised in groups and fed with a unique feed for long periods. Individual pigs within a given population differ in terms of BW, ADG, health, etc. and therefore, differ in the amount of AA they need. Furthermore, there is a large variation in animal responses to AA intake which together with the observed AA interactions highlights the weakness of the ideal protein concept when applied to conventional and unconventional production systems. The inaccuracy of actual models estimating AA and other nutrient requirements, both for individual animals and populations, limits our ability to reduce dietary nutrient supply in growing animals. New mathematical models are needed to precisely estimate nutrient requirements and account for the changes that occur over time and for the variation that exists among animals. These new models should review many of the principles actually used to represent the animals' metabolism and to estimate AA and other nutrient requirements. They also have to integrate the estimation of functional nutrient requirements that will ensure the integrity of the intestinal morphology and microbiota, the immune system, etc.

Predicting nutritive characteristics of heterogeneous dairy pasture in-situ from hyperspectral data

A. Thomson, S. Karunaratne, A. Copland, D. Stayches, E. Morse-Mcnabb and J. Jacobs
Ellinbank Dairy Research Centre, 1301 Hazeldean road, Ellinbank, VIC, 3810, Australia;
anna.thomson@agriculture.vic.gov.au

Measuring dairy pasture regularly offers farmers the ability to make more informed agronomy and grazing decisions in pasture-based systems. This study provides a proof-of-concept that hyperspectral instruments can be used to non-destructively measure changes in nutritional characteristics of dairy pasture *in situ*. The aims were to determine: (1) the degree of accuracy that can be expected when this technology is applied *in situ*; and (2) whether the sensor can perform equally well across a range of growth stages. Predominantly perennial ryegrass pasture samples (n=216) were collected from two contrasting dairy paddocks in Victoria, Australia. Each pasture was sampled weekly over a 4 week period in Spring 2019 to ensure a range of growth stages were represented (weeks 2-5 of regrowth). Reference techniques were used to measure crude protein (CP) and neutral detergent fibre (NDF) concentrations, and calculate metabolisable energy (ME) content, on a dry matter (DM) basis. Spectra were obtained for each sample using an ASD FieldSpec 4 fitted with a custom light shroud before destructive sampling was undertaken. Data pre-treatments were applied based on principle component analysis (PCA), and spectral relationships were modelled using partial least squares regression (PLSR). A method of fusing PLSR outputs with covariate pasture data using a support vector machine regression (SVMR) was also tested. Spectra from week 2 displayed differing characteristics to those from weeks 3-5, likely due to exposed soil, and model accuracies increased when these samples were excluded from the dataset. When independently validated, CP was the best predicted variable with a root mean square error of prediction (RMSEP) of ±26.3 g/kg DM and an R^2 of 0.73. The best models for ME and NDF had an RMSEP of ±0.22 MJ/kg DM and ±24.9 g/kg DM respectively (R^2 of 0.71 and 0.51). Using a fusion model including covariate data moderately improved prediction accuracy in all models to R^2 of 0.74, 0.73 and 0.61 for CP, ME and NDF respectively. In conclusion, hyperspectral sensors are appropriate for use in surveying dairy pastures non-destructively. Further work should investigate lower-cost, automated sensors of this type for pasture-mapping.

Genetic parameters of different feed efficiency traits of Charolais young bulls

S. Taussat[1,2], C. Fossaert[3], G. Cantalapiedra-Hijar[4], L. Griffon[5] and G. Renand[2]
[1]Allice, 149 rue de Bercy, 75012 Paris, France, [2]Université Paris-Saclay, INRAE, AgroParisTech, GABI, Domaine de Vilvert, 78350 Jouy-en-Josas, France, [3]Institut de l'Elevage, Lieu-dit Monvoisin, 35652 Le Rheu, France, [4]INRAE, Université Clermont Auvergne, Vetagro Sup, UMRH, Site de Theix, 63122 Saint-Genès-Champanelle, France, [5]Institut de l'Elevage, 149 rue de Bercy, 75012 Paris, France; sebastien.taussat@inrae.fr

Genetic selection on feed efficiency could increase the profitability of French beef farms. The BEEFALIM 2020 project aims to study the genetic determinism of this trait measured from 2016 to 2019 on 588 young Charolais bulls in four experimental farms. After weaning in commercial farms, the young bulls entered the fattening barns at 303 (±26) days of age and were fed *ad libitum* either with grass or corn silage. During the 29 (±3) weeks of the test period, dry matter intake (DMI) was recorded daily and animals were weighted every 14 days. Average daily gain (ADG) and final body weight (FW) were calculated using regression of body weight on time. Three feed efficiency criteria were used: residual feed intake (RFI), residual gain (RG) and feed conversion efficiency (FCE). Genetic parameters were estimated using WOMBAT software with an animal model containing the contemporary group as fixed effect, age at the start of the test as covariate and farm origins × year, genetic and residual as random effects. During the test period, DMI averaged 9.65 (±1.13) kg/day, ADG 1.51 (±0.29) kg/day and FW 698 (±74) kg. The three traits were closely correlated phenotypically: 0.67 between FI and ADG, 0.77 between FW and ADG and 0.82 between FI and FW. Feed intake and ADG were more heritable (0.34) than FW (0.20). The three feed efficiency traits were moderately heritable (0.22 for RFI and RG; 0.18 for FCE). No genetic correlation was found between RFI and RG or FCE while a strong genetic correlation of 0.99 was estimated between RG and FCE. Residual feed intake was genetically correlated with DMI (0.82), ADG (0.52) and FW (0.49). Both RG and FCE were closely correlated with ADG (0.79 and 0.75, respectively) and moderately with DMI (0.31 and 0.23, respectively) or FW (0.33 and 0.25, respectively). Even if RFI and RG were phenotypically independent with traits used in both regressions, genetic relationships were found between them. Genetic improvement is possible through a selection index combining both feed efficiency and production traits.

Environmental optimisation of diets for genetically selected pigs

T. Soleimani and H. Gilbert
INRAE, INRAE – UMR GenPhySE 24 chemin de Borde Rouge 31326 Castanet Tolosan, France; tara.soleimani@inrae.fr

The environmental impact of pig production is largely dependent on the diet composition, via impacts of the feed production stage and of the pig consumption efficiency stage. Considering these stages jointly for diet formulation seems suitable to improve the environmental impacts of pig production systems. In this study, tailoring the diet formulation precisely to satisfy energy and nutrient requirements was combined with an environmental optimisation of the diet composition. Normalised digestible crude protein (CP) and amino acids (AAs) based on the net energy (NE) were retained to capture the animal growth requirements. An approach was developed to use these indicators to determine for two pig lines selected for different levels of feed efficiency (residual feed intake, RFI) a list of possible tailored candidate diets, to quantify their respective environmental impacts using a life cycle assessment (LCA) approach, and to extract the best environmentally optimised diet. Data from the two pig lines were used to calibrate the nutritional requirements related to high efficient (LRFI) and low efficient (HRFI) animals. The responses of the lines to their corresponding tailored optimised diets were assessed through an LCA method. The environmentally optimised diets tailored for the pig requirements reduced the environmental impacts on average by 8.8% for LRFI and 11.4% for HRFI pigs relative to the conventional diet. Using the optimised diet for HRFI pigs led to even lower impacts than the LRFI pigs fed a conventional diet. In conclusion, the combination of diet formulation tailored to the nutritional requirements of a genetic group of pigs with its environmental optimisation can highly reduce the overall environmental impacts of pig production.

Raw or technologically treated proteaginous seeds as alternatives to soybean meal for dairy cows

S. Mendowski[1,2], P. Nozière[1], A. Ferlay[1], P. Denis[1], G. Chesneau[2] and P. Chapoutot[3]
[1]Université Clermont Auvergne, INRAE, VetAgro Sup, UMR Herbivores, Centre INRA de Theix, 63122 Saint Genès Champanelle, France, [2]Valorex, La Messayais, 35210 Combourtillé, France, [3]Université Paris-Saclay, INRAE, AgroParisTech, UMR Modélisation Systémique Appliquée aux Ruminants, 16 rue Claude Bernard, 75005 Paris, France; s.mendowski@valorex.com

The aim of this study was to quantify the effects on nitrogen (N) utilisation and dairy performance of the substitution of soybean meal (SBM) by either raw or treated proteaginous seeds (PS): faba bean, lupin and pea. Treatments of the PS included heat treatments, thermo-mechanical treatments and tanning. Two databases, gathering results of 31 *in vivo* and 36 *in situ* publications, respectively, were created and N ruminal degradability, N intestinal true digestibility, ruminal parameters, N partitioning and milk production and composition were analysed. Effects of treatments were assessed using t-tests applied to the difference between the tested feedstuffs (i.e. PS or treated PS) and the control ones (i.e. SBM or raw PS, respectively). Despite the limited amounts of available data and the great diversity of feeding practices, some general trends appeared from this quantitative review. Proteins from raw PS were more degradable in the rumen than proteins from SBM (+16 g/100 g). This ruminal protein degradability was lowered by treatments (-13 g/100 g). At similar CP contents of diets, the milk protein content (-0.8 g/kg) and milk N secretion (-4.7 g/d) were decreased with raw PS compared to SBM. In contrast, no significant effect on N partition was observed when treated PS are compared to raw ones, but milk fat content was decreased (-2 g/kg), especially for extruded faba bean and lupin (-3.6 g/kg on average). Finally, N transfer as protein in milk remained lower with treated PS than with SBM but differences were rather low (-2.5 g N/d). Although the variations were repeatable through *in situ* trials, the *in vivo* responses were more variable across studies. This could be partly related to the dietary level of digestible methionine, or to the intensity of the processes, both being poorly described.

Standardised natural citrus extract optimise feed efficiency: a way to reduce carbon footprint

S. Cisse[1,2], H. Bui[1,2], R. Djezzar[3], A. Benarbia[1,2] and D. Guilet[1,4]
[1]Labcom FeedInTech, 42 rue Georges Morel, 49070 Beaucouzé, France, [2]Nor-Feed SAS, 3 rue Amédéo Avogadro, 49070, France, [3]National Higher School of Veterinary, Rue Issad Abbes, Oued Smar, Algiers, Algeria, [4]EA 921 SONAS, 42 rue Georges Morel, 49070 Beaucouzé, France; hoa.bui@norfeed.net

Since the emergence of intensive livestock farming, the environmental impact of animal production is steadily increasing, having bad consequences in global warming. Among the causes of this increase, animal feed is a key element. In order to reduce this environmental impact, decrease the quantity of raw material used in farming seems to be a good solution. There are several solutions to reduce livestock feed inputs. Improving animal's feed efficiency is one of the easiest. in this context, a study has been done in order to assess the effect of a SNCE on feed efficiency. The trial took place at the Technical Institute of Livestock (Algeria). 480 one day old birds (Arbor Acres) were divided into 2 groups. Each group contained 8 replicates of 30 birds. CTRL group: a standard diet without supplementation; SNCE group: a standard diet supplemented with 250 ppm of SNCE. Birds were reared until day 42. Weight, feed intake and mortality from each replicate were recorded weekly. The Feed Conversion Ratio (FCR) was calculated. The intestinal length and villi length were also recorded (1 per replicate). Statistical analyses were performed by Student test. Statistical significance was considered at p <0.05. Results showed a better weight gain and less mortality from SNCE group, compared to CTRL group. In addition, chickens from SNCE group had a lower FCR and higher intestinal and villi length than chickens from CTRL group. No difference was observed on feed intake. According to these data, SNCE supplementation on broilers allow to increase the bodyweight while reducing the FCR of birds. These effects are correlated with longueur intestine and higher villi. We suppose that due to their higher villi and longueur intestine, animal had a better absorption capacity which was translated by higher growth and feed efficiency. Resources becoming scarce, optimising their usage is crucial for sustainable livestock production. According to the results, standardised natural solutions can be useful to reduce the carbon footprint of livestock by improving it efficiency.

Ingestive behaviour and feed conversion of two different Holstein genotypes grazing pasture

D. Talmón[1], M. Garcia-Roche[1], A. Mendoza[2] and M. Carriquiry[1]
[1]Facultad de Agronomía, Universidad de la República, Departamento de Producción Animal y Pasturas, Garzón 780, 12400, Montevideo, Uruguay, [2]Instituto Nacional de Investigación Agropecuaria, Ruta 50 km 11, 70000, Semillero, Colonia, Uruguay; danieltalmon94@gmail.com

Multiparous North American Holstein (NAH; 587±80 kg of body weight (BW) and 3.05±0.35 units of body condition score (BCS); n=11) and New Zealand Holstein (NZH; 530±60 kg BW and 3.23±0.42; n=11) origin dairy cows were used to evaluate the ingestive behaviour and feed conversion grazing pasture during late lactation. A randomised complete block design was used and cows were blocked according to lactation number and days in milk. Cows were milked twice a day and had access to a perennial ryegrass pasture (130 g/kg dry matter (DM) of crude protein and 9.29 MJ/kg DM of metabolisable energy) from 09:00 to 15:00 and from 17:30 to 06:30 h with 36±6 kg DM/cow/d of herbage allowance above 5 cm. During daylight hours grazing time and bite rate were recorded through visual appreciation, daily milk yield was measured and milk samples were collected to analyse milk-solid composition. Pasture DM intake was estimated by energy balance based on retained energy in milk and heat production measurements through the heart rate-O_2 pulse technique. Data were analysed with a generalised linear model including Holstein genotype as fixed effect. There were no differences in milk yield (16.0±1.2 kg/d), protein or lactose percentage (3.49±0.12 and 4.54±0.10%, respectively) but fat percentage was greater (4.63 vs 4.03±0.15%; P<0.05) for NZH than NAH cows. Grazing time, bite rate and bite mass were similar for both genotypes (371±9 min/d, 40±2 bites/min and 1.19 g DM/bite, respectively) as well as pasture DM intake (17.2±0.8 kg DM/d). However, feed conversion expressed as kg of 4%-fat corrected milk per kg DM consumed was greater (1.09 vs 0.90±0.05; P<0.05) in NZH than NAH. Thus, although there were no differences in pasture DM intake and grazing behaviour between genotypes, NZH cows had a better productive performance than NAH cows which was reflected in the greater feed conversion. The results are consistent with previously reported information where NZH cows are able to reach the same milk yield and DM intake than NAH in spite of their lower BW when they are fed with an exclusive based-pasture diet.

Energy efficiency of grazing Holstein dairy cows

D. Talmón[1], M. Garcia-Roche[1], A. Mendoza[2] and M. Carriquiry[1]
[1]Facultad de Agronomía, Universidad de la República, Departamento de Producción Animal y Pasturas, Garzón 780, 12400, Montevideo, Uruguay, [2]Instituto Nacional de Investigación Agropecuaria, Ruta 50 km 11, 70000, Semillero, Colonia, Uruguay; danieltalmon94@gmail.com

Recently, pasture-based systems have increased interest in many temperate regions of the world as consequence of their lower production costs and perceived environment and animal welfare concerns. However, high-producing grazing Holstein cows have reduced energy balance than when they are managed in an indoor-system. The aim of this experiment was to compare the energy partitioning and efficiency of North American Holstein (NAH) and New Zealand Holstein (NZH) origin dairy cows grazing pasture without supplementation. Twenty-two multiparous cows (NAH=11; 587±80 kg of body weight (BW) and 3.05±0.35 units of body condition score (BCS); NZH=11; 530±60 kg BW and 3.23±0.42) in late lactation were grouped based on lactation number and days in milk and then assigned to a randomised block design. Cows were milked twice a day and grazed a perennial ryegrass pasture (crude protein=130 g/kg dry matter (DM) and metabolisable energy=9.29 MJ/kg DM) with 36±6 kg DM/cow/d of herbage allowance above 5cm. Heat production (HP) was measured using the heart rate-O_2 pulse technique and retained energy in milk (RE) was estimated according to NRC (2001) using daily milk yield and its composition. Metabolisable energy intake (MEI) was calculated as HP+RE and residual HP was the difference between measured HP and predicted HP by NRC model. Data were analysed with a generalised linear model with Holstein genotype as fixed effect. No differences in MEI and HP between genotypes (1,394±59 and 942±51 kJ/kgBW$^{0.75}$/d, respectively) were observed but there was a tendency (P=0.08) to greater RE in NZH than NAH cows (497 vs 407±34 kJ/kgBW$^{0.75}$/d) and energy efficiency, expressed as RE/MEI, was greater (P<0.05) for NZH than NAH cows (0.345 vs 0.298±0.015). However, residual HP did not differ between genotypes, but both were greater than zero, representing in average a 5% of MEI. These results may indicate that although HP was similar in both genotypes, NZH cows destined less MEI to maintenance and more to milk production, contrary to NAH cows. In addition, NRC model underestimated the HP of grazing dairy cows.

Feed efficiency evaluation of grazing Hereford heifers

M.F. Marín[1], T. Devicenzi[2], A. Jasinsky[1], E. Navajas[2], A.C. Espasandín[1] and M. Carriquiry[1]
[1]Universidad de la República, Facultad de Agronomía, Producción Animal y Pasturas, Av. Gral. Eugenio Garzón 780, 12900 Montevideo, Uruguay, [2]Instituto Nacional de Investigación Agropecuaria, Andes 1365, 11100 Montevideo, Uruguay; mfedericamarin@gmail.com

Residual feed intake (RFI) is one indicator of feed conversion efficiency, and is defined as the difference between an animal's actual dry matter intake (DMI) and its expected DMI based on the requirements for maintenance and growth over a specified period. Given the difficulties of recording accurately DMI under grazing conditions in large number of animals, RFI is usually assessed in confined conditions using automated feeding systems. The association between RFI and productive performance of grazing animals is still unclear, with evidences of possible genotype × environment interaction. The aim of this work was to evaluate residual heat production (RHP) as a measure of feed efficiency in grazing beef heifers sired by bulls evaluated by RFI. Seventy-one Hereford heifers, progeny of seven sires, were classified in three groups according to their paternal estimated breeding values (EBV) for RFI (percentiles ≤20% for high, 30-50% for medium, ≥95% for low). Heifers (25±0.8 months of age) were managed as a contemporary group on natural grasslands with an average herbage allowance of 3,466±1,264 kg DM/ha and forage allowance of 9.4±1.02 kg DM / kg BW. Heat production (HP) was determined by the heart rate (HR)–O2 pulse (O2P) technique. The HR was measured three times for 4 d each time and the O2P was determined twice over a 60-d period prior to their first insemination. Body weight (BW) and body condition score (BCS) were registered fortnightly and body retained energy (RE) was estimated according to the NRC model. The RHP was defined as the difference between measured and expected HP based on the slope and intercept of a multiple linear regression of HP dependency on heifers' mid-period BW$^{0.75}$, average daily gain (ADG) and average BCS. Data were analysed using a mixed model with paternal RFI group as a fixed effect. No differences (P>0.05) were found between groups in average BW (282±13 kg), BCS (4.25±0.2) or ADG (773±30 g). Neither metabolisable energy intake (70±2.3 MJ/d; MEI=RE+HP), HP (56±2.1 MJ/d) nor RHP (0.08±1.16 MJ/d) between groups were different (P>0.05). However, high RHP heifers (superior quartile; n=18; less efficient) presented 1.2-fold greater (P<0.0001) MEI with no differences in ADG than low RHP heifers (inferior quartile; n=18). These preliminary results indicated that although RHP could be an estimator of feed efficiency in grazing conditions, a greater number of heifers with more divergent paternal EBV for RFI is needed for accurate estimates of the relationship between these variables.

NIRS calibrations for feed intake: a proposal for the definition of reference data

F. Maroto-Molina, D.C. Pérez-Marín, A. Garrido-Varo, J.E. Guerrero-Ginel and C. Riccioli
University of Cordoba, Animal Production, Ctra. Madrid-Cádiz km 396, 14014 Córdoba, Spain; g02mamof@uco.es

Near Infrared Spectroscopy (NIRS) has demonstrated to be a precise and cost-efficient tool for the evaluation of feed composition, even with complex matrices such as a Total Mixed Ration (TMR). However, animal performance is not just about diet composition, as intake, digestibility and other traits are involved. NIRS has also been used to predict feed intake, although most researchers used faecal spectra, which are not appropriate for commercial applications. We designed an experiment to study the feasibility of TMR spectra to predict dairy cow intake. This experiment was selected as part of the Transnational Access Programme of the SmartCow project. It involved four groups of 15 dairy cows each eating four TMR differing in ingredient composition and nutritive value for eight weeks. It was carried out at the Aarhus University experimental farm. NIR spectra were collected for daily samples of each one of the four TMR. Individual cow and average group intake of every experimental group was recorded. TMR intake data showed a large variability between days for the same TMR, which made difficult to use raw intake data to develop NIRS calibrations. The objective of this work was to study several options for intake data processing in order to define reference data suitable to develop NIRS calibrations. Raw intake data was compared to smoothed means and to least squares means considering animal factors (cow weight, days in milk and milk production). The inclusion of animal factors as covariables in NIRS calibrations was also studied. Performance of the different approaches was evaluated in terms of calibration statistics: coefficient of determination (R2) and root-mean-square error (RMSE).

Welfare and production under hot environment: the case of dairy cattle

U. Bernabucci
University of Tuscia-Viterbo, Department of Agricultural and Forests Sciences, Via San Camillo de Lellis, 01100 Viterbo, Italy; bernab@unitus.it

Heat stress (HS) decreases productivity with devastating economic consequences to global animal agriculture. Heat stress can be defined as a physiological condition when the core body temperature of a given species exceeds its range specified for normal activity, which results from a total heat load (internal production and environment) exceeding the capacity for heat dissipation and this prompts physiological and behavioural responses to reduce the strain. Dairy breeds are typically more sensitive to heat stress, and higher-producing animals are more susceptible to heat stress because they generate more metabolic heat. During heat stress ethological changes include decline of feed intake, increase of water ingestion, decrease of urination, greater standing time, and shade or cooling seeking. The increase of respiration and sweating rates reflect the attempt to avoid body hyperthermia, the degree of which is a reliable measure of the severity of HS. Other important physiological changes are linked to the endocrine system, which include decline and increase of hormones regulating basal metabolism and water retention, respectively. The functioning of the immune system is altered in heat stressed dairy cows. Cows giving birth under HS produce colostrum with lower protective value with consequent negative effects on passive immunisation of calves. Several studies documented that HS impairs both innate and acquired immunity. Metabolism of glucose, lipids, proteins and minerals are strongly altered by HS. Also, oxidative status is negatively affected by HS. Productive changes are reduction of milk yield and alteration of milk quality, and reproductive failure. Pathological indicators include those associated to the greater risk of infections and death. Hot periods are characterised by greater incidence of mastitis and death. Consideration of the animal-based indicators described herein may help to adopt early and proper adaptation interventions, which may prevent or mitigate the negative consequences of HS.

Adaptation to heat waves in dairy cows by optimisation of cooling operations

A. Vitali, U. Bernabucci, A. Nardone and N. Lacetera
University of Tuscia, Department of Agriculture and Forest Sciences, Via s. Camillo de Lellis, 01100 Viterbo, Italy;
vitali@unitus.it

Heat waves are increasing in frequency and intensity as result of climate change and they represent a challenge for dairy cows' health and welfare. The study was aimed to assess the effect of 2 daily extra treatments of cooling (EC) added to a standard cooling (SC) program. The SC program was represented by 5 daily treatments, 3 in the waiting and 2 in the feeding lanes. The EC program was based on 1 extra cooling in the morning and 1 in the afternoon. Furthermore, two different lengths of each extra cooling treatment were tested, lasting 45' and 75', respectively. Four distinct trials were carried out in summer 2019, two were performed under moderate heat or non-heat wave (NHW) conditions and two were relative to severe heat or heat wave (HW) conditions. Each trial lasted 72 hours and involved 2 groups of 10 lactating homogeneous cows (milk yield, DIM and parity), exposed to SC or to EC (45' or 75'). Vaginal temperature (VT) in °C and Temperature Humidity Index (THI) were monitored in continuous each 5' by using intravaginal temperature and temperature-humidity loggers, respectively. Individual milk yield was recorded daily at the 3 milking operations. The mean value of THI was significantly higher in HW (77.0 ± 2.7) compared to NHW conditions (74.3 ± 2.3). The EC, both of 45' and 75' did not affect the THI of barns. The VT was significantly higher in HW compared to NHW (39.32 ± 0.36 vs 38.95 ± 0.24). The EC lasting 45' vs SC did not exert effects on VT both in HW and in NHW. The EC lasting 75' vs SC did not affect VT in HW whereas it reduced significantly the VT in NHW (38.80 ± 0.25 vs 38.97 ± 0.28). The EC, both of 45' and 75', did not show any effect on milk yield both in HW and NHW. Under conditions of the present study, cows did not show any relevant sign of severe heat stress indicating that SC was capable to guarantee maintenance of good welfare conditions. The increase in frequency and length of cooling treatments was able to limit body hyperthermia under moderate heat, whereas it did not show any significant effect under the severe heat of HW. Further research is needed to improve effectiveness of cooling operations under HW keeping in mind the evaluation of their sustainability in terms of energy and water consumption.

Technical measures to improve conditions for fattening pigs and environment during warm weather

A.-C. Olsson[1], K.-H. Jeppsson[1] and A. Nasirahmadi[2]
[1]Swedish University of Agricultural Sciences, Department of Biosystems and Technology, P.O. Box 103, 230 53, Sweden,
[2]University of Kassel, Department of Agricultural and Biosystems Engineering, Nordbahnhofstr. 1a, 37213 Witzenhausen,
Germany; anne-charlotte.olsson@slu.se

Due to welfare reasons, fattening pigs in Sweden are kept in pens with partly slatted floors. Partly slatted floors permit usage of straw for bedding, rooting and exploration behaviour as well as allow the pigs to lie down on solid floor surfaces. When clean, pens with partly slatted floor show lower levels of ammonia compared to pens with fully slatted floors. This is partly explained by a reduced slurry pit area. However, impaired pen hygiene is becoming an increasing problem during periods with hot weather and a warmer climate. Heat stressed pigs try to cool themselves by lying on the slatted floor instead of on the solid floor. They also start to urinate and defecate on the solid floor since the slatted floor is occupied and/or since the pigs want to wet their skin to cool down. As a part within the ERA-NET SusAn project PigSys, two technical solutions, for improving pen hygiene in partly slatted pens for growing-finishing pigs, are tested. The pigs are either cooled by sprinkling of low-pressure water on the slatted floor or by convective cooling with increased air velocity on the lying area. The studies are performed in a commercial fattening pig house with 10 identical compartments with 16 pens and about 160 pigs per compartment. Two compartments are filled simultaneously (control versus treatment). The comparison includes 10 batches (2×10 compartments) for two summer periods. The effect, on pig occupation zone in the pens, is evaluated by means of artificial intelligence along with machine vision techniques. The results of how the pigs use the pens are supplemented with observations on pen hygiene (according to a well-defined template) and measurements of ammonia emission (using a photo-acoustic multi-gas analyser). The observed differences are analysed using the SAS software. The preliminary results show that both sprinkling of water on the slatted floor and increased air velocity on lying area have a significant favourable effect on pig occupation zone in the pen, pen hygiene and ammonia emission.

The effect of air temperature, humidity and velocity on respiration rate in gestating sows

P. Brandt[1], B. Bjerg[1], P. Pedersen[2] and G. Zhang[3]
[1]University of Copenhagen, Department of Veterinary and Animal Sciences, Grønnegårdsvej 2, 1870 Frederiksberg, Denmark, [2]Skov A/S, Hedelund 4, Glyngøre, 7870 Roslev, Denmark, [3]Aarhus University, Department of Engineering, Inge Lehmanns Gade 10, 8000 Aarhus C, Denmark; pb@sund.ku.dk

Improved genetics have entailed that sows produce more piglets, more milk and consequently also more heat. Combined with global warming, sows are, particularly in the hot regions of the World, challenged by heat stress. The objective of the present study was to investigate the reactions to heat stress conditions among pregnant sows in order to establish a dataset for development of a thermal index to predict the effects of air temperature, air velocity and air humidity on respiration rate, rectal temperature and skin temperature. The experimental room was equipped with a negative pressure ventilation system with diffuse air inlet through the ceiling, electrical heaters, two steam generators and a dehumidifier. To generate different air velocities the experimental room was equipped with a special air distribution unit to generate an even vertical air velocity 0.8 m above the floor. A total of eight gestating sows were exposed to 6 combinations of three levels of air temperatures (25, 29 & 33 °C), two levels of relative humidity (30 & 70%) and two groups of gestating sows were tested during 12 test days. On each test day three levels of air velocity (0.2, 1 and 2.5 m/s) was tested in a random sequence. Respiration rate (RR) was recorded by counting flank movements for 60 sec. Rectal temperature (RT) was recorded by insertion of a digital thermometer in the rectum of the sow. Skin temperature (ST) was recorded at P2 (7 cm from the spine at the posterior point of the last rib) using an infrared thermometer. RR, RT and ST were recorded every ½ hour during three two-hour test periods. The response variables (RV) RR, RT and ST were fitted in a model by Bjerg *et al.* (2018) with adjustments, including a linear effect of temperature plus a linear effect of humidity×temperature and subtracting air velocity×(constant – temperature). For each RV a trial and error approach was used to search for the combination of the constants that resulted in the best agreement between the predicted and observed values. Modelling RR resulted in an R^2=0.93, but for RT the result was less consistent (R^2=0.82). For ST the results were consistent (R^2=0.98) but showed that the relative humidity had no influence on ST. Thus, RR seems to be the most promising of the investigated RV for predicting the effects of air temperature, air velocity and air humidity on heat stress in sows.

A new approach to combine genetic merit for milk content and heat stress on Spanish goat using PCA

C. Ziadi[1], A. Menéndez-Buxadera[1], M. Sánchez[2], E. Muñoz-Mejías[3], M.D. López[4] and A. Molina[1]
[1]Universidad de Córdoba, Departamento de Genética, Edificio Gregor Mendel. Campus de Rabanales, 14071 Córdoba, Spain, [2]Universidad de Córdoba, Departamento de Producción Animal, Campus de Rabanales, 14071 Córdoba, Spain, [3]Universidad de Las Palmas de Gran Canaria, Departamento de Patología Animal, Producción Animal, Bromatología y Tecnología de los Alimentos, Campus Universitario Cardones de Arucas, 35413 Arucas, Spain, [4]ACRIFLOR, Departamento de Producción Animal, Campus de Rabanales, 14071 Córdoba, Spain; ziadichiraz4@gmail.com

This study was conducted to develop a new approach to combine animal breeding values (EBVs) for test-day total fat plus protein independently estimated along the trajectory of days in milk (DIM) and across the scale of heat stress index quantified by THI. A total of 126.825, 141.856 and 62.834 test-day records collected in Florida, Malagueña, and Murciano-Granadina breeds, respectively were used for this analysis. For that purpose, a random regression model was used including non-genetic effects of herd-test-day, a second order (r) fixed regression for DIM or THI, litter size and female age at kidding with different levels for each breed. A r=2 random regression for DIM or r=1 for THI was used for genetic additive (co)variance and r=1 for permanent environmental effects, the residual variance was homogeneous. Variance components and breeding values were estimated separately for each breed applying Restricted Maximum Likelihood (REML) approach with ASREML3 software. The EBVs for each breed from both models were subjected to a principal component analysis (PCA), showing an antagonism between fat plus protein production and heat stress that varied across lactation curve. The two first eigenvalues explained between 82.5 to 95.0% of genetic variance along DIM and THI, the corresponding eigenvectors were used as weighting factor to combine the original EBVs in a new index. This new index can provide information for breeding programs of these breeds and will permit simultaneous selection of high productive and adapted animals to heat stress.

How does the intestinal metabolism adapt to reduced nutrient intake during heat stress in dairy cow?

F. Koch, S. Görs and B. Kuhla
Leibniz Institute of Farm Animal Biology, Institute od Nutritional Physiology Oskar Kellner, Wilhelm-Stahl Allee 2, 18196 Dummerstorf, Germany; koch@fbn-dummerstorf.de

High ambient temperatures cause heat stress in dairy cows resulting in a reduction of milk yield, feed intake, and alterations in whole-body protein metabolism. Whether the intestine is involved in the adaptation of whole-body amino acid metabolism during heat stress is not known. Glutamine and glucose serve as main energy sources for enterocytes and their utilisation might be modulated during increased ambient temperatures. To discriminate between effects elicited by ambient heat and accompanied reduction in dietary protein and energy intake, we studied markers of the intestinal metabolism in 10 non-pregnant German Holstein dairy cows grouped to heat-stressed (HS) or pair-feeding (PF) at thermoneutrality. Cows in 2nd lactation (245±102 days in milk) were kept in a climate chamber at thermoneutral conditions (15 °C; 63±1% relative humidity (RH) with a temperature-humidity index (THI) of 60) for 6 days. Animals received a total mixed ration twice daily (at 07:00 h and 15:00 h). Thereafter, 5 HS cows were continuously exposed for 4 days to 28 °C (with 52±2% RH; THI=76) with *ad libitum* feeding and access to water, both tempered to 28 °C. The reduction of daily *ad libitum* intake of HS cows was calculated as percentage of the daily mean to provide the same amount of feed energy to PF cows. The 5 PF cows were exposed for 4 days to 15 °C (THI=60). After 4 days of HS or PF, cows were slaughtered to obtain jejunum mucosa samples. Tissue scrapings were analysed for mRNA abundances of *SLC2A2, SCL5A1, PFKL, PKLR* (glucose metabolism), *MCT1, LDHA, LDHB* (lactate metabolism) and *GLS, GLUD1, CSP1* (glutamine metabolism) relative to the reference genes *HPRT1* and *RPL32*. Mucosal amino acid concentrations were measured by HPLC and normalised to tissue protein concentration. The mRNA abundances of *SLC2A2, PFKL, PKLR, MCT1, LDHA, LDHB, GLS, GLUD1* and *CSP1* did not differ between the groups. However, jejunum mucosa of HS cows tended to have higher histidine and anserine concentrations than PF cows (P=0.1). These results suggest that heat stress does not alter glutamine, lactate and glucose catabolism in the intestinal mucosa of dairy cows.

Effect of phytogenic feed additives on performance of dairy cows during heat stress condition

P. Pourazad[1], S. Aditya[2], A. Aulanniam[2], R.L. Balia[2], G.L. Utama[2], T. Aubert[1] and K.R. Wendler[1]
[1]Delacon Biotechnik GmbH, Langwiesen 244209 Engerwitzdorf, Austria, 4209, Austria, [2]Brawijaya University Malang, East Java, Indonesia, Faculty of Veterinary Medicine, Jl. Veteran Malang, Ketawanggede, Kec. Lowokwaru, Kota Malang, Jawa Timur 65145, Indonesia; poulad.pourazad@delacon.com

Lactating dairy cows elevate internal heat loads during the heat stress (HS), which subsequently affects feed intake and performance. Furthermore, the relationship between increased temperature-humidity index (THI) and decreased dry matter intake (DMI)-milk production has been clearly established. A study was conducted to evaluate the effect of phytogenic feed additives (PFA) supplementation on the feed intake and performance in dairy cows during 30 days under HS condition. Twenty mid-lactating Holstein cows were randomly assigned to two different groups, control (CON; n=10) and treatment (TRT; n=10). After 7-d of baseline diet (60% forage:40% concentrates), the TRT group received the PFA at dosage of the (20 g/h/d) for 30 d. The cows in CON group stayed on the baseline diet throughout the experiment. The DMI and milk yield were measured daily while milk samples were taken on d1 (baseline), d10, d20 and d30. The THI was continuously calculated using digital thermo-hygrometer throughout. Data were analysed using MIXED procedure of SAS. TRT group showed an increased DMI (P<0.001) compared to CON group. A higher milk production (l/d) was noticed (P<0.001) in TRT group compared to CON group. A decreased milk fat (P<0.01) and an elevated milk protein (P<0.05) percentage were more pronounced in TRT group (P<0.001). Meanwhile, no difference was found on lactose percentage on both groups. However, the yield (kg) of protein, fat and lactose significantly increased (P<0.05), which could be explained by higher milk production. In conclusion, the data suggest that feeding suitable PFA combination could increase DMI and milk yield as well as alter milk composition during the HS condition.

Effect of heat stress and of feed provision modification on growth and thermoregulation in pigs

A.M. Serviento[1,2], E. Labussière[1], M. Castex[2] and D. Renaudeau[1]

[1] INRAE Agrocampus Ouest, UMR1348 PEGASE, 35590 Saint-Gilles, France, [2]Lallemand SAS, 31702 Blagnac, France; aira-maye.serviento@inrae.fr

With the increasing global mean temperature and more frequent summer heat waves, there is a need to find techniques to mitigate heat stress (HS) consequences on pig production. The aim of the study was to evaluate the effect of HS and of increasing feed provision frequency on performance and thermoregulation responses in finishing pigs. A total of 48 pigs (66.1±1.7 kg) were allocated to 4 experimental groups in 3 replicates. After being housed at thermoneutrality (TN; 22 °C) during 7 days (P1), pigs were subjected to either TN or HS (32 °C) conditions during 20 days (P2). In each environment, the diet (14.9%CP, 9.64 MJ NE/kg) was provided either *ad libitum* (AL; 2 distributions/day) or pair-fed (PF8; 8 distributions/day) with HS pigs fed-AL as the reference group. Thus, there were four experimental groups: TN-AL, HS-AL, TN-PF8, and HS-PF8. In PF8 groups, the calculated daily ration was divided into 8 feed provisions distributed at every 90-min intervals from 09:00 to 19:30. In P2, HS-AL and PF8 pigs had lower ADFI (-19% on average; $P<0.01$), ADG (-25% on average; $P<0.01$), and a reduced final BW (-6.1 kg on average; $P<0.01$) than TN-AL pigs. Rectal temperature was increased by HS (+0.32 °C on average; $P<0.05$) but did not significantly differ within HS nor within TN groups ($P>0.05$). The PF8 groups had lower total viscera percentage than TN-AL pigs (-2.1% on average; $P>0.05$), while HS-AL had intermediate results. Pigs in HS-AL and in PF8 groups had similar slaughter BW ($P<0.05$) but HS-PF8 had more perirenal fat than TN-PF8 (+1.6%; contrast at $P<0.05$). Plasma insulin on d 8 of P2 was higher in HS-PF8 than in TN-PF8 pigs ($P<0.05$). Heat stressed pigs had lower plasma TT3 ($P<0.05$) even compared to TN-PF8 with the same feeding level. Plasma creatinine was higher in HS pigs within AL ($P<0.05$) and within PF8 groups ($P<0.05$). In AL groups, reduced growth performance in HS pigs was mostly due to a reduced feed intake. Increasing feed provision frequency during HS did not seem to improve pig performance. However, the increased fat deposition in internal depot in HS-PF8 could be an adaptation response related to direct effect of HS on pig metabolism.

Effect of supplemental diet with vitamin c and folic acid on broiler chickens under heat stress

A. Gouda[1], S. Tolba[2] and S.A. Amer[3]

[1]National Research Center, Animal production Dept., Dokki, Giza Government, 11865, Egypt, [2]Faculty of Veterinary Medicine, Zagazig University, Zagazig, Department of Nutrition & Clinical Nutrition, Zagazig Government, 44511, Egypt, [3]Faculty of Veterinary Medicine, Zagazig University, Zagazig, Department of Nutrition & Clinical Nutrition, Zagazig Government, 44511, Egypt; shimaamer@yahoo.com

This experiment was conducted to assess the impact of L-ascorbic acid (AA) and folic acid (FA) in mitigating heat stress (HS) deleterious effects when supplemented, either alone or in combination, to broilers' diets. For this aim, the effect of these supplements on the broilers' growth performance, blood parameters, antioxidant and immune status were evaluated. Hatchling Cobb-500 broilers (total=240) were fed either corn and soybean meal-based diet (control group), or basal diet supplemented with 200 mg AA/kg diet, 1.5 mg FA/kg diet, or 200 mg AA plus 1.5 mg FA/kg diet, for 35 days during the summer months (n=6 replicates/group, 10 birds/replicate). The minimum and maximum average temperatures ranged from 84.5 to 96.2 F and relative humidity from 68.5 to 76.5%. The supplemented vitamins either alone and (or) in combination improved ($P<0.01$) broilers' growth performance, levels of thyroid hormones, insulin like growth factor1, blood haemoglobin, total protein, albumin, globulin, heat shock protein70, total antioxidant capacity, catalase enzyme activity, superoxide dismutase enzyme activity, antibody titre against Newcastle disease virus, and decreased ($P<0.01$) heterophil/lymphocytes ratio. The magnitude ($P<0.01$) of vitamins effect on the analysed parameters was higher when they added together. In conclusion, providing a combination of AA and FA at 200 and 1.5 mg/kg, respectively, improved the broilers' antioxidant status with coordinated improvement in the growth performance and health status under heat stress conditions.

Across-generation influence of maternal heat stress during late gestation in dairy cows

C. Kipp[1], K. Brügemann[1], P. Zieger[2], K. Mütze[3], S. Möcklinghoff-Wicke[4] and S. König[1]
[1]Institute of Animal Breeding and Genetics, University of Giessen, Ludwigstraße 21 b, 35390 Giessen, Germany, [2]Diamond V, 2525 60th Ave SW, Cedar Rapids IA 52404, USA, [3]Hessian Association for Performance and Quality Tests in Animal Breeding e. V., An der Hessenhalle 1, 36304 Alsfeld, Germany, [4]Innovation Team Milk Hessen, Lochmühlenweg 3, 61381 Friedrichsdorf, Germany; cordula.kipp@agrar.uni-giessen.de

Heat stress (HS) impairs dairy cows' production, fertility, and health directly. Additionally, negative carry-over effects of HS in utero on offspring performances have been postulated. Therefore, the aim of this study was to quantify the effect of the average temperature-humidity-index (THI) during the last eight weeks of gestation on the offspring of Hessian Holstein dairy cows, considering a comprehensive dataset from calving years 2003 to 2013. Daily THI was merged with fat content measured on first test day (fat%; n=171,150), non-return-rate after 56 days of first insemination (NRR56; n=94,981), calving to first insemination interval after first calving (CFI; n=134,885), productive life (PL; n=119,599) and lifetime productivity (LTP; n=119,634). THI during the dry period significantly influenced all traits (P<0.0001-0.0454), with detrimental effects of increasing THI on offspring performances. For THI 50-59 and THI ≥60 least squares means for fat% were substantially lower (4.21 and 4.20%, respectively) than for THI≤39 and THI 40-49 (4.24 and 4.23%, respectively). Moreover, offspring of mothers exposed to THI≥50 had a lower conception rate than offspring of mothers exposed to THI≤49 (80 and 81%, respectively). Calving to first insemination interval and PL were also negatively influenced by maternal HS (CFI: THI≤39: 88 days, THI≥60: 91 days; PL: THI≤39: 915 days, THI≥60: 889 days). Offspring of mothers exposed to average THI≥60 produced 18,137 kg milk during life. In comparison, offspring of mothers, who experienced THI≤39 during late pregnancy, produced 18,422 kg. The strong detrimental HS effects across generations suggest an optimisation of the HS management in dry cows, and studying mechanism of epigenetics.

Cortisol levels and relative expression of ACTH gene in beef cattle

B.V. Pires[1], N.B. Stafuzza[2], S.B.G.P.N.P. Lima[2] and C.C.P. Paz[1,2]
[1]Ribeirão Preto Medical School (FMRP), University of São Paulo (USP), Departament of Genetics, Avenue Bandeirantes, 3900, Monte Alegre, 14049-900, Brazil, [2]Animal Science Institute, Sertãozinho, SP, Beef Cattle Research Center, Rodovia Carlos Tonani, km 94, Sertãozinho, SP, Caixa Postal 63, CEP 14160-900, Brazil; bianca2510@outlook.com

Cortisol release and adrenocorticotropic hormone in beef cattle are changed by high air temperature, relative humidity and solar radiation in tropical regions. Nelore (Bos indicus) is the most important beef cattle breed in Brazil, which is considered as adapted to tropical climate. This study evaluated the cortisol level and relative expression of ACTH gene in Nelore steers during sun exposure. A total of 37 Nelore steers were evaluated once a month in October and December (2017) and February and March (2018) a period characterised by high temperatures (30 °C to 37 °C). The animals were submitted to three treatments: morning (08:00 to 10:00 h) all animals stayed in a pen with exposure to the sun, shade (11:00 to 13:00 h) half of the animals remained in a shaded pen, and sun (11:00 at 13:00 h) half of the animals remained in a sun pen with absence of shadow. Blood samples were collected before and after treatments from all animals. The cortisol levels and the relative expression of ACTH were evaluated by ELISA and qPCR, respectively. The data were analysed using a MIXED model, residuals were checked for normal distribution and heteroscedasticity and were transformed on a logarithmic scale when null hypotheses were rejected. Treatments did not influenced the relative expression of ACTH (2.23±0.19 morning; 2.08±0.28 shade and 2.14±0.27 sun, P=0.9638) in Nelore steers. The steers had higher cortisol levels in the shade treatment in comparison to morning treatment (13.58±1.23 and 7.21±1.11, respectively). Differences observed between cortisol levels of steers in sun and morning treatment (12.62±1.22 and 7.21±1.11, respectively) were not significant. The cortisol alterations in this study were not sufficient to modify the ACTH gene expression in Nelore steers, which indicates that the Nelore breed is adapted to tropical climate.

A mixed linear model for the analysis of microbiome impact on heat stress in Chinese Holstein cows

J. Szyda[1,2], B. Czech[2,3], K. Wang[4], S. Chen[4] and Y. Wang[4]
[1]National Research Institute of Animal Production, Krakowska 1, 32-083 Balice, Poland, [2]Wroclaw University of Environmental and Life Sciences, Biostatistics group, Kozuchowska 7, 51-631 Wroclaw, Poland, [3]Roche Polska, Domaniewska 39 B, 02-672 Warszawa, Poland, [4]China Agricultural University, College of Animal Science and Technology, 100193 Beijing, China, P.R.; joanna.szyda@upwr.edu.pl

Heat stress poses a serious constrain on production and welfare of livestock. In view of the global climatic changes, the problem is rapidly gaining importance. In our study, faecal samples were collected from 136 Chinese Holstein cows with Estimated breeding Values for three traits related to heat response: rectal temperature, respiratory score and drooling score. Microbiota in each sample were identified by sequencing of a V3-V4 fragment of a gene encoding 16s RNA chain from the 30S ribosomal subunit. The data included 33 single-end and 105 paired-end samples. The bioinformatic analysis comprised standard pre-processing steps such as quality control (FastQC) as well as quality trimming and adapter removing (cutadapt). The Quantitative Insights Into Microbial Ecology 2 (QIIME2) software was used to analyse the 16S rRNA data. The applied analysis pipeline included the following steps: (1) de-noising of errors in reads; (2) de-replication that identifies unique reads; and (3) sequences clustering to collapse similar sequences. Further on, an Amplicon Sequence Variant (ASV) table was created that was used for sequence classification and to create a taxonomy table. Finally, alpha and beta diversity statistics were calculated. Then, in order to assess the interplay between polygenic and microbiotic sources of the variation in heat resistance, a mixed linear model was applied to the data. The model comprised a random additive polygenic effect of each cow with the covariance matrix expressed by the standard numerator relationship matrix and a random microbiome effect with the covariance given by the Bray-Curtis dissimilarity matrix. All computations were carried out at the Poznan Supercomputing and Networking Center.

Plasma antioxidant/oxidant balance of heat stressed ewes fed *Ascophyllum nodosum* and flaxseed

M.G. Ciliberti, M. Soccio, D. Pastore, M. Albenzio, A. Santillo, A. Sevi and M. Caroprese
University of Foggia, Department of the Sciences of Agriculture, Food and Environment (SAFE), Via Napoli 25, 71122, Foggia, Italy; maria.ciliberti@unifg.it

In ruminants, heat stress exposition causes an oxidative imbalance with a production of reactive oxygen species, lipid peroxidation, inactivation of antioxidant defence systems, and activation of heat shock response. Including ingredients such as seaweed or oilseed in the diet can contribute to reduce the negative effects of heat stress on physiological and immunological responses of livestock. The objective of this study was the application of the antioxidant/oxidant balance (AOB) approach for the evaluation of plasma antioxidant status in heat stressed ewes fed seaweed *Ascophyllum nodosum* (AG), whole flaxseed (FS) and their combination (FS+AG). Thirty-two late-lactation Comisana ewes were divided into four balanced groups and exposed to high ambient temperature for 30 days. Each group was individually fed twice daily and received 1.8 kg of oat hay; the experimental groups received 25 g of AG, 250 g of FS, and their combination (FS+AG), respectively. Body condition score (BCS), respiration rate (RR) and rectal temperatures (RT) were registered. Blood samples were collected at 0, 15 and 30 day and the antioxidant capacity (AC), plasma peroxide level (PxL) and antioxidant/oxidant balance values were measured on plasma. In white blood cell lysate, the level of heat shock protein 70 was measured. Data were analysed by ANOVA for repeated measurements using MIXED PROC of SAS. Results showed no differences in BCS and RR among groups; ewes fed FS+AG registered lower RT than ewes fed AG, and the lowest value of heat shock protein 70 concentration. Plasma AC values were lower in ewes fed FS diet than in ewes fed AG and FS+AG diet ($P<0.01$). On average, plasma PxL registered lower values from 15 to 30 day of the experiment ($P<0.01$). When the antioxidant/oxidant balance approach was applied, higher AOB value emerged in ewes fed FS+AG than in ewes fed FS diet both at 15 day and at 30 day of the experiment ($P<0.05$). In summary, the combination of *A. nodosum* and flaxseed in ewes' diet could improve antioxidant defence contributing to protect them from an excessive oxidative imbalance caused by heat stress.

Efficient pig and poultry production: what did we learn from Feed-a-Gene?

J. Van Milgen[1], K.E. Bach Knudsen[2], A. Jansman[3], V. Halas[4], J. Pomar[5], H. Gilbert[6] and G. Garrod[7]
[1]INRAE, Agrocampus Ouest, Le Clos, 35590, France, [2]Aarhus Univ, Postboks 50, 8830 Tjele, Denmark, [3]Wageningen Livestock Research, P.O. Box 338, 6700 AH Wageningen, the Netherlands, [4]Kaposvár Univ, Guba Sándor u. 40, 7400 Kaposvár, Hungary, [5]Univ. Lleida, Alcalde Rovira Roure 191, 25198 Lleida, Spain, [6]INRAE, 24 Chemin de Borde-Rouge, 31326 Castanet Tolosan, France, [7]Newcastle Univ, Centre for Rural Economy, NE1 7RU Newcastle upon Tyne, United Kingdom; jaap.vanmilgen@inrae.fr

The objective of Feed-a-Gene (H2020 grant agreement 633531) was to adapt the feed, the animal and the feeding techniques to improve the efficiency and sustainability of monogastric livestock production systems. Efficiency is a complex trait affecting the components of sustainability. Imported soybean meal is a high-quality protein source, but its use can have environmental and social impacts. Technological treatments applied to European-grown protein sources (i.e. soybean meal, rapeseed meal, and green biomass) allow to obtain fractions with different nutritional characteristics that can be fed to different livestock species. Replacing Brazilian soybean meal with locally-produced protein sources reduced energy cost and impact on climate change, but resulted in a transfer of land-use and in more arable land use. Novel traits indicative for feed efficiency were developed and tested (e.g. individual feed intake in broilers, genomic information on nitrogen efficiency, large-scale measurements of digestibility in individual pigs, blood metabolites, and behavioural traits). These traits can be used in livestock management strategies, and digestibility, microbiota, and blood biomarkers appeared promising for genetic selection. Different mathematic models were developed to assess digestive and metabolic efficiency, animal robustness, and variation among animals. A user-friendly tool was developed allowing users to work with these models. Prototypes of precision feeding systems were developed for broilers, growing pigs, and sows. These systems adjust the type of feed to the daily changing nutritional requirements of individual or groups of animals. Precision feeding in pigs reduced key environmental impacts and increased profitability compared to conventional feeding systems.

Heritability of nitrogen efficiency in Swiss Large White pigs: potential for selective breeding

C. Kasper, I. Ruiz-Ascacibar, P. Stoll and G. Bee
Agroscope, Animal Production Systems and Animal Health, Tioleyre 4, 1725 Posieux, Switzerland; claudia.kasper@agroscope.admin.ch

Nitrogen and phosphorus efficiency are important traits for sustainable pig breeding as pig production contributes to environmental pollution through excretion of these compounds. In addition, 36 million tonnes of soybeans are currently imported annually for European pig production, as domestic plant protein sources often do not provide the required protein quality and are not available in sufficient quantities. Most mineral phosphate sources for animal feed are also imported. The improvement of nitrogen and phosphorus deposition efficiency through selective breeding, without compromising growth performance, is therefore desirable. The aim is to reduce the dietary intake of crude protein and phosphate while maintaining growth rates, carcass compositions and meat quality similar to those currently achieved. Here, we present a preliminary evaluation of the potential of selecting for increased nitrogen and phosphorus deposition efficiency, for which we estimated genetic parameters of nitrogen and phosphorus efficiencies in the dam line of Swiss Large White pigs. We obtained nitrogen and phosphorus efficiency phenotypes from 294 individuals of various live weights by wet-chemistry analyses of empty body and carcasses. To calculate genetic parameters, we fitted Gaussian mixed-effect animal models, which allow the association of variance structures with pedigrees, in a Bayesian framework. The heritability of nitrogen efficiency was estimated at 41 and 36% in the empty body and carcass, respectively, but heritability of phosphorus efficiency was very low (0.3%). However, positive genetic correlations with nitrogen efficiency (r_G=0.73 in the empty body and r_G=0.84 in the carcass) suggest that breeding for nitrogen efficiency could positively affect phosphorus efficiency. A follow-up study, in which we focus on nitrogen and phosphorus efficiency at a live weight of 100 kg, with a larger sample size and a higher number of breeding boars and sows is underway to improve the quality of the estimates.

Comparison of growth curve models for Canadian purebred pigs

L. Maignel[1] and P. Gagnon[2]
[1]Canadian Centre for Swine Improvement, 960 Carling Avenue, Building 75, Ottawa, ON K1A 0C6, Canada, [2]Centre de Développement du Porc du Québec, Tour Belle Cour, 450-2590, Boulevard Laurier, Québec City, QC G1V 4M6, Canada; laurence@ccsi.ca

Hog market weights have been on the rise in North America during the past years. Breeding programs have to adapt to this trend by moving away from traditional approaches of adjusting live growth performance to specific target weights, such as 100 or 120 kg live weight, and towards studying growth patterns at heavier weights. The objective of this study was to compare different approaches to model growth of purebred pigs and explore flexible methods to predict age at any point during the growth phase for an individual animal. A total of 1,500 Canadian pigs including males, females, and castrates from Duroc, Landrace and Yorkshire breeds were weighed seven times between 30 and 160 kg live weight in twelve on-farm trials. Several S-shaped growth curves were tested (Logistic, Gompertz, Richards, von Bertalanffy) with different numbers of parameters (2 to 4). The simplified 2-parameter Gompertz equation was selected since its shape and parameter values make sense from a physiological point of view (maturity weight, inflexion point). To predict age at a target weight for individual animals, different approaches were tested, such as parametric adjustments of reference curves (global, sex-specific and sex within breed-specific curves) and classification methods (e.g. k-nearest neighbours). These were compared to the current sex-specific formulas in use in the Canadian Swine Improvement Program (CSIP). All approaches performed well to predict a target age at 105-120 kg live weight using an observation at a lighter weight. For parametric adjustments, results suggest that adjusting the A parameter of the Gompertz curve is better than adjusting the C parameter when testing is done at a heavier weight (>140 kg). For classification methods, the distance function is critical, and the results were highly dependent on the amount of data available for a given weight class. Strategies for the implementation of classification methods in the CSIP program using machine learning algorithms were also explored.

Effect of the digestible protein reduction on growth performance in fattening pigs fed *ad libitum*

E. Janvier, F. Payola, F. Guillard, C. Launay, E. Schetelat and A. Samson
ADM, Talhouët, Saint Nolff, B.P. 80234, 56006, France; emmanuel.janvier@adm.com

The aim of this study was to investigate the effects of the SID protein reduction in fattening pigs fed *ad libitum*. In total, 72 pigs (68 days of age, 23.7±1.5 kg) were randomly allocated to four dietary strategies. The trial was divided into four periods (d68 to d83, d83 to d104, d104 to d132 and d132 to slaughter). For the first and second period considered as the growing phase, pigs were either fed a high (H; 13.9% SID protein) or a low protein diet (L; 12.4% SID protein) containing 9.70 MJ NE/kg and 0.88% SID Lys. For the third and fourth period considered as the finishing phase, pigs were also either fed a high (H; 12.4% SID protein) or a low protein diet (L; 10.9% SID protein) containing 9.90 MJ NE/kg and 0.80% SID Lys. Then, the dietary treatments compared were HHHH, LLLL, HLHL and LHLH. The results showed that there was no significant difference in the average daily feed intake (ADFI) among the groups for the four periods. No significant difference was observed in the average daily gain (ADG) for the first period (P>0.10), but it tended to differ for the second and third period (P≤0.10) and was significantly different among the groups for the fourth period (P<0.001). The feed conversion ratio (FCR) was significantly different among the groups for the last three periods (P<0.01). When considering the growing and the finishing periods, there was also no significant difference in the ADFI among the groups. The ADG only differed significantly for the finishing period (P=0.01) being the highest for the HHHH and LHLH groups and the lowest for the LLLL and HLHL groups (1,082±78, 1,058±73, 1,005±122 and 956±180 g/d respectively). The FCR tended to differ for the growing period (P=0.10) and was significantly different for the finishing period (P<0.0001), being the lowest for the HHHH and LHLH groups and the highest for the LLLL and HLHL groups (2.58±0.22, 2.62±0.12, 2.74±0.15 and 2.90±0.40 respectively). However, regarding protein efficiency (g protein/kg of growth), the most efficient pigs were those in the LLLL group for both the growing and finishing periods, meaning that they required the least amount of protein to gain 1 kg BW (P<0.0001). Backfat thickness measured at d 146 did not differ significantly among the groups as well as the lean meat content at slaughter (P>0.10). Finally, as the growth performance was not significantly different among the groups for the growing period (d68 to d104), the SID protein concentration can be reduced at 12.4%. However, the reduction of the SID protein concentration at 10.9% for the finishing period (d104 to slaughter) impaired the growth performance. This reduction may be possible at an intermediate level between the H and L diets and would require further investigations.

Effect of a nutritional strategy on fat deposition and lipid-related gene expression in Duroc pigs

E. Jiménez-Moreno[1], P. Aymerich[2], D. Menoyo[3], J. Bonet[2], D. Carrión[1] and J. Coma[2]
[1]Cargill Animal Nutrition, Polígono Industrial Riols, s/n, 50170 Mequinenza, Spain, [2]Vall Companys Group, Polígono Industrial El Segre, 605, 25191 Lleida, Spain, [3]Universidad Politécnica de Madrid, Departamento de Producción Agraria, Calle Senda del Rey, 18, 28040 Madrid, Spain; encarnacion_jimenez@cargill.com

The aim of this study was to evaluate the combined effect of supplementing a feed additive for meat quality in finishing pigs fed a low SID Lys:NE ratio on the *Gluteus medius* adiposity. A total of 380 crossbred pigs sired by Duroc were fed with a common feeding program from 19 to 85 kg live weight (80 days of fattening). On d81, half of the pigs were fed a control diet with 2,500 kcal NE/kg, and 3.20 SID Lys:NE ratio (g/Mcal) and the other half, with a treatment diet containing 2,550 kcal NE/kg, and 2.55 SID Lys:NE ratio and supplemented with a feed additive for meat quality for the last 30 days of fattening. Feed additive increases fat, fat -soluble vitamins and mineral digestibilities improving so, marbling, oxidative stability and sensorial quality of the meat. All pigs were slaughtered on the same day and carcass traits harvest. Hot empty carcasses weighing between 85 and 95 kg were selected for the determination of intramuscular fat (IMF) content, and both ham subcutaneous fat thickness (HFT) and lean percentage were measured. In addition, carcasses with the 6 lowest and 6 highest IMF contents per gender and treatment (48 in total) were randomly chosen for the determination of subcutaneous fat inner layer fatty acid (FA) profile and the analysis of the relative expression of genes encoding lipogenic and lipolytic enzymes in both adipose and muscular tissues. The combined effect of the feed additive and the low SID Lys:NE ratio increased HFT by 1.8 mm (P<0.001) and reduced lean percentage by 1.61 (P<0.01). This nutritional intervention increased the level of expression of FA synthase and stearoyl-CoA desaturase in both adipose and muscular tissues in concordance with a higher oleic acid (P<0.01), as well as total monounsaturated FA content (P<0.05). In conclusion, the supplementation of feed additive for meat quality in finishing feeds with low in SID Lys:NE ratio promoted lipogenesis in the pig *G. medius* increasing ham subcutaneous fat thickness with a higher concentration of oleic acid.

Interaction between fat and fibre level on the nutrient digestibility of pig feed

L. Paternostre[1,2], S. Millet[1,2] and J. De Boever[2]
[1]Ghent University, Faculty of veterinary medicine, Heidestraat 21, 9820 Merelbeke, Belgium, [2]Flanders Research Institute for Agriculture, Fisheries and Food (ILVO), Animal Sciences Unit, Scheldeweg 68, 9090 Melle, Belgium; louis.paternostre@ilvo.vlaanderen.be

In recent years, more and more fibrous feed ingredients are used in pig diets. To ensure a sufficient energy concentration, these diets are usually supplemented with fat. Studies have shown that the fibre content could affect the digestibility of the other components due to an effect on transit time. Other studies have shown that the fat content also could have an impact on nutrient digestibility. The goal of the present study was to investigate the interaction between fat and fibre level on the nutrient digestibility of pig feed. Therefore, two basal diets were formulated, with a low crude fibre (CF) content (LF, 35 g/kg) and with a high CF content (HF, 68 g/kg). To both basal diets 0, 2 or 4% of either animal (pig) fat or soy oil was added, resulting in 10 experimental diets. Acid insoluble ash (0.5%) was added as digestibility marker and each feed was fed *ad libitum* to 6 pens of 3 pigs (averaging between 84 and 110 kg BW). The digestibility of crude protein and crude fat was higher for LF than for HF due to reduced nutrient absorption and/or increased endogenous excretion of the latter. By increasing fat level, irrespective of fat type, apparent fat digestibility increased due to the better digestibility of the added fat and lower endogenous losses in comparison with the fat contained in raw materials. When we calculated the standardised digestibility of both added fats, there was no significant effect of fibre or added fat level nor an interaction. The standardised digestibility of the fat free organic matter was lower in the LF diet, but was not affected by adding fat to the diet. No significant effect of fat type was observed. In conclusion, within the studied ranges no interaction between the fibre and fat level on nutrient digestibility could be seen, supporting the assumption of additivity in diet formulation.

Effects of carbohydrase on ileal digesta, and nutrient and fibre digestibility in growing pigs

G.I. Lee, T.S. Nielsen, H.N. Lærke and K.E. Bach Knudsen
Aarhus University, Animal Science, Blichers Allé 20, 8830 Tjele, Denmark; gilee@anis.au.dk

This study was conducted to investigate the effects of a carbohydrase complex (Rovabio Advance®) in the form of xylanase (Xyl), arabinofuranosidase (Abf) and β-glucanase (BG) on ileal and total tract digestibility of dry matter, energy and nutrients in pigs fed different types of cereal grains and co-products. The experiment was carried out as a 8×4 Youden square design (8 diets and 4 periods by 2 blocks) using 16 growing pigs (33.3±0.8 kg) surgically fitted with a T-cannula in the terminal ileum. The pigs were fed eight experimental diets – four cereal based diets: Maize based (Maize), wheat based (Wheat), rye based (Rye), and wheat and rye mixed diet (WRmix) that were provided without and with enzymes. Apparent ileal digestibility (AID) and apparent total tract digestibility (ATTD) of dry matter (DM), energy, starch, crude protein (CP), fat, and soluble and insoluble fibre were calculated by index method using titanium dioxide as an indigestible marker. Addition of carbohydrase resulted in significant improvement in the AID of high molecular weight arabinoxylan (HMW-AX) ($P=0.044$) and starch ($P=0.042$), whereas none of the other components were affected ($P>0.05$). There was a tendency ($P<0.10$) of a higher AID of non-cellulosic polysaccharide glucose residue (β-glucan) and soluble AX in enzyme treatment diets than in non-enzyme treatment diets. However, there was no interaction between cereal types and the carbohydrase ($P>0.05$). The composition of ileal digesta was influenced by cereal type for total non-digestible carbohydrates, total AX, and HMW-AX ($P<0.05$), whereas there was no enzyme effect on the ileal content ($P>0.05$). Collectively, the results indicate that the carbohydrase is efficient in degrading AX leading to higher ileal digestibility.

A static balance model to quantify and explore mammary nutrient metabolism in lactating sows

U. Krogh[1], H. Quesnel[1], N. Le Floc'h[1], A. Simongiovanni[2] and J. Van Milgen[1]
[1]INRAE, Agrocampus Ouest, PEGASE, 16 Le Clos, 35590 Saint-Gilles, France, [2]Ajinomoto Animal Nutrition Europe, 32 rue Guersant, 75017 Paris, France; uffe.krogh@inrae.fr

Quantitative estimates of mammary nutrient uptake and metabolism in sows are scarce, despite being central for the feeding of lactating sows. The aim was to quantify uptake, metabolism and output of nutrients by the mammary gland and to identify factors regulating mammary metabolism in lactating sows. Results from four studies were integrated into a dataset including data on litter performance, milk composition, and mammary arterial-venous differences (AV–difference) of amino acids (AA) and main energy metabolites. Milk yield (based on litter gain and size) and composition were used to estimate milk nutrient output. Mammary plasma flow (based on Phe+Tyr as flow marker) and AV–differences were used to estimate net mammary fluxes. The dataset constituted the basis for the construction of a static model, in which glucogenic carbon (C), ketogenic C, and AA C were prioritised for synthesis of lactose, fat, and protein, respectively. Surplus C was used for energy processes supporting milk synthesis (for the *de novo* synthesis of fat, lactose, peptide bond formation, milk protein and mammary tissue turnover, and transmembrane transport of glucose and amino acids) and consequently directed to a CO_2 pool. The remaining glucogenic C was used in the carbon skeleton of *de novo* fat. The model suggested that glucogenic C constituted 34% of milk fat C and 88% of net mammary CO_2 release. In addition, 45% of the mammary CO_2 release was used to support *de novo* fat synthesis. Accordingly, 46% of glucogenic C was directly (31%) or indirectly (15%) associated with milk fat synthesis. More than 90% of mammary AA C was secreted in milk protein. Simultaneously, nitrogen balances of individual AA and relationships between AA availability and uptake suggested that the availability of branched-chain AA, Arg, and Phe influence mammary uptake of at least for these AA. In conclusion, milk protein synthesis appears to be the main driver of total mammary AA uptake, but of less importance for some individual AA. Moreover, *de novo* fat synthesis appears to be of great importance for mammary energy efficiency.

Performance, colostrum and milk fatty acid profile in sows fed divergent ω-6:ω-3 fatty acids ratios

T.X. Nguyen[1,2], A. Agazzi[2], M. Comi[2], V. Bontempo[2], G. Invernizzi[2], S. Panseri[2], H. Sauerwein[1], M. Pinerio[1], D. Eckersall[1], R. Burchmore[1] and G. Savoini[1,2]
[1]The MANNA Consortium, University of Glasgow, Glasgow, G12 8QQ, United Kingdom, [2]Università degli Studi di Milano, Department of Health, Animal Science and Food Safety, Via dell'Università 6, 26900 Lodi (LO), Italy; thixuan.nguyen@unimi.it

Dietary ω-6 and ω-3 polyunsaturated fatty acids (PUFAs) influence animal health and performance because they regulate inflammatory and immunomodulatory processes. We tested the hypothesis that decreasing the dietary ω6:ω3 ratio during gestation and lactation sows will improve reproductive performance and the FA profile in colostrum and milk. Sixteen multiparous sows were allotted to two treatments in a completely randomised design from d 28 of gestation onwards. Treatments consisted of dietary ω-6:ω-3 PUFA ratios of 13:1 (HR) and 4:1 (LR) by adding soybean and linseed oils at different amounts to the basal isocaloric diet. Body weight (BW) of sows was measured at d 28, 79 and 108 of gestation and d 26 of lactation. Data were analysed by analysis of variance using the GLM procedures of SAS. The LR increased (P<0.05) BW gain of sows from d 79 to 108 of gestation, the survival rate of piglets at weaning and litter weight gain from d 14 to 26 and from d 0 to 26, compared to HR. Colostrum and milk collected at d 0 (farrowing), d 7 and 26 of lactation showed higher (P<0.05) level of total ω3, C18:3 ω3, C20:3 ω3, C20:5 ω3, C22:6 ω3, and lower (P<0.05) ω6:ω3 ratio in sows fed LR diet. We conclude that decreasing ω6:ω3 ratio in gestation and lactation diets of sow improves reproductive performance and the FA profile in colostrum and milk.

Differences in feeding behaviour of high and low protein-efficient pigs

E.O. Ewaoluwagbemiga, C. Kasper and G. Bee
Agroscope, Animal Production System and Animal Health, Route de la Tioleyre 4, 1725 Posieux, Switzerland; esther.ewaoluwagbemiga@agroscope.admin.ch

Improving protein efficiency (PE) is highly relevant in pig breeding considering its impact on the environment through the excretion of excess nitrogen. Since a large part of dietary protein is excreted, an efficient way to improve protein efficiency would be to reduce dietary crude protein level. However, whether the degree in PE affects their behaviour is unknown. The aim of this study was therefore to explore the differences in feeding behaviour of high protein-efficient (HPE) and low protein-efficient (LPE) pigs when dietary protein is reduced to 80% of the usual recommendation. Pigs (n=302) had *ad libitum* access to feed and water, and individual feeding behaviour was recorded from 20 to 100 kg body weight (BW). Pigs were slaughtered at 100 kg BW and protein content of the carcass measured with Dual-Energy X-Ray Absorptiometry scanner. Protein efficiency was calculated as the ratio of protein in the carcass to total protein consumed. HPE (n=14) and LPE (n=30) pigs were determined as individuals above or below 1 standard deviation of the average PE. Feeding behaviour traits monitored were daily feed intake (DFI; g/day), feed intake per visit (FIV; g/visit), number of daily visits (NDV; visits/day), duration of visits (DUV; min/visit), feeding rate (g/min), and duration at feeder per day (min/day). The average daily gain (ADG) was also calculated. ANOVA was used to analyse the differences in feeding behaviour between these two groups while correcting for effect of sex and experimental series. The average DFI and ADG of HPE and LPE pigs did not differ between both groups. By design, the PE of HPE (45%) was greater than of LPE (33%) and P<0.0001 between the two groups. HPE pigs spent 35% less (P<0.01) time at the feeder per visit, ate 34% less feed (P<0.01) per visit but visited the feeder 27% more (P<0.01) frequently than the LPE pigs. In conclusion, one could speculate that HPE pigs coped with the reduced protein diet by greater feeder visits and lower feed bouts and by doing so, they might have covered amino acid supply for body protein synthesis more constantly compared to LPE pigs.

Dietary interventions and oxidation and energy expenditure in newborn IUGR piglets: a pilot study

J.G. Madsen, M.N. Nielsen and C. Amdi
University of Copenhagen, Department of Veterinary and Animal Sciences, Grønnegårdsvej 2, 1870 Frederiksberg, Denmark; johannes.g.madsen@sund.ku.dk

Piglets exposed to intra uterine growth restriction (IUGR) exhibit lower energy reserves at birth compared with their larger littermates, which increases risk of hypothermia leading to higher mortality rates. The study objective was to investigate two different dietary interventions (DI) for IUGR piglets; Glucose (Glu; Glucose Baxter Viaflo, Denmark) (50 mg/ml) injected subcutaneously (4× 1.5 ml, two in the groin area and two in the neck) and a tube-fed dosage of porcine colostrum (Col; 20 ml warmed to 35 °C) at two different levels of temperature; 22 (T22) and 36 (T36) °C compared with a placebo (P) treatment (20 ml of water) of both IUGR and normal (Norm) piglets (8 groups in total; 48 piglets; n=6). Piglets were weighed prior to insertion into respiratory chambers, where oxygen (O_2), carbon dioxide (CO_2) and methane (CH_4) emissions were measured in 1 h intervals over a 4 h period. Respiratory quotients (RQ) and heat increment (HI) were calculated based on O_2, CO_2, and CH_4 values. Piglets were supplied with the DI 1 h after insertion into the respiratory chambers, and remained for another 3 h. Not until between the 1 h to 2 h after DI in T36 IUGR-Glu and -Col piglets displayed numerically greater HI/kg BW compared with IUGR-P (12.7 and 11.3 vs 7.5%), whereas in T22 IUGR-Col and -P displayed similar increases which were numerically greater than IUGR-Glu piglets (13.0 and 12.8 vs 9.1%). However, 2 h and 3 h after DI, respectively, only IUGR-Col and Norm-P in T22 and IUGR-Glu in T36 displayed HI/kg BW (0.56, 2.04 and 0.81%). In the period from the 1 h to 2 h after DI in T36 RQ changed from 0.34-0.57, 0.28-0.55, 0.32-0.58, 0.52-0.71 for IUGR-P, -Glu, -Col and Norm-P piglets, respectively. In conclusion, only at 36 °C and between 1 h and 2 h after DI IUGR piglets responded positively to Glu and Col treatments compared with P, while RQ values were markedly numerically lower in IUGR than Norm piglets. These preliminary findings underline the potential differences in energy expenditure between IUGR and normal piglets and the need for further investigating the interactions between early DI and housing conditions and its effect of energy metabolism in IUGR piglets.

Poster session

G. Bee[1] and S. Millet[2]
[1]Agroscope, la Tioleyre 4, 1725, Switzerland, [2]Flanders Research Institute for Agriculture, Fisheries and Food (ILVO), Scheldeweg 68, 9090, Belgium; giuseppe.bee@agroscope.admin.ch

Poster session

Reproductive traits and VCL measurement of pigs depending on polymorphisms in the HOXA10 gene

A. Mucha[1], M. Szyndler-Nędza[1], R. Tuz[2], K. Piórkowska[1], M. Małopolska[1] and T. Schwarz[2]
[1]National Research Institute of Animal Production, Sarego 2, 31-047 Kraków, Poland, [2]University of Agriculture in Krakow, Mickiewicza Ave. 24/28, 30-059 Krakow, Poland; aurelia.mucha@izoo.krakow.pl

Homeotic genes (*HOXA*) play an important role in the development of reproductive organs during foetal life and are expressed during the oestrous cycle. The aim of the study was to determine the effect of polymorphisms in the *HOXA10* gene on reproductive performance of sows and vagina-cervix length (VCL) measurement in Polish Large White (PLW) and Polish Landrace (PL) sows and in the genetic reserve breeds of Złotnicka White (ZW) and Złotnicka Spotted (ZS). The study involved 269 PLW and PL pigs as well as 60 ZW and ZS pigs. The analysis included reproductive traits: total number of piglets born (TNB), number of piglets born alive (NBA), number of piglets on day 21 in parity (N21), the age of gilt at first farrowing (FF), average interval between successive litters (IBS), length of pregnancy (LP), and VCL measurement. During insemination, the VCL was evaluated for each sow using a foam tip catheter. Polymorphisms were identified by Sanger sequencing using the GenomeLab DTCS-Quick Start Kit (Beckman Coulter, USA) on a Beckman Coulter sequencer. The analysed breed groups (PLW, PL / ZW, ZS) differed in the frequency of genotypes and alleles of different gene variants. In the PLW and PL breeds, the frequency of genotypes at all three positions exhibited Hardy-Weinberg disequilibrium, while the ZW and ZS breeds did not deviate from Hardy-Weinberg equilibrium. Analysis of the effect of the polymorphism of the different gene variants on litter size traits in PLW and PL sows showed statistically significant differences only for *HOXA10 g.45398714C>G* at TNB and NBA (CG>CC, P≤0.05) and for *HOXA10 g.45398767G>A* at FF (AA>GG, P≤0.05). In the ZW and ZS breeds, *HOXA10 g.45398502C>A* gene polymorphisms had an effect on TNB and LP (respectively: CC>AA, P≤0.01; AC and CC>AA, P≤0.01), *HOXA10 g.45398714C>G* on TNB (CG>GG, P≤0.01), and *HOXA10 g.45398767G>A* on TNB, NBA and VCL (respectively: GG>AA and AG, P≤0.01, P≤0.05; GG>AA, P≤0.05 and GG>AA; P≤0.05). The study showed a greater possibility of predicting reproductive traits and VCL on the basis of *HOXA10* gene polymorphisms in ZW and ZS than in the PLW and PL breeds.

Effects of immunocastration on plasma free amino acid concentration of Iberian pigs

I. Fernández-Fígares, L. Lara, M. Lachica, I. Seiquer, A. Haro and R. Nieto
Estación Experimental del Zaidin, EEZ, CSIC, Profesor Albareda, s/n, 18008 Granada, Spain; rosa.nieto@eez.csic.es

Pig immunocastration -vaccination against gonadotropin releasing hormone (GnRH)- prevents sexual development and boar taint being a feasible alternative to surgical castration (SC). Previous studies indicate that immunocastrated (IC) pigs show better performance than SC pigs. Enhanced animal welfare and increased performance makes immunocastration of particular interest for native pig breeds, with long productive cycles and low capacity for lean growth. In the context of a wider experiment, the purpose of this study was to examine the effects of immunocastration on plasma concentration of free amino acids (PFAA) of Iberian pigs fed different protein content diets. Fifty-four pure Iberian pigs were used (3 sexes: IC males, IC females, SC males; 3 isoenergetic diets: 150, 130 and 110 g CP/kg DM, 6 pigs/treatment combination). The SC pigs were castrated within the first week of life. Vaccination against GnRH was at 4.3 (40 kg) and 6 months of age (70-80 kg). Body weight (BW) was monitored weekly and feed intake daily. Pigs consumed the experimental diets individually housed from 40 to 105 kg-BW. At 105 kg-BW pigs were slaughtered and blood samples taken for plasma separation and PFAA analysis. Effects of immunocastration on post-absorptive PFAA were moderate, and dietary protein had little influence on PFAA profile. Branched-chain amino acids (BCAA; valine, leucine and isoleucine) and lysine were the most abundant within indispensable PFAA; and glycine, glutamic acid, alanine, glutamine and proline among dispensable PFAA. Histidine concentration was greater in IC males than in SC males (32%; P<0.01). Isoleucine was higher in males compared to females (14%; P<0.01), and valine in IC males compared to SC males and IC females (17%; P<0.001). Total BCAA were greater in IC males than in IC females (16%; P<0.01) with SC males in intermediate position. Within indispensable PFAA, proline and taurine were lower in IC males than in the other two groups (0.001<P<0.05). Higher BCAA plasma concentrations were observed in IC male pigs which previously showed increased performance and N retention capacity than the rest of groups. Funded by European project H2020 TREASURE (Grant agreement no. 634476).

Meta-analysis of the effect of a nutritional strategy on fat deposition in Duroc crossbred pigs

E. Jiménez-Moreno[1], J. Aibar[2], J. Mesalles[1], R. Rabadán[1], C. De Blas[2] and D. Carrión[1]
[1]Cargill Animal Nutrition, Poligono Industrial Riols s/n, 50170 Mequinenza, Spain, [2]Universidad Politécnica de Madrid, Departamento de Producción Agraria, Calle Senda del Rey 18, 28040, Spain; encarnacion_jimenez@cargill.com

A meta-analytical approach was used to quantify the effect of a nutritional strategy on backfat thickness, intramuscular fat (IMF) content in longissimus dorsi, and meat quality characteristics in finisher Duroc crossbred pigs. Duroc pig was used as a terminal sired line and crossed with Iberian (D×IB) or with European (D×EU) dam line. Commercial diets met the minimal nutrient requirements for each pig crossbreed over 100 kg live body weight (FEDNA, 2013). The diets with nutritional strategy contained a novel vitamin-mineral premix designed for meat quality and had 2,500 kcal NE/kg and 1.62 SID Lys: NE ratio (g/Mcal) for D×IB- and 2,450 kcal NE/kg and 1.90 SID Lys:NE ratio, for D×EU crossbred pig. Data were obtained from 13 field trials conducted between years 2015 and 2018. Not all meat traits were measured in all trials. Nutritional strategy increased IMF content in longissimus dorsi (4.87 vs 4.26%; $P<0.001$); an effect that was more pronounced in D×EU- than in D×IB crossbred pigs (18.8 vs 3.6% of improvement; $P<0.05$). In Spanish market, D×IB- are slaughtered with heavier hot empty carcasses and have a greater IMF content than D×EU crossbred pigs. No treatment by gender (castrated male vs female) was observed for IMF content ($P>0.1$). Nutritional strategy increased backfat thickness by 11.4% in D×EU crossbred pigs ($P=0.06$). The use of nutritional strategy improved marbling by 0.47 points ($P<0.001$) and meat colour by 0.23 points ($P<0.01$) measured by the methodology of the NPCC scales (1999). In conclusion, this meta-analysis proves the robustness of nutritional strategy to improve the meat quality traits in Duroc crossbred pigs.

Do standard and luxury pig feed really differ in digestibility and growth performance?

L. Paternostre[1,2], J. De Boever[2] and S. Millet[1,2]
[1]Ghent University, Faculty of veterinary medicine, Heidestraat 21, 9820 Merelbeke, Belgium, [2]Flanders Research Institute for Agriculture, Fisheries and Food (ILVO), Animal Sciences Unit, Scheldeweg 68, 9090 Melle, Belgium; louis.paternostre@ilvo.vlaanderen.be

Feed suppliers formulate different types of diets, depending on their clients' needs. In this study, we investigated whether diets with different quality also differ in energy content, nutrient digestibility and whether this translates in better performance. Therefore, we asked 5 feed companies to provide a 'standard' and a 'luxury' feed for pigs in two stages, between 10 and 15 weeks and between 15 and 20 weeks of age, resulting in 10 mutual comparisons. Acid insoluble ash (0.5%) was added as digestibility marker and each feed was fed *ad libitum* to 6 pens of 3 pigs. Pigs were weighed at the beginning and the end of each interval and feed intake was recorded. Feed and faeces were analysed for their chemical composition by accredited ISO methods and the net energy content (NE) was calculated according to CVB (2016). Luxury feeds contained on average more crude protein (179 vs 173 g/kg DM), more starch (444 vs 431 g/kg DM) and more NE (10.91 vs 10.70 MJ/kg DM) than the standard feeds, but similar contents of crude fat (47 vs 46 g/kg DM), sugar (51 vs 50 g/kg DM) and crude fibre (51 vs 55 g/kg DM). In 7 from the 10 intercomparisons, daily gain was better with the luxury feed than with the standard feed, amounting on average to 856 and 824 g/d respectively, but none of the differences was significant ($P>0.05$). A better feed conversion ratio was observed for 8 of the 10 luxury feeds (2.12 vs 2.22 g/g), but the difference was only significant for one feed. The difference in NE between standard and luxury feed was not always reflected in the growth performances. In conclusion, these results show that increasing nutrient density is not always successful for improving feed quality and that differences between standard and luxury feed may be due to other changes like in amino acid composition.

Dietary inosine-5'-monophosphate saved metabolisable energy for finishing pigs
L.P. Bonagurio, C.A. Moreira, L.D. Castilha, A.E. Murakami and P.C. Pozza
State University of Maringá, Animal Science, 5790 Colombo ave., 87020900, Brazil; pcpozza@yahoo.com.br

This study aimed to evaluate the dietary inosine-5'-monophosphate (IMP) associated with reduced metabolisable energy (ME) on the performance and blood plasma parameters of 75- to 100- kg barrows. Fifty-four barrows, averaging 75.62±0.96 kg, were distributed in a completely randomised blocks design with six treatments and nine replicates. Treatments consisted of a positive control diet achieving the ME requirements of the finishing pigs (PC; 3,300 kcal ME/kg), a negative control diet (NC; 3,200 kcal ME/kg), NC supplemented with 0.050, 0.100, 0.150 or 0.200% IMP. Daily feed intake, average daily gain (ADG), feed:gain (F:G), the plasmatic concentrations of triglycerides (TG), cholesterol (CL) and also the high (HDL), medium (LDL) and very low (VLDL) density lipoproteins were evaluated. Data were compared by contrasts and the IMP levels were submitted to regression analysis, fitting linear response plateau (LRP), linear or quadratic regression models. The NC showed a lower ADG than PC, 0.100 and 0.150% IMP. The PC showed only a higher ADG (P=0.019) than 0.050% IMP. Furthermore, the association of the LRP and quadratic models showed a maximum ADG (P<0.001) and F:G (P=0.057) estimated at 0.109 and 0.112% IMP, respectively. All the evaluated IMP levels showed a higher TG than NC and PC (P≤0.05). Supplementing 0.010, 0.150 or 0.200% IMP increased (P=0.019; P<0.001; P=0.001) the CL compared with NC. Moreover, CL concentration linearly increased (P=0.027) according to increasing dietary IMP levels. The PC increased the HDL (P<0.001) compared with NC and all IMP levels. Dietary concentration of 0.050, 0.100 and 0.150% IMP showed lower (P<0.001; P=0.013; P=0.004) LDL than NC, as well as 0.05 and 0.150% IMP were lower (P=0.008; P=0.030) than PC. The NC showed a higher (P≤0.05) VLDL than PC and all the dietary IMP levels. Likewise, the PC provided a higher (P=0.002) VLDL compared with 0.100% IMP. Additionally, a quadratic model was fitted for VLDL (P=0.005) due to the studied IMP levels. In conclusion, 3,200 kcal ME/kg of diet associated with 0.109% IMP may be used for a better ADG of finishing barrows, and the dietary IMP also affected the blood plasma TG, CL, VLDL, HDL and LDL.

Candidate genes related with fatness traits in pigs
K. Ropka-Molik, K. Piórkowska, K. Żukowski, M. Tyra and G. Żak
National Research Institute of Animal Production, Sarego 2, 31-047 Kraków, Poland; grzegorz.zak@izoo.krakow.pl

The fatness characteristic as well as daily gain and feed conversion ratio are one the most important traits in pig production. The previous RNA-seq research allowed us to selected genes which expression was significantly different depend on fatness phenotypes in pigs. Based on whole transcriptome adipocyte tissue profile the SETD2 and TBX2 genes were selected in order to evaluate their potential association with fatness and fattening traits. For both genes polymorphisms were detected (SETD2 – ENSSSCP00000012085.3:p.Thr165Ser; TBX2 ENSSSCP00000021191.1:p. Ala644Val) and genotyped using PCR-RFLP method for 712 pigs (polish landrace – 284; large white – 241 and native pulawska pigs – 134). The genotyping showed that analysed population was not consistent with Hardy-Weinberg equilibrium for both polymorphisms. The association study showed that SETD2 gene was related with feed conversion ratio (kg/kg) and carcass yield (%) (P<0.05). In turn, TBX2 polymorphism was associated with daily gain (g), age at slaughter and average backfat thickness (mm). The obtained results indicated on further researches which should be carried out taking into account large number of pigs and different breeds. The study was sponsored by 01-18-05-21 National Research Institute of Animal Production statutory activity.

Effect of crossbreeds on organically grown pigs growth parameters

I. Jansons[1], V. Sterna[1], L. Degola[2] and D. Jonkus[2]
[1]*Institute of Agricultural Resources and Economics, Department of Plant Breeding and Agroecology, Dizzemes', Dizstende, Libagi parish, Talsi County, 3258, Latvia, [2]Latvia University of Life Sciences and Technologies, Institute of Animal Sciences, Liela street 2, Jelgava, 3001, Latvia; imants.jansons@arei.lv*

In organic farming, it is very important to choose the right crossbreeds of pigs to obtain high-quality pork and improve economic efficiency. The aim of research was to investigate the effect of crossbreeds on fattening pig growth parameters. At the study were used 28 fattening pigs divided in two groups. In the Group 1 were included crossbred pigs Landrace × Large White (LLW) cross to Duroc (D) (LLW × D) with an initial average weight 36.9 kg in the Group 2 crossbred pigs Landrace × Large White cross to Pietrain (P) (LLW × P) with average weight 36.5 kg. Pigs were kept in control observation pens daily recorded individual live weight gain and feed intake. The pigs were provided with all the requirements appropriate to organic farming. In the study organic nutrients were used to prepare diets according to the physiological requirements of the pigs and were prepared two rations grower feed and finisher feed from 0.75 kg until the end of fattening. Diets contained barley, wheat, soybean meal, oil, organic additives. Data were statistically analysed using SAS / STAT 9.22 software package. Obtained results showed the feed conversion was not affected by crossbreed in the Group 1 it was 2.38±0.05, in the Group 2 it was 2.34±0.04. Pigs from Group 1 the final live weight reached at the age 160 days 112.24±1.72 kg the Group 2 by 14 days later 112.45±2.76 kg reached at the age 174 days. Daily live weight gain was 0.86±0.03 kg for pigs in the Group 2 in the Group 1 it was by 15% higher 1.02±0.02 kg between groups were found significant difference (P<0.05). Landrace × Large White cross to Duroc. From the obtained results we can conclude that the crosses pigs (LLW × D) showed better growth rates than (LLW × PJ) crosses.

Time-restricted feeding modifies feed intake and performance of male Mulard ducks

J.E. Rico[1], C.M. Perdomo[2], M. Lehmann[2], L. Delprat[3], M. Lefrançois[3] and D.E. Rico[2]
[1]*Cornell University, 345 Morrison Hall, 14850, USA, [2]CRSAD, 120-A Chemin du Roy, G0A1S0, Deschambault, Canada, [3]Univeristé Laval, Animal Science, 2425 Rue de l'agriculture, G1V0A6, Quebec, Canada; daniel.rico@crsad.qc.ca*

Time-restricted feeding (TRF) is used for the preparation of the crop in ducks destined for foie-gras production. Rodent research suggests TRF can reduce body and liver weight. Our objective was to evaluate the effects of TRF on feed consumption, body and liver weight of ducks. In experiment one, 12 male Mulard ducks (70 d of age, 4.0±0.32 kg of BW) housed in individual cages were blocked by weight and randomly allocated to 1) *ad libitum* feeding (AL; n=6) or 2) two feeding periods of 1 h each (TR; 08:00-09:00 h and 14:00-15:00 h; n=6) during 12 d before slaughtered for liver collection. Feed intake was measured every other day until d 10. Data were analysed using a mixed model including the random effect of duck within block and the fixed effects of treatment, time and their interaction. Feed intake increased progressively (time P<0.001) in both groups, but was not affected by treatment. Body weight was lower in the TR than in the AL group (3.65 vs 3.99 kg respectively; P=0.01), but liver weight was not different (53.8 vs 60.8; P=0.20). In experiment two, 24 male Mulard ducks (52 d of age, 3.89±0.26 kg of BW) were blocked by weight and randomly allocated to 1) AL (n=12) or 2) TR (n=12) for 14 d. Intake was measured daily and animals weighed on d 7 and 14. A subsample of animals was euthanised on d 14 for liver collection (n=4/treatment). Statistical analyses were as described for experiment 1. Feed intake was lower in TR on d 1 and 2, and higher than in AL from d 4 to 9 (P<0.05). Body weight increased over time (P<0.001), but was not different between TR and AL (4.10 and 4.04 kg, respectively), whereas liver weight tended to be higher in TR than in AL (91.8 and 78.3 g, respectively P=0.08). Time-restricted feeding can alter intake and performance of individually housed Mulard ducks, which may depend on age.

Effects of short-term feeding of high-fructose, high-lipid diets on performance of Mulard ducks

L. Delprat[1], M. Lehmann[2], M. Lefrancois[1], J.E. Rico[3] and D.E. Rico[2]
[1]*Université Laval, Animal Science, 2425 Rue de l'Agriculture, G1V0A6, Quebec, QC, Canada, [2]CRSAD, 120-A Chemin du Roy, G0A1S0, Deschambault, QC, Canada, [3]Cornell University, Animal Science, 345 Morrison Hall, 14850, USA; leslie.delprat.1@ulaval.ca*

Type of dietary carbohydrate and lipid are known to alter feed intake and adiposity in rodents. Replacing dietary starch from corn with dietary fructose and saturated lipids may result in a similar phenotype in the duck. 48 male Mulard ducks of (90 d of age, 4.13±0.34 of BW) were blocked by weight and randomly allocated to either 1) Control (75% starch; C) or 2) high-fat, high-fructose diet (11% fat, 42% fructose, 28% starch: FF) in a complete randomised block design. 12 additional ducks from the same cohort were euthanised on d 0 for baseline measures. Intake was measured every 3 days and animals weighed every week until euthanised either on d 14 (n=24) or on d 28 (n=24). Carcass fat content was determined using dual-energy x-ray absorptiometry. Feed intake tended to be higher in control than in FF on 3 and d 16 (P=0.07), whereas energy intake was not different at any time-point. Liver colour was evaluated using a chromameter to determine variation in the green-red (a), white-black (L), and yellow-blue (b) scales. Data were analysed using a mixed model including the random effect of duck within block and the fixed effects of treatment, time and their interaction when appropriate. Body weight and fat content were not different between treatments and averaged 4.07±0.03 kg and 27±1.05%, respectively. Liver weight was 26% higher in C than in FF on d 14 (P<0.001), but not different between treatments on d 28. Livers were more yellow on d 28 for the FF than for the C group (6.6 and 4.8 b values, respectively; P<0.01). Despite its higher energy content, feed intake was not reduced by the FF diet, suggesting a potential alteration of satiety signals, which may impact body liver weight under longer-term interventions in ducks destined for the production of foie gras.

Effects of high energy diets and fructose supplementation in the drinking water on duck performance

L. Delprat[1], M. Lefrançois[1], J.E. Rico[2] and D.E. Rico[3]
[1]*Université Laval, Animal Science, 2425 Rue de l'Agriculture, G1V0A6, Quebec, QC, Canada, [2]Cornell Univeristy, Animal Science, 345 Morrison Hall, 14850, USA, [3]CRSAD, 120-A Chemin du Roy, G0A1S0, Deschambault, QC, Canada; leslie.delprat.1@ulaval.ca*

Long-term feeding of high-fat, high-fructose diets has been shown to alter feed intake and cause fatty liver in rodents. Such diets in combination with the addition of high fructose corn syrup (HFCS) in the drinking water may result in a similar phenotype in ducks destined for the production of foie gras. 50 male Mulard ducks (66 d of age, 4.16±0.37 kg of BW) housed in individual cage were blocked by weight and randomly allocated to 1) Control diet (C; 75% starch); 2) Control diet and sugar water (CSW); 3) high-fat, high-fructose diet (FF; 12% fat, 42% fructose, 26% starch) or 4) high-fat, high-fructose diet and sugar water (FFSW) during 92 d before slaughtered for liver collection. Animals were weighed weekly and feed intake measured every 5 d. 55% HFCS was mixed with tap water (13% w/w) and provided *ad libitum* to the CSW and FFSW groups. Liver colour was evaluated using a chromameter to determine variation in the green-red (a), white-black (L), and yellow-blue (b) scales. Data were analysed using a mixed model including the random effect of duck within block and the fixed effects of treatment, time and their interaction when appropriate. There was a treatment × time interaction (P<0.05) for feed intake as CS and FFS exhibited lower values relative to C and FF (32% reduction on average from d 5 to 90; P<0.05). A treatment × time interaction was observed for body weight as C was higher than all other treatments from d 7 to 90 (P<0.05). Liver weight was not affected by treatment and averaged 63.5±3.2 g. However, Livers were more yellow in FFS than in CS (b scale; P<0.05), and not different between F, FFS and C. Long-term feeding of high high-fat, high-fructose diets did not reduce feed intake, however provision of HFCS reduces feed intake regardless of basal diet composition. Under the conditions of the present trial, high-fat, high-fructose diets seem to have a limited impact on performance of Mulard ducks.

Probiotics in broiler chicks' diet fed with different sources of protein

G. Ciurescu[1] and M. Dumitru[1,2]
[1]*National Research & Development Institute for Biology and Animal Nutrition, Calea Bucuresti no.1, Baloteşti Ilfov, Romania, [2]University of Agronomic Science and Veterinary Medicine, Doctoral School, 59 Marasti Street, Bucharest, Romania; ciurescugeorgeta@yahoo.com*

Probiotics have been provided to be the most effective alternative to antibiotics. The aim of this study was to evaluate the potential of *Bacillus subtilis* ATCC 6051a (Pro), as a probiotic bacterium in broiler diets based of two sources of protein [soybean meal (SBM) and cowpea (CWP; *Vigna unguiculata* [L] Walp). Growth performance (GP), carcass traits, and gut microflora population were investigated using a total of 480 one-d-old mixed sex broiler chicks (Ross 308). The SBM or CWP diets were tested in the presence or absence of Pro (1×10^{11} cfu/kg feed) in a 2×2 factorial arrangement in a completely randomised design. Chicks were randomly assigned to 4 dietary treatments (6 replicate pens with 20 chicks each). Diets of starter, grower, and finisher were formulated to be isocaloric, isonitrogenous with similar content of total lysine, TSAA, Ca and available P, and manufactured in mash form, without inclusion of growth promoters or antibiotics. The results showed that birds fed CWP had comparable GP as those fed the SBM diet. Carcass, breast and legs' yield, and digestive organs size (i.e. gizzard, heart, liver, pancreas, spleen, small intestine, caecum) were not affected by the protein source. Inclusion of Pro in broilers' diets increased BWG during grower (P<0.01), finisher (P<0.01), and overall study period (P<0.001), tended to increase FI (P=0.059; d 0-42) and improved feed efficiency in the grower (P=0.047) and finisher (P=0.043) phase. The addition of Pro significantly decreased abdominal fat (P=0.026), and caecum weight (P=0.034) and tended to increase pancreas (P=0.057), and caecum length (P=0.086) while carcass as well as breast and legs' yield were similar. In addition, Pro was beneficially in modulating gut flora composition. In particular, the cecal *Bacillus* spp. ($\log_{10}$ cfu/g digesta) and *Enterococcus* significantly increased (P=0.016 and P=0.024 respectively) whereas the Coliforms (P<0.001) and *E. coli* count decreased (P=0.040). It is concluded that the *B. subtilis* (ATCC 6051a) as probiotic product significantly improved the GP and beneficially affects the gut bacterial community in SBM and CWP diets.

Metabolic changes in three Iberian pig strains during the finishing period in two feeding systems

N. Garrido[1], F.I. Hernández-García[1], A. González-Bulnes[2], M.A. Perez[1] and M. Izquierdo[1]
[1]*CICYTEX, Animal production, Finca La Orden, 06187 Guadajira, Badajoz, Spain, Spain, [2]INIA, Animal Reproduction, INIA, Avda. Puerta de Hierro s/n, 28040 Madrid, Spain, Spain; ngarridoff@unex.es*

Sixty-five animals of three different Iberian pig strains (Retinto, Lampiño and Torbiscal) were assigned to two different feeding systems, one based in a standard concentrate, and another was the traditional free-range acorn-feeding *montanera*. The aim of this study was to determine blood metabolite levels to characterise physiological parameters for the different strains fed in these different feeding systems and their level changes from the beginning to the end of the finishing period. There were significant differences in blood biochemistry, but none of the blood parameters were within pathological levels. Lampiño was the strain with the highest urea level and the highest urea change. Torbiscal had intermediate values for all parameters and Retinto had the highest values for triglycerides, cholesterol and glucose, as well as for their respective changes. In relation to the feeding system, animals fed with concentrate had the highest values of urea, triglycerides, lactate and glucose, as well as their changes. In contrast, there were no significant differences in cholesterol or fructose levels. In conclusion, there were differences in blood parameters due to genotype, but they also can be modified by the feeding system, with higher values for most parameters in the concentrate-fed pigs than in the acorn-fed animals.

The role of milk and milk products in the diets of children, adolescents, pregnant women and adults

I. Givens
University of Reading, Institute for Food, Nutrition and Health, Earley Gate, RG6 6AR, United Kingdom;
d.i.givens@reading.ac.uk

Different life stages give rise to important nutritional challenges some of which are related to increasingly aged populations in many Western societies. There is however, increasing evidence that diets during childhood and adolescence can impact on health in later adulthood. For example, undernutrition in childhood can lead to stunted growth which is associated with reduced cognitive ability and increased risk of chronic diseases and co-morbidities in adulthood. Despite recent worldwide improvements, stunting in sub-Saharan Africa remains about 40% and some countries have an even higher prevalence. Milk is a key food for reducing stunting with milk proteins having a crucial role. In some Western societies recent reductions in milk consumption have led to sub-optimal intakes of calcium and magnesium by teenage females in particular, at a time when bone growth and mineralisation are at their maximum, and of iodine during pregnancy needed to ensure that supply/production of thyroid hormones to the foetus is adequate. It is of note that the concentration of some key nutrients, particularly iodine is influenced by the diet of the dairy cow. Low intakes of calcium is a particular concern since many populations are also of sub-optimal vitamin D status. This may already have had serious consequences in terms of bone development which may not be apparent until later life. Recent meta-analyses show no evidence of increased risk of cardiovascular diseases from high consumption of milk and dairy foods but increasing evidence of a reduction in the risk of type 2 diabetes associated with fermented dairy foods, yoghurt in particular. With an increasingly aged population there are concerns about the loss of muscle mass and strength in the elderly with associated increased risk of bone breakage. Recent work has shown that milk proteins, and whey protein in particular, have potential to reduce muscle loss. This session aims to examine the role of dairy foods at key life stages in terms of their ability to moderate chronic disease risk.

Advanced monitoring of milk quality to address the demand of added-value dairy products

C. Bastin[1], F. Dehareng[2], N. Gengler[3], M. Sindic[3], E. Piraux[4], H. Soyeurt[3], H. Wilmot[3], C. Bertozzi[1], J. Leblois[1] and F.G. Colinet[3]
[1]Wallon Breeders Association Group, 5590 Ciney, Belgium, [2]Walloon Agricultural Research Centre, 5030 Gembloux, Belgium, [3]Gembloux Agro-Bio Tech, ULiège, 5030 Gembloux, Belgium, [4]Milk Committee, 4651 Battice, Belgium; cbastin@awenet.be

Consumers are seeking for local, healthy and direct-from-producers dairy products. Hence, assessing the suitability of milk to be processed in dairy products either directly in the farms or through local dairy plants is of great interest. Moreover, several studies demonstrated the usefulness of mid-infrared (MIR) spectrometry for the prediction of various traits related to the nutritional and technological properties of milk. This study presents 5 groups of MIR predicted traits related to various aspects of dairy products: (1) milk coagulation traits, (2) cheese and butter yields, (3) nutritional quality, (4) texture and, (5) sensory quality. The MIR prediction equations of these 5 groups of traits were applied on standardised spectra from individual milk samples collected in the frame of the Walloon milk recording scheme. After edits, more than 780,000 records collected between 2017 and 2019 were used. The MIR predictions for coagulation time and curd firmness were combined to define a new trait with 5 levels assessing the overall milk coagulation property, from poorly coagulating milk to optimal milk for coagulation. Casein and calcium contents and titrable acidity were also studied in relation to milk coagulation properties. The nutritional quality of milk was assessed through the content in fat of PUFA and the health promoting index (i.e. UFA / (C12 + C14 × 4 + C16). The spreadability of butter was defined as the ratio of C18:1 cis-9 to C16:0. The sensory quality of milk was assessed through SCC and the content of free fatty acids. Results showed that about 10% of the records were classified as poorly coagulating milk; these records had lower calcium and casein contents with lower MIR predicted cheese yield. Also, 60% of the cows were never classified in the poorly coagulating milk class while 5% of the cow produced poorly coagulating milk at least half of the time. Moreover, cheese yield was highly correlated with the protein and fat contents. The results also showed the influence of days in milk, parity, herd and breed on all traits. Local breeds (i.e. Dual Purpose Belgian Blue and Eastern Red) showed favourable milk fat profile for nutritional quality of milk and texture of dairy products even if these breeds tend to have higher proportion of suboptimal milk for coagulation. These results indicates that dairy farmers have the opportunities to monitor and improve milk quality for the production of added-value dairy products.

The MIQUALAT project: nutraceutical properties of cow milk, preliminary results

C. Marchitelli[1], F. Napolitano[1], M. Contò[1], S. Failla[1], G. Chillemi[2], F. Porcelli[2] and A. Crisà[1]
[1]Consiglio per la ricerca in agricoltura e l'analisi dell'economia agraria (CREA), Research Centre for Animal production and Aquacolture, Via Salaria 31, 00015 Monterotondo, Italy, [2]University of Tuscia, Department for Innovation in Biological, Agro-food and Forest systems (DIBAF), Via S. Camillo de Lellis snc, 01100 Viterbo, Italy; cinzia.marchitelli@crea.gov.it

Milk and dairy products are integral part of human nutrition and they are considered as the carriers of higher biological value proteins, essential fatty acids, minerals, amino acid, antioxidant molecules and several bioactive compounds. The goal of the MIQUALAT project is to evaluate the amount of bioactive prebiotic and protective molecules in milk of different cattle breeds, in two lactation points (60 and 120 day by calving), in order to identify differences between them. Also an investigation both at genomic and transcriptomic levels to identify gene loci, allelic variants and genes differentially expressed associated with phenotypic characteristics will be performed. In the project the experimental herd established at CREA consisted from three groups of lactating cows (purebred Holstein, purebred Simmental and Crosses) and Podolica breed as an outgroup, with twenty-five cows per group. The CREA experimental herd and Podolica breed are differentially reared, intensive vs extensive farming, and experimental herd fed with the same diet (Unifeed) in the analysed lactation points. During the course of the project milk recording data, fatty acid profiles, oligosaccharides, sialic acid, glutathione, thiol compounds, and ^{1}H-NMR spectroscopy-based metabolomics profiles are being determined. Milk recording data analysis by GLM showed significant differences between the three genetic groups of the experimental herd. Multivariate exploratory techniques on milk recording data and NMR profiles discriminate the four breeds and confirmed the Podolica as an outgroup. These differences between CREA herd and Podolica are influenced by feeding dietary factors. This research is supported by MIQUALAT project D.M. 16844/7100/2019 (MIPAAFT).

Mediating role of silence on the issue of animal welfare in the Irish dairy industry

S.E.M. Snijders[1], A. Rieple[1] and L.A. Boyle[2]
[1]University of Westminster, Business School, 35 Marylebone road, NW1 5LS, United Kingdom, [2]Teagasc, Pig Production Department, Animal and Grassland Research and Innovation Centre, Moorepark, Fermoy, Co. Cork, Ireland; s.snijders@westminster.ac.uk

The Irish dairy industry has undergone intensification post quota. This is associated with challenges for the welfare of dairy animals. Given growing competition and demands from consumers for sustainable and ethical practices, this can destabilise the dairy industry. We aimed to elucidate attitudes towards the welfare of dairy animals amongst stakeholders in the industry. Using organisation silence theory as the guiding framework we carried out semi-structured interviews and group discussions with members of the Irish dairy industry, including farmers, policy makers, managers of the co-operatives, as well as university-based animal scientists, vets, dairy researchers and university dairy science students. Analysis was abductive identifying plausible explanations for the attitudes to animal welfare described. Three types of silence were identified, differentially influenced by various characteristics of the industry and its participants: (1) a failure to know that some practices constitute poor cow welfare. This type of silence emerged principally from the contestation of definitions by different stakeholders combined with the lack of hierarchy in the industry which meant that no one voice dominated; (2) a failure to admit that poor welfare exists, either to themselves, colleagues, or to other interested parties such as consumers. The mechanism for this was rooted in the psycho-social aspects of farmer identity and experience, exacerbated by perceived pressures to maintain the grass-based reputation of the industry; (3) a failure to report on poor welfare practices, despite an awareness that these exist. This type of silence emerged to protect the health and welfare of colleagues as well as a fear of scrutiny that would undermine the farmers' sense of self-worth. In conclusion, organisational silence theory can provide a framework to understand the impediments to progress in protecting the welfare of animals in a post quota era and could contribute to improving the sustainability of the Irish dairy industry.

Working towards a socially sustainable dairy industry

M. Von Keyserlingk
Universtity of British Columbia, Animal Welfare Program, 2357 Main Mall, Vancouver BC, V6T 1Z4, Canada;
nina@mail.ubc.ca

In the more than 50 years since Ruth Harrison published her book 'Animal Machines' there has been growing public opposition to many intensive production systems, such as battery cage housing for hens, veal crates for calves, and gestation stalls for sows. This opposition has also led to the bans on the use of battery housing for hens and gestation stalls for sows in some parts of Europe. More recently this opposition has also included many aspects of dairy production, including zero-grazing practices, individual housing of calves, restrictive tie stall housing and cow calf separation. Some stakeholders working within agriculture argue that these intensive systems are beneficial, in part because they perceive these to improve animal health and productivity, and that rather than changing the systems, those working in agriculture should instead seek to better educate regarding these benefits. In this lecture I review work some of our work examining whether efforts to educate the public regarding agricultural practices improves willingness to accept these systems. There is also some evidence showing that although education can improve knowledge it does not automatically improve perceptions of farm animal welfare. Indeed, simply knowing more does little to change our underlying values, and increasing knowledge can result in the development of new concerns. Thus, there is growing evidence indicating that relying on the premise that educating the public will lead to acceptance of farm animal management practices will likely do little to resolve societal concerns about animal welfare on livestock farms, if practices fail to resonate with public values. Research focused on exploring perceptions, knowledge, and values of animal welfare among the public can aid in identifying socially sustainably farming systems. My hope is that by asking the public to tell us how they envision the ideal dairy farm, we can begin to identify areas where current practices are out of step, and consider methods that better align with public values. This type of approach can provide the industry a basis for predicting which factors are likely to come under increasing criticism, and where research efforts should be devoted.

Milk and casein nitrogen use efficiency of dairy farms in the Parmigiano Reggiano production area

M. Simoni[1], R.G. Pitino[1], M.A. Belaid[2], A. Foskolos[3], A. Summer[1], A. Quarantelli[1] and F. Righi[1]
[1]University of Parma, Department of Veterinary Science, via del taglio 10, 43126 Parma, Italy, [2]Università Autonoma de Barcelona, Department of Animal and Food Science, Campus de la UAB, 08193 Bellaterra, Barcelona, Spain, [3]University of Thessaly, Department of Animal Science, Campus Gaiopolis, 41110 Larisa, Greece; marica.simoni@unipr.it

A study was conducted on 19 Holstein dairy farms in the Parmigiano-Reggiano production area with the aim to evaluate on a herd base the milk nitrogen use efficiency (MNE) and casein nitrogen use efficiency (CNE) of dairy cows fed hay-based total mixed ration (TMR). Dry matter intake was calculated based on the amount of diet offered and refused during 2 days. The diets were analysed to estimate nitrogen (N) content and calculate N intake. Data on total milk yield (MY) and milk composition -including casein concentration- were recorded on a farm basis and; milk N yield (N milk), milk casein nitrogen yield (CNY) were determined to calculate MNE and CNE as N milk/N intake and CNY/N intake, respectively. The farms were then categorised based on the daily average MY/cow as low -L- (MY/cow ≤33 kg/day, 9 farms) and high -H- (MY/cow >33 kg/day, 10 farms). MNE and CNE data were normally distributed and analysed as a function of farm categories through the mixed model, with farm as random effect. MNE and CNE ranged from 0.24 to 0.37 and from 0.19 to 0.29, respectively. The L farms (average size: 213±94 cows; 181±20 DIM; average diet: 15.1% CP, 38.7% NDF and 21.5% starch (DM basis)) had an average MY of 30.86±1.60 kg/cow/day. The H farms (average size: 176±95 cows; 174±20 DIM; average diet: 15.3% CP, 37.9% NDF; 22.0% starch diets (DM basis)) had an average MY of 35.86±2.59 kg/cow/day. The H farms showed a trend for higher MNE (0.30 vs 0.27; P=0.068) and CNE (0.23 vs 0.21; P=0.075) in comparison to L farms. Overall, given similar levels of dietary CP, these results seem to confirm the higher MNE of high yielding herds. The CNE could be a further parameter to evaluate the efficiency of N use when milk is processed for cheese production. This project has received funding from the European Union's Horizon 2020 research and innovation programme under the Marie Skłodowska-Curie grant agreement No 777974. We thank the farmers and the volunteer students involved in the study.

Effects of nutritional solutions for dairy cows in reducing nitrogen excretion and ammonia emission
S.A. Salami[1], H. Warren[1] and J. Taylor-Pickard[2]
[1]Alltech (UK) Ltd., Ryhall Road, Stamford, United Kingdom, [2]Alltech Biotechnology Centre, Summerhill Road, Dunboyne, Ireland; saheed.salami@alltech.com

Ruminants excrete 60 to 90% of ingested nitrogen (N) in manure and ammonia (NH_3) volatilisation accounts for the major pathway through which manure N is lost into the environment. Notably, N losses including NH_3 represent important nutrient loss and pollutants that contribute to ecosystem damage, water pollution, climate change and human health problems. To reduce N excretion and NH_3 emissions in dairy production systems, a protein management program can be developed using nutritional technologies such as yeast-derived microbial protein (YMP; DEMP®, Alltech Inc., USA) and a yucca-based extract (YE; De-Odorase®, Alltech Inc., USA). DEMP provides bypass form of microbial protein that allows for a reformulation of the diet for better protein utilisation whereas De-Odorase is a yucca plant extract that binds NH_3 and reduces NH_3 emissions from manure. A simplified simulation was conducted based on static equations to quantify the N excretion and NH_3 emission from the manure of high-producing dairy cows fed YMP and YE (additive scenario). The simulation inputs for the baseline scenario include: number of cows = 1000; average body weight = 650 kg/cow; milk yield = 35 kg/cow/d; dry matter (DM) intake = 24 kg DM/cow/d; crude protein (CP)=16.6% DM diet; and number of lactation days = 305 d. In the additive scenario, the simulation inputs were the same as the baseline scenario except that the diet was reformulated with YMP resulting in a diet CP of 15.6% and cows were fed 3 g YE/cow/d. The effect of YE in reducing NH_3 loss was set at -25% based on results from previous *in vivo* studies that examined YE effect on NH_3 loss in animal housing and manure storage. Compared to the baseline scenario, the additive scenario reduced dietary N intake by -6.0% (637.4 vs 599.0 g N/cow/d), total N excretion by -6.6% (137.92 vs 128.81 tons N), N excretion intensity by -6.6% (12.92 vs 12.07 g N/kg milk) and total NH_3 loss by -30% (83.44 vs 58.45 tons NH_3). Reduction in NH_3-N loss can increase the fertiliser N value of manure (+20.66 tons N), equivalent to an economic value of €17,559. In conclusion, the simulation showed that nutritional strategy combining YMP and YE can reduce N excretion and NH_3 loss in dairy production.

How many cows and how much milk production per cow when the Netherlands goes circular
S. Van Der Beek[1,2] and C. Roozeboom[1]
[1]Wageningen University & Research, Animal Production Systems Group, P.O. Box 338, 6700 AH Wageningen, the Netherlands, [2]CRV BV, P.O. Box 454, 6800 AL Arnhem, the Netherlands; sijne.van.der.beek@crv4all.com

We assumed a circular farming system optimised for food production. In this system land suitable for arable farming is used to grow human food, land only suitable for grassland is used for grazing ruminants, and food waste and by-products from the human food system are used to feed livestock. Such a system deviates from the current system, in which additional by-products are imported and feed that directly competes with human food is used. We studied for the Dutch situation the optimal milk production level of dairy cows when only grassland and by-products were used. Only by-products from food for the Dutch population were available. We optimised the system to maximise the animal protein production. A linear programming model was developed in General Algebraic Modelling System (GAMS) version 24.2. We varied cow weight, energy requirements, protein requirements, feed intake capacity and longevity of dairy cows to study the effect of these parameters on optimal milk production. We also studied the effect of changes in the quality and quantity of available feed. In our optimised base scenario there were 625,261 cows that produced on average 4,939 kg milk per year. In this scenario the animal protein production was 22.4 grams per capita per day. Lowering energy requirements by 10% (simulating a cow that more efficiently converts energy from feed into milk) lead to an increase of average milk production to 6,039 kg milk per cow per year. Cow numbers did not change. Lowering protein requirements had no effect indicating that energy and not protein is the limiting factor in the available diet. Increasing feed intake capacity by 10% lead to an increase in average milk production to 5,775 kg milk per cow per year and number of cows decreased to 587,450. In a situation more comparable to the current Dutch situation with maize land and more and better grassland available, average production increased to 6,470 kg milk per cow per year. The number of cows almost doubled to 1,183,855. In this scenario the animal protein production was 52.5 grams per capita per day. In a situation where three times more by products were available, average production increased to 8,500 kg per cow per year, the number of cows increase to 697,817 and animal protein production was 42.8 grams per capita per day. In a circular food system optimal milk production per cow and number of cows were much lower than in the current production system. Both optimal production level and cow numbers were highly sensitive on the assumptions on quality and quantity of the feed available.

Determinants of calf marketing from dairy production in Baden-Württemberg

C. Reiber, M. Wollmeister and M. Chagunda
Hans-Ruthenberg-Institute of Agricultural Sciences in the Tropics, University of Hohenheim, Animal Husbandry and
Breeding in the Tropics and Subtropics (490h), Garbenstrasse 17, 70599, Germany; c_reiber@uni-hohenheim.de

The objectives of this contribution are to: (1) characterise the calf marketing situation from dairy production in Baden-Württemberg (BW), Germany; (2) to evaluate the effects of breed, sex, age, production system (organic/conventional) and season on calf weight at the time of marketing; and (3) to identify price determinants considering characteristics of a calf and the market situation. A dataset including all calf sales at weekly or monthly livestock markets from 3,753 farms (56% of all dairy farms in BW) in the period from 2014 until 2018 was analysed. Two regression models with the dependent variables price and weight were adjusted for the most important breeds. Since 2015, annual marketing volumes have declined. The most traded breeds were Fleckvieh (FV, 52%), Holstein (HF, 27%), milk-meat crosses (XFM, 7%) and Brown Swiss (BS, 5%). About 8% of these calves came from an organic farm. Compared to conventional farms, organic farms had a lower HF breed share and a higher BS share. The mean age of the marketed calves ranged from 18 days for HF to 38 days for FV. With the exception of HF breed, the average selling age of organic calves was lower. The conventional calves achieved slightly higher average prices across all breeds. The decisive factor for the sales destination was the breed. Almost all FV calves remained in Germany, of which 42% stayed in BW. 45% of HF calves went almost exclusively to Spain whereas only 18% found a buyer in BW. It made no difference whether the calf came from a conventional or an organic farm. Results of the weight model showed that breed and age were most significant determinants. The calves of the breeds FV and XFM showed significantly higher daily weight increases between 14 and 90 days compared to the breeds HF and BV. Slightly lower gains of about 20 grams per day were recorded in female calves and organic calves. The analysis of the price effects showed that the FV and XFM breeds as well as male calves produced significant positive price effects. Significantly price discounts were identified for HF breed, for organic calves (-3.5 €) and for exported calves (-28 €). Results revealed that the low economic value of the HF calves demands for improvement measures, particularly for the organic farms.

Application of random regression models on age for semen volume and concentration in Holstein bulls

I. Fujimoto[1,2], T. Hanamure[2], H. Hayakawa[2], J. Kawakami[3] and K. Hagiya[1]
[1]Obihiro University of Agriculture and Veterinary Medicine, Obihiro city, Hokkaido, 080-8555, Japan, [2]Genetics Hokkaido assoc., Sapporo city, Hokkaido, 060-0004, Japan, [3]Holstein Cattle Association of Japan, Hokkaido Branch, Sapporo city, Hokkaido, 003-0011, Japan; i-fujimoto@gh-assoc.ne.jp

In Japan, the semen from some Holstein bulls is collected at an early age (10 to 13 months) for progeny testing, and it is collected again after progeny testing. In bulls with high genomic enhanced breeding value, semen continues to be collected from an early age into maturity. Semen volume and sperm concentration after progeny testing may differ from when bulls are young. Here, we applied a random regression model on bull age to estimate genetic parameters for semen volume and sperm concentration. A total of 35,294 records collected on 1,284 Holstein bulls by Genetics Hokkaido Association from 2005 to 2013 were used for our analysis. The pedigree file contained data on 4,294 animals traced back five generations from bulls with records. The random regression model contained year–month (107 levels), location (two levels), age (second-order regression), and sampling method (two levels) as fixed effects and additive genetic and permanent environmental components as random effects. For the random regressions a first- or second-order Legendre polynomial was applied. GIBBS3F90 was used to estimate the variance components. A total of 500,000 rounds were completed, with 100,000 rounds discarded for burn-in and samples stored every 10 rounds. In the first-order random regression, the posterior means of heritability for semen volume increased from 0.06 at 10 months to 0.60 at 72 months, and those for sperm concentration increased from 0.09 at 10 months to 0.51 at 72 months. The posterior means of genetic correlations were 0.94 between semen volumes at 12 and 24 months, 0.85 between the volumes at 12 and 48 months, and 0.80 between those at 12 and 72 months. The posterior means of genetic correlations were 0.78 between the concentrations at 12 and 24 months, 0.22 between concentrations at 12 and 48 months, and 0.11 between those at 12 and 72 months. The heritabilities of both traits were higher than those of fertility traits in cows. The posterior means of genetic correlations between semen volume at a young age and older ages were high (more than 0.80).

The role of dairy-beef production in determining the environmental efficiency of dairy farms

A.D. Soteriades[1,2], A. Foskolos[3,4], D. Styles[2,5] and J.M. Gibbons[2]
[1]*Bangor University, Sir William Roberts Centre for Sustainable Land Use, Deiniol Road, LL57 2DG, Bangor, United Kingdom,* [2]*Bangor University, School of Natural Sciences, Deiniol Road, LL57 2UW, Bangor, United Kingdom,* [3]*Aberystwyth University, IBERS, Ceredigion, Aberystwyth, SY23 3EB, United Kingdom,* [4] *University of Thessaly, Department of Animal Science, Campus Gaiopolis, 411 10 Larissa, Greece,* [5]*Ryan Institute, NUI, Plant & Agri-BioSciences Centre, Galway, Ireland; andreasfosk@hotmail.com*

Producing more milk from fewer dairy cows is seen as an effective measure for mitigating greenhouse gas emissions per litre of milk. However, this may come at the expense of higher environmental burdens on the land and water (e.g. land use and eutrophication) per litre of milk. In addition, milk-oriented indicators of dairy farm environmental efficiency ignore the fact that dairy-beef is also a dairy farm output. This study developed an aggregate indicator of dairy farm environmental efficiency (or 'eco-efficiency') using a mathematical method (data envelopment analysis; DEA) able to weight and combine multiple burdens and outputs (milk and dairy-beef production). A sensitivity analysis on the output DEA weights was performed to change the relative weight of dairy-beef relative to milk in the aggregate indicator. The different scenarios were then regressed on indicators of animal- and farm-level farming intensity. For all scenarios, it was found that measures such as increasing milk output per cow and per hectare increased eco-efficiency. However, in scenarios where the weight for dairy-beef relative to milk increased, the relationship between environmental efficiency and farming specialisation (expressed as l milk per kg dairy-beef produced) reversed from positive to negative. In conclusion, dairy-beef production is pivotal in determining the wider environmental efficiency of dairy systems, and its under-representation from efficiency studies give a misleading picture.

The effects of Trp-Phe and acetate combination on milk protein synthesis in Holstein dairy cows

J.R.V. Conejos[1,2], W.S. Kim[1,2], Y.H. Jo[1,2], J.E. Kim[1,2], J.H. Jo[1,2], H.Y. Park[1,2], J.G. Nejad[1,2], J.S. Lee[2], E.J. Kim[3] and H.G. Lee[1,2]
[1]*Konkuk University, Brain Korea 21 Plus Project, 120 Neungdont-ro, Gwangjin-gu, Seoul, 05029, Korea, South,* [2]*Konkuk University, Department of Animal Science and Technology, 120 Neungdont-ro, Gwangjin-gu, Seoul, 05029, Korea, South,* [3]*Kyungpook National University, Department of Animal Science, 80 Daehak-ro, Sangyeok-dong, Buk-gu, Daegu, 41566, Korea, South; hglee66@konkuk.ac.kr*

The objective of this study was to determine the effects of tryptophan (Trp) and phenylalanine (Phe) combination and acetate supplementation on milk protein synthesis in immortalised bovine mammary epithelial (MAC-T) cell line as well in Holstein dairy cows. In *in vitro* study, study 1 was done to determine the optimum dosages for Trp and Phe. Dosage test showed that 0.9 mM Trp and 0.6 mM Phe are the optimum dosages for these two amino amino acids. Then, in study 2, cells were treated with (1) control (2) 0.9 mM Trp; (3) 0.6 mM Phe (4) combination of Trp and Phe (Trp-Phe); and (5) combination of acetate plus Trp and Phe (Acetate + Trp-Phe) with 1:1 ratio of amino acid to acetate for 72 h. Addition of Trp and Trp-Phe combination increased the relative percentage of medium secreted protein concentration (P<0.05). The mRNA relative expression showed that the addition of Trp-Phe and the combination of Acetate+Trp-Phe increased the beta casein relative expression (P<0.05) but not individual amino acids. In *in vivo* study, intravenous administration of (1) saline (control); (2) Trp; (3) Phe; (4) Trp-Phe; and (5) Acetate+Trp-Phe for 5 days with 3 days of adaptation period were done using five lactating cows (587.5±24.01 kg, DIM=122 days) fitted with indwelling jugular vein catheters that were assigned in a 5×5 Latin square design. *In vivo* results showed the potential of amino acid infusion together with energy source (acetate+Trp-Phe) in increasing milk protein synthesis by increasing milk protein yield in the overall period and milk protein concentration in the Trp and Phe group during the last day (day 5) of the experiment (P<0.05). In conclusion, our *in vitro* as well as *in vivo* results showed a potential of amino acid administration together with energy (acetate+Trp-Phe) in increasing milk protein synthesis as shown in the increased milk protein yield.

Evaluation of animal welfare through milk recording in dairy cows

L.M.G. Faria[1], A. Ferreira[2], J.P. Araújo[1,3] and J.L. Cerqueira[1,4]
[1]Escola Superior Agrária do Instituto Politécnico de Viana do Castelo, Ciências Agronómicas e Veterinárias, Refoios, 4990-706 Ponte de Lima, Portugal, [2]Associação para o apoio à Bovinicultura Leiteira do Norte, ABLN, Rua Cidade da Póvoa de Varzim, 55, 4490-295 ARGIVAI, Portugal, [3]Centro de Investigação de Montanha, CIMO, Refoios, 4990-706 Ponte de Lima, Portugal, [4]Centro de Ciência Animal e Veterinária, CECAV, Quinta de Prados, Apartado 1013, 5000-801 Folhadela, Vila Real, Portugal; cerqueira@esa.ipvc.pt

Many authors are coping scientific determination of which parameters are representative of animal welfare according five freedoms. The use of routine herd data, such performance recording, has a high value. Some parameters can be diagnosed through milk recording, such Somatic Cells Count (SCC), β-hydroxybutyrate (BHB) and composition of milk. The purpose of this work was assess the cattle welfare observing some parameters of milk recording. Milk recording data of Associação para o Apoio à Bovinicultura Leiteira do Norte of Portugal from 779 farms were analysed, in triennium 2016/2018 (n=780,958), according ICAR (2017) rules. Effect of parity and lactation phase in milk SCC was analysed by ANOVA and Tukey means comparison test. Statistical analysis was performed with SPSS for Windows v. 22. Milk production (305 d.) was $9,161.0\pm2,535.6$ kg, fat content $3.8\pm0.7\%$, and protein content $3.3\pm0.3\%$. SCC resulted in $150,400\pm214,100$ cells/ml, urea concentration of 259.0 ± 69.5 mg N/l and 0.1 ± 0.1 mmol/l of BHB. Primiparous cows represented 38.3% of total animals, with 83.5% of the cows into first three lactations, and less than 10% with $\geq$4th lactation, which demonstrates the low longevity of cows. 79.3% of the cows presented a SCC-Linear Score $\leq$4 ($\leq$200,000 cells/ml), 8.8% a SCC-Linear Score >5 (>400,000 cells/ml), which expresses very acceptable udder health. An increase in SCC (P<0.001) was observed with parity, revealing 113,370 cells/ml in 1st lactation and 237,340 cells/ml in $\geq$5th lactation. There was an effect (P<0.001) of the lactation phase in SCC, being lower in the 2nd phase (128,100 cells/ml) and the highest value in 4th phase (165,400 cells/ml). 15.5% of the cows had high BHB (>0.1 mmol/l), showing ketosis problems. We conclude that the parameters of milk recording can be a valuable instrument for the assessment of welfare in dairy cows.

Comparison of heat stress indicators based on weather station records in Holstein cows

S. Ishida[1], K. Izumi[1], T. Osawa[2], S. Yamaguchi[3], T. Yamazaki[4] and K. Hagiya[1]
[1]Obihiro University of Agriculture and Veterinary Medicine, Obihiro, 080-8555, Japan, [2]National Livestock Breeding Center, Nishigo, Fukushima, 961-8511, Japan, [3]Livestock Improvement Association of Japan, Tokyo, 135-0041, Japan, [4]Hokkaido Agricultural Research Center, NARO, Sapporo, 062-8555, Japan; 310ka_7x2da@i.softbank.jp

The temperature-humidity index (THI) is used mainly as a heat stress (HS) index for dairy cows. In Japan, only one or two principal weather stations in each prefecture have records of both daily temperature and humidity; many local weather stations have only daily temperature records. Therefore, we used three HS indices to evaluate the heat tolerance of dairy cows. The HS indices were: 1) THI calculated from the daily average temperature (t) and daily relative humidity (rh) at one of the principal weather stations in each prefecture (THI = $1.8 \times t+32(0.55 - 0.0055\times rh) \times (1.8\times t - 26)$); 2) the average daily temperature at the local weather station closest to each herd (TEMP); and 3) the THI calculated from the daily average temperature at the local weather station closest to each herd and the relative humidity at the principal weather station (HTHI). Data included weather records from 532 weather stations and monthly test-day milk yield and somatic cell score (SCS) records for a total of 2,477,162 days from 6 to 305 days after calving in Holstein cows that calved for the first time during 2011 through 2015. The model considered the effects of herd – year, age at calving, days in milk, and THI, TEMP, or HTHI. Permanent environmental effects and residuals were considered as random effects. We used the BLUPF90 program for our analysis. We estimated the mean squared errors (MSEs) in each model and compared accuracies among models for each trait. For both milk yield and SCS, the MSE estimated from models was smaller in order of HTHI, TEMP, and THI. That is, the MSE of the model containing HTHI was slightly smaller than those of the other models. HTHI is the most suitable index of heat tolerance for dairy cows in Japan.

Relationships between type traits and herd life of Holstein cows in different housing in Japan

Y. Nagasaka[1], Y. Nakahori[2], S. Yamaguchi[2], H. Abe[2], S. Nakagawa[2], T. Yamasaki[3], T. Baba[4], J. Kawakami[4], Y. Terawaki[5] and K. Hagiya[1]
[1]Obihiro University of Agricluture and Veterinary Medicine, Obihiro, 080-8555, Japan, [2]Hokkaido Dairy Milk Recording and Testing Association, Sapporo, 060-0004, Japan, [3]NARO Hokkaido Agricultural Research Center, Sapporo, 062-8555, Japan, [4]Holstein Cattle Association of Japan, Hokkaido Branch, Sapporo, 003-0011, Japan, [5]Rakuno Gakuen University, Ebetsu, 069-8501, Japan; sks.ngsk87@gmail.com

Improving longevity increases farm profits. Here, we investigated the relationships between herd life (HL) and type-classification scores with 3 housing types (tie stall (TS), free stall (FS), and grazing (GZ)). Type-classification scores of primiparous cows with 1st calving between 1993 and 2008 and HL data were used. Six classification traits (feet and legs [FL], chest width [CW], angularity, udder support [US], udder depth [UD], and front teat placement [FTP]) were investigated. The FL score was classified into 50 categories (50 to 99) and other traits into nine categories (1 to 9). Housing type was categorised into TS, FS, or GZ from a 2012 farmer survey, resulting in 126,492 data on HL for TS, 88,851 for FS, and 3,989 for GZ. HL was censored at age 84 months. The GLM procedure (R statistics) was used to estimate LSMs of HL on classification scores. Genetic bases were set to 79 in TS for FL and score 1 in TS for linier type traits. LSMs of HL increased (+13.6 to +16.0 for TS, +11.1 to +13.1 for FS, and +15.3 to +26.3 for GZ) with increasing FL score ($\geq$79). LSMs of UD increased (0 to +17.3 for TS, -4.8 to +13.4 for FS, and +0.9 to +21.5 for GZ) in all housing types. With FL scores <79, LSMs of HL decreased linearly with decreasing scores in TS (0 to +13.6) and FS (-2.3 to +11.1). However, for GZ, LSMs of HL did not decrease with decreasing FL scores (<79). For US in all housing types, LSMs of HL were highest at score 6. Intermediate scores for angularity (+9.6 for TS, +7.2 for FS, and +12.6 for GZ) and FTP (+9.5 for TS, +6.9 for FS, and +12.6 for GZ) gave the highest LSMs of HL in all housing types. For CW, the highest LSMs of HL were estimated at score 4 for TS (+2.9), score 3 for FS (+0.5), and score 5 for GZ (+5.1). For all traits, trends of LSMs were similar for TS and FS but not for GZ.

Dynamic modelling for nutritional management of ruminants in the face of climate change

H.C. Dougherty[1,2], M. Evered[1], J.W. Oltjen[3] and V.H. Oddy[1]
[1]New South Wales Department of Primary Industries Livestock Industries Centre, Trevenna Road, Armidale NSW 2351, Australia, [2]University of New England, Animal Science, The Woolshed W049, Armidale NSW 2351, Australia, [3]University of California Davis, Department of Animal Science, 2153 Meyer Hall, Davis, CA 95616, USA; holland.dougherty@une.edu.au

Livestock producers are being asked to become more efficient and more sustainable, while facing higher environmental risk due to climate change. Producers are paid on quality as well as yield, increasing pressure to optimise their systems. However, to optimally manage and feed livestock under increasing risk, producers need tools to assist them in achieving economic and environmental sustainability. Many current nutrition systems attribute variation in performance to feed characteristics and predict energetic requirements additively, first predicting maintenance, then gain, with some adjustment of maintenance as energy intake increases. However, the underlying biology is dynamic and nonlinear-animals may be gaining protein and losing fat, or vice versa, which current systems cannot account for. Rather than predicting energy change and partitioning it into fat and protein, we have developed a model that accounts for whole body gain or loss of fat and protein. Efficiencies are not fixed or defined by feed, as they are in other models, but arise from the interaction of animal, feed, and the animal's nutritional history. This dynamic approach calculates heat production as a function of feed intake and internal pools of protein and fat, and their changes as the animal's nutritional and physiologic states change. This model is dynamic and can reflect both the animal's current state as well as the effects of its previous nutritional state, and therefore is able to capture the variation in body composition due to past and present nutrition. Because of the relatively small number of parameters in this model it is easy to parameterise and to adapt to different situations. By using a dynamic, nonlinear model that reflects not only what the animal is now but what it has been in the past, the effects of previous nutritional circumstances, such as feed restriction from prior management or environmental stressors, may be accounted for in their effects on current body composition and future gain.

The effect of routine practices on the behaviour of beef cows according to their feeding management

K. Orquera[1], I. Ortigues-Marty[2], N. Thollon[3], I. Casasús[1], B. Sepchat[3] and A. De La Torre[2]
[1]Ctr Invest y Tecnol Agroal Aragon (CITA), IA2 (CITA-Universidad de Zaragoza), Montañana 930, 50059 Zaragoza, Spain, [2]INRAE, Université Clermont-Auvergne, Vet-AgroSup, UMRH, 63122 Saint-Genès-Champanelle, France, [3]INRAE, Herbipôle, 63122 Saint-Genès-Champanelle, France; korquera@cita-aragon.es

Animal behaviour can be modified in response to stimuli like weather conditions, social structure and farm management. The aim of this study was to determine if routine management activities such as weighing affected cow daily behaviour. Twelve 4 year-old lactating Charolais beef cows received a diet that met 100% of their nutritional requirements during lactation. From the second month post-calving, cows underwent feed restriction (4 to 10 d), with feed allowance reduced to meet 50% of their energy requirements (Challenge periods, CH), and then returned to full feed (Recovery periods, REC). This was repeated 3 times at monthly intervals. Cows were equipped with Medria® Axel loggers which recorded physical activity continuously during the study at 5-min intervals, providing the most dominant behaviour among five activities (ingestion, rumination, rest, over-activity and other). Cows were moved from their pen to a scale and weighed at 13:30 on some days (BW, n=17 d) but not on others (W0, n=17 d), equally distributed between CH and REC periods. The time devoted to the different daily activities was analysed with a mixed model (R Core Team, 2019) according to weighing (W0 vs BW) and feeding management (CH vs REC). Feeding did not influence ingestion time but affected both rumination (308 vs 473 min/d in CH vs REC, P<0.001) and rest (666 vs 402 min/d in CH vs REC, P<0.001). Only rumination time was longer in W0 than in BW days (406 vs 375 min/d, P<0.05), implying that weighing around midday interfered mostly with the time spent by cows ruminating. The effects of both factors on other and over-activity were less evident. These results should be considered in order to schedule routine management to avoid and/or minimise interference with cattle natural behaviour patterns.

Pasture feeding effects on α-tocopherol content and lipid oxidation of beef from late maturing bulls

S. Siphambili[1,2], A.P. Moloney[1], E.G. O'Riordan[1], M. McGee[1] and F.J. Monahan[2]
[1]Teagasc, Animal & Grassland Research and Innovation Centre, Grange, Dunsany, Co. Meath, Ireland, [2]University College Dublin, School of Agriculture and Food Science, Belfield, Dublin 4, Ireland; sis39@aber.ac.uk

The finishing of late-maturing bulls at pasture offers an opportunity to increase the economic efficiency of beef production. Compared to steers bulls have faster growth rates, higher feed efficiency and higher dressing out percentages, whilst grass is the cheapest feed resource available. The purpose of the study was to investigate the effects of the change from concentrate finishing to pasture finishing on fatty acid profile, α-tocopherol concentration and lipid oxidation of beef from late maturing bulls. 48 Charolais or Limousin sired bulls were assigned to one of four production systems: pasture only (P), pasture plus 25% dietary dry matter(DM) intake as barley-based concentrate (PC25), pasture plus 50% dietary DM intake as barley based concentrate (PC50) or a barley-based concentrate ration (C). Following slaughter at 19 months of age, 14 day aged *M. Longissimus thoracis et lumborum* samples were subjected to simulated retail display (4 °C,1000 lux for 12 h out of 24 h) for 3, 7, 10 and 14 days in modified atmosphere packs (MAP, O_2:CO_2; 80:20). There were higher muscle concentrations of C18:3n-3 (P<0.001), C20:5n-3 (P<0.001), C22:6n-3 (P<0.01), total n-3 polyunsaturated fatty acids (PUFA) (P<0.001) and high highly peroxidisable polyunsaturated fatty acids (HP-PUFA) (P<0.01) in P, PC25 and PC50 compared to C bulls, respectively, but total PUFA content did not differ. There was higher concentration of α-tocopherol (P<0.001) in muscle from P compared to C bulls. α-Tocopherol decreased significantly (P<0.001) in all samples by day 14. Lipid oxidation was higher (P<0.01) in muscle from C compared to P bulls on day 10 and day 14. Finishing bulls on pasture increases the HP-PUFA concentration in muscle from late-maturing bulls but this does not result in increased lipid oxidation due to higher α-tocopherol concentration compared to muscle from concentrate finished bulls. In conclusion, finishing bulls on pasture does not reduce the shelf life of beef compared to beef from concentrate finished bulls.

Impact of animal handling on the stress status of Rubia Gallega beef cattle: study of apoptosis

C. Fuente-Garcia[1,2], N. Aldai[1], E. Sentandreu[2], D. Franco[3] and M.A. Sentandreu[2]
[1]University of the Basque Country (UPV/EHU), Lactiker Research Group, Paseo de la Universidad 7, 01006 Vitoria-Gasteiz, Spain, [2]Instituto de Agroquímica y Tecnología de Alimentos (IATA-CSIC), Avda. Catedrático Agustín Escardino Benlloch 7, 46980 Paterna (Valencia), Spain, [3]Centro Tecnolóxico da Carne (CTC), Avenida de Galicia 4, 32900 San Cibrao das Viñas (Ourense), Spain; claudia.fuente@ehu.eus

Animal handling prior to slaughter is one of the factors majorly affecting the stress status of beef cattle. This is related to 'dark cutting' defect, which implies important economic losses for meat industry. Therefore, the research of biomarkers related to animal stress is a relevant field that needs to be addressed. The aim of this work was to study *post-mortem* apoptosis to understand the impact of animal handling on the stress of Rubia Gallega beef cattle. In this work, 12 calves were reared intensively, while 16 were reared semi-extensively (n=28). Within each production system, half of the animals were mixed with unfamiliar individuals during transport and lairage prior to slaughter (mixed group) while the other half were kept together (unmixed group). After slaughter, the longissimus thoracis et lumborum muscle was sampled at 2, 8 and 24 h. From sarcoplasmic extract, the activities of executioner caspases 3/7 and initiator caspase 9 were measured using a fluorometric assay based on Ac-DEVD-AMC and Ac-LEHD-AMC substrates, respectively. The effect of production system, mixing treatment, *post-mortem* time and their interactions were studied by the GLM of ANOVA. The effect of production system was not significant for any of the caspase activities. The other two factors, mixing treatment and *post-mortem* time were significant ($P \leq 0.05$) but there was an interaction between the two for both caspase activities ($P \leq 0.05$). In this sense, unmixed animals showed decreasing activities over time while a different trend was observed in the mixed group having the highest activities at 8 h *post-mortem*. At 2 h *post-mortem*, caspase activities were of similar value in both mixed and unmixed groups while unmixed group showed lower values at 8 and 24 h compared to mixed group. These results demonstrate an advancement in the understanding of animal stress using apoptosis biomarkers as reliable indicators.

PhenoBR: a model to characterise variations of body reserves in meat ewes

T. Macé[1], D. Hazard[1], F. Carrière[2], S. Douls[2], D. Foulquié[2], E. González-García[3] and M. Taghipoor[4]
[1]INRAE, UMR1388 GENPHYSE, 24 Chemin de Borde Rouge, Auzeville Tolosane, CS52627, 31326 Castanet-Tolosan cedex, France, [2]INRAE, UE321 La Fage, La Fage, 12250 Roquefort-sur-Soulzon, France, [3]INRAE, UMR868 SELMET, 2 Place Pierre Viala, 34060 Montpellier cedex 1, France, [4]INRAE, AgroParisTech, UMR MoSAR, 16 rue Claude Bernard, 75231 Paris Cedex 05, France; tiphaine.mace@inrae.fr

Body reserves (BR) are the main source of energy in ruminants facing negative energy balance challenges e.g. during highly demanding reproductive cycles or feed scarcity periods. This information is of main concern in the context of genetic selection for efficiency and resilience. The objective of this study is to develop a mathematical model to quantify ewes' adaptive response at each reproductive cycle in term of the capacity of mobilisation and accretion of BR. Ewes reared in extensive conditions (n=1,482), from 2002 to 2015, were measured for body condition score (BCS) during one to three parities, through 8 measurements per parity. A system of 7 ordinary differential equations was developed to describe BR dynamic. Using this model, the variation of BCS at each parity is summarised into 4 parameters: tbi and tei associated with the beginning and the end of mobilisation period for parity i, kpi and kbi associated with the capacities of mobilisation and accretion of BR during the parity i, respectively. An algorithm of functional data analysis was applied to estimate the time related parameters (tb and te), and an optimisation algorithm allowed to estimate kb and kp. The model and the parameters estimation procedure allowed to quantify BR mobilisation and accretion capacities for all animals included in the analysis, with low residual errors, indicating the relevance of the model. Estimated individual parameters showed an interesting variability allowing ranking of ewes for their capacity to adapt to physiological and/or environmental challenges. Interesting correlations between parities (e.g. for kp from 0.46 to 0.59) indicated relationships between parities for BR mobilisation and accretion capacities. In conclusion, we demonstrated that our model allows summarising BR dynamic in sheep in few parameters and ranking animals for their BR mobilisation and accretion capacities. These parameters will be further used for genetic analysis of BR dynamics.

HappyMoo project – assessing animal welfare towards milk composition

J. Leblois[1], C. Bastin[1], C. Bertozzi[1], M. Calmels[2], L. Dale[3], C. Grelet[4], C. Lecomte[5], M.-N. Tran[1] and A. Werner[3]
[1]Walloon Breeders Association Group, Rue des Champs Elysées 4, 5590 Ciney, Belgium, [2]Seenovia, Bld des Loges 141, 53940 Saint-Berthevin, France, [3]State Association of Baden-Wuerttemberg, Heinrich-Baumann-Str. 1-3, 70190 Stuttgart, Germany, [4]Walloon Agricultural research Centre, Chée de Namur 24, 5030 Gembloux, Belgium, [5]France Conseil Elevage, Rue de Châteaudun 42, 75009 Paris, France; jleblois@awenet.be

In the context of growing interest in animal welfare, from both consumers, farmers and authorities, it becomes urgent to be able to assess efficiently animal welfare. Protocols to assess welfare at the farm level exist (as Welfare Quality®), but the major issue is the time needed to apply them and the absence of regularly based-assessments. Thus, it is very challenging to find cheap and fast alternatives to assess animal welfare on a regular basis. This is the objective of HappyMoo, a project funded by Interreg NWE. This project comprises 13 partners (8 countries): 8 milk recording organisations, 3 research centres and university, 1 IT company and 1 pilot laboratory. The aim of the project is to roll-out monitoring tools to screen the welfare of dairy cows and specifically to provide alerts about the freedom from disease, hunger and stress. For this, data collected in routine are used to predict the absence of mastitis, lameness, stress and negative energy balance (NEB). These data comprise veterinary diagnosis, phenotype observations but also milk proxy indicators, through the mid-infrared (MIR) spectrum. The first results of the project highlight that milk MIR based predictions can be good indicators for the NEB. Indeed, blood BHB and NEFA predictions in milk, combined with other milk parameters, as citrates and acetone can be a good proxy for NEB and associated physiological imbalance. To obtain information on mastitis, both the direct use of MIR spectrum or MIR predicted mastitis biomarkers, as lactoferrin content in milk are of interest. For lameness and stress proxies, research in on-going. An important challenge is the harmonisation between data coming from 8 countries, as scales to record traits and description of phenotypes can vary considerably. To reach this goal, international standards such as ICAR guidelines for diseases recording are used, together with research to find correspondence between different scales (such as body condition scoring and lameness scales). After harmonisation of data and input into a common transnational database, new machine learning methods will be tested simultaneously in order to release reliable alerts concerning animal welfare. In conclusion, HappyMoo project should end-up with practical on-farm alerts to serve as a management tool for farmers and extension workers.

Estimating energy balance of dairy cows with the GreenFeed system and milk mid-infrared spectroscopy

A. Guinguina and P. Huhtanen
Swedish University of Agricultural Sciences, Department of Agricultural Research for Northern Sweden, Skogsmarkgränd, 901 83, Sweden; abdulai.guinguina@slu.se

The GreenFeed system (GF) is traditionally used to measure enteric methane production from ruminants, but it can also be used to determine energy balance (EB) which is an expensive-to-measure phenotype on a large scale. Moreover, there is a growing interest in using milk mid-infrared spectroscopy (MIRS) fatty acid data to predict EB to facilitate genomic selection purposes. We investigated the relationship between EB determined from the GF (EB_{GF}) and EB estimated from energy requirements for dairy cows using the Finnish feed table (EB_{LUKE}). We also explored the ability of milk fatty acids from MIRS to predict EB_{GF}. Data from 22 early lactating Nordic Red dairy cows in Sweden were used. Cows were fed the same grass silage but supplemented with either cereal grain or fibrous by-product concentrate from week 1 to 18 postpartum. They were milked twice daily and milk was sampled on four consecutive milkings from week 1 to 8 and every other week thereafter for MIRS analysis. Relationship between EB_{GF} and EB_{LUKE} was explored by simple linear regression of EB_{LUKE} on EB_{GF}. The prediction model of EB_{GF} from MIRS was developed using stepwise regression (PROC GLM SELECT in SAS) with k-fold cross-validation by splitting the data into 22 parts based on cow ID. The variation in EB_{GF} explained 76% of the variation in EB_{LUKE} indicating the ability of the GF to measure energy EB with high precision. However, the residuals of EB (EB_{GF}-EB_{LUKE}) were positively related to the metabolisability of the diet suggesting that if the cows were more efficient in converting gross energy to metabolisable energy, their EB_{GF} was higher than that estimated from table values. The accuracy of the model predicting EB_{GF} using milk fatty acid from MIRS was moderate with an R^2 cross-validation (R^2_{cv}) of 0.53. In conclusion, the GF can provide reliable estimates of EB of dairy cows in early lactation. The ability of milk fatty acids from MIRS to predict EB was moderate. A larger data base is needed to improve the accuracy and the robustness of the prediction models.

Effects of electrochemically activated drinking water on bovine milk composition

E. Vargas-Bello-Pérez, R. Dhakal and H.H. Hansen
University of Copenhagen, Department of Veterinary and Animal Sciences, Grønnegårdsvej 3, 1870, Denmark;
evargasb@sund.ku.dk

The objective of this study was to assess the effects of electrochemically activated drinking water (ECW) on milk chlorate, milk perchlorate, milk iodine, milk composition, milk fatty acid profile, and overall performance of dairy cows. Ten Red Danish cows in mid-lactation (203±31 d in milk) were chosen from these 2 groups for intensive sampling. The treated group drank water with 4 ppm of ECW (29 mg/l of chlorate of Neuthox, Danish Clean Water A/S, Sønderborg, Denmark). The treatment lasted 60 consecutive days, with milk and water sampling on d 0, 30, and 60. Additionally, milk samples from both the control group and treated group were taken on d 90 to assess if any carry-over effect was present. Interactions between period and milk yield and somatic cell for the full group and period and milk fat content and milk urea nitrogen in the selected animals occurred. Milk fat, milk fatty acid profile, chlorate, perchlorate, and iodine contents were not significantly different between treatments. Milk urea increased, whereas β-hydroxybutyrate and somatic cell count decreased significantly in the treated groups. Results showed that at a dosing of 4 ppm of ECW, both chlorate and perchlorate concentrations in milk (<0.002 mg/kg) were low, and no deleterious effects on milk production or milk chemical composition were observed. These data can be of use when assessing the effects of ECW on milk and milk powder chlorate and perchlorate levels and provide a context for assessing the potential for influencing human health under the conditions prevailing on a commercial dairy farm.

Cow activity measurements can be used to define new fertility traits for use in genetic evaluation

B. Heringstad[1], K.B. Wethal[2] and K.I. Dragset[2]
[1]Department of Animal and Aquacultural Sciences, Faculty of Biosciences, Norwegian University of Life Sciences, 1432 Ås, Norway, [2]Geno Breeding and A.I. Association, 2317 Hamar, Norway; bjorg.heringstad@nmbu.no

Cow activity measurements are widely used in herd management tools for heat detection. The aim of this study was to examine whether new traits based on activity measures can be useful for genetic evaluation of cow fertility in Norwegian Red. Data were from 284 herds with Lely milking robot and activity tags and included activity measurements from 13,224 lactations of 8,139 Norwegian Red cows. Daily activity measures were available, and we included records from 10 to 150 days in milk for cows with at least 50 records. The trait analysed was interval from calving to first high activity (CFHA). 87% of the cows had at least one episode of high activity recorded. The mean (standard deviation) of CFHA was 42 (28) days. A linear animal repeatability model with fixed effects of month-year of calving, age-parity, and herd, and random animal and permanent environment effects was used for estimation of variance components. The trait CFHA showed significant genetic variation with a heritability of 0.05. This trait reflects the cow's ability to return to oestrus cycle and show heat after calving, which is an important aspect of cow fertility.

Effects of reproductive management on genetic evaluations of fertility traits in dairy cattle

C. Lynch[1], G.A. Oliveira Junior[1], F.S. Schenkel[1], L.R. Schaeffer[1] and C.F. Baes[1,2]
[1]University of Guelph, 50 Stone Road East, N1E 2W1, Canada, [2]University of Bern, Institute of Genetics Vetsuisse Faculty, Bremgartenstrasse 109a, 3012, Switzerland; cbaes@uoguelph.ca

The success of fertility performance on dairy farms, especially for artificial insemination programs, starts with oestrus detection. Detection of oestrus has become more difficult over the years due to decreases in oestrus expression of high-producing dairy cows, with up to 60% of ovulations accompanied by no standing mount. With the use of hormonal protocols which synchronise follicle growth, corpus luteum regression, and ovulation, it is possible to alleviate the pressure of oestrus detection, thus making ovulation time easier to predict and increasing the herds' overall conception rate. This approach is considered precision management of dairy. The use of reproductive management has increased over the past decade, however the effects of these practices on breeding value estimation is not understood. Hormonal synchronisation affects the reproductive physiology and endocrinology involved in the oestrous cycle, allowing for higher rates of conception even for naturally low fertile animals. This leads to genetically inferior cows' performance becoming masked and potentially resulting in similar performance to that of fertile animals. As genetic programs rely on the collection of accurate phenotypic data, phenotypes collected on treated animals likely add bias to genetic evaluations, and can cause changes in genetic and residual variation when compared with non-synchronised herds. In the current study, we used simulation and real data on over 777,000 Canadian dairy cattle to understand the impact of hormonal synchronisation protocols on the accuracy of fertility trait genetic evaluations in dairy cattle. We found that the overall variance of phenotypes was reduced in synchronised animals, and physiological factors related to currently recorded phenotypes varied between herds using heat detection and synchronisation. We show that masking natural phenotypes of animals can have detrimental effects on genetic gain by close to 50%. Finally, we provide recommendations for phenotype collection to account for masked phenotypes.

Relationships of NMR-metabolite profiles with performance, milk and health traits in the bovine

D. Rippel[1], W. Xu[2], A. Kenez[2] and Y. Montanholi[3]
[1]Independent Researcher, Chapada Gaúcha, 38698-000, Brazil, [2]City University of Hong Kong, Kowloon, 83 Tat Chee Ave, NA, China, P.R., [3]North Dakota State University, P.O. Box 6050, Department 7630, 58108, USA; yuri.r.montanholi@gmail.com

The associations of metabolome with key phenotypes could result in the identification of proxies for complex traits in different body matrices (i.e. blood plasma and colostrum whey) with application in cattle husbandry, precision livestock farming and genetic improvement. The objective of this study was to report the associations of the nuclear magnetic resonance spectrum (NMR) across distinct physiological states (late pregnancy, calving and nursing) in relation to performance, milk and health traits. Thirty-six crossbred heifers were productive performance evaluated from weaning (247±17 days of age) to one week before calving (731±26 days of age). Colostrum was collected within 4 h after calving, milk was collected at 47 days post-calving and blood was harvested at the three physiological stages. Total immunoglobulin (Ig) A, IgG and specific IgG anti-ovalbumin (OVA) were evaluated at the three physiological stages in the blood plasma and in the colostrum. Blood plasma and colostrum whey were analysed by 1D ^{1}H NMR using a Bruker Avance III 700 MHz spectrometer. Processed NMR spectral data were analysed in the statistical package R. The difference between physiological stages was visualised by principal component analysis (PCA), and the correlation between NMR spectra and phenotypic traits were tested by Pearson correlation. The PCA scores plot indicated that NMR spectrum presented a substantial overlap among the three physiological stages. Preliminary results reported correlations of untargeted NMR spectrum of blood plasma ranges with traits of interest (i.e. feed efficiency -0.50 to 0.35, OVA -0.27 to 0.27 and milk fat 0.34 to 0.50). Some traits were also correlated with unknown peaks in NMR spectrum (i.e. feed efficiency and 1.88 ppm: 0.50 and, milk protein and 3.33 ppm: -0.42). Correlations of known NMR peaks and measured traits were noted, including: acetate and feed efficiency (0.38), glutamine and OVA (-0.27) and, lactate and milk protein (-0.39). Further investigations are under course to identify other associations across the NMR spectrum of blood plasma and colostrum whey with another 30 plus traits commercially measured in cattle enterprises, as well as to identify and measure the association of specific metabolites associated with performance, milk and health traits of interest.

A validated ELISA method for detecting differences in feather corticosterone between turkey lines

E.M. Leishman[1], N. Freeman[2], A. Newman[2], N. Van Staaveren[1], B.J. Wood[1,3,4], A. Harlander[1] and C.F. Baes[1,5]
[1]University of Guelph, Animal Biosciences, 50 Stone Road E, N1G 2W1, Guelph, Ontario, Canada, [2]University of Guelph, Integrative Biology, 50 Stone Road E, N1G 2W1, Guelph, Ontario, Canada, [3]Hybrid Turkeys, Suite C, 650 Riverbend Drive, N2K 3S2, Kitchener, Ontario, Canada, [4]University of Queensland, School of Veterinary Science, 5391 Warrego Hwy, 4343 Gatton, Queensland, Australia, [5]University of Bern, Institute of Genetics, Vetsuisse Faculty, Hochschulstrasse 6, 3012 Bern, Switzerland; eleishma@uoguelph.ca

Measuring corticosterone levels from feathers is a non-invasive way to assess glucocorticoid levels of birds. This method may contribute to our understanding of physiology in birds; however, it is not often used for poultry species. The objective of this study was to validate a method of detecting corticosterone in turkey feathers using an ELISA and to investigate the differences in corticosterone levels between turkey genetic lines. The ninth primary wing feather was collected from 123 individuals from three genetic lines (lines A (n=46), B (n=24), and C(n=53)). Samples were analysed using an ELISA with validations of specificity, accuracy, and precision. Results showed that feather corticosterone could be measured in the primary wing feathers of domesticated turkeys using an ELISA. There was no significant interaction between the standard curve and serial dilution, indicating adequate specificity of the assay (ANCOVA: F=1.512, P=0.267). The average recovery of corticosterone was 65%, and the intra- and inter-assay CV of the ELISAs were 1.4±2.48 and 12.1±0.02%, respectively. Significant differences were observed in the feather corticosterone concentrations between the three lines (ANOVA: F=12.71, P<0.001). Line C had significantly higher feather corticosterone than Line A (Tukey HSD: P<0.0001) and Line B (Tukey HSD: P=0.036), while no differences were observed between Line A and Line B (Tukey HSD: P=0.366). This is the first report of differences in feather corticosterone between turkey genetic lines. It is possible that the different breeding objectives of each genetic line may indirectly affect the feather corticosterone levels. This relationship will be studied further to investigate the underlying genetics of glucocorticoid levels in domestic turkeys.

Effects of feeding behaviour and physiological stage on the metabolic profile of the local goats

H. Harrabi, M. Abid, M. Fatnassi, S. Bessalah and M. Hammadi
IRA Medenine, LEFS, 4100, Tunisia; hagerharrabi@yahoo.fr

The aim of this work is to study the feeding behaviour of small ruminants in the rangeland of south-eastern Tunisia during two seasons (spring and summer) corresponding to two different physiological stages (lactation and drying up). Indeed, these routes represent an important source of feed for the breeds and their exploitation is done thanks to their mobility which is linked to precipitation or even the state of plant cover. The herds are mainly made up of local goats. Blood samples are taken in the morning and the serum obtained are kept for analyses of glucose, urea, creatinine and total proteins. As well as samples of the fodder plants of the route are taken for analysis of the dry, mineral and nitrogenous matter (NDF and ADF). The chemical composition of the most palatable pastoral plants on the course of the study areas during the course reveals a variation in dry matter ranging from 17.9-87.2% in total nitrogenous matter (<11%). The NDF and ADF rates vary between 40.7 to 85.3 and 19.6 to 59.2% respectively. MM varied from 5 to 20%. During the summer, the pastoral plants studied have higher DM levels that exceed 50%, and poor total nitrogen matter (<11%). NDF rates ranging from 27.4 to up to 82.3% and ADF vary between 16.4 and 61.7%. There is great variability between the levels of MM of pastoral plants ranging from 2.4 to 31.4% MM. The energy metabolism is characterised by higher blood glucose levels in summer than in spring in goats (0.8±0.25 g / l in summer compared to 0.26±0.12 g/l in spring). Lactating goats in the spring season have lower blood glucose levels than those in the summer season. Urea levels in the spring are higher than those in summer (P<0.05). the uraemia increases with the progression of lactation and decreases during the dry period. The creatinine differs (P<0.05) between the spring and summer seasons (10.52±2.95). A highly significant influence of the season on the protein level in goats; this parameter is higher in summer in females than in spring in lactating females. In conclusion, the season (physiological stage) and the type of route induced variations in the concentrations of different classical biochemical parameters in goats living in arid conditions.

The Effect of colostrum lactose content on growth performance of piglets

M. Szyndler-Nędza, A. Mucha and M. Tyra
Instytut Zootechniki, Państwowy Instytut Badawczy, Ul. Sarego 2, 31-047, Poland; magdalena.szyndler@izoo.krakow.pl

Genetic selection with maternal pig breeds has increased the number of piglets born, but has not changed the proportions of major components of sow colostrum, including lactose. Lactose, being the primary carbohydrate present in colostrum, helps to transfer water from blood to colostrum, and together with fat it provides the main source of energy for day-old piglets. This disaccharide is one of the factors that influence the yield of sow's colostrum, which is essential for normal piglet development. The aim of the study was to determine the effect of lactose content in colostrum of sows of maternal breeds on rearing performance of their piglets. The study involved a total of 255 litters of Polish Large White (PLW) and Polish Landrace (PL) sows in their first to third lactation. In each lactation, colostrum was collected 1 h postpartum, and milk was sampled 2 h after morning feeding on day 7 of lactation. Colostrum and milk were analysed for solids, protein, fat, and lactose. Rearing performance of the piglets in 255 parities was determined based on the number of piglets born alive, number of piglets at 7, 14 and 21 days of age, individual weight of piglets at birth and at 7, 14 and 21 days of age. Analysis of data for lactations 1-3 showed that the lower the colostrum lactose content, the higher was the fat, protein and solids content. In the case of milk on day 7, it was observed that sows with the lowest colostrum lactose content (LLAC), compared to the sows with the higher lactose content (HLAC), produced milk with a higher fat (P≤0.05), protein and solids content (P≤0.01). These sows (LLAC) reared 100 g heavier piglets on day 7 (P≤0.05) and 130 g heavier piglets on day 21 of age (P≤0.05) compared to HLAC. Therefore colostrum lactose concentration may serve as a good predictor of growth parameters of the piglets reared by sows. Study financed as part of statutory activity of the National Research Institute of Animal Production, project no. 07-11-02-11.

Fatty acids profile of the reproductive rabbit female: preliminary results

M.L. Garcia[1], R. Muelas[1], I. Agea[1], E. Armero[2] and M.J. Argente[1]
[1]Universidad Miguel Hernández de Elche, Agrofood Technology, Ctra Beniel km 3.2, 03312 Orihuela, Spain, [2]Universidad Politécnica de Cartagena, Department of Agricultural Science and Technology, Paseo Alfonso XIII, 48, 30203, Spain; mariluz.garcia@goumh.umh.es

Fatty acids are an energy source during the process of maturation of oocytes and embryo development before implantation. The aim of this study was to determine the fatty acids profile at mating and 3 d post-coitum in the rabbit female. Blood samples were collected in 8 multiparous females at mating and 3 d post-coitum. All females showed normal embryos, classified as early or compacted morulae. Blood samples were centrifuged and plasma was store at -80 °C. The fatty acids were measured using a gas chromatograph (GC-17A, Shimadzu, Kyoto, Japan) using a Flame ionisation detector (FID), equipped with a CP-Sil 88 for FAME capillary column (100 m × 0.25 mm × 0.36 mm.; 0.20 µm thickness; Agilent technologies, Madrid, Spain). Statistical model included the fixed effect of moment (mating or 3 d post-coitum), line and female as random effect. Data were analysed using Bayesian methodology. The highest concentrations of fatty acids at mating and 3 d post-coitum were linoleic acid (467 ng/ml), palmitic acid (408 ng/ml), oleic acid (345 ng/ml), and stearic acid (185 ng/ml), while myristic acid (26 ng/ml), linolenic acid (18 ng/ml), palmitoleic acid (17 ng/ml), heptadecanoic acid (10 ng/ml) were in fewer concentrations. There was low evidence that concentration of heptadecanoic acid, myristic acid and oleic acid was a 19% (P=0.84), a 40% (P=0.80), and a 16% (P=0.80) higher at 3 d post-coitum than at mating. In conclusion, the same order of fatty acids, according to their concentration, is shown at both reproductive status but several fatty acids concentration seems to increase with early embryo development. These results should be confirmed with a larger database This study is supported by the Spanish Ministry of Economy and Competitiveness (MINECO) with the Project AGL2017-86083 C2-2-P, and Valencia Regional Government with the Project AICO/2019/169.

Assessing feed efficiency in grazing dairy cows through infrared thermography

T. Haak[1,2], A. Münger[2], K.H. Südekum[1] and F. Schori[2]
[1]University Bonn, Institute for Animal Science, Edenicher Allee, 15, 53115 Bonn, Germany, [2]Agroscope, Ruminant Research Group Unit, Route de la Tioleyre, 4, 1725 Posieux, Switzerland; thorsten.haak@agroscope.admin.ch

Genetic selection for feed efficiency is constrained by cost and difficulty of measuring individual feed intake. An alternative is to measure biological proxies of feed efficiency, which can be implemented at farm-level. The objective of this study was to explore the use of infrared thermography (IRT) to predict feed efficiency in dairy cows. A group of 28 dairy cows (14 Holstein and 14 Swiss Fleckvieh) was investigated for two periods during mid and late lactation in a pasture-based system. During a 7-day measuring period, the individual herbage intake of each cow was estimated using the n-alkane marker technique. The surface temperatures (ST, average of surface, maximum and minimum) were recorded (FLIR T620, FLIR Systems, USA) after morning milking, indoors, at multiple body locations (i.e. nose, head, ear, eye, cheek, paralumbar depression, flank, udder, backside, vagina, leg and feet). The ability of the body ST to explain variation in feed conversion efficiency (FCE, energy-corrected milk yield / dry matter intake) and residual feed intake (RFI, effective minus requirements – modelled dry matter intake), was analysed. According to linear models, body ST were positively correlated with FCE (R^2: 0.10-0.50) and negatively with RFI (R^2: 0.06-0.32). At best, FCE and RFI were explained by left flank average ST (R^2=0.50) and flanks average minimum ST (R^2=0.32), respectively. Further, breed and measurement period were included as effects into multiple linear regression models. Predominantly, a significant effect of breed and measurement period was observed. Therefore, when fixed effects were considered, 55% of FCE was explained by backside maximum ST and 43% of RFI was explained by flank average ST and nose average ST. A relationship between surface IRT and feed efficiency was observed. However, ST deviations associated with changes in ambient temperature and/or general conditions could influence the IRT recordings. Standardisation of the IRT process and the conditions during measurement is necessary to accurately assess its potential use to predict feed efficiency.

Relationship between rectal and infrared thermography maximum eye temperature of Arouquesa breed

C. Venâncio[1,2], P. Valentim[1], D. Moreira[1] and S.R. Silva[1,3]
[1]University of Trás-os-Montes e Alto Douro (UTAD), Vila Real, 5001-801, Portugal, [2]Centre for the Research and Technology Agro-Environmental and Biological Sciences (CITAB), VR, 5000, Portugal, [3]Veterinary and Animal Research Centre (CECAV), VR, 5000, Portugal; cvenanci@utad.pt

Monitoring animal health and welfare is a key aspect for cattle producers. One of the most used indicators to assess health and welfare is body temperature. The use of a thermometer to measure rectal temperature is an inexpensive and straightforward way to obtain this indicator. Although it is a simple procedure, it is not of practical application in animals with a more reactive temperament and those that are in extensive free-range systems. Recent research shows that it is possible to estimate the body temperature accurately with techniques that do not require animal restrain. In this sense, the objective of this work is to study the relationship of the infrared thermography (IRT) maximum eye temperature with the rectal temperature in Arouquesa cows. Sixty-three cows were evaluated. In these animals, measurements of the rectal temperature were performed using a digital thermometer (K-jump KG-133), and after this procedure, IRT images of the left head plane were obtained using a FLIR F8 camera (FLIR Systems AB, Sweden). From the eye IRT images and using the FLIR Tools software, the maximum eye temperature data was determined. To ensure exactitude in maximum eye temperature determination, it was used the ellipse tool of the software, and an ellipse fitted to the animal eye. A descriptive statistics and correlation analysis were performed using the JMP-SAS14 software (SAS Institute Inc. Cary, NC, USA). As expected, a small variation is observed for both rectal and eye temperature (coefficient of variation of 1.5 and 2.9% respectively). The results showed a high positive correlation (r=0.934; P<0.01) between rectal and IRT maximum eye temperature. The results show that IRT has a high practical and diagnostic potential to assess the temperature of the cattle without discomfort for the animal. Further research with a larger sample is needed to improve the relationship between the IRT temperature of the eye and the body temperature, making this technique a reliable tool. PDR2020-101-031094.

Towards the identification of nutrition biomarkers in dairy cows

L. Cordeiro[1], H.M. Oliveira[2], A.R.J. Cabrita[1], L. Ferreira[3], F. Castanheira[4], M.R.G. Maia[1], A.J.M. Fonseca[1] and I.M. Valente[1,5]
[1]REQUIMTE/LAQV, ICBAS, Instituto de Ciências Biomédicas de Abel Salazar, Universidade do Porto, R. Jorge Viterbo Ferreira 228, 4050-313 Porto, Portugal, [2]INL- International Iberian Nanotechnology Laboratory, Av. Mestre José Veiga, 4710-057 Braga, Portugal, [3]AGROS UCRL, Lugar de Cassapos, 4490-258 Argivai, Portugal, [4]Alltechaditivos-Alimentação Animal Lda, Av. D. Luís Sá 9 Arm A, 2710-08 Abrunheira, Portugal, [5]REQUIMTE/LAQV, DQB, Faculdade de Ciências, Universidade do Porto, R. Campo Alegre 687, 4169-007 Porto, Portugal; lilianacordeiro@sapo.pt

Digestive diseases are one of the main causes of high economic losses in the livestock sector. Diagnose of these pathologies is based on invasive analysis and in paraclinical parameters. Current concerns on animal health and well-being are driving the research of non-invasive methods for early and effective diagnosis of digestive and metabolic diseases. Therefore, the animal science community is privileging the study of biological fluids as potential test samples for specific diseases through a metabolomics-based approach. This work aimed at the characterisation of the organic composition of biological samples (faeces, urine and rumen fluid) from 3 Holstein cows fed different diets. A non-targeted analysis was performed by GC-FID after liquid-liquid extraction. In rumen fluid the major fraction of the compounds were volatile fatty acids (VFA), ethanal, cresol and 1-phenylethanol. The compounds determined in faeces had a profile similar to the rumen fluid, yet other compounds such as butanal, nona-2,4-dienal, dimethyl sulphide, *tert*-butyl methyl sulphide, 3-ethylphenol and 2-propanol were also identified. The profile of urine samples was less complex, being VFA the most abundant compounds, but at lower levels than those determined in rumen fluid and faeces. The major compounds found in urine were acetic acid, *iso*-butyric acid, *iso*-butanal, *trans*-2-nonenal, 2,6-dimetoxyphenol and eugenol. In all biological samples, positive correlations between the major compounds and dietary protein and starch contents were verified, while negative correlations were obtained for dietary neutral detergent fibre. Fundação para a ciência e tecnologia funding to LC (PD/BDE/136294/2018), IMV and MM (DL 57/2016–Norma transitória) is acknowledge.

Interferon-gamma-dependent mechanisms elicited in bovine macrophages in response to *Neospora caninum*

A.C. Ferreira[1], J. Guimarães[1], R. Pedro[1,2], M. Silva[1], M. Resende[1], M. Vilanova[1,2] and A. Correia[1,2]
[1]i3S – Instituto de Investigação e Inovação em Saúde, Universidade do Porto, Rua Alfredo Allen 208, 4200-135, Porto, Portugal, [2]ICBAS, Instituto de Ciências Biomédicas Abel Salazar, Universidade do Porto, Rua Jorge Viterbo Ferreira, 228, 4050-313 Porto, Portugal; alexandra.correia@ibmc.up.pt

Neospora caninum is an obligate intracellular protozoan that causes neosporosis, a disease responsible for heavy economic losses in dairy and beef industries, for which no commercial vaccine is currently available. Interferon-gamma (IFN-γ) was identified as a key cytokine in mediating protection against *N. caninum*. Since macrophages are privileged host cells in which *N. caninum* proliferates, we aimed to evaluate the impact of IFN-γ signalling in bovine macrophages infected *in vitro* with *N. caninum*. Macrophage cell cultures infected with *N. caninum* were stimulated with different concentrations of bovine IFN-γ. The production of oxygen reactive species (ROS) and nitric oxide (NO) was inhibited with N-acetylcysteine and 1,400 W dihydrochloride, respectively. The expression of genes that could be upregulated in response to IFN-γ was evaluated by qRT-PCR. Parasite killing by macrophages was assessed by quantifying the amount of parasitic DNA by qPCR. Data were analysed by two-way ANOVA with the GraphPad Prism 8.2.1 software. Among the effector mechanisms induced by IFN-γ signalling in macrophages, we identified that genes responsible for the production of ROS and NO, the immunity-related GTPase Cinema and guanylate binding protein (GBP) 4 were upregulated upon IFN-γ stimulation and may constitute major protective mechanisms against *N. caninum* in macrophages. Our results indicate that an immunoprotective approach against bovine neosporosis should potentiate IFN-γ production that will in turn activate macrophages and induce parasite-controlling effector molecules, resulting in protection against this parasitic disease. Funded by FEDER through COMPETE 2020 and FCT – PTDC/CVT-CVT/31020/2017. RP was supported by PD/BDE/135540/2018 FCT grant. AC was supported by FCT Individual CEEC 2017 CEECIND/01514/2017 Grant.

Gut health effects of feeding insect products to livestock

A. Józefiak

Poznan University of Life Sciences, Department of Preclinical Sciences and Infectious Diseases, Wolynska 35, 60-637 Poznan, Poland; agata.jozefiak@up.poznan.pl

Insects at all life stages are rich sources of protein, fat and many other important nutrients. Insects are part of a natural diet for farm animals such as chickens and fish. They can be higher in protein than soy meal and need less food to produce it, use less water, produce less waste and CO_2. Insect proteins can be also a potential source of antimicrobial peptides (AMP) in livestock production. The production of AMP is an early component of an innate immune response toward pathogen infection. Most of insects are also rich in micronutrients such as iron, calcium, vitamins, omega fatty acids, and contain chitin. Chitin acts like probiotic to contribute to a healthy immune system. Therefore, insects are considered to be a potential source for conventional production of protein, and other nutrients, either for direct animal as well as human consumption. Of all regions of the body, the intestinal tract contains the most abundant microbiota. The enteric microbiome is thought to play a significant role in nutrient digestion and uptake, synthesis of vitamins, volatile fatty acids, amino acids, and maintenance of intestinal mucosal integrity and gut peristalsis. It is known that the intestinal microbiota communicates with other organ systems including the brain, lungs, skin and liver. The intestinal microbiome is also influenced by numerous factors, including age, diet, and inflammation process. In animals and humans, alterations in the gut microbiota have been linked to several important diseases and conditions, including obesity, diabetes mellitus, Crohn's disease, ulcerative colitis, and some forms of neoplasia. It has been shown that consuming insects can help support the growth of beneficial gut bacteria, reduce inflammation in the body, improve gastrointestinal function and other parameters of health. A relatively small addition of insects to the broiler and fish diet can modulate microbiome of gastrointestinal tract to help host to absorb nutrients and generate immune responses that can affect host health, making it an important area of current research.

Symbionts, probiotics and pathogens of farmed insects

A. Vilcinskas

Fraunhofer Institute for Molecular Biology and Applied Ecology, Bioresources, Leihgesterner Weg 85, 35392 Giessen, Germany; andreas.vilcinskas@ime.fraunhofer.de

The tremendous success of insects in terms of biodiversity reflects their ability to manage associated microorganisms. The latter determine also the success of insect farming at an industrial scale. The presentation provides an overview about symbiontic microorganisms benefiting farmed insects by assisting enzymatic digestion of the diet, nitrogen-fixation or producing antibiotics. Probiotics are currently developed to strengthen the pathogen resistance of farmed insects in order to prevent the use of antibiotics. We have recently shown that the black soldier fly *Hermetia illucens* has a robust immune system, which relies e.g. on a broad spectrum of antimicrobial peptides exhibiting a diet-dependent expression pattern. However, mass rearing in insect farms promotes inbreeding, allows pathogens to adapt and covert infections can become lethal, if factors such as other pathogens, stress or diet restrictions promote pathogen virulence in an additive, potentiating or synergistic manner. Consequently, advanced tools for the diagnosis of entomopathogens in insects farms and effective strategies that protect the health of farmed insects must be developed.

Effect of seaweed or selenium enriched substrates on live and reproductive performance in BSF

M. Ottoboni[1], F. Defilippo[2], P. Gigante[2], A. Grisendi[2], G. Pupillo[2], P. Bonilauri[2] and L. Pinotti[1]
[1]University of Milan, Department of Health, Animal Science and Food Safety Carlo Cantoni, via Celoria 10, 20133, Italy,
*[2]Istituto Zooprofilattico Sperimentale della Lombardia e dell'Emilia Romagna Bruno Ubertini, Via Pitagora 2, 42124
Reggio Emilia, Italy; matteo.ottoboni@unimi.it*

This study evaluated the effect of inclusion of brown algae (*Ascophyllum nodosum*) or sodium selenite in BSF growing
substrate on selected insect life cycle traits. The larvae were reared on three different substrates: (1) Gainesville diet,
used as control substrate (CTR); (2) *A. nodosum* diet (AN30%), with 30% substitution of the alfalfa meal with brown
algae; and (3) selenium diet (Se), based on Gainesville diet fortified with 0.3 mg/kg of selenium in the form of sodium
selenite. Eggs collected from a stable colony of BSF were placed for hatching on three experimental substrates CTR,
AN30%, and Se. All experiments were carried out under dark condition, at 25 °C with 70% relative humidity, and
repeated for three consecutive generations. Recorded data were: live performance, i.e. percentage of mortality in each
stage, development period (days required to reach pupa stage), larvae weight, percentage of adult emergence; and
Reproductive performance, i.e. number of eggs laid and percentage of hatched eggs. Both mortality and larval weight
did not shown any difference among groups during the three generation study. In contrast, BSF development period
was lower (P<0.05) in CTR than in AN30% and Se group. Further differences were observed in term of percentage of
adult emergence that was higher in CTR (90.4%) and Se (90.4%) compared to AN30% (78.7%). When reproductive
performance were considered, best values for number of eggs laid and percentage of hatched eggs were observed in
the case of CTR. Reproductive performance dramatically decreased in AN30%, and Se group. Combining obtained
results BSF larvae can grown on media containing up to 30% *A. nodosum* or selenium with some implication for both
live and reproductive performance.

***Hermetia illucens* larvae meal: evaluation of an alternative protein source in diet for rainbow trout**

*C. Caimi[1], I. Biasato[1], G. Chemello[1], M. Renna[2], S. Bellezza Oddon[1], M. Prearo[3], A. Schiavone[2], F. Gai[4], C. Trèspeuch[5],
C. Guidou[5] and L. Gasco[1]*
*[1]University of Turin, Department of Agricultural, Forest and Food Sciences, Largo Paolo Braccini, 2, 10095, Grugliasco,
Italy, [2]University of Turin, Department of Veterinary Sciences, Largo Paolo Braccini, 2, 10095, Grugliasco, Italy, [3]Veterinary
Medical Research Institute for Piedmont, Liguria and Aosta Valley, Via Bologna, 148, 10154, Torino, Italy, [4]Institute of
Science of Food Production, National Research Council, Largo Paolo Braccini, 2, 10095, Grugliasco, Italy, [5]MUTATEC,
4 rue Toussaint Flechaire, 84510, Caumont-sur-Durance, France; christian.caimi@unito.it*

Recent investigations highlighted that insect meals can be a valuable alternative to conventional proteins in aquaculture.
This research evaluated the effects of six levels of inclusion of a partially defatted *Hermetia illucens* (HI) meal
on rainbow trout performance. A partially defatted HI meal was used to formulate six isonitrogenous (about 46%
crude protein, as fed), isolipidic (15% ether extract, as fed) and isoenergetic (22 MJ/kg, as fed) diets. The diets were
formulated to have partial replacement of fishmeal (0, 10, 20, 30, 40 and 50%) by increasing levels of HI meal (0%
[HI0], 3% [HI3], 6% [HI6], 9% [HI9], 12% [HI12] and 15% [HI15] on as fed basis, respectively). A total of 576 fish
(initial body weight: 100.1±9.29 g) were randomly allotted to 24 tanks (24 fish/tank, 4 replicates/treatment) and fed
for 121 days. Statistical analysis was performed by means of One-way ANOVA. No differences among the treatments
were observed for all the considered growth performance parameters: individual weight gain, specific growth rate,
feed conversion ratio and protein efficiency ratio. Dietary treatments did not influence the carcass yield (average of
87.67%). Condition factor was higher than 1 in all the treatments but showed the lowest value in the fish fed HI15
(1.08). Hepatosomatic and viscerosomatic indexes showed significant differences among the treatments, with the
highest values being observed in the fish fed HI15 (1.47 and 12.56, respectively). These preliminary results suggest
that a partially defatted HI meal could be included up to 15% with no negative effects on rainbow trout performance.
Research funded by AGER project 'Fine Feed for Fish (4F)' (2016-01-01)

Effect of novel functional ingredient on the performance resistance of pacific white shrimp

A. Richardson[1], J.J. Dantas-Lima[2], M. Walraven[1] and M. Lefranc[1]
[1]INNOVAFEED, 2 Rue de L'Europe, 59231, France, [2]IMAQUA, Pinksterbloemstraat 7, 9090 Melle, Belgium; andrew.richardson@innovafeed.com

This study was performed as part of developing a novel functional feed ingredient for Pacific white shrimp (*Litopenaeus vannamei*). Here we assess the effects of dietary inclusion of a novel functional ingredient (NFI) derived from defatted black soldier fly (*Hermetia illucens*) larvae meal on growth performance, tolerance to salinity stress, and disease resistance when challenged with *Vibrio parahaemolyticus* or a strain of white spot syndrome virus (WSSV). A control diet (CTRL) was used for comparison to three test diets including 4.5, 7.5 and 10.5% of NFI (NFI4.5, NFI7.5 and NFI10.5). After 28 days, all diets with NFI had significantly improved weight gain, feed conversion ratio (FCR) and specific growth rate (SGR) compared to CTRL. The results showed the lowest level inclusion of the NFI (4.5%) created a marked effect, with SGR significantly increasing. During the growth trial the survivability was not affected by diet. NFI in the diets showed a positive pattern towards WSSV resistance, without impacting *V. parahaemolyticus* survival. Overall, we show that there is strong evidence that the use NFI derived from *H. illucens* has potential to improve the performance and health of *L. vannamei* and could generate value for shrimp farmers at low inclusion rates.

Amine metabolism in pigs is influenced by black soldier fly larva meal inclusion in the diet

S. Kar, D. Schokker, M.A. Smits and A.J.M. Jansman
Wageningen Livestock Research, Animal Nutrition, De Elst1, 6708WD, the Netherlands; soumya.kar@wur.nl

Black soldier fly larva (BSF, *Hermetia illucens* L.) meal is considered as an alternative protein source with has the potential to partially replace conventional protein source such as soybean meal (SBM) in livestock species, including pigs. Measuring the profile of systemic amine metabolites by the ultraperformance liquid chromatography-tandem mass spectrometry (UPLC-MS/MS) method reveals the metabolic effects of dietary protein source. The effects include the influence of the intestinal microbiome and amine metabolites as potential biomarker candidates to predict dietary protein-associated phenotypes related to health. The aim of the present study was to investigate the effects on the systemic amine profiles by feeding a BSF based diet relative to a SMB based diet to pigs. Sixteen growing male pigs (average initial body weight 35 kg), housed individually in metabolic cages were either fed with experimental diets containing SBM (n=8; used as reference diet) or BSF (crude protein: 158 g/kg as fed basis; n=8) as single protein source over a period of 4 weeks. Systemic amine levels were measured in blood plasma samples collected at the end of the feeding trial, i.e. d 28 of the experiment. Out of the 68 targeted amine metabolites, we could detect 58 in blood plasma of the pigs. Metabolic pathway analysis considering amine metabolites in pigs fed the BSF based diet, as compared to the SBM based reference diet, showed profound effects on systemic amine metabolism, in particular to the 'glycine, serine and threonine' pathway. In addition, we observed diet specific amine metabolite signature(s) in pigs fed with BSF based diet. For example, sarcosine was significantly (P<0.0001) higher in plasma concentrations in the pigs fed with BSF based diet, that potentially serves as an indicator for increased availability of specific nutrients i.e. choline or amino acids. Moreover, by examining the amine profiles in the blood plasma, we could not detect any significant amine signatures that could compromise the health status of pigs fed with BSF based diet.

Effect of the replacement of soybean by black soldier fly larvae in pig diets – a metabolic assay

D. Murta[1,2,3], R. Antunes[4], A. Pestana[4], I. Vieira[4], R. Nunes[3], A. Sequeira[5], J. Almeida[5] and O. Moreira[5]
[1]Faculty of Veterinary Medicine – ULHT, CBIOS, Campo Grande, 1749-024, Portugal, [2]Faculty of Veterinary Medicine, University of Lisbon, CIISA, Av. Universidade Técnica, 1300-477 Lisboa, Portugal, [3]EntoGreen, Ingredient Odyssey, Santarém, Portugal, Quinta das Cegonhas, Apartado 577, 2001-907, Portugal, [4]Faculty of Veterinary Medicine – ULHT, Campo Grande 376, 1749-024 Lisboa, Portugal, Portugal, [5]INIAV, Estação Zootécnica Nacional, Fonte Boa, Vale de Santarém, 2005-048 Vale de Santarém, Portugal; olga.moreira@iniav.pt

An experiment was conducted with Landrace×Large White×Duroc crossbred pigs aiming to evaluate the effect of the partial and total replacement of soybean meal and soybean oil by black soldier fly larvae meal on the metabolic characteristics, growth performances and carcass traits. Experimental diets (BSF0 – Control, BSF50 – 50% replacement; BSF100 – 100% replacement) were formulated to be isoenergetic and isoproteic (15% CP). Metabolic trials were performed, to study the experimental diets, with finishing pigs. Following a randomised block experimental design twelve non castrated males, with average live body weight of 62±3.6 kg, were individually allocated in metabolism cages for the control of feed intake and separated sampling of faeces and urine, during 7 days periods, in alternate weeks all over the finishing period, after adaptation to the experimental diets. The inclusion of larvae meal in diets was 87.8 and 175.6 kg/ton respectively for diets BSF50 and BSF100. Live body weights (LW), average daily gain (ADG) and feed conversion ratio (FCR) were estimated weekly. Feed and water intake and excreta were measured daily. The animals were slaughtered with 100±6.6 kg LW. Diets and excreta were chemically analysed and apparent digestibility of Dry Matter (DM), Nitrogen (N) and Energy (E) and balance of N, E, Ca and P were calculated. Statistical analysis was performed by SPSS. Mean ADG was 979, 1,083 and 927 g for pigs in diets BSF0, BSF50 and BSF100, respectively. At the time of the submission of this abstract the results were still being analysed, what prevented a discussion and conclusion.

Cuticle-poor black soldier fly meal show casein-matching body protein retention efficiency in mice

R. Gutman[1,2], M. Fialko[2], J. Martinez[1,3] and A. Jonas-Levi[4]
[1]Tel-Hai College, Department of Animal Sciences, Upper Galilee, 12210, Israel, [2]MIGAL – Galilee Research Institute, Laboratory of Integrative Physiology, tarsis 1, 10116, Israel, [3]MIGAL – Galilee Research Institute, Lab. of Animal Ecology and Biodiversity, tarsis 1, 10116, Israel, [4]Tel-Hai College, Department of Food Sciences, Upper Galilee, 12210, Israel; roeeg@migal.org.il

Little is known about the ability of insect protein to support muscle growth in monogastric animals, and the high level of cuticle in insect meal may have adverse effects on insects' meal quality. Hence, we aimed at testing the efficacy of *Hermetia illucens* (black soldier fly larvae – BSF) and *Apis mellifera* (Western honey bee larvae) – based meals, used as the sole source of protein, in supporting muscle growth, and their effect on weight gain, energy intake, and energy and nitrogen digestibility. Also tested was the effect of cuticle reduction on these parameters. C57BL6/J mice were provided standard, nitrogenous-balanced, diets in which different insect-based meals replaced casein. Weight gain under BSF-meal (BSF) matched that under regular diet (Chow) yet was slower compared to casein-based diet (Casein), which matched to of Bees-meal (Bee). BSF-meal with lower cuticle level (BSF-1) resulted in a faster weight gain, compared to BSF and Chow. Further incrementation in protein level on account of cuticle level (BSF-2) did not improve weight gain, which matched that of Bees, even with lower cuticle level (Bee-1). About 15% of the weight difference between BSF and Casein was explained by protein mass difference. Energy intake of BSF-based diets matched that of Casein and was lower than that of Bees-based diets, which matched Chow intake. BSF, Bee, and BSF-1 diets showed lower energy digestibility compared to BSF-2 and Bee-1, which showed Casein-matching digestibility. Absorbed energy did not differ by diet, resulting in a lower weight gain efficiency in Chow and BSF, compared to Casein. These differences may be explained by digestion and absorption costs of cuticle-rich meals or by other means of energy utilisation. BSF and BSF-1 diets showed Chow-matching true nitrogen digestibility that was lower than that of all other diets. Nitrogen absorption was higher under Chow and lowered under BSF and BSF-1 diets, compared to Casein, while Bees-diets showed intermediate levels. Weight and protein gain per nitrogen intake and nitrogen absorption were lower in Chow and BSF diets, compared to all other diets. These results follow the low amino-acid-score of BSF, compared to Casein, while showing that Bees have higher (Casein-matching) dietary benefits, compared to BSF. Decreasing the cuticle level, however, increases the efficiency of BSF-meal to that of Casein.

Quality of the protein from black soldier fly larvae compared to soybean protein in organic broilers

M. Heuel[1], C. Sandrock[2], A. Mathys[3], M. Gold[3], C. Zurbrügg[4], I. Gangnat[1], M. Kreuzer[1] and M. Terranova[1]
[1]ETH Zurich, Universitätsstr. 2, 8092 Zurich, Switzerland, [2]FiBL, Ackerstr. 113, 5070 Frick, Switzerland, [3]ETH Zurich, Schmelbergstr. 9, 8092 Zurich, Switzerland, [4]Eawag/Sandec, Überlandstr. 133, 8600 Dübendorf, Switzerland; christoph.sandrock@fibl.org

Insects in general and especially the larvae of the black soldier fly (BSF) are globally discussed as a possible replacement for soybean in animal nutrition. However, information about their feeding value is scarce. Therefore an experiment was carried out with organic broilers where soybean cake and oil (S) was replaced by defatted meal and fat of BSF larvae grown on two different substrates (A/B). 80 Hubbard S757 broilers were kept in pairs and fattened from day 15 to 63/64 of life with one of the five following diets: a positive control (S/S, protein meal/oil, 21% CP) complying with recommendations for this broiler type, a negative control (SS-, 18% CP), a diet with BSF meal A and fat A (AA, 17% CP), a diet with BSF meal A and BSF fat B (AB, 17% CP) or a diet with BSF meal B (BB, 17% CP, no extra fat). Diets SS-, AA, AB and BB theoretically induced a protein deficiency to determine whether the protein value of the insect material is comparable to that of soybean. One animal per pair was analysed for carcass and meat quality. The average daily gains (g) were 27.1, 20.4, 27.5, 26.4 and 19.9 for SS, SS-, AA, AB and BB (SS- and BB, P<0.05 against the others). The corresponding feed efficiency (g feed/g gain) was 2.7, 3.2, 2.8, 2.9 and 3.1 (SS- and BB, P<0.05 against SS). Dressing percentage was similar at ~68% in all groups. Breast meat proportion was greater (P<0.05) with SS (21%) compared to all other diets (all ~18%). Insect-based diet BB enhanced (P<0.05) the yellowness of skin and meat. Cooking loss differed (P<0.05) between SS (14.3%) and BB (17.2%), and shear force of the meat was also highest with BB. The results indicate that BSF protein meal A has a better protein value than soybean cake as it permitted a better growth than SS- despite the same protein content. Yet, diet AA could not prevent the impairment in breast meat proportion also observed in SS-. BSF protein meal B was widely equivalent to soybean cake. In conclusion, BSF protein meal is a high quality protein source, but its protein value varies between origins.

Evaluation of the suitability of a *Tenebrio molitor* L. meal as ingredient for broiler chicken diets

D. Murawska[1], W. Sobotka[2], T. Bakuła[3], D. Witkowska[4], M. Gesek[5], T. Daszkiewicz[6] and P. Matusewičius[7]
[1]University of Warmia and Mazury in Olsztyn, Department of Commodity Science and Animal Improvement, Oczapowski St. 5, 10-719 Olsztyn, Poland, [2]University of Warmia and Mazury in Olsztyn, Department of Animal Nutrition and Feed Science, Oczapowski St. 5, 10-719 Olsztyn, Poland, [3]University of Warmia and Mazury in Olsztyn, Department of Veterinary Prevention and Feed Hygiene, Oczapowski St. 13, 10-719 Olsztyn, Poland, [4]University of Warmia and Mazury in Olsztyn, Department of Animal and Environmental Hygiene, Oczapowski St. 5, 10-719 Olsztyn, Poland, [5]University of Warmia and Mazury in Olsztyn, Department of Pathological Anatomy, Oczapowski St. 13, 10-719 Olsztyn, Poland, [6]University of Warmia and Mazury in Olsztyn, Department of Commodity Science and Animal Raw Material Processing, Oczapowski St. 5, 10-719 Olsztyn, Poland, [7]Lithuanian University of Health Sciences, Department of Animal Breeding and Nutrition, Tilžės St. 18, 47181 Kaunas, Lithuania; daria.murawska@uwm.edu.pl

Nowadays, insects have gained more and more attention in Europe as an underexploited sustainable protein and nutrient source for poultry feed. The aim of this study was to compare selected growth performance effects and carcass quality characteristics in broiler chickens fed by diets with different *Tenebrio molitor* (TM) processed animal proteins (PAP) meal content. The experiment was run on a total of 800 1-day-old male Ross 308 broilers. At 1-day of age, the chickens were randomly assigned to 4 dietary treatments (10 replications per treatment, 20 birds per pen); TM-0 (0% TM- PAP control group), TM-50 (50% TM- PAP), TM-75 (75% TM- PAP), TM-100 (100% TM-PAP). The birds were raised to 42 d of age and fed *ad libitum* (Starter 1-14 d, Grower 15-35 d, Finisher 36-42 d). The results were processed statistically by one-way ANOVA. The significance of differences between means was determined by Duncan's test (P<0.05). The final body weight (BW) of chicken was respectively: 3,254.0 g (TM-0), 2,741.0 g (TM-50), 2,422.0 g (TM-75), 1,405.0 (TM-100, P≤0.05). The percentage of breast muscles in the total BW decreased from 25.0% (TM-0) to 22.3% (TM-50), 19.1% (TM-75), 15.7% (TM-100, P≤0.05). In conclusion, the replacement above 50% of soybean protein of TM-PAP in the chicken's diet significantly worsens growth performance and carcass quality. Acknowledgements: This work was supported by the National Science Centre, Grant: „Development of a strategy for the use of alternative protein sources in animal nutrition enabling the development of its production on the territory of the Republic of Poland', No Gospostrateg1/ 385141/16 /NCBR /2018.

Effect of diet on the chemical composition of *Tenebrio molitor* larvae

S.F.S. Cardoso[1], T.L. Mateus[1,2,3] and J.C.O. Lopes[1,3]
[1]*Instituto Politécnico de Viana do Castelo, Refoios, 4990-706 Ponte de Lima, Portugal,* [2]*EpiUnit, Unidade de Investigação em Epidemiologia, Porto, Porto, Portugal,* [3]*CISAS, Center for Research and Development in Agrifood Systems and Sustainability, IPVC, 4900-347 Viana do Castelo, Portugal; sarafilipacardoso@hotmail.com*

Tenebrio molitor is a species from the family Tenebrionidae, which has a life cycle characterised by 4 stages of development: egg, larva, pupa and beetle. The aim of this study was to evaluate the effect of diets on the chemical composition of *T. molitor* larvae. The tests were carried out on the ESA-IPVC insectarium (Portugal) under controlled conditions of temperature (°C) and relative humidity (%). Four diets elaborated with conventional cereal usually used to feed other livestock species were tested. It was used 150 larvae per box and 3 repetitions per diet. The tests ended with the appearance of the first pupa in the boxes. At end, the larvae were slaughtered by freezing. The analyses performed (on diets and larvae) consisted in: humidity content, ash content, crude protein content, crude fat content and mineral content (K, Ca, Mg, Na, Fe, Cu, Zn, Mn and P) according to AOAC. The composition and the price of diets used in this study were the following: (A) 80% wheat + 20% soybean meal (0,67€/kg); (B) 40% barley + 40% maize + 20% soybean meal (0,46€/kg); (C) 40% oat + 40% wheat + 20% soybean meal (0,70€/kg) and (D) 40% wheat + 40% wheat + 20% soybean meal (0,49€/kg). The moisture content present in the species *T. molitor*, for four diets, was in agreement with other authors. The ash content of the *T. molitor* larvae were similar to those reported by other authors. Crude protein proved to be the component with the highest expression for larvae fed with the 4 diets, and larvae of diet B showed significantly lower values (42,496±2,221%) than larvae fed with diet D (45,408±1,738%), these results were superior to the results obtained by other authors. In the parameter of crude fat, significant differences were also found between diets, with diet B (35,044±3,265%) getting significantly higher results. The levels of fat presented by the *T. molitor* larvae fed with the four diets were higher than the results obtained by Siemianowska *et al*. To the mineral content, only in calcium levels significant differences were found between diets The studied diets affected the nutritional composition of the two species and this influence of the diet on the larvae of *T. molitor* can be a tool to adapt their chemical composition depending on the purpose of production.

Effect of diet on the chemical composition of *Zophobas morio* larvae

S.F.S. Cardoso[1], T.L. Mateus[1,2,3] and J.C.O. Lopes[1,3]
[1]*Instituto Politécnico de Viana do Castelo, Refoios, 4990-706 Ponte de Lima, Portugal,* [2]*EpiUnit, Unidade de Investigação em Epidemiologia, Porto, Porto, Portugal,* [3]*CISAS, Center for Research and Development in Agrifood Systems and Sustainability, IPVC, 4900-347 Viana do Castelo, Portugal; sarafilipacardoso@hotmail.com*

Zophobas morio is a species popularly known as giant tenebrian commonly produced in Europe and in the U.S.A. The aim of this study was to verify the influence of diet on the chemical composition of Zophobas larvae. The tests were carried out on the ESA-IPVC insectary (Portugal) under controlled conditions of temperature (°C) and relative humidity (%). Were tested 4 diets, used 100 larvae per box and 3 repetitions per diet. Test ended when, after 3 successive weighings, the larvae mass remained the same. At the end, the larvae were slaughtered by freezing. The analyses performed (on diets and larvae) were: humidity, ash, crude protein, crude fat and mineral contents according to AOAC. The composition and the price of diets used in this study were the following: (A) 80% wheat + 20% soybean meal (0.67€/kg); (B) 40% barley + 40% maize + 20% soybean meal (0.46€/kg); (C) 40% oat + 40% wheat + 20% soybean meal (0.70€/kg) and (D) 40% wheat + 40% wheat + 20% soybean meal (0.49€/kg). There were no significant differences in the humidity content or ash content of the larvae fed with the 4 diets. The moisture content is similar to the value of other products created especially for salmon. The ash content of *Z. morio* larvae is in accordance with the levels obtained by Rumpold & Schluter and Finke and these values are slightly higher than those found in conventional meats. Regarding the crude protein content, it was found that there were significant differences between the larvae fed with diet B (42,605±2,222%) and diet D (45,517±1,738%). The protein results obtained were superior to those obtained by Adámková *et al*. Regarding the crude fat parameter, it was found that the larvae of diets B (34,506±5,100%) and C (34,702±3,891%) present content of crude fat significantly higher than those of diet A (29,312±4,082%). The fat content presented by the larvae of the four diets was lower than the results obtained by Araújo *et al*. and Adámková *et al*. but higher than those obtained by Yi *et al*. No significant differences for the elements: Ca, Na, Fe, Cu and P in the *Z. morio* larvae are found. The levels of Ca, Na, Fe and Cu analysed were higher, and that of phosphorus was lower, than obtained by Finke (2015) and Rumpold & Schlüter. The *Z. morio* larvae, fed with the studied diets, contain a high protein and crude fat content, and may be a potential alternative to fish meal and soybean meal in feeds used to feed conventional livestock species.

Use of insect meals in ruminant nutrition: protein evaluation

P.G. Toral[1], P. Frutos[1], M.G. González-Rosales[1], A. Della Badia[1], M. Fondevila[2] and G. Hervás[1]
[1]IGM (CSIC-ULE), Finca Marzanas, 24346, Grulleros, León, Spain, [2]Dpto. Prod. Animal y Ciencia de los Alimentos, IA2, Univ. Zaragoza-CITA, Miguel Servet 177, 50013, Zaragoza, Spain; g.hervas@csic.es

In Europe, most intensive ruminant feeding systems are highly dependent on soybean meal as the source of protein, which fosters the search for alternatives. Although the use of insect meals has been suggested as one of these alternatives, very few reports have characterised this type of feedstuffs in ruminants. In anticipation of a future authorisation of insect meals for use in practical farming, their protein evaluation would be among the first research needs. Nevertheless, concerns exist about the use of methodologies such as batch cultures of ruminal microorganisms or the *in situ* nylon bag technique, and other options must be tested. The objective of this study was to perform a protein evaluation of 4 insect meals (from *Tenebrio molitor*, *Zophobas morio*, *Alphitobius diaperinus* and *Acheta domesticus*) and of soybean meal (as a reference feedstuff). Three different techniques were employed. First, *in vitro* batch cultures of ruminal microorganisms were carried out with increasing levels of starch, and N degradation was estimated by linear regression (ml of gas vs mg of ammonia-N). The second methodology consisted in estimating N disappearance by filtering the *in vitro* incubation residue, whereas the third approach involved the *in situ* nylon bag technique. The crude protein content of insect meals ranged from 38% in *Z. morio* to 70% in *A. domesticus*. The 3 methodologies appeared to establish a similar ranking among feeds, with very similar results in estimations based on regression or *in situ* techniques, but with higher variation in the *in vitro* assessment. Regardless of the method, insects showed lower protein degradations than soybean meal, ranging from 41-50% in *T. molitor* to 70-78% in *A. diaperinus*. Overall, results supported the potential of insects as alternative ingredients to vegetable feedstuffs in ruminant diets. However, further research would be necessary, for example to characterise their amino acid composition or the fibre-bound nitrogen in their chitin exoskeleton, which might contribute to explain their relatively low protein disappearances. Acknowledgements: Ramón y Cajal program, RYC-2015-17230, MINECO/ESF, UE.

The effects of replacing soybean with *Hermetia illucens* on the broiler intestinal tract

J. Catarino[1], M.A. Machado[2], A. Alvarado[1], I. Carvalho[3], R. Matos[1], O. Moreira[4], P. Faísca[1,3,5] and D. Murta[1,5,6,7]
[1]ULHT, Faculty of Veterinary Medicine, Campo Grande 376, 1749-024 Lisboa, Portugal, [2]ULisboa, ISA, Tapada da Ajuda, 1349 – 017 Lisboa, Portugal, [3]DNAtech, Estrada do Paço do Lumiar N.° 22 Edifício E, 1° Andar, 1649-038 Lisboa, Portugal, [4]INIAV, Estação Zootécnica Nacional, Quinta da Fonte Boa, 2005-048 Vale de Santarém, Portugal, [5]ULHT, CBIOS, Campo Grande 376, 1749-024 Lisboa, Portugal, [6]ULisboa, CIISA-FMV, Av Universidade Técnica, 1300-477 Lisboa, Portugal, Portugal, [7]Ingredient Odyssey, EntoGreen, Quinta das Cegonhas, Apartado 577, 2001-907 Santarém, Portugal; daniel.murta@entogreen.com

The increasing demand of broiler production has considerable environmental and economic impact, increasing the urgency to find sustainable alternative feed ingredients. Recent studies indicate that some species of insects have great production potential and may be a source of both protein and lipids, which is the case of black soldier fly (BSF) (*Hermetia illucens*). The objective of the current study was to assess the influence of replacing soy with BSF meal on the intestinal tract morphology. Twenty four, 1-day-old Ross 308 broiler were fed four different diets where soybean meal and soybean oil were replaced by 0, 22, 44 and 66% of dry BSF larvae, for 28 days. Broilers were then slaughtered and the intestinal tract was divided into 5 different portions – duodenum, jejunum, ileum, caecum and colon for histopathology analysis. For each intestinal portion, 6-7 transverse systematic uniform random sections were sampled, and processed for routine H&E staining. Morphometric analysis was performed with the ImageJ software and the colour threshold plug in. For each portion, a pathological assessment was performed and the average total section area, mucosal area, percentage of mucosa and caecal tonsil area were determined. The histopathology assessment did not revealed any histological lesion with pathological significance. The morphometric study didn't disclosed differences in the several parameters measured within the same portion of intestine among the different diet groups (Kruskal-Wallis non-parametric test $P>0.05$). Replacement of soy by BSF has no histological or morphometric effect on broiler intestinal tract. Funding: This study was conducted in the scope of the EntoValor project (POCI-01-0247-FEDER-017675).

Animals in circular food systems and their contribution to healthy diets

H.H.E. Van Zanten[1], B. Van Selm[1,2] and A. Frehner[1,3]
[1]Wageningen University & Research, Animal Production Systems group, P.O. Box 338, 6700 AH Wageningen, the Netherlands, [2]Wageningen University & Research, Plant Production Systems group, P.O. Box 430, 6700 AK Wageningen, the Netherlands, [3]Research Institute of Organic Agriculture FiBL, Department of Socioeconomics, Postfach 219, 5070 Frick, Switzerland; hannah.vanzanten@wur.nl

Do we really need farm animals to secure human food nutrition? Or, can our food system be more sustainable without them? The fact that farm animals provide essential nutrients for humans while causing major environmental problems fuels an often-polarised debate about the role of animals in sustainably feeding an ever-growing population. An increasing body of literature suggests that a transition of animal production towards circularity in the food system might be a solution. Such animals recycle biomass unsuited for human consumption back into the food system and so lower arable land use while contributing to human food supply. But, would such animals also provide the nutrients and foods needed for a healthy diet? We therefore assessed to what extent the recommended intake of animal source food in a healthy diet can be provided in a circular food systems in Europe. More concretely, we assessed the amounts of animal source foods that can be produced from the co-products and food waste derived from healthy diets and the available grass resources. To define a healthy diet we used the recently published healthy reference diet of the EAT-Lancet report. Then, to account for contextual differences, food-based dietary guidelines of five European countries were applied. Our preliminary results show that, in general, protein supply levels can be reached as well as the recommended intake levels of milk, meat, fish and eggs while simultaneously lowering land use. The farm animals that are most efficient in terms of converting biomass unsuited for human consumption into animal proteins are, however, not necessarily the animals that provide the most healthy foods. Our results showed, for example, too high amounts of red meat (mainly a by-product from dairy production) and too low levels of the more often recommended chicken meat. To conclude, animals in circular food systems can contribute to a more sustainable food system but a healthy recommended diet is not necessarily the most sustainable one.

Evaluation of the contribution of 15 European beef production systems to food safety

A. Jarousse[1], P. Veysset[1], P. Madrange[2], S. Hennart[3], A. Mertens[3], G. Pirlo[4], M. Lacurto[4], E. O'Riordan[5], J. Breen[6], C. Pahmeyer[7] and C. Mosnier[1]
[1]INRAE, UMR Herbivores, 63122 Saint-Genès-Champanelle, France, [2]IDELE, Rue de Bercy 149, 75595 Paris Cedex 12, France, [3]CRA-W, Rue du Bordia 4, 5030 Gembloux, Belgium, [4]CREA, Antonio Lombardo, 26900 Lodi, Italy, [5]Teagasc, Oak Park, R93 Carlow, Ireland, [6]UCD, Belfield, Dublin 4, Ireland, [7]University of Bonn, Nussallee 21, 53115 Bonn, Germany; patrick.veysset@inrae.fr

In the framework of the Era-Net project SustainBeef, this study evaluates the contribution of different European beef production systems to food safety through three dimensions: 1) the amount of edible protein and energy produced per hectare of agricultural area (UAA), 2) the efficiency of meat production in terms of human-edible protein and energy and in terms of land used, 3) the capacity of the system to produce food (meat, protein and energy) at low production costs. The analyses are based on 15 case studies representative of beef production systems in Europe (4 French, 4 Belgian, 2 Irish, 2 Italian and 3 German). They cover cow-calf, cow-calf-beef, fattener, dairy and mixed dairy-fattener systems, with or without cash crops. The results show that, at the farm level, systems that also sell milk and cereals have higher protein and energy production per hectare of UAA and lower costs. Beef units are almost all energy consumers due to the low share of energy edible by humans in meat compared to that present in the resources used. Systems using few concentrates, such as cow-calf systems, are mostly net producers of protein. Conversely, cow-calf-beef and fattener systems are net consumers of protein, with the exception of an Italian fattener that uses co-products that are not competitive with human food. Grass-based systems use more area/kg carcass but a major part of this area is non-tilled land, which is not competitive with human food, which is not the case with fattening systems. The systems with the lowest meat production costs are the fattening systems producing the most live-weight/LU/year and dairy systems in lowland which share the costs between milk and meat. In order to complete these results, it will be necessary to couple the cow-calf systems with the fatteners in order to analyse the systems as a whole. The study also plans to propose and model innovations to improve the systems.

Assessing ruminant feed efficiency at world scale: the complex relationship between milk and meat

P. Faverdin[1], J. Agabriel[2], Y. Dronne[3], H. Guyomard[4], E. Marajo-Petitzon[4] and A. Forslund[4]
[1]INRAE, Agrocampus Ouest, PEGASE, 16 le Clos, 35590 Saint-Gilles, France, [2]INRAE, Vetagrosup, UMRH, Theix, 63122 Saint-Genest-Champanelle, France, [3]INRAE, Agrocampus Ouest, Smart-Lereco, rue de Saint-Brieuc, 35000 Rennes, France, [4]INRAE, UAR Departement Ecosocio, rue de Saint-Brieuc, 35000 Rennes, France; philippe.faverdin@inrae.fr

Ruminants are the main bio-converters of fibre-rich biomass. Although animal feed and feed efficiency are key components of the global food system and land use, they are very difficult to assess. The aim of this study was to estimate the feed conversion ratios (FCR) of ruminants for milk and meat production combining statistical data and modelling at the world level divided into 21 regions including 8 regions in Europe. The FCRs (expressed in kg DM/ kg product) necessary to produce one litre of milk or one kg of carcass were calculated using the IPCC's TIER2 methodology which allowed us to estimate dry matter intake (DMI) based on energy needs and diet digestibility. DMI of dairy cows (after their first calving) was attributed almost entirely to milk while DMI of all other ruminants was attributed to meat, the growth of dairy heifers corresponding more or less to the carcass meat of the culled cows. To calculate feed consumption more precisely 8 categories of animals were considered, namely dairy cows, beef cows, females 0-1 year, 1-2 years and + 2 years, males 0-1 year, 1-2 years and + 2 years. A demographic model made it possible to calculate the numbers of animals for each category from comprehensive FAO statistical and technical data. A formulation model was used to allocate feedstuffs to the different productions. The digestibility of the ration was finally estimated for feedstuffs (using feed tables) and for grasslands according to climate data. Results show that FCRs of milk and meat vary greatly between the different regions of the world. They also show that the two FCRs are positively correlated only for very low productivity levels of animals. When the productivity of dairy cows exceeds 4,000-5,000 kg of milk/cow/year, there is a trade-off between milk and meat efficiencies. The increase in milk productivity improves milk FCR but degrades that of meat because more beef cows are needed to compensate for the drop in meat production from the dairy herd. For the same reason, a higher ratio of milk/meat production in a region generally improves the FCR of meat due to a smaller proportion of beef cows.

Exploring European livestock system performance along sustainability dimensions

F. Accatino[1], M. Zehetmeier[2], V. Karger[2], A. Reindl[2], M. Paiva Dos Santos[3], R. Teixeira[3], T. Rodriguez-Ortega[4], J.P. Domingues[1], D. Neumeister[5], E. De Olde[6], A. Van Der Linden[6] and G. Dorfner[2]
[1]INRAE, AgroParisTech, Université Paris Saclay, Paris, 75000, France, [2]LFL, München, 80638, Germany, [3]MARETEC, LARSyS, IST, Universidade de Lisboa, Lisboa, 1049-001, Portugal, [4]CITA, Zaragoza, 50059, Spain, [5]IDELE, Paris, 75000, France, [6]Wageningen University & Research, 6700 AH Wageningen, the Netherlands; francesco.accatino@inrae.fr

Sustainability of livestock systems is a complex concept requiring an evaluation across environmental, economic, and social dimensions. Understanding trade-offs and synergies among those dimensions is vital for informed decision-making, as decisions based on only one dimension might cause unintended consequences on other dimensions. The purpose of this study was to compare the performance of different European livestock system typologies on the sustainability dimensions and to explore correlations between indicators of sustainability (trade-offs and synergies). We collected with a unique protocol farm-level data in 7 European case studies: extensive beef cattle in Portugal; extensive sheep in Spain; extensive beef and dairy cattle in France; mixed crop-livestock dual purpose cattle and pigs in Germany; and laying hens in the Netherlands. We used the data collected to calculate a total of 41 environmental, 11 economic, and 35 social indicators that allowed as much as possible comparability across livestock system typologies. Some indicators were found dependent on the system typologies (e.g. 'Methane emissions' and 'Fraction of subsides in total revenue') whereas other indicators (e.g. 'Stress perceived by the farmer') were found independent on the case study and more linked with other variables. Some correlations were found significant and positive (Pearson correlation coefficient >0.75) like, for example, 'share of organic nitrogen in total nitrogen input' and 'nitrogen feed self-sufficiency'. The protocol for indicator calculation for farms can facilitate the collection of additional data in the future, reinforcing and extending the analysis. The comparison between different livestock system typologies along sustainability dimensions makes it possible, on a larger scale, to make considerations about sustainable land use.

Food Futures: a reporting tool to enhance the sustainability of livestock production systems

A. Aubry[1], R. McGuire[2], J. Birnie[2], N. Scollan[2] and S. Morrison[1]
[1]Agri-Food and Biosciences Institute, Livestock Production Sciences, Large Park, BT366DR, Hillsborough, United Kingdom, [2]Queen's University Belfast, Biological Sciences, 19 Chlorine Gardens, BT9 5DL, Belfast, United Kingdom; aurelie.aubry@afbini.gov.uk

Quantifying and enhancing the sustainability of agri-food production is key to sustain and build markets locally and internationally. The Agri-Food Quest 'Food Futures' project aims to develop a reporting tool to capture, credit and drive positive behavioural change among livestock farmers and their stakeholders, in order to enhance the sustainability of Northern Ireland agri-food production. Under the pillars of Economic, Environmental and Social sustainability, a set of more than 250 indicators, combined into 30 baseline metrics, have been identified following literature reviews and consultation with stakeholders from the food supply chain. The data needed to inform these metrics have been obtained from more than 26 dairy and beef/sheep farms across Northern Ireland, by integrating: (1) existing data that can be automatically obtained through for example existing benchmarking programmes and soil sampling schemes; and (2) new data obtained by surveying the ambassador farms. The data obtained so far indicate that there is a significant potential to enhance further the sustainability at farm level, in particular through behavioural changes. For example, only 45% of the study farms assess and minimise soil compaction, 74% have developed a long term farm business strategy that is reviewed regularly, and 60% feel that they only occasionally have the right work/life balance. The next phase in the project is to develop advanced metrics informed by collecting new, cutting edge data, to better quantify and monitor the effect of activities that have the potential to enhance sustainability at the whole farm level, for example through increased carbon sequestration. Most of the participating farmers are willing to monitor their greenhouse gas emissions through the use of carbon footprint calculators. Through multidisciplinary collaboration, a robust methodology is being developed using machine learning techniques to integrate multiple indicators within one SMART tool that can provide an overall rating of Sustainability for livestock farming systems.

Optimising the contribution of cattle productions to food supply: reconsidering farm type

C. Battheu-Noirfalise, A. Mertens, S. Hennart, M. Mathot and D. Stilmant
Centre wallon de Recherches agronomiques, Durabilité – Systèmes et prospectives, Rue du Serpont, 100, 6800 Libramont, Belgium; c.battheu@cra.wallonie.be

Due to their ability to convert cellulose rich resources into high quality animal products, ruminants can play an important role as net food producers. Nevertheless, in order to (1) increase energy and protein density of the diet of high yielding animals and/or (2) to face limited grassland access under low field accessibility and herd size increase, ruminants diets could contain potentially human-edible feed, such as cereals, leading to feed-food competition. In this work we explore the human-edible feed conversion efficiency (heFCE) and the human-edible food production per hectare of fodder area (heFHA), both under gross energy and gross protein forms, for dairy and meat farms in Wallonia (Belgium). Both milk and meat productions are taken into account. In order to quantify the fraction of human-edible resources in feeds, available literature was used for pure feeds and inquiries were performed near feedstuff producers to take into account the composition of a diversity of concentrates together with their relative importance. Results show that 92,8% of the specialised dairy farms contribute to food security (heFCE greater than 100%). Grass-based farm types show higher heFCE then maize-based types. Results per ha (heFHA) reveal better performances for intensive types (both grass- and maize-based). Indeed, highest heFHA performances were found in systems with dairy cows yielding around 8,000 l of milk per year, enlightening the fact that excluding completely feed-food competition may not be the most efficient way to contribute to food security. On the other hand, intensive dairy calves (Holstein) are not well valorised in the meat sector while fattening's phase of intensive meat calves (Belgian Blue Cattle) are highly competitive for human's food. This work will explore how mixed breeds and terminal crossing of beef breed on dairy breed could reduce the feed-food competition.

Protein quality: implications for sustainability assessments of protein-rich foods

G.A. McAuliffe[1], T. Takahashi[1,2] and M.R.F. Lee[1,2]
[1]Rothamsted Research, North Wyke, EX20 2SB, United Kingdom, [2]University of Bristol, Langford, BS40 5DU, United Kingdom; graham.mcauliffe@rothamsted.ac.uk

Sustainability assessments of protein-rich foods are often carried out under a functional unit (FU) representing protein content, yielding the final metric of environmental burdens per mass of protein. This approach, however, implicitly assumes protein to be a homogenous nutrient that is entirely bioavailable, and, therefore, fails to address complexities surrounding the digestibility and composition of its component amino acids. This study investigated the importance of considering protein quality when evaluating global warming potential (GWP) and land use (LU) associated with animal and plant based foods commonly consumed as protein sources. Baseline GWP and LU per mass of product were acquired from a life cycle assessment (LCA) database for the agri-food sector and recalculated as global weighted averages for individual commodities. Amino acid composition of each food item was taken from USDA's National Nutrient Database. Ileal digestibility of component amino acids was collated by a literature review. For each product, GWP and LU were compared under four distinct functional units: (1) mass of food (baseline: PQ_0); (2) mass of total protein (PQ_1); (3) PQ_1 corrected by protein digestibility corrected amino acid score (PDCAAS: PQ_2); and (4) PQ_1 corrected by digestible indispensable amino acid score (DIAAS: PQ_3). PDCAAS and DIAAS are both commonly-used indices to adjust the mass-based protein value according to a relevant protein's ability to satisfy a body's nitrogen requirements. The results showed that relative GWP between animal and plant based products is substantially affected by the FU employed for the analysis. For example, tofu's GWP was considerably lower than that of pork under PQ_0 and PQ_1 (by 74 and 23%, respectively), while the relationship reversed under PQ_2 and PQ_3 (by 58 and 68%, respectively) due to its relatively low protein scores (PDCAAS=0.56 and DIAAS=0.52). Relative LU between products followed largely similar patterns as GWP. These findings demonstrate the sensitivity of LCA metrics to assumptions on what constitutes the 'function' of food, and thus indicate the importance of addressing the nutritional quality of food according to the research question in hand.

The importance of whole farm nutrient mass balance: a case study at the south of Portugal

L. Martin, D. Santos and F. Miguens
Polytechnic of Coimbra, Coimbra Agriculture School (ESAC), Bencanta, 3045-601 Coimbra, Portugal; luisam@esac.pt

This work has four main objectives. First, apply a whole farm nutrient balance (WFNB) to a large dairy cattle unit of 1987 livestock unit and 560 hectares of arable land for forage. Secondly, use a holistic method as a mean of environmental impact assessment and calculate the NPK farm balance. Then, identify the bad practices of the dairy farming system, that can be controlled in order to improve sustainability and social acceptance. And finally, promote the Whole-Farm Nutrient Balance (WFNB) as a tool of environmental management for dairy farms in Portugal. A two year (2017 and 2018) nutrient dynamics was studied, by the quantification of every input and output of the farm, through the accounting system. Feed, bedding material, water, soil, manure and milk samples were collected and analysed by standard methods for N, P, K and other parameters. A WFNB Software (Version 2.0; February 8, 2012) developed by the Cornell Spear was applied. Based on the soil potential and forage needs a calculation for synthetic and organic fertilisers demands was accomplished. Centred on superficial and ground water nitrogen analysis and on the estimate of the CO_2(eq) balance by the DairyGEM software, a basic environment impact assessment was done. The results revealed an excess of more than 600 kg N per hectare per year at the studied farm and no justification to buy NPK synthetic fertiliser. The low content of soil organic matter can be improved with organic fertilisation available at the farm. With better practices and organic fertilisation, the estimated carbon footprint could be below 0.8 kg CO_2(eq) per kg of fat corrected milk and the ground water protected for nitrates.

Capturing multi-dimensional effects of linking production to area in dairy farming

V. Karger[1], M. Zehetmeier[1], A. Reindl[1], G. Dorfner[1] and J. Halbe[2]
[1]Bavarian Research Center for Agriculture, Agronomy, Menzinger Str. 54, 80538 Munich, Germany, [2]University of Osnabrück, Geography, Seminarstraße 19 a/b, 49074 Osnabrück, Germany; vanessa.karger@lfl.bayern.de

Intensification processes in dairy farming raise questions as to whether these production systems are sustainable. Especially with regards to environmental effects – such as nitrogen leaching due to excessive use of slurry – problems are observed. One option to reduce stress on land and use of resources is by linking the production of milk to the area. This will have an impact on farm management decisions and on a wide variety of sustainability indicators. It is crucial to understand how decision makers are likely to react to restrictions and how the different sustainability indicators are interrelated in order to prevent trade-offs with multi-dimensional sustainability goals. We used a vision design and assessment framework developed by Halbe *et al.* to evaluate the multi-dimensional effects of linking production to area in dairy cattle farming. The participatory approach consisted of 5 one-on-one interviews with experts from the dairy sector in science, practice and politics. During the interviews the experts were asked to identify indicators that would be affected if production was linked to area, to indicate the type of relationship (positive or negative) and to weigh the causal linkages (weak to strong effect). This approach allowed creating individual weighed causal loop diagrams. In a second step individual causal loop diagrams were merged to create a comprehensive diagram. Results showed that linking area to production will put even more pressure on the availability of land, hence is likely to result in further intensification (increasing milk yield per cow, intensive grassland management) and therefore strain on the farmers in high production areas if no boundaries are set. Farmers react with diversification if boundaries are set or suitable options exist. Extensification is likely to occur on less fertile soils. The methodology can facilitate shaping sustainability measures to make best use of synergies and avoid trade-offs between indicators.

The role of animal production in human diets in 2050: effects on nitrogen use and climate change

T.G. Morais[1], R.F.M. Teixeira[1], C. Lauk[2], M.C. Theurl[2], A. Mayer[2], H. Haberl[2], T. Domingos[1] and K.E. Erb[2]
[1]MARETEC – Marine, Environment and Technology Centre, LARSyS, IST, Universidade de Lisboa, Avenida Rovisco Pais, 1049-001, Portugal, [2]Institute of Social Ecology, University of Natural Resources and Life Sciences Vienna, Schottenfeldgasse 29, 1070, Austria; tiago.g.morais@tecnico.ulisboa.pt

Animal products in human diets are an important source of protein and energy. They also play an important role as drivers of climate change (greenhouse gases emissions; GHG), mostly due to enteric methane emissions, and in the nitrogen (N) cycle, due to the production of nitrous oxide emissions from excretion and N leaching. Here, we performed an integrated feasibility assessment of three alternatives to the current ('Business-As-Usual', BAU) diet: reduced meat, vegetarian and vegan diets. For the year 2050, we assessed land availability, N availability, and GHG emissions. We used the BioBaM model to calculate the balance between biomass supply and demand. BioBaM was modified in order to compute, besides C, C N mass balances, which enabled the assessment of the N budget of each diet as well as GHG emissions from nitrous oxide (besides carbon dioxide and methane). The analysis was regionalised, with modelling of 11 regions. The type of animal (ruminant or monogastric) used to obtained animal products is relevant for determining GHG emissions. For example, monogastrics have significantly lower enteric methane emissions when compared with ruminants. However, ruminants use grass in their diet, thus avoiding the GHG emissions in cropland production for feed and using a resource than is neglected in monogastric and plant-based diets because a large fraction of grassland area is unsuited for crop production. Further, organic manure produced by animals is an important source of N in cropland production, avoiding the GHG emissions from manufacturing inorganic fertilisers. So, fully eliminating animal products from human diets requires large yield increases that are only potentially achievable through intensive farming. This integrated assessment shows trade-offs between use of land, GHG emissions and N use of animal products in human diets. Future work is required to explore how these trade-offs can be overcome.

Measurement of patterns of soil organic matter in the Mediterranean Montado ecosystem using NIRS

J. Serrano, S. Shahidian, J. Marques Da Silva, M. Carvalho and A.E. Rato
Universidade de Évora, MED- Mediterranean Institute for Agriculture, Environment and Development, Instituto de Investigação e Formação Avançada, Universidade de Évora, Pólo da Mitra, Ap. 94, 7006-554 Évora, Portugal; jmrs@uevora.pt

From an environmental point of view, animal production is typically recognised as a major contributor to global greenhouse gas emissions. Nevertheless, it is important to add to this equation the positive impact of the Mediterranean Montado ecosystem on the soil, due to the potential for carbon sequestration through an increase in the soil organic matter content (SOM) resulting from the contribution of trees, pasture and grazing animals. This strategy has recognised interest in mitigating and adapting to climate change. However, monitoring SOM to determine carbon sequestration is time-consuming and expensive, as it requires soil sampling with consecutive laboratory analysis. The concept of precision agriculture (PA) can contribute decisively to resolve this situation. In this work, a combination of near infrared spectroscopy (NIRS) technology and multivariate data analysis (PLS regression) was used to estimate and monitor the evolution of SOM in 8 experimental fields in Portugal. Calibration and validation equations were established between SOM predicted and laboratory reference measurements. The determination coefficients (R^2>0.80), the RMSE for prediction (RMSE <0.1%) and the residual prediction deviations (RPD>2.0) demonstrated the use of NIR combined with PLS regression models for efficient estimation of SOM. This results show the practical interest of portable spectrometry, associated with GPS, as expeditious tools for monitoring spatial variability of SOM and, consequently, of soil fertility. Good prospects and opportunities open up for technology-based service providers to develop remote sensing-based computer applications using satellite imagery that enable: (1) dynamic management of animal grazing; and (2) definition of homogeneous management zones for the development of site-specific soil fertilisation/amendment, particularly important because the vast majority of soils that support the Montado ecosystem in Portugal are poor and acidic.

Microalgae supplementation in pigs – muscle microstructure, meat quality, and fatty acid composition

D. Dannenberger[1], C. Kalbe[1], A. Priepke[2] and G. Nuernberg[3]
[1]*Leibniz Institute for Farm Animal Biology (FBN), Institute of Muscle Biology and Growth, Wilhelm-Stahl-Allee 2, 18196 Dummerstorf, Germany,* [2]*State Research Center of Agriculture and Fisheries Mecklenburg-Vorpommern, Institute of Livestock Farming, Wilhelm-Stahl-Allee 2, 18196 Dummerstorf, Germany,* [3]*Leibniz Institute for Farm Animal Biology (FBN), Institute of Genetics and Biometry, Wilhelm-Stahl-Allee 2, 18196 Dummerstorf, Germany; dannenberger@fbn-dummerstorf.de*

The use of microalgae as farm animal diet supplementation provides the opportunity to improve growth and meat quality in poultry, rabbits, ruminants and pigs; however, the results are highly dependent on microalgae species, their proportions and chemical composition. The experimental design of our long-term micro algae pig study involved 31 Landrace sows in two dietary groups, control group (n=15) and micro algae group (n=16). The study was conducted at the experimental pig unit of Leibniz Institute for Farm Animal Biology (FBN), Dummerstorf, Germany. The pigs were fed indoor (group keeping), and the microalgae diets were supplemented by 7% (piglet diet) or 5% (fattening diet) DHA Gold (*Schizochytrium* sp., DSM, Germany), whereas the isoenergetic control diets contained soybean oil plus lard as lipid sources. The pigs of both groups were fed from 33rd to 146th day of age and slaughtered at 100 kg live weight. Carcass traits and meat quality of longissimus thoracis muscle were not affected by the microalgae diet with the exception of reduced drip loss (P=0.01) and increased protein proportion (P=0.04). The major effects of microalgae supplementation were muscle-specific increases in n-3 PUFA concentrations, resulting in increased accumulation of docosahexaenoic acid (DHA) and eicosapentaenoic acid (EPA) and no significant differences for saturated (SFAs) and monounsaturated (MUFAs) fatty acids. The total fibre number (TFN) and the fibre size (given as FCSA) of an average, STO (slow twitch oxidative), FTO (fast twitch oxidative) or FTG (fast twitch glycolytic) fibre did not differ among pigs fed control or microalgae diet in both muscles analysed. The relative fibre type distribution remained unchanged by diet in longissimus thoracis muscle (P≥0.75) but was affected in semitendinosus muscle. In the latter, the microalgae diet resulted in a shift to more oxidative myofibers at the expense of glycolytic myofibres (P≤0.02). In conclusion, microalgae supplementation offers a unique opportunity to enhance essential n-3 PUFA contents in pig meat. The results support small but coordinated changes in skeletal muscle phenotypic appearance and functionality.

Wild boar population management and its impact on local farming, a case report from central Italy

M. Odintsov- Vaintrub[1] and U. Di Nicola[2]
[1]University of Teramo, Faculty of Veterinary Medicine, Localita Piano D'acio, Facolta di Veterinaria, 64100 (TE), Italy,
[2]Ente Parco Nazionale del Gran Sasso e Monti della Laga, L.go Torrione, Isola del Gran Sasso, 64045 (TE), Italy;
modintsovvaintrub@unite.it

Wild boar (Sus scrofa) occupies a controversial role in human-environment interactions as re-introductions resulted in a rapid increase in its population. Consequently, tensions with farmers increased, which promoted the establishment of boar management programs. In the current work, a program launched by the 'Gran Sasso e Monti Della Laga' national park (Italy) is evaluated alongside its effects on local stakeholders. Archived data was paired with direct observations and qualitative interviews (50 n'). Trapping corrals were employed instead of firearms in order to reduce collateral damage, with captured boars being sent to local abattoirs. Since its beginning the program had only minimal technical changes, while its organisation was adjusted several times. The first period (1999-2004) was characterised by low trapping efficiency (0.35) as corrals were managed directly by farmers. The second period (2007-2009) marked an increase in trapping efficiency (0.65) while corral management passed to park authorities. However, this proved to be economically unsustainable and management passed to local private contractors (2009-2019), which also significantly increased trapping efficiency (3.25). Three main groups of stakeholders were identified as extremely invested in the process: (1) farmers; (2) hunters; (3) environmentalists. Farmers expressed a mostly moderate opinion regarding the culling methods, but strongly positive regarding population management and strongly negative regarding other stakeholders. Hunters and environmentalists expressed similar ideas (strongly negative) regarding corral trapping and population control citing animal welfare as a key reason. These opinions resulted in a combined lawsuit that forced the suspension of trapping activity during 2004-2007 (the lawsuit was consequentially archived). In conclusion, wild boar population management plays growing importance as farming-wildlife tension increases. However, the key factor for long terms sustainable management programs appear to be not technical but rather socio-anthropological.

Environmental impacts of applying circularity to a food system: a case study of the Netherlands

B.J. Van Selm[1,2], I.J.M. De Boer[2], M.K. Van Ittersum[1], R. Hijbeek[1] and H.H.E. Van Zanten[2]
[1]Wageningen University & Research, Plant Production Systems group, Droevendaalsesteeg 1, 6708 PB Wageningen, the Netherlands, [2]Wageningen University & Research, Animal Production Systems group, De Elst 1, 6708 WD Wageningen, the Netherlands; ben.vanselm@wur.nl

Applying circularity to food systems is increasingly seen as an important pathway to sustainable food systems. Circular food systems increase resource use efficiency by closing the loop of materials and utilising residual streams (e.g. manure, co-products, food-waste), which might reduce environmental impacts. However, very little is known about how environmental impacts change when circularity is applied to a food system. We created an optimisation model of a circular food system in the Netherlands which integrates crop and livestock production using a food systems approach and thus also taking consumption into account. The model minimises environmental impacts (i.e. greenhouse gas emissions and land-use) while meeting the dietary requirements of the Dutch population. Preliminary results show changes in the crops produced (more protein crops, e.g. beans) and a significant reduction in the number of animals kept; meat producing animals were reduced by at least 50%. Overall, our results suggest that applying circularity to a food system is a promising way to reduce environmental impacts.

Valorisation of effluents from poultry production

A.C.G. Monteiro[1], V. Resende[2] and O. Moreira[2]
[1]IACA, Avenida 5 de Outubro, 21, 2° esquerdo, 1050-047 Lisboa, Portugal, [2]INIAV, Estação Zootécnica Nacional, Quinta da Fonte Boa, 2005-048 Santarém, Portugal; ana.monteiro@iaca.pt

The promotion of an integrated approach to reduce and valorise the different nutrient flows generated within intensive animal production systems is a demand that cannot be ignored by this sector. Livestock production is concentrated in certain regions, some without enough area for land spreading valorisation of effluents. So, in order to be competitive and comply with legal requirements, the sector should promote a circular economy, pursuing new alternatives for effluents management. The objectives of this project are: Optimisation of effluents use as secondary raw materials, recovering energy and nutrients, improving farm nutrient balances and promoting sustainable management; Contribution to sustainable livestock intensification and landscape planning, to face climate change and resources scarcity; A roadmap for effluents management, including technology portfolio, linked to farm characteristics and regional constraints. For the accomplishment of the project objectives questionnaires were applied in the farms (pig, poultry and dairy), the present work will only focus on poultry production. The questionnaires were divided in: (1) identification and general information of the farm and farmer; (2) feed management; (3) water and energy management; (4) type and conditions of animal housing; (5) effluents management. The questionnaire was applied to laying hens and broiler chickens. The laying hens' farms had an average of 3,199 animals with a life cycle of 52 to 55 weeks, and an egg yield of 280 eggs/cycle. The broiler chickens' farms had an average of 697 animals with a life cycle of 17 to 18 weeks. All animals were fed *ad libitum* with compound feed. 70% of the total water consumption was as drinking water. Mean drinking water consumption was 5.44 m3/animal/year. Mean manure production per farm was 41,324.8 tons/year with 33,790.9 kg nitrogen. Sixty percent of the effluents produced are used directly in agriculture (outside poultry farms) and 40% are sent for composting. By adapting its applicability to the regional specificities of the livestock sector in Portugal, the results obtained will contribute to the robustness of overall emissions estimates.

Quality of bulk milk in single- and multi-breed farms of Italian alpine area

T. Zanon[1], A. Costa[2], M. De Marchi[2], M. Penasa[2], S. König[3] and M. Gauly[1]
[1]Free University of Bolzano, Faculty of Science and Technology, Piazza Università 5, 39100 Bolzano (BZ), Italy, [2]University of Padova, Department of Agronomy, Food, Natural Resources, Animals and Environment, Viale dell'Università 16, 35020 Legnaro (PD), Italy, [3]Justus Liebig University, Gießen, Department of Animal Breeding and Pet Genetics, Ludwigstraße 21B, 35390 Gießen, Germany; thomas.zanon@unibz.it

Alpine dairy farms are characterised by small-scale production units, which traditionally host various cattle breeds in different herd combinations. The aim of this study was to investigate the effect of the shift in herd structure on the variation of milk quality. To this purpose, bulk milk traits (gross composition, somatic cells, urea and fatty acid composition) of single- and multi-breed farms of South Tyrol (northern Italy) with either Brown Swiss (BS), Holstein Friesian (HF), Simmental (SI) or combinations of these breeds (BS + HF, BS + SI, HF + SI and BS + HF + SI) were considered. Bulk milk composition was predicted from spectra using mid-infrared spectroscopy. At least three bulk milk samples per month were available for all herds from January 2014 to December 2018. An ANOVA was carried out on milk traits accounting for the fixed effects of herd composition, month and year of sampling, and the interactions of herd composition with month and year of sampling; farm was included as random effect. Orthogonal contrasts were performed for single- vs multi-breed herds, specialised breeds vs dual-purpose breed, and multi-breed herds with SI vs multi-breed herds without SI. Overall, results showed that approximately two thirds of the farms in South Tyrol are single-breed herds and, according to herd size, only farms with more than 20 cows have changed herd structure over time, switching from multi- to single-breed herds. Farms with BS cows produced milk with higher fat, protein, casein and fatty acids contents than farms with SI and HF cows. Milk from single-breed SI herds was characterised by lowest somatic cell counts. Results demonstrated that BS farms were the best in terms of milk composition, and this supports the interest of using this breed in alpine dairy farming in both single- and multi-breed scenarios to improve cheese making properties and farmers' profitability.

The potential of regenerative agriculture on Dutch soils
L. Schreefel[1,2,3], C. Timler[2], R.P.O. Schulte[2], A. Pas Schrijver[2], H.H.E. Van Zanten[1] and I.J.M. De Boer[1]
[1]Wageningen University and Research, Animal Production Systems, De Elst 1, 6708 WD Wageningen, the Netherlands, [2]Wageningen University and Research, Farming Systems Ecology, Droevendaalsesteeg 1, 6708 PB Wageningen, the Netherlands, [3]TiFN, Nieuwe Kanaal 9A, 6709 PA Wageningen, the Netherlands; loekie.schreefel@wur.nl

Agricultural intensification has had a cumulative detrimental impact on the environment. To limit the agricultural related impacts, regenerative agriculture (RA) is increasingly being promoted as a solution for a sustainable food system. RA is described as a mode of agriculture that 'uses soil conservation as the entry point to regenerate and contribute to multiple ecosystem services, with the aspiration that this will enhance not only the environmental, but also the socio-economic dimensions of sustainable food production'. The Netherlands, as the second largest agricultural exporter worldwide, has a large share of these detrimental environmental agri-impacts. In this paper, we will select three Dutch farms as case studies to showcase their potential to limit these detrimental impacts through the use of RA on different soil types (sand, clay, peat). We apply the DEED-cycle (Describe, Explore, Explain and Design) as a framework to further explore regenerative scenarios for the case-study farms. As a first step we use the farm model FarmDESIGN (FD) to provide a baseline assessment of a farming system expressed in a broad set of productive, environmental and socio-economic performance indicators. Secondly, we use the Functional Land Management (FLM) framework, incorporated in the Soil Navigator DSS model, to specifically assess soil functions. The Soil Navigator DSS will, depending on the farmers objectives and assigned improvement options, show the effect of these improvements options on the soil functions. The improvement options will then be integrated back into the (re)design process of FD. Using a multi-objective optimisation algorithm, FD systematically generates 'solution spaces', comprising crop compositions and farm management configurations that would satisfy environmental challenges, while enhancing social welfare and economic prosperity. As such, this study will show the environmental potential of RA on Dutch soils.

Selecting for genetic progress in low-input dairy herds
H. Davis and G. Butler
Newcastle University, School of Natural and Environmental Sciences, Newcastle, Tyne and Wear, NE6 5YL, United Kingdom; h.davis@newcastle.ac.uk

Dairy farmers are scrutinised for their environmental impacts amidst outcry that dairy is not important in the human diet, despite the wide range of fatty acids, protein, vitamins and minerals available in milk. Because grazing cattle often use land where crops don't grow, more and more dairy farmers are turning to low-input and forage-based systems as a response to criticism of intensive dairying. The effect of increased forage on milk quality has been well documented: higher forage increases the concentration of essential omega-3 alpha-linoleic fatty acid and other fatty acids beneficial to human health. By not pushing for high yields but nutritional output, cows are able to maintain health and fertility with very few inputs. In these systems, it is critical that farmers can identify the most suitable and efficient cattle as replacements, but they require alternative indices for measuring efficiency outside of traditional yield focussed indices. This study aimed to identify characteristics that identify 'superior' cows under forage-based management. Twenty-three second and third lactation cows from three forage-based farms were monitored throughout one lactation. The cows selected wore 'Rumiwatch' halters for two weeks each during early, mid and late lactation, recording all grazing and ruminating activities. This also allowed modelling of dry matter intake. Additionally, milk yield and quality (fat, protein, cell count and fatty acid composition) was recorded during each stage. Linear mixed effects models found significant differences in milk quality, ruminating and grazing behaviour, dry matter intake and forage conversion efficiency between farms and between stages of lactation. There were cows on each farm that stood out as 'best performing' (i.e. efficiently converting forage to high quality milk), demonstrating the scope for forage- based farmers to select within herds for genetic improvement. This study shows the large variation in milk quality and cow behaviour between similar systems and suggests that selecting best performing cattle within a low-input dairying system is possible.

Property planning as tools for the design of agricultural integrated systems in fragile ecosystems
A. Conde-Pulgarin, L.C. Bernal-Bechara, R.M. Fajardo-Vergara and D.A. Torres-Chavarro
Universidad de La Salle, Facultad de Ciencias Agropecuarias, Carrera 7 número 179-03, 110141665 Bogotá D.C., Colombia; aconde@unisalle.edu.co

The ecosystem of Paramo is one of the most important in South America for its generation and conservation of water for other ecosystems and for large cities located nearby. The objective of this study was to design and establish, through the direct participation of producers and researchers, integrated production systems of agricultural and livestock, with better indicators of social, economic and environmental sustainability in the locality of Sumapaz, Bogotá Colombia. The rapid agroecological evaluation proposed by Altieri allowed the assessment of sustainability indicators and the analysis of the potentialities and restrictions of the production systems prevailing in the area. Subsequently, the methodology proposed by Mejia *et al.* allowed the design and implementation of new systems and several strategies for the integration between agriculture and animal production systems through property planning workshops, using the Mejía methodology *et al.* In general, due to the impact of large bovines on the soil, and the low productivity of the soil, priority was given to the incorporation of new animal production alternatives, such as cuniculture, based on hydroponic green forage as the main food source. Also, the establishment of production systems with laying hens in grazing, including native genetic lines, and the management of organic waste and animal excreta through composting as a contribution to the recovery of degraded soils.

Use of omics to evaluate the response of dairy sheep and goats to heat and cold stress
A.A.K. Salama, A. Contreras-Jodar, N. Mehaba and G. Caja
Universitat Autonoma de Barcelona, Research Group in Ruminants (G2R), Edifici V, Campus de la UAB, 08193, Spain; ahmed.salama@uab.es

Heat stress (HS) negatively impacts the productivity, health, and welfare of dairy animals, including small ruminants. Despite progress in cooling systems and the implement of several strategies, HS constitutes to be a significant cost for dairy farms. Cold stress (CS), on the other hand, is a significant issue in some regions. Herein results on the omics of blood, milk, and urine of dairy goats and sheep exposed to HS and CS are presented. Blood transcriptomics revealed that HS goats suffer immune-dysfunction due to a decrease in the haematopoiesis and leukocyte diapedesis. In addition, HS results in the disruption of lipid metabolism pathways (i.e. adipocyte and PPAR signalling) of immune cells which compromises innate and the adaptive immune responses. By urine ^{1}H NMR-based metabolomics, several markers of leaky gut under HS conditions are detected. These markers are related to the over-excretion of gut-derived toxic compounds generated by the harmful gastrointestinal microbiota. At the level of mammary gland, HS affects milk RNA-seq transcriptomics, where 699 genes are differentially expressed. Several genes involved in milk fat (ACACA, FASN, SCD, BTN1A1, XDH) and protein (CSN1S1, CSN1S2, CSN3, LALBA) syntheses are downregulated, whereas genes related to cathepsin proteolysis (CTSB, CTSD, CTSS, CTSZ) and cell death are upregulated. Furthermore, milk metabolomics showed that inflammation response to simulated infection is delayed by HS, which is an indicator that the ability to combat pathogens is hindered under high ambient temperatures. Blood metabolomics profile was also altered by low ambient temperatures, where α- and β-glucose increase, which could result from the use of some amino acids for gluconeogenesis. In addition, tyrosine that could be used for the synthesis of catecholamines is also incremented under cold temperatures. In conclusion, omics shed more light on the physiological mechanisms that occur in sheep and goats when exposed to HS or CS. This will help in the establishment of future effective strategies to alleviate the negative effects of HS and CS in dairy small ruminants.

On-farm welfare assessment and its relationship with milk production in dairy sheep farms

G. Caja[1], R. González-González[2], M. Rovai[3], A.A.K. Salama[1] and X. Such[1]
[1]*Universitat Autònoma de Barcelona, Animal and Food Sciences, Group of Research in Ruminants (G2R), Av. dels Turons s/n, 08193, Spain, [2]Grupo Emprendedores Ovino (GEO), Palencia, 34004 Palencia, Spain, [3]South Dakota State University, Dairy and Food Sciences, Alfred Dairy Science Hall, 57007 Brookings SD, USA; gerardo.caja@uab.es*

Welfare is a current hot topic for consumers and animal scientists, while at the same time it is a concern for farmers who are worried about increasing their production costs due to implementation of animal wellbeing measures. With this in mind, currently available animal-based welfare indicators were reviewed looking for valid, reliable and feasible protocols for on-farm wellbeing assessment in dairy ewes. The 5 animal wellbeing principles (Freedoms), despite being internationally recognised, they are ideal states rather than standards for acceptable welfare. Moreover, there are remarkable differences in physiological and behavioural features according to breed, production system and farm size, making necessary specific studies to adapt the key indicators to be used in each case for the assessment protocols under on-farm conditions. A good protocol is a powerful instrument for welfare assessment, but also may be useful for production and health monitoring. Special interest have the available 2-step AWIN welfare assessment protocols for sheep and goats but, until today, they have not been tested in dairy sheep. Use of new technologies (e.g. cameras and sensors) may offer objective welfare indicators but its use in small ruminants is very limited and needs specific adaptations and further research. This review focuses on the state-of-the-art of the wellbeing assessment protocols and measuring technologies currently available for dairy sheep, and summarises the physiological particularities and known impact of some management practices tested in a group (n=35) of intensively managed and high-yielding dairy sheep farms of different breeds (i.e. Assaf, Lacaune and Manchega) in Spain. The use of a simplified indicators index newly proposed (5-F proxy) showed a positive relationship with milk production traits and encouraged the farmers to improve the welfare of their dairy sheep farms.

Performances and plasma redox status in dairy goats fed metabolisable methionine balanced diets

P. Schmidely[1] and L. Bahloul[2]
[1]*Université Paris-Saclay, INRAE, AgroParisTech, UMR Modélisation Systémique Appliquée aux Ruminants, Paris, 75005, France, [2]Centre of Expertise and Research in Nutrition, Adisseo France S.A.S., Commentry, 03600, France; lahlou.bahloul@adisseo.com*

The effect of balancing metabolisable Methionine (MetDi) diets on milk performance and the redox status were studied in 2 trials with dairy goats fed MetDi supplemented diets using either Smartamine M in trial I, or isopropyl ester of 2-hydroxy-4-methylthio butanoic acid in trial II (Adisseo). Goats in trial I (9 wks) were fed either a LP (14% CP) or a HP (16% CP) diets without (M90) or with MetDi (M100) within each diet, to cover 90 and 100% of MetDi requirements. Diets (14% CP) in trial II (10 wks) provided 90 (M90), 100 (M100) and 110% (M110) of MetDi requirements. In both trials, goats were allocated within experimental treatments according to their genetic variant for milk aS1 casein. Milk and blood samples were collected twice and once weekly, respectively. In plasma, concentration of total glutathione (GSt), oxidised glutathione (GSox), Proteins carbonyl (PC), Malondialdehyde (MDA), and enzyme activities of catalase (CAT), superoxydismutase (SOD), and glutathion peroxidase (GSpx) were determined. DMI did not differ between groups in trial I and II. In trial I, M100 goats had higher milk protein content (MPC) than M90 goats with HP diet (32.0 vs 32.1, g/kg) but not with LP diet (30.3 g/kg in both groups). In trial II, MPC was 34.9, 34.0, and 33.g/kg for M100, M110 and M90 goats (P<0.07). The M100 and M110 goats had numerically higher protein yield (128 and 131 g/d) than M90 goats (123 g/d). In trial I, plasma GSt tended (P<0.07) to be higher in M100 goats (83 pM) than in M90 goats (81 pM). No other plasma parameters were affected by MetDi treatment. In trial II, GSt tended also to be higher (P<0.13) in M100 goats (82 pM) than in M90 goats (76 pM), as was GSHox (P<0.05) in M100 goats (15.2 pM) compared to M90 goats (12.8 pM). This was associated with a higher GSpx activity in M100 (94 U/ml) than in M90 goats (89 U/ml). In trial II, catalase was higher (P<0.05) in ME110 goats than in M100 and M90 goats. In conclusion, an increase in milk protein content is associated with an improvement of plasma redox status in goats fed MetDi balanced diets.

Lactational effects of melatonin during spring in 2 breeds of dairy ewes

A. Elhadi, A.A.K. Salama, X. Such and G. Caja
Universitat Autònoma de Barcelona, Animal and Food Sciences, Group of Research in Ruminants (G2R), Av. dels Turons, s/n, 08193 Bellaterra, Spain; abdelaali.elhadi@uab.cat

The effects of melatonin implants were studied in a total of 104 adult dairy ewes of 2 breeds (Manchega, MN; n=51; Lacaune, LC; n=53) in mid-late lactation (120 to 155 DIM). Ewes differed in milk yield (MN, 0.94±0.07; LC, 1.65±0.09 kg/d), but had similar body frame (BW, 80.1±1.1 kg; BCS, 3.40±0.05). They were distributed in 4 balanced groups according to breed (MN, n=25 an 26; LC=26 and 27), fed a TMR *ad libitum* (forage:concentrate = 60:40) and randomly assigned to a 2×2 factorial design (treatment×breed). Treatments were: Control (CO; n=51), that did not receive any treatment; and, Melatonin (MEL, n=53), that were implanted in the ear base (18 mg/ewe; Melovine, Ceva Animal Health, Barcelona, ES) at 125±2 DIM during spring. Milk yield was recorded daily using automatic milk meters (MM25 SG, DeLaval, Tumba, SE) and ruminal bolus transponders (Datamars, Bedano, SW), and milk composition measured on d 1, 7, 21 and 32. Data were analysed by the PROC MIXED for repeated measurements of SAS (v.9.4, SAS Institute Inc., Cary, NC, USA) and LS means separated by PDIFF at P<0.05. On average, no differences were detected on milk yield (1.20±1.1 kg/d; P=0.13) or energy corrected milk (P=0.26), but milk fat (8.11 vs 9.11%; P=0.004) and milk protein (6.23 vs 6.53%; P=0.036) contents increased by MEL treatment. No effects were detected on lactose (P=0.19), urea (P=0.09) and SCC (P=0.11) in milk, as well as in BW (P=0.79) and BCS (P=0.30) of both breeds. MEL treatment increased the level of plasmatic MEL (P<0.001) in both breeds, but no differences in the concentration of PRL (P=0.92) and IGF-I (P=0.98) in plasma were detected throughout the experiment. In conclusion, the use of MEL implants in lactating dairy ewes during mid-late lactation in spring, increased fat and protein milk contents without significant effects on milk yield. These increases may be of interest for improving milk quality and farm income during spring, when milk quality of dairy ewes is usually lower.

Fibre, energy balance, milk production, and milk composition of sheep and goats: where is the link?

A. Cannas[1], A. Nudda[1], A.S. Atzori[1], M.F. Lunesu[1], V. Giovanetti[2] and G. Molle[2]
[1]University of Sassari, Department of Agricultural Sciences, viale Italia 39, Sassari, 07100, Italy, [2]AGRIS Sardegna, Servizio ricerca per la Zootecnia, Olmedo, 07040, Italy; cannas@uniss.it

As milk production level of dairy ewes and goats increases, the nutritional challenges to which they are subjected are also increased. If the ration during lactation is not carefully formulated and monitored, metabolic disorders are likely, especially in highly productive animals. However, while there is a vast research on their energy and protein requirements, very little is known on optimal dietary fibre (NDF) and of non-structural carbohydrates (NSC; sugars and starch) concentrations for ewes and goats. This poses serious difficulties to balance the diets and to predict their dry matter intake (DMI). For this reason, we have been working to refine optimal NDF intake values (i.e. those that do not limit DMI and do not cause rumen acidosis), showing that the optimal level of intake of NDF of lactating ewes is much higher (1.7 to 2.3% of body weight, going from large to small body size ewes) than that reported for lactating dairy cows. In addition, we found a large effect of pasture NDF level on the pasture DMI of ewes. Regarding NSC, we found that their optimal dietary concentrations are similar between lactating ewes and goats in early lactation, while in mid lactation their concentrations should be decreased in ewes but not in goats. Dietary NDF and NSC content also affect the energy balance of the animals and their milk production and composition. On this regard, it appears that the proportion between de novo and preformed milk fatty acids is markedly and quickly affected by the nutritional conditions of ewes and goats and can be used as nutritional indicator of their energy balance and intake of fermentable fibre. Dietary and pasture fibre and NSC are also involved in the milk fat depression phenomena increasingly observed in high yield ewes and goats. Overall, it is clear that carbohydrates, and especially fibre, intake and quality, energy balance, milk production and composition of ewes and goats are fully inter-associated. The aim of this review is to define and untangle these relationships and propose criteria for appropriate dietary balancing.

Effect of condensed tannins of sainfoin on the fatty acid profile of ewe's milk and lamb's meat

S. Lobón, C. Baila, M. Blanco, I. Casasús, J.R. Bertolín and M. Joy
Ctr Invest y Tecnol Agroal Aragon (CITA), Universidad de Zaragoza, Montañana 930, 50059 Zaragoza, Spain;
slobon@cita-aragon.es

Twenty ewe-lamb pairs were used to evaluate the effect of the condensed tannins (CT), through the inclusion of polyethylene glycol (PEG), in the diet of lactating ewes on the fatty acid (FA) profile of their milk and the meat of their suckling lambs. The feeding treatments were: fresh sainfoin (*Onobrychis viciifolia*; SF n=10) and fresh sainfoin plus PEG (SF+PEG, n=10) to bind and deactivate the CT of sainfoin. Ewes were milked weekly and milk samples were collected until the lambs reached the target slaughter weight (11±0.2 kg) with an average age of 27 days. Milk and meat FA were determined by gas chromatography (expressed as % of total FA identified) and sums and ratios of FA were calculated. Regarding milk FA, the inclusion of PEG increased C18:0 and C18:2 9c,11t, and n-6:n-3 ratio during week 1 and 2 of lactation (P<0.05) but not thereafter. Throughout lactation, SF+PEG ewes presented higher percentage of C16:0 and lower of C18:2 n-6, C18:3 n-3, C20:5 n-3 and total polyunsaturated FA (PUFA; 6.5% vs 5.9% for SF and SF+PEG respectively; P<0.05) than SF ewes. Therefore, the CT from sainfoin inhibited the ruminal biohydrogenation. The effect of CT on the FA of the meat of suckling lambs was milder than the effect observed in the milk, due to the *de novo* synthesis of FA in lamb muscle. The inclusion of PEG decreased C18:3 n-3 (2.3% vs 1.7% for SF and SF+PEG respectively; P<0.01) and increased C18:2 9c,11t (0.58% vs 0.71% for SF and SF+PEG respectively; P<0.01). In conclusion, CT from sainfoin improved milk quality because it increased PUFA and decreased n-6:n-3 ratio, and to a lesser extent they improved the meat quality of suckling lamb.

Potentials of milk performance data as indicator for targeted selective treatment in Lacaune sheep

K. Schwarz[1,2], B. Bapst[3], M. Holinger[2], A. Steiner[2], I. Schleip[1] and S. Werne[2]
[1]Eberswalde University for Sustainable Development, Schicklerstrasse 5, 16225 Eberswalde, Germany, [2]Research Institute of Organic Agriculture (FiBL), Ackerstr. 110, 5070 Frick, Switzerland, [3]Qualitas AG, Chamerstrasse 56, 6300 Zug, Switzerland; k.schwarz94@gmx.de

Anthelmintic resistance is a major threat in farming of small ruminants worldwide. One approach to slow down the development of anthelmintic resistance is targeted selective treatment (TST), where a part of animals is left unexposed to anthelmintic treatment and thus providing refugia for susceptible parasites. Closely linked to the successful implementation of TST is the identification of animals in need of treatment. In dairy goats it has been proposed to use milk yield as TST indicator, focussing treatment on high yielding dairy goats. In dairy sheep the relation between milk performance and infection of gastrointestinal nematodes (GIN) is not yet well known. The aim of this study was therefore to investigate the general relation between milk yield and GIN infection in a Swiss Lacaune dairy sheep subpopulation and, based on this, to evaluate milk yield data as a potential TST indicator in dairy sheep. A field study was performed including 1,159 lactating Lacaune dairy ewes on 15 dairy sheep farms in Switzerland. All ewes were sampled once between August and December 2019, when they had exceeded the 70th day of lactation. For each ewe the nematode egg excretion per gram faeces (EPG) was determined and individual milk performance data was obtained closely time-related to the date of faecal sampling. Coprocultures of pooled faecal samples were conducted to determine the proportion of *Haemonchus contortus* on farm level. A linear mixed model revealed that EPG increased significantly with increasing daily milk yield (P<0.01), indicating high yielding ewes to be less resistant to GIN infections than low yielding ewes. The effect was most pronounced in the earlier stage of lactation, but remained within a moderate range. The other included fixed effects milk protein content, lactation day, lactation number and the proportion of *H. contortus* did not show significant relations with EPG. The results suggest the possibility of using milk yield data as TST indicator in dairy sheep. In the frame of H2020 project SMARTER no. 772787.

The pattern of distribution of runs of homozygosity in the Russian local goat breeds

T. Deniskova[1], A. Dotsev[1], M. Selionova[2], H. Reyer[3], K. Wimmers[3], J. Sölkner[4], G. Brem[1,5] and N. Zinovieva[1]
[1]L.K. Ernst Federal Science Center for Animal Husbandry, Moscow, 142132, Russian Federation, [2]RSAU – MAA named after K.A. Timiryazev, Moscow, 127550, Russian Federation, [3]Institute of Genome Biology, Leibniz Institute for Farm Animal Biology, Dummerstorf, 18196, Germany, [4]University of Natural Resources and Life Sciences Vienna, Division of Livestock Sciences, Vienna, 1180, Austria, [5]Institute of Animal Breeding and Genetics, VMU, Vienna, 1210, Austria; horarka@yandex.ru

Russian goat industry includes breeds reared in breeding farms with established pedigree and pastoral goats with no pedigree. An assessment of genomic inbreeding is required to prevent loss of genetic diversity and to provide insight into developmental history of local goats. Thus, our aim was to estimate distribution of the runs of homozygosity (ROH) segments in Russian local goats. Sample included goats from Orenburg (OREF, n=32), Altai Mountain (ALTM, n=33), Soviet Mohair (SOVM, n=30), Dagestan Milk (DAGM, n=14), Dagestan Local (DAGL, n=20), Dagestan Fluff (DAGF, n=14) and Karachaev (KRCH, n=37) breeds. Goats were genotyped using Illumina Goat50KSNPBeadChip. We estimated the ROH segments using consecutive runs method implemented in the R package 'detectRUNS'.We detected ROH in all breeds with mean length ranging from 37.1 in DAGM to 176.1 Mb in KRCH. The maximum ROH length was found in DAGF (645.4 Mb). KRCH had the greatest number of ROH segments (total of 1,468), while 183 ROH segments were found in DAGM. Short ROH segments (1-4 Mb) were predominant in all breeds and varied from 65.8 in KRCH to 86.40% in DAGM. The frequencies of ROH segments of 8-16 Mb ranged from 2.8 in SOVM to 11.2% in KRCH. The highest proportion of the longest ROH segments was detected in DAGL (6.7%), DAGF (5.6%) and ALTM (4.6%), while the minimums were found in SOVM (0.71%) and OREF (0.4%). The largest genome coverage in ROH was identified on CHI1, CHI2 and CHI5 and the lowest was accounted for CHI26 and CHI28. The values F_{ROH} ranged from 0.02 in DAGM to 0.08 in KRCH. The obtained data on the traces of long-term inbreeding in local breeds is relevant for their future sustainable management. The genotyping of 96 goats was funded by RSF No. 19-76-20006. The study was funded by RFBR No.18-316-20006.

Short- and long-term effects of cabergoline at dry-off in 2 breeds of dairy ewes

A. Elhadi, A.A.K. Salama, X. Such and G. Caja
Universitat Autònoma de Barcelona, Animal and Food Sciences, Group of Research in Ruminants (G2R), Av. dels Turons, s/n, 08193 Bellaterra, Spain; abdelaali.elhadi@uab.cat

The effects of cabergoline, a dopamine agonist and inhibitor of prolactin secretion, were studied in 93 dairy ewes of 2 breeds (Manchega, MN, n=48; Lacaune, LC, n=45) in late-lactation. The ewes had similar DIM (208±4 d) and BW (75.5±1.0 kg), but differed in milk yield (MN, 0.47±0.05; LC, 0.97±0.08 kg/d). They were penned during dry-off and fed a total mixed ration (F:C on DM basis, 55:45%) *ad libitum* until dried-off, during peripartum and for the following lactation. Grazing was also allowed after dry-off and the following lactation. Treatments were: CON (control, 1 ml saline; n=47) and CAB (cabergoline, single i.m. injection of 0.56 mg cabergoline/ewe; Velactis, 1.12 mg/ml cabergoline; Ceva, Libourne, FR; n=46). Reactions after injection on the injection site (swelling), udder traits (volume, width and floor distance), udder cistern size (by ultrasonography) and behaviour (eating, lying, standing, drinking and ruminating) were monitored after injection (d 0 to 14). Lambing, suckling (weaning, d 28) and lactational performances (120 DIM) were recorded at lambing (litter and lamb weight) and during the following lactation (daily milk yield, biweekly milk composition). BW and BCS were recorded monthly. Data were analysed by the PROCMIXED for repeated measurements of SAS v.9.4. No reactions or apparent changes of behaviour by CAB were detected in both breeds. Udder volume (P=0.017), width (P=0.012) and cistern area (P=0.040) decreased by CAB during dry-off. No differences were detected in litter size (P=0.18) and lamb weigh at birth (P=0.24), but lambs from LC-CAB treated ewes were 14% heavier at weaning than CO (P=0.009). Apart of expected breed differences on milk production (P<0.001), no CAB effects were detected on milk yield (P=0.90), milk components (P=0.38 and 0.54) and SCC (P=0.28). Only milk lactose content increased in LC-CAB treated ewes (-4%P=0.046). No differences in body reserves were detected by CAB. In conclusion, the use of cabergoline treatment at drying-off improved the udder traits and facilitate the dry-off of dairy ewes, without negative effects on behaviour, lamb performance nor lactational performance during the following lactation.

Effect of proanthocyanidins of sainfoin on secondary compounds in milk and suckling lamb meat

M. Blanco, C. Baila, M. Joy, J.R. Bertolín, I. Casasús and S. Lobón
Ctr Invest y Tecnol Agroal Aragon (CITA), Universidad de Zaragoza, Montañana 930, 50059 Zaragoza, Spain; slobon@cita-aragon.es

Proanthocyanidins may exert a protective effect on polyphenols, carotenoids and liposoluble vitamins in the milk of the ewe and therefore increase their deposition in the tissues of the suckling lamb. The aim of this study was to evaluate the effect of proanthocyanidins of sainfoin (*Onobrychis viciifolia*), with the inclusion of polyethylene glycol (PEG), on the secondary compounds in the milk of the ewe through lactation and the contents in the meat of the suckling lamb. For that, 20 lactating Rasa Aragonesa ewes were individually fed fresh sainfoin *ad libitum* plus 200 g/d of barley. Half of the ewes were orally-dose twice daily PEG diluted in water (50 g PEG/100 ml of PEG 4000) and the other half only water. The suckling lambs had continuous access to their dams until they reached 11 kg of live weight, when they were slaughtered. After 24 h of cooling, the Longissimus thoracis muscle was sampled. The intake of secondary compounds was similar between treatments (P>0.05). Regarding the effect on polyphenols, the inclusion of PEG increased the concentration in milk throughout lactation (43 vs 51 µg/g milk; P<0.01) and their content in meat of the suckling lamb (71 vs 82 µg/g; P<0.05). Similarly, the inclusion of PEG increased retinol concentration in milk (0.7 vs 0.9 µg/g; P<0.05), however, retinol content in meat of the suckling lamb did not reflect this difference (P>0.05). The inclusion of PEG tended to reduce lutein in milk throughout lactation (12 vs 7 ng/ml milk; P<0.10) and in the muscle content of the suckling lamb (12.7 vs 4.6 ng/g; P<0.10). Regarding tocopherols in milk, α- and δ-tocopherols were not affected by the inclusion of PEG (P>0.05). The concentration of γ-tocopherol evolved differently through lactation with the inclusion of PEG (P<0.01), however there were no differences within a week between treatments. The inclusion of PEG did not affect the content of α-, γ- and δ-tocopherol in meat of the suckling lamb.

The investigation of fatty acid profile of Chios sheep during lactation

Z. Basdagianni[1], C. Karaiskou[1], E. Kasapidou[2] and M.A. Karatzia[3]
[1]School of Agriculture, Aristotle University of Thessaloniki, Department of Animal Production, Thessaloniki, 54124, Greece, [2]University of Western Macedonia, Department of Agriculture, Florina, 53100, Greece, [3]Research Institute of Animal Science, HAO-Demeter, Paralimni, 58100, Greece; karatzia@rias.gr

The investigation of fatty acid profile and genetic potential of Chios sheep is vital for animal selection and production of quality dairy products. In the present study, test-day milk yield records of 100 Chios breed sheep from one farm participating in the Chios Sheep Breeders Cooperative 'Macedonia' which implements the genetic improvement program, were analysed for milk quality. Six individual milk samples were collected monthly from February to July for the determination of milk fatty acid profile using gas chromatography. All ewes were at second lactation and were fed under the same conditions without any differences in nutrition or management. Results revealed that lactation stage affected fatty acid profile significantly. Specifically, saturated fatty acids (SFA) concentration was found to be lower at the end of the lactation (1st sampling vs 6th sampling: 70 vs 68%, P≤0.05), while monounsaturated fatty acids (MUFA), polyunsaturated fatty acids (PUFA) and conjugated linoleic acid (CLA) concentrations were higher at the end (24 vs 26%, 4.70 vs 4.79% and 1.23 vs 1.65%, respectively, P≤0.05). Regarding the correlations of grouped fatty acids, a negative one was found between undesirable fatty acids, such as SFA, and beneficial fatty acids for consumer health [unsaturated fatty acids (UFA) (-0.45), PUFA (-0.40), respectively]. Moreover, the results of the study showed that the coefficient of variation (CV%) in most fatty acids was high (>40%), indicating that desired characteristics can be identified and selected for, in Chios sheep. In conclusion, different stages of lactation result in a variable nutritional value, affecting the optimal FA profile of dairy products. Additionally, incorporating milk fatty acid profiling into the sheep selection index, can be an innovative and useful tool in animal selection. The study was funded by the Research Committee of the Aristotle University of Thessaloniki.

Principal component analysis of variation in milk odd- and branched-chain fatty acids in dairy ewes

P.G. Toral[1], G. Hervás[1], M. Plante-Dubé[2], E. Barrio[1], R. Gervais[2] and P. Frutos[1]
[1]Instituto de Ganadería de Montaña (CSIC-Universidad de León), Finca Marzanas, 24346, Grulleros, León, Spain;
[2]Département des Sciences Animales, Université Laval, 2425 rue de l'Agriculture, Québec G1V 0A6, Canada;
pablo.toral@csic.es

Milk odd- and branched-chain fatty acids (OBCFA) are largely derived from bacteria leaving the rumen, which has encouraged research on their use as non-invasive markers of rumen function. In dairy cows, studies have examined relationships between milk OBCFA and dietary components, such as starch, fibre, protein or lipids. However, we are not aware of targeted research on this topic in sheep. Therefore, with the aim of gaining insight into factors controlling milk OBCFA concentrations in dairy ewes, a principal component analysis (PCA) was conducted to examine relationships between diet composition and milk OBCFA profile. A database was compiled using lot observations from 14 trials carried by our team that examined the effects of lipid supplementation on milk fatty acid (FA) profile in dairy ewes (42 dietary conditions). Diets consisted of total mixed rations based on alfalfa hay (in proportions that ranged from 19 to 71%) and concentrates, and contained no additional lipid or lipid supplements from plant or marine origin. The PCA discriminated 2 principal components (PC) that described 46.0 (PC1) and 21.4% (PC2) of the variation in data, and a PC3 that accounted for 9.6% of total variability. The score plot showed that PC1 tended to separate two major groups of lots based on the addition or not of lipids to the ration. The loading plot showed that concentrations of milk 14:0 *iso*, 15:0, 15:0 *anteiso*, 17:0 and *cis*-9 17:1 loaded opposite to dietary unsaturated C18 FA and total FA, and both groups of variables were clearly correlated with PC1, which supports a major role of dietary lipids controlling milk OBCFA. However, milk 13:0 *anteiso* was only correlated, positively, with PC2, which seemed to be influenced by other dietary components, although relationships were less clear. Dietary starch loaded opposite to milk 13:0 *iso* and 15:0 *iso*, and to forage:concentrate ratio and ADF, while crude protein positioned close to the origin of the plot, dismissing its relevance in determining milk OBCFA profile. In conclusion, this PCA would support relevant relationships between diet composition and milk OBCFA in dairy ewes, the strongest influence being that of dietary lipids. Acknowledgements: project AGL2017-87812-R, AEI/FEDER, UE; Ramón y Cajal program, RYC-2015-17230, MINECO/ESF, UE.

Relationships between dietary starch and milk odd- and branched-chain fatty acids in ewes fed oils

P.G. Toral[1], P. Frutos[1], M. Plante-Dubé[2], A.G. Mendoza[1], R. Gervais[2] and G. Hervás[1]
[1]Instituto de Ganadería de Montaña (CSIC-Universidad de León), Finca Marzanas, 24346, Grulleros, León, Spain;
[2]Département des Sciences Animales, Université Laval, 2425 rue de l'Agriculture, Québec G1V 0A6, Canada;
pablo.toral@csic.es

Ruminal bacteria synthesise odd- and branched-chain fatty acids (OBCFA) that can be transferred to milk. Since different bacterial populations have specific enzymes that determine the products of OBCFA synthesis, variations in the concentration of these lipids in milk may reflect changes in the rumen bacterial community. In cows, increasing dietary starch level is known to foster the growth of amylolytic bacteria and limit that of cellulolytic species, reducing milk iso FA concentrations. In a preliminary principal component analysis, we observed a stronger relationship of milk OBCFA with dietary lipids than with starch level in dairy ewes. Therefore, this study examined the relationships between dietary starch and milk OBCFA concentrations in sheep fed lipid supplements with the objective of determining if the presence of additional fat may affect the response of milk OBCFA to starch level. To that aim, a database with lot observations from 14 nutritional trials was used. A total of 47 lots of ewes received lipid supplements, whereas their respective controls (27 lots) were fed the same basal diets with no supplementation. Starch levels were similar in both treatments and ranged from 108 to 257 g/kg diet DM. Relationships between milk OBCFA and starch level were examined using the MIXED procedure of SAS. Prediction models included the fixed effects of the experimental treatment (control vs supplemented), the linear and quadratic effects of starch level, and the interactions between them. In general, prediction models showed different intercepts in the two treatments, whereas relationships between OBCFA and starch remained constant irrespective of the level of the latter. Significant, although moderate, interactions were only observed for the negative linear relationships with 17:0 and 14:0 iso, and for the quadratic positive relationship with 13:0 anteiso. Overall, in dairy ewes, the relationship between dietary starch level and milk OBCFA does not appear to be greatly affected by the presence of supplemental lipids, suggesting potentially independent effects of the two dietary components on the rumen bacterial community. Acknowledgements: project AGL2017-87812-R, AEI/FEDER, UE; Ramón y Cajal program, RYC-2015-17230, MINECO/ESF, UE.

Comparison between rumen bacterial composition of growing and adult Romane sheep
F. Touitou, A. Meynadier, C. Marie-Etancelin, F. Tortereau and C. Moreno-Romieux
INRAE, ENVT, INPT-ENSAT, GenPhySE, Castanet-Tolosan, 31320, France; florian.touitou@envt.fr

Ruminants co-evolved with microorganisms able to degrade plant fibres into absorbable chemical compounds. They are now dependant on the adaptation of their microbiota to the diet. To investigate conservation of the microbiota between lamb stage and adult stage, rumen fluid samples were taken from 48 Romane ewes conventionally raised indoor at the INRAE experimental farm of La Sapinière, both when they were lambs (between 2,5 and 4 month-old) fed only concentrate and at 1-year old fed with a mixed ration of concentrate and forage. Bacterial DNA was sequenced and bioinformatics analyses of the microbiota sequences were implemented with FROGS pipeline. Diversity was estimated using R vegan package and PCA was performed using R mixOmics package. A home–made index (C_x) based on mean relative variation of abundance to maximum abundance corrected by the ratio between the number of OTUs in lambs and ewes was used to evaluate conservation. FROGS pipeline allowed clustering of the 1,178,380 informative sequences into 1,494 OTUs belonging to 150 genera, 48 families, 25 orders, 16 classes and 9 phyla. PCA of the OTUs showed that the samples clustered according to animals' stage indicating that rumen community composition in sheep is likely to be explained by age and most of all diet as previously found. Samples from 1-year old animals were significantly richer and more diverse than those from young animals and all 9 phyla obtained were significantly affected by the animal's age/diet except from Proteobacteria whereas Wang *et al.* showed that this phylum was particularly affected between birth and adult stage. Cumulative abundances of the major one hundred OTUs of lambs represented around 92% of their whole rumen bacteriome while the same OTUs represented less than 6% of ewes'. A mean of 58 OTUs were shared between lambs and ewes. Only 48 OTUs had a $C_x>1$ and within these the ones that were shared by over 40 animals all belonged to the order Clostridiales and most of them belonged to the genus *Lachnospiraceae* NK3A20 group. Despite a good conservation of some OTUs, rumen bacteria do not seem to be conserved between the lamb stage and the adult stage.

Serra-da-Estrela ewe synchronisation: seasonal effects, reproductive performance and lamb production
M.R. Marques, A.T. Belo, J.M. Ribeiro, C.C. Belo and M. Garcia-Herreros
INIAV-IP, Fonte Boa, 2005-048 Vale de Santarém, Portugal; rosario.marques@iniav.pt

Serra-da-Estrela sheep is a Portuguese breed traditionally managed for dairy production. Its Mediterranean location entails seasonality effects affecting productivity due to physiological changes associated to photoperiod. The aim was to determine seasonality effects (mating: autumn vs spring) on reproductive performance and lamb production in FGA/PMSG-synchronised Serra-da-Estrela ewes. 138 ewes were randomly divided into 4 groups: 2 controls [1 autumn (AC) and 1 spring control (SC)] and 2 treated groups [(1 autumn (AT) and 1 spring treated (ST)]. Ewes from AT and ST groups received Flurogestone Acetate (FGA) impregnated intravaginal sponges (40 mg) for 12 d. and one single i.m. PMSG injection (250 U.I.) 24 h before sponge removal. Fertility [Ft =(ewes lambed/ewes mated) × 100], fecundity (Fc=lambs born/ewe mated) and prolificacy (Pr=lambs born/ewe lambed) values were scored in all groups. Birth type (single vs multiple), viability (live:dead ratio till 5th day after lambing) and sex ratio (male vs female) at lambing were scored. Moreover, body weight at lambing (BWL; kg) and body weight at weaning (42 d.) (BWW; kg) were scored. χ^2 and ANOVA test were used to compare variables (season/control/treatment/parameter). There were no significant differences in reproductive parameters (Ft, Fc or Pr), lambing scores (birth type, viability or sex ratio) or BWL/BWW parameters among AC AT and ST groups (P>0.05). When SC vs ST were compared significant differences were observed between both groups in Ft (84 vs 96), Fc (0.9±0.1 vs 1.2±0.2), Pr (1.3±0.2 vs 1.6±0.3), BWL (3.8±0.4 vs 3.4±0.3), BWW (12.7±0.3 vs 11.6±0.4) and birth type (S:M=51:39 vs 49:51) (P<0.05). No significant differences were detected between SC and ST in viability (L:D=94:6 vs 91:9) or sex ratio (F:M=52:48 vs 54:46) (P>0.05). SC derived results showed significant differences regarding Ft, Fc, Pr and birth type compared to AC, AT or ST groups (P<0.05). In conclusion, FGA/PMSG combined protocol during the favourable mating season (autumn) was determined as unnecessary. However, there are interesting advantages in reproductive performance and lamb production when same protocol was used during the unfavourable season (spring) under Mediterranean conditions.

Estimation of genetic parameters and genomic breeding values for AI success in French dairy goats

M. Chassier, A. Piacere and C. Virginie
Institut de l'Elevage, Chemin de Borde Rouge, 31321 Castanet-Tolosan, France; marjorie.chassier@idele.fr

Genetic parameters for artificial insemination (IA) success in French dairy goats were estimated. Data analysis, on a total of 378,269 and 574,477 AI records for Saanen and Alpine breed, respectively, collected from 1981 to 2017, was conducted separately on each breed. The heritability of IA success was estimated using a linear simple repeatability animal model. The model included fixed effects of flock within the year of IA, year within the inseminator, year within the treatment for induction and synchronisation of oestrus, year within the lactation order, year within the interval between previous kidding and IA, year within month of IA within region and year within day of IA within region. Random effects were animal, permanent environmental effect and random residual error. Heritability estimates for IA success were 0.052 (0.003) and 0.045 (0.002) for Saanen and Alpine breed respectively. Despite the low heritability, genetic standard deviations were 9 AI points in Alpine breed and 11 AI point in Saanen breed. A significant progress is feasible if a selection is set up on this trait. 831 Saanen and 1,075 Alpine bucks were genotyped with the Illumina goat SNP50 BeadChip and genomic breeding values were estimated for IA success trait with a within-breed GBLUP (genomic best linear unbiased prediction) model. A low negative correlation (-0.12) between milk yield index and IA success index in Saanen breed, and low positive correlations between fat (0.16) or protein (0.13) index and IA success index in Alpine breed were estimated. In 2020, a genomic evaluation for IA success was developed, this trait will be taken into account to produce elite bucks, by considering the index levels of their sires and dams in mating programs. Early genomic indexes will be produced for young bucks so that they can be selected on this trait before entering in the semen production centre.

Assessment of inbreeding depression on milk production in the two autochthonous Czech goat breeds

L. Vostry[1], H. Vostra-Vydrova[1,2], N. Moravcikova[3], B. Hofmanova[1], R. Kasarda[3], M. Brzakova[1,2] and J. Rychtarova[2]
[1]Czech University of Life Science Prague, Kamycka 129, 16500 Prague, Czech Republic, [2]Institut of Animal Science, Pratelstvi 815, 10400 Prague, Czech Republic, [3]Slovak University of Agriculture in Nitra, Tr. A. Hlinku 2, 94976 Nitra, Slovak Republic; vostry@af.czu.cz

Inbreeding depression, the reduction of fitness caused by inbreeding, is a nearly universal phenomenon that depends on past mutation, selection, and genetic drift. The effect of two different kind of inbreeding (classical inbreeding and new inbreeding) on milk production was analysed in two original Czech goat breeds – White Shorthair Goat and Brown Shorthair Goat. Because these populations are small and closed, there is a concern about the loss of genetic variation which causes the accumulation of deleterious mutations and inbreeding depression. Inbreeding depression for milk production in two autochthonous goat breeds were analysed based on 1,976 animals from White Shorthair goats and 300 animals from Brown Shorthair goats. The effect of tested coefficient on milk performance was investigated using BLUP animal models. The average classical inbreeding and new inbreeding for the reference populations of the two breeds were 2 and 1.7% respectively for White Shorthair Goat and 5.3 and 3.4% for Brown Shorthair Goat. Both tested inbreeding coefficients showed a significantly negative influence on milk performance only in the White Shorthair Goat breed. The statistically non-significant effect of tested inbreeding coefficients in Brown Shorthair Goat may be affected by the small number of analysed individuals with milk performance records. These results suggest that genetic variability has decreased, inbreeding depression has occurred and without changes in the breeding strategy, genetic variability might continue to decline. The study was supported by the projects QK1910156.

Differential gene expression analysis between colour and white skin in the Assaf sheep breed

P.K. Chitneedi, J.J. Arranz, A. Suárez-Vega and B. Gutiérrez-Gil
Universidad de León, Producción Animal, Facultad de Veterinaria-Campus de Vegazana, 24007, León, Spain;
beatriz.gutierrez@unileon.es

In the last years, the Assaf breed, which is a stabilised cross of the Awassi and East Friesian breeds, has become one of the main dairy sheep breeds in Spain. The Spanish Assaf breeders' association (ASSAFE) manages a genetic program for the improvement of milk production and milk quality. Most animals of this breed show white face and white fleece although some of them may show reddish-brown spotting in the face or fleece (associated with its Awassi genetic background). Although coat colour is not considered as a selection criterion in Assaf breed, some breeders may prefer white face and white fleece animals. The objective of this study was to perform a preliminary comparison between two RNA-Seq datasets obtained from white and reddish-brown skin from the same Assaf lamb. Total RNA from skin samples was obtained with the Qiagen Fibrous RNeasy Mini kit and later sequenced using an Illumina HiSeq 2000 sequencer. After the alignment against the ovine reference genome (Oar_v3.1) with STAR_ v2.5.2b and the quantification using HTSeq-count software, an exploratory differential gene expression analysis was performed using EdgeR. A total of 87 differentially expressed genes (DEGs) were identified. The gene ontology (GO) enrichment analyses performed with WebGstalt identified significant (FDR<0.05) GO terms related to melanin and pigment biosynthesis (biological process database) and the melanosome (cellular component database). Moreover, melanogenesis was the only significantly enriched KEGG pathway (FDR<0.05). Some of the highly expressed genes in the reddish-brown skin, such as *PMEL*, *TRPM1*, and *TYR*, are known to be involved in the production of melanin pigmentation in skin and influence natural skin colour variation. This study provides a preliminary analysis of the genes and pathways that might influence skin colour in Spanish Assaf sheep and may help to better understand the genetic basis of coat colour in this species. Project: RTI2018-093535-B-I00 funded by Spanish Ministry of Science.

The effect of maize vitreousness and a starch binder on *in vitro* gas production using rumen fluid

C.W. Cruywagen and J.H.C. Van Zyl
Stellenbosch University, Animal Sciences, Private Bag X1, Matieland, 7602 Stellenbosch, South Africa; cwc@sun.ac.za

The objective of the study was to determine the possibility of binding starch in maize with a commercial starch binder to decrease *in vitro* rumen degradation. One sample each of low vitreous (soft) and high vitreous (hard) maize, were used and treated with either Bioprotect (BP) or distilled water as control (CON) at a rate equivalent to 10 l/tonne. The effect of treatment on starch fermentation was determined in two *in vitro* trials, viz. a gas production trial and a starch disappearance trial. The experimental design in each trial was a randomised block with *in vitro* runs as blocks. The rate (ml/h) and extent (ml/g OM) of gas production of the soft maize (0.037 and 408, respectively) were higher (P<0.05) than that of the hard maize (0.027 and 335). In hard maize, the starch binder did not lower gas production parameters (rate or extent of gas production) significantly. In soft maize, however, total gas production was lower (P<0.05) in the BP treatment (364.9) than in the CON treatment (450.6), but rate was not affected (0.036 for BP and 0.037 for CON). *In vitro* starch disappearance values were determined after 6, 12 and 24 h of incubation for the same four treatments. Mean starch disappearance values were higher (P<0.05) for soft than for hard maize (77.2% vs 66.9%), but the starch binder had no effect on either soft or hard maize. Based on the gas production results, it was concluded that Bioprotect treatment has the potential to reduce ruminal starch fermentation when low vitreouss maize is used in dairy cow diets.

Assessment of antioxidant capacity in dairy ewes after supplementation of feed with milk thistle oil

C. Karaiskou[1], E. Kasapidou[2], S. Makri[3], D. Kouretas[3] and Z. Basdagianni[1]

[1]School of Agriculture, Department of Animal production, Aristotle University of Thessaloniki, 54124 Thessaloniki, Greece, [2]University of Western Macedonia, Department of Agriculture, Terma Kontopoulou, 53100 Florina, Greece, [3]University of Thessaly, Department of Biochemistry and Biotechnology, Viopolis, 41500 Larissa, Greece; ekasapidou@uowm.gr

Milk thistle (*Silybum marianum L.*) is known as an important medicinal herb due to antioxidant activities. The high concentration of flavonoids in milk thistle oil (MTSO) maintains both, animal health and oxidative stability of their food products. Currently, information is lacking in the literature concerning the effect of MTSO supplementation on ruminants' redox status. Therefore, the objective of the current study was to investigate the effect of dietary inclusion of milk thistle oil, on antioxidant capacity in dairy ewes. A total of 20 ewes from the Greek mountain breed, in the second lactation, were allocated in two treatments for 4 months, control [CON (n=10)] fed the basal concentrate diet and MTSO (n=10) fed the concentrate diet supplemented with 3% (w/w) milk thistle oil. Individual blood samples were collected monthly throughout the experimental period for the determination of oxidative stress biomarkers and the potential antioxidant effects of milk thistle oil. The biomarkers assessed in plasma were: total antioxidant capacity (TAC), protein carbonyls (CARB), thiobarbituric acid reactive species (TBARS) and reducing power (RP). The results showed that ewes on the MTSO treatment had significantly lower CARB (10.09 vs 5.00 nmol/mg protein) and TBARS values (10.31 vs 5.00 μmol MDA/l plasma; P≤0.01) compared to CON. Additionally, the concertation of the biomarkers TAC and RP was significantly higher in MTSO group (0.63 vs 0.86 mmol DPPH/l) and (0.091 vs 0.127 mmol potassium ferricyanideml), P<0.01). The results indicate that milk thistle oil supplementation reduced protein oxidation and lipid peroxidation in ewes. Moreover, the observed increase in the concertation of TAC and PR suggests a potentially protective role of milk thistle oil against oxidative stress in Greek mountain breed ewes. In conclusion, supplementation of milk thistle oil could be used for the enhancement of ewes' redox status and for the reduction of oxidative stress.

Pedigree and genomic inbreeding comparison in the Italian Delle Langhe dairy sheep breed

M. Cortellari[1], A. Negro[1], A. Bionda[1], A. Cesarani[2,3], N. Macciotta[3], S. Grande[4], S. Biffani[5] and P. Crepaldi[1]

[1]Univertity of Milan, Department of Agricultural and Environmental Sciences, Production, Landscape and Energy, Via Celoria 2, 20133, Milan, Italy, [2]UGA, Dep Animal and Dairy Science, 425 River Rd, Athens, 30603, GA, USA, [3]University of Sassari, Dep. Agricultural Science, Via De Nicola, 9, 07100 SASSARI, Italy, [4]AssiioNaPa, Via Tomassetti 9, 00161, Rome, Italy, [5]CNR, Istituto di Biologia e biotecnologia agraria, Via Corti 12, 20133, Milan, Italy; matteo.cortellari@unimi.it

The Delle Langhe sheep breed is an autochthonous population of 4,275 heads mainly raised for dairy production in the homonym area of Northern Italy. In this work we compared parentage and inbreeding level (INB) estimated from traditional pedigree data with pairwise mendelian errors (ME) and runs of homozygosity (ROH) inbreeding (FROH) calculated from 50k medium SNPchip data. The objective was to check and improve the reliability of parentage testing and inbreeding estimates. Data for the analysis were from two subsets of Delle Langhe individuals. The first subset (A) included 30 animals sampled from 10 farms (1 male and 2 females per farm). The second subset (B) included 74 animals sampled from only one farm (72 females and 2 males). All data were collected during the Conservation, Health and Efficiency Empowerment of Small Ruminant (CHEESR) project, coordinated by the National Breeders Association. ROH were calculated using PLINK software considering a minimum ROH length of 1 Mb and a sliding window of 20 SNPs; FROH values were calculated as the proportion of the autosomal genome covered by ROH. ME of the pairwise comparisons between all sampled individuals were estimated using an in-house software. Parentage mismatch rates comparing traditional pedigree data and genomic data was 19.5 and 6.6% for subset A and B, respectively. Correlation between ME and pairwise pedigree relationship was -0.62 (P<0.001). A correlation of 0.44 (P<0.001) was observed between FROH and INB. Genomic parameters are indeed efficient tools to identify possible parentage mismatch and to properly estimate genomic relationship among individuals. Both are essential prerequisites for the efficiency of any breeding program and an effective approach to improve animal management also in small populations.

Genomic inbreeding of Nicastrese: conservation of an autochthonous Italian goat breed

M. Cortellari[1], A. Negro[1], A. Bionda[1], A. Cesarani[2,3], N. Macciotta[3], S. Grande[4], S. Biffani[5] and P. Crepaldi[1]
[1]Univertity of Milan, Department of Agricultural and Environmental Sciences, Production, Landscape and Energy, Via Celoria 2, 20133, Milan, Italy, [2]UGA, Dep Animal and Dairy Science, 425 River Rd, Athens, , GA 30602, USA, [3]University of Sassari, Dep of Agricultural Science, Via de Nicola 9, 07100 Sassari, Italy, [4]AssoNaPa, Via Tomassetti 9, 00161, ROme, Italy, [5]CNR, Istituto di biologia e biotecnologia agraria (IBBA), Via Corti 12, 20133, Milan, Italy; matteo.cortellari@unimi.it

Nicastrese is a goat population from Southern Italy. In 2019 there were 6,254 individuals from 105 herds. Recently, within the national CHEESR project, coordinated by the National Breeders Association, a sample of Nicastrese individuals were genotyped. The objective of this work was to analyse breed structure (BS) and calculate inbreeding in Nicastrese goat using Runs of Homozygosity (ROH) estimated from genomic information. Date came from 100 individuals split in 2 subsets. The first subset included 30 unrelated animals sampled from 10 farms. The second subset included 23 and 47 goats sampled from two different and large herds, namely herd A (n=223) and herd B (n=220). The official lab of Italian AIA performed DNA extraction and genotyping with a medium density (50k) Goat SNP chip. ROH were estimated as follows: a ROH length of $\geq$1 Mb, a sliding window of 20 SNP with no heterozygous tolerated. ROH inbreeding coefficients (FROH) were calculated as the proportion of the autosomal genome covered by ROH. ROH analysis allowed to evaluate both the level of inbreeding in the population and the differences among farms. Herd B showed a very high level of outbreeding across all classes of ROH length (1-2, 2-4, 4-8, 8-16, >16 Mb). Herd A had a FROH similar to the one found in BS. In all the subsets, we observed a high ROH number on chromosomes (CHI) 6 and 12. The large number of ROH identified on CHI 6 are consistent with the dairy purpose of this breed and may indicate an artificial selection or a natural selection directed at offspring preservation. CHI 12 contains a syntenic region, both in cattle and sheep, where selection signals were already identified by other studies. The investigation of ROH in different farms can give useful indications for the management of the breed's biodiversity and originality.

Application of multivariate indices in traits of milk of Manchega sheep: preliminary results

E. Angón[1], A. Garzón[1], J. Caballero-Villalobos[1], A. Figueroa[1], R. Arias[2] and J. Perea[1]
[1]Universidad de Córdoba, Animal Production, Campus de Rabanales, 14071, Córdoba, Spain, [2]Centro Regional de Selección y Reproducción Animal de Castilla-La Mancha, Avenida del Vino 10, 13300, Valdepeñas, Spain; eangon@uco.es

Manchega is the most common dairy sheep breed in Spain. The milk from this breed is used to make Manchego cheese, a cured hard cheese manufactured exclusively in the region of Castilla-La Mancha. It is the best-selling cheese variety in Spain and is safeguarded by Protected Designation of Origin (PDO) that guarantees quality. The composition, colorimetry parameters, coagulation properties, and hygienic quality of raw milk are traits of great interest to the dairy industry. However, a complex correlation pattern exists between these variables. In this paper, the multivariate relationship pattern between these variables is explored through factor analysis (AF) of 1,200 individual samples of Manchega sheep milk. This approach was expected to generate new variables with potential technical meaning that could be used for management and breeding purposes. The AF revealed seven common latent factors (F1 to F7) explaining 80.5% of the total variance. F1 showed the positive relationship between the chemical composition of the milk and its cheese yield. F2 showed the negative relationship between syneresis and the rennet clotting time. F3 represented the milk protein content and its inverse relationship with the lactose content and the amount of milk obtained. F4 showed the negative relationship between the hardness and the water content in the curd. F5 was correlated with the luminosity, the colour hue and the red/green colour of the milk, while F6 was correlated with the perceived colour saturation and the blue/yellow colour. F7 was correlated with the somatic cell count and the pH of the milk. The 7 extracted factors showed a well-defined technical meaning. They represent useful indicators of milk characteristics that could be used for management and breeding purposes.

What livestock has to offer to biodiversity & healthy soils: Welcome and introduction

J.L. Peyraud
Animal Task Force, 149 rue de Bercy, 75012 Paris, France; mail@animaltaskforce.eu

Since 2013, the ATF-EAAP Special Session during the EAAP Annual Meeting aims to bring together animal science with practice of animal production and connect researchers, policy-makers, industry representatives and societal organisations. Every year, a different topic is addressed during this half-day session. In 2020, the theme is: 'What livestock has to offer to biodiversity & healthy soils'. This 8th EAAP-ATF Special Session would like to engage discussion with farmers, industries, scientists, policy-makers and with the society. The outcomes of the session will be discussed with a large panel of European stakeholders during the ATF seminar, in Brussels, on 4th Nov. 2020. 'The continuous, accelerating decline in biodiversity is of particular concern as biodiversity provides the fabric of life with a range of ecosystems services which are crucial for human well-being'. 'Main direct drivers of biodiversity loss, in order of their importance, are land use change, overexploitation (through intensive agriculture, forestry and fishing practices), climate change, pollution and invasive species.' 2020 will be the year of biodiversity with the UN aiming to develop a post-2020 framework. Livestock is often blamed for its contribution to biodiversity losses but the reality is more complex. The effects of livestock on biodiversity are variable across farming systems and livestock production can also make a positive contribution to biodiversity objectives and preservation of habitats. Very often, research initiatives on soil fertility do not include animal farming. Still, livestock is a reservoir of solutions to increase soil C-sequestration, biological fertility, organic matter, etc. In some cases, it generates in negative impacts as emission of reactive N and dissemination of medicine residues or antimicrobial resistance. ATF would like to explore the different pathways, needs in cooperation and R&I to support an animal production able to contribute to ecosystems remediation, whatever in soils and on biodiversity.

Assessment of livestock impacts and positive contributions to biodiversity and soil health

F. Teillard
FAO, Viale delle Terme di Caracalla, 00153 Rome, Italy; mail@animaltaskforce.eu

Livestock production is widespread around the world, with terrestrial areas dedicated to rangelands and cropland dedicated to fodder production. Demand for livestock products is projected to grow significantly until 2050, driven by a combination of global population growth and changes in patterns of food consumption due to increasing wealth and urbanisation. The influence of livestock production on biodiversity and soil health needs therefore to be assessed, considering that the exact effects are diverse and depend on the intensity of production, the nature of specific practices, the livestock species used, and local ecological conditions. The quantitative assessment of the impacts of livestock systems on biodiversity and soil health is an area of work to be included in future sustainability assessments.

Trade-offs between livestock, biodiversity and soils health & approaches towards positive impacts
A. Arroyo Schnell (to Be Confirmed)
International Union for Conservation of Nature, Rue Mauverney 28, 1196 Gland, Switzerland; mail@animaltaskforce.eu

A representative of an organisation involved in nature conservation will highlight the relation between the biodiversity crisis and practices in agriculture, fisheries, and forestry on the one hand, and disruption of water flow, inadequate management of waste and discharges, etc. On the other hand, the approach will explore the potential application of positive impact approaches for biodiversity in the commercial agriculture and forestry sectors and drivers for change.

Farmers best practices to mitigate or remediate biodiversity losses
S. Finan
CEJA, 67 Rue de la Loi, 1040 Brussels, Belgium; mail@animaltaskforce.eu

Example of a ruminant farmer's practices towards fostering ordinary biodiversity and soil health by using a diversity of forage species and grassland types, by introducing short term grassland in crop rotation, and by diversification of land use and thus maintaining open habitats and landscape. Introduction of legumes that contribute to nitrogen (fertiliser) and protein (feed) for livestock and contributes to animal health thanks to some of their secondary metabolic compounds.

Animal genetic resources and breeding goals for biodiversity-rich livestock systems

S.J. Hiemstra
WUR, Droevendaalsesteeg 4, 6708 PB Wageningen, the Netherlands; mail@animaltaskforce.eu

To maintain options for livestock systems to adapt to a variety of future environments and contexts it is important to preserve a broad genetic base and livestock gene pool. Complementary strategies are needed to conserve and to properly manage farm animal genetic diversity (local and cosmopolitan breeds), and to exploit the available animal genetic resources in the global context of food and nutrition security and climate change. At the same time, livestock provide a diversity of agroecological, social, cultural and rural economic services, contributing to preservation of biodiversity, diverse landscapes, high-quality food and cultural heritage. Hence, breeding goals and the choice of livestock breeds should be based on characteristics and limitations of different future farming systems and agro-ecological contexts. Knowledge should be mobilised on development of tailor made breeding goals, on better phenotypic and genomic characterisation of breeds, and on genotype × environment interactions and resilience mechanisms.

Good practices of industry driving production

D. Forster
Nestlé, Société des Produits Nestlé SA, 1800 Vevey, France; mail@animaltaskforce.eu

Example of a company involving farmers and developing local and/or labelled products accounting for biodiversity, healthy soils and healthy food products schemes with added value for farmers and appreciated by the consumer.

Questions & answers
J.-L. Peyraud
Animal Task Force, 149 rue de Bercy, 75012 Paris, France; mail@animaltaskforce.eu

Time for questions to the speakers.

Session 12   Theatre 8

Panel discussion: Improving positive contribution of livestock to biodiversity & soil health?
F. O'Mara
Animal Task Force, 149 rue de Bercy, 75012 Paris, France; mail@animaltaskforce.eu

Panel discussion, with speakers and Patrick Worms, president of European Agroforestry Federation (EURAF). Moderated by Vivi H. Nielsen. What are the white spots? Gaps in R&I.

Session 13

Theatre 1

Introduction to the governance of the animal genetic resources challenge session

G. Leroy and R. Baumung

Food and Agriculture Organization, Viale delle Terme di Caracalla, 00153, Rome, Italy; gregoire.leroy@fao.org

Governance can be defined as the way stakeholders organise themselves to make decisions and to ensure that those decisions are enacted in a fair manner. It involves issues either related to power balance, representativeness of actors, information sharing, responsibilities and accountabilities on activities, or rights and benefits. Adequate governance for the sustainable use and conservation of animal genetic resources involves multiple challenges. In this session, we propose to illustrate those challenges regarding a diversity of scales and topics, and discuss how future socio-economic, technological, legal and environmental changes may affect the governance of animal genetic resources in the future. The session will be organised around a small number of interventions by a few selected speakers followed by a round table.

Session 13

Theatre 2

What to preserve, and how to decide?

H. Simianer

University of Goettingen, Center of Integrated Breeding Research, Albrecht-Thaer-Weg 3, 37075 Goettingen, Germany; hsimian@gwdg.de

While 'breed' is generally accepted as the operational unit to be considered in conservation programs, this concept must be challenged from a scientific point of view. First of all, breed is by no means a well-defined category, nor are breeds genetically homogeneous: usually, about 80% of genetic variation is found within and only 20% between breeds within a livestock species. Further, most characteristics of interest have a complex genetic background and are often gradually distributed across breeds or populations. If the aim of a conservation strategy is to preserve with limited resources a wide portfolio of genetic solutions to future challenges, it may be more expedient to preserve a little bit of everything, rather than e.g. focussing on the most endangered breeds. A second aspect that will be addressed is that decisions about livestock conservation are typically made under uncertainty and are characterised by notorious imbalances: when maintaining diversity as a safeguard for future changes, the expected benefit will be on a global level, while conservation activities have to be implemented on a local, national or regional level. Further, there is an imbalance with respect to cost and profit: While the expected profit realised at some point in the future is not quantifiable in form of a predictable monetary return, conservation investments are very concrete, have to be made now and require a substantial financing, which needs to be sustainably provided by some stakeholder. This leads to the challenge of balancing current specific needs, such as deciding what to conserve and who pays the bill, with future potential benefits, where it is not clear if there will be some profit at all, and if so, who will profit when. This, together with the question how this decision making process can be organised leaves us with some major unresolved challenges, which will be discussed and illustrated with some examples.

International genetic evaluations as a technology of governance of the globalised cattle breeding

L. Chavinskaia[1] and V. Ducrocq[2]
[1]Université Gustave Eiffel / INRAE / CNRS, UMR LISIS Laboratoire Interdisciplinaire Science Innovation Société, 3 Allée Jean Renoir, 93160 Noisy-le-Grand, France, [2]INRA, UMR 1313, Génétique Animale et Biologie Intégrative (GABI), Domaine de Vilvert, 78352 Jouy en Josas, France; lidia.chavinskaia@gmail.com

The development of genetic evaluation technologies follows the story of governance of cattle genetic resources by addressing the issue of their values. In the era of industrial breeding, the question of genetic value is not merely a technical question. Put into historical perspective of the period we call 'neoliberal globalisation' (the 1960's to now), technologies of cattle genetic evaluation proved to be technologies of governance. Nationally computed, EBVs accompany genetic products (semen, embryos, live animals) meeting at the global market. The comparability of those values becomes an issue for ordering circulations of cattle genetic resources and for governing the cattle breeding activity on the global level. National interests, scientific objectivity, laws of the neoliberal market and biological nature of the commercialised products come together to challenge the globalisation of the cattle breeding. How to deal with a global pool of resources? Should we account for the diversity of their values? The examples of Interbull (International Bull Evaluation Service), Eurogenomics (the European consortium of semen producers) and of South Africa (as an example of a 'developing' country trying to save its own national cattle breeding industry) helps to understand how different techno-political strategies are competing to address such questions and to set up different solutions suitable for different logics.

Challenges and opportunities in the international governance of animal genetic resources

E. Martyniuk
Institute of Animal Sciences, Department of Animal Genetics and Conservation, ul. Ciszewskiego 8, 02-786 Warszawa, Poland; elzbieta_martyniuk@sggw.pl

Experience have shown that the effective governance policies and procedures for animal genetic resources (AnGR) require sound understanding of the importance and status of these resources. This was not the case for AnGR at the global level until the preparation of the first report on the State of the World's AnGR. Before this report, information was in general limited and dispersed, although breed surveys in the European region and some countries had been conducted. The SoW-AnGR report substantially improved understanding of the status and trends in AnGR paving the way for a regular monitoring system adopted by the Commission on Genetic Resources for Food and Agriculture. The SoW-AnGR led to the adoption of the Global Plan of Action for AnGR (GPA) in 2007, which after review in 2017, received renewed commitment by the FAO Member Countries. The GPA, as an agreed upon international plan of action for AnGR, encourages strengthening of governance policies and procedures for AnGR at the national and regional levels, and promotes development of support tools and measures to manage AnGR by the FAO, the international donor community and FAO member countries. A number of international bodies have acknowledged the importance of AnGR, for example, the CBD Strategic Plan for Biodiversity 2011-2020, Agenda 2030, as well as climate change convention forums. Despite improved understanding of the importance of AnGR, significant questions for the further implementation of the GPA remain. Are AnGR a priority for policy makers? Are adequate financial resources available to ensure their sound management and conservation? Are stakeholders adequately involved in the development of the sector? Is there sufficient AnGR management capacity to develop sustainable strategies and breeding programmes? Is there awareness of new challenges in relation to the Nagoya Protocol and opportunities to include genetic resources for food and agriculture in the Post 2020 Biodiversity Framework?

Sequence-based GWAS for milk production, udder health and morphology traits in French dairy cattle

T. Tribout[1], P. Croiseau[1], R. Lefebvre[1], A. Barbat[1], M. Boussaha[1], S. Fritz[2], D. Boichard[1], C. Hoze[2] and M.P. Sanchez[1]
[1]Université Paris-Saclay, INRAE, AgroParisTech, GABI, 78350, Jouy-en-Josas, France, [2]Allice, MNE, 75012, Paris, France; thierry.tribout@inrae.fr

Genome-wide association studies (GWAS) were performed at the sequence level in Montbéliarde (MO), Normande (NO) and Holstein (HO) dairy cattle breeds for 5 milk production traits, somatic cell scores, clinical mastitis, milking speed and 8 udder morphology traits. The number of bulls considered by trait varied from 1,857 to 2,515 in MO, from 1,427 to 2,203 in NO, and from 4,959 to 6,321 in HO. The variables were the bulls' daughter yield deviations (DYD), derived from the national genetic evaluations. Genotypes of the bulls for around 28 million sequence variants were imputed in 2 steps, using FImpute software: first from 50K level to HD level using 522 MO, 546 NO, and 776 HO HD genotyped bulls as a reference, and then to the sequence level using 1,147 sequenced bulls from the 1,000 bull genomes project (Run 4), including 28 MO, 24 NO and 288 HO bulls. GWAS were done independently within each breed and for each trait, using GCTA software, accounting for the population structure through a HD-based genomic relationship matrix. We detected 27, 14 and 49 significant QTL regions (-logP>8.2) in MO, NO and HO breeds, respectively. Numbers of QTL per trait and breed varied from 0 (protein yield in MO and NO, udder health in NO, 3 and 4 udder morphology traits in MO and NO, respectively) to 11 (protein content in HO). Some QTL were common for several traits within a breed, and/or for same traits in 2 or 3 breeds. Candidate variants (CV) were selected in the QTL, based on the significance of their effect and on their functional annotation. A total of 32,373 MO, 12,316 NO and 52,630 HO cows were genotyped or imputed for a 50K BeadChip augmented with the CV. A second set of GWAS was performed on these cows, considering their YD (ie adjusted own performances) for the same milk production and udder traits, these data being independent from the DYD of the first step bulls. The analyses confirmed 42 and 18 significant (-logP>6) QTL regions for production and udder traits, respectively. Except for 4 production and 4 udder QTL regions, the significance of estimated effect was higher for the CV than for the 50K variants.

Fine resolution CNV catalogue from deeply sequenced cattle genomes

Y.L. Lee[1], H. Takeda[2], G. Costa Monteiro Moreira[2], L. Karim[3], M. Bosse[1], A. Bouwman[1], E. Mullaart[4], W. Coppieters[3], M. Georges[2], T. Druet[2], C. Charlier[2] and GplusE Consortium[5]
[1]Wageningen University & Reserch, Animal Breeding and Genomics, P.O. Box 338, 6700 AH, Wageningen, the Netherlands, [2]University of Liège, Unit of Animal Genomics, GIGA-R & Faculty of Veterinary Medicine, Liège, Belgium, [3]GIGA Institute, GIGA Genomics Platform, Liège, Liège, Belgium, [4]CRV B.V., P.O. Box 454, 6800 AL, Arnhem, the Netherlands, [5]http://www.gpluse.eu/; younglim.lee@wur.nl

Genomes consist of various forms of variations that ultimately contribute to shaping phenotypes. Copy number variations (CNVs) are a form of genetic variation, and arise from gain or loss of DNA. CNV discovery depends on the quality of both sequencing data and reference genome, but in livestock an accurate CNV catalogue and an investigation of the functional impact of CNVs are lacking. We used deeply sequenced cattle genomes from 131 Dutch Friesian Holstein trios (mean coverage: 26x), mapped to the reference genome ARS1.2, to study CNVs. Harnessing the unique pedigree structure in livestock populations, we eliminated spurious CNVs, based on the Mendelian inheritance pattern. Offspring of the 131 probands (~5 animals/proband) enabled us to trace the inheritance of interesting CNVs observed in the probands. Among ~10,000 high quality CNVs, which were ascertained at base pair resolution, ~3,000 overlapped with the coding sequence. Subsequently, we used histone and chromatin modification assay data to investigate whether CNVs are overlapping with gene regulatory elements. Among the ~7,000 CNVs in non-coding regions, ~200 overlapped with putative regulatory elements such as enhancers, promoters, and open chromatin regions. These overlaps imply that CNVs can alter gene expression, either by directly affecting coding sequences or by interrupting gene regulatory elements. A highly interesting CNV in our catalogue is GC gene duplication, overlapping with the last exon of the GC gene. This duplication is located at ~36 kb distance from several known clinical mastitis QTLs. Using RNA seq data generated from liver tissues of Holstein cows (n=178), we confirmed that the duplication is associated with increased expression of GC. This valuable CNV catalogue warrants follow-up research on the functional impact of CNVs.

Genome-wide association studies for iris pigmentation and heterochromia patterns in Large White pigs

G. Moscatelli[1], S. Bovo[1], G. Schiavo[1], G. Mazzoni[2], F. Bertolini[3], S. Dall'Olio[1] and L. Fontanesi[1]
[1]University of Bologna, Department of Agricultural and Food Sciences, Division of Animal Sciences, Viale G. Fanin 46, 40127, Bologna, Italy, [2]Technical University of Denmark (DTU), Department of Health Technology, Lyngby, 2800, Denmark, [3]Technical University of Denmark (DTU), National Institute of Aquatic Resources, Lyngby, 2800, Denmark; giulia.moscatelli8@unibo.it

Coat colour in livestock species is one of the most distinctive traits that characterise many different breeds. Pigmentation does not affect only skin and hairs but also the eyes. Few genomic studies in cattle and horses also investigated iris pigmentation, but no similar studies was conducted in pigs thus far. In this study we analysed eye colour diversity in a Large White pig population (n=897) and reported the results of genome-wide association studies based on several comparisons including pigs having four main eye colour categories (three with both pigmented eyes of different brown grades: pale, 17.9%; medium, 14.8%; and dark, 54.3%; another one with both eyes completely depigmented, 3.8%) and heterochromia patterns (heterochromia iridis, i.e. depigmented iris sectors in pigmented irises, 3.2%; heterochromia iridium, i.e. a whole eye iris of depigmented phenotype and the other eye with the iris completely pigmented, 5.9%). Pigs were genotyped with the Illumina PorcineSNP60 BeadChip and GEMMA was used for the association analyses. The results indicated that the eye pigmented patterns (different grades of brown pigmentation), the total absence of pigmentation in the both eyes, and heterochromia iridis defect were under controlled of SLC45A2 (on chromosome 16, SSC16), EDNRB (SSC11) and KITLG (SSC5), respectively. In addition, to new candidate associations for the eye depigmented patterns were also identified for SNPs on two SSC4 regions (including two candidate genes: NOTCH2 and PREX2) and on SSC6, SSC8 and SSC14 (including COL17A1 as candidate gene). This study provided useful information to understand the genetic mechanisms affecting eye pigmentation in pigs.

Single-SNP and haplotype-based genome scans for teat number in Italian Large White and Landrace pigs

S. Bovo[1], M. Ballan[1], G. Schiavo[1], A. Ribani[1], S. Tinarelli[1,2], V.J. Utzeri[1], G. Galimberti[3], M. Cappelloni[2], M. Gallo[2], S. Dall'Olio[1] and L. Fontanesi[1]
[1]University of Bologna, Department of Agricultural and Food Sciences, Viale G. Fanin 46, 40127 Bologna, Italy, [2]Associazione Nazionale Allevatori Suini (ANAS), Via Nizza 53, 00198 Roma, Italy, [3]University of Bologna, Department of Statistical Sciences 'Paolo Fortunati', Via delle Belle Arti 41, 40126 Bologna, Italy; samuele.bovo@unibo.it

Teat number (TN) is an important parameter for the mothering ability of the sows. Being an economically relevant reproductive trait, TN is the objective of genetic selection which generally impose a minimum threshold value in breeding programmes. Several studies investigated TN in different pig populations, reporting high variability (from 8 to 21 teats), a medium/high heritability (from 0.10 to 0.70) and a polygenic control of the trait. We previously investigated TN in Italian Large White (ILW) pigs and associated its variability to polymorphisms in the VRTN gene. Here, we refined those results in a population of about 4,000 ILW pigs genotyped with the 70K GGP Porcine BeadChip. Moreover, about 2,000 Italian Landrace (ILA) pigs were for the first time investigated. Single-SNP and haplotype-based genome scans were performed via linear mixed model analyses (GEMMA tool). TN for the ILW pigs was in the range 12-20, whereas ILA pigs had 10-18 teats. The two populations had about 53.4 and 43.5% of animals with more than 14 teats, respectively. Genomic heritability of TN was 0.30 (ILW) and 0.38 (ILA). We identified a total of 22 quantitative trait loci (QTL) on 12 porcine chromosomes associated with TN variability in these two populations. Obtained results confirmed the association of the VRTN gene variants in the ILW population and pointed out other genome regions harbouring QTLs for reproductive traits (e.g. litter size, number of vertebra as reported in other populations). ILA pigs showed a completely different set of genome variants affecting TN. Overall, these results provided information that could be used to design breeding programmes to improve TN in Italian pig breeds.

Genome-wide association study for milking speed and temperament in Holstein-Friesian bulls

B. Kosińska-Selbi[1], M. Jakimowicz[1], T. Suchocki[1,2], M. Skarwecka[2], A. Żarnecki[2], W. Jagusiak[2] and J. Szyda[1,2]
[1]Wrocław University of Environmental and Life Sciences, Biostatistics Group, Department of Genetics, Kożuchowska 7, 51-631 Wrocław, Poland, [2]National Research Institute of Animal Production, Krakowska 1, 32-083 Balice, Poland; barbara.kosinska@upwr.edu.pl

The aim of the study was to identify SNPs and KEGG pathways significant for two traits describing workability in Holstein-Friesian bulls, such as milking speed (MSP) and temperament (TEM). MSP and TEM are represented by deregressed breeding values. 28,315 bulls from 22 countries were genotyped with the Illumina Bovine 50K BeadChip, resulting in 54,609 SNPs available for each individual. After quality control for minor allele frequency of at least 0.01 and technical quality of genotyping expressed by a minimum call rate of 99%, 46,216 SNPs were used in further analysis. This data is a part of the EuroGenomics consortium. The first step was to identify SNPs significant for both traits using genome-wide association study (GWAS). Based on the false discovery rate of maximum 5%, 21 SNPs were identified as associated with MSP (located on BTA4, BTA5, BTA6, BTA19, BTA20, BTA29 and BTAX) and 8 as associated with TEM (located on BTAX). In order to link SNPs to coding sequence the Variant Effect Predictor software was used. Further on, the DAVID software was applied to associated genes with KEGG pathways. The second step comprised the estimation of pathway effects using a linear model. Based on the t-test and 1% significance level the most significant KEGG pathway for MSP was prostate cancer: *bta05215* (P=0.0002) and for TEM glycosylphosphatidylinositol (GPI)-anchor biosynthesis: *bta00563* (P=0.008).

Whole-genome sequence association study on birth weaning and carcass traits in Charolais breed

E. Venot[1], S. Taussat[2], A. Govignon-Gion[3], T. Tribout[1], M. Boussaha[1] and C. Hoze[2]
[1]Université Paris-Saclay, INRAE, AgroParisTech, GABI, 78350, Jouy-en-Josas, France, [2]Allice, 149 rue de Bercy, 75012, Paris, France, [3]Idele, 149 rue de Bercy, 75012, Paris, France; chris.hoze@allice.fr

National genomic evaluations have been running twice a year for the Charolais breed in France since 2015. More than 23,000 genotypes associated to phenotypes for traits recorded nationally on field and in slaughterhouses are available for this breed. It provides an opportunity to perform large-scale Sequenced-based Genome-wide association studies (GWAS) on beef traits. Deregressed proofs (DP) from polygenic evaluations were extracted for 2 birth traits: birth weight (BW), calving Ease (CE), 3 weaning traits: adjusted weaning weight at 7 months (AW), muscular (MD) and skeletal (SD) developments, and 3 carcass traits: age at slaughter (AS), carcass weight (CW) and conformation (CC). Maternal genetic effects were also considered for BW, CE, AW weaning traits (BWm, CEm, AWm respectively) in this study. Selection criteria based on minimum r^2 values or minimum number of progeny with performance for animal without performance were applied on these DP: in total 23,000, 2,250 and 3,600 DP were selected for weaning direct, weaning maternal and carcass DP respectively. 50K genotypes were firstly imputed to high-density (777K) using Fimpute3 and a reference population of 633 Charolais bulls and then to whole genome sequence data (WGS) using Minimac4. WGS imputation were based on the RUN7 of 1000 bulls genomes project containing 144 Charolais leading in a set of 25 million variants. GWAS were performed independently for each DP using GCTA, taking into account the population structure through a 50K-based genomic relationship matrix. We detected 109 significant variants (MAF>0.01 and -log10(p)>6). As expected, the number of QTL detected is higher for traits with a greater number of phenotypes. Numbers of QTL per traits varied from 1 (AWm) to 19 (SD) and 19 variants were significant for at least two traits. These results have led to the identification of variants of interest that could be used in the future to improve the accuracy of beef cattle genomic evaluations.

Genes affecting bovine milk mineral composition revealed from sequence-based GWAS

M.-P. Sanchez[1], P. Grosperrin[2], C. Hozé[3], M. Boussaha[1], A. Delacroix-Buchet[1], M. Brochard[4] and D. Boichard[1]
[1]Université Paris-Saclay, INRAE, AgroParisTech, GABI, 78350, Jouy-en-Josas, France, [2]Conseil Elevage 25-90, 25640, Roulans, France, [3]Allice, 75012, Paris, France, [4]Umotest, 01250, Ceyzériat, France; marie-pierre.sanchez@inrae.fr

Although representing a minor fraction of the cow's milk solids, minerals are very important for both technological and nutritional qualities of milk. To identify genomic regions affecting Ca, P, Mg, K and Na in milk, a genome-wide association study (GWAS) was carried out for mineral composition predicted from mid-infrared spectra in Montbéliarde (MO) cows. Data from cows with at least three test-day records during the first lactation (1,506,037 test-day records from 194,934 cows) were adjusted for non-genetic effects and averaged per cow. 50K genotypes, available for a subset of 19,862 cows, were imputed at the HD level (777K genotypes of 522 MO bulls) and then, at the sequence level using 25 millions of sequence variants from the Run7 of the 1000 bull genomes project (1,823 animals including 63 MO bulls). We tested the effect of each sequence variant using a mixed model including also a mean and a random polygenic effect. We found 18, 12, 13, 13, and 17 genomic regions with significant effects ($-\log10(P)\geq8.2$) on milk Ca, P, Mg, K, and Na contents, respectively. All these regions explained around one-third (Ca, P, K, and Na) to two-thirds (Mg) of the genetic variance of the trait and many candidate genes, including SLC37A1, GCK, FSCN3, GRAMD4, SEL1L3, GAK, GID8, RAB6A, FADS6, ANKH, SKA1, and LMAN1, were identified. We found the most significant effects for variants located in the SLC37A1 and ANKH genes, both encoding transmembrane proteins involved in ion transport. For most of these genes, we were able to propose candidate variants that are likely to be either causative or in strong linkage disequilibrium with causative variants. This study, conducted at a very large-scale (19,862 cows with phenotypes and genotypes) on imputed whole-genome sequences, led to the identification of variants located in genes functionally related to milk mineral composition that help to understand its genetic architecture and could be included in genomic prediction to improve the technological and nutritional properties of the bovine milk.

Optimal investment in phenotyping and genotyping in dairy breeding

J. Obšteter[1], J. Jenko[2], J.M. Hickey[3] and G. Gorjanc[3]
[1]Agricultural Institute of Slovenia, Department of Animal Science, Hacquetova ulica 17, 1000 Ljubljana, Slovenia, [2]Geno Breeding and A.I. Association, Storhamargata 44, 2317 Hamar, Norway, [3]The Roslin Institute, University of Edinburgh, Easter Bush, Midlothian, EH259RG, Edinburgh, United Kingdom; jana.obsteter@kis.si

Genomic selection increases genetic gain by reducing generation interval and increases accuracy of early selection. However, breeding programmes have to ensure continuous funding to support it. This first requires a big investment to create an initial training population. An alternative is to use an international training population, if available. In the following years, the breeding programmes have to provide funds for a yearly update of the training population to avoid a drop in prediction accuracy. Here we propose reallocation of a part of the funds from repeated phenotyping in milk recording to genotyping, because repeated milk records increase the accuracy with diminishing return. We evaluated this proposal with a comprehensive simulation of a dairy cattle population under 36 genomic scenarios against a conventional scenario, all at equal costs. The conventional scenario used progeny testing and 11 phenotype records per lactation, while genomic scenarios reallocated some phenotyping resources to genotyping. The genomic scenarios differed in: (1) whether an initial training population was available; (2) the ratio of the price of phenotyping and genotyping; and (3) the number of phenotype records per cow per lactation. We compared the scenarios by genetic gain and underlying accuracy of selection. Despite reduced phenotyping, genomic scenarios with an existing initial training population increased the genetic gain of the conventional scenario up to 143%. Genetic gain increased with increasing investment into genotyping. Genomic selection increased the accuracy for un-phenotyped selection male and female candidates, and dams. In scenarios without an initial training population we first accumulated 2,000 genotypes to start genomic selection. Although this resulted in slightly smaller genetic gain, it did not change the overall trend for genetic gain or prediction accuracy, and increased the genetic gain of the baseline up to 134%. The results will help breeding organisations to optimise resources to implement genomic selection or improve the economic efficiency of their programmes.

Impact of culling reasons on estimated genetic parameters for longevity in North American Angus

H.R. Oliveira[1,2], S.P. Miller[3], L.F. Brito[1] and F.S. Schenkel[2]
[1]*Purdue University, Department of Animal Sciences, 610 Purdue Mall, 47907 West Lafayette, Indiana, USA,* [2]*University of Guelph, Centre for Genetic Improvement of Livestock, 50 Stone Road East, N1G-2W1 Guelph, Ontario, Canada,* [3]*American Angus Association, 3201 Frederick Avenue, 64506 Saint Joseph, Missouri, USA; holivier@uoguelph.ca*

Cattle longevity depends on voluntary and involuntary culling performed by individual farmers. The main goal of this study was to evaluate the impact of different culling reasons on estimated genetic parameters for longevity in North American Angus. Culling reasons (and number of cows; n) considered were: (1) natural death (n=150,229); (2) structural problems (n=24,804); (3) disease (n=4,994); (4) fertility (n=154,419); (5) performance and/or temperament (n=62,005); (6) miscellaneous (n=208,092); and (7) all reasons (n=604,543). Single-trait random regression linear models were used to analyse the binary records (0 = culled and 1 = alive) created for each culling reason dataset. The statistical models used included the systematic effect of embryo transfer, a systematic regression for year-season of birth, and random regressions for herd-year-season of birth, additive genetic, and permanent environmental effects. Fourth order Legendre orthogonal polynomials were used for all systematic and random regressions (cows from 2 to 15 years-old). Bayesian method via Gibbs sampling was used in all analyses. Range and pattern of heritability estimates changed according to the different culling reasons. In general, heritabilities estimated over time for cows culled due to structural problems (0.01 to 0.46) and disease (0.01 to 0.41) were higher than the heritabilities estimated for animals culled due to fertility (0.01 to 0.12), performance and/or temperament (0.01 to 0.17), and miscellaneous (0.01 to 0.09). For natural death, intermediate heritabilities (0.06 to 0.24) were estimated over time. When combining all classes of culling reasons together, heritabilities ranged from 0.07 to 0.12. These results suggest that the different culling reasons affect the heritability estimates for longevity, and they should be taken into account in the genetic evaluation for this trait. The best trait definition in terms of culling reasons will likely depend on the ultimate breeding goal.

Genomic models considering G×E: modelling heterogeneous SNP variances

B. Gredler-Grandl and M.P.L. Calus
Wageningen U&R Animal Breeding & Genomics, Wageningen Livestock Research, Droevendaalsesteeg 1, 6708 PB Wageningen, the Netherlands; birgit.gredler-grandl@wur.nl

One of the aims of the Horizon 2020 project GenTORE (grant agreement no. 727213) is to develop models to improve resilience and efficiency traits in dairy and beef cattle. Resilient genotypes are able to respond to environmental perturbations, therefore being of high importance when genotype by environment interactions (G×E) exist. G×E is typically modelled using a multi-trait approach where the same trait measured in different environments is considered a different, but correlated trait. An alternative is to model G×E with reaction norm models where the breeding values are modelled as a function of the environment defined as a continuous variable. Estimated genetic parameters of both models provide inside scaling (i.e. differences in variances) and re-ranking (i.e. genetic correlations lower than unity) across environments. Genetic parameters for both approaches can be estimated using genomic information, e.g. with GREML or random regression on SNP genotypes (RR-REML) where homogeneous (co)variances are assumed for all SNP across the genome. Since specific regions in the genome may harbour QTL and others may not or loci may have a large effect in one environment and a zero effect in another, the assumption of equal (co)variances across the genome is violated. The objective of this study is to develop genomic prediction models considering G×E allowing for heterogeneous SNP (co)variances across the genome and environments. SNP specific (co)variances allow for varying G×E effects across the whole genome. We propose to make SNP (co)variances heterogeneous by weighing. We suggest to split the data set of interest in two subsets and then follow a two-step approach: (1) estimate SNP effects in the first data set and calculate SNP (co)variances based on the estimated SNP effects and (2) weigh the SNP genotypes using the estimated (co)variances in (1) and simultaneously compute SNP effects and (co)variances in the second data set. We will test the two-step approach for both a genomic multi-trait and reaction norm model implemented in a RR-REML approach. The models will be tested on simulated data first and later will be evaluated on an empirical data set.

Citizen attitudes towards genome editing in farm animals

G. Busch[1], E. Ryan[2], M.A.G. Von Keyserlingk[2] and D.M. Weary[2]
[1]*Georg-August-University Göttingen, Platz der Göttinger Sieben 5, 37073 Göttingen, Germany,* [2]*University of British Columbia, 2357 Main Mall, V6T 1Z4 Vancouver, Canada; gesa.busch@agr.uni-goettingen.de*

The long-term success of new technologies in food production including the genetic modification (GM) of plants and animals depends considerably on how the public responds to these technologies. GM of plants and animals is one of the most promising innovation but also one of the most contentious topics in food production today. Through the emergence of new technologies in the field of GM, especially through the development of genome editing techniques, public concerns will increase as possible areas of application increase. Studies analysing public response to GM and GE mainly focus on the technique whereas studies analysing attitudes towards different applications are scarce. Especially in the field of farm animals, insides into people's attitudes and underlying reasons are missing. Therefore, this study analyses how people from the broader public view different applications in farm animals, namely resistance towards PRRSV in pigs, the production of allergen-free dairy milk and double muscle growth in beef cattle. We further analyse different citizen segments with regard to their perceptions of whether this application should be allowed, risks, benefits and naturalness of GE. To compare attitudes, disease resistance in humans (HIV) and wheat plants (mildew) are added as further scenarios. A total of 3,698 participants from Austria, Canada, Germany, Italy and the US were questioned online. Overall, the highest acceptance of all tested applications gains the HIV-resistance in humans (μ=5.01; scale: 1=very wrong to do over 4=neither wrong nor right to 7=very right to do), followed by mildew resistance in wheat (μ=4.70) and by PRRSV resistance in pigs (4.35). Allergen-free milk in dairy cattle (μ=3.90) and especially the double muscle growth in beef cattle (μ=3.23) show the lowest support. Comparing countries, the Canadian, US and Italian participants show similar attitudes, while Austrian and German attitudes are similar but differ from the others. All in all, four citizen segments could be detected that differ largely in their attitudes towards GE, especially with regard to moral concerns, risks and benefits perceptions.

SNPs selection for differentiation of wild and domestic reindeer populations

V.R. Kharzinova[1], A.V. Dotsev[1], H. Reyer[2], K. Wimmers[2], G. Brem[1,3] and N.A. Zinovieva[1]
[1]*L.K. Ernst Federal Science Center for Animal Husbandry, Moscow region, Dubrovitcy, 60, 142132, Russian Federation,* [2]*Institute of Genome Biology, Leibniz Institute for Farm Animal Biology, Wilhelm-Stahl-Allee 2, Dummerstorf, 18196, Germany,* [3]*Institute of Animal Breeding and Genetics, Veterinärplatz 1 Geb: NA/VIII,Wien, 1210, Austria; veronika0784@mail.ru*

Reindeer (*Rangifer tarandus* L., 1758) is a very important species for the normal functioning of ecosystems and for the life of indigenous peoples. The species is considered to be in an early phase of domestication and often coexists with the wild reindeer. This co-occurrence causes serious problems, namely the inclusion of domestic reindeer to the wild herds and the extermination the wild reindeer as domestic one. The ability to differentiate these forms through molecular genetic approaches would improve the process of domestication, wildlife management and help solving forensic problems. While a clear trend towards the use of single nucleotide polymorphisms for those targets, it is advantageous to select SNPs that might best discriminate studied populations. It is known that genetic markers with the largest genetic distances between populations – determined by applying the genetic distance F_{ST} – are the best candidates for differentiation. Here we aimed at selection a minimum set of SNPs required for distinction of wild and domestic reindeer. A total of 180 (wild, n=46; domestic, n=134) were genotyped with the Illumina Bovine HD BeadChip. Pairwise fixation index (F_{ST}) values were estimated in the R package StAMMP. Candidate SNPs were defined as passing the F_{ST} threshold 0.03. Principal Component Analysis (PCA) based on both the final data set of SNPs and SNP set, selected by F_{ST} values was performed using PLINK v1.90. After filtration, 7,649 SNPs were remained. Further, based on average pairwise F_{ST} values ranged from 0.303 to 0.464, 16 SNPs were detected. Using whole set of polymorphic SNPs allowed clearly differentiating the wild population from domestic reindeer. The similar pattern but with weaker differentiation was observed by the set consisting only of 16 SNPs. A genotyping using reduced amount of SNPs is economically more attractive, and it is expected that the set of 16-SNPs will be robust enough to differentiate reindeer individuals that is the main element of both for reindeer herding and for the stability of the wild population. This work was funded by the Russian Science Foundation Pr. no.16-16-10068.

Preliminary embryo mortality analysis for birth weight homogeneity in mice

N. Formoso-Rafferty[1], J.P. Gutiérrez[2], R. Martín-Amores[2], K.D. Arias[2], M. Arias-Álvarez[2] and I. Cervantes[2]
[1]Universidad Politécnica de Madrid, E.T.S.I.A.A.B. Producción Agraria, C/ Senda del Rey 18, 28040 Madrid, Spain,
[2]Universidad Complutense de Madrid, Facultad de Veterinaria. Producción Animal, Avda. Puerta de hierro s/n, 28040 Madrid, Spain; n.formosorafferty@upm.es

Selecting for litter homogeneity is becoming to be included as one of the selection objectives in genetic breeding programs because it has been shown to be related to higher robustness. A divergent selection experiment for birth weight variability in mice during 23 generations has shown that homogeneous animals are less heavy but having good robustness indicators such as higher litter size, survival and feed efficiency. Differences in litter size between lines could be attributed to the ovulation rate and/or embryo survival. The objective of this work was to compare these components in both divergently selected lines. At day 14 of gestation, 6 females from the high variability line and 10 from the low variability line having a vaginal copulation plug the day after mating, were sacrificed. The genital tract was collected and both horns were measured from the uteral-tubal junction to the oviduct and the number of foetuses were counted. Ovaries were examined in a stereo microscope and the *corpora lutea* were counted (ovulation rate). The ratio between number of foetuses and number of *corpora lutea* were computed (embryo survival). Differences between means in high and low line were -2.07 foetuses, 0.97 hydatidiform moles of the uterus, 0.60 *corpora lutea*, -23.5 mm sum of the horn lengths, -0.2 foetuses/*corpora lutea* ratio, and 2.67 ovules lost from *corpora lutea* to foetuses at 14 days, being the last three significant (P<0.05). Consequently, low variability line had longer uterine horns with higher embryo survival, despite the number of *corpora lutea* was no significant but higher in the high line. Thus, the higher litter size of the low variability line would be due to embryo robustness. The number of hydatidiform moles in the uterus was 2.17 in the high line and 1.20 in the low line on average, suggesting also higher survival in the low variability line. According to these preliminary results, the low variability line presented important reproductive advantages in terms of higher embryo survival, suggesting higher robustness and animal welfare.

Genetic relationship between birth and weaning weight environmental variability in a mice population

N. Formoso-Rafferty[1], J.P. Gutiérrez[2], M.J. Chacón[2], L. Bodin[3] and I. Cervantes[2]
[1]Universidad Politécnica de Madrid, Departamento de Producción Agraria, 28040 Madrid, Spain, [2]Universidad Complutense de Madrid, Departamento de Producción Animal, 28040 Madrid, Spain, [3]INRA, GenPhySE, 31320 Castanet-Tolosan, France; gutgar@vet.ucm.es

A divergent selection experiment for birth weight environmental variability in mice has been successfully performed during 23 generations. Genetic control of the birth weight environmental variability was possible and had direct implications in other interesting traits in livestock. Moreover, it has been revealed that selecting to decrease the environmental variability in the birth weight is beneficial for production, welfare and robustness. But weaning weight is more frequently used as selection criterion in breeding programmes than birth weight. Therefore, the objective of this study was to analyse a possible correlated response in the environmental variability for weaning weight in the divergently selected mice population for birth weight environmental variability. Data included 24,741 birth and weaning weight pairs and 26,764 animals in the pedigree. The model included generation (23 levels), sex (2 levels), parturition number (2 levels) and litter size (16 levels) as fixed effects, and the litter random effect (2,955 levels) besides genetic and residual. The trait was assigned to the mother and same effects were fitted affecting the trait and its variation. The bitrait analysis was done using ASReml 4.1, solving with the double hierarchical generalised linear model. The genetic variance for the environmental weaning weight was 0.10±0.02 being the correlation between mean weaning weight and its variation -0.40±0.08, negative, contrary to the positive correlation found between birth weight and its variation. The correlation between both mean traits was high and positive (0.87±0.03). The genetic correlation between both environmental variabilities was 0.97±0.02 indicating that selection for weaning weight homogeneity would imply also birth weight homogeneity but higher weaning weight.

Biological interpretability of SNP effect size classes in Bayesian genomic prediction models

F. Mollandin, A. Rau and P. Croiseau

INRAE, Allée de Vilvert, 78350, Jouy-en-Josas, France; fanny.mollandin@inrae.fr

Non-linear Bayesian models represent an attractive approach to perform genomic prediction, due in part to their flexibility and ability to perform variable selection. In recent years, a suite of models in the so-called Bayesian alphabet have been proposed to find a compromise between reality, which may correspond to an omnigenic model for complex traits, and computational convenience. In particular, the BayesR model strikes this balance by modelling single nucleotide polymorphisms (SNPs) as a mixture of markers with null, small, medium, or large variance. The BayesRC in turn extends this approach by allowing different a priori categories of SNPs, as defined by previous biological knowledge, to be represented by different mixtures of effects. Although both of these approaches have been shown to achieve improved prediction accuracy in a variety of scenarios, there is still a need to evaluate the extent to which the assignment of SNPs to specific effect size classes (small, medium, large) reflects the true underlying genetic architecture and is meaningful for downstream SNP selection. In this work, we generate several sets of simulated data based on a real set of 50K genotype data in Montbéliarde bulls with a known pedigree, thus preserving the true linkage disequilibrium present in the data. We consider a wide variety of genetic architectures, phenotype heritabilities, and polygenic variances, and we illustrate under which conditions the BayesR and BayesRC models are able to recover known SNPs while also achieving high phenotypic prediction accuracy. Finally, we provide some insight into how the biological interpretation of the SNP classes identified by the BayesR and BayesRC models can be further refined through the incorporation of additional biological a priori information. This work is part of the H2020 project GENE-SWitCH.

Modelling maternal effects on late measured traits in broilers

H. Romé[1], D. Marois[2], C.H. Huang[2] and J. Jensen[1]

[1]Aarhus University, Blichers Alle 20, 8830 Tjele, Denmark, [2]Cobb-Vantress, P.O. Box 1030, 72761-1030 Siloam Springs, USA; helene.rome@mbg.au.dk

The additive genetic effect and the permanent environmental effect of the dam (pem) might affect performances of offspring in broilers. That is particularly true for juvenile performance. Often those effects are ignored in chicken for breeding value estimation (EBV) even if it can affect accuracy of EBV and most likely inflation of those EBV. In this study, we focused on different ways to model maternal effects to predict breeding value of late body weight in a commercial line of broilers. Population consisted of males and females farmed in commercial environment. Body weight (BW) were recorded weekly (week1 to week 5). In a previous study, we showed the importance of fitting fixed effect for dam (i.e. age of the dam) and the pem into the model for genetic evaluation. Adding the maternal additive genetic effect were not significant at later age but was needed at earlier age. In this study, we focused on BW at week 5. Our conventional model ('BW5') considered fixed and random genetic effects for the individuals as well as fixed effects for the age and source of the dam, the pem and the maternal genetic effect. Since maternal effect is bigger at earlier age, we studied the possibility to control maternal effect in week5 by adding information from week 1. Two models were tried. The first ('BW5-BW1') included the fixed and random effect for maternal effect and a regression of BW at week 1 on BW at week 5. The second model ('BW5-EBV1') was similar but instead of BW at week 1, the EBV estimated at week 1 was fitted into the regression. The model 'BW5' showed an accuracy of 0.49 and an inflation of EBV of 0.77. Model 'BW5-BW1' did not show any improvement in term of accuracy and inflation of EBV (accuracy of 0.46 and inflation of 0.74), compare to model 'BW5'. On the contrary, model BW5-EBV1 showed an improvement in term of accuracy and most importantly a reduction of the inflation of EBV (accuracy of 0.53 and inflation of 0.99). This showed that it is possible to account for maternal effect in later BW by considering juvenile trait into the evaluation model. Other ways to improve modelling of maternal effects should be explored.

Predicting sow longevity using pure- and crossbred data

M.W. Iversen[1,2], Ø. Nordbø[2,3], E. Gjerlaug-Enger[2], E. Grindflek[2] and T.H.E. Meuwissen[1]
[1]Norwegian University of Life Sciences, P.O. Box 5003 NMBU, 1432 Ås, Norway, [2]Topigs Norsvin, Storhamargata 44, 2317 Hamar, Norway, [3]GENO SA, Storhamargata 44, 2317 Hamar, Norway; maja.iversen@norsvin.no

Survival and longevity are very important traits in pig breeding. From an economic standpoint it is favourable to keep the sows for another parity instead of replacing them, and from an animal's perspective, better welfare is achieved if they do not experience health problems. In purebred nucleus herds, it is challenging to collect data on longevity, because sows are normally replaced based on breeding value rather than inability to produce, and replacement rates are high. Crossbred sows are however less subjected to high replacement rates and are more likely to be kept in the herd for longer if they can produce large and robust litters. Therefore, the objective of this study was to investigate whether the use of crossbred phenotypes could improve prediction accuracy of longevity in purebreds. In addition, an alternative definition of survival was investigated. The data included phenotypes from two purebred dam lines and their F1 cross. Three traits were evaluated; 1) whether or not the sow got inseminated for a second litter within 85 days of first farrowing (LGY12), 2) how many litters the sow can produce within 570 days of first farrowing (LGY15), and 3) a repeatability trait that indicates whether or not the sow survived until the next parity (Survival). Traits were evaluated both as the same across breeds and as different between breeds. Results indicated that longevity is not the same trait in purebred and crossbred animals (low genetic correlation). In addition, there were differences between the purebreds in terms of which trait definition gave the greatest prediction accuracy. Prediction accuracy for crossbreds was generally poor. The Survival trait is recorded earlier in life than LGY15 and seemed to give a greater prediction accuracy for young animals than LGY15. Thus, for selection of young animals for breeding, Survival would be the preferred trait definition. In addition, results indicated that lots of data were needed to get accurate estimates of breeding values, and that if crossbred performance is the breeding goal, crossbred phenotypes should be used in the genetic evaluation.

EWRAB – a program to calculate economic values for rabbit breeding

Z. Krupová, E. Krupa and M. Wolfová
Institute of Animal Science, Přátelství 815, 10400 Prague Uhříněves, Czech Republic; krupova.zuzana@vuzv.cz

Rabbits are popular source of healthy meat with high nutritional value. A little profit usually generated form rabbitry can be solved by genetic improvement of performance through animal selection. Economic values and economic weights representing the economic importance of the traits and are useful for making the objective decision for selection purposes. Therefore, the comprehensive bio-economic model of the EWRAB software was developed to calculate the economic importance for complex of 20 production, reproduction, health, carcass and feed efficiency traits. In the model, these production systems can be applied: purebreeding system, two-way crossing system (applying crossbreeding for part of doe herd), and three-way crossing system (commercial systems with terminal crossing). Markov chain methodology was applied there to calculate the stationary state of doe herds. To compute the feed costs in certain time periods of animal the energy and protein requirement based on animal weight, growth rate, and doe reproductive status was considered. The economic efficiency of the given production system was expressed as profit per doe per year (calculated as difference between total revenue and total costs). The economic values (expressed per doe per year) calculated for some of the evaluated traits in a typical Czech commercial rabbitry were as follows: -4.29€/kg of live weight of kits at the age of 21 days, 1.20€/kg of daily gain in fattening, 8.37€/kid born alive per litter, 1.14€/% of kids survival from birth till weaning, 8.26€/productive life of doe, 0.60€/% of doe conception rate, 3.88€/% of dressing percentage of slaughtered rabbit and -30.70€/kg of feed conversion during fattening. These values in combination with the traits genetic standard deviations can be further utilised for selection purposes. Moreover, the effect of various production, management, and economic circumstances on the economic efficiency of a rabbit-production system can be evaluated by the software.

Accounting for Manech information in Latxa genetic evaluations

C. Pineda Quiroga[1], A. Legarra[2], J.M. Astruc[3] and E. Ugarte[1]
[1]NEIKER, Department of Animal Production, Campus Agroalimentario de Arkaute, 01080, Spain, [2]INRA, GenPhySE, Castanet Tolosan, 31326, France, [3]Institut de l'Elevage, Castanet Tolosan, 31321, France; cpineda@neiker.eus

Exchange of genetic material between Manech Tête Rouge (MTR) dairy ewes from France, and Latxa Cara Rubia (LCR) ewes from the Basque Country (Spain), has systematically occurred during last ten years. This was possible because of the genetic and phenotypic similarity between breeds, which allows make the best of the MTR superiority. Up to date, genetic differences between both populations have been considered in LCR's genetic evaluation by segregating animals according to their origin country assigning them into genetic groups. To better predict breeding values (EBV), an approach including external MTR information by means of pseudo-records of fictitious daughters has been implemented. For this, milk yield EBV's and its reliability (r) from 395 MTR rams were collected, as well as the genetic variance (σ_a) and heritability (h^2) of milk yield in both populations. Considering h^2 and r values, a fictitious pedigree with 163.546 animals was created, all of which had their respective performance data simulated based on EBV, σ_a from MTR and LCR. All of these generated animals were assigned to a genetic exclusive group. Finally, those simulated data set were added to LCR's pedigree and performance files. Similarly, a conventional LCR data set was prepared. Thus, joint and conventional genetic evaluations, using a single trait repeatability model by BLUP were carried out, and the resulting EBV and r compared. Results showed a significant correlation of EBV's between both evaluations, as well as between ranking of animals (corr>0.95, P<0.05), indicating that the estimated values of both models are highly comparable. A slightly increase of r in the joint evaluation was also found in comparison with the conventional evaluation (0.61±0.18 vs 0.59±0.19). The main change was related with the selection of ewes as mothers of rams to be use for artificial insemination, since the use of the joint evaluation allowed for using 5% more of Latxa mothers, suggesting that classical evaluation overestimate the French genetic. In light of these findings, we can infer that the joint evaluation offers more correctness in the estimation of the genetic results.

Identifying genes and pathways involved in direct calving ease in Holstein Friesian cattle

M. Jakimowicz
Biostatistics Group, Department of Genetics, ul. Kożuchowska 7, 51-631 Wrocław, Poland; michalina.jakimowicz@upwr.edu.pl

Holstein Friesian (HF) cattle is one of the most important cattle breeds in Poland. It is very important to keep cows in good health and enough big size population because of genetic diversity. And it's necessary to know also the genetic background of the most important traits. Knowledge of the genetic background helps to design better breeding programs and this will led to improve polish HF population. Consequently, the main aim behind the study was to check which single nucleotide polymorphisms (SNPs) influence significantly direct calving ease (DCE) in HF cattle. To find genotype-phenotype associations, genome-wide association study and pathway analysis were performed. In the result of the GBLUP model, 2,128 SNPs have a significant influence on DCE, but after false discovery rate (FDR) corrections only 19 of them were considered as significant and therefore used in further analysis. Seven out of 19 were mapped to genes by variant effect predictor (VEP). To four of these genes, functional information in GeneOntology database was available. Also, biological pathways in Kyoto Encyclopedia of Genes and Genomes database were found. From four genes three was found in two KEGG pathways. In KEGG database 289 biological pathways where found. To check if all of these pathways are significant we performed linear regression. In the results, only 31 from 289 KEGG pathways appeared significant.

Investigations of equine chromosome 3 to identify the causal mutation for roan in Icelandic horses

K. Voß[1], I. Blaj[1], D. Becker[2] and G. Thaller[1]
[1]Institute of Animal Breeding and Husbandry, Hermann-Rodewald-Straße 6, 24118 Kiel, Germany, [2] Leibniz Institute for Farm Animal Biology (FBN), Wilhelm-Stahl-Allee 2, 18196 Dummerstorf, Germany; kvoss@tierzucht.uni-kiel.de

Among the variety of different coat colours and patterns in Icelandic horses, roan is characterised by intermingled white and coloured hair on the horse's body while head and distal legs are unaffected. Though a dominant inheritance has been reported the causative mutation is still unknown. Interestingly, if horses are homozygous, an embryonic lethality is presumed based on a survey on Belgian draught horses. Roan was mapped to equine chromosome 3 (ECA 3). More precisely it appears to be part of the equine linkage group II, which consists of two further colour genes (Extension, Tobiano) and genes for albumin, esterase and Vit-D-binding. In addition, roan is associated to *KIT* in different breeds. Polymorphisms in *KIT* produce a multitude of white and spotted phenotypes, therefore, it is a major candidate for roan as well. Similar to Dominant White in horses, an allelic heterogeneity could be responsible for roan. This study aims to decipher the genetic background of roan in the Icelandic breed. Besides descriptive statistical analyses matings of roan animals to investigate the lethality hypotheses for Icelandic horses and an association study verifying the linkage to *KIT*, two roan Icelandic horses (one heterozygous, one marker-tested homozygous for roan) were whole-genome sequenced to identify causal candidate variants. Sequences were aligned to EquCab3.0 and filtered for SNPs, small indels and structural variants. In total, approximately 300,000 SNPs and smaller indels plus 200 structural variants were identified on ECA 3 of which both horses have around 200,000 SNPs and 100 structural mutations in common. The number of potential candidates was further reduced with filtering for zygosity and excluding all breed-specific variants by adding the sequence of two non-roan Icelandic horses as an extra reference. After these filter steps, 5,500 SNPs and 35 structural variants remain as potential candidates for further investigation, e.g. comparison to additional non-roan horses and restriction of the analysis to the genomic region of interest. The investigated region will be reduced to linkage group II, the *KIT* gene and to regions that are homozygous in the homozygous roan horse on ECA3.

TechCare: Integrating innovative TECHnologies along the value Chain to improve small ruminant welfARE

C. Morgan-Davies[1], J.M. Gautier[2], I. Halachmi[3], E. Gonzalez-Garcia[4], L. Grøva[5], G. Molle[6], F. Kenyon[7], C. Dwyer[1], G. Caja[8], A. Rosati[9], E. Sossidou[10] and G. Lagriffoul[2]
[1]SRUC, West Mains Road, EH9 3JG Edinburgh, United Kingdom, [2]IDELE, BP 42118, 31321 Castanet Tolosan Cedex, France, [3]ARO, The Volcani Centre, 7505101 Rishon LeTsiyon, Israel, [4]INRAE, 2 Place Pierre Viala, 34000 Montpellier, France, [5]NIBIO, Gunnars veg 6, 6630 Tingvoll, Norway, [6]AGRIS, Viale Adua, 07100 Sassari, Italy, [7]MRI, Bush Loan, EH26 0PZ Penicuik, United Kingdom, [8]UAB, Plaça Cívica, 08193 Bellaterra, Spain, [9]EAAP, Via Giuseppe Tomassetti, 00161 Roma, Italy, [10]HAO, P.O. Box 60411, 570 01 Thermi, Thessaloniki, Greece; claire.morgan-davies@sruc.ac.uk

This paper presents a multi-actor approach project, *TechCare*, which aims at developing appropriate business models using innovative technologies to improve welfare management for all EU small ruminant (SR) systems, to enable consumers to choose animal welfare friendly products. SRs play a key socio-economic role in Europe, especially in harsh environments where innovative technology is not much implanted. SRs are often managed as a flock/herd level, along the value chain. The paper presents how *TechCare* will encapsulate stakeholders' expectations in terms of welfare and innovative technologies, and will provide them with adapted solutions, in a co-design approach. *TechCare* wll cover all stages of SR production and build novel welfare approaches to develop and validate tools, early warning systems, algorithms and indicators for efficient identification of welfare issues, including positive welfare, to include them in welfare management models. The innovative technologies identified in *TechCare* will pass different stages of validation, from prototyping to large-scale studies. Business models will be constructed and validated with stakeholders, ensuring a wide access to innovative technologies economically sustainable. *TechCare* will offer a durable improvement of SR welfare management using innovative technologies along the whole value chain and across the EU, and will act as an example in terms of approach and solutions for other species and production systems.

Effects of nematode parasitism on activity patterns in first-season grazing lambs

N. Högberg[1], L. Lidfors[2], A. Hessle[2] and J. Höglund[1]
[1]Swedish University of Agricultural Sciences, Department of Biomedical Sciences and Veterinary Public Health, Box 7036, 75007 Uppsala, Sweden, [2]Swedish University of Agricultural Sciences, Department of Animal Environment and Health, Box 234, 53223 SKARA, Sweden; niclas.hogberg@slu.se

We investigated the effects of gastrointestinal nematode (GIN) challenge on activity in first season grazing growing lambs exposed to two different levels of pasture born nematode infections during 7 days prior to and 49 days post weaning. The study consisted of two experimental periods, pre- and post-weaning. The pre-weaning period started when the animals were turned-out to a semi-natural pasture, where ewes and lambs were allocated to one out of two treatment groups (H and L). The pasture was naturally contaminated with nematodes the previous year, exposing both groups to overwintering strongyle larvae. Control groups (L) were dewormed monthly with 0.2 mg ivermectin (Ivomec® vet, oral suspension) per kg bodyweight. Post-weaning, lambs where allocated to one out of four groups based on sex (E = ewe; R = ram) and treatment group (HE, n=15; HR, n=15; LE, n=14; LR, n=14), grazing different but similar non-contaminated ley enclosures. Activity patterns were monitored by fitting animals in each group with IceQube loggers (IceRobotics Ltd). Bodyweight gain (BWG) was monitored weekly, whereas faecal samples were collected every four weeks for nematode faecal egg count (EPG) and parasitological determination using a validated Droplet Digital PCR protocol. All statistical analyses were performed in R (v. 3.6.2), using mixed models (NLME) with repeated measures. EPG was affected by treatment (P<0.0001) and was observed in both groups (L, H) from day -21 from weaning. Parasitological composition results are pending. Furthermore, BWG was affected (P<0.0001) by treatment. Preliminary results show a difference in lying time (P=0.017) 7 days post-weaning, with L animals lying 127 min shorter per day than H animals did. In addition, Motion Index (absolute value of the 3-D acceleration) was affected by treatment, with L animals showing a higher total activity (P=0.017) post weaning. In conclusion, our data supports that changes in activity patterns monitored with accelerometers could contribute to the identification of lambs challenged with GIN.

Rumen temperature of lactating dairy ewes under different ambient conditions by bolus sensors

A. Castro-Costa[1], J. Oliver[2], S. González-Luna[1], B. Chaalia[1], S. Serhan[1], A.A.K. Salama[1], C. Ferrer[2] and G. Caja[1]
[1]Universitat Autònoma de Barcelona, Animal and Food Sciences, Group of Research in Ruminants (G2R), Av. dels Turons s/n, 08193, Spain, [2]Microelectronics and Electronic Systems, Carrer de les Sitges, s/n, 08193 Bellaterra, Spain; gerardo.caja@uab.es

Rumen boluses for small ruminants were developed in the DairyCare Cost Action FA1308. The BioSens2 boluses (60 g and 22×90 mm) had a temperature sensor connected by high RF (433.4 MHz) to a programmable receiver. ATMEL microprocessors were configured for low power operation (Li battery, 3.3 V, 2.1 Ah, 2/3AA) and to collect data every 2 min. Rumen temperature (RT) signal was transmitted to a small transceiver (70×50×30 mm) placed at 5 m of the sheep. Six boluses were administered to 6 Manchega dairy ewes in late-lactation (1.14±0.06 kg/d; 67.5±1.2 kg BW) fed a total mixed ration 1× daily and milked 2× daily. Ewes were penned in a climatic chamber and submitted to thermal-neutral (TN; 20 °C, all the day) or heat-stress (HS; 37 °C day and 30 °C night) conditions, according to a cross-over (2 periods of 28 d each). Relative humidity (50%) and day-night (12-12 h) were maintained throughout the experiment. THI values were: TN (all day, 65) and HS (day, 87; night, 79). Water was at ambient temperature. DM intake decreased 20% (P<0.001) and water consumption increased 35% (P<0.001) in the HS ewes, whereas milk production did not vary (P=0.52), compared to TN. Recorded RT values changed according to ambient temperature and eating and drinking behaviour. Mean and extreme RT values were: TN (39.71±0.01; 38.99 to 40.31 °C) and HS (40.24±0.01; 39.49 to 40.67 °C), the difference being on average 0.53 °C (P<0.001). RT showed an increasing trend during the day, whereas decreased during the night in the HS. On the contrary, RT of the TN ewes decreased during the first 3-h after feeding, increased later and reached a plateau during the night. Dramatic drops of RT were observed in the TN ewes and linked to drinking bouts of cold water which were not observed in HS. In conclusion, the BioSens2 boluses were safely administered to dairy ewes that showed normal eating-drinking behaviour and lactational performances under extreme conditions. Acknowledgement: Ministry of Economy and Competitivity of the Govern of Spain (INIA-RTA2015-00035-C03).

Motivations and barriers for PLF uptake on European sheep farms

J.M. Gautier[1], C. Morgan Davies[2], T. Keady[3], G. Lagriffoul[1], S. Ocack[4], I. Beltrán De Heredia[5], A. Carta[6] and D. Francois[7]
[1]Institut de l'Elevage, PB42118, 31321 Castanet Tolosan Cedex, France, [2]Scotland's Rural College, Hill & Mountain Research Centre, Kirkton, Scotland, United Kingdom, [3]TEAGASc, Animal and Grassland Research and Innovation Centre, Athenry, Ireland, [4]TOGEN, Av. Fatma Gul GUVEN, Gaziantep, Turkey, [5]Neiker Tecnalia, Arkaute Agrifood Campus, E-01080 Vitoria-Gasteiz, Spain, [6]AGRIS-Sardegna, Research Unit: Genetics and Biotechnology, S.S. Sassari-Fertilia km 18,6, 07040 Olmedo, Italy, [7]INRA Occitanie, UMR GenPhySE, 24 chemin de Borde-Rouge, Auzeville CS 52627, 31326 Castanet-Tolosan Cedex, France; jean-marc.gautier@idele.fr

Electronic identification of small ruminants is mandatory since 2010 in Europe. Associated with a context of widespread use of 'connected' tools, the availability of solutions using new technologies to manage livestock and decrease workload, should become relevant for farmers. A survey was undertaken in the seven main EU sheep countries (France, Ireland, UK, Spain, Italy, Romania and Hungary) and Turkey (under two European projects: SheepNet and iSAGE) to determine the use of electronic identification (EID) associated technologies and barriers to the uptake of precision livestock farming (PLF) technologies. A total of 1,148 responses were collected and analysed. Sixty four percent of the respondents believe that EID and PLF are an opportunity for better flock/animal management but only 34% of them use it. This survey also highlighted the type of technologies used and the main motivations and barriers for PLF uptake. To date, in the sheep sector, new technologies are mainly related to drafting, animal location, concentrate feed management and performance testing. This is the first study undertaken at EU level targeting the sheep sector. It identifies the main gaps to tackle and proposes some pathways in order to foster the use of new PLF technologies.

Agroecological livestock farmers and technological innovation

M. Odintsov- Vaintrub, M. Chincarini, I. Fusaro, M. Giammarco and G. Vignola
University of Teramo, Veterinary Medicine, Localita Piano D'acio, Facolta di Medicina Veterinaria, 64100, Italy; modintsovvaintrub@unite.it

Agroecology is an approach to sustainable agriculture which aims to establish farming practices according to ecological principles. Currently, only a small percentage of institutional research activity is dictated to technological development of this field. Therefore, most of the innovation occurs on the farm level, with inventors increasingly sharing information and products within their local sphere of influence. In the current study, the innovators' activity was evaluated using a questionnaire distributed among agroecological communities. Questions included both personal (country, age, sex, etc.) and technical aspects. In total, 52 responses were collected and divided as follows: 14 adopted 'innovative farming techniques' (Farm practices, not technologies), 24 performed small modifications of existing products (fencing and feed), and only 11 introduced original innovations. Those inventions included: 4 regarding cattle (fencing, moving), 3 regarding poultry (light, water supply, and environment control), 2 decision-making aids (software), 1 for beekeeping and one for aquaculture (both regarding feeding). About 80% of the answers came from English speaking countries (the UK, US, and Canada), while the rest from Italy, Spain, Mexico and Kenya. This is an intrinsic bias of the work, as the questionnaire was composed in English and distributed among English speaking groups. A further key objective of the study is to expand the data collection to other languages and geographical areas. In total 6/11 original innovators provided technical detail, with energy supply to timers and actuators being the primary objective. Only one included a computerised ('Arduino') based system for environmental control (temperature/humidity). In conclusion, most of the innovation is decentralised among single individuals resulting in more basic levels of technology, centred mostly around adaptation of existing products and farming techniques. Original innovations are limited by skills, interest, and funds available for each individual farmer. However, with the growing exchange of data and information online, as well as simplified access to technology, a more dedicated product development can emerge as farmers' demand for technology grows.

Tracking individual broiler activity using a passive radio frequency identification system

M. Van Der Sluis[1,2], B. De Klerk[3], Y. De Haas[2], T.B. Rodenburg[1] and E.D. Ellen[2]
[1]Utrecht University, Unit of Animals in Science and Society, Faculty of Veterinary Medicine, P.O. Box 80163, 3508 TD Utrecht, the Netherlands, [2]Wageningen University & Research, Animal Breeding and Genomics, P.O. Box 338, 6700 AH Wageningen, the Netherlands, [3]Cobb Vantress, Koorstraat 2, 5831 GH Boxmeer, the Netherlands; malou.vandersluis@wur.nl

Individual activity levels may be informative for health and welfare of broilers. However, identification and tracking of individual broilers that are group-housed can be a challenge. A possible solution is to implement sensor technology. This study aimed to validate a passive radio frequency identification (RFID) system, that could be implemented from day one of life, for tracking of individual broiler activity. The broilers' pen was fitted with a grid of high frequency antennas and all broilers were fitted with an RFID tag on one of their legs on day 1 of placement. With this setup, the location of broilers could be recorded over time, allowing calculations of approximate distances moved. To validate the RFID system, top-view video recordings of the pen were made, to compare RFID and video in terms of recorded locations, and distances. For the location comparison, four videos taken at different ages, with a duration of approximately seven minutes each, were converted to separate images with a rate of one frame per second. Four colour-marked birds were annotated in these images to determine their location, i.e. at which antenna they were present. These annotations were compared to the RFID recordings. For the distance comparison, approximately one-hour video recordings taken on eight days were analysed using Kinovea to assess the distances moved on video for four colour-marked birds. These distances were compared to RFID-calculated distances. Preliminary results indicate that approximately 62% of the RFID registrations exactly matched with video observations, and that there was a strong rank-correlation of 0.91 between the distances moved on video and as calculated from the RFID data. Overall, it appears that the RFID system is suitable for tracking activity of individual broilers and can facilitate studies of activity of broilers throughout life.

Automatic broiler temperature measuring by IR camera a field trial in a commercial broiler-house

S. Druyan[1], N. Barchilon[1,2] and I. Halachmi[2]
[1]Agricultural Research Organization (ARO), The Volcani Center, Poultry and Aquaculture Department, Animal Sci Inst., 68 HaMakkabbim Road, Rishon Le Ziyyon P.O. Box 15159, 7528809, Israel, [2] Institute of Agricultural Engineering, Agricultural Research Organization (ARO), The Volcani Center, Precision livestock farming (PLF) Lab., 68 HaMakkabbim Road, Rishon Le Ziyyon P.O. Box 15159, 7528809, Israel; shelly.druyan@mail.huji.ac.il

The modern broilers have a low capability to maintain adequate dynamic steady-state mechanisms in the body, and to balance energy expenditure and body water balance under extreme environmental conditions. Moreover, even in temperate-climate, broilers production is negatively affected by heat, due to higher growth rate, which increase heat generation. In order to provide the appropriate micro-climate, modern broiler-houses are equipped with climate control system, with sensors distributed in the broiler-house space measuring environmental parameters around the broilers. Consequently, broiler-house climate control systems might not respond to the actual broiler needs, affecting their performance and thus reducing the efficiency. In this research we tested and validated a new system, which measures the body temperature of an individual broiler, in commercial broiler farms. The system is based on a low-cost IR camera calibrated by a thermistor, algorithm for the IR image processing and the Lasso regression model predicting the actual body temperature. Two thermal systems were placed in two identical houses within a commercial broiler farm. Temperature threshold values were set according to the broiler growth and broilers individual temperatures were monitored from day 11 onward. High consistency in the data acquisition was found between the two systems, which successfully identified an increase in broilers temperature which had crossed the upper critical temperature threshold. The successful identification of heat stress in the broilers has important implications on management of broilers welfare requirements. The results show that this system could be effectively used as a temperature sensor in the climate control loop in broilers houses.

In vivo estimation of body composition: comparison of eight methods in dairy goats

A. De La Torre[1], J. Pires[1], C. Huau[2], M. Monziols[3], C. Xavier[4,5], L. Louis[6], Y. Le Cozler[5], P. Faverdin[5], P. Lamberton[5], I. Chery[7], D. Heimo[4], C. Lonckle[8], P. Schmidely[8] and S. Lerch[4]
[1]INRAE, UMRH, Saint-Genès-Champanelle, 63122, France, [2]INRAE, GenPhySE, Castanet-Tolosan, 31320, France, [3]Ifip, Le Rheu, 35650, France, [4]Agroscope, Posieux, 1725, Switzerland, [5]INRAE, PEGASE, Saint-Gilles, 35590, France, [6]INRAE, UMR SiLVA, Champenoux, 54280, France, [7]Institut Pluridisciplinaire Hubert Curien, UMR7178, Strasbourg, 67037, France, [8]INRAE, MoSAR, Paris, 75005, France; anne.de-la-torre-capitan@inrae.fr

The objective was to test 8 methods for estimation of empty body lipid (EBL) and protein (EBP) mass in dairy goats. The methods tested on 20 Alpine goats (3 ± 0.6 years old; 226 ± 9 DIM; 47 to 72 kg of BW) were: adipose cell size; deuterium oxide dilution space (D2OS); 3-dimension (3D) imaging: whole body 3D-scan and automatic 3D assessment of body condition score (3D-BCS); manual sternal and lumbar BCS and ultrasound imagery; computer tomography and bioelectrical impedance spectroscopy (BIS). Simple and multiple regressions (Proc GLM, SAS) were tested between different variables and EBL and EBP mass measured by chemical analyses after slaughter. Perirenal adipose tissue mass and cell diameter combined with BW provided the most accurate predictive equations for EBL (R^2=0.95, residual coefficient of variation, rCV=12%). Nonetheless, such predictors can only be measured *post-mortem*. The best equations for EBL derived from *in vivo* variables included BW combined with 1/ the volume of fatty tissues measured by computer tomography (R^2=0.92, rCV=17%), 2/ the D2OS (R^2=0.91, rCV=19%), and 3/ the BIS (R^2=0.87, rCV=23%). D2OS combined with BW provided the best equation for EBP (R^2=0.97, rCV=3%), whereas BW alone provided a fair EBP estimate (R^2=0.92, rCV=4%). Manual BCS combined with BW provided good EBL and EBP estimations (R^2=0.80 and 0.94, rCV=28 and 4%, respectively). BCS is a non-invasive technique and does not require particular equipment, but it is subjective and prone to operator bias. Compared to manual BCS, 3D-BCS combined with BW slightly decreased the accuracy of the predictive equation for EBL (R^2=0.74, rCV=32%), and did not improve the estimation of EBP compared with BW alone. Ultrasound and whole body 3D imaging techniques were not satisfactory estimators of EBL and EBP ($R^2\leq0.40$).

Screening for sheep at high risk of fatty liver to reduce the prevalence of pregnancy toxemia

H. Dvir
Institute of Animal Science, ARO – Volcani Center, Rishon LeZion, Israel; haydvir@volcani.agri.gov.il

To address the global increase in demand for food, domestic ruminants have been intensively bred for high productivity, e.g. high prolificacy and milk yields. However, this strong selection has not been suitably matched with the proper metabolic capacity of the dam to cope with the growing energy demands. Prolific sheep and goats exemplify this kind of imbalance between high reproductive potential and limited physiological capacity to sustain it, since during late pregnancy the growing energy demands of their large litter size exceed those they consume in the diet. In severe cases, affected animals develop ketosis and fatty liver (FL) syndrome, which are the underlying metabolic pathologies of pregnancy toxaemia (PT) – the most common and frequently lethal metabolic disease of prolific sheep. For unclear reasons, certain ewes are more resistant to PT. Therefore, developing tools for the early identification of such individuals would help in the selection and culling decisions to improve farming productivity and profitability. In a search for potential biomarkers for fatty liver in sheep, we monitored key physical and biochemical parameters in growing lambs induced for hepatic steatosis. Briefly, lambs grew on a high calorie (HC) concentrate-based diet developed substantial hepatic steatosis compared to lambs grown on the lower calorie (LC) diet (P<0.0001). Consumption of the HC diet induced steady hyperglycaemia and hyperinsulinemia (P<0.0001). Within 4-months of the dietary treatment, the hyperglycaemic lambs presented with insulin resistance (i.e. higher HOMA-IR, P<0.0001), enhanced adiposity (P<0.0001), dyslipidaemia (P<0.05), and higher expression levels of common markers of inflammation. A stepwise regression analysis, with the liver-fat content as the response variable and measured physical and metabolic parameters as the explanatory variables, yielded only two significant effects; average glucose (P<0.0004) and HOMA-IR (P<0.035). Consistently, Pearson correlation analysis showed that blood glucose was the most strongly correlated parameter with steatosis (r=0.77; P<0.0001). The results suggest that a simple point of care measurements of blood glucose in ewe lambs may prove an attractive tool for selection and culling decisions to help minimise the prevalence of FL and PT.

The use of computer tomography to estimate reticulo-rumen content in Alpine goats

J. Pires[1], M. Monziols[2], A. De La Torre[1], P. Lamberton[3], C. Huau[4] and S. Lerch[5]
[1]INRAE, UMRH, Saint-Genès-Champanelle, 63122, France, [2]Ifip, Le Rheu, 35650, France, [3]INRAE, PEGASE, Saint-Gilles, 35590, France, [4]INRAE, GenPhySE, Castanet-Tolosan, 31320, France, [5]Agroscope, Posieux, 1725, Switzerland; jose.pires@inrae.fr

Estimation of reticulo-rumen content (volume and mass) is required in ruminant nutrition to determine effects of diet and environment on gut filling, nutrient turnover and to model digestive processes. Reticulo-rumen content is commonly measured via a rumen cannula. Animal scientists continuously seek to refine experimental procedures by developing less invasive techniques. The objective was to compare reticulo-rumen volume assessed by computer tomography (CT) with *post mortem* measurement of reticulo-rumen content mass in dairy goats. Twenty Alpine dairy goats (3±0.6 years old; 226±9 DIM) with body weight ranging from 47 to 72 kg were used. Goats were housed in a free-stall barn, had free access to hay and water, and were offered 0.75 kg/d of concentrate. Goats were anesthetised and placed in an inflatable mattress before duo CT scan (Siemens, Erlange, Germany) was performed. Between 400 to 500 images were generated per goat and analysed semi-automatically (www.turtleseg.org). The volume of reticulo-rumen and omasum were determined separately. Each goat was slaughtered within 15 min after CT, the reticulo-rumen and omasum were weighed full and after removing the digesta, and content weights were obtained by difference. The SAS GLM procedure was used to test simple regressions between organ volume obtained with CT and digesta mass measured post mortem. Volume of reticulo-rumen and omasum determined by CT were good estimators of digesta mass measured *post mortem* [R^2=0.72 and 0.73, residual standard deviation (rSD)=1.18 and 0.10 kg and residual coefficient of variation (rCV)=11 and 22%, respectively, n=20]. The regression was improved for omasum content when one individual with an extremely low digesta mass (22 g) was excluded (R^2=0.87, rSD=0.06 kg, rCV=12%, n=19). The use of CT may constitute a promising non-invasive method to estimate volume and mass of reticulo-rumem digestive content in small ruminants. Further research is warranted to test the ability of this technique to discriminate dietary treatments that induce differential rumen fill.

Net-zero strategy at the North Wyke Farm Platform: a backcasting analysis for transitional pathways

T. Takahashi, G.A. McAuliffe, P. Harris and M.R.F. Lee
Rothamsted Research, Okehampton, Devon, EX20 2SB, United Kingdom; taro.takahashi@rothamsted.ac.uk

With concerns for global warming greater than ever, determining the feasibility of carbon-neutral ruminant farming is an important task for humanity. While studies have shown that rewilding grassland could potentially offset climate impacts attributable to agriculture, the majority of these analyses are top-down in nature and fail to consider socioeconomic implications arising from such conversion. In particular, the trade-off between carbon sequestration potential of grassland and farm income that must be forgone is seldom investigated, leaving real-world pathways to realising recommended land use rather unclear. The objective of this study, therefore, was to quantify the three-way relationship between: (1) the proportion of land set aside for maximal carbon sequestration; (2) the amount of food producible from the remaining land, and (3) global warming potential (GWP) of the whole farm inclusive of land set aside. Life cycle assessment (LCA) inventory data were collected from the North Wyke Farm Platform, a system-scale cattle grazing trial in Devon, UK, from 2015-2016. Following the development of a baseline model, a fraction of pasture was gradually 'sacrificed' for new woodland and GWP was recalculated under each scenario. Carbon sequestration data for said woodlands were sourced from best available national publications. The analysis was first carried out under current levels of grazing efficiency, and then repeated iteratively to backcast productivity improvements required to achieve carbon neutrality whilst minimising economic impacts. The results showed that ~33% of pasture must be converted to woodland to achieve carbon neutrality under current efficiency. This reduction in farmed area was found to be too large to realistically maintain current food production exclusively by means of intensification. Further examination of the trade-off curve revealed that a combination of 20% land conversion and 20% productivity improvement could enable a GWP reduction by 70% while yielding 96% of current outputs, possibly offering an acceptable short-term balance for most stakeholders. These findings can inform quantitative R&D targets the grazing livestock industry should set into the future.

Milk yield and greenhouse gases emission in mountain dairy cattle systems: an ambiguous relationship

M. Berton[1], S. Bovolenta[2], M. Corazzin[2], L. Gallo[1], M. Ramanzin[1], C. Spigarelli[2], A. Zuliani[2] and E. Sturaro[1]
[1]University of Padova, DAFNAE, Viale dell'Università 16, 35020 Legnaro, Italy, [2]University of Udine, DISAA, Via Sondrio 2, 33100 Udine, Italy; marco.berton.1@unipd.it

The study aimed to analyse the relation between milk production and the emission of greenhouse gases (GHG: methane, CH_4, nitrous oxide, N_2O, carbon dioxide, CO_2) in the north eastern Alps dairy system (Top Value – Interreg ITA-AT). The GHGs emission was computed through a cradle-to-farm gate life cycle assessment. Functional units were 1 kg of fat- and protein-corrected milk (FPCM) and 1 m^2 of land. Herd and manure management, on-farm feedstuffs production, purchased feedstuffs (plus land use change) and materials were included into the system boundaries. Data originated from 75 farms (38 ± 25 LU, 20.9 ± 5.4 kg fat protein corrected milk – FPCM/cow/day), associated to 9 cooperative dairies in the eastern Alps. Mean GHG emission was 1.3 ± 0.2 kg CO_2-eq/kg FPCM and 0.6 ± 0.2 kg CO_2-eq/ m^2. Methane was the first contributor (57%; enteric and manure fermentations), followed by CO_2 (29%; fossil fuels and land use change) and N_2O (14%; fertilisers application). Milk yield was negatively correlated with CH_4 ($r=-0.88$, $P<0.001$) and N_2O ($r=-0.34$, $P<0.01$), but not with CO_2 emission ($r=0.10$, $P=0.38$). Considering instead land surface as functional unit, milk yield was positively correlated with all GHGs emissions (r ranged 0.26-0.46, $P<0.05$). Therefore, high-productive farms mitigated their GHG emission per unit of FPCM with respect to low-productive farms especially through a reduction in biogenic CH_4 emissions, but showed a trade-off in terms of pressure on land surface. In addition, no mitigation was observed for CO_2, an emission that adds new net carbon into the atmosphere. As the share of CO_2 emission on total GHGs emission was nearly 30%, its mitigation could give a notable contribution towards a net zero-carbon system, starting from off-farm feedstuffs (42% of total CO_2 emitted, land use change considered) and on-farm energetic sources (41%). The results of this study highlighted that mitigation strategies aimed to a net zero-carbon balance should consider not only milk yield, but also the origin of the GHG emission, in particular when grassland-based mountain dairy systems are taken into account.

Result-based carbon farming schemes in livestock, a mechanism for boosting low carbon initiatives

J.B. Dolle, C. Brocas and A. Gac
Institut de l'Elevage, Environment department, Avenue Roger Salengro 54-56, BP 80039, 62051, France; jean-baptiste.dolle@idele.fr

The French cattle sector is developing low carbon initiatives to reduce milk and beef carbon intensity by 20% in 2025 compared to 2015. Regarding the 13,000 farmers involved, there is a strong interest to improving methods for measuring, reporting and verifying (MRV) the mitigation actions. There is also need in providing payments for the delivery of climate mitigation benefits, quantified as CO_2 emissions avoided or sequestered. In 2018, the French Government adopted the *Label Bas Carbone* (Carbon Standard), a framework for voluntary carbon reduction project. Complying with this Standard, a MRV process named CARBON AGRI has been developed and certified by the French government. To capture direct and indirect emissions reductions and soil carbon sequestration, CARBON AGRI measures progress from a range of 40 mitigation practices as herd and feed management, crops fertilisation, inputs purchased, grassland management, etc. Based on the national LCA farm tool CAP'2ER®, the reductions are assessed against a baseline level of GHG emissions per unit of product (milk, beef and crop), calculated before the project start. So, each participating farm use its own figures (animal inventory, feed quantity, milk production, manure spreading) to calculate baseline emissions and sequestration. For the five years crediting period, the carbon reduction represents the products 'carbon intensity differences between the baseline and the project, multiplied by the milk and beef productions. An average French cattle farm who reduce carbon footprint from 10 to 15%, could generate 250 to 300 tons of carbon credits for the five years period. These results based emissions reductions are then traded for payment from an external party voluntarily offsetting their emissions (private and public organisations). Thanks to environmental co-benefits (water preservation, contribution to biodiversity, etc.), project developers are selling CO_2 avoided for prices of 30-40 €/t CO_2 which represent 6,000 to 10,000 € per farm. This carbon offset scheme has the potential to play a strategic role in catalysing the transition, providing financial resources but also a rigorous framework to monitor mitigation effort in the sector.

The environmental impacts of beef farms in Alentejo, Portugal, using multiple assessment tools

M.P. Dos Santos[1], T.G. Morais[1], J. Cipriano[1], N.R. Rodrigues[2], T. Domingos[1] and R.F.M. Teixeira[1]
[1]IST, Universidade de Lisboa, MARETEC – Marine, Environment and Technology Centre, LARSyS, Av. Rovisco Pais, Nº 1, 1049-001 Lisboa, Portugal, [2]Terraprima – Serviços Ambientais Lda., Av. Manuel da Maia 36 3º Esq, 1000-201 Lisboa, Portugal; manueldossantos@tecnico.ulisboa.pt

Life Cycle Assessment (LCA) can be used to determine both direct and indirect environmental impacts of any human activity. Given recent progress, it is becoming increasingly more encompassing of biophysical models and hence very helpful to quantify greenhouse emissions (GHG) worldwide. However, LCA results change significantly with methodological choices and tools used. As impacts of meat production are significant contributors to greenhouse gases (GHG) emissions worldwide, here we tested the influence of three different LCA tools assessing the global warming potential of extensive beef production. Farm-level data was collected from 26 farms in the Alentejo region in Portugal, and included fertiliser application, resource utilisation, number and weight of animals per age group, feed consumption and final products. To calculate the carbon footprint (CF) of each farm, we used the tools Coolfarm (CFT), OpenLCA (OPEN) and ModelMeat (unpublished). CFT and OpenLCA assess GHG emissions at a farm level and ModelMeat is an environmental and economic assessment tool developed in Portugal to assess the effects of pasture-based meat production, including also cropland production. Results showed significant differences between the three LCA tools. As expected, the three point out that animal production is responsible for the majority of the total emissions even in multi-purpose farms. There is also a significant contribution from the co-products used for animal feeding. In cropland production, the difference between tools is lower because similar emissions factors are used. Further work should explore the lack of methodological consistency among LCA tools and studies (e.g. database allocation method), take into account that methodological choices are often determined by data availability. For example, IPCC's Tier 1 (used in some of the tools) to calculate enteric fermentation emissions depends only on the number of heads per group, while Tier 2 depends, among other variables, on animal weight. These choices highly influence final outcomes.

Improving the dairy farm efficiency, the first step on the way to the net zero carbon farm

C. Brocas[1], A. Lejard[2], R. Lasbleiz[3] and J.B. Dolle[4]
[1]Institut de l'élevage, Monvoisin, 35652 Le Rheu, France, [2]France Conseil Elevage, 42 rue Chateaudun, 75009 Paris, France, [3]CNIEL, 42 rue Chateaudun, 75009 Paris, France, [4]Institut de l'élevage, 54-56 avenue Roger Salengro, 62051 St Laurent Blangy, France; catherine.brocas@idele.fr

Between 1990 and 2010, the milk carbon footprint has been reduced by 26% in France. Aware of the importance to continue this GHG mitigation, Carbon Dairy project aimed to build the milk carbon footprint reduction road map. A national Life Cycle Assessment tool named CAP'2ER® was developed to measure the milk carbon footprint. We characterised and calculated the carbon footprint of 4,150 dairy farms and performed multicriteria analysis to access the contribution of production parameters to the carbon footprint. The average carbon footprint is 1 kg CO_2e per litre Fat and Protein Corrected Milk. The top-performing farms have carbon footprints 18% lower than the average. Practices with the largest impact are milk yield, age at first calving, quantity of concentrate, N-fertiliser used and fuel consumed. The results also show that the farms can improve economic performance while also reducing carbon footprint. The top-performing group's gross margin was 30€/1000 l higher than the bottom-performing carbon footprint group. In parallel, a sample of 366 farms carried out two CAP'2ER and gross margin analysis during the project. The farms which decrease the most their carbon footprint within three years (-21%) reduce the variable costs by 27€/1000 l as the average decrease by 17€/1000 l. These links between economic and environmental performances encourage now the dairy sector to implement the Low Carbon Dairy Farm initiative at a national level for the 60,000 dairy producers with the objective to engaged all of them in 2028. Behind the efficiency improvement for reducing carbon footprint, the next step to the Net Zero-Carbon farm will be to implemented innovative practices (additives, manure management) and raise carbon storage in soils to compensate the GHG emissions. But, these practices could impact the cost of milk production or the macro-economy of the agriculture sector.

Mitigation strategies and extent of greenhouse gas emissions in beef fattening farms in Italy

S. Carè, G. Pirlo and L. Migliorati
Council for Agricultural Research and Economics, Research Centre for Animal Production and Aquaculture, Via A.
Lombardo, 11, 26900, Italy; sara.care@crea.gov.it

Beef sector is considered a source of GHG, even if it provides a variety of outputs, that can contribute to reduce global warming. For this reason, the sector is responding to this challenge through the 'LIFFE BEEF CARBON' project which aims to reduce greenhouse gas (GHG) emissions by 15% within 10 years across 172 beef farms in France, Ireland, Italy and Spain by adopting mitigation strategies, which improve animal welfare, increase renewable energy, reduce synthetic fertilisers, increase animal performances or increase soil carbon stock. The objective of this study was to evaluate the reduction of GHG emissions in beef production as consequence of the application of a series of mitigation strategies. In Italy, the carbon footprint (CF) of 21 intensive fattening farms was estimated before and after the application of the mitigation strategies. The environmental performance was evaluated using CAP'2ER® (Niveau 2). The functional unit was 1 kg of live weight gain (LWG) and the environmental categories were: global warming (GW), acidification (AC), eutrophication (EU) and energy consumption (EC). The effectiveness of the mitigation strategies was evaluated using Wilcoxon Signed Ranks Test. Factor and discriminant analysis were carried out to show the relationships between the farm's characteristics that best explain the extension of GHG emission reduction. At the beginning of the project, the farms' mean size was 75±74 ha with a capacity of 508±373 animal places; the average LWG was 1.24±0.22 kg/d and the average live weight production was 235±183 T/year. The introduction of the mitigation strategies determined a significant (P<0.05) reduction of 11, 9 and 10% of GW, AC and EC respectively. Farmers that combined more than one mitigation strategies and improved performances of their farms showed the best results in reducing CF. The study showed that it is possible to reduce GHG emissions from specialised fattening beef farms. Farming strategies based on high production intensity showed to be able to mitigate environmental impact and achieve the goal of the project. The activity was funded by European Commission area Environment (LIFE BEEF CARBON project) in collaboration with ASPROCARNE and UNICARVE.

Improving the eco-efficiency of beef production through optimisation of age and weight at slaughter

R.F.M. Teixeira[1], T.G. Morais[1], M.P. Santos[1], N.R. Rodrigues[2], L. Barão[3] and T. Domingos[1]
[1]Instituto Superior Técnico, MARETEC – Marine, Environment and Technology Centre, LARSyS, Av. Rovisco Pais, 1,
1049-001, Portugal, [2]Terraprima, Serviços Ambientais Lda., Centro de Negócios do Porto Alto, Fracção S, Avenida das
Nações Unidas, no. 97, 2135-199 Samora Correia, Portugal, [3]Faculdade de Ciências da Universidade de Lisboa, CE3C,
Campo Grande, 1749-016 Lisboa, Portugal; ricardo.teixeira@tecnico.ulisboa.pt

Animal food products are often assumed to have high environmental impacts, but they are also a crucial source of income for farmers. In policy and business, there is an increasing need to optimise livestock production towards the highest possible eco-efficiency (decreased impacts per value added). Here, we assessed the effect of age at slaughter on the carbon (C) and nitrogen (N) balances of the typical extensive beef cattle production system in Alentejo, Portugal. We used average regional data for two pasture systems: semi-natural pastures (SNP) and sown biodiverse permanent pastures rich in legumes (SBP). We studied potential ages at slaughter between 9 and 18 months. We assumed that during the first 12 months calves may remain in the pasture (with roughage supplementation). After 12 months, calves are confined and fattened using a mix of silage and concentrate feed. We used an extended version of the C and N mass-balance pasture model 'BalSim' that includes C and N flows during confinement and also a whole-system economic balance. Results show that SBP require almost double hours of labour than SNP per hectare. N self-sufficiency is about 90% for SBP and 5% for SNP, and herbage consumption is 80 and 40% for SBP and SNP, respectively. The weight of feed costs on total costs is in the ranges of 29%-51% in SBP 54-65% in SBP. The minimum possible greenhouse gas (GHG) emissions are 17 kg CO_{2e}/kg live weight when the calf is raised in SBP, supplemented with a feed consisting mostly of maize silage and slaughtered at 13 months of age. Any level of high-concentrate feed supplementation always decreases eco-efficiency. After 12 months, farmer income increases because feed costs are lower than the revenue of the marginal weight gain during confinement. However, environmental performance decreases more than proportionally, due to concentrated feed production and manure management emissions.

Effects of different CO_2-eq formulations on optimal R&D strategies to realise carbon-neutral beef

G.A. McAuliffe[1], T. Takahashi[1,2], J.M. Lynch[3] and M.R.F. Lee[1,2]
[1]Rothamsted Research, North Wyke, EX20 2SB, United Kingdom, [2]University of Bristol, Langford, BS40 5DU, United Kingdom, [3]University of Oxford, Oxford, OX1 3PU, United Kingdom; graham.mcauliffe@rothamsted.ac.uk

Life cycle assessments (LCA) evaluating climate impacts of agri-food systems typically employ a single emission metric to represent multiple greenhouse gases (GHGs). In reality, however, each GHG takes a unique pathway to affect the atmospheric environment; therefore, this approach can obscure vital details of the climate response to emissions of different gases. Importantly, such misinformation could also lead to suboptimal transitional policies to achieve net-zero, and in particular areas of mitigation technologies of which advancement should be prioritised. To examine the impact of metric selection on optimal R&D strategies, this paper calculated carbon footprints of a pasture-based beef production system in the UK under both 100-year global warming potential (GWP_{100}) and 100-year global temperature change potential (GTP_{100}) impact assessment methods, which substantially differ in valuation of CH_4 relative to CO_2. A 9×10 factorial virtual experiment was designed to include various combinations of CH_4 and N_2O emission factors (EFs), each representing a distinct advancement scenario focusing on a different set of mitigation technologies. The results showed that the choice of impact assessment method affects reported carbon footprints considerably. Under GWP_{100}, CH_4 arising from enteric fermentation accounted for 39.9% of total emissions in the default scenario, whilst under GTP_{100}, it only accounted for 9.1%. Under GWP_{100}, reductions in both EFs symmetrically resulted in lower mid-point impacts; under GTP_{100}, however, a smaller CH_4 EF did not produce any notable improvement. Finally, when EF reductions were combined with partial conversion of grassland to woodland, the most effective land use option differed depending on the R&D priority due to the nonlinear feedback mechanism between stocking density and emissions intensity. Combined together, the present findings indicate the importance of understanding the climate response assumptions embedded in the model, and the policy risk of solely focusing on the mitigation of CH_4 emissions as often inferred by a GWP_{100} analysis.

Different LFS, same agroecosystem: influence of production model in carbon balance

A. Horrillo, P. Gaspar and M. Escribano
School of Agricultural Engineering. University of Extremadura, Department of Animal Production and Food Science, Avda. Adolfo Suarez, s/n, 06007 Badajoz, Spain; andreshg@unex.es

The management system proposed by organic farming models is based on the simplification and adoption of certain practices leading to improving pastures and soils. These practices can also compensate the greenhouse gases (GHG) emissions derived from livestock husbandry. However, this compensation capacity may depend on the production model adopted. In dehesas' agroecosystem are feasible multiple production models, all of them resulting from management decisions and these determine the final balance on GHG emissions. This study analyses seven production models (ruminants' and Iberian pig's organic farms) located in dehesa with the aim of estimating the balance of GHG emissions and carbon footprint considering their carbon sequestration potential. The method used was Life Cycle Assessment. GHG emissions estimated are those derived from livestock digestion, manure management, soil management, and off-farm inputs (feeding, fuels, and electricity). Carbon sequestration calculations consider carbon fixation due to pasture and crop waste and carbon fixation in soil due to manure fertilisation. The results reveal that the farms producing meat cattle with calves sold at weaning age provide the highest levels of carbon footprint (16.27 kg of carbon dioxide equivalent (CO_2eq) / kg of live weight), whereas the farms with the lowest levels of carbon emissions are Montanera pig and semi-extensive dairy goat farms, i.e. 4.16 and 2.94 kg CO_2eq / kg of live weight and 1.19 CO_2eq / kg of fat and protein corrected milk (FPCM), respectively. Enteric fermentation represents 42.8 and 79.9% of the total emissions of ruminants' farms. However, in pig farms the highest percentage of the emissions derives from manure management (36.5%-42.9%) and animal feed (31%-37.7%). The soil sequestration level has been seen to range between 419.7 and 576.4 kg CO_2eq /ha/year, which represents a considerable compensation of carbon emissions. It should be noted that these systems cannot be compared with other more intensive systems in terms of product units and therefore, the carbon footprint values of dehesa organic systems must always be associated to the territory.

Environmental impact of dairy farms in an Alpine area including the effect on biodiversity
E. Sabia[1], S. Kühl[1], L. Flach[1], C. Lambertz[2] and M. Gauly[1]
[1]Free University of Bolzano, Faculty of Science and Technology, Piazza Università, 39100, Italy, [2]Research Institute of Organic Farming, Kasseler Strasse 1a, 60486, Germany; emilio.sabia@unibz.it

Small-scale dairy farms in Alpine mountain areas play an important and fundamental socio-economical role. The present study aimed to evaluate the environmental impact including the effect on biodiversity of 24 small-scale Brown Suisse dairy farms in South Tyrol. The farms were divided based on the annual milk production per cow in two groups: Low Milk Production (LMP, n=12 farms; ≤5,000 kg/milk/cow/y), and High Milk Production (HMP, n=12 farms, ≥8,000 kg/milk/cow/y). The functional unit was 1 kg of Fat and Protein Corrected Milk (FPCM). The principal difference between LMP and HMP were: pasture area (ha) 1.5 vs 0.4±0.4 (SEM) (P<0.05), meadow area for hay production (ha) 8.3 vs 7.1±1.2, feed concentrate kg/cow/d 3.6 vs 7.9±0.4 (P<0.001). The environmental impact was using the commercial software package SimaPro 8.01, using database of Ecoinvent. The effect on biodiversity has been estimated using two methods: 1. the characterisation factors approach for the estimation of damage score (DS), 2. ReCiPe end-point method expressed in terms of species per year lost. The environmental impact generated in terms of carbon footprint (CF) was greater in LMP than in HMP (1.40 vs 1.01±0.06 kg CO_2-eq pe kg of FPCM, P<0.001). For both groups, the main process involved in terms of CF was enteric emissions followed by the production of concentrate feed. The prevalent chemical compound was CH_4 from biogenic sources followed by CO_2 from fossil fuels. However, when taking the effect on biodiversity into account, the LMP had a lower impact than the HML system for both methodologies used. In LMP and HMP DS per kg of FPCM was 0.39 vs 0.51±0.03 (P<0.01) and ReCiPe species per year per kg of FPCM was 1.59 E-7 vs 2.20 E-7±2.01 E-8 (P<0.05), respectively. A positive correlation was found between the two methods for estimating the effect on biodiversity (R=0.84, P≤0.001). This study shows the importance of monitoring the environmental impact including the effect on biodiversity of dairy farms in Alpine regions.

Enteric methane emission of the Dutch dairy herd
L. Koning and L.B. Šebek
Wageningen Livestock Research, Animal Nutrition, De Elst 1, 6708 WD, the Netherlands; lisanne.koning@wur.nl

Resulting from the Paris Climate Agreement of December 2015, the Netherlands aim to reduce greenhouse gas (GHG) emissions by 49% in 2030 compared to 1990. To achieve this goal the dairy sector needs to reduce methane (CH_4) emissions by 1.0 megaton CO_2-equivalents. Besides the contribution to the worldwide GHG emission, enteric CH_4 emission is also a net loss of energy for the cow. It is valuable to reduce this inefficiency in animal production, resulting in cows that get more metabolisable energy from their feed. Currently the average CH_4 emission of dairy cows in practice are estimated using model calculations which are internationally accepted. It is however unknown if the model estimates the variation in the Dutch dairy sector correctly. The first objective of this study was therefore to gain insight into the average CH_4 emission and variation of the Dutch dairy herd. The second objective was to investigate if variation in enteric CH_4 emission was affected by factors as animal and farm. Enteric CH_4 emission was measured using the Greenfeed (C-lock Inc. Rapid City, SD, US), an adapted feeding station that measures both continually CH_4 concentration and the quantitative airflow. In total 1,718 records of 1,278 dairy cows of 18 farms throughout the Netherlands were collected within a 2 weeks period (plus one week of adaptation) from September 2018 to October 2019. Additionally, individual milk samples were taken to analyse milk composition and feed intake data on herd level (including feed composition and quality) were collected. Linear Mixed Models fitted with Residual Maximum Likelihood were used to analyse the data. The average CH_4 emission was 415±101 g CH_4/cow/day, expressed in fat protein corrected milk 13.9±5.1 g CH_4/kg FPCM. Lactation stage and parity explained the largest part of the variation, approximately 29%, followed by soil type and grazing in relation to season (together 12%), and the content of urea and lactose of the milk plus the lactation value (9%). Feed composition and feed quality components did not show a significant effect on the observed variation. However, there were indications that during grazing the emission was lower and it is advised to focus on that for follow up research. On national level the results are in agreement with model calculations.

Effects of starch level and oil on intake, rumen fermentation, and methane emissions in dairy cows
B. Darabighane, L. Ventto, P. Kairenius, T. Stefański, I. Tapio, H. Leskinen, J. Vilkki and A.R. Bayat
Natural Resources Institute Finland (Luke), Jokioinen, 31600, Finland; alireza.bayat@luke.fi

Four multiparous Nordic Red cows (76 ± 10.4 d in milk) fitted with rumen cannulae were used in a 4×4 Latin square with a 2×2 factorial arrangement of treatments and 35-d periods. Experimental treatments consisted of high (HS) or low (LS) starch level containing 0 or 30 g of a mixture of sunflower and fish oils (2:1 w/w) per kg diet DM. Digestibility was measured by total collection of faeces and CH_4 was measured using SF_6 technique. Rumen microbial community was assessed by 16S/18S rRNA gene amplicon sequencing. Intakes of DM and gross energy (GE) were similar between LS and HS diets. Intakes of CP and starch were greater ($P\leq0.013$) but intake of NDF was lower ($P<0.01$) with HS compared with LS diets. The oil mixture reduced ($P<0.01$) intakes of DM and other nutrients with the exception of increased FA intake. The high starch level increased apparent digestibility of DM, OM and starch while decreased NDF digestibility ($P<0.01$). Both high starch level and oil mixture increased ($P\leq0.02$) the digestibility of GE and tended to increase ($P=0.06$) CP digestibility. Cows receiving HS diets had higher ($P\leq0.056$) levels of total rumen volatile fatty acids (VFA), butyrate, isobutyrate, valerate, isovalerate, and caproate while their acetate was lower ($P<0.01$) and acetate:propionate ratio tended ($P=0.07$) to be lower compared with LS-fed cows. However, starch level did not affect rumen pH and molar propionate ratio, or ruminal CH_4 emissions. Although dietary oil in both diets did not affect rumen pH, total or individual ruminal VFA, decreases or tendencies to decrease ($P\leq0.067$) were observed in ruminal CH_4 emission when calculated as g/day, g/kg milk, and g/kg energy corrected milk. The inclusion of oil in both diets had no impact on the abundance of *Fibrobacter* but increased ($P<0.04$) *Eudiplodinium maggii and* tended ($P<0.09$) to increase *Ruminococcus* and decrease *Methanobrevibacter ruminantium*. The results show that starch level modified rumen fermentation and nutrient digestibility without influencing DM intake or CH_4 emissions. Inclusion of oil mixture reduced DM intake and some of ruminal CH_4 emission indices without influencing rumen fermentation or nutrient digestibility.

Relationship between nourishing performance, carbon footprint and nitrogen balance in Walloon dairy
J. Hurdebise, F. Lessire and I. Dufrasne
University of Liège, Faculty of Veterinary Medicine, Nutrition Unit, Boulevard Colonster 20, 4000 Liège, Belgium; jhurdebise@uliege.be

The objective of this study was to highlight, in 21 dairy farms, the correlations between farm structure and 3 environmental indicators: the nourishing performance index (NPI- number of people fed by animal protein per hectare), net carbon footprint of milk (CF -CO_2 equivalent per litres of energy corrected milk; ECM) and the excess nitrogen balance (NB -kg of nitrogen per hectare). To conduct this study, the CAP'2ER tool developed by the Institut de l'Elevage (IDELE) in France was used. More than 150 data from 2018 were collected during interviews led with 21 dairy farmers from Wallonia (Belgium). The indicators of interest here are NPI, CF and NB. Milk yield (kg) was converted in ECM using the following values: fat%: 4% and protein%: 3.3% milk. The CO_2 conversion factors are estimated at 25 for methane (CH_4) and 298 for nitrous oxide (N_2O). A correlation matrix was run on SAS (SAS 9.3) between the selected environmental indicators and the following parameters: the agricultural surface of the farm (ha), the number of dairy cows/farm and ECM produced per cow. Regarding NPI, we noted a positive correlation with the number of cows per farm, ECM per cow, NB per ha and CF per kg ECM. This latter observation was not confirmed when expressed by ha. The higher NPI, the higher high producing cows on the farm. The highest NPI was observed at highest nitrogen surplus per ha and the highest milk CF. For the excess nitrogen balance, we observed a positive correlation with the number of cows, ECM per cow and milk CF but not with the area. The more there is a nitrogen balance surplus, the more cows produce more on the farm. The more excess of nitrogen per ha there is, the higher is the milk CF. Milk CF did not correlate with area, number of cows per farm or milk production per ha, but did with NPI and excess NB per ha. In conclusion, the NPI of a farm, the excess nitrogen balance per ha and the net carbon footprint of milk demonstrated positive correlation between each other, in these 21 Walloon farms. However, it is interesting to note that the CF per kg ECM is not influenced by farm factors like the surface, the milk yield per cow and the number of cows per farm.

Benefits of low crude protein diets in broilers: beyond CO_2, the importance of N and NH_3 assessment

J. Le Cour Grandmaison[1], B. Méda[2], N. Martin[1], W. Lambert[1] and E. Corrent[1]
[1]Ajinomoto Animal Nutrition Europe, Rue Guersant, 75017 Paris, France, [2]INRAE, Université de Tours, BOA, 37380 Nouzilly, France; lecourgrandmaison_j@eli.ajinomoto.com

The study objectives were to quantify to which extent low crude protein (CP) diets affect broiler performance, greenhouse gas emissions from feed production (GHGf), nitrogen (N) excretion (Nex), ammonia volatilisation (NH_3v) and N entering storage (Nst) per ton of broiler produced. Data from literature were selected: 1) dose response to CP level, 2) published from 2017, 3) presenting performance and diet composition, 4) constant dietary energy and digestible lysine, and 5) indispensable amino acids (AA) adequately supplied. The database included 30 treatments (10 trials from 6 papers), in which CP reduction was performed by replacing soybean meal and oil by cereals and feed-grade AA. For each diet, CO_2-eq value per kg of feed was calculated using feed composition and two databases: GFLI for main feedstuffs, and EcoAlim for micro-ingredients, respectively. This value was then multiplied by daily feed intake (DFI), to calculate GHGf. Nex was estimated by difference between intake and deposition (Nbody=29 g N/kg gain; ITAVI, 2013). NH_3v was estimated using Nex and EMEP volatilisation model (2019), and then modulated according to CP level using equations from Belloir *et al.* (2017). Finally, Nst was estimated by difference between Nex and NH_3v. Data were then analysed by ANOVA and GLM with 'trial' as random factor and 'CP' as covariable. Dependent variables were DFI, average daily gain (ADG), feed conversion ratio (FCR), and environmental parameters: GHGf, Nex, NH_3v and Nst expressed per ton of broiler (assuming a linear growth up to 2.2 kg at 35 d of age). According to linear regression equation, lowering CP increased broiler DFI (P<0.05) and increased ADG (P=0.059) but did not affect FCR (P=0.847). Reducing by 1%-point CP significantly (P<0.001) decreased GHGf by 7% (-226 kg CO_2-eq; R^2=99%), Nex by 11% (-3.6 kg; R^2=97%), NH_3v by 20% (-1.4 kg; R^2=97%), and Nst by 10% (-2.7 kg; R^2=97%) per ton of live broiler. The study demonstrated that low CP strategy in broiler feed is efficient to reduce GHG from feed, N excretion and NH_3 emissions without negatively affecting performance.

Impact of cooling management and anaerobic digester on cows' milk carbon footprint intensity

A. Vitali, G. Grossi, U. Bernabucci, A. Nardone and N. Lacetera
*University of Tuscia, Department of Agriculture and Forest Sciences, via S. Camillo de Lellis, 01100 Viterbo, Italy;
vitali@unitus.it*

The study was aimed to assess the combinate effect of cooling and manure management on the milk carbon footprint (CFP) intensity with a cradle-to-farm-gate approach. Data refer to dairy farm milking 1,200 cows/day and years 2015-2017. Baseline scenario refers to the standard management (BS) that not included neither cooling nor anaerobic manure management. Cooling scenario refers to BS to which was included a cooling program (BS+CS) based on 5 daily treatments (fans + sprinklers) from June to September done in waiting (3) and in the feeding (2) lanes for a total working of 3 h. Anaerobic digester scenario refers to BS+CS to which was included the anaerobic digestion of manure (BS+CS+AD). Enteric CH_4 emissions were calculated for all scenarios using specific diet parameters of DMI, ADF and NDF. In BS and BS+CS scenarios the emissions from manure were calculated for CH_4 and N_2O according to the methodologies adopted in National Inventory Report (NIR) and IPCC guidelines, respectively. In the BS+CS+AD scenario emissions from manure management were not included whereas only those arising from the working of the anaerobic digestion plant were considered. The denitrification-decomposition (DNDC) model was involved in accounting the soil greenhouse gas (GHG) emissions and carbon sinks arising from crops activities for feed production. The SimaPro software was used to esteem emission factor (EF) of the input raw materials and related transports needed for feed production, whereas the EF of energy consumption was from NIR. Finally, the EF of extra-farm feed were extrapolated from bibliography. Total GHG emissions were expressed as kg CO_{2e} for litre of milk. The emission intensity for BS was 1.5 kg CO_{2e}/litre of milk. The adoption of cooling system (BS+CS) improved milk yield efficiency of about 19% and, although fans and higher DMI as consequence of CS have increased overall GHG emissions, the CFP intensity was reduced of 13% (1.32 kg CO_{2e}/litre of milk). The adoption of anaerobic digester (BS+CS+AD) has cut the GHG of a further 16% (1.11 kg CO_{2e}/litre of milk). The sustainability of milk may be increased indirectly by improvement of production efficiency and/or directly by the adoption of mitigation options.

Environmental impact of a dietary supplementation of clays and organic acids in the fattening pigs

I. Skoufos[1], I. Giannenas[2], D.K. Papanastasiou[3], E. Bonos[1], A. Tzora[1], T. Bartzanas[4] and V. Anestis[4]
[1]University of Ioannina, School of Agriculture, School of Agriculture, University of Ioannina, Kostakioi Artas, 47150, Kostakioi Artas, Greece, [2]Aristotle University of Thessaloniki, Laboratory of Nutrition, School of Veterinary Medicine, School of Veterinary Medicine, Aristotle University Campus, 54124, Thessaloniki, Greece, [3]Centre for Research and Technology Hellas (CERTH), Institute for Bio-economy and Agri-technology (IBO), Dimarchou Georgiadou 118, Volos, 38333, Volos, Greece, [4]Agricultural University of Athens, Department of Natural Resources Management and Agricultural Engineering, Iera Odos 75, 11855, Athens, Greece; jskoufos@uoi.gr

A life cycle assessment (LCA) methodology was used to study the environmental impact of the dietary supplementation of attapulgite clay and benzoic acid in the diets of fattening pigs. The *in situ* performance results were examined in two feeding practices: (1) conventional diet (CNVD) used by the farm; (2) dietary addition of attapulgite clay at 4 g/kg and benzoic acid at 5 g/kg (ATTBAD). Functional unit was defined as '1 kg of pig live-weight at the farm gate'. A 'cradle-to-farm-gate' analysis was performed in a Greek fattening pig farm. Emission of CH_4 from enteric fermentation was estimated by the default Tier 1 Intergovernmental Panel of Clime Change emission factor. Default IPCC Tier 2 approach was applied for the emissions of CH_4 and direct N_2O from on-farm slurry management. Indirect N_2O emissions were estimated with the default Tier 1 process. Default European Monitoring and Evaluation Programme Guidebook Tier 2 approach was utilised for the evaluation of NH_3. Modelling of the material and electricity inputs' supply chains was based on secondary LCA Inventory datasets. The ATTBAD system was associated with improved environmental performance, contributing to lower global warming potential (GWP-100), accumulated exceedance-acidification/terrestrial eutrophication. Soybean meal supply was the most important contributor to the GWP-100. Acknowledgements: The project is co-financed by the European Regional Development Fund under the Operational Program 'Competitiveness-Entrepreneurship-Innovation, Bilateral RTD cooperation between Greece-China 2014-2020', EPAnEK. Project Code: T7ΔKI-00313, Acronym: GreenPro.

Methane yield, Ym and emission factor of beef heifers on grasslands

M.S. Orcasberro[1], C. Loza[2], J. Gere[3], P. Soca[1] and L. Astigarraga[1]
[1]Facultad de Agronomía-Universidad de la República, Producción Animal y Pasturas, Garzón 780, 12900 Montevideo, Uruguay, [2]Christian-Albrechts-Universität zu Kiel, Institute of Plant Production and Plant Breeding, Grass and Forage Science / Organic Agriculture, Hermann-Rodewald Straße 9, 24118 Kiel, Germany, [3]Facultad Regional de Buenos Aires, CONICET, Unidad de Investigación y Desarrollo de las ingenierías, UTN.BA, Medrano 951, C117AAQ, C.A.B.A., Argentina; astigarr@fagro.edu.uy

The Campos region is the largest biogeographic unit of native grassland in South America and its environmental conditions make it of high relevance for Livestock, and particularly cow-calf production systems. Despite of the relevance of these systems, the information on methane (CH_4) emissions in pregnant heifers for this region is scarce. Therefore, the aim of the study was to quantify the CH_4 emissions, Methane yield (MY; g CH_4/kg dry-matter intake) or percentage of gross energy in feed converted to CH_4 (Ym), and to estimate the Emission factor (EF – IPCC, 2006; Tier 2) for 40 beef heifers during the last third of gestation and lactation, on continuous stocking in native grasslands. Herbage allowance varied seasonally. In autumn 6, winter 4, spring 10 and summer 6 kg DM/kg live weight. The experiment design was a randomised block with two blocks representing contrasting soils. Data was analysed with the Mixed procedure of SAS (2001), as repeated measures. During May, August and November, CH_4 emission was determined with the SF_6 tracer technique and DM intake was estimated using Cr_2O_3. The results for this trial show significant differences between periods on CH_4 emissions (159, 150, 312 g CH_4/day, on May, August and November respectively; $P \le 0.0001$), with a clear increase in Spring, associated with a higher herbage intake, although MY (23.6 g CH_4/kg DMI) and Ym (7.4) remained similar. The estimated EF from data of the three experimental periods resulted in 74 kg CH_4/head/year. The results, similar to reports for the Bioma Pampa on Brazil, show that CH_4 emissions are affected significantly by measurement period, accompanying the variations on herbage intake. In addition, these results are the first obtained for breeding systems in native grasslands in Uruguay, so they represent an important contribution to the calculations of National GHG Inventory.

In vitro study of a barley-corn mixture system on total gas and methane production

C. Barraso[1], S. Cobos[2], J. Rodríguez[3], M.M. López[1] and P.L. Rodríguez[2]
[1]Finca la Orden-Valdesequera. CICYTEX, Meat Quality Department, Ctra. A-V, km 372, Guadajira, Badajoz, 06187, Spain, [2]University of Extremadura, Department of Animal Production and Food Science, Faculty of Veterinary Medicine, Avenida de la Ciencia s.n, Cáceres, 10003, Spain, [3]COPRECA. Soc. Coop. Ltda, Ctra. Nacional V, km 250, Trujillo, Cáceres, 10200, Spain; cbgil93@gmail.com

Methane is said to be one of the most important gas within greenhouse gas generated by the agricultural sector because its twenty-five times higher global warming potential than CO_2. Spanish agricultural and livestock activity is responsible for almost 11% of GHG emissions with 89% corresponding to methane. Mitigation of this production involves different strategies, including intervention on the rumen microbiota modifying its fermentation pattern. In the case of concentrates, it is conditioned by the nature of the fermented starch. According to this, we proposed an in vitro fermentation test on a barley/corn mixture system (%): 0/100, 25/75, 50/50, 25/75 and 100/0, to assess the total gas and methane production. The barley and maize samples were milled and mixed in the proportions indicated above. In vitro fermentation was carried out following the method of Menke et al., using sheep faeces as a source of inoculum and monitoring gas production at 18 hours. The final gas production was determined from the displacement of the syringe plunger and the accumulated methane was measured using a GMI-PS200 methanometer A total of 8 repetitions/mixing were performed. The results obtained indicate increasing values of gas production as a function of the percentage of barley incorporated, obtaining the maximum for 100% barley and the minimum for 100% corn. As for CH_4 production, we found a decreasing linear response as corn is incorporated. Therefore, incorporating corn as a source of starch in the lamb fattening feed would reduce both gas and methane production.

Mitigation extent of GHG emissions in Sardinia dairy sheep farms by flock fertility improvement

M.F. Lunesu[1], M.G. Serra[2], P. Sau[1], P. Arca[3], S. Contini[2], D. Cossu[3], A. Franca[4], E. Vagnoni[3], M. Decandia[2], A.S. Atzori[1], G. Molle[2] and P. Duce[3]
[1]University of Sassari, Dipartimento di Agraria, Viale Italia 39, 07100 Sassari, Italy, [2]AGRIS Sardegna, Loc. Bonassai, 07100 Sassari, Italy, [3]CNR, IBE, Trav. La Crucca 3, 07100 Sassari, Italy, [4]CNR, ISPAAM, Trav. La Crucca 3, 07100 Sassari, Italy; mflunesu@uniss.it

This study aimed to provide insight on the impact of improving flock fertility on predicted GHG emissions at farm level. For this purpose, two dairy sheep farms located in Sardinia (Italy) were considered as case study. Farmers were surveyed for a complete annual life cycle inventory (LCI) of cradle-to-gate farm production processes. The LCI included information on flock, animal diets, feed purchases, crops, farm stocks, and an energy use audit. For each farm, two scenarios were considered: 1) survey data; 2) after fertility improvement. Total emissions were expressed on kg of fat and protein corrected milk (FPCM) and estimated by the Environmental Footprint Method 2.0 (2010) approach by using the SimaPro 9.0.0.35 software. The two case study farms were of 79 (farm A) and 184 (farm B) ha in size, with flocks of 260 and 695 female heads, and stocking rates of 3.0 and 3.9/head/ha, respectively. Average flock milk production was 114 and 102 kg FPCM/head/year, for A and B farms, respectively. In survey scenario, annual fertility rates were 100 and 90.9% for pluriparous ewes, and 25 and 52% for primiparous ewes, for A and B farms, respectively. Increasing ewe fertility to 100% for both farms, GHG emissions, expressed in kg CO_2eq/kg FPCM, decreased from 4.23 to 3.39 and from 5.81 to 4.99 in the farms A and B, respectively. Reproduction improvement lowered on average 17% of GHG emissions. In addition, the annual lamb sales (kg) and the total milk sold (L) increased by 37 and 25% in the farm A and by 34 and 31% in the farm B, respectively. In conclusion, the predicted mitigation extent (-17%) of GHG emissions observed in this study suggested that the improvement of fertility rate could be considered a good farming practice to reduce GHG emissions from the Sardinian sheep farming sector. The authors gratefully acknowledge the Europe Union for the financial support (Project SheeptoShip LIFE 15 CCM/IT/000123).

GHG emissions and animal welfare in pig production using novel feed ingredients

H.F. Olsen[1], H. Møller[2], S. Samsonstuen[1], M. Øverland[1], L.T. Mydland[1], S. Iakhno[1], H. Sørum[1] and C. Press[1]
[1]Norwegian University of Life Sciences, P.O. Box 5003, 1432 Ås, Norway, [2]Ostfold Research, Stadion, 1671 Kråkerøy, Norway; hanne.fjerdingby@nmbu.no

Over the last 40 years, the intake of meat per capita in Norway has increased by more than 40%, where most of the increase is dominated by pork and poultry meat. The increase in pork and poultry production leads also to increased demand for domestic production of feed, calling for research exploring novel feed ingredients to cover the protein shortage. The project LIVESTOCK is documenting the sustainability of using novel feed ingredients through life cycle analysis (LCA). Based on research within Foods of Norway, a Centre for research-based innovation at NMBU, an LCA is performed to document the impact of replacing a conventional diet for piglets with a yeast-based diet, where yeast partially replaces soybean meal. As a basis for a simulation of possible resilience effects of the gut microbiome, data from a challenge test including 68 piglets were utilised. The piglets, fed two different diets, were challenged with a strain of E. coli known to give diarrhoea. The records included feed intake, weight gain, faecal consistence, behaviour, fever conditions and mortality. The challenge test showed clear, clinical differences between the groups, and the possible effect on growth performance by using a yeast-based feed is compared to a conventional diet and included in the LCA. Incidences of diarrhoea in pig production also affect animal welfare, which is highly valued in Norwegian livestock production and should be included as an impact category of social-LCA (S-LCA), as proposed in recent publications. A wider scope of the S-LCA study will allow a more comprehensive analysis of animal welfare across the value chain and criteria for evaluating animal welfare in S-LCA is needed. The system boundaries are from cradle to the gate of the slaughterhouse, and the LCA-model is developed from a previously developed farm model of pork production. The functional unit is 1 kg of produced pork (slaughter weight). Impact categories included are climate change and animal welfare. The study is ongoing, and results from the study will be included in the final presentation.

Effect of different factors on ammonia emission from pig manure

V. Juskiene, R. Matulaitis and R. Juska
Lithuanian University of Health Sciences, Institute of Animal Science, R. Zebenkos 12, LT -82317, Lithuania; violeta.juskiene@lsmuni.lt

Ammonia emission from manure is one of the biggest environmental problems which generate animal production. NH_3 formation is a result of complex impact of many parameters. Purpose of this study was to investigate the effects of manure surface area, temperature, air velocity, air humidity, atmosphere pressure on ammonia emission from manure as well as to measure the dynamics of emission and to determine the total nitrogen loss from manure. The experiment was carried out at LUHS Institute of Animal Science, using fresh manure from fattening pigs. The study lasted for 3,187 hours (4.4 months), from April to September. Gas emission was measured using the passive chamber method and different size chambers were used to study the effect of manure surface area. NH_3 and CO_2 gas concentrations, temperature of manure, temperature of air, air velocity, air humidity, atmosphere pressure were measured and analysed. The study showed that the highest ammonia emission was determined at the beginning of experiment when manure was fresh, also emissions increased at higher ambient temperatures. NH_3 emission from chambers with different manure surface area differ significantly ($P=0.05$) and was higher from chambers with larger manure surface area. Manure temperature was higher along with the ambient air temperature increases ($P<0.01$). Positive relationship ($P<0.05$) were found between NH_3 gas emission and manure temperature ($r=0.150$), ambient air temperature ($r=0.127$), CO_2 concentration ($r=0.186$). However, when air velocity was higher, ammonia and carbon dioxide concentrations (respectively: $r=-0.181$ and $r=-0.871$), manure temperature ($r=-0.869$), also ambient air temperature ($r=-0.869$) were significantly lower ($P<0.01$). There was no significant relationship between ammonia concentration and relative air humidity. Throughout the study, due to NH_3 emission the loss of total manure nitrogen was about 38% of the initial amount.

Impact of n-3 fatty acid supplementation of sow diets on milk fatty acid and oxylipins profile

E. Llauradó-Calero, R. Lizardo, D. Torrallardona, E. Esteve-Garcia and N. Tous
IRTA, Animal Nutrition, Mas de Bover, crta. Reus-El Morell, km 3,8, E-43120, Spain; eudald.llaurado@irta.cat

Oxylipins are oxygenated compounds formed from n-3 and n-6 polyunsaturated fatty acid (FAs) families via reactions involving at least one oxygenation step. It is well established that oxylipins derived from n-6 FAs have a high proinflammatory potential. In contrast, those derived from n-3 FAs have a direct anti-inflammatory activity, although some also might have a low proinflammatory potential. The objective of this study was to determine the influence of replacing 1.5% (w/w) of animal fat in the sows diet by an n-3 FA source rich in eicosapentanoic acid (EPA; C20:5n-3) and docosahexanoic acid (DHA; C22:6n-3) on the oxylipins and FA profile in sow's milk. Thirty-six sows [Landrace × Large white] from 3rd to 6th parities were divided into groups of two as similar as possible regarding age, parity number and initial body weight, and they were randomly assigned to the control or the supplemented diet. Milk samples were obtained at weaning after i.v. injection of 1.0 ml of oxytocin (20 IU/ml). Fatty acids were quantified by gas chromatography and the oxylipins by liquid chromatography-tandem mass spectrometry. The results were analysed as a randomised block design using a Proc Mixed procedure of SAS. The concentrations of n-3 FAs and oxylipins synthesised from EPA and DHA precursors were increased in the milk of sows fed the experimental diet. Focusing on oxylipins, resolvin E2/3 and 17-HDHA, probably two of the main oxylipins derived from EPA and DHA, respectively, were significantly increased in the supplemented sows. It has been described that both metabolites, resolvin E2/3 and 17-HDHA, possess a very potent anti-inflammatory activity, which might result in an increase of anti-inflammatory cytokine production as well as an inhibition of proinflammatory cytokines synthesis in the offspring. To conclude, supplementation of sow feed with EPA and DHA increase the concentration of these FAs and their anti-inflammatory oxylipin derivatives in milk.

***Nannochloropsis oceanica* inclusion in piglet diets increase n-3 long-chain fatty acids in tissues**

T. Fernandes[1], S.P. Alves[1], D.M. Ribeiro[2], C.F. Martins[2], J. Silva[3], A.M. Almeida[2], R.J.B. Bessa[1] and J.P.B. Freire[2]
[1]Universidade de Lisboa, CIISA- Centro de Investigação Interdisciplinar em Sanidade Animal, Faculdade de Medicina Veterinária, Polo da Ajuda, 1300-477, Portugal, [2]Universidade de Lisboa, Instituto Superior de Agronomia, Tapada da Ajuda, 1349-017, Portugal, [3]Allmicroalgae, Lisboa, 1250-096, Portugal; tatianefernandes@fmv.ulisboa.pt

Microalgae *Nannochloropsis oceanica* (NCO) is rich in eicosapentaenoic acid (EPA). The use of NCO in piglet feeding can result in tissue deposition of EPA that can be converted to docosapentaenoic acid (DPA) and docosahexaenoic acid (DHA). In this study, we evaluate the effect of dietary inclusion of NCO on EPA, DPA and DHA deposition in muscle and subcutaneous fat (SCF). Piglets were divided into 4 groups (n=6): control, 5, 10 and 15% NCO dietary inclusion as a replacement of the basal diet. The trial lasted two weeks and each piglet was kept in an individual metabolic crate. During week 2, faeces were collected for proximal analysis. At the end of the trial, piglets were slaughtered, and muscle, and SCF sampled from the loin area. Fatty acid (FA) methyl esters from muscle, SCF and faeces were extracted and quantified with gas chromatography. Data were analysed using linear and quadratic polynomial contrasts to test the effect of NCO incorporation. The proportion (% of total FA) and the content (mg/g dry matter) of EPA, DPA, and DHA in the muscle increased linearly (P<0.001) with NCO inclusion. However, total FA content decreased linearly in muscle with increasing NCO (P<0.01) contents. For the EPA and DPA in SCF, both the proportions and contents increased linearly with NCO inclusion (P<0.01), whereas DHA was unaffected (P>0.05). Despite the EPA increase in tissues, there was also a linear increase of EPA in faeces (P<0.01). In conclusion, the NCO increased EPA, DPA, and DHA deposition of in both tissues, however, the microalgal cell wall complexity likely reduced the piglets' ability to use the total amount of the available EPA. Work funded ALGAVALOR and UIDP/CVT/276/2020 (CIISA) projects.

Precision feeding optimises efficiency of protein utilisation in pigs

A. Remus[1], J. Van Milgen[2] and C. Pomar[1]
[1]Agriculture and Agri-Food Canada, Sherbrooke, QC, J1M0C8, Canada, [2]INRAe, Saint-Gilles, BRE, 35590, France;
aline.remus@canada.ca

To determine amino acid (AA) requirements for pigs in precision feeding systems, it is important to understand the sources of variation in protein utilisation among pigs. It is possible that using fixed AA ratios to determine requirements limits protein deposition (PD) and contributes to this variation. This study aimed to measure PD in pigs fed with daily tailored diets where either dietary lysine (Lys) or threonine (Thr) were provided at independent variable rates. A total of 95 growing pigs (35 kg body weight (BW)) were used, with electronic ear tags granting them access to automatic feeders. Pigs were randomly assigned to treatments according to an unbalanced 5×5 factorial arrangement with Lys and Thr provided at 5 levels (i.e. 60, 80, 100, 120 and 140% of their estimated individual requirements of Lys and Thr). The pig was considered the experimental unit and the trial lasted 21 days. Daily standardised ileal digestible Lys and SID Thr requirements were estimated using individual daily feed intake and weekly body weight measurements. Total body fat and lean content were measured by dual-energy X-ray absorptiometry on days 1 and 21. Data were analysed using a mixed regression model and means compared by a Tukey test. The variation in the response in PD to Lys and Thr observed was large and similar to those studies using fixed AA ratios, with Lys and Thr intake explaining only 11% of the variation. Results were also analysed by cluster analysis discriminating pigs with low (167 g/d, n=16), medium (191 g/d, n=38) and high (213 g/d, n=37) PD, but with same average Lys and Thr intake among clusters. Low, medium, and high PD clusters differed ($P<0.05$) in gain:feed ratio (0.51, 0.56, and 0.62, respectively), nitrogen retention (50, 58, and 64%, respectively), and in efficiency of Lys (71, 74 and 81%, respectively) and Thr (59, 64 and 70%, respectively) utilisation above maintenance. The efficiency of Lys and Thr utilisation decreased considerably with increasing AA intake level ($P<0.001$), independent of the cluster. Differences in PD were associated to differences in efficiency of nutrient utilisation, and providing Lys and Thr in an independent way did decrease the variability in PD.

Iron supplementation to piglets through a milk cup system compared to parenteral iron administration

N. Jakobsen[1], M.L.M. Pedersen[2] and C. Amdi[1]
[1]University of Copenhagen, Department of Veterinary and Animal Sciences, Grønnegårdsvej 2, 1870 Frederiksberg,
Denmark, [2]SEGES, Danish Pig Research Centre, Axeltorv 3, 1609 Copenhagen, Denmark; nadia_jakobsen@hotmail.com

Piglets are born with limited iron reserves and due to a low amount of iron in sow milk and limited access to natural iron sources, conventional piglets not supplemented with iron after farrowing, are at risk of developing iron deficiency anaemia. An alternative could be to provide iron to piglets through sow milk replacer fed in a milk cup system, which would enable the piglets to control their own iron intake, limit negative side effects associated with parenteral administration and reduce labour. The objective of this study was to investigate if iron can be allocated to piglets through sow milk replacer fed in a milk system with efficacies comparable to intramuscular (IM) administration of 200 mg gleptoferron. This was done by allocating 294 piglets from 21 litters to three different iron treatments. The treatments were Control (C, n=98) providing no supplemental iron, Injected iron (II, n=98) provided 200 mg gleptoferron IM on day 3 postpartum, Milk iron (MI, n=98) provided sow milk replacer 'DanMilk Supreme' added 1% 'Piglet Boozt' *ad libitum* from day 0 until 21 days postpartum. All piglets had access to dry feed from day 6. Start and end weight was registered and haemoglobin (Hb) levels were analysed using the HemoCue 201+ device on day 0, 3, 7, 10, 17 and 21 after farrowing. The results showed that iron treatment had a significant effect on Hb levels ($P<0.001$) that were different between all groups from day 10, resulting in a mean Hb level of 76.2 g/l (C), 120.9 g/l (II) and 105.4 g/l (MI) on day 21. It is concluded that iron can be supplemented to piglets through a milk cup system with efficacies comparable to parenteral administration of 200 mg gleptoferron.

SNAPIG: predicting nutrient digestion kinetics in pigs based on diet and ingredient properties

M. Schop[1,2], A.J.M. Jansman[1], S. De Vries[2], J.L. Ellis[3] and W.J.J. Gerrits[2]
[1]Wageningen Livestock Research, Animal Nutrition Department, P.O. Box 338, 6700 AH Wageningen, the Netherlands,
[2]Wageningen University & Research, Animal Nutrition Group, P.O. Box 338, 6700 AH Wageningen, the Netherlands,
[3]University of Guelph, Centre of Nutrition Modelling, 50 Stone Rd E, Ontario N1G 2W1, Canada; marijke.schop@wur.nl

Currently, the nutritional value of feed ingredients for pigs is based on static values presented in feeding tables. By using such fixed values, however, we ignore the fact, the nutritional value of feed ingredients depends on the kinetics of nutrient digestion, which is affected by diet and animal interactions. To improve future predictions of the nutritional value of feed ingredients and diets, we developed an *in silico* nutrient-based mechanistic digestion model; 'SNAPIG'. The model aims to predict the absorption of nutrients by simulating nutrient digestion kinetics in pigs fed diets varying in feed ingredients and physicochemical properties. By combining *in vitro* (feed ingredients) and *in vivo* (pig) data, the digestion kinetics of starch, protein, fat, and NSP is simulated in the stomach, small intestine, and large intestine, respectively. Subsequently, SNAPIG predicts the absorption kinetics of nutrients, such as glucose and amino acids. Model evaluation against observational data from (portal) blood studies in pigs (12 studies, 32 dietary treatment means), indicated adequate (RMSPE=39%) model predictions regarding the variation in the time of peak (44±15 *v.* 56±20 min after meal) and extent (69±30 vs 63±20% of intake) of glucose absorption after a meal. For protein, the extent of small intestinal digestion was adequate though slightly over-predicted (70±5 vs 78±5%, RMSPE=12%). For amino acid absorption, however, variation in the absorption kinetics of amino acids from various diets and protein sources could not be predicted, despite adequate mean predictions (61±11 vs 58±34 min, RMSPE=60%). These results indicate that SNAPIG is able: to predict nutrient digestion kinetics in pigs fed diets varying in feed ingredient composition and physicochemical properties; and, to identify knowledge gaps in pig nutrition and physiology. As such, it is a promising tool for academia and industry to improve future pig feed evaluation.

Bi-level optimisation of feeding and shipping strategies in pig-fattening units

M. Davoudkhani[1], F. Mahé[2], J.Y. Dourmad[1], A. Gohin[3], E. Darrigrand[2] and F. Garcia-Launay[1]
[1]INRAE, Agrocampus Ouest, PEGASE, 35590 Saint-Gilles, France, [2]Université de Rennes, CNRS, IRMAR – UMR 6625, 35000 Rennes, France, [3]INRAE, Agrocampus Ouest, Smart Lereco, 35011 Rennes, France; mohsen.davoudkhani@inrae.fr

Economic results of pig-fattening systems vary greatly and depend in part on prices of pork and feeds, as well as pig growth performance (e.g. slaughter weight, lean percentage). Previous studies revealed that feeding and shipping strategies are critical factors in the economic outputs of pig production. However, they failed to consider both strategies and the variability in pig growth performance simultaneously. Consequently, the objective of this study was to develop a new procedure to improve the profitability of pig farms by estimating the best compromise among feeding costs, pork price, animal performance, and shipping constraints. We considered a bi-level programming problem in which the upper-level represents a bioeconomic model that simulates the growth of each pig according to its biological traits whereas the lower-level represents a linear least-cost feed formulation. Bioeconomic decisions taken at the upper-level are live weight at diet changes, the percentage of mean amino acid requirement to be covered at the start of each phase, and the target weight for slaughter. Maximising the mean gross margin per fattened pig is the objective function at the upper-level. It depends on pork price and feeds cost (the objective function of the lower-level) which results from the proportion of each feed ingredient (the lower-level decision variables) at the lower-level. The optimisation problem at the upper-level is solved using an evolutionary algorithm. We considered three sets of prices: average pork and feed prices, high pork and low feed prices, and low pork and high feed prices. The changes in pork prices had major impacts on shipping decisions while had minor impacts on feeding decisions. Optimising the shipping strategy at the same time modified the optimal feeding strategy. Considering the bi-level optimisation model improved the gross margin by 1.65 €/pig (5.2%) compared to the situation where each model (the bioeconomic model and least-cost feed formulation) was optimised separately, and by 3.59 €/pig (11.2%) compared to the common practice on farms in France.

Effects of Improvac® on the testicular function of adult male Bísaro pigs

S. Botelho Fontela[1], C. Castelo[2], G. Paixão[1], M.A. Pires[1], R. Payan-Carreira[3] and A. Esteves[1]
[1]*Animal and Veterinary Research Centre (CECAV), Quinta de Prados, 5000-801, Portugal,* [2]*Associação de Criadores de Suínos da Raça Bisara, Edifício da Casa do Povo, Largo do Toural, 5320-311 Vinhais, Portugal,* [3]*MED – Mediterranean Institute for Agriculture, Environment and Development, University of Évora, 7006-554 Évora, Portugal; sbotelho@utad.pt*

Surgical castration of male pigs is a current practice in most European countries. Because of welfare concerns, the present research has been made to find alternatives that provide animal products with the same characteristics as the castrated ones. The main reason for castration is the unpleasant smell and taste of mature pigs, known as boar taint. An alternative to surgical castration is the immunisation of the pigs against gonadotropin-releasing hormone (GnRH), as in Improvac® from Zoetis. This immunisation promotes testicular atrophy, with subsequent reduction of androstenone production – the component responsible for the boar taint. The aim of this study was to assess the testicular function on adult males treated with Improvac. A total of 19 boars selected from two litters were assigned on two experimental groups: boars (n=9) without treatment (control) and immunocastrated males (n=10) who received two doses of Improvac, the first dose at 11 months old and the second dose four weeks after. All pigs were housed in outdoor pens and were fed with commercial diets, acorns and beets. Live weight and scrotal measures were taken in the beginning of the study and every four weeks until slaughter, at 13 months old. Testicles were removed from the carcass and measured post mortem. Histological evaluation was made to assess the testicular function. The small testicles observed in immunocastrated animals suggest an impaired function of these when compared to the testicles of non-castrated boars, which demonstrate full maturity and functioning. The implementation of this kind of protocols can provide a safe alternative to surgical castration in mature male pigs.

Male pig pubertal development and risk of boar taint

C. Dugué[1], S. Ferchaud[2], D. Grivault[2], M.-J. Mercat[3], S. Boulot[3], A. Prunier[4] and C. Larzul[1]
[1]*INRAE, GenPhySE, 24 Chemin de Borderouge, Auzeville CS52627, 31236 Castanet-Tolosan, France,* [2]*INRAE, UE GENESI, La Gouvanière, 86480 Rouillé, France,* [3]*IFIP-Institut du Porc, La Motte au Vicomte, BP35104, 35651 Le Rheu, France,* [4]*INRAE, Pegase, 16, le clos, Domaine de la Prise, 35590 Saint-Gilles, France; claire.dugue@inrae.fr*

Male pigs are commonly surgically castrated, without anaesthesia, to avoid a meat quality defect called boar taint, partially due to accumulation of androstenone in fat. Androstenone is a steroid produced by the testes. To decrease boar taint without using castration, a selection against androstenone should be considered. The purpose of this study is to estimate the relationships between androstenone and reproductive traits in boars. In this study we collected 34 traits for 36 Duroc boars, 40 French Landrace boars and 38 Pietrain Axiom NN boars. Characteristics of the semen (number of spermatozoids, mobility, motility, volume, concentration, duration of ejaculation and viability), libido indicators (number of training sessions before the first successful collection and percentage of later refusals), plasma testosterone and oestradiol levels, backfat levels of androstenone, skatole and indole at 180 and 280 day old, the age at 100 kg, the weight of genital tract and indicators of animal behaviour, health and welfare (human-animal relationship test, tears, health and leg scores). We estimated a high Pearson correlation between plasma oestradiol and androstenone in backfat at around 180 days of age (0.57). This confirms that plasma oestradiol can be used as a proxy for androstenone at the standard slaughter age. From Principal Component Analyses, it seems that fat androstenone at 180 days of age is independent from health score, behaviour toward human, libido and semen characteristics. These results suggest that a selection for lower androstenone in fat or plasma oestradiol in pigs at a standard slaughter age would have very-limited impact on libido or semen production capacities of mature boars.

Evaluation of body development of gilts during gestation and litter characteristics at birth

J.F. Hornstra and J. Zonderland
De Heus Animal Nutrition, R&D, Rubensstraat 175, 6717 VE Ede, the Netherlands; jhornstra@deheus.com

This study was conducted to explore the effect of the development of body weight (BW) and body composition (BC) of gilts during gestation on litter characteristics at birth. A total of 80 Topigs Norsvin (TN70) gilts were used. During gestation, gilts were housed in groups and provided the same diets, with feeding levels increasing gradually. BW and BC were measured at the time of first service and subsequently each three-weeks during gestation and at relocation to the farrowing room. BC was assessed by the amount (mm) of backfat thickness (BFt) and muscle thickness (Mt), measured by ECM Imago.S ultrasound scanner at the P2 position. At farrowing, litter weight (LW), within-litter birth weight variation (LWv), the number of total born (TB), dead born (DB) and mummified (MU) piglets were recorded. BW, BFt and Mt development during gestation were predicted using polynomial regression analysis. Multivariate regression analysis was used to evaluate BW and BC changes during early, mid and late gestation on litter characteristics at birth. At first service gilt age was 247±13 days, BW 157±11 kg, BFt 12.3±1.1 mm and Mt 55.6±3.9 mm. BW increased linearly during gestation (P<0.001) until a BW of 228±15 kg at farrow. A quadratic response was found for BFt (P<0.001) and Mt (P<0.001) development during gestation. BFt and Mt reached a plateau 15.0±1.7 mm and 59.6±3.8 mm respectively at 9 weeks of gestation. An increased BW and Mt during early gestation were associated with a higher number of TB piglets (P≤0.001 and P=0.01, respectively). During mid-gestation, an increase in BFt and Mt was associated with a lower LW (P=0.04 and P=0.02, respectively) and more MU piglets (P=0.025 and P=0.087, respectively) at birth. An increase in Mt also tended to relate with a higher LWv (P=0.08) at birth. Changes in BC during the late gestation were not related to any of the litter characteristics at birth. This study shows that changes in BW and BC of sows during early and mid-gestation have an impact on litter characteristics of gilts at birth.

Impact of straw provision on floor or in racks on tail lesions, hygiene and straw availability

T. Wallgren and S. Gunnarsson
Swedish University of Agricultural Sciences, Dept. of Animal Environment and Health, Box 234, 53223 Skara, Sweden; torun.wallgren@slu.se

Straw enables natural behaviour and reduces tail biting in pigs reared under commercial conditions. Daily straw rations of ~400 g/pig/day is required to fulfil pigs' behavioural needs. Large amounts of straw has however been reported causing poor pen hygiene and blockage of the manure systems. Traditionally, smaller amounts of straw is provided directly on the solid floor in partly slatted pens. This study compared straw provision on the pen floor (C) to straw racks (SR) during the fattening period (30-120 kg LW, 14 weeks in production (WIP)) in a commercial finishing pig farm rearing undocked pigs. The aim was to investigate if straw racks could enable larger straw rations and reduce tail damages, without compromising hygiene. The study comprised of one batch of 459 pigs in 42 pens. Every second pen contained a straw rack holding 44 l straw, remaining pens received a daily straw ration of 25 l straw according to normal routines. Tail damages (TD) were scored weekly (undamaged/swollen/bite marks/wound/ inflamed wound). Amount of clean left over straw and manual cleaning of the pens was scored daily. Although the pigs were provided with straw on the floor since birth, it took ~4 weeks before the pigs managed to empty the straw rack while the pigs in C consumed the majority of the provided straw at all times. Later in production, also the pigs in SR consumed all straw provided. These results may reflect the development of the pigs' spatial cognition rather than their biological need for exploratory material. Preliminary results show that WIP (P<0.0001), Baseline damage (P<0.0001) as well as Treatment×WIP (P<0.01) had significant effect on the percentage of TD. In WIP 2 and 4, pigs in C had significantly more undamaged tail than SR. In WIP 6, SR had significantly more undamaged tails (UT) than C. Baseline damage indicated the TD at the first recording. Pigs with baseline UT were more likely to have UT also in the future compared to other baseline damages. Pigs with baseline inflamed wound were more likely to have damages in the future, compared to other baseline damages. There were no difference in future tail damages between individuals with baseline swelling, bite marks or wounds.

Isoquinoline alkaloids in sow diets reduce body weight loss and improve quality of colostrum

E. Arévalo Sureda[1], V. Artuso-Ponte[2], S.C. Wall[2], B. Li[1], W. Fang[1], J. Uerlings[1], Y. Zhang[1], M. Schroyen[1], C. Grelet[3], F. Dehareng[3], J. Wavreille[4] and N. Everaert[1]
[1]*Gembloux Agro-Bio Tech, University of Liège, Passage des Déportés, 2, 5030 Gembloux, Belgium,* [2]*PhytobioticsFutterzusatzstoffe GmbH, Wallufer Str. 10, 65343 Eltville, Germany,* [3]*Walloon Agricultural Research Centre, Valorisation of Agricultural Products, Chaussée de Namur 24, 5030 Gembloux, Belgium,* [4]*Walloon Agricultural Research Centre, Production and Sectors, Rue de Liroux 8, 5030 Gembloux, Belgium; sc.wall@phytobiotics.com*

The aim of the study was to supplement sow diets with isoquinoline alkaloids (IQ) from plant extracts during gestation to decrease stress at farrowing and improve piglet gut development and growth performance. Twenty-four sows were allocated in three dietary groups: NC (basal diet), IQ1 (90 g/t IQ gestation day (G)80-109 and 150 g/t IQ G110-weaning) and IQ2 (150 g/t IQ G110-weaning). Blood was taken from sows five days before, at farrowing and one week after to measure cortisol, glucose and insulin. Protein, fat, IgA and IgG were analysed in colostrum and milk. Sow's body weight, feed intake, back-fat thickness, and back-muscle thickness were followed. Piglets were monitored for body weight and diarrhoea and on post-weaning day 5, intestinal morphology and gene expression (IL6, IL10, TNFa). Statistical analyses were done by ANOVA, considering sow/litter the experimental unit, the effects of treatment and time were evaluated, and significance was considered when $P<0.05$. The IQ-fed sows tended to lose less body weight. No differences were found in feed intake, back-fat thickness, and back-muscle thickness. Five-days before farrowing, sows' blood glucose and insulin levels were lower in the IQ groups (IQ2, $P<0.05$) compared to the NC group. There were no differences in cortisol between treatments. Colostrum of IQ groups had higher content in protein and IgG ($P<0.05$), but no differences were found in fat content. Piglets showed no effects in the zootechnical parameters nor in the physiological measurements. IQ seems to have their main effect on sow's metabolism, reducing body weight loss during lactation. Providing IQ to sows during the lactation period might suffice to have the desired effects. IQ enhanced the production of protein and IgG rich-colostrum, providing piglets with increased passive immunity.

Poster presentations and discussion

S. Vigors[1] and K. Nilsson[2]
[1]*University College Dublin, Belfield, D4, Ireland,* [2]*Swedish University of Agricultural Sciences, Ulls väg 26, 750 07 Uppsala, Sweden; staffordvigors1@ucd.ie*

Time slot for posters

Effect of the inclusion of probiotics in growing-finishing pig diets: animal performance

D. Villegas Estrada[1], L.E. Robles Jimenez[1], V.L. Villegas Vazquez[1], V. Ambriz Vilchis[2] and M. Gonzalez Ronquillo[1]
[1]Universidad Autonoma del Estado de Mexico, Facultad de Medicina Veterinaria y Zootecnia, Animal Nutrition, Instituto Literario 100, 50000, Toluca, Mexico, [2]Biosimetrics Ltd., Head of Technical Services, Kings Building West Mains Rd, Edimburg, 3H9 3JG, United Kingdom; danielavillegasestrada@gmail.com

The aim of this study was to evaluate published data on the use of probiotics, and its effects on performance in growing-finishing pigs specifically: feed intake, digestibility and N excretion. A database was created with articles that studied the effect of probiotics supplementation on pigs. Publications were obtained from searches on Scopus, Web of Science, Google Academic, Elsevier, ResearchGate, BMC Research, PubMed, and Redalyc. A total of 50 research papers were identified. Of these, 10 articles fitted the selection criteria were included in the final database. Treatments were defined as basal diet (Control), basal diet plus the inclusion of probiotic product with *Bacillus* spp. (*Bacillus* spp.), *Lactobacillus* spp. (*Lactobacillus* spp.) and a mixture of two or more probiotic products (Combo). The variables included weight gain (WG, g/d or kg/d), average daily gain (ADG, kg/d), feed conversion ratio (FCR), average daily feed intake (ADFI, kg/d), sample size (N), breed (n), treatment duration (TD, days), for the diet, metabolisable energy (ME, Kcal/kg of DM), Crude Protein (CP), calcium (Ca) and phosphorus (P, %), Dry Matter digestibility (DMd %), and probiotics dosage (PD, g/kg $LW^{0.75}$). Least-squares means of the control and experimental group and variability measures (standard deviation) were considered. Data were analysed in a completed randomised design. Significant differences (P<0.01) were observed for ME content between *Lactobacillus* spp. and *Bacillus* spp. diets (3,430 and 3,319 kcal / kg DM, respectively). DMd (P<0.05), was higher for *Bacillus* spp. (81.61%) than the rest of the treatments (73.54±0.8% DMd). N intake was higher (P=0.042) for *Bacillus* spp. (6.91 g / kg MBW) than the rest of the treatments (4.14±0.09 g / kg MBW), and N balance was higher (P<0.05) for *Bacillus* spp. than the rest of the treatments. Probiotic supplementation did not affect (P>0.05) ADG and FCR, but it did (P<0.05) in N retention, positively affecting the N balance of the pig. Our data shows that, in the evaluated studies, the inclusion of *Bacillus* spp.(Doses 6.38 g/kg $LW^{0.75}$) in pig diets improve DMd and N balance, compared with *Lactobacillus* spp.(Doses 5.89 g/kg $LW^{0.75}$) and combo probiotics (0.71 g/kg $LW^{0.75}$) inclusion.

The effects of *Nannochloropsis oceanica* on post-weaning pig diets

D.M. Ribeiro[1], C.F. Martins[1], A. Chaves[1], M. Pinho[2], R.J.B. Bessa[2], A.J.M. Fonseca[3], A.R.J. Cabrita[3], J.L. Silva[4], A.M. Almeida[1] and J.P.B. Freire[1]
[1]Instituto Superior de Agronomia, University of Lisbon, LEAF Linking Landscape, Environment, Agriculture and Food, Tapada da Ajuda, 1349-017, Portugal, [2]Faculdade de Medicina Veterinária, Universidade de Lisboa, CIISA – Centro de Investigação Interdisciplinar em Sanidade Animal, Av. da Universidade Técnica, 1300-477 Lisboa, Portugal, [3]Universidade do Porto, LAQV, REQUIMTE, ICBAS, Instituto de Ciências Biomédicas de Abel Salazar, Rua de Jorge Viterbo Ferreira nº228, 4050-313 Porto, Portugal, [4]ALLMICROALGAE, Av. Eng. Duarte Pacheco nº19, 9º Piso, 1070-100 Lisboa, Portugal; davidribeiro@isa.ulisboa.pt

Nannochloropsis oceanica (NCO), an autotrophic microalgae, is highly concentrated in crude protein and long-chain polyunsaturated fatty acids. Weaned piglets are subjected to severe stress and require high-quality feed ingredients. The objective of this study was to evaluate the effect of dietary inclusion of NCO on feed digestibility and growth of weaned piglets. Piglets were divided into 4 groups (n=6): control, 5, 10 and 15% NCO dietary inclusion as a replacement of the basal diet. Each piglet was individually kept in a metabolic cage with *ad libitum* access to water. The trial lasted for two weeks and feed intake was controlled daily. In the 2nd week, faeces were collected for proximal analysis. At the end of the trial, piglets were slaughtered, and samples taken from the small intestine and its contents for histochemical and viscosity analysis, respectively. Results were analysed using the method of polynomial contrasts to test the effect of NCO incorporation levels. Average daily gain and feed:gain ratio were unaffected by NCO inclusion. Dry matter (DM) digestibility linearly decreased with increasing NCO (P<0.01). Ash digestibility increased in parallel with NCO inclusion levels (P<0.001, linear). Viscosity was increased in duodenum+jejunum contents in parallel with NCO inclusion (P<0.01, linear). The ileum had increased villus length as a consequence of increasing NCO inclusion (P<0.05, linear). Incorporating up to 15% NCO in weaned piglet diets had no detrimental effect on growth. The DM digestibility reduction possibly reflects the complexity of the microalgae cell wall and protein.

Suitability of different sampling methods for assessment of sows' cleanliness
S.M. Schmid, S. Terletzki, C. Heinemann, J.J. Hayer and J. Steinhoff-Wagner
Institute of Animal Science, University of Bonn, Katzenburgweg 7-9, 53115 Bonn, Germany; simone.schmid@uni-bonn.de

To evaluate the efficacy of hygiene measures such as the 'sow shower' in pigs or to complement the visual assessment of animals' cleanliness with more objective data, it is necessary to take samples directly of the skin surface. However, there is little information available on endogenous bacterial flora, non-invasive skin sampling methods in pigs and the suitability of sampling methods used to determine skin cleanliness and bacterial load. Therefore, it was the objective of this study to test different sampling methods for the microbiological evaluation of pig skin and evaluate their suitability and reliability. For this, samples were taken at two time points with either low or high contamination. In 34 pigs, adenosine triphosphate (ATP) and protein rapid tests as well as non-selective and selective agar contact plates (ACP) (for total viable count, *Enterobacteriaceae* and *Salmonella*) were applied on 5 different locations (claws, back, thigh, teats, ear tag). ACP were incubated at 36 °C for 48 h and analysed there upon. Furthermore, a sterile moistened cotton glove was swept 3 times along the udder and transferred to 100 ml of sterile saline solution. To analyse the saline solution and produce countable results cotton gloves and solution were blended with a stomacher for 60 s. A serial dilution series was prepared in sterile saline solution with 1% tryptone. All samplings were carried out in a dual approach. Statistical analysis was performed with SAS 9.4 by using linear models with time as fixed effect. Results show that ACP for total viable count were not suitable for use on animal skin as media were completely overgrown and therefore not evaluable. Selective ACP were able to show differences, but should be used on less bristly skin (teats, inner thigh) to avoid damaging the agar surface while pressing it on bristly skin. ATP and protein tests were easily applied and able to reveal differences between low and high groups (P<0.05). On plate count agar, differences were also clearly visible (P<0.05), indicating that glove samples are a suitable method for determining pigs' skin cleanliness.

Can homeopathic treatment prevent mastitis-metritis-agalactia (MMA) syndrome in sows?
I. Czycholl[1], J. Oelkers[1], O. Burfeind[2] and J. Krieter[1]
[1]Christian-Albrechts-University Kiel, Institute of Animal Breeding and Husbandry, Olshausenstrasse 40, 24118 Kiel, Germany, [2]Chamber of Agriculture Schleswig-Holstein, LVZ Futterkamp, Gutshof 1, 24327 Blekendorf, Germany; iczycholl@tierzucht.uni-kiel.de

Mastitis-metritis-agalactia (MMA) syndrome in sows is a common issue after birth causing severe economic losses. Treatment with antiphlogistica and antibiotics is standard veterinarian procedure in the case of disease. Due to the urge to reduce usage of antibiotics, an interesting alternative approach is the prophylactic treatment with homeopathic remedies. However, there is still a lack of reliable scientific studies on the effectiveness of homeopathic treatments. Up to date, so-called complex homeopathics are available that promise, e.g. the prevention of MMA syndrome in sows due to the ingredients which are included according to the similé principle of homeopathic doctrine. However, the usefulness has not been proven scientifically so far. Therefore, the present study aimed at testing whether the prophylactic homeopathic treatment with the complex homeopathic remedies Caulophyllum logoplex and Lachesis logoplex would reduce the occurrence of MMA in sows. In a blind study, 114 sows on two farms received randomly either homeopathics or physiological saline solution via subcutaneous injection. The injections were applied two days before birth (Caulophyllum logoplex/saline solution) and on the day of birth (Lachesis Logoplex/saline solution). Prior to study begin, on farm 1, MMA prevalence was 12.4% and on farm 2 7.0%. The assessment for MMA started for each sow on the day of birth and consisted of the parameters rectal body temperature, overall condition of sow, feed intake, vaginal discharge and agalactia. Moreover, number and weight of live born piglets and number and weight of weaned piglets, was taken into account. For analysis, generalised linear mixed models were calculated. Fixed effects were treatment group, batch, number of litters and day after birth, the random effect was the individual sow. During data analysis, MMA prevalence on farm 1 was 13.6% and on farm 2 6.3%. No significant differences were detected between the treatment groups. Hence, the effectiveness of the application of the complex homeopathic remedies Caulophyllum logoplex and Lachesis logoplex for the prophylactic treatment of MMA syndrome in sows could not be proven.

Formulas to predict cheese-yield traits from Brown Swiss milk to improve dairy chain sustainability

E. Mariani[1], C. Cipolat-Gotet[1], A. Summer[1], M. Malacarne[1], A. Cecchinato[2] and G. Bittante[2]
[1]University of Parma, Department of Veterinary Science, Via del Taglio 10, 43126 Parma, Italy; [2]University of Padova, Department of Agronomy, Food, Natural Resources, Animals and Environment, Viale dell'Università 16, 35020 Legnaro, Italy; elena.mariani@unipr.it

The aim of the study was to develop formulas based on milk composition for predicting cheese-yields (%CY) traits. The %CY traits were measured from individual model-cheeses obtained from 1,264 Brown Swiss' cows reared in 85 herds. Fresh %CY (%CY$_{CURD}$), total solids and water retained in fresh cheese (%CY$_{SOLIDS}$ and %CY$_{WATER}$), and 60 days ripened cheese (%CY$_{CHEESE}$) were considered the 'reference' traits and used as response variables. Training-testing linear regression modelling was adopted: 80% of observations were randomly assigned to training set and 20% to validation set, and the procedure was repeated ten times. Performance was assessed by R$^2_{VAL}$ and RMSE$_{VAL}$. In total 560 models were tested: (1) 80 models were based on fat and protein or casein; (2) in 240 models indirect udder health indicators (UHI; lactose and/or SCS) were added; and (3) in 240 models crude protein was replaced by 4 caseins and/or 2 whey proteins and the non-protein nitrogen (NPN), with or without UHI. The comparison of regression coefficients on %CY$_{CURD}$ with those on %CY$_{SOLIDS}$ and %CY$_{WATER}$ allowed us to disentangle the role of each milk component for the recovery of milk solids or retention of whey in fresh cheese. Moreover, regression coefficients of %CY$_{CHEESE}$ were related to the weight loss of cheese during ripening. Besides the expected effect of milk fat and protein, it was possible to assess the magnitude of different predictors related to casein fractions: β-CN > κ-CN > α$_{S1}$-CN > α$_{S2}$-CN for %CY$_{CURD}$ and %CY$_{WATER}$; κ-CN > β-CN = α$_{S2}$-CN > α$_{S1}$-CN for %CY$_{SOLIDS}$; and α$_{S1}$-CN = β-CN = κ-CN > α$_{S2}$-CN for %CY$_{CHEESE}$. Among whey proteins, α-LA showed an effect more favourable than caseins, whereas β-LG had a negative effect, like NPN and SCS. These formulas provide new insights in the assessment of milk components in a context of processing the milk into cheese. Thus, they may be useful for quantifying the economic weight of milk components in the composite selection indices of the Brown Swiss breed or for the direct genetic improvement of cheese production.

Improvement of genetic evaluation in small cattle populations by considering herd characteristics

J. Herold, K. Brügemann and S. König
Institute of Animal Breeding and Genetics, Ludwigstrasse 21 B, 35390 Giessen, Germany; jonas.herold@agrar.uni-giessen.de

Especially for breeds with small population size and kept in small family farms, it is imperative to improve reliabilities of genetic evaluations through an increase of cow records per contemporary group (CG), and to model the herd environment as detailed as possible. In the present study, we developed a comprehensive survey to monitor 105 herd characteristics in detail, aiming on the definition of specific herd cluster (HC). For herd characterisations regarding feeding, husbandry and management conditions, as well as social-ecological components, we considered 20 dairy farms keeping German Black Pied cattle and 10 farms keeping Holstein-Friesian cows. The 105 herd variables were input parameters for the evaluation of different clustering methods. The evaluation criterion 'average silhouette method' suggested 'a clustering of variables combined with an agglomerative hierarchical clustering' and the creation of four HC. The four HC were included in genetic evaluation models, considering 55,181 test-day milk observations from 5,538 cows from the first three lactations in the recording years 2012 to 2018. The genetic evaluation model includes either the herd test-day (HTD) or herd cluster test-day (HCTD) effect, or alternative herd test-month (HTM) or herd cluster test-month (HCTM). The HCTD modelling implied an increase of cow records per CG (plus 14 records compared to HTD; plus 479 records compared to HCTM). Milk yield heritabilities were 0.23 (HTD), 0.36 (HCTD), 0.23 (HTM) and 0.38 (HCTM). The reliability of estimated breeding values from the test-day model increased from 29.8% (HTD) up to 33.2% (HCTD) for the whole population, and for sires from 65.0% (HTD) up to 70.9% (HCTD). The gain in sire reliabilities depended on the genetic structure, indicating substantial gain for sires with a limited number of progeny. Nevertheless, depicting the herd environment as detailed as possible (HTD and HCTD) was more important for accurate genetic evaluations than increasing the number of cow records per CG. The alternative modelling approach might contribute to improved selection strategies for small populations kept in heterogeneous herd environments.

Pedigree and genomic-based inbreeding in Italian Holstein dairy cows

M. Ablondi[1], A. Sabbioni[1], C. Cipolat-Gotet[1], M. Malacarne[1], C. Dadousis[2], J.B.C.H.M. Van Kaam[3] and A. Summer[1]
[1]University of Parma, Department of Veterinary Science, Via del Taglio 10, 43126 Parma, Italy, [2]University of Firenze, Department of Agriculture, Food, Environment and Forestry, Via G. Donizzetti 6, 50144 Firenze, Italy, [3]ANAFIJ – Associazione Nazionale Allevatori della Razza Frisona e Jersey Italiana, Via Bergamo 292, 26100 Cremona, Italy; michela.ablondi@unipr.it

Dairy cattle breeds have been exposed to intense directional selection for milk production traits over the last fifty years. Intense selection can cause loss of genetic variability and increased inbreeding. Fast selection of breeding candidates via genomic data can boost yearly rate of inbreeding in the population. Inbreeding might in turn result in an excess of deleterious variants and inbreeding depression. Inbreeding is traditionally assessed by pedigree-based estimates (F_{PED}). Nowadays, different measures of inbreeding can be estimated from genomic data. The objective of this study was to assess and compare the level of inbreeding in genotyped Italian Holstein dairy cows from pedigree and SNP data throughout a period of 14 years (2006-2019). A total of 40,921 Italian Holstein dairy cows with imputed high-density genotypes (287,329 SNP after quality control) were available. F_{PED} and the complete generation equivalents (CGE) were computed in ENDOG 4.8. The genomic inbreeding (F_{ROH}) was calculated as the proportion of individual genome size covered by runs of homozygosity (ROH) in PLINK 1.9. Overall, genome-based inbreeding (mean F_{ROH}=0.12) was higher than pedigree-based inbreeding (mean F_{PED}=0.07). As F_{PED} largely relies on pedigree depth, pedigree incompleteness can result in an underestimation of the real inbreeding. Moderate Pearson correlation was observed between F_{ROH} and F_{PED} (r=0.60, P<0.05). The low F_{PED} might be related to the number of CGE registered in the pedigree. The highest correlation was found in the case of animals with highest CGE being equal to 10. Till 2012 the inbreeding per year slightly rose, whereas from 2013 a steep increase was observed. Finally, the rate of inbreeding per generation was equal to 1.67 and 1.58% in the last two generations for F_{PED} and F_{ROH}. The identified increase of inbreeding in the breed has led to in-depth studies to evaluate the effect of inbreeding on milk yield and fertility traits.

Using milk mid-infrared spectra to predict the SARA risk of dairy cows

A. Mensching[1], M. Zschiesche[1], J. Hummel[1], C. Grelet[2], N. Gengler[3] and A.R. Sharifi[1]
[1]University of Goettingen, Department of Animal Sciences, Albrecht-Thaer Weg 3, 37075 Goettingen, Germany, [2]Walloon Agricultural Research Centre, Chée de Namur 24, 5030 Gembloux, Belgium, [3]ULiège GxABT, Passage des Déportés 2, 5030 Gembloux, Belgium; andre.mensching@uni-goettingen.de

Technological progress has facilitated the collection and management of 'Big Data', providing a support for the development of innovative and complex functional traits in animal sciences. Improved monitoring for early detection of subclinical disorders would help to initiate timely management interventions and thus to prevent clinical diseases. In this context the associations between milk mid-infrared (MIR) spectra and sensor technically recorded reference data (reticular pH and temperature, chewing and locomotion behaviour) were investigated in dairy cows. Furthermore, traits derived from the aforementioned reference data were used to develop a subacute rumen acidosis related risk score (SRS). The associations between SRS and the MIR spectra, performance data, blood parameters and the fatty acid profile of the milk were also investigated. For this purpose, data on 100 cows in 10 dairy farms were collected. The sensor data comprise stable climate parameters, pH and temperature from intra-reticular measurement boluses as well as jaw movement and locomotion behaviour recordings of noseband-sensor halters and pedometers. Further records are milk performance, blood parameters and milk main components, MIR predicted fatty acids as well as the milk MIR spectra. Linear mixed models and perceptions derived from a principal component analysis were used to develop the SRS. Partial least squares regression with a stratified 10-fold cross-validation and different spectral pretreatments was used to analyse the associations between MIR and reference data. First results showed moderate associations between milk MIR data and the reticular daily pH range, the SRS, the daily eating and rumination time and the body temperature ($0.24 < R^2 cv < 0.33$). Furthermore, promising was that fatty acid profile changes (lower levels of saturated and short chain fatty acids and particularly higher C17:0 levels) were linked to increased SRS showing potential in the milk composition based detection of conditions where animal welfare is compromised.

Ceramide accumulation in bovine retroperitoneal adipose tissue revealed by sphingolipid profiling

Y.H. Leung[1], S. Baessler[2], C. Koch[3], T. Scheu[3], U. Meyer[4], S. Daenicke[4], Á. Kenéz[1] and K. Huber[2]
[1]City University of Hong Kong, Department of Infectious Diseases and Public Health, Kowloon, Hong Kong, [2]University of Hohenheim, Institute of Animal Science, Stuttgart, Germany, [3]Educational and Research Centre for Animal Husbandry, Hofgut Neumuehle, Münchweiler a.d. Alsenz, Germany, [4]Federal Research Institute for Animal Health, Institute of Animal Nutrition, Braunschweig, Germany; andyleung0201@gmail.com

Sphingolipids are bio-active lipids that can modulate insulin resistance, differentiation and apoptosis in a tissue specific manner. Bovine retroperitoneal adipose tissue (RPAT) was shown to be more active than subcutaneous adipose tissue (SCAT) in terms of insulin response, lipolytic activity and pro-inflammatory signalling. Sphingolipids were discussed to be involved in inflammation, however, their profiles in bovine RPAT and SCAT are currently unknown. We aimed to characterise the sphingolipid profiles using a metabolomics approach, and to assess whether potentially related sphingolipid pathways are different between SCAT and RPAT. Holstein bulls (n=6) were slaughtered, and SCAT and RPAT samples were collected for sphingolipid profiling. 77 sphingolipid species were targeted, including 24 species of ceramides (Cer) and dihydroceramides (DHCer), 19 sphingomyelins (SM) and dihydrosphingomyelins (DHSM), 16 ceramide-1-phosphates (C1P) and sphingosine-1-phosphates (S1P), 10 galactosyl ceramides (GalCer), glucosyl ceramides (GluCer) and lactosyl ceramides (LacCer), and 8 sphinganines (DHSph) and sphingosines (Sph). The concentration of each sphingolipid was quantified by UPLC-MRM/MS. Our results showed that sphingolipids of the de novo synthesis pathway such as DHSph, DHCer, and Cer, were more abundant in RPAT than in SCAT. Sphingolipids of the salvage pathway and of sphingomyelinase pathway such as Sph, S1P, C1P, glycosphingolipid, and SM were more abundant in SCAT. Our results indicate that RPAT had a greater extent of ceramide accumulation, and thereby increased the abundance of further sphingolipid intermediates in the de novo synthesis pathway. This distinctive sphingolipid distribution pattern in RPAT and SCAT can potentially explain the tissue specific activity in insulin response, pro-inflammation, and oxidative stress in RPAT and SCAT.

Body traits and milk productivity indicators of purebred and crossbred dairy cows

M. Piazza, S. Saha, S. Schiavon, N. Amalfitano, G. Bittante and L. Gallo
University of Padova, Department of Agronomy, Food, Natural resources, Animals and Environment (DAFNAE), 35020 Legnaro, (PD), Italy; martina.piazza.1@phd.unipd.it

This study aimed to compare purebred Holstein (HO, no=360) and crossbred cows (CR, no=275) in terms of productivity indicators obtained combining different body measures, body condition score (BCS) and milk yield (MY), composition and cheese-yield traits assessed through individual model cheese-making procedure. Crossbred cows originated from a 3-way rotational crossbreeding scheme starting from HO cows and involving Viking Red, Montebèliarde and HO sires (4 generations, PROCROSS scheme). Cows were kept in 2 herds specialised for typical hard-cheese production. All cows were measured (withers height, hearth girth and body length) and 225 cows were weighed to develop an equation to predict body weight (BW) from body measures. Predicted BW and BCS were used to predict body composition and energy content (NRC) and to estimate the net energy requirements for maintenance (NEm, MJ/d) on the basis of metabolic weight (NRC) or body protein mass (NRC modified). MY and milk composition were used to estimate the energy content of milk (MJ/kg). When compared to HO cows, CR cows yielded (P<0.05) less milk, energy in milk and cheese per day (from -4.5 to -3.5%), whereas differences in terms of daily yield of fat and protein were not significant. Crossbred cows had lower (P<0.01) hearth girth, wither height and body length, greater (+10%) BCS, estimated body fat percentage (+11.5%) and total body energy content (+8%) than HO cows. Conversely, differences in predicted BW (+1.5%) were not significant, likewise those in NEm based on metabolic weight (+1%). Considering the huge differences in body composition, predicted NEm based on body protein mass was on the contrary lower for CR compared to HO cows (-2.5%, P<0.05). Productivity indicators were computed by scaling output traits on BW and on NEm, and CR cows evidenced significantly lower daily yield of milk, milk constituents, milk energy and cheese per unit of BW and unit of NEm based on metabolic weight (from -6 to -3.4%, P<0.05). However, when productivity indicators referred to NEm based on body protein mass as input trait, differences between CR and HO cows ranged between 2 and 3% and were never significant.

G×E for longevity and health in organic and conventional dairy cow herds genetically and genomically

T. Shabalina[1,2], T. Yin[2] and S. König[2]
[1]*Bavarian State Research Center for Agriculture, Institute of Animal Breeding, Prof.-Dürwaechter-Platz 1, 85586 Poing, Germany, [2]Institute of Animal Breeding and Genetics, Justus-Liebig-University of Gießen, Ludwigstraße 21B, 35390 Gießen, Germany; taisiia.shabalina@agrar.uni-giessen.de*

Genotype-by-environment (G×E) interactions indicate different responses of genotypes with environmental changes. Classically, G×E are analysed in multiple-trait animal models, and defining same traits from the different production systems organic and conventional as different traits. Production, longevity and health traits from Holstein cows were recorded in 9 organic (7,915 cows) and 57 conventional German herds (141,778 cows). Longevity was defined as length of productive life (LPL). The disease diagnosis dataset considered three groups of health disorders: mastitis (MAST), ovarian cycle disorders (OCD) including silent estrus, ovarian cysts, corpus luteum persistent, and the claw disorder digital dermatitis (DD). Holstein cows were genotyped with the Illumina BovineSNP50 BeadChip V2 and the Illumina Bovine Eurogenomics 10K low-density chip. Animals with the low density 10K genotypes were imputed to the 50K level. For the estimation of heritabilities and genetic correlations in bivariate animal models, pedigree- and genomic-based relationship matrices were considered. The heritability for LPL using pedigree-based relationships was 0.12 in the organic and 0.08 in the conventional production system. Heritabilities for health traits ranged from 0.02 (OCD) to 0.33 (DD), and were very similar from pedigree and genomic modelling approaches. Genetic correlations for disease traits and LPL between organic and conventional production systems either using pedigree or genomic relationship were lower than 0.80, indicting obvious G×E. In contrast, genetic correlations between moderate heritability test-day production traits form the different systems were larger than 0.80. In an ongoing approach using the genomic marker data, patterns of genome wide associations were different for LPL and health traits in both production systems, and different gene networks and causal pathways were identified. Generally, environmental sensitivity on quantitative-genetic and genomic levels was generally stronger for functional than for production traits.

Exploring the colostrum microbiome in Holstein Friesian and double muscled Belgian Blue cattle

I. Van Hese[1,2], K. Goossens[2], L. Vandaele[2] and G. Opsomer[1]
[1]*Ghent unversity, Reproduction, Obstretrics and Herd Health, Salisburylaan 133, 9820 Merelbeke, Belgium, [2]Flanders Research Institute for Agriculture, Fisheries and Food (ILVO), animal science unit, Scheldeweg 68, 9090 Melle, Belgium; ilke.vanhese@ilvo.vlaanderen.be*

Colostrum, the first milk a cow produces after giving birth, is of vital importance for the neonate. Besides supplying antibodies, referred to as 'passive transfer of immunity', colostrum contains beneficial bacteria as well. In humans, these beneficial bacteria are thought to express immune modulatory function after colonising the infant's gut. The present study investigates colostral microbiome composition from Holstein Friesian (HF) and double muscled Belgian Blue (DMBB) cows. We aimed to find a link with colostrum quality (i.e. IgG concentration) and the transfer of passive immunity in the calf. Colostrum samples were taken from 64 HF and 46 DMBB cows at the ILVO research farm immediately after parturition. Calves received a total of 6 l of colostrum, 3 times 2 l within 2, 6 and 24 h after birth, respectively. At day 3 after birth, blood samples were taken from calves to measure serum IgG levels. Microbial DNA was extracted from colostrum with the Powerfood microbial kit (Qiagen, Germany). Library preparation and amplicon sequencing of the bacterial V3-V4 region of the 16S rRNA gene was performed using the Illumina MiSeq V3-technology (Macrogen, South-Korea). Taxonomy was assigned using the SILVA database with aligned ribosomal RNA gene sequences. Colostrum IgG concentration (mean ± SD) averaged 64.92±12.74 g/l. Serum IgG concentration (mean ± SD) averaged 20.60±9.26 g/l. The most abundant genera were *Acinetobacter, Enterococcus, Pseudomonas, Staphylococcus* and *Streptococcus*. Microbiome composition of colostrum differed significantly between DMBB and HF. In addition, within each breed, there was a significant difference in microbiome composition between colostrum from good versus bad quality, between colostrum from cows that calved in spring versus autumn and between colostrum administered to calves that obtained low versus high serum IgG. However the latter was only observed in colostrum from HF cows. Acknowledgements: Ilke Van Hese is an SB PhD fellow at FWO, Research Foundation – Flanders, project number 1S20220N.

Differential somatic cell count as novel indicator of milk quality of individual cows

M. Mountricha[1], G. Stocco[1], C. Claudio[1], B. Stefanon[2], A. Zecconi[3], M. Franscecutti[4] and A. Summer[1]
[1]*University of Parma, Department of Veterinary Science, Via del Taglio 10, 43126, Parma, Italy,* [2]*University of Udine, Department of Agrifood, Environmental and Animal Science, Via delle Scienze 206, 33100, Udine, Italy,* [3]*University of Milan, Department of Biomedical, Surgical and Dental Science, One Health Unit, Via Pascal 36, 20133, Milan, Italy,* [4]*Associazione Allevatori del Friuli-Venezia Giulia, Via XXIX Ottobre 9, 33033, Italy; maria.mountricha@unipr.it*

A novel understanding of the relationship between Differential Somatic Cell Count (DSCC) and Somatic Cell Count (SCC) has emerged as a new frontier of research for dairy production. The aims of the present study were to investigate the factors affecting the variability of DSCC and SCC and to explore their combined effect on milk yield, composition and coagulation properties. A total of 51,152 individual milk samples were collected from 13,937 dairy cows of 2 breeds (Holstein-Friesian and Simmental) reared in 217 herds in the Northeast of Italy from July to January. Milk samples were analysed using CombiFoss 7 (Foss Electric A/S, Denmark) and the content of fat, protein, casein and lactose, milk coagulation properties, SCC and DSCC (polymorphonuclear leukocytes+lymphocytes, %) were measured. A linear mixed model including individual herd and animal as random factors, herd size, milking type, season, breed, parity and days in milk (DIM) as fixed factors was used to assess DSCC and SCC variability. A second linear mixed model including the same factors together with the fixed effects of DSCC, SCC and their interaction was tested on milk yield, composition and coagulation properties. Animal effect explained the largest part of the variation of DSCC. During lactation, DSCC was high at 60 DIM and then it gradually reduced. Milk yield declined with increasing DSCC and SCC, while the trend was opposite with low SCC and high DSCC. We observed a low % of fat in milk with low SCC and simultaneously an elevated DSCC. Further understanding on the changes in milk yield and composition was provided by the interaction of DSCC and SCC, which likely reflects animal condition. Finally, DSCC could be used to monitor milk quality at herd level and it can provide useful information to include novel traits in the milk payment system.

Reducing the number of critical transitions for dairy cows: effects on milk yield and health

E.E.A. Burgers[1,2], R.M.A. Goselink[1], B. Kemp[2] and A.T.M. Van Knegsel[2]
[1]*Wageningen University & Research, Wageningen Livestock Research, P.O. Box 338, 6700 AH Wageningen, the Netherlands,* [2]*Wageningen University & Research, Adaptation Physiology group, P.O. Box 338, 6700 AH Wageningen, the Netherlands; eline.burgers@wur.nl*

Drying-off, calving, and start of lactation are critical transitions for a dairy cow. By extending the voluntary waiting period for insemination (VWP), calving interval (CI) is extended, reducing the number of transitions per year, with possible positive effects for health. The risk of increased CI is a reduction in milk yield per year, as there are less peaks in milk. This study investigated effects of extended VWP on milk yield, body condition, and disease incidence. Cows (n=150) within one herd were blocked for parity, calving season, expected milk yield and persistency, and 6 weeks after calving within blocks randomly assigned to one of three VWP (50, 125, or 200 days), and monitored for a complete lactation. Fat-and-protein-corrected milk yield per day (FPCM), body weight, and BCS were analysed for the first 40 weeks in lactation and for the last 12 and 6 weeks before dry-off, using a repeated measurements model with fixed effects of VWP, parity (1 or 2+), time, and interactions, and cow as repeated subject. Milk yield was also calculated as FPCM per day of CI. Mean CI was 393, 457, and 503 days, for VWP50, VWP125, and VWP200, respectively. In the first 40 weeks in lactation, FPCM was not affected by VWP for parity 1, while for parity 2+ FPCM was lower in VWP200 compared with VWP125 or VWP50 (32.7 vs 35.6 vs 34.9 kg/day, P<0.10). FPCM per day of CI was not affected by VWP for parity 1, while for parity 2+ FPCM per day of CI was lower in VWP200 compared with VWP125 or VWP50 (23.5 vs 26.5 vs 27.1 kg/day, P<0.05). In the last 6 weeks before dry-off, FPCM was greater for VWP50 compared with VWP125 or VWP200 for parity 2+ (20.3 vs 17.2 vs 14.7 kg/day, P<0.05). In the last 12 weeks before dry-off, for parity 2+, BCS was greatest for VWP200, intermediate for VWP125, and smallest for VWP50 (3.5 vs 3.0 vs 2.6, P<0.05). Extending VWP for heifers did not affect milk yield, but cows with a 200 d VWP had a reduced milk yield per day and less milk before dry off. Extending VWP for cows could increase fattening in late lactation compared with a VWP of 50 days.

On the role of CSN2, CSN3 and BLG genes on the milk protein profile of Brown Swiss cows

N. Amalfitano[1], L.F.M. Mota[1], G.J.M. Rosa[2], A. Cecchinato[1] and G. Bittante[1]
[1]*University of Padova, Department of Agronomy, Food, Natural resources, Animals and Environment, Viale dell'Universita 16, 35020 Legnaro PD, Italy,* [2]*University of Wisconsin, Madison, Department of Animal Sciences, 1675 Observatory Dr, 53706 Madison WI, USA; nicolo.amalfitano@studenti.unipd.it*

A better understanding of the role of the major genes coding for milk protein fractions should be pursued for improving dairy cattle breeding programs. The aim of the present work was to study the effect of the CSN2, CSN3 and BLG genes on the major protein fractions (α_{s1}-CN, α_{s2}-CN, β-CN, κ-CN, β-LG, α-LA), expressed qualitatively as percentage of total nitrogen content (%N) and quantitatively as content in milk (g/l). A total of 1,264 Brown Swiss cows were sampled individually in 85 commercial herds in North-East Italy, and 989 cows were genotyped using the Illumina Bovine SNP50 v.2 BeadChip, from which 37,519 SNP markers were used. The Reversed-Phase High Performance Liquid Chromatography was used for quantifying the content of the six major milk protein fractions, with β-CN, κ-CN and β-LG genetic variants being quantified separately. The protein fractions were analysed using a Bayesian animal model implemented via Gibbs sampling. Effects of days in milk, parity order and the CSN2, CSN3 and BLG genotypes were assigned flat priors; herd and animal additive effects were assigned Gaussian prior distributions. Marginal posterior distributions of parameters of interest were compared between the standard animal model and the mixed inheritance model (i.e. the one with the protein loci included in the model). Results indicated that a considerable portion of the genetic variance was explained by the effect of the three major genes for most of the protein fractions. In particular, the CSN2, CSN3 and BLG genotypes affected not only the proteins they codify for, but also the synthesis of the other protein fractions. For the CSN genes, the B allele was associated with an overexpression of the protein synthesis compared to the κ-CN A and β-CN A^2. Conversely, the β-LG B was associated with a lower concentration of this protein compared with the β-LG A. Overall, the results could be useful for establishing more efficient selection programs aimed to improve the value of milk for dairy industry and also for consumer health.

Minerals affect milk coagulation, cheese yield and recovery of nutrients in the curd

G. Stocco[1], C. Cipolat-Gotet[1], A. Summer[1], M. Malacarne[1], A. Cecchinato[2] and G. Bittante[2]
[1]*University of Parma, Department of Veterinary Science, via del Taglio, 10, 43126, Parma, Italy,* [2]*University of Padova, Department of Agronomy, Food, Natural resources, Animals and Environment, viale dell'Università, 16, 35020, Legnaro, Italy; giorgia.stocco@unipr.it*

The aim of this study was to investigate the effect of Ca, P, K, Mg and Na on traditional milk coagulation properties [MCP: rennet coagulation time (RCT), curd-firming time (k_{20}), and curd firmness (CF) at 30, 45 and 60 min (a_{30}, a_{45}, a_{60})], modelled curd firmness over time (CF_t) parameters [estimated rennet coagulation time (RCT_{eq}), curd-firming (k_{CF}) and syneresis (k_{SR}) instant rate constants, maximum CF (CF_{max}), time at which the CF_{max} value is attained (t_{max}), and potential CF (CF_p)], and cheese-making traits [4 recovery traits (%REC): milk fat, protein, solids and energy in the curd; 3 measures of cheese yield (%CY): fresh, solids and retained water]. Milk samples were collected from 238 cows belonging to 6 breeds, reared in 27 multi-breed herds. Inductively Coupled Plasma – Optical Emission Spectrometry (ICP-OES) was used to measure mineral contents in milk. The lactodynamographic analysis allowed us to measure 240 CF values per sample (60 min analysis; 1 CF value every 15 sec) and model cheese-making procedure was used on 1.5 l milk per cow. Data were analysed using a linear mixed model including DIM, parity, Ca, P, K, Mg and Na as fixed factors, herd and breed as random factors. Milk fat and casein were also included in the model. Results showed that increasing Ca concentrations were associated with reduced coagulation time (about -4 min), fastened curd-firming rate (+2%), lowered time to achieve CF_{max} (-3 min), increased %CY traits and recovery of fat and energy in the curd. An increase in milk P improved especially %CY in solids and %REC traits. High Na content, on the contrary, delayed coagulation time so that much more time was needed to achieve CFmax, and decreased recovery of protein in the curd. The simultaneous inclusion of milk fat, casein and the minerals in the statistical model allowed to ascertain the specific effect of each mineral on both coagulation and cheese-making traits. However, further understanding is needed by considering the role of minerals also in combination with more detailed milk composition (i.e. fatty acids).

Association of mid-infrared-predicted milk & blood constituents with early-lactation adverse events

K.D. Bach[1], D.M. Barbano[2] and J.A.A. McArt[1]
[1]Cornell University, Population Medicine and Diagnostic Sciences, College of Veterinary Medicine, Ithaca, NY 14853, USA, [2]College of Veterinary Medicine, Food Science, College of Agriculture and Life Sciences, Ithaca, NY 14853, USA; kdb23@cornell.edu

Partial least square regression estimates of milk and blood constituents using Fourier transform mid-infrared (FTIR) analysis have shown promise as a tool for monitoring early lactation excessive energy deficit in dairy herds. Our objective was to analyse milk via FTIR to determine the association of early lactation predicted milk β-hydroxybutyrate (BHB) concentrations, predicted blood fatty acid (NEFA) concentrations, and predicted milk de novo fatty acid (FA) percentages relative to total FA concentrations with the risk of disease or removal in early lactation (hyperketonemia, displaced abomasum, metritis, culling, or death) and average daily milk yield during the first 15 wk of lactation. We enrolled 517 multiparous Holstein cows from 2 dairy farms in New York State. Composite milk samples were collected twice weekly from 3 to 18 DIM for a total of 4 timepoints (T1, T2, T3, T4) and analysed using FTIR spectrometry for milk BHB and FA composition and predicted blood NEFA. Blood samples were collected for hyperketonaemia determination (BHB $\geq$1.2 mmol/l) using a handheld meter, and farm-diagnosed occurrence of disease or removal during the first 30 DIM and average daily milk yield during the first 15 wk of lactation were collected from herd management software. The incidence of disease or removal between 3 and 18 DIM was 20.2%. Explanatory models for disease or removal were developed for each predicted constituent of interest at each timepoint using fixed effect multivariable Poisson regression. Repeated measures ANOVA models were developed for each predicted constituent to assess differences in average daily milk yield. For all timepoints, increased risk of disease or removal was associated with higher predicted milk BHB (Relative Risk (RR)$_{T1}$=2.0; RR$_{T2}$=3.4; RR$_{T3}$=5.2; RR$_{T4}$=9.1), higher predicted blood NEFA (RR$_{T1}$=2.7; RR$_{T2}$=2.5; RR$_{T3}$=3.8; RR$_{T4}$=10.0), and lower predicted milk de novo FA relative percentages (RR$_{T1}$=2.9; RR$_{T2}$=3.3; RR$_{T3}$=5.8; RR$_{T4}$=7.2). Average daily milk yield was increased for cows above the cut point for predicted milk BHB (2.1 kg/day) and predicted blood NEFA (3.5 kg/day) and below the cut point for de novo FA relative percentages (2.3 kg/day). Our results suggest that FTIR predicted milk BHB, blood NEFA, and milk de novo FA relative percentages are promising indicators of subsequent disease or removal in early lactation; their positive relationship with milk yield warrants further exploration.

Effects of homeopathic blend on body measurements and weight of weaned Holstein calves

T.H. Silva[1], I.C.S.B. Guimarães[1], R. Melotti[2] and A. Saran Netto[1]
[1]University of Sao Paulo, Department of Animal Science, 225, Duque de Caxias Street, 13635-900 Pirassununga, Brazil, [2]Real H, Technical Manager, 12068, Zila Correa Machado Avenue, 79046-200, Brazil; iulivet3@gmail.com

Homeopathic products have attracted interest during the past few years in order to reduce the abusive use of antibiotics in livestock. The aim of this study was evaluate the effect of homeopathic blend supplementation on body measurements and weight of weaned Holstein calves. A double-blind placebo-controlled trial of an homeopathic blend treatment was performed with 184 weaned Holstein calves (83.01±7.9 days old; 112.5±11.7 kg) which were allocated to 8 paddocks in a completely randomised design experiment. During a 112 days period, the animals received a total mixed ration with the following treatments: control (basal diet + calcium carbonate, top-dressed at 30 g/animal/day: homeopathic vehicle) and homeopathic blend (basal diet + TopVita™ – Real H, top-dressed at 30 g/animal/day: *Sulphur*: 10^{-60} + *Viola tricolor*: 10^{-14} + *Caladium seguinum*: 10^{-30} + *Zincum oxydatum*: 10^{-30} + *Phosphorus*: 10^{-60} + *Cardus marianus*: 10^{-60} + *Colibacillinum*: 10^{-30} + *Podophyllum*: 10^{-30} + Vehicle: calcium carbonate; q.s. 1 kg). Weight and body measures were taken each 28 days until 112 days of supplementation. Data were analysed by a MIXED procedure for repeated measurements of SAS and statistical significance was declared at P≤0.05. No differences were observed on body weight every 28 days period and ADG between treatments during 112 days of the study. Homeopathic blend increased withers height during the initial growing period from d0 to d28 (95.46 vs 96.71 cm) and body depth at d28 (48.2 vs 48.98 cm) and d84 (51.61 vs 52.78 cm) compared to control group, but no differences were detected for the other periods. Pelvic width was higher for homeopathic treatment on d84 (33.3 vs 33.6 cm), but similar results were found for both treatments for the other periods. Homeopathic blend may influence some body measurements of weaned Holstein calves in specific days of supplementation, but there was no difference between treatments at the end of the study.

Homeopathy blend for the control of diseases in weaned Holstein calves immediately after grouping

T.H. Silva[1], I.C.S.B. Guimarães[1], R. Melotti[2] and A. Saran Netto[1]
[1]University of Sao Paulo, Department of Animal Science, 225, Duque de Caxias Street, 13635-900 Pirassununga, Brazil,
[2]Real H, Technical Manager, 12068, Zila Correa Machado Avenue, 79046-200 Campo Grande, Brazil; silvath@usp.br

The aim of the study was to evaluate the effect of homeopathic blend supplementation on tick-borne disease, pneumonia, and diarrhoea in weaned Holstein calves immediately after grouping, in a double-blind placebo-controlled trial. One hundred and eighty-four weaned female calves (83.01±7.9 days old; 112.5±11.7kg) were allocated to 8 paddocks in a completely randomised design experiment. During a 112 days period, animals received a total mixed ration with the following treatments: 1) control (basal diet + calcium carbonate, top-dressed at 30 g/animal – homeopathic vehicle) and 2) homeopathic blend (basal diet + TopVita™-Real H, top-dressed at 30 g/animal – $Sulphur$:10^{-60} + $Viola$ $tricolor$:10^{-14} + $Caladium\ seguinum$:10^{-30} + $Zincum\ oxydatum$:10^{-30} + $Phosphorus$:10^{-60} + $Cardus\ marianus$:10^{-60} + $Colibacillinum$:10^{-30} + $Podophyllum$:10^{-30} + Vehicle: calcium carbonate; q.s. 1kg). All diseases were diagnosed by the farm personnel according to guidelines and training provided by the veterinarians from University of Sao Paulo. Univariate survival analyses were performed according to the Kaplan-Meier method. In addition, three different Cox proportional hazard models were used. The log-rank (Mantel-Cox) analysis detected different survival curves between the homeopathic and control groups (P=0.018) with lower hazard of tick-borne disease diagnostic in calves of the homeopathic blend treatment (hazard ratio, HR=0.73, P=0.036). Furthermore, survival curve comparisons for digestive problems (P=0.006) indicated a lower hazard ratio for homeopathic group (HR=0.77, P=0.078). However, log-rank tests indicated that both groups had similar survival curves for pneumonia (P=0.88) and no difference was detected in the hazard of being diagnosed with pneumonia in homeopathic compared to control group (HR=0.98, P=0.92). These findings suggest that the homeopathic blend aforementioned may reduce the proportion of diarrhoea and tick-borne disease in weaned Holstein calves immediately after grouping.

Mid-infrared (MIR) spectrometry for the quantitative identification of cow's milk in buffalo milk

A.A. Spina[1], N. Costanzo[1], P. Roncada[1], V. Lopreiato[2], F. Trimboli[1], C. Ceniti[1], C. Perri[1] and V.M. Morittu[1]
[1]Magna Græcia University, Department of Health Science, Viale Europa, 88100 Catanzaro, Italy, [2]Università Cattolica del Sacro Cuore, Department of Animal Sciences, Food and Nutrition, Via Emilia Parmense 84, 29122 Piacenza, Italy; aa.spina@unicz.it

In Italy, one of the most consumed dairy products is buffalo mozzarella. The substitution of the buffalo milk with milk of other species, characterised by lower price and greater availability, is a common practice. In current study, FTIR spectroscopy in combination with multivariate calibration of partial least square (PLS) regression was used to quantitatively predict the adulteration of buffalo milk with cow milk. Samples of cow (n=7) and buffalo (n=7) bulk tank milk were collected from various local dairy farms during three years and a total of 119 samples (17 mix for each of the 7 different couple of buffalo-cow milk) were obtained. The best MIR spectrometry model developed showed a good calibration statistics (R2 calib.=0.99861; RMSEC=2.04; R2 valid.=0.99803; RMSEP=2.84; RMSECV=2.44% v/v). Our results show that the proposed method has very good potential for use in the dairy sector as a rapid and inexpensive screening for detecting and quantifying the adulteration of buffalo milk with cow's milk.

Effects of A1 and A2 β-casein variants on milk proteins and technological traits in Holstein cows

V. Bisutti[1], S. Pegolo[1], L.F.M. Mota[1], N. Amalfitano[1], A. Vanzin[1], P. Ajmone[2], M. Brasca[3], G. Bittante[1] and A. Cecchinato[1]
[1]University of Padua, DAFNAE, Viale dell'Università, 16, 35020, Legnaro (PD), Italy, [2]University of Cattolica del Sacro Cuore, Institute of Zootechnics, Via Emilia Parmense, 84, 28122, Piacenza, Italy, [3]CNR, Institute of Sciences of Food Production, Via Celoria 2, 20133 Milano, Italy; vittoria.bisutti@unipd.it

The genetic variants of β-casein (*CSN2*), especially A1 and A2 alleles, have recently risen the attention for their putative human health implications but also for their role in the cheese making process. For this reason, this study aimed at better evaluating the effects of the β-casein (CN) A1 and A2 variants both on milk protein composition and technological traits in a population of 498 Holstein cows reared in 28 herds. A validated reversed phase high performance liquid chromatography method was used to quantify six CN and two whey protein (WP) fractions which were expressed as percentage of total milk nitrogen content. In total, 11 lab-measured milk technological traits were also considered, including milk coagulation properties (MCP), cheese yields and curd nutrients recoveries. Data were analysed with a linear mixed model including the fixed effects of cows' days in milk, parity and *CSN3*, *CSN2* and *BLG* genotypes and the random effect of herd/date. β-CN genotypes had significant effects on all individual CNs and WPs except for κ-CN and on non-protein N fraction. In terms of MCP, β-CN A1A1 had a significantly (P<0.05) lower rennet coagulation time (RCT) than A2A2 (16.4 vs 20.4 min, respectively). Curd firmness at 30 min (a30) was also positively influenced by the A1A1 genotype (43.0 mm for A1A1 vs 35.2 mm for A2A2). Cheese yield and recovery of nutrients in the curd were also slightly higher in A1A1 samples. The positive effect of A1 variant on milk technological traits might be related to the fact that A1A1 genotype was associated to higher proportions of β-CN. These results suggested that, beside κ-CN, also β-CN plays a role on the cheese-making process and in particular the A1 allele seems to have no detrimental effects on the quality of milk destined for cheese making. Acknowledgements. The research was part of the FARM-INN AGER project funded by the Fondazione Cariplo.

Effect of replacing soybean meal by *Lupinus luteus* on meat fatty acids composition of growing lambs

M. Almeida[1], S. Garcia-Santos[2], A. Nunes[3], S. Rito[3], C.M. Guedes[1], L.M. Ferreira[2], V. Santos[1], R. Dominguez[4], J.M. Lorenzo[4], S. Silva[1] and J. Azevedo[1]
[1]Veterinary and Animal Research Centre (CECAV), Quinta de Prados, 5000-801, Vila Real, Portugal, [2]Centre for the Research and Technology Agro-Environmental and Biological Sciences (CITAB), Quinta de Prados, 5000-801, Vila Real, Portugal, [3]University of Trás-os-Montes e Alto Douro, Quinta de Prados, 5000-801, Vila Real, Portugal, [4]Centro Tecnológico de la Carne de Galicia (CTC), Rúa Galicia N ° 4, Parque Tecnológico de Galicia, San Cibrán das Viñas, 32900, Ourense, Spain; mdantas@utad.pt

The present study aimed to evaluate the effect of replacement of soybean meal (SBM) by *Lupinus luteus* (cv. Mister) on meat fatty acid composition of growing male lambs. Two different trials were conducted, trial 1 (T1) and 2 (T2). For both trials the diets were formulated to be nearly isonitrogenous and isoenergetic. Two diets were tested on T1: the control diet (C), containing wheat grain and SBM, and another diet with *L. luteus* replacing 30% of SBM (LL30). For T2, four diets were used: C diet and three diets with SBM replacement of 50, 75 and 100% by *L. luteus* (LL50, LL75 and LL100, respectively). Meadow hay and water were offered *ad libitum*. At weaning (around three months) the lambs were housed in groups of 4, and their growth was controlled throughout the trials. At the end of each trial, the lambs were slaughtered, and fatty acid composition of longissimus thoracis et lomborum was determined (SFA, MUFA, PUFA, n-3, n-6, n-6/n-3). On T1, LL30 showed higher values of MUFA than C (41.18 vs 39.44 mg/g of fat, P<0.05). On T2, diets containing lupine seeds show no effect (P>0.01) on fatty acid composition, although there seems to be a positive trend for LL50, LL75 and LL100 (48.82, 46.16 and 46.25, respectively vs 44.09 for C diet). The results of n-6/n-3 ratio were higher in lambs fed with LL100 and the lowest in the LL50 group (7.08 vs 6.08, respectively). Incorporating lupine seeds in small quantities does not seem to influence the fatty acid profile of the lamb meat. Replacing SBM by *L. luteus* seems to have affected meat fatty acids composition of growing lambs. Further studies are recommended to a more robust understanding of the incorporations of *L. luteus* to replace SBM.

Effect of *Camelina sativa* cake and *Cynara cardunculus* supplements on goat milk functional quality

C. Giromini[1], F. Omodei Zorini[1], S. Sandrini[1], R. Rebucci[1], G. Savoini[1], A. Baldi[1], A. Buccioni[2], A.A.K. Salama[3] and G. Invernizzi[1]
[1]*Università degli studi di Milano, Department of Health, Animal Science and Food Safety, Milano, 20133, Italy, [2]Università degli studi di Firenze, Department of Agriculture, Food, Environment and Forestry, Firenze, 50121, Italy, [3]Universitat Autonoma de Barcelona, Department of Animal and Food Sciences, Bellaterra, 08193, Spain; carlotta.giromini@unimi.it*

The aim of this study was to investigate the effect of *Camelina sativa* cake (Came) and *Cynara cardunculus* (Cardoon) supplementation on total phenolic content (TPC), antioxidant capacity (AOC) and metabolomics profile of goats milk. Eighteen multiparous and primiparous Alpine goats were randomly assigned to three homogenous groups: Came (n=6) receiving 200 g/day of Came, Cardoon (n=6) receiving 200 g/day of Cardoon and control (CT, n=6) receiving the standard diet. The supplementation lasted 21 days. Milk samples and rumen content were collected at d 0 (before supplementation), and 21 for analyses. Milk production and composition were assessed weekly. Milk collected at d21 was *in vitro* digested and further analysed for the TPC using Folin-Ciocalteu assay (mg Gallic Acid Equivalent (GAE)/l) and for AOC (µmol Trolox Equivalent (TE) /ml) by ABTS assay. Data were analysed by one-way ANOVA using GraphPad, Prism. Milk samples collected at d 21 were also analysed by ^{1}H NMR spectroscopy operating at 600 MHz. Metabolomic data were processed by the R (ChemoSpec package) and MetaboAnalyst programs and subjected to partial least square–discriminant analysis to detect differences in the metabolome before and after Came and Cardoon supplementation in the same animals. Results showed that Came and Cardoon milk showed higher (P<0.05) TPC (22.96±6.76 and 32.49±2.08 mg GAE/l) and AOC (207.9±4.25 and 195.4±26.66 µmol TE/ml) compared with CT milk. Metabolomic profile of milk was affected by the experimental day (i.e. d 0 vs 21). Compared to the CT milk, uridine, ethanolamine, hippurate and citrate decreased, whereas lactate, alanine, methionine and choline increased in Came and Cardoon milk. In conclusion, feeding dairy goats with both Came and Cardoon resulted in changes in milk TPC and AOX activities as well as modifications in the milk metabolomic profile.

Effects of dietary dried food waste addition to broiler diets on growth performance

E. Giamouri[1], A.C. Pappas[1], G. Papadomichelakis[1], E. Tsiplakou[1], K. Sotirakoglou[1], K. Feggeros[1], T. Manios[2] and G. Zervas[1]
[1]*Agricultural University of Athens, Nutritional Physiology and Feeding, Iera Odos 75, 11855, Greece, [2]Hellenic Mediterranean University, Agriculture, Stavromenos, GR71410, Crete, Greece; gzervas@aua.gr*

The increasing world population in combination with the improvement of living standards has caused a raising demand for poultry products. In recent years, high cost for animals feeding has led to demand for alternative sources for animal feeds. Food waste may be a potential feed for animals since it is a valuable source of energy, protein, minerals and vitamins. The use of food waste in animal diets has gained considerable attention because of the high prices of conventional arable based animal feeds. The aim of the present study was to investigate the effect of adding dried food waste, collected from hotels in Greece, to the diet of meat type chicken (broilers). Two hundred (200), one-day-old, broilers were used. The broiler chickens were divided in 2 groups which underwent two different experimental treatments for 42 days. There were ten broilers per pen, 100 per treatment. In first treatment, named control (C), the diet did not contain any feed waste but a basal diet based on corn and soybean meal. The diet of second treatment (T) consisted of 15% with dried food waste. Diets were isocaloric and isonitrogenous. At the end of each growing phase body weight were recorded in order to calculate body weight gain, feed intake and feed conversion ratio (FCR). At the age of 38-41 days, a digestibility trial was conducted to determine energy and nutrient digestibility. Two broilers per replicate per treatment were selected for determination of several biochemical and haematological parameters. Carcass yield and breast carcass were also estimated (pH, colour, cooking loss and shear force). All in all, broilers performed well. Chickens in T treatment scored lower body weight (P<0.001) and the feed intake. FCR did not differ between the two groups. No differences in the digestibility of nutrients were observed between the two treatments. Carcass yield was high and did not differ for the two treatments. Biochemical and haematological parameters were similar for two groups. Minor differences on colour traits and shear force were observed. The results of the present study indicate that dried hotel food waste may be an alternative ingredient to be incorporated to poultry diets, and that future studies will determine the optimum inclusion level. The present project was funded by the European LIFE Food4Feed project.

Carbohydrase enzyme complex (CEC) and agro-industrial byproducts in broiler nutrition

W. Al-Izzi and Y. Beckers
University of Liège, Precision Livestock and Nutrition, Passage des Déportés, 2, 5030 Gembloux, Belgium; w.alizzi@uliege.be

Using agro-industrial byproducts in broiler nutrition will lead to a reduction in metabolisable energy content due to the higher fibre content in such feeds, but the supplementation of carbohydrase enzymes could limit the decrease in energy value of the diet. Two experiments were conducted to investigate this hypothesis. In each experiment, 15% of either dried distillers with solubles (DDGS) or wheat bran (WB) were substituted in the diets. The byproducts diets have a lack of 300 kcal/kg in metabolisable energy comparing to control diet. To compensate this shortage, different levels of carbohydrase enzyme complex (CEC) were used (0, 1, 2%) in DDGS experiment and (0, 1, 3.5%) in WB experiment, based on the CEC activity IUs needed to liberate theoretically 300 kcal/kg of energy from byproducts fibres. The enzyme complex was produced at the lab level from the ruminal fluid of 3 cows and the cellulase activities were 11 IU/g DM. In each experiment (3 weeks from day 6 to day 27 of age), 192 one-day old broiler males (Ross 308 strain) were distributed in a completely randomised design on 4 treatments (control diet and 3 byproducts diets) with six repetitions. The results showed a significant higher final weight (7.16 and 5.5%) and daily weight gain (7.64% and 5.5%) of the broilers fed with the DDGS and WB diets supplemented with CEC respectively compared to the control diet, except the 0% CEC in WB diet. Moreover, the results demonstrated that the faecal digestibilities (DM and energy) and the measured AMEn for the DDGS and WB diets were equal or less than that of the control diet. Consequently, higher final weight and daily weight gain with the DDGS and WB diets supplemented with CEC were mainly explained by the increase in feed intake (19% et 3.58%) for DDGS and WB experiments respectively, but the feed conversion ratio was also increased (12%) with the CEC supplementation in DDGS experiment while decreased in WB experiment (2.4%) compared to that of control diet. According to our results, it is possible to increase fibres content of the fast growing broilers diet without penalising the growth performance of the animals when a carbohydrase enzyme is added to the diet.

High-moisture grain meals for swine: proximal analyses, *in vitro* digestibility and estimated energy

C.F. Martins, M. Vasconcelos, D.M. Ribeiro, A.M. Almeida and J.P.B. Freire
Instituto Superior de Agronomia, DCEB – LEAF (Linking Landscape, Environment, Agriculture and Food), Tapada da Ajuda, 1349-017 Lisbon, Portugal; catiamartins@isa.ulisboa.pt

High-moisture (HM) grain meals can be used in swine feeding, nevertheless information is still missing on its nutritional value for this species. In this study, the nutritional quality of eight HM feeds was assessed. A 4×2 factorial study was used with four different seeds (corn, barley, wheat and lupin) and two liquid substrates (water or whey). For each sample, pH, NH_4-N, DM, OM, CP, crude fat, NDF, ADF, ADL, starch, total sugar and caloric value was determined in triplicate. *In vitro* digestibility (IVD) of DM, OM and CP for HM grains were determined for growing pigs (samples in duplicate, in 5 different incubations) using standard techniques. Regression equations were used to predict the digestible (DE) and net (NE) energy from digestible nutrients contents. Metabolisable energy (ME) was calculated multiplying DE by 0.96. Data was analysed using PROC MIXED of the SAS software. HM cereal grains tested had CP content between 7.1 and 43.8% in DM and good storage conditions, highlighting the pH and NH_4-N values of 3.79-4.90 and 4.1-16.7% of N, respectively. IVD of DM was significantly higher to HM wheat, followed by lupin, corn and barley, respectively, with 89.3, 89.0, 83.9 and 83.1%. IVD of OM and of CP had the same pattern to the different seeds studied, respectively, lupin with 93.8 and 96.1%, wheat with 92.2 and 92.8%, barley with 86.5 and 88.1% and corn with 86.1 and 82.3%. The samples preserved in whey exhibited a significant increase on IVD of DM, OM and CP comparing with samples preserved in water (P<0.0007) for all seeds studied. Considering the estimated energy values for HM of seeds, corn had the higher energy values (4,007 and 3,300 kcal/kg DM to DE and NE, respectively) and barley the lower DE and ME values (respectively, 3,457 and 3,319 kcal/kg DM). Moreover, lupin had the lower NE (2,259 kcal/kg DM). In conclusion, the HM lupin had higher IDV of OM and CP, but this was not reflected on the estimated energy, which was higher in HM corn. This occurred due to higher contents of starch, total sugar, and lower fibre content of this cereal vs lupin. Whey was the best fermentation substrate due higher values of IDV of DM, OM and CP for all seeds studied.

Effects of adding Sake lees in milk replacer on growth, faecal traits and blood metabolites in calves

S. Katsumata[1], Y. Hayashi[2], S. Yamanaka[1], A. Obata[2], H. Tsutsumi[3], H. Aoki[1], T. Tsukahara[4], K. Oishi[1], H. Hirooka[1] and H. Kumagai[1]
[1]Kyoto University,Graduate School of Agriculture, Division of Applied Biological Sciences, Kitashirakawa-Oiwake-cho, Sakyo-ku, Kyoto, 606-8502, Japan, [2]Shiga Prefectural Livestock Production Technology Promotion Center, Hino-cho, Gamo-gun, Shiga, 529-1651, Japan, [3]Research Institute, Gekkeikan Sake Co., Ltd, Katahara-cho, Fushimi-ku, Kyoto, 612-8361, Japan, [4]Kyoto Institute of Nutrition and Pathology, Ujitawara, Kyoto, 610-0231, Japan; ksachi93@kais.kyoto-u.ac.jp

Japanese liquefied Sake lees (SL), a by-product of Japanese Sake, is rich in *Saccharomyces cerevisiae* and protein. Previous studies reported that *S. cerevisiae* fermentation products improved health and faecal condition in pre-ruminant calves. The aim of this study was to evaluate the effects of supplementary SL in milk replacer (MR) on growth performance, faecal condition and blood metabolites of calves in the first 90 days of age (d-90). Twenty-four Japanese Black female calves at d-6 were randomly assigned to the following three treatments: no SL supplementation (CON); 100 g/day of SL supplementation (LS); 200 g/day of SL supplementation (HS). All the calves received colostrum and were separated from the cows at d-5. The MR was fed according to their age and birth weight from d-6 to d-90. Calf starter and hay were fed *ad libitum*. Body weight was measured at d-6, d-30, d-60 and d-90. Dry matter (DM) contents, pH and short chain fatty acid (SCFA) concentrations in faeces, and blood metabolite concentrations were measured at d-6, d-14, d-30, d-60 and d-90. Faecal condition was scored daily. Calf starter intake did not differ among the treatments. However, hay intake from d-30 to d-90 in LS was lower than in CON (P<0.05). Faecal pH at d-30 in LS and faecal DM contents at d-60 in HS were lowest (P<0.05). No significant differences were observed in average daily gain, faecal score and faecal SCFA concentrations among the treatments. At d-14, plasma β-hydroxybutyrate (BHB) concentration in LS was higher than in CON (P<0.05) and blood urea nitrogen (BUN) concentrations in LS and HS were higher than in CON (P<0.05). The concentration of BHB at d-60 tended to be higher in LS than in CON (P=0.09). The results suggested that supplementation of SL did not affect the growth performance and faecal condition in pre-ruminant calves. The higher plasma BHB contents only in LS suggested that 100 g/day SL feeding might stimulate the rumen development in calves. The higher plasma BUN contents in SL feeding might reflect the supplementary protein from SL in early nursing stage.

Salty or sweet former food based diets for piglets: effect on growth performance and digestibility

L. Pinotti[1], A. Luciano[1], M. Comi[2], F. Fumagalli[1], N. Rovere[1], M.C. Rulli[3] and M. Ottoboni[1]
[1]Università degli Studi di Milano, Department of Health, Animal Science and Food Safety, Milano, 20134, Italy, [2]Università Telematica San Raffaele, Roma, 00166, Italy, [3]Politecnico di Milano, Milano, 20133, Italy; luciano.pinotti@unimi.it

Former Foodstuffs (FFPs) are products that have lost their commercial value on the human consumption market, due to for example production errors. However, their nutritional value for animal feed purposes is not at all affected. Consequently, biscuits, bread, chocolate bars, pasta, savoury snacks and sweets, high in energy content in the form of sugar, starch, oil or fat can be considered an appealing alternative feed ingredient. Although FFPs composition, may vary to a large extent, they have been indicated as energy sources mainly. Accordingly, in this study, conventional cereal grains have been partially replaced by two category of FFPs (sweet or salty) in post-weaning piglet's diets in order to investigate the effects of these alternative feed ingredients on growth performance and digestibility. Briefly, 36 weaned female pigs (28 d of life, 6.70±1.07 kg) were housed in individual pen. After an adaptation period (7 days), pigs were fed 1 of the 3 experimental diets containing: 1) Control diets (CRT), 0% FFPs; 2) 30% conventional cereals substituted for 30% confectionary FFPs (FFP sweet); 3) 30% conventional cereals substituted for 30% bakery FFPs (FFP salty). The diets were iso-energetic and iso-nitrogenous, and met NRC (2012) requirements. The trial lasted 42 d. Fresh water and feed were available *ad libitum* throughout the whole experimental period. Individual pig body weight (BW) and feed intake (FI) were measured weekly. During the experiment, stool were also collected for determining apparent total tract digestibility (ATTD). In addition, average daily gain (ADG), average daily feed intake (ADFI), and feed conversion ratio (FCR) were be calculated. At the end of the experiment, no differences in BW were observed between groups (P>0.05). The results revealed that ATTD values did not differ between CTR and FFPs based diets. However, ATTD were higher (P<0.05) in salty FFP based diet compared to sweet FFP based one (P<0.05). Overall ADG, ADFI, and FCR were not affected by any dietary treatments. Taken together, these results suggest that the use of FFPs up to a level of 30% in post-weaning diets has no detrimental effects on pig growth performance, although more studies are needed to confirm these results. This study has been done in the frame of the SUSFEED project funded by Cariplo Foundation.

Plasma amino acid profile after feeding a balanced or unbalanced sustainable protein source in pigs

F.A. Eugenio[1,2], J. Van Milgen[2], J. Duperray[1], R. Sergheraert[1] and N. Le Floc'h[2]
[1]*BCF Life Sciences, Boisel, 56140 Pleucadeuc, France,* [2]*INRAE, Agrocampus-Ouest, PEGASE, 16 Le Clos, 35590 Saint-Gilles, France; francisamann.eugenio@inrae.fr*

Metabolic utilisation of amino acids (AA) regarding animal status and feed AA profile can be assessed from postprandial blood profiles. The objective of the experiment was to compare the postprandial plasma AA profile of pigs with contrasting potential for protein deposition (adult Yucatan minipigs vs growing crossbred pigs) after feeding diets either unbalanced (UNB) or balanced (BAL) in AA. For the two diets, the protein source is derived from the extensive hydrolysis of feathers, for which the resulting AA are used in different value chains. The UNB diet had a low Lys content but ratios of Met, Tyr, and His to Lys are close to the ideal profile. The ratios of the other essential AA (Cys, Thr, Val, Ile, Leu, and Phe) were unbalanced. The BAL diet was supplemented with Lys, Met, Ile, and His (plus Tyr in growing pigs) to get AA:Lys near the ideal AA profile. Experimental diets were formulated to provide the same amount of total AA by supplementing non-essential AA (Ala, Asp, Glu, Gly). Four adult and four growing pigs were fitted with a jugular catheter for serial blood collection. After an overnight fast, pigs were fed 75% of their meal allowance and blood was collected during 6 h after the meal. Each pig received each diet three times. In both stages, feeding either diet did not affect (P>0.05) total plasma AA average concentrations. Supplementation with essential AA in BAL resulted in greater (P>0.05) average plasma concentrations of Lys, Met, Ile and His, and of Tyr in growing pigs. However, supplementation of non-essential AA to both diets resulted in lower (P<0.05) plasma concentrations of Ala, Glu, and Gly (and Ser, which was not supplemented) in BAL compared to UNB. Lower Ser and Gly may be due to the use of these AA for Cys synthesis. The concentrations of Thr, Trp, and Val did not differ between diets (P>0.05). With the exception of Arg, Asp, and Tyr, growing and adult pigs responded in a similar way to balancing of the diet. In conclusion, AA supplementation to balance a diet may result in changes in plasma AA profile in growing and adult pigs. Changes in non-essential AA may be due to their roles in carbon and nitrogen metabolism.

Energy value of ensiled lucerne and red clover leaves and their impacts on performance in pigs

S.R. Habit[1,2], L. Stødkilde-Jørgensen[3], S. Krogh Jensen[3] and D. Renaudeau[2]
[1]*Universidade Federal de Minas Gerais, ICA, Montes Claros, 39404547, MG, Brazil,* [2]*INRAE, UMR1348 PEGASE, INRAE Agrocampus Ouest, 35590 St Gilles, France,* [3]*Aarhus University, Department of Animal Science, Foulum, 8830 Tjele, Denmark; david.renaudeau@inrae.fr*

Silages from the leaf fractions of lucerne and red clover may be a valuable energy source for swine. The aim of this study was to measure the energy value of ensiled lucerne (LLS) and red clover leaves (RLS) (trial 1) and the effects of two levels of dietary inclusion (10 and 20%) of LLS and RLS on growth performance (trial 2). In trial 1, total tract digestibility coefficient (TTDC) of energy was measured on a total of 15 pigs (68 kg BW) allotted to 3 different dietary treatments in which silages were given as an additional ingredient with control diet (C), C+22% of LLS (LLS diet), and C+22% of RLS (RLS diet). In LLS and RLS diets,. In trial 2, a total of 60 growing-finishing pigs (97.0 kg on average at the beginning of the 17-d trial) were divided into 5 experimental treatments planned to supply the amount of metabolisable energy (ME, i.e. 37.4 MJ/d) for a fixed amount of DM allowance (i.e. 2.6 kg/d). Pigs from the treatment 1 (T1) were fed a R1 diet formulated from cereals (corn, barley, wheat), wheat bran and soybean meal (12.5 MJ ME/kg and 0.45 g SID lysine/MJ ME). Pigs from T2 and T3 were fed with a mixture of 90% of a R2 diet and 10% of LLS or RLS. Pigs from T4 and T5 received a mixture of 80% of a R3 diet and 20% of LLS or RLS. For compensating the lower ME and protein contents in silages, wheat bran was partially or completely substituted by cereals and soybean meal in R2 and R3 diets, respectively. In the LLS and RLS diets, TTDC of energy was significantly (P<0.05) than in the C diet. On average, DE values of LLS and RLS were respectively 14.0 and 13.8 MJ/kg DM. From the trial 2, ADG and FCR were significantly (P<0.05) influenced by the dietary treatment with depressed performance for the T4 group when compared to the others treatments (796 vs 889 g/d; 3.23 vs 2.90 kg/kg). In conclusion, our results suggest that lucerne and red clover leaf silages have a potential as energy sources in finishing pigs but further studies are needed to better understand why growth performance are reduced in pigs fed 20% Lucerne silage.

Can rumen micro-organisms transformed dietary selenite into elemental selenium?

M.A. Hachemi[1], E. Pinloche[1], M. De Marco[2] and M. Briens[1]
[1]Adisseo France SAS, 6 Route Noire, 03600 Commentry, France, [2]Adisseo France S.A.S., 10, Place du Général de Gaulle, 92160 Antony, France; amine.hachemi@adisseo.com

The lower efficacy of mineral selenium (Se) forms for status improvement compared to organic ones is largely described. However, limited results indicate it could be related to the formation of elemental Se (Se (0)) by rumen microorganisms (RMO). To test this hypothesis, we fed several forms of Se and measured the whole total Se (Tot-Se) as well as the speciation of Se (0) in rumen fluid and plasma of cows. Eleven ruminally cannulated non-lactating Prim Holstein cows were divided into four groups and fed for 23 days either a control diet (hay and protein-energy concentrate) (CON; n=2), or the same control diet supplemented with 10 mg of Se top-dressed as sodium selenite (SS; n=3), zinc-L-selenomethionine (Zn-SeMet; n=3 or hydroxy-selenomethionine (Selisseo® 2% Se; OH-SeMet; n=3). Rumen fluid (RF) was collected on d 22 and 23 and blood for plasma was collected from the tail vein on d 22, both 4 h after morning feeding. Plasma and freeze-dried RF were used for Tot-Se and Se (0) measurement by ICP-MS and HPLC-ICP-MS, respectively. Data were analysed with a Kruskal-Wallis test. Significance was declared at $P \leq 0.05$. Results of plasma Tot-Se were CON: 32; SS: 94; Zn-SeMet: 127; OH-SeMet: 147 µg Se/kg of fresh material with significant differences within groups (P<0.05) except between CON and SS groups (P>0.05). The concentration of Tot-Se in RF was not different between the two sampling days. Concentration of Se in RF was positively correlated with the concentration of Se in plasma (R^2=0.91) and followed the same hierarchy. Se (0) was measured in RF but was not detected in any group except the SS group. The proportion of Tot-Se present as Se (0) in the SS treatment was 42%. We conclude that plasma Tot-Se is representative of rumen Se level and confirmed the higher bioavailability of organic Se forms, particularly OH-SeMet. Those results also confirm a transformation of SS into elemental Se, which could explain the lower bioavailability for SS.

***Hermetia illucens* reared on seaweed or selenium enriched substrates for designing a feed specialty**

M. Ottoboni[1], V. Sele[2], M. Silva[2], A. Grisendi[3], M. Della Porta[1], R. Cazzola[1], F. Selmin[1], R. Ørnsrud[2] and L. Pinotti[1]
[1]University of Milan, via celoria 10, 20133 Milano, Italy, [2]Institute of Marine Research, Nordnesgaten 50, 5005 Bergen, Norway, [3]Istituto Zooprofilattico Sperimentale della Lombardia e dell'Emilia Romagna Bruno Umbertini, Via Pitagora, 2, 42124 Reggio Emilia, Italy; matteo.ottoboni@unimi.it

This study evaluated the effect of inclusion of brown algae (*Ascophyllum nodosum*) or sodium selenite in Black soldier flyers (BSF) growing substrate on the nutritional composition of the deriving prepupae. The larvae were reared on three different substrates: (1) Gainesville diet, used as control substrate (CTR); (2) *A. nodosum* diet (AN30%), with 30% substitution of the alfalfa meal with brown algae; and (3) selenium diet (Se), based on Gainesville diet fortified with 0.3 mg/kg of selenium in the form of sodium selenite. Eggs collected from a stable colony of BSF were placed for hatching on the three experimental substrates, i.e. CTR, AN30% and Se. All experiments were carried out under dark condition, at 25 °C with 70% relative humidity, and repeated for three consecutive generations. Both rearing substrate and deriving prepupae were analysed for proximate and fatty acid composition and total selenium content. Total biomass produced, starting from 300 young larvae under different substrates were: 20.9, 23.4, 20.7 grams of prepupae for CTR, AN30% and Se dietary groups, respectively. The mean protein content was 536 g/kg DM in CTR, 446 g/kg DM in AN30%, and 469 g/kg DM in Se group. All, CTR, AN30%, and Se prepupae contained a high level of saturated fatty acids, namely: 74.8±2.7, 83.0±0.5 and 85.5±0.7% of total lipids, respectively. The inclusion of seaweed in the rearing substrate did not affect the fatty acid profile in the AN30% prepupae compared to CTR and Se dietary groups. By contrast, when selenium content was considered, some differences were observed. Specifically, total selenium content was: 0.41 mg/kg in CTR prepupae, 0.21 mg/kg in AN30% prepupae, and 1.17 mg/kg in Se prepupae. Concluding, it can be proposed that seaweeds and selenium are able to affect BSF larvae composition in term of crude protein and total Se content. However, further investigation are required for optimising BSF meal production for specific feed or food purposes.

Turning fruit and vegetable waste from the wholesale market into feed
D. Tedesco[1], S. Scario[1], E. Biazzi[2], A. Tava[2] and L. Baschenis Begnis[3]
[1]University of Milan, Department of Environmental Science and Policy, Via Celoria 2, 20133 Milano, Italy, [2]Consiglio per la Ricerca in Agricoltura e l'Analisi Dell'economia Agraria, Centro di Ricerca Zootecnia e Acquacoltura, CREA-ZA, Viale Piacenza 29, 26900 Lodi, Italy, [3]H.S.E.Manager So.Ge.Mi. S.p.A. (until July 2019), Via Cesare Lombroso, 54, 20137 Milano, Italy; doriana.tedesco@unimi.it

A large amount of food lost and wasted ends its 'life' with high nutritive elements in it, leading to the squandering of economic, social and environmental resources. Compared to the other food waste categories fruit and vegetable waste (FVW) have a high wastage rate given by their highly perishable nature. The proposal to reallocate FVW into animal feed contribute to the sustainability of livestock production, reducing the impact of feed production for land use. In the assessment of this opportunity, the nutrients composition need to be considered because affected by vegetables seasonal variation. Therefore, to use FVW as a feed component, one of the key aspects to evaluate is the FVW nutritional composition throughout the year. In this study, it has been considered the FVW from the General Wholesale Market of the city of Milan, the largest in Italy, where the fruit and vegetables are sold to retailers. They market more than 1.000.000 tons/year of which about 1.500 tons of unsold products become waste. In the waste-collecting site, samples of each wasted fruit and vegetables were collected monthly in proportion to their amount wasted. Each sample was weighed, cut manually and thoroughly mixed. The FVW samples were analysed for their chemical and nutrients composition. Data were analysed using the GLM procedure of SAS (version 9.4; SAS Institute Inc., Cary, NC). On average the FVW dry matter (DM) was 11.4±1.4%; on DM basis the NDF was 24.0±4.8%, the water-soluble carbohydrates were 33.0±7.5% and the total phenolics were 2.6±0.9%. Although differences in the nutrients composition were found across the year, the results highlighted the significant nutritional value of FVW from the wholesale market and the need for developing appropriate technologies (i.e. heath treatment to reduce the moisture) and safety evaluations, necessary to use FVW as source of feed nutrients.

Performance of 4 handheld NIR spectrometers to predict the dry matter, starch and NDF of corn silage
N. Chamberland, P.H. Vermeulen, B. Lecler, O. Minet, V. Decruyenaere, E. Froidmont and V. Baeten
Walloon Agricultural Research Center, Knowledge and Valorization of Agricultural Products, Chaussée de Namur, 24, 5030 Gembloux, Belgium; n.chamberland@cra.wallonie.be

The Walloon Agricultural Research Center is currently developing analytical methods based on handheld spectrometers for predicting parameters such as dry matter and chemical constituents (starch and neutral detergent fibre (NDF)) of corn silages. Since a few years some handheld spectrometers have been launched on the market to predict the quality of forages *in situ*. The goal of this study is to compare the performance of these devices. Regarding this study, 88 samples of corn silage of 22 varieties were collected between September and October 2019 at four different locations in Wallonia (Belgium). NIR spectra were measured in the lab with four handheld spectrometers: The FieldSpec4 (350-2,500 nm), the Micronir1700 (950-1,650 nm), the Flame-NIR (940-1,665 nm) and the PoliSpec (900-1,680 nm). Each sample was measured on twenty spots representing a surface area of about 20 cm^2. Samples were also measured with a benchtop XDS instrument from FOSS (400-2,498 nm). Reference values were obtained by prediction with a FOSS DS2500 on dried and ground samples for starch and NDF constituents. Dry matter was obtained by drying samples at 60 °C during 48 hours. Database was split into calibration set (68 samples) and validation set (20 samples) using the Kennard-Stone algorithm. Pre-processing was applied using first derivative for dry matter calibration as well as SNV and first derivative for starch and NDF calibrations. Calibrations were built using the Partial Least Square algorithm. All the data analysis was performed using PLS Toolbox 8.5.1 under Matlab R2007b. Results in terms of Standard Error of Cross-Validation (SECV) and Standard Error of Prediction (SEP) for dry matter, starch and NDF are presented in the poster. Two challenges are highlighted concerning this study. Firstly, the high heterogeneity of corn silage samples (stem, leaves and grains) and secondly the narrow range of value for dry matter (average of 35.1% and Standard Deviation (SD) of 5.6%) and for starch (average of 34.3% of DM and SD of 7.8%) and NDF (average of 40.5% of DM and SD of 3.8% of DM).

Ruminal carbohydrase enzyme complex (CEC) vs a commercial cellulolytic enzyme in broiler nutrition

W. Al-Izzi and Y. Beckers
University of Liège, Precision Livestock and Nutrition, Passage des Déportés, 2, 5030 Gembloux, Belgium; w.alizzi@uliege.be

Agricultural by products inclusion in broiler diet is one of the strategies used to reduce production cost. But these byproducts, such as Wheat Bran (WB), known by its high content in crude fibres (9.2%). For this reason, feed experts used cellulolytic enzymes in poultry feed to improve byproducts nutritional value. An experiment was conducted to investigate the efficiency of using two different enzymes on fibres digestibility. The first enzyme used in this experiment is a cellulolytic enzyme complex (CEC) produced form ruminal fluid at laboratory level with 11IU/g DM of cellulase activity and the second is a commercial cellulolytic enzyme with 550 IU/g cellulase activity. Wheat bran diets fabricated with a lack of 350 and 500 kcal/kg ME and supplemented with either CEC or commercial enzyme (11,000 IU/kg diet) to be compared with a control diet in order to investigate broilers growth performance and diets digestibility. In this experiment, 240 broiler males (Ross 308 strain) randomly distributed on 5 feed treatments with six repetitions. The results showed that the CEC increased feed intake, final body weight and daily weight gain 25.1, 30.2 and 30.1% respectively comparing with control and the commercial enzyme diets, resulting in a significant reduction of feed conversion ratio (7.3%). Faecal and ileal digestibility of dry mater, gross energy and AMEn shows a superiority of the control diet comparing with WB diets complemented with CEC and commercial enzymes, but CEC and commercial enzyme supplementation didn't influence DM and energy faecal digestibility. While, ileal digestibility experiment showed that the commercial enzyme improved energy and organic matter digestibility compared to CEC in WB diets, whereas the commercial enzyme has no influence on the growth performance of the birds. Hence, we can say that it is possible to reduce the ME of the diet until 2,500 kcal/kg ME and using 15% of WB rich in fibres without affecting the growth of broiler by using CEC.

The use of banana tree as fibre source to ruminants

S. Teixeira, C. Maduro Dias, C. Vouzela, J. Madruga and A. Borba
University of the Azores, Institute of Agricultural and Environmental Research and Technology (IITAA), INV2MAC (MAC2/4.6d/229), Rua Capitão João d'Ávila, 9700-042 Angra do Heroísmo, Açores, Portugal; alfredo.es.borba@uac.pt

The residues of the banana culture, leaves and pseudostems, could be used as fibre source for feeding ruminants in Banana producing areas, namely in the Macaronesia archipelagos. We intend to first evaluate the nutritive value and envision strategies to improve it. The samples of leaves and pseudostems were harvested in triplicate and dried at 65 °C in an oven with controlled air circulation. The pseudostems were divided in three different portions and chemical composition as *in vitro* digestibility were determined. It was verified that the values of dry matter were low (16.54% in leaves and 6.54% in pseudo-stem), crude protein raging 11.25 DM% in leaves and 7.25% in pseudo-stem. Regarding Fibre values, NDF is higher in leaves (70.07 DM %) than in pseudostems (52.11 DM %), ADL is higher in leaves (9.90 DM %) comparing with pseudostems (6.21 DM %), the *in vitro* DM digestibility is low, (24.42% in leaves and 42.69% in pseudo-stem), supporting the NDF values, although the leaves have higher crude protein values than pseudostems. This by-product can be used in animal feed, however, it will be necessary to carry out tests to improve its nutritional value, namely with NaOH and or with Urea.

Grape pomace intake affects the fatty acids composition of rabbit fat

M. Rolinec, D. Bíro, B. Gálik, M. Šimko, M. Juráček, O. Hanušovský, R. Kolláthová, P. Vašeková, E. Mixtajová and M. Gluchová
Slovak University of Agriculture in Nitra, Department of Animal Nutrition, Trieda A. Hlinku 2, 94976 Nitra, Slovak Republic; michal.rolinec@uniag.sk

Grape pomace is by-product considered as a source of bioactive compounds like antioxidants and polyphenols and is rich in tannin content, which have positive action on health, when fed in optimal amount. In grape pomace are presented mainly unsaturated fatty acids (C18:1cis n9 and C18:2cis n6). Previous researches confirmed possibility fed grape pomace to rabbits. However, there is lac of information describing the effect of grape pomace on fatty acids composition of rabbit fat. Therefore, this was the aim of this study. The experiment included 20 rabbits (*Oryctolagus cuniculus*) randomly divided to two groups (control 'C' and grape pomace 'GP'). Diet in C group consisted from barley grain and hay, both *ad libitum*. In diet of GP group 1% of barley grain was replaced with dried grape pomace. Live weight of rabbits in C and GP group at the start of experiment was 1.15 kg and 1.03 kg, and at the end of experiment was 2.77 kg and 2.89 kg respectively. At the end of experiment, abdominal fat of all rabbits was analysed for fatty acids 'FA' composition using gas chromatography (Agilent technologies, U.S.A.). Results were statistically processed using IBM SPSS v 20.0. The differences of means between groups were tested with T-test. In C and GP group was the most abundant SFA C16:0, MUFA C18:1cis n9 and PUFA C18:2cis n6. Significant ($P<0.05$) difference between C and GP group was detected for C14:1 (0.13% vs 0.21%), C16:0 (35.2% vs 33.6%), C16:1 (2.37% vs 3.63%), C17:0 (1.00% vs 0.86%), C18:0 (7.23% vs 6.34%), C18:1cis n9 (19.7% vs 21.8%), C20:0 (0.20% vs 0.15%), MUFA (22.6% vs 26.0%) and SFA (47.84% vs 45.08%) respectively. Both groups had n6/n3 ration in interval from 4.23 to 4.56. Between groups, significant difference was detected for 7 of 13 total determined FAs. It can be concluded, that the replacing 1% of barley grain with dried grape pomace in rabbit diets affected the FA proportion in abdominal fat, mainly MUFA. Small amount of dried grape pomace can be used in rabbit nutrition as a nutrients source. This experiment was financially supported by The Slovak Research and Development Agency under the contract No. APVV-16-0170.

Nutrients content and microbiological quality of salad crops as feed ingredient for ruminant diets

M. Manoni, N. Rovere, A. Luciano, F. Fumagalli, M. Ottoboni, C.E.M. Bernardi and L. Pinotti
Univestity of Milan, Department of Health Animal Science and Food Safety Carlo Cantoni, Via Celoria 10, 20133 Milano, Italy; michele.manoni@unimi.it

This study evaluated the nutritional composition and microbiological quality coupled with electronic nose odour profile of selected salad crops (SC). Eighteen sample SC deriving from a plant producing salad crops have been collected during the summer season in two sites: at the production site (factory plant) and after transport and one-day storage (to mimic farm site). The samples were then analysed for pH, dry matter (DM), crude protein (CP), neutral detergent fibre (NDF), acid detergent fibre (ADF), and lignin. Eight selected samples were analysed for the enumeration of Total Viable Count (TVC) (ISO 4833), Enterobacteriaceae (ISO 21528-1), Escherichia coli (ISO 16649-1), coagulase-positive Staphylococci (CPS) (ISO 6888), presumptive B. cereus and its spores (ISO 7932), sulphite reducing Clostridia (ISO 7937), Yeasts and Moulds (ISO 21527-1) and the presence in 25 g of Salmonella spp. (ISO 6579). Data obtained were analysed using IBM SPSS Statistics version 21 software (SPSS Inc.). Salad crops at the production site (factory plant) and after transport and one-day storage (to mimic farm site), have shown pH value of 6.0 and 6.4, respectively. For the same material nutrient composition was comparable to that of common fresh grass: the overall mean of CP content was 210 g/kg DM, while average NDF, ADF and ADL contents were 360, 238, and 74 g/kg DM respectively. The main difference between fresh grass and SC was for DM content, that was lower (5.8±0.8 g·kg⁻¹) in SC samples compared to literature grass values. Total Viable Count was limited and Salmonella was always absent. Salad crops has some potential as feed ingredient for ruminant's diets, even though a full verification of their potential need further deep investigation. The present work has been done in the frame of Progetto di Grande Rilavenza IT-RS 'Sustainable animal nutrition' (SUN) funded the Ministero degli Affari Esteri e della Cooperazione Internazionale.

The effect of urea addition on micro mineral profile of grape pomace silage

O. Hanušovský, D. Bíro, M. Juráček, M. Šimko, B. Gálik, M. Rolinec, P. Vašeková and R. Kolláthová
Slovak University of Agriculture in Nitra, Department of Animal Nutrition, Trieda Andreja Hlinku 2, 949 76 Nitra, Slovak Republic; ondrej.hanusovsky@uniag.sk

Last experiments with farm animals showed possibility to use grape by-products as a source of nutrients in animal nutrition, however it is necessary to find the right and economical method for nutrient preservation in grape pomace because of high water content after the wine production in pomace that leads to rapid spoilage. Thus, the main goal of this research was to find the effect of urea addition on micro mineral profile of Pinot Gris (*Vitis vinifera* L.) grape pomace silages. Experiment in the Laboratory of Quality and Nutritive Value was realised. First, fresh matter with dry matter content 45% in laboratory conditions with stabile temperature 22 °C into silos mini bags was ensiled. Control variant was ensiled without addition of urea and experimental variant with urea addition (0.2 kg per 100 kg of fresh matter). After the 35 days of storage in silage the dry matter (gravimetrically) and ash content (sample burning in Muffle furnance at the temperature 550 °C) were determined. Then, for the determination of micro elements chloride extract (ash solution in 3M HCl) from the ash was prepared. In the ash extract the content of micro minerals (Fe, Cu, Mn, Zn) using ContrAA 700 spectrophotometrically by atomic adsorption was determined. The descriptive statistics and differences between means were calculated using IBM SPSS v. 20.0 (Independent samples T-test). In general, the highest concentrations from micro minerals had Fe. However, in the experimental variant (118.48 ± 0.39 mg/kg), higher concentrations of Fe in comparison with control variant (113.57 ± 6.94 mg/kg) were found. On the other side, in control variant higher content of Zn (26.06 ± 3.71 mg/kg) in comparison with experimental variant (25.02 ± 2.18 mg/kg) was observed. Then, the Cu (17.95 ± 0.68 vs 19.48 ± 2.17 mg/kg) and Mn (14.88 ± 0.39 vs 15.10 ± 0.33 mg/kg) content was higher in the experimental variant compared to control variant. Results show that the addition of urea into grape pomace silage does not affect the content of micro elements. This work was supported by the Slovak Research and Development Agency under the contract No. APVV-16-0170.

Changes in the fatty acid profile of milk from ewes fed cauliflower for four weeks

J. Mateo[1], D.E. Carballo[1], I. Mateos, L.M. García-Vázquez[2,3], F.J. Giráldez[2] and M.J. Ranilla[2,3]
[1] Universidad de León, Campus Vegazana s/n, 24, Campus de Vegazana, s/n, 24071 León, Spain, [2]Instituto de Ganadería de Montaña (CSIC-ULE), Dpto. Producción Animal, Finca Marzanas, 24346 León, Spain, [3]Universidad de León, Campus de Vegazana, s/n, 24071 Léon, Spain; mjrang@unileon.es

Brassicas are considered a source of forage in ruminant feeding, and their possible effect on milk quality deserves research. During processing and commercialisation of cauliflower for human consumption, part of the plant biomass becomes a by-product. The aim of this study was to evaluate the changes in the fatty acid profile of ewes' milk due to the use of cauliflower in their diet. Nine Assaf ewes in the middle phase of lactation (2.07 ± 0.20 l) were fed *ad libitum* on a mixture of 1:1 forage:concentrate plus 1.5 kg of cauliflower for 6 weeks (3 for adaptation and 3 for sampling). The animals were milked once daily and 50 ml-milk from each ewe were sampled weekly from the first to the last day of the sampling period (four sampling days). Samples were lyophilised and 0.15 g of lyophylised milk was subjected to transesterification reaction and the fatty acids (FA) were analysed by gas chromatography coupled with mass spectrometry. Detection was performed using the spectral information and standards of FA methyl esters. FA concentrations were expressed as area percentage of the sum of areas of the FA methyl esters identified. Data were analysed by repeated measures ANOVA with time as independent variable and ewe as experimental unit. The main FA detected (accounting for 90% of the total area) in order of abundance were palmitic acid (C16:0), oleic acid (C18:1), myristic acid (C14:0), capric acid (C10:0), stearic acid (C18:0), lauric acid (C12:0), linoleic acid (C18:2), caprylic acid (C8:0), vaccenic acid (C18:1t), pentadecylic acid (C15:0) and linolenic acid (C18:3). C18:1t and C18:3 showed a significant ($P<0.05$) linear increase during the sampling period. The levels at the end of the period were 30% higher than those at the beginning. In contrast, the levels of C18:1 significantly decreased ($P<0.05$). The use of cauliflower in the diet would increase the Ω-6 content in ewes' milk according to a linear pattern.

Chemical composition and nutritional value of almond hulls (*Prunus dulcis*)

C. Costa[1], K. Paulos[1], J.M.S. Costa[1], L. Cachucho[2], P.V. Portugal[1], L.C. Roseiro[1], O.C. Moreira[1,2], E. Jerónimo[3,4] and M.T.P. Dentinho[1,2]
[1]*Instituto Nacional Investigação Agrária e Veterinára, Nutrição Animal, Fonte Boa, 2005-048 Santarém, Portugal,* [2]*Centro Investigação Interdisciplinar em Sanidade Animal, Universidade Técnica, 1330-477 Lisboa, Portugal,* [3]*Mediterranean Institute for Agriculture, Environment and Development, Beja, 7801-908, Portugal,* [4]*Centro Biotecnologia Agrícola e Agro-Alimentar do Alentejo/IPBeja, Beja, 7801-908, Portugal; claudia.dc.costa@gnail.com*

The use of agro-industrial by-products in animal nutrition must be stimulated in order to reduce imports, promote economic and environmental sustainability and reduce human/animal competition for food. In Portugal almonds production has been growing, with new plantations in irrigated areas, following the intensive models, with high number of plants/ha. In 2020 is expected an almond production of 34 thousand tons. Almond hulls are the most important by-product of the almond processing industry. For each 1000 kg of almond nuts produced are available about 500 kg of almond hulls. In this work it was our aim to evaluate for chemical and nutritional value of almond hulls for livestock feed. Almond hulls were analysed for proximate analysis, neutral detergent fibre (NDF), acid detergent fibre (ADF), lignin (ADL), sugar, starch, minerals, fatty acid, amino acids and phenolic compounds. For nutritive evaluation the *in vitro* organic matter digestibility (OMD) was determined. Almond hulls contain high sugar content (29% DM) and moderate amounts of fibre (29% NDF and 18% ADF DM). They are poor in protein (6% DM) and in fat content (<1%) and contain high levels of minerals (10% DM) especially Fe (122 µg/g DM) and K (5% DM). Almond hulls contained 5% DM of phenolic compounds, mostly tannins (4% DM). The OMD is moderate (59%) that may be explained by the negative effect of tannins on digestibility. The results suggest that almond hulls can be a good source of energy for ruminants. However protein supplementation is required when included in animal diets. This work is funded by PRD2020 through the FEADER under the project SubProMais (PDR2020-101-030988) and by National Funds through FCT – Foundation for Science and Technology under the projects UIDB/05183/2020 (MED) and UIDP/CVT/00276/2020 (CIISA).

Effect of *Cistus ladanifer* L. condensed tannins on lamb rumen microbiome and fatty acid composition

O. Guerreiro[1,2], S.P. Alves[3], D. Soldado[2], L. Cachucho[2], A. Francisco[3,4], J. Santos-Silva[3,4], R.J.B. Bessa[3] and E. Jerónimo[1,2]
[1]*Mediterranean Institute for Agriculture, Environment and Development, CEBAL, Beja, Portugal,* [2]*Centro de Biotecnologia Agrícola e Agro-Alimentar do Alentejo, Rua Pedro Soares Apartado 6158, 7801-908 Beja, Portugal,* [3]*Centro de Investigação Interdisciplinar em Sanidade Animal, Av.Universidade Técnica, 1300-477 Lisboa, Portugal,* [4]*Instituto Nacional de Investigação Agrária e Veterinária, Fonte Boa, 2005-048 Vale de Santarém, Portugal; olinda.guerreiro@cebal.pt*

The inclusion of tannins in ruminant diets has been reported to modulate the ruminal biohydrogenation (BH) of unsaturated fatty acids (FA) and to affect the rumen microbiota. *Cistus ladanifer* L. is a shrub rich in condensed tannins (CT), that when fed to lambs has been associated with BH changes. In order to evaluate the effect of three levels of *C. ladanifer* CT (0, 1.25 and 2.5%) and two ways of CT supply (*C. ladanifer* aerial part and *C. ladanifer* CT extract) on FA composition and microbiome diversity of rumen content, thirty-six lambs were fed with a basal diet composed of dehydrated lucerne supplemented with 60 g/kg of soybean oil. Independently from diet, rumen composition was dominated by bacteria belonging to *Firmicutes* (49.7%, average of relative abundance), followed by *Bacteroidetes* (35.8%). At family level, *Clostridiaceae* resulted as dominant group (17.6%), followed by *Prevotellacea*e (13.6%), with *Clostridium* (15.8%) being the main genus found, followed by *Prevotella* (13.5%). *C. ladanifer* CT extract presented lower *Firmicutes* phylum than aerial part. Increasing levels of CT reduced the concentration of total branched chain FA, anteiso-15:0 and iso-16:0 in rumen content. *C. ladanifer* diets induced to higher levels of anteiso-15:0 in rumen than CT extract diets. Dietary treatments also affected the proportion of several BH intermediates in rumen content, particularly the 18:1 isomers, with higher *t*11-18:1 content when 1.25% CT from CT extract was used compared to same CT levels from *C. ladanifer* aerial part. Financed by Alentejo2020 through the FEDER with the project 'CistusRumen' (ALT20-03-0145-FEDER-000023) and by National Funds through FCT – Foundation for Science and Technology under the Projects UIDB/05183/2020 and UIDP/CVT/00276/2020.

Influence of including broccoli in a dairy ewes diet on microbial fermentation in the Rusitec system

A. Martin[1,2], I. Mateos[1,2], C. Saro[1,2], T. De Evan[3], R. Campos[2,4], M.D. Carro[3] and M.J. Ranilla[1,2]
[1]Instituto de Ganadería de Montaña (CSIC-ULE), Finca Marzanas, 24346, León, Spain, [2]Universidad de León, Producción Animal, Campus Vegazana, 24071, León, Spain, [3]Universidad Politécnica de Madrid, Producción Agraria, Ciudad Universitaria, 28040, Madrid, Spain, [4]Universidad Nacional de Colombia, Ciencia Animal, Palmira, 763531, Colombia; imata@unileon.es

Broccoli is a product of the horticultural industry in Spain with high protein and sugars content, and its surplus, wastes and by-products could be used as a substitute of expensive protein feeds in ruminant diets like soybean meal. The objective was to assess the effect of replacing soybean meal and barley grains in the concentrate of a standard dairy sheep diet (CON) by broccoli (BRO) on ruminal fermentation parameters and apparent digestibility in Rusitec fermenters. Diets consisted of alfalfa hay and concentrate 1:1 and had similar protein and NDF contents (17.3 and 29.0% of dry matter, respectively). 4 Rusitec fermenters were used in a cross-over design with 14 days incubation period. Fermenters were given daily 30 g of diets, and in each period 2 fermenters received the diet with no broccoli (CON) and the other 2 received the diet with dried broccoli (14,7% BRO, as feed). On days 10, 11, 12 and 13 of incubation, samples of liquid effluents and gas were collected from the fermenters to analyse the daily production of volatile fatty acid (VFA) and methane, and nylon bags were washed and weighed to determinate the diet apparent digestibility. There were no differences between diets either in the total VFA or in the daily methane production. There were differences between diets in the VFA profile, being greater ($P \leq 0.03$) the acetate and caproate proportion and the acetate:propionate ratio for the BRO diet than for the CON one. Dry matter, crude protein, neutral and acid-detergent fibre apparent digestibility did not differ between diets. Under the present experimental conditions, broccoli successfully replaced soybean meal and barley grains in a dairy sheep without negatively affecting ruminal fermentation or diet digestibility.

Using tomato pulp in dairy sheep diets: nutrition parameters in Rusitec fermenters

I. Mateos[1,2], R. Campos[2,3], A. Martin[1,2], T. De Evan[4], C. Saro[1,2], M.D. Carro[4] and M.J. Ranilla[1,2]
[1]Instituto de Ganadería de Montaña (CSIC-ULE), Finca Marzanas, 24346, León, Spain, [2]Universidad de León, Producción Animal, Campus Vegazana, 24071, Spain, [3]Universidad Nacional de Colombia, Ciencia Animal, Palmira, 763531, Colombia, [4]Universidad Politécnica de Madrid, Producción Agraria, Ciudad Universitaria, 28040, Madrid, Spain; imata@unileon.es

Spain is one of the most important tomato producers in the world. The industrial use of this fruit generates high quantities of by-products, tomato pulp (TP) being one of them. TP is a high-moisture product with a medium protein and energy content that can be used in ruminant feeding. The effect of partially replacing alfalfa hay, soybean meal and sugar beet pulp by TP in a dairy sheep diet on ruminal fermentation parameters and apparent digestibility was tested in Rusitec fermenters. Two dairy diets, with no tomato pulp (CON) and with 17.3% of the by-product (TP) were incubated in a cross-over design in 4 Rusitec fermenters for 14 days. On days 10, 11, 12 and 13 of incubation, samples of liquid effluents and gas were collected from the fermenters to analyse the daily methane and volatile fatty acid (VFA) production, and nylon bags were washed and weighed to determinate the diet apparent digestibility. Daily production of total VFA and methane were higher ($P \leq 0.0037$) in control fermenters than in those receiving the TP diet, but there were no differences between diets either in the VFA profile or in the acetate:propionate ratio. Dry matter, crude protein and neutral detergent fibre apparent digestibility was lower ($P \leq 0.020$) for the TP diet than the control one. In Rusitec fermenters, the inclusion of 17.3% of TP in a milk diet slightly decreased ruminal fermentation compared to a standard diet, but more research is required to determine its optimal inclusion level for dairy sheep feeding.

Ruminal fermentation of a diet including brussels sprouts in a Rusitec system

T. De Evan[1], A. Martin[2,3], I. Mateos[2,3], C. Saro[2,3], R. Campos[3,4], M.D. Carro[1] and M.J. Ranilla[2,3]
[1]*Universidad Politécnica de Madrid, Departamento de Producción Agraria, Ciudad Universitaria, 28040, Madrid, Spain,* [2]*Instituto de Ganadería de Montaña (CSIC-ULE), Finca Marzanas, 24346, León, Spain,* [3]*Universidad de León, Producción Animal, Campus Vegazana, 24071, León, Spain,* [4]*Universidad Nacional de Colombia, Ciencia Animal, Palmira, 763531, Colombia; mjrang@unileon.es*

Brussels sprouts (BS) contain multiple secondary compounds which might modulate ruminal fermentation. As other by-products, wastes of BS coming from industrial horticulture could be used in ruminant feeding after proper evaluation of their nutritive value. The aim of this study was to assess the effect of partially replacing soybean meal and totally replacing barley grains, in the concentrate part, by dried BS in a mixed diet (1:1 alfalfa hay:concentrate) on ruminal fermentation parameters and apparent digestibility in Rusitec fermenters. Two diets with 17.3% protein and 29.0% neutral detergent fibre content were incubated in 4 Rusitec fermenters in a cross-over design in 2 periods of 14 days each. The diets were supplied daily to the fermenters into nylon bags containing 30 g, and in each period 2 fermenters received the diet with no BS (CON) and the other 2 received the diet with dried BS (14,7% BS, as feed). On the last four days of incubation, samples of liquid effluents and gas were collected from the fermenters for determining the daily production of volatile fatty acid (VFA) and methane. In addition, diet digestibility was measured as the disappearance of the diet from the nylon bags. There were no differences between diets either in the total VFA or in the daily methane production. However, caproate proportion was greater ($P \le 0.034$) and valerate and isovalerate proportion was lower ($P \le 0.001$ and $P \le 0.013$) for the BS diet than for the CON one. Dry matter, crude protein, neutral and acid-detergent fibre apparent digestibility did not differ between diets. The results indicate that BS can partially replace conventional feed ingredients, such as soybean meal and barley grains, in mixed diets without negatively affecting ruminal fermentation and diet digestibility and causing only minor changes in VFA profile. *In vivo* research is needed to assess the influence of BS-containing diets on animal performance.

Evaluation of the nutritional value of silages based on agro-industrial by-products

K. Paulos[1], C. Costa[1], J. Costa[1], L. Cachucho[2], P.V. Portugal[1], J. Santos-Silva[1,2], E. Jerónimo[3,4] and M.T.P. Dentinho[1,2]
[1]*Instituto Nacional de Investigação Agrária e Veterinária, Fonte Boa, Santarém, 2005-048, Portugal,* [2]*Centro Investigação Interdisciplinar em Sanidade Animal, Avenida Universidade Técnica, Lisboa, 1049-001, Portugal,* [3]*Mediterranean Institute for Agriculture, Environment and Development, Beja, 7006-554, Portugal,* [4]*Centro de Biotecnologia Agrícola e Agro-Alimentar do Alentejo, IPBeja, Beja, 7801-908, Portugal; katia.paulos@iniav.pt*

With climatic changes, increase in temperature and decrease in precipitation, pastures and forages productions are irregular and increasingly limited to short periods of time, making it necessary to resort to concentrate feeds. In Portugal, concentrate feeds are based in cereals and oilseed cakes, which are high-cost imported feeds. The use of agro-industrial by-products as alternative feed sources must be stimulated in order to reduce imports, promote economic and environmental sustainability and reduce human/animal competition for food. In this study, it was our aim to characterise chemically and nutritionally agro-industrial by-products and mix them in order to produce balanced silages to use in ruminant diets. Potato tubers (PT), carrots (C), wheat bran (WB) and brewers grains (BG) were analysed for chemical and nutritive characterisation and ensiled in bags in order to obtain final balanced silages with 40% of dry matter (DM) and 14% of crude protein (CP). Carrots are rich in sugars (39% DM) and PT are rich in starch (34% DM). Wheat bran and BG are protein sources (29 and 18% DM, respectively). So, two silages were performed: Silage PT: 30% PT + 35% BG + 20% WB + 15% of grass hay and; Silage C: 30% C + 35% BG + 20% WB + 15% of grass hay. The silages obtained were well preserved, with low pH (4.49 and 4.31 in PT and C silages, respectively), NH3-N <10% of total-N and soluble-N <50% of total-N. The CP level, in both silages was 16% DM, the organic matter digestibility was higher in PT than in C silage (56 vs 42%). The concentration of metabolisable energy was estimated at 1,894 and 1,398 kcal/kg in PT and C silages, respectively. This work is funded by PRD2020 through the FEADER, project SubProMais (PDR2020-101-030988) and by National Funds through FCT – Foundation for Science and Technology, projects UIDB/05183/2020 (MED) and UIDP/CVT/00276/2020 (CIISA).

Fruit by-products in Animal Feed: Chemical composition and nutritional value

L. Cachucho[1], K. Paulos[2], C. Costa[2], D. Soldado[1], L. Fialho[1], O. Guerreiro[1,3], J. Santos-Silva[2,4], M.T. Dentinho[2,4] and E. Jerónimo[1,3]
[1] Centro de Biotecnologia Agrícola e Agro-Alimentar do Alentejo, Rua Pedro Soares, s/n, 7801-908 Beja, Portugal, [2]Instituto Nacional de Investigação Agrária e Veterinária, Fonte Boa, 2005-048 Santarém, Portugal, [3]Mediterranean Institute for Agriculture, Environment and Development, CEBAL, Beja, Portugal, [4]Centro Investigação Interdisciplinar em Sanidade Animal, Av Técnica de Lisboa, 1300-477 Lisboa, Portugal; liliana.cachucho@cebal.pt

Edaphoclimatic conditions of Portugal offers opportunity to production of several fruit species. As a result of fruit production, high quantities of products are generated that do not meet the market requirements for human consumption. Fruit by-products are promising sources of nutrients and bioactive compounds, such as phytochemicals compounds and vitamins. So, the application of these by-products in animal feed is an opportunity for their use and valorisation, allowing the reuse of the agricultural by-products and reducing the environmental impact of the food production. The aim of this work was to evaluate the chemical composition, nutritional value and antioxidant activity of fruit by-products highly available in Ribatejo and Alentejo regions of Portugal, specifically apple, pear, watermelon, melon, cantaloupe, blackberry, raspberry and blueberry. Globally, these by-products showed low dry matter (DM) content (5.6-16.5%), low levels of protein (1.9-9.5% DM), low to moderate content of parietal constituents (5.7-21.8% DM of crude fibre), high levels of sugar (50.8-82.1% DM) and high digestibility of organic matter (70.3-89.4%). All fruit by-products are sources of phenolic compounds and natural antioxidants, observing higher antioxidant activity in blackberry, raspberry, blueberry than in other fruit by-products analysed. Results showed that fruit by-products are interesting resources for application in animal nutrition, particularly as functional feed ingredients for improve the animal health and quality of their products. This work is funded by PRD2020 through the FEADER under the project SubProMais – Use of agro-industrial by-products in animal feed (PDR2020-101-030993) and by National Funds through FCT – Foundation for Science and Technology under the projects UIDB/05183/2020 and UIDP/CVT/00276/2020.

Influence of agroindustrial by-products on characteristics and fatty acid profile of lamb meat

T. De Evan[1], H. Pérez[1], A. Cabezas[2], D. Bueno[1], J. De La Fuente[2] and M.D. Carro[1]
[1]Universidad Politécnica de Madrid, Departamento de Producción Agraria (ETSIAAB), Ciudad Universitaria, 28040, Madrid, Spain, [2]Universidad Complutense de Madrid, Departamento de Producción Animal, Facultad de Veterinaria, Avda. Puerta de Hierro, s/n, 28040, Madrid, Spain; t.deevan@alumnos.upm.es

Feeding agroindustrial by-products to ruminants contributes to a sustainable livestock production but can also improve the quality of animal products. The objective of this study was to evaluate the effects of replacing 44% of conventional feeds with corn DDGS, dried citrus pulp (DCP) and exhausted olive cake (EOC) in a concentrate for fattening lambs on meat characteristics and fatty acid (FA) profile. Twenty-four Lacaune lambs (13.8±0.25 kg) were divided in two homogeneous groups and each was fed either a high-cereal concentrate (control) or a concentrate (BYP) including corn DDGS, DCP and EOC (18, 18 and 8%, respectively). Lambs were fed concentrate and barley straw *ad libitum* for 42 days before being slaughtered at about 26 kg body weight. The muscle longissimus dorsi of each lamb was used to determine chemical composition, colour, pH, and FA profile of the meat. Data were analysed as a one-way ANOVA. There were no differences between diets either in hot and cold carcass weight or cold carcass yield. The diet had no effect (P≥0.427) on chemical composition, colour and pH of the longissimus dorsi. Compared with the control-fed lambs, the proportions of C15:0, C16:0, C17:0 were lower (P≤0.034) in the meat from BYP-fed lambs. Feeding the BYP diet also resulted in greater (P=0.030) proportions of total polyunsaturated FA and tended (P=0.069) to increase the proportion of monounsaturated FA. These results would indicate a potentially healthier profile of the meat from BYP-fed lambs, as C16:0 is atherogenic and has been related to an increase in total and low density lipoprotein cholesterol concentrations, whereas monounsaturated and polyunsaturated FA are considered healthier. In conclusion, a mixture of corn DDGS, DCP and EOC can replace 44% of cereal grains and protein feeds in the concentrate for fattening lambs without negatively affecting either growing performance or meat composition, and resulted in a possible healthier meat FA profile.

Antioxidant status and meat oxidative stability of lambs fed *Cistus ladanifer* plant and extract

D. Soldado[1], L. Fialho[1], O. Guerreiro[1,2], L. Cachucho[1], A. Francisco[3,4], J. Santos-Silva[3,4], R.J.B. Bessa[4] and E. Jerónimo[1,2]
[1]*Centro de Biotecnologia Agrícola e Agro-Alimentar do Alentejo, Rua Pedro Soares, s.n, 7801-908 Beja, Portugal,* [2]*Mediterranean Institute for Agriculture, Environment and Development, CEBAL, Beja, Portugal,* [3]*Instituto Nacional de Investigação Agrária e Veterinária, Fonte Boa, 2005-048 Santarém, Portugal,* [4]*Centro de Investigação Interdisciplinar em Sanidade Animal, Av. da Universidade Técnica, 1300-477 Lisboa, Portugal; david.soldado@cebal.pt*

The *Cistus ladanifer* L. (CL) is a very abundant perennial shrub in Mediterranean countries, containing high levels of bioactive compounds with recognised antioxidant activity, including vitamins and phenolic compounds, such as condensed tannins (CT). The aim of this study was to evaluate the effect of inclusion of aerial part and CT extract from CL in lamb diets on muscle antioxidant capacity, lipid oxidation in small intestine and oxidative stability of meat. Six diets were formulated considering three levels of CL CT (0, 1.25 and 2.5% CT) and two ways of CT supply (CL aerial part vs CL CT extract). Basal diet was composed of dehydrated Lucerne supplemented with 60 g/kg of soybean oil. Thirty-six lambs were housed individually, with six lambs per diet. The trial lasted for 35 days. Increasing levels of CL aerial part and CL CT extract in diets increased the phenolic and CT contents and antioxidant activity of diet. Muscle phenolic content and antioxidant activity (ferric reducing antioxidant power assay) were not affected by dietary treatments (P>0.05). Conversely, lambs fed CL aerial part showed lower lipid oxidation (TBARS assay) in small intestine (P=0.048) than lambs fed CL CT extract diets (0.11 vs 0.15 mg malondialdehyde (MDA)/kg). Meat lipid oxidation increased over storage time from 0.09 to 0.40 mg MDA/kg at day 0 and 7 of storage, respectively, and was not affected by the dietary treatment (P>0.05). Present results suggest that CL aerial part exerted antioxidant activity in the gastrointestinal tract. Work funded by Alentejo2020 program through the FEDER under the project CistusRumen (ALT20-03-0145-FEDER-000023) and by National Funds through Foundation for Science and Technology under the individual fellowships to DS (SFRH/BD/145814/2019) and projects UIDB/05183/2020 and UIDP/CVT/00276/2020.

Using agroindustrial by-products in sheep diets: effects on ruminal fermentation and degradability

T. De Evan[1], D. Bueno[1], H. Pérez[1], M.J. Ranilla[2,3] and M.D. Carro[1]
[1]*Universidad Politécnica de Madrid, Departamento de Producción Agraria (ETSIAAB), Ciudad Universitaria, 28040, Madrid, Spain,* [2]*Instituto de Ganadería de Montaña (CSIC-ULE), Finca Marzanas, 24346, León, Spain,* [3]*Universidad de León, Departamento de Producción Animal, Campus de Vegazana, 24071, León, Spain; t.deevan@alumnos.upm.es*

Finding alternatives to conventional feeds, such as agroindustrial by-products, is required for a sustainable livestock production, but nutritional evaluation is needed for an optimal use. This study assessed ruminal fermentation and *in situ* degradability in sheep fed two diets containing 1:1 alfalfa hay:concentrate, with either a high-cereal concentrate (control) or a concentrate (BYP) in which 44% of ingredients were replaced by corn DDGS, dried citrus pulp and exhausted olive oil (18, 18 and 8%, respectively; as fed basis). Four rumen-fistulated sheep were fed the two diets in a cross-over design, ruminal samples were taken at 0, 3 and 6 h after feeding and the *in situ* degradability of the two concentrates were measured. Data were analysed by ANOVA as a cross-over design, and in the analysis of ruminal parameters sampling time was included as a repeated measures factor. Compared to control sheep, BYP-fed sheep had lower ruminal pH (P=0.021; 6.61 vs 6.53; values averaged across time) and NH_3-N concentrations (P=0.016; 104 vs 88.9 mg/l), and tended to have greater concentrations of volatile fatty acids (VFA; P=0.089; 130 vs 142 mmol/l), which would indicate increased ruminal fermentative activity in BYP-fed sheep. The VFA profile in BYP-fed sheep was shifted to greater (P≤0.013) propionate and butyrate proportions at expenses of acetate, resulting in lower acetate:propionate ratio (P=0.001; 3.92 vs 3.54 for control and BYP, respectively). In the *in situ* trial, the control concentrate had lower (P≤0.012) soluble fraction, insoluble but potentially degraded fraction and effective degradability calculated for a rumen passage rate of 6%/h of dry matter (48.6, 38.0 and 71.2%, respectively) than the BYP concentrate (50.5, 42.9 and 74.5%, respectively). The results indicate that the mixture of by-products used in this study stimulated ruminal fermentation and therefore can replace conventional feed ingredients in sheep diets.

The use of NIR Spectroscopy for predicting the chemical composition of dried and undried silage

C. Maduro Dias and A. Borba
University of the Azores, Institute of Agricultural and Environmental Research and Technology (IITAA), Rua Capitão João d'Ávila, 9700-042 Angra do Heroísmo, Açores, Portugal; cristianarodrigues@gmail.com

The goal of this work was to compare the effect of drying corn and grass silage in the results obtained by Near-Infrared Reflectance Spectroscopy (NIRS), when quantifying dry matter (DM), pH, crude protein (CP), neutral and acid detergent fibre (ADF, NDF), and ash. 200 grass silage samples and 200 corn silage samples from the Azores were collected. Each set was further divided into two subsets of the same size. For one of these subsets, specters were collected from the undried samples using a NIR spectrophotometer in reflectance (400-2,500 nm), and for the other from the dried samples. Reference values were determined using traditional methods. Calibration models were then developed using partial least squares regression (PLSR) with full cross validation. These were combined with different spectral pre-treatments: Spectra without treatment (log 1/R), standard normal variate (SNV), combination of detrending and standard normal variate (SNVD), combination of first derivative and SNV, combination of second derivative and SNV, and multiplicative scatter correction (MSC). For each parameter, the optimal calibration was determined through the analysis of statistical information, such as the coefficient of determination in calibration (R^2cal), standard error of cross validation (SECV) and RPD. For grass silage, using undried samples led to more consistent R^2cal values, with 0.92 (undried) vs 0.73 (dried) for dry matter, 0.87 vs 0.52 for pH, 0.92 vs 0.86 for NDF, 0.90 vs 0.91 for ADF, and 0.94 vs 0.90 for ash. Crude protein was the only parameter where the results obtained using dried samples were noticeably better, with 0.95 (dried) vs 0.88 (undried). With corn silage, dry matter showed R^2cal values of 0,90 (undried) vs 0,70 (dried), pH of 0.85 vs 0.56, crude protein of 0.83 vs 0.87, NDF of 0.85 vs 0.87, ADF of 0.86 vs 0.84, and ash of 0.88 vs 0.93. Using NIRS to predict values of undried silage is faster and more practical. According to these results, it may even lead to better overall results when predicting the values of some parameters, particularly with grass silage.

Effect of introducing *Lupinus luteus* seeds to lamb diets on their eating behaviour

M. Almeida[1], A. Nunes[2], S. Rito[2], S. Garcia-Santos[3], C. Guedes[1], L. Ferreira[3], G. Stilwell[4] and S. Silva[1]
[1]Veterinary and Animal Research Centre (CECAV), Quinta de Prados, 5001-801, Vila Real, Portugal, [2]University of Trás-os-Montes and Alto Douro (UTAD), Quinta de Prados, 5001-801, Vila Real, Portugal, [3]Centre for the Research and Technology Agro-Environmental and Biological Sciences (CITAB), Quinta de Prados, 5001-801, Vila Real, Portugal, [4]Centre of Interdisciplinary Investigation in Animal Health (CIISA), Avenida da Universidade Técnica, 1300-477, Lisboa, Portugal; mdantas@utad.pt

The objective of this study was to evaluate the effect of partial or total replacement of soybean meal (SBM) by *Lupinus luteus* grains on lamb eating behaviour. Two different trials (T1 and T2) were conducted. Two diets were tested on T1: the control diet (C), containing wheat grain and SBM, and another diet with *L. luteus* replacing 30% of SBM (LL30). For T2, four diets were used: C diet and three diets with SBM replacement of 50, 75 and 100% by *L. luteus* (LL50, LL75 and LL100, respectively). Meadow hay was offered *ad libitum*. At weaning the 12 weeks old lambs were housed in groups of 4. Eating behaviour was evaluated by registering the frequency of lambs eating hay and the time spent eating the concentrate feeds (TC) on four different days of each trial. Eating hay (H) behaviour was recorded every 10 minutes, between 12 pm and 9 am, using scan sampling. Eye temperature was obtained through thermographic images during the time concentrate feeds were provided. For each diet, the frequency of eating hay behaviour was compared using a Pearson's Chi-square contingency table analysis. The effect of diet on time spent eating the concentrate feeds and eye temperature was examined using an ANOVA procedure. The diet had an effect on H (P<0.05), since animals consuming diets with more *L. luteus* content seem to eat less hay. No differences were found (P>0.01) in TC in T1. However, in T2 animals LL50 and LL75 showed lower values of TC when compared to C and LL100 (9.5 and 10.4 vs 11.6 and 12.5 minutes, respectively). On both trials, eye temperature was lower on lambs fed with diets which had SBM replacement (P<0.01), which can mean a slight aversion to *L. luteus* introduction. Overall, data indicate that replacing soybean meal by *L. luteus* might have a slight effect on lamb eating behaviour.

The f4f feed production process: solar drying as a key component

F. Galliou, C. Bouki, N. Markakis, J. Sampathianakis, M. Sampathianakis, A. Papadaki and T. Manios
Hellenic Mediterranean University, Agriculture, Estavromenos, 71410 Heraklion Crete, Greece; fgalliou@hmu.gr

The LIFE-F4F Project implements in a pilot scale, a solar drying process that provides an innovative, low-tech and low emissions method, for safe transformation of source separated food waste from catering units into animal feed. F4F targets luxury hospitality units, which implement source separation schemes, due to strict quality assurance standards regulations, operating during touristic period. The F4F process involves an efficient food waste collection system, a pretreatment processing unit and solar drying. The pretreatment involves a hand sorting belt, a shredding and pulverising unit, together with a progressing cavity pump. Two drying cells are constructed under a greenhouse, each one served by different bridge turners, a horizontal (rotating drum) and a vertical (screw mixing system) turner. Subfloor heating system is supporting the drying process in both cells, connected with hot water solar panels and a heat pump. The drying cells surface is covered by stainless still, suitable for feed. The final product is sterilised in order to comply with the 142/2011 regulation. Two operational periods have been concluded, the testing period, in summer 2018 and the first full scale operational period, in summer 2019. From the second period, more than 150tn of food waste have been collected and treated in the pilot unit, allowing more solid conclusions to be drawn. Drying on May and October took between 10 to 15 days, on June and September this was reduced to 5 to 8 days and on July and August drying duration dropped into 4 to 6 days. Aim is to reduce moisture levels from 75% (fresh food waste) to 10% dried feed. The quality of the produced feed was high regarding a range of parameters, such as dry matter, crude protein, fat, crude fibre, crude ash, soluble protein, amino acid and mineral content. Regarding pathogens, the final product was analysed for microorganisms with no significant detection. Nuisance, as for example odours and insects was minimum in the pilot unit, as optimisations carried out. Acknowledgments: This research is funded by EU LIFE15 ENV/GR/0002057: 'Food for Feed: An Innovative Process for Transforming Hotels' Food Wastes into Animal Feed'.

The 'Food for Feed' project

M. Georgiou, P. Panteli, K. Borboudaki, I. Giakoumaki and N. Stylianidis
Association of Solid Waste Management in Crete, Arx. Makariou 22, 71202 Heraklion Crete, Greece;
georgiou.maria73@gmail.com

The European Commission has identified Food Waste as one of the most important areas of the European Circular Economy Action Plan. The costs associated with food waste for EU-28 in 2012 are estimated at around 143 billion euros. In this regard, the F4F project implements an innovative, simple technology and low emissions process, which allows the safe transformation of source separated food wastes, from the hospitality industry, into animal feed, utilising an altered solar drying process. The main objectives of the project are: (1) design, operation and evaluation of a separated collection scheme, for hotels' food wastes in a touristic area of Crete (Greece); (2) design, construction, operation and evaluation of a hand sorting, grinding and automated solar drying process that would allow the transformation of the incoming food wastes into animal feed, within the summer period; (3) evaluation of the quality of the produced feed and determination of its commercial; (4) evaluation of the potential role of such a system in EU policies. Acknowledgments: This research is funded by EU LIFE15 ENV/GR/0002057: 'Food for Feed: An Innovative Process for Transforming Hotels' Food Wastes into Animal Feed'.

Salty or sweet former food based diets for piglets: effect on serum metabolites

A. Luciano[1], M. Ottoboni[1], M. Comi[2], F. Fumagalli[1], N. Rovere[1], L. Ferrari[1], D.D. Chiarellli[3] and L. Pinotti[1]
[1]Università degli Studi di Milano, Department of Health, Animal Science and Food Safety, Milano, 20134, Italy, [2]Università Telematica San Raffaele Roma, Italy, Roma, 00166, Italy, [3]Politecnico di Milano, Milano, 20133, Italy; luciano.pinotti@unimi.it

In the present study conventional cereal grains, like corn, wheat and barley, have been partially replaced by sweet or salty former food products (FFPs), in post-weaning piglet's diets in order to investigate the effects of these materials on pig plasma metabolites. Briefly, 36 weaned female pigs (28 d of life, 6.70±1.07 kg) were fed 1 of the 3 experimental diets: 1) Control diet (CRT), 0% FFPs; 2) confectionary FFPs diet (FFP sweet) in which 30% of cereals were substituted by 30% of confectionary FFPs; 3) bakery FFPs diet (FFP salty) in which 30% of cereals were substituted by 30% of bakery/snack FFPs. The diets were iso-energetic (15.3 MJ/kg DM) and iso-nitrogenous (19% CP DM), and met NRC (2012) requirements. The trial lasted 42 d. During the experiment, blood samples were collected on day 0, 21, 42 of the experiment, for determining the following serum metabolites: Total proteins, Albumin, Globulin, Albumin / globulins (A / G), Urea, Alanine aminotransferase (ALT-GPT), Aspartate aminotransferase (AST-GOT), Alkaline phosphatase (ALP), Total bilirubin, Glucose, triglycerides, NEFA, amylase, Total cholesterol, calcium, Phosphorus, Magnesium. Data were analysed by repeated measurement of ANOVA procedure and further explored by PCA. Over the experiment, any treatment effect was observed. Only some time trends have been detected for triglycerides in FFP sweet group on day 21, and for amylase that tended to be lower in both FFP salty and sweet groups over the experiment in comparison with CRT group. Time trend was partially confirmed by the PCA analysis. The results revealed irrespective of the type (salty or sweet), FFP can replace common cereal in post weaning diets, without effects on main serum metabolites. Taken together, these results suggest that the use of FFPs up to a level of 30% in post-weaning diets has no detrimental effects on pig metabolic status. This study has been done in the frame of the SUSFEED project funded by Cariplo Foundation.

Effects of dietary food waste addition on growth performance and meat quality in finishing pigs

E. Giamouri[1], G. Papadomichelakis[1], A.C. Pappas[1], E. Tsiplakou[1], K. Fegeros[1], T. Manios[2] and G. Zervas[1]
[1]Agricultural University of Athens, Nutritional Physiology and Feeding, Iera Odos 75, 11855, Greece, [2]Hellenic Mediterranean University, Agriculture, Stavromenos, GR71410, Crete, Greece; gzervas@aua.gr

Nowadays, the use of food waste as an alternative source of energy in animal diets has gained considerable attention, because of the increasing needs of human population and the high prices of conventional arable based animal feeds. The aim of the present study was to investigate the effect of adding dried food waste (DFW), collected from hotels, to the diet of finishing pigs on growth performance and meat quality traits. Twenty (20) 106-d old castrated male pigs were allotted into two (2) dietary treatments (n=10 pigs/ treatment), namely control (C) and DFW, balanced for body weight (BW; 50.3±2.54 kg). In treatment C, pigs were fed a corn-soybean meal based diet without DFW, whereas in treatment DFW, a diet containing 100 g DFW/kg. Both diets were isocaloric and isonitrogenous and were formulated with similar digestible lysine, methionine+cystine and threonine contents. Feed intake and body weight were recorded at the onset and at the end of the experiment to calculate average daily feed intake (ADFI), daily BW gain (ADWG) and feed conversion ratio (FCR). At the age of 144-147 days, a digestibility trial was conducted to determine energy and nutrient digestibility. At the end (156 days of age), pigs were sacrificed to investigate treatment effects on carcass dressing percentage (CDP), as well as on meat quality indices (pH, colour, cooking loss and shear force). Pigs performed well with a final BW of 98.6 and 94.2 kg for treatments C and DFW, respectively. The ADFI tended (P=0.058) to be lower and ADWG was lower (P=0.027) in DFW compared to C pigs; however, FCR was not affected. No differences in the digestibility of energy and nutrients were observed between treatments. The CDP did not differ between treatments and was 77.0, and 77.7% in C and DFW pigs, respectively. Meat colour traits were also not affected by the dietary treatment. In conclusion, the results support the use of dried food waste in pig feeding. The research was funded by the European LIFE Food4Feed project.

Growth performance, carcass and meat quality of Alentejana bulls fed high-fibre and low-starch diet

J. Santos-Silva[1,2], A. Francisco[1,2], S. Alves[1,3], A. Portugal[2], J. Almeida[2], L. Fialho[4], E. Jerónimo[4,5] and R. Bessa[1,3]
[1]Centro de Investigação Interdisciplinar em Sanidade Animal, Avenida da Universidade Técnica, 1300-477 Lisboa, Portugal, [2]Instituto Nacional de Investigação Agrária e Veterinária, UEISPSA, Quinta da Fonte Boa, Vale de Santarém, 2005-048, Portugal, [3]Faculdade de Medicina Veterinária da Universidade de Lisboa, Avenida da Universidade Técnica, 1300-477 Lisboa, Portugal, [4]Centro de Biotecnologia Agrícola e Agro-Alimentar do Alentejo, Rua Pedro Soares- Campus IPBeja /ESAB, 7801-908 Beja, Portugal, [5]Mediterranean Institute for Agriculture, Environment and Development, Universidade de Évora, Pólo da Mitra, 7006-554 Évora, Portugal; alexandra.francisco@iniav.pt

The present study evaluated the growth performance, carcass and meat quality of 16 Alentejana purebred young bulls fed during 102 days with a conventional (Control) or a high-fibre and low-starch (HFLS) diets. The Control was a conventional cereal-based compound feed diet, while in the HFLS diet 35% of the cereals were replaced by soybean hulls, dehydrated citrus and sugar beet pulps and included 20% alfalfa hay and 4.8% soybean oil. Ryegrass hay was offered to a maximum of 20% of total intake in both diets. Bulls were housed in individual pens, weighted every 14 days and their feed intake was registered daily. Slaughter live weights ranged between 550 to 620 kg and diet did not influence (P>0.05) the average daily weight gain, dry matter intake, carcass and most of meat quality traits. The HFLS diet increased the stability of meat colour and lipids and the deposition of the healthy fatty acids, as vaccenic (t11-18:1), rumenic (c9,t11-18:2) and linolenic (18:3n-3) acids. The financial support was provided through the European Fund for Regional Development (ERDF) (ALT20-03-0145-FEDER-000040 –ValRuMeat- Valorização da carne de ruminantes em sistemas intensivos de produção) and national funds through the Foundation for Science and Technology (FCT) under the UIDB/05183/2020 (MED) and UIDP/CVT/00276/2020 (CIISA) projects.

Digestibility of OM during grazing period and winter feeding depending on the content of crude fibre

P. Kühne[1], H. Scholz[1] and G. Heckenberger[2]
[1]Anhalt University of Applied Sciences, Strenzfelder Allee 28, 06406 Bernburg, Germany, [2]State Institute for Agriculture and Horticulture Saxony-Anhalt, Lindenstraße 18, 39606 Iden, Germany; petra.kuehne@hs-anhalt.de

To achieve a good reproductive performance of suckler cows and a high daily gain of their calves particularly a performance based supply is important. In the EU often there are two opposite interests: Extensive use of grassland versus achieving high animal performance to ensure an economical production. An increased extensivation can lead to a decrease in digestibility of grass during the grazing period or in silages during the winter feeding period. The aim of the investigation was to determine the correlations between the digestibility of organic matter (OM) and content of crude fibre. 7 farms with suckler cows were involved into the investigation from May until December 2018 (monthly sampling of faeces). The farms can be divided into: (1) intensive pasture management with more than 100 kg nitrogen per hectare; or (2) semi-intensive pasture management with less than 100 kg N/ha; and (3) ecologically farms without nitrogen fertiliser on grassland. Faeces were analysed for digestibility of organic matter in the laboratory LKS Lichtenwalde after the method from Lukas *et al*. During the winter period was sampling faeces 2-4 times. In the whole time were sampling feed and analysed for nutrient and energy content. Statistical analysis took place with ANOVA using SPSS Version 25.0 (alpha of 0.05). Digestibility of organic matter during the grazing period 2018 was 71% with variation from average 65 to 74% in the individual farms. Between the content of crude fibre and the digestibility of OM was a correlation of r=- 0.845 (P=0.034) observed. In the winter feeding period was an average digestibility of 65% observed in grass silage, grass hay or Total Mixed Ration. Between the content of crude fibre and the digestibility of OM during the winter feeding period was a correlation of r=- 0.616 (P=0.004) observed. The results of the present study suggest that digestibility of organic matter in suckler cows can be considerably influenced by content of crude fibre in the silages or the ration during the winter feeding period. Getting rations work out of feed quality and body condition score of suckler cows will ensure the cows on correct condition for the calving period.

Phytogenic silica as internal marker for the measurement of feed digestibility in rabbits

E. Fortatos, G. Papadomichelakis, I. Hadjigeorgiou and K. Fegeros
Agricultural University of Athens, Nutritional Physiology and Feeding, 75 Iera Odos street, 11855, Athens, Greece; gpapad@aua.gr

Silicon (Si) assimilated from soil ultimately precipitates into or onto the cell wall structure of plant tissues in an amorphous form (SiO2), called phytogenic silica (PS). The PS is the major fraction of acid-insoluble ash (AIA), which is widely used as internal marker to calculate feed digestibility in animals. Although simple, AIA determination is based on time-consuming gravimetric procedures that require great amounts of sample to produce reliable results, whilst PS can be quantified by more rapid and sensitive methods. The present study aimed to evaluate the dietary PS as alternative internal marker for the calculation of digestibility in rabbits, by comparing to the well-established European reference method (total faecal collection) and the AIA technique. Thirty rabbits weaned at 35 d of age were allotted into 3 groups, namely ERM, AIA and PS (n=10 rabbits/group), balanced for body weight, and were fed the same diet. At 56 d of age, the total tract apparent digestibility (TTAD) of dry matter (DM) and gross energy (GE) was determined using the European reference method (ERM group) and the marker technique (AIA and PS groups). Total faecal collection was also carried out in AIA and PS groups to calculate faecal marker recovery. The AIA marker was quantified in feed and faeces by gravimetry, whereas PS marker using a robust two-phase wet digestion, followed by molybdenum blue colorimetry. Faecal recoveries of AIA and PS markers were 100.3 and 100.4% (SEM 0.02, P=0.939), respectively. The TTAD of DM was 0.589, 0.590 and 0.588 (SEM 0.008, P=0.935), while that of GE was 0.581, 0.585 and 0.584 (SEM 0.009, P=0.910) for ERM, AIA and PS, respectively. In conclusion, PS is a reliable marker for the calculation of feed digestibility in rabbits. It has a complete faecal recovery and yields digestibility values similar to those obtained by the reference method and the AIA technique. Additionally, the determination of PS is faster, less labor intensive and more adaptable to high volume routine analysis, when compared to AIA. The present work is part of a project funded by the State Scholarships Foundation (IKY).

Relationships between in sacco particle losses, feed granulometry and nylon bag pore profiles

O. Dhumez[1], J.C. Motte[2], S. Mendowski[3,4], A. Germain[4], G. Chesneau[4], P. Nozière[3] and P. Chapoutot[1]
[1]Université Paris-Saclay, INRAE, AgroParisTech, UMR Modélisation Systémique Appliquée aux Ruminants, 16, rue Claude Bernard, 75005 Paris, France, [2]Improve, Rue du Fond Lagache, 80480 Dury, France, [3]Université Clermont-Auvergne, INRAE, VetAgro Sup, UMR Herbivores, Theix, 63122 Saint-Genès-Champanelle, France, [4]Valorex, La Messayais, 35210 Combourtillé, France; ophelie.dhumez@agroparistech.fr

Milling proteaginous grains produces significant fractions of fine particle which can influence the rumen degradability estimated by in sacco method. The aim of this study was to quantify particle losses (PL) from nylon bags containing 5 types of feed: soybean meal (S), extruded (E) or raw (R) faba bean (F) and lupine (L), each milled through a 1 mm-sieve at 10K and 15K rpm. Feed PL were assessed by evaluation of the feed residual quantity (RQ, %) in Ankom® bags undergoing a rotated shaking (240 rpm) for up to 1 hour with soft percussion at each rotation. The kinetics of RQ were adjusted using the model: RQ=a+b×exp(-ct), with potential PL calculated as PPL=100-a (%). The pore profiles of the bags were studied by morphogranulometry and the granulometric profiles (GP) of the feeds by laser granulometry; all these profiles were synthetised by classical criteria (d10, d50, d90, SPAN=(d90-d10)/d50). Results were compared by variance analysis to quantify the feed, treatment and milling speed effects and their interactions. Nylon bags had larger and more dispersed pore sizes (d50=75µ, d90=98µ, n=5 bags, n=792 pores) than Ankom specifications (50+10µ). All the tested factors had significant effects: PPL was lower for L than for S or F (26, 30 vs 40%, respectively, P<0.001), decreased with extrusion (29 vs 36% for E vs R, respectively, P<0.001) but increased with milling speed (35 vs 27% for 15K vs 10K, respectively, P<0.001). Extrusion effect was stronger on L than F (-8 vs -5%, respectively, P<0.05) and milling speed had more impact for F than S or L (+9, +8, +6%, respectively, P<0.01) and for E than R feeds (+9 vs +6%, respectively, P<0.05). Most of the hierarchies within factors influencing PPL were maintained for GP criteria, but in a reversed order due to their negative correlation with PPL. However, d50 was not influenced by treatment, and extrusion logically led to higher d10 but reduced d90, highlighting a tighter particle dispersion. Finally, PPL fraction appeared positively related to the cumulated proportion of particles smaller than 98µ (nylon bag d90 value) (Cum98): PPP=13.2+0.60×Cum98 (R²=0.85, ETR=3.5, n=9). So, particle losses from nylon bags proved to be largely dependent on type of feeds and treatments, and potential PL can be predicted by the granulometric profiles of the samples.

Effect of sampling time on apparent digestibility of some selected nutrients in light lambs
J. Pelegrin-Valls, B. Serrano-Pérez, D. Villalba, E. Molina and J. Álvarez-Rodríguez
University of Lleida, Animal Science Department, Av. Rovira Roure 191, 25198 Lleida, Spain; jonathan.pelegrin@udl.cat

Acid insoluble ash (AIA) marker to determine apparent digestibility of nutrients is widely used. This method allows to evaluate the digestibility of feed formulations. However, digestibility results may be affected by diurnal digestion at the time of collection faecal samples. The aim of this study was to evaluate the effect of diurnal sampling time on the dry matter (DM) of faeces and the apparent digestibility of crude protein (CP), organic matter (OM), phosphorus (P) in light lambs fed two CP levels in an ad-libitum concentrate + straw diet. A total of 60 weaned male Ripollesa lambs of 45-60 days old and 15.0±1.5 kg of BW were used. The lambs were submitted to a two-phase feeding program: growing (14 to 19 kg BW) and finishing (19 to 25 kg BW). Pooled faeces samples (approximately 50 g) were collected at 8:00 am, 12:00 am and 16:00 pm by rectal stimulation of at least 3 lambs per pen (6 replicates/ group, 12 pens) at the end of each feeding phase. Concentrate and straw samples were collected the same days and their proportion of each in the ration was estimated using intake on a pen basis. Digestibility estimated from AIA concentration in ration and faeces were analysed through mixed models with repeated measurements that included the effects of time of the day, treatment and their interaction as fixed effects and the pen as random effect. The results showed that the digestibility of OM (69.2±1.5%) and CP (60.7±1.9%) were not affected by the sampling time in either phase. The DM of faeces in growing phase and the P digestibility in finishing were not affected by sampling time either (P>0.05). Nevertheless, P digestibility was higher at 8:00 am than at 12:00 am and 16:00 pm in growing phase (33.7 vs 18.2 and 20.2±3.9%; respectively, P<0.05). The DM of faeces in the finishing phase was lower at 8:00 am than at 16:00 pm (35.7 vs 38.7±1.05%; respectively, P<0.05) while the results at 12:00 am did not differ from the rest (37.2±1.05%; P>0.05). In conclusion, faeces samples to determine apparent digestibility of OM and CP can be sampled independently of the daily time.

Effect of a tannin extract and lactic acid inoculant on silage quality and nutritive value of grass
J.L. De Boever, T. Van Den Bossche, K. Goossens, L. Vandaele and S. De Campeneere
Flanders Research Insitute for Agriculture, Fisheries and Food, Animal Sciences Unit, Scheldeweg 68, 9090 Melle, Belgium; johan.deboever@ilvo.vlaanderen.be

Grass silage is a cheap, home-grown basal feed in many cattle rations. To assure good fermentation and high voluntary intake, wilting grass to at least 35% dry matter (DM) is common practice. This is difficult in unfavourable weather conditions and increases the risk for bad silage quality, particularly when grass is high in protein and low in sugars. Nowadays, inoculants containing *Lactobacillus* strains are frequently used as natural silage additives. Less common are tannins, known to reduce protein degradation in the rumen, but may be able to do so in silages. We compared 4 treatments: 1) control grass without additive, 2) Tanno-SAN L (liquid hydrolysable tannin extract from sweet chestnut wood, Sanluc, Belgium) at 2.5 g/kg grass, 3) Bonsilage Fit G (mixture of *L. buchneri, L. Plantarum and L. Rhamnosus*, Barenburg, Belgium) at 2 mg/kg grass and 4) both additives combined at the same doses. The grass was harvested on 28/10/19 and wilted to 27% DM; it contained 230 g crude protein (CP), 480 g NDF, 66 g sugar and 148 g ash per kg DM. After treatment, the grass was put in PVC micro-silos (H: 35 cm, Ø: 10 cm, 5 per treatment), which were opened 90 d after ensiling. For each silo we determined weight loss, silage quality (pH, NH_3, lactic and acetic acid (AA) and alcohols), chemical composition and *in vitro* digestibility and estimated net energy lactation (NEL), protein digestible in the intestines (PDI) and the degraded protein balance (DPB). The variance in the data was analysed with treatment as fixed factor and means were compared with Tuckey (P<0.05). In comparison to the control grass, tannin resulted in more CP, less ash and a higher NEL, PDI and DPB; the inoculant resulted in more weight loss, a higher pH, more AA and CP, less sugar and more DPB; the combination of additives gave a higher pH, lower NH_3, more AA and CP, less ash and a higher NEL, PDI and DPB. Compared to the inoculant, the tannin resulted in lower loss, less NH_3, AA and alcohols, more sugars and a higher NEL. The studied tannin improved silage quality, energy and protein value and will be further tested for an additional effect in the rumen.

Effects of supplementation VALKALOR® on the zootechnical performances of high-producing dairy cows

P.H. Pomport[1], J.P. Ricaud[2], M. Pondet[2] and D. Tristant[1]
[1]Ferme expérimentale AgroParis Tech, route de la ferme, 78850 Thiverval-Grignon, France, [2]IDENA, 21 rue du moulin, 44880 Sautron, France; pierre-henri.pomport@agroparistech.fr

The aim of this trial conducted at the farm of Grignon is to test the effect of a blend of Hibiscus sabdariffa and essential oils (VALKALOR®) on dairy cow production. Two homogeneous groups of 6 Prim'Holstein dairy cows (70 days in milk) were formed, control (C) and trial (T). The trial was a complete 10-week Latin square design with two 3-week transition periods and two consecutive 2-week periods of individual intake measurement. The basic ration of the two groups consists of 36.1% corn silage, 11.3% alfalfa silage, 4.8% alfalfa hay, 3.2% grass hay, 5.8% overpressed beet pulp, 13.3% crushed grain corn, 5.7% flattened barley, 9.8% oilseed rape cake, 8.4% nitrogen corrector and 2% mineral, salt, bicarbonate. The cows of group T received in addition 15 g of VALKALOR AL881 per day. The nutritional values of the ration were 1,64 Mcal of energy net for lactation/kg of dry matter (DM), 18% crude fibre, 36% neutral detergent fibre, 25.6% Starch, 15% crude protein and 3.3% Fat. A variance-covariance model integrating an animal effect and a PERIOD effect (sampling at the end of period 1 or 2) was performed with the SAS 9.1 software. The intake was not different between the groups (P=0.87). Group T is characterised by a higher raw milk production (P<0.001) and energy-corrected milk production (P<0.001) than group C with 33.6±7.8 and 31.9±8.1 kg/d/cow of raw milk and 32.7±8.0 and 30.7±6.6 kg ECM/d/cow respectively. Therefore feed efficiency is improved (P=0.01) with 1.29±0.25 and 1.21±0.18 kg ECM/kg DM Intake respectively for T and C group. Protein (P=0.001) and fat synthesis (P<0.001) are higher for group T with 998±193 and 1,238±358 g/d/cow respectively compared to 953±220 and 1,146±233 g/d/cow for group C. The lactose level in milk is also 0.3 g/l higher for group T (P=0.004 with T=52.4±1.3 and C=52.1±1.4 g/l), showing higher glucose availability at the udder level. Weight gain was numerically higher (P=0.61) for group T (431 g/d/cow) compared to group C (315 g/d/cow).We can therefore conclude that VALKALOR has a very significant effect with an increase in milk production an average of 2 kg of milk per day.

Diseases and preventive measures in insects produced for food and feed

J. Eilenberg[1], A. Lecocq[1], L. Joosten[2], G. Maciel Vergara[1,3], A.B. Jensen[1] and E. Schmitt[2]
[1]University of Copenhagen, Department of Plant and Environmental Sciences, Thorvaldsensvej 40, 1871 Frederiksberg C, Denmark, [2]Protix B.V, Industriestraat 3, 5107NC Dongen, the Netherlands, [3]Wageningen University, Plant Sciences Group, Laboratory of Virology, Wageningen Campus, RADIX, Droevendaalsesteeg 1, 6708 PB, Wageningen, the Netherlands; jei@plen.ku.dk

Insects produced for food and feed are susceptible to insect pathogens, which can infect and may lower the fitness of the insects, for example by shortening their lifespan or by lowering their fecundity. In the worst case, insect diseases may cause significant losses in the production. To assist the insect industry to be prepared for best prevention and management, joint efforts are needed in order to: (1) obtain an overview on documented real cases of insect diseases in production facilities; (2) prepare diagnostic sheets and simple measurement techniques for use in production facilities; (3) study literature on potential harmful insect pathogens; (4) perform experimental infections in the laboratory to learn about host range of insect pathogens and the pathologies they cause; (5) explore how probiotics can improve insect health. We will give an overview about the current status of efforts on these topics. For example, we have published three diagnostic sheets, free for usage, on diseases in the yellow mealworm *Tenebrio molitor*, the Black soldier fly (BSF) *Hermetia illucens*, and the house cricket *Acheta domesticus*. In addition, we have performed experimental infections in the laboratory and we have looked into the potential of probiotics.

Microorganism-insect interactions during the rearing of insects for food and feed

D. Vandeweyer and L. Van Campenhout
KU Leuven, Department of Microbial and Molecular Systems, Lab4Food, Geel Campus, Kleinhoefstraat 4, 2440 Geel, Belgium; leen.vancampenhout@kuleuven.be

The importance of interactions between insects and their microbiota is increasingly acknowledged in the insect sector. Not only may the microbiota impact health, growth and development of the insect and food safety of the end product, they may also serve as rearing promotor and even contain industrially valuable organisms. This presentation will provide an overview of the current knowledge, in literature and by own research, of the endogenous microbiota of different insect species and emphasise why this should be interpreted in an insect-specific way. Next, the presentation will highlight the potential beneficial effects of microorganisms on insect growth and development as well as the possible adverse effects they may induce. Also their applications in an industrial context (fermentation, waste treatment, etc.) will be considered. Investigating microorganism-insect interactions, however, involves certain challenges that will be discussed in this presentation, e.g. related to culture-dependent and independent analysing techniques. Finally, the presentation will dig into the manipulation of the insect microbiota through inoculation of certain organisms or through sterile rearing. Limitations using those techniques and correct interpretation of results will be addressed.

SUSINCHAIN (SUStainable INsect CHAIN): Introduction to the project

T. Veldkamp[1], H.J. Van Der Fels-Klerx[2], N.P. Meijer[2] and M. Octavia[3]
[1]Wageningen Livestock Research, P.O. Box 338, 6700 AH Wageningen, the Netherlands, [2]Wageningen Food Safety Research, Akkermaalsbos 2, 6708 WB Wageningen, the Netherlands, [3]Sociedade Portuguesa de Inovacao, Av Marechal Gomes da Costa, 1376 POR, 4150-356 Porto, Portugal; teun.veldkamp@wur.nl

The aim of the SUSINCHAIN project is to overcome the remaining barriers for increasing the economic viability of the insect value chain and opening markets. This will be done by combining forces in a comprehensive multi-actor consortium with 18 industry partners and 17 academic partners. SUSINCHAIN considers the entire insect value chain from farm (insect rearing) to fork (insect consumption) based on black soldier fly, housefly, mealworm and crickets. The SUSINCHAIN concept is based on performing all activities needed (innovations, testing, demonstrating, sharing knowledge) to overcome the most important hurdles for scaling up the European insect value chain. Practical experience of consortium industry partners in addressing supply-side barriers as well as related risk management strategies will be evaluated and best practices will be elaborated by the use of living labs. Insect rearing innovations will be focused on organic side streams of vegetable origin that cannot be used directly as animal feed. For improving economic viability, possibilities for marketing of rearing by-products will be evaluated and strategies to avoid insect diseases will be developed. Transport, storage and processing technologies for insects will be optimised and demonstrated at large scale. Microwave and Radio Frequency drying, High Moisture Extrusion and protein recovery from fresh larvae by using enzymatic pre-treatment combined with continuous tricanter centrifugation will be validated and demonstrated. Controlled atmosphere packaging and cold atmosphere preservation for storage and transport of living insects will also be tested. The digestibility of insect meals produced by different processing methods will be assessed and insect meals will be included in feed formulae for large-scale level commercial diets to test optimal inclusion levels to maintain or increase livestock performance and health. Insect based food products will be designed and developed suitable for the domestic preparation of regular dinner meals. The microbiological, chemical and allergenic safety of insects and derived products will be addressed along the entire chain. Hygiene codes, HACCP protocols, and guidelines for safe insect production will be created and distributed among the supply chain actors. Furthermore, a decision support system will be developed to ensure the sustainable growth of the insect value chain in Europe, in addition to the stakeholder's platform provided within the context of the project in order to connect and receive valuable inputs from the relevant actors.

Modelling approaches for sustainable insect production chains

S. Smetana[1], P. Meister[1] and A. Tonda[2]
[1]*German Institute of Food Technologies (DIL e.V.), Prof.-von-Klitzing-str. 7, 49610, Germany, [2]UMR, INRAE, Université Paris-Saclay, 518 MIA, INRAE, Université Paris-Saclay, 147, rue de l'Université – 75338 Paris Cedex 07, France; s.smetana@dil-ev.de*

Insect value chains in Europe are evolving to large-scale industrial systems overcoming economic and environmental challenges. SUSINCHAIN, a H2020 EU-funded project, aims to define the leverages and solutions for sustainable insect value chains from multiple perspectives: economic, environmental, safety, nutritional, etc. Such perspectives have different weights and are often in contradiction to each other at various stages of the value chain. Currently, Life Cycle Assessment (LCA) is the standard for the assessment of environmental impact, allowing to consider multiple system criteria. However, LCA of emerging insect production systems is complicated due to the lack of data and low technology readiness levels. To use LCA efficiently and supply it with data, SUSINCHAIN is relying on modular modelling approaches chosen for ability to rely on already defined set of tasks and logics for time saving and flexibility. Moreover, modelling complexity behind entire insect system is divided into smaller solvable and rather simple tasks (models). For insect production chains, with biological agent in the core of the system, modelling approaches should include the biological nature of insect metabolism and rely on metabolic models. Such approach requires encoding of insect species genome (currently in the scope of a few research groups) and development of metabolic models themselves (in scope of UpWaste project). Metabolic models would provide the foundation for feed conversion models. However, such an approach might take long time and up to full disclosure of metabolic models, LCA of insect chains is relying on energetic, mass flow and environmental fate models. Economic aspect is included in a form of a price model to represent an effect of market influence. These models, representing various modules, are combined with multi-objective optimisation algorithms to define the optimal value chains, satisfying the needs of different actors in the chain. It is predicted that machine learning technique will be applied to adapt modules from similar systems (feed, food and processing) to the case of insect production to tackle the problem of data lack.

Precision insect farming: using sensors to monitor and steer the production process in real time

F. Wouters[1], P. Vanloofsvelt[2] and T. Spranghers[1]
[1]*VIVES University of Applied Sciences, Agro- and biotechnology, Wilgenstraat 32, 8800 Roeselare, Belgium, [2]VIVES University of Applied Sciences, Smart Technologies, Doorniksesteenweg 145, 8500 Kortrijk, Belgium; filip.wouters@vives.be*

Insect products are quite expensive in comparison to other protein rich food/feed sources. This is mostly due to high labour costs (60.9% of all production costs). VIVES developed on automated feeding system for mealworms (*Tenebrio molitor*). Using this system, instead of feeding by hand, labour costs could be reduced by 75%. The feeding robot can be further improved by the implementation of sensors to monitor temperature, moisture, emissions, etc. in the rearing boxes. A new research project at VIVES focuses in the first place on temperature. Our first results showed that fully grown mealworms could increase the temperature in the rearing boxes with 4 °C. When moisture source was added further increases of temperature could be observed. In addition, migration behaviour, shown when larvae are close to pupation, increased the temperature towards the corners of the boxes with 1 extra °C. The monitoring of the temperature in the boxes can reveal certain behavioural patterns. These patterns can be related to feeding state, mortality, developmental stage, etc. Moreover, the growth of the larvae could be monitored using temperature, given that heavier larvae generate more heath. The system could process all the data from the sensors and report to the operator. Consequently, targeted interventions can be made where necessary in order to assure optimal production (e.g. adjust the feeding substrate or regime). Ultimately, the system can learn to recognise the patterns and intervene independently in the production process.

Physical characterisation of black soldier fly rearing diets

L. Frooninckx, S. Goossens, A. Wuyts and S. Van Miert
Thomas More University College of Applied Sciences, Radius, Kleinhoefstraat 4, 2440 Geel, Belgium;
sabine.vanmiert@thomasmore.be

Larvae of the black soldier fly (BSF), *Hermetia illucens*, are voracious feeders that can grow on a wide variety of organic materials. When grown on organic side streams, they are able to convert such low value materials into high value biomass composed of fats, proteins and chitin. These components can be used as a sustainable resource for food, feed and technical applications. The potential of a side stream as an ingredient of a rearing diet for BSF depends among other things on its chemical and physical properties. The influence of chemical parameters such as pH, macronutrient concentration and ratio have already been reported in literature, but the influence of the physical characteristics have been neglected so far. In this study, we investigate the influence of dynamic and kinematic viscosity, syneresis and water holding capacity of semi-artificial BSF rearing diets on growth, biomass conversion and material reduction of BSF larvae. The output of these experiments will give us more insights on the potential of side streams with specific chemical and physical properties as an ingredient of a BSF rearing diet.

Effect of fermentation of nursing diet on hatching, growth and survival of black soldier fly

M. Van Peer, A. Van Gorp, A. Lambrechts, A. Wuyts, L. Frooninckx and S. Van Miert
Thomas More University of Applied Sciences, RADIUS, Kleinhoefstraat 4, 2440, Belgium; meggie.vanpeer@thomasmore.be

Black soldier fly (BSF), *Hermetia illucens*, is a harmless fly species originating from Central- and South-America. Larvae of the BSF are capable of converting low-grade organic waste streams into high-grade raw materials, i.e. protein, fats and chitin. Worldwide, many companies are investing scaling up of the rearing process of this insect species. In order to rear BSF successfully on an industrial scale, all the aspects of the rearing process must be optimised. One of the major bottlenecks during this rearing process is the 'nursing' phase of the eggs and neonate larvae. Using chickenfeed as a nursing diet often results in poor hatching and neonate survival, which is most likely due to fungal development that is caused by the high humidity necessary to keep the substrate moist. Acidification, achieved through fermentation, is a technique that is traditionally used for the conservation of food and feed. Lowering the pH reduces the incidence of spoilage microorganisms. Moreover, fermentation is shown to improve digestibility by enhancing the availability and activity of nutrients. This hypothesis was investigated with the aim to increase larval yield. In this study 3 different lactic acid fermentation starters were used to ferment chickenfeed. Using both eggs and 1 day old larvae, the effect of the fermentation process on growth, survival and feed conversion was investigated. The results are processed using Kruskal-Wallis and Dunn's multiple comparisons test.

Performance of black soldier fly larvae on digestate, pig manure and vegetal waste

T. Veldkamp[1], K. Van Rozen[2], H.J.H. Elissen[2] and R.Y. Van Der Weide[2]
[1]Wageningen Livestock Research, De Elst 1, P.O. Box 338, 6700 AH Wageningen, the Netherlands, [2]Wageningen Research, ACRRES, Edelhertweg 1, 8219 PH Lelystad, the Netherlands; teun.veldkamp@wur.nl

The development of *Hermetia illucens* larvae (black soldier fly; BSF) was assessed in triplicate in seven substrates: chicken feed (CF; reference diet), pig manure solid (PMS), Betafert® solid (BTFS), swill (SW), olive waste pulp (OP), pig manure liquid mixed with chicken feed (PMLCF), and silage grass (SG). On top of each substrate, 2,500 starter (8 days-old) larvae per kilogram (wet) substrate were incubated in 21 plastic containers (75×47×15 cm) (each substrate in triplicate) in a climate chamber (temperature 30.1±0.5 °C, relative humidity (RH) 46.4±3.7% at day 1 and 60.6±1.6% from day 2 to day 6, photoperiod 0:24 hours light:dark). Dry matter content at the start of the experiment was highest in OP substrate, subsequently followed by CF, PMLCF, PMS, BTFS, SG and SW. Substrates at the end of the experiment were much drier. Highest dry matter contents were determined in CF and PMLCF, subsequently followed by OP, PMS, SW, SG and BTFS. Dry matter content of larvae was positively linked to growth rate. Highest dry matter content of the larvae was found on SW, subsequently followed by CF, PMLCF, OP, BTFS, PMS and SG. Highest larval growth rate was found in SW containers. Larval growth rate was even higher than in the containers with the reference substrate CF. Larval growth rate in PMLCF containers did not differ from CF whereas larval growth rate in PMS was lower than in CF. Growth rate of BSF larvae in BTFS, OP and SG was very low. Waste Reduction Index (WRI) was highest in SW containers followed by PMLCF and both were higher than WRI in containers with CF. WRI decreased further in PMS, SG, BTFS and OP, respectively. The Efficiency of Conversion of Ingested substrate (ECI) was highest in SW followed by PMLCF, CF and PMS. The containers with OP, BTFS and SG resulted in a lower ECI than other treatments. Highest NH_3 concentrations were measured above substrates with the lowest larval growth performances (BTFS, PMLCF, OP and SG). Dry matter of substrates, larval growth rate, WRI and ECI were all positively correlated except NH_3.

Pulsed electric filed and sustainable modular insect technologies for food waste processing

S. Smetana[1], S. Ites[1], O. Parniakov[2], K. Aganovic[1], S. Toepfl[2] and V. Heinz[1]
[1]German Institute of Food Technologies (DIL e.V.), Professor-von-klitzing-straße 7, 49610, Germany, [2]Elea Vertriebs- und Vermarktungsgesellschaft mbH, Prof. von Klitzing Str. 9, 49610 Quakenbrück, Germany; s.smetana@dil-ev.de

Insect production for food and feed purposes is rapidly emerging in Europe, in many cases relying on existing processing methods for blanching, freezing, drying, fractionating, but also seeking for more efficient and beneficial biomass treatment methods. The current study is aimed to explore the application of insects in a modular food waste treatment technology with improvement of insect biomass fractionation with Pulsed electric fields (PEF). Selection of insects as bio-transforming agents is associated with their ability to convert low value food waste into biomass and frass which can be sold for other purposes. *Tenebrio molitor* and *Hermetia illucens* are exemplary considered to feed on side streams composed of brewery grains, potato peels and overdate food from retail. Conceptual model of waste treatment technology based on insects was designed and assessed for economic feasibility and environmental impact with Life Cycle Assessment (LCA). Food waste treatment with modular mobile *H. illucens* technology was profitable and more environmentally beneficial than commonly used waste treatment methods in three compared categories for the cases of brewery grains and overdate food treatment. Composting and anaerobic digestion although were better for potato peels treatment in economic effects, land occupation and global warming. Application of *T. molitor* is limited due to longer development time and the need of relatively dry feed material. Application of PEF (1-3 kV/cm and 1-20 kJ/kg) for *H. illucens* larvae pretreatment with following drying and fractionation improved drying rates and fractionation efficiencies. Food waste treatment with designed modular mobile biorefinery relying on *H. illucens* application has the potential to be more profitable and less harmful for the environment than commonly used waste treatment methods for a few types of waste (brewery grains and outdated food). *T. molitor* on the other hand is not suitable for wet waste treatment and might be more beneficial for a batch system dried food waste material.

Soil dynamics and biomass production in short- and long-cycle crops by incorporation of insect frass

R. Menino[1], C. Esteves[2], F. Felizes[1], A. Castelo-Branco[1], P. Fareleira[1], M. Mota[3] and D. Murta[4,5,6]
[1]Instituto Nacional de Investigação Agrária e Veterinária, I.P. (INIAV), Unidade Estratégica de Investigação e Serviços de Sistemas Agrários e Florestais e Sanidade Vegetal, Av. da República, Quinta do Marquês, 2780-159 Oeiras, Portugal, [2]Instituto Superior de Agronomia (ISA), Mestrado em Engenharia Agronómica, Tapada da Ajuda, 1349-017 Lisboa, Portugal, [3]Instituto Superior de Agronomia (ISA), LEAF/DCEB, Tapada da Ajuda, 1349-017 Lisboa, Portugal, [4]University of Lisbon, CIISA, Faculty of Veterinary Medicine, Av. Universidade Técnica, 1300-477 Lisboa, Portugal, [5]Lusófona University of Humanities and Technologies, CBIOS, Faculty of Veterinary Medicine, Campo Grande, 1749-024 Lisboa, Portugal, [6]Ingredient Odyssey, EntoGreen, Quinta das Cegonhas, Apartado 577, 2001-907, Portugal; daniel.murta@entogreen.com

Synthetic fertilisers use, in agricultural soils, contributes to increased production but has detrimental effects in what environment concerns; one alternative to face this conundrum, pointed out as a friendly one, is the use of organic compounds, being the black soldier fly larvae frass (BSFF) a recent candidate. Some studies to evaluate the effect of this organic fertiliser on crop production have been already produced; however there is yet a paucity of information on its environmental impact. The present work is based on studies carried out to evaluate the potential fertilising effect of BSFF in two crops, namely: (1) a 'long-cycle crop' (*Lollium multiflorum* L.), for which it was registered a significant increase of production and soil residual organic matter and mineral components, together with a significant increase of soil residual dehydrogenase activity; (2) a 'short-cycle crop' (*Lactuca sativa* L. var. capitata), for which the effect of BSFF on production was not relevant but the soil residual chemical composition and microbiological activity was significantly increased. The conclusion is that BSFF can be an efficient fertiliser when there is time for nutrients mineralisation (long standing crops) and it is, in any case, a guaranty for the sustainability of soil fertility. Acknowledgements & Funding. The research that yielded these results was funded by PT2020 through the project POCI-01-0247-FEDER-017675: ENTOVALOR – Insects as an opportunity in residues valorisation (2016-2019).

Influence of temperature on selected life-history traits of black soldier fly (*Hermetia illucens*)

M. Shumo[1,2], F.M. Khamis[1], C. Tanga[1], K.M. Fiaboe[1], S. Subramanian[1], S. Ekesi[1], A. Van Huis[3] and C. Borgemeister[2]
[1]ICIPE, Plant Health Unit, Duduville Campus, 00100 Nairobi, Kenya, [2]Center for Development Research (ZEF), University of Bonn, Department of Ecology and Natural Resources Management, Genscherallee 3, 53113, Germany, [3]Wageningen University & Research, Department of Plant Sciences, Laboratory of Entomology, 6700 AA Wageningen, the Netherlands; mshummo@hotmail.com

Insects such as the black soldier fly (BSF) have been identified as potential alternatives to the conventionally used protein sources in livestock feed due to their rich nutrient content and the fact that they can be reared on organic side streams. Substrates derived from organic byproducts are suitable for industrial large-scale production of insect meal. Although efficient in waste management and in feed production, BSF larvae are very sensitive to the external environment such as temperature and rearing medium. Therefore, we studied the effect of temperature and substrate type, i.e. brewers' spent grain (SG) and cow dung (CD), on the development and survival of BSF larvae. Both organic substrates were readily available in Nairobi, Kenya, the location of the experiments. In our experiment, 100 3-5-day-old BSF larvae were placed into containers that contained either SG or CD and further treated at temperatures of 15, 20, 25, 30 and 35 °C. The duration of larval development was recorded, and the prepupae were removed, weighed, and placed individually in separate, labelled, 35-ml plastic cups filled with moist sawdust. The data were subjected to a two-way analysis of variance (ANOVA) using the general linear model procedure. BSF larvae reared on SG developed faster than those reared on CD; the former also favoured higher temperatures for their larval development and emergence into adults. The optimum range was 25-30 °C. With increasing temperatures, the longevity of adult BSF decreased, while the fecundity of females increased. Thus, it is possible to take advantage of the readily available SG waste streams in the urban environments of Kenya to produce BSF larvae-derived livestock feed within a short duration of time and at relatively high temperatures.

Effect of salt availability on growth performance of house crickets (*Acheta domesticus*)
M. Vaga, F. Berthe and A. Jansson
Swedish University of Agricultural Sciences, Department of Anatomy, Physiology and Biochemistry, Ulls väg 26, 75007, Sweden; anna.jansson@slu.se

In mammals, the appetite for sodium is well documented but studies on insects are scarce. The aim of this study was to register voluntary sodium intake in house crickets and its effects on survival, feed intake and growth. Our hypothesis was that salt availability improves house cricket's performance when fed on sodium free diet. About one day old house crickets were collected and placed in 3 l plastic boxes (12 boxes, 36 crickets each) and reared for 57 days. Boxes were equipped with ventilation holes and black plastic hiding tubes inside. Drinking water and low sodium feed, prepared from wheat flower, wheat bran, oats bran, rapeseed meal, calcium and mineral mixture, were available *ad libitum*. Four boxes had Himalayan rock salt and other four had common table salt (99.9% NaCl) available *ad libitum*, and four boxes (control) received no salt. Weight of the crickets (within a box), feed, salt and water and number of crickets was recorded every seven days. By day 57 the survival rate of crickets without salt was only 15% compared (P<0.01) with 31 and 24% in crickets that received table salt and rock salt, respectively. The average cricket weight (59 mg) by the end was lower (P<0.05) in the control groups than in groups offered table and rock salt (185 and 211 mg). None of the crickets in control group reached maturity within the 57 days, but in table salt group 2±1 and in rock salt group 1±1 crickets had matured by the end of the experiment. On average crickets consumed about 0.05% (SD=0.02) of bodyweight of salt daily. It is our recommendation that salt or mineral salt has to be freely available for house crickets when reared on low sodium diets.

CO_2 production and O_2 consumption of black soldier fly larvae grown in different environments
M. Mielenz, G. Das, M. Derno and C.C. Metges
Leibniz Institute for Farm Animal Biology (FBN), Wilhelm-Stahl-Allee 2, 18196 Dummerstorf, Germany; mielenz@fbn-dummerstorf.de

Black soldier fly (BSF) larvae are thought to be efficient in using by-products as nutrient source. However, less information is available on gas emission in this species in relation to larval growth. Aim of our study was to investigate gas exchange in relation to growth when growing larvae were kept in different environments. 150 larvae were added to insect growth boxes (surface area: 23 cm^2) filled with 114 g of 1 day (d) pre-fermented chicken feed (70% water; 30% feed (w/v) and reared in a room at 27.5 °C and 70% humidity for 3 d (room) or for 3 d in a respiratory chamber (RC) with continuous air flow (37.2 l / h) and gas exchange measurement. In parallel, the larvae were kept in the room for a total of 6 d or transferred to the RC from the 4th to the 6th d (2 repetitions; mean larval weight at start: 108.4 mg; 4 groups, n=6 boxes per group). The weight of the larvae was recorded after the end of each 3 d and 6 d period. The continuous production of CO_2 as well the consumption of O_2 was measured at 21 min intervals for 3 d. Larval mass and gas exchange data were analysed by SAS using the procedures GLM or MIXED with repeated measures ANOVA, respectively. After the end of the respective period, about 7% (3 d) or 15% (6 d) of the animals reached the prepupal stage, respectively. At the end of the experiment the mean larval mass was highest in the 3 d RC group (231.3±4.03 mg) and lowest in the 3 d room group (196.7±4.43 mg) (P<0.001). The mean daily body mass gain (BMG) was highest after 3 d in the RC (42.0±1.70 mg) and lowest in the 6 d room group (19.4±1.87 mg). The mean daily BMG after 3 d RC was higher compared to all other groups (P≤0.0001). The CO_2 production per g of larval mass was 45±4 to 60±4 ml per 24 h. The O_2 consumption was lower on d 2 compared to d 3 (41±3 vs 52±3 ml/24 h). In conclusion, RC provided the best growth conditions, likely due to an active air supply on growing larvae inside the available substrate. O_2 consumption appeared to depend on the developmental stage. Further studies are necessary to assess the growth potential in relation to air supply and substrate.

Preliminary study on management risk factors for disease occurrence in *Apis meilifera*

N. Jesus, Y. Vaz and T. Nunes
Faculty of Veterinary Medicine, University of Lisbon, Av. Universidade Técnica, 1300-477 Lisbon, Portugal;
nidia.jesus1995@gmail.com

Beekeeping is a worldwide recognised activity that involves managing honeybees in combs for pollination, honey production, as well as other derived products. Traditionally connected with agriculture, this activity starts to stand out by its own potential, and it requires an adequate management of the apiary and disease surveillance, in order to safeguard high quality products and safe trade. Portugal has a great aptitude for beekeeping practice due to its soil, climate and other environmental conditions. Thereby, this study aimed to do a preliminary analysis of risk factors associated with the occurrence of disease at an apiary level. It was developed at Melbandos, a 'controlled zone' of Portugal, in the region of Lisbon and Tagus Valley. According to Portuguese National Authority for Animal Health the apiaries of these 'controlled zones' are systematically checked for bee diseases and systematic Varroa control is carried out. The study was cross-sectional and applied in 2017 in a sample of 248 apiaries checked for diseases by officially approved laboratories under the Annual Sanitary Apiculture Programme. The study consisted in a sanitary management questionnaire applied to 70 beekeepers and the aggregation of their results of the official anatomopathological tests. The variables were submitted to a descriptive statistical analysis, followed by a univariate analysis and finally culminating in the construction of logistic regression models, using R 3.4.3 software. The models set for 95% confidence (P<0,05) identified as risk factors for the occurrence of varroosis the exchange of frames between colonies (OR=7,0; CI=[1,73-39,0]) and the non-disinfection of the beekeeping equipment with flames (OR=8,7; CI 95%=[2,18-45,8]). Concerning nosemosis the distance between apiaries less than 400 metres (OR=5,0; CI=[1,35-21,5]) and the non-disinfection of frames by boiling (without any positive cases of nosemosis in apiaries submitted to boiling) appeared to be risk factors for the referred disease. Concerning the prevention and management of diseases, one of the main challenges faced by beekeeping sector, this preliminary study highlighted a few sanitary measures that can easily be implemented in apiaries. However, more research and evidence are needed to support the obtained results.

Red clover can be included in the diet of house crickets (*Acheta domesticus*)

M. Vaga, Å. Berggren and A. Jansson
Swedish University of Agricultural Sciences, Dept of Anatomy, Physiology and Biochemistry, Box 7011, 75007, Sweden;
anna.jansson@slu.se

Flowering crops like red clover may offer resources for European insect pollinators in decline. Red clover is commonly used to feed cattle and sheep, but may also be an ingredient in feed for house crickets (*Acheta domesticus*). The aim of this study was to investigate to what extent house crickets choose to consume red clover and if their feed choice affect maturation. Three groups of nymphs (1 day old) were offered three different diets for 62 days; A) a pelleted control diet *ad libitum* (wheat and oat bran, wheat meal, rapeseed meal, $CaCO_3$ and a premix with trace minerals and vitamins), B) free access to early cut red clover hay, wheat meal and rapeseed meal or C) free access to late cut red clover hay, wheat meal and rapeseed meal. All crickets had access to a salt block. Crickets were kept in ventilated plastic boxes with access to shelter and water tubes at 29±1 °C and 50% relative humidity (±10) in a 12 h lighting regime. Crickets were euthanised by freezing at day 62. Choice of feeds on diet B and C was registered (feeds offered were weighed continuously and all left overs by the end of the study). The proportion of nymphs and adult males (AM) and females (AF) were registered for all control animals at day 62 (n=104) and in a random sub sample of crickets offered diet B (133 out of 363 individuals) and C (71 out of 297 individuals) and compared using a Chi-Square test. The proportion of red clover hay, wheat meal and rapeseed meal consumed was 30, 31, 39% and 15, 32, 53% on diet B and C, respectively. On the control diet, the proportions of nymphs, AF and AM were 95, 5 and 0%, respectively. Compared to the control diet the proportions differed on diet B and C, and AF and AM proportions were greater (proportions on diet B: 73, 18 and 9%, proportions on diet C: 69, 21 and 10%, P<0.0001). The study shows that red clover can be used as feed ingredient to house crickets and also indicates that red clover have positive effects on individual maturation. Using late cut red clover crops to crickets also have the additional positive effect that it has greater potential to support wild pollinators than early cut red clover.

Effects of feed deprivation on behaviour of house crickets (*Acheta domesticus*)

E. Roman and A. Jansson

Swedish University of Agricultural Sciences, Dept of Anatomy, Physiology and Biochemistry, Box 7011, 75007, Sweden; erika.roman@slu.se

In commercial cricket rearing, feed deprivation prior to harvest is sometimes practiced in order to enhance food quality. However, feed deprivation can constitute a stressor that may impact cricket behaviour and animal welfare. In this study we compared behaviour in *ad libitum* fed crickets (control) with that of crickets subjected to 24 and 48 h of feed deprivation. Crickets were housed in pairs (8 pairs per treatment, one male and one female) in ventilated transparent plastic boxes ($12 \times 12 \times 4.5$ cm) with yellow lids, containing a black tube-shaped shelter and a water tube. They were kept at 29 °C (± 1 °C) and 50% relative humidity (± 10) in a 12 h lighting regime. Observations (camera recordings) were made at the same time of the day approximately three hours into the light phase, and each box was observed for one minute every 8 minutes during a total of 27 minutes. The observer was blinded to the treatments with exception for the control since feed was visible. Position (outside or inside the shelter), activity level (1 = not altering body position, 2 = small changes in body position in at least one of the crickets, and 3 = at least one of the crickets moving around a lot) and social level (1 = positioned <1 cm from each other, 2 = 1-4 cm from each other and 3 = >4 cm from each other) was registered. Data were analysed with Chi-square analysis and P<0.05 was considered significant. Crickets were inside the shelter to a higher degree during feed deprivation (both at 24 and 48 h) than during control conditions (P<0.0001) and they were also observed close to each other more often at 24 h compared to control conditions (P=0.04). There was no difference in activity pattern compared to control (P>0.18) and signs of cannibalism were not observed in any condition. The study indicates that crickets alter their behaviour in response to feed deprivation but further studies are needed to evaluate the effect on food quality and animal welfare.

Adaptation of the Portuguese insect industry to the restrictions of the 'novel food

R. Nunes[1,2] and J. Gonçalves[2,3]

[1]EntoGreen-Ingredient Odyssey, Santarem, Santarém, Portugal, [2]Portugal Insect, Estação Zootécnica Nacional, Santarém, Portugal, [3]Nutrix, Mourã, Leiria, Portugal; rui.nunes@entogreen.com

With the new regulation on novel food (Reg EU 2015/2283), the Portuguese insect industry cannot place their products on the market, until the EU includes insects in the list of novel foods (after EFSA's positive opinion on the technical dossiers submitted). However, some countries – even without this approval – already allow these products to be placed on their markets. We believe that in order to have a strong and recognised sector it must be safe, and comply with existing regulations, allowing a standardisation of products quality, safety and trust. However, this standardisation must start with the EU regulations and how they are applied in the EU, or the economic and social impact will not be equal and fair. To minimise this unfair scenario, Portugal Insect requested the national authorities, the same type of derogation that was issued by the Italian authorities to the Belgium Pavilion at the 2015 Expo Milano exhibition. Since May 2018, several events have been promoted in schools, stores and supermarkets. Additionally, in 2018 and 2019, the World Edible Insect Day was celebrated in major events, with local authorities attending. These events allow a close contact with consumers, giving the promoters the opportunity to collect valuable feedbacks about their products. But despite this path being followed in close liaison with local authorities, the legal blockade remains, with the Portuguese operators still unable to sell their products. The national insect industry needs to move on to the next phase and be allowed to market products like other countries. After all, the only claim is: One Europe, one rule. On the other hand, assuming the EU will – one day – allow all European players to fairly market insects as food, it is important to know the reason why until then, the market is open for some only. Our journey as members of Portugal Insect during this ramp up of the insect sector, allowed us to conclude that although the unfair legal framework and the fact that authorities are unable to change it, it is of pivotal importance to collaborate, allowing us to raise awareness on the insect sector and its contribution to a sustainable future.

Wet feed distribution on the density and growth rate of *Tenebrio molitor*

D. Deruytter, C.L. Coudron, J. Claeys and S. Teerlinck
Inagro, Insect and aquaculture, Ieperseweg, 87, 8800, Belgium; david.deruytter@inagro.be

When rearing insects, it is important to use a feed that provides an optimal and homogeneous growth. One of the challenges in rearing mealworms is the fact that they need both a dry feed and a wet feed. The dry feed is not a problem as it is a homogeneous mixture that is easily applied evenly in the crate. In contrast, wet feed may result in some problems. In many small-scale rearing facilities it's given daily by hand, but this does not necessarily result in an even distribution of the wet feed. But even automated systems may struggle. Common sense and experience taught us that if the (wet)feed is not distributed evenly, mealworm spatial distribution is not even, and the growth is not optimal. With this experiment we want to quantify the distance to the wet feed before adverse effects start to occur (on density, growth rate and growth rate variability). This was done by placing mealworms in a 60×40 cm crate with *ad libitum* supply of wet feed (Agar) on one side. After one week or multiple weeks, the distribution, average and individual weight of the mealworms was assessed in 7 zones depending on the distance to the wet feed. This was done for 4 replicates. The results indicate that size matters. The smallest assessed size (0.7 mg) seemed too small to move through the dry feed resulting in a limited influence on the spatial distribution. However, the growth was adversely affected when the wet feed was further than 5 cm away. The strongest influence was observed for mealworms with a weight between 3 and 35 mg. This category was mobile enough to crawl to the wet feed but not mobile enough to redistribute afterwards. This results in a densely populated 'active zone' with fast growing mealworms and a virtual 'deadzone' further than 10 to 20 cm from the wet feed (depending on the size). In this deadzone growth was up to 50% lower and with a density of only 10% compared to the active zone. Above an average initial weight of 35 mg, the distance to the wet feed has limited effect. The effects are more pronounced when with increasing duration. In conclusion, homogenous distribution of the wet feed is important, certainly for mealworms below 35 mg. A skewed mealworm distribution within the crate, will result in a skewed dry feed availability per mealworm resulting in a lower growth rate. A rule of thumb that can be used is to multiply the average weight (mg) by 4 and this is the maximum distance between two sources of wet feed if you do not want adverse effects.

The MANNA project, overview and introduction

D. Eckersall[1] and # The MANNA Consortium[2]
[1]University of Glasgow, School of Veterinary Medicine, Institute of III, Jarrett Building, Glasgow G61 1QH, United Kingdom, [2]EJD, http://www.phd4manna.eu/; david.eckersall@glasgow.ac.uk

The European Joint Doctorate in Molecular Animal Nutrition (MANNA) is a EU network whose mission is to provide a Double Doctorate level training programme, valid throughout all Europe, on innovative technologies applied to animal science and nutrition. MANNA will provide future research leaders with the capability to address the needs to improve livestock health, welfare and efficiency. There are 11 PhD projects related to MANNA, partly presenting their results in this session. This project has received funding from the European Union's Horizon 2020 research and innovation programme H2020-MSCA- ITN-2017- EJD: Marie Skłodowska-Curie Innovative Training Networks (European Joint Doctorate) – Grant agreement no. 765423 – MANNA.

Study of the salivary modulation of the rumen microbiome
J.M. Palma-Hidalgo[1], A. Belanche[1], A.I. Martín-García[1], C.J. Newbold[2], S.E. Denman[3] and D.R. Yáñez-Ruiz[1]
[1]EEZ, CSIC, C/ Profesor Albraeda, 1, 18008, Granada, Spain, [2]SRUC, West Mains Road, Edinburgh EH9 3JG, United Kingdom, [3]CSIRO, 306 Carmody Road, St Lucia QLD 4067, Australia; david.yanez@eez.csic.es

Ruminants secret large amounts of saliva which contains salts to maintain rumen pH within physiological range but also numerous immune bioactive molecules that interact with the rumen microbiota. The intrinsic host-derived salivary mechanisms to retain or eliminate rumen commensal microbiota remain elusive. This study was conducted to evaluate *in vitro* the impact of the bioactive components of the saliva on rumen microbiome modulation and the potential host-specificity. A semi-continuous 7 days *in vitro* batch culture (20 ml volume) was used to incubate rumen fluid from four different goats (n=4) with a mix of buffer and either their own saliva (OWN), saliva from a different goat (GOAT) or sheep (SHEEP) or a pool of goats´ saliva autoclaved (AUT) to maintain active only mineral components. One third of the incubation volume was extracted every 12 hours and replaced with the same volume of a mix of saliva:buffer (50:50) plus substrate (66 mg). Samples at days 0, 4, 6 and 7 were collected for quantification of bacteria, protozoa, fungi and archaea and the analysis of the diversity bacterial and archaeal communities using Illumina MiSeq and universal 16rDNA primers. The analysis of the Illumina read sequence data was done using the QIIME software and MixOmics R package. The results showed no differences (P>0.05) across saliva treatments on the abundance of the four microbial groups. The main factor explaining the composition of the microbial community was the animal that provided rumen fluid. No effect of saliva treatment was observed on the alpha diversity indexes; however, the Sparse PLS discriminant analysis identified 100 OTUs that significantly (P<0.05) varied according to one or more treatments. While AUT saliva differed most from the other three treatments containing bioactive molecules, some OTUs also contributed significantly to the differences between OWN and GOAT or SHEEP treatments. These results show that the bioactive components of the saliva (other than minerals) play a role in the modulation of rumen microbiome and that there is an individual host-specificity involved that deserves further research.

Liver proteomics of dairy cows supplied with essential fatty acids and conjugated linoleic acids
A. Veshkini[1,2,3], M. Bonnet[2], L. Vogel[4], A. Tröscher[5], M. Delosière[2], A. Delavaud[2], D. Viala[2], F. Ceciliani[1], H. Hammon[4] and H. Sauerwein[3]
[1]University of Milan, Department of Veterinary Medicine, Via Celoria 10, 20133, Milan, Italy, [2]INRAE, UMR Herbivores, Université Clermont Auvergne, VetAgro Sup, Route de Theix, 63122, Saint-Genès-Champanelle, France, [3]University of Bonn, Institute of Animal Science, Physiology & Hygiene Unit, Katzenburgweg 7-9, 53115, Bonn, Germany, [4]Leibniz Institute for Farm Animal Biology (FBN), Wilhelm-Stahl-Allee 2, 18196 Dummerstorf, Germany, [5]BASF, Chemiestraße 22, 68623 Lampertheim, Germany; veshkini@uni-bonn.de

The liver is a central organ in a cow's ability to adapt to the metabolic demands of lactation. We herein tested whether the liver proteome would differ between cows being supplied either with essential fatty acids (EFA) and conjugated linoleic acid (CLA) or a control fat from 9 wk ante to 9 wk post partum. Late-gestation Holstein cows were abomasally infused with coconut oil (CTRL, n=8, 76 g/d) or a mixture of EFA (78 g linseed + 4 g safflower oil/d) and CLA (38 g/d; Lutalin, BASF) (EFA+CLA). Doses were halved during the dry period (starting 9 wk before calving). Untargeted shotgun proteomics based on liquid chromatography coupled with tandem mass spectrometry was performed on liver biopsies harvested at day 28 post partum. The differentially abundant proteins were analysed by Student's T or Wilcoxon-Mann-Whitney test according to adhesion to the normal distribution and mined using ProteINSIDE (http://www.proteinside.org/). Of the total 1,686 proteins identified, 104 were over- and 30 were under-abundant (P<0.05) in EFA+CLA compared to CTRL livers. Overabundant proteins were annotated by Gene Ontology (GO) terms related to tricarboxylic acid cycle, pyruvate metabolism, glycolysis, and gluconeogenesis pathways, consistent with the improved energy balance of EFA+CLA. GO terms related to oxidative stress and vitamin E metabolism were also recorded. Proteins with a decreased abundance in the EFA+CLA group were annotated by terms related to long-chain FAs uptake and metabolism or vitamin A metabolism. Collectively, enriched pathways annotated by liver proteins revealed that EFA and CLA status in transition cows had an impact on energy, lipid and vitamin metabolism, and oxidative stress balance.

Faecal proteomes of cattle fed on barley treated with propionate or ammonia

Y. Huang, P.D. Eckersall and N.N. Jonsson
Institute of Biodiversity, Animal Health & Comparative Medicine, University of Glasgow, 464 Bearsden Rd, Bearsden, Glasgow, G61 1QH, United Kingdom; 2400204h@student.gla.ac.uk

Freshly harvested cereal grain can be preserved as a cattle feed by acidification with propionate or alkalinisation with ammonia. Ammonia treatment has the added benefit of introducing a source of non-protein nitrogen (NPN), which is used by ruminal microorganisms to synthesise protein that is suitable for subsequent digestion and metabolism by the ruminant. We aimed to develop working methods for bovine faecal proteomic investigations and to differentiate between the faecal proteomes of animals fed on barley that had been treated with propionate or ammonia. The study was an addition to a larger study on the production performance of 217 continental breed beef cattle on two isoenergetic and isonitrogenous total-mixed ration diets after 13-20 d of transition. The calculated ration for cattle on the ammonia-treated diet (AMM) included 13.1 kg (DM) of Maxammon-treated barley (Harbro, Turriff, Scotland), with 0.96 kg molasses and 0.58 kg straw; for propionate-treated diet (PRO) comprised 11.33 kg (DM) Prograin-treated (Harbro, Turriff, Scotland) with 3.45 kg pot-ale syrup and 0.5 kg straw. Both rations included the same pack of vitamins, minerals, blended essential oils (Rumitech, Harbro, Scotland) and yeast (Yea-Sacc, Alltech, UK). We collected 4 fresh faecal samples from the floor (following observation of defaecation) of each of 4 pens (2 pens/treatment), from 07:30 h on one day. Protein was extracted from 16 samples according to standard methods, then subjected to trypsin digestion, followed by liquid chromatography and mass spectrometry (LC-MS/MS) based proteomics. Host proteins such as mucosal pentraxin, Annexin A2, Annexin A4 and polymeric immunoglobulin receptor were identified. Serpin Z4 from barley in the diet was identified in all the samples, while serpin Z7 was found only in the PRO group, which were also noted to have higher faecal starch concentrations and inferior feed conversion efficiency. These are preliminary findings but suggest that faecal proteomic analysis, including quantification such as by tandem mass tag labelling might contribute to the understanding of the role of dietary bioactive peptides and proteins in the digestion and absorption of nutrients by cattle.

Intensity of body condition loss peripartum: implications at the metabolome level

R. Riosa[1,2], M.H. Ghaffari[2], G. Blackburn[1], R. Burchmore[1], D. Suess[3], M. Hoelker[4], M. Drillich[3], H. Sauerwein[2] and M. Iwersen[3]
[1]University of Glasgow, Glasgow Polyomics, Garscube Campus, G61 1BD Bearsden, United Kingdom, [2]University of Bonn, Institute of Animal Science, Physiology and Hygiene Unit, Katzenburgweg 7-9, 53115, Germany, [3]University of Veterinary Medicine Vienna, Clinical Unit for Herd Health Management in Ruminants, University Clinic for Ruminants, Veterinärplatz 1 Geb: MA, 1210 Vienna, Austria, [4]University of Bonn, Institute of Animal Science, Department of Animal Breeding and Husbandry, Endenicher Allee 15, 53175 Bonn, Germany; rriosa@uni-bonn.de

The rapid increase in milk yield after calving is related to substantial mobilisation of body reserves and results in a loss of body condition (BC) in dairy cows. Greater intensities of BC loss are commonly associated with an increased risk for production diseases. We aimed at clustering cows according to the magnitude of body condition score (BCS) loss and reproductive health to identify associations with their serum metabolome in early lactation. Using a database comprising 1,864 cows, we first selected those animals from which production data and blood samples, BCS and BFT (back fat thickness) results (mm), were available; for the remaining 721 cows we then calculated the delta of BFT (ΔBFT) by subtracting BFT after calving (31±3 days in milk; DIM) from BFT before calving (-23±11 d). A X-means clustering procedure of RapidMiner (Studio 9.5; an unsupervised machine learning) was then used to construct the clusters based on their ΔBFT, resulting in: extreme loss (EL, ΔBFT from 17 to 23 mm, n=16), moderate loss (ML, ΔBFT from 9 to 15 mm n=122), small loss (SL, ΔBFT from 4 to 8 mm n=329), no loss (NL, ΔBFT from 0 to 3 mm n=203), and gain (G, ΔBFT from -8 to -1 mm, n=51). We further selected those cows that were inseminated before 70 DIM (not requiring hormonal treatment) obtaining 181 animals grouped as: 7 EL, 26 ML, 88 SL, 46 NL and 14 G. From these animals, an untargeted metabolomics (LC-MS) was carried out on serum samples collected at 31±3 DIM. Results show differences in Lyso-phosphatidylcholine, L-glutamine, taurine, ethanolamine phosphate, L-cysteine, and propionyl-L-carnitine. These results suggest that cows with different patterns of BC loss around calving also differ in their metabolic adaptation. Future analyses, including proteomics, will permit us to study the most important pathways involved in more detail.

Lipidomic profile of milk from dairy cows with subclinical non-aureus staphylococcal mastitis

F. Ceciliani[1], M. Audano[1], M.F. Addis[1], N. Mitro[1], D. Caruso[1], C. Lecchi[1], M.H. Ghaffari[2], M. Albertini[1], F. Tangorra[1], R. Piccinini[1] and V. Bronzo[1]
[1]Università degli Studi di Milano, Via dell'Università 6, 26900, Lodi, Italy, [2]University of Bonn, Institut für Tierwissenschaften, Katzenburgweg 7-9, 53115 Bonn, Germany; fabrizio.ceciliani@unimi.it

Mastitis is an udder inflammation that impacts the dairy industry by reducing milk yield and quality and increasing culling rate. Although many pathogens can cause bovine mastitis, non-aureus staphylococci (NAS) have become the most frequently isolated bacteria from the milk of dairy cows with mastitis, especially in the subclinical form. The molecular mechanisms regulating the mammary gland inflammatory responses to NAS are unclear. The application of system biology approaches to mastitis has provided pivotal information by investigating the transcriptome, proteome, and metabolome. On the contrary, the lipidome lies virtually undiscovered despite its outstanding relevance, being many of the mediators involved in immune defence and inflammation of the mammary gland derived from arachidonic acid metabolites. We aim to cover this gap by determining for the first time the untargeted lipidome of cow milk during NAS subclinical mastitis (SM) with liquid chromatography-quadrupole time-of-flight mass spectrometry (LC-QTOF-MS) approach. The study was carried out on 30 animals. Out of them, 17 were affected by NAS SM, and 13 were healthy (SCC<100×10^3 cells/ml). Sixteen subclasses of lipids were identified in both groups of animals. From 2,506 measured lipids, 372 were significantly changed more than 10-fold (P<0.05) in milk from cows with mastitis as compared to healthy cows. Our results point out the significant influence of NAS on the milk lipidome, contribute to the understanding of inflammatory processes in the bovine udder, and highlight potential novel biomarkers for improving mastitis diagnosis. Part of this work was carried out in 'OMICs', an advanced mass spectrometry platform established by the Università degli Studi di Milano.

Isolation, characterisation and proteomics profiling of porcine milk exosomes

R. Furioso Ferreira[1,2], T. Blees[1], A. Horvatić[2], V. Mrljak[2] and H. Sauerwein[1]
[1]University of Bonn, Institute for Animal Sciences, Physiology and Hygiene Unit, Katzenburgweg 7, 53115 Bonn, Germany, [2]University of Zagreb, Faculty of Veterinary Medicine, Heinzelova 55, 10000 Zagreb, Croatia; rafaelaff.vet@gmail.com

Milk exosomes are gaining interest in research due to their potential as drug vehicles, carriers of presumptive biomarkers, and of mediators involved in various patho-physiological functions. Exosome isolation from milk is particularly troublesome, especially for OMIC analyses, due to its complex nature, the high lipid and protein content, and the presence of milk fat globules. This study aimed at establishing an efficient method for isolation of exosomes from porcine milk, and at the characterisation of their proteome by differential ultracentrifugation coupled with size exclusion chromatography (SEC) and high-throughput LC-MS/MS. Skimmed milk was sequentially centrifuged at 12,000×g for 30 min at 4 °C, at 100,000×g for 1 h at 4 °C and at 150,000×g for 2 h at 4 °C. The exosome pellet was collected, suspended in PBS, and loaded on a SEC column (Izon qEV). After the void volume, 4 fractions of 500 μl were collected. Fractions 2 and 3 contained exosomes, substantiated by verification of the presence of a marker (Western Blotting of TSG101), concentration and size distribution by Nanoparticle Tracking Analysis (NanoSight NS300) and morphology by Transmission Electron Microscopy. In-gel digestion was performed after concentrating exosome fractions (10 kDa cut-off column), resolving the proteins by 12% SDS-gel at 90 V for 15 min, fixing with 25% isopropanol/10% acetic acid and staining with PAGE Blue. Bands were excised and shrunk with acetonitrile (ACN), reduced with 10 mM DTT/50 mM TEAB at 56 °C, alkylated (55 mM iodoacetamide/50 mM TEAB), and digested (trypsin + TEAB/10% ACN; incubated overnight at 37 °C for maximum peptide recovery). Peptides were extracted (5% formic acid/ACN), dried and stored at -80 °C until LC-MS/MS analysis. Proteomic analysis enabled the identification of 1,306 proteins in qEV fraction 2 and 419 in qEV fraction 3 (minimum 2 unique peptides). The proteome of porcine milk exosomes characterised herein provides new information on the composition of this milk protein fraction which is important not only for basic physiology but may also reveal potential industrial applications.

Performance, oxidative status and serum proteome in piglets from sows fed at divergent ω6:ω3 ratios

T.X. Nguyen[1,2], A. Agazzi[2], M. Comi[2], V. Bontempo[2], G. Invernizzi[2], H. Sauerwein[1], M. Pinerio[1], D. Eckersall[1], R. Burchmore[1] and G. Savoini[1,2]
[1]*The MANNA Consortium, University of Glasgow, Glasgow, G12 8QQ, United Kingdom, [2]Università degli Studi di Milano, Department of Health, Animal Science and Food Safety, Via dell'Università 6, 26900 Lodi (LO), Italy; thixuan.nguyen@unimi.it*

This study aimed at assessing the effect of low versus high ratios of ω6:ω3 fatty acids in the gestation and lactation sow diets on their piglets. Sixteen multiparous sows were randomly allocated to two diets with ω-6:ω-3 ratios of 13:1 (C) and 4:1 (T) from d 28 of gestation onwards. Their post-weaning piglets were divided into four groups (10 piglets/group): C-without seaweed supplementation (C0); C-with seaweed (SW) at 4 g/kg feed (CSW); T-without seaweed (T0); and T-with seaweed (TSW). Body weight (BW), plasma and serum were collected at d 0, 7, 15 and 21 post-weaning. Pigs from CSW and T0 had higher ($P<0.05$) BW on d 15 and 21; average daily gain from d 0-21; feed intake from d 0-21; and gain:feed ratio from d 0-15. Protein and lipid oxidation products (AOPP and TBARS) in plasma were not different between groups. However, AOPP and total antioxidants (FRAP) decreased ($P<0.0001$) and TBARS increased ($P<0.0001$) from d 0 to 15. Moreover, FRAP was lower ($P<0.05$) in the SW group compared to the 0 group within the same sow diet on d 7 and FRAP tended to be lower ($P=0.06$) in piglets from T-mothers than those from C-mothers from d 0 to 21. A pilot proteomic experiment on serum showed some variations of the exponentially modified protein abundance index (emPAI) between C0 and CSW groups: albumin ranged from 280 to 604; haptoglobin from 0.47 to 8.16 whereas less difference was found on complement C3 (0.53-0.91) and alpha-2-HS-glycoprotein (0.95-1.62). Further exploiting these results will aid to explain the mechanisms underlying the improved growth performance in piglets from sows receiving a low dietary ω6:ω3 ratio and fed SW post-weaning.

Identifying miRNA-mRNA regulatory relationship on ω-6/ω-3 FA ratio expression profiles in pigs

Y.J.Y. Manaig[1,2], L. Criado-Mesas[1], A. Esteve-Codina[3], E. Mármol-Sánchez[1], A. Castelló[1], J.M. Folch[1,2] and A. Sánchez[1,2]
[1]*Centre for Research in Agricultural Genomics (CRAG) CSOC-IRTA-UAB-UB, Universitat Autònoma de Barcelona, Bellaterra, 08193, Barcelona, Spain, [2]Universitat Autònoma de Barcelona, Department de Ciència Animal i dels Aliments, Bellaterra, 08193, Barcelona, Spain, [3]CNAG-CRG, Centre for Genomic Regulation (CRG), Barcelona Institute of Science and Technology (BIST), Barcelona, 08028, Barcelona, Spain; yronjoseph.manaig@uab.cat*

The present study was conducted to determine differentially expressed (DE) genes, microRNAs (miRNAs), pathways, and correlate their expression profiles in response to ω-6/ω-3 fatty acid (FA) ratio. Porcine *longissimus dorsi* muscle (Iberian × Duroc) with extreme values for ω-6/ω-3 FA ratio (10 high vs 10 low) were used for RNA-Seq and microRNA-Seq and DE analysis. A total of 432 genes were differentially expressed ($P<0.05$), including 226 upregulated and 156 downregulated genes. Gene ontologies and biological pathways ($P<0.05$) were related to muscle structure development (*GO:0061061*), positive regulation of skeletal muscle cell differentiation (*GO:2001016*), SREBP signalling pathway (*GO:0032933*), and adenylate cyclase-activating adrenergic receptor signalling pathway (*GO:0071880*). These pathways are involved in the regulation of glucose and lipid metabolism. The miRNA seed of the DE miRNAs (ssc-miR-15b, ssc-mir30a-3p, ssc-miR-30e-3p, and ssc-miR-7142-3p) was interrogated to the 3' UTR region (7mer-m8 site) of the DE mRNA genes. Out of 432 DE genes, a total of 125 genes to miR-15b, 130 genes to miR-30a/30e-3p, and 54 genes to miR-7142-3p were mapped based on the seed matching on the said region. The expression profiles of DE genes and miRNAs were negatively correlated ($P<0.05$) and filtered according to biological significance (3' UTR match). miR-15b showed correlation with ARRDC3 gene, which interacts with β-adrenergic receptor and is related to lipolysis, whereas, miR-7142-3p was associated with METTL21C gene that is involved in muscle development and protein degradation. Our results have shown genes and miRNAs that are related to the metabolic regulation and muscle development.

Untargeted metabolomics: effects of novel plant extracts in broilers pre and post LPS challenge

F. Riva[1], R. Burchmore[1], D. McKeegan[1], J. Peinado Izaguerri[1], G. Bruggeman[2], D. Hermans[2], M. McLaughlin[1], D. Eckersall[1] and M. Bain[1]
[1]*University of Glasgow, 464 Bearsden, G611QH Glasgow, United Kingdom,* [2]*Nutrition Sciences, Booiebos 5, B9031 Ghent, Belgium; francesca.riva@glasgow.ac.uk*

The need to find good alternatives to antibiotic growth promoters is a major challenge for the broiler industry. Novel plant extracts such as citrus and cucumber, are possible candidates because of their anti-inflammatory, antioxidant and antimicrobial properties. These novel plant extracts contain active molecules such as pectins, limonene and polyphenols that modulate the immune system and metabolism of chickens. In this study a dietary trial (12 birds × 4 replicates × 3 diets) was performed to test the hypothesis that a baseline diet (diet1) supplemented with citrus (diet2) or cucumber extract (diet3) can modulate the immune system, metabolome and performance of broilers. Each diet was fed from day old and all birds were challenged at 15 days old, with *Escherichia coli* lipopolysaccharide (LPS), a bacterial endotoxin known to stimulate the host immune system. Plasma was collected pre and post challenge at 4 time points (T0,12,24,48 h) from the same birds (n=12 per treatment). The plasma samples were extracted and pooled based on pen, treatment and time points and subjected to untargeted metabolomics (MS-LC based) at the Polyomics facilities, Glasgow. Data obtained were analysed using Pimp (polyomics integrated metabolomics pipeline) and Metaboanalyst software. PCA and one way-ANOVA tests were used to compare the effect of time (pre and post challenge) and treatment. The largest change in metabolites across all diets occurred at T12 and T24 post challenge. Preliminary results show that many of the metabolites associated with the Krebs cycle decreased in the first 12 h post challenge and then increased during the recovery period (T24, T48). These metabolites are connected to the production of ATP and therefore act as modulators of the energy metabolism. In conclusion, results from this study will be used to search novel biomarkers of health status and develop new dietary plans.

In vitro impact of citrus pectin (CP) on chicken monocytes' immune response

G. Ávila, G. Grilli, S. Di Mauro, D. De Leonardis, C. Lecchi and F. Ceciliani
Università Degli Studi di Milano, Dipartimento di Medicina Veterinaria, Via Celoria 10, 20133, Milano, Italy; gby31193@gmail.com

Pectin is a dietary fibre primarily composed of repeating units of galacturonic acid. It is found in plant cell walls and is most abundant in citrus fruits. Citrus pectin (CP) was shown to have many beneficial properties in humans and animals like antioxidative, anticancer and anti-inflammatory, having an *in vitro* protective effect on the barrier function of human intestinal epithelial cells as well. Supplementation of broilers' diet with CP has shown to be an effective nutrition strategy, improving energy utilisation, nutrient digestibility and therefore causing an overall positive impact on the animals' growth performance. However, limited information on CP effects on chicken immunity is available. In this study, we sought to assess the *in vitro* impact that CP exerts on both, chicken PBMCs and sorted monocytes' immune response. Cells were isolated from whole blood of healthy chickens and incubated with increasing concentrations (0,0.25,0.5,0.75,1 mg/ml) of CP. To determine CP working concentrations, their effects on cells' apoptosis and viability (lifespan) were assessed. Apoptosis was determined by measuring the activity of caspase-3 and -7, while viability by using an MTT-based assay. CP demonstrated not to have any dose-dependent effect on PBMCs' lifespan, as no differences in apoptosis and viability were observed. However, a decreasing trend in apoptosis with 0.5 mg/ml was seen. We further evaluated the effects of CP (0.5 mg/ml) on chicken monocytes' chemotaxis and phagocytosis. Chemotaxis was determined using transwell migration plates, while phagocytosis by the fluorescein-labelled *Escherichia coli* incorporation assay. Statistical analyses were performed in GraphPad Prism 8.0.2, using repeated measures one-way ANOVA for normally distributed samples. CP inhibited monocytes' chemotaxis and apparently also their phagocytosis. Further analyses on other immune functions (e.g. oxidative burst, killing capability), and integration of system biology approaches should be applied to confirm this anti-inflammatory activity.

MitoCow – interrelationship of milk and plasma metabolite profiles in dairy cows after LPS challenge

W. Xu, Á. Kenéz and Mitocow Consortium
[1]City University of Hong Kong, Department of Infectious Diseases and Public Health, Kowloon, Hong Kong;
wei.xu@cityu.edu.hk

Lipopolysaccharide (LPS) load poses an inflammatory challenge to dairy cows, causing immune system activation and associated metabolic adaptation, reflected by alterations in the plasma metabolome. Alternatively, adaptation processes could be reflected by the milk metabolome, either due to transfer from the plasma or due to altered mammary gland metabolism. We aimed to identify the metabolite markers that can indicate metabolic changes associated with an inflammatory response caused by an intravenous LPS challenge, both in plasma and in milk. Plasma and milk samples of multiparous German Holstein-Friesian cows were collected before and after an intravenous LPS challenge (111 DIM). Cows were either treated without (CON group; n=25) or with 25 g rumen-protected carnitine per day per animal (CAR group; n=27). Both plasma and milk were subjected to a targeted metabolomics analysis using the AbsoluteIDQ p180 Kit (Biocrates Life Science AG, Innsbruck, Austria) by liquid chromatography-mass spectrometry (LC-MS). Metabolite concentration data were analysed in R. Important metabolites to indicate LPS challenge were selected by partial least squares-discriminant analysis and receiver operating characteristic curve (ROC). The correlation between plasma and milk indicators was analysed. Important metabolites in plasma outperformed those in milk, in terms of (1) the number of metabolites indicating LPS challenge, and (2) average ROC area under the curve (AUC). Short-chain acylcarnitines (carbon chain length C2, C3, C4, and C5), long-chain acylcarnitines (C14, C16, and C18), and biogenic amines (sarcosine, glutamine and isoleucine) were identified to have good performance for indicating LPS challenge, either in milk or plasma, with a maximum AUC of 0.93. Valerylcarnitine (C5) was the most powerful metabolite marker both in plasma and in milk, indicating LPS challenge in both CON and CAR. Concentrations of C2, C4, and C5 were positively correlated between plasma and milk (P<0.001). Short-chain acylcarnitines, particularly C5, could reflect metabolic adaptation to an inflammatory signal both in plasma and in milk, highlighting a link between mitochondrial function and immune response.

Liver mRNA abundance of metabolically relevant genes related to inflammation and dietary carnitine

M.H. Ghaffari and Mitocow Consortium
University of Bonn, Institute for Animal Science, Katzenburgweg 7, 53115 Bonn, Germany; morteza1@uni-bonn.de

L-carnitine is essential for fatty acid oxidation (FAO) enabling fatty acid (FA) uptake into the mitochondrial matrix. L-Carnitine has also antioxidant properties which might be particularly relevant during inflammatory reactions when lipid metabolism is dysregulated. With these backgrounds, our objectives were: (1) assessing the hepatic mRNA expression of 4 key genes of carnitine metabolism (CM) and 14 key genes of mitochondrial FAO pathways in dairy cows, and (2) comparing their response to an intravenous challenge with lipopolysaccharide (LPS). A total of 51 multiparous Holstein cows were randomly assigned to two groups of control (CON, n=24) and rumen-protected L-carnitine group (CAR, 25 g/cow/day, n=27) from 6 weeks ante partum (ap) until 18 weeks postpartum (pp). Liver biopsies were collected on d – 42, +100, +110, and +126 relative to calving. On d +110 pp, LPS (*Escherichia coli* O111:B4, Sigma-Aldrich, 0.5 ug/kg BW) was intravenously injected. The expression of targeted genes in liver was assayed by RT-qPCR using microfluidics integrated fluidic circuit chips (96.96 Dynamic Array, Fluidigm, San Francisco, CA, USA). A preamplification of cDNA was performed prior to RT-qPCR. The 3 most stable reference genes were used to normalise the target genes and analysis was done using the MIXED procedure of SAS. The model included the random effect of cow and the fixed effect of the main factors group, time, and their interaction (group × time). The abundance at day 42 ap was considered as covariate. The results showed that the mRNA abundance of 3 of the 4 genes involved in carnitine metabolism and 10/12 in FAO was affected by time after the LPS challenge. From the genes involved in mitochondrial FAO, the mRNA abundance of carnitine palmitoyltransferase 1 (CPT1) and malonyl-CoA decarboxylase (MLYCD) was lower in CAR cows after the LPS challenge. Overall, the expression of most genes involved in lipid metabolism in the liver was affected by the LPS challenge rather than the L-carnitine supplement.

MitoCow – effects of systemic inflammation on telomere length and mitochondrial DNA copy numbers
K.D. Seibt, and Mitocow Consortium
University of Bonn, Institute for Animal Science, Bonn, Germany; katharina.seibt@uni-bonn.de

Telomeres protect chromosomal integrity; they shorten with each cell division; telomere length (TL) may mirror cellular aging processes. The copy number of mitochondrial DNA (mtDNA) reflects the abundance of mitochondria within a cell. Both TL and mtDNA are affected by environmental, physiological, and energy status. In dairy cows the changes in energy balance (EB) during early lactation were demonstrated to be related to TL and mtDNA. Early lactation is accompanied by an increased release of fatty acids (FA) from adipose tissue and also by systemic inflammation. For dissecting systemic inflammation and metabolic adaptation to decreasing EB, we herein compared the effects of inflammation on leukocyte TL and mtDNA both during negative and positive EB in the natural peripartal inflammation and in a standardised inflammatory challenge by administering bacterial lipopolysaccharide (LPS), respectively. To facilitate FA oxidation, half of the cows studied received L-carnitine (*CAR*, 25 g/d; n=26) from d -42 relative to calving onwards, the cows not receiving the supplement served as control (*CON*; n=24). Blood samples were collected at days (d) -42, +42 and +100 relative to calving. All cows were challenged by an injection of LPS (0.5 µg/kg BW) at d +110. Further samples were collected 48 h later. Beside haematology, TL and mtDNA from leukocytes were assessed by multiplex qPCR, i.e. amplifying either TL or mt12s rRNA together with the reference gene β-globin (single copy/cell) and applying strict criteria of validation. Statistical analyses (linear mixed model; SPSS) were performed with time, treatment and their interactions as fixed effects. The TL values as well as the mtDNA values per cell on d -42, +42 and +100 were not different, but the mtDNA values obtained +48 h after LPS were lower than before (P<0.001), irrespective of *CAR*. Taking all TL values into account, an increase of TL +48 h after LPS was detectable (time: P=0.003). In view of the increased total number of leukocytes 48 h after LPS, the LPS-related decrease in mtDNA/cell and the increase in TL might be due to a shift towards younger leukocytes. Neither mtDNA nor TL were altered with stage of lactation and thus EB or by *CAR*.

Statistical analysis of microbiome data
S. Costa-Roura, G. De La Fuente, J. Balcells and D. Villalba
University of Lleida, Av. Alcalde Rovira Roure, 191, 25198, Lleida, Spain; sandra.costa@udl.cat

High throughput sequencing technology results in microbiome abundance datasets that have large proportions of zeros and variable total number of counts per sample along individuals. Prior to statistical analysis, strategies to zero-dealing and normalisation are commonly employed. However, there is an increasing awareness of the compositional nature of microbiome data. When compositionality is ignored, spurious correlations and increase of type I error may arise, so there is a need for applying compositional data analysis (CoDA) to microbiome datasets. The aim of this work was to compare the analysis of ruminal microbiome, applying either the standard statistical procedure (STD) or CoDA methodology. Present microbiome dataset belonged to fattening cattle with different early-life feeding management or treatments: eight calves fed on their dams' milk and with free access to hay and straw until weaning, and ten male calves fed on milk replacer and a starter concentrate until weaning. Ruminal fluid was obtained via oesophagus tube twice, in growing and finishing periods. Rumen bacterial and archaeal community composition was analysed by taxonomic profiling of 16S ribosomal RNA V3-V4 variable regions. Prior to statistical analysis, microbiome dataset was either median-normalised (STD) or carried to a Euclidean space applying the Aitchison's centred log ratio transformation (clr) after replacing zeros by adding 1 to each value (CoDA). To measure differences in microbiome composition between samples, beta diversity was approached through performing either a NMDS based on normalised abundance Bray-Curtis distance (STD) or a PCA based on clr Euclidean distance (CoDA). Multivariate analysis plots showed clear clustering of animals according to period but Adonis test results differed between methodologies (P=0.096 in STD and P=0.001 in CoDA). To decipher which genera were responsible for the differences between treatments, either a LEfSe on normalised abundance (STD) or an ALDEx analysis on clr (CoDA) were conducted over those genera present at least at 50% of the individuals. Genera identified as differentially abundant between treatments were not the same in both procedures, pointing out the main discrepancy between STD and CoDA methodology. Funding by GENTORE H2020 (Project n°727213).

Serum proteomics of dairy cows infused with essential fatty acids and conjugated linoleic acids

A. Veshkini[1,2,3], H. Hammon[4], L. Vogel[4], A. Tröscher[5], M. Delosière[2], A. Delavaud[2], D. Viala[2], F. Ceciliani[1], H. Sauerwein[3] and M. Bonnet[2]
[1]University of Milan, Department of Veterinary Medicine, Via Celoria 10, 20133, Milan, Italy, [2]INRAE, UMR Herbivores, Université Clermont Auvergne, VetAgro Sup, Route de Theix, 63122, Saint-Genès-Champanelle, France, [3]University of Bonn, Institute of Animal Science, Physiology & Hygiene Unit, Katzenburgweg 7-9, 53115, Germany, [4]Leibniz Institute for Farm Animal Biology (FBN), Wilhelm-Stahl-Allee 2, 18196 Dummerstorf, Germany, [5]BASF, Chemiestraße 22, 68623 Lampertheim, Germany; veshkini@uni-bonn.de

Cow's ability to adapt to the metabolic demands of lactation requires numerous endocrine regulations that could be deciphered by screening serum proteome. We aimed to compare the serum protein profiles of dairy cows divergent in essential fatty acids (EFA) and conjugated linoleic acids (CLA) status during the transition from late gestation to early lactation. Holstein cows from 9 wk antepartum to 9 wk postpartum were infused into the abomasum with one of the following treatments: coconut oil (CTRL, n=8, 76 g/d) or a mixture of EFA (78 g linseed + 4 g safflower oil/d) and CLA (38 g/d; Lutalin, BASF) (EFA+CLA). Serum samples collected 4 wk antepartum were analysed using liquid chromatography coupled with tandem mass spectrometry after depleting highly abundant proteins. The groups (log-transformed protein intensities) were compared by the t-test/Wilcoxon-Mann-Whitney test according to adhesion to a normal distribution and mined using ProteINSIDE (http://www.proteinside.org/). Of the 268 proteins identified in the serum, 22 were over- and 12 were under-abundant (P<0.05) in the EFA+CLA group. Overabundant proteins were annotated by Gene Ontology (GO) terms related to triglyceride (TG) homeostasis, retinoid, cholesterol, and lipoprotein metabolisms. The higher abundance of Apolipoprotein C-III (APOC3), a negative regulator of the TG catabolism pathway, may imply reduced TG hydrolysis and uptake by peripheral tissues in EFA+CLA cows in early lactation. Proteins with a decreased abundance in EFA+CLA were annotated by GO terms related to platelet degranulation and positive regulation of I-kappaB kinase/NF-κB signalling. Altogether, enriched pathway annotations revealed molecular signatures related to TG-enriched lipoprotein metabolisms and immune status in transition cows divergent in EFA+CLA status.

Seasonal variation in fatty acid composition of organic milk from cows fed with green fodder

R. Primi[1], A. Cappucci[2], S. Failla[3], M. Contò[3], P.P. Danieli[1], F. Frezza[1], M. Mele[2] and B. Ronchi[1]
[1]Università della Tuscia, Dipartimento di Scienze Agrarie e Forestali (DAFNE), Via San Camillo de Lellis, snc, 01100, Italy, [2]Università di Pisa, Centro di Ricerche Agro-ambientali, Via Vecchia di Marina 6, 5600 Pisa, Italy, [3]Consiglio per la Ricerca in Agricoltura e l'Analisi dell'Economia Agraria, Centro di ricerca Zootecnia e Acquacoltura '02 Monterotondo' (CREA-ZA), Via Salaria 31, 00016 Monterotondo (RM), Italy; ronchi@unitus.it

Feeding affects milk composition, and rearing systems based on grazing and green forage diets are known to improve its healthy fatty components. We conducted a study to investigate the seasonal variation of fatty acid (FA) composition of cow's milk during a complete lactation at farm scale. A homogeneous group of six Brown cows were selected in June 2019 within the herd of an organic dairy farm, in Central Italy. Individual milk samples were collected on June 5 (<100 days in milk, DIM), September 19 (100-200 DIM) and October 15 (>200 DIM) and immediately frozen. Concurrently, single component of the diet and pasture samples were also taken. Chemical and FA profile of the cows' diet components and milk were determined. The relative amounts of saturated (SFA), monounsaturated (MUFA), polyunsaturated (PUFA), short chain (SCFA), medium chain (MCFA) and long chain (LCFA) fatty acids were calculated. The results showed that milk FA composition was influenced by the physiologic phase and the acidic profile of the diet. In comparison with the diets of the cows in mid and late lactation, the diet administered in the first 100 DIM was poorer in total fat, with a lower percentage of SFA and MUFA, but it was richer in PUFA (e.g. +36.3% of the ω-3 fraction). However, there was no a significant decrease of SFA or increase of ω-3 FA in milk from early (<100 DIM) lactation cows. On the other hand, the higher amount of C18:0 and trans-11 C18:1 in early lactation milk suggested a more intense biohydrogenation activity of dietary PUFA in the rumen. In general, this study has shown that, at the farm level, the chemical composition of diet significantly changed with the season, due to the presence of different fodder sources during the whole year. At the same time, the physiologic stage of the cow seemed to be also involved in affecting the fatty acids profile of milk.

Effect of herbage allowances in winter gestation on hepatic gene expression in beef heifers

M. Anzolabehere[1], A. Casal[1], M. Claramunt[2], P. Soca[1], M. Carriquiry[1] and A.L. Astessiano[1]
[1]Facultad de Agronomía, UDELAR, Garzón 780, 12900, Uruguay, [2]Centro Universitario Regional del Este, Ruta 8 km 281, 33000, Uruguay; maite.afigueroa@hotmail.com

The aim of this study was to evaluate the effect of herbage allowance (HA) of grasslands during winter gestation on endocrine/metabolic profile and hepatic gene expression in grazing beef heifers. Forty pregnant Hereford heifers (5.6±0.4 of body condition score (BCS); 130±12 days of gestation (DG)) were used in a randomised block design with two treatments of HA: 4 kg dry matter (DM)/kg body weight (BW) (high; HI) vs 2.5 kg DM/kg BW (low; LO). Plasma and liver biopsies were collected at 190 and 265 DG -fall and winter, respectively- to quantify metabolic and endocrine profile and mRNA expression of gluconeogenesis genes by real time PCR. Data were analysed as repeated measures with a mixed model that included HA, DG as fixed effects and block as a random effect. During winter gestation, all heifers lost (P<0.05) BW and BCS and changes in hormone and metabolic profiles reflected the negative energy balance of winter which was more severe in LO than HI heifers. Insulin and insulin growth factor-1 (IGF1) decreased (P≤0.02) while non-esterified fatty acids (NEFA), beta-hydroxybutyrate and urea increased (P<0.01) from 190 to 265 DG and the decrease in plasma IGF1 was greater for HI than LO heifers. Increase during winter gestation in NEFA and urea was greater (P≤0.05) for LO than HI heifers. Hepatic glycogen decreased (P<0.02) from 190 to 265 DG in all heifers while free glucose in liver decreased only in LO cows. The expression of propionyl-CoA carboxylase mRNA was not modified due to HA or DG. While, glucose 6-phosphatase mRNA increased (P=0.02) and pyruvate carboxylase mRNA tended to increase (P=0.10) during winter gestation for all heifers. From 190 to 265 DG citrate synthase mRNA increased (P=0.03) only for HI heifers. During winter gestation, phosphoenolpyruvate carboxykinase and succinate dehydrogenase mRNA decreased (P<0.01) only for HI heifers. Calf birth weight not differ between HA and averaged 32.0±3.1 kg. Our results indicated that cows adapted their hepatic metabolism in order to supply the increased glucose demands of the gravid uterus and foetus during the last third of gestation when forage mass of grassland decreased due to low temperatures of winter.

MicroRNA expression in mammary gland secretory tissue infected by coagulase-positive staphylococci

E. Bagnicka[1], E. Kawecka-Grochocka[1], K. Pawlina-Tyszko[2], M. Zalewska[3], A. Kapusta[1], S. Marczak[1] and T. Ząbek[2]
[1]Institute of Genetics and Animal Breeding PAS, Postepu 36A St., 05-552, Poland, [2]The National Research Institute of Animal Production, 1 Krakowska St, 32-083 Balice, Poland, [3]University of Warsaw, 1 Miecznikowa St., 02-096 Warsaw, Poland; e.bagnicka@ighz.pl

The aim of the study was the analysis the differences between miRNA expressions in dairy cattle udder secretory tissue infected with coagulase-positive staphylococci (CoNS; n=21) vs non-infected (H; n=6) one. The study was carried out on 27 Polish HF breed dairy cows being between their first and fourth lactation. They were culled due to chronic mastitis after unsuccessful antibiotic therapies (CoPS group) or reproduction problems (H group). The milk samples were microbiologically examined two days before slaughter. The miRNA libraries were prepared from 950 ng of total RNA and then sequenced on HiScanSQ in two technical replicates. We identified 256 known and 260 potentially new miRNAs, and 32 of them were differentially expressed (DE) in CoPS vs H comparison (27 upregulated; 5 downregulated). The most interesting enriched KEGG pathways encompassed bacterial invasion of epithelial cells (hsa05100), endocytosis (hsa04144), focal adhesion (hsa04510), and lysosome (hsa04142). Important GO terms covered cell junction organisation (GO:0034330), cell death (GO:0008219), immune system process (GO:0002376), and signal transduction (GO:0007165). Among DE miRs, we selected that involved in the immune response: miR-99b-5p, miR-145-5p, miR-155-5p, miR-191-5p, and miR-223-5p. In silico analysis, three target genes common for the above-mentioned miRs and involved in the immune response were selected: CRK (CRK proto-oncogene) in the KEGG pathway 'bacterial invasion of epithelial cells', and PELI1 (Pellino E3 Ubiquitin Protein Ligase 1) and B2M (beta-2-microglobulin) in the GO term 'immune system process'. Thus, miRNAs play an essential role in immune system regulation in udder secretory tissue during chronic mastitis. Increased expressions of identified miRNAs could inhibit the action of their targeted genes. Perhaps coagulase-positive staphylococci can modulate the response of the organism during inflammation. Funding: the NSC, Poland, Grant No. 2015/17/B/NZ9/01561.

Feeding lambs with *Nannochloropsis oceanica* induces changes on muscle lipidome

S.P. Alves[1], D. Dannenberger[2], C. Galuska[3], A.C. Vitor[1], A. Godinho[1], A. Francisco[4], J. Santos-Silva[4], R.J.B. Bessa[1] and B. Fuchs[3]
[1]CIISA – Centro de Investigação Interdisciplinar em Sanidade Animal, Faculdade de Medicina Veterinária, ULisboa, Lisboa, Portugal, [2]Leibniz Institute for Farm Animal Biology (FBN Dummerstorf), Institute of Muscle Biology and Growth, Dummerstorf, Germany, [3]Leibniz Institute for Farm Animal Biology (FBN Dummerstorf), Core Facility Metabolomics, Dummerstorf, Germany, [4]INIAV, Santarém, Vale de Santarém, Portugal; susanaalves@fmv.ulisboa.pt

Strategies to increase the content of unsaturated fatty acids (UFA), as the eicosapentaenoic acid (20:5n-3, EPA), in ruminant edible fats include the dietary supplementation with microalgae, but due to the ruminal biohydrogenation (BH), a large range of intermediates are formed in the rumen and deposited in tissues. Our previous studies showed that the cell walls of the microalgae *Nannochloropsis oceanica* can protect EPA from ruminal BH. Thus, this work aims to study the effect of feeding lambs with EPA-rich microalgae in muscle lipidome. We hypothesise that the muscle of lambs fed *N. oceanica* biomass will have more lipids containing EPA while those fed *Nannochloropsis* extracted oil will have a more complex lipid profile. Thus, an untargeted lipidomic approach was applied to lambs muscle from a trial where Merino Branco lambs (7 by group) were fed the diets: (C) Control; (O) C+1.2% of *Nannochloropsis* spp. oil; (L) C+9,2% of *N. oceanica* lyophilised. Diets O and L contained 2.8 g/kg DM of EPA. Lipids from muscle were extracted using a tissues homogeniser (Precellys Evolution, Bertin Instruments, France) and analysed using a Q Exactive Plus (Thermo Scientific). Fifteen lipid classes were identified in muscle covering more than 350 different species. Among diets, the main changes in muscle lipidome were in plasmenylcholine (P-PC), phosphatidylcholine, phosphatidylserine, and triacylglycerols. These changes consisted in a higher proportion of lipids with UFA in both L and O groups than in C. Surprisingly, the P-PC was the lipid class where more than half of the species varied among diets, on the contrary there were no differences in lysophosphatidylglycerol and lysophosphatidylinositol species. The results suggest that P-PC is the muscle lipid class more prone to variation with dietary EPA. This work was funded by Project UIDP/CVT/276/2020 (CIISA).

Gene expression and muscle type in cattle: bond to beef quality

O. Urrutia, A. Arana, J.A. Mendizabal, L. Alfonso and B. Soret
Public University of Navarre, Campus Arrosadía, 31006, Spain; olaia.urrutia@unavarra.es

The knowledge of gene expression underlying the properties and composition of the skeletal muscle is of great interest as it is determinant for beef quality. This work aimed to study differences in the expression of key gene and the metabolic and contractile properties of fibre between *longissimus thoracis* (*LT*) and *masseter* (*MS*) muscles. Sixteen young bulls were used (8 Holstein with 297±3.5 days of age and 230±5.6 kg carcass weight, and 8 Pirenaica with 389±4.4 days of age and 348±4.3 kg carcass weight). Samples were analysed by qPCR, by quantification of enzyme activities and by immuno-histochemical techniques. The expression of genes involved in adipogenesis (*DLK1*, *ZFP423*), fatty acid uptake (*CD36*), transport (*FABP3*) and oxidation (*CPT1B*), triglyceride synthesis (*DGAT2*), lipolysis (*HSL*), fibre type formation (*MYOD, MSTN, MYH1, MYH7*) and encoding cytokines (*ADIPOQ, ADIPOR2, LEP*) were analysed, as well as the activity of glycolytic (LDH and PKF) and oxidative (ICDH and COX) enzymes. Also, the proportion of the main type of muscle fibres (IIX: fast glycolytic; IIA: fast oxidative; and I: slow oxidative) was determined. Data were analysed using different general linear models. The results showed that the expression of *CD36* and *CPT1B* was higher in *MS* muscle than in *LT* (P<0.05) and that *MYOD, MSTN* and *MYH1* expression was higher in *LT* muscle (P<0.01). Glycolytic enzyme activities (LDH and PKF) were also higher in *LT* muscle whereas oxidative enzymes (ICDH and COX) were lower (P<0.001). Finally, higher proportion of fibres types IIX and IIA than type I was found in *LT* (P<0.001), whilst *MS* muscle was composed entirely of slow oxidative fibres. To conclude, *LT* contained mostly type II fast fibres and showed glycolytic metabolism, while *MS* muscle was composed solely of slow oxidative fibres and had oxidative metabolism. This might be related to the greater expression of genes involved in fibre formation (*MYOD, MSTN* and *MYH1*) in *LT* muscle, and the greater expression of genes related to fatty acid transport and metabolism (*CD36* and *CPT1B*) in *MS* muscle. The results provide useful information towards a better understanding of cattle muscle characteristics and the identification of biomarkers that could be valuable to predict beef quality.

Evaluation of different microRNA expression of porcine milk and plasma exosomes influenced by diet

R. Furioso Ferreira[1,2], R. Calogero[3], M. Arigoni[3], V. Mrljak[2], H. Sauerwein[1] and F. Ceciliani[4]
[1]University of Bonn, Institute for Animal Sciences, Physiology and Hygiene Unit, Katzenburgweg 7, 53115 Bonn, Germany, [2]University of Zagreb, Faculty of Veterinary Medicine, Heinzelova 55, 10000 Zagreb, Croatia, [3]University of Torino, Department of Molecular Biotechnology and Health Sciences, Via Nizza 52, 10126 Torino, Italy, [4]University of Milan, Department of Veterinary Medicine, Via Celoria 10, 20133, Italy; rafaelaff.vet@gmail.com

Exosomal microRNA are of increasing interest as a system for the exchange of information between tissues, or from the dam to offspring in the case of milk exosomes. The objective of this study is to evaluate the change in miRNA composition of porcine milk and related piglets plasma exosomes due to different diets. Exosomes were isolated from 24 milk samples from sow's receiving different ratios of n-3:n-6 fatty acids and 24 plasma samples from their piglets in postweaning diets supplemented or not with *Ascophyllum nodosum*, using ExoquickTM Precipitation Solution according to manufacturer's instructions. Nanoparticle tracking analysis confirmed the presence of exosomes, showing particles size of 132 nm. RNAs were purified using the miRNeasy serum/plasma kit in an automated way using the QIAcube instrument. Small RNA libraries were prepared using the Small RNA Library Preparation Kit, starting from 20 ng RNA and using 15 PCR cycles, and each library analysed with High Sensitivity DNA chip using Agilent 2100 Bioanalyzer. Libraries were pooled by 12 each and concentrated with AMPure XP magnetic to 1.6 pM and run on the NextSeq500 sequencer, according to manufacturer instruction, in 75 nts single end sequencing mode. Data analysis was performed as implemented in the docker4seq package in Rstudio. Milk exosomes at day (d) 21 showed a total of 33 differentially expressed miRNAs, but not at d14 of treatment. Plasma exosomes showed a differential expression analysis at d7 after weaning and treatment in two miRNAs, and 7 differentially expressed miRNAs at 21 d of treatment. Enrichment of specific miRNA in the exosomal miRNA asset of plasma and milk show exosomes indicate dietary influences, revealing not only a potential new route for feeding supplements but pointing also to the physiological importance, for understanding the transfer of information from mother to child.

Metabolic fingerprinting of calves' faeces in response to lipopolysaccharide challenge

S. Kamel Oroumieh[1,2], A.A. Naserian[1], R. Valizadeh[1], L. Van Meulebroek[2] and L. Vanhaecke[2]
[1]Ferdowsi University of Mashhad, Department of Animal Science, Mashhad, 91775-1163, Iran, [2]Ghent University, Faculty of Veterinary Medicine, Laboratory of Chemical Analysis, Merelbeke, 9820, Belgium; saeid.kameloroumieh@ugent.be

One of the major causes of economic disadvantages in livestock production is neonatal calves' diseases and mortality. Mortality in neonatal calves has been mostly associated with infectious agents. Information about the causes of diseases and factors influencing mortality is an essential concern in identifying opportunities to enhance the health status of calves. In this context, metabolomics methods have been shown to successfully achieve biomarker identification, response to environmental stressors, etc. In this study, lipopolysaccharide (LPS)-induced acute phase response in calves was employed as a model to investigate the influence of acute inflammatory disease on calves' metabolome. An untargeted mass spectrometry (MS)-based metabolomics approach was used to evaluate healthy versus infected calves associated with metabolic fingerprints following an *in vivo* LPS challenge. A total of 12 male Holstein calves were randomised into two groups: control (n=6) and LPS challenge (n=6). The milk consumption, diet, environment, and other conditions of the experiment were the same between groups. Faeces samples were collected a week before LPS challenge as day 0 and three days after LPS challenge at days 9, 10, and 11. Compound discover 3.0 software generated 9,650 ions for positive and negative ionisation mode. The PCA-X score plots revealed good clustering for the faecal samples between control and LPS groups. Datasets were validated by CV-ANOVA (P<0.05) and the permutation test. The OPLS-DA model for the LPS group between days 0 and 9 was valid: $R^2Y>0.86$, $Q^2>0.57$ and P<0.05. S-plots were created to retain metabolites that were explicitly associated with the LPS challenge, whereby the VIP-value was set at >1.0. In total, 78 LPS challenge associated metabolites were retained, which will be discussed. In conclusion, the used MS-based untargeted metabolomics platform demonstrated to be a powerful platform for the exposure of specific metabolites affected by acute inflammatory diseases in neonatal calves.

In vitro uptake of bovine milk-derived exosomes by monocytes

G. Avila[1], R. Ferreira[2], D. Pravettoni[1], G. Sala[1], S. Di Mauro[1], D. De Leonardis[1], S. Arcuri[1], T.A.L. Brevini[1], V. Mrljak[2], H. Sauerwein[3], C. Lecchi[1] and F. Ceciliani[1]

[1]Università degli Studi di Milano, Via dell'Università 6, 26900, Lodi, Italy, [2]University of Zagreb, Heinzelova 55, 10000 Zagreb, Croatia, [3]Universität Bonn, Katzenburgweg 7-9, 53115 Bonn, Germany; fabrizio.ceciliani@unimi.it

Exosomes are nanovesicles (30-100 nm) with an endosome-derived membrane that modulate cell communication and short and longer-range signalling events by their ability to transfer proteins, lipids, DNA, RNA and metabolites between different cell types. Milk exosomes (ME) can enter circulating immune cells and exert immunomodulatory effects, likely after uptake by their target cells. So far, no information of exosome uptake by bovine immune cells is available. This study aimed to demonstrate the in vitro uptake of bovine ME by bovine monocytes (CD14+). Monocytes were isolated from the blood of 2 healthy animals, using magnetic-activated cell sorting technique (MACS), and 5E5 cells seeded on poly-D-lysine-treated 24-well plates. ME were isolated by ultracentrifugation coupled with size exclusion chromatography and endorsed by the presence of exosomes markers (TSG101 and CD9) in Western Blot. ME membrane was stained with PKH26 red fluorescent cell linker, following manufacturer's instructions. To measure exosome uptake, cells were incubated with 2 concentrations of the stained exosomes (1.2E6 and 1.6E9) or without (negative control) for 12 h and 24 h at 39 °C and 5% CO_2. Cells' nuclei were stained with Hoechst immediately before imaging and observed with an inverted wide-field fluorescence microscope (Nikon Eclipse TE200) at 40×. Fluorescence intensity was measured using ImageJ software. Higher fluorescence intensity was detected at 24 h in cells treated with 1.6E9 exosomes compared to the other two treatments. These results demonstrate that bovine ME are indeed uptaken by bovine monocytes, suggesting an important role in the dam-to-calf transmission of regulatory molecules and immunomodulation. The potential immunomodulatory role on bovine monocytes should be further confirmed by evaluating their in vitro impact on the cells' immune-related activities (e.g. chemotaxis, oxidative burst, phagocytosis).

The effects sub-acute ruminal acidosis on plasma and milk sphingolipids of dairy cows

J.E. Rico[1], E. Sandri[2], J. Levesque[2], A. Kenez[3] and D.E. Rico[2]

[1]Cornell University, 345 Morrison Hall, 14850 Ithaca NY, USA, [2]CRSAD, 120 chemin du Roy, Deschambault, G0A1S0, Canada, [3]Hong Kong City University, Room 505, 852, Hong Kong; jer358@cornell.edu

Bovine milk is a significant source of sphingolipids, dietary compounds that can exert anti-inflammatory actions, and which can modulate the host microbiome. Because sphingolipid synthesis can be modified by diet, we hypothesised that dietary conditions which induce subacute ruminal acidosis may result in simultaneous changes in plasma and milk sphingolipid. Twelve ruminally cannulated cows (120±52 DIM; 35.5±8.9 kg of milk/d; mean ± SD) were randomly assigned to treatment in a Latin square design with 21-d periods. Treatments were 1) SARA induction, 2) control. SARA was induced by feeding a diet containing 29% starch, 24% NDF, and 2.8% fatty acids (FA), whereas the recovery and control diets contained 20% starch, 31% NDF, and 2.3% FA. Plasma and milk samples were obtained on d 21 of each period. A methanol-chloroform extraction was used to separate sphingolipids in plasma and milk. Targeted lipidomics analyses were performed using UPLC-MRM/MS, operated in positive ion mode. Univariate and multivariate analyses of generalised log-transformed and Pareto-scaled data included ANOVA (fixed effects of treatment) and discriminant analysis. Our targeted analysis detected 71 sphingolipids across plasma and milk fat including sphinganine (n=3), dihydro-ceramide (n=8), ceramide (Cer; n=15), sphingomyelin (SM; n=17), and glycosylated ceramides (n=28). Followed by Cer, SM were the most abundant sphingolipids detected in milk and plasma, with a preponderance of 16:0-, 23:0-, and 24:0-carbon sidechains. Although no effects of SARA induction were observed on plasma sphingolipids (all FDR>0.25), we observed consistent reductions in the concentrations of several milk Cer (e.g. 22:0-, and 24:0-Cer; FDR<0.05) and SM (17:0-, 23:0-SM; FDR<0.05) in response to SARA induction. Discriminant analysis revealed distinct metabolite separation of SARA induction and control treatments, with several Cer and SM being distinctively predictive of dietary treatment (i.e. SARA induction or control; VIP scores = 1.5-1.9). We conclude that dietary changes that induce SARA can reduce the secretion of milk Cer and SM, even in the absence of changes in circulating sphingolipids.

In vitro impact of conjugated linoleic acid (CLA) on bovine monocytes' immune response
G. Ávila, C. Catozzi, D. Pravettoni, G. Sala, G. Meroni, P. Martino, C. Lecchi and F. Ceciliani
Università Degli Studi di Milano, Dipartimento di Medicina Veterinaria, Via Celoria 10, 20133 Milano, Italy;
gby31193@gmail.com

CLA are a group of naturally occurring isomers of the essential fatty acid linoleic acid. CLA supplementation in dairy cows' diet has been an important nutrition strategy for mitigating the negative energy balance and improving milk quality and yield. However, there is little information about CLA impact on bovine immune cells. This study aimed to determine the immunomodulatory effects of two CLA isomers, 9c,11t and 10t,12c, on bovine monocytes' (CD14+) immune functions. PBMCs and monocytes isolated from whole blood of healthy cows were incubated with increasing concentrations (10, 50, 100, 500 μM) of each CLA isomer and their effects on cells' apoptosis and viability were assessed. Apoptosis was determined by measuring the activity of caspase-3 and -7, while viability using a MTT-based assay. Statistical analyses were performed in GraphPad Prism 8.0.2, using repeated measures one-way ANOVA for normal distributed samples and repeated measures Friedman test for not normally distributed ones. PBMCs apoptosis was reduced only by 9c,11t-CLA at 10 μM and 50 μM, while viability by both isomers at 500 μM. No effects on sorted monocytes' apoptosis were observed. This study also compared the effects of CLA isomers (50 μM), individually and in the mixture (50:50), with other unsaturated (linoleic acid) and saturated fatty acid controls (stearic acid) on monocytes' apoptosis, chemotaxis, ROS production, phagocytosis and killing capability. Chemotaxis was evaluated using transwell migration plates, while ROS production was determined by cytochrome C reduction assay. Phagocytosis and killing capability were determined by fluorescein-labelled *E.coli* incorporation and intracellular killing of *Escherichia coli*, respectively. Only the CLA mixture reduced monocytes apoptosis and increased ROS production under pro-inflammatory conditions. No effects on monocytes' phagocytosis and killing capability were observed. Remarkably, stearic acid reduced chemotaxis. This study demonstrates that CLA exerts an anti-apoptotic activity, and can increase ROS production in an inflammatory *in vitro* model, suggesting that it may have important roles in modulating some *in vivo* monocyte's immune functions.

Hepatic oxidative status of mature beef cows grazing different herbage allowances of native pastures
A. Casal[1], M. Garcia-Roche[2,3], A. Cassina[3], P. Soca[1] and M. Carriquiry[2]
[1]Facultad de Agronomia, Universidad de la Republica, Departamento de Produccion Animal y Pasturas, Ruta 3 km 363, 60000 Paysandu, Uruguay, [2]Facultad de Agronomia, Universidad de la Republica, Departamento de Produccion Animal y Pasturas, Av Garzon 780, 12300 Montevideo, Uruguay, [3]Facultad de Medicina, Universidad de la Republica, Bioquimica, CEINBIO, Gral Flores 2412, Montevideo, Uruguay; alcas@adinet.com.uy

The aim of this study was to evaluate the effect of two herbage allowances (HA) of rangelands (Campos biome) on hepatic mitochondrial oxidative stress markers of purebred (PU) and the reciprocal F1 crossbred (CR) beef cows. Mature cows (n=32) were used in a complete randomised block design with a factorial arrangement of HA (2.5 vs 4 kg dry matter/day; LO vs HI) and cow genotype (CG). The experiment was conducted during three years and at the end of the third year, cows were slaughtered at 190±10 days postpartum. Liver was dissected, weighed, and samples collected and snap-frozen in liquid nitrogen and stored at -80 °C until analyses. Thiobarbituric acid reactive species (TBARS), protein carbonyls and activity of antioxidant enzymes [glutathione peroxidase (GPX) and superoxide dismutase (SOD)] were determined spectrophotometrically. Relative gene expression of antioxidant enzymes [GPX1, GPX3, GPX4, SOD1, SOD2 mRNA] were determined using SYBR-green real time qPCR. Proteins expressions 4-hydroxynonenal protein adduct (4-HNE) was determined by western blot. Data were analysed in a mixed model using HA, CG, and their interactions fixed effects and block as a random effect. Cow-calf efficiency (g or KJ calf retained energy/MJ cow ME intake) was greater or tended (P≤0.07) to be greater for HI than LO cows and CR than in PU cows. Hepatic TBARS and protein carbonyl concentrations were not affected by HA treatment, CG or their interaction but hepatic 4-HNE expression tended to be greater (P=0.06) for CR than PU cows and tended to be affected by the interaction between HA treatment and CG (P=0.06) as it was greater (P<0.05) in LO-CR than LO-PU cows while it did not differ from HI-CR and HI-PU cows. Neither the hepatic SOD and GPX enzyme activity, protein and gene expression were affected by HA, CG or their interaction. Contrary to what was expected the greater efficiency for CR than PU cows was associated to an increased hepatic oxidative damage, which would probably reflect a greater metabolic activity of this organ in CR than PU cows, particularly in restrictive environments (LO-HA).

The effect of anticoagulant and time and temperature of storage of sheep blood on energy metabolites
F.R. Dunshea
The University of Melbourne, Faculty of Veterinary and Agricultural Sciences, Royal Parade, Parkville 3010, Australia;
fdunshea@unimelb.edu.au

Blood obtained from sheep in the field to assess energy status can't always be processed quickly and may be stored at elevated temperatures for some time. Also, the dogma is that a glycolytic inhibitor is necessary to stop glucose breakdown and that heparin increases triglyceride hydrolysis thereby increasing non-esterified fatty acid (NEFA) concentrations. Therefore, the hypothesis to be tested in the present study was that storage of blood at 4 °C for 24 h would maintain glucose and NEFA concentrations regardless of anti-coagulant. To test this, blood (2× 30 ml) was obtained via jugular cannulae into syringes containing either di-potassium EDTA (1.8 mg/ml blood) or lithium heparin (20 IU/ml blood) from 10 sheep that had been fasted overnight. Five sheep had blood collected into the heparin syringe first followed by the EDTA syringe and this order was reversed for the other sheep. Within anticoagulant, blood was pooled and mixed and maintained on ice before being rapidly allocated to 5 ml tubes for the respective storage conditions. The respective conditions were time (0, 1, 2, 3, 4, 6, 8, 12 and 24 h) and temperature (4, 20 and 37 °C) as whole blood. At the end of each time point blood was centrifuged at 4 °C and 3,000 rpm and plasma harvested then frozen until analysis within a week. For both glucose and NEFA the main and interactive effects of anticoagulant, time and temperature were all significant (P<0.001). Plasma glucose was stable in blood stored at 4 °C (4.25, 4.25, 4.26, 4.38, 4.38, 4.40, 4.35, 4.18 and 4.18 mM at 0, 1, 2, 3, 4, 6, 8, 12 and 24 h, P>0.10) and not different between anticoagulants (4.31 vs 4.27 mM for EDTA and heparin, P>0.10). Glucose was stable for 2 h in blood stored at 20 °C, after which it decreased linearly with time reaching 1.20 mM after 24 h regardless of anticoagulant. Glucose decreased linearly with time in blood stored at 37 °C being exhausted after 12 and 24 h for heparin and EDTA blood, respectively. Plasma NEFA were stable in blood stored at 4 °C (0.27, 0.27, 0.28, 0.28, 0.28, 0.29, 0.29, 0.28 and 0.28 mM at 0, 1, 2, 3, 4, 6, 8, 12 and 24 h, P>0.10) and not different between anticoagulants (0.28 vs 0.29 mM for EDTA and heparin, P>0.10). NEFA were stable for 4 h in blood stored at 20 °C after which they increased linearly with time reaching 0.36 mM after 24 h in heparin but remained stable in EDTA. NEFA increased linearly with time in blood stored at 37 °C with the increase being more rapid in heparin. In conclusion blood can be safely collected with either EDTA or heparin and stored at 4 °C for upto 24 h before processing for glucose and NEFA. Greater care should be taken when plasma is kept at beyond 2 h and 20 °C, particularly if heparin is used.

Biomarkers of inflammation and gut permeability in dairy cows with or without dietary resin acids
N. Qin[1], M. Niku[1], S. Junnikkala[1], J. Vuorenmaa[2] and H. Kettunen[2]
[1]University of Helsinki, Department of Veterinary Biosciences, Agnes Sjöbergin katu 2, 00014 Helsingin yliopisto, Finland,
[2]Hankkija Ltd, Peltokuumolantie 4, 05801 Hyvinkää, Finland; hannele.kettunen@hankkija.fi

During the periparturient period and early lactation, dairy cows are prone to low-grade systemic inflammation which decreases their productive capacity. Inflammatory processes impair intestinal barrier functions and lead to increased gut permeability. The present study investigated the effects of coniferous resin acids on biomarkers of intestinal permeability and systemic inflammation in dairy cows during the first 10 weeks of lactation. Thirty-six Nordic Red cows were assigned to a control diet (CON), resin acid concentrate -supplemented diet (RAC), and a tall oil fatty acid -supplemented diet (TOFA) by a randomised block design from three weeks prior to the predicted parturition until ten weeks postpartum. In both treatments, the daily dose of resin acids was adjusted to 0.65 g per cow. The cows were sampled for blood from the tail vein at wk 2, 3, 6, and 10 postpartum. The following parameters were analysed from blood plasma using commercial ELISA kits: interleukin-1 beta (IL-1β), interleukin-6 (IL-6), interleukin-8 (IL-8), interleukin-10 (IL-10), tumour necrosis factor alpha (TNF-α), intestinal fatty acid-binding protein (I-FABP), lipopolysaccharide-binding protein (LBP), and serum amyloid A (SAA). The log-2 transformed data were analysed with repeated-measures ANOVA. The effect of time was statistically significant for all studied parameters, likely reflecting natural changes in systemic inflammation in dairy cows during the first weeks after calving. Neither of the treatments showed significant effects on IL-1β, TNF-α, LBP, or SAA. The gastrointestinal barrier function biomarker I-FABP tended to be decreased by RAC (P<0.1) but was not affected by TOFA. The two pro-inflammatory cytokines IL-6 and IL-8, as well as the anti-inflammatory cytokine IL-10 were slightly elevated by TOFA (P<0.05) but none of them was affected by RAC. The results suggest different mechanism of action for RAC and TOFA. The used level of TOFA resulted in mild immunomodulating effects. The potential of RAC to positively influence intestinal permeability of lactating dairy cows needs to be verified in future experiments.

Effect of a new mitochondriotropic antioxidant on oocyte maturation and embryo production

C. Teixeira[1], C.C. Marques[1], M.C. Baptista[1], J. Pimenta[1,2], J. Teixeira[3], F. Cagide[4], F. Borges[4], L. Montezinho[5], P. Oliveira[3] and R.M.L.N. Pereira[1,2]
[1]National Institute of Agrarian and Veterinarian Research, INIAV, 2000 Vale de Santarém, Portugal, [2]CIISA, University of Lisbon, Lisboa, Portugal, [3]MitoXT, CNC- University of Coimbra, Coimbra, Portugal, [4]CIQUP, University of Porto, Porto, Portugal, [5]CIVG, Vasco da Gama University School, Coimbra, Portugal; rosa.linoneto@iniav.pt

Reactive oxygen species (ROS) are byproducts of mitochondrial metabolism playing a signalling role under physiological conditions. Under pathological conditions, mitochondrial ROS production increase, leading to oxidative stress, with negative consequences to cells. Gametes and embryos are particularly susceptible to ROS and its regulation is mandatory to the success of assisted reproductive technologies. In this study, we investigated, for the first time, the effect of a new mitochondriotropic antioxidant molecule, AntiOxBEN2, in the prevention of oxidative stress of bovine oocytes and embryos. The new molecule was supplemented to the *in vitro* maturation medium of oocytes (n=1,651) at doses of 0 (control), 10, 20, 50 and 100 µM (6 sessions) during 22 h. Mature oocytes were subjected either to *in vitro* fertilisation with bovine capacitated sperm or to evaluation of nuclear maturation (aceto-lacmoid, nuclear dye) and mitochondrial polarisation (JC1, fluorescent dye). Cleavage and embryo development rates were determined. Data were analysed using the PROC GLIMMIX and PROC MIXED. AntiOxBEN2 dose-dependent induced an improvement in nuclear maturation progression ($P<0.05$) with more advanced nuclear stages and less delayed than control. AntiOxBEN2 supplementation increased the cleavage rate in the concentration of 10 µM ($80.1\pm3.0\%$, $P\leq0.003$) compared to the other groups (control=63.3 ± 3.54, 20 µM=61.8 ± 3.6, 50 µM=58.7 ± 3.6 and 100 µM=$58.4\pm3.5\%$) and doubled blastocyst production rate. In conclusion, the novel mitochondria-directed antioxidant, AntiOxBEN2, improved oocyte maturation and embryo production. Although the results are still preliminary, they suggest that the inclusion of mitochondriotropic antioxidants in the maturation medium should be a strategy to implement in the future. Funded by UID/CVT/276/2019, ALT20-03-0246-FEDER000021, PTDC/BIA-MOL/28607/2017, POCI-01-0145-FEDER-028607, PDR2020-101-03112

MitoCow – effects of systemic inflammation on the ratio of leukocytes & mitochondrial DNA copy number

K.D. Seibt and Mitocow Consortium
University of Bonn, Institute for Animal Science, Physiology & Hygiene, Katzenburgweg 7-9, 53115 Bonn, Germany; katharina.seibt@uni-bonn.de

The copy number (cn) of mitochondrial DNA (mtDNA) within a cell may serve as a marker of the cellular capacity for generating energy. The cn of mtDNA in blood is mainly depending on cell type, its developmental stage and total leukocyte counts. The latter increases during inflammation due to the immune response, resulting mainly from recruiting new leukocytes from the bone marrow. We herein aimed at characterising the effects a standardised inflammatory challenge, i.e. injection of bacterial lipopolysaccharide (LPS), in synopsis with leukocyte counts in mid-lactation dairy cows. Moreover, potential alterations of the response by supplementing L-Caritine (CAR) were tested. Half of the cows studied received L-carnitine (CAR, 25 g/d; n=26) from d -42 relative to calving onwards; the cows without supplement served as a control (CON; n=24). Blood samples were collected at day (d) 100 relative to calving. All cows were challenged by i.v. injection of LPS (E. coli, 0.5 µg/kg BW) at d +110. Further samples were collected 48 h after LPS administration. The mtDNA from leukocytes was assessed by multiplex qPCR by amplifying the mt12s rRNA together with β-globin as reference gene. Strict validation criteria were applied. Total leukocyte counts in blood were assessed by an automatic cell analyser (Celltac, Japan). Statistical analyses (ANOVA, t-test) using SPSS were performed with time, treatment and their interactions as fixed effects. The cn of mtDNA per cell was decreased after LPS, whereas the number of leukocytes was increased ($P<0.001$). When calculating the ratio between mtDNA cn and total number of leukocytes, lesser values were obtained 48 h after LPS challenge than before ($P<0.001$) irrespective of CAR supplementation. This indicates that newly recruited leukocytes have less mitochondria per cell and thus 'dilute' mtDNA cn in blood.

Effect of over-conditioning on hepatic expression of genes involved in fatty acid metabolism of cows

M. Hosseini Ghaffari[1], M.T. Alaedin[1], I. Hofs[1], H. Sadri[1], K. Schuh[1], C. Koch[2] and H. Sauerwein[1]
[1]University of Bonn, Institute for Animal Science, Physiology & Hygiene, Katzenburgweg 7, 53115, Germany, [2]Hofgut Neumuehle, Educational and Research Centre for Animal Husbandry, Münchweiler an der Alsenz, 67728, Germany; morteza1@uni-bonn.de

Our objective was to evaluate the effect of body condition around calving on hepatic mRNA expression of genes involved in fatty acid metabolism of dairy cows. Multiparous Holstein cows were initially pre-selected 15 weeks antepartum from the entire herd based on their previous course of body condition score (BCS) and backfat thickness (BFT) and were classified as either normal-conditioned (NBCS, <3.5 BCS or <1.2 cm BFT) or over-conditioned cows (HBCS, >3.75 BCS or >1.4 cm BFT) at dry-off. Liver (n=5/group) biopsies were collected at d -49, +3, +21, +84 relative to calving. RNA was extracted using the Qiagen reagent (Qiagen, Hilden, Germany) from the liver homogenates. The expression of targeted genes in the liver was assayed by RT-qPCR using microfluidics integrated fluidic circuit chips (Fluidigm, San Francisco, CA) in accordance with MIQE guidelines. The mRNA abundance of 43 genes related to lipid metabolism including carnitine metabolism, fatty acid (FA) uptake, intracellular FA binding, mitochondrial and peroxisomal FA oxidation, and ketogenesis was assessed in the liver tissue. Data were analysed using the MIXED procedure of SAS. The results showed that the gene expression of some of the genes from carnitine metabolism (3 from 4 genes) and FA oxidation (11 from 25 genes) was affected by the day of sampling (transition period). Among all selected genes, the mRNA abundance of acyl-CoA dehydrogenase very-long-chain (ACADVL), hydroxyacyl-CoA dehydrogenase trifunctional multienzyme complex subunit alpha (HADHA), and malonyl-CoA decarboxylase (MLYCD) were greater (at d 21 pp) in HBCS compared to NBCS cows (P<0.05). Overall, over-conditioning around calving would impair the complete β-oxidation in cows in early lactation.

Effect of Ruemanol ML supplementation on carcass and meat quality of White New Zealand rabbits

N. Ivanov
Agricultural Institute – Stara Zagora, Technology in Sheep, Stara Zagora, 6000, Bulgaria; n_t_ivanov@abv.bg

The aim of the study was to determine the effect of the supplement Ruemanol ML in feed on the quality of the meat. Subject of our research were animals from the White New Zealand rabbit breed, divided into experimental and control groups. The animals in the experimental group were fed with granulated feed with Ruemanol supplement and the animals in the control group with granulated feed only. After reaching live weight of 3.0-3.5 kg, we slaughtered 10 male rabbits from both groups. We determined the pre-slaughter live weight, hot carcass weight and cold carcass weight and the carcass yield%. We analysed the technological properties and chemical composition of Musculus longissimus thoracis et lumborum. We found that the animals in the experimental group had a unreliable higher carcass yield% and higher cooling losses than the animals in the control group. The pH24 values in the experimental group were higher than those in the control group (P≤0.05). Results for water-holding capacity (WHC) were higher and cooking losses were lower in the experimental group compared to the control group (P≤0.001). We did not find significant and reliable differences between the groups in determining the chemical composition of the taken meat samples.

Quinoa and Linseed supplementation in lambs' diet: immunological profile and meat tenderisation

R. Marino[1], M. Di Corcia[1], G. Annicchiarico[2], A. Della Malva[1], M. Caroprese[1], A. Santillo[1], A. Sevi[1] and M. Albenzio[1]
[1]University of Foggia, Via Napoli, 25, 71122-Foggia, Italy, [2]CREA, Via Appia, 85054 Bella Scalo (PZ), Italy; antonella.dellamalva@unifg.it

A strong relationship between nutrition and animal welfare has been demonstrated. Particularly, dietary integration of linseeds can modulate the immune response under stress conditions in cow and sheep, but little information are available on seeds supplementation in lamb feeding. The present study investigated the effect of different diet supplementation on welfare and meat tenderisation of Italian Merino Lambs. 32 lambs after weaning were divided into 4 homogeneous groups fed with different diet supplementation: control (CO) with no supplementation, linseeds (LS), quinoa seeds (QS) and a combination of quinoa and linseed diets (QS + LS). Cell-mediated immune response (skin test) and glucocorticoid secretion were assessed *in vivo* while Warner Bratzler shear force (WBSF) and proteomic profile using two-dimensional electrophoresis (2DE) coupled to mass-spectrometry were evaluated on longissimus thoracis muscle. Results highlighted that LS and QS supplementation was able to enhance cell-mediated immune response displaying higher skin fold thickness ($P<0.01$) compared to C and LS+Q lambs; in addition the lowest level ($P<0.01$) of cortisol was found in lambs fed with LS supplementation. Greater meat tenderisation was observed in LS group with lowest values of WBSF ($P<0.001$) and degradation of desmin ($P<0.05$) and TnT proteins complex and the highest presence of 32-25 kDa ($P<0.001$) bands. In addition, protein separation with 2DE revealed a major number of spots and phosphorylation isoforms of fast MLC2 patterns in meat from linseed group and a major number of spots ascribed to sarcoplasmic proteins and fragments of MHC in meat obtained by lamb fed quinoa. Data from the present experiment highlighted that diet can affect differently animal welfare and meat quality. In particular linseed supplementation can help the animal to cope with stressful events and to enhance meat quality producing a better meat tenderness.

Chlorella vulgaris in piglet feeding: effect of an enzyme mixture on the intestinal mucosa proteome

C. Martins[1], A. Dittmann[2], D. Ribeiro[1], W. Wolski[2], M. Pinho[3], J.A.M. Prates[3], J.B. Freire[1] and A.M. Almeida[1]
[1]LEAF, Instituto Superior de Agronomia, Univ. de Lisboa, Tapada da Ajuda, 1349-017 Lisboa, Portugal, [2]Functional Genomics Center, Winterthurerstrasse 190, 8057 Zurich, Switzerland, [3]CIISA, Faculdade de Medicina Veterinária, Av. Univ. Técnica, 1300-477 Lisbon, Portugal; catiamartins@isa.ulisboa.pt

The effect of *Chlorella vulgaris* as an ingredient for piglet's diets and its effects on the intestinal mucosa proteome was studied. Thirty post-weaning piglets were divided into 3 groups (n=10) and fed one of the diets: Control (C; standard diet); Chlorella CH: C diet with 5% Chlorella incorporation) and Chlorella and CAZyme (CZ: – CH diet supplemented with 0.1g/kg CAZyme enzyme mixture). At day 14, animals were sacrificed, intestinal mucosa sampled, and morphological microscopy studies conducted. Intestine proteins were extracted and digested to tryptic peptides. For reversed-phase nano-LC-MS/MS and label-free quantification, peptides were separated on an ACQUITY UPLC M-Class System coupled to an Orbitrap Q Exactive HF, operated in data-dependent acquisition mode. Raw mass spectral data files were processed with Max Quant version 1.6.2.3 and searched against Uniprot pig database. Proteins had differential abundance when $P<0.05$ and fold change above 2.0. No differences were found for growth, ADG and Conversion Rate. Duodenal villi height and the villi height/crypt depth ratio were higher in the CH and CZ animals, indicating a higher maturity of the mucosal development triggered by chlorella inclusion. Proteomics analysis identified over 5,100 proteins. PCA analysis revealed no group clustering. Twenty-five proteins showed differential accumulation between the CZ and C animals, higher in the latter, affecting different pathways: regulation of translation, transcription and cell proliferation and ribosomal biogenesis. Proteomics results point out to a decreased accumulation of proteins involved in protein synthesis in the animals of the CZ group by comparison to control, corroborating microscopy results. In conclusion, the use of the microalgae chlorella and the CAZyme supplementation led to similar productive performances as those of the control animals, although significant changes were detected for intestine development and intestinal proteome profiles.

Decision support tools in cattle – from the cradle to the grave
D.P. Berry
Teagasc, Moorepark, P61 C996 Co Cork, Ireland; donagh.berry@teagasc.ie

This invited presentation describes the developments in genomic management tools that are being put in place through the H2020 GenTORE project. This approach has been enabled by the widespread use of genomics and precision farming technologies to estimate animal worth and thereby provide farmers with tools for enhanced breeding and replacement management.

Quantifying resilience of dairy cows from on-farm time-series measures
Y. De Haas[1] and the GenTORE-WP3-consortium[2]
[1]Wageningen University and Research, Animal Breeding and Genomics, P.O. Box 338, 6700 AH Wageningen, the Netherlands, [2]https://www.gentore.eu/wp31.html; yvette.dehaas@wur.nl

We hypothesise that at-market sensor technologies can be used to develop proxies for complex traits as resilience in dairy cattle. This was tested by comparing variables describing sensor data patterns ('curve-parameters') from resilient with non-resilient cows. Sensor data included data from, for example, weighing scales, activity, rumination activity, and milk production. Curve-parameters were the mean, minimum, maximum, standard deviation, slope, skewness, and the autocorrelation for each sensor for cow for each lactation. Data originated from six partners from France, Italy, the Netherlands, Poland, Switzerland, and the UK. Most partners had data from one or a few herds, only the UK-dataset included 17 farms. Because resilience itself is hard to measure, we have chosen to focus on the ability to re-calve, assuming that impaired resilience will negatively impact reproduction performance. To reflect how good (or bad) this ability was compared to herd mates, a point system was introduced based on a cow's lifetime. This point system included five aspects on top of points for each calving: (1) age at first calving compared to the herd average; (2) calving interval compared to the herd average; (3) number of inseminations; (4) number of events; and (5) 305d milk yield compared to herd peers. We thrived for the same definition for resilience for all partners, but points allocated to each aspect could differ between partners. These weights were, therefore, shared among the partners, and each partner individually ranked the cows according to their own resilience score and to a partner's resilience score. Rank correlations between this score and the ones from the other partners were calculated. These correlations were mostly >0.90, but they do vary between 0.5 and 1.0. Finally, the predictive performance was tested with the 33% most and least resilient lactations. All partners showed that the predictive performance of the models was limited, which implies limited value of individual or a combination of at-market sensor technologies to develop proxies for resilience, with the current trait definition and the current curve-parameters.

A new method to estimate RFI in dairy cattle using time-series data

P. Martin[1], V. Ducrocq[1], D.G.M. Gordo[2], P. Faverdin[3] and N.C. Friggens[4]
[1]Université Paris-Saclay, INRAE, AgroParisTech, GABI, Domaine de Vilvert, 78350 Jouy-en-Josas, France, [2]Aarhus University, Center for Quantitative Genetics and Genomics, Department of Molecular Biology and Genetics, Blichers Alle 20, 8830 Tjele, Denmark, [3]Agrocampus Ouest, INRAE, PEGASE, 16 Le Clos Domaine de La Prise, 35590 Saint-Gilles, France, [4]INRAE, AgroParisTech, Université Paris-Saclay, UMR MoSAR, 16 Rue Claude Bernard, 75005, Paris, France; pauline.martin@inrae.fr

In the current economic and environmental context, improving feed efficiency has become of primary importance. In dairy cattle, the usual way to measure feed efficiency is through residual feed intake (RFI). However, this approach, in its classical form, does not take into account the evolution of the RFI components across the lactation, inducing approximations in the results. We present a new approach that incorporates the dynamic dimension of the data. Using a multi-trait random regression model, daily milk production, live weight, dry mater intake and body condition score were investigated across the lactation. Then, at each time point, the estimated variance-covariance matrix of the animal effects can be used to predict an animal effect for intake. Its difference from the actual animal effect for intake, gives an RFI estimation at each time point. This approach was tested on historical data from the Aarhus University experimental farm (1,469 lactations out of 740 cows) that were shared within the GenTORE project. The newly estimated RFI possessed all the characteristics of a traditionally calculated RFI, with a mean at zero at each time point and a phenotypic independence from its predictors. Moreover, this methodology offers new possibilities for exploring RFI changes over the lactation. For instance, the correlation between the averaged RFI over the lactation and RFI at each time point was found to be always positive and above 0.5, and maximum around mid lactation (>0.9). In addition, the model performed reasonably well in the presence of missing data. This approach allows a dynamic estimation of the traits, free from all time-related issues inherent to the traditional RFI methodology. It can be extended and used in a genetic or genomic selection context. More investigations will be conducted pooling data from different farms or with changing diets.

Genomic prediction using data from multiple pure breeds and crossbreds

E. Karaman[1], G. Su[1], I. Croue[2] and M.S. Lund[1]
[1]Aarhus University, Center for Quantitative Genetics and Genomics, Blichers Alle 20, 8830 Tjele, Denmark, [2]ALLICE, 78350, Jouy-en-Josas, France; emre@mbg.au.dk

Crossbreeding emerges as an efficient strategy for dairy cattle breeding to achieve better productivity and robustness at the animal as well as the system level. In dairy cattle populations, where crossbreeding has been used, animals show large diversity in their origins. In rotational crossbreeding, for instance, crossbred dams are mated with purebred sires from different pure breeds, where the genetic composition of crossbred animals is an admixture of the breeds included in the rotation. How to use the data of such individuals in genomic evaluations is still an open question. In this study, we aimed at providing a methodology for the use of data from crossbred individuals with admixed genetic background together with data from multiple pure breeds, for the purpose of genomic evaluations for both purebred and crossbred animals. Using animals with genomic data from three dairy cattle breeds as base populations, a three-breed rotational crossbreeding system was mimicked. A population of crossbred animals and three purebred populations were simulated for nine generations. Phenotypes were simulated for three correlation levels (1.00, 0.50 or 0.25) of QTL effects between the pure breeds, and two heritability levels representing low (0.05) and high (0.40) heritability traits. Traditional within-, across- and multi-breed predictions were compared with those from a model accommodating data of crossbred individuals by using breed origin of alleles, and estimating breed-specific SNP effects (correlated or uncorrelated). We showed that methods using crossbred data and allowing breed-specific SNP effects can lead to higher accuracies than pooling multiple breeds' data assuming a homogeneous population and thereby common SNP effects. Our findings are relevant for breeding programs where crossing is systematically applied, and also for admixed population involving different subpopulations where exchange of genetic material among those is a routine practice.

Natural 15N abundance of animal proteins: a promising biomarker of feed efficiency in beef cattle

G. Cantalapiedra-Hijar[1], I. Ortigues-Marty[1], C. Martin[1], I. Morel[2] and R.J. Dewhurst[3]
[1]INRAE, Université Clermont Auvergne, VetAgro Sup, UMRH, Saint Genes Champanelle, 63122, France, [2]Agroscope, Posieux, 1725, Switzerland, [3]Scotland's Rural College, Edinburgh, EH9 3JG, United Kingdom; gonzalo.cantalapiedra@inrae.fr

Phenotyping animal feed efficiency, the animal's ability to transform feed into food, is challenging because it is time and labor consuming and not always feasible in livestock systems based on pasture. Thus, biomarkers should be developed and validated for a high-throughput phenotyping of feed efficiency in practical conditions. The natural ^{15}N enrichment of animal proteins over the consumed diet ($\Delta^{15}N = \delta^{15}N_{animal} - \delta^{15}N_{diet}$) has recently been proposed as a biomarker of animal feed efficiency. This study aimed to confirm by meta-analysis the potential of $\Delta^{15}N$ to capture the between-animal variation in feed conversion efficiency (FCE; body weight gain/ dry matter intake) in young beef cattle reared in different European conditions. For this, individual data of $\Delta^{15}N$ measured in plasma and FCE of 468 growing-fattening bulls of different pure and cross continental breeds, from 25 different diets, 8 experiments and 3 countries (France, UK and Switzerland) were evaluated by regression analysis. All animals were tested for at least 60 days for FCE and their blood (7 experiments) or muscle tissue (1 experiment) sampled at the end of the test period. Diets were sampled throughout the feed efficiency test. Diets and animal proteins (plasma or muscle) were analysed for their natural ^{15}N abundance ($\delta^{15}N$) and the $\Delta^{15}N$ was calculated for each animal. Two models were used to assess the relationship between $\Delta^{15}N$ and FCE at the individual level. First, a mixed regression model of FCE on $\Delta^{15}N$ with the experiment and diet as random effects, allowing these two effects to be excluded from the explored relationship. For the second model, residuals were first obtained for FCE and $\Delta^{15}N$ after correcting for the experiment and diet effects and then regressed on each other by simple linear regression. For the first approach, the mixed model confirmed that $\Delta^{15}N$ is significantly (P<0.001) and negatively correlated to FCE within each diet and experiment according to the following equation: FCE=0.27 (se 0.025)-0.030 (se 0.0056) × $\Delta^{15}N$ (RSE=0.017). For the second approach, the significant (P<0.001) correlation between both residuals confirmed the results obtained with the first approach with a similar slope: FCE=-0.031(se 0.0027) × $\Delta^{15}N$ (RSE=0.017). Present data confirm that $\Delta^{15}N$ can be used in future to predict between-animal variation in feed efficiency in beef cattle.

Managing genetic diversity to ensure resilience using the IMAGE multi-species SNP arrays

R.P.M.A. Crooijmans[1], R. Gonzalez Prendes[1], M. Tixier-Boichard[2] and H.2020 Image-Consortium[3]
[1]Wageningen University & Research, Animal Breeding and Genomics, Droevendaalsesteeg 1, 6708 PB Wageningen, the Netherlands, [2]University Paris-Saclay, INRAE, AgroParisTech, UMR GABI, 78350 Jouy-en-Josas, France, [3]EU_H2020_IMAGE, http://www.imageh2020.eu/, 78350 Jouy-en-Josas, France; richard.crooijmans@wur.nl

Monitoring genetic variation of animal collections of ex-situ as well as in-situ collections is important to make decisions on which breeds and animals may need to be stored for the future in genebanks. Inbreeding in every species is of major concern but for small populations it is even more critical. Precision mating to ensure optimal use of the genetic diversity will diminish the chance that deleterious alleles are transmitted which will contribute to ensure resilience. At present, gene banks have few molecular data to characterise their collections and compare them with in-situ populations. Different tools are available according to species. Within the IMAGE project (a European H2020-project) we made two multi species SNP arrays for the major farm animal species represented in gene banks, which include cattle, pig, chicken, horse, goat, sheep for the IMAGE001 array and water buffalo, duck, quail, rabbit, bee, and pigeon for the IMAGE002 array. For each species, on average 10K SNPs were selected for the array. Both arrays can capture biodiversity of traditional breeds for each species on the autosomes and sex chromosomes but also harbour ancestral SNPs, mtDNA SNPs, trait related variation and variation in genes detected in QTL regions. For IMAGE001 we included MHC variation for each species. The Affymetrix IMAGE arrays are worldwide available without restriction at a low cost ($19.50 including genotyping) at the major genotype providers which enhance routine use of these arrays. We validated and tested both arrays with 1,920 and 1,152 DNA samples covering over 300 breeds for IMAGE001 and IMAGE002 respectively. The IMAGE portal has been opened to facilitate exploration of gene bank data (https://www.image2020genebank.eu). The MoBPS software (https://github.com/tpook92/MoBPS) has been developed to facilitate the introduction into breeding programs of genetic diversity from gene banks, using pedigree as well as molecular data.

Increasing duration of feed restriction: performance ranking and variability of beef cows' response

A. De La Torre[1], L. Barreto[1], J. Pires[1], B. Sepchat[2], I. Cassar-Malek[1], I. Ortigues-Marty[1] and F. Blanc[1]
[1]INRAE, Université Clermont-Auvergne, Vet-AgroSup, UMRH, Saint-Genès-Champanelle, 63122, France, [2]INRAE, Herbipôle, Saint-Genès-Champanelle, 63122, France; anne.de-la-torre-capitan@inrae.fr

Animal resilience and robustness are two traits of importance for livestock systems and genetic selection. Their phenotyping is essential to evaluate the adaptation of different genotypes to changing environments. The objective was to characterise productive and metabolic responses of primiparous Charolais cows to two feed restrictions (FR) of different duration. After calving, 13 cows (39±2 mo old, 680±42 kg BW at calving) were fed hay *ad libitum* supplemented with concentrate to fulfil their energy and protein requirements. At 55±6 d post calving, cows underwent a succession of FR (50% of their energy requirements) and *ad libitum* intake: 4 d FR (FR4), followed by 17 d *ad libitum* intake, a second 10 d FR (FR10) followed by 18 d *ad libitum* intake. Individual milk yield (MY; weight-suckle-weight method), BW, BCS and plasma non-esterified fatty acids (NEFA), β-OH butyrate, glucose and urea concentrations were measured before, during and after each FR. Among all measured parameters, MY and NEFA concentrations showed the most pronounced dynamic changes. MY and NEFA returned to their initial levels within 3±1 d, independently of FR duration (SAS Proc Mixed with repeated measures analysis). Functional data analysis was applied to MY and NEFA responses to obtain parameters describing deviation and recovery trajectories in response to FR. FR duration did not change the ranking of animal responses, i.e. responder cows in FR4 remained responders in FR10 (P<0.03 to 0.07) as shown by the linear correlations between deviation and recovery trajectories after FR. Individual variability of MY and NEFA was, however, higher during FR10 than FR4 (Levene test, P<0.05). Altogether 1/ parameters derived from MY and NEFA dynamics accounted for individual differences in responses to FR perturbations and 2/ cow differences appeared to be stable when similar FR are applied for a longer period. Further research is warranted to test these parameters as proxies of animal resilience and robustness and if relevant used to improve future breeding strategies. Gentore (H2020) project funded this study.

Identification of longevity predictors in French dairy cattle

R. Rostellato[1], J. Promp[2], H. Leclerc[3], S. Mattalia[2], N.C. Friggens[4], D. Boichard[1] and V. Ducrocq[1]
[1]Université Paris Saclay, INRAE, AgroParisTech, GABI, Domaine de Vilvert, 78350, Jouy-en-Josas, France, [2]Institut de l'Elevage, 149 Rue de Bercy, 75595, Paris, France, [3]Allice, 149 Rue de Bercy, 75595, Paris, France, [4]Université Paris Saclay, INRAE, AgroParisTech, MOSAR, 16 rue Claude Bernard, 75005, Paris, France; roberta.rostellato@inrae.fr

Resilient animals are able to maintain the normal processes when confronted with environmental perturbations, reducing, in the long term, the risk to be culled. Thus, longevity can be proposed as a proxy of resilience. The length of productive life (LPL), defined as the time between the date of the first calving and the date of culling, is the most popular measure for longevity. It requires a long time to be observed and the identification of its predictors is of high interest. The aim of this study was to investigate the effect of production, reproduction, morphology and health traits on LPL, in order to identify predictors of longevity. Data included 278,237 lactations from 122,470 Holstein Friesian cows reared in 640 herds. Two different longevity traits were investigated: true longevity (TL), defined as the ability to avoid any culling, and functional longevity (FL), that is the aptitude to avoid culling for reasons other than production. Survival analysis was performed using a Weibull proportional hazard model including several non-genetic factors, such as herd-year, year-season, herd size variation, and age at the first calving, with or without adjustment for milk production (i.e. to evaluate FL and TL). The investigated traits were added to this model and tested distinctly from the others. Our study revealed that insemination status, calving ease, mastitis, somatic cell count, body condition score, body depth and udder depth had significant relationships with TL and FL and they can be used as predictors of longevity. In addition, we found differences in estimation of risk of culling between TL and FL, suggesting that milk production plays an important role in culling decisions. The identified longevity predictors can be validated extending the methodology used in this study to other bovine production contexts. In addition, they can be exploited to develop a multi-trait model for genetic and genomic evaluation for resilience.

Between-herd variation in cow resilience and relations to management

M. Poppe, H.A. Mulder, C. Kamphuis and R.F. Veerkamp
Wageningen University & Research, Animal Breeding and Genomics, P.O. Box 338, 6700 AH Wageningen, the Netherlands;
marieke.poppe@wur.nl

Resilient cows are minimally affected by disturbances, such as diseases or heat waves, and when affected they quickly recover. Herd management is expected to affect both resilience and the number of disturbances in a herd. The objectives of this study were (1) to study variation in resilience between herds, and (2) to determine associations between resilience at herd level and management parameters. First, for 227,615 primiparous cows an indicator of their resilience was calculated. This indicator was the variance of daily deviations from expected milk yield (LnVar), where a low LnVar indicated a good cow resilience. The data were divided into 9,917 herd-year classes, and for each class the average LnVar, corrected for genetic and year-season effects (herd LnVar), was computed. For these herd-year classes, also management parameters were derived from milk production recording data, such as average somatic cell count, calving interval, and herd size. Correlations between these management parameters and herd LnVar were then calculated, and a multiple regression of the management parameters on herd LnVar was performed. Herd LnVar differed considerably between herd-years; the highest herd LnVar estimate was more than six times larger than the lowest herd LnVar estimate. The correlation between herd LnVar estimates of subsequent years within the same herd was on average 0.69, indicating that within herds resilience was quite consistent between years. The correlations between the management parameters and herd LnVar showed that high herd LnVar (poor herd resilience) was associated with a high proportion of cows with a rumen acidosis indication (r=0.31), high average somatic cell score (r=0.19), low average fat content (r=-0.18), long calving interval (r=0.14), and low survival to second lactation (r=-0.13). These correlations confirm that herds with high herd LnVar have a high proportion of less resilient cows or many disturbances. According to the multiple regression, the proportion of cows with a rumen acidosis indication explained most of the variation in herd LnVar. In conclusion, herd LnVar differed greatly between herds, and low LnVar seems to indicate good herd resilience or few disturbances.

Simulation of genotype-environment interactions on short and long-term feed efficiency in dairy cows

L. Puillet[1], N.C. Friggens[1], V. Ducrocq[2] and P.R. Amer[3]
[1]Université Paris-Saclay, INRAE, AgroParisTech, MoSAR, 75005 Paris, France, [2]Université Paris-Saclay, INRAE, AgroParisTech, GABI, 78350 Jouy-en-Josas, France, [3]AbacusBio Limited, Morray Place, 9010 Dunedin, New Zealand;
laurence.puillet@inrae.fr

Breeding for more efficient cows is a key challenge for the dairy sector. Future selection strategies have to be sustainable, which means not impairing animal resilience in the face of both environmental trends and increasing fluctuations in environmental conditions. A deeper understanding of the role of genotype by environment interactions (G×E) on feed efficiency at different time scales is therefore relevant when optimising a dairy genetic improvement program. To evaluate the impact of G×E on feed efficiency, an individual cow simulation model was used. It simulates the lifetime phenotypic trajectory of a cow depending on her feeding environment. Each cow is defined by 4 input parameters, representing the genetic scaling of underlying biological processes behind feed efficiency. They define the dynamic capacity of the cow to acquire resource and her strategy of resource allocation among biological functions. The model was used as a building-block to generate a simple paternal half sib population structure for the purpose of estimating genetic parameters for observable traits as driven by the combination of the genetic scaling parameters and the environment. Virtual cows were simulated in 4 contrasted environments: high feed availability, moderate stable feed availability, moderate variable feed availability, and low feed availability. For each cow, 2 efficiency traits (ratio between energy invested in milk production and energy acquired) were calculated: short-term (ST) efficiency (2nd lactation) and long-term (LT) efficiency (lifetime). Genetic variance and covariance components of the 8 traits (2 efficiency definitions × 4 environments) were estimated with a random animal model. Genetic correlations between ST efficiency traits, across the 4 environments, were very high (between 0.92 and 0.99). Genetic correlations between LT efficiency traits decreased as the paired environments became more and more different (from 0.81 down to 0.41). Finally, genetic correlations between ST and LT efficiency traits strongly decreased from 0.49 to -0.68 when feed availability decreased.

Selection for environmental variance in rabbits could modify genomic regions affecting resilience

C. Casto-Rebollo[1], M.J. Argente[2], M.L. García[2] and N. Ibáñez-Escriche[1]
[1]Universitat Politècnica de València, Institute for Animal Science and Technology, C/Camino de Vera S/N, 46022 (València-València), Spain, [2]Universidad Miguel Hernández de Elche, Departamento de Tecnología Agroalimentaria, Ctra de Beniel km 3.2, 03312 Orihuela-Alicante, Spain; 3ccasto@gmail.com

Resilience has recently been related with the environmental variance of traits (V_E). Thus, a greater knowledge of the genetic background of V_E could help to understand better the animal resilience. A successful selection experiment in rabbits for high and low V_E of litter size allowed to identify differences in resilience between animals. The low V_E for litter size line seemed to cope better with the environmental disturbances than the high V_E for litter size line. The aim of this study was to identify genomic regions modified by V_E selection and that could affect the animal resilience. For that, genotypes from 91 does of base population, 142 for high and 134 for low V_E of litter size lines were used to identify signatures of selection. The signatures of selection were identified using three complementary analysis: runs of homozygosity (ROH), extended of linkage disequilibrium (VarLD) and fixation index (F_{ST}). A total of 703 signatures of selection not overlapping between methods were identify across the rabbit genome. The enrichment analysis of the genes in these genomic regions highlighted nine biological process related with the inflammatory and stress response and the social behaviour. The regulation of these processes is related with the ability of the animals to cope with the environmental stimuli. Thus, the selection applied for V_E of litter size could have modify genomic regions that might also affect the animal resilience.

Genetic evaluation for maintenance in dairy cattle – inclusion of carcass weight information

T. Mehtiö, A.-M. Leino and M.H. Lidauer
Natural Resources Institute Finland (Luke), Myllytie 1, 31600 Jokioinen, Finland; terhi.mehtio@luke.fi

Improving of feed efficiency in dairy cattle by animal breeding has started in the Nordic countries. One component of the index for Saved Feed is maintenance cost, which is based on the breeding values for metabolic body weight (MBW). However, body weight recording based on heart girth measurements is decreasing and recording based on scales is increasing only slowly. Therefore, the benefit of including correlated traits, like carcass weight, is of interest. In this study we estimated genetic variation and genetic correlations for five traits describing the maintenance cost in dairy cattle: first, second and third parity MBW based on heart girth measurements, carcass weight (CW) and predicted MBW (pMBW) based on the CW and dressing percentage. The data consisted of 31,109 Finnish Holstein and Ayrshire cows of which 10,184 had CW records from 2007 to 2018. The fitted multivariate animal model for MBW traits included fixed effects of calving year-month, calving age, herd × 5-years period and regression function of lactation trajectory nested within breed, and random effects of herd-year, animal and residual. For CW and pMBW traits fixed effects of slaughter year-month, slaughter age, herd × 5-years period and regression function of lactation trajectory nested within breed × parity interaction, and random effects of herd-year, animal and residual were included in the model. Heritability estimates were 0.45, 0.53, 0.56, 0.51 and 0.59 for first, second, third parity MBW, CW and pMBW, respectively. Estimated genetic correlations among MBW traits were strong (>0.95). Genetic correlation between CW and pMBW was also strong (0.97), and the estimated genetic correlations for CW and pMBW with MBW traits were similar and ranged from 0.77 to 0.86. These results indicate that the inclusion of CW or pMBW as correlated trait into the routine genetic evaluation for MBW in Danish, Finnish and Swedish dairy cattle should be considered.

The analysis of livestock behaviour to concomitantly improve health and performance

L. Canario
INRAE, Animal Genetics, GenPhySE, 24 chemin de Borde Rouge, 31324, France; laurianne.canario@inrae.fr

Selective breeding for high performance has negative repercussions on functional traits but such consequences remain lowly quantified. In pig and rabbit populations, the recording of health and welfare traits is limited. So, few genetic correlations between performance and health have been estimated. With the development of new technology, the automatic and standardised recording of livestock behaviour at large scale could be accessible in the future. Therefore, we investigate the possibility to use sensors in an extended period of time to detect perturbations in animal behaviour as an indicator of health and eventually welfare problems. This can be done by analysing changes in activity patterns at the individual or population level, by comparing several records of the same individual spaced or by comparing activity patterns between individuals at one point in time so as to identify patterns that deviate from the normal pattern. We use different sensors to record sow postural activity and assess associations with identified health problems This approach should enable us to establish connections between behaviour, sow health and piglet survival. In addition, a large concern on reducing the prevalence of tail-biting leads us to consider breeding against this damaging behaviour with use of a genetic strategy. We also quantified changes in maternal and progeny behaviour in response to the genetic improvement of maternal traits in both pig and rabbit species. Associations with progeny survival and growth were analysed. Insights into the genetic determinism of resilience that is related to animal capacity of adaptation to the diversity of systems of production are also considered as a mean for improved breeding.

Selection for social genetic effects in purebreds affects behavioural traits in crossbred progeny

B. Nielsen[1], S.P. Turner[2], L.V. Pedersen[1] and B. Ask[1]
[1]SEGES, Pig Research Centre, Axeltorv 3, 1609 Copenhagen, Denmark, [2]SRUC, Roslin Institute Building, Easter Bush, Midlothian EH25 9RG, United Kingdom; BNi@SEGES.DK

In commercial pig production, reducing expression of harmful social behavioural traits including ear manipulation, tail biting, and skin lesions due to fighting is of major interest. Moreover, farmers like domestic animals to be easy to handle. The aim of this experiment was to identify if selection on social breeding values (SBV) for growth rate in purebred animals affects lesions from ear manipulation and fighting in crossbreds, tail biting, and behaviour in a weigh crate. An experiment was designed to have two treatments, i.e. pigs in pens having either low or high SBV. In total, the experiment included 274 pens of DanBred Landrace-Yorkshire castrates with between 12 and 19 (mean 17.3) pigs per pen. All animals were crossbred offspring of purebred animals from a DanBred nucleus herd in Denmark. Data were analysed by ordinary linear regression using ordered categories of behavioural traits in response to mean SBV of parents. The effect on lesions of the interaction between the variation in start weight and SBV and the environmental effects of pen size, herd, year, and month were also included in the model. The prevalence of pigs with ear lesions, frontal lesions due to fighting or tail bites were 40, 89, and 17%. Preliminary results showed that SBV for growth affect lesions from ear manipulation and lesions to the front of the body from fighting. However, the effect of SBV interacted with the variation in weight of pigs in the pens at the time of transfer to finishing pens. In pens with little variation in start weight, a high SBV resulted in few front lesions compared to pens of low SBV. If the variation in start weight of pigs was high, the opposite effect was found. Furthermore, crossbred animals with high SBV were calmer in the weigh crate compared to those with low SBV. SBV for growth estimated in purebreds affect behavioural traits in crossbreds, but phenotypic outcomes are sensitive to the variation in start weight.

Genetic parameters of antibiotics consumption in pigs

W. Gorssen[1], R. Meyermans[1], N. Buys[1], J. Depuydt[2] and S. Janssens[1]
[1]KU Leuven, Livestock Genetics, Department of Biosystems, Kasteelpark Arenberg 30, box 2472, 3001 Leuven, Belgium,
[2]Vlaamse Piétrain Fokkerij vzw, Deinse Horsweg 1, 9031 Drongen, Belgium; wim.gorssen@kuleuven.be

The administration of antibiotics to livestock animals is under heavy debate, mainly due to the emergence of resistant bacteria threatening human health. Consequently, the European Union's (EU) general guideline is to 'use antibiotics as little as possible, but as often as necessary'. Breeding more robust and healthy pigs may reduce antibiotics consumption. This research aimed to investigate genetic parameters of antibiotics consumption in pigs. Data were collected between 2014 and 2019 on crossbred progeny of Piétrain sires from two experimental fattening units of 'Vlaamse Piétrain Fokkerij' (Belgium). Information on antibiotics consumption was available at a pen level. Per treatment, the number of treated pigs, administered volume (cc), commercial name of antibiotic and cause of treatment were recorded. This allowed us to calculate the total number of treatments (N_treatment), total antibiotics used in mg active compound (AB_mg) and used daily dose (UDD) as mg of active compound per kg of pig. Data were collected on 13,205 crossbred progeny (2,027 full-sib pens) from 619 Piétrain sires, where 627 pens (33%) had at least one registered treatment. Genetic parameters were estimated using an animal model with the R-package BreedR. A low heritability was found for AB_mg (7.3±4.1%), while medium heritabilities were found for N_treatment (25.9±5.3%) and UDD (23.3±6.1%). Genetic correlations were high (r_g=0.89-0.93). Hereafter, predictive ability analysis was performed using 5-fold cross-validation, both within genetic groups (randomly masking 1 in every 5 pens per sire) and across genetic groups (randomly masking all pens for one in every 5 sires). Predictive abilities – Pearson correlation of adjusted phenotype with corresponding EBV – were significantly higher than zero for N_treatment and UDD (respectively: r_{within}=0.18 and 0.20; r_{across}=0.08 and 0.16). Our findings suggest that antibiotics consumption in pigs has a genetic component and is moderately heritable on a pen level. Monitoring antibiotics consumption on an individual level in pigs might further increase selection accuracy.

Genetic correlations between commercial longevity and two stayability traits in pigs

B. Poulsen[1], B. Nielsen[1], T. Ostersen[1] and O.F. Christensen[2]
[1]SEGES Svineproduktion, Breeding & Genetics, Axeltorv 3, 1609, Denmark, [2]Aarhus University, Quantitative Genetics and Genomics, Blichers Alle 20, 8830, Denmark; bgp@seges.dk

This study tested the hypothesis that stayability to service after 1st parity is more strongly genetically correlated with longevity in commercial herds when stayability is measured in commercial herds rather than multiplier herds. The analysis was based on farrowing- and service- records from 470,824 sows (189,263 multiplier; 281,561 commercial) and 300 herds (156 multiplier; 144 commercial). Multiplier sows were either purebred Landrace or Yorkshire and commercial sows were mainly rotationally crossbreds between the two breeds. Commercial longevity was defined as age in days when culled (AgeC), and stayability to service after first parity was defined for both commercial sows (StayC) and multiplier sows (StayM). The genetic correlations between AgeC, StayC, and StayM were estimated by REML using linear mixed models. Genetic parameters were estimated separately for Landrace and Yorkshire. In Landrace, the genetic correlations between AgeC and StayC, AgeC and StayM, and StayC and StayM were 0.86±0.02, 0.24±0.05, and 0.34±0.06, respectively. In Yorkshire, the genetic correlations between AgeC and StayC, AgeC and StayM, and StayC and StayM were 0.81±0.03, 0.17±0.05, and 0.18±0.7, respectively. Conclusively, longevity in commercial herds is more strongly correlated with stayability when stayability is measured in commercial herds rather than multiplier herds.

The Ala610Val substitution in the glucocorticoid receptor enhances endotoxin-triggered immune stress

Z.W. Li[1], E. Kanitz[2], M. Tuchscherer[2], A. Tuchscherer[3], N. Trakooljul[1], K. Wimmers[1] and E. Murani[1]
[1]*Leibniz Institute for Farm Animal Biology, Institute of Genome Biology, Wilhelm-Stahl-Allee 2, 18196 Dummerstorf, Germany,* [2]*Leibniz Institute for Farm Animal Biology, Institute of Behavioural Physiology, Wilhelm-Stahl-Allee 2, 18196 Dummerstorf, Germany,* [3]*Leibniz Institute for Farm Animal Biology, Institute of Genetics and Biometry, Wilhelm-Stahl-Allee 2, 18196 Dummerstorf, Germany; li.zhiwei@fbn-dummerstorf.de*

Glucocorticoid receptor (GR) is a transcription factor that plays a key role in feedback regulation of the hypothalamic-pituitary-adrenal (HPA) axis and participates in many biological activities involved in maintaining homeostasis. Previously, our group discovered a natural Ala610Val substitution in the ligand-binding domain of the porcine GR ($GR_{Ala610Val}$) induced by SNP c.1829C>T. The substitution increases GR sensitivity to glucocorticoids and provokes a compensatory reduction of HPA axis activity, including baseline cortisol levels and adrenal mass, in different pig breeds. The physiological adaptations triggered by the substitution on the one hand allow exploration of HPA axis regulation and function in stress responses, and on the other hand open possibility for genetic improvement of stress resilience. The present study thus aims to determine the effect of the $GR_{Ala610Val}$ substitution on performance and health of pigs in response to immune stress triggered by lipopolysaccharide (LPS) and on feedback regulation in response to dexamethasone (DEX) treatment. After combined DEX/LPS challenge, neuroendocrine, metabolic, haematological and inflammatory parameters were measured and analysed using repeated measures ANOVA. The $GR_{Ala610Val}$ substitution reduced baseline cortisol, ACTH, and triglyceride concentration and increased baseline platelet counts. LPS-induced responses, including sickness behaviours, anorexia, thrombocytopenia, cytokine productions, and metabolic alterations were more intense in $GR_{Ala610Val}$ pigs. In addition, endotoxemic $GR_{Ala610Val}$ pigs were more responsive to DEX therapy than wild types. This study revealed that $GR_{Ala610Val}$ enhances sensitivity of pigs to immune stress, and thus may be used in breeding programs to improve health.

Imputation-based detection of variants affecting the porcine immunocompetence

C.M. Dauben[1], C. Große-Brinkhaus[1], E.M. Heuß[1], M.J. Pröll-Cornelissen[1], H. Henne[2], A.K. Appel[2] and E. Tholen[1]
[1]*Institute of Animal Science, University of Bonn, Endenicher Allee 15, 53115 Bonn, Germany,* [2]*BHZP GmbH, An der Wassermühle 8, 21368 Dahlenburg-Ellringen, Germany; cdau@itw.uni-bonn.de*

Animal welfare and health issues are expected to become an essential part of balanced pig breeding. In literature, several approaches to improve the porcine immune response and immune system are discussed. Therefore, a detailed knowledge about the genetic background is mandatory. Next generation sequencing (NGS) becomes increasingly important, especially through cost-saving imputation from a low marker density to a higher density. In this way, NGS data provides benefits in the basic research as well as in the identification of variants which are mainly involved in the expression of diseases. This study aims to identify and validate variants with a biological relevance for the immunocompetence of piglets and growing pigs. In a first step, univariate GWAS was performed on traits characterising the complete and differential blood count as well as for cytokines and haptoglobin. Phenotypes and genotypes on a 60k level were examined in 535 and 461 piglets from Landrace (LR) and Large White (LW), respectively. Promising regions, especially with across- and within-breed effects and putative pleiotropic effects, were identified on SSC5, SSC9, SSC12, SSC15 and SSC16. In a second step, sequence data of 57 key animals, representing the genetic structure of both populations, was used to impute these regions to a higher density in a set of genotyped animals from the LR (n=2,046) and LW (n=2,000) population. Previously conducted imputation on a 60k level indicates high mean proportions of correctly imputed genotypes for LR (96.22%) and LW (95.93%). First NGS results provide insights into a 500 bp window on SSC15 including a SNP around 56 Mb, indicating putative pleiotropic effects on haemoglobin and haematocrit in LW. After alignment with the reference genome Sscrofa 11.1, variant calling detected five variants within a long non-coding RNA. Imputed genotypes determined a completely homozygous variant in LW, which is divergently identified in LR. In conclusion, this study revealed candidate variants with a putative causal effect on mechanisms interacting in immune system processes to enhance the porcine immunocompetence.

More natural calvings in double muscled beef cattle through selection for a larger pelvic area

L.C.M. De Haer[1], G. De Jong[1], I. Kolkman[2] and S. Scheer[3]
[1]CRV, AEU, P.O. Box 454, 6800 AL Arnhem, the Netherlands, [2]A7 Noord Dierenartsen, Nipkowlaan 17, 9207 JA Drachten, the Netherlands, [3]LTO Noord, Projecten, Zwartewaterallee 14, 8031 DX Zwolle, the Netherlands; lydia.de.haer@crv4all.com

In double muscled beef breeds like Belgian Blue (BB) and Improved Red Pied (IRP) almost 90% of all calvings are performed by a caesarean section, to minimise the risk of stillborn calves. The high percentage of caesarean deliveries may raise consumer concerns about animal welfare. To improve the proportion of natural calvings, a breeding value estimation was set up to select for a minimum pelvic height of 20.5 cm for BB and IRP cattle. Measurements were performed with the Rice Pelvimeter on BB and IRP cattle between the age of 18 months and 6 years. The data set consisted of 2,822 records of 2,137 cows on 105 herds, collected from 2006 to 2015. The percentage repeated measurements was 11% for BB and 38% for IRP. Genetic parameters for pelvic height and width were estimated within and across breeds. The models included the random effect for animal and permanent environment. Herd × measurement date was included as fixed effect. Age at measurement (periods of half a year) and days after calving were included as covariables. In the across model breed was added as fixed effect. Means for pelvic height and width were resp. 180 mm and 147 mm for BB and resp. 184 mm and 153 mm for IRP. Differences in height and width between breeds were small. With increasing age there was a gradual increase in size of pelvic height and width, which slows down till an age of 6 years. Heritabilities across breeds were 0.31 for height and 0.38 for width. Heritabilities for pelvic height and width in IRP cows were respectively 0.28 and 0.36, for BB the heritabilities were 0.26 and 0.28. Repeatabilites were 0.57 and 0.54 for height and width in IRP. Genetic correlations between pelvic height and width were low (0.42 to 0.65), therefore, width and height should be considered as different traits. Including both traits in a multi trait breeding value estimation will increase reliabilities compared to a single trait breeding value estimation. Heritabilities and genetic variation of pelvic height and width show that selection for larger pelvic dimensions is possible. Estimated breeding values are sent to the participating farmers. The BB and IRP herdbooks receive breeding values of their sires.

Genetic analysis of multiple birth events in cattle

S. Widmer[1], P. Von Rohr[2], C. Drögemüller[1] and F.R. Seefried[2]
[1]University Bern, Institute of Genetics, Bremgartenstrasse 109a, 3012 Bern, Switzerland, [2]Qualitas AG, Chamerstrasse 56, 6300 Zug, Switzerland; sarah.widmer@vetsuisse.unibe.ch

Twin and multiple births are rare in cattle and have negative impacts on the cow's performance such as longer calving interval and lower conception rate. In addition, the calves are comparably weak and show a reduced survival rate, whereas the cows have more abortions, dystocia and stillbirths. Similarly, the risk for remained placenta, metabolic disorders, displaced abomasum and ketosis for the dam obviously increases with twin and multiple births. Selection against this trait might be desirable especially for dairy cattle. A genetic analysis was performed in three cattle populations. Data was provided by Swiss national breeding organisations for Brown Swiss (BS), Holstein (HO) and Limousin (LI). Descriptive statistics showed that twinning rates were 4.2% for BS and 3.7% for HO, whereas the multiple birth rate for LI was obviously lower (1.7%). This confirmed previous studies showing a higher prevalence in dairy cattle. Model selection using a backwards elimination strategy based on Akaike Information Criterion (AIC) was used to determine the most influential fixed effects for the trait of interest. Parity and season of birth turned out to have an important impact on twin and multiple births in all datasets. Interestingly, this is the first study showing semen sexing as a significant negative effect for the occurrence of twin and multiple births in the two studied dairy breeds (BS+HO). The identified factors such as parity, season and semen sexing were used as fixed effects in a linear mixed model to estimate variance components. Applying a two-trait (direct and maternal) animal model this resulted in estimated heritabilities of 0.17 (direct) and 0.34 (maternal) for BS, 0.22 (direct) and 0.41 (maternal) for HO and 0.08 (direct) and 0.18 (maternal) for LI. Regardless the obvious breed differences these values will allow to include multiple birth events as a new selection criterion. Currently we aim to use the estimated variance components for breeding value prediction including SNP genotypes as additional source of information. Subsequently, a genome-wide association study to identify quantitative trait loci for twin and multiple births will follow.

Superiority in reproductive longevity linked to homogeneity analysed by survival model in mice

N. Formoso-Rafferty[1], J.P. Gutiérrez[2], A. García-Álvarez[2], T. Pérez[3] and I. Cervantes[2]
[1]*Universidad Politécnica de Madrid, E.T.S.I.A.A.B. Dpto. de Producción Agraria, C/ Senda del Rey, 18, 28040 Madrid, Spain,* [2]*Universidad Complutense de Madrid, Facultad de Veterinaria. Dpto. de Producción Animal, Avda. Puerta de Hierro s/n, 28040 Madrid, Spain,* [3]*Universidad Complutense de Madrid, Facultad de Estudios Estadísticos. Dpto. de Estadística y Ciencia de los Datos, Avda. Puerta de Hierro s/n, 28040 Madrid, Spain; nora.formosorafferty@upm.es*

Selecting for homogeneity has been started to be included as one of the selection objectives in genetic breeding programs because it has improved robustness. A divergent selection experiment for birth weight variability in mice during 23 generations has shown that homogeneous animals are less heavy but more effective in terms of litter size, survival and feed efficiency. The aim of this study was to study the reproductive longevity in both divergent lines, as number of parturitions and as time to the end of the reproductive period. Two generations from both lines with an initial number of 43 females and 43 males were mated one to one, and stayed together from the last birth in order to have consecutives parturitions until the end of the productive life. Females were discarded when the time elapsed from the last parturition was higher than 63 days. The number of parturitions in both lines were compared using a Prentice-Williams-Peterson model adjusting by generation and litter size. Time to the end of the reproductive period between both lines was compared fitting a Cox proportional hazard regression model. Low variability line was associated with higher parturitions rate e.g. adjusted hazard ratio was 2.31 (95%CI 1.77 to 3.01). The Cox model showed that low variability females also presented benefits in terms of time to the end of the reproductive period, adjusted hazard ratio was 0.25 (95%CI 0.17 to 0.38). In a Kaplan-Meier curve, the median of reproductive days was 150 in the high variability line while the median was 260 days for the low variability females. According to these preliminary results, the line selected for low variability presented important reproductive advantages. Higher longevity was found in homogeneous animals suggesting higher robustness and animal welfare.

Potential of DNA pooling to include crossbred slaughterhouse data in genomic prediction of purebreds

J. Marjanovic[1], B. De Klerk[2], Y. De Haas[1], B.W. Dibbits[1] and M.N. Aldridge[1]
[1]*Wageningen University & Research, Animal Breeding and Genomics, P.O. Box 338, 6700AH, the Netherlands,* [2]*Cobb Vantress B.V., Koorstraat 2, 5831 GH Boxmeer, the Netherlands; jovana.marjanovic@wur.nl*

The environments for the purebred and commercial crossbred animals are different, potentially resulting in important traits only being expressed in commercial systems. Commercial animals are generally not phenotyped, however health traits are routinely recorded in slaughterhouses on random subsets of animals for welfare and food safety reasons. Such health data could be linked to the breeding population and may add significant information on the performance of purebreds in commercial environments. As an alternative to individual genotyping, DNA pooling can be used to build a cost-effective reference population to link commercial data and breeding animals. Here we test possible pooling strategies by using footpad dermatitis of broilers as a case study. As a first step, we simulated a typical broiler population in order to test if the DNA pools are representative of individuals used to build the pool. Based on the phenotype, we constructed high and low genotype pools of 25, 10, and 5 animals. Individual and pooled genotypes were used to build a G matrix. Genomic relationship between high pools of different sizes ranged from 0.12 to 0.44, and similar values were found for relationship between low pools. Relationship between high and low pools were ~0. Pools of five or ten individuals, had between 95-100% accuracy when matching individuals to the correct genotype pool, but for pools of 25, accuracy was 0%. Therefore, we could clearly differentiate between high and low pools and for pools of smaller size it was easier to allocate individuals. To validate these results, we constructed DNA pools using real slaughterhouse data. Preliminary results show similar genomic relationships between high (low) pools of different sizes as found in simulation and relationship of ~0 between high-low pools. Identifying individuals that were in the pool was tested for pools of 5 and 10 individuals and had 100% accuracy. As a next step we will investigate how accuracy of allocation of individuals to the pool and size of the pool affect the estimates of SNP effects in crossbred population and selection in purebred population by using simulated data.

Genomic scans of selection signatures in a divergent mice population for birth weight variability
M. More[1], J.P. Gutiérrez[2], N. Formoso-Rafferty[3], C. Ojeda[2], F. Goyache[4] and I. Cervantes[2]
[1]Facultad de Zootecnia, Universidad Nacional Agraria La Molina, Lima, Peru, [2]Departamento de Producción Animal, Universidad Complutense de Madrid, Madrid, Spain, [3]Departamento de Producción Agraria, Universidad Politécnica de Madrid, Madrid, Spain, [4]SERIDA-Deva, Gijón, Asturias, Spain; icervantes@vet.ucm.es

A selection experiment for birth weight variability in mice has created two divergent lines: high (H-line) and low (L-line). It has been described that L-line individuals are more productive and better in terms of robustness and animal welfare. A total of 384 DNA samples, 212 L-line and 172 H-line females were genotyping using a high density mouse genotyping array to identify selection signatures on the genome caused by divergent selection. The quality control was performed using the software PLINK v1.90p. All SNPs mapped on the sex chromosomes, SNPs with a genotyping rate lower than 100% and Minor Allele Frequency lower than 0.05 were removed. A total of 115,612 SNPs located on 19 mice autosomes were retained for the whole sample analysed. Three EHH-based statistics, Cross-population Extended Haplotype Homozygosity (XP-EHH), number of segregating sites by length (nSL) and integrated Haplotype Score (iHS) were used to identify genome-wide signatures of selection. The statistics were calculated using the software Selscan v1.2.0a. The results for each locus were normalised over all chromosomes using the software norm. The normalised values higher than |2| were used to identify SNPs under selection. Genome-wide candidate regions for selection were established from: (1) individual SNPs identified by two or more statistics applied; (2) at least two adjacent SNPs identified by different statistics and overlapping upstream or downstream regions of 75 Kbp. Gene-annotation enrichment and functional annotation analyses were performed with BioMart Software (Ensembl Genes 99 database) and DAVID Bioinformatics Resources 6.8. After normalisation 2,876, 2,898 and 3,903 SNPs were considered to be under selection using the XP-EHH, nSL and iHS test, respectively. A total of 314 candidate selection regions on all autosomes were defined. Gene-annotation enrichment analysis allowed significantly identify 19 functional term clusters, including clusters related with immune system process, potentially involved with animal robustness.

Implementation of a litter vitality index into Austrian maternal pig breeding program
C. Pfeiffer[1,2], B. Fuerst-Waltl[1], P. Knapp[2] and A. Willam[1]
[1]University of Natural Resources and Life Sciences Vienna (BOKU), Gregor-Mendel Str. 33, 1180 Vienna, Austria, [2]PIG Austira GmbH, Waldstr. 4, 4641 Steinhaus, Austria; christina.pfeiffer@pig.at

The strong focus of breeding goals for sows on prolificacy negatively affects welfare of sows and piglets. Therefore, the Austrian pig breeding organisation revised the breeding program (BP) for the dam breeds Large White (LW) and Landrace (L) and implemented a litter vitality index (LVI). The aim of the study was to model various calculations focused on different weighting (in %) of LVI in the total merit index (reference scenario Sc0, and Sc10, Sc15, Sc20) to investigate the effect on the annual genetic gain (AGG = additive genetic S.D. × 100) of all other traits in the BP. A complex deterministic approach was used to model four scenarios using the software ZPLAN. A breeding nucleus of 3,640 Large White and 1,560 Landrace sows, assuming a mean generation interval of 2.3 years each and a selection intensity of 1.40 (boars) and 1.05 (sows). The LVI comprises the three traits mean individual birth weight, standard deviation of birth weight and litter vitality with a weighting of 30:50:20. Other traits considered were number of live born and weaned piglets, functional longevity (LONG) of the sow and several fattening, meat and carcass traits. In total, 17 traits were included, their genetic and phenotypic (co)variances were estimated previously. The mean AGG of LW and L for LVI were 5.30, 7.32 and 9.26 in Sc10, Sc15 and Sc20, respectively. Especially LONG benefited from the inclusion of LVI in all scenarios. AGG for LONG were -0.53 (Sc0), 0.99 (Sc10), 1.60 (Sc15) and 2.18 (Sc20). All fattening, meat and carcass traits showed a slight to moderate loss in AGG, but their AGG remained clearly positive. AGG for number of live born piglets were 8.24 (Sc0), 7.15 (Sc10), 6.55 (Sc15) and 5.87 (Sc20). AGG for weaned piglets were 10.15 (Sc0), 10.44 (Sc10), 10.30 (Sc15) and 10.00 (Sc20). Due to results and focusing on sows and piglet's welfare Sc20 was recommended and finally accepted by the pig breeding organisation. Genetic progress can be expected for litter quality and, simultaneously, for LONG, while that for litter size will be slightly reduced.

Evaluation of inbreeding depression for metritis and ovarian cysts in HF cow's based on ROH analysis

P. Topolski, K. Żukowski and A. Żarnecki
National Research Institute of Animal Production, Balice, Poland, Genetic and Breeding of Cattle, ul. Krakowska 1, 32-083 Balice k. Krakowa, Poland; piotr.topolski@izoo.krakow.pl

Runs of homozygosity (ROH) can be defined as contiguous homozygous regions of the genome where haplotypes inherited from the parents are identical. It has been shown, that the length and frequency of ROH may contribute to the level of inbreeding within the population. In turn, a decrease in the mean degree of heterozygosity within a population is almost always paralleled by greater inbreeding depression, which is manifested by a decrease in the mean genetic and phenotypic value of quantitative traits, specifically related to reproduction of cattle. Metritis (ME) and ovarian cysts (OC) are the most common reproductive disorders in HF cows, generating large economic losses in dairy farming. As a material, we used 1,005 cows genotypes collected on four experimental farms belonging to the National Research Institute of Animal Production in Krakow. The genotypes for low and high-density chips with pedigree information were deposited in Polish cattle SNPdb. Moreover, the veterinary history of ME and OC was collected for each cow during their life. The imputation and LD-based SNP pruning preceded the primary analysis. The final set of 43,772 SNPs was used to identification coefficient of genomic inbreeding (FROH) by the using of KING software. Cows were assigned to 3 groups depending on the level of FROH: FROH <5%; 1st group, FROH≥5% & <10%; 2nd group, FROH≥10%; 3rd group. The results shown that in the analysed population, all cows are inbred, with the mean FROH ranging from 0.03 to 16.59%. Our analysis also showed a relationship between the level of FROH and the frequency of occurrence of both diseases. The lowest frequency of occurrence of ME and OC was found in the 1st group of cows and was equal to 9.91 and 7.78%, respectively. The highest incidence of both diseases was found in the 3rd group of cows (ME=14.12%, OC=15.08%).

From herd to individual decision-making on dairy farms: When and how do we evaluate a decision?

J.O. Lehmann, L. Mogensen and T. Kristensen
Aarhus University, Department of Agroecology, Blichers Alle 20, 8830 Tjele, Denmark; jespero.lehmann@agro.au.dk

New technology in combination with more and better data allow for more individualised management including amount and type of concentrate feeding, prediction of when and how to treat or inseminate a cow as well as differentiated moving of cows between groups. This raises a number of principal questions as to what data to include and how to analyse it at the time of decision, how to evaluate a given decision as well as what criteria that measure a successful decision. Do we only evaluate a decision based on information from the individual animal within the same lactation? Do we include the following lactation? Do herd and farm level characteristics influence this evaluation? To exemplify, we turn to the concept of extended lactations where farmers deliberately delay the time of first insemination after calving. Previous research showed a large individual variation in milk production and persistency through the lactation. Hence, there is a potential for a more targeted approach where only a selection of cows have a delayed time of first insemination. Extended lactation represent a case with incremental data availability as a combination of the previous and the current lactations until the time of decision. Furthermore, previous research showed that extended lactations might lead to increased BCS at dry off, which again may be a risk factor for calving complications, diseases and a concomitant decreased milk yield in the following lactation. An optimal current lactation length may not be optimal if effects in the following lactation are included. In addition, extended lactation leads to fewer young stock with concomitant effects for herd structure and feed supply. These effects may need to be included in the evaluation of the decision taken for the individual animal. Hence, the same set of success criteria for the evaluation of a decision to extend a lactation may need to differ across different farm types. We will attempt to explore many of these questions in the coming years where we aim to test an extended lactation strategy in a randomised experiment on up to 50 dairy farms in Denmark that represent a broad range of farming types.

Indoor cow localisation using Bluetooth Low Energy tags

V. Bloch and M. Pastell
Natural Resources Institute Luke (Finland), Latokartanonkaari 9, 00790 Helsinki, Finland; victor.bloch@luke.fi

Monitoring of cow location inside the barn is important for classification of cow behaviour and recognition of diseases. Usage of commercial positioning systems inside a barn is limited by high cost, low battery life or restrictions in connection with satellites. We propose a system for rough cow locating inside a barn based on tags with (Bluetooth Low Energy (BLE) technology. A low-cost tag (RuuviTag, Ruuvi Innovations, Finland, which is generally used for acceleration measuring and behaviour detection) was attached to cow collar. Signal were sent by tags with frequency of 5 Hz. Ten receiving stations (Raspberry Pi microcontroller, Raspberry Pi foundation, UK) distributed in the barn (42×10 m) received the signal and measured the received signal strength (RSS). The RSS was filtered using a Kalman filter. Since the variability of the RSS depending on the tag orientation could reach up to 20 dB, the dependence of the RSS on the orientation was mapped with the resolution of 45 °. The current tag orientation (except of yaw) was calculated based on tag's accelerometer readings. The barn area was covered by 235 predefined points, where the cow location was calculated. The RSS mapping was calculated for all points and all tag orientations. The map was used to calculate the cow location. The location was filtered using the Viterbi algorithm. To validate the method accuracy, an experiment with 12 tags attached to 40 cows (12 cows at a time) for 2 days was conducted. The reference location of cows was found manually from recorded videos. The preliminary accuracy provided by the system was of 3 m in 42% and 5 m in 60% of samplings. The battery life was estimated to be larger than one year based on power profiling of the tag. In conclusion, the system can be used for detecting queues for milking robots, cows with high motion activity, presence in barn and outside, etc. The accuracy of the system can be improved by applying new technologies for the signal transferring.

Using particle filters for online disease detection in dairy cows

M. Pastell, L. Frondelius and H. Lindeberg
Natural Resources Institute Finland (Luke), Production systems, Latokartanonkaari 9, 00790 Helsinki, Finland; matti.pastell@luke.fi

The aim of this study was to develop an unsupervised model for detecting changes in cow's behaviour and production. The hypothesis was that anomalies in measured time series correspond to disease events. Previous research has mainly focused on supervised learning of specific conditions, e.g. lameness. Supervised approach is difficult to apply for rarely occurring health problems as standard monitoring trials do not yield large enough datasets. We used historical data from the same farm to model normal variability in data and developed a method with 3 steps: (1) forecasting the future values of time series of daily milk yield, lying time and feeding time for each cow using a particle filter with Wilmink function as system equation; (2) applying a group level correction to 1 step ahead prediction error 3) Detecting shifts in prediction error using CUSUM charts. The occurrence of alerts was compared to health data collected in a 6 month animal monitoring trial with 49 cows. A total of 5,075 measurement days of data was collected with 38 separate veterinary treatments and 46 severe hoof lesions identified. The models based on a single sensor gave an alert within 21 days before diagnosis in 74-82% (range between alerts from different sensors) of cases requiring veterinary treatment and for 48-75% of hoof lesions. At least one of the three sensors (combined alert) gave an alert in 97% of cases of veterinary treatments and in 93% of severe hoof lesions. In total, 55% of all alerts occurred within 21 days of either a veterinary treatment or a diagnosis of a severe hoof lesion. With this criterion, 45% of total alerts were considered as false. The number of false positive alerts could be reduced by considering only cases where a change is detected in multiple sensors as correct. However, this also lowers detection sensitivity. In practice, the number of false alerts may be unacceptable for a farmer. On the other hand, according to different studies, farmers are only aware of 25-50% of health problems occurring on their farms. Applying the model developed in this study could give a high confidence that majority of problems are identified early with the cost of potentially extra labour of reacting to false alerts.

Data science algorithms for livestock management based on individual feed efficiency

H. Levit[1], I. Halachmi[1] and I. Shimshoni[2]
[1]Agricultural Research Organization-The Volcani Center (A.R.O), Precision Livestock Farming (PLF) Lab. Institute of Agricultural Engineering, HaMaccabim Road 68, 7528809, Israel, [2]The University of Haifa, Department of Information Systems, 199 Aba Khoushy Ave. Mount Carmel, Haifa, Israel; harelle@volcani.agri.gov.il

In commercial farms, feeding cost is one of the highest among dairy's costs (60%-70%). Therefore, every innovation in this field has a critical impact on farm profits and the industry in general. Farmers try to reduce costs by selecting cheaper ingredients in cow meal, but as for now, the management is done on a group level and not on the individual cows. Cow individual dry matter intake is used for calculating efficiency, today, this data is missing but in the future, this information can be presented in feed intake prediction models, based on feeding behaviour that can be collected on a large scale by wearables sensors (3D accelerometers) already exist in the everyday farm usage. By using data science tools the author wants to suggest dairy farm management based on the individual feed efficiency in order to reduce costs and matching the efficient cow phenotype to the genotype.

Ranking cows for methane emission measured with sniffer and GreenFeed systems

A.E. Van Breukelen[1], M.N. Aldridge[1], L. Koning[2], L.B. Sebek[2] and Y. De Haas[1]
[1]Wageningen University and Research, Animal Breeding and Genomics, P.O. Box 338, 6700 AH, Wageningen, the Netherlands, [2]Wageningen University and Research, Animal Nutrition, P.O. Box 338, 6700 AH, Wageningen, the Netherlands; anouk.vanbreukelen@wur.nl

In the Netherlands there is a target to reduce methane (CH_4) emission in livestock by 2.1 megaton by 2030, and further by 2050. Animal breeding techniques offer perspective to reduce enteric methane emission in dairy cows for which large-scale recording of individual cows is required. Combining records from different methods could help to enlarge data sets, however their agreement in ranking cows should be evaluated first. In this study, we compared two non-invasive spot sampling methods on a dairy farms in the Netherlands. We analysed records taken with the GreenFeed (C-lock Inc. Rapid City, SD, USA), and breath analysers (sniffers) installed in the feed bin of automated milking systems. Both methods record short-term breath concentrations. The sniffer method solely measured CH_4 and CO_2 concentrations in ppm and the flux method (GreenFeed) measured CH_4 and CO_2 concentration, head positioning and airflow rate to provide data in g CH_4/cow/day. Cows were ranked based on their average CH_4 emission, for which Pearson's correlation coefficients were estimated between the two methods. Preliminary analysis on one dairy farm in the Netherlands, for 58 dairy cows, showed a moderate positive correlation (0.31±0.13, P=0.02) between cows measured with the two systems. The low correlation most likely resulted from non-overlapping measurement periods, so individual cows are in different lactation stadia and are fed different rations for the two methods of recording. We are confident that correcting the sniffer data for background gases, and analysing overlapping periods of time (which will be available soon for seven farms), will increase the correlation coefficient between the two methods. Additionally, the coefficient of individual agreement between methods will be estimated, to account for individual agreement between multiple measurements. This will allow us to determine if GreenFeed and sniffer data can be combined in future breeding programs.

Potential auxiliary traits for ketosis based on MIR spectra

B. Fuerst-Waltl[1], A. Koeck[2], A. Werner[3], L.M. Dale[3], J. Sölkner[1] and C. Egger-Danner[2]
[1]University of Natural Resources and Life Sciences Vienna (BOKU), Gregor Mendel-Str. 33, 1180 Vienna, Austria,
[2]ZuchtData EDV-Dienstleistungen GmbH, Dresdner Straße 89/B1/18, 1200 Vienna, Austria, [3]LKV Baden Wuerttemberg,
Heinrich Baumann Str. 1-3, 70190 Stuttgart, Germany; birgit.fuerst-waltl@boku.ac.at

In dairy cattle, the reduction of metabolic diseases by both, management and breeding, has gained in importance. Based on routine milk recording samples, midinfrared (MIR) spectrometry may be used to quantify milk components that are associated with ketosis. The LKV (performance recording organisation) Baden-Wuerttemberg has developed two alternative ketosis risk indices (KetoMIR1 and 2, values between 0 and 1) based on MIR data. Within the COMET-Project D4Dairy, data from a previous project, Efficient Cow, were re-evaluated. Both KetoMIR values were used for a multivariate genetic analysis together with the traits clinical ketosis (KET, yes or no), subclinical ketosis (KTEST, milk keto-test performed twice, 0, 1 or 2 positive tests), fat-protein-ratio on test-day 1 (FPR) and body condition score on test-day 1 (BCS, 1 to 5). Only Fleckvieh cows were considered for this analysis. Depending on trait, number of records ranged from 1,806 (KTEST) to 8,679 (KET). The pedigree comprised 20,313 animals. Heritabilities were 0.01 (KET), 0.07 (KTEST), 0.11 (FPR), 0.16 (BCS), 0.19 and 0.09 (KetoMIR1 and 2) and were in the expected range. Genetic correlations between KetoMIR1 and 2 and KET were rather low (0.15 and 0.09), but the ones to KTEST were moderate (0.41 and 0.45). The results indicate that different information sources and traits can be used to genetically improve the metabolic disease resistance.

PLF poster pitches

I. Adriaens
KU Leuven, Department of Biosystems, Kasteelpark Arenberg 30, box 2456, 3001 Heverlee (Leuven), Belgium;
ines.adriaens@kuleuven.be

During this time-slot, the lead authors of the PLF poster presentations will have the opportunity to pitch their posters and work in a 1 minute talk.

Dynamic cooling strategy based on individual animal's responses mitigated heat stress in dairy cows

H. Levit[1], S. Pinto[2,3], T. Amon[2,3], E. Gershon[4], A.K. Elazary[4], V. Bloch[1], Y.B. Meir[4], A. Arnin[5], J. Miron[4] and I. Halachmi[1]

[1]*Agricultural Research Organization – ARO, Volcani Center, Precision Livestock Farming (PLF) Lab. Institute of Agricultural Engineering, 68 Hamaccabim Road, Rishon LeZion 7505101, Israel,* [2]*Leibniz Institute for Agricultural Engineering and Bioeconomy – ATB, Department of Engineering for Livestock Management, Potsdam 14469, Germany,* [3]*College of Veterinary Medicine, Free University Berlin, Institute of Animal Hygiene and Environmental Health, Berlin 14163, Germany,* [4]*Agricultural Research Organization – ARO, Volcani Center, Department of Ruminant Science, 68 Hamaccabim Road, Rishon LeZion 7505101, Israel,* [5]*Hachaklait Veterinary Service, Caesarea, 38900, Israel; harelle@volcani.agri.gov.il*

Modern technological progress enables measuring an individual cow's temperature in real time by utilising a bolus sensor inserted into the rumen (reticulorumen). However, current cooling systems often work at a constant schedule, based on the ambient environment, rather than on the results of monitoring the animal itself. This study hypothesised that tailoring the cooling methods to the cow's thermal state can mitigate heat stress. The objective is to develop a dynamic cooling system, based on in-vivo temperature sensors (bolus). Thus, cooling can be activated as needed and is thus most efficacious. A total of 30 lactating cows were randomly assigned to one of two groups which received two different evaporative cooling regimes: a control group, which received the common method used in farms, at a preset constant schedule, i.e. 'time-based') TB (cooling; and an experimental group, which received the sensor-based (SB) cooling regime. SB was changed weekly according to changes in the cow's body temperature during the previous week, as measured by reticulorumen boluses. We found that the two treatment groups of cows had similar milk yields (44.7 kg/d), but those in the experimental group had higher milk fat (3.65 vs 3.43%), higher milk protein (3.23 vs 3.13%), higher energy corrected milk (ECM, 42.84 vs 41.48 kg/d), higher fat corrected milk 4% (FCM, 42.76 vs 41.34 kg/d), and shorter heat stress time (5.03 vs 9.46 hours/day) compared to the control group. Dry matter intake was higher in the experimental group. Daily visits to the feed trough were less frequent, with each visit lasting longer. The sensor-based cooling regime may be an effective tool to detect and ease heat stress in high-producing dairy cows under summer heat load in arid and semi-arid zones.

The use of precision livestock farming (PLF) technology to monitor and improve welfare of dairy cows

G.V. Berteselli[1], E. Dalla Costa[1], P. Llonch[2], X. Manteca[2], A.H. Stygar[3], J.K. Niemi[3], M. Pastell[3], R. Wijma[4], S.P. Van Gosliga[5], A. Baxter[6], I. Arbel[7] and E. Canali[1]

[1]*Università degli Studi di Milano, via Celoria 10, 20133, Italy,* [2]*Universitat Autònoma de Barcelona, Cerdanyola del Vallès, Spain, Campus de la UAB Bellaterra, 08193, Spain,* [3]*Natural Resources Institute Finland (Luke), Latokartanonkaari 9, Helsinki, 00790, Finland,* [4]*Cooperativa Ganadera del Valle de los Pedroches (COVAP), Calle Mayor 56, Pozoblanco, 14400, Spain,* [5]*Connecterra, Amsterdam, The Netherlands, Singel 542 Amsterdam, 1017 AZ, the Netherlands,* [6]*Syntesa APS, Strandgade 91 4 SAL Kobenhavn, 1401, Denmark,* [7]*Cattle Watch, Smilanski 15, Rehovot, 76446, Israel; greta.berteselli@unimi.it*

Dairy cow farming is one of the most important European livestock production systems and animal welfare is a key for its sustainability. PLF technology enables farmers to monitor animals automatically and offers a real-time managing system. Through the adoption of electronic data collection, processing and application, PLF differs from other approaches that monitor on-farm dairy welfare by human experts scoring animal-based indicators. PLF technology allows to measure different parameters in dairy farming; wearable and environmental sensors allow to assess indicators for animal welfare such as rumination, activity, feed intake. The current challenge of the use of PLF is to integrate these data into relevant information about animal welfare, in order to return a single output to the farmer. ClearFarm aims at developing a platform integrating welfare indicators from automatic recording technologies to provide a single outcome of animal welfare through a blockchain approach. ClearFarm project is structured in 8 interrelated Work Packages (WP),WP3 will validate data from sensor technology as welfare indicators in dairy cattle systems via the development of a novel algorithm, which will be tested on 8 different farms (organic and intensive) in 4 different European countries. The outcome of selected welfare indicators monitored using PLF technologies will be then compared with alternative assessment methods. These results will help in advancing the understanding of dairy welfare on-farm, potentially assisting farmers to optimise production and inform consumers about the welfare background of animal products.

Genetic analysis on infrared predicted minerals for two Danish dairy cattle breeds

R.M. Zaalberg[1], H. Bovenhuis[2], N. Aargaard Poulsen[3], L.B. Bach Larsen[3], J. Sehested[4] and A.J.B. Buitenhuis[1]
[1]Aarhus University, Quantitative Genetics and Genomics, Blichers Alle 20, 8830 Tjele, Denmark, [2]Wageningen University Research, Animal Breeding and Genomics, P.O. Box 338, 6700 AH Wageningen, the Netherlands, [3]Aarhus University, Food Science, Blichers Alle 20, 8830 Tjele, Denmark, [4]Aarhus University, Animal Science, Blichers Alle 20, 8830 Tjele, Denmark; rooszaalberg1990@gmail.com

Fourier transform infrared (FT-IR) milk spectra have been used for decades to determine milk composition. More recently, many other traits have been predicted with milk spectra. With the rising importance of genomic prediction in animal breeding, defining the genetic architecture of IR predicted phenotypes becomes increasingly important. This study aims at analysing the potential of FT-IR milk spectra to predict milk mineral concentrations in two Danish dairy cattle breeds, and to perform a genetic analysis on FT-IR predicted minerals. A principle least squares prediction model was trained and validated with 251 Danish Holstein cows and 166 Danish Jersey cows. For these cows, the minerals Ca, Cu, Fe, K, Mg, Mn, Na, P, Se, and Zn were extracted from skimmed milk by acid sonication, and identified using inductively coupled plasma mass spectrometry. Absorbance values for wavenumbers were available for wavenumbers from the mid-IR region of 3,008-925 cm-1. The prediction model was applied on a genotyped population of 2,203 Danish Holstein cows with 14,988 milk spectral records and 2,565 Danish Jersey cows with 17,352 milk spectral records. The genotyped population had imputed genotypes for 50k single nucleotide polymorphisms. Infrared predicted minerals were analysed with a repeatability model using Bayz software. For Ca, Na, and P the prediction accuracy of the PLS prediction model was >0.6. The genetic analysis estimated a heritability of 0.19 for Ca, 0.12 for Na, and 0.31 for P. A genome wide association study revealed multiple potential quantitative trait loci (QTL) for FT-IR predicted minerals in genomic regions on chromosome 16 and 17 for Ca, on chromosome 19, 20 and 6 for Na, and on chromosome 1, 6 and 16 for P. In conclusion, this study successfully developed FT-IR based prediction models for milk minerals, and identified potential QTL for IR predicted Ca, Na and P in milk.

The effect of pipelines and databases on the analysis of the faecal microbiota of dairy cattle

B. Czech[1,2], J. Szyda[2,3], K. Wang[4], S. Chen[4] and Y. Wang[4]
[1]Roche, Domaniewska 39B, 02-672 Warsaw, Poland, [2]Wroclaw University of Environmental and Life Sciences, Biostatistics Group, Department of Genetics, Kozuchowska 7, 51-631 Wroclaw, Poland, [3]National Research Institute of Animal Production, Krakowska 1, 32-083 Balice, Poland, [4]China Agricultural University, 17 Qinghua E Rd, Haidian District, 100193 Beijing, China, P.R.; bartosz.czech@upwr.edu.pl

16S rRNA amplicon sequencing is a commonly used molecular technique for quantification, identification and classification of microbes in many ecosystems inhabited by bacteria. Rapid development of next generation sequencing techniques affected dynamic development of bioinformatics tools and therefore choosing the proper tools is not trivial. In this study we performed compositional analysis of 138 samples of cow faecal microbiota using 16S rRNA amplicon sequencing. The aim of this study was to evaluate abundance of bacteria in faecal microbiota to understand diversity of bacteria between and within samples using combination of two pipelines and three taxonomy reference databases. We assessed whether the same biological conclusions could be reached using QIIME2 and mothur pipelines with SILVA, RDP and Greengenes databases. The first steps of the analysis were common for each pipeline and included quality control of raw sequencing data (FastQC) as well as trimming and filtering reads (cutadapt). Then, in the case of QIIME2, reads were: (1) de-noised in order to remove and correct wrong reads; (2) de-replicated to reduce repetitions; and (3) clustered to collapse similar sequences. Afterwards, QIIME2 was applied: to create an Amplicon Sequence Variant (ASV) table, to detect and remove chimeric sequences, and finally to classify sequences and to create a taxonomy table. The mothur workflow first focused on determining unique reads, removing chimeric reads, taxonomic classification, removing non-bacterial sequences, and creating taxonomy that was further used to create an Operational Taxonomic Unit (OTU) table that is classified against the training dataset from the SILVA, RDP and Greengenes databases. The comparison of pipelines and relevant databases concentrated on the phylogeny, alpha and beta diversity, as well as on abundance of detected microbes. Dissimilarity between QIIME2 and mothur pipelines were observed.

Economic evaluation of reproductive performance based on differences in pregnancy rate

L. Krpalkova[1], N. O'Mahony[1], A. Carvalho[1], S. Campbell[1], J. Walsh[1], J. Daly[2] and E. Harty[2]
[1]Institute of Technology Tralee, Lero – the Science Foundation Ireland Research Centre for Software, Institute of Technology Tralee, Clash, Tralee, Co. Kerry, V92 CX88, Ireland, [2]Dairymaster, Causeway, Co. Kerry, V92 NWKO, Ireland; lenka.krpalkova@centrum.cz

The overall objective of this study was to compare the reproductive efficiency of dairy cows and evaluate profit of different managements. The hypothesis is that a dairy farm can substantially improve its economic and environmental performance through increasing pregnancy rate, i.e. increasing the number of eligible cows that become pregnant for a given breeding period. This paper presents a tool which was designed to compare the reproduction efficiency between farms with and without individual cow sensors (MooMonitor+, Dairymaster, Kerry, Ireland). The MooMonitor+ system accurately identifies when the animal is in heat. The tool was developed using dynamic programming in R (Shiny) and shows the changes in costs, revenues and net-return projected for a given change in pregnancy rate. The model calculates from the first day in milk and stops when the last calf was born after successful insemination of each cow. Sensitivity analyses demonstrated that the economic return associated with reproductive performance is greatly affected by the input parameters and therefore real farm and market values are crucial. The average economic gain per percentage point of 21 days (21-d) pregnancy rate (PR) was €14.6 per cow/year. The milk price showed the largest impact on the overall net return. A 10% increase in milk price increased the net return in the average to €250 (10% 21-d PR), €269 (20% 21-d PR) and €275 (30% 21-d PR) per cow/year and both evaluated farms. The resulting cumulative net return across different 21-d PR could be informative to farmers to overview the current situation on the farm and to guide them towards better-informed decisions in the future.

Local sourcing in perspective of the European protein balance

G. Van Duinkerken and T.V. Vellinga
Wageningen Livestock Research, De Elst 1, 6708 WD Wageningen, the Netherlands; gert.vanduinkerken@wur.nl

European animal production chains are increasingly focusing on local sourcing, as a result of societal trends and market demands. This trend of local sourcing does not only concern exclusion of protein sources imported from outside Europe, but it also includes the desire to increase the European self-sufficiency for feed materials in general. Overall, 78% of the total use of feed proteins in the EU is originating from EU countries. This includes grass and forages (45% of total feed protein use). In compound feed, the share of European feed materials is lower. Major categories of compound feed materials used in the EU are oilseed meals and cereals, with 26 and 24% of total feed protein use, respectively. The European self-sufficiency rate for compound feed materials is highly variable. For cereals it is 90%, whereas for high protein (30-50% CP/kg) feed materials it is 29%. In the Netherlands in 2018, 65% of the proteins in compound feeds originated from Europe. There are several pathways to further increase European protein self-sufficiency. Low-protein livestock diets and innovations in novel feed proteins are helpful, as well as reduction of yield gaps in protein crop production. An increased use of crop residues and by-products, reintroduction of processed animal proteins in animal feed and an optimised ratio between animal and plant sourced human food enables both circular agrifood production and local sourcing. All pathways, improving feed protein supply and changing consumers' habits, will have an impact on nutrient recycling, environment, biodiversity and carbon footprint of the European animal production and agrifood system as a whole. Both technical aspects and the effects at European level will be discussed.

Effects of microalgae species and inclusion levels on *in vitro* rumen fermentation parameters

D.J. Meehan[1], M.R.G. Maia[2], J. Silva[3], A.R.J. Cabrita[2] and A.J.M. Fonseca[2]
[1]REQUIMTE, LAQV, Universidade de Trás-os-Montes e Alto Douro, Quinta de Prados, 5000-801 Vila Real, Portugal,
[2]REQUIMTE, LAQV, ICBAS, Instituto de Ciências Biomédicas de Abel Salazar, R. Jorge Viterbo Ferreira 228, 4050-313
Porto, Portugal, [3]Allmicroalgae, Natural Products, SA, R. 25 de Abril, 2445-413 Pataias, Portugal; den2mee2@gmail.com

Strategies to improve the sustainability of ruminant production are a hot topic, in line with the United Nations 2030 Agenda. Dietary modulation is one the most efficient as it directly impacts enteric methane production. In this sense, interest has emerged in microalgae as underexploited feed resource. Impacts of microalgae species and inclusion levels on rumen fermentation and methanogenesis are largely unknown. Therefore, this study aimed to evaluate the effects of incremental levels (2.5, 5 and 10%, dry matter) of *Chlorella vulgaris*, *Nannochloropsis oceanica* and *Tetraselmis* sp. incubated with two substrates (haylage and maize silage) on rumen fermentation parameters after 24-h incubation with rumen inocula *in vitro*. No significant interaction effect between substrate, microalgae species and inclusion level was observed. The interaction between algae species and inclusion level only affected methane production and yield, which were reduced by inclusion of *N. oceanica* at 10% and increased by *C. vulgaris* at 10% ($P \leq 0.010$); *Tetraselmis* sp. inclusion levels had no effect on methane. Most fermentation parameters were affected by the substrate and microalgae species interaction ($P < 0.001$). The lowest gas and methane production were observed with *C. vulgaris* incubated with haylage and *N. oceanica* incubated with maize silage ($P < 0.001$). Haylage promoted the lowest volatile fatty acids production with *C. vulgaris* and the highest with *Tetraselmis* sp. ($P < 0.001$). Fermentation efficiency was higher in haylage with *C. vulgaris* and lower with *Tetraselmis* sp. ($P < 0.001$); no effect of microalgae species was observed in maize silage. These results suggest that microalgae can modulate rumen fermentation but the response depends on the species, inclusion level and substrate. Financial support of Fundação para a Ciência e a Tecnologia to LAQV (UIDB/50006/2020) and MRGM (DL 57/2016 – Norma transitória) and of FEDER through NORTE 2020 to DM (NORTE-08-5369-FSE-000040) is acknowledged.

Can the use of fresh mixed forage in dairy cows' diet improve milk production and CH_4 emissions?

D. Enriquez-Hidalgo[1,2], F.C. Pereira[2,3], D.L. Teixeira[4], L.C.P. Machado Filho[3] and S.R.O. Williams[5]
[1]University of Bristol, Bristol Veterinary School, Langford, United Kingdom, [2]Pontificia Universidad Catolica de Chile,
Facultad de Agronomía e Ingeniería Forestal, Santiago, Chile, [3]Universidade Federal de Santa Catarina, Laboratorio de
Etologia Aplicada, Florianópolis, Brazil, [4]Universidad de O'Higgins, Instituto de Ciencias Agronómicas y Veterinarias,
San Fernando, Chile, [5]Agriculture Victoria Research, Ellinbank, Victoria, Australia; daniel.enriquez@uc.cl

The aim of the study was to evaluate the effect of using fresh ryegrass-berseem clover (MIX) and alfalfa hay-maize silage (CON) as forage basis for Total Mixed Ration (TMR) on dairy cows' performance and methane emissions (CH_4). Lactating Holstein Friesian cows (n=24) were blocked according to productive traits and allocated to each TMR in individual stalls for 10 weeks. TMRs were formulated as a 50:50 forage-to-concentrate ratio. TMR samples (2/week were analysed for dry mater (DM), crude protein (CP) and neutral detergent fibber (NDF) contents and DM digestibility (DMD). DM intake (DMI) and milk yield were estimated daily. CH_4 were estimated using the SF6 technique in weeks 7 and 9 (12 cows/week). Milk composition was estimated once in weeks 7 and 9. Data from weeks 7 and 9 were analysed with mixed models. TMRs had different DM (CON: 523, MIX: 392, SEM: 7.0 g/kg, $P < 0.001$), CP (CON: 166, MIX: 143, SEM: 3.6 g/kg DM, $P < 0.01$) and NDF (CON: 318, MIX: 342, SEM: 6.3, $P < 0.05$), but similar DMD content (803 ± 7.3 g/kg DM, P=0.1). Cows had similar DMI (24.3 ± 1.84 kg DM/d, P=0.8) and milk protein content (35.6 ± 1.25 g/kg, P=0.4), but different milk yield (CON: 34.3, MIX: 28.9, SEM: 1.45 kg/d, $P < 0.01$), milk solid yield (CON: 2.27, MIX: 2.02, SEM: 0.081 kg/d, $P < 0.05$) and milk fat content (CON: 29.9, MIX: 37.2, SEM: 2.23 kg/d, $P < 0.05$). There was no effect of diet type on daily CH_4 (357 ± 31.0 g/cow/d, P=0.2), CH_4 intensity (14.6 ± 0.79 g/kg of DMI, P=0.1) or CH_4 yield (11.4 ± 1.29 g/kg of milk, P=0.8; 163 ± 13.3 g/kg of milk solids, P=0.4; 318 ± 31.6 g/kg of milk protein, P=0.9), but did affected g CH_4 per kg of milk fat (CON: 368, MIX: 317; SEM: 23.9 g/kg of milk fat, P=0.05). We conclude that the inclusion of fresh ryegrass-berseem clover as forage basis for TMR on dairy cows reduced milk yield and did not reduce methane emissions.

A feed additive package and proteolytic extract mitigate heat-induced productivity losses in pigs

M. Fornos[1], E. Jiménez-Moreno[1], Y. Lechevestrier[2], G. Álvaro-Minguito[1], M. Le Gall[2], G. Mantovani[3] and D. Carrión[1]
[1]Cargill Animal Nutrition, Polígono Riols s/n, 50170 Mequinenza, Spain, [2]Provimi France, Parc d'activité Ferchaud, 35320 Crevin, Spain, [3]Cargill Animal Nutrition, Via Ripamonti 89, 20136 Milano, Spain; Encarnacion_jimenez@cargill.com

Heat stress, one of the most stressful events in pig production, is responsible of having detrimental consequences to animal's health and performances. The aim of this study was to evaluate the inclusion of a supplement based on heat stress-mitigating feed additives package combined with a fermentation extract with protease activity on growth performance and carcass parameters of growing pigs under commercial heat stress conditions in Spain. A total of 2,400 [Pietrain × (Landrace × Large White)] pigs (50% intact males:50% females) were used with an average initial body weight of 16.4±1.41 kg and allotted to one of two treatments (Control and Treatment group). Treatment group received the control diet with the supplement from 30 to 134 days of fattening. Pigs fed with treatment diet consumed 10 kg more of feed per pig (P<0.05) and had heavier empty carcasses than those fed with control diet (91.3 vs 90.2 kg per pig, P=0.24). Both groups had similar feed efficiency (2.61 vs 2.59 kg of feed consumed per kg of empty carcass; P=0.63). Treatment group registered an improvement in the uniformity of empty carcass weight that ranged between 75 and 105 kg as compared to control group (87.6 vs 81.6%; P<0.05). In conclusion, the use of the novel heat stress- mitigating feed additives package combined with a proteolytic extract into the diet increased feed intake and improved carcass weight uniformity in growing pigs.

Evaluation of guar meal as alternative to current protein sources for pigs, broilers and rabbits

D. Outor-Monteiro, S. Ferreira, S. Botelho, C. Guedes, M.J. Gomes and V. Pinheiro
UTAD, Animal Science, Quinta de Prados, 5000-801 Vila Real, Portugal; divanildo@utad.pt

Reducing the EU dependency on imported protein sources (PrS) for livestock feeding is an important issue today, being necessary to explore alternatives. Guar meal (GM) is a high protein content (35-50%) is a co-product left after guar gum extraction from cluster bean, an annual legume from the sub-tropical regions but with potential to grow in the European Mediterranean region. This study aimed to evaluate the effect of the partial substitution of conventional PrS by GM korma thermally treated (KoT) or not (Ko) on diets fed to pigs, poultry and rabbits. Four diets were used for each species – a control diet, with soybean meal for pigs and poultry (100 and 261 g/kg of feed, respectively) and with sunflower meal for rabbits (149 g/kg of feed), as the main PrS. The remaining diets having 50 and 100 g of KoT/kg of feed (KoT5 and KoT10, respectively) and 50 g of Ko/kg of feed (Ko5), replacing equivalent proportions of the main PrS. Diets were balanced for essential amino acids, with c.a. the same protein and energy value for each species. Studies involved 32 pigs (Large White × Landrace) × Pietrain in individual pens, between 45 and 105 kg of live weight, with 8 replicates per diet; 256 Ross broilers between 0 and 42 days of age, 8 animals/pen, and 8 replicates per diet, and 160 New Zealand × Californian rabbits, 4 rabbits/pen, between 36 (weaning) and 64 days of age, with 10 replicates per diet. Data were analysed using a one-way ANOVA and the means were compared according to the Tukey test. No significant differences (P>0.05) were observed on daily feed intake, except a 9% improvement in pigs fed Ko5 diet. This group and that fed KoT10 also presented a higher (P<0.05) average daily growth (ADG). ADG of rabbits increased (P<0.05) with all GM diets, in the range of 9-11%. The feed conversion ratio was better with Ko5 and KoT10 diets. No differences on ADG were observed in broilers. In general, GM inclusion did not affect digestibility, except in broilers concerning OM, CP and CF fractions, which were lower for Ko5 diet. The digestibility of CF was also lower for diets KoT5 and KoT10 in broilers and for Ko5 in pigs. The results show that GM co-products may be used as protein sources in EU, reliant on the availability.

Life cycle assessment of dairy farming with the uses of biogas plant and low-protein diet

A. Setoguchi[1], K. Oishi[1], K. Kato[1], H. Kumagai[1], Y. Kimura[2], A. Ogino[3] and H. Hirooka[1]
[1]*Graduate School of Agriculture, Kyoto University, Kitashirakawa Oiwake-cho, Sakyo-ku, Kyoto, 606-8502, Japan,* [2]*Hokkaido Research Organization, Naganuma, Hokkaido, 069-1395, Japan,* [3]*National Agriculture and Food Research Organization, 2 Ikenodai, Tsukuba, Ibaraki, 305-0901, Japan; akira@kais.kyoto-u.ac.jp*

Livestock production needs to mitigate environmental impacts, especially from animal manure management, in response to global demand for sustainable development. Under such circumstance, the uses of biogas plant (BP) and low-protein diet (LP) have been developed as effective mitigation strategies for livestock production. The aim of the present study was to assess environmental impacts of dairy farming with the uses of BP and LP by the life cycle assessment (LCA). The production systems were determined based on a model dairy farm with 250 lactating cows in the northern part of Japan. The system boundary included on-farm crop production (hay and corn silage), purchased feed production, feed transport, animal housing, ruminal fermentation and manure management. The outputs from the production systems were evaluated as a herd basis and the functional unit was defined as 1 kg of fat and protein corrected milk (FPCM). For lactating cows, crude protein (CP) content of conventional diet was set to be 17% and that of LP was 13% with rumen-protected amino acids on a dry matter basis. In manure management, slurry storage was assumed to be a conventional treatment. For the use of BP, power generation from biogas and production of bedding material from dewatered digestate were considered. The slurry and digestate composted in manure management were spread on on-farm crop fields and insufficient nutrients were supplied by chemical fertiliser. As the results, the use of BP reduced environmental impact on climate change due to the reduction of methane emission from manure management. The use of LP reduced environmental impacts on acidification and eutrophication due to the reduction of ammonia emission from spreading slurry and digestate on on-farm crop fields, but slightly increased energy consumption by the increase of chemical fertiliser application corresponding to the decrease of nitrogen in manure from the cows fed LP.

Effect of feeding Birdsfoot trefoil on the faecal appearance of condensed tannins in growing lambs

E. Seoni[1,2], M. Rothacher[2], Y. Arrigo[2], S. Ampuero Kragten[2], G. Bee[2] and F. Dohme-Meier[2]
[1]*Sezione di Scienze Zootecniche, Università di Sassari, Via E. De Nicola, 07100 Sassari, Italy,* [2]*Agroscope, Rte de la Tioleyre 4, 1725 Posieux, Switzerland; frigga.dohme-meier@agroscope.admin.ch*

The aim of the study was to investigate the effect of dietary crude protein (CP) level and active and inactive condensed tannins (CT) from Birdsfoot trefoil on faecal appearance of CT in growing lambs. The study was carried out with 24 White Alpine male lambs (27.2±4.1 kg) and arranged according to a 2×2 factorial design. The basal diet consisted of Birdsfoot trefoil silage and hay. In addition, the lambs were offered either a concentrate to meet the predicted protein requirements (BP) or a concentrate supplemented with soybean meal to achieve a CP level 20% above the requirements (HP). To deactivate the effect of CT, polyethylene glycol (PEG) was used in 2 of the 4 diets: BP+PEG (BP+), HP+PEG (HP+) and BP-, HP-. The diets were offered daily as 2 equal meals at 07:00 and 15:00 for 3 h. After an adaptation period of 14 d, lambs were kept in metabolic cages for 7 d. Feed intake, refusals and faeces were collected daily and analysed for the concentrations of total (T-CT), soluble (S-CT), protein-bound (P-CT) and fibre-bound CT (F-CT) using the HCL butanol method. The data were analysed with the MIXED procedure of SAS using dietary CP level, PEG addition and their interaction as fixed effects. The proportions of the individual CT fractions changed from feed to faeces. Independent of dietary CP level (P=0.08) and addition of PEG (P=0.11), 38% of total ingested CT were excreted with the faeces. The smallest portion of the faecal CT was the S-CT. When expressed as a percentage of total CT intake, the faecal S-CT level was greater with PEG addition than without (P=0.05). The excretion of P-CT, which made up the greatest portion of the faecal T-CT, was neither affected by the dietary CP level nor the PEG addition. The F-CT level tended to be greater in BP+ than HP+, with intermediate values in BP- and HP- (dietary CP × PEG addition interaction, P=0.05). This study shows that only a part of the ingested dietary CT can be found in the faeces and that structural changes occur during the passage through the digestive tract.

Effect of dietary seaweed supplementation on bovine milk yield, basic composition and iodine content

E.E. Newton[1], Á.H. Pétursdóttir[2], G. Ríkharðsson[3], C. Beaumal[2,4], K. Giannakopoulou[1] and S. Stergiadis[1]
[1]University of Reading, School of Agriculture, Policy, and Development, Agriculture Building, Earley Gate, P.O. Box 237, RG6 6AR Reading, United Kingdom, [2]Matís, Vínlandsleid 12, 113 Reykjavik, Iceland, [3]BSSL, Agricultural Society of South Iceland, Austurvegur 1, 800 Selfoss, Iceland, [4]Université de Strasbourg, 4 rue Blaise Pascal, CS, 90032, France; eric.newton@pgr.reading.ac.uk

Seaweed is an underexploited animal feed that has recently gained increased attention due to potential benefits such as reducing methane emissions from cattle, modifying the rumen microbiome including a reduction in pathogens, and improving the nutritional value of milk. The aim of this study was to investigate the effect of seaweed supplementation in dairy cow diets on milk yield, basic composition (fat, protein, lactose, somatic cell count), and iodine concentration. 37 Icelandic cows were split into 3 experimental groups: (1) control (CON, n=11, no seaweed); (2) low-seaweed (LSW, n=13, 0.75% seaweed in concentrate, 36-79 g/cow/day); (3) high-seaweed (HSW, n=13, 1.5% seaweed in concentrate, 80-160 g/cow/day). Groups were balanced for milk yield, fat, protein, somatic cell count and stage of lactation. The seaweed mix used was composed of 9% Laminaria digitata and 91% *Ascophyllum nodosum*. All cows were fed a basal diet of grass silage and concentrates (without seaweed) for 3 weeks, then introduced to the experimental diets for 6 weeks, after which they all return to the original basal diet for 3 weeks. Milk yield was recorded daily and averaged per cow per week. Milk from each cow, silage and concentrate samples (n=439), were collected once a week. Milk was analysed for basic and somatic cell count by CombiFoss 6000 (FOSS Electric, Denmark) and milk iodine by inductively coupled plasma mass spectrometry (Agilent 7900, Agilent, Singapore). The data was analysed by a linear mixed effects model, using dietary treatment, week, and their interaction as fixed factors while milk ID was set as a random factor. Seaweed supplementation did not affect milk yield and contents of fat, protein and somatic cell count. However, there was dose-dependent response in iodine content, which was highest in HSW milk (1,658 ug/kg milk), lowest in CON milk (676 ug/kg milk), and intermediate in LSW milk (1,091 ug/kg milk). The differences in iodine concentrations between CON, LSW and HSW milk peaked 3 weeks in the experimental diets (+1,746 and +4,006 ug/kg milk for LSW and HSW, compared with CON, respectively), but this effect was reduced over the next three weeks of supplementation. This work showed that seaweed can be used to produce milk with higher iodine concentrations; a trait that could be used to reduce iodine deficiency in certain regions and populations.

Feeding value potential of sweetpotato plant in livestock diet: a review

I.F. Olaleru[1,2], O.A. Abu[1] and C.O. Okereke[2]
[1]University of Ibadan, Ibadan Oyo State, Animal Science Department, Ibadan, Oyo State, 440231, Nigeria, [2]National Root Crops Research Institute, Umudike, Abia State, Niger, Farming systems Research Program, Umudike, 100265, Nigeria; olaleru.ibikunle@gmail.com

In recent years, there has been growing interest in the use of sweet potato (Ipomoea batatas) plant as a good alternative energy source in rabbit, poultry, pig and other livestock diets due to their great potential, high energy in the root, high protein and phytochemical in the leaves and vines to mitigate against the high cost of conventional animal feed ingredients which has led to household food and nutrition security soaring prices. In Nigeria, sweet potato production is at an annual output of about 3.92 million tons but however the production is 44% below the potential capacity of 7 million tons. Despite Nigeria being considered the world's second largest producer of sweetpotatoes, their role in food and nutrition security is not yet entirely maximised. By reviewing scientific articles and grey literature we sought the beneficial influence of sweet potato in livestock nutrition towards achieving food security. Dietary recommendations of sweetpotato root, leaf and vines and by-products for ruminants, rabbits, poultry and other livestock have been varied considerably. The major factors limiting the efficient utilisation of these by-products in poultry diets include the high fibre and likely presence of antinutritional factors in some of the sweetpotato varieties. Furthermore, the review provides an overview of the significance and influence of cultivar on the composition and pharmacological activities of sweet potato. Sweetpotato contains a lot of beneficial phytochemicals, some of which are peculiar to certain varieties. It's therefore important to continuously evaluate sweet potato cultivars to identify the appropriate phytochemical composition and bioactivities to be able to fully explore the medicinal value of sweet potato. Researches aimed towards isolation, characterisation and toxicological evaluation of its bioactive compounds may help to strengthen and confirm the possible role of sweet potato as a health promoting food and an alternative remedy for chronic diseases. This review highlights the nutritional value and phytochemical composition of sweetpotato, their recommendations for inclusion and utilisation in livestock diets. It was concluded that efficient inclusion of sweetpotato plants and its by-products in livestock diet will reduce feed cost of livestock production and provide alternative usage for the sweetpotato leaf and vines which are always left to rot on the field after harvesting while also creating additional source of income to sweetpotato farmers and processors.

Opportunities of reducing environmental pollution by locally grown soybean meal pig diets

I. Jansons[1], V. Sterna[1] and L. Degola[2]
[1]Institute of Agricultural Resources and Economics, Department of Plant Breeding and Agroecology, Dizzemes', Dizstende, Libagi parish, Talsi County, 3258, Latvia, [2]Latvia University of Life Sciences and Technologies, Institute of Animal Sciences, Liela street 2, Jelgava, 3001, Latvia; imants.jansons@arei.lv

In order to reduce the negative impact on the environment, animal nutrition should be based on nutrients grown in the local areas. The aim of this study was to determine the effect of feeding local and imported soybean meal to fattening pigs and examining it impact on environmental pollution. In the trial was included 40 pigs divided in two groups 20 in control group and 20 in trial group. The diets were designed to be nutritionally equivalent similar in crude protein content and to be isoenergetic for metabolisable energy. Pigs in the control group received imported soybean meal, pigs in the trial group local farm grown in Latvia extruded soybean meal mixed in the compound feed. The faecal samples were taken at the end of study from the rectum in the slaughterhouse, were determined dry matter, total phosphorus, total nitrogen, organic matter, pH. At the final on the age of 190 days, the pigs which fed soybean meal grown in Latvia, showed 111.88±1.79 kg by 3.3% higher live weight than pigs which fed mixed feed with imported soybean meal. The feed conversation in the control group was 2.39 kg/day in trial group was by 6% better. Diets showed no significant effect on pig feed consumption and feed conversion rates between groups. Were detected chemical composition of faeces dray matter was by 2% higher in control group, pH was the same for both groups. Organic matter was by 1.3% lower in the trial group. The total nitrogen content in samples of faeces was in control group 0.72% in the trial group was by 0,08% less respectively 0.64%.The total phosphorus content in the control group was 0.64% in the trial group was 0.59%. The results of the study showed by using soybean meal grown and processed in Latvia were possible to minimise environmental pollution with nitrogen, phosphorus, organic matter although the difference were not significant. The environmental impact of transportation gas emitted during the import process must be taken into account and, overall, the use of local soybean meal significantly reduced environmental pollution.

A blend of cinnamaldehyde, eugenol and capsicum oleoresin improves rumen fermentation *in vitro*

A. Belanche[1], E. Jimenez[1], P. Romero[1], J.M. Palma-Hidalgo[1], A.I. Martin-Garcia[1], A. Blanchart[2], S. Constantin[2], H. Khelil[2] and D.R. Yáñez-Ruiz[1]
[1]Estación Experimental del Zaidin, CSIC, C/ Profesor Albareda, 1, 18008, Granada, Spain, [2]Pancosma SA, Voie-des-Traz 6, 1218, Le Grand Saconnex, Switzerland; david.yanez@eez.csic.es

This study assessed the effect of using two blends of essential oils (mixture of eugenol 9.5%, cinnamaldehyde 5% and capsicum oleoresin 3,5% -COMBO-, or 20% of capsicum oleoresin (CAPS), provided by Pancosma, Switzerland) on rumen fermentation profile, methane (CH_4), protein metabolism, microbial abundance and enzymatic activity in continuous culture fermenters inoculated with rumen fluid from dairy cattle. The treatments were: control (no supplementation, C), the combination (12.5 mg/day; COMBO) and capsicum (3.75 mg/day; CAPS) in two consecutive periods of 15 days (n=4). The diet supplied consisted (16 g) of a 50:50 mix of grass silage and concentrate. Fermentation vessels were sampled two hours after morning feeding on days 14 and 15 to study microbial fermentation parameters. On day 16 two batch-culture fermenters were inoculated from each vessel to determine total gas production and CH_4 concentration over a 24 hours period. Results were analysed using one-way ANOVA (SPSS). Both additives significantly (P<0.001) increased (+10%) total gas production and reduced CH_4 concentration (1.62, 1.41 and 1.48 ml/10 ml, respectively for C, COMBO and CAPS) and numerically increased (P=0.138) total VFA concentration (+8%). However, no effect (P>0.05) was observed on the proportion of the individual VFAs, the abundance of bacteria, protozoa, fungi and archaea or the activity of carboxymetyl-cellulase, xylanase and amylase. The treatment with the COMBO decreased (P<0.001) the concentration of N-NH3 (-60%) and increased (P<0.01) that of N-AA (+80%), whilst no effect was observed on N-peptides. This suggests an inhibition of the deamination process. The study showed the potential of both additives to improve rumen energy metabolism and of the COMBO to increase protein metabolism efficiency in the rumen. These results highlighted new solutions to improve dairy production while reducing the environmental impact. Further studies are needed to test longer periods of supplementation in *in vivo* conditions.

Lupine grain and sorghum distillers' gains improve performance of weaned calves grazing oat pastures
V. Beretta, A. Simeone, A.I. Bergós, J. Errandonea and J. García Pintos
Universidad de la República, Ruta 3 km 363, 60000, Uruguay; beretta@fagro.edu.uy

Oat pastures during autumn-winter grazing, are characterised for their low DM content, high degradable protein and low soluble carbohydrate concentrations, which limits microbial protein synthesis and total energy intake, impairing liveweight gain (LWG). The present study evaluated the effect of supplementing oat pasture (*Avena bizantina*) with different energy and protein sources, on cattle growth performance and supplement utilisation. Thirty-two Hereford female calves (165.1±3.8 kg) were randomly allotted to 8 grazing paddocks and to 1 of 4 treatments (T): a control without supplement (C), or supplementing 1.0 kg/100 kg LW using: a high starch source as ground sorghum grain (SG, CP 9.6%; fat 2%), sorghum dry distillers grains (DG, CP 34%, 55% RUP, fat 5.7%) or 4), or ground lupine grain (LG, *Lupinus angustifolius*, CP 33%, 30% RUP, fat 5.7%). SG and LG, were both low starch (LS) high fat and digestible fibre sources differing in RUP. Calves grazed during 98 days from June 6, 7-day strips at 5% forage allowance, being supplemented at 9 am. Liveweight was recorded every 14 day. Dry matter intake (DMI) of supplement and forage were measured weekly, and supplement conversion ratio (SCR) was calculated as supplement DMI/ LWG response with respect to C treatment. Data were analysed according to a randomised plot design with repeated measures, and means were compared through orthogonal contrasts (supplemented vs C; SG vs LS; DG vs LG). Average pre-grazing DM biomass did not differ between T (2,855 kg/ha, 44.2 cm, P>0.05). No differences were observed in supplement DMI (1.97 kg/d SE 0.04; P>0.05), but total DMI was higher for supplemented T compared to C (5.44 vs 4.10 kg/d; P=0.08) and for LS grains compared to SG (5.8 vs 4.8 kg/d; P=0.08). This was consistent with LWG response. Supplementation increased LWG in 0.310 kg/d (SE 0.04; P<0.01) while replacing SG for LS grains increased LWG in 0.24 kg/d (SE 0.05; P=0.07). Difference between LG and DG (0.07 kg/d SE 0.06) was not significant (P>0.10). Values of SCR were 13.1, 5.6 and 4.9 for SG, DG and LG (P>0.10). Results suggest that weaned calves grazing oat pastures during autumn-winter benefit form supplementation, with higher response for low starch high protein supplements compared to sorghum grain.

Oat grain as a fibre source in high-grain diets for beef cattle: rumen fermentation and digestion
V. Beretta, A. Simeone, O. Bentancur, M.N. Banchero and V. Martínez
Universidad de la República, Ruta 3 km 363, 60000, Uruguay; beretta@fagro.edu.uy

Given the high fibre content of whole oats (WO, NDF=32%) compared to other cereal grains, it could be used as a fibre source for high grain diets fed to cattle as long as it supplies required physical effective fibre (peNDF) to avoid digestive problems. The present study investigated the effect of replacing, on a weight basis, grass hay (GH, *Setaria italica*) with whole oat grain (WO, *Avena byzantina*) in a total mixed ration, on peNDF supply, rumen fermentation and total tract dry matter digestibility (DMD). Four rumen-cannulated steers were randomly allotted, in a cross over design, to two rations balanced for CP supply (14%): (1) 35% GH/ 65% concentrate; (2) 35% WO/ 65% concentrate (61% sorghum grain, 31% sorghum distiller grain, 8% premix). Experiment consisted of two periods (P, 21 d each), corresponding the first 14 d for transition between diets and the last 7 d for measurements. All animals received the two treatments (T), one in each P. Feed was offered at 2.5 kg/ 100 kg body weight (DM basis) in three meals. *In vivo* DMD was estimated by total faecal collection and intake measurement (d1 to d4). WO grain particles in faeces were estimated in a sub-sample by manual separation. Rumen DM degradability (DEG) of rations was measured *in situ*, placing all bag prior to first meal (d5) and removing replicated pairs at 2, 4, 8, 12, 16, 24, 48, and 72 h. Rumen fluid pH was measured every 3 h on d6. Data was analysed according the following general model: $Y_{ijk}= \mu + A_i + P_j + T_k + e_{ijk}$. and means were compared through Tukey test. Replacing GH for WO did not affect (P>0.10) peNDF supply (21.8 vs 20.4% DM), DM intake (13.7 vs 13.4 kg/d) or DMD (66.8 vs 72.4%), respectively. WO kernels in faeces represented 5.7±2.2% of total WO intake. Mean rumen pH was lower for WO (5.9 vs 6.3, P<0.01) varying within 6.1 and 5.8, below de pH range for GH ration (6.6 to 6.1). DM degradable fraction did not differ between T (GH: 74.0 vs WO:71.2, P>0.10), however fermentation rate was higher for WO (0.35 vs 0.09%/h; P=0.02) so that for a passage rate of 4%/h, effective DEG was higher for WO ration (58.3 vs 65.4%; P=0.048). Results suggest it is viable replacing GH for WO as a fibre source up to 35% of ration DM, without affecting feed DM utilisation.

The effects of replacing soybean meal from dairy sheep diets on their milk yield and quality
A.I. Kalogianni[1], M. Moschovas[1], F. Chrysanthakopoulou[2], G. Katsaganis[1], G. Manessis[1], I. Vastardis[3], I. Bossis[1] and A.I. Gelasakis[1]
[1]Agricultural University of Athens, Animal Science, Iera Odos 75 Str., 11855, Athens, Greece, [2]A.C. Agrinio Union, Papaioannou 24 Str., 30131, Agrinio, Greece, [3]DELTA FOODS S.A., 23rd km Nat.Road Athens-Lamia, 14565, Ag.Stefanos, Attica, Greece; afrokalo@aua.gr

The objective was to assess the effects of replacing soybean meal (SOY) with a mixture of canola meal (CAN), cottonseed meal (COT) and Vicia faba L. minor (VFM) seeds on milk yield (MY) and quality traits (MQT) in dairy sheep. A total of 64 Frizarta and 48 Chios purebred ewes at postweaning, from two intensive farms (A and B, respectively) located in western Greece, were enrolled in the study. In each farm, ewes were randomly separated into two equal groups, according to their lactation stage and MY and assigned to either the control (diet A) or the experimental diet (diet B). In the experimental diet, 3/4 of the control diet SOY was replaced by a mixture of CAN, COT and VFM to produce a diet of similar nutritional value. Dry matter consumption was similar for the two groups. Following a 30-day adaptation period, sampling was performed using ICAR (International Committee of Animal Recording)-approved equipment and protocols. Daily milk-(DMY), fat-(DFY), protein-(DPY), lactose-(DLY), SNF-(DSNFY), TS-(DTSY) yields and the log of somatic cells count (LogSCC) were calculated. Age and body condition score (BCS) were also recorded. SPSSv23 and analysis of covariance was used to assess the fixed effects of farm (2 levels), diet (2 levels), age (3 levels, 2, 3 and ≥4 years) as well as LogSCC and BCS (covariates) on MY and MQT. Diet and age effects were not significant in any case, whereas, farm had a significant effect on DLY (P=0.042). Excluding DFY, a negative association between BCS and the studied traits was observed (P<0.05). A 1-log increase in SCC was associated with about 354, 14, 18, 20, 41 and 53 g decrease on DMY, DFY, DPY, DLY, DSNFY and DTSY, respectively (P<0.05). In conclusion, the partial replacement of SOY with a mixture of CAN, COT and VFM did not adversely affect DMY and MQT of dairy ewes. Research is funded by the Agricultural Cooperative 'Agrinio Union' within the Regional Operational Program 'Western Greece 2014-2020-RIS3', Project code: MIS-5040306.

Effect of yeast (*Cyberlindnera jadinii*) as an alternative protein source on nitrogen balance in mink
M. Kjærulf, C. Marcussen, A.-H. Tauson, A.Ø. Pedersen and C.F. Matthiesen
University of Copenhagen, Department of Veterinary and Animal Sciences, Grønnegårdsvej 3, 1870 Frederiksberg C, Denmark; mak@sund.ku.dk

Mink production is based on protein-rich co-products from the fishing industry and slaughterhouses. An increasing competition for protein sources is expected due to growing animal food production. Thus, there is a need to identify new sustainable alternative protein sources. The amino acid composition of the *Cyberlindnera jadinii* yeast closely resembles that of fishmeal. However, contents of methionine and cysteine are lower in this yeast product than in fishmeal, but the yeast has shown potential as an alternative protein source in diets for piglets and carnivorous fish. The objective of this study was to measure the effect of *C. jadinii* yeast as an alternative protein source on nitrogen (N) balance in growing mink. Thirty-two 10 weeks old male mink kits were divided into four dietary treatment groups. The control diet was based on fishmeal (0Y) and in the three experimental diets 10% (10Y), 20% (20Y) and 40% (40Y) of the digested nitrogen (DN) from fishmeal was replaced with yeast-meal, respectively. The diets were iso-nitrogenous and iso-energetic and fed *ad libitum*, once daily. The kits were housed individually in balance cages during the experiment, which consisted of a three-days adaptation period followed by a four-days collection period. The statistical analysis was carried out in SAS version 9.4 using the proc mixed procedure with dietary treatment as fixed effect. Differences were denoted significant if P≤0.05 and as a tendency when 0.05<P≤0.10. Mink fed the 40Y diet had a significantly lower intake of metabolisable energy (ME) and N compared to the 0Y and 10Y (P<0.01) groups and a tendency towards a lower ME intake than the 20Y (P=0.09) fed mink. The lower ME intake in the 40Y resulted in a significantly lower N retention compared to the other three groups (0Y and 10Y: P<0.001; 20Y: P=0.02). The lower ME intake may be caused by a change in the diet's consistency, when 40% of the DN from fishmeal was replaced with yeast-meal. In conclusion, substituting up to 20% of DN from fishmeal with yeast-meal had no effect on N retention, indicating that this level of yeast-meal can be used as an alternative protein source in diets to growing mink.

Meta-analysis and eco-sustainability of feeding slow-release urea in dairy and beef production

S.A. Salami[1], H. Warren[1] and J. Taylor-Pickard[2]
[1]*Alltech (UK) Ltd., Ryhall Road, Stamford, United Kingdom,* [2]*Alltech Biotechnology Centre, Summerhill Road, Dunboyne, Ireland; saheed.salami@alltech.com*

Slow-release urea (SRU) is a coated non-protein nitrogen source for rumen degradable protein in ruminant nutrition. Feeding SRU can provide sustained availability of ammonia in the rumen for optimal synchronisation with fermentable energy, which in turn increases rumen microbial protein synthesis. Reformulating ruminant diets with SRU has been proposed as an eco-friendly alternative for partial replacement of vegetable protein sources. Two meta-analysis studies were conducted to estimate the effects of replacing vegetable protein sources with SRU (Optigen®, Alltech Inc., USA) in dairy and beef production, respectively. In each study, data were extracted from 17 experiments with 44 and 33 dietary comparisons (control vs SRU) for the dairy and beef studies, respectively. Data on production outcomes in dairy and beef cattle were statistically analysed using a linear mixed model and random-effect model, respectively. For beef cattle, a simulation analysis was conducted using the meta-analysis results to evaluate the economic and environmental impacts of feeding SRU to raise 1000 growing-finishing beef cattle to gain 200 kg live weight. In dairy production, feeding SRU decreased dry matter intake (DMI; -450 g/cow/d) and dietary protein intake (DPI; -113 g/cow/d) with an increase in milk yield (+0.2 kg/cow/d), feed efficiency (FE; +3%) and nitrogen use efficiency for milk production (+4%). The use of SRU replaced about 21% of soybean meal across all dairy diets and decreased the carbon footprint (CFP) of diets (-12.0%; 524.62 vs 461.50 g CO_2/kg diet) and the CFP for the efficiency of feed use for milk production (-14.3%; 370.3 vs 317.28 g CO_2e/kg milk). In beef cattle production, there was no effect of SRU on the DMI and DPI whereas SRU improved the overall live weight gain (LWG; +92 g/d/head) and FE (+12 g LWG/kg DMI/head) of beef cattle. Simulation analysis indicated that the positive effect of SRU in improving LWG and FE reduced feed cost (-6%), *days on feed to slaughter* (-9 days) and the environmental footprint (-2.2%; -111.5 tons CO_2-eq) of beef production. Overall, these results indicate that feeding SRU can contribute to sustainable dairy and beef production.

Effect of chitosan on ruminal fermentation, milk production and quality on dairy cattle

J. Rey, E. Ugarte, A. Garcia-Rodriguez, I. Goiri, R. Ruiz, N. Mandaluniz and R. Atxaerandio
NEIKER, Animal production, Granja Modelo de Arkaute, 01080, Spain; jrey@neiker.eus

Chitosan is a natural polymer obtained from the exoskeleton of crustaceans. It has been reported to have antimicrobial activity, disrupting membrane cells, especially Gram +. It has been reported to modify the ruminal fermentation pattern, being able to decrease the C2/C3 ratio which could result in a improved milk production efficiency. Therefore, the aim of this trial was to study the effect of chitosan on milk production and quality, as well as on ruminal fermentation. Sixteen cows were used (12 Brown Swiss and 4 Holstein breed), divided in 8 blocks of 2 cows each, distributed to control vs chitosan treatment. Blocks were designed according to cows' weight, lactation status and milk yield. This study was carried out during 8 wk, distributed as follows: 10 d to take samples used as covariate, 2 wk of adaptation and the last 5 wk for measurement. Chitosan was included at a dose of 135 mg/kg BW was compared to the control diet, without additive. The chitosan was applied once a day as powder mixed with the concentrate. The control subjects had the same concentrate without chitosan. All cows had roughage fed *ad libitum*. Daily production data were obtained through the milking robot report and milk samples were taken 4 different times in 4 wks. The ruminal liquid was extracted on two days every 12 hours and the VFA were determined by gas chromatography. Milk production and quality was analysed considering repeated measures. VFA data was averaged by cow and ANOVA was used to assess statistical differences. Chitosan significantly increased the propionic acid proportion (16.6 vs 18.4%; P=0.020), and reduced the ratio C2/C3 (3.9 vs 3.5; P=0.029), and the (C2+C4)/C3 ratio (4.9 vs 4.3; P=0.024). There were no significant differences in milk production, or quality aspects of the milk (fat, protein and lactose). In conclusion, although chitosan was able to cause a shift towards a more efficient fermentation pattern, it did not affect either the milk production or quality.

Effects of incorporation of grape stalks in diets for growing rabbits

V.R. Costa-Silva[1], E. Andrade[2], G. Marques[1], V. Pinheiro[1], L. Mendes[1] and M. Rodrigues[1]
[1]UTAD, Animal Science, Quinta de Prados, UTAD, Vila Real, 4740, Portugal, [2]CAPES Fundation, Animal Science, Ministry of Education of Brazil, Brasília, 70040, Brazil; valeriasilva@utad.pt

The aim of this study was to evaluate the effects of inclusion of grape stalks on diets for growing rabbits, as alternative source of fibre. The trial was carried out on 50 crossbred rabbits that were randomly allocated to five groups each with ten animals (five male and five female rabbits each, n=10). Live weight and diets consumed were recorded weekly to determine live weight (g), weight gain (g/d) and feed intake (g/d), during the experimental period. The apparent digestibility of the five diets were determined within 55-60 days of age. Grape stalks were treated for 56 days under controlled conditions with the fungi *Lentinula edodes*. Five diets were developed: a control diet (C) without grape stalk incorporation, two diets with incorporation of different levels of untreated grape stalk (5 and 10%, 5NT and 10NT, respectively); two diets with incorporation of treated grape stalk (5 and 10%, 5T and 10T, respectively). No differences were obtained for the average daily gain, average daily intake and final weight between diets. All the four diets with inclusion of grape stalks showed lower apparent digestibility for neutral detergent fibre (NDF) in comparison to the control diet. Furthermore, for dry matter, organic matter and crude protein digestibility values for the 5NT, 10NT and 10T diets were lower than the control diet. These results show that grape stalks have a potential use for rabbits as a source of fibre.

Impact of red osier dogwood extract on *in vitro* digestibility and fermentation characteristics

W.Z. Yang[1], W.M.S. Gomaaa[2], A.M. Saleema[3], E. McGeough[4], K. Ominski[4] and L.Y. Chen[5]
[1]Agriculture and Agri-Food Canada, Research Centre, 5403 – 1 Ave S., Lethbridge, T1J4B1, Canada, [2]Department of ANCN, Assiut University, Assiut, Egypt, [3]South Valley University, Qena, 83523, Egypt, [4]Department of Animal Science, University of Manitoba, Winnipeg, R3T2N2, Canada, [5]Department of AFNS, University of Alberta, Edmonton, T6G2P5, Canada; wenzhu.yang@canada.ca

Red osier dogwood (ROD) is an abundant native shrub plant in Canada and it is rich in bioactive compounds with total phenolic concentration up to 22% (dry matter (DM) basis). The phenolic compounds include gallic acid, ellagic acid, and quercetin which have antioxidant and antimicrobial properties. The objective of this study was to evaluate the effects of 4 levels of inclusion of ROD extract on gas production (GP), DM disappearance (DMD) and fermentation characteristics in batch cultures with varying media pH. The study was a completely randomised design with 4 inclusion levels of ROD extract (0, 1, 3 and 5% of substrate) × 2 media pH (5.8 and 6.5) + monensin (0.034% of substrate; positive control) in a factorial arrangement. The study was carried out in three replicate experiments. Substrate was a high-grain diet (HG) containing 10% barley silage and 90% barley-based concentrate mix (DM basis). Inoculum was obtained from 2 ruminally fistulated beef heifers offered the HG diet. Substrate (0.5 g DM) ground through a 1-mm sieve was weighed into a filter bag and incubated for 24 h in a culture bottle in 3 replications including each combination of treatments. Data were analysed using mixed procedure of SAS with fixed effects of inclusion level, media pH and their interaction, and the random effect of experiment. There was no interaction between media pH and inclusion level of ROD on GP, DMD and fermentation variables. Increased media pH (5.8 vs 6.5) increased (P<0.01) GP (averaged 198 vs 389 ml/g substrate), DMD (averaged 51.3 vs 64.6%), and total volatile fatty acid production (averaged 74 vs 83 mM). Increasing addition of ROD extract did not affect GP, but linearly (P<0.05) decreased DMD from 56 to 46% at pH 5.8 and from 69 to 61% at pH 6.5. Increasing ROD extract linearly (P<0.01) increased the proportion of acetate from 43 to 47% and 47 to 50% at pH 5.8 and 6.5, respectively. Acetate to propionate ratio increased from 1.68 to 1.93 and from 1.90 to 2.10 at pH 5.8 and 6.5, respectively. Adding monensin did not affect GP and DMD compared with ROD extract. These results indicated that *in vitro* DMD of HG diet varied with dosage of ROD extract and media pH. The decreased DMD along with increased acetate to propionate ratio with addition of ROD extract suggests ROD extract may be beneficial to HG fed cattle for reducing risk of rumen acidosis without negatively impacting fibre digestion.

Lysine and methionine supplementation influence on antler growth of yearling fallow deer (*Dama dama*)
V. Ny[1,2], R. Kotrba[2,3], J. Cappelli[4], D. Bureš[1], M.A. Clar[2,5], A.J. García[4], T. Landete-Castillejos[4], L. Bartoň[1] and F. Ceacero[2]
[1] *Institute of Animal Science, Cattle Breeding, Přátelství 815, 104 00, Praha 22, Czech Republic, [2]Czech University of Life Sciences Prague, Department of Animal Science and Food Processing, Kamýcká 129, 165 00, Praha 6, Czech Republic, [3] Institute of Animal Science, Department of Ethology, Přátelství 815, 104 00, Praha 22, Czech Republic, [4]ETSIAM Universidad de Castilla-La Mancha (UCLM), Departamento de Ciencia y Tecnología Agroforestal y Genética, Campus Universitario s/n 02071, 02071 Albacete, Spain, [5]Mindanao State University-Naawan, College of Agriculture and Forestry, Butuan-Cagayan De Oro-Iligan, 9023 Misamis Oriental, Philippines; nyv@ftz.czu.cz*

For cervid, protein requirement during antler growth is up to 15%. Thus, the requirement of specific amino acids for protein synthesis is also very important. Lysine (Lys) and Methionine (Met) are the two known limiting amino acids for ruminants. And for deer, Lys is very interesting for antler growth, since it is the precursor for collagen production. However, very limiting studies conducting on its effects on antlers. In this study, we show the first trial on a large herd of fallow deer (*Dama dama*) with two experimental designs. Each experiment involved 45 yearling fallow bucks, distributed in three groups based on average weight. The group compositions included Exp. 1: Pasture, Pasture+Barley (0.2 kg/animal per day), and Pasture+Barley+Lys (5 g/animal per day); Exp.2: Pasture+Barley (0.5 kg/animal per day), Pasture+Barley+Lys (9 g/animal per day), and Pasture+Barley+Lys+Met (9 and 3 g/animal per day, respectively). At the first glance, Lys supplementation in the first experiment did not improve any antlers characteristics. However, when the level was increased to 9 g and with the combination of Met in the second experiment, the treatments had good effects on external antler characteristics and antler weight. A very interesting result is the increase in burr perimeter, which is a crucial base for the better and long-term antler growth. The results also suggest a more intense positive effect of amino acid supplementation *in situ*ations when the animals have a lower performance. This result opens opportunity for improving deer farming practices, particularly for velvet production and trophies.

Field pea included up to 45% in bull fattening concentrates does not impair *in vivo* digestibility
S. Lobón, I. Casasús, M. Joy and M. Blanco
Ctr Invest y Tecnol Agroal Aragon (CITA), IA2 (CITA-Universidad de Zaragoza), Montañana 930, 50059 Zaragoza, Spain; icasasus@cita-aragon.es

In the last years, the substitution of soybean by legumes has been promoted in Europe to reduce the dependency on soybean imports. Field pea (*Pisum sativum*), which has high starch and protein content, can be included in the diet of fattening cattle. The aim of this study was to compare the apparent digestibility of dry matter (DM) and nitrogen (N) of the diet and blood metabolites due to the inclusion of pea in concentrates. The main ingredients of the concentrates were: 52% corn, 20% barley, 15% corn gluten feed, 9.7% soya and 0% pea (0% pea); 41% corn, 20% barley, 15% corn gluten feed, 15% pea and 5.5% soya (15% pea); 30% corn, 30% pea, 20% barley, 15% corn gluten feed and 1.3% soya (30% pea); 45% pea, 27% corn, 20% barley and 4% corn gluten feed (45% pea). For that, 4 Parda de Montaña young bulls (initial weight 251±4 kg) received restricted amounts of one of 4 concentrates and straw during 4 consecutive experimental periods. The average daily intake of concentrate and straw was 5.0 (±0.02) and 0.68 (±0.001) kg DM, respectively, and it was not affected by pea inclusion (P>0.05). However, the inclusion of pea in the concentrate tended to increase the apparent DM digestibility (P<0.10) and affected N digestibility (P<0.05). The 30 and 45% pea concentrates had greater N digestibility than the 0 and 15% pea concentrates. The intake of N increased linearly with the rate of pea inclusion, the lowest intake was in the 0% pea treatment and the greatest in the 45% pea treatment (P<0.05). The urinary N excretion was greater in bulls with 30 and 45% pea than in the 0 and 15% pea bulls (P<0.05), whereas the faecal N excretion was not affected (P>0.05). The inclusion of pea also affected the retained N, with higher values in bulls receiving the 45% pea than in those with 0 and 15% pea. The plasma concentrations of IGF-1, total protein, β-hydroxybutyrate and urea at the beginning and at the end of each period were not affected by the inclusion of pea (P>0.05). In conclusion, the total replacement of soybean by pea in the fattening concentrate of bulls did not impair digestibility and improved the N retention.

Assessing the effects of dietary inclusion of *Cistus ladanifer* L. in rumen ciliate protozoa of lambs

A. Francisco[1,2], O. Guerreiro[3,4], E. Jerónimo[3,4], M. Dentinho[1,2], J. Santos-Silva[1,2] and R. Bessa[1,5]
[1]*Centro de Investigação Interdisciplinar em Sanidade Animal, Faculdade de Medicina Veterinária, Avenida da Universidade Técnica, 1300-477 Lisboa, Portugal,* [2]*Instituto Nacional de Investigação Agrária e Veterinária, UEISPSA, Quinta da Fonte Boa, Vale de Santarém, 2005-048, Portugal,* [3]*Mediterranean Institute for Agriculture, Environment and Development, Universidade de Évora, Pólo da Mitra, 7000-083 Évora, Portugal,* [4]*Centro de Biotecnologia Agrícola e Agro-Alimentar do Alentejo, Rua Pedro Soares, Campus IPBeja /ESAB, 7801-908 Beja, Portugal,* [5]*Faculdade de Medicina Veterinária da Universidade de Lisboa, Avenida da Universidade Técnica, 1300-477- Lisboa, Portugal; alexandra.francisco@iniav.pt*

The effects of including in lamb diets *Cistus ladanifer* L. (CL) on rumen ciliate protozoa were studied. Data derived from 3 trials performed by our research team. In the first two trials 5, 10 and 20% DM (trial 1) and 0 and 15% DM (trial 2) of dried and milled CL aerial parts (dCL) were included in the diets. In trial 3 were evaluated 2 forms of CL (dCL and condensed tannins extract (CLCT)) and 3 levels of condensed tannins (CT) (0, 1.25 and 2.5%). The forage:concentrate ratio was 1:1 in trials 1 and 2. In trial 3 the diet was composed by dehydrated lucerne. Rumen protozoa were evaluated by microscopic counting. To each trial, an ANOVA was performed to evaluate the effects in total ciliates (PTZ) and community structure of: 1) the level of dCL inclusion (trials 1 and 2) or; 2) the level of CT and the form of CL inclusion (trial 3). A global regression analysis, including the trial effect, was performed to evaluate the effect of the daily intake of dCL in PTZ and individual genus. The use of dCL increased the PTZ in trial 1 (P=0.001) but had no effect (P>0.05) in trial 2. In trial 3, the dCL did not affect PTZ, but when CLCT was used, PTZ was higher with 1.25% of CLCT and lower with 2.5% (P=0.004), relatively to Control. The use of CLCT resulted in higher proportion of *Isotricha* and *Dasytricha* (P=0.004) than dCL. The level of CT reduced the proportion of *Entodinium* (P=0.007). The PTZ increased linearly with dCL daily intake (P<0.001), but the abundance of *Dasytricha* decreased linearly (P=0.018). Dietary inclusion of dCL and CL tanniferous extract seems to modulate the ciliate protozoa abundance and community structure in the rumen of lambs, but further studies are needed to clarify the effects observed here. Work funded by Alentejo2020 through the FEDER under the project CistusRumen (ALT20-03-0145-FEDER-000023) and by Fundação para a Ciência e Tecnologia (FCT) through PTDC/CVT/103934/2008 and UID/CVT/00276/2013 projects and A. Francisco (SFRH/BD/68773/2010) and O. Guerreiro (SFRH/BD/84406/2012) individual fellowships.

Proteaginous extrusion modifies rumen vs duodenum digestion of N, lysine and Maillard compounds

P. Chapoutot[1], O. Dhumez[1], S. Mendowski[2,3], A. Germain[2], G. Chesneau[2] and P. Nozière[3]
[1]*Université Paris-Saclay, INRAE, AgroParisTech, UMR Modélisation Systémique Appliquée aux Ruminants, 16, rue Claude Bernard, 75005 Paris, France,* [2]*Valorex, La Messayais, 35210 Combourtillé, France,* [3]*Université Clermont Auvergne, INRAE, VetAgro Sup, UMR Herbivores, Theix, 63122 Saint-Genès-Champanelle, France; patrick.chapoutot@agroparistech.fr*

Thermal treatments, especially extrusion, can modify digestion efficiency and increase Maillard compounds in feeds. The aim of this study was to quantify the level of rumen degradability (Deg, %) and duodenum digestibility (Dig, %) of nitrogen (N), available lysine (Lys) and Nε-carboxymethyl-lysine (CML) on 11 feeds: soybean meal (SM) and raw (R) or extruded (E) faba bean (F) and lupine (L). Several conditions were tested for extrusion on F and L: low (LT) or high (HT) temperature, or for maturation on F: with reducing sugars (SUG) or enzyme (ENZ) addition or not (EXTR). Deg was calculated after 16 h rumen incubation (3 cows × 2 replicates) and Dig values were assessed by mobile nylon bag method (3 cows × 7 replicates). Variance analysis were applied on Deg and Dig results to quantify the feed and treatment effects (for N, Lys and CML respectively, RSD=4.9, 6.2 and 22.3% for Deg, and 2.1, 1.2 and 2.0% for Dig). Compared to raw grains (94±1% on average) and SM (85%), extrusion largely decreased N Deg (P<0.001), especially for L vs F (EL: 61 vs EF: 74%) and for SUG vs ENZ (68 vs 78%), but increased N Dig (P<0.001), especially for L (EL: 96, SBM: 95, EF: 91 vs 68% on raw grains), with no influence of temperature and a lower effect for ENZ than SUG (92 vs 94%). The same trends were globally observed with Lys (P<0.001). However, Lys Deg was lower than N Deg (90±1 vs 94±1% and 59±12 vs 70±8% on average for raw and extruded grains respectively) and, at the opposite, Lys Dig appeared higher than N Dig (86±5 vs 68±4% and 97±1 vs 93±3%, respectively). Moreover, positive and linear significant relationships were observed between Lys and N variables: Lys Deg=-34.2+1.33 N Deg; n=11, R^2=0.92, RSD=4.9); Lys Dig=56.9+0.43 N Dig; n=11, R^2=0.84, RSD=2.2). Thus, the flow of metabolisable proteins and available lysine estimated at the duodenum largely improved with extrusion, but with a lower extend for Lys than for N (around ×5 and ×10, respectively). CML Deg largely varied between feeds and treatments (P<0.001) and was lower for raw grains (19% on average) or extruded L (9%) than for SBM (53%), and could even appear negative (-22% for EF). CML Dig had greater values (P<0.001) for extruded grains and SBM than for raw feeds (95% on average, 93% and 74%, respectively), and for F than L (86 vs 82%, respectively). So, extrusion enhanced metabolisable Lys flow but less than for N and most of CML digestion seemed to occur in the intestine.

Evaluation of guar meal and yellow lupine seeds as alternatives to soybean meal for ruminants

C. Guedes, M. Almeida, S. Garcia-Santos, A. Silveira, L. Ferreira, S. Silva, V. Pinheiro, D. Monteiro and M. Gomes
CECAV-UTAD, Animal Science, Quinta de Prados, 5000-801 Vila Real, Portugal; cguedes@utad.pt

In the context of a strong EU dependence on the imported soybean meal (SBM) as a protein source for animal feeding, efforts have been made to find alternatives able to be produced in Europe. Lupines are native European legumes rich in protein and suitable for a sustainable production. Guar meal (GM) has a high crude protein (CP) content and is obtained after the extraction of the guar gum from the seeds of Cluster bean (*Cyamopsis tetragonoloba* L. Taub), a sub-tropic legume that mostly is cultivated in India but with potential to grow in the European Mediterranean basin. Protein degradation in the rumen is an important tool to evaluate the protein value of feeds for ruminants. The objective was to compare rumen CP degradation of GM korma thermally treated (GMKT) or not (GMK) and 4 varieties of yellow lupine seeds (LL; *Lupinus luterus* L.) with SBM using the nylon bags technique. Three non-lactating dairy cows were used. Nylon bags were incubated in the rumen for 2, 4, 8, 16, 24, 48 and 72 h. After removing from the rumen, the bags (including the 0 h bags) were washed with cold water, and dried. Feeds and bag residues were analysed for CP. Feeds were also submitted to ash, crude fat, starch and fibre fractions (i.e. neutral detergent fibre, acid detergent fibre and acid detergent lignin) determinations. The exponential model was used to describe the degradation kinetics of CP. Degradation kinetics (parameters) were analysed using a one-way ANOVA and the means were compared according to the t test. The GMKT presented the highest CP content (50.7%) and LL seeds presented CP values that varied from 38 to 42%. The soluble fraction differed ($P<0.01$) between feeds, with GMKT presenting the highest ($P<0.05$) value (10.2%). Also, differences ($P<0.05$) were observed in the rate of degradation (c), with GMKT and GMK having the highest ($P<0.05$) values (average 0.105 h-1). No differences ($P>0.05$) were observed on c values between LL seeds and SBM (varied from 0.081 to 0.089 h-1). These feeds showed a CP potential (total) degradation close to 100%. The results indicate that guar meal and the yellow lupine seeds studied can be used as sources of degradable protein in the rumen.

Treatment of soybean meal to improve protein utilisation by dairy cows

A. Klop[1], M. Aoun[2], J.P. Ricaud[2] and G. Van Duinkerken[1]
[1]Wageningen Livestock Research, De Elst 1, 6708 WD Wageningen, the Netherlands, [2]Idena, 21, Rue du Moulin, 44880 Sautron, France; gert.vanduinkerken@wur.nl

Soybean meal is a common protein source in dairy cattle diets. By chemical treatment (e.g. with formaldehyde), rumen degradable protein content can be decreased, while increasing the content of intestinal digestible protein. It was evaluated if feed additive Vertan (a blend of Eugenol, Thymol and essentials oils) can be a nature-based alternative for chemical processing of soybean meal. To evaluate the effect of Vertan on performance and nitrogen use efficiency, a feeding trial was conducted as a randomised block design with 3×15 dairy cows. Treatments were a negative control diet (SBM) with soy bean meal (1.4 kg DM/cow/d), a positive control diet (SBM-bp) with formaldehyde treated soybean meal (1.2 kg DM/cow/d) and a treatment diet (SBM-V) with soybean meal supplemented with Vertan (1.4 kg DM/cow/d). Crude protein content of all diets was 15% on a dry matter basis. Parallel to this study, 3 cows fitted with a rumen cannula were used in a Latin square design to measure rumen fermentation characteristics, *in situ* rumen degradation of the diet and digestion. A rumen fermentation model was used to mimic the effect of soybean meal characteristics on fermentation, digestibility and rumen output. Total DM intake was 22.6 kg/cow/d for all treatments. Milk yield did not differ between treatments. Milk protein content was lower ($P=0.037$) for SBM-bp compared to SBM. Blood urea of cows on SBM was higher ($P=0.002$) compared to cows on SBM-bp and SBM-V. Ammonia concentration in rumen fluid tended ($P=0.073$) to be lowest for treatment SBM-V and highest for SBM. Rumen pH tended ($P=0.055$) to be higher for SBM-bp compared to the other treatments. Faecal digestion did not differ between treatments. The output of rumen modelling simulations showed that estimated rumen OM and NDF degradation, VFA concentrations, microbial protein synthesis and enteric methane production were lower on SBM-bp compared to SBM and SBM-V. It was concluded that Vertan has an effect on rumen protein fermentation, resulting in lower rumen ammonia and blood urea.

Oat grown with vetch improves forage production, chemical composition and degradability rate

F.C. Pereira[1], L.C.P. Machado Filho[1], D.C.S. Kazama[1], R. Guimarães Júnior[2] and D. Enriquez-Hidalgo[3,4]
[1]Laboratorio de Etologia Aplicada, Universidade Federal de Santa Catarina, Florianopolis, Brazil, [2]EMBRAPA, Brazilian Agricultural Research Corporation, Brasília, Brazil, [3]University of Bristol, Bristol Veterinary School, Langford, United Kingdom, [4]Pontificia Universidad Catolica de Chile, Facultad de Agronomía e Ingeniería Forestal, Santiago, Chile; daniel.enriquez@uc.cl

This study aimed to evaluate the effect of intercropping common vetch with black oats on the forage production, chemical composition and degradability rate of forage. Two paddocks (2,500 m^2 each) from a Voisin Rational Grazing management system were divided into three blocks each and then into thirds (278 m^2) characterising a randomised block design. Three forage seed compositions were overseeded into third: oats alone (100 kg/ha), vetch alone (60 kg/ha), and oats with vetch (80 + 40 kg/ha). The pasture was fertilised with poultry manure (1.75 t/ha). After 65 days, five samples of forage were cut at soil level into thirds using a 0.5×0.5 m^2 square. Each sample was weighed to measure forage production and then used as a final sample to determine chemical composition. Samples were analysed for dry mater (DM), crude protein (CP), neutral detergent fibre (NDF), acid detergent fibre (ADF) and mineral matter (MM) contents. The 'in vitro' degradability was measured after 6, 24, and 48 hours of incubation. Data were evaluated with linear mixed-effects models including each measured variable as a response, forage categories as explanatory variables, and blocks as random effects. The forage production was higher for vetch alone and oats with vetch compared to oats alone ((1,573.1 vs 686; SEM: 306.6 kg/ha, P<0.01). There was no difference for DM content (17.7; SEM: 2.02%). Vetch grown alone and oats grown with vetch had higher CP (23.9 vs 14.9; SEM: 1.36%, P<0.01) and MM content (8.63 vs 8.31; SEM: 0.27%, P<0.05), but less NDF (53.6 vs 62.9; SEM: 1.90%, P<0.01) and ADF (31.5 vs 34.3; SEM: 0.99%, P<0.01) compared to oats grown alone. The oats grown alone showed less degradability (P<0.01) than the other categories in all the evaluated period of incubation (33.5 vs 45.3; SEM: 7.05%). Intercropping common vetch with black oats increased forage production and improved chemical composition and 'in vitro' degradability.

Influence of maize grain drilling on the water release rate and its nutritional value

M. Grubor[1], S. Zjalic[2], A. Matin[1] and T. Kricka[1]
[1]Faculty of Agronomy, University of Zagreb, Svetosiminska cesta 25, 10000 Zagreb, Croatia, [2]Sveučilište u Zadru, Department for ecology, agronomy and aquaculture, Trg kneza Viseslava 9, 23000 Zadar, Croatia; szjalic@unizd.hr

The most important energy feed is maize grain, and as a consequence of a large number of hybrids, the largest variations in the chemical composition are present. Maize grain must be preserved from one harvest to another. One way of maize kernels conservation is convection drying. The major resistance to the water passage through the kernel during drying is provided by the pericarp. Based on the above, kernel drilling is introduced as a process which breaks down the structure of the pericarp and enables the faster release of water. Since different grain hybrids with different initial humidity, morphological structure and the size of the grain are dried simultaneously in the drying process, different grain behaviour occurs in the drying process. The release of surplus water to the equilibrium humidity (14%) is different for each hybrid, therefore the aim of the paper is to determine the water release rate from the whole and drilled kernels of 3 different FAO maize groups 200, 300 and 400 through exponential equations. Kernel drilling was performed by laboratory „hedgehog', in which the kernels are drilled halfway through by needles. The investigated kernels were dried in a convection dryer at a temperature of about 130 °C and an airflow rate of 2 m/s. Also, the grain fracture after drying by laboratory centrifuge, which represents the simulation of the grain fall in the silage cell, as well as the nutritional value of investigated maize hybrids after the drying process will be determined. From the results, it can be concluded that the drilled kernel dried faster than the whole kernel. Regarding the investigated hybrids, FAO group 200 released the water fastest while FAO group 400 released the water slowest. In the drilled kernel, regardless of the hybrid, a higher percentage of kernel fracture was observed. At the investigated hybrids, FAO group 200 showed more resistance to kernel fracture while kernel fracture values of FAO group 300 and 400 were similar. Hybrids had similar nutritional values, whether it is the whole or drilled kernel.

Mushroom polysaccharides as inhibitors of mycotoxin production

J. Loncar[1,2], A. Parroni[1], A. Motola[1], M. Reverberi[1] and S. Zjalic[2]
[1]Sapienza University of Rome, P.lle Aldo Moro 5, 00185 Roma, Italy, [2]University of Zadar, Mihovila Pavlinovica 1, 23000 Zadar, Croatia; szjalic@unizd.hr

The presence of mycotoxins in feed is among the treats for animal production. Among the treatments with lower environmental impact the mushroom polysaccharides showed to be a promising tool. Tramesan, a polysaccharide isolated from *Trametes versicolor* provided a long lasting inhibition of the synthesis of aflatoxins (AFT) and ochratoxin A (OTA). The characterisation of the polysaccharide showed that it is an α-glucan with presence of some β-bounds. Rather high price, due to the production cost, is the main obstacle for wide scale application of Tramesan. In order to establish the structure of the active polysaccharide it was decided to explore other mushroom polysaccharides active in inhibition of the synthesis of both AFT and OTA and compare the structure with Tramesan. The filtrate extracts of submerged cultures of 96 different mushroom isolates were assayed for their ability to inhibit the two mycotoxins. Only two isolates *T. versicolor* 119, Tramesan producer, and *Schyzophyllum commune* SC-Allianto provided a high inhibition (over 90%) that lasted even after the plateau of fungal growth was reached. The first results indicate that two polysaccharides might have a similar mechanism of activity, by enhancing the antioxidant response in fungal cells. A polysaccharide schyzophyllan, produced by *S. commune*, is known for its biological activity. A commercial schyzophyllan was assayed for inhibition of AFT and OTA. Although an inhibition of around 70% of aflatoxin inhibition and 60% of OTA was observed, it tended to drop down after 5 days of incubation. These results indicate that schyzophyllan could be involved in mycotoxin inhibition, but other molecule(s) play a role in inhibition. The further purifications of filtrates are needed to establish the nature and composition of these molecules.

Effect of the live yeast *Saccharomyces cerevisiae* on feed intake and milk production of dairy cows

K. Goossens[1], L. Vandaele[1], B. Ampe[1], V. Demey[2] and S. De Campeneere[1]
[1]Flanders Research Institute for Agriculture, Fisheries and Food (ILVO), Animal Science unit, Scheldeweg 68, 9090 Melle, Belgium, [2]Lallemand, Animal Nutrition, 19 rue des Briquetiers, 31702 Blagnac, France; karen.goossens@ilvo.vlaanderen.be

The live yeast Saccharomyces cerevisiae CNCM I-1077 (LY, Levucell SC, Lallemand SAS), is expected to stimulate rumen function and nutrient utilisation thus improving the production performance of lactating dairy cattle. A trial with 30 Holstein Friesian cattle in mid-lactation was conducted at the ILVO dairy cattle research stable aiming to measure the effect of LY on feed intake, milk production and nitrogen (N) efficiency. The trial lasted 16 weeks, divided in an adaptation period of 3 weeks and a treatment period of 13 weeks. During the adaptation period, all 30 cows were fed the same mixed ration consisting of 42.5% maize silage, 42.5% pre-wilted grass silage, 12% pressed sugar beet pulp, 1% straw, 1.7% of soy bean meal and 0.3% of feed urea. Each cow received additional amounts of concentrates to meet 105% of the individual requirements for energy and digestible protein. During the treatment period, treated cows (n=15) received LY in a dose of 1×10^{10} cfu/head/day by replacing 1 kg of balanced concentrate by 1 kg of balanced concentrate supplemented with LY. Daily individual roughage and concentrate intakes were recorded. All animals were milked in the milking parlour with recording of individual productions and individual milk sample collection on a monthly basis for analysis of milk composition. The results of the feeding trial were analysed in R 3.6.0 using a linear mixed regression model with treatment and week as fixed effects and a random slope for cow. Dietary supplementation of LY had no significant effect on roughage dry matter intake (P=0.929), total dry matter intake (P=0.652) nor feed efficiency (P=0.609). A significant increase in milk production (P<0.0001) of on average 1.1 kg/day was measured over the 13 weeks treatment period. The milk fat concentration was lower (P=0.045), likely due to a dilution effect. The milk protein concentration was not different (P=0.302) and resulted in a significantly higher daily total milk protein production (P=0.0009). In addition, the milk urea concentration was significantly higher (P=0.022) and the N-efficiency was improved (P=0.002).

Effect of rumen protected lysine on slaughter performance of German Fleckvieh bulls

F. Riepl[1,2], S. Kuenz[1], D. Brugger[3], V. Inhuber[2], W. Windisch[1], H. Spiekers[2] and T. Ettle[2]
[1]Technical University of Munich, Chair of Animal Nutrition, Liesel-Beckmann-Str. 2, 85354 Freising, Germany, [2]Bavarian State Research Center for Agriculture, Institute for Animal Nutrition and Feed Management, Prof. Dürrwächter Platz 3, 85586 Grub, Germany, [3]Institute of Animal Nutrition, University of Zurich, Winterthurerstrasse 270, 8057 Zurich, Switzerland; riepl@wzw.tum.de

Beef production contributes to agricultural ammonia emissions. Reducing environmental impact of cattle farming requires detailed knowledge on amino acid supply. The present study aimed to investigate if rumen-protected (RP) Lys partially compensates for N-deficiency-associated losses in slaughter performance of young Fleckvieh bulls. This experiment comprised 67 German Fleckvieh bulls (age: 156 d, weight: 223 kg). Bulls were randomly allocated to three different diets with 14% CP (CON; n=22), 11% CP (NEG, n=22) and 11% CP + 0.42% Lys (LYS, n=23). All diets were isoenergetic (11.64 MJ ME/ kg DM) and contained rumen-protected Met (0.11% of DM) to avoid potential limitations by this essential AA. The animals were fed *ad libitum* up to d 119. From experimental d 63, subgroups of bulls were slaughtered at weekly intervals until d 119. Data collection comprised daily weight gain (g/d), final live weight at the slaughterhouse (LW, kg), warm carcass weight (kg) and haunch range (cm). Data analysis comprised 3-way ANOVA (treatment group (GRP), days in experiment (DAYS), LW at experimental start, GRP×DAYS). Final LW was significantly increased in the CON group compared to all other groups (P=0.0004). N-restriction reduced warm carcass weight significantly compared to control irrespective of the AA supply regime (P=0.02). Haunch range was significantly different between CON and NEG (P=0.04) but LYS expressed no difference to both controls. Data obtained in LYS group was in any case numerically higher than in NEG. Slaughter performance of N-restricted animals fed RP Lys was not significantly affected compared to NEG. However, values were in any case numerically higher. Such supplements may be able to partially compensate for fluctuations in dietary quality under practical feeding conditions.

Growth performance and health status of veal calves supplemented with hemp meal

L. Bailoni, E. Bacchin, A.P. Berno and N. Guzzo
University of Padua, Department of Comparative Biomedicine and Food Science, Agripolis, Viale dell'Università, 16, 35020, Italy; lucia.bailoni@unipd.it

Recently, interest in hemp (*Cannabis sativa L.*) is increased due to nutritionally beneficial effects for both humans and animals. Hemp seeds are a good source of bioactive compounds, especially PUFAs, mainly n-3 fatty acids. Hemp meal (HM), co-product derived from the mechanical extraction of oil, can be efficiently used as a protein supplement in animal feeding. This study aimed to evaluate the effects of two different levels of HM supplementation on performance and health status of veal calves during the first 91 d of fattening. Forty-eight Belgian Blue calves (35 males, 13 females) were allocated in 12 pens (balanced according to initial BW and sex) and assigned to three dietary treatments: 0% of HM (T0), 3% of HM (T3), and 6% of HM (T6). All groups received a milk replacer diet containing the HM supplementation (30.3% CP, 9.4% lipids, 161 ppm of iron) and additional solid feed based on flaked cereals. The same feeding plan was applied to the three groups. Calves were weighed three times during trial and blood samples were collected by jugular vein at the same time. All data were analysed by GLM of SAS. The average daily gain was similar (P>0.10) for three experimental groups (on average 1,044, 994, and 960 g/d for T0, T3 and T6 group, resp.). The iron concentration in plasma was similar among treatments (67, 72, and 57 mg/dl in the same order) but decreased significantly (P<0.05) from the first to the subsequent samplings (101, 48, and 46 µg/dl, resp. at 0, 43 and 91 d). The pattern of haemoglobin (HGB) concentration in plasma was similar than that of iron. In conclusion, during the first period of fattening, the addition of HM in the diet of veal calves did not affect the growth performances of the animals. The health status was satisfactory, thanks to the absence of pathologies and good plasmatic concentration of iron and HGB. Final consideration of the effect of hemp meal on the calves performance will be obtained by analysing further post-mortem parameters and behavioural data obtained by videorecording system.

Effect of dietary plant phytobiotics and butyrates on broiler blood parameters

I. Skoufos[1], E. Bonos[1], K. Fotou[1], A. Karamoutsios[1], I. Karakousis[2], E. Barka[1], I. Giannenas[3], E. Christaki[3] and A. Tzora[1]
[1]*University of Ioannina, School of Agriculture, School of Agriculture, University of Ioannina, 47150, Kostakioi Artas, Greece,* [2]*Elanco Hellas AEBE, Leoforos Mesogion 335, Xalandri, 15231, Athens, Greece,* [3]*Aristotle University of Thessaloniki, Laboratory of Nutrition, School of Veterinary Medicine, Laboratory of Nutrition, School of Veterinary Medicine, Aristotle University Campus, 54124, Thessaloniki, Greece; jskoufos@uoi.gr*

The aim of this study was to investigate the effect of dietary supplementation with essential oils in combination with butyrates on the haematological and lipid profile parameters of broiler chickens. In this 45 days trial, a total of 360 one-day-old male broiler chicks (Ross-308) were randomly allocated into 4 treatments with 6 replicates (of 15 chicks), all housed in floor pens with rice hulls litter. Treatment A (Control) was fed standard diets in mash form (starter, grower-finisher), which did not contain anticoccidials or antibiotics. The diets of the other treatments were further supplemented: Treatment B with a mixture of phytobiotics based on garlic, oregano and cinnamon (1,5 g/kg); Treatment C with a premix containing butyrates (1 g/kg); Treatment D with both phytobiotics (1,5 g/kg) and butyrates (1 g/kg). Commercial rearing, management and vaccinations procedures were employed. Body weight and feed intake were measured on days 11, 26, 38 and 45. On the last day, blood samples were collected for haematological analyses (Analyser MS4,Melet Schloesing Lab, Osny, France) and lipid analyses (IDEXX VETTEST 8008, IDEXX LAB, USA) and by commercially available kits (Randox, UK). Treatments did not differ (P>0.05) on final body weight or feed conversion ratio. Treatment D showed increased (P≤0.05) blood haemoglobin compared to treatments A and C, whereas no significant (P>0.05) differences were found for blood erythrocytes, haematocrit, leukocytes, lymphocytes, total cholesterol, total triglycerides, high density lipoproteins and low density lipoproteins. Acknowledgement: The project/research is co-financed by the European Regional Development Fund (ERDF) under the Operational Program 'Research-Create-Innovate, 2014-2020', EPAnEK. Project Code: T1EΔK-03856. Acronym: GreenPoultryMeatAntifree.

Looking for the feed autonomy of farms: the case of high density and perennial mulberry forage banks

E. González-García
INRAE, Animal Physiology and Farming Systems Department, UMR SELMET, Bâtiment 22, Campus INRA-Montpellier SupAgro, 34060 Montpellier, France; eliel.gonzalez-garcia@inrae.fr

The current global context of climate change, land pressure and food insecurity issues requires to look for alternative and innovative ways for conceiving farming systems for the future, productive to cover the food requirements of an increasing human population, but at the same time respectful of the already exhausting natural resource base. To achieve these ambitious goals in the design of sustainable animal production systems it is necessary to prioritise the forage and feed autonomy of farms. That's to say, developing alternative technologies to replace the current strong dependency from external feed resources, the most preponderant situation in most farming systems. One of the potential solutions to reduce such dependency from external inputs is to boost important productions of good quality biomasses (feeds) at the farming system level, able to cover the herds' nutritional requirements. The objective of this work is to share our experience in establishing high density (~25,000 plants per ha) and perennial forage banks based on white mulberry (*Morus alba*), a world widely well adapted and highly selected shrub species, traditionally used for silkworm production in different cultures, including Europe. Our experience with the adoption of this technology includes excellent results (i.e. agronomic establishment, high forage yield and excellent forage quality and nutritive value, animal acceptability and productive responses, plantation persistence) in different latitudes i.e. Cuba, Burkina Faso and France. Based in our results during the last decades, we strongly recommend the extensive evaluation of this forage production technology in different agroecological conditions of Europe and other parts of the world.

Genetic relationships between feed efficiency and faecal microbiome in pig lines

A. Aliakbari[1], O. Zemb[1], Y. Billon[2], C. Barilly[1], I. Ahn[1], J. Riquet[1] and H. Gilbert[1]
[1]*Université de Toulouse, INRAE, GenPhySE, 24, chemin de Borde-Rouge, Auzeville Tolosane, 31326, Castanet Tolosan, France,* [2]*INRAE, GenESI, Le Magneraud BP 52, 17700, Surgeres, France; amir.aliakbari@inrae.fr*

Recent advances in bioinformatics and sequencing technologies have made it possible to obtain individual microbiome information for humans, animals and plants. In pigs, as in humans, gut microbiota is an important contributor to the nutrient availability at the gut level. In the present study we aimed to quantify the genetic relationships between two main feed efficiency traits, feed conversion ratio (FCR) and residual feed intake (RFI), and faecal microbial composition in two experimental pig lines divergently selected for RFI (HRFI and LRFI lines). Multivariate linear mixed models of OTUs relative abundancies and performance traits provided heritability (h^2) and additive genetic correlation (r_A) for all traits. Faecal samples were collected at 15 weeks of age in 604 pigs from the G9 and G10 generations of the RFI lines, and about 4,000 FCR and RFI records were available for all generations. From sequencing of the V3-V4 regions of the 16S rRNA gene, a total of 6,792 operational taxonomic units (OTU) were identified in the samples. The 137 OTUs with less than 20% zero abundancies were kept for genetic analyses after log-transformation. A total of 65 OTUs showed a h^2 different from zero (P<0.05), with estimates ranging from 0.13±0.07 to 0.52±0.12. In total, OTUs with significant h^2 were annotated to 13 families and 34 genera. Among those 65 OTUs with genetic background, 10 OTUs had a genetic correlation with FCR different from zero, and 14 OTUs had a significant genetic correlation with RFI. The OTUs with significant r_A with FCR belonged to six families. The OTUs with significant r_A with RFI belonged to four families. Only one OTU, belonging to the Prevotella_9 genus from the Prevotellaceae family, had commonly significant correlation with both traits. Our results showed that some OTU abundancies have a genetic background and significant genetic correlation with feed efficiency traits. These results beside the host genetic effect could deserve more consideration in breeding programs to improve the feed efficiency in pigs.

Microbiability of feed efficiency related traits in pigs

L.M.G. Verschuren[1,2,3], A.J.M. Jansman[3], O. Zemb[4], R. Bergsma[2] and M.P.L. Calus[3]
[1]*Agrocampu Ouest, UMR1348 Pegase, 16 Le Clos Domaine de la Prise, 35590 Saint-Gilles, France,* [2]*Topigs Norsvin Research Center B.V., P.O. Box 43, 6640 AA Beuningen, the Netherlands,* [3]*Wageningen Livestock Research, Droevendaalsesteeg 4, 6708 PB Wageningen, the Netherlands,* [4]*INRA – INPT – ENSAT – Université de Toulouse, GenPhySE, 24 Chemin de Borde-Rouge, 31326 Castanet-Tolosan, France; lisanne.verschuren@wur.nl*

Gut microbiota profiles can be regarded as complex polygenic traits, that are influenced by both the host and the environment. The aim of the present study was to use microbial and genetic relationships to predict feed efficiency related traits (feed intake, body weight gain and feed conversion ratio) in pigs. Data were collected from 538 three-way crossbreed male grower-finisher pigs, all being genotyped at 50k SNPs. The pigs were fed *ad libitum* according to a three-phase feeding program with commercial diets based on wheat, barley and by-products. On the day before slaughter (mean body weight 125 kg) fresh faecal samples were collected per animal and sequenced for the 16S hypervariable region of bacteria (V3/V4) to profile the faecal microbiome, resulting in 2,276 operational taxonomic units after filtering. Using these data we calculated the proportion of variance in feed efficiency related traits explained by variation in the faecal microbiome, i.e. the 'microbiability', and genetic background, i.e. 'heritability'. The microbiability values were 42% for average daily gain, 33% for average daily feed intake, and 46% for feed conversion ratio, whereas heritability values were 24, 39 and 22% respectively. Using leave-one-out cross-validation, we estimated the accuracy of prediction based on the microbiome profile and genotype of the pigs. The faecal microbiome profile had a stronger relationship with variation in average daily gain and feed conversion ratio than data on the genotype of the pig, while the opposite was observed for average daily feed intake. In conclusion, combining information on the genotype of the pig and its faecal microbiota composition could likely improve the accuracy of prediction of phenotypes for feed efficiency related traits.

'Optimum vs maximum': litter size and weight in pigs

E. Sell-Kubiak[1] and E.F. Knol[2]
[1]Poznań University of Life Sciences, Department of Genetics and Animal Breeding, Wołyńska 33, 60-637 Poznań, Poland,
[2]Topigs Norsvin Research Center, Schoenaker 6, 6641 SZ Beuningen, the Netherlands; ewa.sell-kubiak@up.poznan.pl

Litter size and litter weight are traits that require an optimum to serve animal welfare. In addition, there is a need to define a maximal uterine capacity that would provide a healthy litter. In order to investigate 'optimum litter' several traits were defined. First, the highest value out of 5+ records per sow was selected as maximal values for total number born (maxTNB), number born alive (maxNBA), litter weight (maxLW). Second, variability of reproduction traits was measured as log-transformed residuals of all observations per sow (LnVar) per reproductive trait. Lastly, number of stillborn piglets was assigned to each litter with maximal value. In total 23,193 Large White sows from Topigs Norsvin were used for analysis in ASReml 4.1. Maximal values of reproductive traits had much higher heritability than traditionally analysed data, e.g. 0.3 for maxTNB vs 0.1 for TNB, which would enabled better response under selection. Still, at least 5 observations per sow would be required as in traditional breeding approach. All analysed traits had positive genetic correlations with each other ranging from 0.2 to 0.98. The highest correlations were for maxTNB and maxNBA (0.9), LnVarTNB and LnVarNBA (0.98), which should be expected from the biological relationship between TNB and NBA. Also high correlations were present between LnVarLW and maxLW as well as LnVarLW and LnVarTNB (both 0.8), which indicated a high increase in variability of LW with selection for maxLW or when variability of TNB is not controlled. Correlations between stillborn traits and other reproduction traits were low to moderate (0.18-0.64) indicating increase in stillborn piglets with selection for maximal values or not controlled variability of traits. The genetic correlations estimated in this study indicated that selection for any of maximal value traits will also result in an increase in variability of the reproduction traits and in number of stillborn piglets. Selection towards the 'optimum litter' with expectation of using the maximal uterine capacity requires simultaneous consideration of all correlated traits. Research funded by National Science Center grant no. 2016/23/D/NZ9/00029.

Genome-wide association study on maternal behaviour and reproductive traits in two maternal sow lines

K. Schodl[1], B. Fuerst-Waltl[1], N. Khayatzadeh[1], P. Knapp[2], C. Winckler[1] and C. Pfeiffer[1,2]
[1]University of Natural Resources and Life Sciences Vienna (BOKU), Gregor-Mendel Str. 33, 1180 Vienna, Austria, [2]PIG Austria GmbH, Waldstr. 4, 4641, Austria; katharina.schodl@boku.ac.at

Sows expressing normal maternal behaviour may show better performance and have a higher welfare status. However, currently behavioural traits are not included in breeding programs for maternal sow lines in Austria. Such traits are characterised by low heritabilities and thus a genome-wide association study (GWAS) analysis was applied to explore the genomic background of aggressiveness towards stockperson (AS), farrowing behaviour (FB), general maternal abilities (GM), number of live-born piglets (NLB), number of weaned piglets (NWP) and percentage of piglets crushed until weaning (PCW). In total, 595 Large White and 212 Landrace performance-tested sows from 22 Austrian nucleus farms were genotyped using the Illumina PorcineSNP60 BeadChip V2 including 62,621 single nucleotide polymorphism (SNP) markers. Behavioural and reproductive traits were recorded by farmers after training and estimated breeding values (EBV) were derived for all traits. After SNP standard quality control, an association analysis was performed for each of 44,960 SNPs for 807 sows using a univariate linear model in GEMMA software with the EBVs as dependent variable. Due to the large number of hypothesis tests a Bonferroni correction was applied resulting in a significance threshold of $P<10-6$. SNPs with a p-value $<10-4$ were considered suggestive. DNA-regions of ±0.5 mega base pairs next to significant and suggestive SNPs were searched for relevant coding genes. On Chromosome (SSC) 1, 10 and 12 significant SNPs were found for NLB and NWP, respectively. Relevant genes in this area were SMAD2 and SMAD7, which are both involved in embryogenesis and viability of offspring postpartum. On SSC 18 next to the DOCK4 and GPR85 genes, which are involved in the expression of behaviours assessed in animal models for schizophrenia, 4 suggestive SNPs were found for AS. Distinct peaks and relevant adjacent genes could not be clearly identified for the other traits. GWAS identified promising associations, which may be subject to further analyses investigating their potential for implementation in genomic selection for maternal sow lines.

Characterisation of autozygosity in pigs from a three-way crossbreeding

A. Ganteil[1,2], S.T. Rodriguez-Ramilo[2], B. Ligonesche[1] and C. Larzul[2]
[1]SAS NUCLEUS, 7 Rue des Orchidées, 35650 Le Rheu, France, [2]INRAE, Université de Toulouse, ENVT, UMR1388 GenPhySE, 24 chemin de Borde Rouge, Auzeville Tolosane, 31326 Castanet Tolosan, France; audrey.ganteil@inrae.fr

Crossbreeding can be used in livestock animals to increase genetic diversity. This rise of variability is related to the heterozygous status of crossbreds. The evolution of diversity during crossbreeding can be assessed using genomic data. The objective of this study is to describe patterns of Runs of Homozygosity (ROH) in animals from a three-way crossbreeding, from parental pure lines and in their crossbred offspring. The crossbreeding scheme consisted in a first crossbreeding between Pietrain boars and Large White sows, then the Pietrain × Large White sows offspring were crossed with Duroc boars. The offspring of this second crossbreeding were called G0 animals. Finally, offspring of G0 boars and G0 sows were called G1 animals. All animals were genotyped using Illumina SNP60 porcine chip. After applying filtering criteria, analyses were performed with 2,336 animals and 48,579 autosomal SNP. The ROH detection was performed with PLINK software. Results indicated that the mean ROH-based inbreeding coefficients were 0.27±0.05, 0.23±0.04 and 0.26±0.04 for Duroc, Large White and Pietrain, respectively. ROH were detected in the Pietrain × Large White crossbred but the number and the size of these homozygous segments were lower than in their parents. Similar results were observed in the G0 crossbred. However, in G1 crossbred the number and the size of ROH were higher than in G0 parents. Both in G0 and G1 animals, similar ROH hotspots were detected on chromosomes 1, 4, 13 and 15. Moreover, in G1 animals ROH greater than 16 Mb were observed, probably suggesting regions of low recombination rate. The conservation of these homozygous segments in the three populations of crossbred means that some haplotypes were shared between parental breeds. These findings contribute to the understanding of the management of genetic diversity in crossbreed populations.

Towards a pig breeding program aimed at improving animal welfare in Austria

C. Pfeiffer[1,2], K. Schodl[1], B. Fuerst-Waltl[1], C. Winckler[1], A. Willam[1], C. Leeb[1] and P. Knapp[2]
[1]University of Natural Resources and Life Sciences Vienna (BOKU), Gregor-Mendel Str. 33, 1180 Vienna, Austria, [2]PIG Austria GmbH, Waldstr. 4, 4641 Steinhaus, Austria; christina.pfeiffer@pig.at

Commercial pig breeding programs generally emphasise genetic gain in production and efficiency. However, selection for high production may result in co-selection of unfavourable behaviour and health traits. Therefore, the Austrian pig breeding organisation PIG Austria revised the breeding goal for maternal lines and implemented a litter vitality index (LVI) and a maternal ability index (MAI) into routine genetic evaluation. The breeding goal was defined beforehand using a participatory approach involving breeders, staff from the breeding organisations, other industry experts and scientists. Several phenotypic traits regarding piglet vitality and maternal ability were defined and data were collected by the breeders on 23 farms. Breeders were trained twice for data collection, first in group, then individually on-farm. After extensive validation studies, seven traits remained. Traits regarding piglets were mean individual birth weight, standard deviation of birth weight and litter vitality with a weighting of 30:50:20. The estimated heritability (h^2) of LVI was 0.22. Traits regarding MAI were farrowing behaviour, sow's aggressiveness towards piglets, percentage of crushed piglets and overall good maternal ability during suckling period with a weighting of 25:25:25:25. The estimated h2 of MAI was 0.07. Due to the low level of h^2, MAI has not been included in the total merit index (TMI), but can be used as additional selection criterion by breeders and/or multipliers. With the beginning of 2020 LVI has been implemented into the TMI with a weight of 20%, which entailed a re-ranking of breeding boars and sows. In particular pigs with high EBV for number of live born and weaned piglets lost some TMI points. Spearman rank correlations between the former TMI and the TMI since 2020 were 0.95 for the artificial insemination boars and 0.92 for the sows. Including LVI and MAI into routine genetic evaluation can contribute to improved sow and piglet welfare.

Evaluation of the hygiene measure 'sow shower' based on microbial investigations
S.M. Schmid, S. Terletzki, C. Heinemann, J.J. Hayer and J. Steinhoff-Wagner
Institute of Animal Science, University of Bonn, Katzenburgweg 7-9, 53115 Bonn, Germany; simone.schmid@uni-bonn.de

The 'sow shower' is a preventive hygiene measure to remove pathogens and parasites from the sows' bodies before rehousing them to the farrowing pens. Aim is the disruption of infection chains and the reduction of the bacterial load newborn piglets are exposed to. It was the objective of this study to evaluate the sow shower's impact on sows' body hygiene. For this, samples were taken from 34 sows at 6 farms each before and after showering and rehousing, while all farms applied the same detergents (Sauenwaschmittel, THESEO Deutschland GmbH, Wietmarschen, Germany). Sows were randomly allocated to 3 groups: 1. washing with detergent, rinsing after 5 minutes; 2. washing with detergent, no rinsing; 3. washing with water only. Swab samples were used to measure adenosine triphosphate (ATP) and protein content at 4 locations (back, claw, teats, ear tag). Agar contact plates and glove samples taken at thigh, back and teats were microbiologically examined using selective and non-selective media for total viable and coliform count, occurrence of streptococci, staphylococci, *Salmonella* and *Enterobacteriaceae*, as well as Methicillin-resistant *Staphylococcus aureus* (MRSA) and extended spectrum β-lactamase-producing bacteria (ESBL). Additional samples were taken with adhesive tapes to microscopically check for developmental stages of endo- and ectoparasites on skin surface. Statistical analysis was performed with SAS 9.4 by using a linear model with time and group as fixed effects and by calculating correlations. Preliminary results showed that after washing, ATP content was lower at back and ear tag ($P<0.05$), whereas protein content was lower at all sampled locations ($P<0.01$). Furthermore, total viable count was lower at teats and back after washing ($P<0.01$), same accounts for staphylococci count at back and ESBL count at teats ($P<0.05$), as well as streptococci count at teats and back ($P<0.01$). Rinsing off the detergent or allowing it to dry had no significant effect on bacterial load after washing. Further research needs to be carried out with regard to long-term effects on sows' hygiene and piglet health.

Diverging effects of vital or dead probiotics on primary cultured porcine immune cells
S. Kreuzer-Redmer[1], F. Larsberg[2], D. Arends[2] and G.A. Brockmann[2]
[1]University of Veterinary Medicine Vienna, Nutrigenomics, Veterinärplatz 1, 1210 Wien, Austria, [2]Humboldt Univerität zu Berlin, Thaer Institute, Invalidenstraße 42, 10115 Berlin, Germany; susanne.kreuzer-redmer@vetmeduni.ac.at

Probiotics are widely used feed additives for enhancing the health status of pigs. However, the underlying mechanisms have not yet evaluated in detail. In this study, simplified cell culture experiments were conducted with primary cultured porcine immune cells from blood (PBMCs) and from mesenteric lymph nodes (LNLs). PBMC and LNLs were co-cultured with the probiotic *Enterococcus faecium* NCIMB 10415 as vital or UV-treated bacteria in different concentrations and for different time points. After treatment, cells were analysed on cell and transcript level by qPCR, RNA-Sequencing, and flow cytometry. We detected higher relative cell counts of CD8b+ cytotoxic T-cells ($P<0.05$) in the treatment group with vital *E. faecium* after 1, 5 and 20 hours of incubation compared to untreated controls. Treatment with UV-killed bacteria did not result in changes. By analysing B-cells, we observed a different pattern. We found a higher relative cell count of CD21+ B-cells ($P<0.05$) with killed *E. faecium*, while there was no effect with vital *E. faecium*. RNA-Sequencing of primary cultured porcine immune cells revealed 537 significantly differential expressed genes within the groups treated with vital *E. faecium* against control, UV-killed *E. faecium* against control and vital *E. faecium* against UV-killed *E. faecium*. Interestingly, most of the genes which were differentially expressed between the groups treated with vital and killed *E. faecium* (250) were also differentially expressed between UV-treated bacteria and the control without bacteria treatment (198), but not between the treatment group with vital *E. faecium* versus control (9). That hints towards a different mode of action of dead and active bacteria on the immune cells. Taken together, this *in vitro* study demonstrates a diverging effect of vital or dead *E. faecium* on primary cultured porcine immune cells. Furthermore, we suggest, that vital *E. faecium* influences the direction of immune response towards an enhanced response of cytotoxic T-cells at the expense of B-cells, whereas B-cells seem to be stimulated through dead bacteria.

Utilising beef × dairy crossbreds for beef production – Danish experiences

M. Vestergaard[1], P. Spleth[2], R. Stephansen[2], A. Fogh[2], M. Kargo[1] and J. Ettema[3]
[1]Aarhus University, 1 Aarhus University, Foulum, 8830 Tjele, Denmark, [2]SEGES, 2 SEGES, 8200 Aarhus N, Denmark, [3]SimHerd A/S, 8830, Tjele, Denmark; move@seges.dk

In countries with a well-established dairy sector, the most efficient way of producing beef is to raise beef × dairy crossbreds rather than using either purebred dairy animals or a specialised suckler cow system with beef breeds. Many factors contribute to overall efficiency. Results might vary by setting, country or region. In Denmark data from the Central Cattle Database were used to evaluate the consequences of carrying a crossbred foetus for milk yield (current and following lactation), on calving interval, calving ease and calf vigour. Calf birth weight and drinking abilities were assessed. Trials in both research station and private farm settings were conducted with promising beef × dairy crosses slaughtered by 8, and at 12 month of age. Simulations using the SimHerd software showed positive economic value of using beef semen alone and in combination with sexed semen in dairy herds. Results encouraged farmers to increase the use of beef breed semen to around 20% of all inseminations. A reasonable sharing of benefit between dairy farmer and beef producer was needed. A tool based on actual performance and carcass payment was developed and extended to include actual genetic value (for growth, survival rate, health, and carcass value) of the crossbred calf. A 'top-of-the-charge' ranking list of beef sires across breeds was made available. This led to extended use of Belgian Blue semen, now accounting for 80% of the beef breed inseminations. An on-going project is testing over 12,000 crossbred bulls (and heifers) to get genetic values for growth, FCE, GHG emission, carcass value and intramuscular fat. Beef breeds include Belgian Blue, Angus, and Charolais. A chart with different typical beef production systems showing the production economics of the breed combinations (e.g. Holstein bulls, Beef × Holstein bulls, and Beef × Holstein heifers, all slaughtered at 12 months) is available and regularly updated with current input (e.g. feed) and output (e.g. carcass) prices. The presentation will cover the above subjects in a holistic approach based on results from experiments and trials and the tools developed by the Danish extension service over the last 10 years.

In France, a new dairy calf to beef production for the out of home consumers

C. Fossaert[1], D. Le Pichon[2], G. Milet[1], M. Douguet[1], J.J. Bertron[1] and S. Brouard[1]
[1]Institut de l'élevage, 149 rue de Bercy, 75012 Paris, France, [2]Chambre d'agriculture de Bretagne, Station experimentale de Mauron, La Touche Es Bouvier, 56430 Mauron, France; clement.fossaert@idele.fr

Since 2016, the number of dairy calvings in France has been decreasing (-4,4%), and yet the market for French dairy calves remains depressed. Nowadays, 50% of the dairy calves (male of dairy type, male and female crossbred between a dairy dam and a beef sire) available for meat production are raised for veal and 18% are raised as young bulls (<24 months old). On the long run, veal production is decreasing in France, and since the end of milk quotas, young bull slaughterings are also collapsing. This means less and less demand for 14 day old dairy calves, resulting in low prices. As a result, exports have doubled in the past 5 years, now reaching 17% of the calves available for meat production. Among the 3,5 million dairy calves born each year around 20% are crossbred between a dairy dam and a beef sire. Crossbreeding is expanding mostly due to low dairy calves price, lower AI cost, and the development of sexed semen. Sexed semen allows farmers to get their replacement heifers from a restricted number of cows. Often exported, these calves could produce carcasses (~300 kg) well adapted for growing the out of home consumption market, which is currently mostly supplied by imported dairy cow meat. In 2015, following a demand from Interbev Bretagne, some trials have been run on the experimental farm of Mauron by the Chambre d'Agriculture de Bretagne. Young steers and heifers from Holstein cows, mated to Limousin (HoXLim), Angus (HoXAng) or Hereford (HoXHer) bulls have been fattened indoors and their performance compared. These animals were slaughtered at 15-16 month old, with a 300 kg carcass weight. Since the takeover of the experimental farm of Mauron by the Institut de l'Elevage, trials on this system continue with new cross breeds. For the next three years, 336 animals from 7 genetic types (Limousin, Charolais, Blanc Bleu, INRA 95 and Angus bulls mated to Holstein cows and pure Normand and Limousin bulls mated to Normand cows) will be produced as young steers or heifers, on grass based diet. The quality of the carcass produced will be measured, as well as the environmental footprint and animal welfare.

Using genetics to improve dairy-bred & dairy-beef carcases

A. Brown[1], F. Pearston[1], A. Wells[2], K. Matthews[1], M. Winters[1] and M. Coffey[2]
[1]AHDB, Stoneleigh Park, Kenilworth, CV8 2TL, United Kingdom, [2]SRUC, Edinburgh, EH9 3JG, United Kingdom; alex.brown@ahdb.org.uk

Currently 40%-50% of UK beef carcases do not meet market specification, meaning producers do not get maximum value from their stock. Approximately 50% of UK beef originates from the dairy herd, from either purebred dairy cattle & dairy-beef crosses. Genetic improvement for carcase-related traits in dairy & dairy-beef cattle therefore provides an opportunity to add value to both classes of cattle. Carcase trait phenotypes are routinely shared with SRUC from multiple UK processors, with the data representing roughly 40% of the national kill. Alongside pedigree data provided by the British Cattle Movement Service, milk recording organisations & breed societies, we have used a multi-breed & crossbreed genetic evaluation to produce genetic parameters & estimated breeding values (EBVs) for carcase weight, fat & conformation, days to slaughter & average daily carcase gain (ADCG). Heritabilities are moderate to high ($h2$ 0.40-0.63), indicating a good opportunity for genetic improvement. In the dairy sector, as part of their national dairy evaluations, AHDB have published a Dairy Carcase Index based on conformation and ADCG to provide an estimate of genetic merit for purebred dairy sires. Although a useful extra tool for selection, selecting for improved carcases is not a priority for the average dairy farmer, so uptake has been low. Current advice to dairy farmers is to improve their youngstock value through a two-pronged approach; by serving their best females with sexed semen to breed heifer replacements, & serving the rest of the herd with semen from high-EBV beef bulls to improve progeny carcase value. Carcase EBVs for UK beef bulls have been released as the AHDB national beef evaluations, with EBVs published for all five traits. These new evaluations allow dairy producers to select beef bulls that will produce progeny with higher value carcases. An economic carcase merit index that will demonstrate the potential additional profit from the use of high genetic merit beef bulls on both dairy & beef herds is in progress. Alongside current performance recording data, these new evaluations can inform processor premium schemes where producers are financially rewarded for using higher genetic merit bulls.

Performance of dairy × beef bulls and dairy bulls

E. Nadeau, K. Wallin, F. Dahlström, E. Stenberg, M. Johansson and A. Hessle
Swedish University of Agricultural Sciences (SLU), Department of Animal Environment and Health, Box 234, 53223 Skara, Sweden; elisabet.nadeau@slu.se

Carcass value of progeny from dairy cows can be increased by using beef breed semen on cows not suitable for further breeding. Crossbreeding with beef breeds has great potential in Sweden, as the proportion of beef semen to dairy cows is only 13% of all inseminations. This study compared average daily gain (ADG) and carcass traits in bulls, born to a dairy breed dam and a dairy or beef breed sire, when raised indoors at different feed intensities to reach different ages at slaughter. The study included 35 bulls of dairy breed (D; 15 Swedish Red and 20 Swedish Holstein) and 34 crossbred bulls (C; 15 Swedish Red × Angus and 19 Swedish Holstein × Angus). The calves started the experiment at 3 to 4 months of age and were fed either a high-intensity (HI) or a low-intensity (LI) diet *ad libitum* corresponding to slaughter at 15 or 18 months of age, respectively. The HI and LI diets had average grass-clover silage proportions of 35 and 53% of dry matter, respectively. The grass-clover silage was supplemented with rolled barley until slaughter, with rolled peas until 300 kg body weight and with cold-pressed rapeseed cake until 200 kg body weight. Data were analysed in Proc Mixed of SAS with fixed effects of breed (D vs C) and feed intensity (HI vs LI) and random effect of pen. No significant interactions between breed and feed intensity were found. When averaged over feed intensity, the C bulls had higher ADG (1.48 vs 1.33 kg/d, P<0.001), higher carcass weight (389 vs 343 kg, P=0.001), higher dressing percentage (54.2 vs 52.5%, P=0.003) and higher scores of conformation (7.3 vs 5.5, P<0.001) and fatness (9.6 vs 8.1, P<0.001), according to the EUROP classification system with 15 subclasses, compared to the D bulls. The C bulls also had higher visually determined marbling score (2.3 vs 1.5, P<0.001) on a 10-point scale compared to the D bulls. Bulls fed the LI diet had lower ADG (1.34 vs 1.47 kg/d, P<0.001) but higher carcass weight (387 vs 346 kg, P=0.002) and thus higher marbling score (2.1 vs 1.7, P=0.005) than bulls fed the HI diet. In conclusion, the ADG and carcass traits in dairy bulls were improved by crossbreeding with the beef breed Angus.

Antimicrobial drug use and clinical health score of udder-reared dairy calves on Swiss organic farms

R.C. Eppenstein[1], C. Schneider[1], A. Spengler[1], E. Lantinga[2], F. Leiber[1] and M. Walkenhorst[1]
[1]*Research Institute of Organic Agriculture (FiBL), Animal Science, Ackerstrasse 113, 5070, Switzerland,* [2]*Wageningen University and Research, Prof. Lantinga is retired, Droevendaalsesteeg 1, 6708 PB Wageningen, the Netherlands; rennie.eppenstein@fibl.org*

During the recent decade, cow-calf rearing systems, which allow calves to suckle milk from a dairy cow for parts or all of the rearing period are increasingly gaining public interest. To date, no epidemiological multi-herd study exists that evaluates calf health and antimicrobial use for dairy farms using cow-calf rearing systems. This study quantified incidence densities (ID) of veterinary treatment events, antimicrobial treatment incidences (TI) and clinical health scores for dairy calves reared with udder access on 14 organic dairy farms in Switzerland and Southern Germany. Antibacterial drugs were used on 31% of farms and antiparasitic drugs on 15% of farms. Overall, a median ID of 25.35 veterinary treatments per 100 calf-years, a maximum TIUDD of 40 antimicrobial daily doses and a median calf mortality risk of 3% were found. Assessments of the calves' health using clinical health scores, revealed nasal discharge (23.5%) to be the most frequent health impairment, followed by ocular discharge (22.9%) and skin conditions (19.7%). Perianal soiling was observed in 10.1% of calves. Five percent of calves qualified as severely health impaired. Compared to conventional Swiss dairy farmers who rear calves artificially, antimicrobial drug use was substantially lower. This can partly be attributed to the mind-set of organic farmers. In severe cases of calf disease, organic farmers should not be reluctant to resort to antimicrobial drugs, such as to prevent avoidable calf suffering and mortality.

Dam-reared calves: lessons from pioneer farmers for Danish dairy producers

J.O. Lehmann[1], L. Mogensen[1] and M. Vaarst[2]
[1]*Aarhus University, Department of Agroecology, Blichers Alle 20, 8830 Tjele, Denmark,* [2]*Aarhus University, Department of Animal Science, Blichers Alle 20, 8830 Tjele, Denmark; jespero.lehmann@agro.au.dk*

In Denmark, separation of dairy calves from their dam commonly occur within the first 24 to 48 hours whereas weaning occur around either 8 weeks (conventional) or 12 weeks (organic). Consumers increasingly question the practice of early separation, and there is a growing interest among Danish dairy farmers for developing new housing and management systems that incorporate a type of dam-rearing. Danish dairy farms are mostly relatively large with high-yielding cows in capital intensive and high cost systems that sets a tight margin for the milk price, which may challenge the implementation of dam-rearing. Several farmers in northern Europe practice different forms of dam-rearing, and their choices and experience may serve as inspiration for Danish dairy farmers. So far, we have visited 9 farms in Scotland, The Netherlands and Germany that represent a variety of breeds and a herd size ranging from 30 to 130 cows as well as an average of around 5,000 kg milk delivered plus 1,300 kg drunk per calf. The 5 highest yielding herds milk cows twice per day during the dam-rearing period. Six herds combine an initial period of full-time contact with a period of part-time dam-rearing. Separation occur between 4 and 22 weeks after calving, and 6 herds separate cow and calf abruptly whereas 4 herds wean calves abruptly. One case from The Netherlands and 1 from Scotland represent particularly promising systems for Danish dairy farmers. On the Dutch farm, calves are kept in separate deep litter area for the first 120 days, and a milking robot is used to gradually reduce the amount of time that the dams gain access to this area. Weaning occurs 2 weeks after separation with both separation and weaning done gradually. This farm delivers 9,500 kg of milk to the dairy per cow per year. On the Scottish farm, calves and dams are kept full-time in the same area the first 3 months and part-time the following 3 months. Weaning occurs gradually with nose flaps one week before a gradual separation. These case studies will influence the development of different implementation strategies that later will be assessed for various impacts including production economics.

Transition to domestic beef and milk production from dual purposed cattle – effect on climate change

S. Samsonstuen[1], H. Møller[2] and H.F. Olsen[1]
[1]Norwegian University of Life Sciences, Faculty of Biosciences, Aboretveien 6, 1432 Ås, Norway, [2]Ostfold Research, Stadion 4, 1671 Kråkerøy, Norway; stine.samsonstuen@nmbu.no

In Norway, sustainable intensification is an overall political goal to meet the demand of a growing population and to uphold the degree of self-sufficiency. At the same time, the greenhouse gas emissions from agriculture are required to be reduced by 40%, compared to the 1990-level. Only 3% of the domestic area is agricultural land. Livestock plays an important role in the production of agricultural products, as the cultivatable area is mainly used for grass and feed production. Norway is a part of the global discussion on the future role meat should play in the human diet to address future sustainability issues. There are substantial differences in greenhouse gas (GHG) emissions between dairy-based beef and beef from specialised beef breeds, and between production systems in different regions and continents. The scope of the case study is to analyse the effect on GHG emission of altering the domestic meat and milk production. Currently, the Norwegian market for beef and milk is provided by the dual purpose breed Norwegian Red (62%), (several) specialised beef breeds (27%) and imported meat (11%). In the case study, GHG emissions are calculated for the current production level as a baseline. This is compared to a scenario where the demand for beef and milk products are met exclusively by optimising the integrated production of milk and beef through the domestic, dual purpose breed. Such scenario will provide information of the potential to reduce the domestic emissions compared to the baseline situation, the optimal milk yield of the dairy cow in an integrated production system, and how the diet can be changed in the direction of better utilisation of resources in pasture and forage. Results from the study will be included in the final presentation.

Association between morphological classification and productivity of Holstein Friesian cows

J.L. Cerqueira[1,2], J.M.C.T. Sousa[2], L.M.G. Cruz[2], J. Cantalapiedra[3], I. Blanco-Penedo[4] and J.P. Araújo[2,5]
[1]Centro de Ciência Animal e Veterinária (CECAV), UTAD, Quinta de Prados, 5001-801 Vila Real, Portugal, [2]Escola Superior Agrária do Instituto Politécnico de Viana do Castelo, Ciências Agronómicas e Veterinárias, Refoios, 4990-706 Ponte de Lima, Portugal, [3]Servicio de Ganaderia de Lugo, Xunta de Galicia, SGL, Lugo, 27071 Lugo, Spain, [4]Swedish University of Agricultural Sciences, Ciências Clínicas, Almas Allé, 8, 750 07 Uppsala, Sweden, [5]Centro de Investigação de Montanha (CIMO), CIM, Refoios, 4990-706 Ponte de Lima, Portugal; cerqueira@esa.ipvc.pt

The conformation or type of a cow has proven to influence cow productivity and longevity. Morphological classification is an essential tool used to improve the conformation of cows, through the score of each animal. The present study aimed to perform the morphological classification in Holstein Friesian cows. Morphological classification of 1,823 dairy cows was carried out in 30 farms in the Aveiro region in Portugal, according to ICAR (2018) methodology, evaluated on a scale of 1 to 9 points, linear characteristics and on a scale of 100 points for large regions. Statistical analysis was performed with SPSS for Windows version 22. Effect of morphological classification on milk production was analysed by ANOVA and Tukey means comparison test. Average milk yield was $9,633.8 \pm 1,867.2$ kg (305-d), ranging from 6,171.1 kg (farm 29) to 11,643.7 kg (farm 22). The average of the overall classification of all parameters were 81.6 points (Good Plus), being 'capacity' the parameter with the highest score (83.0 points). Mammary gland system was the region with the lowest classification (80.3 points). Farms 14 and 15 revealed highest rankings with 83 points, associated these farms in 'Good Plus' class (80 to 84 points). Farms 2, 17 and 24 showed the lowest morphological classification (<80.5 points). Effect of morphological classification were observed in milk production at 305-d ($P<0.001$). Lowest yields ($7,477.6 \pm 4,005.0$ kg) were registered in animals with 'Fair' classification. In contrast, 'Very good' cows achieved highest production ($10,290.5 \pm 1,694.2$). Therefore, morphological classification is a valuable and reliable instrument for scoring and selecting dairy cows for the milk production trait.

An evaluation of sward age and clover inclusion on the performance of growing steers

L. Chesney[1,2], N. Scollan[1] and F.O. Lively[2]
[1]Queens University Belfast, University Road, Belfast BT71NN, United Kingdom, [2]Agri-Food and Biosciences Institute, Large Park, Hillsborough BT266DR, United Kingdom; Lauren.Chesney@afbini.gov.uk

Improvements in grass breeding has produced modern varieties with improved traits for growth and animal performance. However, in Northern Ireland it is estimated that only 3% of grass swards are reseeded annually which limits the potential of reaping the benefits of this breeding. Inclusion of clover in grass swards has potential to improve livestock performance. The aim of this study was to compare a new and old sward, with and without clover, on forage production and animal performance. This 2 (sward age) × 2 (clover inclusion) factorial study was conducted at AFBI Hillsborough during the 2019 grazing season. There 4 treatments included old sward (OS), old sward plus clover (OS+C), new sward (NS) and new sward plus clover (NS+C). The OS had been last reseeded in 1992 and the NS was reseeded in 2018. The study was carried out in a 7.2 ha field which was sub-divided into 6 blocks, each block containing one replicate of the 4 treatments. This provided 6 paddocks for each treatment randomly located throughout the field. Steers were rotationally grazed around the six paddocks from 19th April until 15th October. Additional cattle were added or removed on a weekly basis to each treatment group to maintain optimal sward height. Sward height was measured weekly throughout the grazing season, with additional measurements taken pre and post grazing. On entry to each paddock grass samples were analysed for nutritional quality and botanical composition. All cattle were weighed fortnightly, with the additional cattle being weighed on entry and exit to the core grazing groups. Data was analysed using REML in GenStat. Average clover content throughout the season was 0, 0, 9 and 13% and average perennial ryegrass content was 61, 87, 61 and 74% for OS, NS, OS+C and NS+C respectively (P<0.001). Total herbage production was 9.8, 10.3, 10.7 and 10.1 t DM/ha for OS, NS, OS+C and NS+C respectively (P>0.05). There was no significant difference in animal performance across the treatments (P>0.05). Results from the study show that sward age and clover inclusion had no effect on animal performance.

Feeding sorghum dry distiller's grains to early-weaned beef calves fed high grain rations

A. Simeone, V. Beretta, M. Anzolabehere and N. Cortazzo
Universidad de la República, Ruta 3 km 363, 60000, Uruguay; asimeone@adinet.com.uy

Replacing sorghum starch for digestible fibre and fat from distillers' grains in rations for growing cattle may reduce ration cost, however it could impair feed conversion. This study evaluated the effect of level of inclusion of sorghum dry distiller's grains plus soluble (DG; 29.8% CP, 50.5% NDF, 13.9% fat) in the diet, on feed utilisation and growth performance of lot-fed early-weaned beef calves. Twenty-four Hereford calves (80±11 kg, 61±12 days old) were randomly allocated to four total mixed rations differing in DG level: 0, 15, 30 or 45% of DM, replacing sorghum grain (SG) and soybean meal (SBM) in the control diet (0% DG, 57% SG, 25% SBM, 0.3% urea, 1.5% molasses, 15% Lucerne hay, 1.5% premix). Chemical composition for control diet was 17.7% CP, 25.5% NDF, 12.4% ADF and 3.2% fat. Rations were balanced for 18% CP across treatments, while NDF and fat contents increased up to 35.5 and 6.6%, respectively in the ration with 45% DG. Animals were fed *ad libitum* in individual pens during 10 weeks. Body weight (BW) was recorded every 14 days, and dry matter intake (DMI) was determined daily. Feed to gain ratio (FG) was calculated based on mean values of DMI and BW gain. In week 5, apparent DM digestibility (DMD) was estimated, and animal behaviour was characterised by visual appraisal during daytime. The experiment was analysed according to a randomised plot design, with repeated measures. Linear and quadratic effects associated to DG level were tested. Increasing DG in the ration did not affect BW gain (1.17 kg/d, SE 0.03; P>0.10) but it linearly increased DMI (kg/100 kg BW, y = 3.47+0.011x, P<0.01) and FG ratio (y = 3.54+0.016x, P=0.01). No differences were observed in DMD (85.3% SE 3.8; P>0.10), but higher DG tended to linearly increase digestible DMI (P=0.06) as well as NDF and fat intake (P<0.01). The probability to find an animal eating, ruminating or idling did no vary between treatments (P>0.10), and no differences (P>0.10) were observed in DMI rate (g/min) after de first (7 to 12 am) and second meal (12 to 5 pm). Although FG ratio worsens as DG level increases, predicted FG=4.3 for DG=45% level is still within the economically viable range. These results constitute a useful tool for decision-making based on price relationships.

Sorghum dry distillers' grains as a supplement for early-weaned calves grazing temperate pastures

A. Simeone, V. Beretta, E. Arduin, S. Purtscher and C.H. Rebollo
Universidad de la República, Ruta 3 km 363, 60000, Uruguay; asimeone@adinet.com.uy

Sorghum dry distillers' grains plus soluble (DG), given their high energy and crude protein content, could be an option for supplementing early-weaned calves grazing summer pastures. Quantifying the response curve to DG in the supplement is relevant to decision support. The present study evaluated increasing levels of substitution of a commercial concentrate for early-weaning (20% CP, 25% NDF, 2.8% fat)] for DG (35% CP, 54% NDF, 4% fat): 0, 33, 66 and 100% on a DM weight basis, as a supplement for calves grazing *Festuca arundinacea* cv Tacuabé (12,5% PC, 54,2% FDN, 32,0% FDA) during summer (Jan 2 to March 28). Forty Hereford calves (81 ± 8 kg, 63 ± 9 days old) were randomly allocated to 8 grazing paddocks to receive 1 of the 4 supplements (n=2 paddocks) offered at 1.0 kg DM/ 100 kg body weight (BW) at 7 am. Pasture was rotative grazed in 7-days strips, with a forage DM allowance of 8% BW. Calves were weighed every 14 days without fasting. Supplement dry matter intake (SDMI) was estimated daily, and pasture utilisation (PU) and forage intake (FDMI) were estimated on even weeks. In weeks 4 and 8, during 3 days, grazing behaviour of calves was characterised by visual appraisal during daytime. The experiment was analysed according to a randomised plot design, with repeated measures and including initial BW as a covariate. Linear and quadratic effects associated to DG level were tested. Pre-grazing biomass did not differ between treatments (2976.4 kg/ha SE 166; P>0.10) as well as SDMI. Refusals were nil for all treatments. DG level in the supplement did not affect (P>0.10) PU (50.2% SE 3.24) or FDMI (4.01%BW SE0.26), but it linearly decreased BW gain (y, g/d= 767 − 1.15x R^2=0.74; P<0.01). Mean bite rate, and the probability to find a calf grazing or ruminating did no vary with treatment (24.8 ± 0.4 bites/min, 0.50 and 0.15, respectively; P>0.05) however, calves with 33% DG in the supplement showed higher resting time (P<0.05). Results evidence lower nutritive value of sorghum DG compared to a ration formulated for early-weaned calves. Nevertheless, as the range of variation in BW gains was low, partial replacement between feeds, based on response curve, could be worthwhile depending of relative feed prices.

Influence of paddock occupation time on the grazing activity of dairy cows in rotational systems

E. Castro Muñoz[1,2], G.F. Da Silva Neto[1,3], A.L.H. Andriamandroso[4], L. Ron[2], C. Montufar[2], O. Debauche[1], F. Lebeau[1] and J. Bindelle[1]
[1]Gembloux Agro-Bio Tech, University of Liège, Precision Livestock and Nutrition/ Precision Agriculture lab/, Passage des Déportés, 2 B, 5030 Gembloux, Belgium, [2]Facultad de Ciencias Agrícolas, Universidad Central del Ecuador, Producción y Salud Animal, Jerónimo Leiton y Gato Sobral, Ciudadela Universitaria. Quito, Ecuador, [3]Federal University of Rio Grande do Sul, Department of Forage Plants and Agrometeorology, Federal University of Rio Grande do Sul, Porto Alegre, Brazil, [4]Université Catholique de Lille, ISA Lille, 48 Boulevard Vauban, 59046 Lille, France; ecastro@uce.edu.ec

A survey was carried out in 42 dairy farms in 2017 in the highlands of Ecuador. It was observed that most of the farms used rotational grazing methods that vary widely in terms of occupation time: from 2 to 3 hours to several days, moving manually an electric fence to open up new paddock areas. Questioning the relevancy of extremely short occupation times, we hypothesised that differences in occupation times impact the eating behaviour of cows and ultimately milk production due to differences in exploration of the pasture and variation along the day in forage allowances. Three *Pennisetum clandestinum*-based pastures were grazed for 7 days each by 4 Holstein cows in-milk using differing in occupation times: long (7 d), medium (1 d) and short (3 h) and replicated in a crossed experimental design for 3 consecutive periods after a 30-d resting time. Two cows per pasture per period were equipped with iPhone S5 to compare grazing activity and movements on the paddocks during day-time. Preliminary results over a single period of measurement seem to indicate that cows grazing with medium term occupation times (1 d) tended to move over longer distances (3.42 ± 1.07 km), than those in long and short occupation times (2.73 ± 0.61 km and 2.37 ± 0.35 km, respectively), while no difference was observed in total grazing time or average speed when grazing (approx. 0.3 km/h). Such a trend in travelled distances might be explained by high pasture exploration requirements in the medium occupation time since animals already know the paddock well for long term and, for short term, the newly offered forage is always presented with a similar structure along a narrow band on the same side of the paddock every three hours.

DGAT1 genotype affects the response of dairy cows to customised dry period management

A. Van Knegsel[1], A. Kok[1], B. Kemp[1], M. Visker[2] and H. Bovenhuis[2]
[1]*Adaptation Physiology group, Wageningen University & Research, De Elst 1, 6708 WD, Wageningen, the Netherlands,* [2]*Animal Breeding and Genomics, Wageningen University & Research, Droevendaalsesteeg 1, 6708 PB, Wageningen, the Netherlands; ariette.vanknegsel@wur.nl*

Short and no dry periods improve metabolic status and fertility, but reduce milk yield in early lactation. Also, they hinder the use of dry cow antibiotics. Customising dry period management for individual cows could balance these benefits and trade-offs. Known cow factors relevant for customising dry period management are parity, milk yield level and somatic cell count. We hypothesise that due to its role in triglyceride synthesis and strong effects on milk production, also the DGAT1 (acyl-CoA: diacylglycerol acyltransferase) genotype affects the response of cows to dry period management. In this study, we randomly assign dairy cows (n=167) to 3 dry period lengths (0, 30 or 60 days). Cows with a dry period were treated with dry cow antibiotics at dry-off. Feed intake (week -8 till 14), body weight and milk yield (week -8 till 44) were recorded daily and averaged per week. Energy balance (EB) was calculated weekly. Lactation persistency was calculated as the decline in milk yield between 100 and 250 DIM. During week 1 till 14 post calving, feed intake was not affected by dry period management or DGAT1 genotype, but milk yield and EB were affected by a genotype by dry period management interaction. No dry period resulted in lower milk yield and greater improvement of EB for cows of the AA and KK genotype than for cows of the AK genotype. During the complete lactation, cows of the AA genotype were less persistent and tended to be heavier after no dry period, compared with cows of the AK or KK genotype. In conclusion, DGAT1 genotype can be a valuable cow characteristic to optimise dry cow management for individual cows in order to minimise milk yield losses and maintain metabolic benefits of a short or no dry period.

Consequences of extending the voluntary waiting period on ovarian cyclicity in dairy cows

J. Ma[1], E. Burgers[1], T. Lam[2], B. Kemp[1] and A. Van Knegsel[1]
[1]*Adaptation Physiology group, Wageningen University & Research, De Elst 1, 6708 WD, Wageningen, the Netherlands,* [2]*Department of Farm Animal Health, Utrecht University, Yalelaan 7, 3584 CL, Utrecht, the Netherlands; junnan.ma@wur.nl*

Extending lactation length in dairy cows is of interest because it reduces the number of calving events per cow per time unit and herewith potentially reduces the risk for health and fertility problems associated with calving and start of lactation. Extending lactation length can be realised by deliberately delaying first insemination, i.e. extending the voluntary waiting period for first insemination (VWP). Moreover, it can be hypothesised that insemination later in lactation is related with improved ovarian cyclicity. The aim of the present study was to evaluate the effect of an extended VWP on ovarian cyclicity and reproductive performance of dairy cows. Holstein-Friesian dairy cows (n=150) were blocked for parity, calving season and expected FPCM. Within blocks, cows were randomly assigned to one of three VWP (50, 125 or 200 days). Cows were artificially inseminated at first oestrous after end of VWP. Milk samples were collected three times a week until pregnancy and were analysed for progesterone concentration. At least two succeeding milk samples with progesterone concentration of 2 ng/ml or greater were used to indicate the onset of luteal activity (OLA) and to classify ovarian cycles. Ovarian cycles were classified as: normal (ovarian cycles of 18 to 24 days in length), short (ovarian cycles <18 days) and prolonged (ovarian cycles >24 days). During the 100 days around the end of the VWP (-50 till 50 days), cows with 200-d VWP had greater percentage of normal cycles (90.6 vs 53.5, P<0.01) and lower percentage of short (0.8 vs 11.2, P=0.02) and prolonged cycles (8.7 vs 35.3, P=0.01) compared with 50-d VWP. Cows with 200-d VWP had less days till pregnancy after end of VWP compared with cows with 125-d or 50-d VWP (31.2 vs 54.8, 58.3 d, P=0.04). In conclusion, extended VWP could improve reproductive performance, which was related with shorter intervals to pregnancy after end of VWP and greater percentage of normal cycles around end of VWP.

GWAS and pathway analysis on *in vivo* performance test traits in a local dual purpose cattle breed

C. Sartori[1], E. Mancin[1], S. Pegolo[1], N. Guzzo[2] and R. Mantovani[1]
[1]University of Padova, Dept. of Agronomy Food Natural resources Animals and Environment, Viale dell'Universita 16, 35020 Legnaro (PD), Italy, [2]University of Padova, Dept. of Comparative Biomedicine and Food Science, Viale dell'Universita 16, 35020 Legnaro (PD), Italy; enrico.mancin@phd.unipd.it

European dual purpose cattle mostly include local populations reared in marginal areas, which selection implies a good balancing between milk and meat traits and a sustainable farming. Meat selection is mainly based on traits recorded at performance testing (PT) stations like average daily gain (ADG), *in vivo* SEUROP fleshiness (FL) and *in vivo* dressing percentage (DP). The present study aimed to perform a genome-wide association study (GWAS) and subsequent post-GWAS analysis on ADG, FL and DP in the local dual purpose Rendena cattle. Phenotypic PT records were collected for about 1,600 young bulls in 34 years of testing, and breeding values were obtained for more than 5,600 animals in pedigree. Genotypic information at a SNP density of 150k was made available for 600 bulls, and 33k genotypes of further 200 were then imputed. BLUPF90 family of programs was used for analysis. The single-step procedure (ssGWAS) allowed to estimate the SNP substitution effects using the whole phenotypic information available. The GWAS results considered the proportion of additive genetic variance in 20-SNPs windows. Alternative software were also detected, like BGLR and EMMAX. The allocation of the SNPs to genes was performed using Ensembl and Genecards databases. Functional categories and biological pathways were detected from GO and KEGG databases. Heritability of traits was 0.36 on average. Genomic regions located on chromosomes (chr) 4, 5, 6, and, most of all, 15 explained more than 10% of the additive genetic variance of ADG. Regions included candidate genes like NCAPG, LCORL, and WIF1, involved in skeletal growth. In addition to regions of chr 6 and 15, most of additive variance for FL and DP was explained by chr 3, 10 and 24, including e.g. the gene FMN1. Biological pathway included e.g. the signalling of wnt proteins responsible for bone development. This study is the first GWAS analysis on *in vivo* FL and DP, and the identification of genomic regions would allow to better understanding and evaluate meat traits in dual purpose cattle.

Effect of castration on fattening and meat quality characteristics of Holstein-Friesian cull cows

T. Teresa Moreno
Centro de Investigaciones Agrarias de Mabegondo-AGACAL, Dpto. Producción Animal, Ctra. de Betanzos a Mesón do Vento km 7, 15318 Abegondo, Spain; teresa.moreno.lopez@xunta.es

Mature beef usually comes from animals in reproductive herds that are culled by age or other management reason, and finished before slaughtering. A complementary technique of fattening cull animals such as cow's castration by surgical removal of ovaries (ovarictomy) was introduced to test whether meat yield and herd profitability can be improved. Twenty Holstein-Friesian cull cows from the CIAM dairy herd were divided into two random groups of 10 cows/group `castrated´ (C) and not castrated´ (NC). The minimally invasive RILU castration method (modified by Ludueña) was used. After castration C and NC animals were both on spring pasture for four months. Afterwards, both groups were fattened for seven additional months till the slaughter time during three consecutive periods (P1, P2 and P3) with concentrate intakes of 4, 6 and 8 kg/cow/d for the following 72, 34 and 106 d, respectively, plus *ad libitum* grass silage. Animals were weighed monthly and the average daily gain (ADG) calculated. Physico-chemical characteristics of loins [% intramuscular fat (IF) and water holding capacity (WHC)] were analysed at the CTC. Results showed that castration had an effect on ADG and beef characteristics. C cows showed lower ADG values than the NC group (0.49 vs 1.14, P<0.0001) during the 30 first days on pasture due to the detrimental surgery effect. This effect on ADG tends to disappear during the following three months after C (1.07 C vs 1.01 NC; NS).There were significant ADG differences between groups at the 212 days of the finishing period, showing C cows the ADG higher values (0.66 C vs 0.44 NC, P<0.05). These differences were more relevant as the fattening period and the concentrate intake increased, (0.56 C vs 0.52 NC, P<0.1) for P1; (0.28 C vs 0.07 NC, P<0.1) for P2; and (0.85 C vs 0.5 NC, P<0.05) for P3. Beef showed better organoleptic characteristics in castrated animals: higher IF (18.3 C vs 13.9 NC; P<0.04) and lower WHC (19.2 C vs 20.8 NC; P<0.08). Castration could be an interesting technique for fattening cull cows because ADG and meat organoleptic characteristics seem to improve.

Genetic parameters for uniformity of harvest weight in white shrimp (*Litopenaeus vannamei*)

S. García-Ballesteros[1], B. Villanueva[1], J. Fernández[1], J.P. Gutiérrez[2] and I. Cervantes[2]
[1]*Instituto Nacional de Investigación y Tecnología Agraria y Alimentaria, Mejora Genética Animal, Ctra. Coruña km 7.5, 28040 Madrid, Spain,* [2]*Fac. Veterinaria (UCM), Producción Animal, Avda. Puerta de Hierro, s/n, 28040 Madrid, Spain; jmj@inia.es*

The uniformity of body weight is a trait of economic importance in the production of white shrimp (*Litopenaeus vannamei*) as the final product is the individual itself. In order to improve this trait through selective breeding, the existence of genetic variability of the environmental variance of body weight is necessary. Although several studies have showed the existence of such variability in other species, to our knowledge, there are no estimates in shrimp. This study aims to estimate the genetic variance for weight uniformity and the genetic correlation between this trait and body weight in a commercial population of this species. As a secondary objective, the genetic correlation between weight uniformity in two environments (selection nucleus and production system) was estimated. The database contained phenotypic records for body weight for 51,346 individuals belonging to the selection nucleus and for 38,297 individuals belonging to the production system. A double hierarchical generalised linear model was used for analysing weight uniformity in both environments. Fixed effects included sex and year for the nucleus data and sex and pond-year combination for the production data. The environmental and additive genetic effects were included as random effects. The estimate of the genetic variance for weight uniformity was greater than zero (around 0.25) in both the nucleus and the production system but the estimated heritabilities for weight uniformity were low (0.04). Despite of this, the high genetic variance coefficient (0.54) indicated a good potential for obtaining response to selection. In addition, the genetic correlation between weight and weight uniformity was close to zero in both environments and so selection for increasing uniformity would not have negative consequences for weight. Finally, the genetic correlation for weight uniformity between both environments was 0.64 indicating that selection in the nucleus will also lead to a response in the production system if this trait is included in the breeding goal.

Defatted *Tenebrio molitor* as protein source for European seabass aquafeeds

A. Basto[1,2], L. Petit[3], J. Pérez-Sánchez[4], J. Calduch[4], E. Matos[5] and L.M.P. Valente[1,2]
[1]*ICBAS, R. Jorge Viterbo Ferreira 228, 4050-313 Porto, Portugal,* [2]*CIIMAR, Av. General Norton de Matos S/N, 4450-208 Matosinhos, Portugal,* [3]*Faculté des Sciences de Montpellier, Pl. E. Bataillon 30, 34095 Montpellier, France,* [4]*IATS-CSIC, Ribera de Cabanes, 12595 Castellón, Spain,* [5]*SORGAL, EN 190-Lugar da Pardala, 3880-728 S. João Ovar, Portugal; anafbasto@gmail.com*

In 2017, the EU authorised insects as protein source for aquafeeds. *Tenebrio molitor* larvae meal (TM) is a rich protein source ($\leq$70% DM) with well-balanced amino acid profile. The success of aquaculture not only depends on fish growth, but also on the ability to produce high quality fish for consumers. Muscle represents the edible part of fish, being flesh firmness one of the most appreciated characteristics. Fillet texture is largely dependent on muscle cellularity which also determines fish growth potential. This study aimed to evaluate the impact of defatted TM (TMd), to substitute fishmeal (FM) in diets for European seabass. Four isonitrogenous diets were formulated to replace 0(CTRL), 40(TM40), 80(TM80) and 100%(TM100) of FM by TMd. Each diet was assigned to triplicate groups of 25 fish (55±5 g) fed 3 times/day, in a RAS (35‰, 22±1 °C). After 10 weeks of feeding, fish were individually weighed and measured. Muscle from 3 fish/tank was collected for evaluation of gene expression, cellularity, colour and texture profile. Overall, fish tripled their weight. Feed intake was lowest in fish fed TM100, but these fish had the best FCR resulting in similar final body weight and final K between treatments. The expression of *myod2* in fish fed TM80 and TM100 was lower than in those fed CTRL, while *mstn* expression was lower in fish fed TM100 compared to CTRL. But *mafbx/atrogenin1* and *mymk* were upregulated in fish fed TM100 compared to the CTRL. Muscle cellularity, texture and colour parameters were similar between fish fed different treatments. These results indicate that the observed gene expression modulation wasn't enough to produce a significant alteration in muscle phenotype, suggesting a high potential of TMd as protein source for aquafeeds. Work supported by ANIMAL4AQUA, funded by Portugal2020, financed by COMPETE-POCI-01-0247-FEDER-017610 and EUH2020 under the TNAprogram-AE090027 within AQUAEXCEL2020-652831. A.B. was financially supported by FCT grant SFRH/BD/138593/2018.

Algae blend: effects on performance of European seabass (*Dicentrarchus labrax*) juveniles

C.S.C. Mota[1,2], O. Pinto[3], T. Sá[1,4], C. Delerue-Matos[3], A.R.J. Cabrita[1,2], H. Abreu[5], J. Silva[6], A.J.M. Fonseca[1,2], L.M.P. Valente[1,4] and M.R.G. Maia[1,2]
[1]ICBAS, Instituto de Ciências Biomédicas Abel Salazar, Universidade do Porto, R. Jorge Viterbo Ferreira 228, 4050-313 Porto, Portugal, [2]REQUIMTE, LAQV, ICBAS, Instituto de Ciências Biomédicas Abel Salazar, Universidade do Porto, R. Jorge Viterbo Ferreira 228, 4050-313 Porto, Portugal, [3]REQUIMTE, Instituto Superior de Engenharia do Porto, R. Dr. António Bernardino de Almeida 431, 4249-015 Porto, Portugal, [4]CIIMAR, Centro Interdisciplinar de Investigação Marinha e Ambiental, Terminal de Cruzeiros de Leixões, Av. General Norton de Matos, 4450-208 Matosinhos, Portugal, [5]ALGAplus, PCI, Via do Conhecimento, 3830-352 Ílhavo, Portugal, [6]Allmicroalgae – Natural Products, SA, R. 25 de Abril, 2445-413 Pataias, Portugal; catiamo4@gmail.com

Macroalgae and microalgae are rich sources of nutrients as protein, lipids, organic minerals, vitamins and bioactive compounds with functional activity. Supplementation of individual species of macro and microalgae were shown to have dose-dependent and species-specific effects; best results being reported at low (<10%) inclusion levels. However, combined inclusion of macro and microalgae are mostly unknown. This study evaluated the effects of increasing inclusion levels (0, 2, 4 and 6%) of a commercial blend of two macro and two microalgae (Algaessence; ALGAplus/Allmicroalgae) on feed intake and growth of European seabass (*Dicentrarchus labrax*) juveniles. Iso-proteic and iso-energetic diets were fed to triplicate groups of juveniles (11.3±2.7 g) for 12 weeks. Fish final body weight increased with Algaessence dietary inclusion in a dose-dependent manner ($P<0.001$). The specific growth rate, voluntary feed intake, viscerosomatic and hepatosomatic indexes increased in fish fed Algaessence compared to the control diet (0%; $P<0.001$). Feed conversion ratio had the lowest values in fish fed either 2 or 4% algae inclusion whilst protein efficiency ratio was highest in fish fed 4% ($P<0.001$). Overall results suggest that performance of seabass juveniles was improved by dietary inclusion of Algaessence. Financial support of Fundação para a Ciência e a Tecnologia to CSCM (PD/BDE/150585/2020), MRGM (DL 57/2016 – Norma transitória), LAQV (UIDB/50006/2020) and MARINALGAE4AQUA project (ERA-NET COFASP/004/2015) is acknowledged.

Ectoparasites of fish farmed in Croatian part of Adriatic Sea

T. Šarić[1], B. Mustać[1], B. Petani[1] and S. Čolak[2]
[1]University of Zadar, Department of ecology, agronomy and aquaculture, Trg kneza Višeslava 9, 23000 Zadar, Croatia, [2]Cromaris d.d., Gaženička cesta 4b, 23000 Zadar, Croatia; tosaric@unizd.hr

Croatian marine aquaculture constitutes significant part of food production in Croatia. The most important species are European seabass (*Dicentrarchus labrax*), Giltheaed seabream (*Sparus aurata*) and Atlantic bluefin tuna (*Tunnus thynnus*). However, in recent years other farmed species such as Meagre (*Argyrosomus regius*) and Common dentex (*Dentex dentex*) are coming in focus of producers. Intensive farming of fish in particular environmental conditions may favour certain parasite species, so that the parasite population increases to a very high level and can cause serious outbreaks of diseases. This is especially a problem because the use of antiparasitic is still limited in fish farming. Point D of Chapter V of Section VIII of Annex III to Regulation (EC) No 853/2004 requires that FBOs (Food Business Operator) ensure that fishery products have been subjected to a visual examination for the purpose of detecting visible parasites before being placed on the market. In this context infestation of fish with ectoparasites are emerging problem for the producers because the restrictions of selling fishery products that are contaminated with ectoparasites for human consumption. In this presentation ectoparasites with significant effect of Croatian mariculture, like *Diplectanum aequans, Sparicotyle chrisophrii, Caligus minimus, Ceratothoa oestroides, Furnestinia echeneis* and *Lernanthropus kroyeri*, will be discussed.

Review on the potential of integrated multi-trophic aquaculture (IMTA) in the Adriatic Sea

I. Zupan[1], T. Saric[1] and M. Peharda Uljevic[2]
[1]University of Zadar, Department of ecology, agronomy and aquaculture, Trg Kneza Viseslava 9, 23000 Zadar, Croatia,
[2]IZOR Split, Šetalište I. Meštrovića 23, 21000 Split, Croatia; zupan@unizd.hr

The Eastern coast of the Adriatic Sea is too oligotrophic to support the commercial aquaculture of bivalves at the open sea. Therefore, most of the bivalve production is settled in areas close to the freshwater inputs, such as river estuaries, lagoons and other more eutrophic sites. These sites are relatively small and with increasing problem of conflict with other users (such as tourism and maritime traffic). IMTA sites in the oligotrophic environment seems like a good alternative for an increase of mussel aquaculture, one of the most consumed species on the EU market. Peharda *et al.* found that the production cycle of the mussels in IMTA conditions in the Adriatic Sea is equal to other sites in the area, while condition index was improved compared to the control sites. Zupan *et al.* (2012) published a review article on the potential and constrains for development of IMTA in the Adriatic, while Zupan *et al.* (2014) again reported a prolonged period of higher condition index in Noah's ark *Arca noae* cultured closer to the sea bass and bream farm. Gvozdenovic *et al.* faster growth and higher condition index of mussels held closer to the fish farms in Boka Kotorska Bay. The need for expanding the bivalve farming area is confirmed by the industry, with several zones for aquaculture being nowadays registered for integrated production of fish and bivalves. Since the Croatian national plan for aquaculture predicts an significant increase in near future, the development of the IMTA practice at the existing fish farm installations could be one of solutions for achieving this goal. The extensive knowledge gained through this period of over 10 years regarding the IMTA in Adriatic should be continued in the support of development of sustainable aquaculture practice.

Saccharomyces cerevisiae-hydrolysate improved performance of Asian sea bass

E. Valkonen[1], H. Kettunen[1], S. Hasan[1], J. Vuorenmaa[1], S. Chumkam[2] and O. Jintasataporn[3]
[1]Hankkija Ltd, Peltokuumolantie 4, 05801 Hyvinkää, Finland, [2]Valaya Alongkorn Rajabhat University, Faculty of Agricultural Technology, 1 Moo 20 Phaholyothin Road, 13180 Pathum Thani, Thailand, [3]Kasetsart University, Department of Aquaculture, Faculty of Fisheries, 50 Ngamwongwan Road, 10900 Bangkok, Thailand; eija.valkonen@hankkija.fi

Asian sea bass (*Lates calcarifer*) is an omnivorous, euryhaline fish with commercial importance especially in Southeast Asian aquaculture. We studied the effect of dietary *Saccharomyces cerevisiae* yeast hydrolysate (YH; Progut®, Hankkija Oy, Finland), on the performance of Asian sea bass. Previously, YH has been shown to improve the performance of farm animals such as chicken, pigs, calves and dairy cows. The present experiment was carried out in earth ponds of 0-2 ppt saline water in Thailand. Juvenile Asian seabass with a mean weight of 13 g were allocated into 12 net cages, 30 fish/cage, at a density of five fish/m3. Aeration was applied into all experimental unit for maintaining DO >5 mg/l, pH 7-8. Commercial-type feed was amended with the yeast at 0% (Control), 0.15%, and 0.30% to create three dietary treatments, each with four replicates. The fish were fed three times/day at 3-5% of body weight for 8 weeks. The feed intake, weight gain, feed conversion ratio (FCR) and mortality of the fish were measured. At the end of study, five fish per test unit were blood sampled for the following immunity-related parameters: white blood cell count, red blood cell count, haemoglobin, haematocrit, immunoglobulin M (IgM), lysozyme activity and glutathione. The study was conducted in completely randomise design and analysed by one-way ANOVA. The Duncan's multiple range test was used to determine the differences between the treatment means. For the time period 0-8 weeks, YH amendment significantly increased average daily gain, specific growth rate and the feed consumption (P<0.05 for all parameters), but did not affect FCR and mortality. The supplement decreased the counts of red and white blood cells (P<0.05), whereas haematocrit was highest in the YH 0.30% group (P<0.05). The other immune parameters were similar in all treatments. In conclusion, the YH supplementation increased feed intake and weight gain of Asian sea bass, thus having a positive effect on fish performance.

Effects of chitosan supplementation on gene expression in external mucosal barriers of common carp

E. Pietrzak[1], A. Dunislawska[1], A. Slawinska[1], J. Mazurkiewicz[2] and M. Siwek[1]
[1]UTP University of Science and Technology, Mazowiecka 28, 85-084 Bydgoszcz, Poland, [2]Poznan University of Life Sciences, Wojska Polskiego 71c, 60-625 Poznań, Poland; siwek@utp.edu.pl

The adaptive immune response of teleost fish is directly related to external mucosal barriers, such as skin and gill mucosa. Mucosa-associated lymphoid tissue (MALT) plays a key role in fish innate immunity. Bioactive compounds, such as chitosan, may have immunomodulatory effects on MALT. We hypothesise that chitosan supplemented in feed modulates the innate immune response in external mucosal barriers of common carp (*Cyprinus carpio*). The aim of this study was to analyse the effects of chitosan on the expression of immune-related genes in skin and gill mucosa of growing common carp. The fish were divided into a control group (fed diet without chitosan), and experimental group (fed diets enriched with 1% of chitosan). The growth test lasted 60 days. After that, the fish was euthanised (n=7) to collect tissues from the lateral line skin mucosa and gills from which RNA was isolated. The analysis of relative gene expression at the mRNA level was carried out using the RT-qPCR for genes associated with innate immune response, i.e. acute-phase protein (*CRP*), antimicrobial proteins (*His2Av* and *GGGT5L*), cytokines (*IL1β, IL4, IL8, IL10*, and *IFNγ*) lectin (*CLEC4M*) lyzosymes (*LyzC* and *LyzG*), mucin (*M5ACL*), peroxidase (*MPO*), proteases (*CTSB* and *CTSD*) and oxidoreductase (*TXNL*). To normalise the data, stability of candidate reference genes was analysed. The geometric mean housekeeping genes *40s s11* and *ACTB* was used for skin mucosa and *ACTB* and *EF1α* for gill mucosa. Relative quantification of the gene expression was calculated with ΔΔCt. In chitosan-fed group, down-regulation in gene expression signatures included: gill mucosa: *IL10* (P<0.001), *INFγ* (P<0.05), *CRP* (P<0.01) and *TXNL* (P<0.01); and skin mucosa: down-regulation of *INFγ* (P<0.05), *CTSB* (P<0.05), and *MPO* (P<0.05), and up-regulation of *His2Av* (P<0.05). In conclusion, chitosan supplemented in feed modulates the innate immune responses in skin and gill mucosa of common carp.

Non-antibiotic interventions: do they work, and can we do more to build our evidence base?

J.M. Sargeant
University of Guelph, Ontario Veterinary College, 50 Stone Rd E., N1G 2W1 Guelph, Ontario, Canada; sargeanj@uoguelph.ca

It is imperative that we develop effective non-antibiotic interventions to maximise the health and welfare of livestock and poultry. It also is important that decisions on preventive or therapeutic interventions are evidence-based. However, although original research studies are the backbone of scientific evidence, the results of a single study are not sufficient for decision-making; instead, it is necessary to build a body of evidence across multiple studies. Newer techniques, such as meta-analysis and network meta-analysis can help to identify effective interventions by summarising all of the available research on the efficacy of two or more interventions intended to prevent or treat the same outcome. For instance, the results from recent network meta-analyses show that teat sealants are efficacious for preventing mastitis in dairy cattle, vaccines for respiratory disease in feedlot cattle do not appear to be efficacious (at least as currently applied), and there is insufficient evidence to determine the efficacy of vaccines for bacterial respiratory disease in swine, despite a relatively large number of trials. Network meta-analysis also is useful for highlighting areas where our research approaches could be enhanced to maximise the value of the research investment. The results from network meta-analyses of management options to reduce illness in livestock will be used as case-studies to illustrate the value of network meta-analytical approaches, and to explore ways to maximise the value of the research that we conduct. Approaches to maximise the value of research include the use of network plots to identify intervention-outcome gaps for targeted research, a consideration of core outcome sets, issues related to trial design and reporting, and ways to enhance research accessibility.

Combining genetic and other approaches to improve disease resilience in animal populations
A. Doeschl-Wilson[1], C. Pooley[1], M. Chase-Topping[1], J. Xie[2], I. Trus[2], C. Bonckaert[2], K. Rediger[2], S. Gueguen[3], R. Bailey[1], V. Bitsouni[1], T. Oppriesnig[1], S. Lycett[1], V. Gerdts[4], S. Walker[4], J. Harding[4], Y. Fang[5], J. Dekkers[6], H. Nauwynck[2] and G. Plastow[7]
[1]University of Edinburgh, The Roslin Institute, Easter Bush, EH25 9RG Scotland, United Kingdom, [2]University of Ghent, Sint-Pietersnieuwstraat 25, B-9000 Ghent, Belgium, [3]Virbac R&D, 13ème Rue, 06510 Carros, France, [4]University of Saskatchewan, 129, 72 Campus Drive, Saskatoon Saskatchewan S7N 5B5, Canada, [5]University of Illinois, 2001 S Lincoln Ave, Urbana, IL 61802, USA, [6]Iowa State University, 513 Farm House Ln., Ames, IA 50011-3091, USA, [7]University of Alberta, 116 St & 85 Ave, Edmonton, AB T6G 2R3, Canada; andrea.wilson@roslin.ed.ac.uk

Biosecurity, vaccination and breeding for disease resistance are the main tools to control infectious diseases in livestock. Yet shockingly little is known how these affect disease transmission. Many veterinary vaccines, as well as genetic resistance, only reduce disease development but don't prevent infection and transmission. This 'leakiness' is concerning as it may accidentally promote disease spread and pathogen evolution to higher virulence. This is particularly pertinent for Porcine Reproductive and Respiratory Syndrome (PRRS) in pigs. PRRS transmission experiments involving 164 and 52 commercial pigs, respectively, were coupled with genetic-epidemiological prediction models to assess the effects of vaccines and host genetics on disease transmission. The models demonstrate that even leaky vaccines, compromised bio-security or partial host genetic resistance can substantially reduce the risk of infection invasion and spread in a population through reduction in host infectivity or the duration of the infectious period. This was confirmed by our PRRSV transmission experiments that found that vaccination did not block virus transmission, but reduced the overall PRRSV transmission potential in vaccinated groups. Analysis of the data from the genetic transmission trial is ongoing. Our results show that even partial vaccination with a leaky vaccine and partial host genetic resistance can have positive consequences in controlling the spread and symptoms of infection. Future evaluation of veterinary vaccines and genetic disease studies would benefit from including transmission experiments coupled with epidemiological models to predict the effectiveness of control strategies in the field.

Dairy cattle ruminal resistome: characterisation and association with productive traits
A. López-Catalina[1], R. Atxaerandio[2], A. García-Rodríguez[2], I. Goiri[2], M. Gutiérrez-Rivas[1], J. Rey[2], J.A. Jiménez[3] and O. González-Recio[1,4]
[1]Instituto Nacional de Investigación y Tecnología Agraria y Alimentaria, Departamento de mejora genética animal, Crta. de la Coruña km 7.5, 28040, Madrid, Spain, [2]Neiker- Basque Institute for Agricultural Research and Development, Department of Animal Production, Campus Agroalimentario de Arkaute, 01192 Arkaute, Spain, [3]Confederación de Asociaciones de Frisona Española, Valdemoro, 28340 Madrid, Spain, [4]Escuela Técnica Superior de Ingeniería Agronómica, Alimentaria y de Biosistemas, Departamento de Producción Agraria, Universidad Politécnica de Madrid, Ciudad Universitaria s/n, 28040 Madrid, Spain; adrian.lopez@alumnos.upm.es

The rumen resistome is the compound of all the antimicrobial resistance genes (ARGs) present in the microbes that inhabit the rumen. The World Health Organization warned about the risk of antimicrobial resistant pathogens, as by 2050 multi-resistant bacteria will kill 10 million people annually, surpassing cancer as our main health concern. Among the 1,461 diseases recognised in humans, 60% of them are caused by multi-host pathogens capable of moving across species. Here lies the importance of characterising the rumen resistome, as ARGs could jump from faeces and saliva within and across species, arriving to humans via direct contact or through the food chain. The ruminal metagenome of 472 Friesian cows from 14 Spanish farms was sequenced using the MinION device from Oxford Nanopore. DNA reads were analysed with the SQMreads tool from SqueezeMeta. We integrated the Comprehensive Antibiotic Resistance Database to align each read to a gene and obtain the number of reads present in the samples. The pipeline was implemented in the CESGA super-computing centre. The 69 most prevalent ARGs were determined. The heritability of their relative abundance (RA) ranged between 0.28(ugd)and 0.52(lnuC). The most remarkable correlations were found between tetQ and methane (-0,75), tetA(58) and milk yield (-0.73) and tetA(58) and protein yield (-0.71).We were able to determine the most prevalent ARGs in the ruminal ecosystem. The RA of these genes showed high heritabilities and genetic correlations with economically important traits. Further studies are needed to gain insights on the role of these genes in the rumen metagenome.

Improving welfare and biosecurity practices increases performance and efficiency of beef farms

F. Righi[1], A. Goi[2], A. Costa[2], E. Pellattiero[2], M. Penasa[2], M. Simoni[1], R.G. Pitino[1] and M. De Marchi[2]
[1]*University of Parma, Department of Veterinary Science, Via del Taglio, 10, 43126 Parma, Italy,* [2]*University of Padova, Department of Agronomy, Food, Natural resources, Animals and Environment, Viale dell'Università, 16, 35020 Legnaro, Padova, Italy; federico.righi@unipr.it*

The aim of the study was to investigate the effect of improving welfare and herd biosecurity practices (30-d quarantine period) on productive performance and efficiency of Italian beef commercial fattening farms. Experimental trials were carried out in 5 commercial fattening farms of a single cooperative (AZoVe, Cittadella, Italy). Charolaise young bulls (n=576) were enrolled in the trials during 4 fattening seasons. At arrival, the animals of each batch (n=48) were randomly divided in 2 theses: (Q) 24 animals were housed for the initial 30 days of fattening in an insulated, independent and sanitised area of the farm, and managed following high standard welfare conditions, and (C) 24 animals were housed with other animals and managed with standard welfare conditions. The animals were individually weighted at the beginning of the trials and after 30 days; the feed consumption was measured during the whole period and diet and faeces samples were collected at day 15 and 30 using the box as experimental unit. Data of individual animals and data collected at box level were analysed through a linear model that included fixed effects of farm, season, and thesis (Q and C) effects on average daily gain, body weight at 30 days, feed conversion rate (FCR) and estimated total tract apparent diet dry matter, neutral detergent fibre and crude protein digestibility. No significant differences were reported between thesis Q and C for average daily gain, whereas body weight at 30 days was greater (469.45 vs 461.03 kg; P≤0.05) and FCR was lower (7.98 vs 9.10; P≤0.05) for Q than C animals, respectively. The estimated digestibility parameters were not statistically different but numerically higher for the Q than the C thesis, partially explaining the better productive performances of the former group. Quarantine and high standard welfare increased beef cattle performance. This study has been conducted within the project 'AntibioticFreeBeef', funded by PSR of Veneto region 2014-2020, grant number 3556074.

Use of antimicrobials in beef cattle in Italy

A. Diana[1], M. Santinello[1], M. Penasa[1], F. Scali[2], E. Magni[2], G.L. Alborali[2], L. Bertocchi[2] and M. De Marchi[1]
[1]*University of Padova, Department of Agronomy, Food, Natural resources, Animals and Environment (DAFNAE), Viale dell'Università 16, 35020, Legnaro, Italy,* [2]*Istituto Zooprofilattico Sperimentale della Lombardia e dell'Emilia Romagna (IZSLER), Via Bianchi 9, 25124, Brescia, Italy; alessiadiana84@gmail.com*

Antimicrobial resistance is a worldwide issue whereby a more prudent use of medications is needed, especially for those antimicrobials (AM) classified by WHO as 'highest priority critically important antimicrobials' (HPCIAs). So far, data on antimicrobial use (AMU) among EU countries have been mostly reported at sales level while information on real use at farm level, particularly in beef cattle, is poor. The most accredited technical unit to measure AMU is the Defined Daily Dose which can be used to calculate the Treatment Incidence 100 (TI100), an indicator to assess the frequency of treatments. Italy is the 2nd EU country in terms of AM sales in livestock production but information on AMU for the Italian beef production is lacking. The aim of this study was to describe AMU in Italian beef sector. Data were collected from January 2016 to April 2019 from specialised fattening farms of a cooperative of beef producers (AZoVe), leading to a final dataset of 1,376 batches. Body weight, starting and finishing days of the fattening cycle and number of animals per batch were included in the dataset, and a TI100 per batch was calculated using Defined Daily Dose Animal for Italy (DDDAit) as technical unit. Data on reason for administration (e.g. type of disease) and number of treated animals were also collected. Data were analysed using SAS 9.4. Results showed a significant reduction in the TI100 as time progressed (P=0.0051) as well as for the TI100 calculated based on the HPCIAs only (P=0.0164) although about 40% of the total amount of treatments administered were still HPCIAs. The most common reason of administration of AM was for respiratory diseases (68.9%) perhaps due to the common practice of keeping beef cattle in groups making more likely the spread of infectious diseases, which was followed by lameness (17.6%). Overall, data showed a general reduction of AMU in beef cattle though a still great use of HPCIAs was found. These findings may be of help in defining EU benchmark criteria for AMU leading in turn to appropriate intervention strategies for the reduction of AM.

Presence of antibiotic-resistant bacteria in pig slurry after storage

M.T. Kleefisch, C. Heinemann, C.D. Leubner and J. Steinhoff-Wagner
University of Bonn, Institute of Animal Science, Katzenburgweg 7-9, 53115 Bonn, Germany; jste@itw.uni-bonn.de

Antibiotic-resistant bacteria can be spread into the environment by the usage of pig slurry as agricultural fertiliser. Due to prohibition of fertilisation during winter, farmers are forced to store the slurry for several weeks. The aim of this study was to investigate how antibiotic-resistant bacteria in fresh pig slurry persist during a storage period. Sampling was performed within 8 weeks after emptying the slurry bunker underneath the stable. Fresh slurry samples of farms with necessary antibiotic treatments were taken from weaning (n=7) and fattening (n=8) production. Temperature and the pH were measured immediately after sampling. Slurry samples were microbiologically analysed within 24 hours. After preparation, the samples were tested for the presence of extended spectrum beta-lactamases producing bacteria (ESBL), vancomycin-resistant *Enterococci* (VRE), methicillin-resistant *S. aureus* (MRSA) with using selective plates and as accompanying background flora for *Pseudomonas spp.* and *S. aureus*. After 11 to 15 weeks of anaerobic storage at 4-5 °C mimicking winter conditions, microbiological analysis was repeated as described above. Data were analysed for time effects using a linear model in SAS 9.4. ESBL E. coli with counts up to 3.7 $\log_{10}$ cfu/ml were found in 8 of 15 slurry samples and other ESBL in 9 samples (1.0-3.8 $\log_{10}$ cfu/ml). MRSA in 3 slurry samples. In 13 of 15 samples *S. aureus* was found. After the storage ESBL *E. coli* was detectable only in 3 of 15 samples with overall lower counts (1.0-2.7 $\log_{10}$ cfu/ml, P=0.03). VRE and MRSA were absent in all samples after storage. *S. aureus* decreased significantly during storage (P<0.001). Our results indicated that storage of pig slurry reduces the spread of antibiotic-resistant bacteria into the environment. M.T. Kleefisch, C. Heinemann, and C. Leubner contributed equally to this work.

Exploring Brazilian pig farmers' attitudes about antimicrobial use

R. Albernaz-Gonçalves[1], G. Olmos[2], M.W. Kowalski[1], R.E. Woodroffe[1] and M.J. Hötzel[1]
[1]UFSC, Laboratório de Etologia Aplicada e Bem-estar Animal, Rodovia Admar Gonzaga 1346. Florianópolis, 88034000, Brazil, [2]Swedish University of Agricultural Sciences, Department of Clinical Sciences, Almas Allé 8, Uppsala, 750 07, Sweden; gabriela.olmosantillon@gmail.com

The aim of this study was to explore the practices and attitudes of pig farmers in southern Brazil (45% breeding farms; 33% farrow-to-finish; 22% growing or fattening farms) regarding antimicrobial use (AMU) and strategies for prudent AMU. We applied semi-structured, face-to-face interviews to a convenient sample 58 pig farmers (62% independent; 26% to integrators; and 12% associated to cooperatives). Interviews covered vet assistance, biosecurity, AMU, access to drugs. Farmers received vet assistance through nutrition companies (48%) and integrators (34%); 18% did not receive constant veterinary assistance. Farmers did not perform vehicle (83%) or visitor (81%) control, water chlorination (70%), quarantine (70%) and rodent control (21%). Most farmers confused the concept of 'vaccine' with use of any injectable drugs, making vaccination information inaccurate. Preventive AMU was reported in feed for sows (88%) and weaned and growing piglets (68%), and orally/injectable for neonatal pigs (51%). The most cited molecules were amoxicillin (40% in feed) and enrofloxacin (78% as injectable). Length of treatment was between 1 to 3 days in 42% of the cases. Farmers purchased AMU from farming stores (57%), from the cooperatives or integrators (21%) and on-site vendors (8%); 14% received from integrators, which provided lists to farmers defining the active principles and treatment courses. Forty-eight percent of participants believed that did a correct AMU on farm. A hypothetical scenario of limiting AMU was rejected (55%), who claimed economic reasons, difficulties in controlling diseases, lack of supervision and producers' awareness. The negative attitudes of farmers towards limiting AMU, added to their belief having a correct AMU, highlights the need of a great effort to gaining a prudent AMU in Brazil by international standards. This includes increased awareness, and overcoming material limitations and psychological barriers; positive encouragement (opposite to punishment) are required for change to happen.

Anthelmintic resistance in gastrointestinal nematodes of sheep raised in mountain farming conditions

I. Poulopoulou[1], C. Lambertz[2], K. Wuthijaree[3] and M. Gauly[1]
[1]Faculty of Science and Technology, Free University of Bolzano, Piazza Università 5, 39100 Bolzano, Italy, [2]Research Institute of Organic Agriculture (FiBL), Kasseler Straße 1a, 60486 Frankfurt am Main, Germany, [3]Natural Resources and Environment Naresuan University, Department of Agricultural Science, Phitsanulok, 65000, Thailand; ioanna.poulopoulou@unibz.it

Endoparasites cause significant losses in sheep production, while, the frequent use of anthelmintics has led to decreased efficacy and the development of resistance. The aim of the present study was to estimate the anthelmintic resistance (AR) against gastrointestinal nematodes (GIN) in sheep raised under mountain farming conditions. Eggs per gram of faeces (EPG) were used for the calculation of faecal egg count reduction (FECR) on eight sheep farms (n=99 animals) after routine anthelmintic treatments using macrocyclic lactone (ML) or benzimidazole (BZ). On 27 sheep farms (n=306 animals) a FECR test was done alike under the controlled application of anthelmintics applying an oral formulation of one of the following drugs: ML, BZ (partly in combination with salicylanilide (SA)) or a combination of imidazothiazole (IT) and SA. AR was assumed if FECR and the upper confidence interval (CI) was P<95% and the 95% confidence level was P<90%. Before the application of the treatments EPG values showed a wide variation (EPG=563, SD=904, min=50, max=8,246). After routine treatment the results revealed that three out of five ML-treated farms showed adequate efficacy while at the other two efficacy was estimated at 75%. From the three BZ-treated flocks only one showed adequate FECR (CI>95%) while two ranged from 68 to 84%. Under controlled conditions, FECR ranged from 77 to 81% indicating AR for all applied anthelmintics. *Trichostrongylus* spp., *Teladorsagia* spp. and *Haemonchus* spp. were identified after ML-treatment, *Teladorsagia* spp. after BZ-treatment and *Trichostrongylus* spp. and *Haemonchus* spp. after combined BZ and SA application. Both routine and controlled application of anthelmintics resulted in a high prevalence of AR, which indicates that immediate actions are needed such as correct dosing and targeted selective treatments to reduce the risk of AR development.

Performance and health of young calves supplemented with a yeast containing probiotic product

M. Thorsteinsson and M. Vestergaard
Aarhus University, Department of Animal Science, Foulum, 8830 Tjele, Denmark; mogens.vestergaard@anis.au.dk

The aim was to investigate the effects of a probiotic product, containing yeast (*Saccharomyces cerevisiae*) and a postbiotic product from *Lactobacillus acidophilus*, on daily gain and health of young rosé veal calves. A total of 120, mainly Holstein, bull calves at a Danish rosé veal production were used. Upon arrival, calves were either allocated to a control diet (CON) (average age and LW: 25.3±1.3 d and 55.1±1.3 kg) or a diet with a probiotic product (PRO) (average age and LW: 25.3±1.0 d and 55.3±0.8 kg) for six weeks. PRO calves received the probiotic product in the milk replacer (MR) and in the concentrate. The probiotic product constituted 0.06% of DM in the concentrate and 0.9% of DM in the MR. All calves were fed MR for 4 weeks and had access to concentrates for all 6 weeks. Water and roughage were offered *ad libitum*. Manure was collected twice, around d 14-17 and 28-31, and blood was sampled on d 3 and 28-35 after arrival. All data were analysed using a mixed procedure of R, except the number of disease treatments that were analysed by using Fisher's exact test. Average daily gain (ADG) was 885 and 949 g/d for CON and PRO calves, respectively (P=0.05). Serum metabolites were similar between the treatments at the first sampling. However, concentrations of total protein (55.6 vs 59.1 g/l, P=0.01) and IgG (11.6 vs15.1 g/l, P=0.03) were significantly higher in serum from PRO compared with CON calves at the second sampling. Supplementing the probiotic product had no effect on the number of disease treatments and DM content in the manure. Thus, it can be concluded that the supplementation of the probiotic product in the diet of young rosé veal calves did not affect the overall health but had a positive effect on average daily gain.

Standardised natural essential oils and their repellent effects on poultry red mites
S. Suor-Cherer, H. Bui, M.A. Benarbia and P. Chicoteau
Nor-Feed SAS, 3 Rue A Avogadro, 49070, France; hoa.bui@norfeed.net

The poultry red mite (PRM), *Dermanyssus gallinae*, is hematophagous ectoparasite. It is a major cause of animal welfare and economic issues. PRM is typically controlled by synthetic acaricidal compounds such as phoxim, fluralaner, etc. The drawbacks of using these products results in the development of resistance, bioaccumulation and public health concern. Hence, a sustainable alternative is needed. Essential oils (EOs) are considered as promising solution for natural PRM management. Orienting hen odour excretion by supplying diet with EOs to repel PRM is emerging subject gained interests. The objectives of this work are: (1) characterise the active compounds of a standardised EO mixture, named Nor-Mite (NM), a complementary feed composed by *Cymbopogon nardus* and *Eugenia caryophyllus*; (2) the capacity to orient odour excretion by supplying NM in laying hen diet to promote PRM repellent. The active constituent of tested EOs was identified by Gas Chromatography-Mass Spectrometry (GC/MS). In order to determine whether or not the modification on hen's odour, characterising constituents of the hen odour were conducted before and after supplying NM in standard diet. Hen's volatiles were trapped using solid-phase microextraction and analysed by GC-MS. PRM repellent capacity of hens fed NM was tested by observing feeding behaviour of red mites. They are free to choose between host fed standard diet and standard diet supplemented with NM. Binomial test was performed to analyse the choice of individuals. Results revealed that NM contain 12 main aromatic compounds which are known and proved for their repellent effect such as eugenol. The chromatograms showed that hen odour was modified and differentiated after the hens consumed the supplemented feed. Behavioural tests showed that hens fed NM were prevented against the assessment of red mites. 70% lower in number of mites chosen to feed on hens ingested Nor-Mite® compared to hens fed standard diet. In conclusion, diet supplemented with Nor-Mite resulted in modifying laying hen odour excretion. Standardised active compounds were showed their repellent effect to limit the assessment of red mites to their host for blood feeding. It results in restricting the proliferation PRM and control the population at acceptable level.

Entry, persistence and dissemination ways of antibiotic resistant bacteria in a chicken farm
C. Heinemann[1], C.D. Leubner[1], E. Sib[2], R.M. Schmithausen[2] and J. Steinhoff-Wagner[1]
[1]University of Bonn, Institute of Animal Science, Katzenburgweg 7-9, 53115 Bonn, Germany, [2]University of Bonn, Institute of Hygiene and Public Health, Sigmund-Freud-Str. 25, 53127 Bonn, Germany; celine.heinemann@uni-bonn.de

A farm manager of a chicken fattening farm with 79,000 fattening places in two stables reported an extensive metaphylactic antibiotic usage of lincomycin, spectinomycin, colistin and amoxicillin to all hatchlings after entry and a reduced susceptibility to antibiotics. Administration of antibiotics was omitted in one production turn and resulted in nearly 20% losses. To identify possible antibiotic resistant bacteria, samples from feeder bowls, inflow of drinking water, drinking lines, pipes and filters from humidification system, dry litter material, discharge pipes and air were taken immediately prior to new entry of the animals and after cleaning and disinfection. Additionally, nasal and rectal swabs from dead arrived hatchlings, without contact to the stable and chickpaper of transport boxes were sampled. In total, 30 samples were investigated for *Pseudomonas aeruginosa*, extended spectrum beta-lactamases producing bacteria (ESBL), namely *P. aeruginosa, Acinetobacter baumannii, Escherichia coli, Enterobacter* spp., *Citrobacter* spp., *Klebsiella* spp. and vancomycin-resistant *Enterococci*, methicillin-resistant *S. aureus* and identified by MALDI-ToF mass spectrometry. Susceptibility to antibiotics was tested via microdilution assay and interpreted according to European Committee on Antimicrobial Susceptibility Testing (EUCAST). High counts for *P. aeruginosa* ($>3.5 \log_{10}$ cfu/ml) were detected in the samples from air, drinking water pipes and the whole humidification system. ESBL *P. aeruginosa* were found in the water pipes, discharge pipes and the humidification system, starting at the filter for incoming water (3 MRGN). ESBL *A. baumannii* were detectable in a water sample (4 MRGN), rectal swabs and on chickpaper. ESBL *Enterobacter* spp. and ESBL *K. pneumoniae* were 3 MRGN. In this case two main entries for resistant bacteria were identified: Hatchlings arrived already burdened with ESBL *Enterobacter* spp. and afterwards sprinkled with *P. aeruginosa*. Targeted cleaning and disinfection measures immediately improved health and lead to reduced antibiotic usage.

Young steers do not benefit from short term sequential grazing with lambs

S. Werne[1], J. Bam[2], M. Holinger[1], A. Steiner[1], S. Thüer[1], M. Leubin[1] and F. Leiber[1]
[1]*Research Institute of Organic Agriculture FiBL, Livestock Sciences, Ackerstrasse 113, 5070, Switzerland,* [2]*ICAR-National Research Centre on Yak Dirang-790101, West Kameng District, Arunachal Pradesh, India, Dirang, West Kameng District, Arunachal Pradesh, 790101, India; mirjam.holiniger@fibl.org*

The IFOAM – Organics International norms impose that ruminants must be grazed throughout the entire grazing season. This approach increases the exposure to pasture borne parasites and eventually increases the number of anthelmintic treatments, especially in young animals. According to the IFOAM norms, the use of synthetic allopathic veterinary drugs is allowed if preventive measures concerning pasture rotation and management have been exploited. Therefore, we examined the sequential grazing of steers and lambs as a preventive measure to control gastrointestinal nematodes in the steers indicated by weight gains and parasite egg excretion. Ten comparable pairs of steers and 5 groups of lambs were used for this study. The 5 lamb groups were assigned to 5 pairs of steers to establish the steer treatment groups. The remaining 5 pairs of steers served as control groups. Two pastures were subdivided into 15 plots. At each pasture change, the treatment steers changed their plots with the lambs, whereas the control steers only changed pasture, but not the assigned plot. The steers were 173±16 days old and had an average live weight of 171±15 kg at the start of the trial. The trial lasted for 105 days during summer 2018. The steers were weighted and sampled (faeces) at regular intervals during the 105 days and faecal egg count was determined at individual levels. Linear and generalised linear mixed effect models were applied using the software R. The model could not reveal significant differences of total body weight and daily weight gains of the steers between treatment and control. Even though gastrointestinal nematode egg excretion rose to a mean of 270±287 eggs per gram faeces in control steers compared to a mean of 140±70 eggs per gram faeces in treatment steers towards the end of the trial, no significant differences of EPG could be shown over the total trial period. We conclude that sequential grazing of steers and lambs does not improve the parasitological status of young steers. In the frame of Core Organic Cofund MIX-ENABLE.

Impact of earthworm or vermicompost supplementation under a challenging condition for broilers

G. Das[1], S.M.M. Seyedalmossavi[1], Z. Li[1], A. Troescher[2] and C.C. Metges[1]
[1]*Leibniz Institute for Farm Animal Biology, Institute of Nutritional Physiology, Wilhelm-Stahl-Allee 2, 18196, Dummerstorf, Germany,* [2]*Chemovator GmbH, Industriestr. 35, 68169, Mannheim, Germany; gdas@fbn-dummerstorf.de*

Earthworms (EW) have a high microbial activity in their gut, consume large amounts of particulate organic matter and excrete holorganic faecal pellets, i.e. vermicompost (VC). We investigated effects of feeding of EW or VC on performance, non-specific immune responses and gut health of broiler chicks under a challenging environmental condition. Material and methods: Starting from the first day (d) of life onwards, male Cobb-500 birds (n=480) were fed either a control diet (CON+) or CON+ supplemented with either 1% EW (on DM basis) (CON+EW) or VC (CON+VC) for 8 d (Period 1; P1). On d-8, half the birds were killed. Half the remaining birds on CON+ diet were either kept on the same diet for further 8 d (P2) or given a challenge diet (CON-) rich in non-starch polysaccharides (NSP). The birds consuming EW and VC in P1 were fed the CON- diet in P2 (i.e. CON-EW and CON-VC, respectively). On d-16, all birds were killed. Pen based average body weight (BW), feed intake (FI) and feed conversion ratio (FCR) were calculated in each P. In the end of P2, the birds were evaluated for the presence of sticky faeces (SF) attached to cloaca. Results: CON+VC improved (P<0.05) BW P1 through an elevated FI (P<0.05), but had no effect on FCR (P>0.05). CON+EW did not differ from the CON+ in terms of growth and FI (P>0.05) in P1. In P2 CON- did not affect growth relative to CON+ (P>0.05). In P2, CON-VC fed birds were still heavier than those fed on CON+. In the end of P2, 10% of CON+ birds had SF. CON- (P<0.05) increased prevalence of SF (40.5%), and VC aggravated this effect (57.9%), whereas CON-EW (18.9%) did not differ from CON+ (P>0.05). As compared with CON-, CON-EW tended to decrease (P=0.072) prevalence of SF. Birds fed on CON-EW had heavier caeca than those birds fed on CON+ (P<0.05). CON+EW tended to increase IgM in P1 (P=0.08). IgY and IgM remained unaffected by dietary treatments in either P. Conclusions: The lower incidence of SF and heavier caeca due to feeding EW may be indicative of a successful inoculation with beneficial microorganisms that might reduce anti-nutritive effects of NSP.

Farmers attitudes towards biosecurity in outdoor pig farms in Spain: results from a questionnaire

P. Obregón, A. Horrillo, P. Gaspar, M. Escribano and F.J. Mesías
School of Agricultural Engineering. University of Extremadura, Avda. Adolfo Suarez, s/n, 06007 Badajoz, Spain;
andreshg@unex.es

The opening up of markets and international trade as a result of increased demand for food, and in particular animal origin food, has in turn led to an increase in the number of regulations and standards relating to livestock health. Among these regulations are those related to the implementation of mandatory biosecurity (BS) measures at farm level. These measures prevent pathogens from entering a farm and reduce the spread of pathogens within a herd. However, the application of these measures is seen with caution by farmers, especially on extensive farms, where the application of BS measures is more complex. In this context, this research aims to know the attitude of extensive pig farmers in Extremadura (SW Spain) on the willingness to adopt and invest in BS measures. The data analysed were obtained from 104 on-line surveys carried out with farmers during 2019. A correspondence analysis and a cluster analysis were performed, and three groups of farmers were identified based on the characteristics of their farms (size, rearing conditions, availability of facilities, etc.) and their willingness to adopt BS measures. The first group (66.3% of the holdings) includes small, closed-herd farms with medium biosecurity levels; the second group (9.6%) consist of small outdoor fattening farms with low biosecurity levels; and the third group (24.1%) ins integrated by large farms with high biosecurity levels. In general, the farms surveyed had the necessary infrastructure to comply with basic BS principles and the farmers applied most of the BS practices. However, an improvement plan must be implemented in order to meet with the regulations in force and thus improve the BS levels of the extensive pig sector mainly in small farms due to the health risk they may represent.

Anthelmintic efficacy in gastrointestinal nematodes of goats raised in South Tyrol

I. Poulopoulou[1], C. Lambertz[2], K. Wuthijaree[3] and M. Gauly[1]
[1]Faculty of Science and Technology, Free University of Bolzano, Piazza Università 5, 39100 Bolzano, Italy,
[2]Research Institute of Organic Agriculture (FiBL), Kasseler Straße 1a, 60486 Frankfurt am Main, Germany, [3]Natural Resources and Environment Naresuan University, Department of Agricultural Science, Phitsanulok, 65000, Thailand;
Ioanna.Poulopoulou@unibz.it

Parasitic infections constitute a serious threat in mountain areas, while the frequent use of anthelmintic drugs and the inappropriate management of grazing land have led to the development of anthelmintic resistance (AR). The aim of the present study was to estimate the AR of gastrointestinal nematodes (GIN) in goats raised under mountain farming conditions in South Tyrol, Northern Italy. The faecal egg count reduction (FECR) was calculated using the eggs per gram of faeces (EPG) on eight goat farms (n=143 animals) after routine anthelmintic treatments using macrocyclic lactone (ML) or benzimidazole (BZ). On five goat farms (n=135 animals) a FECR test was similarly conducted under controlled application of an oral formulation of one of the following anthelmintics: ML, BZ (partly in combination with salicylanilide (SA)) or a combination of imidazothiazole (IT) and SA. AR was assumed if FECR and the upper confidence interval (CI) was P<95% and the 95% confidence level was P<90%. The average EPG number before anthelmintic application showed a large variation (EPG=751, SD=1,090, min=50, max=9,500). After the treatments, three out of the six routinely ML-treated goat flocks showed adequate efficacy. FECR in all others ranged between 64 and 93%. Under controlled anthelmintic treatments, ML had an adequate efficacy on four farms and a FECR of 88% on another one. BZ was effective on all farms. The combination of BZ and SA had a FECR of 99% on the farm. The combination of IT + SA was effective on two farms and had a FECR of 91% on a third farm. Larvae identified after treatment were composed of *Haemonchus* spp. (ML and BZ), *Trichostrongylus* spp. (BZ) and *Teladorsagia* spp. (BZ and SA). Low efficacy observed in goat farms after routine and controlled treatments, which demand immediate actions starting with a correct dosage of available anthelmintics.

The effect of dietary eubiotics on the performance of broilers challenged with *Campylobacter jejuni*

A. Basanta[1], A. Ortiz[2], J.I. Ferrero[2], B. Saldaña[2], C. Millán[1] and J. Peinado[1]
[1]Imasde agroalimentaria, S.L., c/ Napoles 3, 28224, Pozuelo de Alarcón, Madrid, Spain, [2]Nuevas tecnologías de gestión alimentaria, S.L., c/ Marconi 9, 28823, Coslada, Madrid, Spain; abasanta@e-imasde.com

The aim of the trial was to evaluate the antimicrobial efficacy of different composition blends with essential oils, polyphenols and terpenes from Mediterranean plants in broiler chickens experimentally infected with *Campylobacter jejuni* at 14 days of age (1×10^4 colony-forming unit). A total of 240 Ross 308 broilers were randomly allocated into 3 groups with 20 replicates of 4 chickens per replicate. The experimental dietary treatments were formulated by supplementing the basal wheat-soyabean meal diet. The experimental design allocated the groups as follows: 1. an unsupplemented diet; 2. a diet supplemented with phytobiotics (Polifam3®) at a level of 3.0 g/kg feed; 3. a diet supplemented with a new blend of organic acids (Camprotec®) at a level of 6.0 g/kg feed. Throughout the 42 days of the experimental period, performance parameters and mortality were recorded. At 7, 21, and 28 d post-inoculation, the counts of *C. jejuni* in excreta samples were analysed using quantitative real-time PCR. All experimental broilers were infected and the potential effects of infection on these parameters were considered similar for all the treatments. Although no significant differences were found (P>0.10), infected broilers supplemented with phytobiotics and organic acids had greater average daily gain (ADG) and lower counts of *C. jejuni* than infected broilers with an unsupplemented diet. Infected broilers supplemented with 6.0 g/kg of organic acids shower higher European performance efficiency factor (EPEF) and greater reduction in *C. jejuni* counts than infected broilers supplemented with 3.0 g/kg of phytobiotics, although no significant differences were found (P>0.10).

The effect of dietary phytobiotics on the performance of broilers challenged with *Eimeria* oocysts

A. Basanta[1], A. Ortiz[2], J.I. Ferrero[2], B. Saldaña[2], I. Ramirez[1] and J. Peinado[1]
[1]Imasde agroalimentaria, S.L., c/ Napoles 3, 28224, Pozuelo de Alarcón, Madrid, Spain, [2]Nuevas tecnologías de gestión alimentaria, S.L., c/ Marconi 9, 28823, Coslada, Madrid, Spain; abasanta@e-imasde.com

The aim of the trial was to evaluate the anticoccidial efficacy of different supplementing level with a phytobiotic (Polifam3®) in broiler chickens experimentally infected with sporulated oocysts of *Eimeria acervulina* (89,000), *Eimeria maxima* (23,000) and *Eimeria tenella* (17,000) at 14 days of age. A total of 256 Ross 308 broilers were randomly allocated into 4 groups with 16 replicates of 4 chickens per replicate. The experimental dietary treatments were formulated by supplementing the basal wheat-soyabean meal diet. The experimental design allocated the groups as follows: 1. a diet with the addition of Salinomicina coccidiostat at a level of 60 mg/kg feed; 2. an unsupplemented diet; 3. a diet supplemented with the phytobiotics at a level of 1.5 g/kg feed; 4. a diet supplemented with the phytobiotics at a level of 3.0 g/kg feed. Throughout the 35 days of the experimental period, performance parameters, lesion score, and oocyst output were recorded. All experimental broilers were infected and the potential effects of infection on these parameters were considered similar for all the treatments. No significant differences were found between treatments for growth performance and oocyst output for global experimental period (P>0.10). However, infected broilers with Salinomicina coccidiostat and phytobitic supplementation showed lower lesion score than infected broilers with an unsupplemented diet (0.67, 1.04, 0.58, and 0.77 for each experimental treatment, respectively; P<0.01).

A novel approach for development, standardisation, and safety testing of enriched alum-precipitated
M. Farooq
Huazhong Agricultural University, Molecular Nutrition, Key Laboratory of Agricultural Animal Genetics, Breeding and Reproduction of Ministry of Education, 430070, Pakistan; zahid.farooq@uvas.edu.pk

Haemorrhagic septicemia is caused by two serotypes of Gram-negative, nonmotile, coccobacillus bacterium named as Pasteurellamultocida. The two serotypes are B:2 and E:2. The letter denotes the capsular antigen and the numeral 2 stands for the somatic or 0 antigen. In Asia, only serotype B:2 has been reported, while in Africa, the presence of E:2 has been recorded. Haemorrhagic septicemia is a fatal disease of cattle and buffalo all over the world including Pakistan and it causes heavy economic losses every year. The poor farmers cannot bear this loss in the form of less milk production and heavy expenditures on the animal treatment. An enriched alum-precipitated vaccine with reduced dose was prepared and standardised and safety testing of enriched vaccine was performed in Swiss albino mice as well as in natural host. In this experiment, a total of 36 cattle both male and female of different age groups ranging from 4 months to 4 years were used. All these animals belong to all major cattle breeds of Pakistan including Sahiwal, Red Sindhi (crossbred and purebred), Dhani (crossbred and purebred), Lohani (crossbred and purebred), and Cholistani, and exotic breeds including Holstein Friesian and Jersey. These animals were examined for current immune titre prior to vaccination. Animals were vaccinated subcutaneously with 2 and 4 ml dose of new vaccine and were observed for any untoward reaction for 48 h. All the animals were kept under close observation for the next 30 days and all were found safe. The experiment was designed to reduce the dose of the vaccine to 2 ml by using BHI as a growth medium, as well as to increase the number of doses prepared in the same infrastructure, hence reducing the cost of vaccine production. The study proved that vaccine with increased biomass in reduced dose is safe in local as well as in exotic breeds of cattle. Immunity levels have been measured prior to the experiment As it is necessary to check the antibody level against the causative agent prior to vaccination, (which should be non protective i.e. 1:8 according to OIE, if the antibody level is already protective i.e. 1:16 and above it will be difficult to check the efficacy of experimental accine) in order to properly evaluate the effectiveness of experimental vaccine. Actually, the main purpose of study was to check the safety of vaccine prepared with Brain Heart Infusion (BHI) enrichment media in exotic as well as local breeds of cattle in Pakistan as BHI may cause hypersensitivity reactions due to its composition. However, other group of researchers is also working on the above said aspect.

Prevalence of MRSA and ESBL producing bacteria in unweaned dairy calves in western Germany
J.J. Hayer[1], C. Heinemann[1], C. Leubner[1], E. Sib[2] and J. Steinhoff-Wagner[1]
[1]University of Bonn, Institute of Animal Science, Katzenburgweg 7-9, 53115 Bonn, Germany, [2]University of Bonn, Institute for Hygiene and Public Health, Sigmund-Freud-Straße 25, 53105 Bonn, Germany; hayer@uni-bonn.de

Antibiotic resistant bacteria such as the methicillin-resistant *Staphylococcus aureus* (MRSA) is highly prevalent in various livestock production systems (pig, poultry or veal calf). Nevertheless, there is very limited data on the prevalence of antibiotic resistant bacteria in dairy calves. It is likely that they can develop in dairy calves too due to the practice of feeding waste milk or the high disease incidence. Aim of this study was to get an overview of the presence of antibiotic resistant bacteria in unweaned dairy calves and to identify potential risk factors. To achieve this, we visited 23 dairy farms in western Germany (55 to 700 dairy cows) and took nasal and rectal samples from 484 unweaned calves with sterile moistened flocked swabs with 1 ml of liquid amies medium. Samples were analysed on MRSA and on extended spectrum ß-lactamases producing bacteria (ESBL). To detect MRSA amies liquid from each sample was inoculated in pre-enrichment broth (Mueller Hinton broth with 6% NaCl) for 24 h at 37 °C and afterwards plated on selective plates. ESBL were detected through inoculating amies liquid from each sample in pre-enrichment broth (Mossel broth) for 24 h at 37 °C and plating on selective plates. Species identification was conducted by using MALDI-TOF mass spectrometry. Results were descriptively analysed and Spearman's rank correlations were calculated with SAS 9.4. The prevalence of farms with dairy calves carrying EBSL was 83%, with 4 farms that had no calf carrying ESBL. MRSA was found on only 2 farms. 47.3% of all sampled calves carried ESBL and 1.2% MRSA in nose or rectum or in both. Most of the calves tested positive for ESBL carried ESBL *E. coli* (92.1%), followed by ESBL *Pseudomonas* spp. (13.5%) and ESBL coliforms (4.4%). The age of the tested calves correlated negatively with the carriage of ESBL ($r=-0.27$; $P<0.001$), but not with the carriage of MRSA. Furthermore, we were able to detect MRSA in the intestine and ESBL *E. coli* in the noses of dairy calves, which indicates that the usual sampling scheme (nasal sampling for MRSA and rectal sampling for ESBL) may lead to underestimations.

Study and application of ultrafined traditional Chinese veterinary medicine in animal production

S.J. Yu[1], L.L. Chen[2], J.L. Yan[2], K.J. Guo[1] and G.Z. Zhang[2,3]
[1]*Beijing University of Agriculture, College of Animal Science and Technology, No. 7, Beinong Road, Changping District, 102206 Beijing, China, P.R.,* [2]*Henan Kangxing Pharmaceutical Co., Ltd., Kangxing biological science and technology industrial park, zhandian town, 454950 Zhengzhou, China, P.R.,* [3]*Henan University of Animal Husbandry and Economy, College of Pharmaceutical Engineering, No. 6, Longzihong North Road, Zhengdong New District, 450046 Zhengzhou, China, P.R.; kjguo126@126.com*

As living standards have been increasing, Chinese consumers put forward higher requirements on the health, safety and nutrition value of food, therefore organic food has become the first choice for more and more consumers. The superiority of safety and efficacy of Traditional Chinese Veterinary Medicine (TCVM) makes TCVM the first choice in organic farms for raw feed materials and medication. TCVM can enhance animal's immunity and health by multi-target effects mechanisms so as to improve animal performance and product quality. To improve the effectiveness of TCVM, Ultra-fine comminution technology was developed to break the cell wall into fragments and release the active ingredients in order to improve their bioavailability. To test the effects of Dandelion Ultra-fine Powder (DUP) on animal health, 60 mid-lactation dairy cows were randomly chosen in one farm of Mengniu Group. DUP were mixed into TMR, which was fed to cows twice per day. DHI and California Mastitis Test was employed and daily milk production was recorded. The result showed that the average milk production increased from 28.5 kg/d at the beginning to 29.9 kg/d as the average of whole trial period even though the days in milk was about 210 d. The Somatic cell count (SCC) which originally ranged from 180,000 to 2,060,000, decreased by 81% on 7th day after the addition of DUP. Clinic mastitis incidence reduced about 6.3%. In conclusion, DUP had significant effects on reducing SCC of the cow, preventing the cow's clinic and subclinic mastitis, and in turn improving the milk production. The results implied that ultra-fined TCVM was an effective approach to organic production.

Effect of maternal dietary condensed tannins on coccidiosis and gut immunity in suckling lambs

J. Pelegrin-Valls[1], J. Álvarez-Rodríguez[1], B. Aquilue[1], M.J. Martín-Alonso[1], C. Baila[2], S. Lobón[2], M. Joy[2] and B. Serrano-Pérez[1]
[1]*University of Lleida, Animal Science Department, Av. de l'Alcalde Rovira Roure, 191, 25198 Lleida, Spain,* [2]*CITA de Aragón, Av. Montañana 930, 50059 Zaragoza, Spain; jonathan.pelegrin@udl.cat*

In recent years, in animal nutrition there has been an increasing interest in alternative nutrients with antiparasitic properties. However, in Spain, the production system for suckling lambs is characterised by being fed mainly by mother's milk until slaughter. This study hypothesised that the use of sainfoin (*Onobrychis viciifolia*) as a source of condensed tannins (CT) in the diet of ewes may improve the intestinal health of lactating lambs by reducing their faecal coccidian oocyst count. Following lambing, 12 Rasa Aragonesa ewe-lamb pairs were allocated to one of two diets, in individual pens, during 30.2±3.5 days of lactation: SAINFOIN diet (fresh sainfoin *ad libitum* and 200 g/day of barley as a supplement; n=6) and SAINFOIN+PEG diet (fresh sainfoin *ad libitum*, polyethylene-glycol 'PEG' and 200 g/day of barley as a supplement; n=6). On the slaughter day (11.3±0.8 kg of body-weight), blood samples were collected for plasma metabolites analyses and rectal faeces were sampled for oocyst count. Jejunal and ileal tissues were sampled for determining pro-inflammatory and regulatory cytokine gene expression by qPCR. Data were analysed with one-way least square model. The results for urea, creatinine and fructosamine showed no difference between SAIFOIN+PEG and SAINFOIN groups (31.0 vs 29.2±2.6 mg/dl; 0.62 vs 0.6±0.06 mg/dl; 185.1 vs 194.7±8.1 µmol/l; respectively, P>0.05). Nevertheless, maternal SAINFOIN diet tended to reduce the percentage of lambs that shed more than 10 eggs/g oocyst count (83.3 vs 33.3% P=0.07). Furthermore, lambs from the SAINFOIN diet group showed up-regulated mRNA levels of the pro-inflammatory cytokine, tumour necrosis factor-α (TNFA) in jejunal and ileal tissues, along with down-regulation of the T regulatory cytokine transforming growth factor-β1 (TGFB) in jejunal tissue (P<0.05). In conclusion, maternal dietary CT reduce coccidian oocyst excretion probably mediated by a counterbalance between pro-inflammatory and regulatory cytokines in the intestinal tract.

Simulation study for the integration of health traits in horse breeding programs
L. Büttgen, J. Geibel, H. Simianer and T. Pook
University of Goettingen, Department of Animal Sciences, Center for Integrated Breeding Research, Albrecht-Thaer-Weg 3, 37075 Goettingen, Germany; lisa.buettgen@uni-goettingen.de

Osteochondrosis dissecans (OCD) is important in horse breeding both from an animal welfare and an economic perspective. Therefore, a lifelike simulation of the breeding program of German Warmblood horses was performed with the R package MoBPS to assess this problem. Here the complex breeding schemes of horse breeding for both dressage and jumping with different selection steps, such as the stallion licensing, 14-day and 50-day performance test, were taken into account, thus creating an age and selection structure mimicking the German situation. Furthermore, exemplary osseous fragments in fetlock and hock joints were considered in the simulation. Different scenarios which either used threshold selection, index selection or genomic index selection were compared regarding their impact on the health traits as well as the performance traits. A rigorous threshold selection at the stage of stallion licensing performed well in terms of reducing OCD in fetlock by 1.7% and in hock joints by 0.5% within 20 years in comparison to a reference without selection against OCD. However, this would be accompanied by a reduction of the performance traits of about 5%. An inclusion of OCD in the breeding value estimation for approved stallions, which influenced the frequency of use of the stallions, led to lower reduction of OCD occurrence. Depending on the number of phenotype observations considered, this also came along with a minor impact on the riding horse performance traits. The additional use of index selection based on estimated breeding values at the stage of stallion licensing led to a similar reduction of OCD (fetlock 1.5%, hock joints 0.5%), but only with an average loss of 3.5% in the performance traits. Genotyping of stallions improved the accuracy of the breeding value estimation slightly. Genomic gains in regard to OCD and performance traits were basically the same. The increase in inbreeding in all scenarios was neglectable with an average increase in kinship of 0.005 within 20 years.

Genetic evaluation of the Lusitano horse breed in morphology competitions
A. Vicente[1,2,3], R. Serpa[4], J. Ralão[1], L.M. Brito[4,5] and N. Carolino[2,6,7]
[1]Associação Portuguesa de Criadores do Puro Sangue Lusitano (APSL), Estoril, 2765-337, Portugal, [2]CIISA, Faculdade de Medicina Veterinária, ULisboa, 1300-477 Lisboa, Portugal, [3]Escola Superior Agrária, IPSantarém, DCAA, Santarém, 2001-904, Portugal, [4]Escola Superior Agrária, IP Viana do Castelo, Ponte de Lima, 4990-706, Portugal, [5]Centro de Investigação de Montanha (CIMO), ESAPL-IPVC, 4990-706, Portugal, [6]Escola Universitária Vasco da Gama, Coimbra, 3020-210 Lordemão, Portugal, [7]INIAV, Vale Santarém, 2005-048, Portugal; apavicente@gmail.com

The Lusitano horse is the main equine breed in Portugal, often presented in breed morphology competitions in Portugal and abroad. In this study, 7,721 records of model and movements competitions (MM) of the Lusitano breed (1966 to 2017) were used, provided by APSL. In MM, animals are grouped by classes, considering age, gender and type of presentation. Eighteen different competitions from 15 countries were considered and information from the Studbook was used with pedigrees of 76,349 animals. Records of MM evaluation (MME) were converted on a scale of 0-80 points, depending on the sum of points attributed to the classification/ranking, medal and titles awarded. Points from 1st to 7th place were awarded (10 to 2 points, respectively). For gold, silver and bronze medals, 20, 15 and 10 points were given, respectively. As for the titles, 25 points were awarded to the Male or Female Champion and 25 to the Champion of Champions. The genetic parameters of MME were estimated using the BLUP-Animal Model, by REML and univariate analysis, including fixed effects of location×year, breeder, class, birth season and linear effect of individual inbreeding. As random effects, the animal's breeding value, permanent environmental effect and residual effect were considered. MME resulted in a global average score of 18.203±13.166 pts and the environmental effect of location×year and the breeder, justify a considerable variability of MME. MME showed a genetic standard deviation of 3.326 points whereas the estimates of heritability and permanent environmental effect were 0.079±0.020 and 0.106±0.019, respectively (repeatability of 0.185). There were considerable differences between classes, with higher average values for broodmares with foals (+5.5 pts). The degree of Inbreeding depression was estimated in -0.063 pts by 1% inbreeding. Acknowledgement: Project CIISA UID/CVT/00276/2020.

Genetic correlations between gaits of young horse measured by accelerometry and functional longevity

M. Dugué¹, B. Dumont Saint Priest², S. Danvy² and A. Ricard¹
¹GenPhySE, Université de Toulouse, INRAE, ENVT, 31326, Castanet-Tolosan, France, ²IFCE, Pôle DIR, La jumenterie du Pin, Exmes, 61310 Gouffern-en-Auge, France; manon.dugue@inrae.fr

Functional sporting longevity is essential for well-being of horses and satisfaction of riders. Classical selection using longevity breeding values of stallions calculated from competition results is not efficient because reliable information is obtained too late. The goal is therefore to know if gait traits of young horse are genetically correlated with longevity. On one hand, measurements of gaits by accelerometer at walk, trot and canter under saddle were recorded on 1,477 show jumping horses aged 4 to 5 years. Gaits analysis provided 9 principal components (PC) describing walk, trot and canter. On the other hand, longevity estimated breeding values (EBV) for stallions were calculated using a survival analysis from more than a million performances performed by all the 232,861 show jumping horses born from 1981. Random effects were sire and maternal grand sire effects; fixed effects were region of birth, month of birth, year of recording, age at first competition and performance level. Longevity was adjusted by level of performance to avoid confounding with aptitude. EBV were calculated by maximum likelihood using the Survival Kit software. Heritability was 0.12. Longevity EBV were de-regressed to obtained weighted pseudo-performances for 1,968 stallions. Genetic parameters between gait traits and longevity were then carried out with an animal mixed model analysis. Fixed effects for gaits were date/location, sex, age and height at withers and speed as covariates. Relationship matrix was constructed from pedigree over 4 generations (17,644 horses). Parameters were estimated by restricted maximum likelihood estimation method using WOMBAT software. Heritability of gait traits varied from 0.06 (PC linked to rhythm at canter) to 0.50 (PC linked to stride frequency at trot). None of the traits was genetically correlated with longevity. Only one PC, linked to lateral activity at trot, were slightly negatively correlated to longevity: rg=0.49 (s.e. 0.27). Maybe the loss of energy at trot with lateral acceleration is a sign of less effective movement in general affecting the durability of sporting activity.

Implications of genetic correlations between linear gait and jumping traits in the sport horse

K.F. Stock¹, I. Workel², A. Hahn² and W. Schulze-Schleppinghoff²
¹IT Solutions for Animal Production (vit), Heinrich-Schroeder-Weg 1, 27283 Verden (Aller), Germany, ²Oldenburger Pferdezuchtverband e.V., Grafenhorststrasse 5, 49377 Vechta, Germany; friederike.katharina.stock@vit.de

Routine implementation of linear description in the sport horse has given access to detailed and more objective information on specific aspects of both conformation and performance, and the improved individual characterisation of horses with regard to gaits and jumping should benefit breeding applications. However, the larger number of traits implies the need of thorough analyses of correlation patterns in order to optimise the use of linear profiling in the breeding program. The aim of this study was to use the results of the routine genetic evaluation for linear traits of the Oldenburg studbooks for analysing the genetic correlations between distinct aspects of walk, trot, canter and jumping. Based on linear data from 2012 to 2019 which included approximately 15,000 linear profiles of foals and 9,000 linear profiles of adult horses, breeding values were estimated in linear animal models and in a bivariate setting for traits analogously assessed in both age groups (n=2 for walk, n=6 for trot) and univariately for traits assessed only in adult mares and/or stallions (n=4 for canter, n=10 for jumping). Using the genetic proofs of all sires with linearly described progeny (n=2,629 stallions with on average 8.7 progeny), Pearson correlation coefficients between indices and individual breeding values for gaits and jumping were calculated within and across trait groups with SAS software. Correlation analyses revealed very close genetic correlation of 0.92 only between the walk traits referring to reach of front and hind limbs, respectively. The correlations among the other traits within trait group were considerably lower: 0.41 to 0.77 for trot, 0.50 to 0.67 for canter, and 0.42 to 0.71 for jumping. The varying and overall moderate reliabilities of breeding values must be considered when interpreting the results, but there were no indications of redundancies in the system. The correlation patterns indicate that the significantly positive genetic correlations between traits could be favourably used in a larger multivariate setting in which breeding values are jointly estimated for different aspects of gaits and jumping.

Suitability of linear profiling data of Swedish Warmblood foals for genetic evaluation

K. Karlsson and Å. Viklund
Swedish University of Agricultural Sciences, Animal Breeding and Genetics, P.O. Box 7023, 75007 Uppsala, Sweden; asa.viklund@slu.se

Linear profiling, where traits are described on a linear scale between two extremes, was introduced for Swedish Warmblood horses (SWB) in 2013. It complemented the subjective scoring where traits are evaluated in relation to breeding objective. Breeding values for 50 linearly assessed traits have been published since 2017 based on data from 3-year-old horses. In 2014 a linear profiling protocol with 15 traits was implemented at foal inspections. The aim of this study was to investigate the suitability of foal traits for genetic evaluation of SWB by estimating genetic parameters and correlations with traits in 3-year-olds. Data from 2,069 foals linearly assessed between 2014 and 2017, and 4,673 3-year-old horses assessed between 2013 and 2018 were analysed. 400 horses had been assessed both as a foal and 3-year-old. A BLUP animal model with the fixed effects of age in days, event and sex were used to analyse foal data and the model for 3-year-olds included the effects of event and sex. Event was the combination of location and date of assessment. Heritabilities for linearly assessed conformation traits at foal inspections were generally low to moderate, ranging from 0.03 for 'Pastern' (long-short) to 0.28 for 'Forelegs' (toed in–toed out). Heritabilities for gait traits were moderate, ranging from 0.27 for 'Elasticity in canter' (elastic-inelastic) to 0.51 for 'Elasticity in trot' (elastic-inelastic). For subjectively scored traits the heritabilites ranged from 0.36 for 'Walk' to 0.57 for 'General impression'. Very strong genetic correlations (>0.93) were estimated between subjectively scored gait traits and their corresponding linear traits. The subjective score for 'Conformation' was highly correlated (0.84) to the linear trait 'Development' (undeveloped-well developed). Strong genetic correlations were estimated between linear traits assessed at foal inspections and corresponding traits as 3-year-old. For conformation traits the correlations ranged from 0.59 for 'Position of shoulder' (straight-sloping) to 0.93 for 'Type' (refined-heavy), and for gait traits from 0.71 for 'Elasticity in canter' (elastic-inelastic) to 0.92 for 'Stride length in canter' (long-short). Based on the results we conclude that data from foal inspections have potential to be used in genetic evaluation.

Linear scoring in Belgian Warmblood foals: its potential for genetic selection

S. Janssens[1], I. Meurrens[2], M.-A. Van Damme[3], N. Gengler[3] and N. Buys[1]
[1]KU Leuven, Livestock Genetics, Department of Biosystems, Kasteelpark Arenberg 30, 3001 Leuven, Belgium, [2]Belgian Warmblood, Waversebaan 99, 3050 Oud-Heverlee, Belgium, [3]Liège University, Gembloux Agro-Bio Tech, Passage des Déportés 2, 5030 Gembloux, Belgium; steven.janssens@kuleuven.be

Linear scoring data of Belgian warmblood foals were made available for analysis. Data of foals is being collected since 2016, following a simplified scheme derived from the linear profiling of 3 to 4 year old Belgian Warmblood (BWP) mares. The foal's scheme consists of 23 traits (conformation and movements) that are scored on a 9-point scale (using scores from -20 to 20 by steps of 5). In order to determine the potential of the scheme for selecting at an early age, we estimated genetic parameters for 19 traits (4 traits excluded due to many missing values). Observations made on approx. 1,800 foals were edited and 1,612 horses were analysed using a multi-trait animal model. The statistical model included fixed effects for the judge, age in months and a combination of location and date of evaluation. Using canonical transformation, (co)variance components were computed and also applied to generate estimated breeding values (EBV). Estimates of heritabilities ranged from 0.08 to 0.34 and estimated genetic correlations between traits ranged from -0.61 to 0.97, with the majority of traits positively correlated or independent. Heritability estimates in foals seem to be smaller in magnitude than linear traits observed in BWP mares. According to results from approximate genetic correlations between traits of the mares and foals (based on reliability corrected correlations between EBV using Calo's method), genetic evaluation based on linear traits assessed on foals at an early age could be used to predict some of the adult scores in mares.

Investigations of maternal lineages of the Holstein warmblood horse based on mtDNA sequencing
L. Engel[1], D. Becker[2], I. Russ[3], G. Thaller[1] and N. Krattenmacher[1]
[1]Institute of Animal Breeding and Husbandry, Kiel University, Olshausenstr. 40, 24098 Kiel, Germany, [2]Institute of Genome Biology, Leibniz Institute for Farm Animal Biology (FBN), Wilhelm-Stahl-Allee 2, 18196 Dummerstorf, Germany, [3]Tierzuchtforschung e.V. München, Senator-Gerauer-Str. 23, 85586 Grub, Germany; lengel@tierzucht.uni-kiel.de

The Holstein horse is a popular horse breed in Germany, which is especially known for its show jumping ability. Maternal lineages are prominent in the Holstein horse and considered to be important for the breeding success. As the mitochondrial DNA (mtDNA) is strictly inherited maternally and mitochondrial genes are involved in energy metabolism, genetic variation on mtDNA level might lead to lineage specific differences in performance traits. For horses, no corresponding association studies were done so far and genetic diversity of the maternal lineages has not yet been studied. Therefore, the mitochondrial genome of 202 mares belonging to 67 lineages was sequenced i.e. 15.3% of all maternal lineages of the current breeding population. Samples from one to nine mares were available per maternal lineage; the majority of the lineages was only represented by one or two mares. MtDNA sequences were compared to the GenBank reference sequence X79547.1 (Swedish horse). In total, 452 variants were found and in accordance with the literature, about one third of them were non-synonymous and the highest density of variants could be found in the non-coding region. To determine the intra-lineage diversity, mtDNA sequences between mares within a lineage were compared regarding the number of polymorphic sites and showed nearly no differences. The calculation of pairwise genetic distances among all mares revealed a clustering into six groups, with all mares from one lineage belonging to the same group. The mean number of pairwise differences within (π_w) and between (π_b) groups was calculated for further analysis. π_w ranged from 0 to 15.56, displaying a small genetic variation while π_b was higher, ranging from 46.02 to 116.60. These results indicate that the examined lineages may have originated from six ancestral mares and that mtDNA analyses can help to understand the evolution of the Holstein population during tine before pedigrees were recorded.

Towards Genomic Selection in German Warmblood Horses
S. Vosgerau[1], N. Krattenmacher[1], C. Falker-Gieske[2], I. Blaj[1], A. Seidel[1], M. Wobbe[3,4], K.F. Stock[3,4], M. Von Depka Prondzinski[5], R. Reents[3], E. Kalm[1], W. Nolte[6], C. Kühn[6,7], J. Tetens[2,8] and G. Thaller[1]
[1]Institute of Animal Breeding and Husbandry, Kiel University, Olshausenstr. 40, 24098 Kiel, Germany, [2]Department of Animal Science, University of Göttingen, Albrecht-Thaer-Weg 3, 37075 Göttingen, Germany, [3]IT Solutions for Animal Production (vit), Heinrich-Schroeder-Weg 1, 27283 Verden, Germany, [4]Institute for Animal Breeding and Genetics, University of Veterinary Medicine Hannover, Buenteweg 17p, 30559 Hannover, Germany, [5]Werlhof-Institut MVZ, Schillerstr. 23, 30159 Hannover, Germany, [6]Institute of Genome Biology, Leibniz Institute for Farm Animal Biology (FBN), Wilhelm-Stahl-Allee 2, 18196 Dummerstorf, Germany, [7]Faculty of Agricultural and Environmental Sciences, University Rostock, Justus-von-Liebig-Weg 6, 18059 Rostock, Germany, [8]Center for Integrated Breeding Research (CiBreed), University of Göttingen, Albrecht-Thaer-Weg 3, 37075 Göttingen, Germany; svosgerau@tierzucht.uni-kiel.de

Genomic selection requires a large reference population (RP) with informative phenotypes and genotypes. To establish this prerequisite, a joint project of five German warmblood breeding associations with scientific and industrial partners was started in 2017. The aim of this project is to create an appropriate data set across populations comprising 5,000 individuals and to implement genomic selection in sport horses. Currently, 1,984 horses with phenotype and genome-wide SNP genotype information are available. Analyses of the genomic relationship showed low genetic differences between the five populations. Withers height was chosen as reference trait to examine whether the current RP is suitable for genomically-enhanced predictions. Withers height is an appropriate trait for this purpose as it can be measured easily, objectively and has a moderate heritability. A medium accuracy for genomic breeding values of 0.57 was achieved using 1,194 mares in a five-fold cross-validation scheme. The accuracy increased to 0.60 after adding 782 stallions genotyped in a previous project to the RP. These results underline the role of size and structure of the RP and indicate, that the RP across German warmblood populations is suitable for further steps including the use of linear traits as phenotypes.

GWAS and pathway investigation on linear type traits in Italian Heavy Draught Horse breed

E. Mancin[1], C. Sartori[1], G. Pigozzi[2] and R. Mantovani[1]
[1]*University of Padova, Dept. of Agronomy Food Natural resources Animals and Environment, Viale dell'Universita 16, 35020 Legnaro (PD), Italy,* [2]*Italian Heavy Draught Horse Breeders Association, Via Verona 90, 37068 Vigasio (VR), Italy; roberto.mantovani@unipd.it*

The Italian Heavy Draught Horse (IHDH) is now the only native Italian cold blood breed, mainly bred at present for heavy draught and meat. Selection is based on an annual collection of linear type traits in new-born foals, including an evaluation of general aspect, temperament and muscularity. IHDH breed is currently collecting genotypes to include gene information in selection programs. This study aimed to carry out a genome-wide association study (GWAS) and following post-GWAS analysis on IHDH linear type traits. Phenotypic information of about 10,600 young foals collected in 17 years were considered, and breeding values were estimated for more than 17,000 animals in pedigree. Genotypic information at SNP density of 65k was made available for 270 individuals, roughly half males and half females. BLUPF90 family of programs was used for analysis. SNP substitution effects were detected using the single-step procedure (ssGWAS), developed in Bayesian framework and allowing to include all the available phenotypes. The proportion of additive genetic variance was found by considering 20-SNPs windows. Due to the limited sample size of genotypes, other software like the Bayesian BGLR, GBLUP-gwas and EMMAX were used for GWAS. Results of the Bayesian analyses were tested for the presence of QTLs using the Bayesian Factor. SNPs were allocated to genes using Ensembl and Genecards databases. GO and KEGG databases were queried to assess functional categories and biological pathways. Heritability of traits varied from 0.14 to 0.37. GWAS analysis allowed to explain traits variation: e.g. for the first trait, head size and expression, more than 20% of additive genetic variance was explained by regions on chromosomes 7, 9, 10, 13 and 32. Here, candidate genes like ACTN4, involved in cell and muscle motility, and MAP4K1, part of the TGF beta pathway for growth, can be detected. This study is a first evidence of genes associated to both heavy draught and meat aptitude detected via GWAS. Findings would be useful for breeding purposes in IHDH and local cold blood horse breeds.

Genomic evaluation for competition results and gaits measured by accelerometry of jumping horses

H. Crichan[1], B. Dumont Saint Priest[1], S. Danvy[1] and A. Ricard[1,2]
[1]*IFCE, Pole Développement, Innovation et Recherche, La jumenterie du pin, 61310 Exmes, France,* [2]*INRAE, GenPhySe, 24 chemin de Borde-Rouge, 31326 Castanet Tolosan, France; harmony.crichan@inrae.fr*

In order to implement routine genomic evaluation for sport horses in France, we tested genomics evaluation using single step Gblup methodology (BLUPF90) for competition results in jumping and gaits traits measured by accelerometers. Competition data came from official competition results given by the French Federation for Equine sports from 1985 to 2018. There were 296,077 horses performing 1,303,426 annual performances. The performance was measured by the logarithm of sum of points allocated in each event according to rank and technical difficulty of the event. Gaits were measured on 1,477 young horses at age of 4 and 5 years during 2015 and 2016 with accelerometer device. Three principal components per gaits (walk, trot, canter) were extracted from the analysis of this data. Genotyping was performed using successively 3 different beadchips: Illumina 54K, 74K and Affymetrix 670k. A total of 3,658 horses were genotyped. 78% of horses measured for gaits were genotyped. 3,368 genotyped horses had competition results and 1,041 genotyped sires had progeny in competition. We performed imputations using Fimpute3 to harmonise genotype data resulting to 418,037 SNP retained after quality control (MAF>=3%, Hardy Weinberg check, coherence between chips criteria). Genomic evaluations were performed separately for each trait using single step method and adding pedigree up to 4 generations. As results, we calculated genomic breeding value from a truncated file with performances after 2016. To illustrate the efficiency of different selection strategies we compared correlations from different predictors with performances in 2016-2018. The genotype at most important SNP gave a correlation of 0.04, meaning that performance is really a polygenetic trait. The use of judgment in breeding shows gave 0.20, the genealogical breeding value at birth 0.31 and the genomic breeding value at birth 0.34.

Modelling the race effect in genetic evaluations for ranking trait in horses

K.D. Arias, I. Cervantes and J.P. Gutiérrez
Departamento de Producción Animal, Facultad de Veterinaria, Universidad Complutense de Madrid, Madrid, Spain;
icervantes@vet.ucm.es

Genetic evaluation of horse competitions involves difficulties to establish a statistical model when ranking is used as phenotype. In addition to its uniform distribution, under a well conducted genetic selection, there must exist a favourable genetic trend, but it is not observable in the corresponding flat phenotypic trend of the rank trait, and the solutions for other effects in the model has to be affected. To deal with this trait it is usually assumed an underlying continuous variable with Gaussian distribution that explains the horse performances, and the ranking is assumed to be the order of the horses accordingly. The main objective of this study was to find out by simulation what effects were affected in the genetic evaluation to compensate a favourable genetic trend when fitting or not the race effect in the model, as well as assessing the accuracy of predicted breeding values. Four initial discrete generations of 50 males and 50 females were simulated under random mating to create relationships, and ten additional discrete generations were simulated by selecting the best 10 males and 10 females according to their predicted breeding values. An underlying variable with heritability 0.1 to 0.4, affected by an influence of the race and generation was simulated to establish the ranking in 10 or 100 races of 10 random participants each generation, some of the animals having several participations and others none. The genetic evaluation model for the ranking was tested to include the race effect or not. Results showed that race effect estimates collected the competition genetic level, compensating the genetic trend obtained by selection, being the residuals affected when race effect was not fitted. Breeding values accuracy ranged from 0.33 to 0.49 and from 0.65 to 0.86 respectively when 10 or 100 races were performed. True genetic trend was identical when fitting or not the race effect, no difference probably due to the random assignment of horses to races independently of their genetic level. The solutions of the race effect would be a good indicator of the efficiency of the selection process.

Genetic parameters of Quarter Horse racing performance traits in Brazil

R.A.S. Faria[1], A.T. Ospina[1], M.H.V. Oliveira[1], L.Y. Rodrigues[2] and J.A.I.I. Silva[2]
[1]FCAV–Universidade Estadual Paulista–UNESP, Via Prof. Paulo Donato Castellane, 14.884-900 Jaboticabal, Brazil,
[2]FMVZ – Universidade Estadual Paulista, UNESP, Fazenda do Lageado, 18.618-307 Botucatu, Brazil; fariasky@gmail.com

The aim of the study was to evaluate the racing performance of Quarter Horse (QM) in Brazil, through the estimation of the genetic parameters of the earnings trait at two years of age (E2), best time (BT) and time class (TC) at distances 301 m (BT301 and TC301) and 402 m (BT402 and TC402). From a total of 23,482 speed records provided by the Sorocaba Jockey Club (JCS), used 1,900 (E2), 3,365 (BT301), 3,318 (BT402), 6,579 (TC301) and 7,423 (TC402). Data were obtained over a period of 38 years (1978 to 2016) and races were held in five hippodromes, all in the state of São Paulo, Brazil. Uni-variate analyses were performed in order to obtain the components of variance and explore their results compared to the bi-variate analyses, these was carried out considering the E2 characteristic as anchor and changing the other four characteristics one by one. Uni and bi-variate analyses by trait were performed using Bayesian inference to obtain estimates of genetic parameters. The heritability estimates obtained for the traits were low to moderate magnitude, ranging from 0.10 ± 0.04 (E2) to 0.37 ± 0.03 (TC301). The repeatability value of TC301 and TC402 were 0.31 and 0.32, respectively. The heritability values in bi-variate traits analyses were higher than the uni-variate traits, showing moderate to high results, ranging from 0.15 ± 0.05 (E2 with TC301) to 0.76 ± 0.12 (BT402 with E2). The additive genetic, residual and phenotypic correlations showed high, low and moderate estimates, respectively, positive between the E2 trait and the TC traits and negative (as intended) between the E2 and CT traits. This study indicated that E2 presents better results when used in the evaluation associated with other traits, and with the BT trait and at 402 m distance, the genetic information was higher. This suggests that the evaluated traits should be included together in the breeding selection programs for Quarter Horse racing.

Estimation of genetic parameters for fertility criteria in Pura Raza Española mares

D.I. Perdomo-González[1], M.J. Sánchez-Guerrero[1], A. Molina[2], F. Arrebola[3] and M. Valera[1]
[1]Universidad de Sevilla, Dpto. de Ciencias Agroforestales, ETSIA., Ctra. Utrera km 1, 41013 Sevilla, Spain, [2]Universidad de Córdoba, Dpto. de Genética, CN IV km 396, 14071 Córdoba, Spain, [3]Centro IFAPA de Hinojosa del Duque, Ingeniería y Tecnología Agroalimentaria, Ctra. El Viso km 15, 14270 Córdoba, Spain; daviniapergon@gmail.com

Most part of equine breeding profitability depends directly of their reproductive efficiency, result of different biological, environmental and management factors. In general, reproductive and fertility performance in this specie are low in comparison to other domestic species. To the low heritability of fertility's relate traits is added the fact that the equine industry keeps as breeder's animals with a deficient reproductive behaviour but with good sport and morphologic performance. This work pretends to develop evaluation models of different selection criteria on mare's fertility of Pura Raza Española (PRE) horse, which is present in 66 different countries. To do this, reproductive records of 78,576 PRE mares were analysed. These mares were born from 1840 to 2016 with an average of 4.3 offspring in different studbooks. We have evaluated age at first foaling (AFF), total number of foaling (TNF), age at last foaling, interval between first and second foaling, average interval between foaling and three different relations between TNF with the productive life and age of the mare as fertility traits. The model selected as the best included TNF and AFF as fertility traits. Coat colour, breeding stud size, owner stud size, owner stud country and birth year of the mother as fixed factors for both traits. Further, the combination of the season-year for AFF as a fixed factor and the age of mare as a covariate for TFN. Genetic parameters and breeding values were estimated with a Bayesian bivariate model using TM software. The heritabilites ranges from low to medium magnitude, 0.11 (TFN) and 0.28 (AFF) with a negative and high genetic correlation (-0.68). This information implies that AFF and TNF can be included in the PRE breeding program to obtain estimated breeding values for fertility and select more fertile horses.

Mitochondrial D-loop sequence variation in Slovak Warmblood horse

N. Moravčíková, M. Gábor, M. Miluchová, J. Candrák, M. Halo and R. Kasarda
Slovak University of Agriculture in Nitra, Tr. A. Hlinku 2, 94976 Nitra, Slovak Republic; nina.moravcikova1@gmail.com

The aim of this study was to analyse mtDNA D-loop region to assess the level of diversity and phylogenic origin of Slovak Warmblood horse that is in Slovakia recognised as an important animal genetic resource and part of the cultural heritage of small lands. The Slovak Warmblood is a very efficient breed with the excellent characteristics of a sport horse, currently bred for sport as well as work purposes. From a genetic point of view, it is the scion of Austro-Hungarian warmblood breeds and horses which were ennobled from Arabians, English purebreds and half-bloods. A 458 bp D-loop region of 33 horses from different studs spread across Slovakia, and 84 mtDNA sequences of 5 breeds stored in the GeneBank were included in this study. The sequences of English Thoroughbred, Arabian Thoroughbred, Nonius, Trakehner and Hanoverian horses participating in grading-up of Slovak Warmblood horse were used to infer the pattern of mtDNA genetic variation. The consensus sequences aligned using ClustalW method resulted in the final dataset with a length of 387 bp. The polymorphic sites, haplotype numbers, haplotype diversity and nucleotide diversity were calculated to analyse the mtDNA diversity level. Two methods were applied to reconstruct the phylogenetic relationship among D-loop haplotypes in the dataset, Neighbour-Joining method and Median-Joining network. In a total of 59 mtDNA haplotypes and 63 polymorphic sites representing 13.76% of the total mtDNA D-loop sequence under consideration. Obtained average values of haplotype diversity (H_d=0.981±0.004) and nucleotide diversity (P_i=0.027±0.001) pointed to a sufficient level of diversity across sequences analysed. Both the Neighbor-Joining phylogenetic tree and Median-Joining network showed consistent results confirming the breeding history of Slovak Warmblood horse and its multi-matrilineal origin.

Mitochondrial DNA present in the current population of the Lusitano horse breed in Brazil

R.A.S. Faria[1], A.T. Ospina[1], A.A. Vicente[2,3], R. Santos[4,5] and J.A.I.I. Silva[6]
[1]FCAV–Universidade Estadual Paulista–UNESP, Via Prof. Paulo Donato Castellane, 14.884-900 Jaboticabal, Brazil, [2]CIISA–FMV, Universidade de Lisboa, Av. Universidade Técnica, 1300-477 Lisboa, Portugal, [3]Escola Superior Agrária do IP Santarém, Quinta do Galinheiro, 2001-904 Santarém, Portugal, [4]VALORIZA–Centro Investigação Valorização Recursos Endógenos Portalegre, Campus Politécnico, 10, 7300-555 Portalegre, Portugal, [5]IP de Portalegre, ESA de Elvas, Avenida 14 de Janeiro, 21, 7350-903 Elvas, Portugal, [6]FMVZ–Universidade Estadual Paulista–UNESP, Fazenda do Lageado, 18.618-307 Botucatu, Brazil; fariasky@gmail.com

Mitochondrial DNA (mt-DNA) is inherited by maternal origin and is considered an energy generator (ATP) for all cellular activities. The Lusitano Horse Breed (PSL) is originated from hotblood, sensitive and energetic animals, which were important characteristics to athletic performance. The aim of the study was to investigate an indicator of the presence of mt-DNA in the current population of PSL breed in Brazil with pedigree data, serving as a model for other studies. The mt-DNA indicator was related to how many X chromosomes (chrX) from distinct complete founder mares (with absence of father and mother in the pedigree) is still prevalent in the actual population (last generation). Pedigree information (Brazilian Lusitano Horse Breeders Association–ABPSL), including records from 16,511 PSL horses born in Brazil, between the years of 1967 and 2012, and their ancestors, born mostly in Portugal (2,411 horses). The number of chrX was evaluated by transmission analysis along the way of the founder mares to their female progeny who became breeding mares and so on. The average quality of pedigree data was 5.7 ± 1.4 equivalent generations. A total of 223 founder mares transmitted the chrX to the next generation (F1). Dilutions of chrX in progenies were observed over the generations. The 223 chrX from founders were progressively reduced to 129, 100, 84, 81, 70, 55, 33, 17, 7, and 2 chrX over F1 (first generation of breeding mares) to F10 (last generation progeny), respectively. The PSL breed is mainly used in equestrian disciplines like dressage, working equitation, carriage driving and bullfighting, where characteristics of strength and energy are required. The loss of 99.1% of the chrX from the founder mares in the current population limited the transmission of mt-DNA variability (energy) to only two of the founder mares. Directing matings should be considered in live mares with no progenies and good reproductive conditions to discover the presence of mt-DNA from other founders and reduce the loss of the X chromosome from founder animals in the future generations.

Genetic differentiation of the Sztumski and Sokolski coldblooded horses in Poland

A. Gurgul[1], I. Jasielczuk[1], E. Semik-Gurgul[1], K. Pawlina-Tyszko[1], T. Szamotała[1], G.M. Polak[1] and M. Bugno-Poniewierska[2]
[1]National Research Institute of Animal Production, Department of Animal Molecular Biology, Krakowska Str. 1, 31-047 Balice, Poland, [2]University of Agriculture in Krakow, Institute of Veterinary Sciences, Mickiewicza Str. 24/28, 30-059, Poland; grazyna.polak@izoo.krakow.pl

The population of cold-blooded horses in Poland was dramatically reduced during the last decades. To avoid impoverishment of gene pool of a local horse population, a conservation program was established. Two sub-populations: Sztumski and Sokolski coldblooded horses were analysed using SNP genotyping arrays under profile of genetic differentiation, genomic inbreeding and effective population size. Material of study was obtained from Sokolski (n=107; SOK) and Sztumski (n=69; SZTUM) horses. The initial number and validations had 65,157 and the final SNP panel 52,023 markers. Effective population sizes (N_e) were estimated based on 1000 randomly sampled SNP genotypes. The results suggest that despite the low average genetic differentiation of the analysed populations they are characterised by distinct genetic profile which may result from site specific allele frequency differences. The principal components analysis showed that the two populations forms two clusters with some individuals being genetically close in both populations. The average membership coefficient of both horse types ranged from 0.524-0.999 in SOK and 0.539-0.99 in SZTUM. The global genetic differentiation of the analysed populations (F_{ST}) weighted mean 0.012 and 0.014, respectively, suggesting that about 98% of genetic variation can be attributed to inter-individual variation. The genomic inbreeding coefficients were comparable for both populations and depending on the applied measure ranged from 0 for F_{GRM} to about 13% for F_{IS}. Estimated effective population size was high for both populations, but higher for Sokólski (228) than for Sztumski (156.4) horses.

Long-term changes and age variations for G-matrix detected in linear type traits scored in horses
E. Mancin, R. Mantovani, F. Fabio and C.R. Sartori
University of Padova, Dept. of Agronomy Food Natural resources Animals and Environment, Viale dell'Universita 16,
35020 Legnaro (PD), Italy; cristina.sartori@unipd.it

The effectiveness of the quantitative genetic approach to long-term evolution is related to the stability over time of the additive genetic (co)variance matrix G, summarising the relationships among traits of individuals. But, heritability and genetic correlations are likely to change over the years in which a population is selected and a reduction of additive genetic variance is expected for strongly selected traits. It is rarely considered that genetic variance might also change with the age of individuals, thus increasing the complexity of G evolution. This study aimed to evaluate the pattern of G temporal variations by considering the linear type traits measured in a local horse breed, the Italian Heavy Draught Horse. This breed is currently selected for heavy draught and meat, through the evaluation of foals at 6 months for 11 linear type traits, 5 of them directly included in selection index. An evaluation at 30 months is also done for the studbook admission. Genetic parameters for traits scored at the two ages were estimated and compared to understand what is changing in terms of genetic correlations among traits and selection response with age. About 6,700 scores on adults and 11,300 scores on foals were considered for analysis. Moreover, the variance components for 6-months traits were compared between subsets obtained by dividing a 23-years dataset in groups of subsequent years. Heritability measured in 6-months vs 30-months data showed small differences, apart for fleshiness, reducing from 0.36 on foals to 0.25 on adults. Genetic correlations between the same traits measured in youngs and adults ranged between 0.55 to 0.92, and different covariance patterns were found in the two G matrices. In the more recent dataset, additive variances reduced in most of the 11 traits, especially in those included in the selection index as head size-expression, from 0.12 to 0.08. This study provided an empirical example of G 'double variation': over time across population history and during individual life. The knowledge of these variation could provide useful insights for selection purpose.

Partial inbreeding coefficient analysis in Pura Raza Español horse population
D.I. Perdomo-González[1], M.J. Sánchez-Guerrero[1], A. Molina[2] and M. Valera[1]
[1]Universidad de Sevilla, Dpto. de Ciencias Agroforestales, ETSIA, Ctra. Utrera km 1, 41013 Sevilla, Spain, [2]Universidad de Córdoba, Dpto. de Genética, CN IV km 396, 14071 Córdoba, Spain; daviniapergon@gmail.com

The Pura Raza Español (PRE) is an autochthonous Spanish horse population and one of the most important horse breed in terms of census and international impact. Since the establishment of the studbook, the inscription of new PRE horses has been restricted to offspring of horses already registered, becoming a closed population where an increase in the level of inbreeding is inevitable. This work analyses the inbreeding coefficient (F), the relatedness coefficient (AR) and, for the first time, the partial inbreeding coefficient (Fij) of the whole PRE population from common ancestors (founder or non-founders animals that belong to both the maternal and paternal family). The genealogical information of the whole PRE population was analysed (328,706 animals) resulting that the most part of PRE population (47.45%) has a F value between 3.12 and 6.25%, resulting, at the same time, the animals with the higher number of common ancestors. The generational AR values for the whole PRE population and for common ancestors showed similar value and behaviour, first increasing and then slightly constant, indicating that common ancestors are very present in the animals that make up those generations. A total of 324,557 animals (98.74% of PRE population) showed Fij higher than zero (between 0.001 and 25%) due to 6,428 common ancestors (2% of total PRE population). Each horse had between 1 to 99 common ancestors, providing a Fij value to an average of 1,979.4 offspring. Finally, along the decades the number of common ancestors is increasing but the proportion of Fij by each ancestor is getting smaller, the opposite of what would be expected. The reason of this could be due to the census increase, the use of artificial insemination and the increased knowledge about animals breeding values and inbreeding depression.

Inbreeding depression of the occurrence of insect bite hypersensitivity in horse

L. Vostry[1], H. Vostra-Vydrova[1,2], B. Hofmanova[1], J. Citek[3] and I. Curik[4]

[1]Czech University of Life Science Prague, Kamycka 129, 16500 Prague, Czech Republic, [2]IAS, Pratelstvi 815, 10400 Prague, Czech Republic, [3]South Bohemia University, Branisovska 31a, 370 05 Ceske Budejovice, Czech Republic, [4]University of Zagreb, Faculty of Agriculture, Svetosimunska cesta 25, 10000 Zagreb, Croatia; vostry@af.czu.cz

Inbreeding depression, is a nearly universal phenomenon that depends on the past mutation, selection, and genetic drift. Nowadays, pedigree inbreeding coefficient can be replaced by genomic inbreeding coefficient. However, genomic information is not always available for all phenotyped individuals. Recently, a new method has been developed that enables, combining genomic and pedigree information (H matrix), estimation of genomic inbreeding for all individuals in the pedigree. The extensive pedigree and phenotypic information available for Old Kladruber horse provides a unique opportunity to examine the inbreeding effect on Insect bite hypersensitivity (IBH). This analysis was based on 1,295 phenotypic records taken from 536 horses. Not all animals with performance were genotyped (n=47). Thus, in addition to three pedigree inbreeding coefficients (pedigree total, pedigree five generations and Kalinowski new inbreeding coefficient), we have calculated pedigree-genomic inbreeding coefficients for all individuals with IBH records based on related genotyped individuals (n=187). Effect of pedigree and pedigree-genomic inbreeding coefficients were investigated using BLUP animal model for the binary trait. The average values were between 3 and 14% for pedigree inbreeding and between -2% and -18% for pedigree-genomic inbreeding coefficients. The correlation coefficients between pedigree and genomic inbreeding coefficients were between 0.27 and 0.38. Inbreeding depression (significant positive regression coefficients, P<0.05) was observed for all tested models with respect to different inbreeding coefficients. Effect of inbreeding measure by odds of occurrence IBH was estimated between 1.04 for pedigree total inbreeding coefficient and 1.14 for Kalinowski new inbreeding coefficient. We have shown that inbreeding influences the occurrence of IBH while the construction of the pedigree-genomic inbreeding additionally improved the estimation of inbreeding depression. Supported by QK1910156 and MZeRO0718.

Intestinal organoids: innovative *in vitro* tools to study host-microbiota interactions in livestock

M. Beaumont, E. Mussard and S. Combes

INRAE, Castanet Tolosan, 31000 Toulouse, France; martin.beaumont@inrae.fr

Gut microbes play a key role in animal nutrition and health. Therefore, targeting the intestinal microbiota is a promising strategy to improve feed efficiency and resistance to diseases. A better understanding of the mode of action of gut bacteria on animal cells is required to optimise host-microbiota symbiosis. The intestinal epithelium is a critical interface between host and microbes and plays a key role in nutrient absorption and gut barrier function. Thus, experimental models are needed to study host-microbes interactions at the intestinal epithelium level. Recent progress in epithelial stem cell biology allowed the development of intestinal organoid culture *in vitro*. These self-organised 3-dimensional structures are formed by a closed lumen surrounded by a single layer of polarised intestinal epithelial cells. Intestinal organoids contain all epithelial cell types including stem, absorptive and secretory cells. Thus, this innovative *in vitro* model closely replicates epithelial physiology. Initially developed in Humans and rodent models, methods are now available to culture intestinal organoids from several other animal species including pig, rabbit, chicken, cow, sheep and horse. Incubation of intestinal organoids with gut microbes or their products (e.g. bacterial metabolites) has a great potential to reveal how the gut microbiota regulates epithelial barrier function or nutrient absorption. The main limitation is that the apical side of epithelial cells is not accessible in organoids cultured in 3-dimensions while it is the physiological route of exposure of the epithelium to the microbiota. To circumvent this issue, intestinal organoids polarity can be reversed by removal of extracellular matrix or by seeding dissociated cells in 2-dimensional monolayers. In summary, the culture of intestinal organoids *in vitro* has a great potential to improve our understanding of host-microbiota interactions in livestock species while allowing an efficient reduction of *in vivo* animal experiments.

Transcriptomic analysis of pig intestinal organoids divergent for FE and their response to *E. coli*

R.S.C. Rikkers[1,2], O. Madsen[2], J.M. Wells[3], N. Taverne[3], A.J. Taverne-Thiele[3], H. Woelders[1,2], S. Kar[1,2], R. Bergsma[4] and E.D. Ellen[1,2]

[1]*Wageningen University & Research, Wageningen Livestock Research, P.O. Box 338, 6700 AH Wageningen, the Netherlands,* [2]*Wageningen University & Research, Animal Breeding & Genomics, P.O. Box 338, 6700 AH Wageningen, the Netherlands,* [3]*Wageningen University & Research, Host-Microbe Interactomics, P.O. Box 338, 6700 AH Wageningen, the Netherlands,* [4]*Topigs Norsvin, Schoenaker 6, 6641 SZ Beuningen, the Netherlands; roxann1.rikkers@wur.nl*

Feed efficiency (FE) is an important trait in livestock production. It is common to measure and improve FE via the feed conversion ratio (FCR), but the underlying biological mechanisms are still largely unknown and complex. Adult stem-cell derived organoids are powerful experimental models to study mammalian biology, and offer possibilities to advance our molecular understanding of FE. This study investigates differences in molecular phenotype between organoids derived from pigs divergent for FE and their responses when challenged with *Escherichia coli*. Colon and ileum samples were collected from six low FE pigs (FCR=2.61±0.04) and six high FE pigs (FCR=2.19±0.03) and three dimensional (3D) organoids were cultured. These 3D organoids were developed into enhanced, near physiological 2D culture systems. The low and high groups were either challenged with *E. coli* (Le or He) or used as controls (L or H). The results of RNA-seq transcriptomic analysis will be presented and will focus on the differences in gene expression between the low FE and high FE organoids and their responses when challenged with *E. coli*. Thus far, the results suggest a difference in immune response between the low FE and high FE organoids when challenged with *E. coli*. These findings show that organoids could provide better inside into complex biological mechanisms such as FE.

Near complete reduction of rumen methane emission *in vitro* by supplementing high doses of choline

Y. Li, M. Kreuzer, C. Kunz and M. Terranova
ETH Zurich, Institute of Agricultural Sciences, Universitaetstrasse 2, 8092 Zurich, Switzerland; yang.li@usys.ethz.ch

Enteric methane (CH_4) mitigation has often only limited success because of the complexity of the rumen microbiome. With this study, we intended to manipulate and simplify the rumen microbiome by aiding one particular sub-population of rumen methanogens, the Methanomassiliicoccales (MMC) to out-compete other rumen methanogens. This simplified population might be much more vulnerable to CH_4 mitigation strategies. We expected to achieve a dominant MMC population by drastically increasing the methyl-substrate availability for the MMC, especially since MMC has a lower hydrogen threshold than other rumen methanogens. For this goal, an experiment with an 8-fermenter *in vitro* Rumen Simulation Technique (Rusitec) was carried out across 15 days with four replicates. To ensure a constant selection pressure, the artificial saliva of treatment groups were infused with 200 mM of choline chloride (ChCl) or choline bicarbonate ($ChHCO_3$), at the average flowrate of 403 ml/d. This infusion was equivalent to 750 g/kg (ChCl) and 888 g/kg ($ChHCO_3$) diet (consisting of 15 g dry matter of hay: wheat flakes: soybean meal (1:0.7:0.3) and vitamin-mineral mixture). No dominance of MMC was achieved, but a sharp decline of all methanogens and the near complete mitigation (P<0.001) of CH_4 emission at 97.9 and 96.5% was observed in ChCl and $ChHCO_3$ infusion respectively compared to the control. There was a concomitant increase in hydrogen, which was clearly greater with $ChHCO_3$ than with ChCl. The concentration of acetate in the incubation fluid (29.9 mM) was increased (P<0.01) by the treatments (ChCl 57.6 mM; ChHCO3 57.1 mM), the propionate, isovalerate, and valerate concentrations (13.8, 3.5, 5.2 mM) were decreased (P<0.01) (ChCl 7.8, 0.4, 3.3 mM; ChHCO3 4.2, 0.2, 2.3 mM). The proportion of organic matter digested was decreased (P<0.01) by the treatments (ChCl -9%; $ChHCO_3$ -10%). Concomitantly a high level of ammonia was produced owing to the excess of the N-containing choline. An excessive ammonia formation started between 25 and 50 mM of choline as determined in a dose-response run. The next step is performing shotgun metagenomics and metabolomics to decipher the metabolic network and microbial cross-interaction that led to CH_4 mitigation.

In vitro models of oviduct functions, a key for improvement of reproductive biotechnologies

P. Mermillod[1], K. Reynaud[1], Y. Locatelli[1,2], S. Alves[1], L. Cordeiro[1], O. Bernardi[1] and M. Saint Dizier[1]
[1]INRAE, Physiologie de la Reproduction et des Comportements, 37380 Nouzilly, France, [2]MNHN, Réserve de la Haute Touche, 36290 Obterre, France; pascal.mermillod@inrae.fr

A wide range of reproductive biotechnologies are now available to facilitate the genetic management in mammals. The maternal environment offers optimised conditions for fertilisation and embryo development, based on a constant dialogue between somatic tissues and germinal/embryonic cells. More that 50% of pregnancy failures in cattle take place in the oviduct at the time of fertilisation and early embryo development. However, the oviduct is difficult to reach in vivo, making it difficult to study its physiology and functions. To open this black box, an in vitro model of bovine oviduct epithelial cells (BOEC) has been developed and extensively studied. This model allowed to investigate the interactions of BOEC with sperm cells and early embryos, with promising results for improvement of reproductive biotechnologies. However, it has been established that BOEC grown on plastic rapidly dedifferentiate, losing their ability to respond to the steroid hormones known to regulate oviduct activity in vivo. Therefore, OEC culture systems respecting the initial cell differentiation are required. When placed in culture, OEC aggregates spontaneously form vesicles with the apical side of the cells outside and the basal side lining an internal cavity, maintaining better cell differentiation than BOEC on plastic. BOEC can be grown on a semipermeable membrane dipped in a rich basal media and coated by basal lamina proteins (collagen, laminin). The cells grow on this membrane with their apical side up and produce a thin layer of secretions on their top. This 3D culture system provides in vivo like morphology and molecular activities, allowing to mimic a full oestrous cycle by sequential steroids stimulation. This system can be reproduced on 3D printed oviduct scaffolds allowing a microfluidic medium distribution. Finally, the in-gel culture of single OEC allows the proliferation of epithelial stem cells, forming spherical organoids including ciliated and secretory OEC, with the apical side in the internal cavity. These emerging organoids may represent the best in vitro system for the study of oviduct physiology and functions.

Effect of in vitro culture on the stem and migratory properties of primordial germ cells

M. Paradowska, A. Szczerba, A. Dunisławska, K. Stadnicka and M. Bednarczyk
UTP University of Science and Technology in Bydgoszcz, Department of Animal Biotechnology and Genetics, Mazowiecka 28, 85-022 Bydgoszcz, Poland; michelle.paradowska@utp.edu.pl

Primordial germ cells (PGCs) are cells present in developing embryos which contribute to germline and thus, transmit genetic information from one generation to the next generations. PGC circulate in the bloodstream migrating into gonads, where they contribute to sperm and oocytes. There are two types of PGC: cells circulating in the blood (bPGC) and gonadal cells (gPGC). We aim to address the technical issues related to propagation of pure PGC in vitro: 1/ to validate a culture condition so that PGC retain their stem cell characteristics and migratory properties, 2/ to facilitate a fast and non-invasive identification of PGC in the primary material (blood or gonads) using the microscope optics. For in vitro experiments, circulating bPGC were isolated from donors – chicken embryos (green-legged partridge like and Ross308) at stage 13-15 (Hamburger Hamilton stages, H&H) of development and gPGC were isolated at stage 27 (H&H). In vitro tests have involved: short- and long-term culture of PGC in 3 conditions with minimised xenobiotic factors, assessment of proliferation (MTT test) and temporal changes of PGC germline specific markers (SSEA-1,4, CVH, DAZL) and pluripotency genes (cPOUV, NANOG, SOX2). The migratory potential of PGC was confirmed by injecting cultured cells, labelled with a fluorescent dye PKH26, into the blood stream of recipient chicken embryos at stages 14 to 16 (H&H). The gPGCs showed homing ability to the recipient gonads after 5-6 days post-injection. In conclusion, we show to what extent the in vitro propagated PGC retain their stem and migratory/homing properties and provide tools to study the epigenetic modification of PGC by the somatic (gonadal) component of the recipient. The provided simplified methodologies for isolation, cultivation and handling of PGC facilitate their further implementation in strategies to conserve species and produce avian bioreactors. The funding: Polish National Science Centre, no. UMO-2017/27/B/NZ9/01510, research number: BN-52/2019, EcoSET the Polish National Agency for Academic Exchange under Grant No. PPI/APM/2019/1/00003.

Bovine monocyte activation by yeast-derived β-glucans correlates with bovine Dectin-1 expression

A.R.V. Pedro[1,2,3], T. Lima[2,3], I. Ramos[4], E. Martins[2,5,6,7], A.R.J. Cabrita[1,2], A.J.M. Fonseca[1,2], M. Vilanova[2,3], M.R.G. Maia[1,2] and A. Correia[2,3]
[1]REQUIMTE, LAQV, ICBAS, Instituto de Ciências Biomédicas Abel Salazar, Universidade do Porto, Rua Jorge Viterbo Ferreira 228, 4050-313 Porto, Portugal, [2]ICBAS, Instituto de Ciências Biomédicas Abel Salazar, Universidade do Porto, Rua Jorge Viterbo Ferreira 228, 4050-313 Porto, Portugal, [3]i3S, Instituto de Investigação e Inovação em Saúde, Universidade do Porto, R. Alfredo Allen 208, 4200-135 Porto, Portugal, [4]Cooperativa Agrícola de Vila do conde, CRL, R. da Lapa 293, 4480-848 Vila do Conde, Portugal, [5]Escola Universitária Vasco da Gama, Av. José R. Sousa Fernandes 197, 3020-210 Coimbra, Portugal, [6]EPIUnit, Instituto de Saúde Pública, Universidade do Porto, Rua das Taipas 135, 4050-091 Porto, Portugal, [7]ADM Portugal, SA, Zona Industrial de Murtede, 3060-372 Murtede, Portugal; arpedro@i3s.up.pt

Dietary supplements, such as yeast cell wall-based products, have long been used in livestock diets to improve immune function, growth and performance and to reduce negative effects associated with stress and disease. However, mechanistic studies supporting the use of such supplements in ruminants are still scarce. Thus, we aimed to evaluate the effects of yeast-derived products (β-glucans and zymosan) on bovine monocytes and to correlate cytokine production with the expression of the major β-glucan receptor, Dectin-1. Yeast-derived β-glucans were used to *in vitro* stimulate peripheral blood monocytes of Holstein-Friesian cows (n=15). Cytokine and Dectin-1 expression were evaluated 24 h after stimulation by qRT-PCR. Cell surface activation markers were assessed by flow cytometry. Data was analysed using the Mixed model of the GraphPad Prism 8.2.1 software with the fixed effect of treatment and the random effect of the animal. Particulate β-glucans enhanced pro-inflammatory cytokine production and cell activation marker expression. Cytokine production was positively correlated with Dectin-1 expression, suggesting an immunomodulatory effect of yeast-derived products through Dectin-1-signalling in bovine monocytes. Financial support from Fundação para a Ciência e a Tecnologia to RP (PD/BDE/135540/2018), TL (PD/BD/128393), MM (DL 57/2016 – Norma transitória) and AC (Individual CEEC 2017 CEECIND/01514/2017) is acknowledged.

Effects of serums derived from Wagyu calves fed different nutrition on adipocyte differentiation

R. Saneshima, S. Shimamoto, D. Ijiri, A. Ohtsuka, Y. Okamura and T. Gotoh
Kagoshima University, Korimoto 1-21-24, 8900068, Kagoshima, Japan; gotoh@agri.kagoshima-u.ac.jp

Nutrition during early growth stage has potential to alter constitution at adult. It is called 'metabolic imprinting'. The purpose of this study was to investigate how serum derived from Wagyu (Japanese Black) calves of different level of early nutrition influences on adipocyte differentiation and proliferation by using fibroblast cell line *in vitro*. Wagyu calves were randomly allocated into two groups. One is the high-nutrition group (Imp: n=12) received intensified nursing (maximum intake of 1.8 kg/day) until 3 months of age (mo) and then fed a high-concentrate diet from 4 to 10 months of age. Another is the low nutrition group (Cont: n=11) received normal nursing (maximum intake of 0.6 kg / day) until 3 mo and then fed only roughage *ad libitum* until 10 mo. Experiment 1: 3T3-L1 cells were cultured with serums taken from calves of groups of Imp and Cont at 3 mo (groups 3m-Imp and 3m-Cont, respectively) and at 10 mo (groups 10m-Imp and 10m-Cont, respectively) for 2 days as early differentiation period. In each well, the expression of genes related to adipocyte differentiation and proliferation was analysed by real time RT-PCR and DNA concentration was measured. Experiment 2: After Experiment 1, remaining part of the cells was continuously cultured with 10% FBS for 5 days as late differentiation period. Oil red-O staining was carried out to stain fat droplets in cell and the amount of triglyceride (TG) was measured. In Experiment 1, the DNA concentration of well didn't indicate significant differences between 3m-Imp and 3m-Cont. However, they were significantly larger in 10m-Imp than in 10m-Cont (P<0.05). The expression of *Zfp423* as a marker of early differentiation was significantly higher in 3m-Imp than in 3m-Cont (P<0.05). Genes expressions of *C/EBPβ, PPARγ, SREBP1c* and *PCNA* were significantly higher in 10m-Imp than in 10m-Cont (P<0.05). In Experiment 2, larger fat droplets area was observed in 10m-Imp. TG and protein contents/well were significantly larger in 10m-Imp than in 10m-Cont (both, P<0.05). Our study showed that enhancing early nutrition might increase some factors in serum to accrete adipocyte differentiation and proliferation in Wagyu cattle.

Dietary restriction of dairy cows: a lever to study the lipolytic system in milk and mammary gland
C. Hurtaud[1], L. Bernard[2], M. Boutinaud[1] and C. Cebo[3]
[1]INRAE, Agrocampus Ouest, PEGASE, 35590 Saint-Gilles, France, [2]INRAE, Université Clermont Auvergne-Vetagrosup, UMRH, 63122 Saint-Genès-Champanelle, France, [3]INRAE, AgroParisTech, GABI, 78352 Jouy-en-Josas, France; catherine.hurtaud@inrae.fr

Milk lipolysis is defined as the hydrolysis of triglycerides, the major component of milk fat. Short-chain fatty acids released in milk are responsible for rancid flavour. In addition, the presence of partial glycerides impairs milk functional properties such as foaming and creaming abilities. Milk lipolysis represents an important criterion to assess the quality of milk and depends on both animal parameters and breeding factors. Low-energy diets were found to be associated with higher levels of spontaneous lipolysis, particularly in late lactation. In our experiment, dairy cows were fed a diet at 65% of their dry matter ingested *ad libitum* to induce spontaneous lipolysis in milk and for studying milk composition associated with lipolysis. Two groups of 22 cows (165±16 days in milk, half primiparous cows) received either a control diet (100% of the dry matter intake *ad libitum*) or the experimental diet according to a 2×2 cross-over design. Lipolysis was analysed by the copper-soap method. Milk fat globule and casein micelles size distributions were evaluated by laser light scattering. The duration of the restriction was 5 days. The statistical model used was the MIXED procedure of SAS. As expected, feeding restriction caused an increase in spontaneous lipolysis of milk (+0.20 Meq/100 g of fat, P<0.01). At the same time, milk yield and protein content decreased (-5.1 kg/d and -1.3 g/kg respectively, P<0.01) whereas no effect was observed on milk fat content. The increase in spontaneous lipolysis was associated with an increase in the diameter of milk fat globules (+0.17 µm, P<0.05) without effect on the diameter of casein micelles. Dietary restriction induced changes in parameters of the metabolism of dairy cows with an increase in plasma non-esterified fatty acids, triglycerides and urea, indicators of the mobilisation of body fat and the catabolism of proteins associated with dietary restriction. This zootechnical approach will be enriched by proteomics and transcriptomics approaches applied to milk and/or mammary gland on a set of animals selected for being extreme in term of lipolysis.

Effect of omega-3 fatty acids on porcine intestinal *ex vivo* model exposed to stress conditions
R. Rebucci, C. Giromini, T.S. Sundaram, M. Comi, A. Baldi, L. Pinotti and F. Cheli
Università degli studi di Milano, Department of Health, Animal Science and Food Safety, Milano, 20133, Italy; raffaella.rebucci@unimi.it

Many naturally available compounds as n3-polyunsaturated fatty acids (Eicosapentaenoic acid, EPA and Docosahexaenoic acid, DHA) exhibit anti-inflammatory, anti-oxidative and other health beneficial properties in farm animals. The aim of the present study was to demonstrate the anti-inflammatory potential of EPA and DHA using *ex vivo* porcine duodenum tissue explants. Duodenal tissues were obtained from eight pigs at the slaughtering house. Each tissue was dissected in sections of about 1.0 cm^2 and washed using phosphate buffer saline and randomly distributed into 24-well plates containing foam-pads to hold explant in place and 1.5 ml of pre-warmed control (only DMEM) and treatment medium. Explants were co-incubated with 10 µM of EPA:DHA (1:2) and/or lipopolysaccharide (LPS) at 10 µg/ml or Hydrogen peroxide (H_2O_2) at 2 mM (37 °C, 2 hours). Membrane damage was evaluated by measuring LDH release while the secretion of IL-8 was determined by enzyme linked immunosorbent assay (ELISA) in tissue culture supernatants. Data were analysed by one-way ANOVA using GraphPad, Prism. The treatment with H_2O_2 did not induce a significant release of LDH and did not alter IL-8 secretion. After LPS treatment, the LDH release increased up to 22% and the IL-8 secretion increased up to 27% in tissue culture supernatants, compared with control (no treatment). The co-incubation of LPS and EPA:DHA restored (p <0.05) the LDH release and also suppressed the secretion of IL-8 (P<0.05) in LPS challenged tissues compared to control (LPS alone). In conclusion, DHA:EPA consistently elicits strong anti-inflammatory activity. These results support the potential of n3-polyunsaturated fatty acids as anti-inflammatory agents in *ex-vivo* tissues of the gastrointestinal tract.

A complex porcine gut model to assess effects of phytogenics on gut barrier integrity *in vitro*

D. Wendner, T. Schott, V. Rainer, K. Teichmann, E. Mayer and N. Reisinger
BIOMIN Holding GmbH, Technopark 1, 3430 Tulln, Austria; dominik.wendner@biomin.net

An intact gut barrier is essential for gut health in livestock. This barrier can be compromised by disease, toxin exposure, and local as well as systemic inflammation. Measurement of transepithelial electrical resistance (TEER) of cultivated intestinal epithelial cells is commonly used as a model to assess gut barrier integrity *in vitro*. However, such models only account for direct effects on the intestinal cell layer. They do not cover complex interactions in the tissue, for example with blood and immune cells present *in vivo*. These complex interactions might be essential to test substances with assumed beneficial properties, e.g. phytogenics. Thus, the aim of our study was to establish a complex porcine gut model, by indirectly co-cultivating intestinal porcine epithelial cells (IPEC-J2) and porcine peripheral blood mononuclear cells (PBMCs). The two cell types correspond to the apical and basolateral side of the intestinal tissue *in vivo*, respectively. Using this model, we evaluated the effect of oregano oil to counteract negative effects of the plant lectin Concanavalin A (ConA) on the gut barrier. IPEC-J2 were seeded on transwell inserts in 12-well plates and cultivated for 7 days to form a confluent layer in the inserts. PBMCs, isolated from porcine blood, were then seeded into the basolateral compartment and stimulated with ConA (1.25 µg/ml). Simultaneously, oregano oil (12.5 µg/ml) was applied to the test system. TEER was monitored for 72 hours. ConA stimulation decreased TEER by 41.5% compared to unstimulated control (n=6, P<0.001). Application of oregano oil increased TEER by 43.7% compared to the ConA control (n=6, P<0.01). The ConA stimulus activated PBMCs and indirectly compromised IPEC-J2 barrier integrity. Application of oregano oil in addition to ConA increased TEER values. This indicates positive effects on IPEC-J2 barrier integrity, likely by modulating the activity of the immune cells and their cellular interactions with IPEC-J2. The proposed complex gut model allows *in vitro* testing of phytogenic substances, including both direct and indirect effects on barrier integrity.

Comparison of rumen protected lysine products using two *in vitro* procedures

M.M. Mireaux[1], T.P. Pinard[2], F.C. Czernikarz[2] and C.F. Faure[3]
[1]Deltavit, CCPA Group, R&D, ZA du Bois de Teillay, Quartier du Haut-Bois, 35150 Janzé, France, [2]Artémis, ZA du Bois de Teillay, Quartier du Haut-Bois, 35150 Janzé, France, [3]CCPA Group, ZA du Bois de Teillay, Quartier du Haut-Bois, 35150 Janzé, France; mmireaux@ccpa.com

Comparison of commercial rumen protected Lysine products (RPL) is needed to make educated choices between products. To do so trials are necessary but *in vivo* trials can be complicated to set up (availability, timing, and price) whereas *in vitro* procedures are more accessible and they can be independent of cannulated animals. The objective of this study was to propose two *in vitro* methods for evaluating RPL: one with and one without buffered rumen fluid. The two methods where applied on five RPL. In the first *in vitro* method (M1), 0.9 g of RPL (plus free lysine as control) sealed in a nylon bag were incubated in buffered rumen fluid for 5 h simulating the rumen phase. Samples were then digested using an adaptation of the Boisen *in vitro* digestion to mimic the intestinal phase. Dry matter and nitrogen content were analysed before and after incubation and at the end of the intestinal digestion. In the second *in vitro* method (M2), the ruminal phase was replaced by a water soaking: 20 g of RPL were put in 200 ml of water (pH=6.5) in a 39 °C water bath with magnetic stirring for 5 h. Samples were then filtered and analysed for dry matter (DM) and azote (N) content. The two methodologies led to the same hierarchy between products (with one exception). These two in-vitro methods show a good discrimination between products with a ruminal DM degradation of 100% for free lysine and 37, 34, 26, 11 and 9% for RPL in M1. At the end of the intestinal phase, differences between RPL remained for DM degradation (ranging between 48 to 68%) but protein degradation is complete for all products. This study show that *in vitro* methods, even one independent of cannulated animals (M2) allow for a good discrimination between RPL and can help nutritionists choose between products.

Adiponectin in the mandibular glands of cows fed a diet supplemented with essential fatty acids

C. Dall'Aglio[1], F. Mercati[1], P. Scocco[2], E. De Felice[2], A. Tröscher[3], H.M. Hammon[4] and M. Mielenz[4]
[1]*University of Perugia, Department of Veterinary Medicine, Via San Costanzo 4, 06126 Perugia, Italy,* [2]*University of Camerino, School of Biosciences and Veterinary Medicine, Via Pontoni 5, 62032 Camerino, Italy,* [3]*BASF SE, Chemiestraße 22, 68623 Lampertheim, Germany,* [4]*Leibniz Institute for Farm Animal Biology (FBN), Wilhelm-Stahl-Allee 2, 18196 Dummerstorf, Germany; cecilia.dallaglio@unipg.it*

The adipokine adiponectin is mainly expressed in adipose tissue but also found in human and mice salivary glands. Recent studies have shown that there is a relationship between the expression of adipokines and the characteristics of the diet. In particular, the n-3 fatty acids status is positively linked with plasma adiponectin concentrations. In this study, the presence and distribution of adiponectin in the bovine mandibular gland (MG) were investigated to test possible effects from different compositions of dietary fatty acids. Rumen fistulated Holstein cows (n=38, 11.000 kg milk in 2nd lactation) were fitted with abomasal infusion tubes and fed a corn silage based total mixed ration. Cows were infused twice daily with either coconut oil delivering medium-chain fatty acids (CTRL: 76 g/d), linseed-safflower oil mix, delivering mainly n-3 fatty acids (EFA: 78 + 4 g/d, respectively), LUTALIN (content of c9,t11 and t10,c12 conjugated linoleic acids (CLA) in equal amounts; 10 g/d), or EFA + CLA from 9 wk antepartum to 9 wk postpartum. MG were quickly removed after slaughter and processed for routinely tissue preparation. The immunohistochemical reaction was visualised using a polyclonal anti-adiponectin antibody and DAB as chromogen. The software ImageJ, IHC Profiler plugin, was used for the analysis of the immunoreaction. The immunohistochemical study showed a positive reaction for adiponectin in the cytoplasm of ductal epithelial cells and in the serous acinar cells in the CTRL animals. Cows supplemented with EFA or CLA or both showed a positive signal for adiponectin only in the serous cells. These results showed that adiponectin is present in the CTRL bovine MG with a distribution that reflects what has been already described in humans and mice. The differences are linked with the supplements and may suggest a negative effect of fatty acids on local adiponectin production.

Autophagy is required for *in vitro* myogenesis in piglets

A. Vincent, M.H. Perruchot, I. Louveau and F. Dessauge
PEGASE, INRAE, Institut Agro, 35590 Saint-Gilles, France; isabelle.louveau@inrae.fr

Autophagy is a lysosomal degradation pathway with a role in the turnover of cell components via self-digestion. Over the past decade, it has been recognised as an essential process to maintain cellular and energy homeostasis. Nevertheless, little attention has been paid to this process in agronomical animal species. In pigs, the role of autophagy in skeletal muscle homeostasis and more specifically on the formation of multi-nucleated muscle fibres needs to be determined. Primary culture of satellite cells, the resident muscle stem cells, is an appropriate model to investigate macro autophagy (hereafter autophagy), the main autophagy process. The objective of the current study was to evaluate tools to monitor autophagy flux in this cell model and to determine the impact of autophagy on cell differentiation. Samples of longissimus muscle were collected from 4-day-old piglets. After isolation, satellite cells were plated in growth medium, allowed to proliferate up to 80% confluence and then placed in an appropriate culture medium for 6 days to differentiate into myotubes. Proliferating and differentiating cells were explored. Autophagy-related proteins (LC3-I, LC3-II, ULK-1, Beclin, p53, PI3K and p62) were detected by Western blotting in proliferating cells and at day 1, 3 and 6 of differentiation. The two major regulators of metabolism: mammalian target of rapamycin (mTOR), a central hub of nutrient signalling and cell growth, and AMP-activated protein kinase (AMPK), a major sensor for cell energy, were also detected. Then, cells were treated with an inhibitor of autophagy flux, bafilomycin A1 to confirm that autophagy was activated during the conversion of myoblasts into myotubes in our satellite cell culture model. In conclusion, we demonstrated that basal autophagy is required for porcine satellite cell differentiation *in vitro*. The next step will be to demonstrate that autophagosome biogenesis impairment halts cell fusion during stages of differentiation and the fusion step of myocytes.

Identification of the HH2 causal variant and functional validation using CRISPR-Cas9 knockouts

J.B. Cole[1], D.M. Bickhart[2], J.L. Hutchison[1], D.J. Null[1], M.S. Ortega[3] and K. Clark[3]
[1]*Agricultural Research Service, USDA, Animal Genomics and Improvement Laboratory, 10300 Baltimore Avenue, Beltsville, Maryland 20705-2350, USA,* [2]*Agricultural Research Service, USDA, Cell Wall Biology and Utilization Research Laboratory, 1925 Linden Drive, Madison, Wisconsin 53706, USA,* [3]*University of Missouri, Division of Animal Sciences, S109 Animal Sciences Research Center, Columbia, Missouri 65211, USA; john.cole@usda.gov*

A comprehensive framework for identification and validation of genetic defects, including haplotype-based detection of defects, selection of variants from sequence data, and *in vitro* validation using CRISPR-Cas9 knockout embryos, is described. Holstein Haplotype 2 (HH2), which causes early embryonic death, was used to demonstrate the approach. The HH2 haplotype was identified using a deficiency-of-homozygotes approach and confirmed to have undesirable effects on conception rate and stillbirths. Five carriers were present in a group of 183 sequenced Holstein bulls selected to maximise the coverage of unique haplotypes. Three variants concordant with the haplotype calls were found in HH2: a high-priority frameshift mutation resulting from a deletion, and two low-priority variants. The frameshift was confirmed in a separate group of Holsteins from Run7 of the 1000 Bull Genomes Project that shared no animals with the discovery set. Intraflagellar protein 80 (IFT80)-null embryos were generated by truncating the IFT80 transcript at exon 2 using two guide-RNAs annealed to Cas9 mRNA. Abattoir-derived oocytes were fertilised *in vitro* with proven high-fertility semen. Embryos were injected at the one-cell stage either with CRISPR-Cas9 complex (n=100) or Cas9 mRNA (control, n=100) before return to culture. Editing efficiency among three replicates was 26% for biallelic edits and 32% for mono-allelic edits based on visualisation of PCR fragments on an agarose gel. The percentage of putative zygotes that cleaved was determined at 3 d of development (d 0 = day of insemination) and blastocyst rate was estimated at 8 d. IFT80-null embryos arrested at the 8-cell stage of development, consistent with mouse hypomorphs and HH2 carrier-to-carrier matings. A frameshift in IFT80 on chromosome 1 at 107,172,615 bp (p.Leu381fs) disrupts wnt and hedgehog signaling, and is responsible for the death of homozygous embryos.

Genetic parameters for psoroptic mange sensitivity in Belgian Blue cattle

R. Meyermans[1], W. Gorssen[1], E. Heyrman[1], S. Janssens[1], X. Hubin[2], P. Mayeres[2], W. Veulemans[3], S. Van Der Beek[3], E. Claerebout[4], M. Georges[5], C. Charlier[5] and N. Buys[1]
[1]*KU Leuven, Livestock Genetics, Department of Biosystems, Kasteelpark Arenberg 30 box 2472, 3000 Leuven, Belgium,* [2]*Eleveo asbl, Champs Elysées 4, 5590 Ciney, Belgium,* [3]*CRV, Wassenaarweg 20, 6843 NW Arnhem, the Netherlands,* [4]*UGent, Laboratory of Parasitology, Department Virology, Parasitology and Immunology, Salisburylaan 133, 9820 Merelbeke, Belgium,* [5]*GIGA Institute, Medical Genomics-Unit of Animal Genomics, Avenue de l'Hôpital 11, 4000 Liège, Belgium; roel.meyermans@kuleuven.be*

The Belgian Blue cattle breed is known for its high degree of muscling ('double muscling') and low feed conversion ratio. However, this breed is very susceptible to psoroptic mange dermatosis caused by the *Psoroptes ovis* mite. These mite infestations cause severe dermatitis and lesions. Psoroptic mange does not only severely impact animal welfare, but also causes serious economic losses (e.g. due to growth reduction). Within the Belgian Blue breed there are individual differences of sensitivity to psoroptic mange. To study the genetic basis of mange sensitivity in Belgian Blue cattle, 1,436 animals were phenotyped for their mange sensitivity at 55 beef farms during winter housing between 2011 and 2019. Phenotypes were based on the extension (in % of the total body size) and severity of the lesions and mite counts (from three different skin scrapings). Of all phenotyped animals, 921 were genotyped (either using the Illumina BovineSNP50 or the EuroG_MD array). Pedigree records were collected from the respective herdbooks (CRV and awé). Other recorded factors were sex, age, hair colour, skin thickness and farmer. Linear and zero-inflated Poisson models were used to test the significance of these factors. Genetic parameters were estimated using a ssGBLUP via a Gibbs sampler for mixed threshold-linear models (thrgibbsf90). Heritabilities for lesion extent and mite count were estimated at 6.5 and 25%, respectively. The genetic correlation between both mange traits was estimated around 0.70. Both traits seemed to be phenotypes that reflect partially different aspects of the genetic control of mange sensitivity. These estimates confirmed the presence of a heritable component for mange susceptibility.

Identification of novel lncRNA associated with mastitis disease in Holstein dairy cattle

V. Asselstine[1], J.F. Medrano[2], S. Germani[1,3], F. Ceciliani[3], A. Suárez-Vega[1], B. Mallard[1,4], N. Karrow[1], A. Islas-Trejo[2] and A. Cánovas[1]
[1]*University of Guelph, Centre for Genetic Improvement of Livestock, Department of Animal Biosciences, Guelph, ON, N1G 2W1, Canada,* [2]*University of California-Davis, Department of Animal Science, Davis, CA, 95616, USA,* [3]*Università degli Studi di Milano, Department of Veterinary Medicine, Lodi, 26900, Italy,* [4]*Ontario Veterinary College, Department of Pathobiology, Guelph, ON, N1G 2W1, Canada; vasselst@uoguelph.ca*

Bovine mastitis is one of the most common diseases in lactating dairy cows, resulting in milk yield reduction, discarded milk and early culling. Although mastitis has been heavily studied, it remains one of the most challenging diseases dairy producers encounter. With the advent of RNA-Sequencing, the host transcriptome can be studied at a high-throughput level to identify key differences between healthy and mastitic animals. One area of the transcriptome that has not been studied thoroughly in relation to mastitis is long non-coding RNA (lncRNAs). Long non-coding RNAs play a fundamental role in immune regulation and can potentially provide key information regarding mastitis resistance. RNA-Sequencing was performed on six cows using somatic cell (SC) RNA from milk samples from two separate quarters; one quarter classified as healthy (n=6), and one as mastitic (n=6), based on having a SC count <100,000 cells/ml. The goal was to identify lncRNAs that are differentially expressed (DE) between healthy and mastitic samples (FDR<0.05; FC>±2). Preliminary results showed that 244 lncRNAs were DE in the milk SC transcriptome between the healthy and mastitic samples, and these 244 lncRNAs were related to 284 associated mRNA genes. Functional enrichment analysis was performed using the list of 284 associated mRNA genes. Three genes (*Chemokine (C-X-C motif) ligand 2* (*CXCL2*); *Toll-like receptor 2* (*TLR2*); and *Interleukin 10* (*IL-10*)), were significantly involved in metabolic pathways related to host defence against mastitis such as, inflammatory response and defence response. In conclusion, the identification of novel lncRNA associated with host defence to mastitis infection suggests the potential of using this approach to better understand genetic resistance to mastitis in dairy.

Disease susceptibility and resistance in Italian dual-purpose cows: the case of the Valdostana breed

E. De Monte[1], S. Kühl[1], M. Vevey[2] and M. Gauly[1]
[1]*Free University of Bolzano, Faculty of Sciences and Technologies, piazza Università, 5, 39100 Bolzano, Italy,* [2]*National Association Breeders of Valdostana Bovines (A.N.A.Bo.Ra.Va.), Fraz. Favret 5, 11020 Gressan (AO), Italy; erica_dm88@yahoo.it*

Genetic selection, in high yielding dairy breeds, for resistance to specific pathologies is a common practice in many countries worldwide. However, diagnostic data for local breeds are still missing. The aims of this study were: 1) to evaluate the incidence of selected diseases for the year 2018 in the Valdostana dual-purpose cattle, and 2) to identify differences in the susceptibility to specific disorders between the sub breeds composing the Valdostana breed (Red Pied-VRP, Black Pied-VBP, and Chestnut-VC). Pathologies and prescribed treatments, applied to female animals, were registered in an online-database by the veterinarians involved in the project. The recorded diseases were those affecting the udder, reproductive apparatus, locomotor system and parasitic infections. Overall incidence and mean farm incidence (MFI) per sub breed and pathology were identified. In total 12,932 diagnoses were recorded in animals aged 24-99 months in 654 herds (75% of the farms). Mastitis presented a general incidence in Valdostana animals of 10.1%, being higher in VRP cows (11%) than in VC and VBP animals (8.1%) with significantly different MFI (11±0.7% in VPR vs 7.2±0.7% in VC-VBP). Similarly, dry cow therapy was performed at a higher frequency in the VRP breed (incidences 27.4% in VRP vs 24.8% in VC-VBP with MFI in VRP of 23.8±1.2% and of 21±1.1% in VC-VBP; P<0.05). Lack of oxytocin was observed more frequently in VRP animals (2% in VRP vs 1.6% in VC-VBP and MFI was 2.5±0.6% in VRP vs 1.7±0.5% in VC-VBP with P<0.05). Reproductive disorders had an incidence of 10.2% with ovarian cysts recorded at 1.4% in VRP and 7.8% in VC-VBP cows (MFI: 1.4±0.2% in VPR vs 5.9±0.6% in VC-VBP; P<0.05). Disturbances of the locomotor system and parasitoses were recorded at an overall incidence of 2.4 and 9.2% respectively. Laminitis had a MFI of 1.1±0.2% in VC-VBP and of 1.3±0.3% in VRP with P<0.05. In conclusion, the Valdostana cattle is a robust dual purpose breed with low susceptibility to diseases representing the most important challenge in dairy and beef production.

Changes in lactose content of milk as indicators of longevity and udder health in Holstein cows

A. Costa[1], H. Bovenhuis[2] and M. Penasa[1]
[1]University of Padova, Department of Agronomy, Food, Natural resources, Animals and Environment, viale dell'università, 16, Legnaro, 35020, Italy, [2]Wageningen University, Animal Breeding and Genomics Group, P.O. Box 338, Wageningen, 6700 AH Wageningen, the Netherlands; angela.costa.1@phd.unipd.it

Milk lactose percentage (LP) is inversely related with somatic cell count and intramammary inflammation in dairy cows. Moreover, LP decreases with parity, likely as a result of the accumulated effect of udder infections and mammary gland epithelial stress. Thus, changes of LP across parities could be evaluated as potential indicators of longevity (LON) and udder health (UDH). In the present study we defined 6 descriptors of LP changes across the first 3 lactations of 69,586 Holsteins and we estimated their genetic correlations with LON and UDH. At least 5 test-day LP were guaranteed in each lactation, thus $\geq$15 test-day LP were available for each cow. β_{12} was the slope of LP in the first 2 lactations, β_{23} was the slope of LP from lactation 2 to 3, and β_{123} was the overall slope of LP. The 3 lactation means were also calculated for each cow in order to derive LP deltas, i.e. between lactations 1 and 2 (Δ_{12}), 2 and 3 (Δ_{23}), and 1 and 3 (Δ_{13}). Genetic parameters and breeding values were estimated through a linear animal model for the 6 descriptors and the Calo's method was used to assess genetic correlations between traits. Results showed that cows with greater changes of LP (i.e. stronger and negative slopes, and greater and positive deltas) were characterised by impaired LON and worse UDH than cows with smaller changes of LP. The h^2 of descriptors ranged from 0.11 to 0.19 for the slopes and from 0.06 to 0.20 for the deltas. Genetic correlations with LON were 0.19 (β_{12}), 0.08 (β_{23}), 0.52 (β_{123}), -0.19 (Δ_{12}), -0.02 (Δ_{23}), and -0.11 (Δ_{13}), while those with UDH were 0.15 (β_{12}), 0.05 (β_{23}), 0.33 (β_{123}), -0.19 (Δ_{12}), -0.15 (Δ_{23}), and -0.06 (Δ_{13}). Selection responses for the descriptors of LP were also calculated. For comparison, both slopes and deltas of milk yield were investigated. Overall, genetic correlations of milk yield descriptors with LON and UDH were stronger than those of LP descriptors with the same traits, whereas h2 were smaller. Despite this, milk yield descriptors were slightly more efficient than LP descriptors in terms of selection response for both LON and UDH.

Strategy to infer candidate genes, allele substitutions and dominance for endoparasite resistance

K. May[1], C. Scheper[1], C. Weimann[1], C. Strube[2] and S. König[1]
[1]Institute of Animal Breeding and Genetics, Justus-Liebig-University Gießen, Ludwigstraße 21B, 35390 Gießen, Germany, [2]Institute for Parasitology, University of Veterinary Medicine Hannover, Buenteweg 17, 30559 Hanover, Germany; katharina.may@agrar.uni-giessen.de

The study aimed at the identification of candidate genes via genome-wide association studies (GWAS) for endoparasite infection traits and on the estimation of allele substitutions and dominance for SNPs located in the activated leukocyte cell adhesion molecule (*ALCAM*) gene. The phenotypic multi-breed dataset included records from 1,166 cows (German-dual purpose (DSN) and Holstein) for gastrointestinal nematode (INF-GIN) and *Fasciola hepatica* (INF-FH) infections. Endoparasite traits were pre-corrected for environmental effects applying linear mixed models. Pre-corrected phenotypes (residuals) were dependent traits in GWAS, considering 148 DSN cows with imputed genotypes (700k). Associated SNPs were annotated to candidate genes using the ENSEMBL database. The *ALCAM* gene was identified as a candidate gene for endoparasite resistance with one SNP (*c.73+32791A>G*) above the significance threshold (7.47×10^{-6}) on BTA 1. DAVID analyses inferred the 'Cell adhesion molecule pathway' for *ALCAM*, playing a crucial role in immune response during disease. SNPs in *ALCAM* influenced further diseases and production traits. Thus, we focussed on *ALCAM* sequencing in a selective genotyping approach considering a subset of 94 DSN cows, suggesting three SNPs in exon 9 and in intron 9. Estimations of allele substitutions and dominance effects for residuals of endoparasite traits (rINF-FH, rGIN-INF) and the test-day traits milk yield, fat%, protein%, fat-to-protein ratio (FPR), milk urea nitrogen and SCS were carried out via logistic regression analyses, and modelling the probability of the heterozygous genotype as dependent 'trait'. Allele substitutions for rFH-INF were significant for all loci. Associations between rFH-INF with fat%, protein% and FPR were favourable, but unfavourable with rGIN-INF and SCS. Dominance effects were small for the moderately heritable test-day traits (0.0 to 0.47 SD), but quite large and in the direction of desired resistance for rFH-INF (0.51 to 0.82 SD). Results suggest consideration of dominance effects in genomic evaluations for disease resistance.

Genetic analysis of reproduction and metabolic disorders in dairy cattle

E.C. Wiesenekker-Verduijn, J.J. Bouwmeester-Vosman and G. De Jong
CRV u.a., Animal Evaluation Unit, Wassenaarweg 30, 6843NW Arnhem, the Netherlands; lisette.wiesenekker@crv4all.com

Animal health remains an important topic within herd management in the dairy industry. More and more farmers in the Netherlands and Flanders record health issues in their management systems. Some issues they record are reproduction and metabolic disorders. Reproduction disorders have a negative influence on fertility. Metabolic disorders cause high treatment costs and a reduction in milk yield. Farmer recorded data gives the opportunity to develop breeding values to select against these disorders. A breeding value estimation was performed for five different reproduction traits: retained placenta, endometritis, metritis, cystic ovaries and anoestrus. These traits were analysed in a multi-trait model for heifers and cows separately. Furthermore a breeding value estimation for metabolic traits was performed, using a multi-trait model for the traits milk fever and ketosis for heifers, young cows (parity 2) and older cows. For all these traits there were around 200,000 to 400,000 health records available. The prevalence of the reproduction disorders is about 3-5%, with cows having a higher prevalence than heifers. The prevalences of milk fever ranges from around 0.5% for heifers to 17% in older cows. For ketosis this range is 2-8%. Heritabilities of the reproduction disorders are low (1-4%), with a genetic standard deviation of 2-3%. Heritabilities are a bit higher for the metabolic disorders (2-10%), with a genetic standard deviation of 3-6%. The reproduction disorders have large genetic correlations with fertility traits and longevity. A higher incidence relates strongly to a longer calving interval, a lower conception rate and a lower longevity. A higher incidence of the metabolic disorders correlates genetically with a higher milk yield and a lower incidence of other health traits. There is no visible genetic trend for any of the reproduction or metabolic traits. Two index breeding values are calculated based on the reproduction and metabolic disorders respectively. Since April 2020 these indexes and the breeding values of the underlying traits are published in the national evaluation of the Netherlands and Flanders, in order that farmers are able to select bulls to improve the health of their herd.

Genetic relationships between udder health indicators and milk technological traits in Holstein cows

S. Pegolo[1], V. Bisutti[1], L.F.M. Mota[1], R. Tessari[1], P. Ajmone Marsan[2], G. Bittante[1] and A. Cecchinato[1]
[1]University of Padova, DAFNAE, Viale dell'Università 16, 35020, Legnaro (PD), Italy, [2]Università Cattolica del Sacro Cuore, Institute of Zootechnics, via Emilia Parmense 84, 29122, Piacenza, Italy; sara.pegolo@unipd.it

Mastitis is one of the most common and costly diseases in dairy cattle. Beside somatic cell count (SCC) and the more recently introduction of differential somatic cell count (DSCC), new indicators are needed for genetic selection to improve mastitis resistance. Aims of this study were to estimate: (1) variance components and heritability for a plethora of novel phenotypes related to udder health (UH) including SCC (converted in somatic cell score [SCS]), DSCC, lactose, milk pH and 18 lab-measured milk technological traits (MTT) including milk coagulation properties, cheese yields and curd nutrient recoveries; and (2) the genetic relationships between UH and MTT in a cohort of ~1000 Holstein cows reared in Lombardy (Italy). Bayesian animal models were implemented via Gibbs sampling. Posterior means of heritability were relatively high (~0.20) for both SCS and DSCC and high (~0.40) for lactose (LACT) and non-casein nitrogen (NCN). Milk pH had heritability of 0.26. Heritabilities for TT were medium to high (ranging from ~0.11 to ~0.47). Regarding additive genetic correlations, SCS and DSCC had moderate unfavourable effects on curd firmness traits and on rennet coagulation time (RCT) and time to a curd firmness of 20 mm (k_{20}). A negative effect of an increase in milk pH on both RCT and k_{20} was also observed. NCN and SCS had unfavourable effects on %CY traits (estimates from ~-0.20 to ~-0.50) while LACT had favourable (estimates from ~0.20 to ~0.30). Negative effects of DSCC on recovery of protein and fat in curd were also found. The results obtained offer new insights into the relationships between UH and milk technological traits in Holstein cows. This information can be used to set up breeding strategies aimed at the improvement of animal health status and milk cheesemaking aptitude. Acknowledgements. The research was part of the BENELAT project funded by the Lombardy Region under the Rural Development Program 2014-2020.

Genetic and non-genetic influences associated with health and vitality traits in beef calves

T. Condon[1,2], C.P. Murphy[2], R.D. Sleator[2], M.M. Judge[1], S. Ring[3] and D.P. Berry[1]
[1]Teagasc, Moorepark, Co. Cork, Ireland, [2]Cork Institute of Technology, Bishopstown, Co. Cork, Ireland, [3]Irish Cattle and Breeding Federation, Bandon, Co. Cork, Ireland; thomas.condon@teagasc.ie

Awareness and interest in calf health and well-being is growing, prompting change in the management decisions of producers and associated policymakers. The objective of the present study was to use mixed models analyses of a large national database of beef calves to quantify the factors associated with subjectively scored calf vigour, size, scour and pneumonia and to determine the interrelationships among them. The data consisted of health and size data on 236,696 calves born in 10,927 Irish beef herds. Vigour was scored on a scale of 1 (very poor) to 5 (very good). Birth size was scored on a scale of 1 (very small) to 5 (very large). Scour and pneumonia were scored on the number of incidences occurring (0 = no occurrence, 1 = one occurrence or 2 = one or more occurrence). For the purpose of the present study, a scale of 1 to 500 was used for vigour and birth size whereas a scale of 1 to 300 was used for scour and pneumonia. On average, 14.3% of calves were recorded to have had one or more occurrence of scour within the first 5 months of life, whereas 5.9% of calves had one or more occurrence of pneumonia within the first 5 months of life. Vigour score was, on average, 12.01 units lower in male calves than in females. Calves born as twins had an 85.85 unit lower size score and a 1.16 unit greater incidence of pneumonia relative to their singleton contemporaries. Relative to the first period of the calving season (i.e. very early period), the incidence of scour was greater in the third and fourth periods of the calving season (i.e. late and very late period), 1.58 and 1.71 units, respectively. Vigour was moderately heritable, with direct heritability estimates of 0.24±0.35. Birth size was highly heritable, with direct heritability estimates of 0.46±0.43. Direct heritability estimates of scour and pneumonia were 0.04±0.02 and 0.14±0.02, respectively. Breeding for vigorous calves that are less susceptible to scour and pneumonia could provide producers with additional strategies to ensure consumer concerns regarding food quality, safety and calf well-being are being met.

Infectious disease heritability is a poor predictor of response to selection in disease prevalence

A.D. Hulst[1,2], P. Bijma[1] and M.C.M. De Jong[2]
[1]Wageningen University and Research, Animal Breeding and Genomics, P.O. Box 338, 6700AH Wageningen, the Netherlands, [2]Wageningen University and Research, Quantitative Veterinary Epidemiology, P.O. Box 338, 6700AH Wageningen, the Netherlands; dries.hulst@wur.nl

Genetic analysis of infectious disease status (healthy/diseased=0/1) with ordinary linear mixed models typically yields low heritability estimates, suggesting limited response to selection in disease prevalence. Such models, however, ignore that animals can infect each other, which leads to feedback dynamics. To investigate the relationship between the observed heritability of binary disease status and genetic variation in underlying epidemiological variables, we simulated an endemic infectious disease following standard epidemiological models. Results show that an extensive amount of genetic variation in disease susceptibility is needed to reproduce the low heritability estimates as regularly found in field data. To obtain a typical heritability of 0.05, for example, the top and bottom 10% of animals needed to differ by a factor of six in breeding value for susceptibility. However, this substantial difference in breeding value coincides with a much smaller difference in the disease status of the animals. This discrepancy occurs in randomly composed herds, because genetically superior animals are exposed to an average number of infectious herd mates, which masks their genetic superiority. When selecting parents based on their breeding value for susceptibility, such that all herd mates in the next generation have a lower susceptibility, the response to selection in disease status closely corresponds to the selection differential in the breeding value for susceptibility. This response is considerably larger than suggested by the common breeding values for disease status, because animals with lower susceptibility infect fewer herd mates, just because they are less often infected themselves. Our ordinary BV ignore this positive feedback. Thus, genetic variation and potential response to selection in infectious disease status seem considerably larger than currently believed. In contrast to the underlying genetic variation in threshold models, the genetic variation we identify here can be fully used for genetic improvement.

Genetic analysis of resilience indicators based on different lactations and lactation stages
M. Poppe[1], G. Bonekamp[1], M.L. Van Pelt[2] and H.A. Mulder[1]
[1]Wageningen University & Research, Animal Breeding and Genomics, P.O. Box 338, 6700 AH Wageningen, the Netherlands,
[2]Cooperation CRV, Animal Evaluation Unit, P.O. Box 454, 6800 AL Arnhem, the Netherlands; marieke.poppe@wur.nl

Resilience is the ability of a cow to cope with disturbances, such as pathogens. Resilient cows are less affected in their milk yield by disturbances than non-resilient cows. Therefore, a low variance of deviations from expected milk yield (LnVar) could serve as an indicator of good resilience. Resilience may genetically change within and between parities. Therefore, the aim was to perform a genetic analysis on LnVar based on three parts of lactation 1 (day 11-110, 111-210, 211-340) and based on full lactation 1, 2 and 3, and to estimate genetic correlations with health traits. Data of 202,202 first lactations, 194,776 second lactations, and 187,915 third lactations were used. Genetic variance of LnVar increased over the parts of lactation 1 (0.049 to 0.066). Heritability of LnVar was lower based on parts of lactation 1 (0.11-0.15) than based on full lactation 1 (0.21), because of a lower number of records per cow used. LnVar was genetically most similar based on part 2 and 3 (r_g 0.97) and least similar based on part 1 and 3 (r_g 0.81). Genetic correlations of LnVar with udder health and ketosis (adjusted for correlations with average milk yield) were stronger or similar when based on full first lactation (-0.31 and -0.32) than when based on parts of the first lactation (-0.20 to -0.26, and -0.22 to -0.34). Genetic variance and heritability of LnVar decreased over full lactations (0.056 to 0.043, and 0.21 to 0.15). LnVar was genetically most similar based on lactation 2 and 3 (r_g 0.95), and least similar based on lactation 1 and 3 (r_g 0.88). The genetic correlation between LnVar and udder health increased over lactations (-0.31 to -0.52). Our results indicate that resilience is genetically similar throughout life. As a resilience indicator for breeding, LnVar based on the full first lactation is preferred over LnVar based on parts of lactation 1, because of the higher heritability and mostly stronger genetic correlations with health traits. EBV for LnVar based on lactations 1, 2, and 3 could be combined into a multi-trait index to improve resilience throughout the first 3 lactations.

A genetic study of udder health traits with sensor data from automatic milking systems
K.B. Wethal[1], M. Svendsen[2] and B. Heringstad[1,2]
[1]Norwegian University of Life Sciences, Faculty of Biosciences, P.O. 5003, 1432 Ås, Norway, [2]Geno breeding and A.I. association, Storhamargata 44, 2317 Hamar, Norway; karoline.bakke@nmbu.no

The proportion of Norwegian herds with automatic milking systems (AMS) have increased rapidly, and now almost 50% of the dairy cows are milked in AMS. The motivation for the current study was to investigate whether sensors in AMS can contribute genetic information on udder health-traits for Norwegian Red (NR) cows. The specific aim was to estimate genetic parameters of in-line measured electrical conductivity (EC) and online cell count (OCC) in milk. Data from 77 commercial herds equipped with AMS from DeLaval was downloaded using a remote desktop connection. Among these herds, 24 had equipment for measuring OCC. Cows calved in the years from 2015 to 2017; and after editing and quality control 4,714 NR cows with data between 30 and 320 DIM were included in the analyses. For each milking, EC were available on quarter level, and OCC on cow level. Traits analysed were mean EC (ECmean), highest EC value (ECmax) and log-transformed OCC (lnOCC). The traits had one record per milking. In addition, information on lactation average somatic cell score (LSCS) was collected from the Norwegian dairy herd recording system for the same cows. A total of 2,363,928 milkings had information on EC, while 1,490 cows had a total 451,949 records with OCC measurement. Genetic parameters were estimated using bivariate mixed linear animal models. Estimated heritabilities were 0.35, 0.23, and 0.09 for ECmean, ECmax, and lnOCC, respectively. Standard errors of estimates ranged from 0.02 to 0.03. The genetic correlation to LSCS was 0.79 for lnOCC, whereas ECmean and ECmax showed correlations of 0.37 and 0.34, respectively. The genetic correlations between EC-based traits and lnOCC were close to zero. Results showed that AMS-traits can be included in routine genetic evaluations because they are heritable, but a weak genetic correlation for EC-traits to OCC might question the value of EC in a selection program for improved udder health. Further investigation is recommended to fully clarify their potential as indicators of udder health in routine genetic evaluations.

Results of a five years Holstein × Simmental crossbreeding experimentation at CREA

A. Crisà, D. Meo Zilio, R. Steri, C. Marchitelli, G. Catillo and L. Buttazzoni
Consiglio per la ricerca in agricoltura e l'analisi dell'economia agraria (CREA), Research Centre for Animal Production and Acquaculture, Via Salaria 31, 00015 Monterotondo, Italy; cinzia.marchitelli@crea.gov.it

Parallel to the large increase in productivity, dairy bovine breeds show deterioration of functional and performance traits and crossbreeding may help to overcome these problems due to heterosis. A long term research has been established at the Monterotondo experimental barn implementing a rotational breeding scheme between Holstein and Simmental breeds. To date, the experimental herd is made by three groups of lactating cows: purebred Holstein, purebred Simmental and Crosses, in expected relative proportions of 1:1:2. Several productive, economic and reproductive data have been recorded for male and female calves and heifers. Body condition scores (BCS) were registered on heifers at four time points for two reproduction cycles. Results showed that among calves requiring veterinary treatments, crossbred animal sicked less than purebreds (P<0.01). Weight at birth, weight at sale, price at sale and morphological scores for male calves showed intermediate values for crosses (P<0.01). Positive correlations were found between weight at birth and weight at puberty, weight at puberty and age at puberty, age at puberty and age at first insemination (r=0.36, 0.44, 0.52 respectively, P<0.001). Weight at birth, weight at puberty and age at first insemination showed intermediate values for crosses (P<0.01) with heterosis values of 1.57 kg, 17.18 kg and -20 days respectively. Moreover, pregnancy rate at first insemination and the number of services for conception indicated heterosis values of 0,19 and -0,75 respectively. Age at first calving, calving interval, days open and calving rate also showed intermediate values for crosses, with estimated heterosis of -34.9, 4.94, -2.93 days and 0.33 respectively. Simmental heifers showed higher incidence of twin births (18.6%) and calving problems (16.2%) as compared to crosses and Holsteins. BCS showed differences among the three genetic groups at each time point, always recording intermediate values for crossed heifers (P<0.001). Our results confirm crossbreeding as a useful option considering the economic income from male calves and reproductive benefits for heifer.

Genome-wide association study of trypanosome prevalence in cattle of Burkina Faso

B. Yougbaré[1,2], D. Ouédraogo[2,3], A. Soudré[4], B. Zoma-Traoré[2,3], A.S.R. Tapsoba[1], M. Sanou[1], S. Ouédraogo-Koné[3], P. Burger[5], M. Wurzinger[2], N. Khayatzadeh[2], H.H. Tamboura[1], A. Traoré[1], J. Sölkner[2] and G. Mészáros[2]
[1]Institut de l'Environnement et de Recherches Agricoles (INERA), BP 8645, Ouagadougou, Burkina Faso, [2]University of Natural Resources and Life Sciences, Vienna (BOKU), Gregor Mendel Str. 33 A, 1180 Vienna, Austria, [3]Université Nazi Boni de Bobo-Dioulasso, BP 1091, Bobo-Dioulasso, Burkina Faso, [4]Université Norbert Zongo de Koudougou, BP 376, Koudougou, Burkina Faso, [5]University of Veterinary Medicine Vienna, Savoyenstraße 1 A, 1210 Vienna, Austria; yougbernadette@hotmail.fr

A genome-wide association study (GWAS) approach was applied to find SNPs associated with trypanosomosis in purebred and crossbred Baoulé cattle. Trypanosomosis status (positive and negative) was recorded for 387 purebred Baoulé cattle and 350 crossbred Zebu × Baoulé from the Southwest of Burkina Faso, using indirect Elisa. The quality control of the data from the Illumina Bovine SNP50 BeadChip was performed with PLINK 1.9. As the final dataset we used SNPs with MAF> 0.01, with a call rate >95% and those that deviated from HWE with Fisher's exact test with P-value P<10E-6. After applying quality control, 34,346 SNPs and 343 animals for the purebred Baoulé, and 35,404 SNPs and 279 animals for the crossbreds were used for the GWAS analyses with GEMMA software. For purebred Baoulé 6 SNPs showed significant associations with trypanosomosis status on chromosomes (CHR) 8, 9, 16, 22, 24. Four SNPs were significant after Bonferroni correction (P<1.45×10^{-6}), located on CHR16, CHR22, CHR8 and CHR9. For crossbreds, 1 SNP showed significant association with trypanosomosis status on chromosomes CHR 5. These results identify chromosome regions that might harbour the gene(s) that contribute to the trypanosome tolerance in Baoulé cattle. The protein tyrosine phosphatase receptor type G (PTPRG) on CHR 22 (Mb 39.0-39.7) and the alpha-1, 3-mannosyl-glycoprotein 4-beta-N-acetylglucosaminyltransferase C (MGAT4C) on CHR 5(Mb 16.0-16.1) are putative candidate trypanotolerance genes in our study.

Variation of somatic cell count-derived phenotypes for udder health monitoring in Italian buffalo

A. Costa[1], G. Neglia[2], M. De Marchi[1], G. Campanile[2] and M. Penasa[1]
[1]University of Padova, Viale dell'Università 16, 35020 Legnaro PD, Italy, [2]University of Naples Federico II, Via Federico Delpino 1, 80137 Napoli NA, Italy; angela.costa.1@phd.unipd.it

Mastitis data are usually not available on a large scale for all species intended for dairy, thus somatic cell count (SCC) is often used as indicator of udder health. In Italian buffalo, scarce attention has been given to SCC so far and there is need to improve udder health and mastitis resistance in the population to limit direct economic loss due to decrease in milk production and veterinary treatments. Moreover, elevated SCC in buffalo bulk milk is not currently penalise by payment systems. In cattle, high milk SCC is generally associated with a decrease of milk production and poor milk quality, as protein composition, pH and technological properties. Conventionally, the logarithmic transformation of SCC (SCS) is used for statistical analysis as it is normally distributed. Data consisted of 45,312 lactations of 35,623 animals occurred in Southern Italy between 2013 and 2017. Based on recent approach used in cattle, mean (SCS_M) and standard deviation (SCS_SD) of SCS in the first 150 DIM were calculated and severity was defined as the ratio of the n. of test-day SCC above 200,000 cells/ml ('high') to the total n. of test-day SCC (minimum set to 5) in the first 150 DIM. The first test-day SCC was recorded within 35 DIM and the interval between 2 consecutive test-day was lower than 60 days. Fixed effects included in the analysis of variance were parity, season of calving, year of calving and their interactions; cow, herd and residual were the random factors. Both SCS_M and severity increased with parity, whereas SCS_SD decreased. Buffaloes that calved in winter and autumn had significantly greater SCS_M and lower SCS_SD in milk compared to buffaloes that calved in spring and summer. Severity was the greatest (0.23) in lactations that started in autumn. Results indicated that the variation for these traits exists in buffaloes and multiparous animals might be monitored for udder health through SCC trend. Further studies should validate such potential indicators of udder health with clinical mastitis data and future efforts may focus on improving farmers' sensitivity on SCC data interpretation and exploitation.

Polymorphisms of genes related to resistance of cows to mastitis: identification of genetic markers

E. Bagnicka[1], P. Brodowska[1], K. Żukowski[2] and T. Sakowski[1]
[1]Institute of Genetics and Animal Breeding Polish Academy of Sciences, Postepu 36A St., 05-552, Poland, [2]National Research Institute of Animal Production, 1 Krakowska St., 32-083 Balice, Poland; e.bagnicka@ighz.pl

The aim of the study was to identify gene polymorphisms associated with the mammary gland's resistance to mastitis for their potential application/use in MAS. The research was conducted on DNA from hair bulbs collected from 312 cows of the black-and-white PHF variety maintained in four herds. DNA genotyping was performed using Illumina BovineSNP50 v3 BeadChip microarrays. Phenotypic data on daily milk yield and composition (8,090 records) were obtained from the milking control system. The phenotypic data covered information on daily milk yield, fat, total protein, casein, lactose, dry matter, urea, BHB, and acetone contents and SCC, FPD, pH, and fatty acid profile. The SCC value was transformed into a natural logarithm scale before analysis. The GWAS analysis was conducted using the Plink program to choose SNPs significantly associated with the analysed traits, based on the phenotypic and genetic data. The next step was the variance analysis using SAS package with MIXED procedure with a mixed, one-trait repeatability model to estimate the solutions for SNP variants. In GWAS analysis, 487 SNP associated with production traits were identified. Then, 52 SNP were selected to further analysis using analyses of variance method with the MIXED procedure (SAS). It turned out that 35 of them were associated with production and functional traits. From that 35 SNPs, 18 were localised within known genes and three of them in exons. To conclude, the general recommendation to improve MAS could be based on including: (1) CXCR4, IGF1, TLR4, PRRX2, and ADGRB1 genes to increase the cows' resistance to udder inflammation and improve milk quality; (2) DGAT1 and GPIHBP1 genes to the improvement of production traits; (3) MFGE8, FUBP3 and TRAPPC9 genes to improve functional and production features. Funding: This work was supported by the National Center for Research and Development [grant number: SUSAN/I/SusCatt/01/2017] as part of the European research program ERA-NET CO-FUND SUSAN.

Selection on mastitis resistance in Holstein and Normande breeds: genetic and immune responses

R. Lefebvre[1], S. Barbey[2], P. Germon[3], P. Rainard[3], G. Foucras[4] and D. Boichard[1]
[1]Université Paris-Saclay, INRAE, AgroParisTech, GABI, 78350 Jouy en Josas, France, [2]INRAE, UE326 Domaine Expérimental du Pin, 61310 Gouffern en Auge, France, [3]ISP, INRAE, UMR 1282, Université de Tours, 37380 Nouzilly, France, [4]IHAP, Université de Toulouse, Ecole Nationale Vétérinaire de Toulouse (ENVT), INRAE, 31076 Toulouse, France; rachel.lefebvre@inrae.fr

Mastitis is a major issue in dairy cows. Although environmental effects are preponderant, the genetic variability of mastitis resistance is important. A divergent genetic selection experiment on mastitis resistance was carried out at INRA Le Pin experimental farm in Holstein and Normande breeds, yielding females of resistant and control lines, based on their sire breeding values. The aims of this experiment were to evaluate the efficiency of genomic selection on this trait and to better understand the immune background and the relations with other traits. Based on data of 376 cows, overall differences between lines in cell count (-43%), clinical mastitis (-12% affected cows) and udder infection status (-11% positive to a bacteriological test) were favourable to resistant cows, in agreement with genomic predictions. Within-breed differences were similar, with clearer and always significant results in Holstein breed. The whole genome sequence of each cow was imputed from her SNP genotype. Variant frequencies were calculated for each line, then compared between lines to identify regions impacted by selection. For both breeds, 238 regions significantly different between lines were identified, including 14 genes related to immunity. About 40% of these regions are included in QTLs currently used in the French genomic evaluation of clinical mastitis or cell counts. To characterise mechanisms underlying genetic resistance to mastitis, cows were submitted to an intra mammary challenge with lipopolysaccharide (LPS). Inflammation was monitored by cell count and cytokines/chimiokines (IL-6, IL1-ß, IL-8) assays in milk. Even though LPS triggered an inflammation in every infused quarter, resistant cows showed a weaker response at 8 h post injection compared to control cows. Analyses are on-going to determine if these differences are associated with variants located in LPS response genes, and more widely in genes implicated in immunity pathways.

Effects of udder conformation on milk yield and quality traits in Chios and Frizarta ewes

A.I. Kalogianni[1], M. Moschovas[1], G. Manessis[1], I. Vastardis[2], I. Bossis[1] and A.I. Gelasakis[1]
[1]Agricultural University of Athens, Animal Science, Iera odos 75 Str., 11855, Athens, Greece, [2]Delta Foods S.A., 23rd km Nat.Road Athens-Lamia, 14565, Ag.Stefanos, Attica, Greece; gelasakis@aua.gr

The objective of the study was to assess whether udder conformation is related to daily milk yield (DMY) and quality traits (MQT) in Chios and Frizarta dairy ewes. A total of 48 Chios and 64 Frizarta ewes, from two intensive sheep farms rearing purebred animals, were enrolled in the study. Udder conformation, DMY and MQT were recorded within two weeks postweaning and before the morning milking. Udder conformation assessment was performed by a single, trained observer, using 9-degree linear scale for four udder conformation traits (UCT), namely, udder suspension (US, 1=attachment width much smaller than udder height, 9=attachment width much larger than udder height), udder half separation (UHS, 1=no separation, 9=clear separation), udder height (UH, 1=deep udder close to the ground, 9=shallow udder close to abdominal wall) and teat angle (TA, 1=vertically placed teats, 9=horizontally placed teats). ICAR (International Committee of Animal Recording)-approved equipment and protocols were used to estimate MQT, namely, daily milk- (DMY), fat- (DFY), protein- (DPY), lactose- (DLY), solids-non-fat- (DSNFY), total solids- (DTSY) yields and somatic cell counts (SCC). Age and body condition score (BCS) were also recorded. SPSS v23 was used for the statistical analyses. Initially, Spearman's correlation was performed and only one trait from each pair of highly correlated UCT was retained for the subsequent analysis. Then, analysis of covariance was used to estimate the fixed effects of the three retained UCT (US, UC, TA), farm and age, as well as BCS (covariate) on DMY, MQT and the log of SCC (LogSCC). US and UC scores varied from 1 to 7 and TA scores from 2 to 8. US was significantly associated with DMY ($P<0.05$) and all the estimated MQT ($P<0.05$) but not with LogSCC; in particular, US score 7 was associated with the lowest values compared to the rest US scores. No other significant effects of UCT on DMY and MQT were observed. Research is funded by the Agricultural Cooperative 'Agrinio Union' within the Regional Operational Program 'Western Greece 2014-2020-RIS3', Project code: MIS-5040306.

The use of artificial neural networks with radial basis function to predict subclinical ketosis
E.A. Bauer, E. Ptak and W. Jagusiak
University of Agriculture in Krakow, Department of Genetics, Animal Breeding and Ethology, Al. A. Mickiewicza 24/28,
31-120, Poland; e.bauer@ur.krakow.pl

Ketosis is one of the most serious and frequently occurring metabolic diseases in dairy herds all over the world. Due to ketosis, cattle farmers carry large economic losses. The indicator of ketosis is the increased level of ketone bodies in the blood, however, the monthly laboratory verification for lactating cows would be too expensive. Artificial Neural Networks (ANN) are a modelling technique which is capable of projection of extremely complex functional dependencies. ANN networks are characterised by a high accuracy in predicting various occurrences and in processes in which it is difficult to clearly determine cause and the effects of relationships or where there are no principle that allow to logically connect reasons and effect. The aim of the research was to find the relationship between the milk composition and the occurrence of subclinical ketosis in cows with using artificial neural networks model with Radial Basis Function (RBF). The data set used in the study was obtained from Polish Federation of Cattle Breeders and Milk Producers and included information about test day milk yields of 1,085 Polish Holstein-Friesian cows. The ANN simulator implemented in the STATISTICA13 package was used. A classical feedforward ANN between 8 to 15 hidden layers, eight input layers and one output layer, with a backpropagation learning algorithm and sigmoid transfer functions, were applied. A neural model was essentially a regression model used to establish correlations between a vector of independent variables (inputs) and a vector of dependent variables (outputs). The set of milk data composition as: milk yield[kg], protein[%], fat[%], fat-to-protein ratio[%], lactose[%], somatic cell count[cell/ml] and content of urea[mmol/l], β-hydroxybutyric acid (BHBA) $\geq$1.2 mmol/l, and acetone[mmol/l] was used as input variables. Cattle lactation was divided to occurrence of ketosis type I and II. Results show that ANN models were characterised by a sensitivity of 0.647 to 0.682 and specificity ranging from 0.844 to 0.876. For type I ketosis, the best average combination of sensitivity (0.812) and specificity (0.770) and the largest mean AUC under the ROC curve (0.837) were obtained for networks based on 13 neurons in the hidden layer. In the case of ketosis of type II, it was found that ANN with use of 10 or 15 neurons in the hidden layer are were the most efficient for prediction. These type of networks had an average sensitivity of 0.687 and average specificity of 0.853 with an average AUC area of 0.838. The results and conclusions of this study provide new grounds for improvement of ANN to diagnosis ketosis. Acknowledgement: the Research was financed by the National Science Center, Poland no.2017/25/N/NZ5/00793.

Classification of cattle behaviour using convolutional neural networks
D. Pavlovic[1], C. Tachtatzis[2], A.W. Hamilton[2], O. Marko[1], R.C. Atkinson[2], C. Davison[2], C. Michie[2], V. Crnojevic[1]
and I. Andonovic[2]
[1]BioSense Institute, Zorana Đinđića 1, 21101 Novi Sad, Serbia, [2]University of Strathclyde, Department of Electronic and
Electrical Engineering, 99 George Street, G1 1RD Glasgow, United Kingdom; dejan.pavlovic@biosense.rs

The monitoring of cattle behaviour through sensor systems is gaining importance in the improvement of animal health, fertility and management of large herds. Commercial farms commonly implement accelerometer-based systems to monitor the time an animal spends ruminating, eating and overall activity which informs farmers on the health and fertility status of individual cattle. Ill or injured cattle feed and ruminate less, so tracking the duration and frequency of these states provide key indicators of animal health. Activity is used as a metric for the detection of oestrus (heat) which promotes more efficient fertilisation of dairy and beef cattle, reducing operating costs and increasing profits for farmers. The aim of the study was to determine the feasibility of enhancing the accuracy of estimating multiple classifications derived from acceleration-based activity collars can through Convolutional Neural Networks (CNN). CNN models are typically used to classify objects within images, but have been demonstrated to be effective at classifying time-series data across different domains. To evaluate their effectiveness for cattle behaviours classifications, acceleration data was collected from 18 cows across 3 farms using neck-mounted collars which provided 3-axis acceleration values at 10 Hz sampling frequency. Each cow was equipped with pressure sensor halters which provided ground truth data of the animal behavioural state, also at 10 Hz sampling frequency. The ground truth from the halter allowed the CNN model to be trained to predict a number of key cattle behaviours. The model was then tested on separate data to assess performance. The CNN was able to classify the 3 activity states (rumination, eating and other) with an overall F1 score of 82% compared to reported collar classifications with an overall F1 score of 72%.

Machine learning algorithms to predict litter weight gain from birth to weaning in swine
R. Gauthier[1,2], C. Largouët[1] and J.Y. Dourmad[2]
[1]*Univ Rennes, CNRS, Inria, IRISA, UMR 6074, 35000 Rennes, France, [2]PEGASE, INRAE Agrocampus Ouest, UMR 1348, 35590 Saint Gilles, France; raphael.gauthier@inrae.fr*

In lactating sows, high milk production and low voluntary feed intake generally lead to nutrient deficiencies, which in turn may have negative effects on their subsequent reproductive performance and longevity. To accurately handle individual variability, precision feeding systems require to predict nutritional requirements of sows according to their individual milk production. The objective of this study was thus to predict litter weight gain (LWG) from birth to weaning as a proxy variable of milk production. Data were collected on 13,364 lactating sows from 6 farms. Data include information on body weight and parity of sows, litter size and weight at birth, cross-fostering of piglets, lactation duration, litter size and weight at weaning. Five subsets with differences in their attribute's composition were created (S1, S2, S3, S4, and S5). Subsets were randomly split into learning and testing according to an 80:20 ratio. Six supervised machine learning algorithms, namely Linear Regression, Random Forests, k-Nearest-Neighbor, Neural Network, Gradient Tree Boosting and Voting Regressor were selected and trained. The quality of their predictions was assessed in terms of mean absolute error (MAE), and root mean square error in percentage (RMSEP). Average litter weight gain was 67.7 (±20.42) kg. Gradient Tree Boosting algorithm outperformed all other algorithms, on S1, S2, S3, S4, and S5, regarding to every evaluation criterion. Best results were obtained on the subset S5 containing all attributes (r^2=0.80). In details, Gradient Tree Boosting allowed to predict LWG during lactation with a MAE of 8.1 kg, and a RMSEP of 12%. According to this study, Gradient Tree Boosting is a good candidate to accurately predict individual LWG of individual lactating sows and could be embedded in precision feeding decision support system in lactation. This work was supported by the French National Research Agency under the Investments for the Future Program, referred as ANR-16-CONV-0004 (#DigitAg) and was also supported by the European Union's Horizon 2020 research and innovation program (grant agreement no. 633531).

Combining machine learning and remote sensing to characterise Portuguese sown biodiverse pastures
T.G. Morais[1], R.F.M. Teixeira[1], M. Jongen[1], N.R. Rodrigues[2], I. Gama[2], M. Reis[2] and T. Domingos[1]
[1]*MARETEC – Marine, Environment and Technology Centre, Universidade de Lisboa, Avenida Rovisco Pais, 1049-001, Portugal, [2]Terraprima – Serviços Ambientais Lda., Avenida das Nações Unidas, n° 97, 2135-199, Portugal; tiago.g.morais@tecnico.ulisboa.pt*

Grasslands are a crucial ecosystem that supports and provides multiple ecosystem services. In the Mediterranean, sown biodiverse pastures (SBP) rich in legumes are a nature-based, innovative and economically competitive livestock production system. SBP are a mixture of up to 20 species or varieties of high-yield grasses and legumes. They have been shown to accelerate carbon sequestration through soil organic matter (SOM) accumulation. Here, we develop models for estimating standing aboveground biomass (SAGB) and SOM concentration by combining remote sensing (RS) and spectroscopic data processed using machine learning approaches. We used field-measured data from seven different farms collected during two production years (2018 and 2019). Different indirect data sources were then used for SAGB and SOM. Sentinel-2 data was used for SAGB (12 reflectance bands and five vegetative indexes). For SOM, spectrophotometric data (6 reflectance bands) obtained for the soil samples in the laboratory was used as satellite images did not capture soil reflectance. In both cases, additional environmental covariates were included in the model. Both output variables were estimated using a backpropagation artificial neural networks (BP-ANN) model with a single hidden layer. A seven-fold approach was used, where in each fold one farm and year was removed from the training set and used for validation only. In each fold, the optimum number of neurons in the hidden layer was chosen between 1 and 75. Results showed that, in general, the accuracy of the estimation increases with proximity of the sensor, i.e. accuracy is higher in SOM estimation. The overall mean standard error (MSE) of SOM estimation is 6.94 g SOM/kg soil (average SOM content is 20 g SOM/kg soil). The overall MSE of SABG is 509 kg DM/ha (average SAGB is 2,500 kg DM/ha). This work demonstrated the potential of approaches combining remote sensing data with machine learning methods, in this case BP-ANN.

Real-time collection and visualisation of methane emission on dairy farms: new challenges

D. Schokker[1], E.M.M. Van Der Heide[1], G. Seigers[2], Y. De Haas[1], R.F. Veerkamp[1] and C. Kamphuis[1]
[1]Wageningen University & Research, Animal Breeding and Genomics, Droevendaalsesteeg 1, 6708 PB Wageningen, the Netherlands, [2]Wageningen University & Research, Facilities and Services, Akkermaalsbos 12, 6708 WB Wageningen, the Netherlands; dirkjan.schokker@wur.nl

The Dutch government has set an objective to reduce emissions of greenhouse gases (GHG) to 116 Mton CO_2-equivalent in 2030. All sectors have to participate in this reduction, including the livestock sector. A significant contributor to the GHG emissions in the Netherlands is the dairy sector. By making the CH_4 (methane) emission measurements directly available to the dairy farmer, the farmer gets insight in the CH_4 emission of his farm, and can employ different strategies customised to his farm. This will contribute to the reduction of the GHG emissions. To achieve this continuously monitoring of GHG emissions, we have built an infrastructure and this is currently implemented on 17 commercial dairy farms. This infrastructure offers the opportunity to quickly identify issues around the data collection due to for example technical failure of the sensor. We also connected these private CH_4 data with outdoor temperature and wind speed, through API connections with Akkerweb. Data streams are visualised on a web or mobile phone platform, providing real-time information to the farmer. The next challenge is to add individual cow data. To do so, we have performed a feasibility study investigating the use of computer vision for the automatically identification of cows near the CH_4 sensor. This is done by using a camera to capture images of the cows in the milk robot where the CH_4 sensor is located. The video footage is first passed through a blob detection algorithm to select regions of interest, in this case the yellow ear tag of the cow. The regions of interest will then be passed through an optical character recognition algorithm trained to recognise numbers. Both these analyses will be done locally on a Raspberry Pi. The extracted cow identifier number from the optical character recognition model will then be passed on to the Arduino. The Arduino will push this data together with the collected CH_4 measurements to the cloud. This data will flowback to the application platform, where it can provide more detailed information for the farmers by tagging the animal identifier to the CH_4 measurements during milking.

A systematic approach to quantify the impact of feeding and farm management on bovine health

C. Matzhold[1,2], J. Lasser[1,2], C. Egger-Danner[3], F. Steininger[3], B. Fuerst-Waltl[4], S. Thurner[1,2,5,6] and P. Klimek[1,2]
[1]Medical University of Vienna, Spitalg.23, 1090 Vienna, Austria, [2]Complexity Science Hub Vienna, Josefstaedter Str.39, 1080 Vienna, Austria, [3]Zuchtdata EDV-Dienstleistungen GmbH, Dresdnern Str. 89, 1200 Vienna, Austria, [4]University of Natural Resources and Life Sciences, Gregor-Mendel-Str. 33, 1180 Vienna, Austria, [5]Santa Fe Institute, 1399 Hyde Park Road, NM 85701, USA, [6]IIASA, Schlosspl. 1, 2361 Laxenburg, Austria; matzhold@csh.ac.at

Digitalisation in dairy cattle farming holds the promise of substantially improving early detection and prevention of animal disease, in particular through the collection, linkage, and analysis of routine data generated by milking, feeding, and performance recording systems. However, the identification of risk factors from such routine data is complicated by the fact that most frequent diseases are caused by multiple risk factors with strong mutual correlations. Here, we present a systematic framework to disentangle the impacts of feeding and farm management practices on frequent diseases to derive risk profiles for farms. Based on data from 167 Austrian farms and 6,519 cows, we quantify the predictive accuracy of approx. 200 different variables for feeding, housing, bedding and farm management practices in the prediction of diseases by means of multivariate regression models. Individual risk factors can explain up to 60% in the observed variation of disease occurrences, after adjusting for lactation, season, breed, region, and overall farm performance. Regarding feeding practices, we find for instance that intake of grass silage was predictive of strongly reduced risks for ovarian cysts ([OR] 0.39, 95% confidence interval [CI] 0.31-0.50), ketosis (OR 0.42, CI 0.27-0.67) and milk fever (OR 0.69, CI 0.55-0.88); Hay intake showed reduced risk of lameness while concentrated feed intake (OR 15, CI 9.3-24) and maize silages (OR 8.7, CI 6.7-11) showed increased risk. Further, ketosis was strongly reduced in farms with alpine pasturing of heifers (OR 0.28, CI 0.18-0.43) and with long straw litter pens (OR 0.46, CI 0.33-0.64). Lameness increased in farms without alpine pasturing (OR 4.6, CI 3.4-6.2) and with high bed cubicles (OR 3.2, CI 2.8-3.7) whereas low bed cubicles correlate with decreased lameness (OR 0.44, CI 0.40-0.50). A principal component analysis reveals that farms can be grouped into three clusters. We discuss the highly heterogeneous disease risk profiles in these farm clusters and their associated management practices. Farms of the 21st century are becoming digitally integrated sensing systems. We show how the information derived from such systems can be used to increase animal welfare by identifying those management practices whose adoption might lead to the greatest disease risk reductions.

Internet of Things and sensor integration to make herd location affordable

F. Maroto-Molina[1], J. Navarro-García[2], K. Príncipe-Aguirre[2], I. Gómez-Maqueda[2], J.E. Guerrero-Ginel[1], A. Garrido-Varo[1] and D.C. Pérez-Marín[1]
[1]University of Cordoba, Animal Production, Ctra. Madrid-Cádiz km 396, 14014 Córdoba, Spain, [2]Sensowave SL, Avenida de Castilla 1, 28830 San Fernando de Henares, Spain; g02mamof@uco.es

Livestock location technologies has evolved considerably in the last decades. Nowadays, commercial solutions based on Global Positioning Systems (GPS) are available. Nevertheless, most commercial solutions have constraints related to data transmission (2G/3G coverage not available in remote areas) and cost per animal. Both issues make the localisation of all the individuals in a herd not affordable. The objective of this work was to develop a low-cost solution to enable the localisation of a whole herd. We developed a system based on Internet of Things connectivity (Sigfox) and the combination of GPS collars for some animals in the herd and low-cost Bluetooth eartags for the rest of them. The performance of this system was tested in two commercial farms, rearing sheep and beef cattle, by the monitorisation of 50 females in each case. Different collar/eartag ratios, which determine the cost per animal of the location solution, were simulated and evaluated. Results showed that a very low collar/eartag was enough for the monitorisation of a whole sheep herd, mostly due to their gregarious nature. A larger ratio, but still low, was needed for beef cattle. Nevertheless, the optimal ratio strongly depends on the purpose of location data (i.e. farm management or research). Large variability was observed for the number of hourly and daily messages coming from collars and eartags.

Machine learning using transcriptomic data reveals biomarkers to predict pregnancy status in cattle

M.B. Rabaglino and H.N. Kadarmideen
Technical University of Denmark, DTU Compute, Kemitorvet, 2800, Kgs. Lyngby, Denmark; hajak@dtu.dk

Gene expression levels determine uterine receptivity in the cow as soon as around 7 days after ovulation. Several studies applied RNA-seq or microarray technologies to measure the endometrial transcriptome at day 7 in cows that become pregnant (PR) compared to those that did not (non-PR). However, the results were not entirely consistent. The aim of this study was to employ bioinformatics and machine learning (ML) approaches to integrate the transcriptome data from these studies and identify those genes whose expression levels determine pregnancy status in Taurine cattle. Five datasets were obtained from the Gene Expression Omnibus database. The accession numbers and type of animal used were, respectively: GSE115756, Holstein cows (n=17); GSE107741, Japanese cows (n=11); GSE29853, Charolais × Limousine heifers (n=12); GSE36080 and GSE20974, Simmental heifers (n=6 each). In all these studies, endometrial samples were obtained at day 6-7 and animals were classified as PR (n=26) or non-PR (n=26) after ET or AI, depending on the study. The software BioDiscML was employed for gene selection through a combination of ML algorithms. The predictive ability of the potential key genes was evaluated using support vector machine as classifier, using the expression levels of the samples from all the breeds but one, to train the model, and the samples from that one breed, to test it. Finally, the biological meaning of the key genes was explored with the Cytoscape software. Fifty genes were identified as being highly predictive of PR. Using their expression levels in all the samples but a particular breed, they were able to predict pregnancy status with the following accuracy: 100% for the Japanese and Simmental breeds, 94.1% for the Holstein cows and 91.7% for Charolaise × Limousine heifers. Genes with higher expression in the PR cows were related with circadian rhythm, Wnt receptor signalling pathway and embryonic development. The application of ML computational tools allowed the identification of a group of biologically relevant endometrial genes in Taurine cattle. Furthermore, they were able to predict pregnancy status at around 7 days of the oestrous cycle with very high accuracy, despite the animal's breed and category.

A preliminary stock and flow model to support dairy herd management

A.S. Atzori[1] and A. Gallo[2]
[1]*University of Sassari, Department of Agriculture, Viale Itale 39, 07100, Italy, [2]Università Cattolica del Sacro Cuore, Department of Animal Science, Food and Nutrition (DIANA), Via Emilia Parmense 84, 29122 Piacenza, Italy; asatzori@uniss.it*

Dynamic modelling in dairy farming might help to predict future farm performances and support the decision making process. This work defined a preliminary dynamic structure aimed to minimise the system complexity and to outline the modelling of a given Mediterranean dairy herd and its milk deliveries. The modelling process had the objective to accurately predict farm output over time. The modelling process followed a System Dynamics approach and included: (1) a preliminary farm survey to gather technical info and data; (2) the development of a Stock and Flow model on Vensim® (Ventana, Inc), to mathematically and graphically simulate the dynamic connections among variables; (3) the model evaluation against farm records. The survey provided aggregated monthly records, from Jan 2015 to Dec 2018 (4 years), of consistency, feed supply, milk deliveries, reproduction indexes, mortality and health of herd categories. Average farm characteristics in the same period consisted of 1,154±58 milking cows, 205±35.7 dry cows, 1,375±45.2 heifers whereas dry matter intake (DMI) and milk yield were on average equal to 24.2±2.4 and 34.0±3.4 kg/d per head, respectively. The model included 25 variables distributed on a closed aging chain of 4 stocks (replacement heifers and open, pregnant and dry cows) and the respective flow rates of calving, breeding, drying off and culling. Initial settings were based on farm values observed in Jan 2015. Two farm inputs (monthly average of consumed feed from lactating cows and conception rate) were included as exogenous variables resulting of fundamental relevance in model predictions. The simulation showed a good accuracy in predicting the oscillating seasonal pattern of historical farm records of milk deliveries. Cow consistency and milk deliveries were predicted with a RMSPE of 3.5 and 7.2% of observed values (being 57 and 68% due to random variability of data), with high accuracy (Cb=0.90 and 0.97), with good precision (r^2=0.68 and 0.72), and with a satisfactory concordance correlation coefficient (0.74 and 0.82), respectively. Further modelling effort should focus on whether variables affecting intake and reproduction at herd level.

Indicators of body fat mobilisation of lactating beef cows under short nutritional challenges

K. Orquera, G. Ripoll, M. Blanco, J. Ferrer, J.R. Bertolín and I. Casasús
Ctr Invest y Tecnol Agroal Aragon (CITA), IA2 (CITA-Universidad de Zaragoza), Montañana 930, 50059 Zaragoza, Spain; korquera@cita-aragon.es

The aim of this study was to investigate the metabolic adaptation of lactating beef cows to short periods of undernutrition, and assess the accuracy of different indicators of fat mobilisation. The response of 16 Parda de Montaña adult suckler cows to a 4-day energy restriction was analysed in months 2 and 3 post-calving. Prior to restriction and after the challenge, the cows received a diet meeting 100% of their energy requirements (7.0 kg DM hay + 2.7 kg DM concentrate), and during the 4-day challenge the diet met only 55% of requirements (6.2 kg DM hay). With d0 as the start of restriction, on days d-2 (basal phase), d4 (challenge) and d8 (refeeding) several traits were recorded. Cows were body condition scored (BCS, 1-5 scale) and subcutaneous fat thickness was measured by ultrasound in the sacral area (BFT), at the P8 rump site (P8) and at the 13th thoracic vertebra (T13). Plasma was collected for the analysis of β-hydroxybutirate (BHB) and non-esterified fatty acids (NEFA) resulting from fat lipolysis, and malondyaldehide (MDA), a product of lipid peroxidation. A mixed model with month, phase and its interaction as fixed effects and cow as the random effect was used. Thus the BCS and BFT were not affected by month or phase (P>0.05). Both P8 and T13 decreased from month 2 to 3 (P<0.001) and P8 also decreased from the basal to the refeeding phase (P<0.01). Hence, except for P8, external measurements of fat thickness only reflected changes in the long term. The metabolites were affected by the interaction between month and phase. The response to changes was immediate in month 2, but in month 3 the difference among phases was less intense in NEFA (P<0.001) and MDA (P<0.01), and tended to be inexistent for BHB (P=0.06). These are good indicators of lipid mobilisation under a negative energy balance in the short term, especially in early lactation. The individual variation of the adaptive ability of cows remains to be analysed.

The paradox of using residual feed intake or conversion ratios to study feed efficiency in dairy ewe

A. Della Badia[1], G. Hervás[1], P.G. Toral[1], J. Amor[2], A. Belenguer[1], C. Fernández-Díez[1] and P. Frutos[1]
[1]Instituto de Ganadería de Montaña (CSIC-Universidad de León), Finca Marzanas, 24346, Grulleros, León, Spain,
[2]Industrias de Nutrición Animal, S.L.-INATEGA, Ctra. Valdefresno 2, 24228, Corbillos de la Sobarriba, León, Spain;
a.dellabadia@csic.es

Feed efficiency in dairy ruminants is a complex trait that has traditionally been estimated through feed conversion ratios (e.g. the ratio between the amount of feed consumed and the energy corrected milk production – FCR). More recently, the residual feed intake (RFI) is being increasingly used. This index is calculated as the difference between actual feed intake and predicted feed requirements for maintenance and milk production: the lower the RFI, the higher the feed efficiency. Both metrics, FCR and RFI, have been widely used in feed efficiency investigations. However, they probably reflect different mechanisms, which might lead to confusion or apparent contradictions. Therefore, this preliminary work was conducted to compare both indexes in a study carried out with 40 Assaf ewes with the aim of selecting the 20% most efficient and inefficient animals to investigate then the mechanisms underlying the animal-to-animal response. Data of daily intake, milk production, milk fat, protein and lactose, and live weight were collected individually to calculate RFI and FCR. Values of RFI ranged from -0.80 to 0.71 (on average, -0.02 ± 0.063) and those of FCR from 0.61 to 1.52 (on average, 0.93 ± 0.031). Unexpectedly, only 4 out of the 8 ewes classified as the most inefficient according to RFI were also within the 20% most inefficient animals when estimated with FCR. This behaviour was even worse when analysing the most efficient sheep: only 1 out of the 8 animals selected with RFI were confirmed with FCR. In addition, another 2 from the most efficient group according to RFI belonged to the most inefficient when classified by FCR. Only a poor, although significant, linear relationship ($R^2=0.355$; $P<0.01$) was found between both indices. The FCR is nowadays under question due to its correlation with animal size and performance, but still in use. However, many scientists, particularly geneticists, recommend RFI. The poor correlation between them would draw attention to and caution about paradoxical results when either RFI or FCR are utilised to estimate feed efficiency. Acknowledgements: Project CSI276P18, JCyL, FEDER and ESF, UE.

Ruminal biohydrogenation of dietary lipids in dairy sheep that differ in feed efficiency

C. Fernández-Díez[1], G. Hervás[1], P.G. Toral[1], A. Belenguer[1], D.R. Yáñez-Ruiz[2] and P. Frutos[1]
[1]Instituto de Ganadería de Montaña (CSIC-Universidad de León), Finca Marzanas, 24346, Grulleros, León, Spain,
[2]Estación Experimental del Zaidín (CSIC), Profesor Albareda 1, 18008, Granada, Spain; p.frutos@csic.es

It is widely accepted that improving ruminant production requires an enhancement of feed efficiency (FE). However, mechanisms underlying this trait and explaining its large individual variation are still unknown. Some studies have examined the potential link between FE and ruminal microbiota, with certain bacterial populations being more or less abundant in more or less efficient animals. In addition, some of these bacteria might play a role in the biohydrogenation (BH) of dietary lipids and produce bioactive intermediates, which could perhaps be related to FE. To examine this potential relationship, the FE was estimated in 40 lactating Assaf ewes through the residual feed intake to identify the most (n=8) and least (n=8) efficient animals. Rumen samples were collected with a stomach tube and analysed for fatty acid (FA) composition by gas chromatography. An ANOVA was performed to compare the response in both groups of sheep. Results showed a greater proportion of unsaturated fatty acids (UFA) of dietary origin and of several BH intermediates (e.g. cis-9 cis-12 18:2, trans-9 cis-12 18:2, cis-9 18:1; $P<0.10$) in the rumen of less efficient animals. In contrast, an increase of stearic acid (18:0; $P<0.05$) was observed in more efficient sheep, supporting a larger extent of the BH process. Variations in certain odd-chain FA were also found: the content of 17:0 was higher in the most efficient ewes, while that of 13:0 was higher in the least efficient ($P<0.10$). Although these FA are synthesised de novo by ruminal bacteria, available information is too scarce to use them as indicators of changes in specific biohydrogenating groups. In summary, our results show differences in the ruminal BH pattern in dairy ewes that differ in FE, with lower UFA, and greater 17:0 and 18:0 concentrations in the most efficient animals. Further research is needed to understand the role of the rumen microbiota, particularly that involved in BH, and its potential link with feed efficiency [Acknowledgements: Project CSI276P18, JCyL, FEDER and ESF, UE].

Performance and energy requirements for Nellore bulls divergent in residual feed intake

C.D.A. Batalha[1], F.L. De Araújo[2], R.H. Branco[1], L.O. Tedeschi[3] and S.F.M. Bonilha[1]
[1]Instituto de Zootecnia, Centro Avançado de Pesquisa de Bovinos de Corte, Sertãozinho, 14160900, Brazil, [2]Universidade Federal do Recôncavo da Bahia, Centro de Ciências agrárias, ambientais e biológicas, Cruz das Almas, 44380000, Brazil, [3]Texas A&M University, Department of Animal Science, College Station, 778432471, USA; cdabatalha@gmail.com

Dry matter intake (DMI), performance and energy requirements were evaluated in Nellore bulls from divergent classes of residual feed intake (RFI). Thirty-three Nellore bulls (15 classified as low RFI-LRFI and 18 as high RFI-HRFI), from selection and control herds of Instituto de Zootecnia, São Paulo, Brazil Nellore breeding program, were feedlot finished and slaughtered with 400±51 kg of body weight and 542±30 days of age. Eight bulls, three LRFI and five HRFI, were slaughtered at the beginning of the experiment and used as reference bulls. Individual DMI was recorded daily and initial and final body weights were recorded after 16 hours of fasting from total solids. Nine bulls, four LRFI and five HRFI, were fed at maintenance, receiving 65 g of DM/kg$^{0.75}$ BW and 16 bulls (eight LRFI and eight HRFI) were fed *ad libitum*. Diet had 19:81 roughage:concentrate, consisting of *Brachiaria* hay, ground corn, cottonseed, cottonseed meal, citrus pulp and mineral mixture, with 85% of DM and 14% of crude protein. Ultrasound measurements on the *Longissimus* muscle were performed at intervals of 28 days. When two *ad libitum* bulls reached 4 mm of subcutaneous fat thickness, one maintenance bull was randomly chosen and slaughter at the same day. After slaughter, empty body weight (EBW) was measured. Dry matter intake and animal performance were analysed using a random coefficients model. The following variables were included in the model as fixed effects: RFI class, type of feeding and breeding herds. The net and metabolisable energy requirements for maintenance (NEm; MEm) were estimated by exponentially relating heat production with metabolisable energy intake. As expected, the LRFI had lower DMI (5.43 vs 6.36 kg/d; P=0.033) than HRFI, but similar slaughter BW (385 kg; P=0.128) and ADG (0.561 kg/d; P=0.506). The NEm, MEm and the efficiency of metabolisable energy utilisation were 78.0 kcal/d/ kg$^{0.75}$ EBW, 151 kcal/d/ kg$^{0.75}$ EBW, and 51.6%, respectively, for LRFI bulls, and 76.8 kcal/d/ kg$^{0.75}$ EBW, 162 kcal/d/ kg$^{0.75}$ EBW, and 47.4%, respectively, for HRFI bulls. The LRFI bulls produced about 7% less heat than the HRFI ones. The identification and use of LRFI bulls would improve feed efficiency and efficiency of metabolisable energy utilisation without changing performance of beef cattle. Acknowledgments: FAPESP Processes 2017/06709-2, 2018/20080-2 and 2019/17714-2.

Variations in milk fatty acid profile in lactating sheep that differ in feed efficiency

C. Fernández-Diez[1], G. Hervás[1], A. Belenguer[1], J. Amor[2], D.R. Yáñez-Ruiz[3], P. Frutos[1] and P.G. Toral[1]
[1]Instituto de Ganadería de Montaña (CSIC-Universidad de León), Finca Marzanas, 24346, León, Spain, [2]Industrias de Nutrición Animal, S.L.-INATEGA, Ctra. Valdefresno 2, 24228, Corbillos de la Sobarriba, León, Spain, [3]Estación Experimental del Zaidín (CSIC), Profesor Albareda 1, 18008 Granada, Spain; g.hervas@csic.es

Intensive dairy production systems demand high energy inputs, which may compromise their sustainability in the current global context. Increasing feed efficiency (FE) in dairy animals would certainly help improving the competitiveness of production, but the basis of individual differences in this trait remains largely unknown. In the first analyses of an experiment conducted with the 20% most and least efficient ewes (estimated through the residual feed intake, RFI) from a group of 40 lactating Assaf sheep, we observed a greater ruminal biohydrogenation (BH) of unsaturated fatty acids (FA) in the most efficient animals. Since these differences would have an impact on milk FA profile, we hypothesised that milk FAs may also be related to FE and therefore be used as non-invasive biomarkers. Fat in milk samples was extracted and converted to FA methyl esters, which were then analysed by gas chromatography. Differences in milk FA profile between groups were evaluated by ANOVA. The most efficient ewes showed a greater concentration of FA with less than 16-carbons, which suggests enhanced de novo FA synthesis in the mammary gland. However, the milk from the least efficient animals had higher proportions of unsaturated FA, such as *cis*-9 18:1 or some 18:2 isomers, which is consistent with the previously reported differences in the extent of ruminal BH. Similarly, greater milk content of very long-chain n-6 polyunsaturated FA (e.g. 20:2n-6, 20:3n-6, 20:4n-6 or 22:4n-6) was found in these sheep. This greater content would derive from reported apparent decreases in the ruminal disappearance of dietary 18:2n-6, which is the precursor for the synthesis of other n-6 FA in body tissues. In conclusion, milk FA profile from ewes with different RFI values suggest a relationship between FE and lipid metabolism. Milk FA might be used as biomarkers for estimation of FE in dairy ewes, but further research in this field is needed. Acknowledgements: Project CSI276P18, JCyL, FEDER and ESF, UE.

Functional data analysis on MY and NEFA responses of beef cows exposed to feed restrictions

L.B. Mendes, A. De La Torre, J. Pires, I. Cassar-Malek, I. Ortigues-Marty and F. Blanc
Université Clermont Auvergne, INRAE, VetAgro Sup, UMRH, Saint-Genès-Champanelle, 63122, France;
luciano.mendes@inrae.fr

Data used in this study originated from an experiment conducted with 13 primiparous Charolais cows submitted to two consecutive feed restriction (FR) periods, during which feed allowance was limited to meet 50% of NE requirements. Starting at 55 DIM, two consecutive FR of 4 d and 10 d were applied 17 d apart. Milk yield (MY) plasma metabolite concentrations, including non-esterified fatty acids (NEFA), were measured daily during FR, and every 2 to 3 d otherwise. Individual time-series for MY and NEFA were submitted to a functional data analysis (FDA) algorithm (FDA package of R statistical software). Subsequently, we extracted multiple parameters from smoothed functions which describe shape of responses to FR, including initial and final levels (Ppre, Ppost) and amplitude (T) of MY and NEFA deviations from baseline. We estimated areas between response and baseline curves during FR and the following 3 d: A1 (area between tPpre and timeT) and A2 (area between timeT and timePpost). Finally, we defined individual resilience (RFR) as the ability of cows to return to baseline at 3 d after FR. As MY changes over time, MY baseline was smoothed from raw data after excluding measurements collected during FR and the following 3 d. Smoothing time-series for NEFA was achieved for all cows, but MY smoothing was not possible for 2 cows, probably because of their relatively low MY. Comparisons of medians between 4 d and 10 d FR for Ppre and Ppost revealed significant differences (Wilcoxon test, P≤0.05) for MY (Ppre: 6.5 vs 5.7 kg/d; Ppost: 6.1 vs 4.6 kg/d) but not for NEFA (Ppre: 0.07 vs 0.07 mM; Ppost: 0.07 vs 0.08 mM). These results reflect the natural decrease of MY over time, whereas NEFA remained at basal levels before restriction. For T, we observed significant differences for NEFA (0.18 vs 0.26 mM), but not for MY (0.85 vs 0.82 kg/d), indicating that NEFA was potentially more responsive to FR than MY. The variable RFR did not differ significantly between 4 d and 10 d FR because MY and NEFA returned to baseline at 3 d after FR. This approach allows to phenotype individual variability of cow responses to FR and identify potential traits related to animal robustness. Gentore (H2020) project funded this study.

Predicting longevity based on lactation curve, cell count and calving interval in organic cows

F. Moser, A. Bieber, A. Maeschli, A. Spengler Neff and F. Leiber
Research Institute of Organic Agriculture (FiBL), Department of Livestock Sciences, Ackerstrasse 113, 5070 Frick, Switzerland; florian.leiber@fibl.org

Longevity has important impact on lifetime daily milk yields and therefore ecological and economic efficiency of dairy cows. However, actual productive lifespan of dairy cows in Europe is less than four lactations and thus far below the physiological potential. This holds also true for cows in organic systems. Besides management tools, selection criteria at the farmers' hands are needed to improve longevity on farm level. We used data of culled cows from Swiss organic dairy farms to develop prediction models for the number of lactations completed. A dataset comprising 298 cows, which had been in 1st lactation in 2009, and were culled before June 2019, served to fit 32 prediction models initially based on lactation curve parameters (LCP), somatic cell count, age at first calving, calving interval and veterinary events against the binary trait of completing minimum four lactations. The lowest tercile regarding 1st lactation milk yield was excluded in order to separate economic culling decisions from biological reasons. Full and reduced generalised linear models (GLM), generalised mixed effect models, and linear discriminant analyses, respectively with and without inclusion of veterinary events and LCP were applied; all models were based on 1st lactation as well as on 2nd lactation data. In order to rank the models, the resulting scores were applied to a dataset of 7,742 culled cows for predicting their total number of completed lactations as well as lifetime milk yield. Predictability was higher when based on 2nd-compared to 1st-lactation models, and it clearly increased if LCP were excluded. The best prediction for productive lifespan with Rho=14% was reached with a GLM model, excluding LCP. Thus, overall predictability of the best models was still poor. However, for lactation numbers above seven and lifetime yields above 50 tons the rate of cows being false classified by low scores decreased to less than 10%. We conclude that the chosen approach needs to be optimised by larger data for the model development, but might become suitable to support longevity-targeted culling decisions in low-input systems, reliably identifying the poorest percentile.

Assessment of perturbations in lactation to take stock of the existing resilience of dairy cows

A. Ben Abdelkrim[1,2], T. Tribout[2], V. Ducrocq[2], N.C. Friggens[1], D. Boichard[2] and O. Martin[1]
[1]INRAE, UMR MoSAR, Université Paris-Saclay, 75005, Paris, France, [2] INRAE, UMR GABI, Université Paris-Saclay, 78350, Jouy-en-Josas, France; ahmed.benabdelkrim@agroparistech.fr

The robustness of dairy systems is partly related to the robustness of the animals raised under these systems, including their resilience to environmental disturbances. Characterising inter-individual variability can provide the necessary tools to assess the robustness and resilience of animals in a given environment, particularly since modern milk recorders provide new information regarding milking speed or shape of milking curves. We used a model which allows the explicit representation of the perturbations which may occur in the lactation curves, thereby providing a quantification of the variability in perturbations. This model transforms individual time-series data into phenotypic information. The model parameters make it possible to characterise the perturbed and unperturbed profiles of the lactation curve. They also provide estimates of the different perturbations that may exist, the moment when they occur, and their intensity. We applied this model to 6,154 lactation curves from 4,066 Holstein cows recorded in milking parlours in 36 commercial dairy farms. In the process, 4,4024 different perturbations were revealed (with an average of 7.15 perturbations per lactation and a milk loss rate of 6.82%). We used this information as a comparison tool between the different cows studied in order to assess their resilience. Using this type of tools could be an alternative effective, elegant and applicable solution to find a reasonable compromise between optimal production and animal welfare.

Linkage disequilibrium as a signature of positive selection in beef cattle

A. Trakovická[1], N. Moravčíková[1], R. Nádaský[2], P. Polák[3] and R. Kasarda[1]
[1]Slovak University of Agriculture in Nitra, Tr. A. Hlinku 2, 94976 Nitra, Slovak Republic, [2]Poľnohospodárske družstvo Špačince, Družstevná 88, 91951 Špačince, Slovak Republic, [3]Slovak association of beef cattle breeders, Novozámocká 183/408, 95112 Ivanka pri Nitre, Slovak Republic; nina.moravcikova1@gmail.com

This study aimed to assess the distribution of selection signatures in the genome of Charolais and Limousin cattle trough determination of variability in linkage disequilibrium among adjacent syntenic SNPs in the autosomal genome and identify protein-coding genes located in particular genomic areas by available databases. In a total of 85 animals, selected based on the genealogical analysis, were genotyped using International Dairy & Beef Chip for 49,629 SNPs. Only autosomal SNPs and animals with call rate higher than 90% and minimum minor allele frequency 1% in a population were selected to subsequent analysis. The final database consisted of 42362 SNP markers covering 2,503.53 Mbp of the autosomal genome (average SNP spacing 59.14±61.51 kbp). The differences in genome-wide linkage disequilibrium patterns between populations were quantified using VarLD method over sliding windows of 50 SNPs. Selection signatures, identified based on the standardised varLD score, were recognised as regions with top of 1 percentile of signals. Protein-coding genes were screened in the selection signal regions using the cattle genome reference sequence ARS-UCD1.2. Based on the criteria for the detection of selection signals, 15 areas under significant selection pressure were identified across eight autosomes (1, 2, 5, 6, 7, 14, 16 and 17). Overall, 325 genes with different biological functions were located in these regions. The most intense signal was found on the BTA2 chromosome in the area of genes involved in genetic control of double muscle (*MSTN*) and reproduction (*CYP27C1*). The study confirmed that selection pressure within the genome of the Charolais and Limousin breeds was mainly focused on genomic areas controlling animal weight, body frame, carcass yield, meat quality and reproduction.

Long-term effects of selection for prolificacy and kit growth on rabbit performance and behaviour

O. Girardie[1,2], R. Robert[1], M. Maupin[1], J. Hurtaud[1], P. Joly[3], J. Ruesche[2], I. David[2], H. Garreau[2] and L. Canario[2]
[1]Hypharm SAS, La Corbière, 49450, Roussay, France, [2]INRAE, Animal Genetics, GenPhySE, 24 chemin de Borde Rouge, 31324, France, [3]ISARA Lyon, Agrapole, 69364 Lyon, France; lauriane.canario@inrae.fr

An experiment was designed to estimate genetic trends for doe and kit performance and behaviour during lactation in response to 22 generations of selective breeding for litter size and direct and maternal effects for weaning weight. The old-type line (L0) was produced from progeny of frozen embryos of the ancestor population. A cross-fostering design was implemented between L0 and the modern-type line (L22) to quantify contributions of direct and maternal effects to kit performance and behaviour. Does raised kits from a single line. None kit was raised by its biological dam. Data collection was performed from a single batch with 52 L0 and 59 L22 does in parity 1 and among them, 28 L0 and 29 L22 does in parity 2. Litter growth and survival were recorded frequently and milk production was measured with the weigh-nurse-weight method at d21. Nest quality and fur plucking were observed on d0 and d1. Willingness to nurse was assessed from doe behaviour before and after nursing on d5 and d8. Reaction to litter removal was analysed on d6 and d28. Kinetics of exit from the nest was daily evaluated from d12 to d29. Two kits per litter were tested in an emergence test at d22 and d25. Performance data and kit behaviour were analysed with a linear model including doe line, kit line, doe parity, litter size and for kit emergence the day of observation, plus a doe random effect. Maternal behaviour was analysed with a logistic regression. At d21, L22 does produced more milk than L0 does (198 vs 159 g, P=0.002) and produced heavier kits (378 vs 350 g, P=0.0005). L22 does had better maternal abilities than L0 does as referred to nest quality (P=0.06), fur plucking (P<0.0001), milk production (P=0.002) and willingness to nurse (P=0.007). As lactation progressed, L22 kits were observed out of the nest in greater proportion than L0 kits, whether they were raised by L0 or L22 does. L22 kits were bolder than L0 kits in the emergence test. Trends in doe behaviour were favourable to litter performance. Kit genetics influenced its performance and behaviour more than dam genetics in advanced lactation.

Signatures of selection in three indigenous Croatian cattle breeds

V. Brajkovic[1], B. Lukić[2], M. Ferenčaković[1], V. Cubric-Curik[1] and I. Curik[1]
[1]University of Zagreb Faculty of Agriculture, Department of Animal Science, Svetosimunska 25, 10000 Zagreb, Croatia, [2]J.J. Strossmayer University of Osijek Faculty of Agrobiotechnical Sciences Osijek, Department for Animal Production and Biotechnology, Vladimira Preloga 1, 31000 Osijek, Croatia; vbrajkovic@agr.hr

There are three indigenous Croatian cattle breeds, Croatian Busha (HRB), Istrian Cattle – Boškarin (HRI) and Slavonian Syrmian Podolian cattle (HRS). While HRB is classified to the Busha group of cattle, i.e. small shorthorn native breeds, indigenous in countries around Dinaric Alps (Western Balkan Peninsula), HRI and HRS are classified to the group of Podolian cattle breeds, characterised by long horns and grey coat colour and considered to originate from Podolian steppe. We have genotyped populations, representing 40 HRB, 40 HRI and 32 HRS individuals, using the BovineHD Bead Chip (Illumina Inc., San Diego, CA), which contains 777,972 SNPs. To identify selection signals we have applied four different approaches: (1) identification of extremely frequent SNPs in runs of homozygosity (eROHi); (2) Integrated Haplotype Score (iHS); (3) across populations identification of SNPs with extremely high F_{ST} values (ehF$_{ST}$); and (4) across populations Integrated Haplotype Score (Rsb) based on the ratio of extended haplotype homozygosity between populations. We have analysed and compared three indigenous Croatian populations with international and similar breeds, according to the available high-density genomic information. To reduce the number of false-positive results we performed meta-analysis based on fact that minus two times the sum of ln P values is distributed according to chi-square with two times the number of tests as degrees of freedom. Significant genomic regions were further analysed for the identification of candidate genes and their functional properties. The obtained results will contribute to a better understanding of functional value of indigenous Croatian cattle breeds and their adaptation to environmental factors and human needs.

Genomic signatures on differential adaptation in West African livestock

G. Forcina[1], L. Pérez-Pardal[1], I. Fernández[2], I. Álvarez[2], A. Traoré[3], N.A. Menéndez-Arias[2] and F. Goyache[2]
[1]CIBIO-InBIO, Research Centre in Biodiversity and Genetic Resources, University of Porto, Rua Padre Armando Quintas 7, 4485-661 Vairão, Portugal, [2]SERIDA-Deva, Camin de Rioseco 1225, 33394 Gijón, Spain, [3]INERA, 04 BP Boulevard des Tensoba Wam Gobi, 8645 Ouagadougou, Burkina Faso; giovanni.forcina@cibio.up.pt

A total of 237 taurine (106), zebu (89) and sanga (42) cattle from 10 different populations (3 taurine, 3 zebu and 4 sanga) sampled in West Africa were typed using the BovineHD BeadChip of Illumina (777,962 SNPs). After quality control, 543,595 SNPs located on the 29 bovine autosomes were analysed as follows: (1) iHS, nSL and XP-EHH (using zebu cattle as reference population) statistics were used to jointly identify signatures of positive selection (SS) with the software Selscanv1.2; (2) homozygous-by-descent (HBD) segments inherited from ancestors living 1,024 or 2,048 generations ago were identified using the software ZooROH. SS and HBD segments were overlapped within cattle type using the *intersectBed* function of the BedTools software. Only overlapping segments identified in, at least, three populations per cattle type were considered. Genomic segments identified in taurine (196; 123.6 Mb) and sanga cattle (140; 36.6 Mb) that did not overlap with those identified in zebu (9; 4.5 Mb) were considered representative of the genomic uniqueness of each type of West African cattle. Gene- and functional-annotation enrichment analyses allowed to identify 4, 1 and 1 statistically significant (enrichment factor > 1.3) Functional Clusters in taurine (FCt1 to FCt4), sanga (FCs1) and zebu (FCz1) cattle involved a total of 100 candidate genes. distributed as follows: FCt1, 7 immunity-related genes (mainly located on BTA11, coding interleukin-1 and Toll-like receptors); FCt2, 5 genes (located on the same locus on BTA5 involved in bitter taste receptor activity with importance for fitness); FCt3, 68 genes (coding ATP-binding proteins with importance for cell physiology); FCt4, 9 genes involved in the action of the somatotropin hormone; FCs1, 4 immunity-related genes mainly coding oligoadenylate synthase proteins; FCz1, 7 genes located on the Homeobox locus on BTA2 involved in skin ectodermal appendages. Functionality of the clusters identified fits well with the expected differences among the three types of West African cattle.

Investigation of Genotype by Environment interactions in Lacaune dairy sheep in France

D. Buisson[1], J.M. Astruc[1], A. Combasteix[2], C. Gava[3], D. Hazard[2], G. Lagriffoul[1] and H. Larroque[2]
[1]Institut de l'Elevage, BP 42118, 31321 Castanet-Tolosan, France, [2]INRAE, GenPhySE, Université de Toulouse, INRAE, ENVT, 31326, Castanet-Tolosan, France, [3]CNBL, BP 42118, 31321 Castanet-Tolosan, France; diane.buisson@idele.fr

As sustainability has become a major concern in Europe, the iSAGE project has been implemented to investigate how to face the future challenges in sheep and goat sector. Genetic selection in France is a powerful tool to improve profitability and sustainability of dairy sheep production by selecting local breeds able to produce in less favoured areas. So far, genotype by environment interactions (G×E) have not been studied in the French dairy sheep genetic evaluations while selection is used in various environments. If selected animals performed differently depending on the environment, this could have strong negative impacts on the sustainability of many farms. Therefore, G×E was investigated in Lacaune dairy sheep for milk yield (MY), fat and protein contents (FC, PC) and somatic cell counts (SCC). First, flocks have been classified in 4 different clusters according to breeding systems, geographical and meteorological conditions, animal performances and flock effect estimated from genetic evaluation. Then, G×E was studied through both scaling and re-ranking effects. As scaling effect is related to differences in variance of traits across environments, genetic parameters were estimated within each cluster. The re-ranking effect was measured by the estimates of genetic correlations between clusters for each trait. For production traits, estimates of genetic variances are close across clusters and heritabilities ranged for example from 0.22 to 0.27 for MY. The scaling effect is not important for these traits in Lacaune breed. Genetic correlations ranged from 0.86 to 1.00 for MY, FC and PC, indicating few re-ranking between clusters. The re-ranking is more important for SCC since genetic correlations ranged from 0.60 to 0.96. Our results suggested that genetic evaluation in dairy sheep is well adapted for production traits whatever the breeding system while G×E might be taken considered for SCC. These analyses should be extended to other functional traits under selection. This study was supported by the European Union's Horizon 2020 Research and Innovation Action through the project 'Innovation for sustainable sheep and Goat production in Europe (iSAGE)' under grant agreement 679302.

Distribution models unravel drivers of local adaptation in chickens

F. Kebede[1,2], H. Komen[2], S. Alemu[1], O. Hanotte[1,3], T. Dessie[1] and J. Bastiaansen[2]
[1]*International Livestock Research Institute (ILRI), Live-Gene, P.O. Box 5689, Addis Ababa, Ethiopia,* [2]*Wageningen University & Research, Animal Breeding and Genomics, Droevendaalsesteeg 1, 6708 PB Wageningen, the Netherlands,* [3]*University of Nottingham, Cells, Organism and Molecular Genetics, School of Life Sciences, NG7 2UH, Nottingham, United Kingdom; fasilgetachew7@gmail.com*

Environmental heterogeneity is a factor that induces local adaptation. Species distribution models (SDM) rely on presence-only data and are not suited to studying the effect of ecological variables on species' phenotypic variation. We integrated SDM with generalised additive models (GAMs) to identify environmental variables that are important for species and phenotype distribution. A hybrid sampling strategy was designed for Ethiopian indigenous chicken populations, whereby the landscape across each of the four spatial gradients was divided into three environmental clusters. Clusters were distant by 100 km and 2-3 populations were sampled per cluster. Farmers keeping target populations within a cluster visited separate market-sheds. Thirty-four environmental (climatic, soil, vegetation) and 19 phenotypic variables were analysed for 513 birds from 26 populations (6-7 per gradient). R packages (MaxentVariableSelection and ENMeval) optimised model efficiency. Correlated variables ($|r|>0.6$) and those with a relative contribution score of <4% were removed. The best model retained 9 variables (isothermality, temperature seasonality, mean temperature of coldest quarter, precipitation of warmest quarter, precipitation of coldest quarter, solar radiation, water vapor pressure for the month of May, water vapor pressure for the month of August, and soil clay content). SDM identified environmental drivers of local adaptation in indigenous chicken populations. The relationship between environmental predictors and observed phenotypes (LSmeans) will be also be examined. Results will be useful to delineate suitable habitats, predict performance and make recommendations to optimise the design of sustainable chicken breeding and development programmes in Ethiopia.

Putative genes associated with adaptation in the Afrikaner and Brahman breeds from South Africa

S. Mdyogolo[1,2], F.W.C. Neser[1], M.D. Macneil[1,2,3], M.M. Scholtz[1,2] and M.L. Makgahlela[1,2]
[1]*University of the Free State, Animal Science, Bloemfontein, 9300, South Africa,* [2]*ARC – Animal Production, 0062, Irene, South Africa,* [3]*Delta G, Miles City, Montana, USA; neserfw@ufs.ac.za*

Harsh environmental conditions hampered production and livestock in these regions have developed special adaptive mechanisms that includes tolerance to extreme weather conditions, resistance to diseases and the ability to survive on low quality forage. Adaptation and the ability to maintain some degree of production in these conditions develops over generations due to selection pressure on the genome and creates unique genomic signatures. Afrikaner (Afr) (*Bos taurus africanus*) and Brahman (Brah)(*Bos indicus*) are beef breeds suitable for surviving and producing under harsh environmental conditions. The aim of this study was to investigate putative genes that can be associated with adaptation in these breeds. A total of 373 Afr and 573 Brah animals, were genotyped on 150K GeneSeek Genomic Profiler. Markers with minor allele frequency <0.02, marker call rate <0.95 and individuals with call rate <0.90 were excluded. After quality control 330 Afr and 239 Brah animals remained in the study with a total of 83,306 autosomal SNP in common between the two breeds. Runs of homozygosity (ROH) were used to determine chromosomes undergoing selection and containing putative genes. ROH were defined using a sliding window of 150 SNP allowing for 2 heterozygous SNP and a sliding window of 300 SNP and allowing for 3 heterozygous SNP in the Afr and Brah, respectively, over a distance of 1000 Kb. ROH were found on Chromosomes 5, 7, 8, 11, 12, 14 and 15 in Afr, and 5, 6, 10, 14, 20 and 24 in Brah. Genes identified include *PAPPA, PAXBP1, ERAP1* and *SULT1E1* and are associated with immune response, cow fertility and milk production. The distribution of genes in the Afr are mainly associated with adaptation (70%) followed by genes associated with metabolism (17%) and reproduction (13%). In the Brah the association with adaptation is less (46%), followed by metabolism (35%), production (11%) and reproduction (8%) The evidence of adaptive selection signatures in both breeds can be traced back to their historical origin and harsh production environment. This study reveals how selection prioritises adaptation over production in order to ensure survival.

Effect of changes in complex breeding traits on environmental sustainability in the Norwegian Red

T.A. Skjerve, S. Samsonstuen and H.F. Olsen
Norwegian University of Life Sciences, Department of Animal and Aquacultural Sciences, Arboretveien 6, 1433 Ås, Norway;
torgunn.aslaug.skjerve@nmbu.no

In Norway, beef production from the dual-purpose breed, Norwegian Red, covers more than 60% of the national beef consumption. This considerable proportion is achieved through culled cows, surplus heifers and finishing bulls kept to finishing age at the farm. The farm structure development largely follows the same trajectory as in other countries; fewer and larger, with more productive crops, and less, but higher-yielding animals. For the period 1979 to 2015, the number of farms has dropped from 53.793 to 8.800 and the Norwegian Red has had an increase in individual, annual milk yield from 5,600 kg to 7,800 kg. The increase is a result of a long-term, broad breeding goal, but this production efficiency has also led to reduced beef production, as fewer animals are needed to cover the demand for milk products. The shift towards a more intensive production is expected to mitigate the GHG emissions from agriculture, largely due to the lower number of animals and smaller area needed to achieve the production goal. However, this 'sustainable intensification' is not thoroughly documented. These intensive systems require quite intensive inputs through for instance diet composition and amount of feed, which can be challenging in a country dominated by grass- and rangeland. Utilisation of domestic resources in Norway is dependent on ruminants' ability to harvest large parts of the diet from the vast pasture areas of varying quality. It is therefore of interest to study the sustainability of the development of the dual-purpose breed in a Norwegian context. This study will explore sustainability of the development of the dual-purpose Norwegian Red using life cycle analysis (LCA). A baseline scenario for current trait composition is compared to a historic scenario, based on data from the end of the 1970s, to explore the combined effect of changes in the complex breeding traits on the environmental sustainability. The study is on-going, and the results will be included in the final presentation.

Improved protein efficiency for a sustainable pig production for the future

K.H. Martinsen, E.J. Gjerlaug-Enger and E. Grindflek
Topigs Norsvin, P.O. Box 504, 2304 Hamar, Norway; kristine.martinsen@norsvin.no

Protein is a costly nutrient in feed and efficient use of proteins from the feed is therefore important due to global issues such as human population growth, climate changes, greenhouse gas emissions and economics in pork production. In total, emissions from the agricultural sector in Norway represents approximately 9% of the total emissions. From this, 8% is from the Norwegian pork production. Greenhouse gas emissions from pork production can be divided in two parts. One part is related to the feed, which represents 80% of the emissions, whereas a smaller part is related to the animal's digestion and storing of faeces (20%). Breeding companies can contribute to sustainable food production by improving the efficiency of the animals through genetics as a mitigation. Since 1958, Norsvin Landrace has been bred for improved efficiency. This has led to a reduction in carbon dioxide equivalents per kg carcass weight (kg CO_2-eq. kg CW -1) from 4.5 to 3 kg CO_2-eq./kg CW and represents a reduction of 30% from 1960-2019. This reduction is a result of more weaned piglets, lower mortality, shorter growth period, higher feed efficiency and lower mortality. This study investigated protein digestibility to further improve the efficiency of animals. To calculate digestibility, an internal marker from the feed was used. Faeces samples were taken from approximately 500 animals and freeze-dried. All samples were chemically analysed for the marker, whereas 150 samples where chemically analysed for the marker, nitrogen, crude fat and organic matter to use as reference data for development of calibrations for near-infrared spectroscopy (NIRS). All samples were run through NIRS, and individual digestibility was calculated. All animals were genotyped to utilise the genomic relationship between animals and genomic analysis of the trait will be performed when sampling is finished.

Is rumination time a predictor of methane emission of dairy cows?

M. Pszczola[1], R. Mikuła[2], K. Rzewuska[1], S. Mucha[1], W. Nowak[2] and T. Strabel[1]
[1]Poznan University of Life Sciences, Genetics and Animal Breeding, Wolynska 33, 60-637 Poznan, Poland, [2]Poznan University of Life Sciences, Animal Nutrition, Wolynska 33, 60-637 Poznan, Poland; marcin.pszczola@up.poznan.pl

Methane emission (CH_4) is difficult to measure on a large scale. It is, thus, important to search for easily accessible indicator traits. Rumination time (RT) is strongly affected by nutritional management and impacts the whole digestion process including the feed passage rate as well as voluntary feed intake in dairy cows and may impact CH_4 emission. RT is easy to measure and would be a good indicator trait for CH_4 emission in case of a strong relationship between the traits. Therefore, the goal of this study was to assess the level of CH_4 emission at different RT. We analysed 316 Polish Holstein-Friesian multiparous cows that were between 14 to 305 d of lactation. The cows received *ad libitum* partly mixed ration (PMR) which was delivered twice a day and met nutrient requirements for 25 kg of milk yield. RT was assessed based on data obtained from an automatic milking system (AMS, Astronaut, Lely Industries, NV, Maassluis, the Netherlands). CH_4 concentration was measured during milking using sniffer installed in the AMS using an FTIR analyser (GASMET 4030; Gasmet Technologies Oy, Helsinki, Finland). Based on the measured CH_4 and CO_2 concentrations, we have calculated daily methane production (l/d). In total, we collected 14,274 daily observations for CH_4 and RT. Using the RT, the cows were assigned to three groups: low RT below 412 min. (L), medium RT from 412 to 528 min. (M) and high RT above 528 min. (H). We analysed five different phenotypes: 1) daily milk production; 2) daily CH_4 production per metabolic weight; 3) daily CH_4 production per 1 kg milk production; 4) CH_4 production per 1 kg energy corrected milk; and 5) CH_4 production per 1 kg concentrate intake. All the comparisons were made on the phenotypic level. We observed significant differences between the averages of CH_4 phenotypes across the RT groups in most of the cases, except for CH_4/concentrate intake. Although the found differences were significant, the numeric differences were small. Therefore, based on our results obtained with a limited data set, RT seems not sufficient as an indicator trait for CH_4 emission.

A genome-wide recombination map of cattle

S. Qanbari and D. Wittenburg
Leibniz Institute for Farm Animal Biology (FBN), Institute of Genetics and Biometry, Wilhelm-Stahl-Allee 2, 18196 Dummerstorf, Germany; qanbari@fbn-dummerstorf.de

Recombination is a process by which chromosomes are broken and recombine to generate new combinations of alleles, and therefore is a major determinant in shaping genomic variation. Recombination frequencies (r) between markers are used to construct genetic maps, which have important implications in genomic studies. Here, we report a recombination map for 44,696 autosomal SNPs according to the coordinates of the most recent Bovine reference assembly. The recombination frequencies were estimated across 1,052 half-sib families with the minimum number of 30 and a maximum of 4,236 progenies, in a pedigree comprising >360K German Holstein animals. Genome-wide, over 8.9 million paternal recombination events were identified in a pairwise comparison of adjacent markers. The recombination map spans for 24.43 Morgans for a total chromosomal length of 2,489 Mb and an average of ~0.98 cM/Mbp, which concords the available pedigree-based linkage maps. Furthermore, we identified 911 putative recombination hotspots (defined as r>2.5 standard deviations greater than the mean). The recombination hotspots were uniformly distributed as sharp and narrow peaks, corresponding to ~5.8% of the recombination taken place in only ~2.4% of the genome. Given the fact that this map is built on the coordinates of the ARS-UCD1.2 assembly, our results provide the most updated genetic map yet available for the cattle genome.

Genomic evaluation in multi-generation family
T. Okamura, M. Nishio, K. Ishii and O. Sasaki
Institute of Livestock and Grassland Science, NARO, Ikenodai 2, Tsukuba, Ibaraki, 305-0901, Japan; okamut@affrc.go.jp

The improvement in the accuracy of estimated breeding values using multi-generation data is limited by the great distance in the numerator genetic relationship between old ancestors and current subjects. On the other hand, genomic BLUP (GBLUP) with a genomic relationship matrix can be superior to a traditional BLUP using multi-generation data, especially in a small population. The objective of this study was to compare the availability of old ancestors between traditional BLUP and GBLUP. Assuming a small pig population, a non-overlapping population was generated using QMSim for 10 generations, with a breeding population of 10 males and 50 females. From each litter, it was assumed that 1 male and 2 females were candidates for the next parenting and that carcass traits could be obtained and measured from 2 slaughtered individuals. The traits were 1) average daily gain (ADG), which all individuals had; 2) intramuscular fat content (IMF), which only slaughtered animals had; and 3) litter size (LS), which only sows had. Heritability was 0.3, 0.3, and 0.1, respectively. A total of 1,000 QTLs were uniformly generated on 18 chromosomes from a gamma distribution with a shape parameter of 0.4. It was assumed that all individuals had 59,994 SNP markers for ADG and IMF, and that the candidates had it for LS. Ten datasets were created as follows: the 1st dataset included only the 10th generation; the 2nd dataset included the 10th and 9th generations; and the 10th dataset included the 10th, 9th, ..., 1st generation. For each set, the breeding value was estimated using traditional BLUP and GBLUP. The accuracies of each BLUP were correlations between the true breeding value and estimated breeding value in candidates from the 10th generation. One hundred replications were simulated. In BLUP, the accuracies of all traits increased with the addition of a few recent generations, but they were constant with the addition of older generations. On the other hand, in GBLUP, they continued to increase with generations. Although the increments of accuracies with BLUP were 5-23%, they were 14-42% in GBLUP. These results suggest that GBLUP can use information that is too old to estimate the breeding value of current subjects and that GBLUP is superior to BLUP in a small breeding population.

Generalised gametic relationships for the analysis of imprinting effects in Canadian Holstein cattle
I. Blunk[1], G.A. Oliveira[2], C.F. Baes[2,3] and N. Reinsch[1]
[1]Leibniz Institute for Farm Animal Biology (FBN), Institute of Genetics and Biometry, Wilhelm-Stahl-Allee 2, 18196 Dummerstorf, Germany, [2]University of Guelph, Centre for Genetic Improvement of Livestock, Department of Animal Biosciences, 50 Stone Road East, N1G 2W1 Guelph, Canada, [3]University of Bern, Institute of Genetics, Vetsuisse Faculty, Bremgartenstr.109a, 3001 Bern, Switzerland; blunk@fbn-dummerstorf.de

The impact of parent-of-origin effects caused by genomic imprinting can be investigated using a gametic model in an imprinting variance component analysis. This model contains two random gametic effects (one as sire and one as dam); and four equations are required for each individual in the pedigree (including final progeny) as a gametic relationship (G) matrix is used. This quadruples the size of the system of equations compared with that when only one random genetic effect is predicted. As a result, the system of equations may become too large to solve using all available data. By using a numerator relationship matrix, a reduced version of the gametic model requiring a lower number of equations can be implemented. Instead of four gametic effects, two parental transmitting abilities are predicted for each non-phenotyped parent, and final progeny are excluded. In some cases, however, parents may be phenotyped. To combine the advantages of both models, we developed a generalised model that allows for parents with records but excludes final progeny. The model uses a generalised G matrix containing gametic relationships of ancestors with records and numerator relationships of ancestors without records. For the latter, transmitting abilities are predicted. A pedigree of 2,112,986 Canadian Holstein cattle was available, with stature measured as a conformation trait on 1,086,339 cows. Using a gametic model, this dataset would generate 8,451,944 equations. With our generalised model, the size of the system of equations was reduced by more than 60% to 3,372,098 equations, as 756,638 final progeny were excluded and two transmitting abilities were predicted for 1,026,647 breeding animals. The generalised model facilitated an imprinting analysis using all available data and further revealed parent-of-origin effects on stature. This work provides valuable insight to understanding how gametic models can be used to partition imprinting variance components and parent-of-origin effects.

Genomic prediction for crossbred cows by combining marker effects estimated in purebreds

J.H. Eiriksson, E. Karaman, G. Su and O.F. Christensen
Aarhus University, Center for Quantitative Genetics and Genomics, Blichers Alle 20, 8830 Tjele, Denmark; jonh@mbg.au.dk

Genotyping cows for management and selection purposes has become common practice in dairy herds. However, for crossbred cows, genomic predictions have not been widely implemented. Marker effects are expected to differ between breeds because of lower level of linkage disequilibrium between breeds than there is within breeds, marker phase not being completely preserved between breeds and difference in genetic background. Considering breed origins of the alleles of crossbred animals is therefore expected to improve genomic prediction for crossbred animals. Two methods were compared for predicting genetic merit of genotyped crossbred cows with marker effects that were estimated in purebred animals, without considering phenotypes of crossbred animals. The methods were based on combining results from the contributing pure breeds using estimated genomic breed composition (GBC), or estimated breed origin of the alleles (BOA). With the GBC, the genetic merit of a crossbred animal was estimated as a weighted combination of its genomic breeding values computed separately from the SNP effects of the pure breeds, with the weights being GBC. With the BOA, the estimate was the sum of the allele effects for each SNP, with effects taken from the breed origin of its alleles. The two methods were assessed by analysis on simulated data. The simulated genotypes were based on the Danish Holstein, Danish Jersey and Swedish Red cattle and crosses between them. Four different crosses were considered, F1 cross of Holstein and Jersey (H_J), H_J mated with Red cattle (R_HJ), RH_J mated with Holstein (H_RHJ) and H_RHJ mated with Jersey (J_HRHJ). Results of 10 replicates showed that the accuracy measured as correlations between the predicted and the true breeding values for a trait with heritability of 0.40 were 0.69 when applying the GBC method and 0.72 using BOA. The accuracy was highest for H_J, 0.77 and 0.80 for GBC and BOA respectively. For GBC the accuracy was lowest for H_RHJ, 0.66 and for BOA it was lowest for J_HRHJ, 0.68. The results indicate that detecting breed of origin of alleles of crossbred animals for genomic prediction can improve genomic prediction for crossbred cattle when only records of purebred are available.

Comparison of conventional and genomic estimated breeding values in Holstein bulls from Serbia

S. Trivunović[1], D. Glamočić[1], L.J. Štrbac[1], M. Šaran[1], N. Veličković[2] and M. Stefanović[2]
[1]University of Novi Sad, Faculty of Agriculture, Department of Animal Science, Trg D.Obradovića 8, 21102, Serbia,
[2]University of Novi Sad, Faculty of Sciences, Department of Biology and Ecology, Trg D.Obradovića 3, 21102, Serbia;
snezana.trivunovic@stocarstvo.edu.rs

The application of genomic selection in dairy cattle breeding has not been used in Serbia so far. Traditionally, selection criteria are based on phenotypic parameters and estimated breeding value using classic Best Linear Unbiased Prediction (BLUP). Recent year have been worked on implementation of genomic evaluations and for this purpose 36 domestic bulls of Holstein Friesian breed have been genotyped. SNP genotyping was performed using a low-density chip. Estimated breeding values based on genome were obtained through a commercially available marker panel (Igenity, Neogen Corp.). The aim of this paper was comparison of conventionally estimated breeding values (EBV) and genomic estimated breeding values (GEBV) for these bulls. EBV was calculated using the BLUP animal model and this calculation was performed in the software package WOMBAT. The estimation was based on daughter production data for the first and all lactations. The traits evaluated were: milk yield, fat yield, fat percentage, protein yield and protein percentage. The results of EBV and GEBV were compared using Spearman's rank correlation coefficient (r_s). Data comparison showed that compared breeding values for all traits had a weak to moderate positive correlation. The lowest correlation was found between EBV and GEBV (first lactation) for milk yield (r_s=0.179), and the strongest correlation between EBV and GEBV (all lactation) for fat percentage (r_s=0.533). This research is an important step for further development of genomic selection in the cattle breeding programs in Serbia and in the future, we must work on the forming a reference population in Serbia and developing own model for GEBV. Acknowledgements: This paper is part of the project 'Application of genomic selection in cattle breeding', No. 142-451-2480/2019-02, founded by the Government of the AP Vojvodina.

Influence of rare SNP variants on the accuracy of genomic selection in dairy cattle

T. Suchocki[1,2], A. Zarnecki[1], M. Skarwecka[1] and J. Szyda[1,2]
[1]National Research Institute of Animal Production, Krakowska 1, 32-083 Balice, Poland, [2]Wroclaw University of Environemntal and Life Sciences, Department of Animal Genetics, Biostatistics Group, Kozuchowska 7, 51-631 Wroclaw, Poland; tomasz.suchocki@upwr.edu.pl

Predicting of phenotypes from genotype data is widely used in animal breeding since 20 years. Usually, a set of markers included in the genomic selection is edited based on two criterions: a minor allele frequency (MAF) and a quality of genotyping expressed as a call rate. Such filtering leads to the fact that impact of rare genotypes on estimated breeding values is unknown. Recently, rare genetic variants, i.e. polymorphisms with low minor allele frequency, have been used for genetic determination of complex traits. The main reason for this is the phenomenon of so called 'missing heritability' indicated for most of the phenotypes measured in humans. It is expected that variants which represent functional mutations with large effects on complex phenotypes are rare because of the natural (humans) or artificial (livestock) selection pressure. The biological explanation is that since a mutation is functional, it is subjected to selection, which as a consequence, affects the population allele frequency stronger than in the case of a neutral mutation. Dairy cattle seems to be an ideal population to verify this hypothesis. It has undergone a directional selection on production traits for many generations and it has very good records of complex traits and familial relationship. Moreover, recent success of genomic selection provided extensive information on genotypes of single nucleotide polymorphisms distributed all over the genome and available for many individuals. The goal of our study was to verify whether including rare variants into a routine genomic selection model allows for capturing a considerable part of missing heritability underlying production, fertility and type traits in dairy cattle. Especially if using rare variants could increase the reliability of DGV and hence also GEBV.

Comparison of imputation methods in Czech Holstein population

A. Kranjcevicova[1,2], E. Kasna[2], L. Vostry[1,2] and M. Brzakova[1,2]
[1]Czech University of Life Sciences, Faculty of Agrobiology, Food and Natural Resources, Dpt. of Genetics and Breeding, Kamycka 129, 16500 Prague, Czech Republic, [2]Institute of Animal Science, Dpt. of Genetics and Breeding of Farm Animals, Pratelstvi 815, 10400 Prague, Czech Republic; kasna.eva@vuzv.cz

The importance of genomic data in the breeding of dairy cattle is growing. Genomic information is used not only for genomic selection but also for revealing the genetic architecture. Data from DNA genotyping chips of various densities are available. Genotype imputation is the process of using reference population genotyped at a higher density to predict genotypes in population genotyped at a lower density. Our aim was to compare three different methods of genetic imputation with FImpute software and to choose the best one for Holstein population in the Czech Republic. We compared (1) imputation based on pedigree data, (2) population imputation and (3) combination of both approaches. A simulation study was performed on 3,994 animals genotyped on Illumina 50kBeadChip v.2, of which 994 animals were artificially masked on Illumina LD Beadchip v.2. Masked animals were imputed by all three methods. The success rate was based on two parameters: (1) imputation accuracy and (2) the percentage of correctly imputed SNPs. Since the animals to be masked were randomly selected, the whole calculation process was repeated 100 times, and the resulting parameters were averaged. We achieved the highest average accuracy (0.831 ± 0.0004) and the highest percentage of correctly imputed SNPs (0.963 ± 0.0008) with population imputation. For the combined method of imputation, the average accuracy was 0.827 ± 0.0004 and the percentage of correctly imputed SNPs 0.957 ± 0.0009. Since the massive genotyping of animals in the Czech Republic began a few years ago, our database does not contain enough genotyped ancestors. For this reason, pedigree imputation achieved low average accuracy of 0.132 ± 0.0039 and the percentage of correctly imputed SNPs was 0.274 ± 0.0109. The results showed that pedigree imputation is currently not suitable for our population, and we need to focus on the population approach. Supported by the Ministry of Agriculture of the Czech Republic (Projects No. QK1810253, MZE–RO0718) and by the Ministry of Education, Youth and Sports (Project No. LTAUSA19117).

National genomic predictions in relation to international genomic predictions for small dairy breeds

B. Luštrek, M. Štepec, J. Krsnik and K. Potočnik
University of Ljubljana, Biotechnical Faculty, Department of Animal Science, Jamnikarjeva 101, 1000, Slovenia;
barbara.lustrek@bf.uni-lj.si

Routine national dairy cattle genetic evaluation is being upgraded to genomic evaluation. To obtain genomically enhanced breeding values ($GEBV_{IG}$), Slovenian Brown Swiss cattle population has been participating in Brown Swiss cattle international InterGenomics (IG) consortium managed by Interbull Centre (IB) that performs multi-step genomic evaluation for bulls. Multiple Across Country Evaluation (MACE) sire comparison results are used as pseudo phenotypes, while bull genomic information is contributed by member countries. Genomic information is imputed to IB SNP index and each country obtains SNP effects for economically important traits on its own scale, estimated on a common international bull reference population. Only small and chiefly cow part of Slovenian Brown Swiss cattle population is genotyped. Cow genotypes are imputed to the IB SNP index as well, which along with obtained SNP effects enables computation of cow DGV_{IG}. National genomic evaluation resulting in national genomic prediction ($GEBV_N$) has been carried out using modified single-step Genomic Best Linear Unbiased Prediction (ssGBLUP) method that enables integration of external information, avoiding repeated counting of the same information. Input data for milk yield consisted of 1,155,043 test day records and pedigree with 97,989 animals. Genotypes from 2,166 Slovenian animals (1,589 cows and 577 bulls) were included in the evaluation and national reference population was increased by 7,024 foreign sire genotypes obtained from IG consortium with their MACE breeding values. Mean reliability (REL) with standard deviation for $GEBV_N$ of Slovenian genotyped animals was 0.77 (0.16) for bulls and 0.79 (0.11) for cows. Both were similar to 0.78 (0.13) $GEBV_{IG}$ REL for bulls. Estimated genomic prediction correlation coefficient between $GEBV_N$ and $GEBV_{IG}$ was 0.82 (P<0.0001) for bulls. Correlation between $GEBV_N$ and DGV_{IG} for cows was 0.57 (P<0.0001). National genomic evaluation gave satisfactory results that will ease decision making in national selection process especially for cows.

Don't play too much – deep learning to classify true and false positive SNPs in whole genome sequence

K. Kotlarz[1], J. Szyda[1,2], B. Czech[1,3], M. Mielczarek[1,2], T. Suchocki[1,2] and B. Guldbrandsen[4]
[1]Wroclaw University of Environmental and Life Sciences, Biostatistics group, Kozuchowska 7, 51-631 Wroclaw, Poland, [2]National Research Institute of Animal Production, Krakowska 1, 32-083 Balice, Poland, [3]Roche Polska, Domaniewska 39 B, 02-672 Warszawa, Poland, [4]University of Bonn, Institute of Animal Sciences, Endenicher Allee 15, 53115 Bonn, Germany; joanna.szyda@upwr.edu.pl

The data set consisted of whole genome DNA sequences (WGS) of four traditional Danish Red Dairy Cattle bulls, sequenced by the Illumina platform. The average genome sequencing coverage was of 10X. Additionally, the bulls were typed using the Illumina BovineHD BeadArray comprising 777,962 SNPs. After the quality control, 772,532 SNPs were selected for downstream analysis. The first step was SNP classification into true-positives (i.e. a SNP genotype from WGS was concordant with a SNP genotype from a microarray) or false-positives (i.e. a SNP genotype from WGS was discordant with a SNP genotype from a microarray). Out of 2,294,151 considered SNPs pooled across the four bulls, 74,764 were false-positive. In the next step, a deep learning algorithm implemented in Keras was used to build a classifier for false-positive and true-positive SNPs based on standard sequence quality statistics available from the variant calling output, such as QUAL and GQ, read depth at the SNP site expressed by DP and DP2, as well as the three-bases-sequence down-stream and up-stream of a SNP. Three deep learning models were applied: (1) a baseline algorithm; (2) an algorithm in with class weights were added to compute the loss function; and a model (3) in which applies an algorithm that oversamples the false-positive SNPs before parameter estimation. All the models were composed of eight dense layers with five dropout events of 20% of data, implemented after the first five layers. Model parameters were estimated based on the Adam optimisation algorithm from Tensor Flow, implemented via Keras and the binary cross-entropy was used as a loss function. Comparison of the quality of the three models revealed that the best classification was provided by the naïve model, indicating that model regularisation techniques should be treated in caution. All computations were carried out at the Wroclaw Centre for Networking and Supercomputing.

Single-step genomic evaluation for milk and fat yield in buffaloes using random regression models

S.F. Lazaro[1,2], H.R. Oliveira[3], D.F. Cardoso[2,3], A.V. Nascimento[2], H. Tonhati[2] and L.F. Brito[1]
[1]*Purdue University, Department of Animal Sciences, 610, Purdue Mall, 47907, West Lafayette, IN, USA,* [2]*Universidade Estadual Paulista (Unesp), Department of Animal Science, S/N, Via de Acesso Professor Paulo Donato Castelane Castellane, Vila Industrial, 14884-900, Jaboticabal, SP, Brazil,* [3]*University of Guelph, Department of Animal Biosciences, 50 Stone Rd E, N1G 2W, Guelph, ON, Canada; sirlenelazaro@yahoo.com.br*

The application of genomic selection in buffaloes is still incipient compared to other livestock species. In this context, we evaluated the performance of genomic prediction for milk-related traits in Murrah buffaloes, using single-trait random regression models (RRM) and the single-step GBLUP method. The data analysed consisted of 71,048 and 2,428 first-parity records of milk (MY) and fat (FY) yields, respectively. A total of 960 animals were genotyped using the 90K Axiom Buffalo Genotyping array (45,690 markers remained after quality control). The RRM used included the random effects of additive genetic, permanent environment, contemporary group (herd-year-season calving), and residual term. In addition, the fixed effects were milking frequency per day (1 or 2), and age at calving as covariate (linear and quadratic effects). The additive genetic and permanent environment effects were modelled using Legendre Orthogonal polynomials of third and fourth order, respectively. Furthermore, the residual heterogeneity was modelled based on 4 different classes. The predicted RRM coefficients were used to derive Genomic Estimated Breeding Values (GEBVs) for each time point (from 5 to 305 days in milk). Subsequently, the daily GEBVs were compared to the pedigree-based EBV (based on BLUP). The average daily GEBV accuracies (i.e. Pearson correlation coefficient estimated between GEBVs and EBVs) were 0.54 and 0.43 for MY and FY, respectively. The regression coefficients (i.e. b1 from EBV = b0 + b1xGEBV) used to access the average bias in the predictions were 0.60 and 0.73 for MY, and FY, respectively. The moderate accuracy and bias observed for MY and FY indicate that RRM can be used to predict daily GEBVs for this trait. In fact, genomic prediction for MY and FY using RRM and the single-step GBLUP method seems feasible in Murrah buffaloes. However, further studies are needed to validate our findings.

New perspective on intergenerational transmission of characters

I. David[1], A. Aliakbari[1], L. Canario[1], S. Combes[1], J. Demars[1], V. Déru[1,2], H. Garreau[1], H. Gilbert[1] and A. Ricard[3,4]
[1]*GenPhySE, INRAE, Université de Toulouse, INPT, ENVT, 31326 Castanet Tolosan, France,* [2]*France Génétique Porc, la motte au vicomte, 35651 Le Rheu cedex, France,* [3]*Institut Français du Cheval et de l'Equitation, Département Recherche et Innovation, 61310 Exmes, France,* [4]*GABI, INRAE, AgroParisTech, Université Paris Saclay, Département Sciences du Vivant, UMR 1313, 78352 Jouy en Josas, France; ingrid.david@inrae.fr*

In livestock species, animals are selected based on genetic inheritance only. However, other factors than genetics are also inherited and are involved in determining the animal's phenotype, thus playing a role in trait inheritance. These non-genetic sources of inheritance comprise information transmitted *via* physical means, such as epigenetic and microbiota inheritance, and information transmitted *via* learning mechanisms i.e. behavioural inheritance. Disentangling the different sources of inheritance is challenging with only pedigree and phenotypic data. To overcome this problem and still take into account the different sources of inheritance when estimating the transmissible potential of individuals, we developed the transmissibility model that estimates the path coefficients of inherited information from parent to offspring instead of set values of 0.5 that corresponds to genetic inheritance only. Applied to residual feed intake in pigs and rabbits, results of the transmissibility model suggested that phenomena other than genetic explain the phenotypic resemblance between relatives, with a higher transmission from the dam's side than from the sire's side. The environment may influence non-genetic inherited factors. It is thus possible to favour the transmission of positive non-genetic inherited factors *via* the environment in which the animal is raised. Various actions to promote 'positive' non-genetic inherited effects can be implemented at different times during the life of the animal. They can be classified into two main categories: actions that promote transmission of 'positive' non-genetic inherited factors to the target animal *via* other animals, and actions that modulate non-genetic inherited factors via favourable environmental conditions.

Plasma biomarkers of feed efficiency differ across two contrasting diets in growing beef cattle

G. Cantalapiedra-Hijar[1], P. Guarnido[1], S. Taussat[2], G. Renand[2] and I. Ortigues-Marty[1]
[1]INRAE, Université Clermont Auvergne, VetAgro Sup, UMRH, Saint Genes Champanelle, 63122, France, [2]INRAE, AgroParisTech, Université Paris-Saclay, GABI, Jouy-en-Josas, 78350, France; gonzalo.cantalapiedra@inrae.fr

Beef cattle industry could benefit from the identification of individuals with superior feed efficiency. Thus, biomarkers should be developed and validated to contribute to this objective. This study aimed to explore plasma biomarkers of residual feed intake (RFI) in fattening beef cattle through a targeted metabolomics approach. For this, 167 young Charolais bulls were tested for RFI in 3 independent cohorts during at least 6 months. Animals within each cohort were evenly assigned to either a high-starch corn silage diet or a high-fibre grass silage diet. Blood was sampled from each animal one month before the end of the test. Plasma from the 48 most extreme animals in terms of RFI (24 low vs 24 high RFI; balanced within cohorts and diets) were analysed for 630 metabolites by FIA- or LC- tandem mass spectrometry. Metabolites were analysed by ANOVA separately for each diet including the fixed effect of cohort, RFI group and their interaction. In the corn silage diet, plasma concentration of the 3 branched-chain amino acids (BCAA; Leu, Ile and Val) was lower (-15%; $P \leq 0.01$) in low-compared to high-RFI, likely reflecting their higher metabolic use for protein synthesis and muscle accretion. In contrast, creatinine and hydroxyl-proline, markers of muscle and collagen body mass respectively, showed higher plasma concentration (+18-20%; $P \leq 0.01$) in low- compared to high-RFI animals. In the grass silage diet, most of the RFI discriminant metabolites belonged to different classes of triglycerides and their plasma concentrations were always higher (+50% on average; $P=0.003$) in low- compared to high RFI animals, indicating either their lower incorporation into fat cells or their lower use as energy substrate by the muscle. Only creatinine was a common RFI discriminant metabolite in the two diets and Val, though also discriminant, showed opposite trends. Our results suggest that discriminant metabolites of RFI are not the same across two contrasting diets. However, differences in muscle body mass revealed by plasma creatinine concentration seems to be a determinant of RFI regardless the diet.

Economic importance of the number of functional teats in the Czech pig breeds

Z. Krupová, M. Wolfová, E. Krupa and E. Žáková
Institute of Animal Science, Přátelství 815, 10400 Prague Uhříněves, Czech Republic; krupova.zuzana@vuzv.cz

Breeding for higher litter size provided in last decades have had economic and as well welfare consequences. Management interventions are needed in pig farms when the litter size routinely exceed the number of functional teats (NFTs). From the genetic point of view selection for higher NFTs is possible. Therefore, economic weight (EW) of the NFTs was calculated in this study using the enhanced bio-economic model of the EWPIG software. The EW represented the cost differences between naturally and artificially reared piglets until weaning and the cost differences between those animals in finishing period (due to lower growth rate of artificially reared piglets). In total, EW were calculated for the complex of 19 traits of two dam breeds (Czech Large White and Czech Landrace) used in the national breeding system. The evaluated traits were related to growth, prolificacy, carcass, feed efficiency and health. Conception rate of sows and gilts, number of piglets born alive, survival of piglets at birth and until weaning, NFTs and sperm quality traits were considered there as prolificacy traits. Economic effect (EW) of increasing the NFTs by one teat was 8.5 € and 21.1 € per sow of the Czech Large White and Czech Landrace, respectively. The EWs were especially associated to rearing of piglets from birth to weaning (95%) and 39% of these costs were represented by the savings in labour costs. The effect of this trait improvement per sow of the Czech Landrace breed was 2.5 times higher all in accordance with the structure of the evaluated three-way crossing system. Despite the such difference in EWs for NFTs, the relative importance of NFTs (3.7 and 3.5%) was similar in both breeds. Overall economic proportion of prolificacy in dam breeds was 43 and 41%, followed by health traits (27 and 26%), feed efficiency (16 and 18%) and carcass traits (9 and 10%). The NFTs should be incorporated into the breeding goal of the local dam pig breeds to obtaining ideal outcomes with optimal EWs. In both dam breeds, the relative EWs of evaluated traits were similar enough to justify using the same aggregate genotype for selection. The study was supported by project MZE-RO0719 V003 and QK1910217of the Czech Republic.

Evaluation of analytical protocols utilising HTS data aligned tools

A. Bąk[1], D. Bodziony[1], G. Migdałek[2] and K. Żukowski[1]
[1]*National Research Institute of Animal Production, Krakowska 1, 32-083 Balice, Poland,* [2]*Pedagogical University of Cracow, Podchorążych 2, 30-084 Cracow, Poland; kacper.zukowski@izoo.krakow.pl*

The rapid development of techniques and tools used in genetics and genomics has affected new generation sequencing, which has become much more comfortable and cheaper. The result is the generation of a massive amount of data sets, requiring detailed analysis, which becomes impossible without the use of appropriate bioinformatics tools. One of the crucial steps in the analysis of sequencing data is to map readings to a reference sequence. Although the dominance of Illumina SBS technology has been noticeable in recent years, the choice of the tools is hampered and the variety of input data and reference genomes. Moreover, the tool used is crucial for result files and further analysis. The subject of this work is the three most frequently used programs, which have functions that allow working with many platforms: BWA, Bowtie2 and SMALT. The task of the tested aligners is to match short sequences coming from NGS with reference sequences. The most popular: BWA and Bowtie2 use for this purpose the Burrows-Wheeler transformation and SMALT maps the sequences using hashing and dynamic programming. The aim of the work was to compare the quality and efficiency of the programmes under examination, due to three criteria: (1) the quality of the compared sequences of different lengths and from different platforms; (2) coefficient of wrongly compared sequences; (3) the computational resources used. Comparing the results of the mapping analyses for all the programmes used, the least popular SMALT is the best. Obtaining the highest percentage of mapped readings for each platform and maintaining the lowest computational memory usage, turns out to be the most optimal choice. The results presented in this work can be used to verify and rebuild data analysis pipelines from NGS based so far on other tools. By using the tools under appropriate conditions, it is possible to improve the quality of the analyses, speed them up and reduce their cost.

Genetic background of sex determination in Herman's turtle (*Testudo hermanni* Gmelin, 1789)

M. Zorc[1], M. Stvarnik[2], A. Dovc[2] and P. Dovc[1]
[1]*University of Ljubljana, Biotechnical Faculty, Department of Animal Science, Jamnikarjeva 101, 1000, Slovenia,* [2]*University of LJubljana,Veteriary Faculty, Gerbiceva 63, 1000 Ljubljana, Slovenia; peter.dovc@bf.uni-lj.si*

Determination of sex is a trait with very variable genetic background in different taxa. In some taxonomic groups the existence of sex chromosomes defines sex. The other mechanism of genetic sex determination represent so called sex loci which are located on autosomes and those species do not have specific sex chromosomes. The third possibility is predominantly environmental (temperature) dependent sex determination with some genetic elements, which are under environmental control. For *Testudo hermanni* is exact mechanism of sex determination not known yet. However, the breeders know that certain temperature profiles during the hatching period favour development of one or the other sex. In spite of this fact, there is a possibility that in *T. hermanni* also a genetic component is involved in sex differentiation. We performed the whole genome sequencing of *T. hermanni* sex pools, composed of five sexually mature male and female animals in each pool. We obtained 83 Mio reads for the female and 72 Mio reads for the male genome and about 90% of all reads in both sexes could be aligned using BWA-MEM Read Aligner to the *Gotherus evgoodei* (rGopEvg_v1.p) genome assembly which was used as a reference. The sequence data for each pool reached the 4 fold genome coverage of about 90% of the genome. The analysis of both genome assemblies did not reveal any large structural difference among male and female DNA pool, suggesting that in the *T. hermanni* genome no sex chromosomes are present. In the literature we found 37 potential candidate genes associated with sex determination in lizards and turtles (HoxA1 – HoxA13, Dmrt1, Wt1, Sox9, Sf1, Wntr4, NR0B1, MIs12, MIs18A, Amh, Ar, Est1, ZfPM2, Gata4, Igf1, Insr, Lhx9, beta-catenin, Foxl2, Fst, Nf-kB, Crabp2, RARA, Wnt1 and CIRBP). The search for bigger structural differences In the vicinity of these genes and comparison of both sex specific DNA pools revealed putative duplication of the exon 1 in the Foxl2 gene and duplication of the exon 7 in the RARA gene in the male DNA pool. These findings have to be confirmed by resequencing of target genomic regions in both sexes.

Versatile robust low cost genomic DNA extraction solution for use across multiple sample types
Q. Hoang, L. Manley, C. Gunter, M.A. Patil, S. Chadaram, R. Conrad and T. Lei
Thermo Fisher Scientific, AgriBusiness, Life Sciences Solutions, 2130 Woodward Street, 78744, Austin, TX, USA; srinivas.
chadaram@thermofisher.com

Obtaining a low cost and robust method for high throughput sample preparation upstream of genomic platforms is important for laboratories that need to run hundreds or thousands of samples a day. Many laboratories need to run a diverse set of sample types such as blood, hair, semen, and tissue. Low cost workflows often involve a crude lysate that can lead to poor results. The MagMAX™ CORE AgGenomic DNA Extraction Kit was designed to be a robust genetic DNA extraction kit that works with a diverse range of sample types yielding DNA that is suitable across multiple genomic platforms. This kit uses magnetic beads in conjunction with the KingFisher™ Flex Purification System to extract DNA. The total processing time from sample to purified DNA is around 1 hour. Here, we examine DNA isolated from bovine blood, blood cards, raw and extended semen, ear notch, and hair follicles isolated using the MagMAX™ CORE AgGenomic DNA Extraction Kit compared to a more expensive on market magnetic bead-based isolation kit. For down stream applications we tested capillary electrophoresis on an ABI 3500 Genetic Analyzer, Applied Biosystems™ Axiom™ Genotyping Arrays, and targeted GBS with AgriSeqTM HTS Library kits on an Ion GeneStudioTM 5S. The data shows the MagMAX™ CORE AgGenomic DNA Extraction Kit was able to extract DNA from all samples types tested and is compatible with all the genetic platforms tested.

Employing trio information to assess CNV detection performance in array data of Göttingen Minipigs
C. Reimer[1,2], J. Geibel[1,2], T. Pook[1,2], S. Weigend[1,3] and H. Simianer[1,2]
[1]University of Goettingen, Center for Integrated Breeding Research, Albrecht-Thaer-Weg 3, 37075 Göttingen, Germany,
[2]University of Goettingen, Animal Breeding and Genetics Group, Albrecht-Thaer-Weg 3, 37075 Göttingen, Germany,
[3]Friedrich-Loeffler-Insitut, Institute of Farm Animal Genetics, Höltystraße 10, 31535 Neustadt-Mariensee, Germany;
creimer@gwdg.de

Göttingen Minipigs (GMP) are one of smallest existing pig breeds and a prominent model in laboratory and conservation breeding research. To date, several studies using array and sequence derived biallelic SNPs have been conducted, but larger structural variation, such as copy number variations (CNVs) has been neglected so far. Although it is known that such variation plays an important role for the constitution of phenotypical traits, genetic disorders and genetic diversity, the discovery and validation remains challenging, since results might be afflicted with high false-positive rates, which holds especially for non-model species. One strategy to alleviate these difficulties is joint calling of trio samples. We used array data of Göttingen Minipigs originating from the Relliehausen colony, which is owned by the University of Göttingen. The animals were genotyped with the Affymetrix Axiom Porcine Genotyping Array, which contains 658,692 biallelic SNP markers. Individual samples were chosen to gain in total information from 68 Sire-Dam-Offspring trios, with 99 individual pigs involved. There were 12 sires and 21 dams, of which 1 sire and 1 dam served also as offspring in another trio. Basic genotype data were processed using the Affymetrix Power Tools APT 2.11.1. CNVs were called using the PennCNV-Affy (version 1.0.5) 'Joint'-algorithm with population-B-allele frequencies estimated from the signal files and the hhall.hmm model file. Raw autosomal CNVs were filtered for a minimum length of 10 SNPs and 10 Mb maximum length. We detected an average total CNV length of 24.8 Mb per animal and an average CNV length of 125 kb. Based on the known relationships within trios, resemblance between animals could be used as an additional criterion to discard false-positively detected CNVs, resulting in a reliable set of highly specific CNV calls for further analysis.

Genetic and genomic features for Russian Simmental cattle different ancestry derived by SNP data

A.A. Sermyagin[1], L.P. Ignatieva[1], A.V. Dotsev[1], A.A. Zimina[1], G. Mészáros[2], J. Soelkner[2], G. Brem[1,3] and N.A. Zinovieva[1]
[1]Federal Science Center for Animal Husbandry named after Academy Member L.K. Ernst, Podolsk, Dubrovitsy 60, 142132, Moscow, Russian Federation, [2]University of Natural Resources and Life Sciences, Gregor Mendel Str. 33, A-1180 Vienna, Austria, [3]Institute of Animal Breeding and Genetics, University of Veterinary Medicine, Veterinärplatz 1, A-1210 Vienna, Austria; alex_sermyagin85@mail.ru

The population of Simmental cattle in Russia has a third place across country for breeding resources. The aim of this study was to clarify a genetic relatedness for Simmental cattle Russian and Western Europe origin in connection with population structure using SNP data and possibilities of genomic selection. For genotyping we used 50K and 150K BeadCheaps. After quality control were taken into analysis 37,143 SNPs. The training dataset for genomic data included 268 bulls: Russian Simmental ancestry (RUS,35), Western Simmental ancestry (Austrian and German, FLV,130), Sychev breed (SCH,29), Russian Red Spotted breed (RSB, 58) and Holstein as the outgroup (HOL, 16). BLUPF90, Plink 1.90, SplitsTree 4.14.6, R-package StAMPP software were used. By Animal Model approach for pedigree and offspring data, EBVs of 669 sires revealed the highest ranking in HOL (+57.0 kg of milk), followed by RUS/FLV (-15.4 kg) and lowest in RSB (-69.4 kg) for assessing at Simmental herds. The heritability for milk traits ranged from 0.140 to 0.194, for days open 0.061, for body weight 0.073. MDS and Neighbor-Net analyses revealed the clear clustering by ancestry. More distant from HOL was RUS (Fst=0.129), further FLV (Fst=0.127), SCH (Fst=0.112) and RSB (Fst=0.057). Simmental groups of animals formed the common cluster with closest distances between RUS, FLV and SCH (Fst=0.023-0.029). For RSB it was more (Fst=0.040-0.056). GWAS analysis proved significant associations with p-value 10-4-10-6 on chromosomes for milk yield (BTA3, 21, 29), fat (BTA8,9,14,17,20) and protein percentage (BTA 3,5,6,15,24). LD analysis showed (ranged by chromosomes within 0.139 ... 0.212) that the RSB, SCH, FLV and RUS populations can be used as a consolidated reference group to enhance possibilities for predicting genomic EBVs. Supported by RFBR №17-29-08030 and MSHERF №0445-2019-0027

Gene flow in farmed Red deer populations

R. Kasarda, N. Moravčíková and R. Žídek
Slovak University of Agriculture in Nitra, Tr. A. Hlinku 2, 949 76 Nitra, Slovak Republic; radovan.kasarda@uniag.sk

Nowadays, deer breeding has started to play an important role in European agriculture as an alternative to traditional livestock species. This study aimed to analyse the population structure and gene flow dynamics in farmed Red deer based on the cross-species SNP genotyping. The sample of farmed red deer consisted of 99 animals from 8 populations spread across Central and Western Europe. Genomic DNA extracted from hair roots were genotyped by Bovine 150k array. From totally 139,376 SNP markers, 13.98% were successfully genotyped in at least 90% of animals with call rate higher than 10% and minor allele frequency higher than 1% in the whole population. The degree of genetic differentiation was derived from the discriminant analysis of principal components (DAPC) and Wright's statistics. The genetic relationships among populations analysed were tested by inferring the patterns of populations' splits and migrations events in history. The F_{ST}-based matrix of relationships on inter-population level showed strong genetic connectedness between Lithuanian and Poland red deer populations (F_{ST}=0.003) as well as Hungarian and New Zealand (F_{ST}=0.001). The DAPC analysis revealed that even if the allelic frequency significantly varied across populations and each population can be reliably distinguished the intense gene flow among them is evident. The high level of genetic admixture in the Russian population is most likely results of Red deer import from different European counties in the last decades. Obtained results confirmed that the SNP data under consideration are a strong tool to estimate the intensity of migration events among deer farms as well as to predict their membership probability to a particular population reliably.

LESSy – a modern and flexible tool for genetic evaluation
J. Heise
IT Solutions for Animal Production (vit), Heinrich-Schroeder-Weg 1, 27283 Verden, Germany; johannes.heise@vit.de

Large scale genetic evaluations are still the property of highly specialised software. Their development often began in times of very limited computing capacities (CPU, RAM), which made specialised algorithms like 'iteration on data' necessary. Such algorithm implementations are usually relatively costly to maintain or to migrate to new technologies, e.g. from using CPUs to using GPUs. Nowadays, computing capacities are available at much lower cost and are much less restricted. It is possible to build up the coefficient matrix in RAM explicitly even for large equation systems. Ready-to-use general-purpose software libraries like NumPy, SciPy, and TensorFlow (in Python) are Open Source and make management of parallel computation on CPUs and GPUs relatively easy to handle, without the necessity to costly implement specialised custom algorithms. Such Open Source libraries take advantage of the support of powerful communities and undergo rapid progress. To make maximum use of external technological progress and progress in the field of genetics at the same time, LESSy was initialised: a Linear Equation Solving System for use in large scale genetic evaluations. It is written in Python, a widely used and powerful programming language with access to highly optimised libraries like NumPy, SciPy, pandas, and TensorFlow. The base of LESSy's structural design is its modularity: 1) components are easy to replace without changing code in other components 2) components are as self-contained and easy to understand as possible 3) APIs are kept lean to sustain easy maintenance with growing complexity 4) LESSy relies on standard libraries for the representation of the equation system and can thus use readily available algebraic functions. These characteristics make LESSy powerful for large scale genetic evaluations while allowing for rapid implementation of new scientific results at low cost. LESSy is intended to become a collaborative project in the future.

Combining dairy cattle feed intake data from multiple sources
T. Seleguim Chud[1], C. Baes[1,2], F. Miglior[1], K. Houlahan[1], P. Stothard[3], J. Jamrozik[1,4], G. Kistemaker[4] and F. Schenkel[1]
[1]University of Guelph, 50 Stone Rd E, N1G2W1, Guelph, Canada, [2]University of Bern, Hochschulstrasse 4, 3012 Bern, Switzerland, [3]University of Alberta, 1427 College Plaza, T6G2R3, Edmonton, Canada, [4]Lactanet Canada, 660 Speedvale Ave W, N1K1E5, Guelph, Canada; tchud@uoguelph.ca

The ability of a cow to convert feed into milk, fat, and protein is fundamental in dairy cattle production. Measuring traits to evaluate feed efficiency (FE) is expensive, and consequently incorporating FE traits into genetic selection programs has been a challenge. Merging records from different countries would expand the dataset and potentially increase the accuracy of genetic evaluations for FE traits. The Efficient Dairy Genome Project is an international initiative to combine FE records from six different countries. This study aimed to assess a feasible method for combining dry matter intake (DMI) records for a joint genetic analysis. The data set included 45,525 weekly average DMI records for 5 to 305 DIM from 2,293 first lactation Holstein cows across six research herds located in Canada (Ontario and Alberta), Australia, Switzerland, Denmark, and United States. Two traits for DMI were considered: early lactation (5 to 60 DIM; DMI1) and mid-late lactation (61 to 305 DIM; DMI2). Variance components were estimated using four different bivariate repeatability animal models: Model 1 (M1) and Model 2 (M2) were fit to pre-adjusted phenotypes for fixed systematic effects within country (M1) and to unadjusted phenotypes (M2), which, in both cases, were standardised to a normal distribution across countries and then re-scaled to the mean and standard deviation of the Ontario cows. Model 3 (M3) and Model 4 (M4) were fit to the original phenotypes, assuming heterogeneous (M3) and homogeneous (M4) residual variances. The genetic correlation between DMI1 and DMI2 was lower (0.66 ± 0.05) using M1 than M2, M3 and M4 (all $\sim0.85\pm0.04$). The rank correlation of EBV for DMI2 for the top 5% (1,000 animals) between M1 and M2, M3 or M4 ranged from 0.46 to 0.50. The rank correlations between M2, M3 and M4 were all similar (~0.95). Therefore, the methods for combining data from different countries sizably changed the estimated genetic parameters for DMI. Further investigation is required to confirm these findings and determine the best method.

Accordance between measured and predicted dry matter intake in Italian Holstein Friesian cattle

R. Finocchiaro[1], F. Galluzzo[2], F. Omodei Zorini[2], G. Savoini[2], G. Invernizzi[2], P. Ajmone Marsan[3], P. Bani[3], E. Trevisi[3] and M. Cassandro[4]
[1]Italian Holstein and Jersey Association (ANAFIJ, Research and Development, Via Bergamo 292, 26100 Cremona, Italy, [2]University of Milan, Department of Health, Animal Science and Food Safety 'Carlo Cantoni', Via dell'Università 6, 26100 Lodi, Italy, [3]Università Cattolica del Sacro Cuore, Department of Animal science, Food and Nutrition, Via Emilia Parmense 84, 29122 Piacenza, Italy, [4]University of Padova, Department of Agronomy, Food, Natural resources, Animals and Environment, Via delle Università 16, Agripolis, 35020 Legnaro, Padova, Italy; raffaellafinocchiaro@anafi.it

In order to increase the accuracy of the current official predicted DMI (pDMI) in Italian Holstein Friesian (HF) lactating cows, a total of 352 Italian HF with direct measurements for dry matter intake (DMI) have been considered. Individual measured DMI data were compared with the individual pDMI derived from milk yield, fat content, and predicted cow body weight. DMI data of lactating cows were measured using both alkane method (296 lactating cows-UNICATT) and Roughage Intake Control system (Insentec) (54 lactating cows – Animal Production Research and Teaching Centre Lodi (CZDS) – UNIMI). Average DMI was equal to 21.90±4.22 kg/day whereas mean pDMI was 24.28±2.53 kg/day (UNICATT: 22.25±4.26 kg/day; UNIMI: 20.07±3.5). Pearson correlation between DMI and pDMI was 0.52 (P<0.01). Considering groups separately, correlations were 0.54 (P<0.01) and 0.32 (P<0.02) for UNICATT and UNIMI cows, respectively. Difference between the two groups could be related to several circumstances, including sample size, number of herds considered and DMI assessment method. Overall, results are promising, even if other prediction equations can be implemented. However, this approach can be considered a good starting point to estimate the DMI for HF cattle, at population level. Dry matter intake, being a key component of feed efficiency in dairy cattle, is an interesting new trait to consider in selection scheme for dairy cattle populations. In this regard, an Italian consortium has been created in order to build up feed efficiency critical mass data. Furthermore, this study was intended also to set up future objectives and to streamline data-collection process in future experiments.

Development of targeted GBS panels for breeding and parentage applications in cattle and swine

A. Burrell, P. Siddavatam, M. Swimley, C. Willis, M. De Groot, G. Carvalho and R. Conrad
Thermo Fisher Scientific, 2130 Woodward Street, 78744, USA; goncalo.carvalho@thermofisher.com

Parentage testing and genomics-assisted breeding are critical aspects of successful herd management. Due to its highly accurate and reproducible results, targeted GBS is becoming an increasingly favoured technology for SNP genotyping. With the utilisation of next-generation sequencing, labs can test hundreds of samples across thousands of SNPs simultaneously in a simple high throughput workflow starting from either extracted nucleic acid or crude lysis samples. We developed targeted sequencing panels for both cattle parentage, based on 200 SNP markers selected by the International Society of Animal Genetics (ISAG), and swine breeding using a 1,500 SNP imputation panel. Utilising the AgriSeq HTS Library Kit, a high-throughput targeted amplification and re-sequencing workflow, each panel's performance was tested on >96 diverse cattle and swine DNA samples. Libraries were sequenced on the Ion S5™ using an Ion 540™ chip with genotyping calling generated using the Torrent Variant Caller (TVC) plugin The mean genotype call rate of markers across the samples was >98% for the cattle panel and >96% for the swine panel. Concordance across replicate library preparations and independent sequencing runs was >99.9% for both panels. Panel results were compared with results from a DNA array and the genotype call concordance was >99% with the AgriSeq workflows. The cattle panel was also used on field samples by a Netherland service lab to successfully determine the parentage relationships of 45 calves with 48 potential mother cows. The data demonstrates the utility of the AgriSeq targeted GBS approach for cattle and swine SNP genotyping applications.

A flexible automation solution for genotyping by sequencing to maximise sample throughput

M. Swimley, R.C. Willis, A. Burrell, R. Conrad, G. Carvalho and S. Chadaram
Thermo Fisher Scientific, AgriBusiness, 2130 Woodward Street, 78744, USA; srinivas.chadaram@thermofisher.com

Marker assisted breeding using targeted genotyping by sequencing (GBS) is gaining traction as an effective tool for advanced breeding. We have developed and validated a 1,536-barcode set for multiplexed sequencing using AgriSeq™ targeted GBS technology. While 1,536 barcodes provide a tremendous potential sample throughput, the logistics of handling four 384-well plates of barcoded samples can be arduous and time consuming to perform manually. The use of a traditional liquid handler can reduce hands-on-time, however, the number of tips required for processing large numbers of samples can be negatively impactful, both economically and environmentally. To mitigate this impact, we have incorporated a MANTIS® liquid handler from Formulatrix® into the AgriSeq workflow. The Mantis is a positive air-displacement system that precludes the use of disposable tips and delivers the necessary volumes for repetitive dispersals of enzyme mixes, binding solutions, and washes; saving cases of pipet tips and reducing reagent 'dead' volume while easing operator fatigue and potential for technical errors. Capable of delivering a full 384-well plate's worth of reagents in 5-10 minutes, this non-contact dispenser results in significant time savings as well. We have verified that performing the library preparation protocol with the MANTIS provides results that are equivalent to or better than a purely manual workflow or one using a more traditional liquid handler while lessening our environmental impact through reduction of tips used. Sequencing 1,536 unique libraries on a single chip in less than 3 hours on the Ion GeneStudio S5 Prime system makes AgriSeq targeted GBS technology more efficient and affordable.

***In ovo* strategies to improve resilience through gut health in chickens**

A. Dunislawska, M. Siwek, A. Slawinska, K. Stadnicka and M. Bednarczyk
UTP University of Science and Technology in Bydgoszcz, Department of Animal Biotechnology and Genetics, Mazowiecka 28, 85-084 Bydgoszcz, Poland; aleksandra.dunislawska@utp.edu.pl

Intestinal microbiota play a main role in supporting health and improving production parameters in chickens. Intestinal microbiota contain a complex population of microorganisms whose natural habitats are the mucous membranes of the host organism. In young individuals, the intestinal microbiota influences structure of the intestine and immune system development. In adults, it mainly affects nutrient digestion and metabolism. The microbiota composition is determined by environmental conditions (e.g. diet and medication), and the most dynamic colonisation occurs in the perinatal period. To ensure proper stimulation of the intestinal microbiota in chickens, the bioactive compounds (e.g. prebiotics, probiotics or synbiotics) are used as feed additives. However, the most effective method of stimulating the microbiota is to administer the appropriate bioactive compound during embryonic development. This technique, called *in ovo* stimulation, involves delivery of substance into the air cell on day 12 of egg incubation. This way, the beneficial intestinal microbiota is formed inside the egg, without contact with potential environmental pathogens. Our studies showed the significant effect of *in ovo* administration of prebiotics, probiotics or synbiotics on several levels: microbial composition, improvement of physiological and production traits as well as gene expression. The goal of this paper is to present the effects of bioactive compounds delivered *in ovo* on day 12 of egg incubation on the gene expression (at the mRNA and protein level) associated with immune or metabolic traits. Further, the mechanisms of gene expression silencing (epigenetic regulation) will be discussed. The developing intestinal microbiota interacts with the digestive and immune systems of the chicken embryo. The mechanisms of these interactions were determined at transcriptomic, proteomic and epigenetic levels.

The influence of genetics and diet on the microbiome of pigs

S. Vigors, J.V. O'Doherty and T. Sweeney
University College Dublin, Belfield, D4, Ireland; staffordvigors1@ucd.ie

The global population is continuing to grow at a rapid rate leading to increased need for meat and milk products. The pig industry will need to adapt and expand to meet this demand. While this is an opportunity for the industry, it comes with the caveat that expansion must take place with increased financial, environmental and legislative constraints. Improving the productivity of the pig industry is therefore vital in ensuring sustainable growth in the future. The use of dietary additives and the identification of pigs with improved feed efficiency are two means to improve animal productivity. While the mechanisms influencing animal performance through genetic selection and dietary intervention are multi-factorial, one factor that warrants further exploration is the influence of the gut microbiome. The gut microbiome is intrinsically linked to animal productivity, through its impact on nutrient digestion and animal health. By using dietary additives, we potentially can alter the intestinal microbiome to improve animal health and performance. With increased legislative and environmental constraints through the ban on antibiotics and the phasing out of zinc-oxide, there is a need to identify viable natural alternatives to increase productivity particularly in the post weaning period. A wealth of chemodiversity exists in nature in plants and animals developing protective molecules to survive in varying complex biosystems. Our research has established the potential of feeding naturally sustainable bioactives from marine flora and fauna to modify the intestinal bacterial population to improve animal performance and health. While diet is a means to alter the microbiome, the influence of host genetics on the microbiome is not well established. Our research has identified that while the microbiome of pigs is significantly influenced by the environment in which the animal was born, a microbial signature indicative of feed efficiency is still identifiable. These findings suggest that the use of microbial biomarkers could potentially be used a means to identify pigs with improved feed efficiency and improve the productivity of the pig industry.

Improving the robustness of piglets by early neonatal probiotic supplementation

D. Luise, P. Bosi, F. Correa, E. Spinelli, C. Salvarani and P. Trevisi
University of Bologna, Department of Agricultural and Food Science, viale G Fanin 44, 40127, Italy; diana.luise2@unibo.it

The increasing genetic selection for prolific sows decreased the robustness and increased the mortality of their offspring, raising the use of antibiotic use in early life phases. The aim of the present study was to evaluate the effect of an early probiotic supply to improve piglet robustness and their gut health until weaning. Sixty-four litters (820 piglets) were allocated to 4 different groups (205 piglets/group): (1) Control group (CO); (2) *Saccharomyces* group (SA: supplied with 4 ml of 1×10^{10} cfu of *Saccharomyces cerevisiae* var. *boulardii* CNCM-1079; (3) *Enterococcus* group (EF: supplied with 4 ml of 1×10^{10} cfu of *Enterococcus faecium lactiferm* WS200); and (4) a mix of the two probiotics at the same doses (SAEF). Within 48 h from birth, piglets received a single oral dose of the respective probiotic solution or of sterile water. Piglet weights were recorded at the day of the supply (d0), at d7 and d18 of life. Faeces (18 piglets/group) were collected at d7 and d18 for microbial profile analysis by sequencing of v3-v4 regions of 16S rRNA gene using the Illumina MiSeq platform. Data on piglet performance and mortality were arranged in a factorial way with a 2×2 design using a GLM or a GENMOD procedure with a binomial distribution and a Logit link for ADG and mortality rates, respectively. Microbiota sequencing data were analysed using DADA2 pipeline. SA and EC improved the piglet ADG from d7 to d18 ($P<0.0001$) and from d0 to d18 ($P<0.05$). Orthogonal contrasts showed a significant increase of piglet ADG in SA and EF groups compared to CO group (ADG d7-d18 g/day: CO:212; SA:246: EF:251; $P<0.001$; ADG d0-d18 g/day: CO:204; SA:225: EF:218; $P<0.05$) while no effect was observed in the comparison between SAEF and CO groups. Piglet mortality between d0-d7 was not different and tended to be reduced between d7 and d18 in the SA group compared with the CO group (CO: 3.93 vs SA: 1.44, $P=0.08$). These results highlight that early probiotic intervention can support piglet robustness later in life and may contribute to helping piglets face the weaning transition.

Feeding milk replacer with added starch promotes changes in gut maturation in piglets at weaning

C. Amdi[1], M.L.M. Pedersen[2], A.R. Williams[1] and T. Thymann[1]
[1]*University of Copenhagen, Department of Veterinary and Animal Sciences, Grønnegårdsvej 2, 1870 Frederiksberg, Denmark,* [2]*SEGES, Danish Pig Research Centre, Axeltorv 3, 1609 Copenhagen, Denmark; ca@sund.ku.dk*

In Denmark, hyperprolific sows often rear more piglets than they have teats, and in order to accommodate their needs, piglets are often supplemented with milk replacer. The aim of this study was to investigate the effect of a milk replacer containing a high level of wheat flour on growth, gut enzyme activity and immune function of piglets compared to a pure milk-based diet. The hypothesis was that adding wheat flour would induce maturation of the mucosa, as measured by higher digestive activity and improved integrity. To test this hypothesis, piglets were removed from the sow at day 3 and fed either a pure milk replacer diet (MILK) or a milk replacer diet with increasing levels of wheat (WHEAT). The piglets given milk replacer with wheat underwent several changes suggestive of a pre-weaned situation, including increased enzyme activity in the WHEAT group compared with the MILK group. No detrimental effects on the systemic physiology of the pig were observed, with unchanged gut morphology, histopathology and gene expression between the groups. Collectively, the data suggest that the piglets given gradually increasing amounts of wheat in their milk replacer, do not experience detrimental health effects, and instead may have an accelerated production of digestive enzymes. In perspective this may improve their response to weaning. Indeed, the data suggest that the piglets may be undergoing pre-weaning adaptations that, from a practical point of view, could potentially make the weaning transition easier for the animal.

Supplementation of exogenous enzymes in pig diets to improve growth performance and digestibility

E. Aranda[1], L.E. Robles-Jimenez[1], J. Osorio Avalos[1], E. Vargas-Bello-Pérez[2] and M. Gonzalez Ronquillo[1]
[1]*Universidad Autonoma del Estado de Mexico, Facultad de Medicina Veterinaria y Zootecnia, Animal Nutrition, Instituto Literario 100, 50000, Toluca, Mexico,* [2]*University of Copenhagen, Faculty of Health and Medical Sciences, Department of Veterinary and Animal Sciences, Grønnegårdsvej 3, 1870 Frederiksberg C, Denmark; evargasb@sund.ku.dk*

The aim of this study was to perform an analysis comparing different data published on the effects of the use of exogenous enzymes on the performance in growing-finishing pigs, in order to evaluate feed intake, digestibility and N excretion. A database was created from articles where the usage of exogenous enzymes was specified. Publications were obtained from searches on Scopus, Web of Science, Google Academic, Elsevier, ResearchGate, BMC Research, PubMed, and Redalyc. A total of 118 research studies were identified. Treatments were defined as basal diet (Control), basal diet plus the inclusion of an exogenous enzymes product with phytases, proteases, mannanes, xilanases and a mixture of two or more enzymes products (combo). The variables included average daily gain (ADG, kg/d), feed conversion ratio (FCR), average daily feed intake (ADFI, kg/d), dry matter digestibility (DMd %), and enzyme dosage (mg/kg, g/d). Least-squares means of the control and experimental group and variability measures (standard deviation) were considered. Data were analysed in a completed randomised design, the estimation of fixed and random effects was calculated as Dersimonian and Laird (1986), and turkey test was considered ($P<0.05$). ADG (kg/d) was higher with phytase and mannanes supplementation compared with control diet. DMd was higher with phytase supplementation ($85\pm2.38\%$) and mannanes ($83\pm1.44\%$) supplementation compared with control diet ($79\pm1.12\%$). Inclusion of xilanases increased FG, without affects on FCR and DMd. The inclusion of proteases, and combo enzymes did not affect ($P>0.05$) FCR, ADG, and DMd compared with the control diet. Overall, the inclusion of Phytases (2.12 ± 4.56 g/d) and mannanes (1.07 ± 0.49 g/d) improve ADG, and DMd in growing finishing pigs.

Effect of nucleotide supplementation on growth performance and microbiota profile of piglets
F. Correa, D. Luise, C. Salvarani, P. Bosi and P. Trevisi
University of Bologna, Department of Agricultural and Food Science, Viale G. Fanin, 44, 40127, Italy;
federico.correa2@unibo.it

Nucleotides represent a group of bioactive compounds known to be important for the development of the gastrointestinal tract and immune function. Sow colostrum is a good source of nucleotides, but their concentration tends to decrease in milk during lactation. Vaccination is a routine practice in this phase that can cause an increase of nucleotide needs in order to face the immunostimulation. The aim of this study was to evaluate the short-term effect of oral administration of nucleotides before and after weaning on growth performance, health, immunity and the microbiota profile of weaned piglets preliminarily vaccinated against colibacillosis. A total of 98 piglets, genetically susceptible to E.coli F4 infection, were divided in two groups, one orally administrated 4 ml of a solution containing 100 mg of a product based on nucleotides (SwineMOD®, NU) at 10, 15, 18, 21, 27 days of life and a control group (CO). Blood and faecal samples were collected at weaning (26 days of life), and at the end of the trial (12 days post-weaning). In these two time-points, 16 piglets per each time were slaughtered and individual samples of jejunal Peyer's patches were collected. There were no differences in growth performance between the two groups. Piglets supplemented with nucleotides had a higher level of haemoglobin and haematocrit at weaning and at the end of trial. At weaning, the microbiota profile of piglets of the NU group was characterised by a higher abundance of Campylobacteraceae (widely present in faeces of growing pigs), compared to CO. On the other hand, CO was characterised by a higher abundance of Streptococcaceae (predominant in the suckling phase). These differences disappeared at the end of trial. This suggests an earlier maturation of piglet microbiota of the NU group, than in CO. In conclusion the nucleotide supplementation did not influence the growth performance of piglets but could have explicated a positive effect on pig microbiota anticipating its maturation at weaning. There is also an evident effect on the haemoglobin concentration, this data could be better interpreted once the mRNAseq data of blood and Peyer's patches will be available.

Identification of intestinal and faecal microbiome as biomarkers for social stress in pigs
T. Nguyen[1], K. Rutherford[1], M. Auffret[1], G. Simm[2], R. Dewhurst[1], E. Baima[3] and R. Roehe[1]
[1]Scotland's Rural College, Easter Bush Campus, EH25 9RG Edinburgh, United Kingdom, [2]University of Edinburgh, Easter Bush Campus, EH25 9RG Edinburgh, United Kingdom, [3]Zoetis Inc, 10 Sylvan Way, NJ 07054 Parsippany-Troy Hills, USA; rainer.roehe@sruc.ac.uk

Social stress can have a substantial impact on porcine health, wellness and performance with major commercial consequences. Since social stress is difficult to record, biomarkers based on the intestinal or faecal microbiomes are of great interest. A porcine stress model was applied to 19 pigs using repeated re-grouping and reduced space allowance during the last four weeks of the growing finishing period and compared to 19 untreated control pigs. After sedation and slaughter of pigs in the experimental unit, intestinal luminal contents were collected from the caecum, mid-colon and faeces. Total DNA was extracted from intestinal content and faecal samples and amplicons of the V4 region of 16S rRNA were prepared for sequencing using Illumina MiSeq. Raw Illumina reads were annotated with MG-RAST using the SILVA database. The application of the porcine stressors resulted in significant changes in social behaviour (elevated aggression ~450 times the control mean), skin lesions (three-fold higher for the stressed group), salivary cortisol concentration (which increased with time on trial indicating a chronic stress status) and growth performance (16% reduction due to stress) and appetite (12% reduced feed intake under stress). The social stress resulted in numerous significant changes within the microbiota, e.g. enrichment of opportunistic pathogens *Campylobacter* and *Clostridium* and depletion of *Faecalibacteria* and *Prevotella* (which are also reduced in humans with mental disorders and are abundant in healthy microbiomes). The microbiota in the colon and faeces were substantially more informative than those in the caecum as biomarkers associated with stress. The results agree with the concept of the microbiota-gut-brain axis as indicated by increased hypothalamic–pituitary–adrenal axis activity (cortisol concentration) due to social stress and an associated change in microbiota composition which besides other objectives, such as identification of resilience to pathogens, can be used as an effective biomarker for stress.

Effects of enriched housing on gut microbiome in piglets

D. Schokker[1], C. Wen[2], H. Woelders[1], H. Smidt[2] and I. Van Dixhoorn[3]
[1]Wageningen Livestock Research, Animal Breeding and Genomics, Droevendaalsesteeg 1, 6708 PB Wageningen, the Netherlands, [2]Wageningen University, Microbiology, Stippeneng 4, 6708 WE Wageningen, the Netherlands, [3]Wageningen Livestock Research, Animal Health & Welfare, De Elst 1, 6708 WD Wageningen, the Netherlands; dirkjan.schokker@wur.nl

A previous study showed that enriched housing increased disease resilience of piglets. The present study assessed whether disease resilience could be associated with housing-induced changes in faecal microbiota composition. We studied immunological parameters and faecal microbiota of piglets at weeks 2, 4, 5, 7, and 9 after birth, and of the lumen of jejunum, ileum, and colon at week 9 by 16S rRNA gene sequencing. Piglets were either housed in an enriched (E) or standard barren (B) pen, where E provided double space and rooting substrate of peat and wood shavings. Piglets were weaned after sampling at week 4. A significant difference (P<0.01) in faecal microbiota composition was observed at week 2 using principal response curve analysis (PRC) based on pairwise Unweighted UniFrac distances between samples at genus level for the evolution of microbial composition under the influence of enriched housing during the investigated period. The PRC model revealed a significantly higher relative abundance of several genus-level taxa in E housed piglets, including: at week 2 *[Ruminococcus]_gauvreauii_group* (P<0.01) and *Christensenellaceae_R-7_group* (P=0.0035), at week 4 *Lachnospiraceae_g_uncultured* (P<0.01), at week 5 *Catenibacterium* (P=0.01). Whereas in B housed piglets a significantly higher relative abundance of *Escherichia-Shigella* (P<0.05) was observed at week 5 and *Enterococcus* at week 2 (P=0.05). At week 9, the ileal microbiota showed distinct grouping of B and E pigs when using Principal Coordinate Analysis of weighted UniFrac and Bray-Curtis distances (P<0.05). Moreover, inter-individual variation for ileum and colon communities was clearly lower in E than in B (P<0.01). However, no difference was observed between E and B piglets in alpha-diversity at any of the time points for faeces and gut luminal contents. We conclude that there are differences in gut microbiome between housing groups, but future work is needed to describe possible causality and the link of the microbiome with performance, behavioural, and immunological variables.

Supplementation of SCFA and MCFA to the diet of growing pigs stimulates growth performance

M.M.J. Van Riet[1,2], P. Bikker[3], S. Vermaut[1] and I. Peeters[1]
[1]Eastman Chemical Company, Technologiepark 21, Zwijnaarde, Belgium, [2]Ghent University, Heidestraat 19, Merelbeke, Belgium, [3]Wageningen University & Research, De Elst 1, Wageningen, the Netherlands; sabien.vermaut@eastman.com

The combination of short chain fatty acids (SCFA) and medium chain fatty acids (MCFA) supplemented to the feed may positively contribute to the growth performance of weaned pigs, as this combination covers a broad antimicrobial spectrum by targeting both pathogenic gram-negative bacteria (by SCFA) and gram-positive bacteria (by MCFA). However, the effects are less well known for growing pigs. The objective of this in-vivo study was to assess the efficacy and acid regulatory effects of SCFA (formic + lactic acid) and SCFA+MCFA (C8/C10) on performance of growing pigs. A total of 432 growing pigs (entire male and female, Tempo × (Dutch landrace × Great Yorkshire)), 9 weeks of age (21.9±3.2 kg bodyweight), were selected from three batches and randomly allocated to 3 dietary treatments; negative control (NC), SCFA blend, SCFA+MCFA blend, all with 0.5% inclusion level. Pigs were housed in a conventional housing system with 12 pigs per pen and in total 12 replicates (pens) per treatment. A cereal grain-soybean meal-based pelleted grower diet was fed for 40 days according to the nutrient requirements of pigs in this period. The bodyweight and feed intake were recorded, as well as faecal consistency, and ADG and FCR were calculated accordingly. The data were analysed using a general mixed model for growth performance and a generalised linear mixed model with binomial distribution for faecal consistency. Inclusion of 0.5% SCFA+MCFA positively supported the ADG and ADFI in the first three weeks of the study compared to the negative control diet. Growth performance for the SCFA blend was intermediate between the NC and SCFA+MCFA diets. The effect of SCFA+MCFA was most consistent throughout the study and this may suggest a beneficial effect of the additional MCFA supplementation. Faecal consistency was not significantly different between dietary treatments groups. In conclusion, dietary SCFA+MCFA supplementation contributed most to the growth performance of growing pigs.

Poster presentations and discussion

S. Vigors[1] and P. Trevisi[2]
[1]University College Dublin, Belfield, D4, Ireland, [2]University of Bologna, Dipartimento di Scienze e Tecnologie Agro-Alimentari Viale Fanin 46, Bologna, 40126, Italy; staffordvigors1@ucd.ie

Time slot for posters

In vitro large intestinal fermentation of growing Iberian pigs under heat stress

A. Francisco[1], I. Fernández-Fígares[2], R. Campos[3], Z. Pardo[2], A. Martín[1,4], A. Costas[1], M. Lachica[2], I. Mateos[1,4], F.J. Giráldéz[1,4] and M.J. Ranilla[4]
[1]Universidad de León, Producción Animal, Campus Vegazana, 24007 Leon, Spain, [2]CSIC, Estacion Experimental del Zaidin, Profesor Albareda 1, 18008 Granada, Spain, [3]Universidad Nacional de Colombia, Dep. Ciencia Animal, Carrera 45, 111321 Bogota, Colombia, [4]CSIC, Instituto de Ganadería de Montaña, Ctra. Leon-Vega de Infanzones, 24346 Grulleros, Spain; ifigares@eez.csic.es

Changes in large intestine fermentation may be expected in heat stressed (HS) animals. The aim of the investigation was to study *in vitro* intestine fermentation differences in growing Iberian pigs under control (22 °C, 30 d; TN) or HS (30 °C, 30 d) conditions. Sixteen Iberian barrows (44 kg) were assigned to HS or TN pair-fed group. The diet was barley-soy bean meal based covering all nutrient requirements. Pigs were slaughtered at 60 kg and rectum content collected and kept at -80 °C until the fermentation experiment. We set up an *in vitro* batch culture method to assess fermentability of ingredients for pig diets. Pectin and starch were fermented *in vitro* in a faecal slurry consisting of an anaerobic culture medium, salts and faeces (5%, $P<0.05$) from pigs under HS or TN conditions. After 24 h of fermentation, production of gas, VFA and NH_3 concentrations were measured. Heat stress increased total VFA, propionate, butyrate production (13%-17%, $P<0.05$) and gas production (8%, $P<0.05$) compared to TN pigs when starch was used as substrate. Furthermore, there was a decrease in acetate production (17%, $P<0.05$), acetate:propionate ratio (32%) and valerate molar proportion (31%, $P<0.05$). When pectin was fermented, heat stress similarly increased total VFA, acetate and propionate production (12, 11 and 47%, $P<0.05$), propionate molar proportion (32%, $P<0.05$), gas production and NH_3 concentration (10 and 18%, $P<0.05$). Additionally, there was a decrease in valerate production and molar proportion (51 and 56%, $P<0.05$) and butyrate and valerate molar proportions (17 and 56%, $P<0.05$). Moreover, acetate:propionate ratio and valerate molar proportion were decreased (33 and 31%, $P<0.05$). Large intestine fermentation of Iberian pigs under HS seems to perform better than in TN conditions using starch or pectin as substrates.

Omics techniques for the analysis of the gut-liver axis after *in ovo* delivery of synbiotic

A. Dunislawska[1], A. Herosimczyk[2], A. Lepczynski[2], M. Ozgo[2] and M. Siwek[1]
[1]UTP University of Science and Technology in Bydgoszcz, Department of Animal Biotechnology and Genetics, Mazowiecka 28, 85-084 Bydgoszcz, Poland, [2]West Pomeranian University in Szczecin, Department of Physiology, Cytobiology and Proteomics, Janickiego 29, 71-270 Szczecin, Poland; aleksandra.dunislawska@utp.edu.pl

The gut-liver axis is defined as the anatomical and functional interaction between the gastrointestinal tract and liver. This axis strongly mediates the interaction between intestinal microbiota and host metabolism. It is estimated that up to 30% of metabolites in peripheral blood are of bacterial origin. The host microbiota produces various metabolites that are absorbed through the intestinal tissue and into circulation and to the liver. The aim of study was a multi-stage analysis of liver response after earlier stimulation of the chicken embryo with a synbiotic. Eggs on 12 day of incubation were injected with *Lactobacillus plantarum* with raffinose family oligosaccharides. Liver was collected from adult chickens for RNA, DNA and protein isolation. Gene expression at the mRNA level was estimated by microarray analysis. Analysis at the protein level was performed using the 2-DE and MALDI-TOF methods. For the silenced genes, the effect of delivered substance on changes of DNA methylation was determined. Microarray analyses showed that the synbiotic strongly activates the metabolic, signal transduction and cell communication genes. In proteome analysis, 7 proteins were found to be up-regulated and 2 down-regulated in response to stimulation. Literature data suggest that microbiota itself as well as their primary fermentation products may have impact on mitochondrial functions. Based on the microarray data, down-regulated genes were selected for methylation analysis. There was a significant increase in *ANGPTL4* methylation relative to the control group, while the change in the level of mRNA expression was -10.5. ANGPTL4 gene plays an important role in maintaining energetic homeostasis via the regulation of lipoprotein lipase activity. Results explain the mechanism of action of synbiotic delivered *in ovo* directly on the relationship between the intestinal microbiota and the liver, and indirectly on the host organism.

Effect of cowpea and probiotic on broiler performance and caecal microflora populations

M. Dumitru[1,2] and G. Ciurescu[2]
[1]University of Agronomic Science and Veterinary Medicine, Department of Biotechnology, Doctoral School, Blvd. Marasti, no. 59, Sector 1, 011464, Romania, [2]National Research & Development Institute for Animal Biology and Nutrition, Animal nutrition, Calea Bucuresti, no. 1, 077015, Ilfov, Romania; mihaela.dumitru22@yahoo.com

Price and availability of soybean meal (SBM) on global markets is constantly changing, thereby stimulating interest in maximising the use of locally produced protein sources. This study investigated the effect of different levels (0, 10 and 20%) of raw cowpea (CWP; *Vigna unguiculata* [L] Walp) seeds, as a replacement of SBM, with and without probiotic addition on growth performance, digestive organ sizes and caecal pH (0 to 42-d). The impact of the treatments on the caecal microflora population at 42-d was evaluated. A total of 720, unsexed 1-d-old Cobb 500 broilers were divided into 6 groups with 4 replicate pens (30 birds/replicate pen). Data were analysed as a 3×2 factorial arrangement with 3 levels of CWP with and without probiotic (3×10^8 cfu/g feed). The probiotic used was based on *Lactobacillus plantarum* ATCC 8014. Urease activity (pH change) in CWP was not detected. The results showed that CWP at 10 or 20% without probiotic in an optimised diet on digestible amino acid contents maintained broiler growth performance (2712.7-2,696.2 vs 2,718.5 g in the SBM group; P>0.05). The digestive organ sizes (i.e. gizzard, heart, liver, pancreas, small intestine, caecum) and pH of the caecal digesta were not affected (P>0.05) by treatments. Inclusion of LP as a probiotic in broiler diets increased BWG during the overall study period (P<0.001) and improved FCR (P=0.045). LP addition, also was beneficial in modulating gut microflora composition. In particular, the caecal LAB significantly increased (P=0.030; $\log_{10}$ cfu/g of wet digesta), whereas the coliform count decreased (P<0.001) vs with those without probiotic. It is concluded that CWP and probiotic addition had a beneficial effect on broiler growth responses and caecal microflora populations.

N-glycan profiling of chicken intestine as a tool for designing improved feed supplements

H. Kettunen[1], A. Heiskanen[2], A. Olonen[2], T. Satomaa[2], J. Saarinen[2] and J. Vuorenmaa[1]
[1]Hankkija Ltd, Peltokuumolantie 4, 05801 Hyvinkää, Finland, [2]Glykos Finland Ltd, Viikinkaari 6, 00790 Helsinki, Finland; hannele.kettunen@hankkija.fi

Cells of all vertebrates communicate through interactions between short carbohydrate chains referred to as glycans. Host-microbiota interplay in the intestine is also based on these sugar structures. Study of the carbohydrate language of the body is called glycomics. This yet rather uncommon technology will likely increase in importance with the emerging understanding of host-microbiota interactions to human and animal health. Here we used N-glycan profiling for analysing various parts of the intestinal mucosa of broiler chickens with the aim of identifying target structures for pathogens. Three broiler chickens were sampled for tissue from proximal jejunum, distal jejunum, distal ileum and caecum. The samples were thoroughly rinsed with phosphate buffered saline and scraped for mucus and epithelial cells. The protein-bound N-glycans in mucosal lysates were liberated with glycosidase F from *Chryseobacterium meningosepticum* (Calbiochem, USA). The detached glycans were purified and analysed by MALDI-TOF mass spectrometry. The intensity of all N-glycan signals in the mass spectra was measured. Based on their molecular structures, the N-glycans can be divided into neutral and acidic, and further into several types. Both neutral and acidic N-glycans showed similarities and differences between the intestinal compartments. In all samples, the most common neutral N-glycans corresponded to high-mannose type N-glycans, many of which can be seen as housekeeping glycans produced by all tissues. Jejunal and ileal profiles differed especially in the abundance of the high- and low-mannose glycans, while the caecal samples were rich in complex-type glycans. Compared with neutral N-glycans, the acidic N-glycans showed more variation and presence of sialylated structures which may act as pathogen receptors. The mapping of the N-glycan profile of chicken intestine will serve as a foundation for rational design of novel, functional feed supplements from the brewer's yeast *Saccharomyces cerevisiae* or other sustainable raw materials.

Replacement of long straw in fattening lambs

L. Sarramia[1], F. Touitou[1], P. Lamy[2], A. Vermeil[3], F. Enjalbert[1] and A. Meynadier[1]
[1]GenPhySE, INRAE, INPT, ENVT, Université de Toulouse, Castanet-Tolosan, France, [2]OVI-PC, Route Albi, 81340 Valence-d'Albigeois, France, [3]RAGT, Rue Emile Singla, 12000 Rodez, France; annabelle.meynadier@envt.fr

Lamb fattening is a breeding activity with low gross margin. In order to increase this margin, it is of interest to generate the highest value from its co-product: transformation of the skin into luxury leather. That was the aim of the 'Peau'Lux' project regrouping partners from Roquefort lambs sector including OVI-PC and RAGT, and research institutes including INRAE. The present work focused on the replacement of long straw in fattening lamb diets and bedding to reduce skin damage caused by straw, without altering growth. 36 lambs raised indoor and were divided in 3 equal groups: control (C) fattened on straw bedding with long straw and a 8% CF and 36% starch concentrate, and 2 test groups T1 and T2 fattened on plant granule bedding with straw pellets and a 15% CF and 24% starch concentrate, supplemented with yeast for T1 and with yeast+yeast cell walls for T2 (T1 and T2 received the same dose of life yeast). Lambs were weighed and for 30 of them rumen fluid was sampled for ruminal bacteria, fatty acids (FA) and volatile FA (VFA) analysis. FA and VFA analysis were performed with GC. Bacterial DNA was extracted and sequenced using 16s rRNA gene with Miseq technology. Bioinformatic analysis of the sequences were implemented with FROGS pipeline and R Phyloseq package was used to estimate biodiversity indices. Lamb growth was not different between the 3 groups (P=0.88): 357 g/d on average. For VFA, iC4 and iC5 were 78% higher (P<0.01), and for FA, C13:0, C14:0, C15:0ai and C18:1c11+t15 more than 30% higher (P≤0.01) and C18:2n-6 and C18:3n-3 28% lower (P<0.05) in C than in T1 and T2 groups. Concerning bacterial community, α and β diversities were different (P<0.05) in T1, with a higher diversity, compared to C and T2, suggesting that addition of yeast+yeast cell walls was less efficient to prevent the loss of bacterial diversity with acidogenic diet than yeast alone. Moreover our results confirmed a negative link between C2/C3 and final lamb weight, and showed positive links between some bacterial families, like Prevotellaceae and Ruminococcaceae. Our results show that lambs could be fattened without long straw by adapting the concentrate.

Effects of gallic acid on nutrient uptake and intestinal epithelial cell integrity in IPEC-J2

M. Tretola, G. Bee and P. Silacci
Agroscope, Institute for Livestock Sciences, Department of Economic Affairs, Education and Research EAER, La Tioleyre
4, 1725 Posieux, Switzerland; marco.tretola@agroscope.admin.ch

Gallic acid (GA) is a natural phenolic compound found in several fruits and medicinal plants and because of its antioxidant, antimicrobial and anti-inflammatory properties also widely used as dietary supplement. However, little is known about how GA can influence intestinal transport and epithelial cell integrity. Therefore, the aim of this study was to investigate the effects of GA on both intestinal nutrient uptake and tight junction integrity in an *in-vitro* intestinal model. To determine trans-epithelial resistance (TEER) and the absorption of L-arginine and L-lysine, differentiated porcine intestinal columnar epithelial cells (IPEC-J2) were incubated with 5, 25 or 50 µM GA for 24 h in filter inserts and then mounted in Ussing chambers. In addition, IPEC-J2 cells were either lysed to evaluate tight junction integrity determining Claudin-1, Occludin and Zonula Occludens-1 mRNA and protein expression, and to evaluate the cationic amino acid transporter 1 (CAT-1) protein expression. Compared to the untreated IPEC-J2 cells, TEER and Claudin-1 protein abundance were lower (-20% and -40%, respectively; $P<0.05$) in the 50 µM GA treatment. Claudin-1 mRNA expression tended (-20%; $P=0.07$) to be lower in the 50 µM compared to the 5 and 25 µM GA treatment. The L-arginine and L-lysine uptake was greater ($P<0.05$) in the 5 then in the 25 and 50 µM GA treatments. This study showed that low concentrations of GA influenced intestinal uptake of the tested cationic amino acids without detrimental effects on intestinal cell barrier function, whereas at higher concentrations GA tended to permeabilise intestinal epithelium affecting the expression of tight junctions proteins. Further studies will be carried out to investigate the mechanisms involved and to clarify the link with the health-promoting properties exerted by the GA.

Gallic acid affects glutamate and arginine uptake and intestinal epithelial cell integrity *ex vivo*

M. Tretola, P. Silacci and G. Bee
Agroscope, Institute for Livestock Sciences, Department of Economic Affairs, Education and Research EAER, La Tioleyre
4, 1725 Posieux, Switzerland; marco.tretola@agroscope.admin.ch

Beneficial effects of gallic acid (GA), one of the most abundant phenolic compound found in several fruits and medicinal plants, have been already demonstrated. However, how it affects luminal membrane integrity and nutrient uptake across the small intestinal epithelial membrane is not fully investigated. This study aimed to evaluate the acute effects of GA on intestinal epithelial integrity and amino acid uptake across the brush border of porcine intestine. Trans-epithelial resistance (TEER), together with L-glutamate and L-arginine uptake were studied in porcine jejunum segments obtained from 100 kg Swiss Large White pigs and mounted in Ussing chambers in the presence (5, 25 and 50 µM GA for 30 min) or absence of GA. In addition, the jejunum segments were lysed to determine Claudin-1, Occludin, Zonula Occludens-1 and cationic amino acid transporter 1 (CAT-1) protein expression. The 50 µM GA treatment decreased ($P<0.05$) TEER values (-20%) and the Claudin-1 (-30%), Occludin (-60%) and Zonula Occludens-1(-50%) protein abundance, while the 5 µM and 25 µM GA treatment only decreased Claudin-1 protein abundance (-30% and -50%, respectively; $P<0.05$), compared to the untreated segments. In addition, the 25 µM GA treatment increased ($P<0.05$) L-glutamate and L-arginine uptake, the latter confirmed by an increased (+30%; $P<0.05$) protein abundance of CAT-1. This experiment implied that GA improves intestinal uptake of glutamate and arginine at the concentration of 25 µM without detrimental effects on intestinal barrier permeability, but decreases the integrity of the tight junction at higher concentrations *ex-vivo*. Correlations between the observed results and health-related properties of GA will be further investigated.

A randomised control trial of housed dairy cows to explore the effect of overall space allowance

J.S. Thompson[1], R. Robinson[1], C. Hudson[1], N. Bollard[1], K. Woad[1], J. Huxley[2], J. Kaler[1] and M.J. Green[1]
[1]University of Nottingham, School of Veterinary Medicine and Science, LE12 5RD, United Kingdom, [2]Massey University, School of Veterinary Science, Palmerston North, 4442, New Zealand; martin.green@nottingham.ac.uk

Current literature does not provide a clear understanding of the relationship between cow health, welfare productivity, and the housed environment. Of particular note, is the dearth of evidence on living space requirements. To evaluate the impact of living space on cow production and reproduction, including underlying physiological changes. A randomised controlled, long term (364d) study was designed to evaluate a spatial intervention. Cows were matched based on parity and days in milk and randomly allocated to a control group (total: 9 m^2 per cow, GB median) or 'treatment' group (total: 14 m^2 per cow, GB upper 95%ile). All other aspects of the environment were matched between groups and both groups experienced identical management routines. Data collection included all key reproductive events, ultrasound scanning to quantify follicular development, blood sampling to assay anti-Mullerian hormone and milk sampling to assay progesterone. Production data included daily yield per cow, rumination time and bodyweight. Data analysis comprised time to event and mixed effects modelling, conducted using R statistical software. The group with more living space produced significantly more milk; an increase of 366 litres per 305 d lactation. Rumination times were 15 minutes longer per day in cows given more space. There was a reduced time to conception in the control compared to the treatment group (hazard ratio 0.60, P=0.03). Commencement of luteal activity showed no significant difference between groups and no differences were found in time to first service, AMH concentrations, endometritis scores or antral follicle counts. Housed cows produced substantially more milk when provided with the same feed but greater living space. However, the cows with greater living space took significantly longer to conceive than cows in the control group. This is the first study to identify alterations in reproductive and production performance solely attributable to a change in living space allowance in dairy cows.

Automated detection of social behaviour at the feed bunk in dairy cows – the influence of regrouping

N. Melzer[1], J. Langbein[2] and B. Foris[1,3]
[1]Leibniz Institute for Farm Animal Biology, Institute of Genetics and Biometry, Wilhelm-Stahl- Allee 2, 18196 Dummerstorf, Germany, [2]Leibniz Institute of Farm Animal Biology, Institute of Behavioural Physiology, Wilhelm-Stahl-Allee 2, 18196 Dummerstorf, Germany, [3]University of Britisch Columbia, Animal Welfare Program, 2357 Main Mall, V6T 1Z4 Vancouver, Canada; melzer@fbn-dummerstorf.de

Previous work has shown that an algorithm based on electronic feeder data can be used to automatically detect agonistic interactions and the dominance hierarchy in dairy cow groups. However, it is not clear if this approach is applicable when the group composition changes. Our aim was to investigate the reliability of the algorithm when regrouping lactating Holstein cows. We used three established groups (n=11, 12, and 14) and formed two new groups (n=14) by choosing 4, 4, and 6 cows from the established groups. We performed 48 h continuous video analysis considering all agonistic interactions in the whole pen before (T1), directly after (T2), and one week after (T3) regrouping. Additionally, the corresponding data from electronic bins (feeder and drinker) were analysed using the algorithm to detect agonistic replacements. In T1 most agonistic interactions were observed at the feed bunk (>58%) and a minority at the walking alley (<25%). In contrast, in T2 the proportion of agonistic interactions at the walking alley increased to a similar percentage as at the feed bunk, and this pattern was still present in T3. Dominance was calculated using the normalised David's score and Spearman rank correlation was used to determine the associations between dominance hierarchies obtained from video observation and electronic feeder data. The correlation between hierarchies were higher in T1 in contrast to T2 and T3. Moreover, the hierarchies based on electronic feeder data were flat, with small variability between cows, especially during T2. Our results underpin the importance of investigating differences between pen areas for detecting social behaviour. In summary, care has to be taken when calculating dominance after regrouping based on agonistic interactions solely at the feed bunk.

Assessing the thermal comfort of dairy cows housed indoor during high heat periods

B. Fagoo and J. Capdeville
Institut de l'Elevage, 56, avenue Roger Salengro, 62051 Saint Laurent Blangy, France; bertrand.fagoo@idele.fr

The objectives were to develop a decision tool based on heat indexes to characterise the thermal comfort of dairy cows inside the building during hot periods. The HLI (Heat Loaded Index) which combines humidity, air velocity and the impact of solar radiation (black globe temperature) was use to allow an assessment of the atmosphere conditions inside the buildings. Fifteen buildings were observed once either. The visit was planned only if the weather forecast was hot and stable. The data were always collected on the hottest hours of the day. The measurements of four variables were collected: temperature, humidity, air speed and black globe temperature every two meters in length and width. Each parameter was linear interpolated to have values every 0.50 m. Humidity, air speed and black globe temperature were used to calculate HLI. The HLI values were codes in five colour scale (Neutral conditions, light stress, moderate stress, severe stress and extreme stress). Then, a 'thermal' mapping of each monitored buildings was carried out. In the observed buildings, temperature averaged 32.7±0.8 °C, and humidity averaged 43.8±3.1%. The air speed averaged 0.58±0.51 m/s. With fans, we recorded 0.81±0.71 m/s and without fans 0.26±0.21 m/s. The black globe temperature averaged 33.5±1.1 °C, 0.8 °C warmer than the temperature. Depending on the building, the difference was from 0.1 to 1.8 °C. Thermal mapping within each building highlighted a high heterogeneity of climatic comfort zones. The HLI averaged 84.2 with a standard deviation of 3.1, a minimum of 1.12 in one farm and a maximum of 5.04 in a farm equipped by fans. The diagnostic method provides clues on the optimisation of existing buildings. Thermal mapping also helps to objectify the efficiency of specific equipment (fans) meant to improve the thermal comfort in the building. This method is to be popularised with livestock consultants in order to better appreciate the operation of buildings in the summer and then to provide the appropriate advice. Nowadays, reduction of solar radiation is becoming increasingly necessary and when the farmers invest in fans, they have to choose the good equipment to have a climatic comfort in each part of the building.

The effect of age and housing on assessment of animal welfare for unweaned dairy calves

J.J. Hayer, D. Nysar, C. Leubner, C. Heinemann and J. Steinhoff-Wagner
University of Bonn, Institute of Animal Science, Katzenburgweg 7-9, 53115 Bonn, Germany; hayer@uni-bonn.de

While animal welfare assessment protocols for adult cattle have extensively researched, little is currently known about the assessment of animal welfare of dairy calves. One of the most important questions in developing an animal welfare assessment protocol for dairy calves is, if it is possible to use the same protocol for calves independent of age since they may differ greatly in regard to aspects such as health status, physiology of digestion or housing. Therefore, the aim of this study was to determine if selected welfare indicators in unweaned dairy calves are influenced by age and housing. For this purpose, we recorded management factors and the evaluation of farmers by survey, the conditions on site as well as various animal related welfare indicators. A total of 25 indicators (such as health issues, feeding or cleanliness) from 42 dairy farms and 878 calves in western Germany were evaluated. Results were descriptively analysed. Spearman's rank correlations and two linear models with housing and age respectively as fixed factor were conducted with SAS 9.4. The age of the calves correlated highly with housing (r=0.68, P<0.001). Calves were kept in single housing for up to two weeks by half of the farms (50%), while 27.5% of the farms kept them in single housing for two to three weeks and 22.5% for more than three weeks. Overall, 33% of the assessed calves were held in single housings and 67% in groups. The availability of solid feed and water was highly dependent on housing (P<0.001) as 59% of the farmers just provide it in groups. Furthermore, calves in single housing showed significantly more often symptoms of diarrhoea, inflammations around the ear tags and navel as well as less coughing than calves housed in groups (P<0.001, resp.). The bedding (P<0.001) in single housing was cleaner than those in groups as well as claws (P<0.001) and carpal joints (P<0.001). On the other hand, the rating of cleanliness of the whole calf, number of runts and presence of complications after disbudding were not affected by housing or age. Overall, the results of this study indicated that age and housing form of dairy calves needed to be considered in welfare assessment protocols for dairy calves.

Dairy cattle management and housing for warm environments

A. De Vries and I.M.M. Toledo
University of Florida, IFAS Dairy Extension, 2250 Shealy Drive, Gainesville, Florida 32608, USA; devries@ufl.edu

Heat stress negatively affects the performance, health and welfare of dairy cattle. Increasing milk yields are lowering the minimum temperature-humidity index at which heat stress starts to depress milk production. The rise of global temperatures exacerbate the problem. The average dairy cow in the USA experiences 96 heat stress days during the year if not cooled. Experience and research has shown that management and housing may alleviate the effects of warm environments on dairy cattle. For example, dietary interventions such as feeding melatonin and chromium supplementation may reduce the detrimental effects of heat stress on fertility. An effective strategy is altering the timing of feeding to take advantage of the lower night-time temperatures. Some cattle are more heat tolerant than others, so breeding values for thermotolerance are being calculated. Inclusion of the slick-hair gene into high producing Holstein cattle is another approach of making dairy cattle more tolerant of high temperatures. In-vitro produced embryo transfer improves conception rates during heat stress, but cost may be prohibitive. Housing design, including cooling systems, is a primary way to reduce heat stress. Fans and water sprinklers or misters increase evaporative cooling. In the USA, there is continuous development in size, number, and position of fans, as well as schedules for sprinklers. Newer systems are designed to be more energy and water efficient. Tunnel barns and cross-ventilated barns are constructed to help direct airflow to cool cows. Cooling of heifers and dry cows has been receiving more attention in the last decade. Recent research has shown that cows provided relief from thermal stress during the dry period give birth to heavier calves, produce more milk during the next lactation and have better immune response during the transition period. Late gestation heat stress of dairy cows also depresses milk production of their daughters and possibly granddaughters. In the USA, economic losses are estimated to be $87 per cow per year if dry cows were not cooled. We observe that heat abatement strategies including management of feeding, reproduction, genetic selection and housing design are varied and continue to receive greater emphasis in dairy production across the USA.

Farmer and consumer attitudes towards utilising composting bedding materials from cattle barns

M. Klopčič[1], M.E. Waldrop[2], A. Kuipers[3] and J. Roosen[2]
[1]UL, Biotechnical Faculty, Domžale, Slovenia, [2]TUM, School of Management, Freising, Germany, [3]WUR, Livestock Research, Wageningen, the Netherlands; marija.klopcic@bf.uni-lj.si

The lying and walking area in freewalk cattle housing systems consists of organic bedding material, such as wood chips and straw, which is cultivated and aerated to stimulate the composting process. The compost area can also be utilised for food production during the grazing period of the herd. However, it is important to understand how farmers and consumers perceive this multi-functional use of the barn in order to gauge its market potential. The multi-functional use of the freewalk system was demonstrated on the experimental farm Logatec by utilising the compost bedding for fattening pigs and growing vegetables during the grazing period. In a survey across six EU countries, 80 farmers who are familiar with freewalk (FW) and cubicle (CB) housing were questioned about various technical and marketing aspects of the FW housing with composting material and CB housing with slurry using a 7-point scale. Results from consumer focus groups conducted in Austria, Germany, and Slovenia and a corresponding quantitative survey in eight EU countries (n=3,693) were used to assess consumer perception of re-using the compost material from the FW system for food production. Farmers found the composted bedding material to be a significantly better soil improver than slurry; however, the cost of bedding material was seen as the biggest disadvantage of this system. The FW system was also seen as having a higher market potential, delivering high quality manure and dairy products. Most consumer focus group participants had a positive view of re-using the compost for other products, but some concerns were raised regarding using the compost to grow produce that is usually un-cooked such as salad. The quantitative survey results indicate consumers mostly feel safe consuming food products grown with the compost despite participants stating to be somewhat worried about the safety of the food they eat in general. However, more than half of consumers think the compost can be used for non-edible products (e.g. soil, energy) compared to around 25% thinking the compost is suitable for growing produce that is normally un-cooked and for raising animals.

Algorithms for dairy barn layout – the base of calculation models for labour requirement estimation

J. Mačuhová and S. Thurner
Institute for Agricultural Engineering and Animal Husbandry, Vöttinger Str. 36, 85354, Germany;
juliana.macuhova@lfl.bayern.de

Time studies are applied to determine standard times and to create calculation models for estimation of labour requirement for individual dairy farming tasks using particular work routines and technical equipment in dependence of different influencing variables (e.g. herd size and barn design (number of stall rows, stocking density, and number of special need pens)). However, the proper functionality of calculations can be assured only when barn dimensions needed for this (e.g. length of feed bunk and dimensions of forage storage for feeding tasks and dimensions of milking parlour and holding area and distances between pens and parlour for milking tasks) are also calculated appropriately in the calculation models if they are unknown. Therefore, layout algorithms (with decision tree and equations) were developed in MS-Excel for calculation of dairy barn dimensions. The layout algorithms require as an input the herd size and various barn layout parameters (e.g. number of stall rows, free stall dimensions, width of feed alley and alleys between rows, and width of crossovers). The barn layout parameters can be specified in the input fields, or they base on minimum requirements or empirical values from known practical solutions already included in the algorithms (e.g. 1.25 m and 2.5 m for free stall dimensions). As an output, the algorithms calculate at first the number of animals in individual barn areas (according to parameters such as calving interval, the length of dry period, and minimum cows to build a separate group) and then the barn dimensions. So, for example, for a herd size of 100 cows (plus for heifers 3 weeks before calving), the barn area with stalls would be 81.25, 60.00, and 48.75 m long and 11.25, 13.25, and 18.75 m width for 2-, 3-, and 4-row design, resp. Algorithms are primarily used to estimate the size of barn and other areas of interest to calculate the influencing variables for calculation models in order to estimate labour requirement (e.g. the necessary distances to walk or drive during work operations). However, they determine a barn layout with the size of all areas. Therefore, they can be also useful tools for farmers and advisors for planning of a dairy barn.

Livestock integration improves resilience of crop systems

P. Carvalho[1], P. Nunes[1], C. Peterson[2], G. Farias[1], F. Moojen[1], L. Szymczak[1], W. Filho[1], T. Kunrath[1], A. Martins[1], J. Trindade[3], E. De Souza[1], A. De Moraes[1], C. Bremn[1], G. Lemaire[1] and J. Ryschawy[4]
[1]UFRGS, UFRGS, Porto Alegre, Brazil, [2]MAD Agriculture, Bolder, Bolder, USA, [3]IRGA, IRGA, Porto Alegre, Brazil, [4]INP- ENSAT, UMR AGIR, Castanet Tolosan, France; paulocfc@ufrgs.br

Decades of decoupling livestock and crops since the Green Revolution produced very specialised systems that were successful in the intensification of food production but failed to protect the environment. Lack of diversity has been seen as one of the main causes of this failure. Integrated crop-livestock systems (ICLS) are experiencing a renewed interest in many countries not only because they are more diverse than specialised systems, but because they are a rare example of reconciliation between system intensification and environmental quality. However, those qualities alone are no longer enough. The global agenda of future food demand develops in a scenario of uncertainty. Climate models forecast weather variability over the coming decades, so the characteristic of resilience of a food production system will be more crucial than ever. In this context, can ICLS face a future scenario of uncertainty? More specifically, are ICLS more resilient than specialised systems to economic and climatic variability? We use data from long-term experiments in Southern Brazil to investigate the effects of recoupling grazing animals to specialised crop systems. Beef and sheep grazing temperate and tropical pastures in rotation with soybean, maize, rice and occasionally with trees, otherwise known as on-farm or co-located ICLS, are among the diversity of long-term crop-livestock systems we consider. Results in these systems have long been demonstrating that ICLS increase whole-system yield (e.g. in terms of human-digestible protein), increase profitability, and reduce negative impacts on the environment (e.g. by soil and nutrient cycling indicators). Recent data provides evidence that adding a whole new trophic level (i.e. grazing animals) to specialised cropping systems improves resilience of nutrient cycling functions and economic indicators to climatic and market stressors, and promotes the long-term stability of the whole system. We conclude that moderate grazing reduces the chance of crop failure in unfavourable years and benefits the stability of pasture production.

What internal mechanisms make sheep farms more resilient to technical and economic hazards?

M. Benoit[1], F. Joly[1], F. Blanc[1], B. Dumont[1], R. Sabatier[2] and C. Mosnier[1]
[1]INRAE, Ecosocio, UMR Herbivores, 63122 Saint Genes-Champanelle, France, [2]INRAE, TERRA, UR 767, 84000 Avignon, France; frederic.joly@inrae.fr

The sustainability of livestock systems can be assessed through their productive and environmental performances, including the ecosystem services they provide and their resilience to hazards. We used a modelling approach to assess how key performance indicators respond to technical and market hazards in five contrasted meat-sheep farms in France and Ireland. Hazards were associated to three key technical (ewe fertility, prolificacy, lamb mortality) and four economic variables (prices of two types of lambs, concentrate use, energy use). We used a mechanistic model to simulate farm functioning and calculate farm performance over 3,000 iterations (simultaneous random draws with hazards on previously-mentioned seven variables), which allowed us to quantify: (1) the compensatory effects of different types of technical and economic mechanisms that lead to more stable economic performance; and (2) the probability of economic collapse of meat-sheep farms through a diachronic analysis. We showed that the most resilient systems in terms of net income coefficient of variation are those combining low level of inputs with at least two lambing periods per year. The most stable income per worker were observed in the accelerated reproduction system with three lambing periods per year, but its high demand in inputs (concentrates) reduced farm net income, increasing in the same time its coefficient of variation. Overall, variations in technical variables have the larger effects on income variability than those of economic prices. This was related to the specific characteristics of sheep production, with a very high potential prolificacy and highly variable lamb mortality. The short gestation duration of the species is quite compatible with the possibility to implement multi periods lambing systems which buffers the variability of the three technical variables, and enhances farmer's adaptive capability to adjust to fluctuations in ewe fertility. The diachronic analysis confirms the results of income sensitivity to hazards with a 18% probability of succession of 3 years with an average loss of income higher than 3,700 € for the one lambing a year farming system vs a 4% probability for the accelerated farming system (three lambing periods).

A thirty-year analysis of trajectories of evolution of cattle farming systems in Central Pyrenees

E. Muñoz-Ulecia[1], A. Bernués[1], I. Casasús[1], S. Lobón[1], A. Olaizola[2] and D. Martín-Collado[1]
[1]Ctr Invest y Tecnol Agroal Aragon (CITA), IA2 (CITA-Universidad de Zaragoza), Unidad de Producción y Sanidad Animal, Avda. Montañana 930, 50059, Zaragoza, Spain, [2]Universidad de Zaragoza, Departamento de Ciencias Agrarias y del Medio Natural, Miguel Servet 117, 50013, Zaragoza, Spain; emunnozul@cita-aragon.es

Mountain regions are characterised by their complex interrelations between human and environmental systems. In this coupled system, changes induced by political or socioeconomic transformations at different levels generate consequences for mountain livestock systems functioning, resulting in changes in their structure, management and economic performance which may alter their resilience. The analysis of the diversity of current states and past trajectories helps understanding how socioeconomic drivers at different scales affect mountain farm resilience which informs the development of optimal agricultural strategies for future changes. In this study we aimed to determine: (1) the main changes occurred during the last three decades in cattle farming systems in Central Pyrenees in Spain; (2) the main trajectories followed by farms during this period; (3) the socioeconomic drivers of such evolution paths. A constant sample of 50 beef cattle farms of three valleys with different economic development pathways was surveyed in 1990, 2004 and 2018. Trajectories of evolution were analysed using multivariate statistical. Globally, main production orientation in the area changed from mixed dairy-weaned calves to specialised beef systems with on-farm fattening in the first period, and to sucker cattle farms in the second period. Other changes were related to: increase of herd size and grazing period and dependence of subsidies, decrease of the labour input and increase of gross margin per work unit. Besides this general trend, four different trajectories of evolution were identified: three of them specific to each valley and one that was common to all of them. A preliminary discriminant analysis showed that farm trajectories were driven by household size, farmer age and education level, farm dynamism and dependence of subsidies. Our results show that farm resilience is influenced both by internal social drivers at the farm scale and external policy drivers at the EU level.

Learning from traditional societies for the agroecological transition of Western livestock systems
F. Joly and G. Brunschwig
Université Clermont Auvergne, INRAE, VetAgro Sup, UMR Herbivores, 63122 Saint-Genès-Champanelle, France;
gilles.brunschwig@vetagro-sup.fr

The application of the agroecology framework to livestock farming is based on 5 principles: integrated management of animal health (P1), reduction of inputs using ecological processes (P2), closure of cycles to reduce pollution (P3), use agronomical diversity to increase resilience (P4), and preserve biodiversity (environmental and agronomical) by adapting practices (P5). We examine here to which extent traditional livestock farming societies (TLFS) complies with these principles, through 2 contrasted examples. Andeans systems use cattle, alpacas, llamas and sheep to exploit a variety of natural forage resources, including crop residues, along altitudinal gradient from 2,000 to 4,500 m a.s.l. Mongolian systems use goats, sheep, horses, cattle and camels, quasi exclusively fed on natural rangelands. Both systems produce a variety of products, for markets and for subsistence, and both benefit from their diversity to build resilience strategies. In the Andes, manure fertilise crops and exchange can take place with crop farmers. In Mongolia, herders sell some of their resistant camels or horses to rebuild their goat herd, when this latter has suffered severe winter losses. These TLFS comply with 3 agroecological principles as they use few inputs (P2), aims at closing cycles through crop livestock integration (Andes) (P3) and use species diversity to increase resilience (P4). This compliance is however imposed by the difficult climatic conditions, and the low availability of resources. It is also labour-intensive and exposes herders to climate uncertainty. To inspire agro-ecological transition in Western countries, research should address TLFS weaknesses through 4 questions: How could the labour burden needed to exploit the diversity of resources be reduced? Could it integrate farmers traditional knowledge? Which production optimisation should be targeted to preserve resilience? What importance should be given to the breed phenotypes? How could systems anticipate hazard and not only react to them? Could agro-climate prediction models help? What is the attitude of farmers towards such systems? Could they consider it as examples? Can farmers representation of modernity be an obstacle to the replication of such systems in Western context?

Territorial resources mobilisation shapes agroecological transitions in crop-livestock systems
G. Martel[1], V. Thenard[2], J.P. Choisis[3] and M. Moraine[4]
[1]INRAE, UMR 0980 BAGAP, 55 rue Rabelais, 49000 Angers, France, [2]INRAE, UMR AGIR, 24, chemin de Borde-Rouge, 31326 Castanet Tolosan, France, [3]INRAE, UMR SELMET, Avenue Agropolis, 34398 Montpellier, France, [4]INRAE, UMR Innovation, Avenue Agropolis, 34398 Montpellier, France; vincent.thenard@inrae.fr

Recent advances on agroecological transitions are mainly based on concepts and theories and are not grounded to a specific territory, while transitions need specific resources to occur. By exploring 8 farming systems over 4 territories we aimed at underline the role of the diversity of resources in the sustainability of farming systems. We defined four kind of territorial resources: natural (soils, water, ecosystems), technical (equipment, infrastructures, factories), social (knowledge sharing, training, farmers' groups), and economic (supply chains, public policies, local community support). Among the diversity of farming systems within each territory, we select two that are far apart in term of agroecology mobilisation. Both systems combine different resources. But, when agroecological systems rely on resources linked to a proximity area and based on biodiversity, other systems look at resources allowing global exchanges and needing inputs. Farming systems try also to mobilise resources that are connected between them. As an example in Brittany, grassland dairy cow system refer to natural resource 'permanent grassland', on social resources delivering knowledge on permanent grassland management and on economic resource valorising labelled products. In the other hand the diversity of resources mobilised allows duplication of certain knowledge increasing its resilience to resources changes in the territory. As an example in Aveyron grassland dairy sheep systems combine technical knowledge on natural breeding but also integrate the network of local breeders and share their experiment on natural breeding. We conclude that, the larger the diversity of resources is, the easier will be the possibility to find a path to change while remaining resilient.

Assessing sustainability and resilience of a French extensive beef-cattle system

F. Accatino[1], D. Neumeister[2], C. Pinsard[1] and C. Pineau[2]
[1]INRAE, AgroParisTech, Université Paris-Saclay, 16, rue Claude Bernard, 75000, France, [2]French Livestock Institute, 149, rue de Bercy, 75595, France; francesco.accatino@inrae.fr

European livestock systems are increasingly subject to challenges of environmental, social, economic, and institutional type. We applied a framework, developed in the European research project 'Sustainable and Resilience EU Farming Systems (SURE-farm), to assess the resilience of an extensive grassland-based beef cattle system in the Bourbonnais region (France). According to the framework, resilience has three dimensions: robustness (capacity to withstand shocks without changing configuration), adaptability (capacity to change configuration in response to challenges), transformability (capacity to significantly change structure). We performed a series of interviews, participatory workshops and focus groups with local stakeholders with the purpose of making an inventory of challenges and of the strategies applied by actors to face them: these made it possible to assess the performance of the system in the three dimensions of resilience. The main challenges were related to low profitability (economic), difficulty to find successors and public distrust of farming practices (social), recurrent droughts (environmental), and a non-flexible policy (institutional). The strategies applied by actors were mostly focused on enhancing robustness (e.g. insurance schemes and storing feed in face of droughts), less strategies were dedicated to enhancing adaptability (e.g. developing cooperatives and changing practices to fulfil social expectations) and only a few strategies, rarely applied, were dedicated to transformability (e.g. developing grass fattening in the region). Strategies suggested by stakeholders as to be applied in the future were mostly focused on adaptability (e.g. improve the coordination of actors along the value chain). Results of this series of workshops will be used for promoting collective reflection to increase the resilience of the region.

Economic perspectives of small beef cattle farms in mountain regions

V. Angerer[1], S. Kühl[1], U. König Von Borstel[2] and M. Gauly[1]
[1]Free University of Bozen, Universtitätsplatz 1, 39100 Bozen, Italy, [2]Justus-Liebig-Universität Gießen, Department of Animal Breeding and Genetics, Leihgesterner Weg 52, 35392 Gießen, Germany; verena.angerer@unibz.it

For many farmers in South Tyrol, the most Northern Province of the Italian Alps, the production of beef seems to be an interesting income alternative to milk production: especially small-scale mountain farms are often managed as a sideline and therefore require a time-extensive farming system. Moreover, South Tyrol still has a considerably unexploited potential for beef production: currently only about 20% of the total beef demand of the region is met by local producers. The aim of this study are the economic assessment of the two prevailing beef-production systems suckler cow farming and heifer/ox fattening. In a first step, economic data of 68 South Tyrolean beef cattle farms were recorded. In a second step, a detailed economic analysis of 34 farms (20 suckler cow and 14 heifer/ox fattening farms) was performed. All costs and revenues were divided by the average number of beef cattle per farm during 2017. The results reveal that suckler cow farms earn higher mean (±SD) profit per animal (suckler cow farms:1,432±1,824 €/animal; heifer/ox fattening farms: 435±524 €/animal) (P<0.05), but only if subsidies are included in the calculation. If subsidies are not considered, both groups exhibit a negative mean economic outcome per animal (suckler cow farms: -783±2,300 €; heifer/ox fattening farms: -538±519 €). Only 10% of the suckler cow farms and only 5% of the heifer/ox fattening farms earn a positive profit without subsidies. The monetary result further decreases drastically, if also depreciations and forage production are regarded as cost factors. The study proves that currently, beef production is not profitable for small-scale mountain farms since revenues for regionally produced beef are far too low to cover the high production costs. Therefore, cost saving strategies (e.g. collaboration between farmers) and marketing programs are needed to decrease costs on the one hand and to increase prices on the other hand.

Balance between dairy cows' production and supporting ecosystem services in highlands pastures

S. Raniolo, M. Ramanzin, A. Squartini, G. Concheri and E. Sturaro
University of Padova, DAFNAE, viale dell'università 16, 35020 Legnaro PD, Italy; salvatore.raniolo@phd.unipd.it

Mountain pastures are agro-ecosystems depending on livestock farming and providing a large spectrum of services. Nowadays, these systems are facing social, economic and climatic challenges that threaten their equilibrium and externalities. This study aimed at characterising the relationships between grazing livestock, land morphology, and the microbial community of mountain pastures. Four different mountain pastures of Italian eastern Alps, grazed by dairy cows producing milk for PDO cheese, were considered during multiple summer transhumance (from mid-June to mid-September). Herd size (average ± SD) was 74±56 dairy cows, pasture surface 165±77 ha, and elevation 1,990±81 m asl. The study was developed at different spatial scales, with a particular focus on the supporting ecosystem services linked to the nitrogen cycle. At a macro scale, cattle movement was GPS-tracked to estimate individual grazing patterns and map the local intensity of grazing. Daily distance travelled by cows was 8.0±3.8 km. At a micro scale, pasture soils were characterised with a molecular approach to obtain the nitrification (copies of bacterial and archaeal amoA genes) and denitrification potentials (copies of nosZ genes) genes. Slope was the most important morphological variable at both spatial scales, determining animal movement patterns, also between breeds, and the nitrification potential of soils, which increased at lower slopes. In contrast, the denitrification potential (nosZ gene copies) of grazed areas was unaffected by land morphology. In addition, amoA Archaea showed a much higher residual variability than nosZ. The knowledge of how grazing patterns and environmental conditions determine individual productivity and soil supporting ecosystem services can be used to develop good practices for the sustainable management of mountain livestock-grassland systems. Identifying multi-purpose pasture management practices may increase the resilience of grazing agro-ecosystems while maintaining the productivity of local livestock systems.

Using animal diversity to build resilient trajectories: a viability approach

F. Joly[1], R. Sabatier[2], M. Benoit[1] and C. Mosnier[1]
[1]Université Clermont Auvergne, INRAE, VetAgro Sup, UMR Herbivores, Theix, 63122 Saint-Genès-Champanelle, France,
[2]INRAE, UR Ecodéveloppement, 84000 Avignon, France; frederic.joly@inrae.fr

Traditional pastoral societies commonly use a diversity of livestock species to face climate hazards. They use it as portfolios comprising slow growing resistant species, such as camels, that provide a safety net in case of hazards, and less resistant species, such as sheep, with high growth potential. They stabilise their production over time by adjusting the numbers of both types of animal. For this reason, multispecies herding is common in subsistence-based systems exposed to hazards in Asia, Africa or South America. Here we present outputs of a model parametrised for a Mongolian case study. In this country, when forage is scarce, livestock gets vulnerable to cold and snowy winters, which causes severe peaks of mortality. The model built in the framework of the viability theory is able to scan available management options, to identify those ensuring the long-term respect of predefined constraints (income and subsistence consumption of livestock products). To meet this aim, after a mortality peak, the model suggests to sell camels or horses to buy goats. Goats can this way compensate losses quickly and once the herd has recovered, the model will promote the opposite exchange. It thus will reconstitute the pastoralist's safety net and through continuous species adjustments, it will stabilise the multi-species herd production. This management of portfolio of diverse animal traits can inspire agro-ecological transitions in western countries. The modelling framework used in our Mongolian example could help design resilient systems, based on a diversity of animal types. It could be used to assess how the management of different robustness traits of dairy cows can improve the long-term stability of milk production. Robustness can concern the resistance of various biological functions (milk production, reproduction and survival) to diverse hazards (drought or heat wave). Cows with different robustness profiles would replace this way the species of the Mongolian example. Such investigation could contribute to the implementation of nature-based solutions for climate hazard mitigation in a European context.

Added value of local sheep breeds in alpine agroecosystems

M. Orsi, G. Bittante, L. Gallo, M. Ramanzin, S. Raniolo and E. Sturaro
University of Padova, DAFNAE, viale dell'università 16, 35020 Legnaro PD, Italy; enrico.sturaro@unipd.it

Mountain livestock farming systems are facing climatic, social and economic changes, which cause abandonment of marginal and less favoured areas and intensification in the most productive areas. This study is part of a project (Sheep Al.L. Chain, RDP Veneto Region) aiming to improve the competitiveness of local sheep breed farms through the valorisation of their link with mountain agroecosystems. We considered two local sheep breeds of Eastern Italian Alps, 'Lamon 'and 'Alpagota', which have a population of 400 and 2,600 heads, respectively, and are used to produce lambs for typical products. A total of 40 farms (21 for Alpagota, with a total of 1,900 heads; 19 for Lamon, with a total of 395 heads) were surveyed to collect data on farm organisation, flock structure and management (on farm questionnaire), land use management (GIS approach), and value chain organisation (participatory processes). The farms rearing Alpagota sheep have an average flock size of 90 heads, ranging from 4 to 400 sheep, and farmers are associated in a cooperative with a Slow Food Presidium. The farmers with Lamon sheep are mainly smallholders with very small flocks (around 20 sheep per farm), and the breed is at risk of extinction. The link between the two local sheep breeds and mountain agroecosystems is very strong: land use is characterised by a large number of small patches of grasslands (on average 102 patches with an average size of 0.2 ha/farm) with steep slopes (on average 35% of slope for highland pastures). The results of this study were used to develop strategies aiming at improving the competitiveness of these systems. In particular, an application for smartphone has been developed to support mating plans, with the aim to limit the risk of inbreeding. To valorise the link between local breeds and landscape, a 'territorial marketing' strategy has been tested on the sampled farms. Communication to the consumers and to the relevant stakeholders of the added value of local sheep breeds in marginal mountain agroecosystems can contribute to favour the resilience of small ruminant farms. The involvement of farmers (in particular smallholders) in cooperative/multiactor approaches is fundamental for the conservation of Alpine sheep breeds.

Complementary approaches for local breed adaptation: the cases of several French Mediterranean sheep

A. Lauvie[1], M.O. Nozières-Petit[1], L. Perucho[2] and C.H. Moulin[1]
[1]UMR SELMET INRAE CIRAD Supagro Univ Montpellier, France, 2 Place Viala, 34060 Montpellier Cedex 01, France,
[2]INRA (now INRAE) LRDE, Quartier Grossetti, 20250 Corte, France; anne.lauvie@inrae.fr

Local breeds have been long promoted for their adaptation to various constraints and to specific farming systems. Adaptation is particularly important while dealing with resilience of Mediterranean livestock farming systems. The same adaptation notion includes several aspects: (1) specific properties of animals or animal populations that allow them to maintain themselves while exposed to constraints or specific environment, (2) dynamic processes that help animals or animal populations to maintain themselves in an environment. These approaches and their methodologies are rarely put in perspective when dealing with local breeds. Based on several case studies of local sheep breeds in French Mediterranean area (the Corsican, Raïole, Caussenarde des Garrigues and Rouge du Roussillon sheep breeds) we contribute to such a perspective. We show that the main objectives of both approaches are different: animal biological characteristics are at the core of the first approach while human practices and their interrelations with animal behaviour and characteristics are at the core of the second. We also show that hardiness is a core notion of both approaches, but that the first approach involves questioning the multiple characteristics associated to it and the modalities of their objectivisation, while the second approach shows that the notion can be used at different scales including the breed scale (e.g. in farmers discourses to compare breeds, or to mention characteristics taken for granted for an animal of a given breed). Finally, we show that the risk of the first approach is to limit the question to biological characteristics while other dimensions of the local breeds are at play (e.g. role of collective human dynamics around the breed as a criterion of choice of this breed). A difficulty of the second approach is to comprehend how actors deal with this adaptation at a collective level and to consider the diversity of levels at play (not only the genetic selection). We discuss the complementarity of both approaches and underline the need for interdisciplinary and participative approaches to allow such a complementarity.

The environment and the biometry in the Northwest Portuguese sheep breeds

N.V. Brito[1], J.C. Lopes[1], V. Ribeiro[2], R. Dantas[2] and J.V.Leite[2]
[1]IPVC, CISAS, Quinta do Mosteiro, Refoios, 4990-706 Ponte de Lima, Portugal, [2]AMIBA, R. Domingos Marques 40, 4730-264 Lanhas, Portugal; nunobrito@esa.ipvc.pt

The northwest region of Portugal has undergone a profound change in its agricultural structure, with the abandonment of agricultural activity and the increase of the use of autochthonous sheep breeds. These sheep exploited for meat production can belong to Bordaleira (crosswool) Entre Douro e Minho (BEDM breed), usually accompany cattle, whether at work or in the pasture, well-formed and moderately developed or Churra do Minho, small sheep with coarse and long wool, grazed exclusively in herds, which are traditionally kept in turn (Vezeira) by the farmers. The characterisation of this population began in 2001, detecting a morphology and wood characteristics diversity and originating, in 2007, the two breeds Genealogic Books (GB). Studies are proceeding, regarding the genetic, productive and reproductive parameters and the adaptability to different ecosystems. The aim of this study is to evaluate the influence of the biometry and the environment, in particular altitude, in the breed production system. All registered animals, born after 2003, were analysed, in a total of 1,394 females and 51 males of BEDM and 908 females and 16 males of the Churra. Conformation, productivity and mobility measures were estimated, according to the Breeds Manual and data were analysed using the SPSS22.0. Our results present significant differences ($P \leq 0.05$) between sex and breed, with higher values in males and BEDM, relatively to parameters of anterior (height at withers, chest width, body length) and posterior structure (rump length and width). Significant differences were found, regarding breed in functionality (hock alignment, mobility) and sex in posterior structure (rump angle). No significant differences were observed in functionality (lateral hock curvature) and reproductive characteristics (udder insertion height and supernumerary teats). These biometric differences contribute to explain the geographic distribution of breeds, occupying BEDM the valleys, rarely exceeding 700m in altitude and Churra the highest mountains where poor soils predominate, with rocky outcrops and sporadic agricultural crops- scrub, low grass and shrub plants, with low energy levels and poor digestibility.

Camel herd management under pastoral system in southern of Tunisia

N. Letaief[1] and S. Bedhiaf[2]
[1]National Agronomic Institute of Tunisia (INAT), Animal sciences, 43, Charles Nicolle Avenue, 1082 -Tunis- Mahrajene, Tunisia, [2]National Agricultural Research Institute of Tunisia, Animal and Forage Production, Hedi Karray street, 2049 Ariana, Tunisia; neyrouz2008@gmail.com

Camels played an important role in the life of southern Tunisia population with 90% of camel flocks raised in this region. The aim of this study is to describe husbandry and breeding practices, herd structure, traditional veterinary practices and animal production parameters for camels raised under pastoral system in the region. Data were collected from a total of 35 camel owners field surveys within 12 delegations of the governorates of Kebili, Tozeur and Medenine. Result surveys highlighted that the only raised breed camel is the 'Maghrebi' breed which is represented by seven different local ecotypes showing different body measurements and coat colours. According to farmers, animals are mainly kept for meat production while camel milk is rarely commercialised (14%) with an average daily yield of 2.5 litres. Pastoralists practice mobile animal husbandry where camels moved to the desert for grazing. After returning, animals are grouped and identified, vaccinated and received feed supplementation from November to next March which coincide with oestral season 'El hdad'. The she-camel reaches sexual maturity at 3 to 4 years while males at the age of 4 to 5 years. Generally, one stallion can cover 20 to 40 she camels in one season. Regarding the economic importance of camel for southern regions inhabitants, a national strategy was established, and many efforts are made in order to improve camel flocks' production in Tunisia.

Global warming: reconciling biodiversity maintenance with mountain farming economic sustainability

P. Scocco[1], C. Dall'Aglio[2], F.M. Tardella[1], E. De Felice[1], F. Mercati[2] and A. Catorci[1]
[1]*University of Camerino, Biosciences and Veterinary Medicine School, Via Pontoni 5, 62032 Camerino, Italy,* [2]*University of Perugia, Department of Veterinary Medicine, Via San Costanzo 4, 06126 Perugia, Italy; cecilia.dallaglio@unipg.it*

The global warming is one of the factors threatening the sub-Mediterranean dry grasslands by both socio-economic and environmental view. This research aims to deep our knowledge on how counteract such threat. We attempted to understand the drivers and mechanism behind the species assemblage modification, and experienced adaptative management to face climate changes and ensure a high level of animal welfare and food value. We analysed the dry grasslands functional variation under different stressors, disturbance intensities and environmental gradients to achieve predictive models of grassland modifications. We performed a trial, mainly based on animal feed supplementation, analysing the interplay between sheep and forage features, comparing milk/cheese chemical composition under different management conditions. Sensory and consumer tests were performed on cheeses to evaluate the consumer willing to pay. Findings showed the effects of increasing summer aridity on trait-related species assemblages lead to a shift from acquisitive to retentive traits and to regeneration strategies enabling plants to cope with unpredictability of climate variability. Reduction of disturbance fosters the spread of dominant tall grasses leading to decreased nutrient value of pastures and a loss of plant diversity. Productive grasslands were high negatively influenced by increasing summer aridity. Forage worsening caused the increase of rumen keratinisation degree and a decline of sheep welfare and productivity. The adaptive management experimented proved to foster the production quality. Milk and cheese showed an increase of poly-unsaturated fatty acids and Vitamins A and E. These features reflected in the cheese taste and consumer appreciation, which led to the theoretical availability of the most of consumers to pay more than the current market price for the cheese. Proper management practices proved to be a key tool in maintaining grasslands ecosystem services and to increase the quality of productions and satisfy customers' demand, respecting the animal welfare and reducing the risks of mountain farming abandonment by increasing the farm income.

Revitalising the sheep and goat sector in Greece: the strategy of workforce development

G. Papadomichelakis[1], G. Zervas[1], K. Tsiboukas[1], D. Voloudakis[2] and E. Lazaridou[2]
[1]*Agricultural University of Athens, 75 Iera Odos street, 11855, Athens, Greece,* [2]*Rutgers State University Office, 2 I. Kranidioti street, 57001, Thessaloniki, Greece; gpapad@aua.gr*

Sheep and goat dairy products are key components of the agri-food sector in Greece and hold a unique position within Europe. However, the complex interplay between market prices, national or international policies and consumer preferences, in combination with the socio-economic impact of the recession over the last decade have created an uncertain environment. To address this challenge, the sector must adopt effective solutions. A SWOT analysis indicated that low educational background of farmers, slow technological penetration, high production costs and inexistent strategic planning, are the main weaknesses of the sheep and goat sector in Greece. On the other hand, there are important opportunities for the new generations, mainly due to the limited restrictions and competition from other European countries, and the growing demand for sheep and goat milk. This paper will present one of the main strategies adopted to revitalise the sheep and goat sector; the Livestock Farming and Dairy Technology (LFDT) training program. The LFDT was launched in 2018 within the framework of the innovative multiyear 'New Agriculture for a New Generation' project (NeAGeN) that aims to create career opportunities and entrepreneurship for youth in the agri-food sector in Greece. Rutgers University is leading this program, in partnership with the Agricultural University of Athens and the American Farm School. The LFDT performs targeted training programs to meet specific needs of the sheep and goat sector by combining experiential education, scientific counselling, mentoring and business consulting. Its objective is to develop a new workforce of educated and skilled farmers; consequently, farms resilient to internal changes and external disturbances, and capable of exploiting emerging opportunities. The presentation will explore the key aspects of the activity and will emphasise on the impact of this strategy after 3 years (2018-2020) of implementation, including financial outcomes and future perspectives. The NeAGeN program is implemented through an exclusive grant from the Stavros Niarchos foundation.

Effect of maize silage replacement on the health status of two beef cattle breed categories

G. Riuzzi, S. Segato, B. Contiero and F. Gottardo
University of Padova, Dept. of Animal Medicine, Production and Health, Viale dell'Università 16, 35020, Italy;
giorgia.riuzzi@unipd.it

Meat consumers have been paying more attention to the way cattle are fed. The main concern is replacing maize silage (MS) as roughage source to make beef cattle diets more suitable for animal health and less competitive towards human nutrition. The study wanted to evaluate the impact of diets including decreasing amounts of MS on animal health and rumination. The study involved 11 specialised beef farms located in the Veneto region (Italy). Over 12 months, 4 visits per farm were carried out to identify animals with lameness, cough, nasal and ocular discharge, hampered respiration, diarrhoea, bloated rumen. At each visit, 20 animals were also assessed in terms of number of chewing acts per bolus. The study was an experimental two-factorial design where three levels (% on DM basis) of MS (high maize silage, HMS≥28%; low maize silage, LMS<28%; no maize silage, NMS=0%) were fed to French meat breed (FMB) vs dual purpose crossbreed (CSB) animals. In case of change of diet, the visit took place after an adaptation period. The number of animals affected by one of the health parameters on total number of farm animals (%) was analysed by a chi-square test. Number of ruminating acts per kg of NDF was analysed using an ANOVA that included the farm nested within breed, diet and their interaction. Overall, breed category has the main effect on the health status as the FMB animals seemed to be significantly more affected by lameness, nasal and ocular discharge, hampered respiration, diarrhoea, bloated rumen. They also seemed to ruminate less. The diet effect proved to be significant mainly within the FMB group. When less amount of MS was included in the TMR, there was an improvement in animals' health conditions and rumination: less nasal and ocular discharge and hampered respiration and higher rumination. Acknowledgements This research was made possible by funding from MIPAAF and SusAn, an ERA-Net co-funded under European Union's Horizon 2020 research and innovation programme (www.era-susan.eu), under Grant Agreement n °69623.

Standardisation of BSF feed experiments: the EAAP working group actions and Ring test protocol

L. Gasco[1], G. Vandenberg[2] and D. Deruytter[3]
[1]University of Turin, Department of Agricultural, Forest and Food Sciences, largo P. Braccini 2, 10095, Italy, [2]University of Laval, Department of Animal Sciences, Pavillon Paul Comtois Ste-Foy, Québec G1V 0A6, Canada, [3]INAGRO, Ieperseweg 87, 8800 Rumbeke-Beitem, Belgium; laura.gasco@unito.it

The increasing interest for the insect sector has led to an unprecedented rise of publications and a large number of papers have been published on insect rearing using various substrates and methodologies. This reduces the possible scientific and industrial value as comparisons between papers is difficult. During the EAAP conference in Ghent (2019), the importance to have harmonised criteria for designing trials in order to compare results, was discussed and a Working Group (WG) to standardise insect feeding studies was set up. This WG merged with the SUSINCHAIN (H2020) and was expanded with several other experts. The aim is to draw up a set of recommendations for trials in relation to aspects such as the control diet, the rearing densities, the minimum size of insect containers and number of replicate in feeding trials, and the measurements to be done. As a first step, the WG decided to focus on black soldier fly (BSF) protocol standardisation. It was agreed that a Ring test would be performed by some members. The goal of this test would be to assess the inter-laboratory variability when using the same substrate and environmental conditions. In order to have the same homogeneous feed, a single batch of a basic diet will be prepared and sent to the partners involved. Upon arrival, a sample of diet will be analysed to assess inter-laboratory differences. The BSF neonate larvae rearing will be performed under agreed conditions and protocol. A feeding experiment will then be conducted on 5 day-old larvae following the same standardised protocol. In particular, an agreement has been reached for: substrate humidity, and environment temperature and lightings conditions; quantity of feed/container cm²; number of larvae/kg of feed, sampling and end of trial times, parameters to be recorded during the trial, how to collect and to process larvae as well as analyses to perform. The WG comprises about 20 persons and it was impossible to list all of them in this abstract. Authors are grateful to all of them for the important inputs in the WG.

Insects as an international feed ingredient – the standardisation challenge
D. Murta[1,2,3]
[1]University of Lisbon, CIISA, FMV, Av. Universidade Técnica, 1300-477 Lisboa, Portugal, [2]CBIOS, FMV-ULHT, Campo Grande, 1749-024, Portugal, [3]EntoGreen, Ingredient Odyssey, Quinta das Cegonhas, Apartado 577, 2001-907 Santarém, Portugal; daniel.murta@entogreen.com

Insects has been pointed as a viable large-scale solution for feeding production animal during the last ten years. Since then, many private and public research have been done in order to identify the most suitable insect species to be produced and to determine production conditions and viable substrates. Several of these researches resulted in parallel developments and different production procedures, in which both the biological part of the production and the processing methods are different. This situation results in variability in the production performance and final product characteristics between the different companies that are being stablished around the world, even when producing the same insect species. This is a challenge to insects' recognition as a feed ingredient to be obtained in the international market with a known quality and composition. In order to promote a stronger sector, insect producers and researcher must start by coordinating the standardisation of terminology of this new industry, ensuring that everyone is understood. Besides that, a standardisation of final products quality and definition is also needed, as it is pivotal to recognise that products may have different nutritional characteristics when obtained using different production methods. Besides that, although researchers have been supporting the development of this sector, transferring the research developments from the lab to the industry results in many changes and the outcomes may no longer be repeatable, decreasing the impact that such research could have. This implies a stronger connection between research centres and companies, and a better understanding of the challenges associated with large scale production. Such understanding relies on the ability of the researchers to replicate the large-scale production conditions, however, in this point, it is also important to know to what extent can companies disclose their production methods in order to allow this standardisation. It may be concluded that in order to allow a better connection between research and production, basic terms and production methods must be standardised, allowing the comparison and repeatability of different research lines to be translated into full-scale production.

The science-policy interface and its role in the development of the European insect sector
C. Muraru, C. Derrien and J. Mathew
International Platform of Insects for Food and Feed, A. Lacomble 59, 1030, Belgium; info@ipiff.org

The European insect sector is a new agricultural industry, gradually growing into a reputable actor in terms of products' quality and sustainability. Across the EU, insects and their derived products have diverse uses in products intended for the food, feed or pet industries. According to the estimations of IPIFF members, the sector is expected to produce circa three million tonnes of insect protein by 2030, bringing a valuable contribution towards improving EU's self-sufficiency in terms of protein-rich ingredients. On the other hand, an optimistic forecast indicates that up to five million tonnes could be produced. To reach these targets, representatives from the industry, academia, as well as policymakers, made a public call (i.e. at the IPIFF International Workshop organised in December 2019) for a stronger collaboration between all actors from across the insect production value chain – with the aim to facilitate the development of science-based evidence relevant for the advancement of the regulatory context. This call – initiated by IPIFF and its more than 50 members – aims at improving synergies between stakeholders, in line with the FAO multi-sectorial approach – private sectors, regulators and scientists (the so-called 'Golden Triangle Concept'). As part of this presentation, IPIFF describes the regulatory roadmap of the European insect sector, the possibilities opened by the EU legislation with regards to the use of insects in animal feed, as well as the potential of the sector in contributing to the objectives of the 'Farm to Fork' Strategy launched by the European Commission in March 2020.

The research priorities of the European insect sector

A. Paul

International Platform of Insects for Food and Feed, A. Lacomble 59, 1030, Belgium; aman.paul@protix.eu

According to the FAO, the global meat consumption in 2050 will be 73% higher than in 2010. Feeding a growing population while also respecting the planetary boundaries requires a comprehensive multi-sectorial approach in terms of how food is produced. To this end, the contribution of innovative sectors, such as insect farming, is considered a viable solution to the increasing demand for protein-rich ingredients. Used in both human food and animal feed, such products (1). have a low environmental footprint – reducing the pressure on limited resources; (2). are an interesting source of nutrients; and (3). contain functional components that could promote animal and human health. Research conducted in a European context played an essential role in the advancement of the EU legislation for farmed insects. To facilitate the development of science-based evidence, the International Platform of Insects for Food and Feed recently published a report that summarises the key research priorities of the European insect sector. These topics are, on the one hand, focusing on elements that have been previously investigated and provided promising conclusions, but also on subjects that would require a closer attention from the scientific community. In this talk, IPIFF presents the main research priorities of the European insect sector, while also highlighting the key research gaps. The talk will cover matters related to circular agricultural practices (the use of new substrates in insect farming), agricultural fertilisers (the application of insect frass) and health (the nutritional and health benefits of insects in food and feed).

Equidae production: roundtrip in the era of globalisation

N. Miraglia and E. Salimei

Molise University, Agriculture, Environment and Food Sciences, Via De Sanctis, 86100 Campobasso, Italy; miraglia@unimol.it

The background of the present situation of Equidae production lies on a strong and well documented scientific activity that started in the years '70 in Europe and US with fundamental research on nutrition and genetic. In horse populations there has been a great concern both pedigree and digestion physiology and feeding evaluation systems. These studies implemented considerably and were shared by all the research institutes of the world, generating a perfect globalisation concept. In this context, consultation and discussion with representatives of European horse breeding industry and scientists have been carried out to state experts in Equine science: key challenges, knowledge gaps and the opportunities to meet the needs of the horse breeding in Europe. As a consequence, the role of sponsor became more and more relevant and was linked to a progressive limitation of funds destined to research and a strong demand coming from the 'Equine industry'. Horses became more popular in the growing industry of sports, leisure, and hobby farming. The booming development of leisure riding and the diversification of usage of horses, and the increasing role of horses in the use of territories became of high concern for European and local socio-economy. Special attention was given to: breeding performance horses for race and sport using most of the modern tools to maximise the efficiency of breeding and husbandry systems and to prevent the risk of health disorders; breeding leisure horses, ponies and donkeys for riding education, hobby, tourism and therapy implementing relevant management systems to reduce the cost of breeding and to fit the social need of the end-users; preservation of endangered breeds to maintain biodiversity using new modern tools in the scope of strategic selection plan; breeding horses for alternative animal products. Based on this evolution, the paper aims to point out strengths and weaknesses of the Equidae production and the privileged points of development in the next future.

Green assets of equines in Europe

A. Rzekęć[1], C. Vial[1,2] and G. Bigot[3]
[1]French National Research Institute for Agriculture, Food and Environment INRAE, MOISA, 2 place Pierre Viala, 34060, France, [2]French Institute for Horse and Horse Riding IFCE, Pôle innovation, développement, recherche, Jumenterie du Pin, 61310 Exmes, France, [3]French National Research Institute for Agriculture, Food and Environment INRAE, Territoires, Université Clermont Auvergne, 63000 Clermont-Ferrand, France; celine.vial@inrae.fr

Environmental awareness is increasing in agriculture and society. In this context, equines can be active actors in the ecological transition in Europe, being not only animal producers, but also ecosystem service providers. The aim of our study is to highlight the most important services provided by equines for the environment at the European level – which we called 'green assets'. In order to answer public and professional stakeholders' questions about the inclusion of equines in public policies according to their priorities, we first met stakeholders to understand their main issues. Then, a literature review of the available knowledge about environmental impacts of equines was conducted. Finally, our work examines the environmental assets (and limits) of equines that appear to be most important in the context of European policies about agriculture and rural development. The five major green assets identified by stakeholders are linked to: (1) equines specificity as grazing and domestic biodiversity; (2) their geographical distribution and their land use; and (3) to some uses by human beings as tourism and work. Equines impact landscape, biodiversity, greenhouse gas emissions, soil and water quality in many ways: creating ecosystem niches in pastures, producing renewable energy through traction maintaining and creating trails for their use in tourism, maintaining sensitive areas thanks to the adaptation of local breeds to environmental conditions. Practical recommendations are proposed and rely on literature review to help owners and stakeholders to promote these assets into their projects and debates. The next step is to quantify these green assets on a national level, for every country in Europe. To this end, a data collection application is being implemented in order to improve knowledge exchanges within Europe and find a consensus to promote equines in the European ecological transition. It is being tested on five countries presenting a dynamic equine industry and a proactive attitude towards ecological transition (France, Belgium, Poland, Sweden, Croatia) with the will to extend the tests to 10 countries in 2020.

Ecosystem function of semi-feral horses with specific focus on orthopteran density

C.-G. Thulin and A. Jansson
Swedish University of Agricultural Sciences, Anatomy, Physiology and Biochemistry, Box 7011, 750 07 Uppsala, Sweden; carl-gustaf.thulin@slu.se

Up until a few hundred years ago, wild horses roamed the European subcontinent. To test the ecosystem function once upheld by these wild horses we used a Swedish, native horse breed, the Gotland Russ, in a field experiment outside Uppsala, Sweden. The horses were kept in three enclosures (~10 hectares each) year around without supplementary feeding (four horses/enclosure). Their impact on landscape and biodiversity were assessed in an enclosure/exclosure design (three exclosures per enclosure) with and without treatment (i.e. horse grazing). Previous results show that year around grazing haltered the loss of floral diversity, benefitted pollinating insects and prevented shrubification (forest regeneration). Here we address how the horses impact orthopteran density. The total number of orthopterans observed within one square meter were counted at four positions in all exclosures, at four positions in the enclosures which were clearly grazed by the horses and at four positions in the enclosures that were left ungrazed by the horses and with a subjectively comparatively high number of droppings ('toilets'). All registrations were made during July under sunny weather conditions and data was analysed using ANOVA and Tukey grouping. There were more orthopterans in toilets compared to exclosures[a] and enclosures[b] (1.4 ± 0.2 vs 0.7 ± 0.2[a] and 0.4 ± 0.2[b], [a]$P<0.01$ and [b]$P<0.001$) but there was no difference between exclosures and enclosures. Our results show that the differential grazing pressure and faecal droppings by the horses created microhabitats where density of grasshoppers was higher than intensively grazed- and non-grazed areas. Thus, feral horse keeping seems beneficial for orthopterans. The study as a whole indicate that feral horse keeping can benefit flora and insect species diversity of the open landscape, and that horse grazing prevent reforestation of pastures.

How do grazing horses shape biodiversity – stressing the lack of evidence in mesophile grasslands

G. Fleurance[1,2] and B. Dumont[1]
[1]Université Clermont Auvergne, INRAE, VetAgro Sup, UMRH, Centre INRAE Clermont Auvergne, Rhône-Alpes, 63122 Saint-Genès-Champanelle, France, [2]IFCE, Pôle Développement Innovation et Recherche, 61310 Exmes, France; geraldine.fleurance@inrae.fr

In an agricultural context where farmers and citizens have growing environmental concerns, horses – whose are more than 6 millions across Europe – have a significant role to play for land management and biodiversity conservation. We review the state of art on the effects of horse grazing on plant and fauna diversity in different types of grasslands and heathlands across Europe. Among 40 references found in the WoS in February 2020, 35 were surveys from areas of high nature value. Horses are less affected than ruminants by digestive constraints: their high voluntary intake on roughages enables them to control competitive grasses as well as reeds and sedges in wet areas. This maintains open areas and promotes the coexistence of many plant and animal species in pastures, especially when these are grazed at a lenient stocking rate. Mixed grazing with cattle appears effective for controlling shrubs that are usually avoided by horses unable to detoxify secondary metabolites from dicotyledons. Their consumption of forbs and legumes is lesser than that of ruminants, with positive consequences on flowering plants and flower-visiting insects compared to sheep grazing or no grazing. Horses with their two sets of incisors create stable short patches of grass within a matrix of taller vegetation. Increase in the structural heterogeneity of pastures favours grassland biodiversity. Under low to moderate stocking rate, mixed grazing between cattle and horses would maximise botanical diversity compared to horse or cattle grazing alone, as cattle increase diversity in patches of tall grasses avoided by horses. Horse trampling under high stocking rate could have detrimental effects on plant and animal species. In mesophile grasslands which are commonly used in horse production systems in Europe, a number of studies have improved our understanding of factors (body size, sward height, supplementation) affecting voluntary intake and feeding choices by horses. Only five references report how grazing horses impact grassland biodiversity, revealing an increase in legume abundance in stable grazed patches and some effects on insect diversity. Further research is needed in mesophile grasslands to propose management options that jointly meet production goals and ecosystem integrity.

Distances walked in an open stable system by long established and newcomer horses

F. Hildebrandt, J. Krieter, K. Büttner, J. Salau and I. Czycholl
Institute of Animal Breeding and Husbandry, Christian-Albrechts-University, Olshausenstr. 40, 24098 Kiel, Germany; fhildebrandt@tierzucht.uni-kiel.de

Horses are flight animals that travel large distances in the wilderness. Modern group holding is supposed to enable species-appropriate locomotion. Thus, the aim of the study was to investigate the daily walking distances of horses in a 'HIT active stable'. This study included 53 horses (23 mares & 30 geldings) held in one group in Northern Germany. Each horse was equipped with a GPS sensor (sample frequency: 0.1 Hz) appended to a collar over an observation period of 9 months (June 2018 – February 2019). Firstly, the daily walking distances for the whole group were analysed with a generalised linear mixed model including the fixed effects sex (male, female), season (winter, summer), age class (young: <18 years, old: ≥18 years) and the random effect animal. No significant differences were found in sex (6.0±0.1 km; P>0.05). The wintertime resulted in significantly smaller walking distances of 5.0±0.1 km in comparison to summer (7.2±0.1 km) (P≤0.05). Travel distances were greater in summer because of pasture usage and longer activity periods. Furthermore, younger horses walked significantly more (6.6±0.1 km) than older ones (5.4±0.1 km) (P≤0.05). Younger horses tended to show more play and exploration behaviour, resulting in larger distances. Secondly, a similar model was used to analyse the effects of the inclusion of new horses in the herd on walking distances. This compared the walking distances of day 1-5 directly after inclusion of the new horse to days 31-35 (after acclimatisation period) and how each newcomer (14 horses) affected walking activity in the herd. Here, the interaction between newcomer and inclusion status (day 1-5 vs 31-35 after inclusion) was included additionally to sex and age class. Animal was used as random effect. The interaction between newcomer and inclusion status was significant (P≤0.05). Thus, for 8 out of the 14 newcomers, significantly higher walking distances were found for the herd on the 5 days directly after their inclusions. Additionally, four horses showed a corresponding tendency. The integration of new horses resulted in an increased unrest shown by higher distances.

Effect of ageing time on consumer preference, sensory in-mouth description and texture of horse meat

L.R. Beldarrain[1], I. Etaio[1,2], L. Moran[1], M.A. Sentandreu[3], K. Insausti[4], L.J.R. Barron[1] and N. Aldai[1]
[1]University of the Basque Country, Lactiker Research Group. Department of Pharmacy and Food Science, Miguel de Unamuno 3, 01006 Vitoria-Gasteiz, Spain, [2]University of the Basque Country, Laboratorio de Análisis Sensorial Euskal Herriko Unibertsitatea, Miguel de Unamuno 3, 01006 Vitoria-Gasteiz, Spain, [3]Instituto de Agroquímica y Tecnología de Alimentos, Carrer del Catedratic Agustin Escardino Benlloch 7, 46980 Paterna, Spain, [4]Public University of Navarre, IS-FOOD, Calle Cataluña AGR, 31006 Pamplona, Spain; lorea.rivera@ehu.eus

Horse meat is not very popular but its consumption is slowly increasing in several countries due to its recognised nutritional and environmental benefits. In terms of consumer acceptability of meat, tenderness is a key factor that can be considerably improved during the ageing process. The objective of this study was to assess the in-mouth acceptability (IMA), sensory drivers and texture of aged horse meat (0, 7, 14, 21 d). Ten Hispano-Bretón foals (5 females, 5 males) were reared under grazing and natural suckling conditions. At 11-13 months of age, animals were finished on *ad libitum* concentrate (100-120 d). Horses slaughtered at 15-17 months provided carcasses of 246.2±14.0 kg. After 48 h, rib joints (n=20) were transported to the lab, loins excised and sliced into 1.5 cm steaks. Vacuum packed steaks were randomly assigned to 0, 7, 14 or 21 d of ageing (4 °C, no illumination). A consumer study was performed with 120 volunteers and grilled (71 °C internal) horse meat samples. IMA using a hedonic scale and sensory description using Check-all-that-apply (CATA) were measured. Texture was also recorded using a texturometer with a Warner Bratzler Shear Force (WBSF) device. For data analysis, ANOVA, Tukey's, Cochran´s Q and MacNemar´s tests were applied. Correspondence Analysis was also performed on selected CATA terms. Overall, meat aged for 7 d obtained higher IMA scores than unaged meat; other aged meats showed intermediate scores. WBSF decreased significantly from 0 to 14 d and, in fact, IMA and WBSF were inversely related between 0 and 7 d. After 7 d, other attributes seem to affect IMA. Among CATA terms, texture attributes were the most discriminant. In aged meat, 'juicy', 'tender' and 'easily dissolving' were the most cited while 'dry', 'tough' and 'chewy' were the least cited terms. Odour/aroma related terms were also affected, as ageing influences Maillard reaction compounds. This approach was able to describe and discriminate among meats. The best acceptability results were obtained at 7 d of ageing while longer periods would require an economic investment not translated into higher consumer acceptability.

How training can reduce stress in loading procedure in meat donkeys?

F. Dai[1], E. Dalla Costa[1], E.U.L. Heinzl[2], S. Cannas[1] and M. Minero[1]
[1]Università degli Studi di Milano, Dipartimento di Medicina Veterinaria, via Celoria 10, 20133 Milano, Italy, [2]Università degli Studi di Milano, Direzione Sicurezza, Sostenibilità e Ambiente, Via S. Sofia 9, 20122 Milano, Italy; francesca.dai@unimi.it

Transport procedures (loading, travel and unloading) are known to be stressful for animals, consequently having both short term and prolonged effect on animal welfare. In particular, loading is considered to be one of the most stressful components of transport. To reduce transport-related stress, habituation could play a crucial role decreasing behavioural reactions to a previously novel situation. Donkeys kept for meat production are generally transported to the slaughterhouse without any training. The aim of this study was to evaluate the effect of habituation to transport procedures on stress related behaviours during loading in meat donkeys. Fourteen Romagnolo donkeys (M=10; F=4; 1.2±0.4 years) were included in the study and randomly divided in two groups: Control (C; n=7) and Habituation (H; n=7). Foals in the H group were gradually habituated to be transported for short distances (from one pasture to another), travelling together with adult habituated donkeys. Short transports were conducted once a month to maintain their memory until the transport to the slaughterhouse. All donkeys were used to human contact. Donkeys were transported to the slaughterhouse in five different days, using the same truck, in small groups (two to four donkeys per transport). The loading phase was video-recorded; time for loading and donkey behaviour were analysed using Solomon Coder, applying a focal animal continuous recording method. Loading time was significantly shorter for H donkeys (mean 7.97±4.62 sec) than C donkeys (mean 83.23±143.84 sec) (Mann-Whitney test; P=0.004), and H donkeys showed more forward locomotion toward the truck than C donkeys (Mann-Whitney test; P=0.026). C donkeys showed significantly more stress-related behaviours such as turning back, moving backwards, refusing to proceed and rearing (Mann-Whitney test; P=0.026), and required a higher number of human intervention to load (H: mean 1.29±0.55; mean C: 5.43±10.11). These results, although preliminary, suggest that habituation to transport could reduce stress during loading in meat donkeys, reducing time needed and decreasing the need of human intervention for loading.

Oxidative profile in fattening horses fed with different feeding strategies

F. Raspa[1], F.R. Dinardo[2], I. Vervuert[3], D. Bergero[1], E. Valvassori[4], P. De Palo[2] and E. Valle[1]
[1]University of Turin, Largo Braccini, 2, 10095 Grugliasco (TO), Italy, [2]University of Bari A. Moro, Piazza Umberto I, 70121 Bari (Italy), Italy, [3]Leipzig University, An den Tierkliniken 9, 04103 Leipzig, Germany, [4]Public Veterinary Service, Via Ferrero, 28, 10022 Carmagnola (TO), Italy; federica.raspa@unito.it

The study investigates the effect of two feeding strategies on the oxidative balance in horses. The trial was carried out in a horses fattening farm for 100 days (fattening period). 19 Bardigiano horses (age 14.3±0.7 months) were randomly divided in two groups. One group was fed with high amounts of concentrate (HCG; n=9): 6 kg/animal/day of hay and 7 kg/animal/day of a cereal-based pelleted feed (starch 55% as fed), in 2 meals/day. The second group was fed with high amounts of fibre (HFG; n=10), with the same hay *ad libitum*, and 3.5 kg/animal/day of a fibrous-pelleted feed (starch 20% as fed) in 2 meals/day. At slaughter, blood and liver samples were collected. *L. lumborum* muscle was sampled 24 hours after slaughter. All samples were analysed for glutathione peroxidase (GPx), catalase (CAT), and superoxide dismutase (SOD); plasma and muscle for TBARs, hydroperoxides, and carbonylated proteins. Student's *t*-test and Mann Whitney U-test were applied (P<0.05). Muscular GPx was higher in HCG than in HFG (P=0.02). Plasmatic CAT had higher activity in HFG than HCG (P=0.03). Muscular TBARs were significantly higher in HFG than in HCG (P=0.04). Different feeding strategies affect oxidative system. The higher levels of CAT in plasma of HFG suggest that animals tend to be highly protected by oxidative damages, as it is one of the most rapid/effective antioxidant enzyme. Muscle samples were stored 24 h at refrigerated conditions before sampling. This could explain the higher TBARs concentration in HFG: higher intake of fibre is correlated to higher unsaturation of intramuscular fat. The lack of significant differences in liver could be due to its high capacity in reducing high amounts of oxidised metabolites, so the effect of the feeding system was insufficient to modify the investigated enzymes. Concluding, different feeding strategies influence oxidative status of horses, and plasmatic CAT activity could be a useful mean for revealing effects of the feeding system on oxidative balance in horses.

The effect of mare's milk on the viability of Caco-2 cells

J. Založnik, M. Narat and K. Potočnik
University of Ljubljana, Biotechnical Faculty, Department of Animal Science, Jamnikarjeva 101, 1000, Slovenia; klemen.potocnik@bf.uni-lj.si

The Caco-2 continuous cell line is a well-known human intestinal epithelium model, which has been used in several research studies for testing the absorption of different bioactive substances, their digestion, distribution, and release of metabolites. In recent years, mare's milk has become more and more popular in Europe. Reasons include the health benefits known from regions of the Eurasian steep where consumption of mare's milk has been a long-standing tradition. Despite the growing interest, mare's milk remains relatively unresearched. In fact, only a few experimentally supported studies of the effects of mare's milk on the human immune system are available. Seeing that mare's milk is becoming increasingly important among the general public, this paper focuses on researching whether or not Caco-2 cells could be used as a model for studying the effects of mare's milk. We used the cell model to study how different concentrations of mare's milk affect the viability of Caco-2 cells. To perform the experiment, the cells were incubated for 3, 6, 24, 48, and 72 hours in 5, 10, 20, and 40% concentrations of mare's milk. We monitored cell viability using the trypan blue cell counting method. Based on our results, we found that mare's milk had a negative effect on the viability of Caco-2 cells. With the increase of mare's milk concentration, the viability of the cells showed a statistically significant decrease. The duration of incubation also had a significant effect on cell viability. The highest growth was observed for the 24-hour incubation and demonstrated a statistically significant increase in viability compared to the 3 hour incubation period. Following 24 hours of incubation, cell viability decreased; however, the differences were not statistically significant. Our research showed that this experiment could not be performed with high concentrations of mare's milk. On the other hand, the present study yielded ideas for further research using different cell lines, where viability would not be affected in the same way as it was for the Caco-2 cell line.

A method to assess the valorisation of grass by horses in grassland

M. Valleix[1], T. Jousset[2], E. Keller[3] and L. Wimel[1]

[1]The French Horse and Riding Institute (IFCE), Experimentation research center of Chamberet, 1 Impasse des Haras, 19370 Chamberet, France, [2]/, La maison blanche, 14100 Saint Germain de Livet, France, [3]AgroSup Dijon, 1025 esplanade Erasme, 21000 Dijon, France; marianne.valleix@ifce.fr

Currently, agriculture intends to optimise its economy by decreasing financial charges and trying to develop sustainable systems. The valorisation of grassland serves these two purposes. Thus, a tool was created to help farmers to manage their land: Herb'Valo. It assesses the quantity of grass consumed by animals in pastures. Already functional for cattle and goats, the adaptation for horses and sheep is in progress. The model for horses was developed at the experimental centre of Chamberet over two years, 2018 and 2019. The grazing of three plots was studied for three and four herds (growing horses, mares not in reproduction, pregnant mares and mares nursing their foals) in 2018 and 2019 respectively. The first year, 99 mares were studied and 75 the second year. Using these data sets, a version of Herb'Valo for horses was adapted from the cattle version and was tested at the experimental centre and at four commercial horse farms. The Herb'Valo results were analysed with a Multiple Factor Analysis (MFA). This showed that grass height is inversely correlated with the progress of the grazing season. Moreover, the density of grass decreases while its height increases. These two results are in accordance with the literature. It also shows that the number of passage in the plot the quality of grass and the year have a strong influence on the predicted quantity of consumed grass. In contrast, the plot and the herd seem to have little influence on this result. So, physiological characteristics of animals not appear to be determinant in the efficiency of grazing. So, by taking into account plots' (area, type, interventions, etc.) and animals' (physiological characteristics, body weight, etc.) characteristics, Herb'Valo can predict the performance of the farm in terms of animal weight gain, the quantity of consumed grass and several elements about horses and pastures capacities (body condition, intake capacity or grazing pressure for example).

The potential role of horses in territory sustainability

A.S. Santos[1,2] and L.F. Lopes[2,3]

[1]CITAB-UTAD, University of Trás-os-Montes and Alto Douro, Vila Real, 5001-801, Portugal, [2]Ruralidade Verde, Lda, Vila Real, 5000-773, Portugal, [3]CIFAP-UTAD, Department of Forestry and Landscape Architecture, Vila Real, 5001-801, Portugal; ruralidadeverde@gmail.com

The Garrano breed is one four autochthonous horse breeds in Portugal. These animals have been mostly used as working animals and recently have become an important part of the rural development and promotion in the Northern inland of Portugal, they are recognised as Portuguese socio-cultural heritage, and its present uses are mainly related to leisure. In present days the Garrano mainly exists in a free ranging system, some of the herds are completely wild, representing a very important link in the mountain ecosystem. Herds walk freely in the National Peneda-Gerês Park (PNPG), the only Portuguese National Park, located in the northwest of Portugal. The main source of feed of these animals is supplied by natural pasture areas called 'lameiros', mountain agro-ecosystems that have a good potential for animal feeding due to their high botanical diversity, and also mountain areas. Several studies conducted in Spain and Portugal in similar breeding systems showed that equines are able to include up to 30% of woody species in their diet and refer that equines can constitute a biological tool for controlling shrub encroachment and the re-growth of invasive plants, enhancing ecosystem biodiversity, and help keeping forest areas cleaned reducing the occurrence of forest fires that may cause serious environmental and economical losses. Observational data show the effectiveness of these animals in 'cleaning' schrub areas. These results indicate that these animals have an important part of maintaining biodiversity in these agro-ecosystems, due to their feeding behaviour and preferences and can be useful in controlling and/or reducing schrub encroachment in mountain areas, playing an important role in the maintenance of biodiversity and, simultaneously, producing environmental and socio-economic benefits to rural populations. These advantages can/should be further explored and seem as an alternative usage for these animals in areas that are characterised by low demographic density and rural abandonment.

Macro minerals distribution in different fractions of donkey milk

F. Fantuz[1], S. Ferraro[1], L. Todini[1], L. Cimarelli[1], A. Fatica[2], F. Marcantoni[1] and E. Salimei[2]
[1]Università degli Studi di Camerino, via Sant'Agostino 1, 62032 Camerino (MC), Italy, [2]Università degli Studi del Molise, via De Sanctis, 86100 Campobasso (CB), Italy; salimei@unimol.it

It is known that nutritional and technological properties of milk are also influenced by mineral composition and distribution. The aim of this trial was to study the concentration of Ca, P, S, Mg, K, and Na and their distribution in fat, casein, whey proteins and aqueous phase of donkey milk. Sixteen lactating donkeys (6 Amiata and 10 Ragusana breed) were used to provide individual milk samples, collected mechanically. Milk samples were analysed for casein content. Fat, casein and whey proteins were removed from whole milk samples by subsequent centrifugation, ultracentrifugation and ultrafiltration, respectively. Skimmed milk, a supernatant whey (soluble) fraction and the aqueous phase of donkey milk were therefore obtained. The concentration of the aforementioned elements was measured by inductively coupled plasma-mass spectrometry in whole milk and fractions. The concentration of elements associated with fat, casein (colloidal) and whey proteins was then calculated. The effect of fat, casein, and whey proteins removal was determined by analysis of variance for repeated measures. Results on casein percentage (6.2 g/l), and on milk concentration of Ca (799 mg/l), P (484 mg/l), Mg (78 mg/l), K (707 mg/l), and Na (143 mg/l) were within the range available in literature. The concentration of elements in whole and skimmed milk did not differ, the fat fraction carrying no or very little amount of the investigated elements. Sodium was entirely contained in the aqueous phase of milk. Approximately 63% of total Ca, 53% P, 33% Mg and no K were associated with the casein fraction. The molar ratio of colloidal Ca, P and Mg to casein was 2.04, 1.36 and 1.07 mmol/g casein, respectively. A small but significant amount ($P<5\%$) of Ca, P and K, and 9% of Mg were associated with whey proteins. The aqueous phase of milk contained 32% of total Ca, 43% P, 58% Mg and 97% K. The milk concentration of S was 142 mg/l, of which 64% was associated with whey proteins and only 25% with casein, indicating a higher content of sulphur-containing amino acids in donkey whey proteins than casein.

Brakes and levers for horsemeat consumption in France in 2019

A. Lamy[1], C. Vial[1,2] and S. Costa[1]
[1]MOISA, INRAE, CIHEAM, CIRAD, Montpellier SupAgro, Univ Montpellier, 2 place Pierre Viala, 34000 Montpellier, France, [2]IFCE, Pôle développement innovation et recherche, La Jumenterie du Pin, 61310 Exmes, France; celine.vial@inra.fr

Morally taboo today in different parts of the world, horsemeat consumption has also been taboo for a long time in France, until the second half of the 19th century. It really became a specialised market at the beginning of the 20th century. However, during the past fifty years, it has been collapsing, representing in 2019 1% of butcher's meat volume purchased in France. Two-thirds of the French population do not consume any horsemeat. In this context, our research aims to identify the factors that can encourage or limit horsemeat consumption. To answer our study questions, we use literature review, questionnaire survey and in-depth interviews with consumers and non-consumers. The survey and the qualitative interviews show four main clusters of determinants. Firstly, the relationship between man and horse emerges most easily and quickly from the interviewees. The nature of this relationship presents many variations between purely emotional and strictly functional positions. Different strategies operate, such as the dissociation between a horse intended for meat production and another devoted to racing or leisure activities. Secondly, we note the anchoring of the product in eating habits, in connection with educational routines and food socialisation during childhood and later life. Indeed, opportunities to discover horsemeat are usually limited to family circles. Thirdly, individual taste preferences condition and divide consumers from non-consumers according to distinct appreciations. Finally, we find the last factors in the marketing conditions of horsemeat supply, and more particularly in the modalities of access in terms of price and distribution. In terms of managerial implications, it seems important to redefine commercial strategies for professionals. A reflection must be carried out both on consumer targeting and on the characteristics of the proposed offer in terms of products, prices, distribution and communication.

Effect of breed and finishing diet on foal growth parameters and foal meat quality

M.V. Sarriés[1], A. Cittadini[1], J.M. Lorenzo[2], B. Adrian[1], J.L. Saéz[3], K. Insausti[1], M.J. Beriain[1] and R.M. Canals[1]
[1]Universidad Pública de Navarra, ISFOOD, Campus de Arrosadía, 31006, Pamplona, Spain, [2]CTC, Rua Galicia, 4, Parque Tecnológico de Galicia, 32900, San Cibrao das Viñas, Spain, [3]INTIA, Av. Serapio Huici, 22, 31610, Villava, Spain; vsarries@unavarra.es

The aim of this work was to study the effect of breed and type of finishing diet in 24 foals from Jaca Navarra (JN) and Burguete (BU) breeds on the growth parameters (Average Daily Gain (ADG) and Dressing Percentage (DP)) and on the foal meat quality (colour coordinates; L*, a*, b*, C* and H*; Water Holding Capacity (WHC) and pH). The animals were raised under extensive conditions and they were slaughtered at 21+2 months of age. Before that, one group of 12 foals (6 from JN and 6 from BU) were supplemented with conventional fodder (CF) (4- 15 kg/foal/day) and the other group of 12 foals (6 from JN and 6 from BU) was supplemented with organic feed (OF) (1-5 kg/foal/day) during 3-4 months. There were significant differences due to the effect of breed (P<0.001) and the type of finishing diet (P<0.05) on the ADG and on the DP (P<0.05) values, since the BU-CF reached the highest ADG (BU-CF: 1.24+0.28; BU-OF: 0.93+0.17; JN-CF: 0.73+0.17; JN-OF: 0.63+0.14). In addition, this type of supplementation produced the highest DP values in JN followed by BU (JN-CF: 63.57+1.61; BU-CF: 61.95+0.71; JN-OF: 60.09+0.63; BU-OF: 59.30+1.2) breeds. Regarding meat quality (aged 4 days), only breed showed significant differences in the b* and C* colorimetric coordinates (P<0.005; P<0.001, respectively) and in the WHC (P<0.001). BU foals reached the highest values of b* and C* and the lowest values of WHC compared to JN foals. Finishing diet and breed affected significantly the pH values (P<0.005; P<0.001, respectively), since the highest values were found in the JN-OF group, followed by the JN-CF, BU-OF and BU-CF groups respectively. In conclusion, the type of fodder supplementation had a significant effect on the growth parameters where CF supplementation improved DP by 5%. Finally the breed affected to some parameters of meat quality, showing the JN breed meat reached the lowest values of b* and C* and the highest values of WHC and pH. These results will be useful for future organoleptic studies.

Partition of essential trace elements in donkey milk fractions

F. Fantuz[1], S. Ferraro[1], L. Todini[1], L. Cimarelli[1], A. Fatica[2], F. Marcantoni[1] and E. Salimei[2]
[1]Università degli Studi di Camerino, via S.Agostino 1, 62032 Camerino (MC), Italy, [2]Università degli Studi del Molise, via De Sanctis 1, 86100 Campobasso (CB), Italy; salimei@unimol.it

The aim of this trial was to study the concentration of Zn, Cu, Mn, Se, Mo and Co in donkey milk and their partition among fat, casein, whey proteins and aqueous phase. Individual milk samples were collected mechanically from 16 lactating donkeys (6 Amiata and 10 Ragusana breed). Subsequent centrifugation, ultracentrifugation, and ultrafiltration (cut-off 3 kDa) were carried out to remove fat, casein and whey protein, to obtain skimmed milk, a supernatant whey fraction and the aqueous phase of milk, respectively. The concentration of the mentioned elements was measured by inductively coupled plasma-mass spectrometry. The concentration of elements associated with fat, casein and whey proteins was then calculated. The effect of fat, casein, and whey proteins removal was determined by analysis of variance for repeated measures. The milk concentration of Zn (2,729 µg/l), Cu (77.2 µg/l), Mn (4.7 µg/l), Se (4.1 µg/l), Mo (3.0 µg/l) and Co (0.37 µg/l) was within the range previously published for donkey milk. The effect of milk fractionation was significant for all the investigated elements. The percentage of samples below the limit of detection was between 60 and 80% for Mn in ultracentrifuged, and for Mn and Se in ultrafiltered samples (aqueous phase), therefore the lower and the upper bound were calculated for such elements in these fractions. The effect of fat removal was not significant for Zn, Cu, Mn and Se but a significant, although small, amount of Mo (approximately 14%) and of Co (7%) was associated with fat. The large majority of milk Zn (95%) and Cu (82%) was associated with casein, as well as the majority of Mn (min. 73%) and Mo (46%). Significant amount of Se (38%) and Co (28%) were also associated with casein. The majority of Co (37%) and of Se (min. 41%) was associated with whey proteins, as well as 13% of Cu and 8% of Mo. The aqueous phase contained 33% of Mo, 27% of Co and between 10% (lower bound) and 25% (upper bound) of Mn and Se.

Effect of the finishing diet on carcass characteristics of Jaca Navarra foal using image analysis
A. Cittadini[1], M.V. Sarriés[1], J.M. Lorenzo[2], A. Gamboa[1], J.L. Saéz[3], K. Insausti[1], M.J. Beriáin[1] and R.M. Canals[1]
[1]IS-FOOD. Universidad Pública de Navarra, Campus de Arrosadía, 31006, Pamplona, Spain, [2]CTC, Rúa Galicia, 4,
Parque Tecnológico de Galicia, 32900, San Cibrao das Viñas, Spain, [3]INTIA, Avenida Serapio Huici, 22, 31610, Villava,
Spain; aurora.cittadini@unavarra.es

The purpose of this work was to study the effect of the finishing diet on carcass traits of twenty-four Jaca Navarra foals using video image analysis. Foals were obtained from local farms after weaning and reared at pasture until slaughter. Previous to it, the animals at a mean of 17 months of age were divided in two groups: one group of twelve foals was supplemented with conventional fodders (CF) (11 kg/foal/day) and hay (2.5 kg/foal/day); whereas the other group of twelve foals was supplemented with an organic fodder (OF) (4 kg/foal/day) and silage (5-6 kg/foal/day). Both groups of foals were gradually introduced to the commercial feeds using oat and silage over 3-4 weeks period before fattening. In both cases, the finishing diet was of 3-4 months. Images from the left side of the carcasses were captured using a high-resolution digital camera (Olympus, E300, Tokio, Japan). Video image analysis software (Imagej 1.52t) was employed to obtain the following morphometric measures: carcass length (CL), compactness index (CI), chest depth (CD), leg depth (LD), perimeter and area. The data were examined using a one-way ANOVA with IBM SPSS Statistics 25.0 program software package. The finishing diet significantly affected the foal growth parameters. In particular, the organic supplementation showed significantly lower ($P<0.01$) average daily gain for the OF group compared to the other one. As a result, the dressing percentage was also significantly ($P<0.001$) affected with values from 63.29 ± 1.54 to $59.62\pm0.79\%$, for CF and OF groups, respectively. Regarding the zoometric measures, the OF group had significantly lower LD values ($P<0.05$) compared to the other one. In conclusion, the type of finishing diet had a significant effect on the foal carcass characteristics. Considering the lack of information about the carcass quality of this autochthonous breed in danger of extinction, this study could be a starting point for producers of these animals.

Extensive equid milk as an alternative and sustainable dairy product
A. Blanco-Doval, M. Díez-Ramos, L. Moran, L.J.R. Barron and N. Aldai
University of the Basque Country (UPV/EHU), Lactiker Research Group, Department of Pharmacy & Food Sciences,
Miguel de Unamuno, 3, 01006 Vitoria-Gasteiz, Spain; ana.blancod@ehu.eus

In the last two decades, horse meat production has increased in Spain. Although most of live animals are located in the southern and western regions, 93% of the meat producing farms are sited in the northern regions. Horses are fed in the nearby valley and mountain grasslands and forests, and after weaning, foals are usually moved to eastern regions to be fattened and slaughtered. In order to mitigate rural depopulation by improving economy and social farming, local foal meat and mare milk production and commercialisation could be alternative and sustainable options to retain value in northern regions. Mare milk production is novel and, therefore, this work evaluates its advantages and disadvantages. In terms of advantages, extensive management of equids has several environmental and socio-economic benefits. It creates a mosaic landscape simultaneously increasing the biodiversity of the area. Besides, since they consume gorse, the coverage of shrub is diminished and wildfires can be reduced. Equids, as non-ruminant herbivores, contribute little to greenhouse gas emissions, and their meat and milk are rich in forage-derived polyunsaturated fatty acids, especially ω-3, with enhanced nutritional value. Moreover, extensive farming systems help preserving culture and tradition, including genetic conservation of native heavy and multipurpose horse breeds. However, some concerns need consideration. When grazing together in grasslands, cattle, sheep and horse have similar diet preferences and compete for quality forage. Equids are less efficient in fibre digestion and need to consume more to reach their nutritional requirements. Regarding milk production, mares might not be well adapted to milking with lower yields compared to other dairy species. Overall, mare milk could be a sustainable novel product retaining the added-value in the northern Spanish regions and providing several parallel benefits for the environment, micro-economy and social status of equine breeders. Moreover, it seems a suitable dairy alternative for consumers with cow's milk protein allergy and a potential source of bioactive compounds.

Proteome analysis of sarcoplasmic protein fraction associated with meat organoleptic characteristics

A. Della Malva[1], M. Albenzio[1], M. Caroprese[1], P. De Palo[2], A. Maggiolino[2], A. Santillo[1], A. Sevi[1] and R. Marino[1]
[1]university of Foggia, Agricultural Food and Environmental Sciences, via Napoli, 25, 71122, Italy, [2] University of Bari, Department of Veterinary Medicine, S.P. per Casamassima, km 3, 70010, Valenzano (BA), Italy; antonella.dellamalva@unifg.it

Although sarcoplasmic proteins not directly impact muscle tenderness, their evaluation is useful to identify protein biomarkers related to meat quality parameters during post-mortem aging. Therefore, the aim of this research was to study the effect of aging on meat organoleptic properties and sarcoplasmic protein changes in 3 different horse muscles. 36 samples excised from longissimus lumborum (LL), semitendinosus (ST) and semimembranosus (SM) muscles of 12 Italian Heavy Draft Horse carcasses were analysed at 1, 3, 6, 9 and 14 days of aging. At each time, colour, texture profile analysis and changes of sarcoplasmic proteins with SDS-PAGE and Two-Dimensional Gel Electrophoresis (2DE) were estimated. ST muscle showed higher values of lightness and redness (P<0.001), while, LL muscle showed lower values (P<0.05) of hardness and chewiness parameters. During aging, a progressively decrease of hardness was observed in ST muscle reaching the lowest value at 14 days. Aging affected the intensities of several sarcoplasmic protein bands, particularly, band corresponding to glycogen phosphorylase b kinase decreased in intensity after 6 days of aging in LL muscle, and, after 14 days of aging, was almost disappeared. Additionally, LL muscle showed some polypeptides in the 28-20 kDa area starting from 6 days of aging. 2DE image analyses results showed a decrease of sarcoplasmic protein spots during aging reaching the lowest value in ST muscle at 14 days of aging. Proteins separation also revealed, at 14 days of aging, an increase of 2 spots of tropomyosin alpha and beta chains in the LL muscle. In addition, in ST spots ascribed to glycerol 3-phosphate dehydrogenase, superoxide dismutase and phosphoglucomutase-1 also revealed differences in abundance at 14 days. Data highlight that aging affect post-mortem biochemical processes with different intensities in each muscle and revealed potential protein biomarkers to monitor meat quality characteristics.

Updating dietary strategies for high yielding dairy sheep according to the INRA2018 feeding system

G. Caja
Universitat Autònoma de Barcelona, Animal and Food Sciences, Group of Research in Ruminants (G2R), Av. dels Turons, s/n, 08193, Spain; gerardo.caja@uab.es

The new INRA Feeding System for Ruminants was recently published to face the challenges of predicting productive responses (i.e. quantity and quality), improving animal health (i.e. nutritive risks) and reducing environmental emissions (i.e. CH_4 and N). Its framework was built (Systali Project) using meta-analyses of databases and modelling. Because sheep data was lacking, its model was less developed than others. Basically, the system uses net energy (1 UF=1,760 kcal NE), intake capacity (IC) and fill units (UE, encombrement), and truly digestible protein (PDI) values, which were updated. INRA2018 accounts for digestive interactions and flow of nutrients, and therefore, feed values depend on the final ration. Moreover, concentrates have fill values and nutritive requirements vary according to metabolic efficiencies and allowances. The PDI-N is not more used, being replaced by rumen protein balance (RPB). With regard to dairy ewes, although IC predictions for lactating ewes were maintained, current equations include the negative effect of body condition score. Energy requirements also consider displacement costs. Overall, energy requirements for a 70 kg BW are increased for maintenance (5%), flushing (12%) and growth (8%) and decreased for body gain (-59%) and lactation (-4%). Similarly, PDI requirements increased by efficiency (PDIeff=0.58, on average) and wool requirements (maintenance, 52%; pregnancy, 10%) but decreased for body gain (-50%). New requirements for a high-yielding dairy ewe producing 500 kg standard milk (fat, 7.6%; protein, 5.5%) in 270 DIM (1.85 kg/d, on average) are 596 UF_L and 69 kg PDI for whole lactation. Nutrient concentration during early and late lactation result to be 130 and 110 $gPDI/UF_L$. Late pregnancy is also critical (110 $gPDI/UF_L$). When PDI allowances increase to 130%, as usually done in high-yielding dairy ewes, efficiency decreases (PDIeff=0.50) and PDI requirements increase (maintenance, 76%; pregnancy, 20%, lactation, 16%) although it decreases for body gain (-41%); nutrient concentration being 150 and 130 $gPDI/UF_L$. In practice, high BCS (>3.5) and a milk yield objective (MYo) satisfying near 80% of the ewes of a group are recommended (MYo = Mean + 0.8 SD).

Divergent selection on residual feed intake in Romane meat sheep breed to dissect biological process

F. Touitou[1], C. Marie-Etancelin[1], J.-L. Weisbecker[1], D. Marcon[2], D. François[1], R. Bessa[3], A. Meynadier[1] and F. Tortereau[1]
[1]INRAE, INP-ENVT, INPT-ENSAT, GenPhySE, Castanet-Tolosan, 31320, France, [2]INRAE, La Sapinière Experimental Unit, Bourges, 18000, France, [3]University of Lisbon, Faculdade de Medicina Veterinária, Animal Production Systems, Lisbon, 1300-477, Portugal; florian.touitou@envt.fr

Feed efficiency is a trait of major interest for all livestock species because breeding efficient animals is not only cost-saving but also leads to a decrease of environmental impacts. The biological processes underlying feed efficiency are poorly described in sheep. In our INRAE experimental farm of La Sapinière we have developed divergent lines of Romane meat sheep, the divergence being established on their breeding values for residual feed intake. The phenotype of male lambs (fed *ad libitum* with low-energy concentrate pellets) is obtained from the recording of daily feed intakes, body weights and body composition traits during a 6-week period starting from 90 days of age. After this first period of phenotyping, rumen and blood samples were taken from each individual. Lambs were then controlled during an additional 6 weeks period for their intake of forage, and after this period, the same sampling was performed. All the samples were analysed for metabolomics through NMR. The spectra were then analysed using the ASICS R-package. Analysis of ruminal fatty acids were performed through GC equipped with FID and a 100 m column after identification of peaks using GC coupled to Mass Spectrometer. Relative abundances of ruminal bacteria were obtained after 16s rRNA gene Miseq sequencing and FROGS pipeline analysis. Multivariate and discriminant analysis (PLS-DA) methods implemented in the MixOmics R-package were applied. After two generations of selection, efficient lambs ate 130 g less concentrates during the first period, and 100 g less forage during the second period than less efficient lambs. After each period of control, we identified metabolites that discriminate the efficient lambs from the less efficient ones, such as butyrate and choline chloride in plasma of lambs fed with concentrate and L-lysine and L-leucine in plasma of lambs fed with forage. These results from blood metabolomics will be combined with ruminal metabolomics (including fatty acids) and microbiota data to identify pathways involved in feed efficiency in sheep fed with different diets.

Milk traits as predictors of the milk coagulability in Assaf sheep

H. Marina[1], B. Gutiérrez-Gil[1], R. Pelayo[1], A. Suárez-Vega[1], C. Esteban-Blanco[1], A. Reverter[2] and J.J. Arranz[1]
[1]Universidad de León, Departamento de producción animal, Campus de Vegazana, León 24007, Spain, [2]CSIRO Agriculture & Food, St Lucia, QLD 4067, Australia; hmarg@unileon.es

Sheep milk is mainly used to manufacture high-quality cheeses. For many dairy sheep breeds, the economic relevance of cheese yield has encouraged the development of breeding programs aiming the increase of milk total solids. Cheese-making ability is also related to milk coagulation properties (MCP), which are studied through parameters such as rennet coagulation time, or curd firmness at 30 or 60 minutes, etc. At the practical level, non-coagulation of milk samples has been identified as an important problem for the industry. However, the routine measurement of MCP in commercial populations for inclusion as selection criteria in sheep breeding programs is expensive and not practical. A previous study of our research group in Spanish Assaf sheep identified about 13% of non-coagulating samples within 60 min from the addition of the clotting enzyme. The present work aimed to predict the coagulation behaviour of individual milk samples, as a binary trait (coagulating or non-coagulating samples), based on the milk trait information. For that, considering milk samples from 1,145 ewes with available records from the official milk control and analysed for MCP traits, we implemented a linear regression model to calculate a coagulability score (CS) for each milk sample. The model included milk traits from the official recording system (fat, lactose and protein percentages, dry extract, milk yield, somatic cell count, urea), the pH of milk, and days in milk as covariates. In contrast, the age of the ewes, the number of births, and the number of born lambs were included as fixed factors. The results of the analysis showed that the differences in the CS averages between the coagulating and non-coagulating milk samples were statistically significant (P<0.001). For several intervals of the distribution of predicted values, the CS reliability reached values greater than 90% for the two types of samples. These results support the implementation of the approach here described to predict the coagulability potential of milk samples without the need to measure MCP traits specifically. Future studies should try to improve the prediction efficiency of the applied model and assess if it can be used in different dairy sheep breeds.

Effects of condensed tannins from sainfoin in ewe milk production and suckling lamb growth

C. Baila, S. Lobón, M. Blanco, I. Casasús, J.R. Bertolín and M. Joy
Ctr Invest y Tecnol Agroal Aragon (CITA), IA2 (CITA-Universidad de Zaragoza), Animal Production, Avda. Montañana, 930, 50059 Zaragoza, Spain; cbaila@cita-aragon.es

There is an increased interest in using sainfoin (*Onobrychis viciifolia*) due to the presence of condensed tannins (CT), however their effects are not well known. The aim of the study was to evaluate the effect CT from fresh sainfoin on productive performance during 4 weeks of lactation. For that, 20 lactating ewes and their lambs were individually fed fresh sainfoin *ad libitum* plus 200 g/d of barley. Randomly, half of the ewes were daily orally-dosed 100 g of PEG 4000 /200 ml water per ewe as tannin-binding agent (SF+PEG) and the other half received no PEG (SF). Sucking lambs were permanently with their dams and fed maternal milk until they reached the target slaughter weight of 10-12 kg. The intake of sainfoin was recorded daily and its chemical composition was analysed. The body weight (BW) of ewes and lambs, body condition score (BCS) of ewes and the milk yield were recorded weekly. Milk was analysed for chemical composition, polyphenol content (Folin-Ciocalteu method) and antioxidant capacity (ABTS method). The presence of CT did not affect the intake of dry matter, protein, fibre and energy (P>0.05). The BW and BCS of dams and BW of lambs were similar between the groups (P>0.05). Milk yield was affected by the interaction between treatment and the week of lactation (P<0.001), SF ewes peaked at 2 week, whereas the SF+PEG ewes peaked at the 4th week. Milk protein and fat contents were similar between treatments (P>0.05) but urea tended to decrease (P<0.1) in SF group. The content of polyphenols was affected by both presence of CT (P<0.001) and week of lactation (P<0.05), with greater content in SF+PEG than SF ewes, and higher values on the first week of lactation. However, this greater content was not reflected in the antioxidant capacity of milk (P>0.05). In conclusion, the CT from sainfoin did not affect the intake and BW of animals but varied the pattern of evolution of milk yield and decreased the polyphenols and urea content in milk.

Quantifying and genotyping protein fractions in milk of different goat breeds by RP-HPLC

G. Secchi[1], N. Amalfitano[1], S. Pegolo[1], A. Cecchinato[1], M.L. Dettori[2], M. Pazzola[2], G.M. Vacca[2] and G. Bittante[1]
[1]University of Padova, DAFNAE, Viale dell'Università 16, 35020 Legnaro (PD), Italy, [2]University of Sassari, Department of Veterinary Medicine, Via Vienna 2, 07100 Sassari (SS), Italy; giorgia.secchi@phd.unip.it

The knowledge of detailed composition of milk protein fractions is important for dairy goat industry because of its impact on technological properties of milk and the nutritional value of milk and dairy products. The caseins (α_{s1}-, α_{s2}-, β- e κ-casein) and whey proteins (α-lactalbumin, β-lactoglobulin) are the same as in bovine milk but their genetic variants are different because of different polymorphisms at these loci. The aim of this study was to quantify the protein fractions and concurrently to identify their genetic variants developing a method based on Reverse-Phase High Performance Liquid Chromatography (RP-HPLC). This project involved 1,272 goats from 6 different breeds reared in 35 farms located in Sardinia (Italy). Individual milk samples were analysed for milk composition, coagulation, curd firming and syneresis, cheese yield and milk nutrients recovery in curd. To quantify the content of protein fractions of different genetic variants, freeze-dried protein samples of milk previously genotyped for genetic variants were used for calibration of HPLC since commercial standards are not available for goat milk. Five variants of α_{s1}-casein (A, B, E, F and null; *CSN1S1* gene; 11 genotypes), 2 variants of α_{s2}-casein (A/F and C; *CSN1S2* gene; 3 genotypes), 3 variants for β-casein (A, C, and null; *CSN2* gene; 4 genotypes) and 5 variants for κ-casein (A, B, C, D, and null; *CSN3* gene; 8 genotypes) were identified and quantified by HPLC. The statistical analysis of preliminary data revealed that polymorphisms in goat milk proteins have strong effects on the concentration of different caseins and on milk coagulation, curd firming, cheese yield and nutrient recovery traits. Moreover, we confirmed that different goat breeds are characterised by different genotypes frequencies, which are associated to different amounts and proportions of caseins in milk. These results provided useful information for defining the value of dairy goat industry and for breeding programs aimed at improving goat milk quality and technological characteristics.

Monitoring metabolic parameters in dairy sheep at the beginning of lactation

A.T. Belo, M.R. Marques, J.M.B.F. Ribeiro and C.C. Belo
Instituto Nacional de investigação Agrária e Veterinária, INIAV, IP, Qta Fonte Boa, EZN, 2005-048 Vale de Santarém, Portugal; anateresa.belo@iniav.pt

The importance of selectivity of pasture species by Assaf dairy sheep was evaluated during the first 6 weeks of lactation. Ewes were given a choice between ryegrass and fescue (AF); ryegrass, fescue and white clover (AFTB); lucerne, fescue and white clover (LFTB), given individually. A 4th group received pasture including the 4 species (PAST). Sheep were supplemented with 1,137 Mcal of metabolisable energy and 20% crude protein (CP) until day 21 of lactation. Forage and supplement were provided twice a day. The ewes suckled the lambs until 21 days and were milked afterwards. Data was subjected to analysis of variance (PROC MIXED) and means were compared by Tukey test, at 5% significance level. Metabolic indicator data was log transformed before analysis. Productive aspects were evaluated. Ewes showed preference for legumes that reached 85% of the DM consumed in the AFTB, and 89% in LFTB. Intake of DM in AFTB and LFTB was 78 and 57% higher than AF respectively. Diets consumed presented CP contents of 18% (AF), 23.5% (AFTB) and 22.6% (LFTB). Average daily gain of lambs was higher in LFTB (408 g.day^{-1}) and lower in AF (314 g.day^{-1}). Milk production for LFTB ewes was also higher than AF (1,579±226.1 vs 917±159.8 ml/day). Blood indicators – glucose, non-esterified fatty acids (NEFA), insulin (INS) and urea-N – were evaluated to establish the ewes' metabolic performance. Determinations were made on 4 occasions during lactation, throughout the day. Glucose values did not vary over the period, being higher in LFTB and PAST (average of 3.203±0.1369 mmol/l) and lower in AF (2.645±0.1031 mmol/l) while NEFA levels were only significant in AF in the first 4 weeks of lactation (0.496±0.0280 mmol/l). Insulin levels were higher in AFTB and LFTB (9.685±1.0842 µU/ml) and increased throughout the day. Urea-N levels were high in the AFTB and LFTB treatments (mean 36.535±1.028 mg/dl) compared to AF and PAST (23.860±1.033 mg/dl) and did not vary over the period. The indicator NEFA proved to be important in defining the amount of energy to be supplied to the ewes while urea-N levels indicate that supplementation of grazing ewes should consider the excess degradable protein obtained from pastures.

A crosstalk between small ruminant's nutrition and quality of dairy products

E. Vargas-Bello-Pérez[1] and M. González-Ronquillo[2]
[1]University of Copenhagen, Department of Veterinary and Animal Sciences, Grønnegårdsvej 3, 1870, Denmark, [2]Universidad Autónoma Del Estado de México, Facultad de Medicina Veterinaria y Zootecnia, Instituto Literario 100 Ote., 50000, Toluca, Estado de México, Mexico; evargasb@sund.ku.dk

Animal nutrition and animal physiology are important research fields that contribute to the quality of final products. Understanding the crosstalk between those animal disciplines and food sciences is often seen as divergent fields but actually, they complement each other. Our research approach is based on the use of dietary lipids as a nutritional strategy to modulate specific changes in dairy products at different levels: proximate composition, fatty acid profile, and sensory characteristics. Another important aspect from our approach is that all dietary interventions are designed to either improve or maintain overall animal performance. In order to favour changes in the physicochemical and sensory characteristics of dairy products it is important to understand the different metabolic pathways that dietary lipids undergo in the animal's system. This is the cornerstone for our research, which promotes the formation of bioactive fatty acids without compromising intrinsic attributes from dairy products. Those fatty acids could be C18:1 t11 (vaccenic), C18:2 c9, c12 (linoleic), C18:2 ci9, t11 (rumenic), and C18:3 c9, c12, c15 (linolenic). In the case of sheep trials, we have had the opportunity to use local by-products from the olive oil industry whereas for goats, we have used by-pass oils or whole seeds from sunflower and linseed. We have noted that the chemical configuration (i.e. degree of saturation of fatty acids, chain length, geometry and location of double bonds) of dietary lipids will promote differential changes on animal performance or final product quality. This is challenging, as each lipid source will provoke specific changes in the physicochemical and sensory characteristics of dairy products. Overall, dairy products from small ruminants can be naturally improved by supplementation of different lipid sources. This research approach could be useful for the dairy industry, especially for those players looking to provide an added value for their products via animal's nutrition and more importantly, without removing fat contents.

Methionine balanced diet improves performances and biosynthetic traffic in mammary epithelial cell

E. Chanat[1], S. Lemosquet[1], A. Leduc[1], L. Bahloul[2] and M. Boutinaud[1]
[1]INRAE, Agrocampus Ouest, PEGASE, 16, Le Clos, 35590 Saint Gilles, France, [2]Adisseo France S.A.S., CERN, 6, route noire, 03600 Commentry, France; eric.chanat@inrae.fr

Methionine (Met) or energy supplementation increase milk protein and fat yields in cow. Here we investigated whether this could be explained by increasing flow of milk components in the secretory pathways of mammary epithelial cells. Multiparous Alpine goats (mid lactation, n=47), grouped by levels of expression of *CSN1S1*, were assigned to 4 treatments in a randomised complete block design. Treatments were: LE (Low Energy), LEMet (LE, balanced Met), AE (Adequate Energy) and AEMet (AE, balanced Met) for 5 weeks. Diets consisted in a fixed amount of hay per group with individual amount of energy concentrates (LE vs AE: 1.47 vs 1.54 Mcal/kg DM). Isopropyl ester of 2-hydroxy-4-methylthio butanoic acid (HMBi 0.24% concentrate DM) was incorporated in the LEMet and AEMet concentrates based on 100% cow Met requirement. Goats (n=23) were slaughtered and mammary tissue was processed for Western blotting using secretory compartment specific markers. In goats fed Met balanced diets, milk protein yield and casein content significantly increased (P=0.01), as was previously observed in cow. The amount of the endoplasmic reticulum (ER) markers Calnexin and ERLIN2 decreased (20%, P≤0.05) in goats fed the LE diet. Met balanced diets had the opposite effect on both markers (20%, P≤0.05) and on protein disulphide isomerase (45%, P≤0.05). These observations are in agreement with a positive effect of Met on the activity of the ER, the site where milk specific proteins and lipids are synthesised. On the other hand, a specific marker of the exit site of the Golgi apparatus and secretory vesicles formation (AP1) decreased with the LE diet (25%, P≤0.05) and its highest level was found in goats fed Met balanced diet at AE supply. Concerning ßCOP, a marker of intra Golgi transport, its variation clearly evoked a decrease in membrane transport at LE diets. These data show that energy level has a direct impact on membrane traffic in the secretory pathway of mammary epithelial cell while suggesting that Met improves ER activity and has the tendency to further promote intracellular transport of milk components and, ultimately, their secretion.

Responses of dairy ewes to heat stress according to their phenotypical heat tolerance

B. Chaalia, S. Serhan, S. Gonzalez-Luna, X. Such, A.A.K. Salama and G. Caja
Universitat Autonoma de Barcelona, Research Group in Ruminants (G2R), Edifici V, Campus de la UAB, 08193, Spain; ahmed.salama@uab.es

Effects of heat stress (HS) on Manchega dairy ewes were evaluated using animals that differ in their phenotypical heat tolerance. Rectal temperature (RT) and respiratory rate (RR) were recorded in 24 ewes (158±5 DIM) before and after 2-h heat challenge (36 °C, 43% humidity). Ewes were classed by the RT and RR change ratio (CR = after/before HS). The CR values averaged 5.2±0.2 and distributed normally. In a second trial, 5 ewes from each phenotype with similar milk yield but different (P<0.05) CR (Tolerant: A, 4.6±0.2; Sensitive: B, 5.7±0.1) were used to evaluate their productive and metabolic responses to HS. The design was crossover of 2 periods (3 wk each) and 2 climatic conditions: 1) thermo-neutral (TN; 15-20 °C day-night), and 2) HS (day, 37 °C; night, 30 °C). Humidity (50%) and dark-light (12-12 h) were constant. RT, RR, milk yield, and feed and water intakes were recorded daily. Milk and blood samples were collected and analysed weekly. At d 19, response to glucose tolerance test (GTT; 0.25 g/kg BW) was evaluated by collecting blood samples at 10 time-points (min −15 to 120) to analyse glucose and insulin. No significant HS × phenotype interaction was detected for RT, RR, feed intake, water consumption, milk yield, or milk composition. HS ewes had greater (P<0.001) RT (+0.54 °C), RR (+177%), water consumption (+35%), but lower feed intake (-20%; P<0.001) than TN. Milk yield (0.63±0.1 kg/d) did not vary between treatments, but milk fat (-14%) and milk protein (-17%) contents decreased (P<0.01) in HS. Blood glucose, insulin and NEFA did not vary between TN and HS, but blood prolactin (+415%) and creatinine (+10%) were greater (P<0.01) in HS ewes. HS ewes had similar insulin response to GTT (P>0.10), but numerically greater (P=0.19) glucose disposal than TN. The A ewes under HS tended to have greater insulin secretion during GTT, but similar glucose disposal compared to B ewes. In conclusion, late lactating Manchega ewes were relatively tolerant to HS conditions. Few metabolic differences between A and B heat tolerance phenotypes were detected under high ambient temperatures. Acknowledgement: Project funded by the Spanish Ministry of Economy and Competitively (INIA-RTA2015-00035-C03).

Singular versus composite trait: what difference for maternal traits selection in meat sheep?

E. Cobo[1], J. Raoul[1,2] and L. Bodin[1]
[1]INRAE, Animal Genetic, 24 Chemin de Borde Rouge – Auzeville Tolosane, CS 52627, 31326 Castanet Tolosan cedex, France,
[2]Idele, Campus INRA, Chemin de Borde Rouge, BP 42118, 31321 Castanet Tolosan cedex, France; emilie.cobo@inrae.fr

In France, selection of maternal traits for meat breeds is mainly based on few traits: litter size, lamb viability and pre-weaning weight. Each trait is independently evaluated through a BLUP animal model and selection is based on a linear combination of estimated breeding values (EBV) according to the breeding goal of each breed. Is the selection based on a composite trait, the 'litter weight', which encompasses current selected traits, be more adapted and comprehensive to breeders? We first estimated the genetic parameters of both singular traits and the composite trait. We then analysed changes on potentially selected animals that occurs when moving from singular traits to composite trait selection. Official records from 2006 to 2019 of two meat sheep breeds, Ile de France (80,484 ewes and 261,069 litters) and Blanche du Massif Central (87,768 ewes and 298,605 litters), were analysed using SAS and Asreml software. For singular traits, the genetic parameters and genetic values were estimated according to the models routinely used in meat sheep: a two traits model for litter size after natural and hormonal induced oestrus and a direct and maternal effects model for pre-weaning weight (weight of the lambs at 30 days of age). For the composite trait, we first computed the sum of the weights of lambs per litter, pre-corrected or not by the sex in two situations. The first situation included only litters with no lamb dead before weaning. The second situation included litters with and without dead lambs. The genetic parameters and genetic values were then estimated using different models. The routinely used maternal index, computed from a linear combination of EBVs for singular traits, was compared to the EBVs for the composite trait. Correlation coefficients between 'singular' and 'composite' EBV have been calculated in the two situations.

The purebred-crossbred correlation in broilers and layers: a review

M.P.L. Calus, J.A. Bos, P. Duenk and Y.C.J. Wientjes
Wageningen University & Research, Animal Breeding and Genomics, P.O. Box 338, 6700 AH Wageningen, the Netherlands;
yvonne.wientjes@wur.nl

Commercial broilers and layers are typically 4-way crossbred animals that are housed under commercial conditions, implying that the breeding goal in poultry breeding is to improve crossbred performance. Selection, however, takes place in the parental purebred lines based on performance measurements of purebreds that were housed under nucleus conditions. The response to selection in crossbred performance is therefore bounded by the purebred-crossbred genetic correlation (r_{pc}). The aim of our study was to review published r_{pc} values in poultry. In total, results of 14 different studies were summarised, of which 4 on broilers and 10 on layers. Six studies (5 on layers and 1 on broilers) compared purebred and crossbred animals housed in the same environment, 5 studies (all on layers) compared purebred and crossbred animals housed in different environments, while another 3 studies on broilers compared two groups of purebred animals housed in a nucleus versus commercial environment. In total, 128 correlations have been reported, with an average value of 0.71. Average values in broilers ranged from 0.63 for foot pad dermatitis, 0.70 for body weight, to 0.86 for meat amount. Average values in layers were 0.61 for laying rate, 0.68 for maturity, 0.69 for egg quality, 0.70 for mortality, 0.83 for body weight, 0.84 for egg weight, and 0.86 for egg colour. Due to differences in experimental set-ups, the reported correlations for layers all included the contribution of genotype by genotype interactions to the r_{pc}, while two-thirds also included the contribution of genotype by environment interactions. Most reported correlations for broilers provide an estimate of the contribution of genotype by environment interactions. Of the studies using crossbred animals, almost all used 2-way crossbreds, and one used 3-way crossbreds, but none considered 4-way crossbreds. Thus, the values that apply in practice for the 4-way crossbreds are likely to be lower than the average reported values, also because 9 out of 14 studies included either only genotype by environment or only genotype by genotype interactions. Nevertheless, the reported correlations suggest that measuring crossbred performance in commercial environments may be especially relevant for laying rate, egg quality, maturity and mortality in layers and foot pad dermatitis and body weight in broilers (average correlations ≤0.7), while it may provide limited benefit for body weight, egg weight, and egg colour in layers, and meat amount in broilers (average correlations ≥0.83).

Predicting the purebred-crossbred genetic correlation from variance components in parental lines

P. Duenk, P. Bijma, Y.C.J. Wientjes and M.P.L. Calus
Wageningen University and Research, Animal Breeding and Genomics, Droevendaalsesteeg 1, 6708PB, the Netherlands;
pascal.duenk@wur.nl

Breeders benefit from heterosis and breed complementarity by mating individuals from genetically distinct purebred lines to produce crossbred production animals. While the aim of such breeding programs is to improve crossbred (i.e. hybrid) performance, selection usually takes place in the purebred lines based on purebred performance. The response to selection in crossbred performance therefore depends on the genetic correlation between purebred and crossbred performance (r_{pc}). The r_{pc} can be lower than one due to non-additive effects in combination with differences in allele frequencies between parental lines. This suggests that r_{pc} can be expressed as a function of parameters in the parental lines. In this study, we derive expressions for r_{pc} based on genetic variances within, and the genetic covariance between parental lines. These expressions were derived for a genetic model with only additive and dominance effects (D), and only additive and epistatic additive by additive effects (E_{AA}). We validated our expressions using simulations of purebred parental lines and their crosses, where the parental lines were either selected or not. Finally, using these simulations, we investigated the value of r_{pc} for other genetic models, for which expressions could not be derived. Results show that our expressions provide exact estimates of r_{pc} for models D and E_{AA}, and accurate upper and lower bounds of r_{pc} for two other genetic models, using information of parental lines only. In conclusion, our work demonstrates how the r_{pc} depends on non-additive effects, and aids in determining bounds of r_{pc} for different genetic models.

Swedish dairy farmers' preferences for different breeding tools

J.B. Clasen[1], C. Bengtsson[1,2], H.N. Källström[3], E. Strandberg[1], W.F. Fikse[4], S. Østergaard[5], M. Kargo[6] and L. Rydhmer[1]
[1]Swedish University of Agricultural Sciences, Animal Breeding and Genetics, Box 7023, 75007 Uppsala, Sweden, [2]VikingGenetics, VikingGenetics Sweden AB, 53294 Skara, Sweden, [3]Swedish University of Agricultural Sciences, Urban and Rural Developments, Box 7012, 75007 Uppsala, Sweden, [4]Växä Sverige, Ulls väg 26, 75651 Uppsala, Sweden, [5]Aarhus University, Animal Science, Blichers Allé 20, 8830 Tjele, Denmark, [6]Aarhus University, Molecular Biology and Genetics, Blichers Allé 20, 8830 Tjele, Denmark; julie.clasen@slu.se

Genomic testing, sexed semen, beef semen, crossbreeding, and embryo transfer are breeding tools that have shown potential for genetic and economic improvements in dairy cattle production. However, despite the potential of these tools are well known, only a small fraction of the dairy producers use these tools systematically. In-depth interviews with Swedish farmers revealed their attitudes towards different breeding tools. The largest concern among the interviewed farmers was the economic uncertainty in future dairy production. The expected fluctuating future milk prices made most of them hesitant to implement long term breeding strategies, such as crossbreeding. The unstable market for live animals made some farmers not use genomic selection and sexed semen for production of extra higher genetic merit heifers. Some practical concerns were lower conception rates, more difficult calvings and stillbirths, and risk of disease transmission using sexed semen, beef semen, and embryo transfer as well as a concern of these tools to be time consuming. The increased speed of genetic progress that some of these breeding tools provide were of concern for some farmers; they thought that 'too fast' genetic progress indirectly could lead to poor animal welfare. Based on these interviews, a new hypothesis was formed: farmers prefer breeding tools that are easy to implement, are less time consuming, do not compromise animal welfare (such as calving difficulties and diseases), and provide economic benefit. To test the hypothesis, 3,000 dairy farmers in Sweden are invited to participate in a web-based survey. Results from this preference study will show farmers' obstacles and motives for using different breeding tools.

Exploration of production conditions: a step towards the development of a community breeding program

E.A.M. Omer, S. Addo and D. Hinrichs
University of Kassel, FB 11, Organic Agricultural Sciences, Nordbahnhof 1a, 37213 Witzenhausen, Germany;
uk061369@uni-kassel.de

Butana cattle is a Sudanese breed highly adapted to high temperatures and it plays an essential role in the supply of milk in rural communities. Milk yield of the breed is low and unable to cope with the rising demand for milk. Therefore, many flock owners cross Butana cows with Holstein Friesian bulls indiscriminately, as organised breeding programs are lacking. This study explored the current production conditions of Butana cattle to identify breeding objectives from farmers´ perspectives using a survey. Data on 202 respondents and on farmer-group discussions were collected between October 2018 to January 2019 from 17 villages in the Butana region of central Sudan. The data were analysed using a Chi-squared test, multiple response analysis, and binary logistic regression. From our results, all respondents kept Butana cattle primarily for milk production. Other functions included meat (42%), draught power (9%) and insurance (8%). The average milk yield per/cow/day was 6.6 kg and lactation length was 6.7 months. On a five-point scale (5=most important), milk yield (4.55±0.05), growth rate (4.04±0.07) and lactation length (3.92±0.08) were highly preferred for future developments. Grazing ability and disease resistance were already above optimal levels and less important. On average, 4 crossbred animals per farm were observed, while 63% of farmers had small herds size (<11). Nearly two-thirds (68%) of respondents were willing to adopt crossbreeding using exotic breeds to increase milk performance and about the same proportion were willing to exchange breeding bulls and establish a breeders' association. None of the respondents kept records and 81.7% claimed to have abilities to memorise all relevant information. Interestingly, highly educated farmers were 2 times (P<0.05) more likely to adopt record keeping. Starting with farmer education on the relevance of record-keeping and by exploiting small herd sizes coupled with the willingness to engage, the setup of a community-based breeding program is a milestone in achieving the outlined breeding objectives.

Breeding programs in the South American Creole cattle

G. Mészáros[1], R. Martínez[2], C. Lucero[2], W.O. Burgoz Paz[2], M. Naves[3], H. Doekes[4], J. Windig[4], T. Pook[5] and H. Simianer[5]
[1]University of natural resources and life sciences, Vienna, Gregor Mendel Str. 33, 1180 Vienna, Austria, [2]Corporación colombiana de investigación agropecuaria, AGROSAVIA, km 14 Vía Mosquera, 250047 Bogotá, Colombia, [3]Institut national de recherche pour l'agriculture, l'alimentation et l'environnement, INRAE, Route de Godet 97131, Petit-Bourg, Guadeloupe, [4]Wageningen University and Research, P.O. Box 338, 6700 AH, Wageningen, the Netherlands, [5]University of Göttingen, Albrecht-Thaer-Weg 3, 37075 Göttingen, Germany; gabor.meszaros@boku.ac.at

The IMAGE project aims to explore novel ways to manage animal genetic resources. Such management in a data driven society is more efficient when using newly developed or improved software tools, closely linked with a breeding program design. The objective of our work was to enhance the breeding programs of the Creole cattle populations in Colombia and Guadeloupe in two steps. The first step was a simulation of the breeding program and the possible increase in genetic gain, given the country specific characteristics. In this step, the MoBPS (Modular Breeding Program Simulator) software was used. The estimated breeding values for weaning weight were used to demonstrate the projected changes within the Colombian Blanco Orejinegro (BON) population, including the bulls from gene banks. The second step followed, with the optimisation of the breeding program to constrain inbreeding levels. This step was done utilising the optimal contribution methodology with GENCONT software. With the implementation of selection in the BON population we achieve the expected increase in production levels. In addition, the average heterozygosity also decreases, thus the inbreeding levels increase. Such one-sided selection might lead to unsustainable breeding practices. In a simulation we show that the inbreeding levels of the population could be improved by using the gene bank bulls from the conservation program. The follow up optimal contribution selection approach was implemented to suggest suitable mating plans for simultaneous increase of production level, while putting constraints on inbreeding levels. The use of gene bank bulls had a clear added value, and enabled an even better management of genetic diversity in the population.

Breeding strategies for resilience of dairy production to weather changes in small ruminants

M. Ramon[1], V. Kapsona[2], M.J. Carabaño[3], C. Díaz[3], G. Banos[2] and E. Sánchez-Molano[4]
[1]IRIAF, Av Vino 10, 13300 Valdepeñas, Spain, [2]SRUC, The Roslin Institute. Easter Bush Campus, Midlothian, EH25 9RG Edinburgh, United Kingdom, [3]INIA, Crta de La Coruña km. 7,5, 28040 Madrid, Spain, [4]The Roslin Institute and R(D) SVS, The Roslin Institute. Easter Bush Campus, Midlothian, EH25 9RG Edinburgh, United Kingdom; mramon@jccm.es

A potential strategy to address the impact of climate change on sustainability of livestock production is the use of novel animal phenotypes for resilience in breeding schemes. However, antagonistic correlations between resilience and production level may undermine the potential benefits. The objective of this study was to develop breeding strategies aiming to enhance animal resilience to weather changes and assess the effect on other breeding traits. Weather and animal data from Spain (dairy sheep) and UK (dairy goats) were used to derive individual phenotypes for resilience and estimate genetic parameters and correlations with other traits. Respective populations with animal records for performance and resilience traits were simulated assuming a polygenic model and non-overlapping generations, and estimated breeding values for production, climate resilience and functionality traits were combined into a selection index in each generation. Different scenarios were based on the varying emphasis of the traits included in the breeding goal. For dairy sheep, traits considered were daily milk, fat and protein yields, conception rate and, as resilience trait, protein yield to daily temperature change. A 10-20% relative emphasis on resilience resulted in animals maintaining a high production regardless of weather conditions. For dairy goats, traits were daily milk yield, longevity, mastitis incidence, age at fist kidding and as resilience trait, milk yield to daily temperature change. Despite the antagonistic correlation between milk yield and resilience, an index comprising milk production and functional traits remained unaffected when a 10% emphasis on resilience was considered in the breeding goal. In conclusion, not considering resilience led to performance losses due to weather changes. Therefore, breeding strategies placing a low emphasis on resilience are recommended to reduce losses due to weather change without compromising other traits in the breeding goals.

Stochastic simulation of breeding schemes for Red dairy cattle breeds across Europe

C. Schmidtmann[1], M. Kargo[2], A.C. Sørensen[2] and G. Thaller[1]
[1]Kiel University, Institute of Animal Breeding and Husbandry, Hermann-Rodewald-Str. 6, 24118 Kiel, Germany, [2]Aarhus University, Center for Quantitative Genetics and Genomics, Department of Molecular Biology and Genetics, Blichers Allé 20, 8830 Tjele, Denmark; cschmidtmann@tierzucht.uni-kiel.de

Breeding programs of Red dairy breeds in Europe (ERDB) are primarily organised at national level. However, some countries have already recognised the potential of collaboration and implemented common breeding strategies (e.g. VikingRed: Denmark, Sweden, Finland). One way to strengthen the cooperation between ERDB is to establish common breeding lines across these breeds, meaning that the populations develop in directions specialised for particular production systems. On the one hand, this offers the possibility to produce animals, which are well adapted to specific production conditions. On the other hand, this might be advantageous especially for smaller sized populations in terms of genetic progress and inbreeding as they can make use of genetic material provided by other breeds. The aim of this study is to investigate the consequences of common breeding concepts for ERDB by stochastic simulation. In a first step, two production systems were defined reflecting the most prevailing production conditions for Red breeds in Northern Europe. These production environments differed, for instance, regarding the level of input factors required for milk production and the application of management and reproductive technologies such as heat observation and MOET. For both production systems economic values were derived by means of a stochastic bio-economic model in order to set up appropriate breeding goals. Based on these breeding goals, most optimal animals for both production environments were selected from a shared reference population consisting of ERDB and, subsequently, breeding programs were simulated over a time horizon of 30 years. Consequences of these breeding programs will be evaluated and judged on genetic gain for a given rate of inbreeding. Furthermore, the number of bulls selected for both breeding goals will be studied in the course of time, depending on changing correlations between the defined breeding goals.

Prediction of gametic variance and its use in breeding programs

C. Hoze[1], A. Baur[1], S. Fritz[1] and D. Boichard[2]
[1]Allice, 149 rue de Bercy, 75595 Paris, France, [2]Université Paris-Saclay, INRAE, AgroParisTech, GABI, Domaine de Vilvert, 78350 Jouy-en-Josas, France; didier.boichard@inrae.fr

Within-parent variability of progeny breeding values is variable across parents. This gametic variance is related to the level of heterozygozity of the QTLs, to linkage disequilibrium and coupling/repulsion between QTLs. In breeding programs, the use of parents with the highest breeding values but also the highest gametic variance maximises the probability of obtaining and selecting extreme elite progeny and thus generating high genetic gain. Although the detection of the parents with high gametic variance is of high interest in selection, there is no obvious predictor of this parameter. In this study, we estimated this gametic variance by simulation and compared it to various predictors. First, for each proposed couple of genotyped sire and dam, a large number of offspring were generated by simulating meiosis and fertilisation, and their genetic merit was estimated by computing their direct genomic value (DGV) for all evaluated traits and for total merit index by using the allelic effects obtained in the official genomic evaluation. In addition, their genotype was predicted for all known genetic defects. To obtain a good prediction of the probability of getting an exceptional progeny, at least 500 progeny must be simulated, generating a large computing load (50 million progeny per 100 sires and 1000 dams). The gametic variance of the couple was estimated by the variance of the progeny DGV. This estimate was compared to various indicators calculated from the genotypes of the parents: heterozygosity, length of runs of homozygosity, sum of of squares of the contrasts between allelic effects, or gametic variance estimated from allelic effects and linkage disequilibrium. This latter parameter was found to be the best estimator of the gametic variance observed in the progeny, while marker homozygosity and runs of homozygosity have limited prediction value. In practice, the service has been offered to the breeding companies since 2019 by GenEval. This work was carried out within H2020 GenTORE project.

Using SNP genotypes for automatic ploidy screening of cattle embryos to improve pregnancy rates

A.C. Bouwman[1], E. Mullaart[2], J.M. Henshall[3], R. Hawken[3], I. Hulsegge[1], D. Schokker[1], C. Kamphuis[1] and R.F. Veerkamp[1]
[1]Wageningen University & Research, Animal Breeding and Genomics, P.O. Box 338, 6700 AH, the Netherlands, [2]CRV B.V., P.O. Box 454, 6800 AL Arnhem, the Netherlands, [3]Cobb-Vantress Inc., Siloam Springs AR, 72761-1030, USA; aniek.bouwman@wur.nl

Embryos are routinely used in current cattle nucleus breeding programs, but pregnancy success rate is still rather low, especially for *in vitro* produced embryos. This could in part be caused by increased incidence of aneuploidy (e.g. trisomy and monosomy) and polyploidy observed in embryos. The cattle embryos for the nucleus breeding program are routinely genotyped with SNP-arrays before placement for genomic selection purpose. Abnormalities in ploidy can be detected using intensity data resulting from genotyping. Hence we can exploit these genotypes to investigate the incidence in cattle embryos and their relation to pregnancy success rate. Genotype data from 150 embryos is available with 10K or 50K SNP nicely distributed over the different chromosomes. The B-allele frequency, Log2 R ratio, X and Y normalised intensity signals will be used to identify abnormalities. For the embryos that were placed in recipients we know whether there was a successful pregnancy or not, therefore we can investigate to which extent the detected abnormalities caused pregnancy failure. A pipeline will be built in Microsoft Azure that will automatically detect and classify specific abnormalities, (e.g. monosomy or triploid) using machine learning. The developed pipeline can be used to pre-screen genotyped embryos routinely and only place embryos without observable abnormalities in recipients.

The breeder's equation in the era of genomic selection

H.H. Swalve and M. Wensch-Dorendorf
Martin Luther University Halle-Wittenberg, Institute of Agricultural and Nutritional Sciences, Theodor-Lieser-Str. 11, 06120 Halle, Germany; hermann.swalve@landw.uni-halle.de

Following Jay L. Lush, the breeder's equation, $R = h^2 S$, is widely known. In this formula, R denotes the response to selection, h^2 the heritability in the narrow sense and S is the selection differential. If standardised selection differentials are used, the formula becomes $R = i h^2 \sigma_P$, where i is the selection intensity and σ_P is the phenotypic standard deviation of the trait in question. This formulation is equivalent to $R = i h \sigma_A$, where σ_A is the additive genetic standard deviation and h is equivalent to the accuracy of selection based on own performance, often in general denoted as r_{TI}. The breeder's equation then can be written as $R = i r_{TI} \sigma_A$. This latter expression of the breeders' equation greatly aids in understanding that while the expression given as $R = h^2 S$ appears to suggest that for a trait with low heritability not much genetic progress can be expected, $R = i r_{TI} \sigma_A$ much more clearly demonstrates that substantial progress by genetic selection can be achieved if the genetic variation is large and the accuracy of selection is at least moderate. Traditionally, as estimates of heritabilities for functional traits and health traits are commonly low, it has been assumed that the expected genetic progress from genetic selection for these traits also is marginal. In the era of genomics, now, after phenotyping schemes have been put into place that enable deriving genomic breeding values from large calibration samples, the accuracy of selection, r_{TI}, for gEBV often is of substantial magnitude. Reliabilities for gEBV for health traits in the magnitude of 50 to 70% (equivalent to $r_{TI} = 0.70$ and 0.83, respectively) are not unrealistic any more. At the same time, accuracies for gEBV of health traits often show to be substantially higher than theoretical expectations and many health traits exhibit fairly large coefficients of variation on the phenotypic as well as on the genetic scale. Assuming equal selection intensities, the parameter $R \% = (r_{TI} \sigma_A / mean) \times 100$ can be defined and applied for comparisons. Using model calculations, this presentation will demonstrate what has to be expected for genetic progress in health vs traditional traits in today's era of genomics.

Genotyping strategies in Latxa dairy sheep breed

I. Granado-Tajada[1], L. Varona[2] and E. Ugarte[1]
[1]Neiker, Produccion Animal, Campus Agroalimentario de Arkaute, 01080, Spain, [2]Instituto Agroalimentario de Aragon (IA2), Universidad de Zaragoza, Unidad de Genetica Cuantitativa y Mejora Animal, Calle de Miguel Servet, 177, 50013 Zaragoza, Spain; igranado@neiker.eus

The inclusion of genomic information into genetic evaluations has been shown to be beneficial for big and well-connected populations, such as dairy cattle. However, for small populations with lower use of artificial insemination, it is necessary to consider their specific characteristics to study the possible advantage of using genotyped individuals. So, the aim of this work is to compare the effect of different genotyping strategies in terms of the accuracy of the predictions, based on the data structure of two dairy sheep breeds, Latxa Cara Rubia (LCR) and Latxa Cara Negra from Euskadi (LCNEUS). To do so, a population was simulated using a base of real pedigree and genotypes of 374 LCNEUS and 451 LCR animals genotyped with the Illumina OvineSNP50 BeadChip. The obtained phenotypic and genotypic data were used into genetic (BLUP) and genomic (ssGBLUP) evaluations for milk yield. For each breed, the genotypes of 1000 males with progeny data were simulated and young rams without phenotypic data were used as validation group. No gain in accuracy was found, resulting in almost the same accuracy into both methodologies (0.46 LCNEUS and 0.53 LCR), which is probably conditioned by lacking genealogy. So, more studies are being done to understand if the percentage of missing genealogy or the data structure are affecting these results. Also, the benefits of natural service rams and females genotyping will be studied.

Design of community-based breeding strategies for llamas in the Central Andes of Peru

G. Gutiérrez[1], J. Mendoza[1], M.J. Arias Coronel[2] and M. Wurzinger[1,3]
[1]Universidad Nacional Agraria La Molina, Av. La Molina s/n, 15024 Lima, Peru, [2]Direccion Regional de Agricultura Pasco, Edificio Estatal no. 2, San Juan, 19001 Cerro de Pasco, Peru, [3]BOKU-University of Natural Resources and Life Sciences, Department of Sustainable Agricultural Systems, Gregor-Mendel-Strasse 33, 1180 Vienna, Austria; maria.wurzinger@boku.ac.at

The llama production system in Peru can be characterised as an extensive, low-input, pasture-based system. Llamas are usually kept by smallholder farmers at altitudes between 3,800 m and up to 5,000 m above sea level. In recent years some farmers in the Central region of the Peruvian Andes showed increased interest in improving their herds as llamas seem to be better adapted to the changing weather conditions. So far, no national breeding program has been established by the government. Therefore, farmers in the region of Cerro de Pasco founded the association ProLlama with the support of the university with the aim to establish community-based breeding strategies. The majority of the 600 registered members keep the K´ara type (used for meat), but some also keep the Chaqu type (used for fibre). Thus, different breeding strategies were developed using a participatory approach with individual interviews, consultation workshops and field visits. Communal cooperatives play a key role in the breeding program as they can operate as nucleus herds and redistribute proofed males to smaller herds. The K´ara males are selected based on growth performance, whereas for Chaqu type fibre traits such as fibre diameter and fleece colour will be considered. For K´ara llamas a dispersed nucleus consisting of 7 local nucleus herds is proposed. Through a rotation scheme of selected males these herds will be genetically connected. For the smaller Chaqu population a central nucleus will be established at one communal cooperative and linked with 7 multiplier herds. Selected males will be kept in a central mating centre, but also lent to farmers, who live in very distant locations and cannot afford transport of their female animals. These different breeding strategies take into account the existing institutional settings, but also geographical aspects of widespread and distant living families.

Genomic analysis of the inbreeding depression for litter size in two varieties of Iberian pig

C. Hervás-Rivero[1], J. Caselllas[2], J.L. Noguera[3], N. Ibañez-Escriche[4], M. Martín De Hijas-Villalba[2], M. Vázquez-Gómez[2], S. Negro[5], J.P. Rosas[5] and L. Varona[1]
[1]Universidad de Zaragoza, Instituto Agrolimentario de Aragón (IA2), 50013.Zaragoza, Spain, [2]Universitat Autònoma de Barcelona, Dept. Ciència Animal i dels Aliments, 08193, Spain, [3]Institut de Recerca i Tecnologia Agroalimentàries (IRTA), Genètica i Millora Animal, 25198 Lleida, Spain, [4]Universitat Politècnica de València, Instituto Universitario de Ciencia y Tecnología Animal, 46022 Valencia, Spain, [5]Programa de Mejora Genética 'CASTUA', INGA FOOD S.A. (Nutreco group), 06200 Almendralejo (Badajoz), Spain; lvarona@unizar.es

Inbreeding depression is expected to be more noticeable in fitness-related traits, such as pig litter size. Recent studies have suggested that the genetic determinism of inbreeding depression may be heterogeneous along the genome. Therefore, the objective of this study was to perform a genomic scan along the pig autosomal genome to detect the genomic regions that control the inbreeding depression for litter size in two varieties of Iberian pig (Entrepelado and Retinto). The datasets consisted of a total of 1,575 (326 sows) and 1,727 (310 sows) records for litter size related traits (Total Number Born and Number Born Alive) for the Entrepelado and Retinto varieties, respectively. All sows were genotyped with the Geneseek GGP Porcine HD 70K (Illumina), and, after appropriate filtering, 35,190 markers were used. The genomic scan was performed by solving single-step mixed model equations for each genomic segment of 20 SNP. The model included two covariates, the first with the heterozygosity of the specific region of 20 SNP markers and the second with the heterozygosity of the remaining autosomal genome. The results of the first covariate were used to determine the genomic regions that contribute to inbreeding depression. Several genomic regions that may have a more substantial influence in the inbreeding depression of litter size in SSC6 (at Mb 132), SSC8 (Mb 141) and SSC14 (Mb 84) for the Entrepelado variety and in SSC1 (Mb 13), SSC6 (Mb 15), SS7 (Mb 114), SSC9 (Mb 15), SSC9 (Mb 37) and SSC11 (Mb 81) and SSC14 (Mb 141) for the Retinto variety were detected. Moreover, no genomic regions were shared between varieties.

Evaluation of the breeding value of workability traits in Polish Holstein-Friesian cattle population

B. Szymik[1], W. Jagusiak[1,2], P. Topolski[1] and K. Żukowski[1]
[1]National Research Institute of Animal Production, 1, Krakowska Street, 32-083 Balice, Poland, [2]University of Agriculture in Krakow, 21, Mickiewicza alley, 31-120 Krakow, Poland; bartosz.szymik@izoo.krakow.pl

Workability traits (WT) are included in breeding programmes in most European countries. In Poland, those traits are not yet covered in the routine breeding value estimation. The WT include the milking speed (MS) measured on a scale of 1 to 5 points and the temperament (MT) measured on a scale of 1 to 3 points. Since 2006, cows are scored for WT on the second test-day of the first lactation. These data are collected in the SYMLEK database. The study was carried out on the 102,834 records of Polish Holstein-Friesian cows. First, the heritability of both traits as well as the genetic and phenotypic correlation between them were estimated. A bivariate linear model was applied to estimate (co)variance components. The model included the fixed effects of herd-year-season (HYS), lactation stage, fixed regressions on a percent of Holstein-Friesian genes and age of calving and random genetic animal effect. The herds for which no variability of MS or MT was found were removed from the basic data set. In other herds, standardisation has been carried out to a common variance. Additionally, cows that did not have at least 9 half-sisters in the HYS subclass were excluded. After the restrictions were imposed, the data set consisted of 20,695 cows assigned to 908 HYS subclasses. These cows were daughters of 1,399 bulls. Variance components were estimated using the Bayesian method with the Gibbs sampling algorithm. The estimated heritability was 0.120 for MS and 0.08 for MT. Genetic and phenotypic correlations between MS and MT were 0.28 and 0.003, respectively. The breeding values of cows were estimated using a bivariate animal model. As a result, genetic and phenotypic trends were found for both traits. Regression coefficients of mean breeding values per year of calving cows ranged from 0.27 for sires (MS) to 0.72 for cows (MT). The results of our research confirm that it is possible to estimate the breeding values of bulls for both traits in the Polish HF cattle population and to incorporate the developed methodology into the system of routine breeding value estimation in Poland.

Nanotechnology as a key approach to improve animal feed science and production

S. Reis[1], J. Albuquerque[1,2], I. Van Dorpe[3], A.J.M. Fonseca[2] and A.R.J. Cabrita[2]
[1]University of Porto, LAQV, Requimte, FFUP, Porto, 4099-002, Portugal, [2]University of Porto, LAQV, Requimte, ICBAS, 4099-002, Porto, Portugal, [3]PREMIX-Especialidades Agrícolas e Pecuárias. Lda, 4900-934, Viana do Castelo, Portugal; shreis@ff.up.pt

Nanoparticles are nowadays widely used in several sectors, as nutrition, therapy, and diagnosis. Nanotechnology is a promising and emerging field with a great potential to improve the nutritional value of animal feeds. Indeed, the development of nanodelivery systems revealed to be a promising strategy for enhancing the efficacy of bioactive compounds by protect them from degradation, enhancing its bioavailability, and modifying feed functionality. The different nanomaterials that have the potential to be used and the key parameters that should be taken in account during the nanoparticle optimisation for feed industry applications will be highlighted. As a case study, the development of a lipid nanocarrier to improve lysine bioavailability, capable of protecting it across the bovine digestive tract and delivering it into the bloodstream where it could be used by the animal, will be presented. The rationale that guided the planning and development of this study will be described as well as the most important technical concerns when designing a nanoparticle-based approach.

Effects of grain processing and undegradable fibre on rumen pH and fermentation of high grain diets

W.Z. Yang[1], T. Ran[1,2], A.M. Saleem[1,3], K.A. Beauchemin[1] and G. Penne[4]
[1]Agriculture and Agri-Food Canada, Research Centre, Lethbridge, T1J4B1, Canada, [2]Faculty of Veterinary Medicine, University of Calgary, Calgary, T2N1N4, Canada, [3]Department of APP, South Valley University, Qena, 83523, Egypt, [4]Department of APS, University of Saskatchewan, Saskatoon, S7N5A8, Canada; wenzhu.yang@canada.ca

Fibre is required by feedlot cattle fed high-grain diets to reduce the risk of rumen acidosis and optimise growth rate and feed efficiency. It was hypothesised that altering ruminally fermentable carbohydrate and undegradable fibre (uNDF) levels in finishing cattle diets would affect the risk of acidosis. The objective of this study was to investigate the effects of processing index (PI, weight after processing/weight before processing × 100) of barley grain and dietary uNDF level on dry matter intake (DMI), ruminal pH and fermentation characteristics of finishing beef cattle. Six ruminally cannulated beef heifers (body weight, 715±29 kg) were used in a 6×6 Latin square design with 3 PI (65, 75 and 85%; fine, medium, coarse, respectively) × 2 uNDF levels (low and high; 4.1 vs 5.0% of DM) factorial arrangement. The heifers were fed *ad libitum* a total mixed ration consisting of 10% barley silage (low uNDF), or 5% silage and 5% straw (high uNDF), 87% dry-rolled barley grain, and 3% vitamin and mineral supplement. Data were analysed using the mixed procedure of SAS with a model that included fixed effects of PI, uNDF and interactions, and random effects of period and heifer. Sampling day and sampling time were considered as repeated measurements. An interaction of PI with uNDF occurred (P<0.01) for DMI (kg/d); DMI was greater (P<0.05) for high (12.7) versus low (12.1) uNDF diets with 85% PI, with no difference in DMI (average, 12.1) between low and high uNDF diets with 65 or 75% PI. There was no interaction between PI and uNDF for mean rumen pH or duration of pH<5.8, <5.6 and <5.2 over a 24-h period. However, duration of pH<5.8 (14.6 vs 13.3 h) and <5.6 (10.8 vs 8.6 h) was reduced (P<0.05) by feeding high vs low uNDF diets. The PI did not affect total rumen volatile fatty acid (VFA) concentration or molar proportions, but VFA concentration was less (172 and 162 mM; P<0.01) and acetate proportion was greater (46.0 and 48.2%; P<0.01) with high vs low uNDF. The propionate proportion was less (P<0.04) with high (35.0%) than low uNDF (40.4%) for 75% PI, but it was not different for PI of 65 or 85%. Ratio of acetate to propionate was not impacted by PI or uNDF level. These results demonstrate that increasing dietary uNDF level is an effective strategy to improve rumen pH status in feedlot cattle, regardless of extent of grain processing. In contrast, minimising the extent of grain processing did not reduce the risk of rumen acidosis.

Requirement-oriented supply of organic broilers with riboflavin from fermentation of *Ashbya gossypii*

C. Lambertz[1], J. Leopold[1], S. Ammer[2], B. Thesing[2], C. Wild[3], K. Damme[3] and F. Leiber[4]
[1]Research Institute of Organic Agriculture (FiBL), Kasseler Strasse 1a, 60486 Frankfurt am Main, Germany, [2]Georg-August-University, Department of Animal Sciences, Albrecht-Thaer-Weg 3, 37075 Göttingen, Germany, [3]Poultry Competence Centre of the Bavarian Institute for Agriculture, Mainbernheimer Strasse 101, 97318 Kitzingen, Germany, [4]Research Institute of Organic Agriculture (FiBL), Ackerstrasse 113, 5070 Frick, Switzerland; christian.lambertz@fibl.org

The required riboflavin (vitamin B2) demand of poultry is hardly met by the native contents of grains and legumes. A supplementation is consequently necessary to avoid deficiency in this vitamin, which has detrimentous effects on performance, health and welfare. In organic production riboflavin supplementation must origin from GMO-free sources. A dried product from fermentation of the yeast-like fungi *Ashbya gossypii* with a high native content of riboflavin was tested in two trials with 800 mixed-sex one-day old Ranger Gold™ chicken. Four different riboflavin concentrations were tested in each run. In the first trial, a control diet contained only native riboflavin contents without supplementation (2.9 mg riboflavin/kg), the other diets were supplemented at graded dosages up to contents of 9.2 mg/kg. In the second trial, starter feeds contained 4.0 mg and 5.6 mg riboflavin/kg, whereas contents at the later fattening stages were reduced. Body weight and feed consumption were measured weekly. Additionally, slaughter parameters were assessed (dressing, proportion of valuable cuts, abdominal fat, liver, heart and gizzard). In the first trial, the native riboflavin contents did not meet the requirements of the slow-growing broilers and resulted in deficiency symptoms. High supplementations, however, did not have positive effects on performance and slaughter parameters. Consequently, a reduction of the riboflavin supplementation particularly during the later fattening period seems to be possible to compensate higher production costs of the tested product compared to conventional vitamin production by GMO.

Effect of feeding press cake silage as replacement for high quality grass silage on milk production

E. Serra[1], M.B. Lynch[1], M.H. Bock[1], J. Gaffey[2], J.P.M. Sanders[3], B. Koompans[4] and K.M. Pierce[1]
[1]University College Dublin, School of Agriculture and Food Science, Belfiled, Dublin, Ireland, [2]Institute of Technology Tralee, Clash Road, Tralee, Ireland, [3]Wageningen University, Valorisation of plant production chains, Droevendaalsesteeg, Wageningen, the Netherlands, [4]GRASSA! BV, Sint Jansweg, Venlo, the Netherlands; eleonora.serra@ucdconnect.ie

The biorefinery process represents an opportunity to sustainably diversify the product base of grassland and reduce the impact of agriculture on the environment. In this biorefinery process, press cake silage is obtained by ensiling the dry material remaining after mechanical separation of the liquid juice from fresh grass. The objective of this experiment was to evaluate the effect of direct replacement of traditional high quality grass silage (25% DM, 16% CP, 77% DMD) with press cake silage (37% DM, 10% CP, 62% DMD) in the diet of dairy cows on milk yield and composition. A randomised complete block design experiment involving 30 early lactation dairy cows (70 DIM) was conducted over a 56-d period with two dietary treatments (T; n=15) as follows: T1 (GS) consisted of grass silage (14 kg DM) plus concentrate (C) (7.4 kg DM) and soya bean meal (SBM) (0.44 kg DM) and T2 (PC) had the same quantities of C and SBM but 2/3 of the grass silage was replaced with press cake silage (5 kg DM grass silage; 9 kg DM press cake silage). 4 kg of C was offered in the parlour at milking twice daily while the remaining feed was offered as a TMR once daily. Data were analysed using the proc mixed procedure of SAS (Version 9.4). There were no significant differences in milk yield and composition between dietary treatments in this study. Milk yield for GS was 31.14 kg/d (±1.27) and 29.29 kg/d (±1.16) for PC (P=0.18); milk fat concentration was 4.42% (±0.16) GS vs 4.26% (±0.17)PC; P=0.40 and milk protein concentration was 3.35% (±0.08) GS vs 3.34% (±0.08) PC; P=0.89. Results show that press cake silage can act as a partial replacement for high quality grass silage in the diet of early lactation autumn calving dairy cows without a negative impact on milk yield and composition.

Physiological adaption of broiler chickens challenged with varying qualities of soybean cake

S. Kuenz[1], D. Hoffmann[1], S. Thurner[2], K. Damme[3], W. Windisch[1] and D. Brugger[1]
[1]Technical University of Munich, Chair of Animal Nutrition, Liesel-Beckmann-Str. 2, 85354 Freising, Germany, [2]Bavarian State Research Center for Agriculture, Insitute for Agricultural Engineering and Animal Hunsbandry, Voettinger Str. 36, 85354 Freising, Germany, [3]Bavarian State Research Center for Agriculture, Department for Education and Poultry Research, Mainbernheimer Str. 101, 97318 Kitzingen, Germany; sylvia.kuenz@wzw.tum.de

Soybeans can express high amounts of trypsin inhibitor activity (TIA), which impairs protein utilisation and thus animal performance. Hence, TIA has to be reduced by heat and pressure. Different processing techniques may result in vast differences in qualities. The present study used differently processed soybean cakes (SBC) to test the physiological adaption of broiler chickens in a fattening as well as a digestibility study. For this project, two feeding trials were conducted. For both trials several SBC with a wide range of TIA (0.3- 23.6 mg/g) were created using four processing techniques (thermal, hydrothermal, pressure and kilning). For trial 1, 1,680 broiler chickens (Ross 308) were used to test 34 SBC variants in a performance trial. On d 1, 11, 24 and 35 weight were recorded. On d 35 birds were slaughtered. For trial 2, 5,490 birds were used in total to test amino acid digestibility of 45 SBC. From d1 to d14 birds were fed, with a commercial starter diet. From d15 onwards, birds were fed with experimental diets (diets containing 0.5% TiO_2 as indigestible marker). On d21 birds were euthanised with CO_2 and digesta from the terminal ileum was collected. Data collection comprised zootechnical performance, pancreas weight and standardised ileal amino acid digestibility Data analyses occurred as linear regression applying the SAS 9.4 package. Life weight and feed efficiency were downgraded with rising dietary TIA significantly in a linear way (P<0.001), whereas pancreas weight significantly increased (P<0.001). Additionally, we demonstrated that each mg of dietary TIA decreased amino acid digestibility. Cysteine was showing the worst response to TIA (4.95% Cys digested at 8.7 mg/g TIA). Broiler organism adapted to increasing levels of TIA with pancreatic hypertrophy to compensate for the TIA effect on protein digestion in the small intestine.

Effects of increasing levels of insect fat supplementation on growth performance of nursery pigs

G. Hosotani[1], J. De Laat[2], J. De Oliveira[1], P. Ringard[3] and D. Melchior[4]
[1]Cargill, Research and Development Centre Europe, Vilvoorde, 1800, Belgium, [2]Cargill, Animal Nutrition Innovation Centre, Velddriel, 5334, the Netherlands, [3]InnovaFeed, Nesle, 80190, France, [4]Cargill, Animal Nutrition, Crevin, 35320, France; guilherme_hosotani@cargill.com

The objective of the experiment was to determine the dose-response effects of dietary levels of insect fat on growth performance of nursery pigs. A 39-d trial was conducted with a total of 120 pigs (initial BW 7.0±0.96 kg; d22) randomly assigned to 1 of 4 dietary treatments with 4 increasing levels of insect fat replacing soy oil; 0, 1, 3, and 4% in phase I diets (d0 – 14), and 0, 0.5, 1.5, and 2% in phase II diets (d14 – 39). Each treatment consisted of 10 replicates with 3 pigs per pen (mixed sex) and blocked by initial weight. At d 8, 14, 26, and 39, pigs were weighed and feed disappearance was measured. Faecal score was evaluated at d 4, 8, and 14. Data were analysed as a randomised complete block design using R software with pen as experimental unit and contrasts were used to determine linear and quadratic effects. Faecal score data were analysed as binomial variable (0 = normal and 1 = soft). Differences were significant at P≤0.05 and a trend between P>0.05 to P≤0.10. No differences in ADG and ADFI were observed among treatments in all measurements. At d 8, there was linear improvement in feed efficiency (G:F) with increasing levels of insect fat (P=0.03) with G:F at 0.79, 0.80, 0.90, and 0.89 of pigs fed 0, 1, 3, and 4% insect fat. During week 1, there were numerical linear increases in ADG and ADFI. ADG were 114, 123, 123, and 125 g/d and ADFI were 92, 104, 112, and 112 g/d with insect fat at 0, 1, 3, and 4%, respectively. Effects in ADG and ADFI remained until the end of experiment (d39), resulting in BW at 27.03, 27.30, 27.58, and 27.85 kg, respectively. There were no differences in probability of having more watery stool among treatments, however, there was a tendency of linear improvement at d 14 with increasing levels of insect fat (P=0.06). In conclusion, feeding insect fat does not affect overall growth performance of nursery pigs, however, feeding insect fat at levels up to 4% of insect fat during first week post-weaning can improve feed efficiency.

Effects of fresh forage species on *in vitro* total gas and methane production: preliminary results

S. Contini[1], G. Molle[1], G. Scanu[1], M. Decandia[1], P. Bani[2], V. Giovanetti[1], M. Acciaro[1], G. Serra[1] and A. Cabiddu[1]
[1]Agris Sardegna, Animal Science Department, Loc. Bonassai, Sassari, 07040, Italy, [2]Università Cattolica del Sacro Cuore, Department of Animal Sciences, Food and Nutrition, Via Emilia Parmense, 84, 29100, Italy; acabiddu@agrisricerca.it

The objective of this study was to investigate the effect of different inclusion levels of fresh forage species (grasses: oat or ryegrass; legumes: sulla or berseem clover) on *in vitro* total gas and methane production of dairy sheep diets. The diets were previously tested in two digestibility trials. In trial 1 oat and ryegrass were evaluated at high, medium and low level of inclusion (57, 46 and 32% of total diet DM); in trial 2 sulla and berseem clover were evaluated at high, medium and low level of inclusion (48, 34 and 21% of total diet DM diet). The total diets included also hay and concentrate. Overall we tested 12 diets during three fermentation runs: samples (440±0.2 mg) were incubated in triplicate with 40 ml of medium and 20 ml of rumen fluid. Total gas (GP) and CH_4 production were measured at 2 and 24 h of incubation. Data were analysed by GLM considering diet, sampling hour and their interaction as fixed effects. Data of GP and CH_4 were expressed as ml/g of digestible DM (DDM). Legume-based diets showed higher CP and EE content than grass-based ones (15.27 vs 14.00% DM, for CP and 3.55 vs 2.82% DM, for EE), while NDF was higher in grass than in legume diets (41.06 vs 36.40% DM). The content of total polyphenols in the diets was similar between legumes and grasses (1.19 vs 1.22% DM). Overall diet and sampling hour affected both GP and CH_4 production. Legumes increased GP and CH_4 compared to grasses, in particular at 24 h (P<0.01). No effect of inclusion level (high, medium and low) was detected on GP and CH_4. In general, sulla showed the highest value of GP (221 ml vs oat, ryegrass and berseem clover (197, 197 and 211 ml; P<0.01, respectively) and CH_4 (47 vs 37, 38 and 43 ml P<0.01; for sulla, ryegrass, oat and berseem clover respectively). Moreover, *in vitro* CH_4 production at 2 and 24 hour of incubation increased significantly (P<0.05) with the level of sulla in the diet.

Ketosis and energy balance milk MIR spectral predictions – practical use

L.M. Dale[1], F. Onken[2], H. Spiekers[3], P. Hertel-Böhnke[3], E. Stamer[4], F.J. Auer[5], A. Köck[6], C. Egger-Danner[6], K. Drössler[1], F. Gollé-Leidreiter[1] and A. Werner[1]
[1]*State Association for Performance and Quality Inspection in Animal Breeding of Baden Wuerttemberg, Heinrich-Baumann-Str. 1-3, 70190 Stuttgart, Germany,* [2]*German Association for Quality and Performance Testing e.V., Irmintrudisstraße 15, 53111 Bonn, Germany,* [3]*Bavarian State Research Centre for Agriculture, Prof.-Dürrwaechter-Platz, 85586 Grub-Poing, Germany,* [4]*TiDa, Animal and Data GmbH, Bosseer Str. 4c, 24259, Westensee, Germany,* [5]*State Association for Performance and Quality Inspection in Animal Breeding of Austria, Dresdner Straße 89/19, A-1200 Wien, Austria,* [6]*ZuchtData, Dresdner Straße 89/B1/18, 1200 Wien, Austria; awerner@lkvbw.de*

Ketosis, a metabolic disorder associated with milk characteristics, can be measured by MIR spectra at individual level. Ketosis causes a further decrease in feed intake and milk yield, a more pronounced negative energy balance while milk fat content increases by body fat mobilisation. The aim was to use the dairy cows' veterinary ketosis diagnosis, milk MIR spectra from routine milk recording and energy balance data for identifying meaningful variables associated with ketosis and negative energy balance. Farm management and services of the milk recording organisation and its feed advisors should benefit from improving the established KetoMIR model and validating two ways of calculating energy balance by MIR spectra. The KetoMIR model was based on 810,496 spectra from 10,079 herds in D4Dairy project. The energy balance models calculated according GfE (2001) in NEL (EB-NEL) and Susenbeth (2018) in ME (EB-ME) were developed based on ~30,000 spectra from German research farms from the optiKuh project. KetoMIR included DIM 5 to 120, while the energy balance models have included the whole lactation period. Statistical analysis was made in R software using 'glmnet' library. Spectra were first standardised, were applied first derivative and worked on 212 wavelengths. The strong correlation with ketosis indicators justified the construction of a multi class scheme based on thresholds. These were applied to ketosis probability in order to overcome the restrictions of a binary classification in the central region of probability. The KetoMIR model probability showed high correlation with EB-NEL and ME, energy deficit, blood BHB and milk yield. The moderately endangered class is a signal for the farmer to consult the veterinary for examining the identified cow in order to prevent ketosis disease in time. KetoMIR index could be used in herd management to detect general feeding deficiencies in the late and early lactation transition period at herd level. Within the projekt D4Dairy models were evaluated and optimised for the use in routine herd management and breeding evaluation.

Evaluation of replacing Dacron bags with ANKOM bags for indigestible NDF determination

M. Rønn and M.R. Weisbjerg
Aarhus University, Department of Animal Science, AU-Foulum, 8830 Tjele, Denmark; martin.weisbjerg@anis.au.dk

Indigestible NDF (iNDF) is an important feed value. The standard procedure applied in Nordic Feed Evaluation System (NorFor) is the NDF residue after 288 h *in situ* incubation in Dacron bags. The aim was to study if the standard method could be replaced with less resource demanding methods and if a finer ground sample could be used. Five combinations were included; ANKOM F57 and F58 bags both with 1.0 mm and 1.5 mm screen size used for milling and the standard method with 1.5 mm screen size were tested on 16 samples of grass silage, whole crop-maize and maize stem. One additional treatment with ANKOM F57 bags and 1 mm screen size with NDF determination before rumen incubation was included, to test if the two analyses could be combined. Three replicates of each sample within treatment were distributed between three cows and incubated in the rumen for 288 h. NDF residues from standard procedure were analysed using FibertecTM 2010 and ANKOM bag residues using ANKOM 2000 Fibre Analyzer. The results for NDF and ash-corrected iNDF (iNDFom) from the standard procedure were 47.4, 46.1 and 64.0 g NDF/kg DM and 6.3, 8.6, and 13.8 g iNDF/kg DM for grass silage, whole crop-maize and maize stems, respectively. A two-way ANOVA showed no difference (P=0.63) and a high correlation (R=0.98-0.99, P≤0.001) between the five combinations. An analysis on 1 mm screen size samples in F57 bags with or without NDF determination before rumen incubation showed significant lower iNDF (P<0.01) with NDF determination before rumen incubation with no difference for grass silage (6.4 and 6.2 g iNDF/g DM), but major differences for maize samples, as iNDF was reduced with 2.3 and 8.2 g iNDF/kg DM for whole crop-maize and maize stem. Therefore, ANKOM bags can be used to determine iNDF with both milling screen sizes. However, F58 bags showed greater risk of rupturing as five bags with low or no amount of NDF left after incubation were excluded from the analysis. Furthermore, it is not possible to run iNDF sequentially after NDF analysis in the F57 bags.

SESAM sensor technology for milk producers

Z. Vassilev[1], M. Hoenen[2], K. Drössler[1], A. Werner[1] and F. Gollé-Leidreiter[1]
[1]State Association for Performance and Quality Inspection in Animal Breeding of Baden Wuerttemberg, Heinrich-Baumann-Str. 1-3, 70190 Stuttgart, Germany, [2]Chamber of Agriculture Alsace, rue de Rome 2, 97013, Schiltingheim, France; ZVassilev@lkvbw.de

Sensors dominate modern agricultural technology more and more, also could milk recording organisation benefit from this. Sensor assisted Alpine milk production (SESAM) project is an Interreg Alpine Space project between State Association for Performance and Quality Inspection in Animal Breeding from Austria, Germany (Bavaria and Baden-Württemberg), Association of the South Tyrolean Breeders (Italy), the Chamber of Agriculture in Slovenia and Alsace (France), the IT service provider Qualitas AG (Switzerland), the scientific partners: University of Ljubljana (Slovenia) and the Hahn-Schickard-Gesellschaft (Germany). The main aim was to promote in the Alpine region intelligent, sustainable and integrative growth. The objective was to develop a tracking system for dairy cows based on an existing system that integrates the data from the milk performance test and the health monitoring systems. The SESAM tracking system at final stage will monitor: running, lying, eating and chewing dairy cows and can localise the animals. By integrating animal health data, it will be possible to identify emerging health problems in dairy cows (ketoses, acidosis, mastitis). The system has to be developed in multi-lingual and specially tailored to small and medium-sized businesses in the Alpine region. The sensor units were attached to cows' neck by a belt, level 1 algorithm was developed, and the infrastructure elements base station (antenna) and gateway, hardware and software could already be used. A central, multi-lingual server will record the sensor information and display it to users. In early 2020, the system has been installed in 40 farms in Alpine region from SESAM project. The system has to be tested, and in addition, all events relating to animal health will be recorded in order to identify behaviour patterns in cow's health events. These behavioural patterns will be the basis for the development of level 2 algorithms. In order to start the extensive test in 40 dairy farms, the sub-projects: casing, infrastructure, sensor and software development must be successfully merged.

From an open-access computer image analysis method to a smartphone app for beef marbling evaluation

M. Bonnet, B. Albouy-Kissi, B. Meunier, P. Tisseur, D. Huguel, M. El Jabri and J. Normand
INRAE, UMR Herbivores, Université Clermont Auvergne, VetAgro Sup, Route de Theix, 63122 Saint-Genès-Champanelle, France; muriel.bonnet@inrae.fr

Marbling is one of the most important attribute related with beef quality, consumer satisfaction and purchasing decisions. A rapid, non-destructive and non-invasive evaluation of beef marbling from fresh product is desired in laboratories and industries dealing with beef analysis and processing. The present study, thus reports an open-access computer image analysis method based on a fresh beef rib image captured under non-standardised and uncontrolled conditions, for determining intramuscular, intermuscular and total fat content. For this purpose, 7 dairy (mean slaughter age 20.4 months) and 31 beef (mean slaughter age 16.9 months) young bulls as well as 85 dairy (mean slaughter age 65.6 months) and 7 beefs (mean slaughter age 73.0 months) cull cows, heifers or steers were slaughtered. This beef population was deemed to represent the marbling of the beef market in Europe. Cross section images of the 6th rib were captured with a smartphone Galaxy S8 using the default settings except for the strobe which was set always active. A dedicated program was developed in macro language using the ImageJ v1.52i image processing open source software. The obtained 17 processed image features were mined relatively to gold standard measures, namely intermuscular fat, total fat and muscles dissected from a rib and weighed, and intramuscular fat content (IMF) determined by the Soxhlet method. The best predictions were obtained by Sparse Partial Least Squares (SPLS) for both IMF percent and rib composition parameters, and by combining animal and image analysis features obtained from the caudal face of the 6th rib captured on a red table. These predictions were more accurate than those based on animal and image analysis features issued from the caudal face of the 5th rib captured on the hanged carcasses. The external-validated prediction precision were of 90% for IMF, and ranged from 71 to 86% for the total fat, intermuscular and muscle rib weight ratios. Both gold standard results, binary derived processed images animal and image analysis features were used to develop neural network implemented in a smartphone app. The external-validated prediction accuracies of neural network were of 76% for IMF, and 67% for intermuscular fat. Therefore, an easy, low cost, eco-friendly and rapid method based on a smartphone picture from the 6th rib of bovine carcasses provides an accurate method for fat content determination.

CO_2 tracer method disrupts the ranking of cows according to methane emissions

P. Huhtanen[1], P. Lund[2], M.R. Weisbjerg[2] and A.R. Bayat[3]
[1]Swedish University of Agricultural Sciences, Skogsmarksgränd, Umeå, Sweden, [2]Aarhus UNiversity, AU Foulum, P.O. Box 50, Tjele, 8830, Denmark, [3]National Reseach Institute Finland, Jokioinen, 31600 Jokioinen, Finland; pekka.huhtanen@slu.se

There has been an increasing interest to reduce methane (CH_4) emissions from dairy cows by selecting low-emitting animals. This requires reliable phenotypic on-farm measurements of individual cow's CH_4 emissions from large number of animals with a reasonable cost. A large numbers of records on individual animals under commercial conditions can be obtained by 'sniffers' installed in the feed bin of automated milking stations or concentrate feeders that measure CH_4:CO_2 ratio in exhaled air. Predicted CO_2 is used as a tracer gas to estimate daily CH_4 production and is predicted from metabolic body weight (MBW), ECM yield and days in pregnancy, i.e. assuming constant efficiency of energy utilisation for different metabolic functions. Data from respiration chamber studies (307 cow/period observations) were used to evaluate this concept. Residual feed intake (RFI) and residual ECM production (RECM) were estimated using observed ECM yield, MBW and energy balance as energy sinks. The cows were classified to three efficiency groups (low, medium and high) of equal sizes. Model 1 underestimated CO_2 production and consequently CH_4 emissions compared to observed values or Model 2. The models ranked the cows similarly ($R^2>0.99$). Relative to low efficiency cows, CO_2 production of high efficiency cows was overestimated by approximately 2 kg/d irrespective of model or efficiency definition (RECM vs RFI). At given feed intake and MBW, efficient cows produce more milk than inefficient cows and thereby have a higher predicted CO_2 production. However, actual CO_2 production in efficient cows is less than in inefficient cows at similar intake and MBW. Because of biases in predicted CO_2 production, CH_4 emissions were overestimated by approximately 60 g/d (15%) for efficient compared with inefficient cows. It is concluded that ranking the cows according to CH_4 emissions by the 'sniffer' and tracer gas method favours inefficient cows compared with efficient cows.

Residual CO_2 production as a trait of feed efficiency in dairy cows

P. Huhtanen[1], A. Guinguina[1], P. Lund[2] and A.R. Baya[3]
[1]Swedish University of Agricultural Sciences, Skogsmarksgränd, 90193 Umeå, Sweden, [2]Aarhus University, P.O. Box 50, 8830 Tjele, Denmark, [3]Natural Resource Institute Finalnd, 31600 Jokioinen, Finland; pekka.huhtanen@slu.se

The productivity of dairy cattle has risen considerably due to advances in nutrition, genetics, and management. Feed efficiency (FE) is an important trait under practical conditions having a major influence on profitability and environmental efficiency in the dairy industry. Until now, main improvements in FE have been derived from dilution of maintenance requirement to greater amount milk. However, further potential is limited due to already high production level, increased body weight (BW) and reduced diet digestibility with increased intake. Improvement in FE is also hampered by lack of on-farm determinations of individual feed intake and energy balance. The objective of this study was to study if individual animal CO_2 production can be used as a proxy of FE. Data from respiration chamber studies (289 cow/period observations) were used to evaluate relationships between CO_2 and heat productions, and between residual CO_2 production (RCO$_2$) and residual feed intake (RFI). Residual CO_2 was estimated as differences between actual and CO_2 production predicted from ECM yield and metabolic BW (MBW). Residual feed intake was estimated as the difference between actual DM intake (DMI) and intake predicted from observed ECM yield, MBW and energy balance. Residual CO_2 was closely related to heat production (adjusted RMSE 2.0 MJ/d) and CO_2 predicted DMI better than DMI = ECM + BW –model (Adj. RMSE 1.04 vs 1.39 kg/d) indicating that CO_2 measurements can detect differences in FE derived from improved efficiency of ME utilisation. RCO_2 was closely related to RFI: RFI (kg/d) = $0.0\pm0.15 + 0.88\pm0.039 \times RCO_2$ (kg/d) (Adj. RMSE=0.43 kg/d). Including energy digestibility (ED) to the model reduced adj. RMSE of RFI prediction to 0.23 kg/d with a bivariate model: RFI = (kg/d) = $15.3\pm0.70 + 0.83\pm0.023 \times RCO_2 - 0.021\pm0.001 \times ED$. It is concluded that the cows can be ranked according to RFI predicted from RCO_2 and that predictions can further be improved by including estimates of digestibility in the model without measuring feed intake. If CO_2 could be reliably measured on-farm conditions, selection of cows according to feed efficiency will be possible.

Expert locomotion score prediction based on IMU data with machine learning in turkeys

J.E. Doornweerd[1], A. Savchuk[2,3], B. Visser[2] and A.C. Bouwman[1]
[1]Wageningen University & Research, Animal Breeding and Genomics, Droevendaalsesteeg 1, 6708 PB Wageningen, the Netherlands, [2]Hendrix Genetics, Spoorstraat 69, 5831 CK Boxmeer, the Netherlands, [3]Jheronimus Academy of Data Science (JADS), Sint Janssingel 92, 5211 DA 's-Hertogenbosch, the Netherlands; janerik.doornweerd@wur.nl

In livestock production, locomotion is an important health & welfare trait. In turkeys, locomotion scores are heritable, repeatable, and valuable for selection but the process to acquire them is laborious, invasive and subjective. Motion sensor application in poultry is largely underdeveloped but could be used for the automated and objective scoring of locomotion. The aim of this study was to predict expert locomotion scores based on inertial measurement unit (IMU) data. Currently, breeding candidates are scored by a human expert as a one-off subjective snapshot whilst walking through a corridor. Each bird is individually scored from poor (1) to good (6). During this routine procedure, the birds were recorded. Three IMUs were attached to each bird (n=83) with Velcro straps, one on each leg and one on the neck. The IMUs provided 3D accelerometer, gyroscope & magnetometer data. A video camera provided information on the morphology of the bird. Gradient boosting was used for step recognition based on leg IMU data (F-score: 0.82, on an allowed distance of 0.2s). The steps (n=1,736 after quality control) served as input for feature extraction. IMU features were mainly concerned with the descriptive statistics of each IMU variable, whereas the video provided the angle of the legs during double support. The model had a mean per class error of 0.29 on the test set. The current approach shows promise in providing objective locomotion scoring. Knowledge gained could enhance the application of motion sensor technology in other livestock species.

Development of image analysis pipeline to predict body weight in pigs

H. Yu, K. Lee and G. Morota
Virginia Polytechnic Institute and State University, Department of Animal and Poultry Sciences, 3695 Litton Reaves Hall, 175 West Campus Drive, Virginia Tech, Blacksburg, Virginia 24061, USA; haipeng.u@gmail.com

Average daily gain can reflect the growth rates, diet efficiency, and current health status of livestock species such as pigs. However, labour-based measurement of body weights (BW) is intensive and may induce stress to pigs. Therefore, we developed an automatic 3D image-based computer vision system to predict BW. We employed the Intel RealSense depth camera D435 to capture the RGB and depth images of eight pigs from nursery to finishing phases across two months. During the experiment, each pig was video recorded for around 3 mins per day, with six frames per second, and manually weighed using an electronic scale. The recording resulted in around 1,080 images for each pig per day. We developed an image processing pipeline using OpenCV-Python. Specifically, each pig within the image was segmented by a thresholding algorithm, and a contour box of the pig was identified to extract width and length. The depth of the pig was captured by an active infrared stereo sensor using the Intel RealSense software development kit 2.0. The volume of the pig was derived by multiplying length, width, and depth. We processed frame by frame, and the third quantile of these morphological image descriptors was used as the image-derived measures for each pig. The Pearson correlation coefficients between scale-based BW records and morphological image descriptors were 0.314 (length), 0.512 (width), 0.881 (depth), and 0.579 (volume). A goodness of fit obtained by fitting multiple linear regression by regressing the BW records from the electronic scale on the morphological image descriptors was 0.8 in R^2. Overall, depth, width, and volume seem to be more correlated with BW records. We conclude that our proposed 3D computer vision system could potentially provide an effective way to predict the BW of pigs.

ATP1A1 and NPY genes expression associated with heat stress and residual feed intake in beef cattle

B.V. Pires[1], N.B. Stafuzza[2] and C.C.P. Paz[1,2]
[1]Ribeirão Preto Medical School (FMRP), University of São Paulo (USP), Departament of Genetics, Avenue Bandeirantes, 3900, Monte Alegre, Ribeirão Preto, 14049-900, Brazil, [2]Animal Science Institute, Sertãozinho, SP, Beef Cattle Research Center, Rodovia Carlos Tonani, km 94, Sertãozinho, SP, 14160-900, Brazil; bianca2510@outlook.com

Heat tolerance is an important trait in beef cattle and can influence feed efficiency. The *ATP1A1* and *NPY* genes are associated with residual feed intake (RFI) and heat tolerance in cattle. This study compared the expression of *ATP1A1* and *NPY* gene expression in Nelore (*Bos indicus*) and Caracu (*Bos taurus*) in the sun and shade environment. The study was carried in Sertãozinho, SP, Brazil during October 2017 (28.5 °C and 53.5% relative humidity), and evaluated 35 Caracu and 30 Nelore steers (12-15 months of age). Three different environments were evaluated: in the morning, all animals remained in a pen with some trees in the sun, in the afternoon the steers were separated and evaluated in shade and sun. Feed intake was measured during 88 days for each animal using automated feeding stations. The RFI for each animal was calculated based on a linear regression model of feed intake on mean metabolic live weight and average daily gain. All steers were classified into the most efficient (RFI<0) and least efficient (RFI>0). The blood samples were collected in the morning (07:30 am) and the afternoon (03:00 pm) from all animals and the relative expression of *ATP1A1* and *NPY* genes were identified by qPCR. The relative change in mRNA expression levels was calculated using the delta Ct method using the *GAPDH* housekeeping gene. The data were analysed by the GLM procedure. A total of 33 steers showed RFI>0, and 32 steers showed RFI<0. In sun treatment, Nelore steers obtained higher *NPY* expression than Caracu steers (3.15 and 0.55, respectively). The *NPY* expression in three treatments was not different (P>0.05) in Nelore breed, while in Caracu breed the shade treatment resulted in the highest expression (2.75). The treatments and breeds did not influenced the *ATP1A1* expression (P>0.05). Animals' most efficient showed lower *ATP1A1* expression than animals least efficient (2.20 and 2.77, respectively). *ATP1A1* and *NPY* expression in Caracu is more likely to have alterations in comparison to Nelore breed in the hot climate.

Polymorphism in β-casein and κ-casein genes and milk performance of Polish Holstein-Friesian cows

B. Sitkowska[1], M. Kolenda[1], D. Kamola[2], Z. Lach[3] and Ł. Lewicz[4]
[1]UTP University of Science and Technology, Department of Animal Biotechnology and Genetics, Mazowiecka 28, 85-084 Bydgoszcz, Poland, [2]Polish Federation of Cattle Breeders and Dairy Farmers, Żurawia 22, 00-515 Warszawa, Poland, [3]Breeding Center of Osięciny, Włocławska 2, 88-220 Osięciny, Poland, [4]Agricultural Cooperative Bądecz, Bądecz 2, 89-320 Bądecz, Poland; sitkowskabeata@gmail.com

The aim of the study was to determine polymorphism within two milk protein genes: β-casein (CSN2) and κ-casein (CSN3) in the population of 897 Polish Holstein-Friesian cows of the black and white variety kept in three herds in northern Poland. The relationship between genetic variants (SNPs) and milk yield per day of milking, milk yield per full lactation, fat, protein and lactose percentages in milk and dry matter content in milk was studied using one-way analysis of variance. Total of 22 SNPs in the β-casein (10 SNP) and κ-casein (12 SNP) genes were tested revealing polymorphism only within two SNPs of the CSN2 gene and four in CSN3. Within β-casein, two polymorphic sites were identified: g.8101A>C (later called CSN3_8101) and g.8178A>C (CSN3_8178). In the case of CSN3_8101 SNP the largest share of individuals (49%) was heterozygous, while in the case of CSN3_8178 the BB genotype was clearly dominant (frequency of 89%). Within κ-casein, four SNPs were analysed: g.13068C>T (CSN2_13068), g.13104A>C (CSN2_13104), g.13124A>G (CSN2_13124) and g.13165A>G (CSN2_13165). The highest share of heterozygotes (42-43%, respectively) was found for CSN2_13068 and CSN2_13165 SNPs, while in the case of CSN2_13124 polymorphism, 22.5% of cows was heterozygous and 1% was homozygous for BB genotype. Significant differences in the level of analysed milk yields were observed between genotypes. Cows with the BB genotype for CSN3_8101 SNP, compared to AA homozygotes, produced over 850 kg more milk in full lactation, had a lower dry matter (by 0.20%) and fat content (by 0.20%). With regard to SNP CSN2_13165, it was found that animals with the BB genotype produced significantly (P≤0.05) less milk with a higher concentration of dry matter and its components compared to individuals with the AA genotype. AA genotype for SNP CSN2_13165 as well as BB genotype for CSN3_8101 were found to be associated with a better milk performance in the studied population.

Frequency of selected genes associated with functional traits in Polish Holstein-Friesian cows

M. Kolenda[1], B. Sitkowska[1], D. Kamola[2], D. Piwczyński[1], Z. Lach[3] and Ł. Lewicz[4]
[1]UTP University of Science and Technology, Department of Animal Biotechnology and Genetics, Mazowiecka 28, 85-084 Bydgoszcz, Poland, [2]Polish Federation of Cattle Breeders and Dairy Farmers, Żurawia 22, 00-515 Warszawa, Poland, [3]Breeding Center of Osięciny, Włocławska 2, 88-220 Osięciny, Poland, [4]Agricultural Cooperative Bądecz, Bądecz 2, 89-320 Bądecz, Poland; kolenda@utp.edu.pl

The aim of the study was to identify polymorphism in 27 polymorphic sites in 21 genes (namely APAF1, EVC2, F11, F8, FBN1, FECH, GART, GH1, GLRA1, ITGB2, KRT5, LTF, LYST, MAN2B1, MC1R, MITF, MRC2, NAGLU, OPA3, PMEL, UMPS genes) associated with various functional features. Genotypic data obtained from 1,900 Polish Holstein-Friesian cows of the black and white variety was analysed. The genetic structure of the studied population in terms of genotype frequency was determined. The obtained results indicate that for 21 polymorphic sites (SNPs) studied population was monomorphic. Only 6 of the analysed SNPs showed genetic diversity (APAF1, GART, ITGB2, LTF, MC1R, and MITF genes). In the case of APAF1 (apoptotic peptidase activating factor 1), only 4% of all individuals were heterozygous and 96% were BB homozygotes. For mutations in the GART (phosphoribosylglycinamide formyltransferase) gene 98.3% of the population was AA homozygous, while 1.7% was heterozygous. Mutations in APAF1 and GART genes are important as they are associated with abortion of the foetus. Only 4 individuals (0.2% of the tested population) were heterozygous for mutation in the ITGB2 gene (integrin subunit beta 2), for which the association with Leukocyte adhesion deficiency, type I (BLAD) is well known. The remainder of the population was homozygous for the A allele. Three genotypes (45% AA, 44% AB and 11% BB) were detected in the study population for the LTF (lactotransferrin) gene. For two genes associated with coat colour, MC1R (melanocortin 1 receptor) and MITF (melanocyte inducing transcription factor), 3 possible genotypes were detected, however, in both cases BB genotypes predominated (94 and 95%, respectively). The low genetic diversity in the studied genes shows that over the years a selection aimed at improving various functional features was carried out, which in effect reduced the occurrence of many undesirable mutations responsible for various genetic disorders in cows.

Polymorphism in β-lactoglobulin gene and milk performance of Polish Holstein-Friesian cows

M. Kolenda[1], B. Sitkowska[1], D. Kamola[2], Z. Lach[3] and Ł. Lewicz[4]
[1]UTP University of Science and Technology, Department of Animal Biotechnology and Genetics, Mazowiecka 28, 85-084 Bydgoszcz, Poland, [2]Polish Federation of Cattle Breeders and Dairy Farmers, Żurawia 22, 00-515 Warszawa, Poland, [3]Breeding Center of Osięciny, Włocławska 2, 88-220 Osięciny, Poland, [4]Agricultural Cooperative Bądecz, Bądecz 2, 89-320 Bądecz, Poland; kolenda@utp.edu.pl

The aim of the study was to determine polymorphism within the β-lactoglobulin (LGB) gene in a population of 894 Polish Holstein-Friesian cows. In total, 10 point mutations (SNPs) were analysed. The relationship of genetic variants of this protein with the milk yield in full lactation and chemical composition of milk (protein, lactose and dry matter contents) was statistically examined with the use of one-way analysis of variance. Out of 10 analysed SNPs only two localised in exon 3 and 4, respectively, proved to be polymorphic: g.3982 TGA>CGG (later called LGB_3982) and g.5263T>C (LGB_5263). Most cows in the tested population (49%) in terms of LGB_3982 and LGB_5263 SNPs were heterozygotes. Cows with genotype BB for LGB_3982, compared to AA homozygotes, produced over 715 kg more milk in full lactation, had a higher percentage of dry matter (by 0.20%) and protein (by 0.10%). These differences were proven to be statistically significant ($P \leq 0.05$). In relation to LGB_5263 SNP, homozygous individuals for the BB variant gave more milk (milk yield in full lactation was over 10,800 kg), than heterozygotes (over 10,700 kg of milk) and homozygotes AA (10,286 kg). The analysis of this SNP revealed statistically significant ($P \leq 0.05$) differences between the dry matter, lactose and protein contents in milk. Highest levels of dry matter and protein contents were noted in the milk of individuals with the BB genotype, while lactose content was greater in the AA genotype. In relation to both LGB_3982 and LGB_5263 SNPs, milk yield in full lactation of heterozygous cows was similar to BB homozygotes, which in turn had statistically better milk yield than animals of AA genotype. In the studied population, 8 SNP sites within the β-lactoglobulin gene were monomorphic and only two sites revealed genetic diversity. An association between BB genotype for LGB_3982 and LGB_5263 SNPs and an improved milk performance was found in the tested population.

Frequency of milk protein genes in Polish Holstein-Friesian cows

B. Sitkowska[1], M. Kolenda[1], D. Kamola[2], D. Piwczyński[1], Z. Lach[3] and Ł. Lewicz[4]
[1]UTP University of Science and Technology, Department of Animal Biotechnology and Genetics, Mazowiecka 28, 85-084 Bydgoszcz, Poland, [2]Polish Federation of Cattle Breeders and Dairy Farmers, Żurawia 22, 00-515 Warszawa, Poland, [3]Breeding Center of Osięciny, Włocławska 2, 88-220 Osięciny, Poland, [4]Agricultural Cooperative Bądecz, Bądecz 2, 89-320 Bądecz, Poland; sitkowskabeata@gmail.com

The aim of the study was to determine the genetic variability within β-casein (CSN2), κ-casein (CSN3) and β-lactoglobulin (LGB) genes in the Polish dairy cattle population. A population of 1,900 Holstein-Friesian cows of the black and white variety was examined. 10 polymorphic sites within CSN2, 12 within CSN3 and 10 in LGB gene were selected for the study. The genotype frequency was calculated for a total of 34 polymorphic sites. If genetic monomorphism of a given SNP was excluded, population was examined in terms of Hardy-Weinberg equilibrium. Within the CSN2 gene, only 5 sites showed genetic diversity, this sites being the following: g.8101A>C (later called CSN2_8101), g.8178A>C (CSN2_8178), g.8219A>C (CSN2_98219), g.8267C>G (CSN2_8267), g.8356T>C (CSN2_8356). A total of 6 SNPs was detected in the CSN3 gene: g.12971T>G (CSN3_12971), g.13065C>T (CSN3_13065), g.13068C>T (CSN3_13068), g.13104A>C (CSN3_13104), g.13124A>G (CSN3_13124), g.13165A>G (CSN3_13165), while in LGB 3 SNPs were identified: g.3982TGA>CGG (LGB_3982), g.5174T>C (LGB_5174T), g.5263T>C (LGB_5263). In the case of polymorphism in CSN2 gene, only one SNP (CSN2_8101) showed greater differentiation (14% of individuals had AA genotype, 48% had AB and 38% BB genotype), in the other cases one genotype dominated in the population (CSN2_8178 AA; 88% of the population; CSN2_8219 BB; 99.8%, CSN2_8267 AA; 93%, CSN2_8356 BB; 99.8%). The CSN3 gene showed greater variability – a high number of heterozygotes was identified in the case of the following polymorphisms: CSN3_13068 (47% of the population), CSN3_13104 (45%), and CSN3_13165 (47%). The LGB gene was also diverse in terms of studied polymorphisms. 50-52% of all individuals were heterozygous in terms of LGB_3982, LGB_5174 and LGB_5263 SNPs. Out of 32 selected SNPs for three genes 18 were monomorphic, which may suggest that in previous years a selection unified cows in terms of many polymorphic sites.

Digital claw analysis as part of precision health management at dairy farm

R. Kasarda[1], N. Moravčíková[1] and J. Tomka[2]
[1]Slovak University of Agriculture in Nitra, Tr. A. Hlinku 2, 949 76 Nitra, Slovak Republic, [2]NPPC-Research Institute of Animal Production in Nitra, Hlohovecká 2, 951 41 Lužianky, Slovak Republic; radovan.kasarda@uniag.sk

The aim of the study was the assessment of morphometric traits of high producing Holstein cow claws farmed at the University farm in high input, high-intensity production system. Average production level of in total 512 dairy cows was 10,307.7/9,791 kg (1st lact.) milk with 3.86% fat; 3.16% protein. Average number of 2.16 lactation, length of lactation 389 days, age at first calving 25.8 months, days open 421 days and average functional length of production life (right censored) 800.5 days. The farm is considered in TOP Slovakia Dairy farms in 34th place. Because majority of claw disorders is present of hind claws, evaluation of claw formation was made based on digital images taken on individual cow from right hind claw (distal, lateral) allowing to measure claw angle, heel index, claw height, claw diagonal and claw width using NIS-Elements image analysing platform. Additionally, two additional traits were introduced as part of precision health management: total claw area and functional claw, calculated based on claw distal digital images. Claw management routine at the farm is as follows: 2 standard, (5-month routine) trimming of whole herd during the year by external service and individual emergency trimming in acute stage by farm MVD. In total, 164 cow claw digital images were analysed. Average claw angle was 48.6 °, heel index 4.02 cm, claw height 6.01 cm, claw diagonal 11.75 cm and claw width 5.51 cm. Total area and functional area were 43.97 and 31.36 cm^2, respectively. Sole ulcer was observed in 36 individuals (21.95%), digital dermatitis in 3 (1.83%) and interdigital dermatitis and heel erosion in 1 case (0.006%). It can be concluded that based on regular evaluation of claw trimming in the herd as part of precision management is resulting in considerable improvement of claws reaching overall expectation in claw angle (45-50 °) and heel index over 3 cm. In total 75.61% of cows were evaluated not lame which is in accordance with overall consideration herd as healthy with only 1.23% (2) cases evaluated as moderate and 0.006% (1) as severe.

Dairy-4-Future: propagating innovations for more resilient dairy farming in the Atlantic area
S. Foray and A. Le Gall
Institut de l'Elevage, Monvoisin BP 85225, 35650 Le Rheu Cedex, France; sylvain.foray@idele.fr

Funded by the Interreg Atlantic Area Program, the Dairy-4-Future project (2018-2021) aims to increase the competitiveness, sustainability and resilience of dairy farms in the Atlantic Area. The project involves 5 countries (Ireland, United Kingdom, France, Spain and Portugal) and covers, from Scotland to Azores, 12 Atlantic regions that together represent 20% of EU-28 milk production and 100,000 farmers working in a wide diversity of milk production systems. Dairy-4-Future focusses on four key issues: analysing strengths and weaknesses of the dairy sector in Atlantic area, fostering dairy sector economic resilience, improving resource use efficiency, and determining sustainable dairy systems for the future. The main objective is to identify, evaluate and then widely propagate innovative practices to European dairy technicians and breeders, through transnational seminars or farm open days and technical tools: publications, videos, training tools, etc. The project puts innovative farmers at the centre of practice-based research work, and combine several methods to adapt and develop scientific knowledge, which will lead on technical solutions and recommendations to be shared across the network. With 11 technical partners and 21 associated partners, Dairy-4-Future relies on a strengthened scientific and technical network, completed by 10 experimental farms and 100 pilot farms.

SWOT analysis of the North Portugal dairy sector
J. Almeida[1], M. Gomes[1], S. Silva[1,2] and H. Trindade[1,3]
[1]Universidade de Trás-os-Montes e Alto Douro (UTAD), Quinta de prados, 5001-801 Vila Real, Portugal, [2]CECAV, Quinta de prados, 5000-801, Portugal, [3]CITAB, Quinta de prados, 5000-801, Portugal; htrindad@utad.pt

This work presents the results of a SWOT analysis conducted to evaluate the status of the North Portugal dairy sector. The information was collected in consultation with farmers and various dairy sector intervenients from public (ministerial, municipal and academic), private (industry) and cooperative institutions, under the scope of the 'Dairy4Future' INTERREG project. During a workshop, the participants were invited to build a shared diagnosis of the strengths, weaknesses, opportunities and threats linked to dairy farming in the region. Additionally, participants were involved on the identification of services and disservices or constraints associated to the North Portugal dairy sector regarding economic, social, environmental and agricultural issues. The identification of innovative practices to enhance the services and limit or fix the disservices and the identification of factors that prevent the adoption of innovations reducing the most controversial disservices were also discussed. Analysis of the data collected showed that main positive points of North Portugal Dairy sector are the existence of a dairy cluster relatively well organised that supplies a well-dimensioned industry, which guarantees production flow at contracted prices, although low. In addition, the region benefits from favourable edaphoclimatic conditions with high productivity, that sustain a maize silage based diet, milk is produced with high quality control (food safety) and total traceability. The most negative issues are the low and volatile milk prices at production, the instability of public policies (lack of coherence of policies between ministries and uncertainty in the future), heavy dependency on the import of commodities (especially protein-rich materials) for concentrates, the difficulty to comply with environmental rules especially on manure management, the scarcity and high price of land and the absence by the dairy industry of a policy for the development of innovative products. The future of the Portuguese dairy sector can also be affected by the negative perception on milk from some consumer segments.

Industry stakeholder assessment of the Scottish dairy sector and its future growth
H.J. Ferguson, P.R. Hargreaves and M.D. March
SRUC, Dairy Research and Innovation Centre, Hestan House, The Crichton, Dumfries, DG1 4TA, United Kingdom; holly.ferguson@sruc.ac.uk

Scotland is the fifth largest milk-producing region in the UK, with 77% of its milking herd located in Southern Scotland. Despite decreasing herd numbers, Scottish herd sizes have grown, with the number of dairy cows increasing to 180,293 in 2019. The EU Interreg project Dairy-4-Future (D4F) aims to increase the competitiveness, sustainability and resilience of dairying in the Atlantic Area region of the EU. D4F aims to identify, evaluate and propagate innovative on farm practices to the European dairy industry and beyond. A case study and evaluation of the dairy sector in each D4F member region has been carried out, including a SWOT (strengths, weaknesses, opportunities, threats) analysis of milk markets, enterprises and farming practices. The Scottish analysis gathered data from government, research and private bodies, as well as interviews with key individuals and an interactive SWOT analysis with dairy stakeholders at two large agricultural events. Participants were asked to rank issues in order of importance to the Scottish dairying community. This work has resulted in a sizeable and unique dataset with detailed feedback from Scottish stakeholders. Overall, Scotland had a positive outlook within its dairy industry, in comparison with surveys from other countries in the project, and recognised the opportunities that 'Scotland' as a brand provided. The SWOT results showed that many of the threats and weaknesses ranked highest in Scotland are issues which can be acted upon. The top four weaknesses identified were: 1. Supermarkets have too much power in the market 2. Poor transport links 3. Little encouragement for new entrants to the sector 4. Poor signal and internet capacity, which impacts upon isolation. Of the threats identified, some could be addressed through public engagement, such as the public's perception of dairying and its role in climate change. Opinions and attitudes within the dataset provide a strong direction for development of a plan to improve resilience and sustainability of dairying in Scotland. Issues identified by the Scottish dairying community are important and should shape future research and government policy.

Rendered services and dysservices of dairy farming to the territories – a bottom-up approach
C. Perrot
Institut de l'Elevage, Economie, 149 rue de Bercy, 75012 Paris, France; christophe.perrot@idele.fr

The Dairy-4-Future project focuses on the improvement of dairy farming's sustainability in the UK, Ireland, France, Spain and Portugal. This can be achieved by fostering the services that dairy farming offers to the territories and by reducing the dysservices. Services are defined as positive impacts of the dairy sector on three dimensions of sustainability (economic, environmental and social issues). On the contrary, dysservices, with dys- like dysfunctioning, are defined as negative impact on these issues. In some extent, with dedicated actions, these dysfunctional services can be reduced or fixed. Some scientific literature exists on the positive and negative impacts of farming or livestock farming on the territories with local or global impacts. But neither dedicated for dairy farming, nor with an exhaustive list of detailed types of impacts. Anyway, the relevant list of impacts and the ranking of these are strongly related to specific territories, with specific natural and socio-economic contexts. To get a qualitative identification (without any quantitative or monetary assessment) of relevant, ranked list of items or issues, interactive workshop with local stakeholders (farming, dairy processors, advisors and researchers, GNO, local authorities, etc.) were organised in each regional case study of the project. Stakeholders were told to identify positive or negative impacts of local dairy farming on their territories in four category of functions: provisioning (food, etc.), rural vitality, environmental quality. cultural heritage and quality of life. 165 services and 135 dysservices have been identified, with a quite balanced repartition between functions. A word cloud's analysis has been carried out to get a representation of the most important issues at stake for dairy farming in each category. The combination of services and dysservices at local scale is meaningful. Identification of several types of bundle of services/dysservices has been achieved with a Bertin's matrix analysis. It revealed proximities between items and/or territories (like Northern Ireland and Cornwall; Southern Ireland and Normandy; Galicia and Brittany).

Financial effects of innovative dairy farming practices in the Atlantic Area

M.D. March[1], P.R. Hargreaves[1] and M. Topliff[2]
[1]SRUC, Dairy Research & Innovation Centre, Hestan House, The Crichton, Dumfries, DG1 4TA, Scotland, United Kingdom, [2]AHDB, AHDB Dairy, Stoneleigh, Kenilworth, CV8 2TL, England, United Kingdom; paul.hargreaves@sruc.ac.uk

A favourable climate for forage allows dairy farms situated in Atlantic Area regions of Europe to supply around 20% of all milk produced in the EU28. Agriculture is an important economic activity in the Atlantic Area, however the dairy sector faces challenges surrounding economics and the environment, such as volatile milk prices, and the need to reduce carbon emissions. Through the Interreg Dairy-4-Future project, 100 innovative producers form a group of pilot farms, and detailed economic data were captured annually. Farms were recruited in Atlantic Area regions of Portugal, Spain, France, Ireland and the UK, and various innovative practises were categorised by type, as having mainly financial, environmental or social benefits or consequences to the dairy management system. We assessed farm financial performance in terms of profitability, operating costs and physical performance using indicators such as labour efficiency, fixed costs ratios, and outputs retained, and individual farms were benchmarked against regional averages. One objective was to analyse data so as to illuminate any financial effect of farm innovations and to cluster innovative practises by category, such as animal health or environment. Case studies of specialist dairies were prepared to describe economic performance and any financial effect of a range of on farm innovations. Results will be used as a guide for dairy farmers across Europe who may want to implement an innovative farming techniques or novel production methods on their holding. Progressive dairy farming techniques were wide ranging, examples include the use of robotics technologies for feeding and milking and alternative management styles included once a day milking and 'cow with calf' systems. Some farms focussed on financial outcomes such as low cost feed inputs and efficient labour use, while environmental innovations included organic or antibiotic free, products to reduce ammonia emissions, mixed sward grasslands, minimum till systems and reduced water use.

Development and test of a new platform for sustainability assessment of pasture-based animal systems

R.F.M. Teixeira[1], N.R. Rodrigues[2], I. Gama[2], M. Reis[2], J. Palma[2], T. Valada[2], T.G. Morais[1] and T. Domingos[1]
[1]Instituto Superior Técnico, MARETEC/LARSyS, Av. Rovisco Pais, 1, 1049-001, Portugal, [2]Terraprima, Serviços Ambientais Lda., Avenida das Nações Unidas, nº 97, 2135-199 Samora Correia, Portugal; ricardo.teixeira@tecnico.ulisboa.pt

Grassland farmers have difficulty finding easily usable tools to help them assess and improve the sustainability of their practices. Access to quality data from farmers is also a problem for researchers as dedicated surveys are time-consuming and frequently seen as a burden by farm managers. To help address these problems, we present here a new farm-level platform specifically tailored for grassland sustainability management of pasture-based meat, milk and dual-purpose animal production systems. The 'ModelMeat' platform, developed by Terraprima in Portugal, is an online tool that receives: (1) spatial data, namely the location and contour of the farm or parcels within a farm; and (2) agronomic and economic declarative data from farmers regarding land use, type, breed and age of animals, stocking rates, management operations, among other variables. To facilitate and expedite data input, the tool provides defaults to assist the farmers (e.g. the farmer selects the use of a given parcel and an automated fact sheet with a list of standard operations is automatically generated for editing by the farmer). Using those data, the system returns: (1) the greenhouse gas emissions (GHG) emissions of the farm, including farm-level emissions from soils and animals, as well as life cycle emissions due to the use of materials and energy; (2) the GHG balance of the farm including also biogenic carbon (i.e. carbon sequestration); and (3) the costs and revenues of the activities. Farmers can create scenarios (e.g. different stocking rates or feed) and estimate their environmental and economic effects. Due to the geospatial nature of the analysis, farmers also access external data sources in the platform related to their farm such as vegetation indices calculated using satellite data. The data aggregated in the platform can be shared by the farmer and used by third parties such as researchers, public administrators or certification companies. We present here also a demonstration of how the tool was used by a pool of 40 farmers in Portugal and the results obtained.

Birth weight and early nutrition affect performance of Holstein heifers only until 15 months of age

J. Jurquet[1], A. Philibert[1], D. Plouzin[2] and Y. Le Cozler[3]

[1]*Institut de l'Elevage, Maison Nationale des Eleveurs, 149 rue de Bercy, 75595 Paris cedex 12, France, Metropolitan,* [2]*Ferme expérimentale des Trinottières, La Futaie, 49140 Montreuil sur Loir, France, Metropolitan,* [3]*INRA, Agrocampus Ouest, UMR PEGASE, Domaine de la Prise, 35590 Saint Gilles, France, Metropolitan; julien.jurquet@idele.fr*

Decreasing age at first calving (AFC) is an efficient way to decrease Green House Gazes (GHG) emissions in dairy farms. In Holstein heifers, 24 months (mo) of age is generally recommended, with a body weight (BW) around 550-600 kg just after calving. High growth rate during rearing is generally associated with such a strategy. In present study, we hypothesised that BW at birth and intake of both milk and concentrate before weaning had limited effect on such strategy, when sufficient feed is available after weaning. A dataset from an experimental dairy farm in western France was used, originating from 402 Holstein heifers born between 2001 and 2015. For each heifer, BW at birth, at weaning, 4, 6, 15 mo of age (BW_b, BW_w, BW_4, BW_6, BW_{15} respectively) and milk yield during the 1st lactation on a 305 days basis (MY) were available. For the 281 culled animals, lifetime milk production (LMP) was calculated. As feed and milk intake before weaning were registered, dry matter intake (DMI) and metabolisable energy intake (MEI) before weaning were determined. At birth, calves weighted 41 (±5) kg. Age and BW_w averaged 63 (±8) days and 89 kg (±13), respectively. Average DMI and EMI/calf during rearing were 72.9 (±16.7) kg and 356 (±70) MJ, respectively. BW_w was highly correlated with DMI (Pearson correlation coefficient, r=0.89) and EMI (r=0.85) at weaning, but the strength of this correlation decreased over growth and disappeared at calving (0.15 and 0.12 respectively). 305 d MY was neither correlated with DMI and EMI. A complementary analysis based on 251 heifers with a similar rearing period length (57±2 d) indicated that BW_b (41±5 kg) was correlated with BW_w (0.57), BW_4 (0.54) and BW_6 (0.49), but not with AFC (-0.19), 305 d MY (0.10) or LMP (0.07). Results from this study indicated that calves with low intake before weaning due to a short milk feeding period and/or with low BW at birth were able to first calved at 24 mo, if an adequate feeding strategy is used after weaning. AFC at 24 mo appeared to be achievable for most Holstein heifers, resulting in a decrease of GHG emission in most dairy farms.

Agricultural knowledge and innovation systems (AKIS) in the Atlantic area: a focus on dairy farmers

M. Dorigo[1] and S. Foray[2]

[1]*Agriculture and Horticulture Development Board (AHDB), Stoneleigh Park, Kenilworth Warwickshire CV8 2TL, United Kingdom,* [2]*Institut de l'Elevage (IDELE), Monvoisin BP 85225, 35650 Le Rheu cedex, France; martina.dorigo@ahdb.org.uk*

Feeding the world population in the long term in a sustainable way is the main challenge faced by agriculture. In this perspective, innovation will be increasingly critical to support the competitiveness of the sector, making it also resilient and sustainable at the same time. The success of innovation, however, depends on how effectively the different AKIS (Agriculture and Innovation Systems) players operate in creating and delivering it to the end user. Whilst being AKIS players themselves, dairy farmers are also the final users of any knowledge and innovation generated and delivered by other AKIS players. Therefore, the AKIS in any country or region should operate with the goal of effectively delivering innovation to the farmer and enhancing its implementation. The current study provides a description of the end user population (dairy farmers) in the Atlantic Area, with a focus on current and future generations: it also proposes some suggestions to increase the effectiveness of the AKIS in implementing innovation in dairy farming.

Fatty acids profile in milk of different species after *in vitro* digestion: effect on adipocyte

A. Santillo[1], L. Figliola[1], M.G. Ciliberti[1], M. Caroprese[1], R. Marino[1], M. D'Apolito[2], I. Giardino[2] and M. Albenzio[1]
[1]University of Foggia, Department of the Sciences of Agriculture, Food and Environment (SAFE), Via Napoli, 25, 71122 Foggia, Italy, [2]University of Foggia, Department of Medical and Surgical Sciences, Via Napoli, 71122, Italy; antonella.santillo@unifg.it

The study aimed to evaluate the fatty acid (FA) profile of raw milk and of the corresponding digested milk from different sources (human, liquid formula, donkey, bovine, ovine and caprine milk) to gain information on the nutritional quality in infant nutrition; moreover, aimed to study the effect of digested milk on human adipose tissue, using an *in vitro* culture of mature adipocytes 3T3-L1. All milk sources were subjected to *in vitro* digestion; afterwards total FA in raw milk sources and free FA in digested milk sources, were analysed by gas chromatography. Murine 3T3-L1 fibroblasts were propagated, differentiated and treated with DMEM medium add BSA conjugated with 2% of different digested milks. On mature adipocytes 3T3-L1 treated with digested milks, cellular viability, apoptosis, oxidative response and gene expression levels of NF-kB p65, HMGB1, SREBP-1c and FAS were evaluated. Data were processed using ANOVA for repeated measures. In all the milk sources studied, the percentage distribution of FA liberated upon gastrointestinal digestion did not reflect the patterns found in the corresponding milk sources. Differences in the free FA of digested milk sources could be attributed to multiple factors (the interaction between milk substrate and digestion process; the presence of endogenous enzymes to milk; the different structure and organisation of the milk fat globules; the position of bounds in TAG). All digested milks decreased cell viability in mature adipocytes and induced cell death, partially due to apoptic event, although no differences were observed among milk sources. Each digested milks exerted pro-inflammatory effect in adipocytes through over expression of HMGB1 and NF-kB p65 although lower gene expression was found in human milk and milk from ruminant species. Results suggest that milk from minor species may be exploited by the dairy industry for the design of novel formulations able to promote a correct development and functioning of adipose tissue to sustain human health.

Ammonia production from chitosan fed diet derived slurry

J. Lizarrañde, E. Rosa, A. Doblas, H. Arriaga, A. Garcia and P. Merino
NEIKER-Basque Institute for Agricultural Research and Development (BRTA), Department of Conservation of Natural Resources, Berreaga kalea, 1, 48160 Derio, Bizkaia, Spain; erosa@neiker.eus

Chitosan is a natural polysaccharide polymer which has been tested for different applications. Used as feed additive in ruminants, it was related to increasing the rumen fermentation efficiency. Less is known about the side-effect of chitosan on animal N balance, and the subsequent NH_3 emission from chitosan fed diet derived slurry. This work aimed to study at lab-scale the NH_3 emissions at barn stage from chitosan diet derived dairy cattle slurry. Sixteen dairy cows were split into 2 groups for 45 days. One group was supplied with chitosan (CHI) and the other was set as control (CTR). All cows were fed with maize and grass silage based diets. Urine and faeces were separately collected from each cow for a week. Composite samples were produced with the faeces and the urine, which were stored at 4 °C. Slurry was produced at the lab taking into account the proportion of 377 ml excreted faeces per 222 ml urine. 500 ml slurry were incubated in jars (1 L) at 4, 20 and 30 °C for 7 days. Ammonia concentrations were measured with a photoacoustic gas analyser (INNOVA 1412) every 24 h. A multifactorial variance analysis followed by a Tukey's HSD test was carried out (significance at $P<0.05$). Initial urea contents of 32 individual urine samples were measured. There were no differences between treatments at 4 °C (mean values: CTR=139 mg/m^3, CHI=169 mg/m^3). However, significant differences were found between CTR and CHI treatments at 20 and 30 °C, with significantly higher NH_3 production in CHI treatment (mean values: 20 °C CTR=237 mg/m^3, 20 °C CHI=342 mg/m^3; 30 °C CTR=321 mg/m^3, 30 °C CHI=407 mg/m^3). Ammonia production of CHI treatment at 20 °C equalled NH_3 production of CTR treatment at 30 °C. Significant differences were observed in urea values between treatments (mean values: CTR=12.4 g/l, CHI=17.0 g/l). This fact would explain the observed NH_3 concentration differences in jars incubations. We conclude that the addition of chitosan to the diet increased NH_3 production from the slurry because of the higher urea content of the urine. The magnitude of such NH_3 production was modelled by air temperature.

Genetics of gut microbiota of growing pigs in three contrasted environments

H. Gilbert[1], Y. Labrune[1], M. Le Sciellour[2], Y. Billon[3], M. Giorgi[4], I. Hochu[1], J.L. Gourdine[5], O. Zemb[1], D. Renaudeau[2] and J. Riquet[1]
[1]GenPhySE, Université de Toulouse, INRAE, ENVT, 31326, Castanet-Tolosan, France, [2]PEGASE, INRAE, Agrocampus Ouest, 35590, Saint-Gilles, France, [3]GenESI, INRAE, 17700, Surgères, France, [4]PTEA, INRAE, 97170, Petit-Bourg, France, [5]URZ, INRAE, 97170, Petit-Bourg, France; helene.gilbert@inrae.fr

Faecal samples for 1,600 pigs were collected in a temperate farm, in a tropical farm, and during an acute heat stress (i.e. 3 weeks at 30 °C at 23 weeks of age) in the temperate farm, in a genetically connected design between all environments. The V3-V4 regions of the 16SRNA were sequenced and 1,688 operational taxonomic units (OTU) were obtained. All animals were genotyped with a 60K SNP-chip. Feed had the same nutritional composition between environments, and was identical in the two periods tested in the temperate farm. Earlier studies showed that feed intake was reduced in the tropical environment, as was growth rate and fatness, and some G×E were evidenced on these traits. Linear mixed models were used to run association studies in each environment at the level of OTU and at different taxonomic levels. About 10% of the OTU had heritability close to zero in one of the environments and higher than 0.2 in another environment. In addition, associations between SNPs and some microbiota components were identified: which most strongly differed between the three environments. Altogether, our results suggest different genomic control of the faecal microbiota composition of growing pigs in temperate, acute heat stress and tropical chronic heat stress conditions. These potential G×E on microbiota could result from different feed resources, feed intake or other parameters directly related to breeding environment differences. They question the potential of selection for microbiota composition given the large range of changing environments in which pigs are now produced worldwide. This study is part of the Feed-a-Gene Project, funded from the European Union's H2020 Programme under grant agreement no 633531.

Directional selection of Large White pigs for gut microbiota enterotypes

C. Rogel-Gaillard[1], J. Estellé[1], Y. Billon[2], J.J. Leplat[1], G. Lemonnier[1], J. Lecardonnel[1] and C. Larzul[3]
[1]INRAE, GABI, Vilvert, 78530 Jouy-en-Josas, France, [2]INRAE, GenESI, Le Magneraud, 17700, Surgères, France, [3]INRAE, GenPhySE, Borde Rouge, 31320 Castanet-Tolosan, France; claire.rogel-gaillard@inrae.fr

Studying the influence of host genetics on gut microbiota composition is challenging because it is difficult to dissociate the variability due to genetic effects from the variability due to the environment, notably the maternal microbial environment at birth. Thus, we launched a study to demonstrate coevolution of the host and its gut microbiota by directional selection over two generations. We have shown the existence of two enterotypes on populations of 60-day-old piglets (d60) by sequencing the 16S rRNA gene from faecal DNA. These enterotypes are characterised by differential abundances of the genera *Prevotella* and *Mitsuokella* (PM enterotype) or *Ruminococcus* and *Treponema* (RT enterotype). We have started a selection experiment by generating two pig lines contrasting for their enterotype. In the initial population (generation G0), we studied 317 piglets and confirmed a stratification according to the PM or RT enterotype at d60. The founders for the next generation (G1) were selected among the G0 population for their higher abundance in Prevotella and Treponema for the lines PM and RT, respectively (6 males and 30 females per line). Each male inseminated five females from the same line, and four piglets were sampled at d60 for faecal microbiota composition analysis. We showed significant differences between the two lines in the relative abundance of the four bacterial genera that characterise the enterotypes (P<0.001, from 0.6 genetic standard deviation for *Treponema* to 1.3 for *Prevotella*). The observed contrasts were in the expected direction, thus demonstrating a response to selection consistent with the expected genetic response. The heritabilities of the relative abundance of the bacterial genera were estimated by combining G0 and G1 animals (588 piglets): h^2=0.30 for *Prevotella*, h^2=0.31 for *Ruminococcus*, h^2=0.39 for *Treponema*, h^2=0.35 for *Mitsuokella*. All these results demonstrated a significant influence of host genetics on the composition of gut microbiota at d60 in pigs and a capacity of directional selection over generations.

Modelling the impact of the macroalgae *Asparagopsis taxiformis* on rumen microbial fermentation

R. Muñoz-Tamayo[1], J.C. Chagas[2], M. Ramin[2] and S.J. Krizsan[2]
[1]*Université Paris-Saclay, INRAE, AgroParisTech, UMR Modélisation Systémique Appliquée aux Ruminants, 75005, Paris, France,* [2]*Department of Agricultural Research for Northern Sweden, Swedish University of Agricultural Sciences (SLU), Skogsmarksgränd, 90183 Umeå, Sweden; rafael.munoz-tamayo@inrae.fr*

The red macroalgae *Asparagopsis taxiformis* is a potent natural supplement for reducing methane production from cattle. *A. taxiformis* contains several antimethanogenic compounds including bromoform that inhibits directly methanogenesis. The antimicrobial activity of *A. taxiformis* impacts as well other microbial groups of the rumen microbiota producing changes in the fermentation profile that may have detrimental effects on animal health and productivity. The positive and adverse effects of *A. taxiformis* on the rumen microbiota are dose-dependent and operate in a dynamic fashion. It is therefore key to characterise the dynamic response of the rumen microbial fermentation for identifying optimal conditions on the use of *A. taxiformis* as a dietary supplement for methane mitigation. Accordingly, the objective of this work was to model the effect of *A. taxiformis* supplementation on the rumen microbial fermentation under *in vitro* conditions. We used a published mathematical model of rumen fermentation, where the rumen microbiota is represented by three functional microbial groups namely sugars utilisers, amino acids utilisers and hydrogen utilisers (methanogens). Model parameters associated to the kinetic rates of microbial conversion and the flux distribution of substrate utilisation were estimated by calibration using experimental data from an *in vitro* study assessing the dose-response impact of *A. taxiformis* on fermentation and methane production. In such a study, *in vitro* fermentation was carried out with rumen inoculum from two lactating Swedish Red cows during an incubation period of 48 h. *A. taxiformis* was supplemented at six treatment levels. Methane production, volatile fatty acids (VFAs), and ammonia were measured. The model captured efficiently the dose-dependent effect of *A. taxiformis* on the dynamic pattern of VFAs and methane. For methane, average determination coefficients (R^2) and concordance correlation coefficients (CCC) were 0.87 and 0.93 respectively. For VFAs, R^2=0.79, CCC=0.91. For ammonia, R^2=0.79, CCC=0.88. These results show the potential of our model as a tool for predicting rumen microbial fermentation. We are working on model extensions to account for *in vivo* conditions. We expect that our model developments can be useful for designing methane mitigation strategies maintaining animal health and productivity.

Effects of Lacaune ewes selection for somatic cells score and milk persistency on rumen bacteria

G. Martinez Boggio[1], A. Meynadier[1], R. Rupp[1], H. Larroque[1], C. Allain[2] and C. Marie-Etancelin[1]
[1]*INRAE, GenPhySE, Castanet Tolosan, 31320, France,* [2]*INRAE, Experimental Unit of La Fage, Saint-Jean et Saint-Paul, 12250, France; guillermo.martinez-boggio@inrae.fr*

The ruminal microbiota plays a central role in the nutrition of its host, directly affecting milk production and animal health. There are no such studies in sheep that describes this link, thus the aim of our study was to determine differences in ruminal bacteria between divergent lines for somatic cells score and milk production persistency of Lacaune ewes. We had rumen fluid samples of 700 dairy Lacaune ewes raised at the INRAE Experimental Farm of La Fage sampling between 2015 and 2019. These ewes were adult animals, mostly in their second lactation, and belonged to two divergent lines selected for somatic cells score (SCS) or milk production persistency during lactation (PERS). The phenotypes of ewes obtained from milk recording controls over the five years were somatic cells count, milk production, milk fat content, milk protein content and milk dry matter content. Ruminal metagenome was sequenced using 16s rRNA gene with Illumina Miseq technology. We analysed microbiota sequences with FROGS pipeline to obtain relative abundances of OTUs (in proportion of total sequences per sample) and R Phyloseq package to estimate biodiversity indices. Thereafter we grouped OTUs according to phylum, family and genus levels. Fixed effects would be included in the variance analyses were the lines (SCS+, SCS-, PERS+, PERS-), run sequencing effect, date of rumen fluid sampling, lactation number (2 up to 7 lactations), stage of lactation (45 to 133 DIM) and litter size (1, 2 or more lambs), moreover we tested also suckling method of the ewe (maternal or artificial). ANOVA of α-diversity and relative abundance of OTUs or taxa at different taxonomic levels were estimated by including the significant effects. Further, a discriminant analysis (with sPLS-DA) was applied on OTUs with divergent lines as discriminant factor, and some links were researched between phenotypes and OTUs (with sPLS), using MixOmics R-package. With these analyses, we found several links between ewes' performances and rumen bacteria and we characterised the ruminal microbiota of divergent lines.

Non-invasive method to evaluate the occurrence of trans-10 shifted rumen biohydrogenation in lambs

S.P. Alves[1], A. Francisco[2], A.P.V. Portugal[2], E. Jerónimo[3], J. Santos-Silva[2] and R.J.B. Bessa[1]
[1]CIISA, Faculdade de Medicina Veterinária, Universidade de Lisboa, 1300-477 Lisboa, Portugal, [2]Instituto Nacional de Investigação Agrária e Veterinária (INIAV), Fonte Boa, 2005-760 Vale de Santarem, Portugal, [3]CEBAL, Centro de Biotecnologia Agrícola e Agro-Alimentar do Alentejo, ICAM, IPBeja, 7801-908 Beja, Portugal; rjbbessa@fmv.ulisboa.pt

Ruminal microbiota isomerise and hydrogenate the double bonds of the dietary C18 polyunsaturated fatty acids (FA), through a process known as biohydrogenation (BH). The normal BH pathways produce 18:1t11 as the main intermediate, but also CLA-c9t11 and other minor isomers. In tissues, the 18:1t11 can be converted into CLA-c9t11 by the action of the enzyme delta 9-desaturase, and both 18:1t11 and CLA-c9t11 have been considered as health-promoting. However, when ruminants are fed with low-fibre, high-starch diets, an undesirable modification on BH pathways often occur, where the 18:1t11 is replaced by 18:1t10 as the major intermediate formed (i.e. trans-10 shift). The 18:1t10/18:1t11 ratio, calculated after analysing the FA composition of digestive contents, tissues or milk is used indicator of the occurrence of the trans-10 shift. In this study, we aimed to develop an *in vivo* non-invasive method to evaluate the trans-10 shift in lambs. Hence, we analysed the FA composition of rumen contents and faeces from 85 Merino Branco lambs. Samples were freeze-dried and transesterified into FA methyl esters using a basic followed by acid catalyst in methanol, and then analysed by gas-chromatography with flame ionisation detection. Linear regression equations ($Y=aX+b$) were applied to the data to evaluate the accuracy of faeces as marker of the FA in rumen contents. We confirmed significant linear regressions for the proportion of 18:1t10 ($Y = 0.86x + 0.58$; $r^2=0.916$), 18:1t11 ($Y = 1.27x + 0.86$; $r^2=0.703$) and t10/t11-18:1 ($Y = 1.63x – 0.18$; $r^2=0.885$). Thus, we propose using faeces FA composition as marker of the proportion of 18:1t10, 18:1t11 and 18:1t10/18:1t11 ratio in the rumen. This is a non-invasive method, which causes no discomfort to the lambs and that is very accurate to estimate the ruminal function. Financial support was provided by the ALT20-03-0145-FEDER-000040 (ValRuMeat) and UIDP/CVT/00276/2020 projects.

Characterising the microbiota of traditional cattle breeds in the Netherlands and Finland

R.P.M.A. Crooijmans[1], R. Gomez Exposito[2], R. Gonzalez-Prendes[1], T. Reilas[3], M. Makgahlela[4], J. Kantanen[3], C. Ginja[5], D.R. Kugonza[6], N. Ghanem[7] and H. Smidt[2]
[1]Wageningen University & Research, Animal Breeding and Genomics, Droevendaalsesteeg 1, 6708 PB Wageningen, the Netherlands, [2]Wageningen University & Research, Laboratory of Microbiology, Stippeneng 4, 6708 WE Wageningen, the Netherlands, [3]Natural Resources Institute Finland, Myllytie 1, 31600 Jokioinen, Finland, [4]Agricultural Research Council, Animal Breeding and Genetics, P/Bag X2, Irene, 0062 Pretoria, South Africa, [5]Universidade do Porto, CIBIO-InBIO, Rua Padre Armando Quintas 7, 4485-661 Vairão, Portugal, [6]Makerere University, Department of Agricultural Production, P.O. Box 7062, Kampala, Uganda, [7]Cairo University, Animal Production Department, El-Gammaa, 12613 Cairo, Egypt; richard.crooijmans@wur.nl

The ruminant gut microbiota has been shown to be affected by numerous factors, including host diet, medications, ecosystem and the host genetic background. Depending on the type and chemical composition of animal feed (amount of fibres, sugars and fat) differences in microbiota composition have been identified in different species. As ruminants, they have four-chambered stomachs for efficient breakdown of indigestible plant material via fermentation, which further proceeds in the intestine, and this process is performed by a complex microbiota along the digestive tract that converts food into energy. Gut microorganisms play an important role in maintaining the physiology of the host. However, gut microbiota diversity over breeds is poorly studied. We investigated the genetic variability of indigenous breeds on adaptation and the options to improve their performance, longevity and disease resistance within the OPTIBOV-project. Faecal microbiota composition is measured through 16S rRNA gene sequencing from 200 animals of eight traditional cattle breeds from the Netherlands and Finland representing Deep Red (24), Groningen White Headed (20), MRY (23), Dutch Belted (23), Dutch Friesian (24), Northern Finncattle (25), Eastern Finncattle (25) and Western Finn (25), and ten commercial Holstein Friesian animals from two countries (5 from the Netherlands and 5 from Finland). Impact of genotype on the variation in microbial composition is explored through multivariate analyses.

Effect of rearing system on the microbial communities in rumen of Maremmana and Aubrac steers

M. Daghio[1], F. Ciucci[2], G. Conte[2], A. Serra[2], B. McAmmond[3], C. Viti[1], J.D. Van Hamme[3], A. Buccioni[1] and M. Mele[2]
[1]University of Florence, Department of Agriculture, Food, Environment and Forestry, Piazzale delle Cascine 18, 50144 Firenze, Italy, [2]University of Pisa, Dipartimento di Scienze Agrarie, Alimentari e Agro-ambientali, Via del Borghetto 80, 56124 Pisa, Italy, [3]Thompson Rivers University, Department of Biological Sciences, 805 TRU Way, V2C 0C8 Kamloops, BC, Canada; matteo.daghio@unifi.it

Rumen microbiota plays a key role in the conversion of plant biomass and in lipid metabolism in ruminants. A deep characterisation of the microbial communities can help the development of feeding strategies to improve the quality of animal products and to increase the sustainability of livestock productions. Our aim was to evaluate the effect of the rearing system (grazing or feedlot) on the microbial communities in rumen liquor (RL) of steers from two different breeds well suited for extensive farming conditions: Aubrac (AU) and Maremmana (MA). Forty 6-month-old AU and MA steers (average body weight of 250 kg) were allotted into 2 experimental groups: grazing (10 animals for each breed) and feedlot (10 animals for each breed). The steers were slaughtered (between 20-22 months of age and around 600 kg) and samples of RL were collected. All the experiments of this study were performed according to the approved guidelines from the European Directive 2010/63/UE and DL 4/03/2014 n 26. Fatty acids (FAs) composition was determined by GC-FID. DNA was extracted from each sample and the microbial communities were characterised by high-throughput sequencing of the partial 16S rRNA gene. Saturated FAs (SFAs) were the most abundant FAs (78%), mainly composed by stearic acid (SA, C18:0), followed by palmitic acid (PA, C16:0), which showed different concentration in the two breeds: RL of grazing MA steers showed lower content of SA and higher content of PA. The most abundant unsaturated fatty acids were oleic acid (C18:1 *cis* 9) and linoleic acid (C18:2 *cis* 9 *cis* 12). The most abundant bacterial families were *Lachnospiraceae*, *Prevotellaceae*, *Rikenellaceae* and *Ruminococcaceae*. The families *Rikenellaceae* and *Ruminococcaceae* were more abundant in the RL of MA breed, conversely the family *Prevotellaceae* was more abundant in the RL of AU breed.

Poster presentations and discussion

S. Vigors[1] and P. Trevisi[2]
[1]University College Dublin, Belfield, D4, Ireland, [2]University of Bologna, Dipartimento di Scienze e Tecnologie Agro-Alimentari Viale Fanin 46, Bologna, 40126, Italy; staffordvigors1@ucd.ie

Time slot for posters

Effect of weaning and antioxidant compound supplementation on oxidative status in piglets

Y. Martel-Kennes[1,2], A. Caldwell[1], C. Robert[1], J.P. Laforest[1] and J. Lapointe[3]
[1]Université Laval, Sciences Animales, Ville de Québec, Québec, G1V 0A6, Canada, [2]Centre de Recherche en Sciences Animales de Deschambault, Deschambault, Québec, G0A 1S0, Canada, [3]Agriculture and Agri-Food Canada, Sherbrooke R & D Centre, Sherbrooke, Québec, J1M 0C8, Canada; yan.martel-kennes@crsad.qc.ca

The objective of this project was to characterise the effects of a nutritional supplementation of antioxidant compounds on energy and oxidative status during the post-weaning period. In this experiment, 96 piglets selected on d19 were allocated to three groups. They received different concentrations of antioxidant compounds (low dose of vitamin E, LE: 4,8 mg of α-tocopherol; high doses of vitamin E and C, HEC: 30 mg of α-tocopherol and 150 mg of vitamin C; high doses of vitamin E, C and a commercial melon concentrate rich in antioxidants, HECM: HEC and 15 mg of the melon concentrate per day and per pig), starting on d19. Piglets were weaned on d21, thereafter fed a basal diet deprived of vitamin E and were euthanised on d21, d23, d26 and d29 (n=8/treatment/day) for liver, intestinal mucosa and blood sample collection. Plasma levels of oxidative damage to DNA (8-hydroxy-2'-deoxyguanosine, 8-OHdG) were higher on d21, d23 and d26 than on d29 (P<0.01). Plasma glutathione peroxidase (GPx) activity gradually increased following weaning to reach its highest concentration on d29 (P<0.01). This suggests an increase in systemic oxidative stress and was also related to significant increases in superoxide dismutase (SOD) and GPx enzymatic activities in other tissues. Cellular and mitochondrial SOD activities in the liver as well as cellular and mitochondrial GPx activities in the liver and intestinal mucosa increased significantly following weaning (P<0.05). Administration of HECM reduced cellular GPx activity in the liver compared to the LE group during the overall period (P=0.02). No effects of weaning or antioxidant treatments have been observed on liver adenosine triphosphate (ATP) concentrations (P>0.05). These results show that weaning of piglets has significant consequences on oxidative stress that occur very quickly after this event. In addition, some antioxidant combinations administered to piglets reduced liver GPx activity during the post-weaning period, which could be related to a positive effect on oxidative status.

Organic acid based product supplementation in pigs challenged with the enterotoxin *E. coli*

V. OcElova[1], M.J. Jimenéz[2], R. Berrios[1], S. Stelzhammer[1] and A.P.F.R.L. Bracarense[2]
[1]BIOMIN Holding GmbH, Erber Campus 1, 3131, Austria, [2]Universidade Estadual de Londrina, Laboratory of Animal Pathology, Campus Universitário, Rodovia Celso Garcia Cid, km 380, Londrina, 86057-970 Paraná, Brazil; vladimira.ocelova@biomin.net

Weaning transition has several dietary, environmental and social factors that predispose piglets to gastrointestinal disorders and enteric infections by pathogens. Ongoing research has focused on finding feed additives to improve animal performance and to prevent pathogen challenges to reduce antibiotic use in livestock production. An organic acid based product (OABP) was used to evaluate its ability to prevent gastrointestinal tissue, liver and lymph node damage during a challenge study with *Escherichia coli*. A total of 90 healthy 21 day-old piglets were fed a basal diet for 7 days and later assigned to 3 treatments (10 replicates each). At 28 days of age a control group of pigs (CON) was fed with a basal diet, while the group COL was supplemented with colistin 200 mg/kg feed. The OABP group was supplemented with a blend of propionic, formic and acetic acids together with cinnamaldehyde and permeabilising substances (Biotronic® Top3, BIOMIN, Austria) at 1 kg/t feed. One week after the randomisation, all animals were orally challenged with an enterotoxic *E. coli* (K88; 5 ml 1.1×10^9 cfu/ml). Six animals per treatment were euthanised 5 days after induced challenge. Tissues of small intestine, liver and mesenteric lymph nodes were collected and subjected to histological analysis with determination of lesion score. The performance of the rest of animals was recorded for a total period of 5 weeks. Lesion score of jejunum and liver tissue was significantly improved (P<0.05) in OABP group when comparing to CON. Numerical improvement (P>0.05) after OABP supplementation was observed for lesion score of ileum (compared to CON) and lymph nodes (compared to CON and COL group). Feed conversion rate in OABP group was numerically improved in comparison to CON and COL group (P>0.05). In conclusion, the administration of an organic acid based product reduced the lesion score of intestinal tissue, liver and lymph nodes influencing positively growth performance parameters of piglets in OABP group.

Yeast hydrolysate prevented *E. coli*-induced increases in piglet gut tissue conductance *ex vivo*

E. Valkonen[1], H. Kettunen[1], S. Hasan[1], J. Vuorenmaa[1] and G. Breves[2]
[1]Hankkija Ltd, Peltokuumolantie 4, 05801 Hyvinkää, Finland, [2]University of Veterinary Medicine Hannover, Institute for Physiology and Cell Biology, Bischofsholer Damm 15, 30173 Hannover, Germany; eija.valkonen@hankkija.fi

The enterotoxigenic *Escherichia coli* with F4 fimbriae (ETEC F4) is known to increase intestinal permeability in pigs – a process which may lead to loss of homeostasis, diarrhoea, and reduced performance. The cell wall of *Saccharomyces cerevisiae* contains macromolecules that bind to the fimbriae of F4-ETEC and may thus inhibit its ability to bind to intestinal mucosa. Here we compared two *S. cerevisiae*-products: a hydrolysate (HYD) and a cell wall product (CW) in preventing *E. coli*-induced increases of intestinal permeability in an *ex vivo* model. A section of middle jejunum was removed immediately after the euthanasia of a piglet. The tissue was carefully rinsed with ice-cold physiological saline solution. The mucous tunic was mechanically separated from the muscle and serosa layers. The epithelium was cut and placed into sixteen pairs of Ussing chambers. Krebs-Ringer solution at pH 7.4 was used for a 30 min preincubation. Four replicate chamber pairs were then assigned to the following treatments: (1) untreated control; (2) negative control: fresh culture of F4-ETEC; (3) F4-ETEC + HYD; and (4) F4-ETEC + CW. Tissue conductance, an indirect measure of epithelial permeability, was continuously recorded for two hours. During the two-hours incubation conductivity between 18 and 35 mS/cm^2 was measured in all of the sixteen chambers, representing normal values for the given test conditions. The conductance remained at the same level throughout the two hours in the untreated control and in the F4-ETEC + HYD-treatment. Instead, the F4-ETEC-group and F4-ETEC + CW showed steadily increasing conductance values. As the reciprocal value of conductance is electrical resistance, it means that HYD but not CW can compensate for the reduction of resistance caused by F4-ETEC. Further research is needed to identify which cellular mechanisms cause the conductance changes activated by F4-ETEC, how HYD counteracts these events *ex vivo*, and whether HYD also inhibits pathogen-associated permeability increase in live pigs.

A dose of symbiotics and vitamins at birth has lasting effects on piglet microbiota and performance

M. Girard, M. Tretola and G. Bee
Agroscope, Tioleyre 4, 1725 Posieux, Switzerland; marion.girard@agroscope.admin.ch

Early-life microbial colonisation plays a central role in the development and the maturation of the gut, improving development and disease resistance of piglets. This study aimed to measure whether a single dose supplement given at birth modified faecal microbiota of piglets and improved their performance. The supplement was a mixture of probiotics, prebiotics, immunoglobulins, vitamins and selenium. At birth, piglets received either 2 ml of this supplement (S+) or 2 ml of water (S-). In order to study faecal microbiota, 29 female piglets originating from 14 litters, 8 and 6 litters assigned to S+ and S-, respectively, were monitored from birth to two weeks post-weaning. Their faeces were collected at 16 d of age and 2 weeks post-weaning to measure volatile fatty acid levels and to investigate the gut microbiota composition by high throughput sequencing. The growth and the occurrence of diarrhoea from 1 week before weaning to 2 weeks post-weaning was also monitored. The S+ female piglets had a greater (P<0.05) average daily gain during the whole experiment compared to S- female piglets. It was mainly due to a 72 g/d greater (P<0.05) growth in the post-weaning period for the S+ compared to the S- piglets. At 16 d of age but not 2 weeks post-weaning, S+ piglets had 40% greater (P<0.05) faecal volatile fatty acid levels, mainly acetate and propionate, than S- piglets. The supplementation reduced (P<0.05) diarrhoea occurrence, especially in the first week post-weaning where the occurrence was of 12.5% in the S+ piglets and 26% in the S- piglets. Not surprisingly, α- and β-diversity indexes differed between 16 d of age and 2 weeks post-weaning. *Ruminococcaceae* was the main family at 16 d of age while *Prevotellaceae* family dominated the bacteria population at 2 weeks post-weaning. At 16 d of age, among many other, the genus *Candidatus saccharimonas*, belonging to the phylum *Saccharibacteria*, was the most enriched (P<0.05) in S+ piglets while the relative abundance of the genus *Butyricimonas* was significantly increased (P<0.05) in S- piglets. This study highlighted the importance of early-life interventions on the growth performance and health in the 2 first weeks post-weaning.

Effect of betaine on *in vitro* large intestinal fermentation of Iberian pigs under heat stress

Z. Pardo[1], M.J. Ranilla[2,3], A. Francisco[3], R. Campos[4], A. Martín[3], I. Mateos[2,3], M. Lachica[1] and I. Fernández-Fígares[1]
[1]CSIC, Estacion Experimental del Zaidin, Profesor Albareda 1, 18008 Granada, Spain, [2]CSIC, Instituto Ganadería Montaña, Ctra. Leon-Vega de Infanzones, 24346 Grulleros, Spain, [3]Universidad de León, Producción Animal, Campus de Vegazana, 24007 Leon, Spain, [4]Universidad Nacional de Colombia, Dep. Ciencia Animal, Carrera 45 # 26-85, 111321 Bogota, Colombia; ifigares@eez.csic.es

Betaine could help to cope with osmotic stress of intestinal cells and bacteria under heat stress (HS). There is lack of information regarding the effect of betaine on intestinal microbial activity in pigs under heat exposure. The aim of the investigation was to study the effect of betaine (5 g/kg) on *in vitro* intestinal fermentation capacity in growing Iberian pigs under HS (30 °C, 30 d). Sixteen Iberian barrows (43 kg) were assigned to control or betaine group. The diet covered all nutrient requirements and was based on barley and soy bean meal supplemented or not with betaine. At 60 kg, rectum content was collected and kept at -80 °C until the fermentation trial. We set up an *in vitro* batch culture method to assess fermentability of ingredients for pig diets. Pectin and starch were fermented in a faecal slurry consisting of an anaerobic culture medium and faeces from pigs fed the control or betaine diet under HS. After 24 h of fermentation, production of gas and VFA and NH_3 concentrations were measured. Total VFA production was unchanged by betaine. However, betaine decreased propionate production and propionate molar proportion (25%-26%, $P<0.05$) while increasing valerate production (70%, $P<0.05$), acetate:propionate ratio (22%), molar proportions of butyrate, isoacids and valerate (25, 53, 106, $P<0.05$) and NH_3 production (9%, $P<0.05$) when starch was used as substrate. When pectin was the substrate, acetate and valerate production (6 and 78%, $P<0.05$) and molar proportion (4%, 69%), acetate:propionate ratio (16%) and NH_3 production (4%, $P<0.05$) increased while propionate and butyrate production (12%, $P<0.05$) and molar proportion (15.5%, $P<0.05$) and gas production (7%, $P<0.05$) decreased. In conclusion, betaine modified *in vitro* large intestine fermentation profile of Iberian pigs under HS, depending upon substrates.

Rumen microbiome of ruminants from extreme environments

G. Forcina[1], A. Dos Santos[1], L. Pérez-Pardal[1], J.G.V. Carvalheira[1,2] and A. Beja-Pereira[1]
[1]CIBIO-InBIO, Research Centre in Biodiversity and Genetic Resources, University of Porto, Rua Padre Armando Quintas 7, 4485-661 Vairão, Portugal, [2]Abel Salazar Institute of Biomedical Sciences, University of Porto, Rua de Jorge Viterbo Ferreira 228, 4050-313 Porto, Portugal; giovanni.forcina@cibio.up.pt

The study of microbiome composition is a novel branch of metagenomics which has gained momentum over the last few years boosted by the massive decrease of sequencing costs and the increasingly powerful NGS techniques. This enables unprecedented insights into microbial ecologies, which in turn bear huge applicative potential in research areas as wide ranging as human medicine to animal health and production. Driven by commercial interests, a number of studies has been devoted to investigate the microbiome composition in domestic livestock over the last decade. However, aside from these ongoing efforts for investigating the microbial communities hosted in commercial breeds which are delivering insights of major socioeconomic importance, much less is known about local breeds of high cultural other than economic value. These resources deserve major interest also in terms of adaptation to a range of peculiar – and sometimes even extreme – environmental conditions resulting from the interplay among several major evolutionary forces such as genetic drift, gene-flow, mutation-selection balance and various forms of natural and human-induced positive selection. Deciphering the role played by microbiome composition in this process is of utmost relevance, especially in an epoch of increasing global climate change. At a local level, these studies are of key importance to protect and strengthen the production strategies of rural marginal areas by promoting the sustainable use of agrobiodiversity resources while enhancing food security, in compliance with Goals 2, 3, 12, 13 and 15 of the United Nations 2030 Agenda. The aim of this study is to explore the rumen microbial composition s inhabiting remote areas of Angola with extreme ecological conditions (hot and dry) combining lab work on faecal samples with in-silico analyses. The results will be compared with those obtained in commercial breeds from Portugal to evidence patterns of similarity/dissimilarity in terms of presence/absence and abundance of microbial taxa.

Dehydrated olive oil wastewater alleviates the incidence of post weaning diarrhoea

M.R. Mellino[1,2], C. Ollagnier[1], G. Bee[1], N. Pradervand[1] and G. Battacone[2]
[1]*Agroscope, Swine Research Unit, Animal Production Systems and Animal Health, Rte de la Tioleyre 4, 1725 Posieux, Switzerland,* [2]*University of Sassari, Department of Agricultural Sciences, viale Italia, 39, 07100, Italy; battacon@uniss.it*

Post weaning diarrhoea (PWD) is one of the health issues where most of the antimicrobials are used in pig production. With multidrug-resistant pathogens increasingly spreading, finding solutions to diminish the severity of PWD without antibiotics becomes even more critical. Plant-based bioactive compounds like phenolic compounds (such as hydroxytirosol, tirosol and verbascoside) have been shown to have antimicrobial activities *in vitro*. These compounds can be found in residues after the process of refining vegetable oils. The aim of this study was to test whether substances present in dehydrated olive oil wastewater (OOWW) can be used in starter diets for weaned piglets to reduce the PWD incidence without negatively impacting the growth performance. At 25±1.05 d of age and with average body weight of 7.41±1.16 kg, 64 weaned pigs (4 piglets per pen) susceptible to enterotoxigenic *Escherichia coli* (ETEC) F4ac were assigned within litter to 4 treatments: control (C), OOWW1 (C + 0.5% OOWW), OOWW2 (C + 1% OOWW) and OOWW3 (C + 2% OWW). During the 19-d of trial piglets had *ad libitum* access to the diet and water. The body weight was determined at d 0 (weaning), 7, 14 and 19 and feed intake per pen was determined daily. Faecal scores (1, normal, to 4, watery diarrhoea) were monitored on d 0, 3, 4, 5, 7, 10, 14 and 19. In the mixed linear model, dietary treatment, the experimental day and the 2-way interaction was used as fixed effects and animals within pen as random effect. Regardless of the OOWW inclusion levels, average daily gain, feed intake and feed conversion ratio was not (P<0.05) affected. However, the faecal score was lower (P<0.06) in OOWW1 and OOWW3 than C piglets. In conclusion, the OOWW inclusion had no detrimental impact on growth performance but alleviated the incidence of PWD at an inclusion level of 0.5 and 2%.

Climate resilience in dairy sheep production in Europe

M.J. Carabaño[1], G. Arsenos[2], D. Buisson[3], C. Díaz[1], D. Hazard[4], H. Larroque[4], C. Pineda-Quiroga[5], M. Serrano[1], A. Tryantafillidis[2], V. Tsartsianidou[2], E. Ugarte[5] and M. Ramón[6]
[1]*INIA, Ctra de La Coruña km 7, 28040 Madrid, Spain,* [2]*Aristotle University of Thessaloniki, Thessaloniki, 54124 Thessaloniki, Greece,* [3]*Institut de l'Elevage, BP 42118, 31321 Castanet-Tolosan, France,* [4]*INRAE, Université de Toulouse, 31326, Castanet Tolosan, France,* [5]*Neiker, Arkaute, 01080 Vitoria-Gasteiz, Spain,* [6]*IRIAF-CERSYRA, Avda del vino, 10, 13300 Valdepeñas, Spain; mjc@inia.es*

Dairy sheep production is one of the traditional livestock productions in Europe, mainly in the Mediterranean region, which is one of the most responsive regions to global climate change. Animal resilience to extreme weather is considered one of the key elements for sustainability of dairy sheep production in this region. Productive records of 644,863 dairy ewes of 5 breeds (Assaf, Chios, Lacaune, Latxa and Manchega) and meteorological information from weather stations closest to farms were used to characterise overall and individual response to extreme weather and produce novel phenotypes within the iSAGE project. Heat stress thresholds in terms of average daily temperature for dairy sheep ranged between 17 °C in the Latxa breed in Northern Spain to 26 °C in the Greek breed Chios. Productive decay due to heat stress was largest in the highly selected breed, Assaf, and largest for fat and protein yields than for milk yield. Although declines in productive and reproductive parameters were evident in these breeds, the estimated loss in productivity associated with milk traits was relatively small at the population level for all breeds; only around 1-2% of the current average production per degree of temperature above the maximum heat load. For AI fertility, the local breed Manchega showed a larger negative impact associated to high temperatures than the more selected Assaf breed. Slopes of decay were up to 1% at the highest heat load. For milk traits, individual response showed substantial variability. For example, for the Assaf breed, the 0.1% of the more susceptible ewes showed differences in slopes of around 70 g/d in milk yield compared with the 0.1% more tolerant. The Chios and Latxa breeds showed the largest variation in response to heat. The observed variation of resilience phenotypes across sheep breeds enables for downstream decisions on breeding selection process.

The Karagouniko Sheep: response of the breed to heat stress

M.A. Karatzia[1], E.N. Sossidou[2], D. Tsiokos[1] and C. Ligda[2]
[1]Research Institute of Animal Science, HAO-Demeter, Paralimni, 58100, Giannitsa, Greece, [2]Veterinary Research Institute, HAO-Demeter, HAO Campus, 57001, Thessaloniki, Greece; karatzia@rias.gr

The purpose of the present study was to investigate the response of sheep of the indigenous Greek breed Karagouniko, reared semi-extensively in Thessaly, during heat stress (H.S.) and to assess H.S. effects on their health and welfare. The survey was carried out from May to September 2019. Ambient temperature and relative humidity values were recorded at 5-minute intervals by digital sensors during grazing and on-farm. At the same time, using collars fitted with GPS positioning trackers, the daily grazing route (distance, duration and resting points) was recorded. Upon return from grazing, measurements of physiological parameters [body/rectal temperature (BT/RT), respiratory rate, salivation, etc.] on 10% of the flock were recorded. Additionally, behavioural observations at individual and flock level attributed to heat stress, as well as the welfare level (AWIN protocol) of the flock were estimated. Temperature Humidity Index (THI) was calculated based on well-established methodology. The duration of H.S.-initiating conditions was estimated for May at 6h24m/24 h, of which 4h54min relate to T.H.I.$\geq$23.3 and 1 h to T.H.I.$\geq$25.6. In June, the duration of H.S. was calculated at 15h10m/24 h, with 12h18min referring to T.H.I.$\geq$23.3 and 10h28min T.H.I.$\geq$25.6. Furthermore, in August, the duration of H.S. was determined to be 17h53m/24 h, with 11h40min referring to T.H.I.$\geq$23.3 and 9h22min T.H.I.$\geq$25.6. It appears that sheep of the Karagouniko breed possess the ability of adapting their grazing behaviour according to H.S. intensity by decreasing grazing route distance (8.4 km in June/ 6.1 km in August) and duration of grazing (4 h in June/3.5 h in August). Despite the significant increase of mean BT (June:32.59$\pm$0.157 °C/August:34.45$\pm$0.151 °C, P$\leq$0.005), mean RT increased within normal range (June:38.65$\pm$0.078 °C/August:38.93$\pm$0.061 °C, P$\geq$0.005). This could be attributed to a potential genetic adaptability of the breed to H.S. and merits further investigation. The research was carried out in the framework of the ARIMNet2 project 'PeRFORM – Breeding and management practices for locally adapted breeding and productive sheep and goat breeding systems'.

Assessing the ancestry of American Creole cattle using genetic markers

C. Ginja[1], L.T. Gama[2], O. Cortés[2], P. Zaragoza[2], I. Martín-Burriel[2], J.L. Vega-Pla[2], M.C.T. Penedo[2], P. Sponenberg[2], J. Cañón[2], A. Sanz[2], A.A. Egito[2], L.A. Alvarez[2], G. Giovambattista[2], S. Agha[2], A. Rogberg-Muñoz[2], M.A. Lara[2], J.V. Delgado[2] and A. Martínez[2]
[1]Universidade do Porto, CIBIO/InBIO, Centro de Investigação em Biodiversidade e Recursos Genéticos, R. Padre Armando Quintas, 7, 4485-661 Vairão, Portugal, [2]BioBovis Consortium, Departamento de Genética, Universidad de Córdoba, Edificio Gregor Mendel, Planta Baja. Campus Universitario de Rabanales, 14071 Córdoba, Spain; catarinaginja@cibio.up.pt

The origins of the American Creole cattle and major genetic contributions to their development remain unclear. We analysed uniparental genetic markers and autosomal microsatellites in 4,658 DNA samples from 114 cattle breeds distributed worldwide, including: 40 Creole breeds representing the American continent; 39 Iberian breeds; 18 African breeds; 6 British breeds; 6 Continental European breeds; and also 5 Indicine breeds sampled in the Americas. The information on the various genetic markers was combined to investigate: (1) the diversity, identity and genetic structure of Creole cattle; (2) the evidence for a direct African influence in the development of Creoles; (3) the signals of an Iberian genetic signature in current Creole populations, (4) the extent of *Bos indicus* introgression in the gene pool of Creole cattle. The genetic legacy of Iberian cattle is still represented in Creoles, but other influences could also be detected, even though in most cases Creoles remain well differentiated. The African contribution to the genetic composition of Creoles is clear in our work, and while this may have occurred by an indirect path through Iberian breeds, the direct influence of African breeds on Creole cattle is undoubtedly demonstrated by their sharing of unique maternal and paternal lineages. Programs for the genetic management of Creole breeds of cattle are urgently needed, aimed at the conservation of these unique genetic resources. With this goal, efforts must be made to overcome the gap existing between the state-of-the-art genomic tools currently available and their application to undervalued breeds, such as Creoles, kept in marginal regions.

Combining categories of crossbred females to improve the overall performance of a dairy cattle herd
J. Quénon, M.-A. Magne and S. Ingrand
INRAE, 24 chemin de Borderouge, 31320 Castanet-Tolosan, France; julien.quenon@inrae.fr

There is renewed interest in rotational crossbreeding in Western countries to improve reproductive performance of dairy cattle cows, especially for the Holstein (HO) breed. To our knowledge, few studies are interested in the impact on the overall performance at herd scale of genetic diversification, which is involved in the transition of a herd towards rotational crossbreeding. Our hypothesis is that there are herd compositions of crossbred categories that optimise its multi-criteria performance. The objective of the present study was twofold: (1) to define and compare performance of various categories of crossbred females; and (2) to assess the best combinations of these categories within a herd to optimise its performance. We used milk record data of 19 French commercial dairy crossbred herds from 2008 to 2018. The HO breed was the initial breed from which sampled dairy farmers managed rotational crossbreeding. The dataset was made up of 5,110 lactations from 2,448 cows, for which milk yield and contents and fertility performance data were available. First, we defined categories of crossbred females [a; b] according to two indicators: a, the percentage of Holstein genes from 25 to 100% and b, the theoretical value of heterosis from 0 to 100%. Secondly, we used linear models to compare performance profiles of each category of crossbred females. We showed for example that 305-days milk yield was greatest for HO (7,823 kg) and lower for F1 (7,748 kg), 3-way G2 (7,394 kg) and 2-way G2 (6,211 kg). In a third step, based on the mean performance we built multi-criteria performance profiles for each category. We simulated all herd compositions as combinations of the different categories following a design of experiment approach and modelled the resulting multi-criteria performance for each herd composition. Then we assessed the best herd composition according to three sets of constraints: (1) 'Milk yield ≥8,000 kg/cow'; (2) 'Milk yield ≥6,000 kg/cow'; and (3) 'CCI≤60 days'. These results bring new evidence of the benefits of rotational crossbreeding on the whole herd performance rather than solely on individual performance. It allows us to discuss the implications for managing the transition from purebred HO to crossbred herds.

Characterisation of traditional Dutch cattle breed specific copy number variations
R. Gonzalez-Prendes[1], M. Makgahlela[2], J. Kantanen[3], C. Ginja[4], D.R. Kugonza[5], N. Ghanem[6] and R.P.M.A. Crooijmans[1]
[1]Wageningen University & Research, Animal Breeding and Genomics, Droevendaalsesteeg 1, 6708 PB, Wageningen, the Netherlands, [2]Agricultural Research Council, Animal Breeding and Genetics, P/Bag X2, Irene, 0062 Pretoria, South Africa, [3]Natural Resources Institute Finland, Animal Breeding and Genetics, Myllytie 1, 31600 Jokioinen, Finland, [4]Universidade do Porto, CIBIO-InBIO, Rua Padro Armando Quintas 7, 4485-661 Vairão, Portugal, [5]Makerere University, Department of Agricultural Production, 7062 Kampala, Uganda, [6]Cairo University, Animal Production, El-Gammaa, 12613, Cairo, Egypt; rayner.gonzalezprendes@wur.nl

The OPTIBOV project aims to investigate the genetic variability of traditional/indigenous breeds with respect to adaptation and the options to improve their performance, longevity and disease resistance. Here we identified and characterised copy number variants (CNV) from 120 whole genome sequences of five traditional Dutch cattle breeds representing Deep Red (24), Groningen White headed (20), MRY (23), Dutch Belted (23), Dutch Friesian (24) and five commercial Holstein Friesian animals. The number of sequence reads per animals varied between 203,644,434 (Dutch Friesian) and 377,428,034 (MRY). All reads were aligned to the ARS-UCD1.2 reference genome. The genome coverage after alignment varied from 7× to 13× and the CNV calling was performed based on a read depth algorithm implemented in CNVnator tool. The total number of CNVs detected after filtering was 63,320 covering a total sequence length varying between 1.8 to 4.4 Mb. Deletions represented 60% (38,033) of the total CNVs whereas 25,287 (40%) were duplications. To detect the structural genomic variants affecting local breeds the breed specific CNVs were studied. The number of breed specific CNVs ranged from 39 in Holstein Friesian to 1,554 in the Dutch Belted breed. The total sequence length of these breed specific CNVs varied from 1 Mb in Holstein Friesian to 4.4 Mb in the Dutch belted breed. Variation in genome structure was observed among breeds, and further study on genes underlying the breed specific CNVs will be performed.

Gene banking strategies to ensure resilience of farming systems in less favoured areas

M. Tixier-Boichard[1] and Image Consortium[2]
[1]*University Paris-Saclay, INRAE, AgroParisTech, GABI, CRJ, 78350 Jouy-en-Josas, France,* [2]*www.imageh2020.eu, Belgium; michele.boichard@inrae.fr*

To be resilient, farming systems need to keep options for the future. Gene banks keep options for livestock by being a back-up, supporting *in situ* conservation, supporting research to assess adaptation and monitor trends. Less favoured areas have harsh climatic conditions, limited disease control and low support to farmers. What can gene banks offer here? The IMAGE H2020 project has addressed the different dimensions of a gene banking strategy. This requires a multi-actor approach with a governance supported by a public policy. Economic optimisation showed that coordination of gene banking strategies across countries can save money or make possible to preserve more breeds at the same cost. A gap analysis showed that local breeds are less present in gene banks than commercial transboundary breeds, where collection of biological material is easier. Yet, local breeds are likely to harbour original genetic variants useful for adaptation. Thus, a special effort has to be directed towards local breeds before they become endangered and too much inbred, which severely impairs reproduction and collection of germplasm. Gene banks must trigger developments in reproductive biotechnologies, in order to improve the fertility of cryopreserved material and match expectation of farmers. Collecting biological materials other than semen is needed in order to speed up the recovery of desired genotypes. IMAGE has made several achievements in this respect for mammal embryos and for gonads or germ cells of chickens. The health status of local populations can be a severe limit to gene banking, and public policies should define measures to organise collection in safe conditions, even if not at the level required for international trade. Finally, documenting the material by genomic and phenotypic data is key to promote the future use of gene bank collections. IMAGE has developed: (1) two multi-species SNP arrays to assess diversity of animal genetic resources, (2) the MoBPS software to simulate conservation or breeding programs using gene bank material with associated data and (3) a portal to connect data. Training staff and users of gene banks is now a priority to benefit from the progress made in IMAGE.

Benefits from recent and on-going projects on adaptation & resilience in French dairy sheep & goats

J.M. Astruc[1], D. Buisson[1], V. Clement[1], G. Lagriffoul[1], H. Larroque[2], A. Legarra[2], I. Palhiere[2] and C. Moreno-Romieux[2]
[1]*Institut de l'Elevage, BP 42118, 31321 Castanet-Tolosan, France,* [2]*INRAE GenPhySE, Université de Toulouse, INRAE, ENVT, CS 52627, 31326 Castanet-Tolosan, France; Jean-michel.astruc@idele.fr*

In France, dairy sheep and goats have developed efficient breeding schemes, whose annual genetic gain ranges from 0.10 to 0.27 genetic standard deviation of the total merit index, according to the breeds. From 2015 to 2020, the 5 dairy sheep breeds and 2 main dairy goat breeds have shifted to full genomic programs, based on reference population constituted of AI progeny-tested males. Currently, the breeding goals are similar for almost all the selected breeds and may be per se qualified as balanced, since they include efficiency criteria (milk yield and content), resilience criteria (udder health, resistance to scrapie), and adaptation to mechanic milking in order to mitigate the workload of the farmer in the milking parlour (udder morphology). Expectations for improving sustainability, robustness, resilience while still increasing efficiency are more and more pressing. These new objectives are emphasised by the different stakeholders: the farmers, the industry, but also the consumers, the citizens. Opportunities occur to cope with these requests: a larger efficiency offered by genomic programs, new technologies and methods (e.g. use of MIR spectra), awareness and consent of all stakeholders to bring new challenges at the agenda (climate change, growing resistance to antibiotics or anthelmintics). Various recent or on-going projects (e.g. iSAGE, RUSTIC, ARDI, SMARTER) are addressing the topics of adaptation, sustainability, resilience in small ruminants. They bring or intend to bring responses by activating new levers, following new paths. Through the French dairy sheep and goat illustration, we proposes to explore the following ways, that are the core objectives of the on-going SMARTER project: breeding for new efficiency and resilience traits in a context of more agro-ecological systems and search of self-sufficiency in the feeding systems; benefiting from international harmonisation and cooperation; tackling the genotype × environment interactions; sharing the ideas and solutions across the wider range of stakeholders.

Documenting animal genetic resources in the Mediterranean: interaction and cooperation in the region

C. Ligda[1], G. Hadjipavlou[2], M. Djemali[3], M. Fadili[4], E. Charvolin-Lemaire[5] and E. Sturaro[6]
[1]Hellenic Agricultural Organization, VRI, P.O. Box 60272, 57001 Thessaloniki, Greece, [2]Agricultural Research Institute, P.O. Box 22016, 1516 Lefkosia, Cyprus, [3]INAT, University of Carthage, Animal Genetic and Feed Resources Research, 43, av. Charles Nicole, 1082,Tunis, Tunisia, [4]INRA Morocco, 25 Avenue Ennasr, BP 415 RP Raba, Morocco, [5]INRAE UMR GABI 1313, Domaine de Vilvert, Bat 211, 78352 Jouy en Josas, France, [6]University of Padova, DAFNAE, Via dell'Università 16, 35020 Legnaro (PD), Italy; chligda@otenet.gr

This study aims to present the status of animal genetic resources documentation in the Mediterranean region through specific country examples and discuss the links between recorded data and the actual situation, the activities carried out and their effect on the status of local farm animal genetic resources. For this purpose, the cases of France, Italy, Greece, Cyprus, Morocco and Tunisia were examined. The relevant data from the Domestic Animal Diversity Information System were exported and analysed. The key figures refer to the number of breeds registered, breeds per risk status, population trends, degree of data completeness, etc. According to the analysed data, the completeness of updated population figures varied from 20 to 100%. The SDG 2.5.2 indicator in the countries was widely diverse, potentially indicating the extent to which DAD-IS was used for breed population data monitoring in each country. In addition, complementary data were collected through a questionnaire to understand the organisational structure and the relevant policies in each country. Previous work carried out in the frame of the European Regional Focal Point on farm animal genetic resources (ERFP) has been used and extended to Mediterranean countries not previously included in that survey. Means to stimulate data recording and to productively use the DAD-IS system are also discussed, such as developing specific tools to analyse groups of breeds, enhancing the communication and interaction with researchers working in the field in order to increase data availability and improving data flow from institutional to national, to regional and international databases. This ongoing work reflects the primary aims and activities of the ERFP WG Documentation and Information and the EAAP Mediterranean WG.

The AGRICYGEN project: genomic studies of animals, plants and microbes under Cyprus conditions

G. Hadjipavlou, D. Fasoula, M. Omirou and I.M. Ioannides
Agricultural Research Institute, 1516, Lefkosia, Cyprus; georgiah@ari.gov.cy

The Project AGRICYGEN (CYprus AGRIcultural Genomics CENtre) was funded in 2017-2018 by the Teaming call phase 1 of the EC Horizon 2020 Framework Programme and is since then fully supported by national funds. The project employs a unified approach, through genomics, to improve the productivity and sustainability of the whole system (animals, plants and microbes), and produce novel products, services and recommendations for improving livestock output, crops used for feed and soils cultivated for feed production. Genomic characterisation of local sheep and goat breeds, crops and microbial communities will facilitate advanced genomic research, relevant to local and native agro-biodiversity management, traceability of related products, and mitigation of climate change effects on agricultural production. The initial focus of the project is on employing research approaches for fulfilling the halloumi cheese PDO requirements in small ruminant milk and feed production, within the harsh and fluctuating environmental and climatic conditions of Cyprus. The project outcomes will have direct impact on the Cyprus economy and society, and will significantly contribute to sustainable development of agriculture and animal production on the island. AGRICYGEN also seeks to contribute to mitigating and adapting to climate change effects on crop and animal production in Cyprus. Project findings and strategic processes developed will be highly relevant to the entire eastern Mediterranean, Middle East and North Africa.

Whole-genome scan of sequence data for selective sweeps in South African indigenous cattle breeds

M.L. Makgahlela[1], A.A. Zwane[1], K.S. Nxumalo[1], J. Kantanen[2], C. Ginja[3], D.R. Kugonza[4], N. Ghanem[5] and R.P.M.A. Crooijmans[6]
[1]Agricultural Research Council, Animal Breeding and Genetics, Pretoria, South Africa, [2]Natural Resources Institute Finland, 31600 Jokioinen, Finland, [3]CIBIO-InBIO, Centro de Investigação em Biodiversidade e Recursos Genéticos, Universidade do Porto, Vairão, Portugal, [4]Makerere University, School of Agricultural Sciences, College of Agricultural and Environmental Sciences, Makerere, Uganda, [5]Cairo University, Animal Production Department, Faculty of Agriculture, Giza, Egypt, [6]Wageningen University & Research, Animal Breeding and Genomics Group, Droevendaalsesteeg 1, Wageningen, the Netherlands; tumi2002@gmail.com

South African (SA) cattle breeds are genetically diverse, and have been subjected to natural and artificial selection leading to their adaptation to local environments, and changes in their morphology and physiology. Within the OPTIBOV project, a whole-genome scan of selective sweeps in indigenous Afrikaner (16), Nguni (24) and Tuli (10) cattle was carried out to identify genes that were targeted by selection. This information will be useful to develop appropriate selection and breed improvement programs for optimised production in local environments. DNA samples were sequenced using Illumina PE150 at 10x coverage. Runs of homozygosity (ROH), fixation index (FST), haplotype and marker differentiation (hapFLK and FLK respectively) were computed to identify putative selective sweeps. Annotation was performed using the animal QTL database to identify genes. About 688 candidate selective sweeps with Z(Fst) values between 3.5 and 9.0 were identified across the breeds with 223 putative regions (Z(Fst)≥6.0). Gene ontology identified KIT and MITF genes associated with skin pigmentation in cattle. MLANA and SYT10 genes have been associated with coat colour, while ADAMS3 gene has been associated with fertility in cattle. Enrichment analysis in Nguni identified 58 genes associated with phenotypes. NOD2 and IL21R genes were associated with inflammatory bowel diseases in cattle. The study provides insight into the genetic mechanisms that influenced selection, and some processes that shaped the genomes of these breeds.

Imputation accuracy of crossbred dairy cows

L.H. Maugan[1], I. Croué[2], S. Fritz[2], R. Lefebvre[3], P. Croiseau[3] and C. Hoze[2]
[1]Institut de l'Élevage, MNE, 149 rue de Bercy, 75012 Paris, France, [2]ALLICE, MNE, 149 rue de Bercy, 75012 Paris, France, [3]Université Paris-Saclay, INRAE, AgroParisTech, GABI, 78350 Jouy-en-Josas, France; laure-helene.maugan@agroparistech.fr

Crossbreeding is a growing strategy among dairy farmers, usually chosen to compensate weaknesses of a breed, in order to obtain robust cows, to facilitate the transition to a different farming system (organic farming, specifications of a PDO, etc.) or to combine various breed strengths. One of the objectives of the GenTORE project is the implementation of genomic evaluation on crossbred cows, which requires the imputation of crossbred genotypes with sufficient precision. In this study, we used genotypes from Normande × Holstein crossbreds, bred on the French experimental farm Le Pin au Haras. 1,208 crossbred animals with known pedigree were genotyped with the Illumina 50k SNP chip. Out of them, 243 were used as a validation population and their genotypes were reduced to the markers of the Illumina 10k SNP chip. The genotypes of the validation animals were imputed to 50k, using the other crossbred animals and the purebred ancestors as a training population. FImpute 2.2, Beagle 5, and Eagle 2.4.1 were used for imputation. Imputed genotypes and true genotypes of the validation population were compared and an imputation error rate was measured for each software. The three software were compared based on imputation error rate and memory requirements. FImpute always performed as good or better than the other two for all criteria and hence was determined the most suitable one for the imputation of crossbred animals. For FImpute, error rates were lower than 0.1%, probably partly thanks to the very small and very well known pedigree, and had low memory requirements. Hence, imputation of crossbred dairy cattle seems feasible for the purpose of routine genomic evaluations, even based on small datasets.

A SNP-BLUP genomic evaluation with breed-specific SNP effects for three-way dairy crossbreds

I. Croué[1], D. Boichard[2] and P. Croiseau[2]
[1]Allice, 149 rue de Bercy, 75012 Paris, France, [2]Université Paris-Saclay, INRAE, AgroParisTech, GABI, Domaine de Vilvert, 78350, Jouy-en-Josas, France; iola.croue@inrae.fr

Crossbreeding is a growing strategy among dairy farmers, usually chosen to compensate for weaknesses of a breed, in order to obtain robust cows, to facilitate the transition to a different farming system (organic farming, specifications of a PDO, etc.) or to combine various breed strengths. However, in many countries, genetic evaluation is only available for purebred cattle and farmers who choose to implement a crossbreeding scheme lose access to genetic evaluation information. This study proposed to investigate the prediction accuracy of a SNP-BLUP genomic evaluation with breed-specific SNP effects, in the case of a three-(or n-) breed rotational crossbreeding scheme, in the frame of the GenTORE European project. This approach requires the breed of origin of alleles to be known. A first software was developed to determine the breed of origin of alleles, based on haplotype identification, and tested on simulated data, in a three-breed rotational crossbreeding scheme. Estimated breeds of origin were compared to the true breeds of origin and the error rate was 1.5%. Hence, this approach was accurate enough to be used for the purpose of crossbred genomic evaluation. A second software developed for genomic evaluation was validated through simulation. Data were simulated based on 50K SNP, for different reference population sizes and trait architectures. Prediction accuracy and bias were estimated using a validation study. They were compared to the ones obtained using a classical multiracial SNP-BLUP, in which all animals were evaluated as part of a single population, with no distinction between breeds. Attention was paid to computing time and memory requirements. This approach will be used to analyse real datasets including purebred and crossbred animals. To this end, more than 8,500 genotyped crossbred animals from different types of crosses (2 or 3-way crosses) have been collected in the EVAGENOC project funded by Apis-Gene.

OPTIBOV-PT: genomic characterisation of Portuguese native cattle for optimal performance

C. Ginja[1], S. Guimarães[1], L. Blaschikoff[1], C. Bruno De Sousa[1], R. Fonseca[1], D. Gaspar[1], A.E. Pires[1], M. Makgahlela[2], J. Kantanen[3], D.R. Kugonza[4], N. Ghanem[5] and R. Croojimans[6]
[1]CIBIO-InBIO, Universidade do Porto, Campus Agrário de Vairão, R. Padre Armando Quintas, 7, 4485-661, Portugal, [2]Agricultural Research Council-Animal Production Institute, Pretoria, Pretoria, South Africa, [3]Natural Resources Institute Finland, Myllytie 1, 31600 Jokioinen, Finland, [4]Department of Agricultural Production, Makerere University, P.O. Box 7062, Kampala, Uganda, [5]Animal Production Department, Cairo University, Giza, Egypt, [6]Animal Breeding and Genomics Group, Wageningen University & Research, Droevendaalsesteeg 1, 6708 PB Wageningen, the Netherlands; catarinaginja@cibio.up.pt

Iberian native cattle retain high genetic diversity relative to that of their European counterparts, including: taurine Y1 and Y2 unique patrilines; distinct Q-maternal lineages and a strong influence of T1-matrilines of African origin; as well as African taurine cattle ancestry in their autosomal genomes. This renders Iberian breeds a great model to study the genomic impact of cattle diversification. The OPTIBOV international project aims to study bovine populations from 6 biogeographic regions, from north to south, in Europe and Africa. Here, the genomic variability of the Barrosã, Mirandesa and Mertolenga Portuguese native breeds will be investigated to identify markers for (positive) adaptation traits and for (negative) traits to be improved. Recommendations of The Food and Agriculture Organization of The United Nations were followed for phenotyping 30 animals of each of these breeds. Whole-genome resequencing data was generated to carry out association studies of well-defined phenotypes under local conditions. We will study inbreeding in detail and genomic regions which have been under natural (environmental adaptation) as well as artificial (human-mediated) selection. An important deliverable of OPTIBOV is a genotyping assay specific for more diverse traditional cattle to include markers associated with specific traits. OPTIBOV results will be used to improve breeding programs. Also, some unique adaptive features of local cattle to harsh environments can be useful for adjusting mainstream breeds to climate change.

Genomics and epigenetics of farmed insects

A. Vilcinskas
Fraunhofer Institute for Molecular Biology and Applied Ecology, Bioresources, Leihgesterner Weg 85, 35392 Giessen, Germany; Andreas.Vilcinskas@agrar.uni-giessen.de

Insect farming has become a worldwide growing agribusiness focusing on the bioconversion of organic side streams for the sustainable production of proteins, lipids, chitin and other valuable compounds. The genomes and transcriptomes of farmed insects such as the black soldier fly *Hermetia illucens* and the mealworm *Tenebrio molitor* have recently become available, providing the opportunity to design tailor-made industrial strains using CRISP/CAS. Such modern genome editing approaches enable the optimisation of growth rates, the adaptation to industrial organic side streams as diets and boosting the immune system of farmed insects. The future options for insect farming can be expanded beyond the genetic tools to encompass epigenetic approaches, given that transcriptional reprogramming associated with insect growth, metamorphosis, fecundity and pathogen resistance is mediated by epigenetic mechanisms such as DNA-methylation, histone acetylation and microRNAs. The presentation will provide an overview about our knowledge from model insects including the red flour beetle *Tribolium castaneum*, which belongs to the same family as the mealworm. Of particular interest are epigenetic mechanisms, which can path the way for the transfer to farmed insect species in order to address challenges resulting from mass rearing of insects at an industrial scale.

Exploiting genetic variation in insect populations

E.D. Ellen[1], T. Veldkamp[2], P. Bijma[1] and A.C. Bouwman[1]
[1]Wageningen University & Research, Animal Breeding and Genomics, Droevendaalsesteeg 1, 6708 PB, the Netherlands, [2]Wageningen University & Research, Animal Nutrition, Droevendaalsesteeg 1, 6708 PB, the Netherlands; esther.ellen@wur.nl

With the current (and future) demand of high-quality protein sources for food and feed, and the current food waste of at least 30%, a major role can be foreseen of using insects as bio-waste converters into high-grade food and feed ingredients. However, a major challenge in insect mass production is the large-scale efficient production of healthy populations. Tools developed in livestock breeding could enable insect producers to capture variation between individuals, to genetically select for specific populations with distinct traits. A previous study in flour beetles (*Tribolium castaneum*) was focused on estimating genetic parameters for three life history traits in two populations: pupal body mass, development time and growth rate. For all three traits, the estimated additive genetic variation ranged from 38 to49% of the phenotypic variation, indicating substantial genetic variation. Furthermore, it was found that flour beetles that developed more slowly achieved a higher pupal body mass. In another study, in black soldier flies, it was found that separated populations can differ substantially in their efficiency to convert a certain waste stream, suggesting genetic differences among populations. Both studies suggest a substantial role for breeding to improve traits important for insect populations (e.g. body weight and development time). In this study, we will show the impact of genetic selection on traits important for insect production, by quantifying potential selection responses. Utilising genetic variation within and between populations will enable insect producers to select for efficient and healthy populations.

Genome of the yellow mealworm *Tenebrio molitor*: high throughput sequencing and possible applications

K. Labadie[1], E. Eleftheriou[1], B. Vacherie[1], J.-M. Aury[1], F. Berro[2], A. Rios[2], T. Lefebvre[2] and M.-A. Madoui[1]
[1]CEA, Genoscope, Institut François Jacob, Génomique Métabolique, 2 rue Gaston Crémieux, 91000 Evry, France, [2]Ÿnsect, R&D dpt, Genopole, Campus 3, Bât.2, 1, rue Pierre Fontaine, 91000 Evry, France; tle@ynsect.com

Tenebrionids beetles have been studied for decades as model in genetics. One of them, the yellow mealworm, *Tenebrio molitor*, is also the most promising insect to be industrially farmed as a new source of proteins for both feed and food. Its use in livestock showed a growing interest for ecological and sustainable production system with competitive yields. However, *T. molitor* lacks molecular characterisation, therefore the development of genomic resource is of great interest as a prior to molecular-based improvement. The European BBI-JU Project FarmŸng aims to establish a genomic reference of this mealworm species and to define breeding strategies for both the preservation and the exploitation of its genetic diversity in the context of an industrial farm. CEA-Genoscope prepared high molecular weight DNA from *T. molitor* males from Ynsect's beetles populations and prepared libraries to combine Oxford Nanopore Technology long reads with Illumina short reads. This produces a first genome draft that was scaffolded with Hi-C data resulting in a high-quality assembly. The genome was annotated using *T. molitor* transcriptomes from different development stages and other beetle proteomes available. The yellow mealworm genome offers new opportunity for the identification of links between genotypes and agronomical traits such as growth or feed conversion indicators, protein and lipids content or pathogen resistance. This will enable the development of new *T. molitor* strains for industry.

Use of the Affymetrix microarray in a cross-species approach to determine *T. molitor* gene expression

C. Lopez-Viso[1], M. Castellanos-Uribe[2], S.T. May[2], T. Parr[1] and A. Salter[1]
[1]University of Nottingham, Future Food Beacon (Protein Platform), Sutton Bonington Campus, LE12 5RD. School of Biosciences, United Kingdom, [2]University of Nottingham, Nottingham Arabidopsis Stock Centre (NASC), School of Biosciences, LE12 5RD, United Kingdom; carlos.lopezviso@nottingham.ac.uk

Insects are of interest as a sustainable source of protein which could be included in the food production systems. Due to its high protein content, the larvae of *Tenebrio molitor* (yellow mealworm) has a potential application as an ingredient in animal feed and human food. Their nutritional composition is determined by the environmental conditions and the genetic background. Therefore, the genetic knowledge could provide the understanding of metabolic pathways and mechanisms that are controlled by the regulation of gene expression. However, the whole genome of *T. molitor* has yet to be sequenced, and there are only a few annotated genes. Consequently, targeted improvements in growth, feed utilisation and nutrient composition of this organism using genetics selection tools are currently limited. Affymetrix GeneChip microarrays allow transcript expression profiles to be compared. Although the commercial microarrays are provided only for a limited number of species, robust results have been published using cross-species transcriptome arrays. We have analysed *T. molitor* transcripts using an Affymetrix Drosophila GeneChip. The hybridisation of cDNA generated from mRNA expressed at five developmental stages (larvae and pupae) showed significant ($P<0.05$) differential genes expression. Genes were clustered by gene ontology categories and potential orthologs have been identified. This work shows a novel genetic approach which will allow to broaden the knowledge of *T. molitor* gene expression. It will give an insight into the regulation of metabolic pathways, and potentially allow the identification of those that could be selected to improve mealworm growth efficiency and nutritional composition. Furthermore, this method appears to be an approach which can be used to study gene expression in other insect species of whose genomes have yet to be characterised.

Developing genotyping-by-sequencing method for black soldier fly (*Hermetia illucens*) genome studies

M. Tapio, D. Fischer, O. Bitz, T. Iso-Touru and I. Tapio
Natural Resources Institute Finland (Luke), Myllytie 1, 31600 Jokioinen, Finland; miika.tapio@luke.fi

The black soldier fly (*Hermetia illucens*) is an emerging insect-based solution for circular bio-economy. A genotyping method would make dissecting or improving genetic traits more accessible. Genotyping by sequencing (GBS) using the double digest restriction-site associated DNA sequencing (ddRAD) is a flexible way to generate genotype data even for species with limited previous knowledge on their genome. The objectives here were 1) to create a pipeline for generating genotype data useful for studies on both dissecting the genetic architecture of traits and for advanced breeding operations, and 2) to evaluate the validity and performance of the pipeline in black soldier fly. The choice of restriction enzymes for library preparation was based on in silico analysis that defined the expected number and locations of the sequenced genome fragments, library characteristics, and the mock genome size. We created ddRAD libraries for a sample-set of 10 black soldier flies using SphI and EcoRI restriction enzymes and generated paired-end sequence data using Illumina NextSeq550. For the data processing, we adjusted the existing GBS-SNP-CROP pipeline and implemented it as a snakemake workflow. To evaluate the recovery ratios, the repeatability was evaluated between the samples and between the mock genome and the reference genomes. Subsampling of the dataset was done to estimate the minimum required sequencing depth to obtain a reliable set of variants. We estimated that using EcoRI and SphI, the sequencing library would have 18-40k fragments and would represent ~1% of the genome, based on the 1.1 GB size of the recent reference genome (ASM983516v1). Assuming that approximately 0.5% of sites are variable, this could detect ~50k variants. Sequencing generated 0.35-1.8 million paired-end reads per sample. The total length of the mock genome exceeded the expected length. We observed 250k variants initially. This reduced to 8,000 variants with the strict quality filtering. The variants can be used as genetic markers in biological research and making breeding decisions. The generated pipeline is evolving further and allows for detailed genomic studies of black soldier fly.

Breeding programme design for black soldier fly and practical implementation

L.A.C. Lara[1], I. Pocrnic[1], T. Farrugia[2] and G. Gorjanc[1]
[1]The Roslin Institute and Royal (Dick) School of Veterinary Studies, University of Edinburgh, Easter Bush Campus, EH25 9RG, United Kingdom, [2]Beta Bugs Limited, Edinburgh, EH25 9RG, United Kingdom; llara@ed.ac.uk

This study will describe the design and implementation of a black soldier fly breeding programme with the main goal of enabling continuous improvement while striking operational simplicity and maintenance of genetic variability. To this end we have conducted a life-cycle analysis of black soldier fly and used these parameters for deterministic and stochastic modelling of different breeding programme scenarios. The scenarios are reminiscent of breeding programmes in aquaculture and cross-pollinating plants. In this context we are modelling the effect of different population sizes, size and number of groups, phenotyping strategies and selection intensities. In parallel to the modelling work we are testing the operational applicability and performance of a subset of the scenarios in real populations. This work will guide the development of state-of-the-art breeding programmes for black soldier fly and other insects, both for the insect-as-fed and insect-as-food sectors.

Group specific CNVs in scouts and recruits of honeybee

M. Mielczarek¹, S.L. Rodriguez-Zas², J. Szyda¹ and B. Southey²
¹Wroclaw University of Environmental and Life Sciences, Kożuchowska 7, 51-631 Wroclaw, Poland, ²University of Illinois at Urbana-Champaign, 1207 W. Gregory Dr. Urbana, IL 61801, USA; magda.a.mielczarek@gmail.com

Copy Number Variations (CNV) encompass duplications and deletions of long DNA fragments that can span whole genes and regulatory regions and disrupt phenotypes. The present study investigated CNV differences between scout and recruit honeybees that present different behaviours. Whole genome sequences from 22 scouts and 22 recruits were assessed for quality control, and poor-quality reads were removed. Two implementations of a bioinformatics pipeline that use either the 3.1 (GCA_003254395.2), or the 4.5 honeybee genome reference assembly (GCA_000002195.1) were used. CNV were detected using the software Lumpy in the multi-sample mode. To minimise false positive CNV, polymorphisms of length <50 bp or >1 Mbp, and located in ambiguous or homopolymeric regions of the reference assembly were removed. Also, CNVs were deemed group specific if present in at least 11 honeybees in a group and absent in the other group. For the 3.1/4.5 assemblies, the number of duplications were 3,225/3,093, while the number of deletions were 5,348/4,409. The duplication length varied from 82-920,100 bp (2,410±23,415) and from 82-48,601 bp (772.835±2,036) for the 3.1 and 4.5 assemblies, respectively. The deletion length varied from 50-920,055 bp (1,524±17,334) and from 50-30,925 bp (510±1,161) for 3.1 and 4.5, respectively. No scout-specific CNV was identified, whereas recruit-specific fragments were deleted in the overlapping genes LOC408840 (5'-3' exoribonuclease 2 homolog) and LOC411116 (nephrin). For the 3.1/4.5 assemblies, the deletion on chromosome 5 spanned the positions 9,737,471-9,737,608/11,508,552-11,508,689. LOC408840 promotes termination of transcription, while LOC411116 participates in multiple biological processes (e.g. cell adhesion, regulation of excretion). The results suggest that LOC408840 and LOC411116 may play a role on the behaviour differences between scouts and recruits. Project was supported by Polish National Agency for Academic Exchange and Poznan Supercomputing and Networking Centre.

Whole genome sequence analysis of buckfast honeybees

G. Minozzi¹, B. Lazzari², M.G. De Iorio¹, A. Talenti³, P. Crepaldi⁴, R. Rizzi¹, E. Facchini¹, A. Stella², G. Gandini¹ and G. Pagnacco²
¹University of Milan, Department of Veterinary Sciences (DIMEVET), Via Celoria 10, 20133, Italy, ²CNR-IBBA, Via Bassini 15, 20133 Milan, Italy, ³The Roslin Institute, Easter Bush Campus, Midlothian, EH25 9RG, United Kingdom, ⁴Univeristy of Milan, Dipartimento di Scienze Agrarie e Ambientali – Produzione, Territorio, Agroenergia (DISAA), Via Celoria 2, 20133 Milano, Italy; giulietta.minozzi@unimi.it

At the end of the last glaciation numerous subspecies of *Apis mellifera* were established in Europe. In Italy, the *Ligustica* variety of *A. mellifera* has settled and is now well adapted to the mild climate and to the rich floristic biodiversity of the territory. With the spread of Varroa destructor and the increasingly dramatic use of pesticides in agriculture, the Italian bee populations are increasingly dependent on human action for its conservation, both for the control of the parasites and for the threat posed by pesticides. At mean time, these effects have favoured in the Italian territory other genetic types that are more performing in terms of productivity and resistance to pathogens, as the Buckfast bees. The availability of the honeybee genome allows using next-generation sequencing methods to define subtypes/subpopulations more precisely. In this study, whole genome sequencing has been performed on 40 Buckfast bees, sampled in Italy, by Illumina technology at a 30× coverage. Preliminary analyses of their genomic structure indicate that the Buckfast bees differ from the *Ligustica* bees present in Italy and that within the Buckfast bees sampled, several subpopulations are present that reflect their original selection origin/breeder.

Simulation of a breeding scheme accounting for inbreeding and maternal lines in *Apis mellifera*

F. Phocas[1], T. Kistler[1] and B. Basso[2]
[1]Université Paris-Saclay, INRAE, AgroParisTech, GABI, Domaine de Vilvert, 78350 Jouy-en-Josas, France, [2]INRAE, A & E UR406, Site Agroparc, 84914 Avignon cedex 9, France; florence.phocas@inrae.fr

Implementing an efficient breeding program for *Apis mellifera* is difficult due to genetic (haplo-diploïdy) and reproductive (polyandry) specificities, as well as due to the complexity of traits of interest in this species. Phenotypes such as honey production, aggressiveness or swarming behaviour are measured in the hive at the colony scale and are explained by the genes of both the queen (maternal effect) and her numerous worker daughters (direct effect). To help bee breeders design breeding schemes that remain efficient in the mid or long-term (20 years), a study was performed to explore through stochastic simulation the effects of various selection methods (mass or pedigree-based BLUP selection, with or without restrictions on maternal origin), population sizes (6, 12 or 24 dam queens) and polyandry levels (each queen was mated to either 1, 8 or 16 drones) on direct and maternal breeding values and the inbreeding level of the population after 20 years of selection. Based on the assumptions of the infinitesimal genetic model, traits were described by an animal model considering haploid drones individually in the derivation of the relationship coefficients among individuals. Choosing the best queens of the next generation independently of their maternal origin increased the expected genetic gains by 25% after 20 years of selection, but also increased inbreeding by 50% compared to a within-maternal line selection. These gains were all the more important as the selection was based on BLUP estimated breeding values and not only on the own performance of the candidates for selection. A doubling of the population size corresponded to an increased genetic gain of 20% after 20 years for a level of consanguinity decreased by more than 30%. A breeding program that would use single-drone insemination for 20 years would increase average inbreeding levels of queens by 50% over a similar breeding scheme that retains polyandry as a mode of reproduction, while no significant extra-genetic gain would be obtained.

Historical changes of the Swedish horse population, its use and breeding methods

J. Philipsson
SLU, Dept. of Animal Breeding and Genetics, Box 7023, 75007 Uppsala, Sweden; jan.philipsson@slu.se

Since the domestication of the horse a close relationship has developed between horse and man. The diversified use of the horse has led to its unique position among domesticated animals in serving needs of mankind. As late as in the beginning of the last century horses were of vital importance for transportation of people as well as of post, goods and rescue vehicles, for army purposes, agricultural and forestry work, for food, and finally, for sport and leisure. Over time these roles of the horse have changed dramatically, especially since World War II. The Swedish horse population decreased from about 700 thousand at that time to about 80 thousand in 1970, mostly due to increased use of automobiles and mechanisation of agriculture, forestry and of the army. After 1970 a remarkable upward change in numbers and use of the horse, and consequently changed breeding objectives, took place following the positive economic development in Europe. In 2010 the number of horses had increased to about 360 thousand of which a large majority were horses for sport and leisure. In 1960 draft horse stallions were used to 87% of covered mares while it dropped to 7% in the year 2000. The horse is still a product of agriculture, but the type of horse owners has changed and it serves other needs of the society than before. Riding has become the next largest sport for young people in the country, and plays an important social role. The changed use of horses means that we now have about 30 breeds used for sport and leisure, Standardbred trotters and Warmblood riding horses being the most common. The Gotland pony is the only breed of domestic origin. The market of breeding, especially of Standardbred trotters and Warmblood riding horses, has become truly global, not least due to use of artificial insemination, which today comprises more than 90% of the coverings in these two breeds. While only sport results are used for genetic evaluation of trotters, various complementary performance tests for young riding horses have been developed to shorten generation intervals in breeding of Warmbloods. BLUP-methods for genetic evaluation were applied already in 1986. Today molecular gene testing for specific traits is practised and international evaluation of Warmblood stallions is being developed.

Evaluation of economic effects of the equine industry in two French rural areas

D. Vollet[1,2] and G. Bigot[1]
[1]INRAe, UMR Territoires, Centre de Clermont-Auvergne-Rhône-Alpes, 9 avenue Blaise Pascal, CS20085, 63178 Aubiere, France, [2]Agro-Campus Ouest, 2 rue André le Notre, 49000 Angers, France; genevieve.bigot@inrae.fr

Public decision-makers often question the weight of the horse industry in the economics of rural areas. In order to answer this question, an estimation of the direct, indirect and induced economic effects of the horse industry has been implemented in two French rural areas (NUTS3): Ain and Allier, known for the attractiveness of their horse industry and the diversity of representative sectors: racing, sport and leisure, meat production. The model used was a hybrid between the economic or export base and the Keynesian theory (mixed multiplier). The first wave of expenditure (or direct effects) was obtained by a random direct survey of all types of structures (breeders, equestrian centres, race trainers), in order to collect precise data on their characteristics: nature (supplies, services, farriers, veterinarians, etc.), volume and location. Indirect effects were estimated by the economic base model. In Allier, the percentage of basic jobs from the equine industry was 0.76% of the total number of jobs. This relative role was much lower in Ain (0.18%) and closely related to equestrian centres and saddle horses breeders. The first wave of spending was significantly higher in Ain (0.32) than in Allier (0.23). The weakness of the indirect and induced effects in Allier was explained both by the diversity of structures: breeders of all sectors (race, leisure, sport and meat production), similar number of equestrian centres and race trainers and the limited local supply of the race sector which induced moreover, the higher turnover. In addition, the share of external demand in total demand for each segment of the horse industry is all the higher as the area is 'specialised' in this segment. For example, nearly 40% of direct employments in equestrian centres was driven by external demand in Ain, while 90% of them satisfied a local demand in Allier. The equine sectors thus seem to be at the origin of a 'consumption base' effect from anchored territorial resources (races in Allier) or from a growing external demand (Switzerland and the town of Lyon for equestrian centres in Ain) by diverting some of the external consumption to their profit.

Social network analysis for a better understanding of the location of horses in France

J. Tapprest[1], R. Di Biagio[1,2], H. Farchati[1,2,3], A. Merlin[1], J.P. Amat[2], M. Saussac[2] and C. Sala[2]
[1]ANSES, Laboratory for Animal Health in Normandy, 14430 Goustranville, France, [2]ANSES, Lyon, 69364, France, [3]University of Caen Normandy, 14000 Caen, France; jackie.tapprest@anses.fr

French regulations require the declaration of the location of every equine premises but not of every equine. Nevertheless, this obligation is not fully followed. This imperfect knowledge of the spatial distribution of the equine population is thus a limiting factor for the protection of equine health and welfare. In France, several sources of data of varying quality are available. The French equine census database SIRE, managed by the French Horse and Riding Institute (IFCE), collects individual equine data information as well as data about the location of all owners and some holders. Data about dead equines and their location are centralised in the Fallen Stock Data Interchange database managed by the French Ministry of Agriculture. Besides, other databases focus on specific populations such as racehorses or equines controlled for compliance with the regulations by IFCE. The objective of our study was to evaluate whether the combined use of these databases could provide a better knowledge of the location of equines in France. A dataset was built from all the available data sources to provide for a sample of equines with their holder and owner and their locations. Spatial social networks of holders and owners were built according to the type of equine (saddle horses, draft horses, ponies, etc.). Data management was performed with mySQL and the analysis with the R software. The features of the sub-networks (size, centrality, cohesion, spatial distribution, etc.) were very different according to the breed category. The results are of particular interest for the improvement of demographic knowledge in identifying the holding places and the number of equines held by individuals, but also for monitoring and management of health events by identifying connected individuals and cut-off points deconstructing the network in the event of an epidemic and providing knowledge about locations to facilitate controls. Acknowledgements to the French Ministry of Agriculture, the French Horse and Riding Institute, the French fallen stock companies, the Regional Council of Normandy and the Fonds Eperon.

Brakes and levers of draught horses farming in France

G. Bigot, M. Valleix, G. Brétière and J. Veslot
INRAe, UMR Territoires, Centre de Clermont-Auvergne-Rhône-Alpes, 9 avenue Blaise Pascal, CS 20085, 63178 Aubiere, France; genevieve.bigot@inrae.fr

In France, 9 natives breeds of draught horses were preserved until the beginning of the 21st century thanks to the technical and financial assistance provided by the Ministry of agriculture, in particular for reproduction. The phasing out of this aid since 2010 resulted in a halving of the number of mares and breeders between 2007 and 2017, which have since remained at 17,000 mares and 6,000 breeders nationwide. Currently, French breeders raise mainly small equine herds in association with cattle on grassland, particularly in highlands. To understand the reasons and the limits of farmers to still raise draught horses today, a survey was carried out in Auvergne (NUTS2), a region gathering 18% of French breeders of draught horses. Located at an altitude of around 900 m, 50 farmers were surveyed. They managed an average of 3 mares and 60 cows and their young, on 100 ha of grasslands. In particular, 43% of farms held less than 4 broodmares, 38% had from 4 to 10 and 19% held more than 10 broodmares, unrelated to the number of cows and the size of the agricultural area. 85% of breeders interviewed declared that they owned horses out of passion. The main constraint mentioned by farmers was the extra work induced by horses. Almost a quarter of respondents quoted problems related to reproduction (natural or artificial). However, farmers with more than 25 mares, or having at least 2 stallions, did not mention this problem. Farmers with less than 4 mares mentioned also monitoring foaling. They quoted in addition, specific problems related to horses (compared to cattle) such as more lively behaviour or a need for specific care. Regardless of the size of the herd, only 20% of breeders mentioned the low market price for foals and the too few outlets for harnessing or traction. It emerges from this survey that main problems encountered in the draught horse farming concerned reproduction and specific work for horses, especially for breeders with less than 4 mares. Thus, to maintain a draught horse farming in uplands (and more generally in France), reproduction assistance services and practical training would certainly be useful, in addition to subsides currently proposed by local and national policies.

Growth and development of the Lusitano horse: application of non-linear growth functions

M.J. Fradinho[1], D. Assunção[2], A.L. Costa[2,3], C. Maerten[2], V. Gonçalves[2], P. Abreu[2], M. Bliebernicht[2,3] and A. Vicente[1,4]
[1]CIISA, Centro de Investigação Interdisciplinar em Sanidade Animal, Faculdade de Medicina Veterinária, Universidade de Lisboa, Lisboa, Portugal, [2]Pôle Reproduction–Haras de la Gesse, Boulogne-sur-Gesse, Boulogne-sur-Gesse, France, [3]Embriovet, Lda., Muge, Muge, Portugal, [4]Escola Superior Agrária de Santarém, IPS, Santarém, Santarém, Portugal; mjoaofradinho@fmv.ulisboa.pt

The present study aimed to characterise the growth patterns of the Lusitano horse using non-linear models. A total of 4,008 records for body weight (BW) and height at withers (HW) were obtained from 134 Lusitano foals (70 colts and 64 fillies) born and raised in a reference stud-farm in the south of France. Data were regularly collected between birth and 56 months of age, when horses were already on regular work. Several sigmoid growth functions (Brody, Logistic, Gompertz, von Bertalanffy and Richards) were adjusted using the NLIN procedures of SAS. However, the Richards equation $y = A(1 - b.exp(-kt))M$ was chosen for further analysis because it was the best fit model for both variables. Growth rates (ADG, kg/d or cm/d) were obtained from the first derivative of the equations and the effect of sex was also evaluated. The mean mature BW was 579.9±9.4 kg and the average mature size was 163.2±0.6 cm for HW. Considering the Lusitano breed standard, the proportions (%) of mature BW at 6, 12, 24, 36 and 48 months of age were, respectively, 44, 62, 83, 95 and 102%. The proportions (%) of mature HW for the same ages were 83, 90, 97, 99 and 101%. Sexual dimorphism was observed on both variables (P<0.001), with higher maturity values observed for males. Growth rates for BW and HW were similar to those proposed for other sport breeds. ADG were higher than those observed for the Lusitano horse managed on extensive systems, on different geographical regions. The present study clearly shows an influence of environmental factors that may modulate the potential growth response of this breed under different production systems. Acknowledgement: Project CIISA UIDP/CVT/00276/2020.

Biometric and ponderal evaluation of neonate foals: influence of maternal factors

A.R. Teixeira[1,2], M.T. Leão[2,3], A.L. Costa[2,3], C. Maerten[2], P. Abreu[2], M. Bliebernicht[2,3], A.S. Santos[1,4] and M.J. Fradinho[5]
[1]Universidade de Trás-os-Montes e Alto Douro, Quinta dos Prados, 5000 Vila Real, Portugal, [2]Haras de la Gesse, Pôle de Reproduction, Chemin du Château de la Gesse, 31350 Boulogne sur Gesse, France, [3]Embriovet, Lda, Rua Fonte Costa, 2125-348 Muge, Portugal, [4]CITAB-UTAD, Quinta dos Prados, 5000 Vila Real, Portugal, [5]CIISA-FMV, Universidade de Lisboa, Avenida da Universidade Técnica, 1300-477 Lisboa, Portugal; ateixeira004@gmail.com

The growth and development of foetus during pregnancy are strongly conditioned by maternal factors. Knowledge of these factors is essential to ensure and promote healthy growth of the newborn foal. This study aimed to identify the influence of some maternal factors, namely parity, weight and area of the placenta, the age and weight of the mother and the duration of pregnancy, on the weight and biometric measurements of the foals at birth. During the breeding season of 2019, 41 Lusitano mares of the same stud were monitored. Placenta was assessed, measured and weighed, within 24 hours of delivery. Foals were also weighed and measured (withers height; WH, thoracic perimeter-TP and cannon bone perimeter; CP) in the first 24 hours after birth. Average gestation time was 339 ± 8 days, weight of placenta was 5.46 ± 1.1 kg and area was $5,411\pm803$ cm^2. Average foal weigh and measurements at birth where 49.5 ± 6.3 kg, WH of 98.7 ± 5.1 cm, TP of 80.6 ± 11.7 cm and CP of 12.5 ± 0.8 cm. The duration of pregnancy did not influence foals' weight or measurements at birth. The height at withers was influenced by parity, with primiparous mares, given birth smaller foals (95.4 ± 1.5 vs 99.7 ± 0.9 cm; $P<0.05$). The weight and area of the placenta had a significant effect on the foals' measurements. Larger and heavier placentas were associated with larger and heavier foals at birth ($P<0.05$). The mothers 'age only influenced foals' weight and CP, with younger mothers having lighter foals and with lower CP at birth ($P<0.05$). There was also a positive correlation between the weight of the mothers and the weight of the foals (0.36; $P<0.05$), although there was no correlation between the weight of the mare and the weight of the placenta.

Individual coat traits in a native horse breed kept under semi-wild Nordic conditions

S. Ringmark, K. Dahlborn and A. Jansson
Swedish University of Agricultural Sciences, Dept. of Anatomy, Physiology and Biochemistry, Box 7011, 75007, Sweden; anna.jansson@slu.se

Year round grazing by horses during semi-wild conditions can have several positive effects on biological diversity. To ensure that horse welfare is acceptable the breed, but most important the individuals used, must possess traits that make them hardy and healthy. We have studied 12 stallions of the ancient native Swedish breed Gotlandruss for 2.5 years when they were kept year-round grazing without supplementary feeding in three pasture-forest enclosures (10 ha each, four horses/enclosure) outside Uppsala in Sweden. The study started when the horses were 1 year old. Daily and weekly registrations of traits related to hardiness were made throughout the study. The aim of this part of the project was to study individual effects on traits related to coat function. Coat length at withers was measured weekly and length of coat was analysed using a mixed model including effects of horse, month and year. The number of horses observed shivering the first year and, at any time, affected by *Bovicola equi* was registered. Differences in coat length in January as 1.5-2 year old between horses observed shivering and horses that did not shiver were analysed using a T-test. The number of swellings from insect bites were also registered in each horse and summarised per season (summer start: >5 °C for 5 days, winter start: <5 °C for 5 days) and year and analysed in a mixed model including effect of horse, year and season. Horse and year had an effect on coat length and in September to May coat length at withers was longer than in June to August ($1.3-3.5\pm0.2$ vs $1.0-1.1\pm0.2$ cm, $P<0.05$). Coat length did not differ between individuals affected by *B. equi* and unaffected individuals ($P>0.05$). Coat length was also not different in the individuals where shivering was observed compared to in those that did not shiver ($P>0.05$). The number of insect bites were affected by year and season but not by individual horse. The study indicates that coat length varies within individuals of this breed and that there are individuals that show hardy traits like no shivering and no *B. equi* infestation. Further studies are needed on which coat traits that are of importance during semi-wild Nordic conditions.

Genetic diversity of Connemara pony populations in New Zealand, Australia, and Canada

A.K. Michler[1], S. Ramsay[2], C. Castaneda[3], E.G. Cothran[3] and M.L. Cox[4]
[1]HfWU Nürtingen-Geislingen, Equine Management, Neckarsteige 6-10, 72622 Nürtingen, Germany, [2]Southern Institute of Technology, School of Veterinary Nursing, 133 Tay Street, 9840 Invercargill, New Zealand, [3]Texas A&M University, CVM, VIBS, 664 Raymond Stotzer Pkwy, College Station, 77843, USA, [4]Generatio GmbH, Blumenstr. 49, 69115 Heidelberg, Germany; melissa.cox@centerforanimalgenetics.com

The preservation of genetic diversity is an important consideration in native breeds. The Irish native Connemara Pony faces this challenge as it has gained popularity around the world. To assess genetic diversity, breed societies in New Zealand (NZ), Australia (AUS) and Canada (CA) provided DNA profiles of animals in their studbooks which represent full populations: 593 AUS, 150 CA, 175 NZ. Pre-existing data was also included from 69 Connemara ponies from the USA (not representing the full population). Using 15 STR loci, factors including expected and observed heterozygosity (H_E, H_O), inbreeding coefficients (F_{IS}), and genetics distances (PCA, FCA) were calculated. PCA analysis revealed that the countries have four distinct populations. FCA indicated that ponies from NZ and CA are least related while those from the USA and AUS are most closely related. The highest mean number of alleles per locus (MNA) was found in AUS (7.83) and the lowest in the US (7.25). The H_E ranged from 0.774 in AUS and CA to 0.765 in NZ, while the H_O ranged from 0.789 in CA to 0.779 in NZ and the US. F_{IS} ranged from -0.019 in NZ to -0.011 in the US. A recent study focusing on Connemara ponies in Ireland and the UK found an F_{IS} of 0.033; the highest F_{IS} in this study is -0.011. The H_O and the MNA in the four distant populations are also higher than in Ireland and the UK (H_O=0.718, MNA=6.58). Our data indicate that geographical distance has not had a negative impact on the genetic diversity of the populations in this study. In contrast, they may serve as a source of diversity for each other and for the native populations in Ireland and the UK, helping breeders maximise and sustain diversity while maintaining the type and breed character that marks the Connemara Pony.

Foal suckling bouts frequency and duration from 3 days to 3 months old

L. Wimel[1], C. Chardon[2], P. Wehbe[3] and J. Auclair-Ronzaud[1]
[1]The French Horse and Riding Institute (IFCE), Experimentation research center of Chamberet, 1 Impasse des Haras, 19370 Chamberet, France, [2]IUT Brest-Morlaix, 12 rue de Kergoat, 29200 Brest, France, [3]ENSAT, Avenue de l'Agrobiopole, 31326 Auzeville-Tolosane, France; laurence.wimel@ifce.fr

In horses, many studies focused on milk quality and some on milk yield but information regarding suckling frequency and bout duration is scarce. These data are useful to determine foal's time budget and milk intakes. Thus, it may allow estimating energy intakes knowing milk quality during the two first months of the foal's life, when he relies mainly on milk for his feeding. A better understanding of the foal energy intakes will allow adequate management regarding his feeding. Experiment was conducted in 2019 on 26 Anglo-Arab multiparous mares and their foals. Foalings occurred between the 1st of April and the 4th of May. All mares and foals are kept together on pasture from 3 days after foaling to weaning. Individuals were observed at four periods after birth (a few days, 1 month, 2 months and 3 months) at 8 to 12 am and 2 to 5 pm. Measurements were repeated, with two observation days at each period. Direct observations and video recordings were used to estimate the frequency and duration of suckling bouts. Each couple was observed for 15 minutes during each hour of observation. Thus, 6 mares and foals were observed for 15 minutes and the 26 in an hour. Individuals were randomly assigned to the groups, which changed at each studied period. Two ANOVA were computed to study the impact of different factors on suckling bout frequency and duration. In both cases were assessed as explicative variables: age of the foal, sex of the foal, observation date and observation time. Age of the foal has a significant effect on both frequency and bout duration (P<0.01). Correlation curves were drawn with r^2 of 0.52 and 0.68 respectively. Observation time has a significant effect on bout duration (P<0.01). No effect of the foal sex was observed. This first experimentation gives information regarding suckling behaviour for saddle breed horses kept outdoor. Nonetheless, more has to be done regarding the characterisation of both frequency and bout duration at night.

Live weight prediction of Miranda donkey from body measurements using image analysis

S.R. Silva[1], C. Guedes[1], D. Bacellar[2], B. Leiva[2] and M. Quaresma[1]
[1]CECAV, University of Trás-os-Montes e Alto Douro, Quinta de Prados, 5000-801 Vila Real, Portugal, [2]Associação para o Estudo e Proteção do Gado Asinino, Rua da Igreja, 5225-011 Atenor, Miranda do Douro, Portugal; ssilva@utad.pt

Using a scale is the best method to access live weight (LW). However, this approach is time-consuming and stressful. The use of an image analysis system (IAS) could be a safer alternative to predict LW. The objective of the present work is to predict donkey LW using IAS. Sixty-three donkeys of the Miranda breed (6.6±4.4 years old; 280±104 kg LW) were used. The height at the withers (HW), measured at the highest point of the withers, and two body length, BL1- measured from the point of the shoulder to the ischial tuberosity, and BL2, measured from the elbow to the head of the femur, were taken. The animals were weighted on a platform scale (Salter Brecknell® PS-3000HD, USA). Images from lateral view were capture using a digital camera (Nikon D3100) featuring a 14.2-megapixel sensor, placed 4 m from the animals. For scale purposes, two red lasers with a 650 nm wavelength, were mounted on a frame with a 32 cm distance between them. From the captured images and using Fiji software (ImageJ 1.49u) body measurements equivalent of those described previously were obtained. A correlation analysis between IAS and body measurements was performed. Multiple linear regression stepwise combined with k-fold cross-validation were used to predict LW. The coefficient of determination (R^2), the root mean square error of the cross-validation (RMSE) were used to evaluate the model. Statistical analyses were done with JMP-SAS14 (SAS Institute Inc. Cary, NC, USA). A significant correlation between IAS and body measurements was found ($r>0.96$; $P<0.01$). The prediction model performed by multiple linear regression using k-fold cross-validation shows that IAS can explain 86% of the LW variation (RMSE=39.6 kg). The results show that the IAS can be an accurate non-restraint approach to predict the donkey LW.

Neonate foal viability evaluation: Apgar index application

A.R. Teixeira[1,2], M.T. Leão[2], A.L. Costa[2,3], C. Maerten[2], M. Bliebernicht[2,3], A.S. Santos[1,4] and M.J. Fradinho[5]
[1]Universidade de Trás-os-Montes e Alto Douro, Quinta dos Prados, 5000 Vila Real, Portugal, [2]Haras de la Gesse, Chemin du Château de la Gesse, 31350 Boulogne sur Gesse, France, [3]Embriovet, Lda, Rua Fonte Costa, 2125-348 Muge, Portugal, [4]CITAB-UTAD, Quinta dos Prados, 5000 Vila Real, Portugal, [5]CIISA-FMV, Universidade de Lisboa, Avenida da Universidade Técnica, 1300-477 Lisboa, Portugal; ateixeira004@gmail.com

The early assessment of the newborn foal's vitality is essential to reduce its mortality in the postpartum period. This study aimed to assess the vitality of newborn foals through the application of the Apgar index. For this purpose, 31 births were monitored in the 2019 season, in a stud farm located in the south of France. A total of 26 Lusitano foals and 5 warmblood foals where monitored. In this evaluation a modified Apgar score was used, considering seven parameters: heart rate-HR, respiratory rate-RR, muscle tone-MT, reflexes of irritability (nose-NO, ears-EO and back-BC) and mucous colour; MC. Each parameter was assigned to a rating of 0 to 2 points, with the global sum ranging from 0 to 14. All foals were evaluated immediately after birth (first minute of life-T0), with a new evaluation after 15 minutes (T15). The global score obtained at time T0 was lower than the score obtained at time T15 (11.97±0.17 vs 12.74±0.17; $P<0.01$), with no differences between breeds. Considering the scores of each parameter between the two moments of assessment, there was a significant improvement in HR, BC and MC and a decrease in RR. According to what is described in the literature, it can be concluded that the foals included in this study presented a stable situation, not requiring any type of intervention after birth. In cases where the foals were born a little more debilitated, there was a recovery during the first 15 minutes after delivery. The application of the Apgar score in neonatal foals can thus be considered as a useful tool in the assessment of their postnatal viability.

Portuguese Equidae native breeds: main characteristics, census and demographic parameters

A. Vicente[1,2,3], N. Carolino[1,2,4,5] and M.J. Fradinho[2]
[1]Sociedade Portuguesa Recursos Genéticos Animais, SPREGA, Santarém, 2005-048, Portugal, [2]CIISA, Faculdade de Medicina Veterinária, Lisboa, 1300-477, Portugal, [3]Escola Superior Agrária, IPSantarém, Santarém, 2001-904, Portugal, [4]INIAV IP, Santarém, 2005-048, Portugal, [5]Escola Universitária Vasco Gama, Coimbra, 3020-210, Portugal; apavicente@gmail.com

The aim of this study was to summarise the main characteristics of the Portuguese *Equidae* native breeds. Portugal is a small country from the southwest of Europe with 50 recognised domestic native breeds. Among them there are 6 *Equidae* breeds, 4 equine (Lusitano, Sorraia, Garrano and Terceira pony) and 2 donkey populations (Miranda and Graciosa). In this review we present an overall census of each population, average body weight, height at withers, skills and uses, dynamics in the studbook and some demographic parameters such as known generations, generation interval and inbreeding. Lusitano horse is the main equine breed, with a close studbook and small census of 3,623 breeding females (half outside Portugal), spread around the globe with more than 30 countries breeding it. The Garrano pony is a small horse from the northwest, with a close studbook, with 1,994 breeding females, 162 stallions, and 435 breeders. Sorraia horse is a primitive southern European breed regarded as an universal genetic resource, very resistant and versatile, managed as a close population since 1937 with only 18 breeders, ~150 breeding females and a total of ~350 animals worldwide. Terceira Pony, from Azores, with an open studbook, was the last to be recognised (2014), with a total of ~100 animals descending from 14 founders. The Miranda donkey from the northeast is a long bay coat animal, with a calm temperament suited for agriculture, milk production and leisure, with a total of 756 females, 60 stallions and 460 breeders. The Graciosa donkey, from Azores, is the most recently recognised donkey breed (2015). With a very small size, a scarce census of 92 dams, of which only 22 are in production, 13 stallions and 63 breeders, are bred for conservation, tourism and milk production. Beside the opportunities for research and conservation programs, all these breeds represent important socio-economic and ecological values, that will enhance the sustainability of Mediterranean production systems. Acknowledgement: Project CIISA UID/CVT/00276/2020.

The use of estimated breeding values (EBVs) to improve lamb performance and carcase composition

N. Rutherford[1], B. Waters[1], J. Rankin[2] and A. Aubry[1]
[1]Agri-Food and Biosciences Institute, Agriculture Branch, Hillsborough, BT266NH, United Kingdom, [2]AgriSearch, Innovation Centre, Hillsborough, BT266NH, United Kingdom; naomi.rutherford@afbini.gov.uk

Sire and dam genetics are both significant determinants of lamb performance and meat quality. Genetic improvement on commercial sheep farms is largely realised through the purchase of breeding males. The adoption of estimated breeding values (EBVs) into flock breeding programmes can increase both animal performance and the rate of genetic gain within sheep populations. However, at present it is estimated that only 40% of sheep holdings in the UK use EBVs to select new breeding sires. The objective of this study was to evaluate the effect of sire EBV on the performance, carcase characteristics and meat quality of their progeny. This study involved 637 ewes over two breeding years. Ewes were artificially inseminated with 23 different terminal sires from two breeds (Suffolk and Texel), with either a high or low EBV for muscle. A total of 1,171 lambs were born, weighed at birth, and then fortnightly until slaughter. All lambs were ultrasound scanned at a mean age of 14 weeks, with muscle depth and fat depth being measured. In addition, a subset of 160 lambs were scanned using computer tomography (CT) at a mean age of 15 weeks. CT scanning was used to determine the relative proportions of muscle, fat and bone within each third of the body (fore, saddle and hind). Lambs were drafted for slaughter at approximately 45 kg live weight (LW). At slaughter, carcase weight, conformation and fat classification were recorded for all lambs. The saleable meat yield proportion of 277 carcases were calculated. In addition, a meat sample from the loin was retained from each of these carcases for instrumental meat quality measurements (colour, ultimate pH, cooking loss and Warner Bratzler shear force). Results indicate that lambs from the high EBV sire group had a higher muscle depth and muscle:fat ratio and reached target slaughter weight earlier than lambs from the low EBV sire group. Thus, this study provides an evaluation of the potential to improve lamb performance through the use of EBVs.

Microbial deterioration of lamb meat of Portuguese origin as affected by its intrinsic properties

U. Gonzales-Barron[1], S.C. Coelho-Fernandes[1], G. Santos-Rodrigues[1], J.M. Lorenzo[2], R. Bermúdez-Piedra[2] and V.A.P. Cadavez[1]
[1]Centro de Investigação de Montanha, Instituto Politécnico de Bragança, Animal Science, Campus de Santa Apolónia, 5300-253 Bragança, Portugal, [2]Centro Tecnológico de la Carne de Galicia, Avenida de Galicia 4, Parque Tecnolóxico de Galicia, 32900 San Cibrao das Viñas, Ourense, Spain; ubarron@ipb.pt

The study aimed to elucidate how intrinsic properties affected the evolution of spoilage microorganisms in refrigerated vacuum-packed (VP) lamb meat from BEDM and CGB breeds. Thirty CGB and 30 BEDM lambs were four months old when slaughtered. Physicochemical analyses (i.e. pH, water activity and proximate composition) were performed on meat at day 1 after slaughter, while microbial counts (mesophiles, psychrotrophic, lactic acid bacteria and Pseudomonas spp.) were essayed on VP meat at days 3, 9 and 15. Mixed-effects models were adjusted to each of the microbial groups to assess the effect of breed, time of maturation and each of the intrinsic properties of meat, in separate. Breed had a significant effect on all microbial groups ($P<0.0001$): meat from BEDM breed presented higher ($P<0.0001$) populations of mesophiles, lactic acid bacteria, Pseudomonas spp. and psychrotrophic bacteria, since its higher ultimate pH (means: 5.77 for BEDM vs 5.58 for CGB) accelerated spoilage rate ($P<0.0001$). While water activity and protein content were not found to modulate microbial deterioration ($P>0.05$), the growth of spoilage bacteria was found to be exacerbated by higher moisture ($P<0.0001$) and higher ashes content ($P<0.001$). By contrast, a higher fat content retarded ($P<0.0001$) the growth of spoilage bacteria in VP lamb meat. In order to extend the shelf-life of Portuguese-origin lamb meat, animal handling must be enhanced to minimise pre-slaughter stress, and a carcass classification system should be adopted towards the selection of fatter animals and chilled carcasses of optimal ultimate pH.

Meat quality attributes from two Portuguese autochthonous lamb breeds

G. Santos-Rodrigues[1], U. Gonzales-Barron[1], S.C. Coelho-Fernandes[1], J.M. Lorenzo[2], R. Bermúdez-Piedra[2] and V.A.P. Cadavez[1]
[1]Centro de Investigação de Montanha, Instituto Politécnico de Bragança, Animal Science, Campus de Santa Apolónia, 5300-253 Bragança, Portugal, [2]Centro Tecnológico de la Carne de Galicia, Avenida de Galicia 4, Parque Tecnolóxico de Galicia, 32900 San Cibrao das Viñas, Ourense, Spain; giselar0608@hotmail.com

The objective of this study was to evaluate the effects of maturation time on the meat quality attributes of two Portuguese autochthonous lamb breeds. A total of 30 Bordaleira-de-Entre-Douro-e-Minho (BDM) and 30 Churra-Galega-Bragançana (CGB) lambs were reared in a semi-intensive system in the cities of Ponte de Lima and Bragança, respectively. Lambs were slaughtered at 4 months of age under identical conditions at the slaughterhouse. Carcasses were refrigerated for 24 h, and samples of L. dorsi muscle were cut, vacuum-packed (VP) and kept at 4 °C for 15 days. The meat quality attributes, quantified on days 3, 9 and 15 after slaughter, were: colour in L*, a*, b* space, lipid oxidation (TBARs), cooking loss (CL) and meat tenderness by Warner-Bratzler test (slope, work and maximum force). A series of mixed models were adjusted to assess the effects of lamb breed, carcass cold weight (CCW) and maturation time on the meat quality attributes. All meat quality attributes were affected by maturation time ($P<0.0001$) excepting L * ($P=0.156$). During maturation, higher CCW ($P=0.049$) and higher pH24 ($P=0.002$) tended to produce higher CL. The active tenderisation of VP lamb meat for both breeds occurred until 9 or 10 days of maturation. However, regardless of breed, the main properties determining the tenderness attributes of lamb meat were pH24 ($P<0.0001$) and CCW ($P<0.0001$). However, the breed effect should be interpreted with caution, since breed was strongly confounded with pH24 (5.58-5.64 for CGB meat and 5.77-5.83 for BDM meat) and with CCW (13.7 to 14.8 kg for CGB and 6.5-7.6 for BDM).

Interaction between age and live weight at mating of ewe lambs on their reproductive success
A.N. Thompson[1], E. Bowen[1], J. Keiller[2], D. Pegler[3], G.A. Kearney[4] and C.A. Rosales Nieto[5]
[1]Murdoch University, College of Science, Health, Engineering and Education, 90 South Street, Murdoch, WA, 6150, Australia, [2]Cashmore Park, 114 Wilmots Road, Cashmore, VIC, 3305, Australia, [3]Oaklea Genetics, P.O. Box 1018, Mount Gambier, SA, 5290, Australia, [4] GK Biometrical Services, 36 Payne Road, Hamilton, VIC, 3300, Australia, [5]Universidad Autónoma de San Luis Potosí, Facultad de Agronomía y Veterinaria, San Luis Potosí, 78321, Mexico; andrew.thompson@murdoch.edu.au

The younger ewe lambs can be mated successfully the easier they can be integrated with the mating of the adult ewe flock the following year. In this paper we tested the hypothesis that both liveweight and age of ewe lambs at mating would influence their reproductive rate and the survival of their progeny. To test this hypothesis, we analysed data from more than 10,000 maternal ewe lambs collected from 2010 to 2017 by ram breeders. The ewe lambs had full pedigree records including birth type, age and liveweight at mating plus records of the birthweight and survival of their progeny. The average liveweight and age at mating was 40.1 kg and 228 days. The reproductive rate and weaning rate responses to liveweight at mating were curvilinear (P<0.001) and if ewe lambs achieved 45 kg by mating their reproductive rate and weaning rate were within 5% of their maximum. There was also a quadratic (P<0.01) effect of age at mating on reproductive rate which increased only marginally when ewe lambs were older than 8 months at mating and there was no effect on reproductive rate beyond 8.5 months. By contract, the effects of age at mating on weaning rate were linear up to 10 months of age. Liveweight (P<0.001) and age (P<0.001) at mating both had significant positive effects on progeny birth weight when included in the same statistical model. The model predicted that an extra 10 kg of liveweight or one-month of age at mating independently increased the birth weight of their progeny by 0.16 kg. Surprisingly, liveweight at mating had no significant effect (P<0.1) on progeny survival whereas age at mating had a significant positive effect (P<0.001) that remained (P=0.05) even when birth weight was included. This data has contributed to whole farm modelling to determine the optimal age and liveweight for mating maternal ewe lambs.

Birth difficulty affects the quality of maternal care given by a ewe immediately after birth
A. Redfearn[1,2], E. Janodet[3], J. McNally[2], H. Brewer[2], E. Doyle[1], R. Doyle[4] and S. Schmoelzl[1,2]
[1]University of New England, Armidale, 2350, NSW, Australia, [2]CSIRO, FD McMaster Laboratory, New England Highway, Armidale, 2350, NSW, Australia, [3]National School of Agricultural Engineering, Bordeaux, Bordeaux, 33170, France, [4]University of Melbourne, Parkville, 5511, VIC, Australia; amellia.redfearn@csiro.au

Birth is a stressful event, and any deviation from 'normal' can have serious effects on the dam and offspring. In sheep, the critical period for ewe-lamb bonding is 2 h immediately after birth. If bonding is disrupted, or insufficient, the lamb is at much higher risk of death. A difficult labour increases the risk of poor bonding, as the ewe may be too exhausted to care for the lamb, or the lamb may be compromised and unable to perform the correct behaviours. In this experiment, 15 video records of ewes lambing outdoors were selected based on clarity, completeness and labour difficulty (normal (N) n=9; difficult (D) n=6). Videos were continuously annotated with BORIS (Behavioural Observation Research Interactive Software) for 2 h immediately after full expulsion of the lamb. Four behavioural metrics were used; behavioural transition frequency (BTF)- the number of behaviour changes per hour; grooming, including sniffing, bonding and licking; care, including vocalisation, nudging and suckling; and locomotion, including standing, lying, walking, scratching, stepping and circling. Ewes that experienced a difficult labour had a lower BTF than normal ewes (P=0.02). Difficult ewes performed fewer grooming behaviour bouts (P=0.005), and spent less time performing grooming behaviours than normal ewes (P=0.006). Additionally, ewes with a difficult labour had shorter bouts of care behaviours compared to ewes with a normal labour (P=0.03). Finally, ewes with a difficult labour spent significantly more time performing locomotion behaviours, specifically standing (P=0.03) than ewes with a normal labour (P=0.009). Labour difficulty has an obvious and important effect on the ability of the ewe to provide adequate care to the lamb immediately after birth. Identification of difficult labour events may assist with intervention and management decisions, improving lamb survival.

Merino ewes selected for carcase traits are more energy efficient than ewes selected for wool traits

S. Blumer[1], J. Young[2] and A. Thompson[1]
[1]Murdoch University, 90 South Street, Murdoch, 6150, Australia, [2]Farming Systems Analysis Service, 476 Tindale Rd, Denmark, 6333, Australia; s.blumer@murdoch.edu.au

Fat tissue depth in ruminants is negatively associated with feed efficiency during growth. Fat tissue is more energy dense per kilogram than muscle, hence gaining fat is more energy expensive than gaining lean tissue. However, the converse is also true. During negative energy balance adult animals with proportionally more fat have greater energy reserves to buffer seasonal deficits in pasture quality and quantity. Previous work has demonstrated that leptin is a significant predictor of residual feed intake (RFI) in adult Merino ewes, explaining 27% of RFI while sub-cutaneous fat depth explained 11%. Composition is an important consideration for efficiency in the adult flock and leptin may be a useful and easily measured predictor of whole-body fat stores. We used 20 adult Merino ewes from distinct genotypes (1: selection pressure on CARCASE traits – fat and muscle; 2: selection pressure on WOOL traits – primarily fleece weight) to assess leptin as a proxy for whole body fatness measured using CT and investigated differences in composition and feed efficiency during a 42-day housed experiment. Leptin was a good predictor of whole-body fat as measured by CT (R^2=0.74; P<0.001), however was significantly different between the genotypes. WOOL ewes had more leptin than CARCASE ewes, despite being leaner (0.91±0.09 vs 0.65±0.10 µg/ml; P<0.05). The genotypes were not different for RFI with similar liveweight change and intake (P>0.05). However, CARCASE ewes carried proportionally more fat at the start and finish of the feeding period, and under *ad libitum* feeding proportionally more of the liveweight gain was fat tissue. Given that fat tissue is more energy expensive to deposit, the CARCASE line ewes stored more energy per MJ of ME consumed than the WOOL line ewes.

Impacts of the lifetime ewe management training program for farmers on the Australian sheep industry

A.N. Thompson[1], D. Gordon[2], E. King[3], M. Scott[3] and J. Trompf[4]
[1]Murdoch University, College of Science, Health, Engineering and Education, 90 South Street, Murdoch, WA, 6150, Australia, [2]Livestock Logic, 60 Portland Road, Hamilton, VIC, 3300, Australia, [3]Australian Wool Innovation Limited, Level 6, 68 Harrington Street, Sydney, NSW, 2000, Australia, [4]J.T Agri-Source Pty Ltd, Wangaratta, VIC, 3677, Australia; andrew.thompson@murdoch.edu.au

Practical guidelines for managing Merino ewes to improve lamb marking rates, whole farm profit and animal welfare were developed in Australia about 15 years ago and have been delivered via the Lifetime Ewe Management (LTEM) training program to more than 4,000 producers that manage 12 million ewes. Eight hundred producers that graduated from LTEM between 2008 and 2018 were evaluated using a telephone survey to capture data relating to attitudes, skills, practices and productivity preceding their commencement of LTEM and again after completing LTEM. Farm characteristics, production data, practices, skills and attitudes were analysed separately using General Linear Mixed Models with a logit-transformation, where appropriate, and linear regression was used to model various relationships either singly or with multiple variates for the mean values of graduation years. LTEM participants increased their whole-farm stocking rate from 8.5 to 9.3 DSE/ha, increased lamb-marking rate from 97.3 to 104.3% and reduced ewe mortality from 4.1 to 3.0%. The single most important management practice relating to the impacts of LTEM on lamb marking rates was the rate of adoption of pregnancy scanning for multiples and differential management of twin ewes both pre-LTEM and due to LTEM. Adoption of this practice increased from 25% pre-LTEM to 65% post-LTEM. Pre-LTEM adoption of pregnancy scanning and differential management of twin ewes, the change adoption of this practice due to LTEM and the number of ewes explained 86% of the changes in lamb marking rate achieved by the different cohorts of graduates. This model predicted that adoption of pregnancy scanning for multiples and differential management of twin bearing ewes increased lamb-marking rates by 14%. The scale of the productivity and welfare gains achieved by participants of LTEM are unprecedented by previous extension efforts in Australia.

Phytoestrogen profile of retail goat and cow milk in the UK

S. Stergiadis[1], N.P. Nørskov[2], S. Purup[2], D.I. Givens[3] and M.R.F. Lee[4,5]
[1]University of Reading, School of Agriculture Policy and Development, Agriculture Building, Earley Gate, P.O. Box 237, RG6 6AR, Reading, United Kingdom, [2]Aarhus University, Department of Animal Science, AU-Foulum, Blichers Alle 20, P.O Box 50, 8830 Tjele, Denmark, [3]University of Reading, Institute for Food, Nutrition and Health, P.O Box 217, RG6 6AH, Reading, United Kingdom, [4]University of Bristol, Bristol Veterinary School, BS40 5DU, Langford, Somerset, United Kingdom, [5]Rothamsted Research, North Wyke, EX20 2SB, Okenhampton, United Kingdom; s.stergiadis@reading.ac.uk

The dairy goat industry in the UK has been growing, and currently accounts for approximately 100,000 goats, producing 34 million litres of milk annually. This provides a market value of £70 million. Consumers, especially those with milk intolerances/allergies, perceive goat milk as a healthier alternative but the nutrient profiling of goat milk in the UK is not widely known. This study compared UK retail goat and cow milk for the concentrations of phytoestrogens (lignans, isoflavones and coumestans), which are associated with benefits in human health. Monthly milk samples (n=84) from 3 goat and 4 cow milk retailer brands were collected over 12 months. Phytoestrogen profile was assessed using LC-MS/MS (microLC 200, Eksigent/AB Sciex, USA; QTrap 5500 MS, AB Sciex, USA). Phytoestrogen concentrations data were analysed by a linear mixed effects model (Genstat 17th edition), using species, month and their interaction as fixed factors and milk ID as random factor. Compared with cow milk, goat milk contained less secoisolariciresinol (−62.7%), matairesinol (−50.0%), lariciresinol (−47.4%), hydroxymatairesinol (−56.2%), enterolactone (−67.6%), plant lignans (−52.5%), mammalian lignans (−66.6%) and total lignans (−66.5%), and more daidzein (+747.3%), glycitein (+167.5%), naringenin (+185%), equol (+985.8%) and isoflavones (+964.3%). In the UK, goat dairying is more intensive (typically all-year-round housing) than conventional dairy cow production. Certain intensive dairy practices (e.g. increased dietary soyabean meal instead of wheat, barley and oat) may reduce contents of lignans and increase contents of isoflavones in goat milk. In addition rumen microbiome or genetic differences between species may not be excluded. Consuming goat milk may increase intakes of isoflavones and reduce intakes of lignans, but the potential effect on human health was not assessed in the present study.

Meta-analysis on the effect of biopreservatives on *Staphylococcus aureus* inactivation in cheese

B.N. Silva[1,2], V.A.P. Cadavez[2], J.A. Teixeira[1] and U. Gonzales-Barron[2]
[1]Centre of Biological Engineering, University of Minho, Campus Gualtar, 4710-057 Braga, Portugal, Portugal, [2]Centro de Investigação de Montanha, Instituto Politécnico de Bragança, Animal Science, Campus de Santa Apolónia, 5300-253 Bragança, Portugal; ubarron@ipb.pt

Biopreservation methods based on the use of natural extracts and starter cultures have been proposed as hurdles to increase the microbiological safety of many food products, including cheese. The objective of this study was to construct two separate meta-regression models on the inactivation of Staphylococcus aureus (SA) in cheese containing essential oils (EOs) and added lactic acid bacteria (LAB). Twenty challenge studies were considered appropriate for inclusion in the meta-analyses (n=299), and the following information was extracted: antimicrobial class (EO or LAB) and name, mean log reduction, storage temperature, exposure time, antimicrobial application (i.e. cheese mixture, milk or film), and antimicrobial concentration. The EOs model revealed the significant impact of application type (P<0.0001), storage temperature (P<0.0001) and inoculum concentration (P=0.019) on SA microbial reduction. The effects of exposure time and antimicrobial concentration were dependent on the type of application (P<0.0001). The application of EOs in milk promoted the highest microbial reduction, whereas incorporation in films presented the lowest inhibitory effect. Among the types of EOs meta-analysed, lemon balm and sage produced the greatest bactericidal effects. The LAB model did not show differences (P=0.091) in the inhibitory effect achieved by different applications (milk or cheese mixture), but revealed the interaction between this term and exposure time (P=0.040). This meta-analysis has emphasised that the experimental practice of inoculating the antimicrobial in shredded cheese ('cheese mixture') should no longer be employed in challenge studies, since this application method biases the actual results. The inverse relationship between pathogen's inoculum size and microbial reduction (P=0.025) should be further investigated.

New selection criteria for female fertility in Florida dairy goat using random regression models

C. Ziadi[1], E. Muñoz-Mejías[2], M. Sánchez[3], M.D. López[4], F. Arrebola[5], A. Menéndez-Buxadera[1] and A. Molina[1]
[1]Universidad de Córdoba, Genética, Edificio Gregor Mendel. Campus de Rabanales, 14071 Córdoba, Spain, [2]Universidad de Las Palmas de Gran Canaria, Departamento de Patología Animal, Producción Animal, Bromatología y Tecnología de los Alimentos, Campus Universitario Cardones de Arucas, 35413 Arucas, Spain, [3]Universidad de Córdoba, Departamento de Producción Animal, Campus de Rabanales, 14071, Córdoba, Spain, [4]ACRIFLOR, Edificio de Producción Animal, Campus de Rabanales, 14071, Córdoba, Spain, [5]IFAPA, IFAPA, Hinojosa del Duque, 14270, Hinojosa del Duque, Córdoba, Spain; ziadichiraz4@gmail.com

This study was conducted to estimate genetic parameters of new selection criteria for female fertility across the age trajectory in Spanish Florida dairy goat using random regression models (RRM). Traits analysed (univariate) in this analysis were: parity number (PN) treated as a character and as a deviation between optimal and real parity number at each age considered as a continuous trait (DORPN). Using 136.583 reproductive records of 50.803 females over a period of 34 years (1986-2019), genetic parameters were estimated with Bayesian inference using the GIBBS3F90 software. Estimates of heritabilities were moderate to high, ranging from 0.28 to 0.41 for PN and from 0.25 to 0.38 for DORPN, respectively. Phenotypic correlations for PN across ages were close to unity and for DORPN, they varied from -0.41 to 0.84. The genetic correlations for PN and DORPN at different age points were positive in all cases ranging from 0.61 to 1.00 and from 0.20 to 0.98 for PN and DORPN, respectively, decreasing their magnitude when increasing the female age. This results indicate that selecting for these trait separately or through a selection index could lead to improve goat female fertility.

Crude protein level in post-weaned lambs diet did not affect productive and meat quality parameters

C. Baila[1], M. Blanco[1], G. Ripoll[1], J. Pelegrín-Valls[2], J. Álvarez-Rodríguez[2] and M. Joy[1]
[1]Centro de Investigación y Tecnología Agroalimentaria de Aragón, Animal Production, Av. Montañana, 930, 50059 Zaragoza, Spain, [2]Universitat de Lleida, Animal Production, Plaça de Víctor Siurana, 1, 25003, Lleida, Spain; cbaila@cita-aragon.es

In the Mediterranean area, the light lamb production is based on flocks of medium-small frame sized autochthonous breeds. Fattening lambs are concentrate-fed indoors until 20-25 kg live weight (LW). In this phase, the crude protein (CP) in the concentrate ranges between 15-21% on dry matter basis (DM). Optimising the level of CP in the lamb's diet reduces feed costs and contributes to the mitigation of environmental emissions of ammonia and nitrous oxide, and could improve the production efficiency. The aim of this study was to evaluate the effect of the reduction of the CP level (Control vs Low) in the fattening concentrates of light lambs of Ripollesa breed. The iso-energetic concentrates and straw were fed *ad libitum*. Half of the lambs received a concentrate with 20% CP from 15 to 19 kg LW and a concentrate with 19% CP until 25 kg LW (Control). The other half received a concentrate with 18% CP and 17% CP in the same periods (Low). Weekly, intake and LW were recorded. After slaughter and cooling for 24 h, carcass characteristics (weight, carcass yield, fatness score and colour of Rectus abdominis muscle) and meat quality parameters of Longissimus thoracis et lumborum muscle (pH, colour, haeminic pigments, lipid oxidation, chemical composition and fatty acid profile) were evaluated. The reduction of 2% CP tended to increase the daily concentrate intake ($p<0.10$) without affecting the feed *conversion rate* ($p>0.05$). Regarding the effect on carcass quality, the reduction decreased the carcass weight (11.9 vs 11.3 kg, $p<0.05$) and the carcass yield (49 vs 46%, $p<0.01$), but no effect was observed on the fatness score and the colour of Rectus abdominis muscle ($p>0.05$). The reduction of 2% of CP had minor effects on meat quality. Minor changes in fatty acids and an increase of metmyoglobin content at 3 and 6 days of air exposure were detected ($p<0.05$). It is advisable to reduce the CP of the concentrate during fattening of light lambs of medium-small framed breeds, although the price of the lamb and the cost of protein ingredients has to be taken into account. However, more studies should be carried out to evaluate possible effects of CP level on meat oxidation.

Genome-wide association studies for growth traits and body size in backcross sheep family

T. Deniskova[1], S. Petrov[1], A. Sermyagin[1], A. Dotsev[1], M. Fornara[1], H. Reyer[2], K. Wimmers[2], V. Bagirov[1], G. Brem[1,3] and N. Zinovieva[1]
[1]L.K. Ernst Federal Science Center for Animal Husbandry, Moscow, 142132, Russian Federation, [2]Institute of Genome Biology, Leibniz Institute for Farm Animal Biology, Dummerstorf, 18196, Germany, [3]Institute of Animal Breeding and Genetics, VMU, Vienna, 1210, Austria; horarka@yandex.ru

Despite high population demand for lamb, sheep meat industry is not economically beneficial in Russia because local breeds are slowly growing and low meat productive. Thus, implementation of genomic selection to overcome disadvantages of Russian local breeds has a great value. The aim of our study was to identify QTLs for growth traits and for body size in the (Katahdin× Romanov) × Romanov backcross family. Sheep were genotyped with Ovine Infinium HD SNP BeadChip (Illumina, USA). After quality control a total of 459,868 SNPs were left for the GWAS performed in PLINK v.1.9. Threshold p-value for Bonferroni genome-wide significance was set to $P=1.09\times10^{-7}$ (0.05/459868). The data was visualised in R package qqman. Phenotypic data included body weight (BW), withers height (WH), rump height (RH), back height (BH), chest depth (CD), chest width (CW), body length (BL), oblique body length (OBL). The measurements were recorded at the age of 6, 42 and 180 days. Significant SNPs, including oar3_OAR6_18252318 ($P=7.29\times10^{-7}$), oar3_OAR3_160072113 ($P=8.84\times10^{-7}$), oar3_OAR7_37679053 ($P=1.70\times10^{-7}$), oar3_OAR7_43279978 ($P=5.45\times10^{-9}$), were found for OBL at six months (59.52±0.70 cm) and for BL at six months (57.95±0.75 cm). Besides a significant SNP (oar3_OAR15_58479722, $P=9.35\times10^{-7}$) was identified for CW at six months (11.58±0.29 cm). GWAS identified genes CNOT2, CNOT6, PPARD, ELOVL6, PITX2, CAPN3 and MYOD1 which were previously reported in sheep and involved in regulation of energy and lipid metabolism in the skeletal muscles, growth and development, carcass traits, myogenic differentiation. In addition, GWAS detected suggestive SNPs located nearby genes KLHDC1, KLHDC2, DGAT2, DGKH that are responsible for muscle cell migration and growth processes in cattle and other mammals. Currently, obtaining of additional backcross families is in process to address more fully the mechanisms underlying the genetics of growth traits. The study was funded by RFBR,No.17-29-08015.

LR-validation of genomic selection in Norwegian White sheep

J.H. Jakobsen[1], X. Yu[2], T. Blichfelct[1], J.C. McEwan[3], H. Baird[3] and T.H.E. Meuwissen[2]
[1]The Norwegian Association of Sheep and Goat Breeders, Box 104, 1431 Ås, Norway, [2]Norwegian University of Life Sciences, Aboretveien 6, 1433 Ås, Norway, [3]AgResearch Limited, Invermay Agricultural Centre, Private Bag 50034, 9053 Mosgiel, New Zealand; jj@nsg.no

Genomic selection is under implementation in the population of Norwegian White sheep. The objective of this study was to evaluate if genomic selection outperforms pedigree selection using the LR-validation procedure. The current reference population is composed of 4,887 lambing ewes born 2017, and 8,115 test-, elite and AI rams born during the years 2013 through 2019. Animals have been genotyped on various Illumina SNP chip panels (8K, 18K, 18K_v2) and imputed to the Illumina SNP chip 18K_v2 genotyping density using the Beagle 5.0 imputation software. The VanRaden Method I was used for setting up the G-matrix from the imputed genotypes. Eighteen years of phenotypic information of the three trait groups; lamb growth and carcass traits (Nrec=4,911,832); litter size (Nrec=2,525,074); and lamb fleece weight and quality (Nrec=645,593); collected in commercial breeding flocks and from abattoirs were extracted from the Norwegian Sheep Recording System. A joint pedigree was traced as far back as possible on animals with phenotypes. GEBVs and AEBVs were predicted using ssGBLUP and pedigree BLUP, respectively, in the DMU package. The value of including genomic information into the breeding program of Norwegian White sheep was computed using the LR-validation procedure by correlating GEBVs and AEBVs, and in addition computing the regression slope of cov(AEBV,GEBV) on var(AEBV) by trait and genotyping cohort. Across traits, the LR-validation showed the largest benefit of genomic selection for the youngest selection candidates and for the maternal traits. However, the regression slope was less than unity and the benefits are likely to have been overestimated. Future sampling will be targeted towards both improving genomic breeding value accuracy and addressing the causes related to this bias.

A variant in intron 11 of RXFP2 gene is associated with the polled phenotype in North Aegean sheep

A. Kominakis, A. Hager, E. Tarsani and I. Hadjigeorgiou
Agricultural University of Athens, Animal Science, Iera Odos 75, 18855 Athens, Greece; acom@aua.gr

Aim of the present study was to identify genetic variants and candidate genes for the horned/polled phenotype in a local sheep population kept on three North Aegean islands (Lemnos, Lesvos and Agios Efstratios) of Greece. A total number of 256 ewes genotyped with the 50K SNP array were used. Application of quality control criteria at a marker level resulted in a number of 37,201 autosomal markers retained for further analysis. Genome wide association analyses with application of additive and dominant multi-locus mixed models followed. Marker-based heritability was as high as 0.26 and 0.16 in the additive and dominant model, respectively. Marker-trait association analyses revealed a highly significant marker OAR10_29511510.1 (C>T) (additive p-value=4.82e-24, dominant p-value=1.52e-30). The marker explained a large proportion of the phenotypic variance of the trait i.e. 0.37 and 0.43 in the additive and the dominant genetic model, respectively. Polled animals (n=174) were either homozygous CC (n=167) or heterozygous CT (n=7). No homozygous TT polled animals were recorded. The majority of horned animals (n=82) were either heterozygous CT (n=55) or homozygous CC (n=20) while there were only n=7 homozygous TT ewes. OAR10_29511510.1 is located in intron 11 of *RXFP2* (relaxin/insulin-like family peptide receptor 2) gene that has been related to horn development and types in various sheep breeds. Current findings are indicative of dominant gene action and confirmed the strong candidacy of the *RXFP2* gene in the genetic control of presence/absence of horns in sheep.

Effects of cryopreservation on CASA parameters: preliminary study in Portuguese sheep breeds semen

P. Cordeiro[1], J.P. Barbas[1,2], M.C. Baptista[1], C.C. Marques[1], R.M.L.N. Pereira[1,2] and J. Pimenta[1,2]
[1]Instituto Nacional de Investigação Agrária e Veterinária, Q.Fonte Boa, 2000 Vale de Santarém, Portugal, [2]CIISA-Universidade de Lisboa, Av.UTL, 1300 Lisboa, Portugal; rosa.linoneto@iniav.pt

Studies on computer-assisted evaluation of semen (CASA) from Portuguese native sheep breeds, are meager. This study was planned to quantitatively compare the kinematics of fresh and frozen-thawed semen, using CASA and to study their interrelationships. Data from visual observation parameters (motility, morphology and vitality), were also used. A total of 46 good quality semen ejaculates from 6 rams belonging to 4 Portuguese native breeds (Churra Algarvia, Churra do Campo, Churra Galega Bragançana and Saloia), were studied for CASA kinematic parameters and visual observation, in fresh and frozen-thawed spermatozoa. Data were analysed with Paired-samples T-test, Pearson product-moment correlation and Cohen's d. Present findings on CASA and visual analysis found a significant (P<0.05) decline in mean values of post-thawed sperm when compared to fresh semen. With respect to effect size of cryopreservation, we observed a large effect (d≥0.8) in the following variables: (1) visual (vitality and motility); and (2) CASA [beat cross frequency (BCF), straightness (STR) and linearity (LIN) of track, total motility (TM), slow, medium, rapid, curvilinear (VCL), average path (VAP) and straight line (VSL) velocity]. VAP, VSL and VCL were the CASA variables where cryopreservation elicited the highest (d≥1.7) effect size. Overall, cryopreservation had a large (d≥0.8; 59%), medium (d≥0.5; 23%) and small (d≥0.2; 18%) effect size in the observed variables (CASA and Visual). Moreover, among the main CASA parameters, the highest (positive) correlations were observed between VAP and VSL (r=0.94; r=0.94; P<0.01), and between VAP and VCL (r=0.82; r=0.90; P<0.01) in fresh and frozen-thawed semen respectively. Noteworthy that VAP and VCL were previously described by others to present significant positive correlations with the ability to migrate in sheep cervical mucus. This preliminary study provided fundamental descriptions on both visual and CASA characteristics, that will be essential to improve post-thawed sperm quality of native sheep breeds. (Funded by UID/CVT/276/2019,ALT20-03-0246-FEDER000021).

Effects of dietary inclusion of an EPA-rich microalgae on lamb meat quality and fatty acid profile

A.C.M. Vitor[1], A.M. Godinho[1], A. Francisco[2], J. Silva[3], J. Santos-Silva[2], R.J.B. Bessa[1] and S.P. Alves[1]
[1]CIISA – Centro de Investigação Interdisciplinar em Sanidade Animal, Faculdade de Medicina Veterinária, ULisboa, Avenida da Universidade Técnica, 1300-477, Portugal, [2]INIAV, Instituto Nacional de Investigação Agrária e Veterinária, Quinta da Fonte Boa, Vale de Santarém, 2005-048 Santarém, Portugal, [3]Allmicroalgae, Fábrica Cibra, 2445-287 Pataias, Portugal; anacrisvitor@fmv.ulisboa.pt

Ruminant meat is characterised by having a low content in n-3 long-chain polyunsaturated fatty acids (n-3 LC-PUFA), which is linked to the extensive ruminal biohydrogenation (RBH) of the dietary unsaturated fatty acids (UFA) and low endogenous elongation and desaturation of 18:3n-3. To overcome the RBH, the n-3 LC-PUFA can be subject to ruminal protection vehicles, some of which could be considered natural, like microalgae. *Nannochloropsis oceanica* (NO), is an EPA-rich microalgae that thanks to its thick cell wall could partially protect EPA from RBH. This study aimed to evaluate the fatty acid (FA) composition of meat from lambs fed NO dried in two ways (freeze-dried, FD and spray-dried, SD). Twenty-eight Merino Branco ram lambs with an initial live weight of 18.9±3.9 kg were housed and randomly assigned to individual pens and one of four diets: control (C); NO-oil (O); NO-FD (FD) and NO-SD (SD). All NO diets contained 2.8 g EPA/kg dry matter (DM). DM intake, average daily gain and carcass traits were not affected ($P>0.05$) by the diets but carcass yield was increased ($P<0.05$) by diets containing microalgae. Sensory characteristics of the meat ($P>0.05$) were not affected by diets with microalgae except for the presence of off flavours ($P<0.05$) that increased with the NO-FD diet. Neither total lipid nor total FA contents were affected by diet ($P>0.05$). The 20:5n-3 (EPA), 22:5n-3 (DPA) and total n-3 PUFA in meat increased with all NO diets but were highest ($P<0.05$) with the NO-FD and NO-SD diets. All diets containing NO decreased the n-6/n-3 ratio and increased the 18:1t11 in the meat. This work was funded by Project UIDP/CVT/276/2020 (CIISA).

Pregnant ewe activity increases perinatally with a diet including fodder beet (*Beta vulgaris*)

A. Redfearn[1,2], S. McCoard[3], C. Erichsen[3,4], C. Dwyer[4] and S. Schmoelzl[1,2]
[1]University of New England, Armidale, 2350, NSW, Australia, [2]CSIRO, FD McMaster Laboratory, New England Highway, Armidale, 2350, NSW, Australia, [3]AgResearch, Grasslands Research Centre, Palmerston North, 4442, New Zealand, [4]SRUC, Roslin Institute, Edinburgh, EH25 9RG, United Kingdom; amellia.redfearn@csiro.au

Maintaining a positive energy balance during pregnancy is essential for the development of the foetus, and for maintaining the health of the dam. Providing access to feed of sufficient energy content can be a challenge, and this has led producers to investigate novel feed types that contain highly metabolisable energy, such as fodder beet. This experiment investigated whether a diet containing fodder beet would lead to changed activity levels of pregnant ewes. 56 ewes were randomly assigned to two treatments; feed containing chaff and fodder beet (FB; n=32) or chaff only (CTRL; n=24). Ewes were gradually introduced to fodder beet (FB only) and transitioned to single pen housing with 24 h surveillance by day 100 of gestation. A week before ewes were due to lamb, accelerometers were deployed on the neck, and were removed around 24 h after lambing. In total, 22 activity records were collected. Accelerometer data were compiled and analysed using ActiLife software (ActiGraph, Pensacola, FL). Three levels of activity were used for this analysis; sedentary (S), light (L) and moderate (M). On average, FB ewes spent significantly less time sedentary than CTRL ewes (P=0.007). In addition, FB ewes spent significantly more time in light activity (P=0.02) and moderate activity (P=0.03) than CTRL ewes. Both treatment groups spent significantly more time sedentary than active (L and M combined; P<0.0001). Both treatments spent significantly more time sedentary than active, which is likely a reflection of the single-pen housing. FB ewes spent more time active than CTRL ewes. The observed difference in activity may be due to differences in available energy from the different feeds. Alternatively, the difference may be reflective of differential feeding behaviour related to the different nature of the feed. To help answer this question, video footage should be analysed. Further and more detailed analysis of accelerometer data may give insights as to whether the diet had any effect on behaviour during the birth process itself.

Influence of maternal GH2-Z genotypes and feeding regime on Serra da Estrela lambs' body weight

M.R. Marques[1,2], J.R. Ribeiro[2], A.T. Belo[2] and C.C. Belo[2]
[1]CIISA, FMV, Avenida da Universidade Técnica, 1300-477 Lisboa, Portugal, [2]INIAV-IP, Fonte Boa, 2005-048 Vale de Santarém, Portugal; rosario.marques@iniav.pt

Growth hormone (GH) is one of the main hormones involved both in puberty and gestation. Directly or indirectly, it influences the homeorhetic control of female metabolism during gestation and lactation. Two alleles of the GH gene have been described in ovine. The *Gh1* allele contains a single gene copy (GH1), whereas in the *Gh2* allele the gene is duplicated (GH2-N and GH2-Z copies. *GH2-N* and *GH2-Z* copies are expressed in the ovine placenta in a temporal-specific way. The objective of this work is to access how Serra da Estrela ewes *GH2-Z* gene copy genotypes [AA (R9R/S63S), AB (R9C/S63S), and AE (R9R/S63G)], and ewes' feeding regime during pre-pubertal phase (group R; growth rate restricted to 79 g/day; or group N – growth rate of 106 g/day) influences their lambs' weight at lambing (LWL), lambs (body weight gain; BWG) and ewes estimated milk production (by the BWG of the lambs; PL) until the 21st day of lactation. Ewes were genotyped at one month of age and evenly distributed by the feeding regimes. Lambs were weighed at lambing and weekly until the 21 days of age, throughout six lambing seasons. Data was analysed using a mixed model. Serra da Estrela lambs weighted on average 4.1±0.05 kg at lambing. Single lambs were significantly heavier than twins (4.5±0.05 vs 3.7±0.05; P<0.0001) and males were significantly heavier than females (4.3±0.05 vs 4.0±0.05; P<0.0001). Genotype × feeding regime interaction affected significantly the LWL: lambs from AB ewes at R group (4.3±0.08 kg) were significantly heavier (P<0.05) than lambs from AA or AE ewes (4.0±0.07 kg and 4.1±0.11; respectively; P>0.05), and lambs from AB ewes at N group (4.0±0.08 kg). The weight at lambing of the lambs from group N ewes was not affected by genotypes. Considering the lambing type, single lambs for AA ewes were significantly lighter than the other lambs (P<0.001), for both feeding regimes. Twin LWL were significantly influenced by the genotype and feeding regime of their dams (P<0.05). The results suggest the existence of an epistatic regulation of the LWL of the lambs in Serra da Estrela ovine breed. Project PTDC/CVT/112054/2009 funded by Foundation for Science and Technology (FCT)

PRNP genetic variability in Portuguese Merino da Beira Baixa sheep breed

M.F. Santos-Silva[1], C. Oliveira E Sousa[1], I. Carolino[1], P. Jacob[1], P. Cardoso[2] and N. Carolino[1]
[1]Instituto Nacional de Investigação Agrária e Veterinária, INIAV I.P., EZN, Fonte Boa, 2005-048 Vale de Santarém, Portugal, [2]Associação de Produtores Agropecuários, OVIBEIRA, R. José Cifuentes 11D/E, 6000-244 Castelo Branco, Portugal; nuno.carolino@iniav.pt

Scrapie is a transmissible disease of sheep and goat. Prion protein gene (*PRNP*) polymorphism at codons 136 (A/V), 154 (R/H) and 171 (R/H/Q) were related to different levels of sheep genetic resistance to classical *scrapie*. Resultant genotypes were classified into five degrees of resistance, ARR/ARR being the more resistant and VRQ/VRQ the more susceptible. This work developed in collaboration with the breed Herd Book, aimed to analyse the present pattern of variability of *PRNP* locus at Merino da Beira Baixa Portuguese sheep breed, as a contribution to the defence of animal autochthonous genetic resources following the directives and support of the Portuguese PDR2020 program. Merino da Beira Baixa is a local endangered breed (5,800 females, 262 males registered, 40 breeders) from the centre of Portugal raised in a tri-purpose function milk, meat and wool, perfectly adapted to his environment conditions with an important role to farmers sustainability. Blood samples of 471 animals from 14 farms were analysed using a primer extension procedure designed to detect point mutations in codons 136, 154 and 171. Five alleles and nine genotypes were identified in the global population. The ancestral ARQ was the more frequent allele (0.77), followed by ARR (0.21). AHQ, VRQ and ARH were present at low frequencies of 0.04, 0.02 and 0.003 respectively. The more predominant genotype was ARQ/ARQ (0.55), followed by ARR/ARQ (0.29). The more resistant genotype ARR/ARR showed low frequencies of 0.06 similar to ARQ/AHQ. All the other genotypes found had frequencies under 0.03 and the more susceptible allele VRQ/VRQ was not present. This results, very similar to our previous findings (2003), don't reveal increase of the more resistant genotype in the population studied, showing that only 7% of this population can be included in the first two classes of resistance to classical scrapie, while the remaining has only an intermediate (34.4) to low (58.6) degree of resistance.

Factors associated to failure of passive immunity transfer in Latxa dairy sheep

R. Atxaerandio[1], X. Averós[1], I. Beltrán De Heredia[1], J. Arranz[1], N. Elguezabal[2], M.A. Haro[1], A. García-Rodríguez[1] and R. Ruiz[1]
[1]Neiker, Animal Production, CA de Arkaute, 01192 Arkaute, Spain, [2]Neiker, Animal Health, PT de Bizkaia, 48160 Derio, Spain; ratxaerandio@neiker.eus

Current practices in veterinary medicine promote a more rational antibiotic use in livestock farming to face antimicrobial resistance. Improving animal management and biosecurity, especially around birth, is crucial. Immediately after birth, lamb survival depends on colostrum management and intake to acquire adequate passive immunity transfer (PIT). The aim of this study is to assess the prevalence of failure of PIT (FPIT) in the sheep Latxa breed by exploring factors affecting the success of passive immunisation, which should help to enhance disease resistance. Preliminary FPTI estimates were already obtained from a sample of lambs from the Neiker sheep flock born during the 2018 and 2019 lambing seasons. Lambs were bottle fed a colostrum amount equal to 10% BW at birth (Strategy A; 2018) or 15% BW (Strategy B; 2019). Lambs were blood sampled at 24 and 48 h of age to quantify blood immunoglobulin G (IgG) concentration using RID. Strategy B nearly doubled the apparent absorption efficiency (AAE) rate in comparison to strategy A (16.5% vs 28.0%). The prevalence of FPIT in 2018 was 40 or 73%, depending whether the cut-off value considered was 10 or 15 g/l; in 2019, FPIT was 45 or 80% for the same cut-off values. Colostrum immune quality and hygiene, as well as the impact of environment, animal management and welfare related issues, among others, may explain the high prevalence of FPIT and the low AAE. However, this scenario needs to be verified in commercial flocks. To do so, 14 commercial flocks were monitored during the 2019/2020 lambing season. Colostrum samples (13.3±4.1 samples/flock), and serum samples were collected from lambs between 24-60 h after birth (20.1±6.5 lambs/flock), and at 21 d of age (17.1±7.3 lambs/flock). Also information to characterise farms between pre-birth and weaning was collected using questionnaires. Results on colostrum immunological quality, the prevalence of FPIT, and factors associated to an adequate PIT and maintenance of IgG immunity will be presented, and will serve to enhance management strategies to improve colostrum quality and PIT, to reduce perinatal mortality and to increase disease resistance.

Young EAAP: jobs of the future

C. Lambertz
Research Institute of Organic Agriculture (FiBL), Kasseler Strasse 1a, 60486 Frankfurt am Main, Germany; christian.lambertz@fibl.org

This debate on the pros and cons of staying in academia or changing to companies working in the livestock sector targets young scientists during their early stages of their scientific career independent whether at master's, PhD or Postdoctoral level. Whenever graduations approach, important decisions with impact on our future career have to be taken. For young scientists a variety of options arise with job opportunities becoming more and more diverse. Similarly, job requirements change constantly and may vary even more dramatically in the future, for example in view of developments such as digital farming. In this session, several speakers will highlight their career paths, including both changing from academia to the livestock industry, and the other way round. The first is for most of the young and early career scientists the usual way. However, also the way back to universities or research institutions might interest people who left academia for a certain time. Accordingly, livestock scientists with a variety of career pathways moving from research to livestock industries or vice versa will present us their motivations and discuss skills and expertise that are helpful and needed for successful job changes in the livestock sector.

Long term response to selection in the genomic era

P. Bijma
Wageningen University and Research, Animal Breeding and Genomics, Droevendaalsesteeg 1, 6708 PB Wageningen, the Netherlands; piter.bijma@wur.nl

The history of livestock genetic improvement demonstrates that response to traditional selection, such as mass or pedigree-BLUP selection, can continue for a very long time. Other than restricting the rate of increase in mean kinship (or 'inbreeding'), there is no trade-off between short and long term response with traditional selection. In the genomics era, however, there appear to be considerable trade-offs, which questions the sustainability of genomic selection. For example: (1) allele frequency change with GS is far from optimal. This happens because optimum genomic prediction methods strongly regress the effects of rare alleles and of alleles of small effect, which is precisely opposite to the optimum long-term selection strategy; (2) In genomic selection schemes, parents are often selected before own performance records are available, so that de novo mutations in the selection candidate are not targeted. (3) Even when own-performance records are included in the GEBV, genomic prediction puts extremely little emphasis on new mutations, particularly compared to mass selection. (4) Empirical results suggest that genomic selection generates much stronger selection signatures in the genome than, e.g. pedigree BLUP. Here I will identify and discuss several short vs long term trade-offs in the genomics era, and discuss potential solutions. This will include at least the following topics: In our pedigree-based thinking, additive genetic relatedness, which measures the similarity of breeding values of individuals, is closely related to genetic diversity and inbreeding. I will argue that this is no longer the case in GRM. In contrast to pedigree models, genomic models allow optimisation of response over multiple generations, aiming for example at bringing together favourable alleles. Such methods have a history in plant breeding, but have received little attention in livestock. I will discuss potential merits of such approaches for livestock. I will conclude by proposing a research agenda to address open questions.

Long-term selection of layers in the presence of dominance

I. Pocrnic[1], L.A. De C. Lara[1], C. Gaynor[1], J.M. Hickey[1], A. Wolc[2,3], J. Arango[2], D. Lubritz[2] and G. Gorjanc[1]
[1]The University of Edinburgh, The Roslin Institute, Easter Bush Campus, EH25 9RG Edinburgh, United Kingdom, [2]Hy-Line International, P.O. Box 310, Dallas Center, IA 50063, USA, [3]Iowa State University, Department of Animal Science, Kildee Hall, Ames, IA 50011, USA; ivan.pocrnic@roslin.ed.ac.uk

This contribution evaluates long-term genetic gain and variance in a layer breeding programme in the presence of dominance. Most nucleus breeding operations in animal breeding are based solely on additive genetic architecture, including the development of new breeding methods such as the optimal contribution selection (OCS). However, non-additive genetic effects such as dominance can have a significant effect on genetic gain. Here we explore the role of dominance in long-term selection of layers without and with optimisation. To this end, we have stochastically modelled a breeding program over 40 years in line with genetic parameters from the real data. We compared scenarios where the three correlated traits were influenced by either additive effects solely, or by additive and dominance effects. Each scenario included 20 initial years of traditional (BLUP) selection, followed by 20 years of either continuation of BLUP selection or switch to genomic (single-step GBLUP) selection. Furthermore, for each scenario, we compared truncation selection without and with minimum inbreeding mating and OCS. As expected, genomic selection delivered higher gain than the traditional selection, but also reduced more variance. Their efficiency of converting variance into gain was similar. In the long-term OCS maintained higher efficiency of converting variance into gain than the truncation or minimum inbreeding mating. We will further explore comparisons between additive and dominance scenarios to unravel the mechanisms that are driving the gain, how they influence the long-term variability, with the goal to provide general considerations for sustainable breeding of layers.

Evolution of genetic variance due to selection in Manech Tête Rousse dairy sheep

F.L. Macedo[1,2], O. Christensen[3] and A. Legarra[2]
[1]UdelaR, F. Veterinaria, A. Lasplaces 1620, 11200 Montevideo, Uruguay, [2]INRAE, France, GenPhySE, Castanet Tolosan, 31326, France, [3]Aarhus University, QGG, Blichers Allé 20, 8830 Tjele, Denmark; fernando.macedo@inrae.fr

Genetic variance for a trait under selection is reduced due to Bulmer effect. However, few studies describe empirically this evolution over time of genetic variance in real populations. In this work we estimated the genetic variance for milk yield of each yearly cohort of progeny-tested sires (σ_{sy}^2) in Manech Tête Rousse dairy sheep, a breed that has been under selection since the 80s for milk yield and since the 2000s for other traits as well. We work with 540,999 individuals and 1,842,295 records of milk yield. We used a BLUP animal model with repeated measures and unknown parent groups. Heterogeneity of variance was pre-corrected. Estimates of σ_{sy}^2 were obtained for the tested rams for each year 'y' from 1981 to 2014 following Sorensen *et al.* (2001) using Gibbs sampling. Briefly, σ_{sy}^2 were computed for each group of rams, obtaining in this way a posterior distribution of σ_{sy}^2 for each cohort of sires born in year 'y'. The genetic variance at the base population was 499.15±5.21, permanent environmental variance was 409.29±3.58, the residual variance was 857.25±1.15 and the h^2 was 0.28±0.0026. The genetic variance of the selected rams σ_{sy}^2 decreases (irregularly) since the beginning of the selection program in 1981 (σ_{sy}^2=515) reaching values of 288 and 282 in 2001 and 2008, respectively. In the last years (2009 to 2010) the σ_{sy}^2 increases up to values ranging between 360 and 380. On the other hand, the average relationship within each group of rams increases steadily up to 0.09 in 2014, but this only accounts for a small reduction of the genetic variance – most reduction is therefore due to Bulmer effect. In 2010 the milk composition was added to the selection criteria, possibly explaining the release of genetic variance and an increase in σ_{sy}^2 in the last period of the study. As a conclusion we confirm reduction in the genetic variance due to (mostly) Bulmer effect and (marginally) increased relationship in the population, and we also confirm that changing the selection objective releases genetic gain for the previously selected traits. Acknowledgments: ARDI (Poctefa program), European Union's Horizon 2020 (G. A. no. 772787).

Similarities between Mendelian sampling distributions and their possible application in breeding

A.A. Musa and N. Reinsch
Leibniz Institute for Farm Animal Biology, Institute for Genetics and Biometry, Wilhelm-Stahl-Allee 2, 18196 Dummerstorf, Germany; musa@fbn-dummerstorf.de

Mendelian sampling variance (MSV) is the genetic variability among progeny due to the inheritance of random samples of alleles from both parents. MSV quantifies the deviation of the realised breeding values of progeny from expected values. Consequently, it influences mate selection and allocation when small or large MSV is desired, depending on the user's specific goals. However, similarities among the MSV of individuals are required to make optimum decisions, which have not been derived in previous studies. In this study, we derived a special kind of similarity matrix S that defines the Mendelian segregation patterns of zygotes among mate-pairs and demonstrated its potential application in visualisation, mate allocation, and optimum mating decisions. We derived S using prior knowledge of parental diplotypes, recombination rates, and estimated marker effects. The diagonal elements of S give the MSV of zygotes from each potential mate-pair, while the off-diagonal elements compare the Mendelian sampling distributions among mate-pairs. We found that this matrix is useful not only for visualisation purposes but also for genetic interpretations because their values depend on linkage, heterozygosity, and trait-genetic architecture. Mate-pairs with common heterozygosity of markers with large effects for a trait in coupling appear more similar than those with small effects. To optimise mate selection and allocation, we applied principles of portfolio theory – known from asset management. We derived an efficient frontier using the probability to breed top-ranking individuals and S, which provided a diversifying effect. The efficient frontier provided a set of mate-pairs/matings that offer the maximum genetic gain with minimum risk (considered as MSV) while maintaining diversity by keeping different heterozygous chromosome regions active. Like genomic truncation selection, our approach for optimising mate selection and allocation uses genomic estimated breeding values to maximise genetic gain, but our approach also considers the MSV and determines the optimal set of mate-pairs that are required to produce the next generation. The matrix S is potentially important for visualising trait-specific haplotype diversity in a population and in genetic improvement or conservation programs.

The long-term effects of selection on the genetic architecture of traits

Y.C.J. Wientjes[1], P. Bijma[1], M.P.L. Calus[1], B.J. Zwaan[2], Z.G. Vitezica[3] and J. Van Den Heuvel[2]
[1]Wageningen University & Research, Animal Breeding and Genomics, P.O. Box 338, 6700 AH Wageningen, the Netherlands,
[2]Wageningen University & Research, Laboratory of Genetics, P.O. Box 338, 6700 AH Wageningen, the Netherlands,
[3]INRAE, GenPhySE, 24 Chemin Borde Rouge, 31326 Castanet-Tolosan, France; yvonne.wientjes@wur.nl

With non-additive gene action, the average effect of a causal allele depends on its allele frequency and the allele frequencies at interacting loci. Since drift and selection change allele frequencies, they also change the average effects of alleles. Genomic selection has likely accelerated this process, because it is more accurate and focusses more on genes of large effect. We aimed to quantify the changes in genetic architecture and genetic variation under genomic vs traditional selection. We simulated a population for 50 generations with selection for a single trait, controlled by either only additive effects, or additionally also dominance and epistatic effects. When additive, dominance and epistatic effects were simulated, 50% of the variation was epistatic at the functional level, although at the statistical level this was only 5%, with 62% of the variation being additive, and 33% dominance. Results show that genomic selection always outcompeted other selection strategies for short-term gain. For long-term gain, mass selection was close to genomic selection, and even outcompeted it when epistasis was present. Loss in statistical additive genetic variance was comparable across functional modes of gene action, and around 75% for genomic selection, slightly less for pedigree BLUP selection, and only 50% for mass selection. Surprisingly, dominance variance increased as a result of selection when non-additive effects were present. Over 50 generations of genomic selection, the genetic correlation between generation 0 and 50 was around 0.7 with only additive effects due to a difference in the subset of genes affecting the trait. This genetic correlation was 0.5 when dominance was present, and only 0.1 when also epistatic effects were present. The genetic correlation between generations decreased only slightly faster with genomic than with pedigree BLUP selection, but considerably slower with mass selection. We conclude that the presence of non-additive effects and method of selection have a large impact on the long-term effects of selection.

Genomic and temporal analysis of genetic covariation between traits

M. Spehar[1], C. Edel[2], R. Emmerling[2], K.U. Götz[2], J. Hickey[3], I. Curik[4] and G. Gorjanc[3,5]
[1]Ministry of Agriculture, Vukovarska 78, 10000 Zagreb, Croatia, [2]Bavarian State Research Centre for Agriculture, Prof.-Dürrwaechter-Platz 1, 85586 Poing-Grub, Germany, [3]University of Edinburgh, The Roslin Institute, Easter Bush Campus, Midlothian EH25 9RG, United Kingdom, [4]University of Zagreb, Faculty of Agriculture, Svetošimunska 25, 10000 Zagreb, Croatia, [5]University of Ljubljana, Biotehnical Faculty, Groblje 3, 1230 Domžale, Slovenia; marija.spehar@hapih.hr

Understanding genetic variation is a fundamental task in quantitative genetics. Traditionally we used the pedigree-based model to estimate genetic covariances at the pedigree base. We can estimate temporal changes in genetic covariances by summarising sampled realisations of genetic values from the fitted model. Here we extend this approach to a marker-based model for temporal and genomic analysis. We demonstrate the approach by analysing dairy (milk yield; MY and fat yield; FY) and beef traits (net daily gain; NG and carcass grading; CG) in ~9K progeny-tested dual purpose Fleckvieh bulls genotyped with ~50K markers. First, we fitted a multivariate marker-model and saved samples from posterior distribution of allele substitution effects. Second, we used these samples to calculate realisations of genetic covariation based on obtained samples from posterior distribution of breeding values, which were in turn summarised to obtain samples from posterior distribution of genetic covariances and correlations among traits encompassing the whole genome or genome regions. We then summarised these genetic values to analyse genomic and temporal trends in genetic covariation between the traits. Allele substitution effects were positively correlated among most of trait combinations (rMY:FY=0.65, rMY:NG=0.07, rFY:NG=0.03, and rNG:CG=0.52) with an exception of rMY:CG=-0.10 and rFY:CG=-0.12. Overall correlations within dairy and beef traits were high and positive (rMY:FY=0.81, rNG:CG=0.52), while they were low to moderate between these two groups of traits (rMY:NG=0.23, rMY:CG=-0.05, rFY:NG=0.18, and rFY:CG=-0.10). A detailed analysis of the trends shows that most of the changes are driven by the build-up of negative linkage-disequilibrium (the Bulmer effect) for individual traits as well as for pairs of traits, while allele frequencies and corresponding genic covariances changed very little. These observations are consistent with the directional selection process in the breeding programme and near infinitesimal genetic architecture of the analysed traits.

Signatures of selection and environmental adaptation across East Adriatic sheep breeds

B. Lukic[1], V. Brajkovic[2], I. Drzaic[2], M. Ferenčaković[2], V. Cubric-Curik[2] and I. Curik[2]
[1]Faculty of Agrobiotechnical Sciences Osijek, University of J.J. Strossmayer of Osijek, Department for Animal Production and Biotechnology, Vladimira Preloga 1, 31000, Croatia, [2]Faculty of Agriculture, University of Zagreb, Department of Animal Science, Svetošimunska 25, 10000, Croatia; blukic@fazos.hr

Sheep are one of the most important livestock species in Croatia, found predominantly in the coastal and mountain areas along the East Adriatic, with farming that is primarily extensive. In the present study, we have sampled 200 individuals belonging to eight sheep breeds (Istria sheep, Krk Island sheep, Cres Island sheep, Rab Island sheep, Lika sheep, Pag Island sheep, Dalmatian Pramenka, Dubrovnik Ruda sheep). East Adriatic sheep are the closest populations to a world-wide distinctive group of Balkan sheep populations, which has never been analysed for selection signals and environmental adaptation. We have genotyped all sampled animals with Ovine Infinium® HD SNP BeadChip (606,006 SNPs). After quality control, four different approaches that identify selection signatures were performed; a) identification of extremely frequent SNPs in runs of homozygosity (eROHi), c) Integrated Haplotype Score (iHS), c) across populations identification of SNPs with extremely high F_{ST} values (ehF$_{ST}$) and d) across populations Integrated Haplotype Score (Rsb) based on the ratio of extended haplotype homozygosities between populations. While the eROHi and IHS analyses were performed within each population, ehF$_{ST}$ and Rsb analyses were performed in contrast to internationally known sheep breeds. Subsequently, we have analysed and compared all significant SNPs (regions) with respect to breed and used methodologies, while potential candidate genes were identified and classified according to their biological function. This study contributes to the better understanding of the environmental adaptation and genomic architecture of the East Adriatic sheep populations.

Genetic diversity and signature of selection for temperature stress in Chinese cattle breeds and yak

P.H.F. Freitas[1], Q. Xu[2], Y. Wang[3], Y. Zhang[3], P. Yan[4], H.R. Oliveira[1,5], F.S. Schenkel[5] and L.F. Brito[1]
[1]Purdue University, Animal Science, 270 S. Russell St, 47907, West Lafayette, USA, [2]Beijing Jiaotong University, 8212 Zhixing Building, 100044, Beijing, China, P.R., [3]China Agricultural University, 17 Qinghua E Rd, 100083, Beijing, China, P.R., [4]Lanzhou Institute of Husbandry and Pharmaceutical Sciences of CAAS, 335 Jiangouyan, 730050, Gansu, China, P.R., [5]University of Guelph, 50 Stone Rd E, N1G 2W1, Ontario, Canada; pfreita@purdue.edu

Climate change has become a major concern around the world as it negatively impacts the sustainability of livestock production and, consequently, food safety in a rapidly growing population. Understanding the adaptation of different cattle breeds to local environments is of paramount importance for genetic improvement and conservation of genetic resources. The aim of this study was to investigate genetic diversity and unravel genomic regions potentially under selection for heat and cold tolerance in Chinese cattle breeds adapted to divergent climatic conditions and in domesticated yak (*Bos grunniens*). A total of 37,358 SNP were included in this study. Signatures of selection were identified by grouping the cattle populations in divergent clusters (defined as heat tolerant, n=109; and cold tolerant, n=203), and yak (adapted to cold temperatures and high altitudes, n=45). In general, moderate to high levels of genetic diversity were observed in all populations. The proportion of polymorphic SNP ranged from 0.197 (yak) to 0.992 (Mongolian cattle). The H$_O$ and H$_E$ ranged from 0.023 (yak) to 0.366 (Sanhe cattle), and from 0.021 (yak) to 0.358 (Sanhe cattle), respectively. The inbreeding coefficients estimated based on the excess of homozygosity and runs of homozygosity ranged from -0.291 (Sanhe) to 0.937 (yak), and from 0.000 (Yunnan Humped Hazake) to 0.087 (yak), respectively. Signatures of selection analysis (F_{ST} and HapFLK) revealed important genomic regions on BTA8, BTA10, BTA14 and BTA15. Interesting positional candidate genes were found in these regions, which were related to the expression of heat shock responsive genes (*EEF1D*) and DNA repair and maintenance (*TONSL*). These novel findings provided essential candidate genes to explore genetic mechanism related to thermal tolerance and created new angle in the conservation of local genetic resources.

Selection signatures in sheep breeds of Middle East and European origin

S. Eydivandi[1,2], M. Amiri Roudbar[3], S. Salek Ardestani[4] and G. Sahana[2]
[1]*Islamic Azad University, Department of Animal Science, Behbahan Branch, Islamic Azad University, Behbahan, 13198-63617, Iran,* [2]*Aarhus University, Center for Quantitative Genetics and Genomics, Blichers Allé 20 building G20 8830 Tjele Denmark, 8830, Denmark,* [3]*Agricultural Research, Education & Extension Organization (AREEO, Department of Animal Science, Safiabad-Dezful Agricultural and Natural Resources Research and Educat, 333, Iran,* [4]*Dalhousie University, Department of Animal Science and Aquaculture, Dalhousie University, Truro, NS, B2N 5E3, Canada Depar, B2N 5E3, Canada; sirous.eidivandi@mbg.au.dk*

Both natural and artificial selection leave patterns on the genome during the domestication of animals and lead to changes in allele frequencies between populations. Therefore, detecting genomic regions expose to selection in livestock may assist to better understand the involved processes in genome evolution and discovery of genomic regions that can be related to economic traits and ecological interests. Genetic diversity analyses were conducted in fifteen sheep breeds including six indigenous breeds from the Middle East: Iranian Balouchi, Afshari, Moghani, Qezel, Karakas, Norduz, and nine modern commercial breeds from Europe, namely East Friesian Sheep, Ile De France, Moureous, Romane, Swiss Mirror, Spael-White, Suffolk, Comisana, and Engadine Red Sheep. We used genome-wide data generated by the Illumina OvineSNP50 Genotyping BeadChip array. We applied two complementary statistical analyses; FST (fixation index) and XP-EHH (cross-population extended haplotype homozygosity) to detect selection signatures in Afshari and East Friesian Sheep. The Z(FST) and XP-EHH detected 495 and 226 genomic regions indicating signatures of selection, respectively. Using the XP-EHH approach, we found the gene CAMKMT which was reported earlier to influence a number of economic traits. Both Z(FST) and XP-EHH approaches identified 11 shared genomic regions as signatures of selection including eight candidate genes that they were enriched for 33 significant Gene Ontology (GO) terms. Several of the GO terms were involved in forebrain development, male gonad development, heart development, and aorta development.

Danish dairy breeds are economically similar

M. Kargo[1,2], L. Hein[1], A. Fogh[1], J.B. Clasen[3] and R.B. Davis[1,4]
[1]*SEGES, Cattle, Agro Food Park 15, 8200, Denmark,* [2]*Aarhus University, Department of Molecular Biology and Genetics, Blickers alle 20, 8830, Denmark,* [3]*Swedish university of Agricultural Sciences, Animal Breeding and Genetics, Ultuna, 75007, Sweden,* [4]*Simherd A/S, Niels Pedersensves Alle 2, 8830, Denmark; morten.kargo@mbg.au.dk*

Systematic crossbreeding is gaining growing interest in the dairy sector. But in order to develop superior systematic crossbreeding programs it is a prerequisite that the purebred breeds that are used are economically similar. Different breeds have different strengths and weaknesses, it is therefore of importance to know the genetic level for different traits as well as for total economic performance in a given production system, to evaluate if breeds are economically similar. To evaluate the potential of crossbreeding in Denmark, we have therefore investigated the economic performance of Holstein, Red Dairy Cattle (RDC) and Jersey under Danish production circumstances. The analysis was conducted on phenotypes from purebred herds with either Holstein, RDC or Jersey cows. The phenotypic levels were corrected for regional differences within Denmark. For each herd, the average production results for yield, fertility, disease (udder-, claw- and general health), calving, conformation and longevity were calculated. The economic results for these differences were estimated using economic values from the 2018 revision of the Nordic Total Merit index. The total economic result showed that on average, there was no significant difference between the three purebred breeds. For cows born in 2011 and 2012, RDC and Jersey cows were 6 and 3 € inferior to Holstein cows respectively in conventional herds. Based on phenotypes from organic herds and using the same economic values, the figures were 2 (RDC) and 17 (Jersey) € less than Holstein. However, using organic economic values, the economic result shifted to RDC being 19 € and Jersey 16 € superior to Holstein. Regarding individual traits, Holstein were superior with regards to milk production and conformation compared to RDC and Jersey, but RDC and Jersey cows had fewer diseases, easier calvings and superior fertility. In conclusion, this study found that the three main Danish dairy breeds are economically similar and are therefore suitable in a systematic crossbreeding program.

Genetic parameters and genetic trends for female fertility in Icelandic dairy cattle

T. Thorarinsdottir[1], E. Albertsdóttir[2] and S. Eriksson[3]
[1]*Agricultural University of Iceland, Hvanneyri, 311, Iceland,* [2]*The Icelandic Agricultural Advisory Centre, Hagatorgi 1, 107 Reykjavik, Iceland,* [3]*Swedish University of Agricultural Sciences, Animal Breeding and Genetics, P.O. Box 7023, 75007, Uppsala, Sweden; tordis95@gmail.com*

Fertility, represented by calving interval, has been included in the genetic evaluation of Icelandic cattle since 1993. In spite of this an unfavourable genetic trend is seen and, recent implementation of test-day models for the genetic evaluation of milk production has changed the premise of using calving interval to represent fertility. The aim of this study was to estimate genetic parameters and genetic trends of different female fertility traits in Icelandic dairy cattle and suggest new traits for the genetic evaluation. Insemination records for the first three lactations and the heifer period of 52,951 Icelandic cows were used to analyse the traits: conception rate at first insemination (CR), number of inseminations per service period (AIS), interval first to last insemination(IFL), interval calving to first insemination (ICF), interval calving to last insemination (ICL) and calving interval (CI). Correlations between fertility and production traits were also estimated. Five different linear animal models were used to estimate (co) variance components. Breeding values were compared, and genetic trends were investigated. Estimated heritabilities for the fertility traits were low, ranging from 0.01 (IFL) to 0.08 (ICF). Genetic correlations between heifer and cow performances within traits ranged from 0.23 to 0.81. Between lactations within traits, the genetic correlations ranged from 0.36 to 1.00. Genetic correlations between different heifer traits were strong, and between different cow fertility traits they ranged from weak (-0.17) to very strong (0.97). Genetic correlations between the fertility and production traits were generally unfavourable, but more pronouncedly so for AIS, IFL, ICL and CI than for ICF and CR. Estimated genetic trends seemed unfavourable for the traits AIS, IFL and CI and favourable for ICF and ICL. In a revised genetic evaluation for fertility in Icelandic cattle, the traits ICF and IFL should be included to represent cow fertility, and CR to represent heifer fertility.

Single step GWAS study of classical and endocrine fertility traits in Swedish Red and Holstein cows

G.M. Tarekegn[1,2], E. Strandberg[2], R. Båge[3,4], P. Gullstrand[2], E. Rius-Vilarrasa[3], S. Andonov[2], J.M. Christensen[5] and B. Berglund[2]
[1]*Bahir Dar University, Department of Animal Production and Technology, Bahir Dar, 79, Ethiopia,* [2]*Swedish University of Agricultural Sciences, Department of Animal Breeding and Genetics, Ulls Väg 26, Uppsala, 75007, Sweden,* [3]*Växa Sverige, Ulls Väg 26, Uppsala, 75007, Sweden,* [4]*Swedish University of Agricultural Sciences, Department of Clinical Sciences, Ulls Väg 26, Uppsala, 75007, Sweden,* [5]*Lattec, Slangerupsgade 69, 3400 Hillerod, Denmark; britt.berglund@slu.se*

The study aimed to identify candidate genomic regions associated with classical and endocrine fertility traits in Swedish Red (SR) and Holstein cows. A total of 3,955 lactations from 1,164 SR and 1,672 Holstein cows were included in the study. Automatically collected and analysed in-line milk progesterone (P4) samples from 14 herds with the DeLaval Herd Navigator™ were used. Endocrine traits studied were days from calving to commencement of luteal activity (C-LA), first luteal phase length (LPL), length of inter-luteal interval, length of first inter-ovulatory interval (IOI), luteal activity during the first 60 DIM, and proportion of samples with luteal activity during the first 60 DIM. In addition, classical fertility traits were investigated. A total of 180 SR and 312 Holstein cows were genotyped with a low density SNP chip panel and imputed to 50K SNP chip. We employed a single step GWAS approach to explore candidate genomic regions associated with the fertility traits. The result revealed 990 and 415 SNPs above the threshold for SR and Holstein cows, respectively. The annotation analysis resulted in 281 SNPs located in 241 genes. The functional enrichment analysis, in DAVID tools, further reduced the genes to 80 that were mediated in various biological processes and KEGG pathways in a wide range of functions, which include folliculogenesis, embryogenesis, uterine growth and development, immune response and ovarian cysts. Overall, 67 of these associations were detected in SR cows and 13 in Holsteins. Most of the genes were associated with LPL and IOI in SR cows, but in Holstein no association with any of the endocrine traits were found, except for C-LA. For the classical traits, some genes were found to be associated with calving interval. SR cows are known to have better fertility than Holsteins. However, a challenge for future research will be to understand why different regions seem relevant for different traits and breeds and the practical implications of this for genomic selection.

Genotype by environment interaction in South African Holsteins under two production systems

M. Van Niekerk[1,2], F.W.C. Neser[2], J.B. Van Wyk[2] and V. Ducrocq[1,2]
[1]Université Paris-Saclay, INRAE, AgroParisTech, UMR GABI, Jouy-en-Josas, 78350, France, [2]University of the Free State, Department of Animal, Wildlife and Grassland Sciences, 205 Nelson Mandela Drive, Bloemfontein 9300, South Africa; michiel@ganna.co.za

Varying long-term, average annual rainfall between regions in South Africa (SA) lead to dairy herds choosing to mostly utilise either a total mixed ration (TMR) or pasture based (PAST) production system (PS), depending on the rainfall area the herds are situated in. Accurate selection of sires to produce daughters that perform optimally within their applicable PS is of importance in the presence of genotype by environment interaction (G×E). Evidence of such a G×E effect in SA Holsteins (SAHST) were shown in a previous study where a significant difference was found between genetic parameters for TMR and PAST PS. An alternative random regression model (aRRM) that accommodate random effects to continuously change over the lactation, has been proposed for genetic evaluation of SAHST. This allows for the calculation of an additive genetic persistency (AGP) effect over the lactation, in addition to the additive genetic average production effect (AGA). The aim of the study was to use the aRRM to assess the magnitude of the G×E effect. The dataset consisted of first lactation, test-day records from herds utilising either a PAST or TMR PS. Milk (MP), butterfat (BFP) and protein (PRP) production were each considered as two separate traits according to the production system from where the records originated (PAST or TMR). Three separate, multivariate analyses (MP, BFP and PRP) were carried out to attain estimates using REML. Heritability estimates for MP in the PAST PS (0.29) were significantly higher than for the TMR PS (0.18), especially during early lactation. Within lactation genetic correlations for MP between AGA and AGP were negative and stronger for the TMR PS (-0.36) than for the PAST PS (0.03). BFP and PRP followed the same tendencies in these regards. Genetic correlations between the AGA effects for MP, BFP and PRP were positive and very strong (0.93, 0.98 and 0.92, respectively) whilst being moderate for the AGP effects (0.56, 0.65 and 0.52, respectively). Re-ranking of sires between the PAST and TMR PS are likely to take place because of the AGP and not the AGA effect.

Estimation of genetic parameters and GWAS for semen production traits in Belgian Blue cattle

J.L. Gualdron[1], M. Solé[1], J. Zhang[1], A.S. Gori[2], X. Hubin[2], E. Henrotte[2] and T. Druet[1]
[1]GIGA-R, University of Liège, Unit of Animal Genomics, 11 Avenue de l'Hôpital (B34), 4000, Belgium, [2]Association Wallonne de l'Elevage, Ciney, 5590, Belgium; jlgualdron@uliege.be

In the Belgian Blue cattle breed, artificial insemination (AI) accounts for approximately 50% of mating and has an important role in dissemination of elite sires. The breed is also frequently used for cross-breeding with dairy breeds. Semen production is therefore essential, both in terms of quantity and quality. Several related traits are routinely measured in AI stations, allowing the implementation of a genetic evaluation. We herein estimated the genetic parameters for semen production traits measured on fresh semen including semen production, concentration, motility score and % viable. We subsequently performed genome-wide association studies (GWAS) with these traits. To that end we relied on 98,196 records available for 770 bulls, including 546 individuals genotyped with the Illumina Bovine 50K or the BovineHD genotyping array. We selected only records of first ejaculates and from bulls 1 to 8 years old. Genetic parameters estimated with univariate or multivariate models indicated that although repeatabilities were high (ranging from 0.37 to 0.55), the heritabilities were relatively modest (ranging from 0.07 for semen production to 0.22 for sperm concentration). Several traits presented strong genetic correlations such as motility score with %viable (0.98) or semen concentration with all traits (-0.62 with production, 0.71 with motility and 0.64 with % viable), suggesting that a multiple-trait evaluation model might increase prediction accuracies. GWAS were performed using an haplotype-based approach that resulted in the absence of significant association with the four traits. Multiple-trait QTL mapping approaches will be applied to increase the power of our study. In addition, we are now studying the contribution of the X-chromosome to the genetic variance. We will also test whether there is evidence for global or regional (locus-specific) inbreeding depression associated with these traits. Finally, additional traits measured on frozen semen will also be explored.

Inbreeding and pedigree analysis of the European red dairy Cattle

S. Nyman[1], A.M. Johansson[1], V. Palucci[2], A. Schönherz[3], B. Guldbrandtsen[4], A. Roozen[2], D. Hinrich[5] and D.J. De Koning[1]
[1]Swedish University of Agricultural Sciences, Department of Animal Breeding and Genetics, Box 7023, 75007 Uppsala, Sweden, [2]Interbull Centre, Department of Animal Breeding and Genetics, Swedish University of Agricultural Sciences, Box 7023, 75007 Uppsala, Sweden, [3]Aarhus University, Department of Animal Science, Immunology and microbiology, Blichers Allé 20, 8830 Tjele, Denmark, [4]University of Bonn, Department of Animal Sciences, Endenicher Allee 15, 53115 Bonn, Germany, [5]University of Kassel, Department of Animal Breeding, 37213, Witzenhausen, Germany; sofia.nyman@slu.se

Pedigrees and inbreeding levels of the European Red Dairy Breed were analysed. The depth of known pedigree, average generation intervals, effective population size, the effective number of founders and ancestors were determined. The animals most influential within the current population were identified. Indicators were based on pedigree data including 9,073,403 animals born between early 1900 to 2018 from Denmark, Finland, Germany, Latvia, Lithuania, the Netherlands, Norway, Poland and Sweden. Thirty-two breeds defined according to Interbull standards, were included in the pedigree. The most represented breed was the Red Dairy Cattle (RDC) breed while the second largest one has been identified with the Dutch breed Meuse Rhine Yssel (MRY). The deepest complete generation equivalent was 9.39 found for RDC in 2017. Mean pedigree completeness ranged between 0.6 for Finnish Cattle (FIC) and 7.51 for RDC. Effective population size for the whole pedigree was estimated to 166 animals, and varied between 35 (Rotes Höhenvieh; RHV) and 226 (RDC). All breeds with more than 500 animals displayed a generation interval between 5 and 7 years except from FIC with a generation interval of 12 years. The mean inbreeding coefficient for all animals in the pedigree was 1.5%. The highest mean inbreeding coefficients were observed for RHV and RDC, 2.5 and 1.6% respectively, while the smallest was observed in FIC (0.1%). The top 20 animals born between 1960 and 2018 explained 29% of the total marginal contribution. The most influential animal was a Dutch MRY bull born in 1960. There is a need to adopt strategies aimed at minimising inbreeding to avoid losses on genetic diversity. Especially for smaller breeds, such as e.g. RHV, with high inbreeding levels and low effective populations sizes.

Life cycle assessment of cattle production in the South West of Spain

C. Reyes-Palomo[1], E. Aguilera[1,2], M. Llorente[3], C. Díaz-Gaona[1], G. Moreno[3] and V. Rodriguez-Estévez[1]
[1]Universidad de Córdoba, Cátedra Ganaderia Ecologica, Campus Universitario de Rabanales Ctra. Madrid-Cádiz km 396, 14071 Cordoba, Spain, [2]Universidad Politécnica de Madrid, CEIGRAM, P° Juan XXIII, 11, 28040 Madrid, Spain, [3]Universidad de Extremadura, INDEHESA, Calle Virgen del Puerto 2, 10600, Spain; ganaderiaecologica@uco.es

In the south West of Spain, the cattle production is linked with extensive farms in an agroforestry system, known as dehesa, in which cows and calves are in pastures until weaning. Then calves usually are transported to fattening facilities. This study has been conducted collecting data from 15 dehesa farms; some of these with the fattening period of calves in the same farm. However, the most common system was sale of calves at weaning (6 months old and 230 kg of live weight). Some data have been collected from fattening farms to calculate the environmental impact until the slaughtering moment. Life Cycle Assessment has been carried out to perform this environmental analysis. The enteric fermentation and manure handling emissions has been calculated using Tier 2 equations on IPPC guidelines and Spanish National GHG Inventories. The not differentiable emissions have been assessed by economic approach and environmental impacts have been calculated with Ecoinvent 3, Agri-footprint and ELCD databases. The limit of the system has been 'cradle to gate'. The GHG emissions obtained for calves raised in the dehesa system have been 31.96 kg CO_2eq/kg of live weight at slaughtering; where supplementary feed has been the main cause of GHG emissions. These results are consistent with the data shown in Weiss and Leip for beef production in Spain, with 30 kg CO_2eq/kg beef. However, the result of the current study is slightly higher in comparison with Eldesouky et al.; who showed 17.74 kg CO_2eq/kg of live weight for calves produced in a single dehesa farm. In conclusion, this is the first study dealing with a representative sample of dehesa cattle farms in a Mediterranean region. Besides, it is expected an important carbon sequestration linked to this grassland system with Quercus trees; hence, this carbon sequestration should be added in a future research to have a higher accuracy in the carbon footprint calculation.

Effects of milk PTA and supplementary CP on milk quality in late lactation grazing dairy cows

M.J. Doran[1], F.J. Mulligan[2], M.B. Lynch[1], A.G. Fahey[1], M. O'Sullivan[3], H. Ryan[3] and K.M. Pierce[1]
[1]*University College Dublin Lyons Farm, School of Agriculture and Food Science, University College Dublin Lyons Farm, Celbridge, Co. Kildare, W23 ENY2, Ireland,* [2]*University College Dublin Lyons Farm, School of Veterinary Medicine, University College Dublin Lyons Farm, Celbridge, Co. Kildare, W23 ENY2, Ireland,* [3]*Institute of Food and Health, School of Agriculture and Food Science, University College Dublin, Belfield, Dublin 4, D04 V1W8, Ireland; michael.doran@ucdconnect.ie*

Irish dairy production is seasonal and grass based. Supplement is offered when grass growth and quality are not sufficient to meet herd demand. Late-lactation presents problems with milk quality (MQ) and processability (MP), and studies show MQ and MP can be altered through nutrition. Genetics may play a role but this is not well defined. The aim of this study was to investigate effects of 305-d milk predicted transmitting ability (PTA), supplementary crude protein (CP) concentration, and their potential interaction on selected MQ (composition), and MP (ethanol stability (ES) and rennet coagulation time (RCT)) parameters in late lactation spring-calving grazing dairy cows. A factorial experiment with two milk PTA groups (low milk (LM) (-60 kg) versus high milk (HM) (+130 kg)), and two supplementary CP levels (12% versus 18% with equal PDI supply) offered at 3.5 kg DM/cow/day, was conducted over 53 days. Forty-eight Holstein Friesian (24 LM and 24 HM PTA) dairy cows, blocked on parity and balanced on DIM, BCS, and Economic Breeding Index, were assigned to one of two feeding strategies resulting in four treatments (n=12): 1) LM PTA + 12% CP (L12); 2) LM PTA + 18% CP (L18); 3) HM PTA + 12% CP (H12) and 4) HM PTA + 18% CP (H18). Cows grazed full time and were offered 15 kg DM grass/cow/d. Repeated measures ANOVA (MIXED procedure) on SAS 9.4 was used to analyse MQ and MP parameters. Differences between means were determined by F-tests using Type III sums of squares. Milk yield was greater in H18 than H12 (+2.58 kg; P=0.049). Fat, protein, casein, and lactose concentration did not differ between treatments. HM PTA and 18% CP reduced fat (-0.16%; P=0.04 and -0.14%; P=0.04 respectively). The 18% CP increased milk urea (+0.0035 g/100 g milk; P=0.04). RCT and ES did not differ between treatments. In conclusion, milk PTA and CP% did not affect the MP studied, suggesting that offering low supplementary CP in late lactation does not impact on milk processability.

Heritability coefficients of the main fatty acids of bovine milk

T. Sakowski, P. Brodowska and E. Bagnicka
Instytut Genetyki i Hodowli Zwierząt PAN w Jastrzębcu, Department of Animal Improvent and Biotechnology, ul. Postępu 36A, Jastrzębiec, 05-552, Poland; t.sakowski@ighz.pl

The aim of the study was to estimate the heritability coefficients of the main fatty acids contained in bovine milk. The research was conducted 312 black and white PHF cows of the black-and-white variety maintained in four herds (n=109, 64, 75, and 52 animals). The average milk yield was almost 10,000 l per lactation with 4.2% of fat and 3.5% of protein contents. Phenotypic data on daily milk yield and composition (8,090 records) and pedigree information on 7,036 individuals, including 2,047 sires and 4,634 dams, were obtained from the official milking control system. The analysis using DMU4 package with AI-REML method was conducted with one-trait, mixed animal model with herd-year-season of calving and parity as fixed effects, and with animal, herd-year-month of milking, and error as random effects. Legendre polynomials up to 5th power were also considered. The phenotypic data covered, in addition to basic information on the milk composition, information on fatty acid profile (C14:0, C16:0, C18:0, C18:1, MUFA, PUFA, SFA, TFA, and TUFA contents). The highest contents were stated for SFA (2.55%), and TUFA (1.11%). The average contents of MUFA, PUFA, C18:1, C14:0, C16:0 C18:0 in bovine milk were 1.49, 1.54, 0.06, 0.45, 1.13 and 0.40%, respectively. However, the coefficients of variation (CV) were rather high and ranged between 20-40%. The estimated heritability coefficients were as follow: C14:0 – 0.33, C16:0 – 0.36, C18:0 – 0.20, C18:1 – 0.08, MUFA – 0.07, PUFA – 0.15, SFA – 0.38, TUFA – 0.10. Summing up h^2 of SFA and PUFA were estimated at an average level, therefore an effective selection towards the reduction of the content of saturated fatty acids and elevation of unsaturated fatty acids in milk is possible. Funding: This work was supported by the National Center for Research and Development [grant number: SUSAN/I/SusCatt/01/2017] as part of the European research program ERA-NET CO-FUND SUSAN.

Heritability of pregnancy losses estimated from in-line milk progesterone profiles in Swedish dairy

P. Ask-Gullstrand[1], E. Strandberg[1], R. Båge[2], J.M. Christensen[3] and B. Berglund[1]
[1]Swedish University of Agricultural Sciences, Department of Animal Breeding and Genetics, Box 7023, 750 07 Uppsala, Sweden, [2]Swedish University of Agricultural Sciences, Department of Clinical Sciences, Box 7054, 750 07 Uppsala, Sweden, [3]Lattec, Blytækkervej 10, 3400 Hillerød, Denmark; patricia.gullstrand@slu.se

The aim of this study was to assess the extent of, and genetic parameters for, pregnancy losses in dairy cattle using in-line milk progesterone records collected by DeLaval's Herd Navigator™ in 14 Swedish herds. A total of 337,693 progesterone samples were connected to 10,385 inseminations (AI) from 5,315 lactations of 1,468 Swedish Red (SR) and 1,876 Holstein (H) cows. Pregnancy loss traits were defined as early embryonic loss (1-24 days after AI), late embryonic loss (25-41 days after AI), foetal loss (42 days after AI to calving), and total pregnancy loss (from day 1 after AI to calving). The following classical fertility traits were also analysed: interval from calving to first service, interval from calving to last service, interval between first and last service (FLS), and calving interval. Fixed effects included breed, parity (1, 2, ≥3) and oestrus cycle number when the AI took place. Herd by year and season of AI, cow, and permanent environmental effect were considered as random effects. Milk yield from 305d lactation was included as a fixed regression. Least square means with standard error, and heritabilities were estimated in a mixed linear model. Extensive early embryonic loss was found (approx. 50%), however, there was no difference (P>0.05) between breeds. SR was significantly superior (P<0.0001) to H in the remaining pregnancy loss traits, with late embryonic loss at 6.9±0.9% and 14.5±0.9%, foetal loss at 7.7±0.9% and 12.3±0.9%, and total pregnancy loss at 57.3±1.1% and 63.5±1.1%, respectively. SR also performed significantly better than H in all traditional fertility traits, except for FLS. Heritabilities were estimated at 0.03, 0.05, and 0.01, for early and late embryonic loss, and foetal loss, respectively. In conclusion, SR cows had significantly lower pregnancy loss during late embryonic stage, foetal stage and in total, and had better fertility than H cows. The heritability estimates for pregnancy loss traits were of the same order of magnitude as for classical fertility traits.

Feeding high-grain diets to cattle: effect on further grazing intake, rumen kinetics and digestion

M.V. Burjel, A. Simeone, O. Bentancur, N. Zabalveytia and V. Beretta
Facultad de Agronomía, Universidad de la República, Ruta 3 km 363, 60000, Paysandú, Uruguay; victoriaburjel@gmail.com

Lot-feeding high grain diet to growing cattle during winter is recommended as strategy to overcome low pasture growth during this season. However, depending on roughage level in the ration, this could affect forage utilisation when returning to grazing in spring. This study aimed to evaluate the effect of winter feedlot ration roughage level (RL) on subsequent spring pasture dry matter intake (DMI), the *in vivo* DM digestibility (DMD), ruminal ammonia-nitrogen concentration (N-NH3), ruminal pH and the *in situ* degradability (DEG) of forage dry matter (DM). Four rumen-cannulated steers were random allocated to two winter dietary treatments (T) differing in ration RL (*Setaria italica*): 0% or 35% DM. Rations were formulated for equal protein supply (14% DM) and fed at 2.5 kg DM/100 kg LW for 62 days. In spring (73 days) both T strip-grazed a mixed grass-legume pastures (3,533±1000 kg DM/ha; 6 kg DM/100 kg LW). Measurements were performed during weeks (W) 1, 3 and 5 after changing diets from feedlot to grazing. DMD was estimated using the acid insoluble ash content as an internal marker, and DMI as the difference between pre and post-grazing herbage mass. On day 1 of each W, hand-clipped forage samples were incubated in the rumen for 0, 6, 24 and 48 hours. Following, on day 3, rumen fluid samples were taken at 0, 3, 6 and 9 hours after entering the new daily strip and analysed for pH and N-NH3 concentration. Data were analysed according to a complete randomised design with repeated measures. General model included T, W and T×W effects; for pH, N-NH3 and DEG the effects of sampling time and its interactions were also considered. Tukey test was used to compare means. None of the evaluated variables were affected by T or T×W (P> 0.05). Spring pasture DMI was significantly lower in W1 (P<0.01) compared to W3 and W5. Both T showed higher (P<0.001) DEG in W5 (50%) than in W1 and W3 (36 and 39% respectively). Results evidence a transition period when moving cattle from a high-grain diet to grazing, during which the animal adjusts pasture DMI and rumen DEG.

Estimation of genetic parameters for reproduction traits in South African Brahman cattle

E.D. Cason, F.W.C. Neser, P.D. Vermeulen and J.B. Van Wyk
University of the Free State, Animal Sciences, 205 Nelson Mandela Road, Park West, 9301, Bloemfontein, South Africa;
casoned@ufs.ac.za

In any cattle production system, reproductive performance is one of the main economic aspects that influence profitability. Therefore, accurate genetic parameter estimates are essential to obtain accurate breeding values in order to maximise the response to selection. The objective of the present study was to obtain genetic parameters for age at first calving (AFC) and calving interval (CI) for Brahman cattle from South Africa. In this breed, selection has focused mostly on production traits which can result in reduced reproductive performance. Data on South African Brahman reproduction performance was collected by the Brahman Cattle Breeders Society of South Africa and after editing consists of 51,263 records covering a period from 1953 (the introduction of the breed in South Africa) to 2019. A combination of multiple trait and repeatability models was used to estimate genetic parameters which in turn was used estimate breeding values (EBVs) for each of the reproductive traits. Preliminary results indicate that the heritability estimates are well within the results obtain within the literature. However, very little genetic progress were made in the two traits tested. The low heritability for CI indicate that good practical management will be essential for improving fertility efficiency. The study shows that greater emphasis should be placed on reproductive traits in the selection program, as a decrease in age at first calving and in calving intervals would improve productivity.

Estimation of breeding values for milk yield in Iranian Holstein using random regression models

Y. Salimiyekta[1], J. Jensen[2] and N. Emam Jomeh Kashan[1]
[1]Azad University, Science and Research Branch, Animal Science, Shahid Hesarak Boulevard, Daneshgah Square, 1477893855, Tehran, Iran, [2]Aarhus University, Molecular Biology and Genetics, Blichers Allé 20, Foulum, 8830 Tjele, Denmark; yasamin.sa@mbg.au.dk

The objective of this study was the estimation of genetic parameters for milk yield and use of this to develop genetic parameters for lactation persistency in Iranian Holstein dairy cows. Data consisted of 1,449,910 records from 144,991 cows in first lactation between 1993 and 2017. (Co) variance components were estimated using random regression test day model and the Average Information RML method. Legendre polynomial functions of order 3 were chosen to fit the additive genetic and permanent environmental effects and homogenous residual variance was assumed. Three different persistency measures were defined; P1: the sum of estimated breeding value from day 61 to day 280 as a deviation from the estimated breeding value of day 60. P2: the sum of estimated breeding value from day 30 to day 280 as a deviation from the sum of estimated breeding value from day 5 to day 30 and P3: the average of estimated breeding values for test day milk yields from day 255 to day 305 as a deviation from the average of estimated breeding values from day 50 to day 70. The heritability of milk yield_(305) was estimated to be 0.36. The heritability of P1, P2 and P3 were estimated to be 0.12, 0.13 and 0.18 and genetic correlations between P1, P2 and P3 and milk yield_(305) were 0.78, 0.78 and 0.10, respectively. Response to selection for P1, P2 and P3 was estimated to be 23.60, 25.27 and 58.90. Besides, correlated response to selection for different criterions and milk yield_(305) was estimated to be 31.24, 32.54 and 8.25, respectively. These results show that although P3 has higher heritability and lower genetic correlation with milk yield_(305) it does not have a high correlated response to selection and can not improve milk yield_(305) in next generation. However, P2 that has higher correlated response to selection with milk yield_(305) is a better criterion for lactation persistency in dairy cattle.

Effect of age, weight and score condition in the ovarian cyclicity of native heifers

R.M.L.N. Pereira[1,2], R. Romão[3], J. Pimenta[1,2], C.C. Marques[2], E. Bettencourt[3], L. Capela[1,2], M.C. D'abreu[3], J. Pais[4], P. Espadinha[5], N. Carolino[1,2] and C. Bovmais[2,3,4,5]
[1]CIISA, Universidade de Lisboa, Lisboa, Portugal, [2]Instituto Nacional de Investigação Agrária e Veterinária, UEISBRG, 2000 Vale de Santarém, Portugal, [3]Universidade de Évora, Polo Mitra, Évora, Portugal, [4]Associação de Criadores de Bovinos Mertolengos, ACBM, Évora, Portugal, [5]Associação de Criadores de Bovinos de Raça Alentejana, ACBRA, Assumar, Portugal; nuno.carolino@iniav.pt

Age at onset of puberty is a major determinant of lifetime reproductive efficiency and productivity in beef cows. The objective of the present work was to identify ovarian cyclicity and development of the reproductive system and their relationship with age, weight and score condition in Mertolenga (MERT) and Alentejana (AL) heifers raised extensively. This work was carried out in four herds in the Alentejo (2017-2019). The MERT (n=50) and Al (n=43) heifers were evaluated between the ages of 308 and 613 days. To identify the beginning of regular ovarian activity, gynaecological examinations of the uterus and ovaries were monthly performed by transrectal palpation and ultrasound. Heifer's body condition (scale 1-5) and weight were also assessed. Data were analysed by the PROC FREQ and PROC LOGISTICA. AL and MERT heifers had a mean condition score of 3.40±0.71 and 3.77±0.70, weighing 362.13±70.24 and 290.1±30.14 kg, respectively. The mean dimensions of the ovaries were: left (AL=22.88±6.87 and MERT=21.86±6.49 mm) and right (AL=23.30±8.50 and MERT=22.42±7.64 mm) sides, presenting more developing follicles (right, AL=9.69±4.12 and MERT=8.47±5.00 mm; left, AL=8.76±4.66 and MERT=10.5±6.15 mm) than mature corpus luteum (right, 22.29±3.95 and left, 16.6±6.23 mm). In both breeds, the uterus was classified more often as medium size (AL=84 and MERT=76%, P>0.05) and the probability of the onset of ovarian cyclicity depends on age and body condition (P<0.001). The age for the onset of ovarian cyclicity had a greater variability in the AL breed. The present study identified large differences in the onset of ovarian cyclicity among herds, breeds and body condition of heifers from the Portuguese native breeds, opening new windows to a future implementation of measures to improve their productive efficiency and longevity. Funded by PDR2020-101-3112, UID/CVT/276/2019

On-farm mortality of Mertolenga progeny in Alentejo beef cattle farms

R. Santos[1,2], L.A. Conceição[1,2], M.C. Caetano[3], M. Minas[2], L. Pereira[2], M.G. Carvalho[2] and L. Costa[2]
[1]VALORIZA – Research Center for Endogenous Resources Valorization, Research Centre for Endogenous Resource Valorization, Polytechnic Institute of Portalegre, Portugal, [2]Polytechnic Institute of Portalegre, Av. 14 de Janeiro, 7350-092 Elvas, Portugal, [3]General Directorate for Food and Veterinary, Regional Directorate for the Alentejo, Rua D. Isabel, 7000-880 Évora, Portugal; luis_conceicao@ipportalegre.pt

On-farm mortality is an important cattle welfare indicator and has a negative impact on farmers' revenue. Mertolenga is an indigenous breed raised in beef cattle farms in the Alentejo region, South Portugal, and known by its rusticity and maternal traits. The goal of this study was to assess on-farm mortality on progeny of Mertolenga dams (both pure-bred and cross-bred) on the Alentejo region. On-farm mortality was assessed during the years of 2016 to 2018 in the Alentejo districts of Beja, Évora and Portalegre. Meteorological data (average monthly air temperature and monthly accumulated rainfall) were obtained from records of meteorological stations in the three districts. Data from 47,367 births were collected from the official national animal registry (SNIRA). From these, 1971 on-farm deaths were recorded (an overall mortality rate of 4.2% of births). Pre-weaning (0 to 180 days) mortality rate was 2.9% of births. Statistical analysis was performed using non parametrical tests, as data showed non normal distribution. Both overall and pre-weaning mortality rates were significantly higher in the Évora district. Odds ratio for on-farm deaths in the Évora district was 1.2919 (P<0.0001). There were no significant differences between the three analysed years. Although the number of births differed significantly between months, overall and pre-weaning monthly mortality rates exhibited no significant differences. Pre-weaning mortality was more prevalent in months with lower air temperatures and post-weaning mortality was more prevalent in months with higher rainfall values. Overall and pre-weaning on-farm mortality in the Mertolenga progeny was low to moderate, when compared to previous studies in other breeds. Further work should focus on the influence of environmental conditions on on-farm mortality, considering predicted climate change scenarios.

Coconut oil and peppermint oil as phytotherapeutic treatments of sub-clinical mastitis in dairy cows

T.W. Furness and A. Westland
Myerscough University Centre, Agriculture, Myerscough Hall, St Michaels Road, Bilsborrow, PR3 0RY, Preston, United Kingdom; awestland@myerscough.ac.uk

Clinical mastitis costs industry £149-£250 per case. The effect of sub-clinical mastitis on milk production is estimated to cost between £294 -£634 per cow per lactation in addition to subsequent effects on fertility. As antimicrobial resistance threatens global health and food security the need to develop alternatives has led to research into Phytotherapy. Peppermint oil (PO) and coconut oil (CO) have proven antimicrobial properties both *in vitro* and *in vivo*. Their effects on mastitic pathogens when applied as cutaneous udder creams have not been widely studied, despite, PO being routinely used in organic farming. Milk leukocyte differential (MLD) analysis was used to assess sub-clinical status of infected (n=21) and uninfected (n=47) quarters at pre-trial and at 24 hour intervals throughout three days of treatment. Two groups of nine cows and eight cows, were treated on all four quarters with CO and PO respectively. At pre-trial, 24, 48 and 72 hour intervals cows were sampled using Q-Scout Milk Leukocyte Differential (MLD) Lab. In addition, aseptic samples were taken pre and post-trial for bacteriology to determine colony forming units (cfu). A significant reduction (P≤0.01) in cfu of gram-positive bacteria was found in cows 72 hours after treatment with PO with no significant reduction in CO treated cows. MLD analysis in infected quarters treated with CO had a significant decrease (P≤0.05) in Total Leukocyte Count 72 hours after treatment. Neutrophils counts of uninfected quarters significantly (P≤0.05) increased after 24 hours then significantly decreased (P≤0.05) after 48 hours with PO treatment indicating a possible increase in vasodilation assisting innate immune response. Coconut oil, however, appeared to inhibit vasodilation after 72 hours due to decreases in total leukocyte counts suggesting reduced innate immune response thus making it potentially less effective as a phytotherapeutic tool than PO for sub-clinical mastitis. A larger trial could further assess mode of action of PO and thus aid the uptake of commercially available peppermint oil liniment creams to reduce antibiotic use.

Available and missing information on production factors in Parmigiano Reggiano area dairy farms

R.G. Pitino[1], M. Simoni[1], A. Foskolos[2], E. Tsiplakou[3], M. Ablondi[1], A. Quarantelli[1] and F. Righi[1]
[1]University of Parma, Department of Veterinary Science, via del Taglio,10, 43126, Parma, Italy, [2]University of Thessaly, Department of Animal Science, Campus Gaiopolis, 41110, Larisa, Greece, [3]Agricultural University of Athens, Department of Nutritional Physiology and Feeding, Iera odos 75, 11855, Athens, Greece; marica.simoni@unipr.it

A survey was conducted on 59 farms in the Parmigiano Reggiano cheese production area to understand the awareness level of breeders on the availability and use of production factors in their farms. An interview including questions on animal capital, productivity and management, feedstuff inventory and origin, animal nutrition and rations per animal category, facilities and structures characteristics, land management and use, mechanical equipment and work allocation was submitted to the breeders. Almost all breeders knew the number of animals per category in their farm and, for the lactating cows, milk production and quality were well known on a herd basis. The 91.5% of the breeders knew the average age at 1st calving whereas the number of calves weaned for reproduction (84.7%) or sold for meat production (52.5%) were less known. The 92.2% of farmers was able to estimate the amount and origin of the feedstuffs available in the farm, but only 44.1% of them was aware of the exact formula of the feedstuffs supplied to the different categories of animals. Buildings size and stoking capacity data were available for the 70.8% of the cases and a similar proportion of breeders (69.5%) was able to provide information about land usage and crops productivity. A complete mechanical equipment list was provided in the 79.7% of the cases while the 55.9% of the breeders were able to provide an estimation of labour allocation in animal activities. Results show that a sensible proportion of farmers are missing key data related to herd management, nutrition, facilities and crops productivity; moreover, an accurate traceability of work allocation is unavailable in a large number of dairy enterprises. Lacking of those information can have detrimental effects on their efficiency and profitability. This project has received funding from the European Union's Horizon 2020 research and innovation programme, under the grant agreement No 777974.

Effect of early lactation feeding treatment on the milk performance of dairy cows of divergent EBI

O. Quigley[1,2], L. Delaby[3], K.M. Pierce[1] and F. Buckley[2]
[1]UNiversity College Dublin, School of Agriculture and Food, Belfield, Dublin, D4, Ireland, [2]Teagasc, Animal & Bioscience, Animal & Grassland Research and Innovation Centre, Moorepark, Fermoy, Co. Cork, P61 P302, Ireland, [3]INRA, Physiologie, Environnement et Génétique pour l'Animal et les Systèmes d'Elevage, UMR1348 Pegase, 35590, Saint-Gilles, France; orlaith.quigley@teagasc.ie

The objective was to evaluate the implications of early lactation feeding treatment (FT) on the production performance of two genetic groups (GG) of Holstein-Friesian cows based on the Irish national dairy breeding index, the EBI (Economic Breeding Index); ELITE representing cows in the top 1% nationally (EBI €210) and National Average (NA) representing the mean genetic merit (EBI €108). A total of 201 individual cows with 344 lactation records were available over 3 years (2017-2019). For the purpose of this analysis the grazing season was divided into period 1 (P1 – 18 weeks) and period 2 (P2-16 weeks). Three contrasting FT were implemented in P1 and a single FT implemented in P2. The FT in P1 were Control (CTL), Lower Grass Allowance (LGA) and High Concentrate (HC), with a target post-grazing sward heights of 4.5-5, 3.5-4 and 4.5-5 cm, respectively. Concentrate supplementation of 3 kg/cow/day was offered to LGA and CTL, while 7 kg/cow/day was offered to HC. During P2 all cows were offered a grass only diet grazing to 4.5-5 cm. In both P1 and P2 GG had no effect on daily milk yield (MY) but Elite cows had higher ($P\leq0.001$) milk fat (FP) and protein (PP) content and higher ($P\leq0.072$) milk solids (MS; fat + protein) yield. Feeding treatment during P1 had a significant effect eliciting the predicted effects of HC increasing MY and MS while depressing FP, and LGA decreasing MY, PP and MS. A carryover effect was observed for MY and MS during P2 ($P<0.001$) associated with HC producing greater MY than the CTL and LGA during P2. Furthermore, there was a tendency for a GG×FT ($P=0.0551$) during P2, a more positive carry over effect was expressed by National Average compared to ELITE.

Farmer's application of the breeding value predictions in the Pirenaica beef cattle population

D. López-Carbonell, J. Altarriba, C.M. Then-Rodríguez, C. Hervás-Rivero, C. Moreno and L. Varona
Instituto Agrolimentario de Aragón (IA2), Departamento de Anatomía, Embriología y Genética Animal, Calle Miguel Servet, 177, Universidad de Zaragoza, 50013 Zaragoza, Spain; 767339@unizar.es

The breeding scheme of the Pirenaica Beef Cattle population has been implemented since the 80s, and it involves the prediction of the breeding values for five traits. Two of them recorded on live animals (BW: birth weight and WW: weaning weight), and the remaining provided by the slaughterhouse (CCC: cold carcass weight, FAT: fatness under the SEUROP scale and CONF: conformation). The breeding values obtained are then combined to create two different selection indexes, one for the selection of live animals (ICV) and another for meat production (ICC). However, it could be possible the farmer breeding strategies do not follow the recommendations provided done by prediction. Being so, the objective of this study was the evaluation of those breeding strategies. We split the total number of individuals born each year, from 2010 to 2012, in five clusters: bulls, cows, sires with progeny (SP), dams with progeny (DP) and sires with progeny higher than 30 (SP30). SP and DP represents between 1.74 to 1.91% and between 9.79 and 14.03% of the bulls and cows born each year, respectively. We compared the average predicted breeding values for SP, DP and SP30 with respect to the distribution of the breeding values for the five traits and two indexes, for each analysed year. For dam selection, there is not a clear average tendency, as their average breeding values did not differ significantly from the average of the population. In contrast, sires with the highest breeding values for traits CONF, ICC, and ICV are systematically selected, respectively. The preliminary conclusions of this study are that farmers follow more intensively the breeding recommendations for sire selection than for dam selection, and that meat production traits have the most weight in their breeding decisions. Nonetheless, further research needs to be carried out to evaluate the breeding strategies at the individual farm level.

The effect of dietary Hydroxy-selenomethionine on selenium bioavailability in milk and plasma

M.A. Hachemi[1], N.L. Whitehouse[2], J.R. Sexton[2], S.M. Hollister[2], L.H.P. Silva[2] and M. Briens[1]
[1]Adisseo France SAS, 6 Route Noire, 03600 Commentry, France, [2]University of New Hampshire, 105 Main St, Durham, NH 03824, USA; amine.hachemi@adisseo.com

Selenium (Se) is an essential trace mineral that if deficient in the soil will therefore be deficient in diets fed to dairy cows. Selenium supplements exist in inorganic and organic forms, with the organic form being seleno-yeasts (SY) or pure forms such as selenomethionine or hydroxy-selenomethionine (OH-SeMet). The objective was to determine the amount of Se that was transferred to milk and blood of mid to late lactation dairy cows when supplemental Se from a OH-SeMet (Selisseo® 2% Se, Adisseo France SAS) was fed compared to an unsupplemented group and a group supplemented with a SY. Twenty-four lactating Holstein cows, 12 multiparous and 12 primiparous (178±43 days in milk) were used in a randomised complete block design for 91 days (7 days for covariate and 84 days for treatments). Treatments were 1) basal diet with an expected Se background of 0.2 mg Se per kg as-fed (negative control; NC), 2) basal diet + 0.3 mg Se per kg as-fed from SY (positive control; SY-0.3), 3) basal diet + 0.1 mg per kg as-fed from OH-SeMet (OH-SeMet-0.1), and 4) basal diet + 0.3 mg per kg as-fed from OH-SeMet (OH-SeMet-0.3). Blood (from the tail vein at 10:30 h) and milk samples (am and pm milking) were obtained from each on the last 3 days of the covariate week and weeks 1, 2, 3, 4, 6, 8, 10 and 12 after supplementation to determine plasma and milk Se concentration and plasma glutathione peroxidase (GPx) activity. Data were analysed using the PROC MIXED of SAS with REPEATED measures. Significance was declared at $P \leq 0.05$. The mean plasma Se concentrations were: 120, 134, 122 and 142 ng/ml to NC, SY-0.3, OH-SeMet-0.1 and OH-SeMet-03 respectively ($P < 0.001$). The milk Se concentrations on average were: 57, 97, 78 and 116 ng/ml to NC, SY-0.3, OH-SeMet-0.1 and OH-SeMet-0.3 respectively ($P < 0.0001$). In terms of redox status, there was no differences on the plasma GPx activity between groups. Those results are a confirmation that organic Se forms can increase milk and plasma Se concentrations. Moreover, when administered at the same level of supplementation, OH-SeMet showed to be more efficient than SY to improve those Se concentrations.

Effects of a combination of rumen-protected amino acids prototypes on performance of Holstein cows

G. Kim[1], T.D. Marbun[1], J. Park[1], H.G. Lee[2], J.O. Moon[3], J.S. Park[3] and E.J. Kim[1]
[1]Kyungpook National University, Department of Animal Science, Sangju, 37224, Korea, South, [2]Konkuk University, Department of Animal Science and Technology, Seoul, 05029, Korea, South, [3]CJ CheilJedang Research Institute of Biotechnology, Suwon, 16495, Korea, South; ejkim2011@knu.ac.kr

Methionine (Met) is one of the first-limiting amino acids in dairy cattle fed with typical corn- and soybean-based diets. Moreover, limited studies have suggested that tryptophan (Trp) and/or histidine (His) are considered as limiting amino acids in the diets of high-producing dairy cows. We recently chemically synthesised prototypes of rumen-protected Trp, His, and Met and examined their effects on bovine mammary tissue cell culture *in vitro* (published elsewhere), both individually and in combinations. Here, we evaluated the effects of a combination of rumen-protected Trp, Met, and His prototypes on the performance of Holstein dairy cows. Forty cattle were used, and a randomised block design was followed; the cattle were blocked according to their parity, milk yield, days in milk, and body weight and were allocated as either control (no supplementation) or total-mixed ration (TMR) with 30 g supplementation (TMR-30). A combination of rumen-protected amino acids (Trp:Met:His = 17.25:8.4:4.35 g) was top-dressed on TMR, with 30 g per cow for 3 weeks. Data were analysed using Student's t-test with the SAS software. Compared with control, TMR-30 fed rumen-protected amino acids had greater dry matter intake ($P < 0.05$). Milk yield was not different between control and TMR-30. Although not significantly different, milk protein content of TMR-30 was numerically higher than that of control. No differences were observed in terms of milk fat, lactose, and milk urea nitrogen concentrations. Collectively, supplementing rumen-protected Trp, Met, and His prototypes can improve the performance of dairy cows. Further studies to examine the effects of using different levels of the prototypes are warranted.

GWAS for growth and carcass traits in the Rubia Gallega beef cattle population

C. Then-Rodríguez[1], J. Altarriba[1], C. Hervás-Rivero[1], C. Moreno[1], M. Martínez-Castillero[2], D. López-Carbonell[1] and L. Varona[1]
[1]Universidad de Zaragoza, Instituto Agrolimentario de Aragón (IA2), 50013 Zaragoza, Spain, [2]University of Padova, Department of Agronomy, Food, Natural Resources, Animals and Environment (DAFNAE), 35020 Legnaro PD, Italy; lvarona@unizar.es

We carried out a series of single-step genomic selection analyses using a dataset that included 288,960 records for trait birth weight (BW), 75,631 for weaning weight (WW), 76,868 for cold carcass weight (CCW), 76,859 for conformation using the SEUROP scale (CONF) and 76,723 for fatness (FAT). The pedigree included 417,903 individual-sire-dam entries, were 713 were genotyped with the Affymetrix Axiom Bovine chip. After standard filtering, genotypes for 45,279 SNP markers were available. From the results obtained by the single-step genomic analyses, we used the equivalence between the GBLUP and SNPBLUP models to calculate the solutions for SNP effects, which were then used to determine the percentage of variance explained by genomic regions ranging from 0.5 to 3 Mb and between 10 to 60 SNP. The results allowed us to identify 10, 9, 6, 5, and 6 genomic regions for BW, WW, CCW, CONF, and FAT, respectively. Some of the genomic regions have pleiotropic effects such as BTA1 (131-132 Mb) – BW, FAT, CCW-, BTA2 (1-11) -FAT and CONF-, BTA6 (36-38) – BW, CCW, FAT-, BTA11 (69) -CONF and CCW, BTA15 (23-24) -BW and FAT-, BTA16 (24-26) -BW, CONF, WW- and BTA 20 (20-22) -BW, WW, and CCW-. These genomic regions contain appealing candidate genes: NCK1: Cytoplasmatic protein NCK1 at BTA1, MSTN: Myostatin at BTA2, LCORL: Ligand Dependent Nuclear Receptor Corepressor Like and NCAPG: Non-SMC Condensin I Complex Subunit G at BTA6, BMP10: Bone morphogenetic protein 10 at BTA11, DRD2: dopamine receptor 2 at BTA15, BPNT1: 3'(2'),5'-bisphosphate nucleotidase 1 at BTA16 and ACTBL2: Beta-actin-like protein 2 at BTA20.

Bull ejaculates got on the field: storing temperature and extender vs raw semen on semen quality

A. Fernandez-Novo[1], S. Santos-Lopez[2], C. Barrajon-Masa[3], P. Mozas[3], E. Caceres[2], M. Gomez[2], A. Garrafa[2], J.V. Gonzalez-Martin[2], A. Oliet[3], S. Astiz[4] and S.S. Perez-Garnelo[4]
[1]CEU-UCH, Valencia, 46115, Spain, [2]UCM, Madrid, 28040, Spain, [3]CENSYRA, Madrid, 28794, Spain, [4]INIA, Madrid, 28040, Spain; aitorfn@gmail.com

Breeding bull soundness (BBS) evaluation is designed to detect infertile and subfertile bulls. Semen quality assessment is an important part of BBS, which can be affected by ejaculate management until evaluation. Our objective was to determine the effect of storing temperature (5 °C vs Ambient Temperature; AT), of the extender used and of the no use of extender (Raw Semen stored at 37 °C; RS) on beef bulls semen quality, until evaluation at a laboratory (up to 2 h), in terms of total and progressive motility (TM and PM), sperm viability (SV) and sperm morphology (SM). Ejaculates from 10 bulls from commercial farms were immediately divided into seven aliquots (RS, AndroMed®, Bioxcell® and INRA®; two aliquots/extender). One aliquot was chilled at 5 °C(-0.2 °C/min); the other kept at AT (22-25 °C) until analysis. TM and PM were measured with Computer-Assisted Sperm Analysis (CASA; SCA-Microptic®), SV and SM were assessed through eosin-nigrosin stain-slides count. Results were analysed with Sigma-Plot 12®. TM tended to be higher when semen was diluted with any extender (lowest value in RS: 56.6±19.28; the highest result with Bioxcell at 5 °C: 76.1±16.19% P=0.074). Bioxcell preserved PM independently of storing temperature vs RS (RS: 36.2±28.47 vs Bioxcell at 5 °C: 70.3±16.22% and at AT: 67.7±16.09%; P<0.05); AndroMed differed significantly from RS only when stored at 5 °C (RS: 36.23±28.47 vs AndroMed: 365.18±17.84; P<0.05). By contrast, INRA at 5 °C did not induce any difference in PM vs RS (P=0.565) but it was better at AT (RS: 36.2±28.47 vs INRA: 64.8±18.64% P<0.05). PM did not differ among extenders and storing temperature (P>0.5). SV (P=0.815) and SM (P=0.914) did not vary regardless of extender (including RS), nor by storing temperature. In conclusion, total and progressive motility were better preserved when using extenders, especially Bioxcell or AndroMed. Sperm viability and morphology were similar regardless of the use of extender or not, and of the storing temperature. Therefore, it is important to use commercial semen extenders, even when analysing ejaculates <2 h after collection.

Bull ejaculates got on the field: effect of time lapse, storing temperature and extender used on pH

A. Fernandez-Novo[1], S. Santos-Lopez[2], C. Barrajon-Masa[3], P. Mozas[3], E. Caceres[2], M. Gomez[2], A. Garrafa[2], J.V. Gonzalez-Martin[2], A. Oliet[3], S. Astiz[4] and S.S. Perez-Garnelo[4]
[1]UCH-CEU, Valencia, 46115, Spain, [2]UCM, Madrid, 28040, Spain, [3]CENSYRA, Madrid, 28794, Spain, [4]INIA, Madrid, 28040, Spain; sgarnelo@inia.es

Breeding bull soundness (BBS) evaluation is designed to detect subfertile bulls. Seminal exam is part of the BBS, with different methodologies described to analyse ejaculates immediately after collection. However, these evaluations cannot be always performed immediately. Therefore, the objective was to determine the effect of storing factors (temperature, extender and time lapse) on pH of ejaculates from beef bulls, collected at commercial farms. Ejaculates from 25 bulls were divided into six aliquots (Andromed®, Bioxcell® and INRA®; two aliquots/extender). One aliquot was chilled at 5 °C (-0.2 °C/min), the other kept at ambient temperature (AT; 22-25 °C) until analysis. Values of pH was assessed at <2, 4 and 24 h post-ejaculation with indicator paper strips (Whatman®). A 3×3×2 experiment design (3 times lapses × 3 extenders × 2 temperatures) was implemented. Results were analysed with a 3-way ANOVA (Sigma-Plot 12®). Values of pH revealed no statistical difference between time lapse (6.48 at 0 h, 6.45 at 4 h and 6.43±0.027 at 24 h; P=0.379) nor extender used at any time (P=0.212). However, storing temperature influenced pH values (P<0.001), and also the interaction between time and temperature (P<0.001). When storing at 5 °C average pH values were 6.53±0.022 and at AT 6.38±0.022 (means difference of 0.15; P<0.001). At 5 °C pH values increased from 6.49±0.038 (<2 h) to 6.61±0.038 (24 h; P>0.05), however, at AT, pH decreased from 6.47±0.038 (<2 h) to 6.25±0.038 (24 h; P<0.05). The pH difference according to storage temperature was significant after 24 h: 6.61±0.038 for 5 °C and 6.25±0.038 at AT (difference of 0.361; 5.76% of reduction; P<0.001); but there were no statistical differences at <2 h (P=0.639) or at 4 h (P=0.298) between storage temperatures. In conclusion, in terms of pH, seminal samples should be stored at 5 °C when it is not possible to analyse them within 4 h after collection. When evaluated up to 4 h after collection, pH values do not differ by storing temperature nor by extender used.

Growth and reproduction traits in Portuguese Preta cattle carrying the nt821 myostatin-gene mutation

M.C. Feliciano[1], E. Bettencourt[1], A. Amaral[2], F. Ferreira[3] and L.T. Gama[2]
[1]MED. University of Évora, Polo da Mitra, 7006-554 Évora, Portugal, [2]CIISA. Faculty of Veterinary Medicine. University of Lisbon, Avenida da Universidade Técnica, 1300-477 Lisboa, Portugal, [3]ACBRP, E.N. 10, 2139-909 Samora Correia, Portugal; mcarmo.vet@gmail.com

The effects of the nt821 myostatin-gene mutation on growth and reproduction traits in Preta cattle were assessed by evaluating animals from two herds with the genotypes wild type (+/+), heterozygous (mh/+) and homozygous (mh/mh). The number of animals evaluated ranged from 105 for dressing percentage to 298 for calving interval. Adjusted weight at 120 days was about 139 kg in mh/+ and mh/mh, and 125 kg in +/+ (P<0.05). However, differences between genotypes at 210 days were minor (P>0.1) and at 365 d the lowest mean weight was observed in mh/mh individuals (P<0.05). The lowest average daily gain from 120 to 210 d was observed in mh/mh, but the means for the various genotypes were similar for daily gain between 210 and 365 d (P>0.05). Mean carcass daily gain up to slaughter was higher in mh/+ by about 40 g (P<0.05), but did not differ between mh/mh and +/+ (P>0.05). Carcass dressing was higher in mh/mh by about 3% (P<0.01), but did not differ between the other genotypes. The means for age at first calving were 34.0, 37.0 and 40.0 months in +/+, mh/+ and mh/mh heifers, while the mean calving interval for the same genotypes was 432.3, 445.7 and 523.8 days. The means for mature weight were 574.7, 542.7 and 506.0 kg in +/+, mh/+ and mh/mh cows, respectively (P<0.01). Taken together, our results indicate that, in Preta cattle, the mh mutation has a detrimental effect on age at first calving and on calving interval, and leads to a reduction in cow mature weight. The effect of mh on calf growth depended on the genotype and age considered, with lower growth rate in mh/mh calves before weaning, whereas carcass growth by day of age was higher in heterozygous animals and dressing percentage was higher in the homozygous group.

Dietary trace mineral supply vs recommendations: cross-sectional study in Canadian dairy herds

M. Duplessis[1], D. Pellerin[2] and C.L. Girard[1]
[1]Agriculture & Agri-Food Canada, 2000 College St, J1M0C8 Sherbrooke, Canada, [2]Université Laval, 2425 rue de l'Agriculture, G1V0A6 Québec, Canada; melissa.duplessis@canada.ca

The aim of this trial was to determine concentrations of Co, Cu, Mn and Zn in the diets of lactating cows in commercial herds and to compare the supply vs the requirement based on the National Research Council. A total of 100 dairy herds located in Quebec, Canada were enrolled. Each ingredient in the diet was analysed for trace minerals (TM) using inductively coupled plasma spectrometer. On a dry matter basis, Co, Cu, Mn and Zn (mg/kg) of diets were calculated by multiplying mineral composition of each ingredient by their respective dry matter percentage fed to each cow (n=4,432). Mineral recommendations were computed for each cow according to the NRC (2001) using the factorial approach (except for Co, one recommendation regardless of the level of production). Cows were divided into 3 groups: <21 days in milk (DIM); Between 22 and 190 DIM; >190 DIM. Proc MIXED and UNIVARIATE of SAS were used. Tukey test was used when the result was significant. Co in diet averaged 0.64±0.03 mg/kg among DIM groups and exceeded the recommendation by 479±26 (min: 30, max: 1,276)% (P=0.28). Cows <21 DIM were fed a diet containing greater Cu, Mn, and Zn, but closer to the recommendations, than cows with DIM>21 DIM (P<0.05). Dietary Cu, Mn, and Zn averaged 20.1±0.7, 81.3±3.3, and 90.7±3.1 mg/kg, respectively, within the first 21 DIM whereas Cu, Mn, and Zn averaged 17.6±0.6, 69.3±3.0, and 78.7±2.4 mg/kg, respectively, when DIM>21. There was no significant difference for cows between 22 and 190 DIM and for those with DIM>190 (P>0.05). Before 21 DIM, dietary supply in Cu, Mn, Zn was greater than the recommendation by 33±6 (min: -55, max: 282), 320±22 (min: -22, max: 1,215), 36±6 (min: -63, max: 555)%, respectively, whereas after 21 DIM they exceeded the recommendation by 61±5 (min: -47, max: 464), 422±20 (min: 37, max: 1,915), 80±6 (min: 53, max: 1,044)%. These results showed that most diets exceeded TM recommendations. Supplying more of those TM in the diet probably increases their concentrations in manure, then representing an environmental issue when manure is spread in the field. This could lead to a TM build-up in the soil, affecting soil microbiota and causing plant phytotoxicity.

Estimation of economical values for Polish Holstein, Polish Red and Scandinavian Red

B. Kosińska-Selbi[1], C. Schmidtmann[2], M. Kargo[3], J.F. Ettema[3] and J. Szyda[1,4]
[1]Wroclaw University of Environmental and Life Sciences, Biostatistics Group, Department of Genetics, Kożuchowska 7, 51-631 Wroclaw, Poland, [2]Kiel University, Institute of Animal Breeding and Husbandry, Hermann-Rodewald-Str. 6, Hermann-Rodewald-Str. 6, 24118 Kiel, Germany, [3]Aarhus University, Center for Quantitative Genetics and Genomics, Department of Molecular Biology and Genetics, Blichers Allé 20, 8830 Tjele, Denmark, [4]National Research Institute of Animal Production, Krakowska 1, 32-083 Balice, Poland; barbara.kosinska@upwr.edu.pl

Economical values (EV) are key elements in derivation of breeding goals for dairy cattle. They are defined as expected monetary gain per unit improvement of a breeding trait, while keeping the remaining traits constants. The objective of the present study was to calculate EV of breeding traits (production, reproduction, health and calving traits) for the Polish dairy breeds such as Polish Holstein (PH), Polish Red (PR) and Scandinavian Red dairy breeds (ScR). ScR are a combination of Norwegian Red (NR) and Swedish Red (SR), because both populations are not numerous and the assumptions similar they are analysed as one breed. PR is one of the local small breeds, which is covered by the national program for the protection of genetic resources. According to the Polish Federation of Cattle Breeders and Milk Producers current population of PR cows is estimated to 3,000. PR has very low average milk yield equal to 3,787 kg. Average milk yield for PH is 8,050 kg and for ScR 7,311 kg. For deriving EV, the stochastic bio-economical model SimHerd was used. SimHerd allows to derive EV based on biological parameters of each animal and management herd parameters. Simulations run stochastically, by simulating 40 years of economic consequences, with 1000 replications. For each trait, EV were estimated based on the differences between two marginal scenarios. IIn order to avoid double counting, the EV from SimHerd were corrected using models with mediator variables. The results have shown the largest differences for the traits of conception rate among the considered dairy breeds: $EV_{PH}=197$ €, $EV_{PR}= 84,3$ €, $EV_{ScR}=91,3$ €.

Mapping of genomic features related to mastitis resistance in dairy cattle

T. Iso-Touru, D. Fischer and J. Vilkki
Natural Resources Institute Finland (Luke), Myllytie 1, 31600 Jokioinen, Finland; terhi.iso-touru@luke.fi

Mastitis is the costliest disease in dairy cattle and also a major animal welfare concern. Mastitis resistance is a complex trait, with low to moderate heritabilities and thus breeding for resistance has not been effective. The aim of our project is to characterise genomic features related to mastitis resistance to improve functional annotation of the bovine genome and generate better tools for reducing mastitis in dairy cattle. We have extracted primary bovine mammary epithelial cells (pbMEC) from milk samples from two groups of cows: 5 with predicted good resistance to mastitis (high GEBV, no history of clinical mastitis) and 5 with low predicted and proven resistance to mastitis (low GEBV, recorded mastitis infections); and have started challenge experiments with Finnish *Escherichia coli* and *Staphylococcus aureus* strains. Samples from different time points will be analysed by RNASeq and miRNASeq, and samples with significant differences in gene expression will also be analysed by ATACSeq to reveal differences in chromatin accessibility. First results on differences in response to *S. aureus* challenge on gene expression will be presented. All protocols follow FAANG guidelines, and the data will be submitted to the FAANG databases. Identified candidate regions for regulation of immunological reactions in mastitis will be compared with candidate regions and variations from previous GWAS for udder health, and whole genome sequences from all samples to identify potential causative variations. The project is funded by a grant from the Academy of Finland (No. 317998).

Effects of milk PTA and supplementary CP on milk production in late lactation grazing dairy cows

M.J. Doran[1], M.B. Lynch[1], F.J. Mulligan[2], A.G. Fahey[1] and K.M. Pierce[1]
[1]University College Dublin Lyons Farm, School of Agriculture and Food Science, University College Dublin Lyons Farm, Celbridge, Co. Kildare, W23 ENY2, Ireland, [2]University College Dublin Lyons Farm, School of Veterinary Medicine, University College Dublin Lyons Farm, Celbridge, Co. Kildare, W23 ENY2, Ireland; michael.doran@ucdconnect.ie

Irish dairy systems are predominantly grass-based and spring-calving to coincide with the onset of the grass growing season. Studies show nitrogen utilisation efficiency is low, particularly in late lactation. Grazed grass is naturally high in crude protein (CP). Therefore, there may be potential to reduce supplementary CP in feed. The aim of this experiment was to investigate effects of 305-d milk predicted transmitting ability (PTA), supplementary CP concentration, and their potential interaction on milk production and composition in late lactation (+200 DIM) spring-calving grazing dairy cows. A factorial experiment with two milk PTA groups (low milk (LM) (-60 kg) versus high milk (HM) (+130 kg)), and two supplementary CP levels (12% versus 18% with equal PDI supply) offered at 3.5 kg DM/cow/day, was conducted over 53 days. Forty-eight Holstein Friesian (24 LM and 24 HM PTA) dairy cows, blocked on parity and balanced on DIM, BCS and Economic Breeding Index, were assigned to one of two feeding strategies resulting in four treatments (n=12): 1) LM PTA + 12% CP (L12); 2) LM PTA + 18% CP (L18); 3) HM PTA + 12% CP (H12) and 4) HM PTA + 18% CP (H18). Cows grazed full time and were offered 15 kg DM grass/cow/d. Repeated measures ANOVA (MIXED procedure) on SAS 9.4 was used to analyse milk production and composition parameters. Differences between means were determined by F-tests using Type III sums of squares. Milk yield was greater in H18 than H12 (+2.58 kg; P=0.05), while fat, protein, casein, fat + protein, and lactose kg did not differ between treatments. The 18% CP supplement increased milk yield (+1.90 kg; P=0.009), protein (+0.07 kg: P=0.004), casein (+0.06 kg; P=0.005), fat + protein (+0.12 kg; P=0.03), lactose (+0.08 kg; P=0.02), and urea (+0.002 g/100 g milk; P=0.04) compared with the 12% CP. Constituent concentrations did not differ between treatments. In summary, milk PTA and supplementary CP% did not alter milk composition. Furthermore, increased CP% led to increased milk and fat + protein yield, suggesting offering an 18% CP is beneficial for improving production in late lactation.

Histidine dose effects on lactational performance and plasma amino acid concentrations in dairy cows
S.E. Räisänen and A.N. Hristov
The Pennsylvania State University, University Park, PA, 16802, USA; ser45@psu.edu

Two experiments (E1 & E2) were conducted to determine dose effect of His on milk production and composition and plasma amino acid (AA) concentrations in lactating dairy cows fed a metabolisable protein (MP) adequate (E1) or deficient (E2) diets (NRC, 2001). Mid-lactation Holstein cows (E1: n=16 and E2: n=20) were used in replicated 4×4 Latin square design experiments with 4, 28-d periods. Treatments for E1 were total mixed ration (TMR) with 1.8% digestible (d)His of MP (His1.8; TMR1), TMR with 2.2% dHis (His2.2; TMR2) and TMR2 supplemented with rumen-protected (RP)His to supply 3.0% or 3.2% dHis of MP (His3.0 and His3.2). Treatments in E2 were a basal TMR with 1.8% dHis of MP (His1.8) and the basal TMR top-dressed with RPHis to supply 2.4, 2.8 or 3.3% dHis of MP (His2.4, His2.8 and His3.3). Dry matter intake (DMI) and milk yield (MY) were recorded daily and milk and blood samples collected on the last wk of each period. Data were analysed using PROC MIXED of SAS with treatment and period in the model. Square and cow within square were random effects. Linear and quadratic effects of His dose were tested. His did not affect DMI (P≥0.52; average 23.4 kg/d; SEM=1.42 and 21.0 kg/d; SEM=1.95, for E1 and 2, respectively), whereas MY tended to increase quadratically in E1 (P=0.08; 41.5, 42.7, 43.3, and 42.3 kg/d, respectively), and increased linearly in E2 (P=0.001; 37.1, 38.9, 39.2 and 39.1 kg/d). Milk true protein concentration (TPC) increased linearly (P=0.004; 2.62, 2.83 and 2.78) and TP yield tended to increase (P=0.10) with dHis in E1. Milk TPC was not affected (P>0.10) in Exp2, but TP yield increased linearly (P<0.01; by 80 g/d; His1.8 vs His3.3). Plasma His increased linearly (P<0.001) with His dose in E2 (21.0, 27.0, 35.1 and 40.0 µM, respectively). Milk and milk TP yields were optimised with dHis supply at around 3.0% of MP on both MP-adequate and deficient diets, with the effect of supplemental His being more pronounced with the MP-deficient diet.

Dairy cow response to lysine and methionine when reducing metabolisable protein through soybean meal
S. Lemosquet[1], C. Cirot[2], L. Bahloul[3] and P. Lamberton[1]
[1]INRAE, Agrocampus Ouest, PEGASE, 35590 Saint-Gilles, France, [2]Ajinomoto Animal Nutrition Europe, 32 rue Guersant, 75017 Paris, France, [3]Adisseo France S.A.S, Centre of Expertise and Research in Nutrition, 03600 Commentry, France; sophie.lemosquet@inrae.fr

Reducing soybean meal (SBM) in dairy diets could be a way to increase protein autonomy of French dairy farms. However, decreasing metabolisable protein supply (PDI in INRA, 2018) through SBM in corn silage based diets, that are known to be deficient in 2 amino acids (AA), lysine (Lys) and methionine (Met), decreased dry matter intake (DMI), milk yield (MY) and protein yield (MPY). The effect of balancing Lys (AjiPro®-L) and Met (Smartamine M®) supplies at Low (LP) or High (HP) levels of Protein (124 vs 148 g/kg DM of CP; 84.7 vs 94.2 g/kg DM of PDI) were investigated according a 2×2 factorial design on 40 Holstein dairy cows at 84±20 days in milk. They were assigned to a randomised complete block design in 4 groups of 10 cows for 5 weeks. Four corn silage diets (58.7% of DM) at 1.44 MCal/kg DM of net energy were offered *ad libitum* in which the SBM were incorporated at 2.5 vs 8.9% in HP vs LP diets, respectively. The levels of Lys and Met in % of PDI were at 6.1 and 2.0 in unbalanced AA (LM-) and at 7.0 and 2.3 in balanced AA (LM+) diets, respectively. Results were submitted to variance covariance analyses using the MIXED procedures of SAS including the effects of parity, PDI, AA profiles and interactions on the mean of weeks 3 to 5 (n=37). Increasing Lys and Met supply tended (P<0.1) to increase DMI (24.3 vs 24.9 kg of DM in LM- vs LM+ diets). Milk protein yield were at 1,032, 1,116, 1,155, 1,151 g/d and MY were at 34.7, 36.8, 37.5, 37.1 kg/d in LPLM-, LPLM+, HPLM- and HPLM+, respectively. They both decreased with decreasing PDI supply (P<0.01) and increased in response to AA only at LP diets (P × AA interaction: P<0.03). On a repeated measurements analysis, MY tended to increase with AA supplementation only in multiparous cows (40.3 vs 40.8 kg/d; n=6 per diet). Body weight tended to be lower at LPLM- (677 kg; P × AA: P=0.06) than in other diets (from 687, 693, 690 kg in LPLM+, HPLM-, HPLM+, respectively). This study confirmed that better balancing Lys and Met in low PDI diets with low SBM supply increased dairy cows' performances.

Can xylanase addition increase *in vitro* enzyme solubility of organic matter from maize stem?

M. Rønn[1], K.E.B. Knudsen[1], N.B. Kristensen[2] and M.R. Weisbjerg[1]
[1]Aarhus University, Department of Animal Science, AU-Foulum, 8830 Tjele, Denmark, [2]SEGES, Agro Food Park, 8200 Aarhus, Denmark; martin.weisbjerg@anis.au.dk

An accurate determination of organic matter digestibility is of great importance to estimate the energy value of animal feeds. The method to determine the enzymatic digestibility of organic matter (EDOM) is an *in vitro* multi-enzymatic method used in Denmark for feed value estimations for cattle. Former EDOM estimations have shown low solubility of maize stem compared to *in vitro* rumen liquor solubilities. The aim of this study was to test if addition of different xylanase enzymes in the EDOM procedure would increase the solubility of maize stem. Seven procedures were compared; addition of three xylanase compounds in two concentrations and the standard solution without xylanase were tested with four maize stem samples in two replicates. The tested xylanase compounds were endo-1,4-beta-xylanase (purified from rumen microorganism; Megazyme) used in concentrations of 1.5 and 3 ml/l, NS 39163 (Novozymes) and Rovabio Advance (Adisseo) both used in concentrations of 5 and 10 ml/l added to the standard fibre degrading enzyme mix. With standard enzyme solution without xylanase the average EDOM solubility of the four maize stem samples were 51.3% and a two-way ANOVA revealed that Rovabio Advance (10 ml/l) and NS 39163 (5 and 10 ml/l) were higher (<0.05) in solubility than the standard procedure with EDOM values of 53.7, 53.7 and 54.1%. The average EDOM values for Adisseo (5 ml/l) and Megazyme (1.5 and 3 ml/l) were 51.6, 51.4 and 50.7%. All six treatments showed high correlation with the standard procedure ($R=0.99$-1.0). Even though two tested xylanases resulted in a statistically significant increase in solubility of maize stem, the magnitude of the increase is not considered to be sufficient to revise the standard procedure.

Wide-scale associations for workability traits in the Polish HF dairy cattle

K. Żukowski[1], W. Jagusiak[1,2], P. Topolski[1] and B. Szymik[1]
[1]National Research Institute of Animal Production, Krakowska 1, 32-083 Balice, Poland, [2]University of Agriculture in Cracow, Mickiewicza 21, 31-120 Cracow, Poland; kacper.zukowski@izoo.krakow.pl

One of the most critical traits for new, automated dairy farming with great economic importance is milking temperament and milking speed, both so-called workability traits. Milking speed (MS) can be defined as cow ability to entirely give in a short time the milk produced by the udder gland. The milking temperament (MT) is 'type and degree of reaction of the animal in its overall surrounding conditions'. The disorders for the optimal MS could increase the risk for mastitis or affect to the high level of somatic cells level in udder gland, and many cases lead to dairy cow disposal. According to the high economic importance, the genetic and further, genomic evaluation of dairy cattle could lead to better management of the herds and aiming breeding goals. Moreover, the knowledge of the genetic background could help to understand processes associated with workability traits. This study aimed to wide-scale associations based on SNP genotype data of workability traits in the Polish HF dairy cattle. The analysed dataset covered more than 1.3 million cows born between 2006 and 2018 and phenotypically evaluated for workability traits (MS and MT, both with reliability). Moreover, the pedigree information was limited to three generations, and each ancestor has evaluated breeding value for MT and MS. Finally, in the analysed dataset we had more than 2.1 animals including almost 17K sires, 450K dams and 400K grandparents. More than 2,000 genotypes of bulls and cows were obtained from the low and high-density Illumina chips. A mixed linear model analysis assessed the association between each SNP and each of the traits and indicate candidate genes. Additionally, the pleiotropy character of genes with other milk production and functional traits was analysed. The most significant association signals were identified for genes located on chromosome 7, 10, 14, and 18, and on 11, 15 and 17, respectively for MS and MT. However, the additional analysis is needed to incorporate identified regions which QTLs and genes.

Semen production and quality from bulls of seven Portuguese cattle breeds

A. Priester[1,2], P. Martins[3], R. Dantas[3], R. Valentim[2] and V.A.P. Cadavez[2]
[1]*Universidade Tecnológica Federal do Paraná, Av. Sete de Setembro, 3165 Rebouças, 80230-901 Curitiba-PR, Brazil,*
[2]*Centro de Investigação de Montanha, Instituto Politécnico de Bragança, Animal Science, Campus de Santa Apolónia,*
5300-253 Bragança, Portugal, [3]*Federação Nacional das Associações de Raças Autóctones, Rua Franca, nº 534 S. Torcato,*
4800-875 Guimarães, Portugal; fera.storcato@net.novis.pt

Bull fertility has a great impact on herd performance and producer's profits. Thus, the aim of this study was to determine the effect of age and breed on semen production and quality from seven Portuguese cattle breeds: Alentejana, Arouquesa, Barrosa, Mertolenga, Minhota, Mirandesa and Preta. The data collection was conducted at the Centre for Collection and Analysis of Semen from the National Federation of Autochtonous Breed Associations, located in São Torcato, Guimarães. Data consisted of 1,128 semen samples from 186 bulls aged 14 to 36 months, and the semen properties studied were: ejaculate volume (VO, ml), sperm cells concentration (SPZC, million/ml), total number of sperm cells (SPZN, million) per ejaculate. Data were analysed using a linear mixed model, with bull age and breed as fixed effects, and bull within breed as a nested random effect, using nlme package from the R Software. The bull's age at collection had no effect ($P>0.05$) on the three semen traits studied. The VO and SPZC were affected ($P<0.05$) by bull's breed, and the bull random effect explained 47% of the variance in VO. It was concluded that the high between-bull variation observed makes it possible to select individual bulls on the basis of some traits.

High-throughput phenotyping of intakes in small ruminants

J.L. Weisbecker[1], C. Huau[1], J.F. Bompa[2], D. Marcon[3], L. Estivalet[3], C. Marie-Etancelin[1], F. Tortereau[1], T. Heirman[2],
F. Laperruque[2], D. Francois[1] and E. Ricard[2]
[1]*INRAE, GenPhySE, Castanet Tolosan, 31320, France,* [2]*INRAE, CATI SICPA, Castanet Tolosan, 31326, France,* [3]*INRAE,*
La Sapinière Experimental Unit, Osmoy, 18390, France; christel.marie-etancelin@inrae.fr

Feed efficiency is a trait of major interest for all livestock species because breeding efficient animals is not only cost-saving but also leads to a decrease of environmental impacts. The genetic improvement of this trait requires feed intakes to be recorded individually. For small ruminants which are gregarious animals, feeding is mainly managed at the group-level and information on individual intakes is difficult to get. Since the 90s', automatic concentrate feeders have been used in an *ad libitum* version for meat sheep in our experimental farm. Regarding forage intakes, no devices did exist until recently to record individual intakes within groups, whereas forage is the main feed for ruminants. At INRAE, the CATI SICPA team developed automatic feeders adapted to sheep and goats bred in pens. Three devices were developed, each of them recording individual intakes of concentrate, forage or water. A weighing pan has been coupled to the water automatic feeder to record animal body weights at each visit. All these devices can handle a large number of animals, each visit (in duration and quantity intake) being assigned to an individual through electronic identification. We have equipped our experimental farm with these different devices since 2015. We evidenced that efficient lambs (tested under a concentrate diet) eat less forage later in life than less efficient lambs. We also highlighted that daily intakes of forage and water were correlated (+0.35) and that the number of visits to the different devices were also correlated (from +0.31 to +0.45). The automatic weighing system led to weights highly correlated (> 0.98) with weights from handled weighing. All these high-throughput phenotypes will be analysed jointly with other biological, physiological and genomic traits in order to study phenotypes of interest (feed efficiency, longevity, social interactions) and to quantify response to nutritional or infectious challenges. Finally, these data can also help in identifying major events (technical, meteorological, health, etc.) and then to test the resilience of animals under these stressful events.

Session 54 — Theatre 2

The concept and perspectives of precision mineral supplementation
H.L. Foged

Organe Institute, Skødstrupbakken 64, 8541, Denmark; henning.lyngsoe.foged@gmail.com

Precision Mineral Supplementation is an IoT technology system consisting of sensors, data, software and communication technology apart from conventional hardware. The relevance of Precision Mineral Supplementation is caused by increasing productivity and use of Total Mixed Ration (TMR) feeding in dairy production. In TMR, minerals and vitamins are mixed in a constant ratio to other feed ingredients. This creates a situation of undersupply with minerals and vitamins in the mobilisation phase and in the entire critical transition period where cows' intake of their ration is averagely about 20% under their nutritional needs, estimated on basis of their typical weight loss in that period. This undersupply causes a dysfunctional immune response, oxidative stress and an imbalanced energy metabolism among the cows in the critical period of the lactation and increases risks for loss-giving health and reproduction problems. Precision Mineral Supplementation is the practical and economic method to meet these dysfunctions, via ensuring a more balanced mineral supplementation of the cows and also allowing an economic and practical use of advanced fed additives that are cost-effective for transition cows. Other methods include group feeding and overdosing. Separate TMR feeding of transition cows with an increased ratio between minerals and energy is rather laborious, it requires investment in cow separation gates and causes larger social stress due to cows' more frequent group change, not to mention the higher administration. General overdosing with mineral feeds in physical amounts or by using chelated minerals or natural forms of vitamins and minerals is relatively expensive and leads to even larger oversupply in the mid and late lactation phases, which is problematic as several micro minerals are unwanted heavy metals that pollutes the nature via excretion through faeces. Precision Mineral Supplementation meets consumers demand for food of high ethical quality and society's demand for improved resource economy.

Session 54 — Theatre 3

Precision mineral supplementation and its impacts on dairy cow performance
H.L. Foged
Organe Institute, Skødstrupbakken 64, 8541, Denmark; henning.lyngsoe.foged@gmail.com

Analyses of testbed registrations shows positive effects of Precision Mineral Supplementation on dairy cow performance. Test cows that had access to extra mineral feed supplements in the critical transition period from 21 days before calving and until 98 days after calving had 21% lower somatic cell counts, 39% more pregnancies, 32% more of them showed heat, they had 12% less disease incidences requiring involvement of a veterinarian, and 35% fewer were culled. All mentioned figures concerns performance within 98 days from calving. In addition, good effects were seen on the milk yields in the herds where the cows were eating averagely 60-100 grams extra minerals per cow per day in the critical transition period. Every other cow that calved in the period from August 2019 to July 2020 was allowed to eat extra mineral feed supplements from Pitstop+ feeders during the critical transition period from 21 days before calving and until 98 days after calving. The multi-purpose testbed activities were carried out at large dairy farms in Germany, Latvia, Lithuania and Denmark within the frames of the Internet of Food and Farm 2020 mega project. Precision Mineral Supplementation has emerged as an IoT technology to challenge the adverse development in dairy cow health and fertility, to decouple this from the productivity trend, and to solve the Gordian knot of correct supplementation of TMR fed dairy cows with minerals and vitamins, given an average gap of 20% between need and supply in the critical transition period, and an average 10% oversupply in the rest two thirds of the lactation. The mentioned average gap of about 20% corresponds to an undersupply of averagely 50-70 grams of mineral feed supplements per cow per day over the about 120 days long transition period. Similarly, cows typically receive 25-35 grams of mineral feed supplements more per day than they need in mid and late lactation. It is expected that the positive effect of Precision Mineral Supplementation is caused by the supplementation of the small body stores of micro-minerals of importance for the transition cow's immune status and thus for its health, fertility and productivity.

Influence of the social hierarchy within the herd on the individual drinking behaviour of dairy cows

A. Boudon, C. Gérard, M.-C. Meunier-Salaün, V. Lollivier and J. Lassalas
INRAE, Agrocampus Ouest, PEGASE, 35590, Saint-Gilles, France, 16, Le Clos, 35590 St-Gilles, France;
anne.boudon@inrae.fr

The absence of thirst is a fundamental criterion of animal welfare but it is very heterogeneously respected in dairy farming. Recommendations of the number and the location of drinkers exists but few studies have quantified the variability in drinking behaviour between cows from the same herd in relation to social hierarchy. A herd of 22 mid lactating cows was observed for 3 weeks. The cows were housed in a free stall and fed a total mixed ration distributed in individual troughs specific to each animal, with 2 meals a day. After each meal distribution, the cows were blocked at the feed fence for 1 hour. The drinking behaviour was monitored using 6 connected drinking troughs that recorded, for each drinking bout, the start time, duration, volume and identity of the cow. Six cameras also recorded the agonistic behaviours of the cows in the stall. Nine days of recording were analysed to build a matrix of agonistic interactions between individuals. 828 agonistic interactions were observed. They occurred mainly in the areas of the stall with drinking troughs and in the hours following the meals (61% of the agonistic interactions were observed within 2 hours following the morning meal). Drinking bouts were also very concentrated on the time after meals and some drinking troughs were clearly preferred over others depending on their distance from the troughs or their accessibility. The quantities of water drunk by drinking trough varied from 150.6 to 420.2 l/d. Four hierarchical groups were defined. The most subordinate groups had lower visit numbers and lower drunk volumes at the drinking troughs that were more used by the herd and the most subordinate group drank more than the other groups at the drinking troughs that were least used by the herd. The number of available drinking troughs was far higher than the most severe recommendations. It seems relevant, in the continuation of this work, to question the satisfaction of the needs for watering of the subordinate animals of a herd in systems strictly respecting the recommendations.

Blood parameters predictions based on milk MIR spectral data

L.M. Dale[1], K. Drössler[1], F. Gollé-Leidreiter[1], H. Spiekers[2], P. Hertel-Böhnke[2], E. Stamer[3], F. Onken[4] and A. Werner[1]
[1]State Association for Performance and Quality Inspection in Animal Breeding of Baden Wuerttemberg, Heinrich-Baumann-Str. 1-3, 70190 Stuttgart, Germany, [2]Bavarian State Research Centre for Agriculture, Prof.-Dürrwaechter-Platz, 85586 Grub-Poing, Germany, [3]TiDa, Animal and Data GmbH, Bosseer Str. 4c, 24259, Westensee, Germany, [4]German Association for Quality and Performance Testing e.V., Irmintrudisstraße 15, 53111 Bonn, Germany; ldale@lkvbw.de

Different metabolites such as nonesterified fatty acids (NEFA), β-hydroxybutyrate (BHB), glucose, insulin, IGF1, calcium and adiponectin are commonly used as parameters of negative energy balance in dairy cows. The objective of this study was to develop a mid-infrared (MIR) calibration based on blood parameters and on milk production in pre and postpartum Holstein and Simmental cows. Blood samples were collected from 893 animals in the prepartum (-14 to -1 day) and postpartum (+7 to +9 DIM, 28 DIM and 100 DIM) from 12 research farms of different German states such as Baden Württemberg, North Rhine-Westphalia, Bavaria, Schleswig-Holstein, Rhineland-Palatinate, Lower Saxony and Mecklenburg-Western Pomerania in between 2014 and 2017. The present work is part of the collaborative project optiKuh, funded by the German Federal Ministry of Food and Agriculture. Approx. 3,500 milk samples with NEFA, BHB, glucose, insulin, IGF1, calcium, adiponectin information were available for developing the individual model calibrations. The individual models were based on standardised and unstandardised spectral data. To detect the outliers a difference between fat determined by the FOSS or Bentley spectrometers and the RobustMilk equation was calculated. A difference lower than 2% was accepted, thus considering 30% of the data set as outliers. To identify animal variables that were positively or negatively associated with cow energy status, the spectral data set was first pre-processed by Savitzky-Golay first derivative to remove the offset differences between samples for baseline correction. Then three models were performed using the novel multivariate method Canonical Powered Partial Least Squares (CPPLS) by the 'pls' package in R. The 1st model was based on spectral data random selection, the 2nd on random selection of animals and the 3rd was a global model with a cross-validation trial. The global NEFA, BHB, glucose, insulin, IGF1, calcium and adiponectin calibration models showed high coefficients of determination (R^2=0.80, 0.80, 0.84, 0.80, 0.85, 0.79, respectively 0.85) and poor RPD (2.24, 2.23, 2.49, 2.22, 2.56, 2.20 respectively 2.59). The RPD is the ratio of standard deviation to standard error of cross validation. We have to underline that a poor model allows only rough screening.

Methane, energy balance and feed efficiency MIR predictions evaluation on dairy cows population

L.M. Dale[1], A. Werner[1], K. Drössler[1], F. Gollé-Leidreiter[1], H. Spiekers[2], P. Hertel-Böhnke[2], E. Stamer[3], B. Khula[4], F. Onken[5] and A. Vanlierde[6]
[1]State Association for Performance and Quality Inspection in Animal Breeding of Baden Wuerttemberg, Heinrich-Baumann-Str. 1-3, 70190 Stuttgart, Germany, [2]Bavarian State Research Centre for Agriculture, Prof.Dürrwaechter-Platz, 85586 Grub-Poing, Germany, [3]TiDa, Animal and Data GmbH, Bosseerstr. 4c, 24259, Westensee, Germany, [4]Leibniz Institute for Farm Animal Biology, Wilhelm-Stahl-Allee 2, 18196 Dummerstorf, Germany, [5]German Association for Quality and Performance Testing e.V., Irmintrudisstr. 15, 53111 Bonn, Germany, [6]Walloon Agricultural Research Centre, Chée de Namur 24, 5030 Gembloux, Belgium; awerner@lkvbw.de

Methane (CH_4) is one of the major greenhouse gas (GHG) emissions. Along with other sources it is naturally produced during the rumen fermentation process of ruminants. Different factors could have an impact on CH_4 quantity emitted i.e. feed, herd management or genetic selection for efficient animals. The aim was to compare milk MIR CH_4 predictions relevance obtained with different equations based on same reference dataset but including or not fix effects (1,203 records). The CH_4 models were built on SF6 tracer gas method and respiratory chambers measurements with PLS, CPPLS and GLMNET methods and took into account the stage of lactation. The models performances have been statistical validated in yearly dairy data production. A Pearson correlation analysis has been made with milk MIR indicators. For feed efficiency were found negative correlation with PLS model and positive correlation with CPPLS and GLMNET. Negative correlations were found between all CH_4 models and blood NEFA, C18-1Cis9, Energy-Detect, KetoMIR, INSAT and positive correlation showed SCFA and MCFA and SAT. The PLS, CPPLS and GLMNET models showed positive correlation with EB NEL and ME. The correlations fit well with the usual metabolic effects of extreme body fat mobilisation e.g. an increased concentration of NEFA, ketone bodies, long chained unsaturated fatty acids, a decreased concentration of MCFA and an extreme negative EB. The effect can be explained by reduced feed intake and rumen activity in these situations. The limits of applicability must still be defined in order to ensure the relevance of the predictions obtained before the models can be used in practical conditions.

Use of precision livestock farming tools to evaluate the performance of growing-finishing pigs

V.F.C. Cruz, R. Charneca, F. Baptista, T. Morgado and J. Rico
MED – Mediterranean Institute for Agriculture, Environment and Development, University of Évora, Ap 94, 7006-554 Évora, Portugal; jcsr@uevora.pt

The environmental conditions have an impact in animals' welfare and productivity. The use of PLF tools for environmental control and for animal feeding, as the electronic feeding stations (EFS) may help to optimise the production process. Already widely used in gestation phase, the EFS use is not so common in the growing/fattening phases. Within the AWARTECH (Animal Welfare Adjusted Real Time Environmental Conditions of Housing) project, three trials were carried out in an environmental controlled room where the environmental temperature (T) and relative humidity (RH) were permanently monitored and the feed intake and the pigs' weights were recorded by the EFS. Performance of growing-finishing pigs was tested in 3 different conditions: Winter (W) – T mean=13 °C and RH mean=74%; Thermoneutrality (TN) – T mean=21 °C and RH mean=74%; and Summer (S) – T mean=29 °C and RH mean=63%. In each trail, after a 15 d habituation period in TN conditions, 8 females (Piétrain × TN60) were fed *ad libitum* in a EFS (Schauer Compident MLP II) until a commercial slaughter weight. Recorded data was analysed by ANOVA using trial as fixed effect. The BW at the beginning of each trial was similar averaging 48±3 kg. Slaughter BW was higher in TN than in W (103±2 vs 96±2 kg, P<0.05) and in both cases no different than S (99±2 kg). The average daily feed intake (ADFI) wasn't different between trials although it was numerically lower in S (2.3±0.8 kg/d). The average daily gain (ADG) was higher in TN than in W (883±22 vs 778±22 g, P<0.01) and both were not different than S (851±22 g). The feed conversion rates (FCR) were, respectively for W, TN and S of 3.1±0.1, 2.9±0.1 and 2.8±0.1 with significant differences between W and TN (P<0.05) and W and S (P<0.001). The pigs were influenced by the environmental conditions with poor performance in W conditions. The S situation had no impact in the tested conditions. The use of PLF technologies allowed to record permanently and automatically the environment and pigs' data giving increased information in a shorter time and can be useful for an optimised production, reducing the welfare problems of group feeding.

Faecal-NIRS for predicting animal-to-animal variation in feed organic matter digestibility in cattle

D. Andueza[1], F. Picard[1], J. Pourrat[1], A. De La Torre[1], M. Devant[2], C.K. Reynolds[3], E. Froidmont[4], L. Bernard[1], C. Martin[1], P. Nozière[1] and G. Cantalapiedra-Hijar[1]
[1]INRAE, Centre de Theix, 63122 Saint-Gènes Champanelle, France, [2]IRTA, Torre Miramon, 08140 Caldes de Montbui, Spain, [3]University of Reading, School of Agriculture, Policy and Development, Reading, United Kingdom, [4]Walloon Agricultural Research Center, Rue de Liroux, 8, 5030 Gembloux, Belgium; donato.andueza@inrae.fr

Ruminant feed utilisation efficiency is determined in part by organic matter digestibility (OMD). However, OMD is expensive and time consuming to measure it in routine. Faecal near infrared (faecal-NIR) spectroscopy has been developed for predicting OMD in cattle across, but its potential for estimating the between-animal variability of OMD has not been reported. The objective of this study was to evaluate the use of of faecal-NIR for assessing the between-animal variability of OMD across different dietary and experimental conditions in cattle. Faecal samples from individual beef and dairy cows associated (n=454) to individual measured OMD values were used. Values ranged from 55.1 and 85.0% (mean±SD; 70.6±4.20%) for 9 experiments at 4 EU sites using 50 different diets. A NIR calibration model for predicting OMD was obtained and validated by cross-validation. Relationship between cross-validated OMD predictions and measured OMD values were analysed by regression analysis with faecal-NIR cross-validated predictions as fixed factor and diet within experiment as random factor. The faecal-NIR prediction was positive and correlated (P<0.001) to measured OMD. The model residual standard error was 2.05%. The effect of diet within experiment was significant (P<0.05) but minor since it representing 25% of error. For a given diet, the correlation coefficients between measured and faecal-NIR predictions when higher when measured OMD variability between animals was high. Our data highlight the potential of faecal-VIS/NIR to discriminate individuals in terms of OMD within a same diet. It could be used as an economical high-throughput tool for exploring between animal variability of OMD.

Better control of the sow health using a novel sensor embedded in an ear tag

A. Herlin[1], S. Dasen[2], P. Liechti[2], A. Peñaranda[3], I. Cuevas Martinez[3], A. Jara[3], A. Bugueiro[4], I. Gomez[4], C. Ollagnier[5], P. Renevey[2] and C. Verjus[2]
[1]Swedish University for Agricultural Sciences, Biosystems and technology, P.O. Box 103, 24355 Alnarp, Sweden, [2]Centre Suisse d'Electronique et de Microtechnique, Rue Jacccquet-Droz 1, 2002 Neuchatel, Switzerland, [3]HOP Ubiquitous S.L., Louis Bunuel 6, 30562 Ceuti, Spain, [4]Digitanimal S.L., C/Fredico Cantero VIllamil 2B, Mostoles (Madrid), Spain, [5]Agroscope, Rte de la Tioleyre 4, 1725 Posieux, Switzerland; anders.herlin@slu.se

The challenges in pig production involve improving competitiveness by better efficiency and control of production. A reduction of working hours and human presence in the pig barn, will make it difficult to manually, carefully monitor the pigs, and as, often subtle, disease symptoms can be unnoticed. There are critical points in piglet production with focus on the sow. The duration of the farrowing is important to establish as a prolonged farrowing can reduce the future fertility of the sow. The health of the sow around farrowing is critical. The post-partum dysgalactia syndrome (PPDS) is a complex of diseases where symptoms are vague but leading to reduced milk production, which has serious consequences for newborn piglets. The novel sensor includes a photoplethysmography sensor (PPG) and an accelerometer, which are integrated into an ear tag. It monitors physiological parameters of the sow (level of activity and resting heart rate). Changes in resting heart rate are linked to health issues (like fever) and the accelerometer will monitor the level of activity of the animal. We have developed and tested the system, including the capture of signals from the sensor, the collection of reliable data and the connectivity under practical conditions. The PPG sensor is sensitive to movement and we identified factors that could interfere with the PPG signal like heavy breathing, which was observed in sedated animals and the impact of small ear movements. The signal was acceptable on resting animals. The technical testing included also strategies for saving energy to optimise battery life with respect to data collection and intervals for data transfer. Further development includes data analysis and information management and application layers where processed information will be provided to the farmer in near real-time.

Thermal imaging cameras vs mobility scoring for early detection of digital dermatitis in dairy cows

T. Gerrard, J. Dodgson and A. Westland
Myerscough University Centre, Agriculture, Myerscough Hall, St Michaels Road, Bilsborrow, PR3 0RY Preston, United Kingdom; awestland@myerscough.ac.uk

Lameness in dairy cows is the most persistent challenge in both UK and global production systems. The most common cause of lameness is Digital Dermatitis (DD), which affects the soft tissue region of a cow's foot. Currently, the full impact of DD on farms is not properly understood with a prevalence rate of 31.6% and cases costing on average £50-£100 in UK production systems. An investigation comparing a high-tech thermal imaging camera (Flir© T420bx) with a mobile phone thermal imaging camera (Seek© compact XR 299 USD) was carried out over a five-week period to determine if thermal imaging (TI) can detect DD earlier when compared to mobility scoring. Temperature differences between the two cameras when comparing subjects with healthy feet versus lame animals was also investigated using three anatomical areas the coronary band (CB), interdigital gland (IG) and interdigital space (IS). Mobility scores, milk yields, parlour temperature and lying times were measured weekly to determine if TI temperature was affected by these factors. The 66 Holstein-Frisian cows were split into two groups, heifers and late lactation cows, to determine whether TI could detect DD in heifers just entering the milking herd. Significant differences between camera temperatures were identified (P≤0.05) in the CB and IS; however, there was no significant difference (P≥0.05) in the IG. A mobility score of 'two' presented a significantly higher temperature (P≤0.01) compared to 'zero' and 'one' for both cameras. Lame animals with higher foot temperatures had significantly (P≤0.05) lower milk yields and higher lying times compared to cows with healthy feet. Parlour temperature (°C) had no significant effect on TI temperatures (P≥0.05). Heifers displayed lower maximum TI temperatures in both lame and non-lame feet compared to cows. This study supports previous work that temperatures of 32 °C and above is a key indicator of the presence of DD in lame animals. Both cameras displayed higher TI temperatures for cows identified as lame compared to cows with healthy feet, indicating the potential of using TI cameras for quick effective detection of digital dermatitis.

Using a pecking detection system to test for correlations between pecking and injuries in turkeys

J.J. Gonzalez[1], A. Nasirahmadi[2] and U. Knierim[1]
[1]University of Kassel, Farm Animal Behaviour and Husbandry, Nordbahnhofstraße 1a, 37213 Witzenhausen, Germany, [2]University of Kassel, Agricultural and Biosystems Engineering, Nordbahnhofstraße 1a, 37213 Witzenhausen, Germany; jennifer.gonzalez@gast.uni-kassel.de

Turkey farmers often report an increasing birds' activity before a cannibalistic outbreak. A considerable proportion of this activity is likely beak related behaviour which consists of feeding, drinking, preening, and environmental and bird pecking. In order to test possible relations between changes in environmental pecking activity (PA) and abnormal bird pecking, a pecking detection system was newly developed based on AI. The aim of this study was to determine its detection performance and to analyse correlations between PA and injuries. This was carried out on a near-commercial German research farm in two batches of each 2,170 female turkeys (B.U.T. Big 6, intact beaks). Two pecking objects (PO1, PO2) were equipped with a microphone as automatic detection system, and above each a camera was mounted. Detection performance was ascertained in the first batch over a whole fattening period (13 weeks: week 2 to 14 of life) using randomly selected 300 s of audio and video data per week. In the second batch, PA was detected continuously during the rearing phase (n=27 days) for 12 hours during the light period. Number of birds separated or killed due to pecking was documented daily. Overall detection performance was very good, with an accuracy of 96.8%, sensitivity of 92%, specificity of 97.8%, precision of 89.6%, and F1-score of 90.8%. Daily means for PA during rearing correlated highly between PO1 and PO2 ($R_{Sperman}$=0.759, P≤0.000001, n=27). Therefore, PA at both objects was merged to PA_{total}. PA_{total} showed high standard deviations in the first week of detection, and PA levels started to decrease in week 3 from 592 to 32 pecks per day on day 19 of detection which was also the day with the highest number of injuries. PA_{total} showed a moderate negative correlation with the number of injured birds ($R_{Sperman}$=-0.487, P=0.009, n=27). The tested pecking detection system is promising for the valid recording of continuous PA, although it should be further tested under different farm conditions. The first hints on potential relations between object pecking and cannibalism need further investigation.

PLF to better understand the behaviour of young beef cattle at fattening

A. Philibert, B. Mounaix, E. Vanbergue and A. Aupiais
Institut de l'Elevage, DATA'STAT, 149 rue de Bercy, 75595 Paris Cedex 12, France; aurore.philibert@idele.fr

The behaviour of fattening cattle can be difficult to observe due to the density of animals in pens and to the scarcity of contacts with human (making the animals more reactive to observers and giving less opportunity for farmer to observe his animals). Behaviour is though a meaningful welfare indicator, especially in constrained farming systems. In 2018, 36 young Charolais cattle (MW 316 kg) were equipped in two farms with pedometers during the first month at fattening to explore their behavioural adaptation. Statistical analysis were carried out to better calculate behavioural indicators and their variability. Two peaks of activity by day were observed in relation to feeding procedures, with no significant effect of farm or pen. Inter-individual variability occurred and may be interpreted as a consequence of competition for feeding or drinking. A circadian rhythm was showed, with less time spent standing between 9:00 pm and 6:00 am. On average, cattle spent 14 hours lying, either standing or ruminating. Posture changes (standing to lying or reverse) were rare even during daytime: on average, very few cattle showed more than 1 change/hour. No significant (P>0.01) evolution of those behavioural indicators was observed between the day of arrival and 30 days later. No significant effect (P>0.01) was observed when cattle had been pre-conditioned in their farm of origin before entering the fattening farm. This first insight of the behaviour of cattle in French fattening systems (straw bedding <15 cattle pens) may indicate that this system does not affect aversely the time-budget of cattle. It suggests some rapid adaptation of most animals to housing conditions, but a possible impact of competition for some animals. Precision livestock farming proves to be efficient to better measure the behaviour of fattening cattle, provided the development of technologies accessible to farmers.

Long-term BCS monitoring by 3D camera assistant system

V. Bloch, M. Lidauer, P. Mäntysaari, T. Mehtiö, A. Kemppainen and M. Pastell
Natural Resources Institute Luke (Finland), Latokartanonkaari 9, 00790 Helsinki, Finland; victor.bloch@luke.fi

Continuous monitoring of body condition score (BCS) is important for effective management of dairy cows. Though, manual evaluation of the BCS is time consuming and cannot be performed with sufficient frequency, while commercial automatic systems are still not in common usage. The aim of this study was to develop a system suitable for longitudinal BCS monitoring. We developed a system based on 3D cameras (RealSense D435, Intel) installed on a scale after the exit from the milking parlour. The cameras were installed on 2.5 m height above the rear part of the cow back, symmetrically on both sides of the cow distanced by 0.5 m from each other, providing the top-side views parallel to the cow back. We developed a 3D image processing algorithm classifies the body condition according to the criteria commonly used by classifiers The algorithm replicates the analysis of shapes and curvatures of cow body parts characterising the BCS, including the shape of the hook bones, short ribs, vertebrae at the middle of the back, tail head and the line between the hook and the pin bones. Hence, the algorithms used the same BCS grading table, which is used by classifiers, with scores ranging from 1 to 5. The table of grading is once defined by a herd classifier according to the features of a specific breed. With the developed system, change in BCS is being monitored at a research herd with 127 Ayrshire cows (2.4±1.6 lactation) during a one-year period. To verify the developed BCS evaluation system, a manual scoring was performed by a classifier once every second week. In the first three months of the experiment 476 manual BCS scores classifications from the 127 cows were compared with the automatic scoring. The error was 0.36±0.31 (mean ± STD) units. In conclusion, the automatic BCS system with manually tuned grading table can provide a tool for constant monitoring of BCS considering the specialties of the herd.

Mask R-CNN based detection of group housed cows and automated definition of social contacts
J. Salau and J. Krieter
Kiel University, Institute of Animal Breeding and Husbandry, Hermann-Rodewaldt-Straße 6, 24118 Kiel, Germany;
jsalau@tierzucht.uni-kiel.de

Network analysis provides multiple parameters to analyse herd structure and animal interactions. In social networks the animals are regarded as nodes while animal contacts define the edges. In this study, video recordings from 8 cameras AXIS M3046-V were used to monitor the contacts within a herd of 36 Holstein Friesian cows housed in Northern Germany. Cows were recorded daily between morning and afternoon milking, and all cameras were installed 3.5 m above the centre line of the barn. The barn dimensions are 12×26 m and 37 lying cubicles and 18 feeding places are provided. Centrality parameters like degree and betweenness are associated with individual nodes of the network. The number of edges connected to a node is called the degree, and betweenness is the number of shortest paths that pass through a node. Parameters as diameter and density are associated to the whole network. Diameter is defined by the longest of all shortest paths between two nodes, and density is the ratio of the number of edges to the number of possible edges. Fully connected sub networks are called cliques. Convolutional neural networks (CNN) are a deep learning method in image analysis that uses a sliding dot product between an image and a much smaller kernel to extract detailed features. The Mask R-CNN framework for the segmentation of individual objects combines a deep CNN with a Region Proposal neural network to find objects in the images which is followed by an object classification network. On top of drawing bounding boxes around the found objects, Mask R-CNN provides an additional CNN to generate segmentation masks. This state of the art framework was trained on the recorded video material and applied to detect the animals. From these, contacts between cows were determined automatically. For 15 consecutive days, one undirected network per day was build. Degree and normalised betweenness ranged from 1 to 20, respectively, 0 to 0.147, implicating significant differences regarding the contact behaviour of the cows under analysis. Density and diameter of the daily networks varied between 0.19 to 0.31, respectively, 3 to 5. The maximal number of cliques larger than 3 cows and the size of the largest clique ranged from 49 to 255 and 3 to 6. As a stable herd was analysed, only moderate day to day variations of the social structure were reflected by the results. Future work focuses on parameter deviations due to replacements of animals in the group.

Cattle activity recognition using Recurrent Neural Networks and TensorFlow on Android Smartphones
J.F. Ramirez-Agudelo[1,2], S. Bedoya-Mazo[1], R. Muñoz-Tamayo[2], S.L. Posada-Ochoa[1] and J.R. Rosero-Noguera[1]
[1]Universidad de Antioquia, Grupo de Investigación en Ciencias Agrarias, GRICA, Calle 67 No. 53-108, 1226 Medellín, Colombia, [2]Université Paris-Saclay, INRAE, AgroParisTech, UMR Modélisation Systémique Appliquée aux Ruminants, 16 rue Claude Bernard, 75005 Paris, France; johnfredy.ramirezagudelo@inrae.fr

The use of pedometers or activity tags is expensive to record cattle's behaviour for short-time periods (e.g. 24 h). Under this particular situation, the development of low-cost and easy-to-use technologies is relevant. Similar to smartphone apps for human activity recognition, which analyses data from embedded triaxial accelerometer sensors, we develop an Android app to record activity changes in cattle. We used the Venelin Valkov's codes (https://github. com/curiousily), whom proposes a Recurrent Neural Network with Long-Short-Term Memory units to build this kind of projects. For our app development, three main steps were followed: data acquisition, model training, and app deploy. For data acquisition, we developed a system in which three components are used: two smartphones and a Google Firebase account for data storage. In this system, one phone is used by the human observer to record the actual cattle activity (walking, grazing, supplement eating, resting or ruminating) into Firebase. At the same time, the second phone, placed on the backside of the cattle's head, records the triaxial accelerometer values and the actual activity registered in Firebase into a text file every 100 milliseconds. During 30 days, 20 grazing dairy cows were individually monitored for the database building. For model training, the database (n=9,972,000) was reshaped in fixed-length sequences for each activity (80% for train and 20% for test). The model training was performed with 100 Epoch number, after which it was obtained a Test accuracy of 0.98 and Loss of 0.14. The performance of our training was assessed by the construction of the confusion matrix. For all actual activities, the trained model provided a positive high prediction (> 97%). The TensorFlow API was used to store learned parameters (history, predictions, and checkpoint), and to freeze and save the graph weights into a single protobuf file. Finally, Valkov's Android project was used as a template for our app deploy.

Air sample as a welfare friendly method to detect swine influenza A virus in pigs?

C. Fablet, S. Herve, V. Dorenlor, F. Eono, E. Eveno, N. Barbier, S. Gorin, A. Keita, G. Simon and N. Rose
ANSES, B.P. 53, 22440 Ploufgragan, France; christelle.fablet@anses.fr

This study aimed at assessing the ability to detect swine influenza A virus (swIAV) in bioaerosols collected at the pig's breathing zone compared to samples taken from live animals. It was carried out in five batches of pigs with influenza-like syndrome (rectal temperature $\geq$40.5 °C, coughs, sneezes). In each batch, nasal swabs were taken from a sample of six pigs with hyperthermia. Then, three groups of two pigs were formed. Bioaerosol samples were collected at the breathing zone (5 to 100 cm from the pig) of the two animals using a portable, silent and self-contained air sampler based on electrostatic precipitation (BIODOSI®, CEA-Tech, France). Dust particles were collected on a stainless steel ring placed in the sampler. The ring was then removed and placed into a transport medium. Three sampling times (5, 10 and 15 min) and two volumes (2 ml and 12.5 ml) of minimum essential medium (MEM) for transport were tested. All samples were analysed by M-gene real time RT-PCR. In the three batches where swIAV genome was detected in nasal swab supernatants, 4-6 pigs tested positive with Ct values ranging from 17.2 to 38.1. In those batches, at least one air sample also tested positive, with the 5 min×12.5 ml MEM sampling conditions having the lowest detection rate (5/9 positive samples) and the 15 min sampling time the highest (9/9 positive samples with 2 or 12.5 ml MEM). Ct values of air samples ranged from 31.4 to 39.5. This field study showed that swIAV may be detected in the air from infected pigs using handheld electrostatic precipitator device. Although the low apparent genomic loads in air samples (Ct >30) did not permit to investigate the virus infectivity in these samples, air sampling at the pig's breathing could be a promising way to conduct non-invasive viral surveillance among swine farms.

Automatic body position and behaviour detection of sows based on video images

S. Küster[1], P. Nolte[2], B. Stock[2] and I. Traulsen[1]
[1]*Georg-August-University, Department of Animal Sciences, Albrecht-Thaer-Weg 3, 37075, Goettingen, Germany,*
[2]*University of Applied Sciences and Arts Hildesheim/Holzminden/Goettingen, Faculty of Natural Sciences and Technology, Von Ossietzky-Str. 99, 37085, Goettingen, Germany; steffen.kuester@uni-goettingen.de*

The aim of this study was to determine if computer vision can be used to automatically recognise the behaviour and the body position of sows during final gestation. For this purpose, a two-step solution including object detection (1st step) and behaviour/body position classification (2nd step) was developed. The automatic detection of sow body parts and pen equipment (objects) was trained on labelled random video images (950) of sows crated in free-farrowing pens using a pre-trained supervised Convolutional Neural Network (YOLO v3). The average precision (AP) of the trained objects on the validation set (550) were 0.97 (straw trough), 0.95 (feeding trough), 0.97 (head), 0.86 (jute bag), 0.78 (tail), 0.78 (legs), 0.66 (teats). To classify behaviours and body positions based on the detected objects a deterministic algorithm was developed, which automatically classifies and outputs a body position or behaviour of the sow under consideration of context, temporal and geometric values of the detected objects. Finally, 57% of the defined body positions/behaviours (lying laterally (right, left), lying ventrally, standing, eating/drinking, nest-building behaviour and sitting) were correctly detected on the validation set and 18% were partially correctly detected (e.g. actual position: lying ventrally; determined position: lying laterally). Especially problems with the separation of the positions standing, lying ventrally and sitting increased the amount of fail detections since nearly the same body parts (objects) are visible for these positions. Videos/Images with depth information could improve the separation of these positions. In conclusion, an algorithm using an implemented Convolutional Neural Networks can successfully detect single body parts as well as pen equipment while behaviour classification algorithm needs to be improved.

Impact of immunocastration of gilts on instrumental and chemical traits of Teruel dry-cured hams

L. Pérez-Ciria[1], G. Ripoll[2], D. Allueva[1], F.J. Miana-Mena[1], M. Blanco[2] and M.A. Latorre[1]
[1]Universidad de Zaragoza-IA2, C/ Miguel Servet 177, 50013 Zaragoza, Spain, [2]CITA de Aragón-IA2, Avda. Montañana 930, 50059, Spain; leticiapcgm@gmail.com

In a previous trial carried out with gilts intended for the Protected Designation of Origin (PDO) 'Teruel dry-cured ham', a positive impact of immunisation against GnRH was detected on carcass fat thickness. According to the Regulation of this PDO, more than 16 mm of fat depth over the gluteus medius muscle are required to favour salting and avoid excessive drying of pieces. Nowadays, a study was carried out with 32 dry-cured hams of Duroc × (Landrace × Large White) gilts to evaluate the effects of immunocastration on weight losses during the dry-curing process and on instrumental and chemical characteristics. Half of them belonged to intact gilts (IG) and the other half to immunocastrated gilts (IMG). All pigs received the same management in the farm and in the abattoir. Hams were weighted several times during the dry-curing process (lasted 19 months). Once cured, colour, texture and chemical composition by near-infrared spectroscopy (NIR) and by classical analyses were determined in the biceps femoris muscle of 10 hams per treatment chosen at random. Data were analysed using the GLM procedure of SAS. The fresh-ham weight and the dry-cured-ham weight were used, when significant, as covariates for weight losses and for instrumental and chemical characteristics, respectively. Total ham weight losses during the dry-curing process tended to be lower (P=0.058) in IMG than in IG, leading to hams from IMG tended to be heavier (P=0.057). Yellowness (P=0.055) and chroma (P=0.017) were lower in IMG than in IG. Hams from IMG presented lower moisture (P=0.001) and water activity (P=0.015), higher intramuscular fat content (P=0.049 by classical analyses and P=0.077 by NIR) and tended to show higher ashes proportion (P<0.10) than those from IG. Under our experimental conditions, it can be concluded that immunocastration improves some parameters related to the quality of Teruel dry-cured hams of gilts. This work was funded by MINECO (Project AGL2016-78532-R) and by FITE and FEDER, through the operative programs 'Construyendo Europa desde Aragón'.

Sensory differences of Vinhais (Portugal) meat sausages with 3 ripening times

S.S.Q. Rodrigues, L. Vasconcelos, E. Pereira, A. Teixeira and F. Sousa
CIMO, ESA Instituto Politécnico Bragança, Campus Sta Apolónia, 5300-253, Portugal; srodrigues@ipb.pt

This work aimed to evaluate the effect of 3 (6, 9 and 12 days) ripening times on the sensory characteristics of Vinhais meat sausages produced according to the PGI specifications, to investigate the possibility of creating a new PGI product: Vinhais meat sausage to roast. Twenty-four appearance, odour, texture and taste attributes were evaluated by a qualified taste panel. The data were submitted to the product characterisation procedure. The results show that 13 of the 24 assessed attributes have significant discriminatory power over the sausages with different ripening times. The sausages with 12 days of ageing were firmer and harder, had a darker colour, more difficult chewability, and had a higher flavour intensity and persistence, also presented more spots in the exterior. Sausages with nine days of curing showed higher brightness, and also, not significantly different from 12 days sausages, darker colour after cooked, and not significantly different from 6 days sausages less firmness. Sausages with six days of ripening showed smaller values for most of the attributes. They were more tender, more homogeneous, and presented the lowest values of firmness, colour, chewability, brightness, flavour persistence, lighter muscle colour, more opaque aspect of the fat, fewer spots, more pink muscle colour and more intact wrapping.

Dietary net energy affects pork quality of finishing pigs

C.A. Moreira, L.P. Bonagúrio, N.Y. Sitanaka, L.A.C. Esteves and P.C. Pozza
State University of Maringa, Animal Science, Colombo Ave 5790, 87020900, Brazil; pcpozza@yahoo.com.br

Two experiments were carried out to evaluate the increasing dietary net energy (NE) levels, for 70- to 100- kg barrows, on the performance and pork quality. Experiment I: Twelve crossbred barrows, averaging 85.75±6.80 kg of initial body weight, were distributed in a randomised blocks design with two treatments, six replicates and one animal per experimental unit, in order to determine the metabolisable energy (ME) of the experimental diets with the higher and lower calculated ME content. The prediction equation used to estimate the NE was proposed as follow: NE = 0.730ME + 1.31EE + 0.37Starch – 0.67CP – 0.97CF. Experiment II: average daily feed intake (ADFI), NE efficiency, average daily gain (ADG) and feed:gain (F:G) were determined with 45 barrows, averaging 70.10±1.26 kg of initial body weight, distributed in a randomised blocks design, with five NE levels (2,345, 2,425, 2,505, 2,582, 2,665 kcal/kg), nine replicates and one animal per experiment unit. The diets with higher and lower NE content were mixed and after that were diluted one another to provide the intermediate NE levels (2,425, 2,505, 2,582 kcal/kg). Drip loss (DL) and colour parameters as L*(luminosity), a*(red – green) and b*(yellow – blue) were determined in the Longissimus lumborum (LL) muscle. The degrees of freedom related to NE levels were deployed in orthogonal polynomials to obtain the regression equations. The studied dietary NE levels provided a linear reduction (P=0.00039) of the ADFI. Quadratic effects were observed for NE efficiency (P=0.0027), ADG (P=0.0352) and F: G (P=0.0024), estimating the optimum NE levels at 2,485, 2,493 and 2,533 kcal/kg, respectively. It was also observed a linear reduction of DL (P=0.0001) and Minolta colour (+a*) parameter (P=0.0462) of the LL due to increasing dietary NE levels. In conclusion, the dietary NE levels affected the performance and pork quality of finishing pigs, and the level of 2,493 kcal/kg provided the higher ADG.

Post-mortem meat inspection using live-video transmission with remote guidance

V. Almqvist[1], C. Berg[1], A.H. Kautto[2] and J. Hultgren[1]
[1]Swedish University of Agricultural Sciences, Department of Animal Environment and Health, Box 234, 53223 Skara, Sweden, [2]National Food Agency, Control Support, Box 622, 75126 Uppsala, Sweden; viktor.almqvist@gmail.com

Consumers' demand for locally produced food increases. Small-scale abattoirs are often situated in the countryside, and often slaughter a low numbers of animals, which entails substantial costs for authorities that perform mandatory veterinary inspections. With current environmental concerns, and travel costs expected to rise, there are strong incentives to streamline veterinary inspections. Advances in telemedicine and streaming video techniques might make inspections via remote video a viable alternative to on-site inspections. A centrally located inspecting veterinarian could guide a non-trained assistant at the abattoir during inspections using video and audio equipment. We compared traditional on-site inspection with remote inspection using live-video software designed for remote guidance using an augmented-reality overlay. The study was conducted at a Swedish large-scale pig abattoir during the spring of 2019. During 28 days, two trained veterinarians inspected 400 carcasses (including organs), using one of the described methods each, recording the presence of 31 types of lesions in accordance with governmental instructions. Following routine inspection by regular control staff, the studied carcasses were arrested for additional veterinary inspection. To include healthy controls, carcasses without any indication of lesions were also arrested, in random order. The veterinarians shifted method repeatedly and used both methods an equal amount of times. Most common lesions were other pneumonia (neither enzootic pneumonia nor porcine pleuropneumonia), tail lesions and pleuritis/pericarditis, at an estimated prevalence of 48.9, 43.9 and 30.6%, respectively. The highest Cohen's kappa values were obtained for tail lesions (0.77), abscess (0.76) and other pneumonia (0.74), and the lowest for old injury (-0.01), porcine pleuropneumonia (0.21) and systemic disease (0.24). After merging lesion types into 13 logical categories, fixed-effects logistic regression revealed no effect of method on the probability of recording any specific lesion category. In conclusion, remote post-mortem meat inspection using video transmission and guidance by augmented reality appears promising.

Growth modulation of Sasso C44 roosters in the 'Galo de Barcelos' production system

P.S. Vaz[1,2], J.P. Araújo[2,3], J.L. Cerqueira[2,4] and J. Oliveira[5,6]
[1]Quinta de Eira Vedra, Ardegão, Ponte de Lima, 4990-535, Ponte de Lima, Portugal, [2]Escola Superior Agrária do Instituto Politécnico de Viana do Castelo, Refóios do Lima, 4990-706 Ponte de Lima, Portugal, [3]Centro de Investigação da Montanha (CIMO), Instituto Politécnico de Viana do Castelo, 4900-347, Viana do Castelo, Portugal, [4]Centro de Ciência Animal e Veterinária (CECAV), UTAD, UTAD, 5000-801, Vila Real, Portugal, [5]Centro de Investigação e de Tecnologias Agroambientais e Biológicas (CITAB), UTAD, 5000-801, Vila Real, Portugal, [6]Escola Superior Agrária de Viseu, Instituto Politécnico de Viseu, 3500-606 Viseu, Portugal; psantosvaz@gmail.com

Specifications of 'Galo de Barcelos' production system imply the use of autochthonous or slow growth roosters and a well-defined pattern of feeding, housing and sanitary rules. These characteristics limit the potential growth of the animals, conditioned by the purpose of improving the meat quality and organoleptic attributes, creating an exclusive product. Every animal is identified with a wing tag with a serial number. The present work aims to characterise growth behaviour of the Sasso C44 roosters and identified the best age to slaughter the animals using different nonlinear models. The non-linear models of Gompertz, von Bertalanffy, Brody and Logistic were used. Data editing was performed in the excel spreadsheet and descriptive analysis and modelling was obtained through SPSS v25. There were monitored 379 roosters produced on a farm associated to 'Confraria G. O Galo de Barcelos', from april 2017 to october 2019, with 4.0 ± 1.6 weights per animal, comprising a minimum of 2 and a maximum of 9 records (total of 1,522 weights). For all roosters, it was considered, according to the strain standard reference, the weight of 39 g for the day-old chick. The last weight for some roosters was obtained at the 193 days-old. The Gompertz model was the one with the best fit to the data, with a determination coefficient of 0.887, followed very closely by von Bertalanffy model, with an $R^2=0.884$. The Brody and Logistic models presented a R2 of 0.715 and 0.518, respectively. In conclusion, Gompertz model is adequate to estimate the growth of the Sasso C44 roosters in this production system, allowing to define the best moment for slaughtering the animals.

Poster discussion Part 1: Pig, poultry and rabbit husbandry for improved product quality

S. Millet[1] and K. Stadnicka[2]
[1]Flanders research institute for Agriculture, Fisheries and Food, Animal Sciences Unit, Scheldeweg 68, 9090 Melle, Belgium, [2]UTP University of Science and Technology, Animal Breeding and Biology, Mazowiecka 28, 85-084 Bydgoszcz, Poland; katarzyna.stadnicka@utp.edu.pl

This interactive discussion will allow grasping the latest knowledge as to how various aspects of animal management affect product quality. The presented research mainly involves (but is not restricted to) to dietary strategies, effects of immunocastration, and farming management.

Poster discussion Part 2: Pig, poultry and rabbit husbandry for improved product quality

S. Millet[1] and K. Stadnicka[2]
[1]*Flanders research institute for Agriculture, Fisheries and Food, Animal Breeding and Biology, Scheldeweg 68, 9090 Melle, Belgium,* [2]*UTP University of Science and Technology, Animal Breeding and Biology, Mazowiecka 28, 85-084 Bydgoszcz, Poland; katarzyna.stadnicka@utp.edu.pl*

This interactive discussion will allow grasping the latest knowledge as to how various aspects of animal management affect product quality. The presented research mainly involves (but is not restricted to) to dietary strategies, effects of immunocastration, and farming management.

Oleuropein extract supplementation in pig diet improves drip loss and stability of meat

A.I. Rey[1], A. De-Cara[1], L. Calvo[2], P. Puig[3] and T. Hechavarria[3]
[1]*Faculty of Veterinary Science, Animal Production, Avda. Puerta de Hierro s/n, 28040 Madrid, Spain,* [2]*Incarlopsa, i+D department, Ctra. N-400 km. 95400, 16400 Tarancón. Cuenca, Spain,* [3]*Andres Pintaluba, S.A., i+D department, Polígono Industrial Agro-Reus Prudenci Bertrana, 5, 43206 Reus, Spain; anarey@ucm.es*

There is scarce information of the specific supplementation of oleuropein extract on stability of pork. The aim of this investigation was to study the effect of dietary oleuropein extract supplementation in comparison with vitamin E and selenium or their combination on some meat quality characteristics such as drip loss, and thiobarbituric acid reactive substances (TBARS). Pigs (Large white × Landrace) were distributed into four groups (n=12/per group) and fed the experimental diets ((1) control: C; (2) α-tocopheryl acetate + 0,26 mg/kg selenium: VE; (3) 96 mg oleuropein/kg diet: OLE; and (4) the combination of 192 mg oleuropein/kg + 100 mg α-tocopheryl acetate/kg + 0.26 mg selenium /kg: VEOLE) during the last 35 days of fattening from 60.2±1.5 kg until final averaged weight of 121.4±2.3 kg. Fresh samples from the longissimus lumborum muscle at the level of the last rib (15 cm in size) were used for measurements. Drip loss percentage was determined by difference between weight loss after 72 hours of storage in refrigerated conditions at 4 °C. TBARS were quantified spectrophotometrically in meat on day 0, 3, 5 and 8 of refrigerated display under fluorescent light after heating sample extracts in presence of thiobarbituric acid. The statistical analysis of data was carried out using the statistical program SAS v. 9.3. The drip loss of muscle was lower in those groups supplemented with antioxidants when compared to control (P=0.05). No differences were detected in drip loss percentage between different groups supplemented with antioxidants. TBARS averaged values reached the lowest production in VEOLE whereas C group had the highest when compared to the other groups (P<0.05); being the averaged production of TBARS intermediated and similar in VE and OLE groups. In conclusion, dietary supplementation of oleuropein at 96 mg/kg as extract presentation could be incorporated into diets in substitution of vitamin E to improve stability and other quality characteristics such as drip loss of pork in Mediterranean countries.

Effect of dried rose petals on performance, meat quality and boar taint of non-castrated male pigs

S. Ivanova[1], T. Stoyanchev[2], T. Nikolova[1] and I. Penchev[2]
[1]Agricultural Academy, Agricultural Institute, Animal Nutrition and Technology, 3 Simeon Veliki blvd, 9700 Shumen, Bulgaria, [2]Trakia University, Faculty of Veterinary medicine, Student Campus, Rectorat, 6015 Stara Zagora, Bulgaria; ivanovapeneva@gmail.com

The residue of distilled rose (*Rosa damascena*), which is obtained after the production of rose oil and rose water in Bulgaria, is rich in flavonoids and could alter the mechanism of development of boar taint in the liver. The aim of this study was to investigate the effect of dried rose petals on the performance, meat quality and boar taint in pork from non-castrated male pigs. A total of 30 male pigs from Danube white breed with an average body weight of 67 kg were used in the experiment. Pigs were equalised by origin, age, and weight and were placed in group pens and fed *ad libitum*. Pigs were allocated to one of the three groups – control, castrated pigs (CON); non-castrated pigs without any additive (NCAS) and non-castrated pigs with addition of 5 g of dried rose petals in 1 kg of feed (NCAS+R). At the end of the experiment at 186 days, after 40 days, they were slaughtered at about 108-109 kg live weight. Before slaughter, the analyses of backfat thickness were made and blood samples were taken. Sensory evaluation of boar taint after heating of all carcasses was made by two independent trained experts. Parameters of pork quality – carcass weight, pH, the area of MLD, percentage of lean meat, water holding capacity and the meat colour were measured. There were no statistically significant differences in weight development, blood parameters and meat quality in general between groups. The statistically significant differences were measured in backfat thickness (P<0.001) between castrated and the two groups of non-castrated pigs as well as in lean meat percentage (P<0.01), but not between NCAS and NCAS+R pigs. The testosterone level was about twice smaller (P<0.05) in NCAS+R pigs in comparison to NCAS pigs. The percentage of smell was higher (40% against 30%) in the carcasses of NCAS pigs in comparison with the carcasses of NCAS+R pigs. The colour of their meat was characterised by more reddish coloration (P<0.05).

Preliminary evaluation of local pig breed Sarda and its Duroc crossbred for carcass and meat quality

A. Di Gianvito[1], G. Spanu[1], A. Fenu[1], M. Acciaro[2], S. Ampuero Kragten[3] and G. Battacone[1]
[1]University of Sassari, Department of Agricultural Sciences, viale Italia, 39, 07100, Italy, [2]Agenzia Agris Sardegna, Department of Research for Animal Production, Loc. Bonassai, 07040, Olmedo (SS), Italy, [3]Agroscope, Tioleyre 4, 1725, Posieux, Switzerland; battacon@uniss.it

The Sarda pig is the native breed of Sardinia (Italy) with low growth rate a high genetic potential for deposition of adipose-tissue which is often too high and not always suitable for the requests of consumers and the pork industry. Therefore, the crossbreeding of the Sarda breed with a leaner breed, such as Duroc, could be an interesting way to sustain and protect the local breed, as the essential core to obtain F1 pigs having features suitable for traditional meat industry. The aim of this research was to study the effects of genotype, Sarda (SA) and Sarda × Duroc crossbreed (S×D), and gender (immunocastrates IM and females F) on carcass and meat quality traits. All males were vaccinated with Improvac® at the age of 155 and 200 days. A total of 15 carcasses (9 IM and 6 F), from animal reared in the same experimental conditions, were considered. From the left half carcass, the backfat thickness, the weight of primal cuts for traditional curing processing were recorded. Concentrations of skatole and androstenone were quantified in backfat of IM. The ANOVA was carried out to assess significant differences. Carcass and primal cuts weight of S×D were significantly higher (P≤0.001). No effects of breed were detected in lean meat and fat thickness. Sex had no influence on carcass or primal cuts. The proportion of primal cuts was not affected by genotype and sex with the exception of belly (P≤0.05) resulted with higher yield for SA. Trimmed cuts, ham cheek and belly, intended for traditional dry-curing process resulted higher in weight (P≤0.001) for S×D. In backfat of IM the mean concentration of androstenone was 0.17 and 0.18 01 µg/g in S×D and SA animals, respectively; while that of the skatole were 0.01 and 0.02 µg/g. Although preliminary, this study showed how the crossbreeding with Duroc could balance some limits of Sarda pure breed, providing heavier carcasses and cuts that are more suitable for curing and processing.

Red wine in cured Bísaro pig ham to minimise salt content and improve physico-chemical atributes

J.P. Araújo[1,2], J.L. Cerqueira[2], M. Barros[3], E. Fernandes[3], R. Rodrigues[2], B. Lebret[4] and P. Preciosa[3,5]
[1]*Centro Investigação Montanha (CIMO), Instituto Politécnico de Viana do Castelo, Viana Castelo, 4900-347, Portugal,*
[2]*Escola Superior Agrária, Instituto Politécnico de Viana do Castelo, Refóios do Lima, 4990-706 Ponte Lima, Portugal,*
[3]*Escola Superior de Tecnologia e Gestão, Instituto Politécnico de Viana do Castelo, Avenida do Atlântico, 4900-348
Viana Castelo, Portugal,* [4]*INRA, UMR PEGASE, 35590 Saint-Gilles, France,* [5]*CISAS, Instituto Politécnico de Viana do
Castelo, Rua Escola Industrial e Comercial de Nun'Álvares, 4900-347 Viana do Castelo, Portugal; pedropi@esa.ipvc.pt*

Bísaro pig is a local breed located mainly in the North of Portugal. For obtaining high quality traditional products from this breed, the present work aims to develop a pork ham with low salt content, evaluate chemical, colorimetric and texture indicators and compare with the traditional ham manufacture. For this study, 30 Bísaro animals were subjected to the same semi-extensive production system. All animals were slaughtered with live weights of 120.53±13.12 kg at 287.86±14.97 days of age. The hams were removed from the left side of each carcass and were divided into two groups. In the control group, the pork hams were manufactured by a traditional method (n=14). In a second group (n=14) red wine was added during the manufacturing process. After the manufacture period of thirteen months all hams were analysed. There was no significant difference in pH, moisture, protein, fat and minerals. However the salt content (% NaCl w/w), TBARS (mg MDA/kg) and water activity values of pork ham manufactured with the addition of red wine were reduced compared to the control group (P<0.017 and P<0.001).Addition of red wine increased the redness value in the hams from the innovative process (P<0.001) compared to the control group. In texture profile values of shear force, brittleness, gumminess and adhesiveness, there was no significant difference. In conclusion, the addition of red wine during ham manufacture showed the effect of increasing the quality of product, mainly the parameters related to low salt content and oxidation index, with focus on anti-oxidant activity of red wine. This innovative manufactured ham may represent an alternative to the traditional form of ham production.

Effect of NaCl replacement by KCl and SubSalt4 on the mineral content of pork meat sausages

A. Teixeira[1], I. Ferreira[1], E. Pereira[1], R. Dominguez[2], J.M. Lorenzo[2] and S.S.Q. Rodrigues[1]
[1]*CIMO, ESA Instituto Politécnico Bragança, Campus Sta Apolónia, 5300-253, Portugal,* [2]*CTC, Centro Tecnologico de
la Carne de Galicia, Rúa Galicia N° 4, Parque Tecnologico de Galicia, San Cibran das Viñas, 32900 Ourense, Spain;
srodrigues@ipb.pt*

This work aimed to evaluate the effect of NaCl reduction and substitution by KCl and SubSalt4 on the mineral content of pork meat sausages with 4 different salt formulations (Form1: 2% NaCl, Form3: 1.5% NaCl+0.5% KCl, Form4: 1.5% Sub4 salt+0.5% NaCl, Form5: 0.5% Sub4 salt+1.5% NaCl) and two ripening times. The quantification of mineral elements (Na, K, Ca, Cu, Zn, Fe, P and Mn) was performed by inductively coupled plasma-optical emission spectroscopy (ICP-OES). Three repetitions and two replicas of the experiment were made for each type of sausage. Statistical analysis was performed using the statistical package JMP Pro 11.1.1 by Copyright © 2013 SAS Institute. Main effects (salts treatment, ripening time) and interaction were tested as fixed effects and the random effects of repeated measurements (3) of individual sausage and two replications. Results show a significant increase in the minerals' content with ripening time increase. The NaCl reduction and substitution by KCl promoted a higher Na reduction than SubSalt4, and the total substitution of NaCl by KCl and SubSalt4 had the highest reduction in this mineral. Also, significant differences were found in Ca, K and P. The partial or total Na replacement did not affect the ripening process once there are no significant differences between treatment for water activity (a_w), it was around 0,91 and 0.84-0.87 for the ripening periods. The ashes content, between 4 and 6% of the sample, was not affected by the different treatments.

Effect of immunocastration of male pigs on instrumental and chemical traits of Teruel dry-cured hams

L. Pérez-Ciria[1], G. Ripoll[2], D. Allueva[1], M. Blanco[2], F.J. Miana-Mena[1] and M.A. Latorre[1]
[1]*Universidad de Zaragoza-IA2, C/ Miguel Servet 177, 50013 Zaragoza, Spain, [2]CITA de Aragón-IA2, Avda. Montañana 930, 50059 Zaragoza, Spain; leticiapcgm@gmail.com*

Pigs intended for Teruel dry-cured ham elaboration must reach a high slaughter weight (around 130 kg) to meet the desirable quality. Therefore, castration of male pigs is necessary to avoid boar taint. Traditionally, it has been carried out surgically, but considering piglet welfare, alternatives are being researched; among them, immunisation against GnRH stands out. Thus, a study was carried out with 14 dry-cured hams of Duroc × (Landrace × Large White) male pigs to evaluate the impact of immunocastration on weight losses during the dry-curing process and on instrumental and chemical characteristics. Half of them belonged to surgical castrated males (SCM) -during the first week of life- and the other half to immunocastrated males (IM) -by three doses of Improvac® at approximately 20, 60 and 80 kg of body weight-. All animals received the same management in the farm and in the abattoir. Hams were weighted several times during the dry-curing process (lasted 19 months). Once cured, colour, texture and chemical composition by near-infrared spectroscopy (NIR) and by classical analyses were determined in the *Biceps femoris* muscle. Data were analysed using the GLM procedure of SAS. The fresh-ham weight and the dry-cured-ham weight were used, when significant, as covariates for weight losses and for instrumental and chemical characteristics, respectively. There were no differences in weight losses during the dry-curing process (salting, post-salting, drying, aging and total period), neither in chemical composition (moisture, ash, protein and intramuscular fat proportions) nor in Warner-Bratzler shear force (hardness) between hams from IM and those from SCM (P>0.10). Only some colour variables were affected; yellowness (P=0.045) and chroma (P=0.015) were lower in IM than in SCM. Under our experimental conditions, it can be concluded that the type of castration has scarce influence on the quality of Teruel dry-cured hams of male pigs. This work was funded by MINECO (Project AGL2016-78532-R) and by FITE and FEDER, through the operative programs 'Construyendo Europa desde Aragón'.

Scrotal morphometry of Bísaro pig males treated with two or three administrations of Improvac®

G. Paixão[1], J. Marques[2], S. Botelho[1], A. Esteves[1], R. Charneca[2] and R. Payan-Carreira[2]
[1]*CECAV, Animal and Veterinary Research Centre, University of Trás-os-Montes e Alto Douro, Quinta dos Prados, 5000-801 Vila Real, Portugal, [2]MED – Mediterranean Institute for Agriculture, Environment and Development, University of Évora, Ap 94, 7006-554 Évora, Portugal; rmcc@uevora.pt*

Seventeen male Bísaro (BI) pigs were treated from 13 weeks of age with two (n=6; Grp1) or three injections (n=11; Grp2) of Improvac® (Zoetis), 4-weeks apart. Animals were reared in intensive conditions and fed *ad libitum*, until 29 (Grp1) and 33 (Grp2) weeks of age. Scrotal dimensions (total width; average height and width of the left and right scrotal sac) were measured every 4 weeks from the first injection (FI) until the 16th week post-administration, and the variation between every measurement were calculated in order to surveil the immunosuppressive effects of the vaccination protocols. Variation were also compared with the normal scrotal growth in non-treated, age-matched animals (GrpC; n=24). Variation of scrotal dimensions did not differ significantly between treated and control groups, from the first to the 2nd administration of Improvac. After the 2nd injection, however, the variation and the absolute scrotal dimensions differ between treated and control groups (P≤0.001). When comparing between treated groups, the decrease was higher in Grp1 for total width (P=0.074) and average height (P=0.0025). The decrease is reverted between the 8th and the 12th week post FI for Grp1, even though the overall dimensions of the scrotum remained significantly lower than in Grp2 and GrpC, and below the initial values for the group. Animals in Grp2 continuously registered a decrease in the scrotal width until the last observation (16 weeks after FI). The size of the scrotum for both treatment groups remained significantly lower than the control group by the end of the evaluation. This study showed that in prepubertal Bísaro pigs, the immunocastration protocol with three administrations of Improvac allows a more consistent and prolonged suppression of the reproductive axis, as reflected by the subsequent decrease in the scrotal dimensions. This protocol seems to better match the foreseen for the Bísaro system, where producers sought to attain one-year old animals at the time of slaughter.

Non-destructive technologies to evaluate hams' characteristics from immunocastrated Iberian pigs

J. García-Gudiño[1,2], M. Gispert[1], A. Brun[1], I. Blanco-Penedo[3], F.I. Hernández-García[2], M. Izquierdo[2] and M. Font-I-Furnols[1]
[1]*IRTA, Product Quality, Finca Camps i Armet, 17121 Monells, Spain,* [2]*CICYTEX, Animal Production, Finca La Orden, 06187 Guadajira, Spain,* [3]*SLU, Department of Clinical Sciences, P.O Box 7054, 75007 Uppsala, Sweden; javier.garciag@juntaex.es*

Imaging technologies such as computed tomography and ultrasound can be used in livestock. All heavy pigs need to be castrated and immunocastration is an alternative to surgical castration. The objectives of the present work were: (1) to determine the relationship between fat thickness (measured *in vivo* by ultrasounds and post mortem by a ruler) and the composition of the ham determined with computed tomography; (2) to evaluate the feasibility of ultrasounds used *in vivo* to find out differences in fatness between animals from different treatments; and (3) to evaluate the effect of early (pre-pubertal) and late (pre-finishing) immunocastration on carcass fatness and ham tissue composition of the Iberian pigs evaluated with non-destructive technologies (ultrasound scanning and computed tomography). Twenty pure Iberian pigs were used, 10 were early immunocastrated (4.5, 5.5 and 9 months) and 10 were late immunocastrated (11, 12 and 14 months). Ultrasounds were used to measure fat thickness in the loin and ham regions. Pigs were slaughtered at 17 months of age and loin fat thickness was measured with a ruler. Ham was computed tomography scanned and their composition was determined. Results show a good correlation between measurements obtained *in vivo*, in the carcass and in the ham with the different technologies and that ultrasounds are useful to determine differences in fatness in live animals. Carcasses from early immunocastrated pigs were heavier and fatter than those from late immunocastrated pigs. Thus, depending on the requirements of the final product, vaccination protocol needs to be adapted.

Performance of immunocastrated or surgical castrated Bísaro pig breed males

R. Charneca[1,2], J. Marques[1], G. Paixão[3] and R. Payan-Carreira[1,2]
[1]*Universidade de Évora, Ap 94, 7006-554 Évora, Portugal,* [2]*MED – Mediterranean Institute for Agriculture, Environment and Development, Ap 94, 7006-554 Évora, Portugal,* [3]*Universidade de Trás-os-Montes e Alto Douro, CECAV, Quinta dos Prados, 5000-801 Vila Real, Portugal; rmcc@uevora.pt*

Twelve male pigs of the Portuguese breed Bísaro (BI) raised in intensive conditions and fed *ad libitum* were used to test the effects of imunocastration on growth performance. Pigs were reared until about 140 kg body weight (BW) and 8.5 months of age. In the facilities, two temperature environments were observed during the trial: thermoneutrality (TNE; 15-25 °C average temperature) until ~97 kg BW and cold (CE; 10-15 °C), from 97 to 140 kg BW. Two shots of Improvac vaccine were administrated, 4-week apart, to 6 pigs (immunocastrated group; IMC) at 83±5 d of age (35±2 kg BW) and at 111±5 d (58±2 kg BW). Another 6 pigs (SC group) were surgically castrated at an average age of 94±4 d and 38±2 kg BW. Starting at 125±4 d and 67±7 kg BW, ultrasound measures of backfat thickness (BF – at P2 point) and *Longissimus dorsi* depth (LD; loin eye) were performed monthly. Considering the 2nd Improvac doses or surgical castration as effective sex neutralisation date, no differences were observed in average daily gain (g/d, ADG) between groups (581±30 in IMC and 570±30 in SC, P=0.806) until slaughter, even when adjusted to the same initial weight. There were no significant ADG differences between groups in both temperature conditions, although ADG was lower in CE vs TNE (449±25 vs 764±25, P<0.001). When adjusted to BW at first measure date, IMC pigs showed lower BF than SC pigs (9.7±0.8 vs 15.1±0.8 mm; P=0.003) but similar LD depth. When adjusted to the final weight, LD depth remained similar in both groups but BF tend to be higher in CC pigs (31.5±1.7 vs 25.9±1.7 mm, P=0.064). LD depth gain (%) was not different between groups but tend to be higher in TNE than in CE conditions (P=0.064). BF gain was also no different between groups but was significantly higher in TNE (89 vs 26%, P<0.001). Even though contrary to the reported in the literature, this study shows no or minor effects of immunocastration on the performance of Bísaro pigs raised in intensive conditions, despite the high impact of thermal environmental conditions both on growth and fat deposition.

Distinctive natural solutions to finishing heavier male pigs without castration and low boar taint

L. Martin, R.P.R. Da Costa, M.A.P. Conceição and A. Ramos
Polytechnic of Coimbra, Coimbra Agriculture School (ESAC) Bencanta, 3045-601 Coimbra, Portugal; luisam@esac.pt

Advance the knowledge on raising male pigs for heavy carcasses with decreased boar taint by natural means like feeding and housing is the main objective of this work. We tested the effective inclusion rates of sugar beet pulp and chicory root prebiotics in finishing diets as well as space allowance and enrichment to provide better health and welfare. The aim is to study solutions to raise heavy male pigs without surgical or chemical castration. A factorial (3×2) design was implemented to estimate main effects and interactions of 3 diets and 2 housing treatments, on pigs plasma indole, skatole, androstenone and cortisol through ANOVA. 60 pigs starting at 5 months and slaughtered at 7 months of age, randomly allocated to 6 parks for different treatments were housed at EU welfare requirements or better. Plasma from jugular blood collected at the beginning middle and end of the experiment was analysed for androstenone, skatole and indole by a UHPLC-HR-Orbitrap-MS validated method and for cortisol by ELISA. Skin lesions were quantified based on the welfare quality® protocol and health accessed at post-mortem and by hemogram results interpretation (blood samples collected 2 days before slaughter). The results showed that both diets with sugar beet pulp or sugar beet pulp and chicory had a significant effect on reducing plasma androstenone. Plasma skatole and indole were for most samples bellow quantification limit. Improved housing had a significant effect on reducing plasma androstenone and cortisol. A positive interaction was found between diet and housing with a decrease of androstenone accentuated by the two factors together. Better health and welfare indicators could be found for animals under improved housing.

Herd structure and production efficiency in Iberian pig farms

S. Sanz-Fernández[1], C. Díaz-Gaona[1], P. López Romero[2], R. Quintanilla[2], N. Alòs Saiz[2] and V. Rodríguez-Estévez[1]
[1]University of Cordoba, Animal Production, Campus Universitario de Rabanales, 14071. Córdoba, Spain, [2]Institut de Recerca i Tecnologia Agroalimentàries, Torre Marimon, 08140. Caldes de Montbui (Barcelona), Spain; v22safes@uco.es

The production efficiency of pig farm depends on its herd structure and the productivity of each parity. To maintain a steady production rate (weaned piglets per sow and year, WPSY), farms must have a balanced herd structure with the highest percentage of sows in the first parity and a progressive reduction throughout the following cycles. The aim of this study is to evaluate the herd structure of the 50 Iberian pig farms included in the database BDPorc-I managed by IRTA. This has been evaluated according two levels of productivity in 2019: 25% of farms with the highest productivity and rest of the farms. Besides, these farms have been classified into 3 groups of herd structure: G1, census structure in triangle with few sows of ≥8 parities; G2, census structure in triangle but with more sows of ≥8 parities than in 7th parity; G3, irregular census structure. In addition, an analysis of variance to compare means and Pearson's linear correlation coefficient are performed. The most productive farms tend to have a higher percentage of sows in the first parities (with a triangle structure) than the other farms: 1st parity (P), 18.11 vs 15.88; 2nd P, 16.18 vs 15.00; 3rd P, 14.58 vs 12.94; 4th P, 12.49 vs 10.77; 5th P, 11.10 vs 9.82; 6th P, 10.05 vs 9.82; 7th P, 8.06 vs 9.31. However, between both productivity groups there only are significant differences ($P<0.05$) in $\geq8^{th}$ P, 9.43 vs 16.39, with lower percentage in the most productive farms. Besides, there is a negative Pearson's correlation between $\geq8^{th}$ P percentage and number of farrowings per sow and year ($r=-0.439$, $P\leq0.05$). Average productivity comparison of farms according to their structure does not show statistically significant differences, but the trend is higher annual productivity for G1 and G2 farms (18.3 and 17.35 PWSY, respectively) compared to G3 farms (16.83 PWSY). It is confirmed that the ideal herd structure is a triangle. In addition, although Iberian sows tend to be slaughtered with a greater number of parities, the most productive farms maintain a percentage of $\geq8^{th}$ P sows lower than of 7th P sows.

Alternative pig farms: a diversity of housing types and health statuses

C. Fablet, E. Dubarry, N. Rose and F. Pol
ANSES, B.P.53, 22440 PLOUFRAGAN, France; christelle.fablet@anses.fr

An online survey was conducted to describe the housing systems and health statuses of alternative pig herds. In this study, an alternative farm was defined as having at least one housing system different from closed confined building with fully slatted floor in at least one part of the production stage or with at least 30% of the pigs housed in alternative system. The questionnaire was filled in by 102 French farmers. Of this population, 56% were farrow-to-finish herds, 20% were wean-to-finish herds, 16% were grow-finish herds, 4% were breeding herds and 4% were breeding-to-post-weaning herds. Farrow-to-finish herds had a lower proportion of pigs reared in alternative systems with on average 43% of their pigs housed in these systems (vs >70% of the pigs from the other herd types). Gilts, sows and growing-finishing pigs were more frequently housed in alternative systems than the other types of pigs with respectively 77, 56 and 61% of the herds with alternative housing at these stages. Deep litter systems were found at every production stage. Free-range systems were mainly used for the breeding herds (gilts to lactating sows). Finishers were more frequently reared in buildings with outdoor access than the other pigs. Different health profiles were identified by hierarchical clustering. In breeding herds, four health clusters were found ranging from a cluster of herds without frequent health troubles to a cluster of herds with sows having reproductive disorders, gilts suffering from respiratory troubles and sows and sucklers suffering from leg disorders. After weaning, four health clusters were also found from herds without frequent health problems, herds reporting lameness and respiratory diseases in finishers, herds experiencing diarrhoea in weaners and respiratory troubles in finishers to herds facing leg disorders in weaners and respiratory diseases as well as bites in weaners and finishers. These findings show a diversity of housing systems with few free-range rearing for weaners and finishers. They indicate that a diversity of health troubles are encountered in alternative systems and that room from improvement need to be considered in some herds. Identifying factors associated with these profiles may help optimise herd health management programmes.

Body weight evolution, from birth to adult weight, in pigs of Sarda breed and its Duroc crossbred

A. Cesarani[1], M. Acciaro[2], M.R. Mellino[1], A. Fenu[1], G. Spanu[1], A. Di Gianvito[1] and G. Battacone[1]
[1]University of Sassari, Department of Agricultural Sciences, viale Italia, 39, 07100 Sassari, Italy, [2]Agenzia Agris Sardegna, Department of Research for Animal Production, Loc. Bonassai, 07040 Olmedo (SS), Italy; battacon@uniss.it

Sarda pig breed is a local breed from Sardinia (Italy) usually reared in extensive conditions. The maintenance of this local breed is important because of its rusticity, strong relationship with native area and biodiversity safeguard. Recently, farmers of Sarda pigs seek more efficient alternatives to obtain animals with highest productive efficiency. A possible alternative could be crossbreeding the Sarda sows with commercial boars to produce crossbreed animals (F1). Aim of this study was to compare growth traits of animals belonging to two pig breeds: Sarda purebred (SA) and Sarda × Duroc crossbreed (S×D). For the experimental trial, 18 piglets (9 SA and 9 S×D) born within 2 days were used. Individual body weigh has been recorded at born and: twice per week until 60 days of age; every 2 weeks until six months of age, and monthly until their slaughter at 8 months. A linear model considering group (SA and S×D), sex (immunocastrates and females) and their interaction was applied to birth and final weights. Animals growth in the two groups was modelled using Logistic and Gompertz models. Group was highly significant (P<0.001) both at birth at trial end, whereas sex was not significant in both cases. Interaction group/sex was significant only at the end of the trial (P<0.05). Average weights at birth were 1.34±0.24 and 1.88±0.11 for SA and S×D, respectively. Lowest birth weight was registered in SA group (1.06 kg), while the largest one in S×D group (2.04 kg). Animals of S×D group were significantly heavier also at the end of the trial: 160±16 (S×D) and 117±11 (SA). Logistic model showed better fitting criteria (AIC and BIC), and therefore growth curve was estimated using this model. As expected, larger estimated weight at adult (about 225 days) was estimated in the S×D group. Females tended to be heavier than castrated males. Results of the present study, even if based on small number of animals, showed that using Duroc as crossing breed can improve weights both at birth at adult stages.

Growth performance in Bísaro pigs under different feeding regimes

S. Botelho Fontela[1], C. Castelo[2], G. Paixão[1], R. Payan-Carreira[3] and A. Esteves[1]
[1]Animal and Veterinary Research Centre (CECAV), Quinta de Prados, 5000-801, Portugal, [2]Associação de Criadores de Suínos da Raça Bísara (ANCSUB), Edifício da Casa do Povo, Largo do Toural, 5320-311 Vinhais, Portugal, [3]MED – Mediterranean Institute for Agriculture, Environment and Development, University of Évora, 7006-554 Évora, Portugal; sbotelho@utad.pt

Bísaro pig is a Portuguese breed presenting the largest expression in the north of the country. The high industrialisation of the sector almost led this breed into extinction, and to be raised in mainly small family farms. With the implementation of a conservation program this rustic breed was recovered and due to the excellent meat and meat products quality is nowadays a prized asset in the region. The aim of this study was to evaluate the live weight of the piglets after weaning in three farms with different feeding regime. A total of 36 piglets were studied across the farms, starting at the age of 8 weeks old until 17 weeks old. All the animals were fed with growth commercial diets according to the farm's routine management and had water *had libitum*. The pigs were housed in indoor group pens, separated from the females albeit cohabiting the same room. The live weight records were taken every 15 days, and the amount of feed was recorded each time it was placed in the feeders. Average daily gain was significantly different in all three farms (P<0.001) as well as the feed intake (P<0.001). Data shows that Farm 3 was not feeding the animals enough to promote ideal growth, as shown by the feed conversion rate (P<0.001). In conclusion, the three different farms had very different feeding managements, and some did not promote optimal development of the animals. This results are important to the farmers as they can improve the animal production, obtaining larger and healthier pigs and, therefore, more profitable. This work was supported by the project Icas- Bísaro (reference nº. PDR 2020-101-031029) and the project UIDB/CVT/00772/2020 funded by the Fundação para a Ciência e Tecnologia (FCT).

Managing the herdbook of an endangered Portuguese swine population: the Malhado de Alcobaça pig

A. Vicente[1,2,3,4], A. Roque[4], J. Bastos[2] and N. Carolino[1,3,5]
[1]Sociedade Portuguesa Recursos Genéticos Animais, SPREGA, Santarém, 2005-048, Portugal, [2]Federação Portuguesa Associações Suinicultores, FPAS, Montijo, 2870-219, Portugal, [3]CIISA, Faculdade Medicina Veterinária, Lisboa, 1300-477, Portugal, [4]Escola Superior Agrária, IPSantarém, Santarém, 2001-904, Portugal, [5]Instituto Nacional Investigação Agrária Veterinária, INIAV, Santarém, 2005-048, Portugal; carolinonuno@hotmail.com

Malhado de Alcobaça breed is the 3rd Portuguese native swine breed and represents a population from the centre west of Portugal and was officially recognised in 2003. The herdbook has information of more than 11,000 animals (1985-2020) and the last animals have more than 10 known generations on their pedigree. It´s a very endangered breed with only 211 breeding sows, 12 boars and 9 active breeders. Since 2014 all candidates for breeding must be submitted to a grading process by analysing their morphology under 5 marks and, at the same time, placing an ear tag ID, collecting, simultaneously, tissue for DNA analysis, for parentage testing and genetic characterisation. All data from the gradings was analysed to obtain the descriptive statistics and the linear and quadratic effect of age at grading, besides considering the effects of breeder, gender and year. Since 2014, 448 pigs where graded (415 sows and 33 boars) with average age of 15.87±8.88 months (15.96±8.99♀; 14.61±7.35♂). Only 7 different breeders produced breeding sows and boars, and 4 of them are responsible for 83% of the breeding stock in use. For the grading grid (scale with maximum of 10 pts.) the average scores for morphology were 8.54±0.72 pts for type and development; 8.35±0.68 pts for back, loin and croup; 8.24±0.64 pts for shoulders, chest, belly and flanks; 7.99±0.81 pts for legs, stances and gaits; and 8.43±0.66 pts for sexual characteristics, genitals and nipples, obtaining and overall sum of 82.58±4.45 pts. Globally boars obtain more 1.5 points than sows and the score for legs was the lowest with higher coefficient of variation. Breeder and year of grading were significant (P<0.05) for all morphological traits analysed and gender was only significant (P<0.001) for back, loin and croup. Age at grading had a significant linear effect (P<0.05) in Overall sum and for some partial morphological scores. Acknowledgement: Project CIISA UID/CVT/00276/2020.

Consumers' willingness-to-pay for information on PLF technology and welfare in pork
E. Maes, T. Van De Gucht, C. Vandenbussche, J. Maselyne and S. Van Weyenberg
ILVO, Burg. Van Gansberghelaan 115 bus 1, 9820 Merelbeke, Belgium; jarissa.maselyne@ilvo.vlaanderen.be

Optimal use of data can improve the management of the entire pork value chain. This is what we are working on in the Pig Farm Management use case in the IOF2020 project (H2020, grant nr. 731884) via the use of Internet of Things and Precision Livestock Farming (PLF) technologies such as early warning systems and intelligent dashboards. The pig farming practices and innovative developments are most often unknown by the general public. Where technological developments (in this case automated monitoring systems) are aimed at a more efficient production but also an improved animal health and welfare, we believe this creates added value for the animal and the end product, as more (objective) data is gathered to maintain and validate good farm management. But the use of technology on livestock farms can in fact be perceived as negative by the general public. A consumer survey was designed with general (demographic and preference) questions and a discrete choice experiment of filet mignons with different levels in price, animal welfare label and farm technology level. Of 454 respondents, 204 did not buy filet mignons and 250 did. Age, gender, consumption patterns and education level was quite evenly spread amongst respondents. However, half of them were in some way related to agriculture. Price, animal welfare label, gender and some specific preferences had a very significant effect on consumer's choices. The utility of the level of technology for health monitoring was only significant for individual level monitoring and not at group level. The latter, however, did not result in a significant willingness-to-pay for pork originating from farms with health monitoring technology if compared with a reference price of € 12/kg. These results confirm that price and welfare are important attributes for consumers, but technology is either not known enough or not valued highly enough for consumers to pay extra at the moment. We should therefore communicate more towards consumers about on-going developments in automated monitoring systems so they become more familiar with the possibilities. In addition, we should work towards objective impact assessments of these systems on animal health and welfare.

Association between FTO gene polymorphisms with growth, carcass and meat quality traits
D. Polasik[1], M. Tyra[2], G. Żak[2] and A. Terman[1]
[1]West Pomeranian University of Technology, Al. Piastów 17, 70-310 Szczecin, Poland, [2]National Research Institute of Animal Production, Ul. Sarego 2, 31-047 Kraków, Poland; grzegorz.zak@izoo.krakow.pl

Recent study showed that gene encoding alpha-ketoglutarate-dependent dioxygenase (FTO) is associated with body mass index (BMI) and the risk of obesity in humans. In pigs, FTO gene expression is significantly higher in backfat, whereas m. longissimus dorsi shows the second highest level. FTO gene has been proposed as a marker for fat deposition, but it may be also considered as a marker for growth traits in pigs due to its high expression in muscles. Therefore, the aim of this study was to investigate specific polymorphisms (g.400C>G SNP) in porcine FTO gene in relation to six growth, nine carcass and nine meat quality traits. The study included 578 sows that belong to the following breeds: Polish Landrace (n=269), Polish Large White (n=189), Puławska (n=68), Pietrain (n=31), Duroc (n=14) and Hampshire (n=7). FTO genotypes were determined by use of PCR-RFLP method. Association analyses were performed by means of GLM procedure for first three breeds separately as well as for all animals covering six breeds. The results show that FTO genotypes were correlated with some growth and carcass traits only in Polish Large White pigs. Sows with CC genotype were characterised by highest values of test daily gain and backfat thickness from 5 measurements (P≤0.05), however those with GG genotypes by highest values of weight of loin without backfat and skin, loin eye area and meat percentage (P≤0.05). Regarding to meat quality traits, we found higher values of pH24, measured in m. longissimus dorsi in pigs with GG genotype (P≤0.05). It was observed for Polish Large White breed and the joined group. GG genotype was also associated with highest value of water holding capacity (P≤0.01) in Puławska breed.

Organic acids in plasma at early rabbit gestation

M.L. Garcia[1], R. Muelas[1], I. Agea[1], E. Armero[2] and M.J. Argente[1]
[1]*Universidad Miguel Hernández de Elche, Agrofood Technology, Ctra Beniel km 3.2, 03312 Orihuela, Spain, [2]Universidad Politécnica de Cartagena, Department of Agricultural Science and Technology, Paseo Alfonso XIII, 48, 30203 Murcia, Spain; mariluz.garcia@goumh.umh.es*

The early embryo development needs high levels of organic acids as energy substrates that can be obtained by diffusion from the blood plasma to the lumen of the oviduct or metabolised in the epithelium of the oviduct. The aim of this study was to determine lactic acid and citric acid concentrations at mating and 3 d post-coitum in plasma. Relationship between lactic acid and citric acid at 3 d post-coitum with ovulation rate was analysed. A total of 18 multiparous rabbit females were used. Blood samples were collected at mating and 3 d post-coitum. Blood samples were centrifuged and plasma was stored at -80 °C. Females were slaughtered at 3 d post-coitum and ovulation rate was estimated as the number of corpora lutea. Lactic acid and citric acid concentrations were analysed by High-Performance Liquid Chromatography (Agilent 1100 series HPLC System) with UV detection, using Agilent Hi-Plex H column (7.7×300 mm, 8 µm particle size). Concentrations were calculated from analytical standard calibration lines (Organic Acids Kit, Supelco, Spain). Data were analysed using Bayesian methods. The statistical model included the reproductive status effect (mating and 3 d post-coitum), line and female as random effect. Lactic acid concentration was similar at mating (78.2 µg/ml) and at 3 d post-coitum (72.3 µg/ml, P=0.63). Citric acid concentration was higher (17.8 µg/ml at mating vs 14.0 µg/ml at 3 d post-coitum, P=0.92). The relationship between ovulation rate (y) and lactic acid measured at 3 d post-coitum (x) was lineal (y = 9.001 + 0.017x, R^2=0.41). Citric acid was not related to ovulation rate. In conclusion, lactic acid concentration increased with ovulation rate, and citric acid concentration was higher at mating than at 3 d post-coitum. These results should be confirmed with a higher database, and more organic acids should be identified not only in plasma but also in oviductal fluid. This study is supported by the Spanish Ministry of Economy and Competitiveness (MINECO) with the Project AGL2017-86083 C2-2-P, and Valencia Regional Government with the Project AICO/2019/169.

Effect of wheat malting on slaughter performance of slow growing broiler chickens

S. Kuenz[1], K. Damme[2], W. Windisch[1] and D. Brugger[3]
[1]*Technical Universtiy of Munich, Chair of Animal Nutrition, Liesel-Beckmann-Str. 2, 85354 Freising, Germany, [2]Bavarian State Research Center for Agriculture, Department for Education and Poultry Research, Mainbernheimer Str. 101, 97318 Kitzingen, Germany, [3]University of Zurich, Institute of Animal Nutrition, Winterthurerstrasse 270, 8057 Zurich, Switzerland; sylvia.kuenz@wzw.tum.de*

Exogenous phytase is often used in conventional production systems as feed additive to increase phosphorous (P) utilisation in the presence of high dietary concentrations of phytate. However, this is not allowed in European organic farming systems. Previous studies suggest that malting is accompanied by significant hydrolysis of phytic acid. The aim of this experiment was therefore to investigate the effect of malted versus raw wheat on slaughter performance of slowly growing meat chickens under terms of practical feeding. Four different diets were formulated for this experiment containing either wheat or malted wheat in combination with and without phosphorus (P) supplementation from Monocalcium-phosphate, respectively (diet A: malted wheat + low P; diet B: malted wheat + high P; diet C: raw wheat + low P; diet D: raw wheat + high P). 2.000 1-day-old Ranger Classic broiler chickens (both sexes) were used for this experiment. From d1 to d16 the birds were fed with an organic starter diet. On d17 the equal number of birds from both sexes were randomly allocated to the four experimental diets (five replicates per diet, 100 birds per pen) comprising a completely randomised design with balanced sex ratio. On d57 birds were slaughtered and weights of carcass, cuts (wings, thighs, breast) and edible viscera (heart, gizzard, liver) were documented. Statistical analyses comprised ANOVA (wheat variant, phosphorus and interactions). Animals fed with untreated wheat had significantly heavier carcasses (P<0.001) and cuts (P<0.001). Surprisingly, P content had no significant effect on these parameters. Either wheat variant, nor P content had significant impact on weights of edible viscera. Malting had no beneficiary effect on slaughter performance. Interestingly, slowly growing broiler chickens appear to be less sensitive to phytic acid than high performance breeds under the present experimental conditions.

Effect of supplementation with organic selenium on the microbiological status of broilers meat

D. Konkol[1], E. Popiela[1], M. Korzeniowska[1], M. Korczyński[1], H. Iwański[2] and H. Różański[2]
[1]Wrocław University of Environmental and Life Sciences, Chełmońskiego 38C, 51-642 Wrocław, Poland, [2]AdiFeed, Opaczewska 43, 02-201 Warsaw, Poland; damian.konkol@upwr.edu.pl

In recent years, poultry meat producers are focused on producing meat with functional features that prevent the risk of disease. This effect can be achieved by using antimicrobial and antioxidant compounds in animal nutrition, such as phytobiotics, vitamin E, polyunsaturated fatty acids and selenium. There is little information available in the literature on the effect of selenium on microbiological contamination of chickens meat. Therefore, the aim of this experiment was to evaluate the effect of organic selenium used in broiler nutrition on the microbiological status of the chickens meat. The research was carried out in a commercial broiler house. Cobb 500 line broilers were used in the experiment. Chickens were divided into two experimental groups consisting of four replicates with 50,000 birds per replicate. Birds from the experimental group received basal feed with 100g of selenium yeast preparation per ton of feed mixture. Meat samples (30 breast muscles and legs from each replication) were collected on the 28 day of the experiment and at the end of the fattening period (on the 42 day). Then the temperature measurement of the obtained samples was evaluated and the total number of microorganisms (according to ISO 4833-1: 2013-12) and *Enterobacteriaceae* (according to ISO 21528-2: 2005) were determined. The obtained data was transformed to log10 cfu / g before performing the analyses. The results were statistically analysed using Statistica ver. 13.1. The differences between the groups were assessed using the t-Student test. The differences were statistically significant when P<0.05. There were no statistically significant differences in the temperature between meat samples. At the end of fattening period, a higher (P<0.05) total number of microorganisms and *Enterobacteriaceae* was observed in meat from group receiving selenium in the feed. Based on the obtained results of the present study, it can be concluded that the addition of selenium in the broilers feed does not improve the microbiological status of broilers meat.

Dietary protected versus non-protected essential oils on chickens' performance and meat quality

E. Bonos[1], I. Giannenas[2], A. Tzora[1], E. Gouva[1], I. Anastasiou[1], E. Barka[1], L. Chatzizisis[1], E. Christaki[2] and I. Skoufos[1]
[1]University of Ioannina, School of Agriculture, School of Agriculture, University of Ioannina, Kostakioi Artas, 47150, Arta, Greece, [2]Aristotle University of Thessaloniki, Laboratory of Nutrition, School of Veterinary Medicine, School of Veterinary Medicine, Aristotle University Campus, 54124, Thessaloniki, Greece; ebonos@uoi.gr

Aromatic plants and their essential oils are natural sources of bioactive ingredients that can benefit chicken health and performance. It can be hypothesised that the encapsulation of such substances using can protect them and improve their activity when used as feed additives. In a feeding trial (35 days) 576 male one-day-old chicks (Ross 308) were randomly allocated to 3 treatments (12 pens of 16 chickens). Control (A) was fed standard maize and soybean meal diets. Treatments B and C were fed the same diets, further supplemented with either a mixture of unprotected essential oils (50 g/ton) or protected essential oils (40 g/ton), respectively, with carbacrol as main active substance. Body weight was recorded on days 10, 24, and 35; feed consumption and mortality recorded daily. At the last day of the trial, meat samples were collected and analysed by FoodScan™. Treatment C had higher (P<0.05) final body weight, higher body weight gain, reduced feed intake, and better feed conversion index, compared to the other two treatments. Furthermore, meat protein was higher (P<0.001) and meat fat was lower in treatments B & C, compared to the control. Based on the results, the use of encapsulated essential oils seems to work better than the unprotected use. Acknowledgments: This research has been co-financed by Greece and the European Union (European Regional Development Fund) in context 'Research–Create–Innovate' within the Operational Program (Competitiveness, Entrepreneurship and Innovation (ΕΠΑΝΕΚ) of the NSRF 2014-2020. Project Code: T1EΔK-03856. Acronym 'GREEN POULTRY MEAT ANTIFREE'

Muscle amino acids of poultry genetic lines with distinct growth (and moving) rate

S. Failla, D. Meo Zilio, M. Contò, F. Cenci and M. Guarino Amato
CREA, via salaria 31, 00015, Monterotondo, Rome, Italy; david.meozilio@crea.gov.it

Chickens in organic farming have outdoor access, but growth speed and body development may counteract moving. Foraging behavior of fast-growing strains can be influenced by the rapid weight gain. Moreover, high final liveweight often leads to gait problems. It is interesting to characterise commercial hybrids according to moving potential. Physical exercise modifies muscle amino acids (AA) in humans and rats, but few researches are available for chickens. Assuming a different catabolism depending on physical activity, modifications in the relative distribution of specific AA is a sign of activity that may be used as welfare and management quality check. The Aim was to compare AA of breast and thigh in 6 lines, with different growth rate (Hubbard RedJA (C), CY5XJA87 (CY), M22XJA87 (M), Ranger Classic (RC), Ranger Gold (RG), Rowan Ranger (RR)), 15 animals for genotype and to group them accordingly. Animals received the same diet, according to European Union Regulations on Organic Farming, *ad libitum*, and had access to pasture starting from 21 d until slaughter, at 81 d. Outdoor activity was observed and recorded. Animals were categorised in three clusters: high movement (HM); medium movement (MM); low movement (LM). Thigh and breast samples were used for AA determination by acid hydrolysis (HCl 6N). Analysis was performed in HPLC using the AccQ-Fluor reagent kit (Waters). Distribution of AA in thigh was significantly different according to group (P<0.001). Branched chain AA were 15.27 vs 15.64%, in LM (M and RC) when compared to HM (A and RR). Threonine and phenylalanine were higher in HM (4.15 vs 3.96 and 3.52 vs 3.67%). Sulphur AA were lower in HM when compared to LM (e.g. 8.55 vs 8.94%, for methionine). Low differences resulted in breast. Therefore, a higher physical exercise may affect the AA composition of chicken, at least the thigh. As free ranging condition is a pre-requisite in organic poultry production, this approach would allow to evaluate physical activity from metabolic point of view and could be used to develop an index of movement for welfare and management evaluation. A multi-trait index, including movement and exploring aptitude markers, could be useful to check suitability of commercial hybrids for organic farming.

Layer HACCP – critical control points to prevent untrimmed hens from feather pecking and cannibalism

A. Riedel[1], A. Nording[2], P. Hiller[2], N. Kemper[1] and B. Spindler[1]
[1]University of Veterinary Medicine Hannover, Foundation, Institute for Animal Hygiene, Animal Welfare and Farm Animal Behaviour, Bischofsholer Damm 15, 30173 Hannover, Germany, [2]Chamber of Agriculture of Lower Saxony, Oldenburg, Mars-la-Tour Straße 6, 26121 Oldenburg, Germany; anna.katharina.riedel@tiho-hannover.de

Since beak-trimming in laying hens has been abolished in Germany, avoiding feather pecking and cannibalism is of utmost importance. The aim of this project is to develop and test critical control points on farm to minimise such behavioural problems in rearing and laying hens. The steps to prevent these problems are complex. Parameters as weight, uniformity, laying performance and animal behaviour can help to detect early signs of a beginning problem. Furthermore, factors as feed structure and composition, additives and feeding times help to maintain healthy hens. Light management, the barn climate and an exact monitoring of the flock as well as well-trained caretakers also play a role to decrease the risk of feather pecking and cannibalism. Ten flocks of laying hens with different genetics are currently monitored in three different housing systems. Flocks sized 9.000 to 40.000 hens are kept in conventional aviaries, free range or following organic standards. At least ten visits over the hen's lifespan take place and data about weight, plumage damage, health status and barn climate is collected. Assessments of feed composition and parasitological examinations are carried out, and management and performance parameters are documented after interviews with the animal caretaker. Based on this data, critical control points are defined to propose a tangible package of measures to the animal caretaker, which is implemented till the next visit. To date no flock has developed cannibalism, while first ones reached the age of 70 weeks. Identifying these individual, specific control points supports the farmer to optimise the husbandry of the hens. Factors as suitable feed, enrichment materials and adapted management help to provide healthy, long living and highly productive hens. This work is funded by the German Federal Ministry of Food and Agriculture (BMEL) based on a decision of the Parliament of the Federal Republic of Germany, granted by the Federal Office for Agriculture and Food (BLE; grant number 2817MDT200/201).

Selected growth parameters of farm-raised mallard (*Anas platyrhynchos* L.)

D. Murawska[1], V. Hanzal[2,3], P. Janiszewski[4], P. Matusewičius[5] and D. Witkowska[6]
[1]University of Warmia and Mazury in Olsztyn, Department of Commodity Science and Animal Improvement, Oczapowski St. 5, 10-719 Olsztyn, Poland, [2]Czech University of Life Sciences Prague, Department of Game Management and Wildlife Biology, Kamýcká St. 129, 165 00 Praha 6 – Suchdol, Czech Republic, [3]University of South Bohemia in Ceske Budejovice, Department of Landscape Management, Studentska St. 13, 370-05 Ceske Budejovice, Czech Republic, [4]University of Warmia and Mazury in Olsztyn, Department of Fur-Bearing Animal Breeding and Game Management, Oczapowski St. 5, 10-719 Olsztyn, Poland, [5]Lithuanian University of Health Sciences, Lithuania, Department of Animal Breeding and Nutrition, Tilžės St. 18, LT-47181 Kaunas, Lithuania, [6]University of Warmia and Mazury in Olsztyn, Department of Animal and Environmental Hygiene, Oczapowski St. 5, 10-719 Olsztyn, Poland; daria.murawska@uwm.edu.pl

The number of consumers seeking new taste experiences has been growing steadily, and there has been an increased interest in meat from wild-living animals, including game birds. Captive-reared mallards are released from farms into natural habitats for hunting purposes. Mallard meat is considered a delicacy, in a many countries game meat from free-ranging wild animals can be legally sold and distributed. The objective of this study was to analyse selected growth parameters of farm-raised mallard. The experimental materials comprised 210 mallard ducklings (sex ratio 1:1), raised from hatch to 154 d of age. Starting from the first day of age, at two-wk intervals, 20 ducklings (5 males and 5 females) were selected randomly for slaughter and were used for future analysis. Data were subjected to crossed two-way analysis of variance. The significance of differences in mean values between age groups was determined by Duncan's D test. Between one day and twenty two weeks of age, the average body weight (BW) of mallards, increased nearly 38.5-fold of male (from 35.4 g to 1,363.5 g), and 34.5-fold of female (from 35.7 g to 1,230.5 g, P≤0.01). Over the analysed period, total lean meat weight of male increased 91.1-fold (from 5.41 g to 493,6 g, P≤0.01), of female 84.9-fold (from 5.29 g to 449.0 g). Lean meat content increased significantly in the breast, wings and back of growing mallard, whereas a significant decrease in lean meat content was observed in the legs and neck. In conclusion, age has a significant effect on body growth rate and distribution of lean meat of carcass of farm-raised mallard.

Morphological characterisation of Portuguese autochthonous chicken breeds

V. Ribeiro[1], P.S. Vaz[2], R. Dantas[1], J.V. Leite[1], J.C. Lopes[3] and N.V. Brito[3]
[1]AMIBA, Associação de Criadores de Raça Bovina Barrosã, Vila Verde, 4730-264 Lanhas, Portugal, [2]Quinta da Eira Vedra, Ardegão, 4990-535, Portugal, [3]Polytechnic Institute of Viana do Castelo, CISAS, Center for Research and Development in Agrifood Systems and Sustainability, Polytechnic Inst, Quinta do Mosteiro-Refoios, 4990-706 Ponte de Lima, Portugal; virginia.ribeiro@amiba.pt

The biometric study of the breeds is fundamental in the knowledge of the morphological characteristics and the productive potentialities, especially in the autochthonous populations. The first reference to Portuguese poultry breeds is recent, in the 30's of the last century, despite their relevant role in the more traditional production systems and the strong contribution to the domestic economy. The aim of this study is to phenotypically characterise the four Portuguese poultry breeds (Pedrês Portuguesa, Preta Lusitânica, Amarela and Branca), using different biometric measures and live weight and to evaluate, under production conditions specific to the artisanal system, the effect of several factors in each of the studied breeds. For the present analysis, 5 biometric measures and the live weight of 436 animals (372 Females and 64 Males) of the four breeds: Amarela, 122 animals (103 F, 19 M); Branca, 64 animals (53 F, 11 M); Preta Lusitânica, 127 animals (107 F, 20 M) and Pedrês Portuguesa, 123 animals (109 F, 14 M) were collected. The measures were: body length (CC), circumference of the chest (CirP), shank length (CT), circumference of the shank (PT) and wing span (EA). For the statistical analysis the animals were grouped in 3 productive cycles (less than 1 year, from 1 to 2 years, more than 2 years) and the effects of the breed, sex, farm and age were evaluated and correlations between the different variables were estimated. The Branca breed stands out in all the biometric measures, while the other breeds are more similar. The males are bigger and heavier, and between all the breeds are significant differences for all the variables, with the exception of CT. The influence of farm is significant for the PT and the correlations between variables were superior to those related to body size. The present study is important to highlight the differences between the breeds, and to guide the breeding strategies that may contribute for the valorisation of these breeds.

Interdigital hyperplasia in Holstein cows: a highly underestimated and strongly hereditary disease

H.H. Swalve[1], X. Zhang[2], R. Pijl[3], F. Rosner[1], M. Wensch-Dorendorf[1], I. Gauggel[1] and B. Brenig[2]
[1]Martin Luther University Halle-Wittenberg, Institute of Agricultural and Nutritional Sciences, Theodor-Lieser-Str. 11, 06120 Halle, Germany, [2]University of Goettingen, Institute of Veterinary Medicine, Burckhardtweg 2, 37077 Göttingen, Germany, [3]Independent Researcher, Fischershäuser 1, 26441 Jever, Germany; hermann.swalve@landw.uni-halle.de

Interdigital hyperplasia (IH) is a disease of the bovine hoof that manifests in excessive growth of tissue in the interdigital space between the paired claws. Commonly, incidence rates between 3 and 6% are reported. In the analysis of the database of observations at time of hoof trimming solely collected by the third author in his practice of hoof trimming, an incidence rate of around 10% was found. This elevated finding may be attributed to the fact that the practice records contain many long-term clients and thus cows are followed throughout their lives. To exploit this finding even further, the database was restricted to cows that were indeed followed from their first observation in first lactation in herds that had been visited >10 times. The resulting data set contained 70,810 observations from 15,388 cows showing a lifetime incidence of 16.6% for IH. Within the data, a single herd with a highly elevated incidence rate for IH of >50% was identified and subsequently used for further studies. A total of 94 cows were genotyped using a 50K SNP array for a case control study of healthy vs affected animals. Highly significant associations were found for SNP on BTA8 in and near the tyrosine kinase-like orphan receptor 2 (ROR2) gene. Subsequent sequencing of the region yielded a functional variant highly associated with the status for IH. As ROR2 plays a key role in ossification of the distal limbs and is associated with brachydactylies in humans, its role as a candidate for IH is very likely. Findings were underpinned from gene expression and protein analysis based on fine-needle biopsies of tissue. Based on definition of phenotype, the proportion of the total variance accounted for by the mutation was 20 to 32%.

Recording lameness in dairy cattle

A.M. Christen[1], C. Egger-Danner[2], N. Capion[3], N. Charfeddine[4], J. Cole[5], G. Cramer[6], A. Fiedler[7], T. Fjeldås[8], N. Gengler[9], M. Haskell[10], M. Holzhauer[11], G. De Jong[12], A. Koeck[2], J. Kofler[13], K. Müller[14], J. Pryce[15], A.A.M. Sogstad[16], K.F. Stock[17], G. Thomas[18], E. Vasseur[19] and B. Heringstad[8]
[1]Lactanet, Canada, [2]ZuchtData, Vienna, Austria, [3]University of Copenhagen, Denmark, [4]CONAFE, Spain, [5]USDA, USA, [6]University of Minnesota, USA, [7]Hoof Health Practice, Germany, 8Norwegian University of Life Sciences, Aas, Norway, [9]Gembloux Agro-Bio Tech, Université de Liège, Belgium, [10]Scotland Rural College, United Kingdom, [11]GD Animal Health, Netherlands, [12]CRV, Netherlands, [13]University of Veterinary Medicine, Austria, [14]Freie Universität, Germany, [15]La Trobe University, Australia, [16]Animalia, Norway, [17]IT Solutions for Animal Production (vit), Germany, [18]Institut de l'Élevage, France, [19]Mc Gill University, Canada; bjorg.heringstad@nmbu.no

Routine recording of lameness has been established in several countries. It is known that regular lameness scoring allows earlier detection of lame cows, which enhances the treatment effectiveness. The ICAR working group for functional traits created guidelines for recording of lameness in dairy cattle. The aim was to harmonize traits and to standardize data recording. The guidelines include considerations and recommendations for improved lameness recording in the context of herd health management, animal welfare, benchmarking and genetic evaluation. The guidelines describe common lameness scoring methods used worldwide and developed ICAR recommendations for a harmonized scheme. It includes practical recommendations for recording of lameness in free-stall and tie-stall barns as well as aspects of training of assessors and suggestions about how many cows should be assessed, how often and when. Collected data can be used to evaluate the herd's lameness control and prevention strategies, for further analyses and research, and also for breeding purposes. The use of lameness scores as a valuable animal-based welfare indicator is discussed as well. We acknowledge the other representatives of the ICAR WGFT and claw health experts contributing to this work.

Routine claw trimming and cow behavioural variations as tools for early detection of claw disorders

L. Magrin, G. Cozzi, I. Lora, P. Prevedello and F. Gottardo
University of Padova, Department of Animal Medicine, Production and Health, Viale dell'Università 16, 35020 Legnaro, Padova, Italy; luisa.magrin@unipd.it

Claw disorders are still perceived by dairy farmers as a relevant welfare and economic problem. Often, they are subclinical and not easily visible, and thus their prevalence may be underestimated by farmers. During their development, claw disorders may affect cow behaviour, however. This study aimed at monitoring claw disorders in a dairy herd with an automatic milking system during the routine twice-yearly claw trimming, and investigating their impact on the individual cow activity and rumination. A group of about 60 lactating cows was trimmed during 4 sessions by a veterinarian, who evaluated the claw condition identifying specific lesions. Cows were classified as healthy or affected. Data on individual activity and rumination time performed by cows during 5 weeks before and after each trimming session were continuously registered by neck collars (composed of acceleration sensor, microphone and microprocessor). These behaviours were statistically analysed using a mixed model that considered the fixed effects of the pathological condition (healthy/affected), week/day from trimming, and their interaction, with the random effect of week/day and cow as subject. Affected cows by claw disorders were 36.7%. Among infectious disorders, digital dermatitis was the most frequent, affecting 15.4% of the total cows; while among non-infectious, ulcers were detected in 13.8% of the cows. Activity of affected cows started decreasing progressively (<390 bits/d) 6 days before disorders' identification and it remained lower than that of healthy ones for up to 3 weeks after trimming. Rumination of healthy and affected cows were statistically similar in each week before and after trimming, and in particular after claw disorders' identification and treatment, affected cows' rumination reached the same values of healthy ones. However, focusing on a specific 10-days period before trimming, it can be noticed that rumination decreased below 440 min/d only for affected cows 6 days prior trimming. A routine claw trimming along with a regular control of cow activity and rumination variations should be considered good tools to detect and treat claw disorders at early stages.

Reliability of lameness scoring and related welfare measures during test-day recording visits

M. Suntinger[1,2], B. Ruthner[1], L. Maurer[1], C. Egger-Danner[2] and C. Winckler[1]
[1]Division of Livestock Sciences, University of Natural Resources and Life Sciences (BOKU), Gregor-Mendel-Straße 33, 1180 Vienna, Austria, [2]ZuchtData EDV Dienstleistungen GmbH, Dresdnerstraße 89/18, 1200 Vienna, Austria; suntinger@zuchtdata.at

Lameness and related measures such as skin alterations have repeatedly been suggested for on-farm welfare assessment protocols. However, to support herd management on farm level or to genetically improve animal health on population level, reliable data recording is needed. To achieve standardised recording the assessment might be carried out in the course of milk performance recording visits. The objectives of this study were thus to determine (1) the inter-observer reliability of assessing lameness and related welfare indicators (2) the agreement between gait scoring and a standing score for lameness assessment. For this purpose, on 22 dairy farms using milking parlours, five trained persons of performance recording organisations recorded the standing lameness score (e.g. resting a foot), claw position as well as cleanliness of the lower hind legs and hairless spots, injuries and swellings of the hock on two consecutive test days. The assessment was done in the milking parlour parallel to test-day sampling. In a timely manner, an external person (silver standard) and the farmers themselves assessed the same traits while cows were at the feed bunk. They additionally scored locomotion, cleanliness and skin alterations of the upper hind legs. Agreement with the silver standard ranged from PABAK 0.32 to 0.93 across farmers and from 0.24 to 0.87 across control assistants. It was high (>0.6) for gait score and injuries, moderate (0.4-0.6) for claw position and hairless spots and low (<0.4) for cleanliness. Severely lame cows were reliably detected in the milking parlour, but slightly lame cows remained largely undetected. The study showed that trained farmers were able to score lameness and animal welfare traits mostly with medium to high agreement. In-parlour lameness scoring within test-day recording offers potential to identify problem cows, but seems insufficient for early detection; additionally, milking parlour type considerably affected feasibility of the assessment. Regular recording by a trained external person would provide most reliable data.

Interpretation of data on claw health and from repeated locomotion scoring of Holstein dairy cattle

K.F. Stock[1], A.M. Choucair[2] and K.E. Müller[2]
[1]*IT Solutions for Animal Production (vit), Heinrich-Schroeder-Weg 1, 27283 Verden (Aller), Germany,* [2]*Freie Universität Berlin, Dept. of Vet. Med., Ruminant and swine clinic, Koenigsweg 65, 14163 Berlin, Germany; friederike.katharina.stock@vit.de*

Routine screening and documentation protocols for claw health have been proposed for systematic use in herd health management. However, feasibility of additional scorings and comprehensive data recording at hoof trimming differs considerably between farms complicating data interpretation for targeted support. The aim of this study was to compare usability of claw health records from routine trimming and results from additional locomotion scorings to reflect the claw health status of cows and quantify the potential impact of its impairment on lactation performance. Data for the statistical analyses were collected in four commercial dairy farms in 2015/2016 in the course of a research project. For 2,415 lactations of 2,170 Holstein dairy cows, standardised claw health records from routine trimmings, available through herd management software, and locomotion scores from repeated scorings (LS scale 1 to 5) were available and merged with information on milk yields and disposal from milk recording. All farms were visited biweekly for data collection, and all scorings were performed by the same trained assessor. For multiple analyses of variance, performed in general linear models using SAS software, binary coding was used to distinguish between lactations without and with certain claw disorders, clinical signs of lameness (LS>3) and health related disposal. In addition, highest LS within lactation and lactation milk yield were analysed. Herd, parity and year of calving were considered as fixed effects. Modelling lameness, yield and disposal in dependence of claw health, significantly increased probability of lameness and disposal was found for claw ulcers and white line disease/abscess, but only of lameness for others. In cows with completed lactations, the same disorders showed significant reduction of milk yield. These findings were consistent with significance of highest LS and lameness for both yield and disposal. Claw health data and LS records can be seen as complementary sources of information in herd health management, with the former allowing more targeted measures for long term improvement.

Risk factors associated with milk fever in Austrian Fleckvieh

J. Hintringer[1], C. Fuerst[2], C. Egger-Danner[2] and B. Fuerst-Waltl[1]
[1]*University of Natural Resources and Life Sciences Vienna (BOKU), Gregor Mendel-Str. 33, 1180 Vienna, Austria,* [2]*ZuchtData EDV-Dienstleistungen GmbH, Dresdner Straße 89/B1/18, 1200 Vienna, Austria; birgit.fuerst-waltl@boku.ac.at*

Risk factors associated with milk fever (hypocalcaemia) occurrence were studied based on records from Austrian Fleckvieh cows in the years 2006 to 2017. In 15,329 of the 317,295 lactations (=4.83%) milk fever was recorded based on veterinarian diagnoses or farmers' observations around calving. However, the milk fever frequency was only 0.3% for first lactating cows and increased up to 12.9% in the seventh lactation. For the analyses, two logistic regression models for the observations of the first (n=63,308) and second to sixth lactations (n=37,365, 30,458, 23,357, 16,929 and 11,046) were applied including 5 and 13 fixed effects, respectively, as well as the random effect of herd. The following characteristics (ranked in descending order of influence) were identified as significant risk factors in at least one lactation: milk fever diagnosed in an earlier lactation, difficult calving, higher energy-corrected-milk-yield in a prior lactation, longer calving interval, stillbirth, higher milk protein content at drying off, lower milk fat content at drying off as well as higher somatic cell count at drying off. While calving year was not found to be a significant risk factor in any lactation, calving month partly was. However, no tendencies for a higher risk in certain seasons could be observed. In the management of milk fever prophylaxis, it is essential to pay attention to the risk factors that were identified.

Genetic relationship between body reserves mobilisation and feed intake in Holstein cows?

R. Lefebvre[1], P. Faverdin[2], S. Barbey[3], T. Tribout[1], P. Martin[1] and D. Boichard[1]
[1]Université Paris-Saclay, INRAE, AgroParisTech, GABI, 78350 Jouy en Josas, France, [2]INRAE, AgroCampus Ouest, UMR 1348 PEGASE, 35590 Saint Gilles, France, [3]INRAE, UE326 Domaine Expérimental du Pin, 61310 Gouffern en Auge, France; rachel.lefebvre@inrae.fr

The dairy cow faces large changes in nutritional requirements after calving. As feed intake adjusts only gradually to the strong increase in nutritional needs due to milk production, body mobilisation cannot be avoided in the beginning of the lactation. Excessive body mobilisation increases the risk of low fertility, of several diseases, and of poorer longevity, decreasing the animal robustness. Although body mobilisation is associated to milk yield, the potential for uncoupling production from body mobilisation remains largely unknown. The aim of this study was to estimate the correlations between feed intake, milk production and body mobilisation trajectories according to body condition score breeding value. This study involved 330 Holstein cows from three experimental facilities. Cows were genotyped and characterised by their direct genomic value for body condition. Their milk production, feed intake and body weight were recorded daily all along their lactation and body condition was scored monthly. Blood metabolites were also measured. Trajectories were found to be dependent on body condition breeding values (BCg). In the beginning of lactation, BCg did not affect feed intake but milk production was lower (-1.5 kg), and body weight and body condition were higher for animal with high BCg (+19.7 kg and +0.2 points per BCg genetic standard deviation, respectively). Contrarily, after 4 months of lactation, cows with high BCg showed more daily feed intake (+4 kg), production (+1.4 kg milk) and less body weight (-13 kg). These trajectories were therefore quite different: low BCg seems to be more favourable short term but at the expense of the mid-term feed intake and persistency. This study is part of the Deffilait project funded by ANR (ANR-15-CE20-0014-03) and ApisGene.

Effects of dietary CLA or a milk fat depressing diet on energy balance of early lactating cows

A.R. Bayat[1], A. Razzaghi[1,2], M. Sari[3], P. Kairenius[1] and J. Vilkki[1]
1Natural Resources Institute Finland (Luke), Animale, 31600 Jokioinen, Finland, [2]Beihagh Nutri-Paya, Ferdowsi University of Mashhad, 91775 Mashhad, Iran, [3]Ramin Agriculture and Natural Resources University, 1234 Ahvaz, Iran; razzaghi@um.ac.ir

Thirty-five multiparous Nordic Red dairy cows from d 1 to 112 postpartum were used in a complete randomized block design. Experimental treatments consisted of: (1) a basal diet based on grass silage (CON); (2) the same basal diet supplemented with 100 g/d conjugated linoleic acid supplement (CLA); and (3) a grass silage-based diet with a higher-starch content and supplemented with 40 g/kg of DM of a mixture of sunflower oil and fish oil (2:1 w/w) (MFD). Energy balance measurements were conducted using 4 open-circuit respiratory chambers over a 5-d period that was repeated at 28-d intervals. Feed intake and milk yield were measured daily and digestibility was measured by total fecal and urinary collection. Feeding the MFD diet decreased (P<0.01) intake of DM, OM, and GE in cows from wk 3 to 15 compared to CON and CLA groups. Overall, nutrient intake increased (P<0.01) from wk 3 to wk 15 of lactation in all diets while that was lower for cows fed MFD diet. However, MFD diet had greater (P<0.01) DM, OM and GE digestibility compared with CON during the experiment. Milk yield was not affected by treatments over the first 11 weeks of lactation but different patterns over time resulted in lower (P<0.05) milk yield in the MFD group compared with the CLA group at wk 15. The milk fat content was effectively depressed (P<0.01) in the CLA and MFD groups at wk 3 to 15 relative to the CON group while the MFD group showed even lower (P<0.01) milk fat content than the CLA group. The energy balance of the CON group was lower than the MFD group and it was numerically lower than that of the CLA group from wk 3 to wk 15 of lactation. However, the energy balance of the CON and CLA groups increased at wk 11 and 15 compared to that at wk 3 and 7. The results show that the MFD diet considerably lowered nutrient and energy intakes compared with CON group during early lactation period while it improved energy balance without any adverse effect on milk yield. While both CLA and MFD depressed milk fat content effectively, the MFD diet was more effective in improving energy balance than CLA group.

Use of mid-infrared spectroscopy to predict lameness related traits for dairy cows

M.-N. Tran[1], C. Bertozzi[1], L. Dale[2], N. Gengler[3], C. Lecomte[4], A. Werner[2] and J. Leblois[1]
[1]*Walloon Breeders Association Group, Rue des Champs Elysées 4, 5590 Ciney, Belgium, [2]State Association of Baden-Wuerttemberg, Heinrich-Baumann-Str. 1-3, 70190 Stuttgart, Germany, [3]ULiège, Gembloux Agro-Bio Tech, Passage des Déportés 2, 5030 Gembloux, Belgium, [4]France Conseil Elevage, Maison du Lait, Rue de Châteaudun 42, 75009 Paris, France; mntran@awenet.be*

As lameness is a major welfare issue and culling reason for dairy cows, being able to objectively record it is very important. However, even though it is one of the most costly health problem (together with mastitis and infertility), many lameness cases are not detected, also due to difficulties to assess this phenotype. Practically, lameness phenotype can be recorded using the gait scoring, but it is done in a small amount of farms. Thus, objective detection of lameness on a regular basis is one of the objectives of the Interreg NWE HappyMoo project. This project focusses on 3 of the 5 freedoms defining animal welfare: absence of hunger, stress and disease. Concretely, gold standard will be compared to data collected in routine and mostly milk composition. The first step was to define gold standard, with the challenge of data coming from 8 different countries, therefore with different methods and formats. It was chosen to work on both lameness phenotype (acquired with gait scoring) and hoof diseases. To harmonise data, lameness phenotype, which can be recorded on different scales: 1 to 3, to 5 or a binary trait (presence/absence), will be registered under a unique scale. Using recently published ICAR guidelines, bridges between the different existing scales will be developed. For hoof disorders, it was chosen to work with the 8 most recorded pathologies, again using ICAR format and codes in order to be as universal as possible (white line disease, sole ulcer, interdigital phlegmon, interdigital hyperplasia, digital dermatitis, sole haemorrhage, heel horn erosion and double sole). Ultimately, harmonised reference data will be compared to milk composition obtained via regular milk recording and in particular the milk mid-infrared spectrum. Work will be challenging, hoping that innovative machine learning methods could help HappyMoo to develop tools allowing the detection of lameness at early stages and on a regular basis.

Relationships between the energy balance and fertility or health disorders in Holsteins in Japan

A. Nishiura[1], O. Sasaki[1], T. Tanigawa[2] and H. Takeda[1]
[1]*Institute of Livestock and Grassland Science NARO, Tsukuba, Ibaraki, 3050901, Japan, [2]Hokkaido Research Organization, Nakashibetsu, Hokkaido, 0861135, Japan; akinishi@affrc.go.jp*

It is important to improve the energy balance (EB) of dairy cows in lactation to prevent the deterioration of fertility and health. Our objective was to investigate relationships between EB predicted using milk traits and fertility or health disorders. The data to create the equation for prediction of EB consisted of records of 156 lactations of 102 Holstein cows which were dried off in 2016-2019 in Hokkaido Research Organization. We measured milk yield and dry matter intake daily and milk components and body weight weekly. We calculated the average EB per 10 DIM, then the number of EB values was 30 per one lactation. Multiple regression model was built to predict EB. Independent variables were DIM, milk yield (MY), fat % (F), fat yield (FY), protein % (P), protein yield (PY), lactose % (L), lactose yield (LY), fat to protein ratio (FPR), fat to lactose ratio (FLR) and protein to lactose ratio (PLR). These variables except DIM were together with 'd' variables, which were the current minus the previous value. Model reduction was carried out by stepwise regression. The averages of predicted EB (PEB) from 6 to 105 DIM were divided into 3 classes: low (under -10 MJ/day), middle (from -10 to 0 MJ/day), and high (over 0 MJ/day). The averages of days to first AI, days to conception, days to first ovulation, days to first oestrus and first conception rate in each PEB classes were calculated. Also, the average numbers of treatments of disorders per cow in each PEB classes were calculated. DIM, LY, FPR and dPY were included in the equation to predict EB ($R2=0.49$). The averages of days to first AI, conception, first ovulation and first oestrus were longer in lower PEB class. The averages of first conception rate were lower in lower PEB class. The average numbers of treatments of digestive, fertility, udder health, locomotion, endocrine and metabolic disorders were more in lower PEB class. It would be possible to improve fertility and health performance by improving EB in the early lactation stage and PEB could be used as the index of EB.

Metabolites in blood serum of suckler cows during the grazing period

H. Scholz[1], P. Kühne[1] and G. Heckenberger[2]
[1]Anhalt University of Applied Sciences, Faculty LOEL, Strenzfelder Allee 28, 06406 Bernburg, Germany, [2]State Institute for Agriculture and Horticulture Saxony-Anhalt, Lindenstraße 18, 39606 Iden, Germany; heiko.scholz@hs-anhalt.de

The objective of the studies reported was to examine the concentration of different blood serum metabolites in grazing suckler cows during the grazing season. Evidently, more knowledge about metabolism of beef cattle and suckler cows is required in order to improve health and productivity in this area of livestock production under German conditions. Blood samples were taken from cows every month from the jugular vein during the grazing period from May until September (years 2015 until 2019; 3 farms) and were analysed for the content of urea (UREA), total protein (TP), ß-hydroxybutyrate (BHB), cholesterol (CHOL) and the activity of ASAT and GLDH. Feed samples for chemical analysis were taken every month on each farm during the grazing season. For statistical analysis within each farm were estimated the average concentration of blood serum concentrations per month to minimise the effect of different herd sizes. Statistical analysis took place with ANOVA with fixed effects of farms (1-3) and month (May until September) using SPSS Version 25.0. The average serum TP concentration of the cows was 73 g/l with a large variability within the grazing period. Concentration of BHB and CHOL in blood serum of suckler cows was documented within the reference values. The serum ASAT activity was significant higher ($P<0.05$) in May and June than in the other part of the grazing period. Increased activities of ASAT and GLDH can indicate a damage of the liver. In the investigation, there was a tendency of increasing blood urea concentration from onset to the end of grazing season. If the crude protein content decreases during the grazing period, this increase can only be justified by a decrease in the milk yield and thus a lower excretion of nitrogen from the milk. The results indicated that at the onset of the grazing period the high amount of crude protein in the grass often results in high activities of ASAT and GLDH. An increase of blood urea content during the grazing season was found. Options for optimising protein and energy supply are advisable to keep the suckler cows healthy.

DCAB of feed and NSBA in urine of suckler cows during the grazing period

H. Scholz[1], P. Kühne[1] and G. Heckenberger[2]
[1]Anhalt University of Applied Sciences, Faculty LOEL, Strenzfelder Allee 28, 06406 Bernburg, Germany, [2]State Institute for Agriculture and Horticulture Saxony-Anhalt, Lindenstraße 18, 39606 Iden, Germany; heiko.scholz@hs-anhalt.de

Dietary Anion-Cation Balance (DCAB) in grazing systems under German condition show a tendency to decrease from May until September and often are measured DCAB lower than 100 meq per kg dry matter. Lower DCAB in grass feeding system can change the metabolic status of suckler cows and often are results in acidotic metabolism. The hypothesis was that metabolic imbalances could be identified by urine measurement in suckler cows. The farm study was conducted during the grazing seasons 2017 and 2018 and involved 7 suckler cow farms in Germany. Cows had free access to water and salt block and free access to minerals (loose). The dry matter of the grass was determine at 60 °C and were than analysed for energy and nutrient content and for the Dietary Cation-Anion Balance (DCAB). Urine was collected in 50 ml-glasses and analysed for net acid-base excretion (NSBA) and the concentration of creatinine and urea in the laboratory. Statistical analysis took place with ANOVA with fixed effects of farms (1-7), month (May until September) and number of lactations (1, 2 and ≥3 lactations) using SPSS Version 25.0 for Microsoft Windows. An alpha of 0.05 was used for all statistical tests. During the grazing periods of the years 2017 and 2018 was observed an average DCAB in the grass of 167 meq per kg DM. A very high variability could be determined and decrease to the end of the grazing period. Between the DCAB of the grass and the NSBA in urine of suckler cows was a correlation according to PEARSON of r=0.478 (P≤0.001) or after SPEARMAN of r=0.601 (P≤0.001) observed. The results obtained show that up a decrease in the DCAB is related to a decrease in NSBA in urine of suckler cows. Monitoring of metabolic disturbances should include analysis of urine, blood, milk and ruminal fluid.

Genetic relationship between foot and claw health traits and lameness in Czech Holstein cattle

L. Zavadilová, E. Kašná and Z. Krupová
Institute of Animal Science, Genetics and Breeding of Farm Animals, Přátelství 815, 104 00 Prague 10, Czech Republic; zavadilova.ludmila@vuzv.cz

The heritabilities and genetic correlations for foot and claw disorders and lameness (L) were estimated in the Czech Holstein cattle by repeatability linear animal models, optionally with random herd_year (hy) effects. The disorders and L, as 0/1 occurrence from calving to 305 d of lactation, 47,174 lactations of 31,750 Holstein cows were recorded on 52 farms in the Czech Republic from 2017 to 2019 using the national web application 'The Diary of Diseases and Medication' designed for the continuous recording and evaluation of health traits based on the farmers. Three groups of foot and claw disorders were defined: skin diseases (SD), including digital and interdigital dermatitis, and interdigital phlegmon; claw disorders (CD) including ulcers, white line disease, horn fissures, and double sole; and overall foot and claw disorders (OFCD) comprising all the recorded disorders and L. Digital dermatitis (DD), interdigital phlegmon (IP), ulcers (U) and lameness (L) were analysed as separate traits. Lactational incidence rates (number of affected lactations/number of lactations at risk × 100) for CD, SD, and L were 9.9, 5.7, and 4.8%, respectively. Estimates of heritability were from 0.008 for IP, to 0.042 for CD and 0.043 for L or U when no random hy was employed. If random hy was included in the model heritability decreased due to the increase of the total explained variance. Phenotypic correlations of foot and claw disorders to lameness were lower than 0.069 (IP). On the contrary, those between foot and claw disorder traits were high; U and CD 0.963, DD and SD 0.823. Genotypic correlations between L and foot and claw disorder traits were low and nonsignificant, from 0.036 OFCD to 0.236 SD except for IP 0.507 (P<0.05). However, genetic correlations between three groups of foot and claw disorders and DD, IP and U were strong and significant, from 0.536 (DD and CD) to 0.971 (U and OFCD). We conclude that lameness is not applicable for OFCD definition and has to be used as a separate selection trait. This work was supported by the Ministry of Agriculture of the Czech Republic, institutional support MZE-RO0718 and by the National Agency for Agricultural Research, Project No. QK1910320.

Metabolites in blood serum of suckler cows during winter feeding period in varied feeding situations

P. Kühne[1], H. Scholz[1] and G. Heckenberger[2]
[1]Anhalt University of Applied Sciences, Strenzfelder Allee 28, 06406 Bernburg, Germany, [2]State Institute for Agriculture and Horticulture Saxony-Anhalt, Lindenstraße 18, 39606 Iden, Germany; petra.kuehne@hs-anhalt.de

The objective of the studies reported was to examine the concentration of different blood serum metabolites in suckler cows in the transition period. In the last 8-10 weeks ante partum suckler cows were often fed with silage or hay with low energy and a higher content of crude fibre to control body condition. This can result in a significant reduction of dry matter feed intake during the hole transition period and a higher challenge of metabolism. Suckler cows were fed a Total Mixed Ration (TMR) in the last 8 weeks before calving and grass silage after calving. Blood samples were taken from cows from the jugular vein during the transition period from 6 weeks ante partum and 4 weeks post partum (years 2009 to 2019) and were analysed for the content of UREA, Total Protein (TP), ß-hydroxybutyrate (BHB), cholesterol and the activity of ASAT and GLDH. Feed samples for chemical analysis were taken every month. Dry matter feed intake of suckler cows were measured during the transition period by each cows and day. For statistical analysis were estimated the average concentration of blood serum concentrations per month to minimise the effect of different amount of cows. Statistical analysis took place with ANOVA with fixed effects of ration and week using SPSS Version 25.0 (alpha 0.05). The average BHB concentration was 0.3 mmol per litre and shows no tendency for energy imbalance. Activity of ASAT and GLDH was on average within the reference values, however sometimes too high. A high variation could be determined in the concentration in blood serum for UREA and Total Protein. Metabolites related to the crude fibre content in TMR show significant differences for ASAT, GLDH and TP. Activity of ASAT was above the reference values and GLDH shows a significant difference between groups of crude fibre content in the rations. Concentration of cholesterol was decreased in ration with more than 30% crude fibre. Lower concentration of cholesterol in blood serum of suckler cows was associated with a lower feed intake ante partum. Blood serum metabolites can be used as an indicator of energy status and health of suckler cows in the transition period.

Colostrum: The magic potion to ensure newborn survival in ungulates

R.M. Bruckmaier and J.J. Gross
Vetsuisse Faculty, University of Bern, Veterinary Physiology, Bremgartenstr. 109a, 3012 Bern, Switzerland; rupert.
bruckmaier@vetsuisse.unibe.ch

Ungulates like bovine, equine, and porcine species cannot transfer complex proteins such as immunoglobulins (IG) from the maternal blood to the foetus during pregnancy. Thus, the newborn does not have any immune defence at birth. Instead of a materno-foetal transfer during pregnancy (as e.g. in rodents and humans), ungulates depend on the immunisation via the oral uptake of colostrum to bridge the weeks until the immune system of the offspring has developed. The passive immunisation via colostrum requires its immediate availability. The intestinal absorption of intact IG is only possible for a few hours before the secretion of proteolytic enzymes in the intestine develops. The uptake of colostrum shortly after birth requires an early initiation of lactogenesis besides an efficient active transfer of IG into mammary secretions during the final stage of pregnancy. Early lactogenesis requires an early decline of blood progesterone, which is common to all ungulates. The concentration of IG is considered the most important factor to determine colostrum quality and its immune protective effect for the newborn. Their concentrations vary considerably between animals, but also between individual glands within an animal. However, the IG concentration of individual animals as well as individual glands of an animal is highly repeatable in successive lactations. Therefore, a considerable effect of genetics as well as epigenetics on colostrum quality is assumed. Beyond IG, colostrum contains a number of bioactive substances such as hormones, growth factors, and enzymes that have been demonstrated to be crucial for a fast development of the gastro-intestinal tract but also for mammary gland function itself during the final stage of pregnancy. In addition, colostrum is characterised by high amounts of galacto-oligosaccharides in addition to free lactose. Because these lactose-based oligosaccharides are not cleaved by beta-galactosidase in the small intestine, they serve as prebiotic nutrients for bifidobacteria and lactobacilli in the large intestine. The proliferation of these microorganisms causes a considerable decline of the faeces pH, which inhibits the growth of pH-sensitive intestinal pathogens.

Effect of early life rumen microbial programming on feed utilisation and gut microbiome in goats

A. Belanche, J.M. Palma-Hidalgo, A.I. Martín-García and D.R. Yáñez-Ruiz
Estación Experimental del Zaidín, EEZ, Profesor Albareda 1, 18160, Granada, Spain; a.belanche@csic.es

This study explored the concept of rumen microbial programming using early life nutritional interventions which potentially modify the rumen colonisation pattern towards a desirable microbiota providing adaptive advantages later in life. A total of 32 newborn goat kids were randomly distributed into 4 groups and orally inoculated (daily during 2.5 months) with autoclaved rumen fluid (AUT), fresh rumen fluid from adult animals adapted to forage (RLC) or concentrate diets (RFC) or without inoculation (CTL). Experimental groups were always kept physically separated. At 6 months of age animals were fed with forage or concentrate-rich diets in two consecutive periods and finally euthanised. Rumen samples were collected through the experiment to study the microbiome. Inoculation with fresh rumen fluid accelerated the rumen microbial colonisation in early life facilitating the transition from liquid to solid feed. Most of these effects persisted at 6 months of age and RFF and RFC animals had higher bacterial (+64 OTUs), methanogens (+4 OTUs) and protozoal diversity across diets, whereas CTL animals remained protozoa-free. This greater microbial complexity had a positive effect in the rumen energy metabolism such as higher VFA concentration (+18%), butyrate molar proportion (+48%) and digestibility of the forage diet (+8%), resulting in higher *in vitro* methane emissions (+41%). The higher rumen fermentative activity observed in RFF and RFC animals was accompanied with an up-regulation of genes related with the epithelial growth (IGFR1), cell proliferation (Cyclin A), VFA absorption (MCT1) and VFA metabolism (HMGCL) in the rumen wall, as well as a greater papillae width (+43%). However, inoculation with fresh rumen fluid had a negative impact on the rumen protein metabolism resulting on higher ammonia concentration and lower microbial protein synthesis (-55%) when fed the forage diet. Control animals had higher bacterial and fungal concentrations in the hindgut, possibly as a compensatory mechanism. In conclusion, early life inoculation with fresh rumen fluid can be considered, to some extent, as a rumen microbial programming strategy, but the suitability of its use in practical farming would depend on the diet consumed by the animals later in life.

Calcium and P absorption and feed efficiency in weaned pigs is reduced by low P intake of the sows

P. Bikker[1], A. Liesegang[2] and J. Van Baal[3]
[1]Wageningen University & Research, Wageningen Livestock Research, P.O. Box 338, 6700 AH Wageningen, the Netherlands,
[2]University of Zurich, Institute of Animal Nutrition, Vetsuisse Faculty, Winterthurerstr. 270, 8057 Zürich, Switzerland,
[3]Wageningen University & Research, Animal Nutrition Group, P.O. Box 338, 6700 AH Wageningen, the Netherlands;
paul.bikker@wur.nl

Insight in calcium (Ca) and phosphorous (P) metabolism of sows is required to optimise P utilisation in sows and their offspring. This study was conducted to determine the early nutritional programming of P and Ca metabolism in weaned pigs via maternal dietary P supply during gestation and lactation. A 2×2 factorial arrangement was used with 2×14 sows on low (LP) and high P (HP) diets (50 and 100% of recommendations) in gestation and lactation and LP and HP nursery diets (60 and 110% of recommendations) during 5 weeks post weaning. Faeces, urine and blood of sows was collected in early and late gestation and lactation. Piglets were sacrificed at birth, 21 days of age and 35 days post-weaning, and serum, mucosal scrapings from the jejunum, and kidney cortex were harvested to determine expression of Ca and P transporter genes with qPCR. Faeces were collected from d 32-35 post weaning to determine total tract digestibility. The reduction in P-content of sow diets did not influence birth weight, growth rate of piglets in lactation, body weight at weaning, body composition or bone characteristics of piglets at birth and weaning. Hence, the sows appeared to largely buffer the reduction in P intake with their body reserves. The reduction in dietary P content enhanced apparent Ca-digestibility ($P<0.001$) in sows, but not in nursery pigs. This effect was associated with a drastic increase in serum 1,25-OH2-D3, a decrease in PTH, and an increase in expression of calcium binding protein CaBP9K, Ca channel TRPV6 and phosphate transporter NPT2c, as observed in nursery pigs. Jejunal PIT1 expression was increased ($P=0.069$) in nursery pigs originating from LP sows, suggesting an effect on the capacity of intestinal P uptake. Digestibility of Ca ($P=0.003$) and P ($P=0.02$) in weaned pigs was lower in offspring of LP sows irrespective of the P-content of the nursery diets. In addition, the P reduction in sows diets also reduced the feed efficiency (higher FCR) in nursery pigs. The reduction in P content in nursery diets reduced growth performance, bone mineralisation, bone strength and carcass mineral content in weaned pigs. In conclusion, the results indicate differential Ca-P interactions in the digestive tract of sows and growing pigs, and a long term programming effect of the P content of sow diets on nutrient utilisation in the offspring. Nonetheless, the immediate effect of P content in nursery diets is greater than the effect of the early nutritional programming.

Neonatal glutamine supplementation effects on jejunal development in low birth weight piglets

J. Schregel, Q. Sciascia, Z. Li, A. Tuchscherer and C.C. Metges
Leibniz Institute for Farm Animal Biology, Wilhelm-Stahl-Allee 2, 18196 Dummerstorf, Germany;
schregel@fbn-dummerstorf.de

Intestinal development is delayed in low (L) birth weight (BiW) piglets. Glutamine (Gln) is a primary energy source for enterocytes and beneficially affects piglet growth rate and intestinal health. We explored the effect of oral Gln supplementation during the first 12 days (d) of life, on jejunal biochemical indices (BI) and mRNA abundance encoding anti-oxidative defence, tight junction and cell function proteins. At birth (d0), male L (0.8-1.2 kg, n=48) and normal-BiW (N; 1.4-1.8 kg, n=48) littermates born to German Landrace gilts were selected. Litter size was standardised to 12 piglets; piglets nursed by their dams were randomly assigned to oral supplementation groups (1 g Gln or 1.22 g alanine (Ala)/kg body weight×d; isonitrogenous): Gln-L, Gln-N, Ala-L or Ala-N. At 5 and 12 d jejunal RNA, DNA and protein were isolated. BI of total RNA, DNA, protein (µg/mg tissue), cell size (total protein/DNA), translational capacity (total RNA/DNA) and efficiency (total protein/RNA) were assessed. The relative mRNA abundance of GPX1, GPX2, GSRa, CGL (glutathione metabolism), Claudin-4, ZO-1, ZO-2, Occludin (tight junctions), glutaminase, PSMC3 (ATPase subunit proteasome) and PCNA (DNA replication) were assessed by rt-qPCR. Data were analysed with the MIXED procedure and Tukey-Kramer test ($P<0.05$) of SAS. At 5 d, total RNA ($P=0.09$) and translational capacity ($P=0.04$) were higher, whilst translational efficiency was lower in Ala-N compared to Gln-N piglets. Total RNA was lower ($P=0.09$) and total protein ($P=0.1$), cell size and translational efficiency ($P=0.04$) were higher in Ala-L compared to Ala-N piglets, while other groups did not differ. At 12 d, no differences in BCI were found. At 5 d, the mRNA abundance of Gpx1 ($P=0.06$), Occludin ($P=0.09$), Psmc3 ($P=0.01$) and Zo-1 ($P=0.06$) were higher in Ala-L compared with Ala-N. At d 12, only two differences in mRNA abundance were observed, Pcna was lower in Ala-L vs Ala-N ($P=0.05$) and Cgl was higher in Ala-L vs Gln-L ($P=0.06$). In conclusion, the differences in jejunal BCI and mRNA between Ala-L and Ala-N were not observed between Gln supplemented littermates, and may be associated with the improved growth of Gln-L piglets we have previously reported in this cohort.

Effect of dietary energy source replacement in milk replacer for calves

J.N.R. Wilms[1,2], M.A. Steele[1], J. Martín-Tereso[2] and L.N. Leal[2]
[1]Department of Animal Bioscience, University of Guelph, 553 Gordon St #501, Guelph, ON N1G 1Y2, Canada, [2]Trouw Nutrition, Stationsstraat 77, 3811 MH, the Netherlands; juliette.wilms@trouwnutrition.com

Compared with whole milk (WM), milk replacers (MR) for calves usually contain higher levels of lactose and lower levels of fat. Implications of MR macronutrient composition on performance and metabolic parameters is not well known. In this light, the effect of three MR formulations and WM powder were evaluated on intakes, growth and health in 96 Holstein male calves fed *ad libitum*. Calves were blocked based on day of arrival and within each block, calves were randomly assigned to one of 4 treatments including: a high lactose MR (44% lactose; 19.7 MJ/kg; HL; n=24), a high protein MR (26% protein; 20.0 MJ/kg; HP; n=24), a high fat MR (25% fat; 21.3 MJ/kg; HF; n=24), and WM powder (26% fat; 21.6 MJ/kg; WM; n=24). In the first 2 weeks (phase 1; P1), calves were individually housed and were fed 3 l three times daily (135 g/l). Calves were then moved to group housing and were fed *ad libitum* from week 3 to week 6 (P2). Weaning was gradual and took place between week 7 to 10 after arrival (P3). In weeks 11 and 12, calves were fed solids only (P4). Concentrates, chopped straw, and water were available *ad libitum* throughout the study. Body weight was measured weekly and blood was sampled in weeks 1, 2, 3, and 4. In P2, MR intakes were higher for HL (11.5 l/d) than HF and WM (10.0 l/d; P<0.05) and for HP (10.9 l/d) than WM (P<0.05). Starter intakes were higher in HP (3.78 kg/d) compared to HF and WM in P4 (3.35 kg/d; P<0.05). Growth over the entire experimental period was higher in HL than WM (P<0.05). Similarly, ADG was higher in HL (1.28 kg/d) than HF and WM calves (1.10 kg/d; P<0.05) in P2. Blood pH, blood base excess, blood HCO_3, blood tCO_2, pCO_2 were lower in WM fed calves (P<0.05). Similarly, haematocrit and blood Na^+ were lower in WM (P<0.01). Results suggest that the lower energy density of the HL and HP resulted in higher milk intakes and that the high glucose signal from the HL treatment may explain the higher BW. Moreover, MR feeding affected acid-base balance and blood haematology when compared to WM powder. Whether these differences are biologically relevant requires further investigation.

Gestational undernutrition in extensive grazing conditions affects muscle traits in newborn lambs

J. Ithurralde[1], R. Perez-Clariget[1], A.L. Astessiano[1], M.J. Abud[1], A. Casal[1] and A. Bielli[2]
[1]Facultad de Agronomía, Udelar, Producción Animal y Pasturas, Garzon 780, 12900, Montevideo, Uruguay, [2]Facultad de Veterinaria, Udelar, Morfología y Desarrollo, A. Lasplaces 1550, 11600, Montevideo, Uruguay; javiithu@gmail.com

The aim of this study was to evaluate the effects of maternal undernutrition -induced by a lower natural pasture allowance- from 23 days before conception until 123 days of gestation on muscle gene expression of newborn lambs. Multiparous single-bearing Corriedale ewes (n=335, 48.9±0.3 kg body weight (BW) and 2.74±0.02 body condition score (BCS)) were used in a randomised block design (three repetitions) and assigned to two natural pasture allowances: High-pasture-allowance (HPA) ewes grazed on natural grasslands at 10-12 kg of dry matter (DM)/100 kg of BW/day; and Low-pasture-allowance (LPA) ewes grazed on 5-8 kg of DM/100 kg of BW/day. After day 123 of gestation all ewes grazed on Festuca arundinacea at 2.8 kg DM/kg BW and were supplemented with rice bran (200 g/animal/day) and 50 ml of crude glycerine/animal/day. Twelve hours after birth muscle samples were collected from the first eighteen (9/treatment; 3/repetition) male newborn lambs. Muscle gene expression was evaluated in the Longissimus lumborum of newborn lambs by SYBR-Green real time RT-PCR. Data were analysed using a mixed model including the effect of treatment as the fixed effect and the block as the random effect. Maternal undernutrition increased (P=0.05) the expression of MyHCI mRNA, tended to increase (P=0.07) the expression of GLUT 4 and PAX 7 mRNA, while it reduced (P=0.04) the expression of IGF-I (P=0.04). The higher expression of MyHCI and GLUT 4 mRNA as well as the lower expression of IGF-I mRNA in muscles from in utero undernourished newborn lambs might be due to secondary myogenesis impairment and could be associated with an early phase of enhanced postnatal insulin sensitivity and glucose uptake followed by the later emergence of insulin resistance. In consequence, our results suggest that a lower maternal pasture allowance affects muscle characteristics of newborn lambs with possible future effects over animal health, growth and meat-producing performance.

Calf metabolic profiles and performance until weaning from beef heifers on grasslands

M.S. Orcasberro and A. Alvarez-Oxiley
Facultad de Agronomía-Universidad de la República, Producción Animal y Pasturas, Garzón 780, 12900 Montevideo, Uruguay; alvarez-oxiley@fagro.edu.uy

The aim of the study was to evaluate the effect of two native pasture allowances in beef heifers on continuous grazing, during the last third of gestation and lactation, on metabolic profiles and calf growth from birth to weaning. Two treatments of forage allowance (HA) were evaluated (8 and 5 kg DM/kg LW, annual average) with 20 pregnant heifers (10 per treatment, Hereford and Aberdeen Angus), in second third of gestation at the beginning of the trial (May). At calving (September), calves were weighed. Within 24 hours postpartum and monthly until weaning (April), blood samples and body measurements were made: length of head (H); distance between shoulders (SH); rump length (RL); body length (BL); thoracic circumference (TC); long front leg (LFL); front leg circumference (FLC); long hind leg (LHL); circumference of hind leg (CHL). To estimate the growth pattern and potential mature size of calves, the Frame Score (FS; scale 2-9 from small to large frame animals) was calculated from LHL at weaning. The experiment was a completely random design. Data were analysed with the Mixed procedure of SAS (2001), as repeated measures. Cows LW (343 kg), BCS (3.9, scale 1-8) and calf birth weight (31.5 kg), did not differ between HA treatments (P>0.001). At weaning, calf LW was higher in HA5, explained by a higher daily gain on the second half stage from birth to weaning (January-April; HA8: 0.61 vs HA5: 0.87 kg/d; P<0.0001). Plasma glucose (4.49 vs 4.41 g/l; P=0.744), urea (21.9 vs 20.5 g/l; P=0.179) and total protein (56.5 vs 57.7 g/l; P=0.338), HA8 and HA5 respectively, did not differ between treatments. However, TC (124 vs 132 cm, P=0.0139) and LHL (100 vs 106 cm, P=0.0015) were higher in HA5. FS was 4 (large size) and 3 (medium size) for HA5 and HA8, correspondingly. The results of this study are still preliminary, but they suggest that beef heifers submitted to a 5 kg DM/kg LW during the last third of gestation, might reprogrammed the postnatal growth trajectory affecting calf performance to tall and large size animals with relatively heavy body weights.

Effect of poor maternal nutrition on skeletal muscle microRNA expression of foetal calves

S. Muroya[1], Y. Zhang[2], A. Kinoshita[2], K. Oshima[3], Y. Gotoh[3], I. Oshima[2], K. Otomaru[2], M. Sano[4], S. Roh[5], M. Futohashi[5], M. Oe[1], K. Ojima[1] and T. Gotoh[2]
[1]*NARO Inst Livestock and Grassland Science, Tsukuba, Ibaraki 305-0901, Japan,* [2]*Kagoshima University, Kagoshima, Kagoshima 890-8580, Japan,* [3]*Western Region Agricultural Research Center, NARO, Ohda-shi, Shimane 694-0013, Japan,* [4]*University of Shiga Prefecture, Hikone, Shiga 522-8533, Japan,* [5]*Tohoku University, Aoba-ku, Sendai 980-8578, Japan; muros@affrc.go.jp*

Maternal nutrition during gestation has influences on development of offspring in rodents, however, little is known about bovine foetal development. The objective of this study was to investigate influences of maternal cow nutrition during gestation on skeletal muscle microRNA (miRNA) expression in Japanese Black cattle. Eight cattle were allocated to high (HN; n=4) and low nutrition (LN; n=4) treatments that met 120 and 60%, respectively, of the nutritional requirements. The foetuses were removed by Caesarean section at 8.5 mo after conception, and a portion of *longissimus thoracis* (LT) muscle was collected. Total RNAs were extracted from the LT muscles, and miRNA expression was comprehensively analysed using miRNA microarray (Agilent). The expression levels of miRNAs of interest was validated by RT-qPCR. Then the predicted target genes of miRNAs of interest were applied to gene ontology (GO) analysis to gain molecular biological events associated with the miRNAs. The maternal feeding treatment resulted in lower carcass and LT weight in LN than in HN foetuses (P<0.05). The miRNA transcriptomic results combined with statistical analysis showed that a total of 13 miRNAs were different in expression between HN and LN foetuses (P<0.05). In the qPCR results, expression of 3 miRNAs including miR-33a (P=0.027) was lower in the LN than in the HN. According to the results of GO analysis, the predicted target genes of the 3 miRNAs of interest were found to be associated with in utero embryonic development and both positive and negative transcriptional regulation from RNA polymerase II (RNAPII) promotor. These results suggested that the three downregulated miRNAs were involved in regulation of genes associated with LT muscle development and RNAPII-associated transcriptional regulation in foetal calves under reduced maternal nutrition.

Supplemental liquid feeding of suckling piglets on growth, sow condition and milk composition

Z. Li[1], A. Winkler[2], Q. Sciascia[1], S. Goers[1], A. Tuchscherer[1], J. Schregel[1], G. Dusel[2] and C.C. Metges[1]
[1]Leibniz Institute for Farm Animal Biology, Wilhelm-Stahl-Allee 2, 18196 Dummerstorf, Germany, [2]Bingen Technical University of Applied Sciences, Berlinstraße 109, 55411 Bingen, Germany; metges@fbn-dummerstorf.de

The genetic selection towards high sows' prolificacy has increased litter size where milk supply is limited for piglets. We investigated the effect of supplemental liquid feeding during the suckling period (SLF) on piglet growth, sow condition (SC) and milk macronutrient and amino acid composition (MC). At a commercial farm 35 sows (DanBred, parity 1-5) and their offspring were randomly assigned to treatment (T, n=17) or control (C, n=18) groups, resulting in 226 T and 205 C piglets. Group T piglets received supplemental milk replacer (1-15 d) and liquid feed (16-27 d), while Group C only received dry creep feeds (6-21 d). Both groups received the same weaning (22-27 d) and post-weaning (POW, 28-69 d) feeds. Piglet body weight (BW) was measured at 0-27, 40, 44 and 69 d. Sow BW, back-fat thickness and SC were measured at 7 d antepartum and 26 d postpartum, and MC was determined at 1, 7 and 21 d of lactation. Data was analysed by SAS using the MIXED procedure and the Tukey test (P<0.05). Litter sizes were 16.6 (T) and 15.7 (C) (P>0.2). SLF reduced piglet pre-weaning (PRW) mortality (-30.7%, P<0.1), increased piglet BW at 44 and 69 d (+11.2% and +8.8%, respectively, P<0.05), and average daily weight gain (ADWG) during PRW (+4.0%, P<0.1) and POW (+7.3%, P<0.05) periods. With consideration of piglet sex, SLF increased BW of male and ADWG of both sexes during POW periods (P<0.05). In Group T, males were heavier than females during PRW and POW periods (+2.5%~+9.4%, P<0.05). In parities 1-3, piglets of both sexes in Group T had higher PRW and POW BW (P<0.1) and PRW ADWG (P<0.05) than Group C. In parities 4 and 5, SLF improved BW and ADWG during POW period only. Additionally, male and female piglets with SLF had higher PRW (P<0.1) and POW (P<0.05) ADWG in parities 1-3 than in 4 or 5. No effect of SLF on SC was observed, whilst the effect on MC was minor. In conclusion, SLF improved survival and growth of piglets, and the effect seems to be greater in males and piglets of younger sows (parities 1-3), whilst SLF did not affect MC or SC.

Glutamine supplementation stimulates cell proliferation in the muscle of low birth weight piglets

Y. Zhao, E. Albrecht, Z. Li, J. Schregel, S. Görs, Q. Sciascia, C.C. Metges and S. Maak
Leibniz Institute for Farm Animal Biology, Wilhelm-Stahl-Allee 2, 18196 Dummerstorf, Germany;
elke.albrecht@fbn-dummerstorf.de

Low birth weight (LBW) piglets may profit from adapted nutrition during the early postnatal phase improving their survival rate and growth performance. This study aimed at elucidation of the effect of supplementation of newborn piglets with glutamine (GLN) on the cellular muscle development. A total of 144 low and normal birthweight (NBW) male piglets were either supplemented with 1 g GLN/kg body weight or an isonitrogeneous amount of alanine (ALA) (4 groups: NBW-GLN, NBW-ALA, LBW-GLN, LBW-ALA). The supplementation was provided between postnatal days 1 and 11 (dpn). Twelve piglets per group were slaughtered at 5, 12 and 26 dpn, one hour after injection with Bromodeoxyuridine (BrdU, 12 mg/kg). Samples of M. longissimus were immediately frozen in liquid N_2 and stored. Free amino acid (AA) concentrations were measured by HPLC. Proliferating cells, that incorporated BrdU, were detected by immunofluorescence with an anti-BrdU antibody and were quantified with image analysis. Results were statistically analysed with the MIXED model in SAS (v. 9.4) with fixed factors group and age and their interaction, as well as sow as random factor. The results indicate that supplementation increased the free GLN and ALA content in muscle tissue of respective piglets at 5 dpn (P<0.01). The concentrations were similar after the end of supplementation in all groups at 26 dpn (P>0.8). Proliferating, BrdU-positive cells were detected close to muscle fibres, but also within connective tissue and in blood vessels. The number of BrdU-positive nuclei per mm^2 decreased with age (P<0.001) and was greater in LBW-GLN than LBW-ALA piglets (P=0.01). Furthermore, LBW-GLN piglets had more total nuclei per area unit than LBW-ALA piglets (P=0.02). First attempts were made to identify the cell type of proliferating cells. Further differences among the groups were not detected. The results indicate that GLN supplementation increased intramuscular AA concentration and promoted cell proliferation particularly within M. longissimus of LBW piglets at 5 dpn, thus contributing to compensation of growth retardation.

Perilipins and the intramuscular lipid deposition in glutamine supplemented, low birth weight piglet

Y. Zhao, E. Albrecht, Z. Li, J. Schregel, Q. Sciascia, C.C. Metges and S. Maak
Leibniz Institute for Farm Animal Biology, Wilhelm-Stahl-Allee 2, 18196 Dummerstorf, Germany;
elke.albrecht@fbn-dummerstorf.de

Piglets with low birth weight (LBW) can partially compensate the growth retardation with adapted nutrition. The current study investigated the influence of glutamine (GLN) supplementation during the early suckling period on lipid deposition in the M. longissimus. Four groups were generated consisting of male LBW piglets and their normal birth weight (NBW) littermates. Piglets were supplemented with either 1 g GLN/kg body weight or an isonitrogenous amount of alanine (ALA) between postnatal days (dpn) 1 and 11. Twelve piglets per group were slaughtered at 5, 12 and 26 dpn and samples of M. longissimus were collected. Cryostat sections were used for Oil Red O staining, to quantify adipocytes and intramyocellular lipid droplets, and for immunohistochemistry of PLIN1-5. The mRNA abundance of PLIN1, 2, 4, 5 was determined by RT-qPCR. The protein abundance of PLIN1-5 was quantified with western blots. Results were statistically analysed with the MIXED model in SAS (v. 9.4) with fixed factors group and age and their interaction, as well as sow as random factor. The results indicate significant effects of age and BW on intramyocellular lipid droplets (P<0.01), but minor supplementation effects. Adipocyte development increased with age, but was not influenced by birth weight class nor by supplementation. PLIN1 was localised around lipid droplets in mature and developing adipocytes, while PLIN2 was localised around intramyocellular lipid droplets, PLIN3 and 4 at cell membranes of muscle fibres and adipocytes, and PLIN5 in undefined cells between muscle fibres. PLIN2 mRNA tended to be expressed higher in LBW piglets (P=0.09), corresponding to more intramyocellular lipid droplets. PLIN4 mRNA level was higher in piglets at 26 dpn compared to piglets at 5 and 12 dpn (P<0.01). There was a decrease of PLIN1, 2, 4 and 5 relative protein abundance with age (P<0.01) and a higher protein level of PLIN3 and 5 in LBW piglets (P<0.05). The results indicate importance of birth weight and age for intramuscular lipid deposition and different roles of PLIN family members in this process, but no clear GLN supplementation effect.

Maternal and direct dietary polyphenols affect meat and carcass quality in lambs and goat kids

K. Giller[1], S. Sinz[1], M. Kreuzer[1] and S. Marquardt[1,2]
[1]ETH Zurich, Institute of Agricultural Sciences, Animal Nutrition, Univesitätstrasse 2, 8092 Zurich, Switzerland,
[2]International Livestock Research Institute (ILRI), Mazingira Centre, Nairobi, Kenya; katrin.giller@usys.ethz.ch

Antimicrobial and antioxidant properties of polyphenols may affect rumen and host metabolism and thus animal growth and meat quality. This study aimed at assessing the effects of maternal and direct polyphenol supplementation of sheep and goats on growth, slaughter performance and meat quality. Lactating East Friesian Dairy sheep (n=11) and Saanen goats (n=9) received a hay-based diet including concentrate (55:45), which was supplemented either with (P) or without (C) 7.4% grape seed extract (GSE) rich in polyphenols. The offspring suckled from their dams. After weaning, lambs (n=16) and kids (n=13) received a hay-based diet including concentrate (50:50) supplemented either with (P) or without (C) 5.6% GSE. This resulted in a total of four groups per species defined as CC (dam and offspring fed C), CP (dam fed C, offspring fed P), PC (dam fed P, offspring fed C) and PP (dam and offspring fed P). At 9 to 12 weeks of age, offspring was slaughtered to obtain hot carcass weight and collect samples of the M. longissimus thoracis (LT) at 24 h post mortem for meat quality analysis after 21 days of aging. Data were analysed using a Mixed Model with species, maternal and offspring diet and their interactions as fixed and sex as random factors. Maternal GSE supplementation increased average daily gain in lambs but not in kids (lambs: 360 vs 284 g; kids: 263 vs 251; P=0.017). Maternal and offspring GSE supplementation hardly affected carcass and meat quality. Redness (a*, lambs: 10.2 vs 10.5; kids: 11.2 vs 9.86; P=0.035) and yellowness (b*, lambs: 1.97 vs 2.37; kids: 3.18 vs 1.36; P=0.034, all P vsC) of the LT were lower in GSE-supplemented lambs but higher in GSE-supplemented kids whereas maternal diet was ineffective. Drip loss, shear force and gross composition of the LT differed between species but were unaffected by diet. A diet interaction was, however, observed for the ash content of the LT (CC: 16.0, CP: 14.2; PC: 12.4, PP: 15.4 g/kg; P=0.008). No interaction was observed for all three fixed factors. The reasons for the observed species-specific responses to maternal polyphenol supplementation need to be further evaluated.

Early feed restriction of ewe lambs and effects on male offspring during the fattening period
S. Andrés, A. Santos, C. Valdés and F.J. Giráldez
Consejo Superior de Investigaciones Científicas (CSIC), Instituto de Ganadería de Montaña (CSIC-Universidad de León), Finca Marzanas s/n, 24346 (Grulleros, León), Spain; sonia.andres@eae.csic.es

Nutritional programming caused by feed restriction during the early life of ewe lambs may affect the feed efficiency traits and biochemical profile of the offspring. In this study, the replacement ewe lambs (F0) raised in a previous project (a group of ewes fed milk replacer *ad libitum* – ADL – vs a group of ewes restricted – RES – to 62.5% the intake level of milk replacer during the suckling period) were mated to obtain the progeny (20 male lambs first generation, F1). These lambs were fed *ad libitum* a complete pelleted diet during the fattening period, weighed weekly to measure feed efficiency (residual feed intake, RFI) and slaughtered being 85 days-old. All the animals were blood sampled 3 times along the experiment from the jugular vein (vacuum tubes with no anticoagulant, 10 ml) to obtain serum and measure biochemical profile (BA400 Biosystems). The preliminary results revealed that lower numerical values of daily dry matter intake (DMI) were obtained for the male lambs during the fattening phase (780 vs 885 g/day for lambs born from ADL and RES dams, respectively; P=0.075). This trend towards significant differences in DMI could not be attributed to difference in the digestibility or in the weight of different parts of the gut. In any case, the level of significance was not reached for RFI (-17.21 vs 15.50 for the offspring born from ADL and RES ewe lambs, respectively; P=0.261). When biochemical parameters were measured in the serum samples lower significant values (P<0.05) were observed for AST-GOT, urea and Mg for the lambs born from the RES group, whereas insulin was increased (P<0.05) when compared to the ADL-born lambs. The higher levels of insulin might be indicating insulin resistance in the lambs born from RES dams. The higher numerical values for the carcass (perirenal) and non-carcass (mesenteric, omental) fat depots in these animals would support this hypothesis because insulin is a lipogenic hormone. Regarding the AST-GOT and urea levels, the differences observed in the concentration of these parameters might be indicators of differences in the liver functionality of both groups of lambs. In any case, these alterations were not enough to cause significant variations in feed efficiency traits of lambs born from early feed restricted ewes.

Responses of chicken (*Gallus gallus domesticus*) embryos to pre-hatch bromide exposure
H.L. Lucht, N.H. Casey and R.J. Coertze
University of Pretoria, Department of Animal and Wildlife Science, Pretoria, 0028, South Africa; norman.casey@up.ac.za

In previous research, the no observed adverse effect level (NOAEL) of 0.01 mg/l for bromide was validated using the chicken embryo model. The proportions of heart mass to whole embryo mass changed throughout embryonic development and interference with this process at any stage may have adverse effects on chick survivability. The aim of this study was to investigate the effect of different concentrations of bromide administered by *in ovo* injection into the albumen on embryo development in the pre-hatch period of chicken (*Gallus gallus domesticus*) embryos. 1,280 fertilised Ross-308 eggs were randomly assigned to 4 treatments (T): T0mg/l (Control), T0.01mg/l, T0.5mg/l and T1mg/l. A volume of 0.2 ml of NaBr of each treatment was injected into each egg except those assigned to the Control group. Eggs were incubated at standard temperature and humidity until hatch. At sampling points (SP), embryonic day (ED) 14 and ED20 and on hatch day (HD), wet body mass was measured and the heart collected and weighed. At ED 14 and 20 and at HD, only viable embryos and chicks were chosen for sampling. The relative heart mass to whole body mass was calculated (%). Data were analysed by a one-way ANOVA and the REML procedures using Statistical Analysis Software (SAS). A one-way ANOVA revealed significant differences (P≤0.1) in relative heart mass (%) between treatments on HD where T0mg/l (Control)=0.88±0.03, T0.01mg/l=0.72±0.03, T0.5mg/l=0.68±0.03 and T1mg/l 0.68±0.03. An analysis using REML showed no significant differences between treatments within sampling points (P≤0.1) and the interaction T×SP was NS (P≤0.1080). The conclusion is that embryonic exposure to bromide concentrations 0.01, 0.5 and 1 mg/l resulted in decreased relative heart mass over the 21-day incubation period in chicken embryos. This may have implications for post-hatch growth and sustainability of chicks to maturity.

Effect of early life rumen microbial programming on productivity in dairy goats
A. Belanche, J.M. Palma-Hidalgo, A.I. Martín-García, E. Jimenenz and D.R. Yáñez-Ruiz
Estación Experimental del Zaidín, CSIC, Profesor Albareda, 1, 18160, Spain; a.belanche@csic.es

Rumen microbial transplant from adult animals to young animals in early life has been described as an effective method to accelerate the rumen microbial development, however little is known about the persistency of the effects and the impact on the productivity of the adult animal. In this study 32 newborn goat females were randomly distributed into 4 groups and orally inoculated (daily during 2.5 months) with either autoclaved rumen fluid (AUT), fresh rumen fluid from adult animals adapted to forage (RLC) or concentrate diets (RFC) or no inoculation (CTL). Experimental groups were always kept physically separated. At 9 months of age goats were artificially inseminated and milk yield was daily monitored during the first lactation. The entire lactation (6 months) was divided into 3 consecutive periods in which proportion of forage in the diet was increased. Diets consisted on alfalfa hay and commercial concentrate at 30/70 (LowF), 60/40 (MidF) and 100/0 (HighF) forage to concentrate ratios during the first, second and third period, respectively. Feed intake, rumen digesta, blood and milk samples were collected at the end of each period to monitor feed utilisation. Rumen digesta was incubated in batch cultures for 24 h to estimate methane production. Results indicated that inoculation with fresh rumen fluid (RFF and RFC) promoted a more diverse rumen microbiota which resulted in an improved energy uptake by the animal as noted by the higher forage intake (+9%), butyrate molar proportion (+25%) and plasma concentrations of beta-hydroxybutyrate (+28%) and glucose (+7%). Although no differences were noted in milk yield and overall feed efficiency, inoculation with fresh rumen fluid increased milk fat content (+9%) and concentrations of poly-unsaturated fatty acids such as linolenic (C18:3), linoleic (C18:2) and omega-3 acids. On the contrary, CTL animals remained protozoa-free and had increased concentration of rumen lactate (+13%), milk oleic acid (C18:1) and higher methane emissions (+31%) without affecting feed utilisation. Despite the limited number of animals used, this study demonstrated that inoculation of rumen fluid in early life had not negative impact later in life on milk yield and seems to improve energy utilisation in dairy goats.

DNA methylation of metabolic genes in the liver induced by *in ovo* stimulation of the chicken embryo
A. Dunislawska, A. Slawinska and M. Siwek
UTP, University of Science and Technology, Mazowiecka 28, 85-084 Bydgoszcz, Poland; siwek@utp.edu.pl

DNA methylation inhibits DNA transcription by addition of the methyl residues to the cysteine within the CpG islands of the gene promoters. The process of the DNA methylation can be modulated by environmental factors, such as intestinal microbiota. In poultry, composition of the intestinal microbiota can be stimulated by *in ovo* delivered bioactive compounds, such as prebiotic, probiotic, or synbiotic. The administration of bioactive compounds has been demonstrated to influence mRNA gene expression in immune, intestinal, and metabolic tissues. Liver, as the main metabolic organ, plays a key role in many vital processes, including nutrient metabolism, fat digestion, blood protein synthesis, and endocrine management. The aim of this study was to determine the effect of prebiotic, probiotic or synbiotic delivered *in ovo* on the level of hepatic DNA methylation in chicken. *In ovo* stimulation was performed on day 12 of egg incubation in broiler chickens (ROSS) and Green-legged partidgelike (GP). Bioactive compounds delivered *in ovo* included probiotic (pro)– *Lactococcus lactis* subsp. *cremoris*, prebiotic (pre)– galactooligosaccharides (GOS), or synbiotic (syn)– *L. lactis* subsp. *cremoris* with GOS. Liver samples were collected from six randomly selected individuals from each group on day 42 post-hatching. DNA methylation of 6 metabolic genes (down-regulated at mRNA level) was analysed using quantitative methylation specific PCR (qMSP). Oligonucleotides for qMSP analysis were designed using MethPrimer Significant changes in DNA methylation of metabolic genes in liver included *KLHL6* (pro), *SYK* (pre), *ANGPTL4* (pre) and *NR4A3* (syn) in ROSS, and *SYK* (pro) and *ANGPTL4* (pre) in GP. The obtained results confirm that changes in methylation levels vary depending on the chicken genotype and the bioactive compound delivered *in ovo*. The down-regulation of the metabolic gene expression in the liver mediated by *in ovo* stimulation had epigenetic character. This research was financed by the grant UMO-2017/25/N/NZ9/01822 funded by the National Science Centre (Poland).

Effect of nanocurcumin in response to a lipopolysaccharide-induced inflammation in Holstein calves

S. Kamel Oroumieh[1,2], L. Van Meulebroek[2], R. Valizadeh[1], L. Vanhaecke[2] and A.A. Naserian[1]
[1]*Ferdowsi University of Mashhad, Department of Animal Science, Mashhad, 91775-1163, Iran,* [2]*Ghent University, Faculty of Veterinary Medicine, Laboratory of Chemical Analysis, Merelbeke, 9820, Belgium; saeid.kameloroumieh@ugent.be*

Curcumin (CUR) is the primary active component of turmeric, a yellow compound originally isolated from the plant *Curcuma longa*. CUR has been known to own anti-inflammatory activity in experimental animals. The relatively poor bioavailability of CUR has been reported as the main problem. Most studies so far showed that formulated CUR (i.e. nanoparticles) has better bioavailability and biological activity. The present study is the first experiment that evaluated the effect of nanocurcumin on calves. The aim of this study is to evaluate the immunomodulatory properties of nanocurcumin to response to an LPS inflammation model in neonatal Holstein calves. A total of 24 male Holstein calves were randomised into four groups: (1) negative control (CON); (2) positive control (LPS); (3) nanocurcumin, 2 mg/kg BW (CUR2); (4) nanocurcumin, 4 mg/kg BW (CUR4). With the exception of the CON group, others were intravenously challenged with 0.5 µg/kg BW ultrapure LPS. Nanocurcumin was mixed with whole milk and used from a week before the LPS challenge to the end of experiments. Blood collection and clinical scoring were conducted at regular time points until 72 h post LPS challenge. Data were analysed by analysis of variance with repeated measures using JMP (13.2) software. There was no significant difference between LPS, CUR2, and CUR4 in clinical signs, cytokines, and acute-phase proteins; however, CUR4 recovered faster and demonstrated a more moderate level in cytokines. In conclusion, although nanocurcumin did not exert any significant effects on acute phase response in neonatal Holstein calves, it seems that using nanocurcumin could improve the health status of calves.

Gestational undernutrition in extensive grazing conditions: effects on muscle traits in heavy lambs

R. Perez-Clariget[1], J. Ithurralde[1], P. Genovese[2], A. Saadoun[1], C. Cabrera[1] and A. Bielli[2]
[1]*Facultad de Agronomia, Udelar, Producción Animal y Pasturas, Garzon 780, 12900, Montevideo, Uruguay,* [2]*Facultad de Veterinaria, Udelar, Morfología y Desarrollo, A. Lasplaces 1550, 11600, Montevideo, Uruguay; raquelperezclariget@gmai.com*

We evaluated the effects of maternal undernutrition induced by a lower natural pasture allowance between gestation days 30 and 143 on muscle characteristics of lambs. Multiparous single-bearing Corriedale ewes were randomly assigned to two natural incremental pasture allowances: (1) high pasture allowance [HPA, n=16 (eight carrying male and eight carrying female foetus) grazed on 14-20 kg of dry matter (DM)/100 kg of body weight (BW)/day; (2) Low pasture allowance [LPA; n=17 (seven carrying male and ten carrying female foetus) grazed on 6-10 kg of DM/100 kg BW/day. From gestation day 143 until weaning (90 days) all animals grazed on natural pasture with unlimited availability. After weaning, lambs were individually *ad libitum* fed with a mixed diet until slaughter at 200 days. Histochemical fibre types, total glycogen and lipid content were evaluated in samples from the Supraspinatus muscle. Data were analysed in a randomised block design using a mixed model including the effects of treatment, sex, and their interactions as fixed effects and the block as random effect. Maternal undernutrition increased the proportion of oxidative fibres (P<0.0001) and the proportion of PAS strongly stained fibres (P=0.0007) while it tended to increase muscle glycogen content (P=0.07). Maternal undernutrition reduced the proportion of glycolitic fibres (P<0.0001) and the mean diameters of both glycolytic (P=0.008) and oxidative fibres (P=0.03). Furthermore, maternal undernutrition also affected in a sex-dependent way total lipid content (treatment×sex: P=0.0013) and the mean diameter of oxidative fibres (treatment×sex: P=0.02). Oxidative fibre diameter was reduced in male LPA lambs, while lipid content was increased in muscles from female LPA lambs. Our results suggest that maternal undernutrition affects muscle characteristics with possible effects over meat-producing performance and meat quality, and that some of these effects could depend on the offspring's sex.

A novel functional protein effectively replaces fishmeal in creep feed and starter diets for piglets

M.A. Ton Nu[1], I. Lupatsch[2] and H. Schulze[2]
[1]Agilia, Skjernvej 42, Videbaek, Denmark, [2]AB Agri, 64 Innovation Way, Peterborough, United Kingdom; matn@agiliaglobal.com

This study aimed to test a novel functional protein (Cellpro – 52% CP, co-processed soy and yeast protein) as fishmeal (FM) alternative for piglets in 2 experiments. Experiment 1 studied effect of replacing FM by Cellpro in starter diets on weaned piglet's performance. 108 weaned pigs (9 reps/group) were fed one of 2 diets from weaning at D 26-40 (7% FM vs 10% Cellpro), from D 40-54 (3% FM vs 7.5% Cellpro) and a common grower diet from D 54-80. Pigs fed Cellpro diets showed similar daily intake and gain as pigs fed the FM diets from D 26-54 (P>0.05). Feeding Cellpro in starter diets generated a carry-over effect on performance by a 10% better daily gain and a 6% improved FCR in grower phase (P=0.09). Experiment 2 studied effect of feeding Cellpro instead of FM in creep feed and starter diets. 20 litters were divided and fed one of 2 creep feeds for 10 days before weaning (8% FM vs 11.4% Cellpro). Each treatment was divided into 2 sub-groups, each fed 8% FM or 11.4% Cellpro diet from weaning at D 28-34, and 5.5% FM or 7.5% Cellpro from D 34-56 and a common grower diet from D 56-71. This led to 4 groups in total (9 reps/group). Replacing FM by Cellpro in creep feed resulted in similar daily gain and weaning weight (P>0.05). There was no interaction between creep feed × starter treatment but fixed effect of creep feed on post-weaning performance. Feeding Cellpro in creep feed resulted in up to 4.5% better FCR in all post-weaning phases and 40 g higher daily gain in grower phase (P<0.05). Health status and faeces scores were comparable between pigs fed FM or Cellpro. The carry-over effect of feeding Cellpro in creep feed on post-weaning phase is due to its nutritional and functional benefits that supports a healthy gut and strong immune system in piglets. As observed in an ex-vivo piglet model, Cellpro efficiently inhibited adhesion of pathogenic *E.coli* to mucosa of pig's intestinal epithelium cells. This means fewer pathogenic bacteria colonising the gut, less microbial fermentation and more available nutrient for the host. In conclusion, Cellpro is effective in replacing FM in creep feed and starter diets and enables piglets to better exploit their genetic growth potential.

Core-microbiome development in calves is affected by the oral inoculation with mature rumen liquid

H. Huuki[1,2], S. Ahvenjärvi[1], P. Lidauer[1], J. Vilkki[1], A. Vanhatalo[2] and I. Tapio[1]
[1]Natural Resources Institute Finland (Luke), Myllytie 1, 31600 Jokioinen, Finland, [2]University of Helsinki, P.O. Box 28, 00014 Helsinki, Finland; ilma.tapio@luke.fi

Improving the microbial composition of cow rumen might improve animal performance. We studied the extent to which mature rumen microbiome can influence the establishment of the core microbiome in calves during the pre-weaning period. At birth, twin calves were randomly split into treatment (T) and control (C) groups (n=6, each). They were kept in individual pens for 8 weeks. T-group was given an oral dose of fresh mature rumen liquid 3× week starting from week-2. C-group was not treated. Rumen microbiome was determined using 16S/18S rRNA and ITS1 amplicon sequencing. A core microbiome was defined as a set of microbial OTUs, present in all animals of a group on a particular week. The number of bacterial OTUs over time increased in both groups, but the between animal variation was large. At week-2, 9 OTUs among the T-group core bacteria had donor origin. Five of them, representing *Prevotella* sp., *Selenmonas ruminantium*, *Methanobrevibacter* sp. and *Bibersteinia* sp., remained in the core until week-8, indicating successful seeding. OTU composition in C-group bacterial core fluctuated more on weekly basis. At the start of inoculation, T-group calves acquired 22 ciliate protozoa OTUs from the donor. Many of them (21% of the total core) persisted until week-8. C-group calves remained defaunated from ciliate protozoa until week-6 but on week-8 they shared 10 out of 14 their core ciliates with T-group and donor. Among ciliates, *Epidinium caudatum* dominated T-group rumen on week-2, but *Isotricha* sp. was predominant on week-8. C-group at week-8 was dominated by *Eremoplastron diblobum*. The number of fungal OTUs was highest at week-2 and decreased over time, but no significant differences between the groups were observed. The fungal community had 7 OTUs, representing *Caecomyces* 1 and *Neocallimastix* 1, shared between the donor and both groups at week-2. Additional 16 OTUs were shared only between donor and T-group but none between the donor and C-group. At week-8, *Piromyces* 2 and *Buwchfawromyces/SK2* were the predominant fungal taxa in both groups. Our results suggest that mature rumen inoculum influences microbial colonisation in calves.

Early feed restriction of ewe lambs and effects on DNA methylation of the filial generation 1 (F1)

S. Andrés[1], O. Madsen[2], A. Santos[1] and F.J. Giráldez[1]
[1]Consejo Superior de Investigaciones Científicas, Instituto de Ganadería de Montaña (CSIC-Universidad de León), Finca Marzanas s/n, 24346 (Grulleros, León), Spain, [2]Wageningen University and Research, Animal Breeding and Genomics Centre, P.O. Box 338, Wageningen, 6700 AH, the Netherlands; sonia.andres@eae.csic.es

Milk or milk replacer restriction during the early life of ewe lambs (F0) might program both, health status and feed efficiency during post-weaning phases, and transfer epigenetic marks to the next generation (F1). In this study, the replacement ewe lambs (F0) obtained in a previous project (a group of ewes fed milk replacer *ad libitum* –ADL– vs a group of ewes restricted –RES– to 62.5% the intake level of milk replacer during the suckling period) were raised under similar post-weaning conditions, and mated to obtain progeny (24 female lambs first generation, F1). Both, F0 and F1 were fed *ad libitum* with a complete pelleted diet between 7.5 to 9.5 months old to measure feed efficiency traits (residual feed intake). Animals were blood sampled at the age of 15 months (F0) or 35 days old (F1) from the jugular vein for DNA isolation. DNA methylation was assessed by Reduced Representation Bisulfite Sequencing (RRBS). Results of the RRBS data analysis will be presented to evaluate the effect of feed restriction during the suckling period of ewe lambs on DNA methylation patterns (F0) and the effects on the next generation (F1). The first preliminary results indicate that there are no global differences in the level of methylation between the different groups of lambs, which is in agreement with the lack of differences in feed efficiency (residual feed intake data) of both, F0 and F1. We do however observed a larger difference in significant methylated sites in pairwise comparison of the F1 (lambs born from ADL vs RES dams) compared to the F0 suggesting that the feed restriction during the suckling phase of ewe lambs do have a larger effect on next generation. These variations in the methylation patterns also might be related to the different age of blood sampling for F0 (15 months-old) and F1 (35 days-old). A detailed analysis of the genes related to the significant differential methylated sites will be presented. In any case, the feed restriction of ewe lambs during the suckling period (62.5% intake level) was not enough to cause clear differences in feed efficiency traits of F0 and F1.

Effect of yeast supplementation in sow diet on sow and piglet microbiota, health and performance

N. Le Floch[1], F.A. Eugenio[1], C. Achard[2], E. Apper[2], S. Combes[3] and H. Quesnel[1]
[1]INRAE, Agrocampus-Ouest, PEGASE, 35590 Saint Gilles, France, [2]Lallemand SAS, 19 rue des Briquetiers, BP59, 31702 Blagnac, France, [3]INRAE, Université de Toulouse, ENVT, GenPhySE, 31326 Castanet Tolosan, France; nathalie.lefloch@inrae.fr

Feeding probiotics like live *Saccharomyces cerevisiae* var. *boulardii* (SB) in swine diets is recommended to promote a better health and reduce antibiotic use during critical periods like weaning. This study was conducted to determine if SB added in the diet of sows during the last 2 months of gestation and the 4 weeks of lactation may contribute to supporting health and performance of the piglets before and after weaning. Crossbred sows (n=45) from parity one to nine were allocated to two dietary treatments, Control (n=23) and SB (n=22). Sows in the SB group were fed the same standard gestation then lactation diet as the Control sows but with the addition of SB at 1×10^9 cfu/kg of feed. Piglets were weaned in challenging conditions (mixing of litters, no pen cleaning). SB supplementation in sow diets influenced faecal microbiota of the sows and their piglets. Five days after weaning, the alpha-diversity was lower (P<0.05) in piglets from SB sows than in piglets from Control sows. Analysis of microbiota with Partial Least Square Discriminant Analysis discriminated faeces from SB sows from that of Control sows at 110 days of gestation (37% error rate). Piglet samples could also be discriminated according to the diet of their mother, with a better discrimination early after birth (day 6 of lactation) than after weaning (day 5 post-weaning, 11% vs 34% error rate). Five days after weaning, piglets had greater white blood cell count, plasma haptoglobin and oxidative stress than before weaning (P<0.001). Nevertheless, SB supplementation in sow diets had no effect (P>0.05) on most of health criteria measured in blood and growth performance of piglets during lactation and the post-weaning period. Moreover, dietary supplementation of SB to sows did not elicit any changes (P>0.05) on their reproductive performance, metabolic and health status, nor in the immunoglobulin and nutrient content of colostrum and milk. In the present experimental conditions, feeding SB to sows influenced sow and piglet microbiota with no consequences on health and performance.

Epigenetic regulation of the gene expression by early chicken microbiome reprogramming
A. Dunislawska, A. Slawinska and M. Siwek
UTP University of Science and Technology in Bydgoszcz, Department of Animal Biotechnology and Genetics, Mazowiecka 28, 85-084 Bydgoszcz, Poland; aleksandra.dunislawska@utp.edu.pl

Epigenetic regulation of the gene expression results from interaction between the external environment and transcription of the genetic information encoded in DNA. It includes heritable modifications within the DNA structure, such as CpG methylation. Methylated CpG regions within the gene promoters lead to silencing of the gene expression. Factors contributing to epigenetic regulation include intestinal microbiota, which in chicken can be potently modified by *in ovo* stimulation. The aim of this study was to estimate the methylation level of CpG islands of candidate genes in chickens stimulated *in ovo* with bioactive compounds. Incubated eggs were *in ovo* stimulated on day 12 of incubation in two chicken genotypes: broilers (ROSS 308) and Green-legged Partidgelike (GP). The injected compounds included: probiotic (pro) – *Lactococcus lactis* subsp. *cremoris*, prebiotic (pre) – galactooligosaccharides (GOS), and synbiotic (syn) – *L. lactis* subsp. *cremoris* with GOS. Six randomly selected individuals from each group were sacrificed on day 42 post-hatching. Immune-related tissues (spleen and caecal tonsils) were collected and DNA was isolated. Methylation analysis was performed using quantitative methylation specific PCR (qMSP) of 11 target genes, which were silenced at gene expression level. Oligonucleotides were designed by using MethPrimer. qPCR was done on LightCycler480 using chemistry with SYBR Green. In caecal tonsils, significant changes of DNA methylation were determined in *ANGPTL4* (all stimulants) and *CYR61* (syn) in ROSS, and in *SERPING* (pro, syn) and *TNFRSF14* and *IKZF1* (syn) in GP. In spleen, significant changes DNA methylation was determined in *NR4A3* (all stimulants), *TNFRSF14* (pre), *IKZF4* (pro, pre), and *NFATC1* (pro) in ROSS, and in *ANGPTL4* (pre), *TNFRSF14* (pro, pre) and *IKZF1* (pro) in GP. The obtained results confirm epigenetic character of the gene expression modulation in chickens mediated by bioactive compounds delivered *in ovo* on day 12 of egg incubation. Epigenetic silencing of the gene expression depends on the delivered compound and the chicken genotype. This research was financed by the grant UMO-2017/25/N/NZ9/01822 funded by the National Science Centre (Poland).

Measuring the impact of transport on livestock welfare
T. Collins, E. Dunston-Clarke, R. Willis, D. Miller, A. Barnes and T. Fleming
Murdoch University, College of Science Health Engineering and Education, Murdoch WA 6156, Australia; t.collins@murdoch.edu.au

The welfare of animals during long distance transport presents significant challenges to livestock, particularly when transport by sea across several climatic zones. Stressors include physical and physiological factors, such as fatigue and thermal extremes, and the effects of these hazards increase with journey time. Additional stressors on psychological wellbeing include novelty, mixing and poor handling, and are evident when loading and unloading. Notably, commercial preparation of shipments involve multiple livestock handling events. While studies have investigated the effect of some resources, there are no protocols that capture the animal experience and the cumulative effect of these stressors. Behavioural changes are important welfare indicators and may provide early signals when aspects of the journey become aversive. Behaviours that have been correlated to physiological outputs during land transport events have been described. A comprehensive welfare framework based on the principles of Welfare Quality is proposed to monitor livestock during transit by the collection of numerous animal-based outcomes. A practical list of animal-based measures is described and designed to be collected through pen-side observations by stockpersons at multiple time points along the supply chain. Measures include aspects of health and behaviour, such as the percent of individuals lame or injured, and those exhibiting resting or panting behaviour. In addition, measures of animal demeanour, using qualitative descriptors that can capture animal body language, can be analysed. The numbers of animals showing an increased flight zone or those being described as agitated and restless can be quantified and compared to different pens or stock type. A pilot study describing 240 sheep transported from on-farm, until vessel embarkment shows the demeanour of animals change throughout the journey. The framework requires monitoring and reporting of stock prior to loading, during transit and after discharge, and encourages self-audit for continuous improvement. Thus, the development of a welfare protocol suitable for use in livestock across the supply chain could help mitigate welfare risks and promote improved conditions for transport.

Prevalence and causes of skin, tail lesions and ham defects assessed at slaughter

M. Vitali, E. Santacroce and P. Trevisi
University of Bologna, Department of Agriculture and Agri-food Sciences and Technologies, University of Bologna. Viale Fani, Viale Fanin 46, 40127 Bologna (BO), Italy; marika.vitali4@unibo.it

Meat inspection is of great interest for welfare monitoring. Aggressive behaviours, rough handling or stress pre-slaughter can result in skin lesions, fractures and deep hematomas and affect muscle metabolism, as in PSE meat. In countries with typical production of PDO hams, skin must be unblemished, and hams must not present defects. Recording ham defects can identify new suitable welfare indicators or know the effect of pre-slaughter conditions on ham quality. The study aimed to investigate prevalence of skin and tail lesions and ham defects in pigs carcasses and to identify relationships between skin and tail lesions and ham defects with pre-slaughter conditions (transport distance, lairage duration, and season) and carcass quality (carcase weight and lean meat percentage). 10,070 carcases from 79 batch were assessed. Skin and tail lesions were recorded in 5 areas of the body on a 0-2 scale. A final score for each area was calculated considering both the number and severity of the lesions. Prevalence of ham defect was recorded the day after. Transport length, lairage, season, carcass weight and lean meat percentage were recorded. GLMM procedure assessed relationship between the variables. Skin lesions, veining and fractures in the ham resulted lower in winter than in summer (P<0.0001), pulled muscle increased (P<0.0001). Long-distance transport (LT) decreased middle lesion (P=0.005) and low-fat cover defect (P<0.03) compared to medium and short transport (MT and ST). A significant reduction of lean meat percentage in LT compared to MT and ST (P<0.0001) was observed. Long lairage (LL) increased lesions in the front area of the body compared to short and medium lairage (SL, P=0.025 and ML, P=0.004). Carcass weight was lower with LL compared to SL and ML (P=0.04; P=0.10). Tail biting was not affected by any of the tested parameters, but it influenced lean meat percentage, which increased at the increasing of tail lesions. Results confirm the need to improve pre-slaughter conditions in both animal welfare and cost-benefit perspectives. Ham defect were promising as welfare indicators. Tail lesions affected carcass quality confirming its multifactorial origin and its importance in the pork chain.

Effect of gilts' breed and social experience on skin lesions at mixing into sow groups

L.M. Hannius, L. Keeling, C. Anderson, D. De Oliveira and A. Wallenbeck
Swedish University of Agricultural Sciences, Department of Animal Environment and Health, P.O. Box 7068, 750 07 Uppsala, Sweden; linda.marie.hannius@slu.se

The transition from individually stalled to group housed gestation sows is an ongoing process within pig production in Europe. Group housing offers the sows freedom of movement and possibilities to display important species-specific behaviours. However experience in Sweden, where gestation sows have been group housed since the 1980-ies, shows that it also carries an increased risk of damaging behaviour. The aim of this study was to assess effects of gilts' breed and social experience on skin lesions at mixing into sow groups after weaning. We assessed 83 first parity sows of two lines of Yorkshire indirectly selected for single (Dutch Yorkshire, DY, 45 sows) or group housed systems (Swedish Yorkshire, SY, 38 sows). These sows, balanced across lines, were reared in two different social environments to give four different combinations of social experience (1) from 2 to 5 weeks of age – half of the litters had access to the piglets and sows in the neighbouring pen (access pen, 41 sows) (2) from 10 weeks to farrowing – half of the groups were mixed with two unfamiliar gilts (mixed groups, 41 sows). Lameness and the occurrence of skin lesions were recorded before and after a paired interaction test, performed just before mixing into the sow group, and then again 4 days after introduction to the sow group. Data was analysed using PROC GLIMMIX in SAS, using bin-distribution l and logit link. Preliminary results show an increase in proportion of gilts with skin lesions from the assessment before the paired interaction test (37.5±3.77%, LSM±STERR) to the assessment after (91.6±3.77%) followed by further increase after mixing into the sow group (99.0±4.01), P<0.001 for all. A higher proportion of SY gilts compared to DY gilts had skin lesions at the assessments before the paired interactions (P<0.05), but there were no breed differences at the assessments in the sow groups. Moreover, a higher proportion of SY than DY gilts were lame (P<0.05). These preliminary results indicate that breed differences in the prevalence of skin damages and lameness at mixing with unfamiliar sows exist, but that the social experiences of the gilts have less effect.

Artificial vs conventional rearing of pigs: tail lesions and losses, skin lesions and performance

M. Gentz[1], A. Lange[1], S. Ammer[1], O. Burfeind[2] and I. Traulsen[1]
[1]Georg-August-University Göttingen, Livestock Science, Albrecht-Thaer-Weg 3, 37077 Göttingen, Germany, [2]Chamber of Agriculture of Schleswig-Holstein, Gutshof 1, 24327 Blekendorf, Germany; maria.gentz@uni-goettingen.de

This study aims to investigate the effects of artificial rearing during suckling period on piglets' tail lesions and losses as well as skin lesions and performance during rearing and fattening period. After weaning, 27-days old pigs (50% docked, 50% undocked) from 1) artificial rearing (AR), 2) group housing of lactating sows (GH) and 3) conventional farrowing crate (FC) were housed in a conventional rearing system. After 40 days of rearing they were regrouped and rehoused to fattening. Weekly, the status of tail lesions as well as tail losses was assessed individually (rearing n=889, fattening n=321). Additionally, the skin lesions were recorded after weaning and after regrouping for fattening period and the daily weight gain during suckling, rearing and fattening period was documented. The incidence of tail lesions was generally lower in AR than in GH and FC (P<0.01). The highest percentage of tail lesions occurred in GH in week 7 (undocked=74.1%) while the AR pigs reached their maximum in week 18 (32.7%). The farrowing systems had a significant effect on tail losses at the end of fattening (P<0.01). The AR showed the lowest incidences of tail losses (AR=8.0%, GH=72.0%, FC=57.1%) at the end of fattening. Pigs raised in AR had significantly fewer skin lesions after weaning (AR=19.0%, GH=35.0%, FC=82.4%). After regrouping to fattening there were no significant differences in skin lesions. There were significant differences in the daily weight gain during suckling (AR=140 g, GH=228 g, FC=260 g) and rearing period (AR=471 g, GH=423 g, FC=430 g) (P<0.01). There were no differences in daily weight gain during fattening. In conclusion, the frequency of tail lesions and losses as well as skin lesions was significantly lower in AR pigs. The daily weight gain was only different during suckling and rearing. The AR stands out positively, however, it must be emphasised that the beneficial effect of reduced regrouping cannot be clearly separated from the housing system during suckling period. Reduced animal welfare and additional economic costs must also be considered.

The HealthyLivestock approach: welfare and behaviour of broilers hatching in different systems

M.F. Giersberg[1,2], R. Molenaar[2], I.C. De Jong[3], C. Souza Da Silva[3], H. Van Den Brand[2], B. Kemp[2] and T.B. Rodenburg[1,2]
[1]Utrecht University, Faculty of Veterinary Medicine, Animals in Science and Society, Yalelaan 2, 3584 CM Utrecht, the Netherlands, [2]Wageningen University & Research, Adaptation Physiology Group, De Elst 1, 6708 WD Wageningen, the Netherlands, [3]Wageningen University & Research, Wageningen Livestock Research, De Elst 1, 6708 WD Wageningen, the Netherlands; m.f.giersberg@uu.nl

The peri-hatching environment can affect health, resilience and welfare of broiler chickens in later life. Optimal early and later life conditions will likely contribute to a reduction in the use of antimicrobials in broiler production. The aim of this study was to investigate effects of recently developed hatching systems on the well-being of broiler chickens in early and later life. Therefore, a grow-out experiment was performed with chickens that hatched either conventionally (HH, hatchery hatched, no light, feed or water in the hatcher), in a system which provided feed and water in the hatcher (HF, hatchery fed) or on-farm (OH, on-farm hatched, where feed and water were available after hatch and, in addition, transport of day-old chickens from hatchery to farm was not necessary). The animals were reared in three batches, in 12 floor pens/batch (1,150 animals/pen) with a total of 12 replicates of each treatment. Several animal-based indicators were assessed following standard protocols: plumage cleanliness, footpad dermatitis (FPD), hock burn, skin lesion (at d21 and 35 of age), and gait score (d35). Furthermore, a set of behavioural tests was carried out: novel environment (d1 and 21), tonic immobility, novel object, and avoidance distance test (d4 and 35). Plumage cleanliness, hock burn and skin lesion were affected by age but not by hatching system, with older broilers scoring worse than younger ones (P<0.05). An effect of hatching system was only found for FPD, with HH chickens having more frequently and more severe lesions compared to HF and OH chickens (P<0.05). All responses measured in the behavioural tests were again affected by age but not by hatching system. In later life, chickens acted significantly less fearful than during the first days of life. The results indicate that conventionally hatched chickens scored significantly worse for the key indicator FPD, whereas, in general, hatching system seemed to have minor effects on welfare and behaviour of broilers.

Animal health and welfare in alpine dairy cattle farms

K. Katzenberger[1], E. Rauch[2], M. Erhard[2], S. Reese[1] and M. Gauly[1]
[1]Free University of Bolzano, Faculty of Science and Technology, Piazza Università 5, 39100 Bolzano, Italy, [2]Ludwig-Maximilians-University Munich, Department of Veterinary Sciences, Faculty of Veterinary Medicine, Veterinärstraße 13/R, 80539 Munich, Germany; katja.katzenberger@unibz.it

Small-scale farm structures characterise alpine milk production. Because of topographical constraints and herd size mainly tie-stalls are used, although they are associated with a critical welfare situation due to the restriction of movement. The present study aims to evaluate health and welfare of dairy cows by developing a monitoring system. Therefore, animal-based (e.g. BCS, avoidance distance, integument alterations, claw conformation, lameness) and resource-based indicators (e.g. comfort around resting) were recorded in 204 dairy farms (93 tie-stalls and 111 free stalls) in North and South Tyrol from March till October 2019 including a dataset of 1,891 cows. First analyses revealed the following results: BCS was too lean in 34.2% of the animals. Cows in free stalls showed significantly more frequently avoidance behaviour when compared with cows in tie-stalls (54.7% vs 38.8%). The presence of integument alterations (hairless patch areas, swellings, lesions) was also significantly affected by the type of stall. Integument alterations localised on the neck were recorded in 45.3% of the animals observed in tie-stalls and only in 7.4% of the cows kept in free stalls. Furthermore, they were most frequently found in the carpus (48.0%) and knee regions (44.8%) and there were significant differences between tie- and free stalls (65.5% vs 35.1% in the carpus region, 70.3% vs 26.1% in the knee region, respectively). Cows kept in tie-stalls showed significantly less alterations in the presence of mattresses when compared with the absence of mattresses in the lying down area (in the carpus region 26.5% vs 67.6%, in the knee region 41.5% vs 75.0%). Overgrown claws were recorded in 44.4% of the cows observed in tie-stalls, while the percentage was 31.8% in free stalls. No significant differences were found between the systems regarding the prevalence of lameness (14.8% vs 11.9% in free stalls). The values are clearly showing the urgent need for improvements. Regular monitoring of the indicators could be the first step towards improving welfare levels by data-based derivation of recommendations and controlling long-term effects.

Attitudes and barriers towards footrot in sheep flocks of south-central France: a qualitative study

M. Doucet[1,2], E. Souil[2,3], F. Corbière[1,3] and E. Royer[1,4]
[1]UMT Pilotage de la Santé des Ruminants, 23 chemin des Capelles, 31076 Toulouse, France, [2]Idele-institut de l'élevage, bd des arcades, 87060 Limoges, France, [3]Ecole Nationale Vétérinaire de Toulouse, UMR 1225 IHAP, 23 chemin des Capelles, 31076 Toulouse, France, [4]Idele-institut de l'élevage, BP 42118, 31321 Castanet-Tolosan, France; eric.royer@idele.fr

Footrot caused by *Dichelobacter nodosus* is a contagious bacterial disease altering sheep welfare and performance, and inducing heavy economic losses. In France, various control practices have been recommended without significant results on the incidence of the disease. To investigate the motivators and barriers of farmers to apply the best management practices, a total of 30 selected French sheep farmers were interviewed using a semi structured questionnaire. Farmers were recruited accounting for farmer age and herd size from three sub-groups, identified by veterinarians, of herds with prevalence of footrot in three geographical areas (Lot, Vienne, Haute-Vienne). Results show that farmers consider footrot as a difficult pathology. The concern is low to medium for half of the farmers (14/30), but high for others (9/30). Impacts on working time and mood are particularly perceived, less those on health, performance and benefits. Farmers are able to diagnose footrot but do not score it methodically. Risk factors linked to environment (grassland and litter humidity, housing and grazing management) are spontaneously considered as important, whereas those linked to internal and external biosecurity are not spontaneously mentioned. The means of control considered as important are trimming (25/30), antibiotics (18/30), footbath (17/30), culling of affected animals (16/30) and vaccination (14/30). Affected ewes are mainly trimmed (27/30) and foot soaked (15/30) while for disease prevention one or two methods are used (22/30): footbath, trimming, vaccination and culling. Only 6 out of 30 sheep farmers indicated a strategy combining several methods, and 12 farmers spontaneously reported that their implementation of control measures was not successful. Farmers expected more solutions proposed by research as well as so-called natural methods. The results of this work will be used to design achievable control plans and communication strategies for reducing lameness at farm level.

Alternative dairy production systems with higher ethical attributes as perceived by veterinarians

M. Brscic
University of Padova, Department of Animal Medicine, Production and Health, Viale dell'Università 16, 35020, Italy;
marta.brscic@unipd.it

This study aimed at assessing the perspective and attitude of bovine veterinarians in regards to alternative dairy production systems to the early cow-calf separation with higher ethical milk attributes. The study was carried out in North-eastern Italy using a short questionnaire distributed to 89 bovine veterinarians during a national congress. The questionnaire was organised in a first part on the characterisation of the responder; the ranking (1 most important to 6 less important) of six triggering issues around modern dairy production; specific questions investigating respondent opinion on the major pulling factors driving the change of the conventional dairy production system towards alternatives with higher ethical attributes and perspective regarding cow-calf systems (scored on an visual analogue scale from 0 min to 12 cm max). Thirty-five questionnaires were collected back and these preliminary results were submitted to descriptive statistics. Respondents were prevalently male (26) in their forties (mean age 39 ± standard deviation 13.2 years) and they were mainly veterinarians (26). Respondents ranked differently the different challenges of the dairy sector: 14 of them ranked animal welfare as the most important (rank 1) and 14 of them ranked societal aspects and demands as the less important issues (rank 6). Alternatives to the early cow-calf separation were marked as an opportunity with a mean score of 6.3±3.7 (0 minimum to 12 maximum) but hardly feasible with a mean score of 5.6±3.4 on the same scale. Respondent opinion on the potential factors determining major driving or pulling forces towards higher ethical attributes of the dairy production were market demand (9.0±2.4), higher milk and dairy product prices (8.8±2.6), consumer awareness (8.4±2.9), specific legislation (7.3±3.2), and milk quality and shelf life (7.3±2.9). Results of this study, although sample size does not allow statistical inference, highlight that veterinarians are aware that early cow-calf separation is under debate although they see it hardly feasible and they might not be ready to promote cow-calf systems on dairy farms, waiting for market demand, consumer awareness, or the coming in force of species-specific legislation.

Extent of transport of cattle and pigs between EU-member states 2014-2018

K. Dahl-Pedersen[1] and M.S. Herskin[2]
[1]University of Copenhagen, Veterinary Clinical Sciences, Hojbakkegaard Allé 5A, 2630 Taastrup, Denmark, [2]Aarhus University, Animal Science, Blichers Allé 20, 8830 Tjele, Denmark; kdap@sund.ku.dk

Background: Animal transport includes different phases and stressors and may be a strenuous and exhausting experience for the animals. Within the European Union, transport of animals is regulated by Council Regulation (EC) No 1/2005, stating that 'For reasons of animal welfare the transport of animals over long journeys, including animals for slaughter, should be limited as far as possible'. The aim of this study was to create an overview of the intra EU movements of cattle and pigs, and possible identify trends. Materials and methods: The study was based on data from the official database TRACES, the European Commission's online management tool for intra EU trade with e.g. live animals, feed and plants. Results: Overall, the number of cattle and pigs transported between EU member states has increased during the period of study. Transport of pigs (sum of breeding, production, and slaughter) amounted to 33.4 million animals in 2018, an increase of 12% since 2014. In the same time, pigs transported for breeding and slaughter have decreased by 75 and 25%, respectively, but pigs transported for production has increased by 29% to 24.3 million. Transport of cattle (sum of breeding, production and slaughter) amounted to 4.3 million animals in 2018, an increase of 7% since 2014. Transport of cattle for breeding has decreased by 34%, but transport of cattle for production and slaughter have increased by 22 and 9%, respectively. Cattle and pigs for production refer primarily to unweaned calves and pigs weighing less than 30 kg. Several countries can be found on the list of largest exporters as well as importers of animals of the same category, e.g. in 2018 Belgium was the largest importer of cattle for slaughter and at the same time second largest exporter of cattle for slaughter. Conclusions: Live pigs and cattle are transported in large numbers between EU member states, suggesting a complex underlying trading pattern. The intent of Council Regulation (EC) No 1/2005 to limit long journeys seems in many cases not to be complied with as millions of animals are transported through several countries before reaching their destination.

Improvement of calf behaviour and veal quality using rearing at foster cows

G. Grodkowski[1,2], M. Stachelek[1], K. Puppel[2], P. Solarczyk[2], K. Cypko[2] and T. Sakowski[1]
[1]*Institute of Genetics and Animal Breeding PAS Jastrzębiec, Postępu 36A, 05-552 Magdalenka, Poland,* [2]*Institute of Animal Sciences, Department of Animal Breeding, Ciszewskiego 8, 02-786 Warszawa, Poland; t.sakowski@ighz.pl*

The experiment was carried out at the ecological farm Ekofarma Kaszubska in Poland. Ten Holstein-Friesian bulls of the same age were selected for the experiment and divided into two experimental groups. Five calves were raised in a pen without access to their mothers. They were taken with whole milk served from a bucket equipped with a teat. The remaining five bulls were reared at two foster cows. The rearing lasted six months, after which the animals were slaughtered. During the rearing, behavioural observations of each of the experimental groups were carried out every month for 3 hours. During the observation, the number of 'licking cases' of the pen equipment or other calf was counted. The study was conducted in three identical replications one after the other. The collected experimental data were statistically analysed using IBM SPSS Statistics. It was found that calves reared with suckler cows gained weight faster and were characterised by better muscle class (assessed in the EUROP system) compared to the control group. Behavioural assessments showed less adverse behaviours, such as licking other calves or pen equipment, in the suckling group compared to the control group. This difference was most evident in relation to calves up to 3 months of age. No significant differences were found between the groups of older calves. This is due to the increased need for calves to suck in the first weeks of life. Strong urine drinking tendencies were demonstrated in both groups, so this type of behaviour would not be related to the rearing system, but rather to mineral deficiencies in the diet. Based on the obtained results, it can be concluded that the increased availability of milk, and the frequent natural intake of milk directly from the udder, have a positive effect on calves' growth and well-being. 'The authors acknowledge the financial support from SusAn, an ERA-Net, co-funded under European Union's Horizon 2020 research and innovation programme (www.era-susan.eu), Grant n °696231 and funded by the National Center for Research and Development'.

Cleaning of pipes in pig liquid feeding system for improved animal health and performance

A. Lind[1], C. Lindahl[1] and M. Åkerfeldt[2]
[1]*RISE, Research Institutes of Sweden, Bioeconomic and health, Box 7033, 75007 Uppsala, Sweden,* [2]*Swedish University of Agricultural Sciences, Department of Animal Nutrition and Management, Box 7024, 750 07 Uppsala, Sweden; ann-kristina.lind@ri.se*

Poor hygienic quality of feed can cause diarrhoea in pigs and result in slower growth. Disease spread due to lack of hygiene reduces animal welfare and production capacity. Therefore, cleaning of the stable is necessary for successful production. The aim of this study was to investigate the effects of cleaning of the discharge pipes in liquid feeding systems with a special cleaning tool (Envirologic Pipe Cleaner) on the discharge pipe cleanliness, feed hygiene quality and pig health and growth. A case-control study was performed on a commercial pig farm with All-in All-out batch wise production. At weaning, 473 pigs in existing litters were moved to a growing unit where the feed discharge pipes in half of the pens was cleaned prior to the study. The litters were evenly distributed to either cleaned (CL) or uncleaned (UCL) pens, thus 22 pens/treatment, 10-12 pigs/pen. Six UCL and six CL discharge pipes were randomly selected for visual assessment of cleanliness and collection of feed samples, which were performed at three times. Daily, all pigs were individually monitored for diseases and injuries by the staff and the pigs were weighed two times during the study period. At the first sample occasion, UCL pipes contained mould, yeast fungi and enterobacteria, in some pipes higher than the recommended threshold values. There was no significant difference in pH between UCL and CL pipes (P=0.951) or between sampling occasions (P=0.246). Enterobacteria were found in all samples and varied between 3.6-6.0 log cfu/g where UCL pipes had higher levels (P=0.013). UCL pipes also tended to contain slightly more mould sponges (P=0.052). The CL pigs had a lower daily weight gain compared with the UCL pigs (P=0.011). Cleaning the feed pipes in pig stables has the potential to improve animal health by improving the hygienic quality of the feed, but more research is required in the field, in order to be able to investigate the long-term effect in the entire herd of cleaning the pipes.

A survey on hygienic challenges and animal welfare in mobile houses for laying hens

L. Rieke[1], M. Corbach[2], F. Kaufmann[2], H. Pieper[3], J. Van Der Linde[4] and N. Kemper[1]
[1]*Institute for Animal Hygiene, Animal Welfare and Farm Animal Behaviour, University of Veterinary Medicine Hannover, Foundation, Germany, Bischofsholer Damm 15, 30173 Hannover, Germany,* [2]*Faculty of Agricultural Sciences and Landscape Architecture, University of Applied Sciences, Osnabrueck, Germany, Albrechtstr. 30, 49076 Osnabrück, Germany,* [3]*Animal Husbandry, Chamber of Agriculture Lower Saxony, Hameln, Germany, Klütstr. 10, 31787 Hameln, Germany,* [4]*Animal Production and Livestock Breeding, Chamber of Agriculture North Rhine Westphalia, Viersen, G, Gereonstr. 80, 41747 Viersen, Germany; lorena.rieke@tiho-hannover.de*

Mobile houses for laying hens have gained increasing interest over the last years as an animal friendly and environmental sound alternative to static housing systems. However, with regard to biosecurity, and especially in the event of an outbreak of an epizootic disease such as Avian Influenza, mobile houses have deficits in the possibilities to establish hygienic measures. Therefore, the aim of the presented survey was to provide first insights into the main challenges and to reveal potential improvement approaches. Questionnaires were designed and 20 farmers with mobile houses in Lower Saxony and North Rhine Westphalia were asked to give information about experienced situations in case of the last Highly Pathogenic Avian Influenza outbreak in 2016/2017. In particular, data on legal requirements and control points by the local veterinary offices were collected. The herd size of the questioned farmers ranged from 225 to 1,200 animals, and they kept their animals in fully mobile trailer-like systems from the same manufacturer, but of two different sizes, either for 225 or for 300 laying hens. The main uncertainties and problems were revealed and the farmers assessed hygienic management as well as animal health as most challenging. During the outbreak, the local veterinary offices regulated that laying hens had to be kept indoors for average 3.5 months, resulting in limited space due to no free range. Thereupon, in 87% of the interviewed farms, the behaviour of the laying hens changed resulting in feather pecking and cannibalism. None of the farmers had a hygiene barrier in the mobile house. In 60%, the hygienic measures consisted in changing clothes and shoes, but the implementation was difficult, and concrete guidelines were missing, both on the part of the famers and of the veterinary offices. The outcome of this study is used to develop a concept consisting of a modified mobile house and a management package focusing on general and specific hygienic aspects. The study is part of the project *Hyg-MobiLe*, which is supported by funds of the German Government's Special Purpose Fund held at Landwirtschaftliche Rentenbank.

Prevalence of pecking injuries and footpad dermatitis in turkey flocks

N. Van Staaveren[1,2], E.M. Leishman[2], S. Adams[2], B.J. Wood[2,3,4], A. Harlander[1] and C.F. Baes[2,5]
[1]*University of Guelph, The Campbell Centre for the Study of Animal Welfare, Department of Animal Biosciences, Guelph, Ontario, N1G 2W1, Canada,* [2]*University of Guelph, Centre for the Genetic Improvement of Livestock, Department of Animal Biosciences, Guelph, Ontario, N1G 2W1, Canada,* [3]*Hybrid Turkeys, Suite C, 650 Riverbend Drive, Kitchener, Ontario, N2K 3S2, Canada,* [4]*University of Queensland, School of Veterinary Science, Gatton, Queensland, 4343, Australia,* [5]*University of Bern, Institute of Genetics, Vetsuisse Faculty, Bern, 3001, Switzerland; nvanstaa@uoguelph.ca*

Pecking-related injuries and footpad dermatitis are important welfare issues in turkeys which can lead to mortality or culling. Little is known about pecking injuries and footpad dermatitis in Canadian turkey flocks. Turkey farmers were asked to assess the prevalence of pecking-related injuries and footpad dermatitis in their flocks in a cross-sectional study. Farmers inspected thirty birds and recorded the severity of footpad dermatitis, pecking injuries to the head/neck area, and back/tail area, on a 0 – 2 scale. Information on flock sex and age were recorded and a final 63 flocks were included in the study. Flock prevalence was calculated as the percentage of birds affected by score >0 for each category. Generalised linear mixed models were used to explore relationships between prevalence of the different welfare issues and season, sex, and age of the flock. The median prevalence of head injuries was 3% (IQR: 0 – 10%), 7% (IQR: 0 – 13%) for back injuries, and 27% (10 – 67%) for footpad dermatitis. A higher prevalence of head injuries within a flock was correlated to a higher prevalence of back injuries (r=0.54, P<0.001). Though neither of the pecking injuries were correlated with the prevalence of footpad dermatitis (P>0.05), birds with back injuries were more likely to have footpad dermatitis (OR=1.5, 95%CI 1.09-2.01) and head injuries (OR=4.1, 95%CI 2.67-6.44). While the average prevalence of pecking injuries and footpad dermatitis was numerically higher in tom flocks, there was no significant difference between tom and hen flocks when adjusting for age. The large variation between flocks and relationships between the different welfare issues highlights a need for proper management strategies that address these concurrently.

Using infrared thermography as a tool to assess acute stress of donkeys during weighing procedure

S.R. Silva[1], M. Almeida[1], D. Bacellar[2], B. Leiva[2] and M. Quaresma[1]
[1]*CECAV, University of Trás-os-Montes e Alto Douro, Quinta de Prados, 5000-801 Vila Real, Portugal,* [2]*Associação para o Estudo e Proteção do Gado Asinino, Rua da Igreja, 5225-011 Atenor, Miranda do Douro, Portugal; ssilva@utad.pt*

In response to the growing concern regarding animal welfare, measuring stress caused by husbandry practices is crucial. Some techniques to measure stress involve invasive procedures, such as blood sampling, which can cause a stress response. To overcome this problem, non-invasive or minimally invasive methods have been developed, including the use of infrared thermography (IRT). Recent research has shown that IRT can be a useful tool to assess acute sympathetic and hypothalamic, pituitary and adrenocortical responses. The present study aimed to investigate the use of IRT to assess the maximum temperature of the eye in donkeys during weighing on a platform scale. Thirty-two Miranda breed donkeys (8.3±3.7 years, 333±62 kg) were studied. The animals were divided into two categories: (1) collaborative (n=21); and (2) non-collaborative (n=11) according to the difficulty in placing the animals on a platform scale (Salter Brecknell® PS-3000HD, USA). After LW record, IRT images of the donkey faces were obtained with an infrared camera Flir F4 (Flir Systems AB, Sweden). The eye IRT temperature data was extracted from thermal images using the FLIR Tools software. To ensure that the maximum eye temperature was measured, the software ellipse tool was used and an ellipse fitted to the animal eye. Data were analysed using an ANOVA, following least significant difference Student's t-test as comparisons test. All statistical analyses were performed using the JMP-SAS14 software (SAS Institute Inc. Cary, NC, USA). The results showed significant differences (P<0.05) in IRT eye temperature between collaborative and non-collaborative animals (36.31 vs 35.49 °C, respectively). The results show a reduction in the eye temperature of the non-collaborative group, probably due to stress activated vasoconstriction. Further research is needed to fully understand the relationship between IRT with other stress indicators to establish the value of this technique as a reliable tool for welfare assessment.

The use of euthanasia in French equines in the context of the strengthening of the legislation

A. Merlin[1], N. Foucher[1], M. Linster[1,2], G. Cazeau[3], M. Delerue[4], M. Marsot[5], C. Sala[3], J. Schneider[4], B. Ferry[4], J.P. Amat[3] and J. Tapprest[1]
[1]*ANSES, Laboratory for Animal Health in Normandy, Physiopathology and Epidemiology of Equine Disease Unit, RD675, 14430 Goustranville, France,* [2]*Pathological Anatomy Unit, National Veterinary School of Alfort (ENVA), 7 Avenue du Général de Gaulle, 94700 Maisons-Alfort, France,* [3]*University of Lyon-ANSES, Laboratory of Lyon, Epidemiology and support to Surveillance Unit, Cedex 07, 69364 Lyon, France,* [4]*French horse and riding institute (IFCE), Le Château, 19230 Arnac Pompadour, France,* [5]*Unité EPI, Animal Health Laboratory, INRA, ANSES, ENVA, Université Paris-Est, 14 Rue Pierre et Marie Curie, 94701 Maisons-Alfort, France; aurelie.merlin@anses.fr*

One of the consequences of the enhancement of the equine identification legislation is the increase of the number of equines excluded from the food chain. The question arises then of what is the becoming of these potentially unwanted equines in terms of well-being and health. Among possibilities of disposal, euthanasia appears. However, little knowledge is available on this option despite its major issues. In this context, the aim of this study was to estimate the rate of euthanasia in French equines and identify factors associated with it. A web-based survey was performed and distributed by the French horse and riding institute (IFCE) to the 13% of owners (n=5,158) throughout France who used the web service of the IFCE's ATM-équidés ANGEE association to organise the removal of their dead equine between April 2017 and April 2018. Through the survey, information about the equine characteristics, its activity and environment, the mode of death (natural or by euthanasia) and the cause of death were collected. Factors associated with euthanasia were identified using a logistic regression. The response rate was 11% (n=548). In our sample, 71% of equines were euthanised and others were naturally dead. The percentage of euthanasia increased significantly with the age of the animal, reaching 80% for equines above 20 years old. Health-related factors associated with the higher rates of euthanasia were generally poor state, tumoral process, senescence, traumatic accident and colic. Conversely, heart problems and unknown causes were associated with the lowest percentage of euthanasia. The sex, the breed, the activity and the environment did not influence the percentage of euthanasia. To conclude, these results allowed us to estimate the rate of euthanasia in French equines and to identify factors associated with it. This study will be strengthened and expanded to more owners and to veterinarians and regularly repeated to measure over time the impact of the enhancement of the traceability regulation on the use of these unslaughtered equines and their health and well-being.

Evaluation of the quality of the equine information system (SIRE) database for equine traceability

H. Farchati[1,2], A. Merlin[2], M. Saussac[2], X. Dornier[3], M. Dhollande[3], D. Garon[1], J. Tapprest[2] and C. Sala[2]
[1]University of Caen Normandy, Caen, 14000, France, [2]Anses, Maisons-Alfort, 94700, France, [3]French horse and riding institute, Arnac Pompadour, 19230, France; halifa.farchati.ext@anses.fr

Created in 1976 and managed by the French Horse and Riding Institute (IFCE), the SIRE database is the national database for equines in France. SIRE collects, on a declarative but mandatory basis, individual data related to every equine living in France as well as data about their owners and keepers. The objective of our study was to evaluate the SIRE database quality regarding the updating and completeness of information recorded by owners and keepers in order to evaluate the quality of traceability of equines in France and evidence possible improvements. An online survey was conducted via the Sphinx® software and included 6,244 keepers and 13,869 owners registered in SIRE who has agreed to be contacted by email. After six months of survey and two follow-ups of mailings, the final response rate was 20.1% (n=2,788) for the owners and 19.5% (n=1,217) for the keepers. Results indicated that some information was not always declared to IFCE. Indeed, 0.8% of 2,004* owners had declared no equine in SIRE and 2.6% of 2,004* owners had declared only some equines; 32.0% of the 50 owners whose all equines were registered dead in the SIRE had, in fact, at least one equine alive; and 68.6% of the 51 keepers whose equine premises were all registered as closed in SIRE had, in fact, at least one equine premise open. Besides, some information was not updated regularly in SIRE. In particular, the following changes were not always registered: the castration of 32.6% of 1,038 geldings was not declared; the death of 5.9% of 2,738 equines registered alive in SIRE; the change of address of 33% of 2,303* owners; the closure of all equine premises of 4.7% of 1,166* keepers; and the discontinuation of equine ownership of 3.6% of 2,738* owners. In conclusion, the general quality of the SIRE database seems satisfactory. However, an increase in people's awareness of the regulations and the importance of regularly updating information seems necessary to improve the traceability of equines. This is particularly true for the change of ownership or address and for the opening and closing of equine premises. * Concerned by this item.

Genetic resistance to a gastro-intestinal parasite in young sheep weakly effective during peripartum

F. Douhard[1], S. Aguerre[1], A. Corbishley[2], T.N. McNeilly[3], D. Marcon[4], J.-L. Weisbecker[1], L. Bordes[5], P. Jacquiet[5] and C. Moreno-Romieux[1]
[1]GenPhySE, Université de Toulouse, INRAE, ENVT, 31326 Castanet-Tolosan, France, [2]The Roslin Institute and Royal (Dick) School of Veterinary Studies, University of Edinburgh, Edinburgh, United Kingdom, [3]Moredun Research Institute, Bush Loan, Penicuik, United Kingdom, [4]INRAE, Unité Expérimentale INRAE, Domaine de La Sapinière, 18390 Osmoy, France, [5]Ecole Nationale Vétérinaire de Toulouse, UMR INRA/ENVT 1225 IHAP, UMT Santé des Petits Ruminants, BP 87614, 31076, Toulouse Cedex 03, France; frederic.douhard@inrae.fr

Gastrointestinal nematode (GIN) infections are a major hindrance to the development of sustainable grazing systems as they cause significant production losses. The current GIN control strategy (anthelmintics) is unsustainable due to the emergence of drug resistant parasites. Although genetic selection for resistance to GIN is a promising alternative, its effectiveness across different physiological stages remains unclear. To look into this issue, we studied the responses to GIN artificial infection in genetically resistant and susceptible female meat sheep at four different stages: during growth (n=91), during the periparturient period of 1st lambing (n=48 out of the 91), in 1st third of 2nd pregnancy (n=81 out of the 91), and around 2nd lambing (n=55 out of the 91). Females were from the 2nd generation of a divergent selection experiment on GIN fecal egg count (FEC) measured at 6 months of age (between-line difference >3 genetic standard deviations). The parasitic challenges consisted of a single-dose of 10,000 third-stage *Hæmonchus contortus* larvae, except during 1st peripartum where a trickle infection was applied (1,000 larvae per week during 9 weeks). Each infection was ended by drenching. A large and consistent between-line difference was observed in FEC and haematocrit loss during growth, but also later during early pregnancy. However, the difference was largely reduced during the peripartum period, regardless of the infection protocol. Overall, our results suggest that selection based on experimental infection of young sheep during growth promotes adult resistance, but is temporally limited during periparturient relaxation of immunity. This study was in the frame of H2020 project SMARTER n °772787.

Genetic evaluation systems and breeding programs in sheep and goats: an international perspective

L.F. Brito[1], D. Berry[2], H. Larroque[3], F.S. Schenkel[1], G. Ciappesoni[4], A. O'Brien[2], F. Tortereau[3], E. Ugarte[5], I. Palhiere[3], B. Bapst[6], J. Jakobsen[7], G. Antonakos[8], A. Kominakis[8], V. Clement[9], G. Bruni[10], V. Loywyck[9], E. Massender[1], H.R. Oliveira[1], J. Posta[10] and J.M. Astruc[9]
[1]University of Guelph, Guelph, N1G2W1, Canada, [2]TEAGASC, Fermoy, P61 C996, Ireland, [3]INRAE, Toulouse, 31326, France, [4]INIA-UY, Canelones, 90000, Uruguay, [5]NEIKER, Vizcaya, 48160, Spain, [6]Qualitas AG, Zug, 6300, Switzerland, [7]NSG, As, 1431, Norway, [8]Frizarta, Agrinio, 30100, Greece, [9]IDELE, Toulouse, 31326, France, [10]ARAL, Crema, 26013, Italy; britol@purdue.edu

Genetic selection has been a key tool for improving productive performance in small ruminant populations around the world, especially in Europe. This has been achieved through selective breeding for numerous traits, using diverse genetic evaluation systems and breeding schemes. Knowledge of the alternative approaches taken is paramount to the design of efficient and integrated genomic breeding programs. In this study, we summarised information on 48 sheep and goat breeding programs, genetic and genomic evaluation systems and resources available in 12 countries involved in the SMARTER project. This was done using published reports and surveys distributed to all partners. Responses to the surveys reveals information for more than 9, 16, and 20 dairy goat, dairy sheep, and meat sheep breeds involved in genetic schemes, respectively, with ~3,083,562 animals included in data collection schemes. The main groups of traits recorded across countries are: (1) milk yield and composition, mastitis indicators, udder and body conformation, and reproduction in dairy sheep and dairy goats; and (2) growth, reproduction, health, ultrasound, wool, and carcass in meat sheep. Seven countries have progeny testing schemes, but only 5 use artificial insemination. There are numerous challenges to be addressed (e.g. disparity of trait recording, SNP panels, statistical models used, joining pedigrees across countries as well as grouping breeds based on genetic similarity, and an average of ~30% of animals with unknown sires). However, there are many opportunities to use the current resources and develop collaborative approaches to optimise selection for novel breeding goals such as resilience and efficiency in small ruminants across countries.

Milk transcriptome analysis identifies genes and pathways affecting feed efficiency in dairy ewes

M. Franco[1], P.G. Toral[2], C. Esteban[1], B. Gutiérrez-Gil[1], G. Hervás[2], J.J. Arranz[1], P. Frutos[2] and A. Suárez-Vega[1]
[1]Universidad de León, Producción Animal, Facultad de Veterinaria-Campus de Vegazana, 24007, León, Spain, [2]Instituto de Ganadería de Montaña (CSIC-Universidad de León), Finca Marzanas, 24346, Grulleros, León, Spain; asuav@unileon.es

Improving feed efficiency (FE) reverberates in the profitability and sustainability of dairy sheep production by reducing feed costs and environmental footprint. FE in dairy sheep has been defined as the capability of a ewe to transform the feed nutrient consumed into milk. One of the currently established indexes to evaluate FE is the residual feed intake (RFI). The specific evaluation of the milk transcriptomic profile of ewes with extreme FE phenotypes may be of particular interest to identify genes and metabolic pathways related to this trait. This study aimed to identify differentially expressed genes (DEGs) between sheep with extreme RFI values through the analysis of milk somatic cell (MSC) transcriptome. For that, RNA-sequencing was performed on RNA extracted from 14 milk samples collected from high (n=7) and low (n=7) efficient ewes. An average of 43 million paired-end reads were generated per sample. Samples were aligned against the Ovine genome assembly (Oar_V3.1) using STAR and the quantification was performed with RSEM software. The differential expression analysis, performed with DESeq2, identified 23 DEGs (FDR<0.05) between high and low feed-efficient ewes. Among them, 3 genes were upregulated (logFC$\geq$2) and 20 genes were downregulated (logFC$\leq$-2) in the high feed-efficient group compared with the low feed-efficient one. Interestingly, upregulated genes in high feed-efficiency sheep were associated with metabolic pathways involving lipid biosynthesis and protein secretion. In contrast, genes downregulated in high feed-efficient ewes were related to apoptotic processes, which agrees with previous results in cattle and pigs. In summary, results from this research improve our knowledge of the metabolic pathways underlying milk production-related feed efficiency in dairy ewes, which could be a first step towards improving the favourable phenotype through genetic selection. Projects EU-SMARTER & JCyL CSI276P18, FEDER and ESF.

Identification of homozygous haplotype compromising fertility traits in dairy sheep

M. Ben Braiek[1], S. Fabre[1], C. Hozé[2], J.M. Astruc[3] and C. Moreno-Romieux[1]
[1]INRAE, 24 chemin de Borde-Rouge, 31326 Castanet-Tolosan, France, [2]Allice, 149 rue de Bercy, 75595 Paris, France, [3]Institut de l'Elevage, 24 chemin de Borde-Rouge, 31326 Castanet-Tolosan, France; maxime.ben-braiek@inrae.fr

In the frame of H2020 project SMARTER no. 772787 In livestock population under selection, the small effective size and inbreeding can cause the emergence of recessive deleterious mutations. When homozygous, these mutations may be responsible for embryo or neonate lethality, or genetic defects, affecting female fertility and animal welfare. To detect such mutations, a reverse genetic screen was applied on phased 50k SNP genotypes and pedigree data to identify Homozygous Haplotype Deficiency (HHD) in Lacaune (LAC, n=19,102) and Manech Tête Rousse (MTR, n=4,900) dairy populations. We detected 10 significant HHD, 5 presenting a complete deficit of homozygous animals (3 in LAC, 2 in MTR), and 5, a partial deficit of 79 to 96% compared to the expected (4 in LAC, 1 in MTR). These haplotypes spanned regions from 1.2 to 3.0 Mb with a frequency of heterozygous carriers between 4.4 to 17.4%. Then, we defined risky mating as mating between heterozygous rams at a HHD and females coming from heterozygous rams at the same HHD. Using logistic binary models, we tested the effect of risky mating for each HHD on two fertility traits: conception at AI and stillbirth rates. HHD in complete deficit were mainly associated with a decrease in conception rate in LAC breed and an increase of stillbirth rate in MTR breed. Only 2 HHD in partial deficit in the LAC breed were associated with an increase of stillbirth rate. Finally, we tested a putative selective advantage of heterozygous rams at the 10 HHD. The daughter yield deviation of 4 main traits (milk, fat and protein yields, somatic cell score) selected in dairy sheep in France were tested by variance analyses comparing carrier and non-carrier rams. Among the 10 HHD, 4 detected in LAC breed had significant positive impact on at least one of the selected traits. Thereafter, using available and newly generated sequence data of HHD carriers and non-carriers, we will try to identify putative causal mutations. The further management of these mutations in the LAC and MTR selection schemes will allow to improve the overall fertility and lamb viability.

Identification of QTL for body reserves in meat sheep

T. Macé[1], E. González-García[2], J. Pradel[3], C. Durand[3], S. Douls[3], S. Parisot[3] and D. Hazard[1]
[1]INRAE, UMR1388 GENPHYSE, 24 Chemin de Borde Rouge, Auzeville Tolosane, CS52627, 31326 Castanet-Tolosan cedex, France, [2]INRAE, UMR868 SELMET, 2 Place Pierre Viala, 34060 Montpellier cedex 1, France, [3]INRAE, UE321 La Fage, La Fage, 12250 Roquefort-sur-Soulzon, France; dominique.hazard@inrae.fr

In ruminants, body fat reserves (BR) is a relevant phenotype that could be used to improve animal's adaptive capacities to challenging environments. Changes in BR consist in the alternation of accretion and mobilisation periods, depending on physiological stages and/or environmental factors. In a previous study, we reported BR levels and BR changes over time to be heritable traits. In the present study, a genome wide association analysis was carried out to research QTL (Quantitative Trait Loci) associated with such BR trait in meat sheep. Romane ewes (n=490) reared under extensive and harsh conditions were phenotyped for BR at three or four physiological stages (Mating, Pregnancy, Lambing and Weaning) during one to three productive cycles. Measurements included body condition score (BCS), body weight (BW) and key biomarkers for BR i.e. non-esterified fatty acids (NEFA), beta-hydroxybutyrate (β-OHB) and triiodothyronine (T3). Ewes were genotyped using ovine 50K or 15K beadchips. After SNP quality checking and imputation, approximately 44K SNP were used in the analyses. GWAS were performed using single SNP or haplotypes analyses. Among the many QTL associated with BCS or BW mapped on 13 chromosomes, three main genomic regions on chromosomes 1 and 12 were of particular interest for BCS due to high level of significance and QTL overlapping for correlated traits. These regions harbour several interested genes previously described to be involved in biological functions linked to body fat metabolism. Many QTL associated with biomarkers for BR were mapped on 19 chromosomes but only one reached the genome wise threshold. This QTL was associated with β-OHB level at mating and mapped on chromosome 12 but in a different region from this associated with BCS. In conclusion, present results contribute to a better knowledge of genetic variability of BR in sheep. The SNPs polymorphisms found in this study may offer opportunities for improving BR management in sheep. This study was supported by the European Union's Horizon 2020 through grant agreement 679302.

On the accuracy of estimating resilience

M. Ghaderi-Zefreh, V. Riggio, O. Matika, A. Doeschl-Wilson and R. Pong-Wong
Roslin Insitute, Easter Bush campus, EH25 9RG; Midlothian, United Kingdom; mghaderi@ed.ac.uk

Resilience is a desirable trait in livestock species. Resilient animals will perform well even in challenging environments. This trait can be measured as the inverse of the slope of a fitted line in a reaction norm model for an animal. The success of selection for resilience depends on accurately estimating breeding values for this trait. The difficulty to obtain good estimates is because it requires the individual or their close relatives being measured on a range of environments. Genomic prediction may improve the accuracy of resilience estimates as the evaluation uses information from related and non-related individuals. Our aim was to quantify the benefit of genomic prediction on the accuracy of the estimated breeding values for resilience in sheep using simulation in the frame of H2020 project SMARTER no. 772787. We also assessed the effect of the distribution of the families on the accuracy. Data were simulated assuming a population of sheep. For each individual, the phenotypic performance values associated with different environments were generated based on an assumed linear reaction-norm model for resilience. In addition, the phenotype was parameterised with a heterogeneous residual term. Breeding values for slope and intercept were assumed to be polygenic traits that were affected by a number of pleotropic loci. Our results indicate that the use of genomic data improves the accuracy up to 24% compared with standard evaluation using pedigree information This benefit increased when the heritability of slope is lower. In addition, the spread of the offspring of a given sire across different environments has an impact on the accuracy. When the individuals of a family are in a similar environment the accuracy can drop up to 40% compared to when family members are measured across the whole range of environments, but the benefit of genomic is still at least 10%. The results of this simulation study suggest genomic prediction can improve the genetic evaluation of resilience traits and increase the potential benefit of selection for such trait.

Development of a genomic test-day model and correlations with robustness traits in French goats

M. Arnal[1,2], V. Ducrocq[3], C. Robert-Granié[1] and H. Larroque[1]
[1]GenPhySE, Université de Toulouse, INRAE, ENVT, 24 chemin de Borde Rouge, 31326, Castanet Tolosan, France, [2]Institut de l'élevage, Campus INRA, Chemin de Borde Rouge, BP 42118, 31321 Castanet Tolosan Cedex, France, [3]Université Paris-Saclay, INRAE, AgroParisTech, UMR GABI, Domaine de Vilvert, 78350 Jouy-en-Josas, France; mathieu.arnal@idele.fr

The objectives were to develop a single-step genomic test-day model (ssTDM) for French dairy goat for milk yield (MY) and to study the correlations between the new estimated breeding values (EBVs) generated by the ssTDM and robustness traits such as artificial insemination (AI) fertility and longevity. The study was based on goat test-day, measured from first to third parity between 2000 and 2017 (1,306,704 Saanen and 1,698,032 Alpine goats). 1,242 Saanen and 1,925 Alpine animals were genotyped with the Illumina goat SNP50 BeadChip. The best ssTDM to fit MY test-day was obtained with a quadratic Legendre polynomial to model genetic and permanent environmental effects. A rank reduction of variance covariance matrix was performed by eigenvalues decomposition in order to reduce computing time and complexity. The ssTDM developed generated four EBVs, for production level and persistency, for both primiparous and multiparous. EBVs obtained with the ssTDM for production level were closed to those obtained with the classical lactation model used for official genetic evaluations. The use of genomic information increased the accuracy of the EBVs of young males and reduced biases. Correlations between daily MY EBVs and EBVs for AI fertility or longevity were studied for bucks with at least 25 progeny (555 Alpine bucks, 430 Saanen bucks). High daily MY EBVs at the time of AI had an unfavourable relationship with fertility. Milk persistency was negatively correlated with fertility. For longevity, a high daily MY EBVs in primiparous was associated with a lower longevity, whereas high daily MY EBVs at the end of lactation in multiparous was positively correlated with longevity. Persistency was positively correlated with longevity in both primiparous and multiparous. In the context of looking for robust animals, selection based on multiparous EBVs seems more relevant than a selection based on current EBVs mixing primiparous and multiparous data. EBVs for milk persistency could be used as an additional selection tool according to the objectives of breeders: favouring longevity and very long lactations or on the contrary preferring grouped kiddings for a seasonal production.

Meta-analysis of genetic parameters for resilience and efficiency traits in goats and sheep

F. Tortereau[1], S. Mucha[2], R. Rupp[1] and J. Conington[2]
[1]INRAE, INPT-ENVT, INPT-ENSAT, GenPhySE, 31326 Castanet-Tolosan, France, [2]SRUC, Animal & Veterinary Sciences, Easter Bush, Midlothian EH25 9RG, United Kingdom; sebastian.mucha@sruc.ac.uk

Genetic selection has been successful in increasing animal efficiency. However this can deteriorate due to changing environmental conditions. It is therefore important to improve resilience of the animals. The aim of this study was to obtain a better insight into the relationship between efficiency (i.e. milk production, growth, wool) and resilience traits (i.e. mastitis, parasite resistance, longevity), in order to properly account for it in breeding programmes. Meta-analysis was performed to obtain pooled estimates of heritability. The dataset comprised of genetic parameters collated from 13 partners of the H2020 SMARTER project (no. 772787) coming from 7 countries. This included 43 published papers together with some unpublished results. The dataset included 1,150 estimates of heritability and genetic correlation for 20 sheep and 4 goat breeds. Pooled genetic parameters were estimated in dairy goats for milk yield (MY), protein yield (PY), fat yield (FY), somatic cell score (SCS), parasite resistance (PR), and longevity, and in meat sheep for adult body weights (ABW), body condition scores (BCS), mastitis phenotypes (MAS) and PR. The analysis was divided into two steps: (1) hierarchical Ward clustering using Gower distance as a measure of dissimilarity between estimates from the analysed studies. Clustering took into account variables such as breed, country, method of analysis, and sample size; (2) meta-analysis to obtain pooled estimates of heritability and genetic correlations. The model used in the meta-analysis included a random effect of clusters obtained in step (1). Estimation was done using REML algorithm. Pooled heritability estimates (min and max h2 of individual studies in brackets) for the analysed traits were: 0.27 (0.11 to 0.35), 0.29 (0.25 to 0.34), 0.30 (0.25 to 0.35), 0.21 (0.19 to 0.24), 0.16 (0.04 to 0.22), 0.10 (0.08 to 0.14) for MY, PY, FY, SCS, PR, and longevity, respectively in dairy goats. In meat sheep, pooled heritability estimates (min and max h2 of individual studies in brackets) were: 0.40 (0.24 to 0.57), 0.25 (0.12 to 0.35), 0.07 (0.02 to 0.11) and 0.16 (0.08 to 0.27) for ABW, BCS, MAS and PR respectively.

Potential exists to improve ewe mothering ability and lamb vigour through breeding

A.C. O'Brien[1], N. McHugh[1], T. Pabiou[2] and D.P. Berry[1]
[1]Teagasc, AGRIP, Teagasc, Moorepark, Fermoy, P61 P302, Co. Cork, Ireland, [2]Sheep Ireland, Bandon, P72 X050, Co. Cork, Ireland; aine.obrien@teagasc.ie

The objective of the present study was to establish the risk factors, including the extent of genetic variability, associated with ewe mothering ability and lamb vigour in a multi-breed Irish sheep population. Ewe mothering ability was subjectively scored on a five-point scale for each lambing event (regardless of litter size) where 1 = very poor (ewe has no interest in her lamb(s)) and 5 = very good (ewe is very protective, licks her lamb(s) immediately). Lamb vigour was also scored on a five-point scale where 1 = very poor (lamb is still not standing after 60 minutes) and 5 = very good (lamb is standing within five minutes). Following edits, 21,826 ewe mothering ability records from 15,324 ewes as well as 44,619 lamb vigour records remained. Risk factors associated with both lamb vigour and ewe mothering ability were determined using multinomial logistic regression and variance components were estimated using linear animal mixed models. Fixed effects considered for both traits included contemporary group, breed proportion, and coefficients of heterosis and recombination loss. For lamb vigour, dam breed proportion and the dam coefficients of heterosis and recombination loss were also considered as risk factors. Ewes in their first parity had a greater probability of having a poorer mothering ability score compared older ewes (P<0.01). Similarly, ewes that had a no lambing difficulty had a greater probability of having a good mothering ability score compared to ewes that experienced lambing difficulty (P<0.01). Lambs that experienced a difficult birth were more likely to have a poorer lamb vigour score compared to those that had an easy birth (P<0.01). The direct heritability of mothering ability was 0.06 (0.01). The risk factors identified for both ewe mothering ability and lamb vigour will enable better on-farm decisions by grouping specific at-risk groups for extra monitoring. Ample genetic variation exists for both lambing-related traits investigated indicating that genetic improvement in ewe mothering ability and lamb vigour is possible.

Genetic parameters for VIA-derived muscle, bone and fat weights in Texel-sired crossbred lambs

A. Tolkamp[1], J. Conington[1], J. Yates[2], E. Smith[2], J. Draper[3], N. Clelland[1] and N. Lambe[1]
[1]SRUC – Scotland's Rural College, AHES, Roslin Institute Building, SRUC, United Kingdom, EH25 9RG, United Kingdom, [2]Texel Sheep Society, 4th Street, Stoneleigh Park, Kenilworth, CV8 2LG, United Kingdom, [3]ABP food group, Edison Road, Coleshill, Birmingham, B46 1DA, United Kingdom; arjan.tolkamp@sruc.ac.uk

In the UK, 27% of the UK's 13M breeding ewes are mated to a Texel ram, thereby genetic improvement in this breed has a significant impact on the industry. However genetic improvement is largely undertaken in purebred flocks with a disconnect from the information generated from their crossbred progeny performance and carcass data in the abattoir. Meat yield and tissue distribution are 'hard to measure traits', however, recently calibrated Video Image Analysis (VIA) technology has been shown to better predict meat yield from primal regions (r~0.8-0.96) compared to existing EUROP carcass classification. The aim of this work is to quantify genetic and phenotypic properties of VIA-derived muscle, fat and bone weights in the shoulder, saddle and hind leg primal regions, and growth data from 2,340 Texel X Lleyn lambs to inform future genetic improvement for these traits. New data on these traits (2017-2019) were collected from seven Texel Sheep Society commercial phenotype farms across the UK, with all lambs slaughtered at one UK abattoir fitted with VIA technology. The pedigree consisted of 16,309 individuals including 2,340 lambs with 75 sires, 1,205 dams, 53 grand-sires, 71 grand-dams, 101 great grand-sires and 120 great grand-dams. Lleyn pedigree was not available. Lambs and sires were genotyped with a 17k and 50k single nucleotide polymorphism (SNP) bead chip respectively and genotypes were used for parentage verification. Variance components for X-bred lamb carcass traits were estimated using a multi-trait sire model including farm, year, slaughter batch, sex and adjusted for carcass weight. Variance components for growth traits were estimated using a single trait sire model including dam age, birth rank, sex, farm, and year. Heritability estimates ranged between 0.10-0.46 for muscle, 0.21-0.53 for bone, 0.19-0.34 for fat and 0.24-0.40 for growth traits. These results demonstrate the potential for VIA technology to enable genetic selection for hard to measure carcass traits.

Association between feed efficiency and methane emissions, performance and health in Merino sheep

I. De Barbieri, E.A. Navajas, D. Giorello, J.I. Velazco, G. Banchero, B. Rodríguez, F. Rovira and G. Ciappesoni
Instituto Nacional de Investigación Agropecuaria, Ruta 5 km 386, 45000, Uruguay; idebarbieri@inia.org.uy

Decreasing feed intake without negative consequences on animal performance and health provides an opportunity to increase profitability of grazing sheep systems. Selection by residual feed intake (RFI) increases feed conversion efficiency by reducing dry matter intake (DMI) at constant body weight (BW) and bodyweight gain (BWG). The impact of selection by RFI on several performance traits was investigated (in the frame of H2020 project SMARTER no. 772787) in 278 Merino lambs (ewe and rams), sired by 12 rams. One-year old lambs were allotted to one of three RFI tests based on sex, birth type and age. In each test, animals were allocated to one of five automated feeding systems in accordance to BW and sire. During the 56-day RFI tests lambs were fed *ad libitum* with Lucerne haylage (DM 53.8%, CP 21.8%, FDA 29.3%, FDN 36.0%, EE 2.4%). Extreme 25% highest (148 g DMI/d/an) and lowest (-135 g DMI/d/an) RFI animals were selected for the analysis. The effect of RFI group was analysed by a general linear model including RFI group, dam age, birth type, contemporary group, sex and lamb age as fixed effects. RFI, DMI, BWG, fat depth, CH_4 and CO_2 emission and O_2 consumption during RFI test, clean fleece weight (CFW), fibre diameter, staple length, wool production potential (WPP=CFW/BW×100) at one-year old and faecal egg count after weaning were measured. RFI group affected (P<0.05) DMI, RFI, CFW, WPP, CH_4 (g/d, g/kg DM, g/kg BW) and CO_2 emission and O_2 consumption. Less efficient animals ate more (1.449 vs 1.157 kg DM), produced a heavier fleece (3.27 vs 3.14 kg), had a higher WPP (6.9 vs 6.5%), CH_4 (24.3 vs 22.8 g/d) and CO_2 emission (1,028 vs 958 g/d), methane yield (0.55 vs 0.52 g/kg BW) and O_2 consumption (941 vs 894 g/d), while their methane intensity was lower (15.5 vs 17.8 g/kg DM). Other traits were not affected (P>0.05) by RFI group. In agreement with other studies, improving RFI reduces DMI, and CH_4 (total and yield) emissions, without modifying BWG or wool quality. No significant effect of RFI on health traits was found in our study. These favourable results must be outweighed with a potential reduction on fleece weight or wool/BW in a fine-wool production system.

Detection of unrecorded environmental challenges and genetic determinism of resilience in lambs

C.A. Garcia-Baccino[1,2], C. Marie-Etancelin[2], F. Tortereau[2], D. Marcon[3], J.L. Weisbecker[2] and A. Legarra[2]
[1]*Universidad de Buenos Aires, Facultad de Agronomía, Av San Martín 4453, 1417, Buenos Aires, Argentina,* [2]*GenPhySE, Université de Toulouse, INRAE, ENVT, 31326, Castanet-Tolosan, France,* [3]*Unité Expérimentale INRAE, Domaine de La Sapinière, 18390, Osmoy, France; carolina.garciabaccino@inrae.fr*

Resilient animals are capable of remaining productive under different environmental challenges. Rearing in increasingly heterogeneous environmental conditions increases the need of selecting resilient animals. Under normal productive rearing conditions, challenge events are sometimes unrecorded and from unknown source. Therefore, being able to identify periods in which a potential environmental challenge occurred provides an opportunity to quantify variability among animals and select those more resilient. In this study, we present a simple and practical data-driven approach to identify unrecorded environmental challenges to evaluate genetic determinism of resilience to these events. A total of 951 Romane male lambs were phenotyped for feed intake over an 8-year period (from 2009 to 2016) at the INRAE La Sapinière experimental farm. Feed intake was automatically recorded using automatic concentrate feeders during an 8-week period each year. A total of 51,832 daily feed intake (DFI) records were available. We fitted a mixture of two Gaussian distributions on the natural log-transformed coefficient of variation (CV) of DFI and computed the posterior probabilities of pertaining to the second component (high values of CV) for each day. Consequently, we were able to differentiate 'low CV days' (with low probability of having a high CV) and 'high CV days', with increased variability in DFI probably due to the occurrence of an environmental challenge. These probabilities were included as a covariate in a reaction norm animal model to evaluate genetic determinism of resilience to unrecorded environmental challenges. Variance components were estimated using Gibbs sampling and REML. The genetic correlation between the level and sensitivity to inferred environmental challenge was -0.73±0.05, showing that a hypothetical selection for increased DFI would result in decreased environmental sensitivity (increased resilience). In the frame of H2020 project SMARTER no. 772787.

SMARTER EU project: SMAll RuminanTs breeding for efficiency and resilience

C. Moreno-Romieux[1], J.J. Arranz[2], J.M. Astruc[3], D. Berry[4], T. Byrne[5], J. Conington[6], A. Doeschl-Wilson[7], P. Frutos[2], A. Legarra[1], A. Meynadier[1], C. Mosconi[8], C. Paul-Victor[1], R. Pong-Wong[7], A. Rosati[9], R. Rupp[1], B. Servin[1], C. Soulas[3], A. Stella[10], V. Thenard[1] and The Smarter Consortium[1]
[1]*INRAe Toulouse, CS 52627, 31326 Castanet-Tolosan France, France,* [2]*Universidad de Leon CSIC, Campus de Vegazana s/n, 24071 León, Spain,* [3]*IDELE, 149 rue de Bercy, Paris, France,* [4]*TEAGASC, Oak Park, R93 XE12 Carlow, Ireland,* [5]*AbacusBio, 10 John Street, WC1N 2EB London, United Kingdom,* [6]*SRUC, King's Buildings, West Mains Road, EH9 3JG Edinburgh, United Kingdom,* [7]*University of Edinburgh, Easter Bush, EH25 9RG Midlothian, United Kingdom,* [8]*SERVICE ICAR, Via Savoia 78, 00198 Rome, Italy,* [9]*EAAP, Via G. Tomassetti 3A, 100161 Rome, Italy,* [10]*CNR, Via Alfonso Corti 12, 20133 Milano, Italy; carole.moreno-romieux@inrae.fr*

In the frame of H2020 project SMARTER no. 772787, we are developing and will deploy innovative strategies to improve Resilience and Efficiency (R&E) traits for sheep and goat. Several R&E traits including feed efficiency, mobilisation of body reserves, disease resistance, survival and welfare are being investigated. Experimental populations are used to identify new predictors of these R&E traits and the trade-offs the animal faces to overcome external challenges. We aim to characterise the underlying genetic and genomic variability governing these R&E traits and trade-offs. The genotype-by-environment interactions (G×E) of R&E traits will be estimated in large commercial populations. The adaptation genes will be also characterised comparing under-utilised breeds that are well-adapted to their environments. New methods of genetic analysis and genomic selection are proposed to take into account R&E traits and G×E, aiming at estimating accurate R&E genomic predictions in different environments across different populations. We will improve R&E predictions by performing genomic selection across countries thanks to a new cooperative EU and international initiative in this project. We are also developing new models at both animal and farm levels to predict trade-offs between productivity and R&E traits. Finally, we will propose new breeding strategies that utilise R&E traits and trade-offs and that also balance economic, social and environmental challenges from their implementation.

Benefits of genomic selection for meat sheep according to the available pedigree and the level of AI

J. Raoul[1,2] and J.M. Elsen[1]
[1]GenPhySE, Université de Toulouse, INRAE, ENVT, Campus INRA, 31326, Castanet-Tolosan, France, [2]Institut de l'Elevage, Campus INRAE, 31320 Castanet-Tolosan, France; jerome.raoul@idele.fr

Many studies reported additional genetic gain when genomic selection is implemented. However the available pedigree information and the reproductive mode (artificial insemination vs natural mating) potentially affect the benefit of genomic selection, especially for the selection of a late-in-life trait. All over the world, numerous meat sheep breeding programs diverge according to their pedigree structure and the level of inseminated ewes. To assess the effect of both parameters on the genetic gain for a maternal trait, we developed a stochastic simulation software that mimics meat sheep breeding programs. Depending on the strategy, selection was based on EBVs or GEBVs computed with a BLUP or single step GBLUP animal model using blupf90 software. Natural mating and artificial insemination based designs, inspired by the current diversity of designs used for French meat sheep breeds, were modelled and two genomic strategies were tested and compared to a conventional selection strategy for a 25 year-period: genomic selection based on a male (all sires) or a male (all sires) and female reference population. Genetic gain was computed as the regression slope of the average true breeding value of first parity dams over a time interval between years 10 and 25. Genomic selection based on a male reference population did not always outperform conventional selection, especially when sire information was missing for a part of dams. Regardless of the design, genomic selection based on a male and female reference population was always favourable and genotyping 25% of the females in the nucleus resulted in a significant increase in gain compared with conventional designs. The project SMARTER (H2020) no. 772787 funds the study.

What is the energetic cost of resistance to gastro-intestinal parasite? A modelling approach

F. Douhard[1], C. Moreno-Romieux[1] and A.B. Doeschl-Wilson[2]
[1]GenPhySE, Université de Toulouse, INRAE, ENVT, F-31326, Castanet-Tolosan, France, [2]The Roslin Institute and Royal (Dick) School of Veterinary Studies, University of Edinburgh, Edinburgh, United Kingdom; frederic.douhard@inrae.fr

Selection for resistance to gastro-intestinal nematode (GIN) is a promising strategy for the sustainable control of parasitism in grazing systems, but the consequences on production traits are still not fully known. While promoting a strong immunity to GIN should alleviate parasite-induced damage, a metabolic cost through the diversion of nutrients away from productive functions may incur. Accordingly, there may be an optimal level of immunity to select on, which depends on the nutrient allocation and on the costs of immunity. These would need to be estimated in order to accurately predict selection responses. Our aim was to estimate trade-offs and the energetic cost of mounting an immune response against GIN infection by fitting a mechanistic host-parasite interaction model to data from female lambs that were experimentally infected with *Haemonchus contortus*. Individual data related to 42 lambs from two lines divergently selected on resistance to this parasite. At 5 months of age those lambs were infected during 5 weeks with a single-dose of 10,000 third-stage larvae while being fed *ad libitum* a high-protein, low-energy diet. The model was based on an energy balance approach to predict changes in body weight (BW), body fat thickness (BFT) and immune response cost from observed feed intake. The immune response cost was included as an unknown component to estimate. Immune response was part of an interaction with the within-host parasite dynamics and this determined faecal egg count (FEC) and haematocrit over the course of infection. The model was fitted individually using a normalised root mean square error calculated over the repeated measures of BFT, BW, haematocrit, and FEC. Our results showed that the broad variation in observed responses to GIN infection can be largely explained by two parameters controlling immunity. They also indicate a small energy cost of immunity associated with reduced feed intake, which suggests a trade-off between GIN resistance and body fatness during growth. This study was in the frame of H2020 project SMARTER no. 772787.

Divergent selection for functional longevity in Alpine goats as a model to assess animal resilience

C. Huau[1], J.A. Torres-Penna[1], T. Fassier[2], J. Pires[3], M. Tourret[3], I. Palhière[1], N.C. Friggens[4] and R. Rupp[1]
[1]INRAE, INPT-ENVT, INPT-ENSAT, GenPhySE, chemin de borde rouge, 313265 Castanet-Tolosan, France, [2]INRAE, Domaine de la Sapinière, 18390 Osmoy, France, [3]INRAE, Univ. Clermont Auvergne, VetAgro Sup, UMR Herbivores, 63122 Saint-Genès-Champanelle, France, [4]INRAE, AgroParisTech, Univ. Paris-Saclay, MoSAR, 16 rue Claude Bernard, 75231 Paris, France; nicolas.friggens@agroparistech.fr

Given the anticipated long-term change in climate and the evolution towards more extensive production systems, ruminants will find themselves in increasingly variable and challenging environments. Therefore, there is a need for better knowledge on animal resilience, that is the ability to cope with a wide range of environmental disturbance, and to develop new breeding tools and objectives. To address this issue, we created and analysed data from goats under divergent selection for functional longevity (LGV+ and LGV- lines), which undergo physiological, nutritional and inflammatory challenges. The first 3 cohorts comprise 281 Alpine goats born since 2017. They were sired by 12 LGV+ and 11 LGV- bucks, with breeding values for longevity equal to +171($\pm$73) days and -156 ($\pm$50) days, respectively. The preliminary survival analysis of these goats showed a reduced longevity in LGV- goats after 2 years. Resilience was assessed by studying adaptation to peripartum (a spontaneous physiological challenge) and to induced nutritional challenges (2-d of straw diet) during early and midlactation, i.e. catabolic and anabolic stages. The catabolic challenge resulted in an average loss of production of 62% with rapid recovery. Both milk and blood metabolite profiles during the latter challenge showed large animal variability, with a difference (P<0.05) for beta hydroxybutyrate (BHB) between genetic lines. Furthermore, available data from 34 goats showed greater (P<0.05) prepartum plasma NEFA and BHB for LGV- compared to LGV+, suggesting greater reliance on body reserve mobilisation and metabolic stress for LGV- to support late gestation. Ongoing research will include more animals, an inflammatory challenge with lipopolysaccharide and comprehensive modelling to identify the path that resilient animals take through the multi-dimensional space. SMARTER (H2020), ACTIVE-GOAT and RESILAIT (Apisgene) projects fund the study.

Can a domestic sheep industry make genetic benefits from using foreign sires?

N. Fetherstone[1], F.S. Hely[2], N. McHugh[1] and P.R. Amer[2]
[1]Teagasc, Mellows Campus, Athenry, Co.Galway, H65R718, Ireland, [2]AbacusBio, 442 Moray Place, Dunedin, 9016, New Zealand; nicola.fetherstone@teagasc.ie

The widespread use of superior foreign sires has contributed to economic and genetic gains in many species. The objective of this study was to create a gene flow model which could use sheep industry data to evaluate whether a domestic sheep population would benefit from the widespread use of foreign genetics. An Irish-New Zealand case study was derived. Data was retrieved from the respective national maternal breeding objectives; the Sheep Ireland Replacement Index and the New Zealand Maternal Worth. The gene flow model predicts the genetic improvement in economic merit of the maternal index for future generations of commercial (COM) sheep as a consequence of the widespread implementation of different breeding strategies, i.e. the incorporation or exclusion of foreign genetics in the domestic industry. Subpopulations used in the model range from conservative (CON) and progressive (PRO) breeders who used only domestic genetics, to foreign (FOR) breeders that supplied rams directly into domestic flocks, or PROFOR who used a combination of PRO and FOR genetics. Benefits were expressed relative to a base situation, worth €49.18m after twenty years, whereby COM flocks sourced 87 and 13% of rams from CON and PRO breeders respectively, and never used FOR genetics. Scenario 1 shifted the market share away from CON towards PRO breeders, at a rate of 5% per annum, with increased selection intensity to model farmers retaining only the top ranked 20% of animals for replacements, contributing an extra €132.94m to the industry after twenty years. Scenario 2 modelled an optimised PROFOR scenario, with five years of FOR sires sourced directly by COM flocks, followed by shifting half of the market share of PRO to PROFOR, accumulating an extra €97.30m after twenty years. Results show that large gains can be achieved by the industry without the use of foreign genetics, through shifting the market share away from CON towards PRO domestic breeders. However, the short term tactical use of superior FOR genetics may increase the proportion and rate at which the shift occurs. The authors would like to acknowledge the funding provided through the frame of H2020 project SMARTER no. 772787.

Comparison of sheep genotype metrics across breeds and countries

A.C. O'Brien[1], J.M. Astruc[2], A. Tolkamp[3] and D.P. Berry[1]
[1]Teagasc, AGRIP, Teagasc, Moorepark, Fermoy, P61 P302, Co. Cork, Ireland, [2]IDELE, BP 42118-31321, Castanet-Tolosan, France, [3]SRUC, Edinburgh, EH9 3JG, United Kingdom; aine.obrien@teagasc.ie

The objective of the present study was to compare the allele frequency of different meat sheep breeds in Ireland and the United Kingdom as well as different dairy sheep breeds from France. The frequency of each allele per single nucleotide polymorphism (SNP) was available for five meat sheep breeds from Ireland (i.e. Belclare, Charollais, Suffolk, Texel and Vendeen), two meat sheep breeds from the UK (i.e. Scottish Blackface and Texel), and five French dairy sheep breeds (i.e. Basco-Béarnaise, Black-faced Manech, Corse, Lacaune, and Red-faced Manech). Allele frequency data were available on 44,040, 577,400 and 48,059 SNPs from the sheep in Ireland, the UK, and France, respectively. A total of 38,883 SNPs were common to all sheep populations. All genotyping had been undertaken using Illumina platforms and the allele frequencies of all SNPs were subsequently aligned to the Illumina 'Allele A/B' format; the frequency of the 'A' allele was subsequently calculated in each of the 12 populations. An informative SNP was identified where the frequency of the 'A' allele was between 0.2 and ≤0.8 in each of two pair-wise breeds compared. A non-informative SNP was defined as when the two breeds were compared, the frequency of the 'A' allele in either breed was either <0.2 or >0.8. On average, 49.33% of all 38,883 common SNPs were informative between each of the 66 pairwise comparisons. The pair of breeds with the maximum number of informative SNPs was, as expected, the Irish Texel and Scottish Texel (59.89% of SNPs were informative) breeds; the minimum number of informative SNPs between the Scottish Texel and Irish Suffolk (40.49% of SNPs). The number of SNPs that were informative in one breed but non-informative in another ranged from 1,081 (informative in the Scottish Texel but not in the Irish Texel breed) to 9,350 (informative in the Corse but not in the Suffolk breed). The present study identified both informative and non-informative SNPs across both meat and dairy sheep. The results from the present study will contribute to the development of future genotyping panels for sheep to maximise the applicability of the genotyping panel across multiple sheep populations.

Study of rumen microbiota in dairy sheep with different feed efficiency using nanopore sequencing

C. Esteban-Blanco[1], P.G. Toral[2], C. Fernández-Díez[2], A. Suarez Vega[1], B. Gutiérrez-Gil[1], O. González-Recio[3], G. Hervás[2], P. Frutos[2] and J.J. Arranz[1]
[1]Universidad de León, Producción animal, Facultad de Veterinaria, Campus de Vegazana, 24007 León, Spain, [2]Instituto de Ganadería de Montaña (CSIC-Universidad de León), Finca Marzanas, 24346 Grulleros, León, Spain, [3]Instituto Nacional de Investigación y Tecnología Agraria y Alimentaria (INIA), Mejora Genética Animal, Crta. A Coruña km 7.5, 28040 Madrid, Spain; jjarrs@unileon.es

Ruminants can obtain high-quality products using low-quality forages, which are mainly indigestible for humans. Until now, second-generation massive parallel sequencing has been the most used technology in the majority of studies of microbiota in livestock species using two approaches; 16S rRNA gene sequencing and shotgun metagenomics. However, the launching of third-generation single-molecule technologies offers long-read output with low cost and rapid real-time analysis. In this preliminary study, we aim to assess the complete end-to-end microbial composition of rumen using the MinION sequencer (Oxford Nanopore Technology) in two groups of ewes with divergent residual feed intake (RFI) values. From a total of 40 lactating Assaf ewes, the 20% most efficient (RFI<-0.40) and inefficient (RFI>0.39) sheep were selected for DNA sequencing of ruminal fluid, using Nanopore technology. The fluid was collected with a stomach tube. After quality control and removing host DNA, an average of 380.000 reads per sample with a mean length of 714 bases were aligned against the NCBI-nr protein database performing taxonomic and functional assignments on individual reads rather than contigs. No significant differences were detected in the microbial community between the feed efficiency groups studied here. Predominant genera among samples were *Prevotella*, *Methanobrevibacter* and *Fibrobacter*, as usually found in other rumen studies. Also, some of the functions observed in this research, such as methane metabolism, peptidoglycan biosynthesis, energy production and conversion, protein degradation and inorganic ion transport, could be related with higher feed efficiency in dairy ewes. These preliminary results suggest a more relevant role of microbiota function than taxonomy in explaining differences in feed efficiency in dairy sheep. Projects: EU-SMARTER & JCyL CSI276P18, FEDER and ESF.

Genomic solutions for improved gastrointestinal parasite resistance in Scottish Blackface sheep

A. Pacheco[1], T.N. McNeilly[2], G. Banos[1] and J. Conington[1]
[1]Scotland's Rural College, Roslin Institute Building, Easter Bush, EH25 9RG Midlothian, United Kingdom, [2]Moredun Research Institute, Pentlands Science Park, Bush Loan, EH26 0PZ Midlothian, United Kingdom; antonio.pacheco@sruc.ac.uk

Global sheep production incurs important economic losses due to gastrointestinal parasitism. The aims of this study were to estimate genetic parameters for and identify genomic regions associated with new disease and immunological traits to underpin genetic improvement programmes for disease resistance in sheep. Faecal egg and oocyst counts (strongyles (FEC$_S$), *Nematodirus* (FEC$_N$) and coccidia (FOC)), a 5-point faecal soiling score (DAG) and live weight (LWT) data were collected from 3,731 Scottish Blackface lambs on an extensive hill farm (2011 to 2017). Cellular immune traits reflecting T-helper type 1 (Th1), Th2 and regulatory T cell (Treg) responses were determined by quantifying cytokines interferon gamma (IFNγ), interleukin (IL)-4 and IL-10, respectively, from *ex vivo* stimulated whole blood. Humoral immunity was quantified via circulating nematode-specific immunoglobulin (Ig)A. Genetic parameters (heritability (h^2) and genetic correlations (r_g)) were estimated with ASReml v3.0. Animals were genotyped using OvineSNP50 Bead Chip level (50k SNP chip). Final dataset contained 45,827 SNP markers. Genome-wide association studies (GWAS) were performed using GEMMA. Significant genetic variation was found in all traits which could be integrated in breeding programmes aimed at improving resistance (h^2 between 0.14±0.03 and 0.77±0.09). Increased genetic resistance to FEC$_S$ did not appear to compromise resistance to FEC$_N$ and FOC (r_g=0.74±0.09 and 0.17±0.03, respectively). There was an antagonistic r_g of -0.33±0.15 between DAG and LWT, as well as between Th1 responses and LWT. Th1 and Th2 immune responses were positively correlated at a genetic level, suggesting a lack of counter-regulation between Th1 and Th2 immunity, and were partially under the same genetic control (positive r_g). Finally, IgA levels were positively correlated with Th2 and Treg responses (r_g=0.32±017 and 0.85±0.17, respectively). Several SNP markers reached significance at chromosome-wide level for both disease and immunological traits, with some genes identified around these SNPs being of interest.

A novel method to minimise labour costs of herbage monitoring based on spatial heterogeneity

A.G. Jones, T. Takahashi, P. Harris and M.R.F. Lee
Rothamsted Research, North Wyke, EX20 2SB, United Kingdom; andy.jones@rothamsted.ac.uk

Continuous monitoring of herbage quantity available on grazing fields can contribute to effective decision-making for both animal and pasture management. While a range of precision farming tools are available to assist accurate collection of this information, adoption rates are generally low, with many farmers prioritising cost, convenience and lifestyle over absolute accuracy. Manufacturers of rising plate meters (RPM), which measure compressed sward height and convert it to herbage mass, typically advise a 'pasture walk' along a W-shaped transect using the device to account for within-field spatial variability. However, anecdotal evidence suggests that this guideline is often perceived as excessively labour intensive relative to the perceived benefit, likely reducing current uptake of this technology. The objective of this study was therefore to investigate the loss of accuracy associated with a straight diagonal walk, and identify conditions under which this can replace the existing protocol. The field component of the study was conducted at the North Wyke Farm Platform, a farm-scale grazing trial in Devon, UK. Herbage data were recorded using a Jenquip EC20 Bluetooth Electronic RPM during the 2019 grazing season. Measurements were taken weekly under both 'W' and diagonal patterns, recorded using the same equipment, operator and sample size. Furthermore, to evaluate the universality of the results, a spatial simulation model was developed to synthesise herbage maps for alternative soil, weather and topographical conditions, and a 'virtual walk' was carried out under both protocols. Across all real fields and realistic virtual spatial patterns, no statistically significant difference was observed between the two sampling strategies regarding either estimated total pasture cover or within-paddock variation. On the other hand, the time required for diagonal walks was 63% shorter on average. These findings suggest that efficacy of herbage mass monitoring is comparable under both sampling techniques, while the cost difference is substantial. Current RPM sampling guidelines are therefore economically sub-optimal and recommending an alternative sampling pattern could dramatically increase the likelihood of technological adoption.

Trade-offs between economic performance and environmental impact of dairy farming in Central Norway
H. Steinshamn, M. Koesling, F. Walland and O. Flaten
Norwegian Institute of Bioeconomy Research, Gunnars veg 6, 6630, Norway; havard.steinshamn@nibio.no

The objectives were to quantify indicators of environmental impact of milk production and economic performance of dairy farms in Central Norway, to assess the association between economic returns and environmental impact the effect of management factors. Data were from 200 dairy farms in Central Norway that participated for three consecutive years (2014-2016) in the National dairy herd recording system. Standard life cycle assessment, input-output calculations and economic analysis methods were used to quantify the indicators. Correlation analysis were used to test relationship between indicators and multiple linear regression to identify farm characteristics and management factors associated with the indicators. Sale of animal as live animal or for slaughter were converted to kg energy corrected milk (ECM) equivalent. The gross margin per kg ECM sale averaged NOK 5.94 but ranged from NOK 4.07 to 8.91 across farms. Carbon footprint (CF) varied from 0.87 to 2.55 CO_2-eq/kg ECM, fossil energy use (Energy-cost) from 2.35 to 7.51 MJ/kg ECM, and nitrogen cost (N-cost) from 3.30 to 12.18 kg N input/kg N output in milk and meat sale. Gross margin was positively correlated with CF (0.34), Energy-cost (0.39), and N-cost (0.20), suggesting that there are trade-offs between economic performance and environmental impact. However, multiple linear regression revealed that for some farm management factors and characteristics a reduction in environmental impact of milk production can be achieved and at the same time improving the gross margin. Grassland yield and pasture proportion of the total dietary intake of dairy cows were both associated with reduced CF and higher gross margin. The use of purchased concentrate per cow were associated with reduced gross margin and increased Energy-cost. Increasing milk production per cow were associated with reduced CF, Energy-cost, and N-cost, without impact on gross margin. Our results suggest that there is a trade-off between economic and environmental sustainability measures, but dairy farmers in Central Norway may both improve their economic performance and reduce their environmental impact by improving grassland yield and increasing the use of pasture to dairy cows.

Adaptation to climate change of extensive livestock farming in the Iberian Peninsula: LIFE LiveAdapt
S. Sanz-Fernández[1], J.A. Salatti[1], C. Díaz-Gaona[1], A. Madrid[2], P.M. Herrera[3] and V. Rodríguez-Estévez[1]
[1]University of Cordoba, Animal Production, Campus Universitario de Rabanales. Edificio de Producción Animal. Cátedra de Ganadería Ecológica, 14071, Córdoba, Spain, [2]Institut de l'Élevage (IDELE), Service Fourrages et Pastoralisme, Campus INRA, Chemin de Borde Rouge, BP 42118, 31321 Castanet-Tolosan Cedex, France, [3]Fundación Entretantos, Calle Antonio Lorenzo Hurtado,1, E47014 Valladolid, Spain; liveadapt@uco.es

The LIFE LiveAdapt project (co-funded by the European Union through the LIFE17 program, reference LIFE17 CCA/ES/000035) seeks solutions for the adaptation to climate change (CC) of extensive livestock farming in Southern Europe. CC problems and possible solutions have been determined through a questionnaire for farmers. This work analyses the answers of 103 farmers. 83% of them already observe CC impacts in their region. Firstly, they are concerned about the availability of water due to changes in the distribution of rainfall, followed by an increase of temperatures, with a longer summer season. They also indicate a changes in grass and fodder production (more grass in winter and less in spring-summer), with a general decrease of production; as well as variations in the phenology of crops, trees, grasses and other weeds. At the zootechnical level, they detect reproductive problems and emerging diseases and health problems from interactions between wildlife and livestock, besides competing for resources (pastures and water). A lack of water for drinking is showed in Mediterranean regions. Finally, health problems in crops and trees (Quercus sp.) are also mentioned. The farmers' measures for adaptation to the reduction of pasture production are: transhumance, rotational grazing and foraging of summer stubbles. Besides, some farmers have reduced their stocking rates. What is more, as a result of the decrease in production, they recognise the need to reduce fixed production costs and increase sales prices. As a result of the decrease of production, these farmers recognise the need to reduce fixed production costs and increase prices of sale. Finally, they highlight the need for training on the different CC adaptation measures and higher level of transfer for technology and knowledge.

Analysis of workload and livestock monitoring on alpine pastures

J. Maxa and S. Thurner
Bavarian State Research Centre for Agriculture, Institute of Agricultural Engineering and Animal Husbandry, Voettingerstr. 36, 85354 Freising, Germany; jan.maxa@lfl.bayern.de

High labour workload together with difficult relief conditions are one of the reasons of decreasing livestock units during the last decades in many regions of the Alps. Nevertheless, modern techniques for tracking of the animals such as GPS were introduced in the last years. Those can help to reduce daily workload, improve pasture management as well as monitor livestock grazing behaviour on alpine pastures. Therefore, the main aim of this study was to: (1) investigate the workload on selected alpine farms; and (2) to test and evaluate the usage of the tracking system by the farmer in such areas. The workload of herdsmen was registered for 34 activities divided into four categories on a total of 9 alpine farms located mainly in Bavaria, Germany. Furthermore, every herdsman carried a GPS data-logger in order to estimate daily walked distances needed to search the animals on the pasture. The evaluation of the tracking system was conducted on one selected farm comparing the workload registered by the herdsman for periods with and without utilisation of the tracking system. The results of the workload analysis showed that 65% of the total workload was spent by the herdsmen for the category 'animal', followed by the categories 'management' (21%), 'pasture' (10%) and 'stable' (4%). The majority of the time spent for the category 'animal' consisted of the activities related to daily control and search of animals on the pasture. On average 5.9 km and 1,004 m altitude were passed daily by the herdsmen for those activities. The evaluation of the usage of the tracking system resulted in a significant ($P<0,05$) decrease of time needed to control the animals on the pasture with a tracking system. Furthermore, usage of such system omits the activity of livestock searching. Livestock tracking systems, if additionally equipped with other motion sensors, can further reveal beside workload reduction also the opportunity to monitor behaviour of grazing animals.

Meat production potential of impala (*Aepyceros melampus*) under intensified systems

T. Needham[1], R. Engels[2], D. Bureš[3,4], R. Kotrba[1,5] and L. Hoffman[2,6]
[1]Czech University of Life Sciences Prague, Department of Animal Science and Food Processing, Kamýcká 129, 16500 Suchdol, Czech Republic, [2]University of Stellenbosch, Department of Animal Science, Private Bag X1, Matieland, 7602, Stellenbosch, South Africa, [3]Czech University of Life Sciences Prague, Department of Food Quality, Faculty of Agrobiology, Food and Natural Sciences, Kamýcká 129, 16500 Suchdol, Czech Republic, [4]Institute of Animal Science, Department of Cattle Breeding, Přátelství 815, 104 00, Uhříněves, Czech Republic, [5]Institute of Animal Science, Department of Ethology, Přátelství 815, 104 00, Uhříněves, Czech Republic, [6]University of Queensland, Centre for Nutrition and Food Sciences, Queensland Alliance for Agriculture and Food Innovation, 39 Kessels Rd, 4108, Coopers Plains, Australia; needham@ftz.czu.cz

Wildlife farming is the fastest growing agricultural sector in South Africa and has begun to intensify to optimise animal production. However, little information is available on the meat production potential of important antelope species, such as impala. The aims of this study were to compare the effect of sex and production system on the slaughter performance of impala. Sub-adult impala (n=35) were culled from intensive (12 males and 11 females; 200-ha grazing and minimal supplementary feeding) and semi-extensive (12 males; 0.25-ha boma/paddock with *ad libitum* total mixed ration) production systems within the same game farm. The full body, external and internal offal, and warm carcass was weighed to determine the proportionate yield of each item. After 24 h of cooling (4 °C) six commercially important muscles were removed from the semi-extensive male and female impala only, and weighed. Data was analysed using General Linear Models and univariate ANOVAs in SAS (SAS Institute Inc.) with both sex and production system as fixed effects in their respective analyses, and animal as the random effect. Shapiro-Wilk test was used to test normality and Fisher's LSD test was used to compare means ($P<0.05$). Male impala had a higher dressing percentage than females (59.1±0.76 vs 55.6±0.76%), indicating a higher meat production potential. Minor differences in offal yields provides little motivation for these factors to be considered when processing sub-adult impala carcasses. Furthermore, intensification of feeding of male impala contributed no benefit in carcass or offal yield and thus management thereof should be evaluated. Overall, impala showed higher dressing percentages than livestock, indicative of advantageous meat production potential.

Milk authentication according to the main dietary roughage source by using DART-HRMS

G. Riuzzi[1], I. Lanza[1], M. Bragolusi[2], B. Contiero[1], R. Piro[2], A. Massaro[2], V. Bisutti[1], S. Segato[1] and F. Gottardo[1]
[1]*University of Padova, Dept. of Animal Medicine, Production and Health, Viale dell'Università 16, 35020, Italy,* [2]*Istituto Zooprofilattico Sperimentale delle Venezie, Viale Fiume 78, 36100, Italy; giorgia.riuzzi@unipd.it*

To avoid frauds or misperceptions, dairy sector would benefit from using accurate techniques to authenticate milk and dairy products. Among these, direct analysis in real time-high resolution mass spectrometry (DART–HRMS) is an emerging tool to identify a molecular fingerprinting. The study aimed at evaluating DART reliability to discriminate among milk samples from 3 farming systems relying on different main roughage sources. 14 specialised Italian dairy farms (Veneto, Italy) were grouped in 3 experimental dietary thesis based on: grass-hay (GH, 3 farms), crop silage-hay (SH, 5 farms) and maize silage (MS, 6 farms). Over 2018, 70 raw bulk milk samples have been collected (5 samples/farm) and analysed by using a DART-HRMS system. The DART variables were normalised and submitted to a statistical data fusion procedure. In the data fusion matrix, a PLS-DA was applied to graphically separate the 3 dietary groups within their 0.95-confidence interval. The statistical approach identified the most informative chemical variables, such as organic acids and mono- and diglycerides, able to accurately discriminate the 3 thesis. The data fusion matrix was also split into a train set, used to carry out a linear discriminant analysis (LDA), and a test set, to build a confusion matrix. The outcomes of the confusion matrix related to the LDA showed very high values ($\geq$0.90) of sensitivity and specificity for all groups, confirming the powerful capacity of DART technique to authenticate milk samples according to the feeding management. In conclusion, our findings revealed that DART profile is a powerful metabolomics fingerprint to perform a complete discrimination among milk obtained from cows fed with increasing permanent-meadow hay's inclusion levels as replacement of maize silage. This research was made possible by funding from a SAFIL project, funded by FONDAZIONE CARIVERONA, and from MIPAAF and SusAn, an ERA-Net co-funded under European Union's Horizon 2020 research and innovation programme (www.era-susan.eu), under Grant Agreement no. 69623.

Peer community in animal science: a free publication model for transparent and open science

R. Muñoz-Tamayo[1], M. Hess[2], S.A. Huws[3], B.L. Nielsen[1], E. Norberg[4,5], M. Pastell[6], L.O. Tedeschi[7], P. Trevisi[8] and M. Vayssier-Taussat[9]
[1]*Université Paris-Saclay, INRAE, AgroParisTech, UMR Modélisation Systémique Appliquée aux Ruminants, 75005, Paris, France,* [2]*Department of Animal Science, University of California, 2251 Meyer Hall, Davis, CA 95616, USA,* [3]*Institute for Global Food Security, Queen's University of Belfast, University Rd, Belfast, BT7 1NN, United Kingdom,* [4]*Department of Animal and Aquacultural Sciences, Faculty of Biosciences, Norwegian University of Life Sciences, Arboretveien 6, Ås 1432, Norway,* [5]*Department of Molecular Biology and Genetics, Aarhus University, Blichers, Allé 20, Tjele 8830, Denmark,* [6]*Resources Institute Finland (Luke), Production Systems Unit, Latokartanonkaari 9, P.O. Box 2, 00791 Helsinki, Finland,* [7]*Department of Animal Science, Texas A&M University, College Station, TX 77843-2471, USA,* [8]*Department of Agricultural and Food Sciences (DISTAL), Alma Mater Studiorum-University of Bologna, Viale Fanin 46, 40127 Bologna, Italy,* [9]*UMR BIPAR, Animal Health Laboratory, INRAE, ANSES, Ecole Nationale Vétérinaire d'Alfort, Université Paris-Est, 94700 Maisons-Alfort, France; rafael.munoz-tamayo@inrae.fr*

The scientific publication system is urged to evolve practices that enhance free dissemination and access to research findings, and ensure reproducibility and transparency while safeguarding scientific integrity from the detrimental effects of the *publish and perish* race. The evolution should introduce new publication models at low or no costs for readers and authors. Currently, a method for free research dissemination is the use of preprint servers such as bioRxiv (https://www.biorxiv.org/) that allow researchers to share their work immediately and before publication in a traditional journal. However, the main drawback of preprints is the lack of a formal peer review that guarantee the scientific quality of the posted manuscripts. The adoption of preprints in biology and the life sciences is rising, while in the animal science domain the preprint practice is still marginal. In this contribution, we present the international open science initiative Peer Community In (PCI) Animal Science (https://animsci.peercommunityin.org/) that overcomes the putative quality problem of the preprint publication model by proposing a free and transparent evaluation process of scientific papers based on peer reviews. PCI Animal Science operates similarly to a journal with editors and reviewers. PCI Animal Science belongs to the parental project Peer Community In (https://peercommunityin.org/) and is managed by scientists without the intermediation of any commercial publisher. We have currently 40 editors, and this number is increasing. The managing board (signatory of this abstract) invites the EAAP community to support our initiative and to join us to embrace open and transparent science.

Reproductive and economic results after 5dCosynch with or without P4-device in Holstein heifers

P. Pallares[1], J. Heras[1], J.L. Pesaantez-Pacheco[2] and S. Astiz[3]
[1]DairyProffesionals, Girona, 17820, Spain, [2]Univ. Cuenca, Cuenca, 010220, Ecuador, [3]INIA, Madrid, 28040, Spain; astiz.susana@inia.es

Our aim was to compare the economic efficiency of the 5dCosynch protocol for Fixed Timed Artificial Insemination (FTAI) with intravaginal progesterone device (5dCO-P4) or not (5dCO). The study lasted 18 mo. at a commercial heifer farm. Conception rate (CR), AI/pregnancy (AI/P) and Age at Conception (AC) was recorded after AI 811 heifers randomly distributed into both groups up to 3 AI. Progesterone (P4) and ultrasound exams were performed on 34% of heifers (276/811) at days 0, 5, 8 and 15 to verify synchronisation rate (SR) depending on P4 and ovarian structures. THI values were daily measured and THI$\geq$69 at AI considered heat stress (HS). A total of 173 and 638 AIs were during HS and without HS, respectively; 503 were first AI, 205 2nd and 103 3rd AI. From all AIs, 679 were FTAI and 132 after oestrus detection. AIs were analysed together (IBM SPSS v.25.0) after verifying no differences with logistic regression (Wald stepwise forward method with P>0.10). Economic estimations included results of AC, CR and AI/P depending on THI, a cost of 2.2€/heifer/day and actual costs of each hormonal protocol (9.8€ and 21.5€, respectively), estimating the cost/pregnancy (CP). With HS at AI, SR was 54.6 and 90.3% for 5dCO and 5dCO-P4, respectively (OR=0.13; 95%CI, 0.02-0.63; P=0.01), while SR was 68.9 vs 69.2% without HS (P>0.05). Similarly, CR was lower in 5dCO (52.4 vs 67.4%; OR=0.35; 95%CI, 0.18-0.70; P$\leq$0.001) with a significant interaction group × THI (P=0.03) and group × SR (P=0.03), with 5dCo CR=49.0% vs 5dCO-P4 CR=74.0% with HS and no difference without HS. During HS 5dCO-P4 shortened the AC in 17d and reduced the CP in 4.9€. Without HS 5dCO-P4 reduced AC in 5d and increased CP in 11.8€. Therefore, economic efficiency depended on THI at AI with 5dCO working better than 5dCO-P4 during the cool season (1.58€ CP-difference between protocols) and with 5dCO-P4 inducing better results than 5dCO with heat stress (43.3€ CP-difference). In conclusion, optimal economic efficiency will be achieved when adapting protocols to heat stress, implementing 5dCO-P4 in summer and 5dCO during the rest of the year, reducing use of hormones and management, without worsening benefits. Funded by Virbac.

Measure to manage forage crops by remote sensing

L.A. Conceição[1,2], R. Santos[1,2], B. Maças[3] and J.R. Da Silva[4]
[1]Valoriza, Research Centre for Endogenous Resource Valorization, Polytechnic Institute of Portalegre, Portugal, [2]Polytechnic Institute of Portalegre, Av 14 de Janeiro, 7350-092 Elvas, Portugal, [3]Unidade investigação Recursos Genéticos e Biotecnologia (INIAV), Estrada de Gil Vaz, Apartado 6, 7351-901 Elvas, Portugal, [4]MED Mediterranean Institute for Agriculture, Environment and Development, Universidade de Évora Pólo da Mitra, Apartado 94, 7006-554 Évora, Portugal, Portugal; luis_conceicao@ipportalegre.pt

In Alentejo region, under Mediterranean climate, forage crops are determinant for the food sustainability of livestock of ruminants in extensive production systems given the periods of scarcity of pasture in both winter and summer, boosted by the extreme climatic phenomena. Currently, sensory technologies are increasingly available to the farmer to monitor crops and support decision-making. The objective of this study was to evaluate the correlation between NDVI, NDWI vegetative indices and the productivity of a mix forage crop under no-till, through the use of images by remote sensing of the Sentinel-2 constellation. The trial took place in 2019, on a 30 ha plot at Herdade da Comenda, in Caia (coordinates 38 ° 53 '39" N; 7 ° 03' 03" W), were the assessment of the biomass productivity NDVI and NDWI indices was carried out in a georeferenced grid in 32 places of the plot at grazing, cut for hay-silage, second cut for hay-silage and cut for hay. At any of monitored moments, significant correlations were explained for both the NDVI index and the NDWI index and biomass production, explained by linear adjustment models for grazing (R^2 0.92; P<0.01; 0.70; P<0.01), hay silage (R^2 0.75; P<0.01; R^2 0.59) and hay (R^2 0.53; P<0.01; R^2 0, 33; P<0.01). The results obtained demonstrate the interest that remote sensing technologies through satellite images may have in evaluating the production of forage crops and in decision-making processes for delineation of management zones.

Multivariate indices in technical-economic and structural traits of Iberian pig: preliminary results

J. García-Gudiño[1], I. Blanco-Penedo[2], E. Angón[3] and J. Perea[3]
[1]CICYTEX, Animal Production, Finca La Orden, 06187 Guadajira, Spain, [2]SLU, Department of Clinical Sciences,
P.O Box 7054, 75007 Uppsala, Sweden, [3]UCO, Animal Production, Campus Rabanales, 14071 Córdoba, Spain;
javier.garciag@juntaex.es

Iberian traditional pig production is located in the southwest of the Iberian Peninsula. This traditional livestock production is linked to the ecosystem of the *dehesa*. Therefore, the farming activity contributes to *dehesa* conservation. The size of farm, intensification of system, productivity, or profitability of pig production are variables of great interest that determine to characteristics of this ecosystem. However, a complex correlation pattern exists between these variables. In this work, the multivariate relationship pattern between these variables is explored through factor analysis (AF) of 70 farms. This approach was expected to generate new variables with potential technical meaning that could be used for management purposes. Three factors resulted from AF. F1 was correlated mainly with the kilograms of live weight produced, stocking density (swine) and land use referred to the *dehesa* (percentage used in pig production). This factor (F1) indicates that Iberian farms with more extensive management have greater use of natural resources in the *dehesa*. F1 was named as 'management'. F2 was correlated with number of sows, number of animals produced (piglets, growers or fatteners) and income per annual work unit. In this case, F2 indicates that greater intensive management generates a higher economic benefit. F2 was called 'profitability'. Finally, F3 was correlated with kilograms of live weight produced and farm size (hectares). F3 indicates that Iberian farms produce less kilograms of live weight per hectare when they have a greater area. This factor (F3) was named 'land yield'. Preliminary results show that Iberian traditional pig farms are more cost-effective when the production management is more intensive. Considering economic and environmental components in the analysis is necessary to obtain a sustainable development in the Iberian pig farms. Thus, the next step is to characterise Iberian pig farms economic and environmentally through the factors (F1, F2 and F3) generated.

Milk fatty acid composition and lipid quality nutritional indices in different sheep farming systems

E. Kasapidou[1], Z. Basdagianni[2], C. Karaiskou[2], V. Papadopoulos[1], A. Kesidis[1] and A. Tsiotsias[1]
[1]University of Western Macedonia, Department of Agriculture, Terma Kontopoulou, 53100 Florina, Greece, [2]School of Agriculture, Department of Animal production, Aristotle University of Thessaloniki, 54124 Thessaloniki, Greece;
ekasapidou@uowm.gr

Information on the nutritional indices of sheep milk fat is very important for the production of quality dairy products due to the increased consumer demand for healthier products. Sheep milk fatty acid composition can be modified via nutritional management such as grazing. This research studies lipid nutritional quality indices of milk produced in different farming systems. Bulk tank milk samples were collected from commercial semi-intensive (n=20) and intensive (n=20) farms and were analysed for fat content, fatty acid composition and lipid nutritional quality indices [hypocholesterolaemic and hypercholesterolaemic fatty acid ratio (h/H), atherogenic index (AI), thrombogenic index (TI), desirable fatty acids (DFA), and n-6/n-3 ratio]. One-factor variance analysis was employed for the comparisons between farming systems (P≤0.05). There were no significant differences in the fat content between the two farming systems (6.63 and 6.46 for milk from semi-intensive and intensive farms, respectively). The h/H fatty acid ratio was significantly higher (P≤0.01) in milk produced at the semi-intensive farms (0.64) in comparison to that from intensive farms (0.55). Significant differences were also observed in the AI and TI indices between the two systems farms (2.31 vs 2.64, P≤0.05 for AI and 1.63 vs 1.82, P≤0.01 for TI for semi-intensive and intensive respectively). However, both indices did not exceed the acceptable value of 3 in milk from both farming systems. The content of DFA was also significantly higher in the milk from the semi-intensive farms (17.46 vs 15.35 P≤0.05). The ratio of n-6/n-3 exceeded the desirable value of 4 in both farming systems but it was significantly lower in milk produced at the semi-intensive system in relation to the intensive one (4.84 vs 5.59 P≤0.05). In conclusion, the improved lipid quality indices observed in milk produced under semi-intensive system can be attributed to the nutritional management which includes grazing of local plant species.

Biometric characterisation of Sasso C44 roosters in the 'Galo de Barcelos' production system

P.S. Vaz[1,2], J.L. Cerqueira[2,3], J. Oliveira[4,5] and J.P. Araújo[2,6]
[1]Quinta de Eira Vedra, Ardegão, 4990-535, Ponte de Lima, Portugal, [2]Escola Superior Agrária do Instituto Politécnico de Viana do Castelo, Refóios do Lima, 4990-706 Ponte de Lima, Portugal, [3]Centro de Ciência Animal e Veterinária, UTAD, 5000-801, Vila Real, Portugal, [4]Centro de Investigação e de Tecnologias Agroambientais e Biológicas, UTAD, Vila Real, 5000-801, Portugal, [5]Escola Superior Agrária de Viseu, IPV Viseu, 3500-606 Viseu, Portugal, [6]Centro de Investigação da Montanha, Instituto Politécnico de Viana do Castelo, 4900-347, Viana do Castelo, Portugal; psantosvaz@gmail.com

'Galo de Barcelos' specification's production system involves the use of slow growth or autochthonous roosters. Biometric traits are used to characterise body conformation, compare growth in different animals, describe a population or breed and define ideal time for slaughter. The present work aims to characterise biometry and liveweight of Sasso C44 roosters at slaughter. For this biometric study, 88 Sasso C44 roosters were subjected to the same semi-extensive production system, carried on a farm associated to 'Confraria Gastronómica O Galo de Barcelos'. Liveweight and slaughter age were recorded, and five linear measurements were taken according FAO guidelines. Liveweight was 4,213.5±400.8 g corresponding to a slaughter age of 153.4±21.7 days. Linear measurements (cm) obtained were: wing span, 57.7±2.86; chest circumference, 36.7±2.18; body length, 51.2±2.92; shank length, 9.8±1.14 and shank circumference, 6.3±0.45. The slaughter age presented the highest coefficient of variation, 14.2% and wing span the lowest (4.95%). Most of the correlations are not significant (P>0.05), being the highest's values between liveweight/chest circumference (0,55; P<0.001) and liveweight/wing span (0,35; P<0.001). The principal component analysis indicate that the first two components were responsible for 43.2% of total variation. Communalities with highest extraction values were liveweight (0.71) and chest circumference (0.73). First component accounted to 25.7% and second one to 17.5% of the total variation. Liveweight and chest circumference presents the greatest correlations with first component while body length, shank length and shank circumference with the second component. To robust these results, further studies are required including more records.

Breeding strategies for sustainable beef cattle production in the tropics

R. Carvalheiro[1,2] and L.G. Albuquerque[1,2]
[1]National Council for Scientific and Technological Development (CNPq), SHIS QI 01, Conj. B, Lago Sul, 71605-001, Brasília, DF, Brazil, [2]São Paulo State University (UNESP), Via de Acesso Prof. Paulo Donato Castelane, s/n, 14884-900, Jaboticabal, SP, Brazil; roberto.carvalheiro@unesp.br

Beef cattle production plays an important role in food security and supplying high-quality protein for human consumption. Around 2/3 of beef cattle are raised on tropical grasslands, where animals usually face constrained management and nutrition conditions, especially during the dry seasons due to reduction of pasture quality and availability. Climate changes have been magnifying this constraint, as environmental variation has been more pronounced and more difficult to predict, adversely impacting pasture and livestock production. This highlights an increasing necessity for beef cattle farms in the tropics to focus more on efficiency and sustainability rather than total productivity. Breeding strategies play a crucial role in increasing adaptability of beef cattle to constrained conditions and also to mitigate the impact of cattle on climate changes, ultimately affecting efficiency and sustainability of beef production. In our presentation, we will discuss the effectiveness of different breeding strategies to cope with the effects of climate changes, such as: crossbreeding indicine (*Bos indicus*) and taurine (*Bos taurus*) breeds; selection for traits directly associated with efficiency/adaptability (e.g. residual feed intake, methane emission, heat tolerance); selection based on reaction norm (response of each animal to environmental variation); selection for increasing uniformity of production, among others. The impact of genomics in some of these breeding strategies will also be discussed. In addition, we will present the main challenges and opportunities for sustainable beef cattle production in the tropics, under a livestock breeding perspective.

Phenotypic and genetic characterisation of dairy sheep production resilience to climate fluctuations

V. Tsartsianidou[1], G. Banos[1,2,3], Z. Basdagianni[1], D. Chatziplis[4], V. Kapsona[3], E. Sánchez-Molano[2], K. Gkagkavouzis[1], N. Karaiskou[1], G. Arsenos[1] and A. Triantafyllidis[1]
[1]Aristotle University of Thessaloniki, University Campus, 54124, Thessaloniki, Greece, [2]The Roslin Institute, Easter Bush Campus, EH25 9RG, Edinburgh, United Kingdom, [3]Scotland's Rural College, Easter Bush Campus, EH25 9RG, Edinburgh, United Kingdom, [4]International Hellenic University, Sindos Campus, 57400, Thessaloniki, Greece; tsarvale@bio.auth.gr

Small ruminant milk production is of major importance for the Greek livestock sector and is challenged by climate change affecting the Mediterranean region. Breeding for resilience to climate change is a possible mitigation strategy. The resilience of milk yield to weather fluctuation was studied in one of the most productive Greek sheep breed, the Chios breed. Data included 364,173 test-day milk records of 29,331 ewes raised in 91 flocks. These records were matched with weather data from the nearest meteorological stations to the farms. Reaction norm functions were fitted to random regression models for the statistical analysis of the joint data to derive novel resilience phenotypes reflecting milk production response to weather fluctuations. Substantial phenotypic variation was observed among individuals. Significant heritability was estimated for the resilience phenotypes (0.19). A downstream genome-wide association study was conducted on 664 individuals genotyped with the Ovine SNP50KBead Chip. One genome and two chromosome-wise significant SNP markers were identified, indicating candidate genomic regions located at chromosome five associated with animal resilience. Subsequent functional annotation of the significant SNP markers revealed closely located genes related to animal adaptation to climate biological processes. Examination of the genetic and phenotypic correlations of the novel trait with other production and functional traits and incorporation in future breeding goals constitute the next steps.

Genome-wide association of heat tolerance in dairy sheep

M. Ramon[1], M.J. Carabaño[2], C. Diaz[2] and M. Serrano[2]
[1]IRIAF. Instituto Regional de Investigación y Desarrollo Agroalimentario y Forestal, CERSYRA, Av del Vino 10, 13300 Valdepeñas, Spain, [2]INIA. Instituto Nacional de Investigación y Tecnología Agraria y Alimentaria, Mejora Genética Animal, Crta de la Coruña km 7.5, 28040, Spain; mramon@jccm.es

The rise of global temperatures and the increased frequency of extreme weather events associated with climate change are already having a negative impact on agriculture and livestock. Understanding how animals respond to thermal stress is key to developing strategies that improve adaptation to climate change. Production (milk, fat and protein yields) and fertility (AI success) records provided by the breeders' associations of the Manchega (AGRAMA) and Assaf (ASSAFE) Spanish dairy sheep breeds and weather data from the Spanish Meteorological Agency (AEMET) and SIAR were used to obtain individual measures of thermotolerance. For that, a random regression mixed model approach fitting an animal effect was used to obtain individual slopes under heat stress as a pseudo-phenotype of thermotolerance for the GWA studies. From these two populations, a total of 768 Manchega ewes and 1,456 Assaf ewes and rams were genotyped using the 50K Axiom(TM) Ovine Genotyping Array from ThermoFisher and a 50K customised Affimetrix array from INATEGA (Spain), respectively. HD Illumina genotypes (700K) were also available for 300 Manchega and 170 Assaf, and used to impute 50K genotypes to the HD platform. Quality controls of genotypes included removing samples and SNPs with call rates below 95 and 98% respectively, as well as those SNPs with a MAF bellow 0.05 or presenting mendelian errors. Association studies were carried out using a linear mixed model analysis. A significant number of genome regions associated with thermotolerance fall into regions that contain genes related to productive traits. Other regions point to genes involved in the regulation of systemic arterial blood pressure, multicellular organism growth and cell differentiation, regulation of lipolysis and thermogenesis or spermatogenesis among others. Results from this study reveal the existence of sound regions associated with the individual response to thermal stress. This information may help to establish efficient strategies to face thermal challenges and therefore enhance the improvement in the sustainability of the livestock sector.

Effects of heat stress on prolificacy in Iberian pigs

M. Muñoz[1], C. Caraballo[1,2,3], P. Palma-Granados[1,3], G. Gómez[4], G. Matos[4], F. Sánchez-Esquiliche[4] and J.M. García-Casco[1,3]
[1]INIA, Departamento de Mejora Genética Animal, Carretera de la Coruña, km 7,5, 28040, Madrid, Spain, [2]Imasde Agroalimentaria S.L, Calle Nápoles, 3, 28224 Pozuelo de Alarcón, Madrid, Spain, [3]Centro de I+D en Cerdo Ibérico, INIA, Departamento de Mejora Genética Animal, Ctra. Santos, 1, 06300 Zafra, Badajoz, 06300, Spain, [4]Sánchez Romero Carvajal, Carretera San Juan del Puerto, s/n, 21290, Jabugo, Huelva, Spain; mariamm@inia.es

The increase of temperature could trigger heat stress on livestock altering the physiology of animals through a decrease in their prolificacy. Although Iberian pigs are well adapted to the high temperatures, the effects of heat stress on prolificacy traits should be analysed to be considered in selection programs. The objective of the present study was to analyse the effects of heat stress on the number of piglets born alive (NBA) and estimate genetic parameters for NBA taking into account the heat stress. Data included records of NBA from 8,595 litters born in 177 farrowing batches of 2,441 dams and 29 sires. Temperature-Humidity index (THI) was built using the data obtained from the closest Public Weather station. Heat stress was quantified through the definition of different heat loads, being the daily heat load the number of degrees surpassing a particular threshold (THI_0). According to this, different heat stress thresholds and the sum of heat load from 1 to 7 days were evaluated. In addition to this, genetic parameters were estimated using different models. Akaike information criterion (AIC) values revealed that the heat load function using a THI_0 equal to 22 °C and a heat load of two days previous to the birth date was the most optimum, under these conditions the estimate of NBA decreases in -0.066 (SE=0.025) per °C. Regarding the genetic component estimation, AIC values also showed better perform of a univariate model including heat load as a covariate than other one disregarding this parameter. On the other hand, the multivariate analyses did not reveal NBA can be considered as an independent trait during heat stress and not heat stress periods. Although this study reveals interesting finding about the effects of heat stress on prolificacy of Iberian pigs, further analyses considering other models as random regression should be carried out to improve the knowledge in this field.

Identification of genomic regions in Large White pigs associated with climatic conditions

L. Getmantseva, S. Bakoev, O. Kostyunina, A. Traspov, N. Bakoev and Y. Prytkov
Federal Science Center for Animal Husbandry named after Academy Member L.K. Ernst, Moscow region, Podolsk urban district, Dubrovitsy village, 60, 142132, Russian Federation; ilonaluba@mail.ru

In the end of the nineteenth century in the Soviet Union (the territory of the Russian Federation with adjacent republics), a huge selection work was conducted, which was aimed to acclimating the imported population of Large White pigs. As a result, new breeds (based on Large-White breed) or types of Large-White pigs adapted to different climatic conditions were obtained. The aim of our work was to study of long sections of ROH, which could be associated with pig acclimatisation. Samples (ear plucks) of these pigs were early collected and stored in the unique collection of biomaterials of the Federal Science Center for Animal Husbandry named after Academy Member L. K. Ernst. 210 samples of pigs belonging to breeds were selected for analysis: Civil (n=33), Achinsk type of Large White (n=20), Belarusian Large White (n=32), Large-White of Moscow region (n=32), Large White of Tyumen region (n=39), Ukrainian White Steppe (n=30) and Urzhumskaya (n=24), which, depending on their geographical location, were divided into three groups: 'Center', 'North' and 'South'. GeneSeek GGP Porcine HD chip (San Diego, USA) was used. The analysis of homozygous regions was performed using the DetectRuns 0.95 package of the R software environment. As a result, total of 24,883 homozygous segments were identified. The largest number of SNPs in the ROH regions was detected in the SSC14 chromosome in all three groups, as well as in the SSC1 and SSC4 chromosomes in the South group. For further analysis, long ROH (more than 10 Mb) were selected, which are found in more than 40% of the animals in the group. Such sections were defined only in the 'South' and 'North' groups. Among these sites, a homozygous region on SSC1 can be allocated in positions 163762514 ... 180054118, occurring in more than 40% of pigs in both the 'South' and 'North' groups. In this area, the zinc-finger family genes TSHZ1, ZNF407, and ZNF516 can be selected, which are actively involved in the regulation of transcription, and, according to Perdomo-Sabogal and Nowick (2019), are responsible for regulatory changes associated with human adaptation. This research was supported by the Russian Scientific Foundation (RSF) within Project No. 19-16-00109.

The environmental impact of breeding for resilient laying hens
M. Schop[1], I.J.M. De Boer[1] and H.A. Mulder[2]
[1]Wageningen University & Research, Animal Production Systems group, P.O. Box 338, 6700 AH Wageningen, the Netherlands, [2]Animal Breeding & Genomics, Wageningen University & Research, P.O. Box 338, 6700 AH Wageningen, the Netherlands; marijke.schop@wur.nl

Mitigating the environmental impact of animal production has mainly been focused on increasing of animal productivity. A disturbance of the animal through its environment, e.g. through heat-stress or disease, often results in lower productivity. To what extent animals can cope with disturbances, is also known as resilience. Recent analysis of egg production data in pure line laying hens shows evidence for genetic variation in resilience-indicators. This variation offers the opportunity to genetically improve the resilience of laying hens. However, trade-offs between resilience-indicators and production traits exist, potentially affecting the environmental impact of egg production when selecting for resilience. This study aimed to quantify the trade-offs between breeding for resilient laying hens, and the environmental impact of egg production. A life cycle assessment was conducted to study effects on green-house gas emissions (GHG) and land use (LU) of egg production. First, the environmental impact of one genetic standard deviation-change in single production traits was quantified. Improving egg production (+10.3 eggs/hen/round), survival (-1.2%-point mortality) or feed efficiency (-0.076 kg feed/kg eggs), decreased GHG (baseline: 1.8 kg CO_2-eq/kg egg) and LU (baseline: 3.2 m^2/kg eggs) by 0.5, 0.4, 1.9%, and 0.3, 0.1, and 2.1%, respectively. Secondly, the effect of breeding for resilient laying hens was quantified based on single-trait selection, using genetic correlations between production traits and resilience-indicators (i.e. variance, skewness, and autocorrelation in deviations of egg production) to estimate hen productivity. Selecting hens for lower variance and autocorrelation, and higher skewness, altered GHG by +0.8 (CI:+0.4 to +1.2), -0.1 (CI:-0.6 to +0.5), and 0.1 (CI:-0.7 to +0.5) %, and LU by +0.6 (CI:+0.2 to +1.0), 0 (CI:-0.5 to +0.5), and -0.2 (CI:-0.4 to +0.7) %, respectively. These results indicate that, depending on the resilience-indicator, single-trait selection for resilient laying hens causes an increase (i.e. trade-off) or no effect on the environmental impact of egg production.

Estimation of genetic parameters for milk urea content
R.E. Jahnel[1], I. Blunk[1], D. Wittenburg[1], M. Mayer[1], H. Täubert[2] and N. Reinsch[1]
[1]Leibniz Institute for Farm Animal Biology, Institute of Genetics and Biometry, Wilhelm-Stahl-Allee 2, 18196 Dummerstorf, Germany, [2]Vereinigte Informationssysteme Tierhaltung w. V., Heinrich-Schröder-Weg 1, 27283 Verden / Aller, Germany; jahnel@fbn-dummerstorf.de

The dairy cattle industry emits enormous amounts of nitrogen, which pollutes the environment. Therefore, breeding and selection of cattle with outstanding nitrogen efficiency could minimise environmental pollution. Since milk urea content (MUC) correlates with the amount of nitrogen emitted, this study aimed at estimating genetic parameters of MUC using a Holstein Friesian dairy cattle population in Mecklenburg Western Pomerania. In this study, data from first, second and third lactations of 287,843 dairy cows, collected between January 2008 and June 2019, were analysed. Average milk urea levels ranged from 226±68.6 ppm to 234±67.6 ppm across lactations. Statistical analysis using univariate and bivariate random regression models was performed with ASREML 4.1. The results revealed moderate heritabilities ranging from 0.21±0.01 to 0.24±0.01, which decreased with increasing lactation number. Repeatability estimates ranged from 0.41±0.00 to 0.42±0.00. Genetic correlations between MUC and milk yield (0.13), fat percentage (0.15) and protein yield (0.17) showed the highest outcomes, while correlations between MUC and fat yield (0.03), protein percentage (0.04) and SCS (0.01) were nearly zero. Moderate heritabilities show a possibility on reducing MUC by considering this trait in the breeding goal. Moreover, estimates for correlations with other milk production characteristics resulted in low values, so that disadvantageous selection effects are not expected from breeding against high nitrogen excretions.

Long term recording of breath methane concentration as a potential trait for dairy cattle breeding

M.N. Aldridge[1], H.J.C. Van Dooren[2], I. Hulsegge[1] and Y. De Haas[1]
[1]Wageningen University and Research, Animal Breeding and Genomics, P.O. Box 338, 6700 AH Wageningen, the Netherlands, [2]Wageningen University and Research, Livestock and Environment, P.O. Box 338, 6700 AH Wageningen, the Netherlands; michael.aldridge@wur.nl

To date, studies that measure breath methane concentration (CH_4) using multi-gas analysers 'sniffers', recorded for a limited time period and a limited number of farms. We explored the potential for a trait defined from continuously measured CH_4 within milking robots. After data cleaning and filtering (background barn CH_4 removed), records from 12 farms and 1,412 cows, covering a period of three to nine months, was analysed. The mean CH_4 for each milking robot visit per cow was calculated, and for this analysis the trait was defined as the total mean CH_4 for all visits. Phenotypically there were clear differences between cows and between farms. The mean CH_4 between farms ranged between 296 ppm and 978 ppm, with all farms having a normal distribution. The phenotypic standard deviation ranged between 23 ppm and 272 ppm across farms. When combining data from all farms the mean cow CH_4 was 551 ppm with a standard deviation of 213 ppm, and slightly positively skewed. Most importantly we demonstrated there is variation within the trait, and ranking cows based on long term methane production is not only possible but practical. All animals within the analysis have been genotyped and there is some linkage between farms, the next step will be to estimate the genetic variance of the trait. Based on these results fitting farm will likely be an important fixed effect. Accuracy of estimates could be improved by including information such as the lactation curve, time of day during milking event (background barn CH_4 increases during the day, and feeding times effect cow produced CH_4), and time of year (pasture and barn effect). The trait definition can also be refined. For example, we are currently exploring methods that remove CH_4 dips (caused by the cow's head moving away from the sample air intake), testing different lengths of recording windows and how to combine them (this analysis used a single window using all available data), and treating it as a ratio trait with CO_2. Finally, CH_4 will continue to be recorded on these farms, more cows and farms are also being added, which will improve the accuracy of genetic parameter estimation.

The impact of crossbreeding and feed efficiency on methane emission

R.B. Davis[1], J.B. Clasen[2], M. Kargo[3,4], J.F. Ettema[1] and S. Østergaard[1,5]
[1]SimHerd, Niels Pedersens Allé 2, 8830 Tjele, Denmark, [2]Swedish University of Agricultural Sciences, Ulls väg 26, 756 51 Uppsala, Sweden, [3]Aarhus University, Dept. of Molecular Biology and Genetics, Blichers Allé 20, 8830 Tjele, Denmark, [4]SEGES, Agro Food Park 15, 8200 Århus N, Denmark, [5]Aarhus University, Dept. of Animal Science, Blichers Allé 20, 8830 Tjele, Denmark; rbd@simherd.com

There is a growing focus on the climate impact of dairy production, and several countries have ambitious goals to reduce emissions. Crossbreeding has shown potential to improve herd fertility and health, and hereby improving longevity, which reduces the number of replacement heifers needed in the herd. Several studies have found a positive effect of crossbreeding on the herd net return, but just a few studies have investigated the effect on methane emission. An important aspect of methane emission is feed efficiency. Some studies suggest that crossbred animals are more feed efficient, but it is unknown if it is due to heterosis or breed effects. In this study, we investigated if crossbreeding can be a tool to reduce methane emission. A sensitivity analysis of feed efficiency in crossbred animals versus purebred animals was conducted to quantify the effect of the potential improved feed efficiency in crossbred animals on methane emissions. Three scenarios based on average Danish dairy herds were simulated in the SimHerd Crossbred program: purebred Holstein, two-way rotational crossbreeding between Holstein and Red Dairy Cattle, and three-way rotational crossbreeding using Montbéliarde as the third breed. Feed efficiency was defined as feed intake per 305-day ECM. For the crossbred herds four scenarios were simulated were crossbreds had 2.5, 5, 7.5 and 10% less feed intake.

Mitigation of greenhouse gas emissions beef cattle production systems

S. Samsonstuen, L. Aass and B.A. Åby
Norwegian University of Life Sicences, Faculty of Biosciences, Aboretveien 6, 1432 Ås, Norway; stine.samsonstuen@
nmbu.no

Mitigation options to reduce the greenhouse gas emissions from beef cattle production were investigated using the whole-farm model HolosNorBeef. HolosNorBeef is an empirical model based on the HolosNor model and the methodology of the Intergovernmental Panel on Climate Change with modifications to Norwegian conditions. The model estimates direct emissions of methane (CH_4), nitrous oxide (N_2O), and carbon dioxide (CO_2) from on-farm livestock production, and indirect emissions of N_2O and CO_2 associated with inputs used on the farm. Mitigation scenarios involving female reproductive performance (i.e. calf mortality rate and number of produced calves per cow per year), young bull production efficiency (i.e. age at slaughter and carcass weight), supplementation of the inhibitor 3-nitrooxypropanol (3-NOP), and various combinations of strategies were investigated. Baseline scenarios were typical herds of British and Continental breeds with emission intensities of 30.8 kg carbon dioxide equivalents (CO_2 eq/kg carcass and 29.2 kg CO_2 eq/kg carcass, respectively. Improving calf survival reduced emission intensities 3% across breeds. Continental breeds showed largest potential for reducing emission intensities from improved carcass production (-6.6%). Combining mitigation options in a best case scenario reduced emission intensities 11.7% across breeds. The inhibitor 3-NOP reduced the net GHG emissions dependent on application level. Assuming a 33% reduction in enteric CH_4 emissions during housing period (Sept 15 to May 31) in the best case scenario reduced the emission intensities further by 8.3%.

Ruminal microbial gene abundances as a selection criteria to reduce methane emissions in beef cattle

M. Martínez-Álvaro[1], M. Auffret[1,2], R. Stewart[3], J. Lima[1], R.J. Dewhurst[1], C.A. Duthie[1], M.A. Cleveland[4], M. Watson[3]
and R. Roehe[1]
[1]SRUC, Easter Bush, EH25 9RG, United Kingdom, [2]Danone Nutricia Research, 128 Av. Vauve, 91767, France, [3]Edinburgh
Genomics, Easter Bush, EH25 9RG, United Kingdom, [4]Genus plc, 202 Moravian Valley, WI53597, USA; malvaro@sruc.ac.uk

Genetic selection for reduced methane emissions (CH_4) is hampered by the high costs of recording, which led us to investigate abundances of rumen microbial genes as proxy traits for breeding. The aim of this study was to estimate the heritabilities (h^2) and genetic correlations (r_g) between host-releasing CH_4 and the abundances of ruminal microbial genes. Data were available from 359 beef cattle of various breeds (Aberdeen Angus, Limousin, Charolais and Luing) and fed two basal diets (forage and concentrate). We detected 1,141 microbial genes present in all animals by aligning whole metagenomic sequence reads of microbial DNA extracted from post-mortem rumen samples to the KEGG database. The ratio of each microbial gene abundance by the 16S rRNA microbial gene (K01977) was used to avoid statistical artefacts due to the compositional nature of the data. Genetic parameters were estimated using Bayesian methodology by fitting CH_4 and the abundance of each microbial gene in multiple SNP-BLUP uni- and bi-variate models. Heritability of CH_4 was 0.39±0.10, and most microbial genes (731/1,141) showed h^2 >0.25±0.06-0.13. Genetic correlations between 276 microbial genes and CH_4 were >|0.6| at a probability >0.85 of being different from zero. These microbial genes were involved in interesting metabolic pathways, including glycine metabolism (n=64), which provides substrates for methanogens, or synthesis of Coenzyme B (n=31), a molecule necessary for the last step of methanogenesis. As an example for breeding, 31 microbial genes showed negative r_g with CH_4 (from -0.61 to -0.86), and favourable properties as selection candidates (h^2 from 0.21 to 0.33, and CV ranging from 34-118%). This is the first study showing host h^2 of ruminal microbial genes abundances and their r_g with CH_4 in cattle. We found microbial genes explaining CH_4 that have great potential to be integrated in future breeding programs and as selection criteria to exploit novel sources of genetic variance in order to mitigate CH_4 in cattle.

The correlation of substitution effects across populations and across generations

A. Legarra[1], Y. Wientjes[2], C.A. Garcia-Baccino[1,3] and Z.G. Vitezica[4]
[1]INRA, GenPhySE, 31326 Castanet Tolosan, France, [2]Wageningen University & Research, P.O. Box 338, 6700 AH Wageningen, the Netherlands, [3]Universidad de Buenos Aires, Facultad de Agronomia, 1417 Buenos Aires, Argentina, [4]INP-ENSAT, GenPhySE, 31326 Castanet Tolosan, France; andres.legarra@inra.fr

Selection, and Genomic Prediction in particular, is sensitive to additive (substitution) effects or 'alphas' of causal genes to vary across time (generations) or space (populations). Even if causal genes were exactly known, change in substitution effects hamper use of 'old' generations to predict 'new' generations, or across-population prediction. The change of substitution effects from population (generation) b to population (generation) b' is due to deviations from functional additivity, i.e. G×E and G×G interactions (dominance, epistasis). We propose a model in quantitative genetics to predict the change of substitution effects due to G × G interactions that are a result of a change in allele frequencies. No explicit *functional* dominance or epistatic mechanism is assumed, so the model is very general. This model approximates small changes in substitution effects from b to b' using Taylor series expansions. Finally, the magnitude of the change in substitution effects is a function of change in allele frequencies (which can be computed or forecasted) and of magnitudes of *statistical* dominance and epistatic variances (which can be estimated). As a final result, the correlation of substitution effects is a function of the following. First, the variances of (*statistical*) additive, dominance, and additive by additive epistatic variances. Second, the distribution of allele frequencies at causal genes. Third, the relationship (or the genetic distance) of the two populations b and b'. Based on literature estimates of variance components, of relationships across populations or generations, and on different guesses of the distribution of allele frequencies, we expect correlations of substitution effects across breeds around 0.40-0.70, and the expected correlations across generations are roughly 0.95 for distances up to 5 generations and 0.90 for 10 generations apart.

Understanding impact of causative SNP on genomic predictions

I. Misztal[1], I. Pocrnic[1,2], M. Perez-Enciso[3] and D.A.L. Lourenco[1]
[1]University of Georgia, Animal and Dairy Science, 30602, USA, [2]The Roslin Institute, Midlothian, EH25 9RG, United Kingdom, [3]CRAG, Cerdanyola del Vallès, 08193, Spain; ignacy@uga.edu

Most of accuracy of the genomic selection is due to estimation of chromosome segments, with no need to identify causative SNP nested in those segments. On the other hand, if only the causative SNP were used in a fixed SNP model, the accuracy could be close to 100%. As gains with incorporation of potential causative SNP are limited in practice, the purpose of this study was to 1) determine the impact of causative SNPs on solutions of SNP effects, 2) develop recommendations for incorporation of causative SNP. A population was simulated assuming effective population size 60 and 100 equidistant causative SNP with identical substitutions effects. Causative SNP were included in 50 k SNP genotypes. Ten generations of 2,000 animals were simulated, with the last 3 genotyped. Analyses were by single step GBLUP, with solutions converted to SNP values and subsequently to p-values for each SNP; p-values account for the population structure. Manhattan plots for standardised SNP solutions showed large values only for few of the 100 causative SNP and were very noisy. Manhattan plots for p-values were similar to those for SNP solutions, indicating little impact of population structure, as is typical in simulated data. P-values over the statistical threshold pointed to only 8 out of 100 causative SNP. SNP profiles were creating by averaging SNP solutions +-100 SNP around causative SNP. The profiles showed a normal-like response around each causative SNP, with a distinct peak for the causative SNP. Each causative SNP influenced about 50 adjacent SNPs. Simulation results explain current results with the use of sequence data. Causative SNP are accounted for by many adjacent SNP, and selection of markers close to causative SNP is not likely to improve accuracy. The accuracy could be improved if causative SNP are identified. Then, their successful incorporation requires removal of nearby SNP to avoid double counting and decreasing shrinkage for causative SNP. Such a strategy is likely less successful with multiple causative SNP per chromosome segment.

When is sequence data going to help increasing accuracy of genomic predictions in livestock?

D. Lourenco[1], S. Tsuruta[1], S. Jang[1], B. Fragomeni[2] and I. Misztal[1]
[1]*University of Georgia, 425, River Rd, 30602, Athens GA, USA, [2]University of Connecticut, 17 Manter Rd, 06269, Storrs CT, USA; danilino@uga.edu*

One of the greatest benefits of using the common medium-density SNP chips is the addition of non-redundant information to the genetic evaluation system that complements phenotypes and pedigree, which considerably increases accuracy. As sequence data is becoming available for many livestock species, there is a question on whether this information can help to boost the accuracy of genomic predictions beyond what has already been achieved with SNP chips. Several studies have been conducted by our group using simulated and real dairy and beef cattle populations that included from 1,000 to almost 30,000 animals with full or imputed sequence information. In the simulated data, up to 1,000 causative variants were generated and used to construct genomic relationship matrices with or without SNP. For the real datasets, the potential causative variants were identified based on genome-wide association (GWA) and were added to the current SNP chips. Additional scenarios included the use of only causative variants and the use of all sequence SNP. Genomic predictions were obtained based on genomic BLUP (GBLUP) or single-step GBLUP (ssGBLUP). Overall, in real datasets, we observed no significant increase in accuracy by using all sequence SNP, causative variants alone or combined with SNP currently used for genomic prediction. However, an increase in accuracy of almost 100% was observed in simulated datasets when the causative variants were added to the 60k SNP panel and their simulated variances were accounted for by the prediction model. When the variances were set to neighbour SNP instead of to causative variants, a decay in accuracy was observed. Our results show that if true causative variants are identified, together with their position and the variance explained, a boost in accuracy can be observed. This raises a question on the effectiveness of the methods used to select causative variants in real data. Possibly, the combination of sequence and other layers of omics data (e.g. functional data) can help to identify some of the true causative variants in real livestock populations.

Genetic evaluation including an intermediate omics trait

O.F. Christensen[1], A. Legarra[2], L. Varona[3] and V. Boerner[1]
[1]*Aarhus University, Center for Quantitative Genetics and Genomics, Blichers Alle 20, 8830 Tjele, Denmark, [2]INRAE, GenPhySE, BP52627, 31325 Castanet Tolosan, France, [3]Departamento de Anatomía, Embriología y Genética, Departamento de Anatomía, Embriología y Genética, Miguel Servet, 177, 50013 Zaragoza, Spain; olef.christensen@mbg.au.dk*

In plant and animal breeding and genetics there has been an increasing interest in intermediate omics traits that mediate the effects of genetics on the phenotype of interest. Such traits are for example transcriptomics, metabolomics, or proteomics, and they are increasingly becoming commonly available in larger quantities at decreasing costs. For inclusion of such intermediate traits into a genetic evaluation system, there is a need for statistical models and methods that integrate phenotypes, genotypes, pedigree and such omics traits, and provide estimated breeding values. Here, we develop a model and associated inference methods for genetic evaluation including intermediate omics data. The model consist of two model components, a model for the phenotypes given the observed omics expression data, and a model for the omics expression data as a function of genetic and environmental effects. A formula for breeding values as a function of effects in the model is derived. A procedure for best linear unbiased prediction is presented, which consists of first solving a mixed model equation for omics prediction of the phenotypes, and second solving a mixed model equation for estimating breeding values given omics predictions of phenotypes. In this model omics traits can partially or totally determine the genetic effects of the trait of interest. Both complete and incomplete omics data are considered.

Genomic prediction with single-step genomic BLUP using a subset of genotypes in US Holstein

Y. Masuda, S. Tsuruta and I. Misztal
University of Georgia, 425 River Road, Athens, GA 30605, USA; yutaka@uga.edu

As of January 2020, the US dairy database includes more than 3.8 million genotypes. Most of the genotypes are for heifers, and only a fraction of them have phenotypes. Although the use of all genotypes in genomic prediction is an ideal strategy, the same prediction-accuracy can be available using a subset of genotypes with a decrease of computing cost in single-step genomic BLUP (ssGBLUP). We validated genomic predictions of young bulls between all genotypes and a subset of genotypes in US Holstein. We calculated the benchmarks using the full data set, provided by the Council of Dairy Cattle Breeding, including 61M phenotypes of 305-d protein yield, 36M pedigrees, and 2.3M genotypes. The benchmarks included daughter-yield-deviation (DYD) from pedigree BLUP, DYD from ssGBLUP, and GPTA from ssGBLUP. We cut off the last 4 years from the full data, and the truncated set included 841K genotypes up to 2014 (841K; ALL). Out of it, we created two sets of genotypes including bulls only (142K; BULL) and bulls and cows with records and with both parents known (256K; BULLCOW). For validation, we chose 3,250 bulls that had at least 50 phenotyped-daughters in 2018, but that had no daughters in 2014. The coefficient of determination (R^2) and the slope coefficient (b_1) were calculated from a linear regression of the benchmark on GPTA. For GEBV-ssGBLUP as the benchmark, R^2 was 0.82 for ALL, 0.76 for BULL, and 0.82 for BULLCOW. Whereas b_1 was around 0.90 in ALL and BULLCOW, BULL showed a lower value (0.83). We observed the same tendency in the other benchmarks. Using bull genotypes alone drop the accuracy probably because of limited information. The use of cow genotypes increases accuracy and reduces bias. The inclusion of genotyped heifers does not improve the accuracy of bull predictions. Statistics for validations based on GPTA-ssGBLUP indicate good stability of genomic predictions. Routine analyses by ssGBLUP can include only genotypes for bulls with daughters and cows, with the remaining animals predicted indirectly.

International single-step genomic evaluations in beef cattle

R. Bonifazi[1], J. Vandenplas[1], J. Ten Napel[1], A. Michenet[2], A. Cromie[3], R.F. Veerkamp[1] and M.P.L. Calus[1]
[1]Wageningen University & Research, Animal Breeding and Genomics, Droevendaalsesteeg 1, Radix building 107, Wageningen Campus, 6700 AH, Wageningen, the Netherlands, [2]Interbull Centre, Department of Animal Breeding and Genetics, SLU, Almas Allé 8, Box 7023, 75007, Uppsala, Sweden, [3]Irish Cattle Breeding Federation, Highfield house, Bandon, P72 X050, Co Cork, Ireland; renzo.bonifazi@wur.nl

Cattle international evaluations allow the comparison of animals' estimated breeding values (EBV) across different countries. Interbeef international beef cattle evaluations involve up to 11 countries, 5 breeds (Limousin, Charolais, Beef Simmental, Angus, Hereford), and 2 trait groups (weight and calving traits). Several national evaluations in beef cattle already combine genomic information with pedigree and phenotypic information. However, it is currently unknown how much Interbeef international evaluations could benefit from genomic information. The objective of this study was to investigate the feasibility and the potential benefits of an international multi-trait single-step approach, where countries are modelled as different traits. Age-adjusted weaning weight phenotypes for 481,067 Limousin males and females, born between 1975 and 2018, were available for seven European countries. Genotypes from four countries were available for about 17,000 animals. The number of genotypes per country varied from about 300 to 11,000. After imputing all genotypes to a common reference panel, analyses include: (1) a pedigree BLUP as currently run at the national level ($PBLUP_{nat}$); (2) an international multi-trait pedigree BLUP, as currently run by Interbeef ($PBLUP_{int}$); (3) a single-step SNP-BLUP using only national genotypes (if any) and performances ($ssSNPBLUP_{nat}$); and (4) an international multi-trait single-step SNP-BLUP ($ssSNPBLUP_{int}$). The results will quantify for each country the potential benefits of moving from a national evaluation ($PBLUP_{nat}$ or $ssSNPBLUP_{nat}$) to an international evaluation ($ssSNPBLUP_{int}$), and of including genomic information at the international level ($PBLUP_{int}$ vs $ssSNPBLUP_{int}$). This research will highlight the benefits and challenges of a single-step genomic international evaluation for beef cattle populations, both in countries with and without national genomic evaluations already in place.

Benefits from cow genotyping in the genomic evaluation for conformation in Fleckvieh

E.C.G. Pimentel, C. Edel, M. Erbe, D. Krogmeier, R. Emmerling and K.-U. Götz
Institute of Animal Breeding, Bavarian State Research Center for Agriculture, Prof.-Dürrwaechter-Platz 1, 85586 Poing-Grub, Germany; eduardo.pimentel@lfl.bayern.de

The German-Austrian official genomic evaluation for conformation traits in Fleckvieh has used information from bull genotypes in a calibration set of a two-step system since its introduction in 2011. Because the increase in the amount of information at the male side is limited, in the last years females have started to be genotyped. In August 2019 the first official genomic breeding values for conformation traits from a single-step system including cow genotypes were published. The objective of this work is to present an overview of the benefits from the new system with additional cow genotypes. Data from the December 2019 run of the German-Austrian official genomic evaluation for 24 conformation traits in Fleckvieh were used. The dataset included 193,507 genotyped animals, from which 89,962 were female and 22,436 had phenotypes for conformation traits. A further dataset reduced to the state of December 2015 was used to conduct a validation study. This reduced dataset included 50,168 genotypes, 10,060 of which from females and 2,937 with phenotypes. Theoretical reliabilities of genomic breeding values from the 2019 runs and realised reliabilities from the validation runs were compared to the ones from sets of runs, in which all genotypes of females were excluded from the datasets. Comparisons of theoretical reliabilities were made for different groups of male selection candidates depending on the degree of relatedness to the genotyped cows. Average absolute gains in theoretical reliabilities for udder score ranged from 3 to 4 percentage points depending on the group. Validation reliabilities also increased with the inclusion of cow genotypes. For example, in udder score it went from 0.507 to 0.528, despite the relatively low number of cows with both genotypes and phenotypes in December 2015. In conclusion, the new single-step system including cow genotypes resulted in better validation statistics and theoretical reliabilities. With the cow genotyping initiative recently started, a further improvement in validation reliability is expected in the next years.

Potential of multi-breed genomic prediction based on effective number of chromosome segments

J. Marjanovic[1], X. Yu[2], S.E. Wallen[2], T.H.E. Meuwissen[2], D. Hinrichs[3] and M.P.L. Calus[1]
[1]Wageningen University & Research, Animal Breeding and Genomics, P.O. Box 338, 6700AH Wageningen, the Netherlands, [2]Norwegian University of Life Sciences, Dep. of Animal and Aquacultural Sciences, P.O. Box 5003, 1432 Ås, Norway, [3]University of Kassel, Department of Animal Breeding, Nordbahnhofstr. 1a, 37213 Witzenhausen, Germany; jovana.marjanovic@wur.nl

Genomic selection (GS) has become the method of choice in many breeding programs. In dairy cattle, the main appeal of selection based on genomic breeding values, is the possibility to increase genetic gain by reducing generation interval. This is especially beneficial for numerically small cattle breeds, where accelerated genetic improvement may improve competitiveness and long-term perspectives of those breeds. However, GS has not been effective in small breeds due to difficulties to obtain sufficiently large breed-specific reference population for genomic prediction (GP). This may be overcome by adding individuals from other breeds to the reference population, where the added benefit strongly relies on relatedness between the breeds. Effective number of chromosome segments (Me) can be used as an indicator of relatedness between individuals from different breeds and to directly predict expected accuracy of multi-breed GP. Using 50k SNP data of three Red cattle breeds, MRY from the Netherlands (387 individuals), Red and White dual purpose from Germany (816 individuals), and Norwegian Red from Norway (~20,000 individuals), our aim is to assess the benefit of multi-breed GP for each of those breeds. We will estimate Me within and between populations, which will enable to predict the expected accuracy of GP within each of the breeds, and the expected added accuracy when combining different breeds in the reference population. In all applications so far, Me is computed on a genome-wide basis. Here, we will estimate local Me and evaluate its pattern across the genome. Multi-breed GP may be strongly driven by tagging large effect QTL and assigning appropriate variances to them. Variation in local Me values may provide insight into the expected benefits of variable selection models over regular GBLUP. The Me will be computed as the reciprocal of the variance of the difference between genomic and pedigree relationships. This study is part of the ERA-NET SusAn project ReDiverse, that aims to increase resilient and competitive use of European Red Dairy Breeds.

Easy implementation of QP transformation in ssGTBLUP

M. Koivula[1], I. Strandén[1], G.P. Aamand[2] and E.A. Mäntysaari[1]
[1]Natural Resources Institute Finland (Luke), Animal Genetics, Myllytie 1, 31600 Jokioinen, Finland, [2]NAV Nordic Cattle Genetic Evaluation, Agro Food Park 15, 8200 Aarhus N, Denmark; minna.koivula@luke.fi

The increasing amount of genomic information in the single-step evaluations has caused problems in the convergence of iterative solving. This can be due to incorrectly accounting genetic groups in the computations if genetic groups are included in the mixed model equations (MME) through QP transformation in full pedigree relationship matrix (A) but not in the pedigree (A_{22}) and/or genomic (G) relationship matrices for the genotyped animals. Often this problem can be solved by properly accounting for the contributions of the genotyped animals to the genetic groups. In practice, this means that for each genetic group, the elements in the row of $H_+^{22}=G^{-1}-(A_{22})^{-1}$ corresponding to an animal are weighted by the proportions of genes the animals received from the group (Q-matrix), i.e. product $H_+^{22}Q$. With ssGTBLUP, the QP transformation is possible to do by including new columns (TQ) into the original T matrix. The contribution due to the A_{22}^{-1} matrix is as easy by including the 'phantom parents' to the set of genotyped animals. We applied single-step test-day (TD) model to Nordic Holstein data where unknown parent group coefficients were accounted in: (1) ssGTBLUP A^{-1}, (2) ssGBLUP with H_+^{22}; (3) ssGTBLUP with H_+^{22}; and (4) ssGTBLUP A_{22}^{-1}. The TD data included 8.4 million cows with records, 10.4 million animals in the pedigree, and 178,177 genotyped animals. To reduce over-dispersion, 30% of the residual polygenic effect was included in G. All MME were solved with MiX99 software. Methods (2), (3) and (4) gave the same results: correlations between GEBVs of both genotyped and non-genotyped bulls were 0.999. Also, the genetic trends, as well as standard deviations of the GEBVs by birth year, were the same. The central observation was that case a) i.e. single-step ignoring QP transformation both in G and A_{22} did not converge in a reasonable time. Thus, we conclude that: (1) QP transformation is needed in G and/or A_{22} for good convergence of the single-step model; (2) different methods to make QP transformation to genotyped animals give similar results; and (3) QP transformation is easy to implement also in the ssGTBLUP which with large genomic data is computationally efficient.

Compatibility of pedigree and genomic relationships in single-step evaluations for crossbreeding

J. Vandenplas[1], R. Bergsma[2], C.A. Sevillano[1,2] and M.P.L. Calus[1]
[1]Wageningen University and Research, Animal Breeding and Genomics, P.O. Box 338, 6700 AH Wageningen, the Netherlands, [2]Topigs Norsvin Research Center B.V., P.O. Box 43, 6640 AA Beuningen, the Netherlands; jeremie.vandenplas@wur.nl

To ensure the compatibility of pedigree and genomic relationships, single-step evaluations require the centring of the genotypes using the allele frequencies in the unselected founder population of the pedigree. When multiple populations are involved, like in three-way crossbreeding, base population allele frequencies corresponding to each purebred population are required. In practice, these base population allele frequencies are rarely available because genotypes are usually only available for animals selected in the current breeding population and a few preceding generations. This issue can be solved by including one metafounder per purebred population in the pedigree and by centring all genotypes with allele frequencies equal to 0.5. Metafounders can be considered as pseudo-individuals with relationships among them estimated based on genotyped descendants. In our experience, estimating these relationships may be difficult, e.g. due to a lack of genotypes in one of the populations. Another solution is to fit an additional covariable to model the mean breeding value of the genotyped animals, while centring all genotypes using the observed allele frequencies. In this study, we extended this concept to single-step evaluations for three-way crossbreeding. In this case, an additional covariable is fitted for each of the three purebred populations, and crossbred animals are related to these additional covariables through their breed compositions. We tested the two approaches using an 11-trait dataset that includes phenotypes and genotypes of three commercial pig lines and of their three-way crossbred descendants. A single-step SNPBLUP approach was used for the analyses. Genetic groups (one per line) were also fitted with the additional covariables that model mean breeding value of the genotyped animals. Accuracy and bias of the different approaches were assessed by using the average three-way crossbred offspring performance of recent boar line sires as independent variable. The results show similar accuracies and bias across the different approaches. Computational performances will be also reported.

What information does ssGBLUP need to give correct evaluations in the presence of preselection?

I. Jibrila, J. Vandenplas, J. Ten Napel, R.F. Veerkamp and M.P.L. Calus
Wageningen University and Research, Animal Breeding and Genomics, Droevendaalsesteeg 1, 6708 PB Wageningen, the Netherlands; ibrahim.jibrila@wur.nl

Our previous study has shown that in a two-stage genetic evaluation, using ssGBLUP in the second stage evaluation takes care of the impact of preselection (first stage selection) on bias of the dispersion of breeding values of selection candidates and on correctness of estimated genetic gain (measured by the difference between true and estimated realised genetic gain). As a follow-up, this study investigated what information ssGBLUP needs in the second stage evaluation to give unbiased breeding values and to correctly estimate genetic gain. We simulated a breeding programme with a single-trait breeding goal with a heritability of 0.1. The simulated population included 15 generations in which parents were selected based on BLUP, with no preselection. To select the parents of the potential generation 16, 10% of males and 15% of females in generation 15 were pre-selected based on their GEBVs, and these GEBVs were obtained using ssGBLUP with complete pedigree (generations 0 to 15), phenotypes of all the animals in the five most recent generations, and genotypes of all the animals in the three most recent generations. In the second stage evaluation, several scenarios with different amounts of pedigree, genotypic and phenotypic information were implemented. We observed that bias was completely absent and genetic gain was correctly estimated only when: (1) all the information included as reference in the preselection model was also included in the second stage evaluation model; and (2) genotypes of preselected candidates were included in the second stage evaluation model. We also observed that including the genotypes of the preculled animals in the second stage evaluation model did not affect the results. We conclude that to give unbiased breeding values and correctly estimate genetic gain, in the presence of preselection, it is sufficient to supply ssGBLUP with: (1) information representing the data of the reference population used to compute the breeding values used for preselection; and (2) the genotypes of the preselected animals, which are the main source of information that informs ssGBLUP that the preselected animals are not a random sample of the offspring from their parents.

Integration of MACE breeding values into domestic multi-trait test-day model evaluations

T.J. Pitkänen[1], M. Koivula[1], I. Strandén[1], G.P. Aamand[2] and E.A. Mäntysaari[1]
[1]Natural Resources Institute Finland, Myllytie 1, 31500 Jokioinen, Finland, [2]NAV Nordic Cattle Genetic Evaluation, Aarhus N, 8200 Aarhus N, Denmark; timo.j.pitkanen@luke.fi

The integration of information for bulls from Multiple Across Country Evaluation (MACE) into the domestic test-day model (TDM) is demonstrated by blending MACE yield indices into the Nordic TDM. The 9-trait TDM includes simultaneously test-day records of milk, protein and fat from the first three lactations. The official 305 d lactation total yield indices for milk, protein, and fat are calculated from the TDM solutions. Each yield index is a weighted average of 305 d breeding values of the three parities. Yield indices and corresponding reliabilities are submitted to Interbull which performs MACE and returns MACE yield indices and reliabilities. In this study, we included MACE information for EUROGENOMICS bulls into the Nordic TDM. A bull was considered to have additional information in MACE if its reliability for milk, protein and fat indices in MACE were at least 0.01 units higher compared to the Nordic TDM reliability. The TDM gained additional information for almost 28,000 bulls. The integration process had three steps: (1) The multitrait reversed reliability approximation was used to obtain effective record contributions (ERC) for the selected 28,000 bulls. The ERC was approximated first using reliabilities from the Nordic TDM, and then using reliabilities from MACE. (2) Yield indices and ERC were used to calculate multitrait deregressed proofs (DRPs) separately using the Nordic TDM and MACE data. To avoid double counting of information, correlations between the evaluated milk, protein and fat indices were accounted during the ERC and DRP calculations. (3) Based on the two DRPs and two ERCs, pseudo-observations were calculated for the selected 28,000 bulls. Pseudo observation approximates the external record information in MACE. Original TDM models breeding values of milk, fat and protein in three lactations by 15 random regression coefficients. Three new covariable sets were added to model which describes the three new pseudo-observations. After including MACE information, the correlations between MACE and Nordic yield indices were 0.99, 0.99, and 0.99, prior inclusion they were 0.72, 0.77, and 0.67, for milk, protein and fat, respectively.

Approximate individual animal reliabilities in single-step genomic model

H. Ben Zaabza[1], M. Taskinen[1], T. Pitkänen[1], G.P. Aamand[2], E.A. Mäntysaari[1] and I. Strandén[1]
[1]Natural Resources Institute Finland (Luke), 31600 Jokioinen, Finland, [2]NAV Nordic Cattle Genetic Evaluation, 8200 Aarhus, Denmark; hafedh.benzaabza@luke.fi

Calculation of individual animal reliabilities of genomic breeding values from single-step genomic BLUP (ssGBLUP) requires elements in the inverse of the coefficient matrix of the mixed model equations (MME). The inversion is often computationally unfeasible due to the large size of the MME. We used a simple approximation. Approximate reliabilities for the genotyped animals were computed using weighted SNP-BLUP/GBLUP (wSNP/GBLUP) where the weights were effective record contributions (ERC). The ERC for the genotyped animals were calculated using reversed reliability approximation from an animal model without genomic information. The method was tested with a dataset including about 20,000 genotyped animals from Finnish Red dairy cattle population with 136,593 pedigree animals. Correlation (MSE) between the reliabilities from the correct ssGBLUP and the wSNP/GBLUP was 0.988 (0.0003). The regression coefficient (slope) and intercept of reliabilities from ssGBLUP on wSNP/GBLUP were 0.93 and 0.04, respectively. Correlation between the reliabilities in ssGBLUP and animal model (without genomics) was 0.945. Both the slope and the intercept of the ssGBLUP on the animal model reliability indicated increased reliability due to genomic information. The method can be extended for non-genotyped animals by blending the increased genomic information in animal model. Preliminary results suggest that the wSNP/GBLUP and blending can be used to approximate ssGBLUP reliabilities for all animals even for large data sets.

MastiMIR-Mastitis detection with the help of veterinary diagnosis and MIR spectral data

L.M. Dale[1], A. Werner[1], K. Drössler[1], F. Gollé-Leidreiter[1], F.J. Auer[2], C. Egger-Danner[3], A. Köck[3], J. Sölkner[4] and L.M. Rienesl[4]
[1]State Association for Performance and Quality Inspection in Animal Breeding of Baden Wuerttemberg, Research and Development, Heinrich-Baumann-Str. 1-3, 70190 Stuttgart, Germany, [2]State Association for Performance and Quality Inspection in Animal Breeding of Austria, Certification Authority, Dresdner Straße 89/19, 1200 Wien, Austria, [3]ZuchtData, Genetics, Dresdner Straße 89/B1/18, 1200 Wien, Austria, [4]University of Natural Resources and Life Sciences Wien, Livestock Sciences, Gregor-Mendel-Straße 33, 1180 Wien, Austria; ldale@lkvbw.de

Mastitis disease appearance at farm level had not just decreased the milk production, also produced veterinary costs, welfare issues, increased culling rate or caused lower milk payment. Mastitis had been lately associated with a wide range of characteristics that can be measured in milk, and milk components could be determinate true mid infrared (MIR) spectrometry. The objective of this study was to build a spectrometric tool, such as MastiMIR and to see if it is possible to determinate true the milk quality the animal healthy status. The aim was to evaluate the diagnosis usability and MIR indicators for the improvement of early mastitis prediction at LKV Austria as was similar in LKV Baden Württemberg done. The dataset contains 635,614 spectral data from LKV Austria GMON herds. The first trial is composed from 70% of data in calibration and 30% in validation, while the second trial 70% and respectively 30% of farms in calibration and validation datasets. To identify animal variables that were positively or negatively associated with mastitis, the spectral data was first standardised, then pre-processed by first derivative and the legendre polynom model was applied for days in milk correction. For the non-healthy class, spectral data with mastitis diagnosis for a given cow within -21 to 30 days was chosen, while for healthy class no diagnosis data was registered. As fix effects were considered the sampling moment, lactation stage, breed and legendre polynomial spectra. No SCC edits were performed for this model. The MastiMIR calibration model showed 60.7% sensitivity and 71.3% specificity for the period chosen. If the days within spectral date and diagnosis is -7 to 0 the sensitivity is increased to 74.5% and for the same period if the SCC edit is done, 400,000 SCC for non-healthy data and 50,000 SCC for healthy data the sensitivity and specificity is increased to 86.3 and 77.7% respectively. The MastiMIR model provides four classes of mastitis warning such as not, moderately, significantly and severely endangered. The moderately endangered class is signal for farmers and could contact the vets and a control could be made in order to prevent the mastitis diseases.

Classification of sickness behaviour using XGBoost with sensor data and on-farm health recordings

M. Gertz[1], K. Große-Butenuth[2], W. Junge[1], B. Maassen-Francke[3] and J. Krieter[1]
[1]Institute of Animal Breeding and Husbandry, Kiel University, Olshausenstraße 40, 24098 Kiel, Germany, [2]365FarmNet Group GmbH & Co. KG, Hausvogteiplatz 10, 10117 Berlin, Germany, [3]GEA Farm Technologies GmbH, Siemensstraße 25, 59199 Bönen, Germany; mgertz@tierzucht.uni-kiel.de

The objective was to analyse the feasibility of an automated detection system for locomotor-related diseases under strictly practical conditions. For this purpose, motion sensor data and on-farm health recordings of locomotor-, respiratory- and udder-health issues were recorded of 397 cows on a commercial farm in Germany, employing two sensors at the leg and neck simultaneously. Sensor Data and health recordings were evaluated with the XGBoost algorithm to predict 'sick' and 'healthy' cows. Classification examples were constructed by dividing sensor data on basis of eight different window lengths and by applying four different features to produce the resulting segments. Results suggest that sensor information and health recordings can be well utilised to learn sickness-behaviour patterns, which enable a classification of data excerpts to 'sick' and 'healthy' by achieving 86% (± 2%) AUROC, 81% (± 2%) F-Measure as well as relatively balanced specificity (78%) and sensitivity (81%) levels when using all variables and features available. Results further indicate that sensible feature selection can reduce computing time greatly by only minor losses in classification performance. XGBoost classification seems to be a powerful, easy-to-use and efficient method for identification of sickness behaviour under practical conditions. Nevertheless, the results also indicate that more research is necessary into the choice of features, the ideal window lenght(s) or segmentation strategy, interaction of feature variants and variable preselection. Over-all, the applied approach indicates a substantial potential for the development of an automated detection tool in the future, capable of an on-going evaluation of locomotor-associated diseases.

Early disease detection based on feeding behaviour traits using machine learning methods in pigs

A.T. Kavlak[1], M. Pastell[2] and P. Uimari[1]
[1]University of Helsinki, Agricultural Sciences, Koetilantie 5, 00014 Helsinki, Finland, [2]Natural Resources Institute Finland (Luke), Latokartanonkaari 9, 00790 Helsinki, Finland; alper.kavlak@helsinki.fi

Early detection of health and welfare compromises is essential for timely intervention to increase treatment success, reduce the impact on welfare, and encourage sustainable pig production. The objective of this study is to develop a method to monitor changes in behaviour patterns that can be used for detection of pigs that may need medical treatment or extra management, thus avoiding possible situations where diseases occur as untreatable. In this study, the raw data, provided by the Figen Oy, included feeding behaviour and disease information. The data included records of 10,336 pigs from Finnish Yorkshire and Landrace and their F1 crosses. Feeding behaviour traits were measured as a number of visits per day, time spent in feeding per day, daily feed intake, time spent in feeding per visit, feed intake per visit, and feed intake rate. Health information of the pigs was based on daily monitoring of pigs by the test station staff. The combined data included 869,689 observations (healthy: 537,697; cough: 663; limp: 5,116; loss of appetite: 815; skin damage: 749; tail bitten: 2,112). A machine-learning algorithm was applied to predict the individual health state of each animal based on the features (slope, variance, skewness, kurtosis of each feeding behaviour trait) calculated from the feeding data. The features were based on a 10-day sliding window from the beginning of the test period until the end of the test period (95±3 days). A software package eXtreme Gradient Boosting (Xgboost) was used for building the model and making predictions. The data were split into learning (70%) and testing (30%) sets. The model achieved balanced accuracy ranging from 60 to 76% between classes. The initial analyses indicated that the variance and slope of the traits have higher importance score as features than skewness and kurtosis. The positive predicted values were quite low for the sick diagnosis of animals. The sensitivity was predicted as moderate (e.g. healthy and loss of appetite (57%), skin damage (41%)) whereas specificity was high (e.g. cough (95%), loss of appetite (96%), limp (90%)) between the classes. More features will be extracted, and different machine learning methods will be implemented in this data to improve the accuracy of the prediction.

Automated sickness detection in dairy cattle using multivariate cumulative sum control charts

I. Dittrich[1], M. Gertz[1], B. Maassen-Francke[2], K. Große-Butenuth[3], W. Junge[1] and J. Krieter[1]
[1]Christian-Albrechts-University, Institute of Animal Breeding and Husbandry, Olshausenstr. 40, 24098 Kiel, Germany, [2]GEA Farm Technologies GmbH, Siemensstr. 25, 59199 Bönen, Germany, [3]365FarmNet Group GmbH & Co. KG, Hausvogteiplatz 10, 10117 Berlin, Germany; idittrich@tierzucht.uni-kiel.de

Sensor systems and related detection-algorithms provide potential aid within the field of early sickness detection, whereby various useful variables are provided, e.g. physical activity or rumination. Characteristics like autocorrelations appear and are likely to inhibit algorithm accuracy. Hence, the aim of this study was the selection of variables by using the principal component loadings (PCLoad) from principal component analysis and the variable importance in projection (VIP) from partial least squares. Furthermore, the selected variables were employed in multivariate cumulative sum control charts (MC), enabling an online monitoring of cows at risk for disease. Data was collected between September 2018 and April 2019 of 480 cows wearing two sensor systems (Neck, Leg). After the selection of 154 healthy (no sickness events) and 144 sick cows (300 sickness events), two data sets (NM, LM) were generated combining both sensors with milk yield and conductivity (M). Variables showing a high PCLoad ($\geq|0.7|$) and those that achieved a VIP of ≥0.8 were chosen for the use in MC control charts. From the NM data set, activity and rumination were selected by the PCLoad, while from data set LM activity and resting variables were selected. The VIP selected both performance parameters and resting variables from the LM data set and conductivity and feeding information from the NM data set. The MC of the PCLoad detected 76% (LM) and 72% (NM) of the sickness events, respectively, with 20 false positive cows per day in both data sets. The VIP based variable selection showed 71% detected sickness events in both data sets. Hence, 20 (LM) or rather 19 (NM) cows a days were counted as false positive. The area under the ROC-curve was determined 0.9 for all calculated MC control charts. Combining the selected variables (PCLoad, VIP) with MC control charts shows a good and consistent sickness detection rate, however the generated false positives could increase farmers' workload and therefore inhibit the overall performance of an algorithm like this.

Indicating resilience or reflecting disease? Relating milk fluctuations and mastitis in dairy cows

A. Kok[1], G. Tsousis[2], G. Niozas[3], M. Kaske[4], B. Kemp[1] and A. Van Knegsel[1]
[1]Wageningen University & Research, Department of Animal Sciences, De Elst 1, 6708WD Wageningen, the Netherlands, [2]Clinic of Farm Animals, Aristotle University of Thessaloniki, Thessaloniki, 54627 Thessaloniki, Greece, [3]Clinic for Cattle, University for Veterinary Medicine, Bischofsholer Damm 15, 30173 Hannover, Germany, [4]Department for Farm Animals, Vetsuisse Faculty, University of Zurich, Winterthurerstrasse 260, 8057 Zurich, Switzerland; akke.kok@wur.nl

Dairy cows are exposed to pathogens, management transitions and other challenges throughout the lactation cycle. Resilience is the capacity to be minimally affected by such disturbances. Ultimate resilience would be absence of disease and easy transitions, whereas lack of resilience can result in an increase in incidence or severity of disease. Fluctuations in milk yield throughout lactation have been used as indicator of resilience and were associated with breeding values for disease. However, it is unclear whether fluctuations in milk yield indicate resilience in the absence of disease, or whether these fluctuations reflect the reduction in milk yield upon disease. In this study, we related fluctuations in milk yield in early lactation to the occurrence of mastitis later in lactation. The dataset included daily milk yield of 414 cows, of which 249 had no mastitis in the 305 days lactation (NO); 29 had mastitis in the first 30 days in milk (EARLY); and 136 had mastitis later in lactation (LATE). Individual lactation curves were fit using quantile regression. Fluctuations were defined as deviations of realised daily milk yields from the fitted lactation curve; and fluctuations in early lactation were summarised as natural logarithm of the variance between 10 and 30 days in milk (lnVAR). A linear model was used to assess the difference in lnVAR between cows with no, early and late mastitis. Milk yield level was included as a covariate, as it was strongly correlated with lnVAR. Early mastitis was associated with a higher lnVAR in early lactation, i.e. greater fluctuations in milk yield, than no or late mastitis (lnVAR of 2.9, 2.0 and 2.1 for EARLY, NO and LATE, respectively). No difference was detected in lnVAR in early lactation between cows that remained healthy versus cows with mastitis later in lactation. However, mastitis was preceded by a drop in milk yield 2 days in advance. In this study, fluctuations in milk yield reflected disease, but did not indicate resilience in healthy cows.

Classification of cow behaviour with deep neural networks using accelerometer and tracking data

L.P.J.J. Noldus[1,2], S. Van Bruggen[2], H. Heuer[2], E.A. Van Dam[2,3], B.J. Loke[2], A. Van Gijssel[2] and E.K. Visser[4]
[1]Radboud University, Department of Biophysics, Donders Institute for Brain, Cognition and Behavior, Heyendaalseweg 135, 6525 AJ Nijmegen, the Netherlands, [2]Noldus Information Technology BV, Nieuwe Kanaal 5, 6709 PA Wageningen, the Netherlands, [3]Radboud University, Department of Artificial Intelligence, Donders Institute for Brain, Cognition and Behavior, Montessorilaan 3, 6525 HR Nijmegen, the Netherlands, [4]Aeres University of Applied Sciences, De Drieslag 4, 8251 JZ Dronten, the Netherlands; lucas.noldus@noldus.nl

Monitoring individual behaviour plays an important role in assessing animal health and welfare, and is becoming increasingly consequential with the trend towards larger group housing systems for cattle. Automatic behaviour recognition with machine learning can aid this monitoring in a less laborious and more consistent manner than manual scoring by human observers, especially for large-scale data collection. However, machine learning approaches for behaviour classification traditionally rely heavily on extensive manual feature engineering and expert knowledge. Deep learning approaches have shown to improve upon these models by outperforming them while reducing the need for handcrafted features. We present a deep learning model for cattle behaviour classification. The neural network was trained on a small dataset consisting of 3D accelerometer and ultra-wideband tracking data from a neck-mounted sensor on two cows for 9 hours each, recorded in a cow barn. This data was obtained using TrackLab, a software package for data acquisition of group-housed animals, and manually annotated. Our model is able to accurately classify feeding behaviour (with 95% accuracy for classifying eating and ruminating) and shows promising results for locomotion (with 87 and 81% accuracy for classifying lying and standing, respectively), which is an improvement over earlier efforts to classify behaviour using accelerometer data only. Due to the size of the dataset, behaviours that occur less often, such as walking, still prove a challenge for the precision of the model. For future work, we plan to extend our approach to different and larger datasets. We hope that applications of automatic behaviour classification can contribute to further developments in livestock research and precision livestock farming.

Automated recognition of postures and drinking behaviour for the detection of compromised health pig

A. Alameer[1,2], I. Kyriazakis[3] and J. Bacardit[2]
[1]Newcastle University, School of Natural and Environmental Sciences, Newcastle upon Tyne, NE1 7RU, United Kingdom, [2]Newcastle University, School of Computing, Newcastle upon Tyne, NE4 5TG, United Kingdom, [3]Queen's University Belfast, Institute for Global Food Security, Belfast, BT9 5DL, United Kingdom; ali.alameer@newcastle.ac.uk

Changes in pig behaviours may be used to detect early signs of problems, such as in animal health. Automated vision-based early warning systems have been developed to detect behavioural changes in groups of pigs to monitor their health and welfare status. In commercial settings, automatic detection of pig postures and drinking behaviour remains a challenge. Here, we developed a system that automatically identifies pig postures (standing, sitting, lying lateral and lying sternal) and drinking behaviour. Pigs were monitored by top view RGB cameras that covered a large area of the pen including the drinking area, and animal behaviours were detected using deep learning-based methods. Our first objective was to demonstrate the ability of this automated method to identify behaviours of individual animals with high precision. We then tested the system ability to detect changes in group-level behaviours due to a food restriction protocol. Two deep learning-based detector methods, including faster regions with convolutional neural network features (Faster R-CNN) and you only look once (YOLO) combined with Residual Network (ResNet-50), were developed to precisely identify pig postures and drinking behaviours of group-housed pigs. We evaluated our method using routine data recorded at a commercial pig farming environment. Our experiments show that our system could recognise the postures and drinking behaviour of individual pigs with a mean average precision (mAP) of 0.9888 ± 0.0094. When the pig feeding regime was disrupted, we observed significant deviations from the daily *ad-libitum* routine in the standing, lateral lying and drinking behaviours. These experiments demonstrate this method is capable of robustly and accurately monitoring pig behaviours under commercial conditions without the need for additional sensors or individual markings.

Millimetre-wave radars for the automatic recording of sow postural activity

A. Dore[1], M. Lihoreau[2], Y. Billon[3], L. Ravon[3], J. Bailly[3], J.F. Bompa[4], E. Ricard[4], H. Aubert[1], D. Henry[1] and L. Canario[4]

[1]LAAS-CNRS, MINC, 7, avenue du Colonel Roche, Toulouse, 31000, France, [2]Research Center on Animal Cognition CNRS, 118, route de Narbonne Toulouse, 31000, France, [3]INRAE, UE GenESI, fief de magnereau, 17700, France, [4]INRAE, GenphySE, Chemin de Borde Rouge, Castanet-Tolosan, 31320, France; alexandre.dore@laas.fr

The global objective is to analyse the variations in the pattern of activity in a Large White population and estimate associations with piglet survival. The aim of the study is to evaluate the efficiency of millimetre-wave radars based on frequency modulation to monitor the postural activity of lactating sows kept in a crate. The monitoring is performed with two radars operating at a frequency of 122 GHz positioned on a metal structure 35 cm above the crate at two strategic locations: above the head and above the sow's back. The acquisition speed was 20 obs/sec. For validation of the positions predicted from radar sensor data, sow positions were recorded with a digital camera and video records were analysed by observers trained in the same way. Preliminary results were obtained on two sows analysed over a period of 24 h and two other sows over a period of 12 h and using information collected each second. First trials focused on the ability of radars to distinguish three positions: standing (ST), sitting (SI) and lying (LY). The prediction ability was tested with machine learning applied to random forests and deep learning. The model used is derived from VGG16, which is a state of art model to analyse images. Analyses were carried out individually for each sow, using 40% of the beginning of the data base as training data set and the remaining 60% as validation data set. On the validation set, the global statistical sensitivity was 93% (94% for ST and between 80 and 93% for both SI and LY). The 3 positions were also detected by random forest with similar sensitivity but the data coming from the radar have to be preprocessed before. More sows will be monitored for validating the use of millimetre-wave radars as an accurate and non-invasive innovative technology for the study of the sow pattern of activity. The analysis of more sows is in progress and additional behavioural traits of interest might be identified from radar signals.

Relationship between reproductive cycle and body temperature in the mare

J. Auclair-Ronzaud[1], K. Varona[2], C. Julé[3], T. Jousset[1,4] and L. Wimel[1]

[1]The French Horse and Riding Institute (IFCE), Experimentation research center of Chamberet, 1 Impasse des Haras, 19370 Chamberet, France, [2]Montpellier SupAgro, 2 Place Pierre Viala, 34060 Montpellier, France, [3]AgroParisTech, 16 rue Claure Bernard, 75231 Paris, France, [4]Current address, La maison blanche, 14100 Saint-Germain de Livet, France; juliette.auclair-ronzaud@ifce.fr

Body temperature appears to be a good indicator of oestrus and/or ovulation detection in several farm species such as beef, dairy cows and sheep. Thus, tools using this information to send on-time alerts to the farmer are available on the market, allowing a better management of herd reproduction. In horses, few studies have been conducted on the subject. Especially due to the lack of device allowing automated and continuous recording of the body temperature. Indeed, for cows vaginal or ruminal devices have been developed and can be used routinely. In horses, however, these tools are not available. A preliminary study has been conducted on 17 mares. Mares were kept in group in pasture, under natural light, with grass and water available *ad libitum*. Body temperature was measured using a mandatory identification microchip inserted in mares' neckline. Temperature was automatically recorded every 10 minutes, even though the reader device was still under development and points are missing for each studied individual. Phase of the reproductive cycle was determined by ultrasound once a week and hCG (human chorionic gonadotropin) was administrated to induce ovulation at a dose of 750 or 1,500 UI. Three phases were distinguished: luteal phase, ovulation and follicular phase. Temperature variations were studied using a mixed linear model with a quadratic function to take into account temperature 24 hours variations. Circadian rhythm has been observed with a nadir of 36.7±0.02 °C at 7 am and a peak of 37.6±0.02 °C around 5 pm. In addition, mean decrease of 0.15±0.01 °C is detected during ovulation. This preliminary study shows variations of body temperature during the reproductive cycle of the mare using a non-invasive technic to measure temperature. Further work has to be done to be more precise regarding body temperature variations and to relate it to hormonal fluctuations.

Assessment of two shearing practices on animal welfare in alpaca using infrared thermography

A. Cruz[1], S. Quispe-Potosino[1], I. Cervantes[2], A. Burgos[1] and J.P. Gutiérrez[2]
[1]Fundo Pacomarca, INCATOPS S.A., Miguel Forga 348, Arequipa, Arequipa, Peru, [2]Universidad Complutense de Madrid, Facultad de Veterinaria, Avda Puerta de hierro s/n, 28035, Spain; gutgar@ucm.es

Shearing affects the alpacas` welfare and causes stress which can lead to an increased abortion rate. Traditional shearing (T) is usually carried out with scissors, while a shearing machine is used in the mechanised (M), being much faster and supposedly less stressful. Animal stress has been reported to be indirectly assessed by increased cortisol levels, decreased blood glucose levels, increased heart rate and increased rectal temperature, but invasive manipulations also influencing the animal welfare are needed to obtain these values. Therefore, this study aims to compare the stress produced by both shearing methods using the eye temperature as an indicator measuring it before (B), during (D) and after (A) shearing, by using infrared thermography as a non-invasive method. Temperatures were measured from one meter distance from the lacrimal caruncle, with a Flir E6xt infrared thermographic camera, adjusting for the environment temperature and emissivity to the value of 0.98. A total of 212 alpacas from two herds (180 and 32) in southwestern Peru were registered for at least one of the three temperatures in each animal. Shearing time was also registered. A total number of records of 65, 185 and 16 temperatures for B, D and A the shearing, and records of 38, 11 and 15 temperature differences D-B, A-B and A-D were obtained, respectively. The six variables were analysed using a multivariate generalised linear model including herd and method of shearing as effects. All the temperatures were also jointly analysed fitting also the stage (B, D or A). Mean values were 35.77 °C, 37.13 °C, 36.16 °C for B, D, and A temperatures and 1.18, 0.55 and -0.45 for D-B, A-B and A-D differences respectively, showing significant differences in temperatures between D and the other stages of the shearing (p <0.05). The differences D-B, A-B and A-D were respectively for T and M, 1.50 and 1.13, 0.74 and 0.38, and -0.76 and -0.30. There seemed to be relevant differences between shearing methods being M less stressful, but they were not significant and more records are needed to confirm this hypothesis. In addition, mean shearing time was respectively 1,065 and 230 seconds for T and M, showing longer exposure to stress in the first one.

Which feeding behaviours of dairy calves are associated with BRD bouts?

M.C. Cantor, M.M. Woodrum Sester and J.H.C. Costa
University of Kentucky, Dairy Science Program, 404 W.P. Garrigus Building, 325 Cooper Drive, Lexington, KY, 40546, USA; costa@uky.edu

Bovine Respiratory Disease Complex (BRD) affects calf welfare and productivity. The aim of this study was to investigate if feeding behaviours changed during the 7 d proceeding clinical BRD diagnosis in preweaned calves. A cohort of calves were health scored daily, lung ultrasound scored and weighed twice weekly from birth until 2 weeks post-weaning (90 d). Two automatic feeders (milk and starter) recorded daily intake, drinking speed, rewarded and unrewarded visits. Milk allowance was 10 l/d of milk replacer for 50 d. Clinical BRD was defined as UW-calf (>4) and positive lobar (>2.99 cm^2) consolidation on ultrasound. Only BRD prior to weaning was included: 50 BRD calves were paired-to-healthy calves by season, age, and weight. The effect of BRD on feeding behaviour for 7 d proceeding BRD was assessed in a linear model, with health status (BRD yes/no) as a fixed effect, grouped by pair, and time was a repeated measure with calf as the subject. Calves were on average, 32.0±12.5 d of age at clinical BRD, and weighed 55.7±10.0 kg. There was a significant day by health status interaction for milk intake, unrewarded visits and starter intake. Milk intake was lower for BRD calves on d -7, -6, -5, -4, and on d -3 (P=0.01), but not on d -2 and d -1. Milk intake tended differ d 0 (healthy 9.3±0.3 vs BRD 8.6±0.3 l/d; P=0.06). Drinking speed and rewarded visits were not different. Yet, unrewarded visits tended to be lower for BRD on d -6 and d -5; (P=0.05), and were lower on d -4, -3, -2, and -1 (-2.2±0.8 visits; P=0.01). Unrewarded visits were also different on d 0 and higher for healthy calves (healthy 4.1±0.6 vs BRD 2.4±0.6 visits; P=0.03). Starter intake was lower for BRD calves on d -7, -6, -5, -4, -3, and -2 (P=0.001), but not on d -1. Starter intake differed on d 0 (healthy 0.32±0.04 vs BRD 0.14±0.04 kg/d; P=0.001). Results suggest that milk and starter intake are associated with clinical BRD as early as -7 d, but feeding behaviour the day prior to clinical BRD is not different from the day of diagnosis. Unrewarded visits remained lower than healthy calves closer to the day of BRD diagnosis (-4 d to d 0). Results suggest feeding behaviours are associated with clinical BRD days before diagnosis and may be used in a predictive model.

Comparison of methods for short-term milk production forecasting in dairy cows

I. Adriaens[1,2], B. Aernouts[1] and M. Pastell[3]
[1]KU Leuven, Department of Biosystems, Kleinhoefstraat 4, 2440 Geel, Belgium, [2]KU Leuven, Department of Biosystems, Kasteelpark 30, box 2456, 3001 Heverlee (Leuven), Belgium, [3]Natural Resources Institute Finland (Luke), Latokartanonkaari 9, 00790 Helsinki, Finland; ines.adriaens@kuleuven.be

Diseases and metabolic problems are often reflected in the cow's production dynamics as milk yield perturbations. Detection and characterisation of these perturbations in a cows' milk production in real time can help the farmer to (1) identify cows at risk that require specific management actions (detection), (2) follow-up on the recovery and cure to evaluate treatments and (3) characterise the milk losses caused by a specific event in order to chart impact of perturbations both at individual and herd level. A common method for revealing perturbations in milk yield is to use a model that does short-term forecasts of expected yield, after which the changes in the forecast residuals can be monitored. Several methods for solving the forecasting problem have been proposed, however model performance is rarely compared on standardised datasets. The aim of this research is to compare two recently developed methods for online, short-term prediction of milk yield using a lactation curve model on a common dataset. The methods included in the study are: (1) A nonlinear state space model with Wilmink function as lactation curve model fitted using SIR particle filter; (2) An online updated, linearised quarter-level mixed model with Wood's function as lactation curve model using Bayes estimates for the short-term predictions. The prediction performance of both models is compared on approximately 250 unperturbed lactation curves to show the differences in performance of predicting the reference milk yield when no perturbations are present. Furthermore, we (1) introduce the estimation method for both models; (2) discuss the advantages and disadvantages of both methods from computational and application-specific point of view.

Kinetics of calcium and phosphorus absorption and utilisation in pigs and poultry

P. Bikker[1], Y. Hu[1,2], J. Van Baal[2], A. Mens[1], J.W. Resink[3], M. Rodehutscord[4] and M.M. Van Krimpen[1]
[1]Wageningen University & Research, Wageningen Livestock Research, P.O. Box 338, 6700 AH Wageningen, the Netherlands, [2]Wageningen University & Research, Animal Nutrition Group, P.O. Box 338, 6700 AH Wageningen, the Netherlands, [3]Trouw Nutrition, Research and Development, Stationstraat 77, 3811 MH, Amersfoort, the Netherlands, [4]University of Hohenheim, Institute of Animal Science, Emil-Wolff-Str. 6-10, 70599 Stuttgart, Germany; paul.bikker@wur.nl

Calcium and phosphorus are essential nutrients for bone development, cell structure and numerous biochemical processes, in pigs and poultry. These minerals need to be absorbed in an adequate ratio for optimal post-absorptive utilisation. It has been demonstrated in many studies that a high dietary calcium content may reduce the absorption of phosphorus and our data suggest that phosphorus supply may influence calcium absorption as well. It is a matter of debate to what extent interactions with phytate and phytase play a role. This presentation aims to further elucidate calcium to phosphorus interactions in different segments of the digestive tract, the role of microbial phytase, the influence of limestone particle size, and the regulation via active and passive absorption mechanisms in pigs and broilers. Several studies in pigs and broilers were conducted in our lab, with diets varying in calcium, phosphorus and phytase concentration and differences in limestone particle size. Precaecal (broilers) or total tract (pigs) digestibility was determined, animals were sacrificed to collect digesta and mucosa from different segments of the digestive tract and plasma concentration, nutrient balance and bone characteristics were determined as markers of post-absorptive mineral utilisation. Briefly, phosphorus was largely passively absorbed for the proximal and distal small intestine and absorption was enhanced in the proximal small intestine by use of microbial phytase. Calcium was absorbed from the proximal small intestine and the large intestine in pigs and seemed to be more actively regulated by up- and down regulation of transporter gene expression. The impact of calcium on phosphorus absorption was mediated by phosphorus solubility and ileal phytate degradation, with a bigger effect in phytase supplemented diets. Unlike earlier studies, the results indicate that limestone particle size plays a significant role in pigs and poultry, not only with respect to digestibility but also on post absorptive utilisation. The potential role of synchronisation of calcium and phosphorus will be discussed.

Assessing the responses of growing pigs to deficient dietary phosphorous

M.M. Misiura[1], J.A.N. Filipe[1,2], C.L. Walk[3] and I. Kyriazakis[4]
[1]Newcastle University, Agriculture Building, NE1 7RU, Newcastle upon Tyne, United Kingdom, [2]University of Aberdeen,
Ashgrove Rd W, AB25 2ZD, Aberdeen, United Kingdom, [3]AB Vista, Woodstock Ct, SN8 4AN, Marlborough, United Kingdom,
[4]Queen's University, Biological Sciences Building, BT9 5DL, Belfast, United Kingdom; m.m.misiura@ncl.ac.uk

While phosphorus (P) has generated plenty of research interest in pig nutrition due to its economic and environmental importance, the following two central issues warrant further investigation: (1) how intake is regulated on foods containing different P levels; (2) how limited nutrients are partitioned within the body. We addressed these knowledge gaps in the context of growing-finishing pigs by conducing meta-analyses of published literature data and by developing a predictive, mechanistic model of P utilisation and retention incorporating this evidence. Our statistical analyses of the literature data suggest: (1) there is no change in FI response in pigs given lower P feed contents; (2) the ash-protein allometric relationship in the body is influenced by feed composition, with an isometric relationship only holding for pigs given access to balanced feeds; (3) priority is given towards P retention in soft tissue over P retention in bones. Subsequent results of the developed mechanistic model of P retention incorporating the above data-based evidence, suggest that a potential reduction in P feeding recommendations could be possible without compromising average daily gain; however, such a reduction would impact P deposition in bones. The outcomes are expected to enhance our understanding of P nutrition on pig growth and body composition, and could contribute towards issuing more accurate P feeding guidelines in the future that reduce environmental impact of pig production.

Liquid and soaked feed improve Ca and P digestibility of fattening pigs

R. Lizardo, B. Villca and D. Torrallardona
Irta, Animal Nutrition, Ctra REUS-Morell, km 3.8, 43120, Spain; rosil.lizardo@irta.es

Liquid feed is extensively used in most of the EU countries to raise fattening pigs. However, few research was carried out to compare this technique with to the traditional pelleted feed usage. The aim of this study was to evaluate nutrient digestibility of pigs raised on dry, fresh and soaked liquid diets. Data from two trials, with 144 male and female Pi×(LR×LW) pigs in each, were used for this purpose. Pigs were housed by sex in 36 pens and received a 2-phase feeding regimens from 25 kg to slaughter weight. On the 1st trial, 3 treatments corresponding to a pelleted diet distributed *ad libitum* on dry (T1) or liquid (T2) form, or as liquid after 10 h soaking (T3) were used. On the 2nd trial, 3 other treatments corresponding to a diet without (T1) or with phytase inclusion (T2) distributed both in liquid form or after 10 h soaking (T3) were used. A ratio of feed to water of 1:2.5 l was used for all liquid diets, and they were distributed after 5 minutes mixing twice a day. Fresh faeces samples were collected from pens by 60 kg BW, and before slaughter for nutrient analysis at the lab. Major nutrient digestibility was not affected (NS) whereas mineral digestibility was significantly improved in both trials. Phosphorus digestibility improved with liquid (P<0.001) and soaked (P<0.001) feeds, as well as with phytase inclusion into the diet (P<0.01). Calcium digestibility also improved (P<0.01) with liquid feeds after soaking. In conclusion, liquid and soaking diets increase mineral utilisation, and may be beneficial to improve health and growth performance of fattening pigs.

Kinetics of bone mineralisation in replacement gilts determined by a non-invasive method

P. Schlegel[1], P. Floradin[1,2], C. Pomar[2] and M.P. Létourneau-Montminy[2]
[1]Agroscope, Tioleyre 4, Posieux, Switzerland, [2]Laval University, Rue de l'Agriculture 2425, Quebec, Canada; patrick.
schlegel@agroscope.admin.ch

To limit the use of mineral phosphates, there is an increased interest to feed phosphorus levels allowing maximised growth, without fulfilling the higher requirement for maximised bone mineralisation to growing pigs. In some contexts of production, replacement gilts are raised with fattening pigs until 80-100 kg BW. Considering that bone mineralisation of replacement gilts should be maximised to expect a high lifetime breeding performance, the application of such a feeding strategy requires a better knowledge about the animal's capacity to mobilise and store bone mineral reserves of calcium and phosphorus. Temporal kinetics for bone mineralisation require non-invasive methods, such as dual energy X-ray absorptiometry to allow repeated measures on the same animal. This presentation aims: (1) to verify the accuracy of dual energy X-ray absorptiometry to replace classical invasive measurements for calcium and phosphorus status, such as chemical and mechanical bone traits using data from our piglet and growing pig experiments; (2) to present new data about the modulation of bone mineral content in replacement gilts between 60 and 140 kg body weight in response to a depletion-repletion period of dietary calcium and phosphorus. The response of regions of interests reflecting different bone types such as spongious vertebrae, longitudinal bones from the foot or the head will also be highlighted. Briefly: (1) the accuracy of DXA to estimate chemical and mechanical traits was higher in data from dissected bones than from non-dissected feet; (2) replacement gilts were able to recover their bone mineralisation by 140 kg body weight after having received a diet low in calcium and phosphorus between 60 and 100 kg BW and the depleted diet did not result in any detrimental effects on growth performance. Dietary strategies to limit the use of mineral phosphates in growing-finishing pigs thus seem to be compatible with raising replacement gilts thanks to their high homeostatic regulation capacity.

Effect of dietary phytase inclusion on phosphorus digestibility and bone strength of pigs fed diets

R. Lizardo[1], B. Villca[1], P. Wilcock[2] and G. Cordero[2]
[1]IRTA, Animal Nutrition, Reus-El Morell road, km 3.8, 43120 Constantí, Spain, [2]AB Vista, Technical, Marlborough, SN8
4AN Wiltshire, United Kingdom; rosil.lizardo@irta.cat

An experiment was conducted to determine the effect of phytase supplementation on mineral digestibility and bone strength in growing pigs. One hundred eighty male and female Pietrain×(Landrace×Large White) pigs of 20 kg liveweight were used. They were distributed into 9 blocks according to body weight and allocated at 4/pen in 45 pens for a 42-d trial. Intra-block pigs were randomly distributed to the 5 treatments corresponding to positive (PC: 0.35% digestible P) and negative (NC: 0.11% digestible P) control diets or the NC supplemented with 300, 600 or 1,200 FTU/kg of phytase (Quantum Blue, AB Vista), respectively. After 6 weeks of trial, fresh faeces were collected from pens for determination of total tract digestibility. One pig per pen was euthanised at the end and left metatarsal bones collected for bone breaking strength and stiffness determinations. Regardless the diet or the dose of phytase used no statistical differences were observed for dry and organic matter, and energy digestibility. On the contrary, ash, P and Ca digestibility were significantly improved (P<0.01) due to phytase inclusion and the dose used. Treatments containing 600 FTU/kg and 1,200 FTU/kg showed better P and Ca digestibility (P<0.01) than the PC. Metatarsal bones from pigs fed the NC diet were lighter (P<0.01) than those from the other treatments in full agreement with body weight measurements (P<0.01). Bone breaking force and stiffness of metatarsals from pigs fed the PC and 1,200 FTU/kg of phytase were higher (P<0.001) than those from 300 and 600 FTU/kg, and they were all higher than the NC treatment (P<0.001). It can be concluded that inclusion of diets with phytase improved phytic-P utilisation and bone strength of pigs fed diets without inorganic P supplementation, and that the response is dose dependant.

Does mineral supplementation reduce the negative effects of heat stress on pig's performance?

M. Rodríguez[1], A. Monteiro[2], G. Montalvo[1], A. Piñon[2] and J. Morales[1]
[1]PigCHAMP Pro Europa, Dámaso Alonso 14, 40006, Spain, [2]Animine, 10 Rue Léon Rey Grange, 74960, France; amonteiro@animine.eu

Heat stress (HS) is one of main reasons of the pork industry losses via its effects on intestinal integrity and consequent reduced growth performance. Zinc (Zn) and copper (Cu) improve gut health. They can be supplemented from oxide (Oxd) or sulphate (Sph) sources. Oxd may present lower interaction with other nutrients in the gut; thus be more bioavailable to exert their functions than Sph. Therefore, the aim of the present study (Project E! 11780) was to determine the effect of Sph and Oxd ($ZnSO_4$ + $CuSO_4$ vs Animine, ZnO + Cu_2O) supplemented at 30 and 15 mg/kg feed, respectively, on performance of growing pigs during hot summer time. A total of 120 Topigs × Pietrain entire males and females (31.9±2.51 kg) were randomly allotted to 24 pens (2 treatments; 12 replicates/treatment; 5 pigs/pen) in three different fattening rooms (8 pens/room). The trial lasted 50 days, then performance was measured. The MIXED procedure of SAS software was used including 'Room' and 'Sex' as random effects and initial body weight (BW) as co-variable. Moreover, a correlation analysis was performed to quantify the relationship between performance and the room temperature (T). Oxd source numerically improved average daily gain (ADG; 0.78 vs 0.76 kg/d) and final BW (63.8 vs 63.2 kg) compared with Sph source, but differences were not significant (P>0.05). However, a trend was observed for feed conversion ratio (FCR; 2.06 vs 2.14 kg/kg for Oxd and Sph, respectively; P=0.09). Mean values of T during the experimental period was different between rooms (range 26.0-27.5 °C). Correlation analyses showed that when the mean of daily maximum T in the room exceeded 33.5 °C, all performance parameters were affected. In the case of ADG and final BW, pigs from both experimental groups were affected but with different correlation (r=-0.42 vs r=-0.99 for Oxd and Sph, respectively). It means that T>33.5 °C affected more Sph than Oxd group, associated with the differences observed in performance (+2.6% in ADG, +0.9% in final BW and -2.8% in FCR). In conclusion, supplementing Zn and Cu as Oxd may improve ADG, final BW and FCR of growing pigs under HS.

Effect of different sources of zinc on productive performance and ileal microbiota in broilers

A. Forouzandeh[1], L. Blavi[1], M. Rodríguez[2], A. Monteiro[3], J.F. Pérez[1] and D. Solà-Oriol[1]
[1]Universitat Autònoma de Barcelona, Animal and Food Sciences, Bellaterra, 08193 Barcelona, Spain, [2]PigCHAMP Pro Europa, 40006, Segovia, Spain, [3]Animine, 74960, Annecy, France; asal.forouzandeh@gmail.com

An experiment (Project E! 11780) was performed to determine the effects of two sources of Zn as zinc sulphate ($ZnSO_4$) and potentiated zinc oxide (pZnO; HiZox®, Animine, France) at two levels of inclusion (40 and 80 mg/kg) on growth performance and ileal microbiota. A total of 450 one-day-old male chickens (Ross 308) were randomly allocated to 5 dietary treatments: a basal diet without Zn addition (NC) and the basal diet supplemented with 40 or 80 mg Zn/kg as $ZnSO_4$ or pZnO (9 pens/ treatment, 10 chicks/ pen). Broilers were challenged by reusing an old litter with high concentrations in *Clostridium* and *Coccidia* to promote necrotic enteritis. Individual body weight (BW) and average daily feed intake (ADFI) was recorded at d 21, 35 and 42. At d 43, one broiler per pen was euthanised to obtain ileum content for microbial characterisation. Performance data were analysed with ANOVA using the GLM procedure of SAS. Alpha and Beta diversity were analysed using Vegan package and taxa differences with MetagenomeSeq in RStudio v.3.5.1. Birds fed pZnO at 80 mg/kg had higher (P<0.05) BW at d 42 (2.50±0.19 kg) compared with the other treatments (2.33±0.17 kg for $ZnSO_4$ and 2.37±0.13 kg for NC). ADFI was significantly increased (P=0.003) in broilers fed pZnO at 80 mg/kg, however, no differences were observed for FCR (overall of 1.68±0.08) on d 42. The effects of Zn supplementation at 80 mg/kg for both sources on the microbial characterisation of ileum content were determined by using 16S rRNA gene sequence and compared to NC. Supplementation of pZnO increased (P<0.05) α diversity at Genus level compared with the NC and the $ZnSO_4$ group (Shannon index = 0.858, 0.238, and 0.639, respectively; and Simpson index = 0.353, 0.082, and 0.296, respectively). There were no differences in β diversity among the treatments. It is concluded that supplementation of 80 mg/kg of Zn as potentiated pZnO increased microbial diversity of ileum content in broilers challenged with *Coccidia* and *Clostridium* which might explain the positive impact of HiZox on growth performance of broilers.

Riboflavin requirements in organic poultry: graded supplementation to layers and parent hens

F. Leiber[1], Z. Amsler[1], M. Leubin[1], C. Baki[1], R. Eppenstein[1], C. Lambertz[2], V. Maurer[1] and H. Ayrle[1]
[1]Research Institute of Organic Agriculture (FiBL), Ackerstrasse 113, 5070 Frick, Switzerland, [2]Research Institute of Organic Agriculture (FiBL), Kasseler Straße 1a, 60486 Frankfurt, Germany; florian.leiber@fibl.org

Added riboflavin (vitamin B2) in organic animal feeding needs to originate from GMO-free fermentation. This makes the process much more expensive compared to GMO-based riboflavin. Therefore, the level of supplementation matters from an economic point of view. Requirement definitions for B-vitamins are generally based on older empirical data and have never been defined for specific organic conditions. Two experiments (LAYERS and PARENTS) were conducted to test graded dosages of riboflavin in feeds for hens, in order to define safe lower thresholds of supplementation. The experiment LAYERS included 135 Lohmann Brown Classic laying hens. They were allocated to nine groups and supplemented with either 1.5, 3.0, or 4.5 mg riboflavin/kg fresh matter (FM) feed (L1.5, L3.0, L4.5). In PARENTS, 10 groups of 10 parent Hubbard JA 57 hens and one Hubbard S77 cockerel each were allocated to riboflavin supplementation of either 2.5 or 4.0 mg/kg FM feed (P2.5, P4.0). LAYERS lasted for 18 weeks; data for PARENTS are from 14 weeks (ongoing). Hens were fed *ad libitum* and housing conditions were according to organic standards. Realised total riboflavin concentrations in the feeds (including native riboflavin from feed components) were 5.0, 4.5, and 3.0 mg/kg FM for L4.5, L3.0, and L1.5, respectively, and 8.0 and 6.0 mg/kg FM for P4.0 and P2.5. In both experiments (18 weeks for LAYERS, 14 weeks for PARENTS), no treatment effects on feed consumption, body weight, laying performance, eggshell strength and yolk colour as well as plumage and keel bone integrity scores occurred. Symptoms of lameness were not observed at all. Riboflavin concentration in egg yolk was in the range of 0.55 to 0.65 mg/100 g yolk for all treatments in both experiments without treatment effects, except for L1.5, where it dropped to 0.4 in week 18 (P<0.05). In conclusion, no clinical symptoms of riboflavin deficiency were observed in any of the treatments of both experiments. However, lower riboflavin concentrations in yolk of L1.5 indicated a lower metabolic level. This level of supplementation is therefore not recommended.

Modulation of zinc, copper and iron transporters in weaned pigs fed dietary zinc oxide

D.B. Dalto[1], A.K. Novais[1], F. Guay[2], Y. Martel-Kennes[2,3], G. Talbot[1], M. Lessard[1], J.J. Matte[1] and J. Lapointe[1]
[1]Agriculture and Agri-Food Canada, Sherbrooke R&D Centre, 2000, College street, J1G 5H9, Sherbrooke, Quebec, Canada, [2]Université Laval, Département de Sciences Animales, 1506-2325, rue de L'Universite, G1V 4G2, Quebec, Quebec, Canada, [3]Deschambault Research Centre in Animal Science, 120, chemin du Roy, G0A 1S0, Deschambault, Quebec, Canada; jacques.matte2@canada.ca

This study evaluated the impact of dietary zinc oxide (ZnO) on the gene expression of zinc (Zn), copper (Cu), and iron (Fe) transporters in jejunum and liver of weaned pigs. From weaning (d21), 120 piglets were fed a basal post-weaning diet (without antibiotics) supplemented (n=40/treatment) with 100 mg/kg (LZn), 1000 mg/kg (MZn) or 3,000 mg/kg (HZn) of ZnO until d42. Dietary Cu (130 mg/kg) and Fe (260 mg/kg) were constant throughout the experiment. Piglets were slaughtered at d21 (before treatments), d23, d35, and d42 (n=10/treatment/day) for samples collection. mRNA expression of Zn, Cu, and Fe transporters were evaluated by quantitative real-time PCR. No effect of treatments (P≥0.11) was observed at d21 and d23 in jejunum and liver for any gene. For Zn-related genes, jejunum *Zip4* expression was lowest for HZn, intermediary for MZn and greatest for LZn at d35 and d42 (treatment × age interaction; P≤0.01). In contrast, *Znt1* was greatest for HZn at d35 and d42 in jejunum and liver and did not differ between MZn and LZn, except at d35 in jejunum (LZn<MZn) (treatment × age interaction; P≤0.01). For Cu-related genes, no treatment effect (P≥0.26) was detected for *Crt1* and *Atox1* in jejunum and liver. In both tissues, *Mt3* expression was greatest for HZn, intermediary for MZn and lowest for LZn at d35 and d42 (treatment × age interaction; P≤0.01). A tendency for treatment × age interaction was detected (P=0.06) for *Atp7a* expression in jejunum where MZn was lowest and HZn and LZn were greatest at d35. Values for *Atp7b* in liver were greatest for HZn and did not differ between MZn and LZn at d35 and d42 (treatment × age interaction; P=0.03). For Fe-related genes, no treatment effect (P≥0.12) was detected for *Dmt1* in liver and *Heph*, *Fpn1* and *Tfrc* in jejunum and liver. A tendency for treatment × age interaction (P≥0.06) was detected for *Dmt1* in jejunum where LZn was highest at d35 and HZn was lowest at d42. The expression of *Fth1* was (or tended) to be affected in jejunum and liver (treatment × age interaction; P≤0.07) where values were highest for HZn at d35 and d42, intermediary for MZn and lowest for LZn in jejunum but MZn and LZn did not differ in liver. In conclusion, intestinal and hepatic Zn, Cu and Fe transporters were affected by dietary ZnO levels in post-weaning piglets.

Poster discussion 'Mineral and vitamin nutrition in pigs and poultry'

S. Millet[1] and G. Bee[2]
[1]Flanders Research Institute for Agriculture, Fisheries and Food (ILVO), Scheldeweg, 68, 9090 Melle, Belgium, [2]Agroscope, la Tioleyre 4, 1725 Posieux, Switzerland; sam.millet@ilvo.vlaanderen.be

In this session we will discuss the accepted posters of the session 'Mineral and vitamin nutrition in pigs and poultry'

Reduced dietary P and Zn supply to laying hens affects egg weight and egg Zn

D. Brugger[1], P. Urban[2], J. Pokorny[3], R. Puntigam[3], J. Slama[3], K. Schedle[3], M. Schlattl[2] and W. Windisch[2]
[1]Institute of Animal Nutrition, University of Zurich, Winterthurerstrasse 270, 8057 Zurich, Switzerland, [2]Chair of Animal Nutrition, Technichal University of Munich, Liesel-Beckmann-Strasse 2, 85354 Freising, Germany, [3]Institute of Animal Nutrition, Livestock Products and Nutrition Physiology, University of Natural Resources and Life Sciences Vienna, Muthgasse 11, 1190 Vienna, Austria; dbrugger@nutrivet.uzh.ch

Reduction in Zn supply promoted subclinical Zn deficiency in laying hens after 8 d. The present study investigated how this affects the weight, shell strength and concentration of Ca, P and Zn of eggs. 48 fully grown laying hens (50% Lohmann Brown-Classic, 50% Lohmann Selected White-Classic) were kept in 24 boxes housing each 2 hens (average body weight 2,134±222 g and 1,809±82.1 g). Laying performance was 0.93 egg/hen/day. The experiment comprised a 14 d acclimatisation and 8 d experimental phase. During acclimatisation, animals were fed according to recommendations except for P (low P: 3.67 g/kg; high P: 8.42 g/kg by adding 4.5 g P/kg from mono-Ca-phosphate). During the total experimental phase of 8 d, the P feeding regime continued and the Zn supply was additionally split in two dosages (low Zn: 27.3 mg/kg, high Zn: 130 mg/kg by adding 100 mg/kg from ZnSO4×7H$_2$O)). Egg weight, egg shell strength and Ca, P and Zn concentrations in whole eggs were measured in the end of each phase. Statistical analysis comprised 3-way ANOVA (P, Zn, Genotype; SAS 9.4; P≤0.05). Egg weight was significantly lower with lower dietary P at the end of the 14 d acclimatisation (P=0.01) but the effect was no longer evident in the end of the experimental phase. Egg shell strength was not affected. Reduced P feeding increased Zn in whole eggs at the end of pre-feeding (P=0.03). Alimentary Zn deficiency significantly decreased egg Zn at the end of the experimental phase (P=0.006). Concentrations of P and Ca in eggs were not affected. Our recent data highlighted increased efflux of Zn from bone under the terms of reduced P-feeding. This seemed to increased Zn deposition into the eggs at the end of the pre-feeding. Irrespective of the P feeding regime, dietary Zn deficiency reduced egg Zn after only 8 d, indicating it to be a quite sensitive marker of body Zn status under experimental conditions.

Effect of zinc-methionine supplementation on the intestinal morphology of pigs
J.M. Romo, R. Barajas, I. Enrriquez-Verdugo, G. Silva-Hidalgo, J.A. Romo, H.R. Guemez and J.M. Uriarte
Facultad de Medicina Veterinaria y Zootecnia de la Universidad Autónoma de Sinaloa, Boulevard San Ángel s/n, Colonia San Benito, CP 80246, Culiacán de Rosales, Sinaloa, Mexico; jumanul@uas.edu.mx

With the objective of determine the effect of zinc-methionine supplementation during the period of gestation-lactation (GL) and development-finalisation (DF) in the morphology of the intestinal epithelium of pigs under warm or cool environment, were realised two experiments. 192 pigs (Large White × Landrace × Large White × Pietrain, with 96 per Exp.) with an average age of 79 d and 26.39 kg of weight were used, to which one of four treatments was assigned in a complete randomised block experimental design, with 2×2 factorial arrangement. The treatments were: T1 (Control; n=24), mothers not supplemented-pigs not supplemented; T2 (Zn DF; n=24), non-supplemented mothers-pigs supplemented with 100 mg of Zn/kg DM; T3 (Zn GL; n=24), mothers supplemented with 100 mg of Zn/kg DM-pigs not supplemented, and T4 (Zn GL + Zn DF; n=24), supplemented mothers + supplemented pigs. During the warm season (Exp. 1) the ratio villus height: depth crypt was greater (P<0.01) in pigs supplemented with Zn (3.36 vs 2.77) during the hot season. During the cool season the treatment did not change the productive yield (villus height 436 vs 446 μm, and 113 μm of depth crypt); however, supplementation during GL tended to reduce (P=0.06) mortality. The results allow us to conclude that the addition of Zn to the diet improves the morphology of the intestinal epithelium of the pig during the development-finalisation stage, under conditions of caloric stress.

Vitamin D3 in drinking water of fasted pigs increases α-tocopherol concentration and meat stability
A.I. Rey[1], J. Segura[1], R. Escudero[1], A.I. Rodriguez[2] and L. Calvo[2]
[1]Faculty of Veterinary Science. University Complutense of Madrid, Animal Production, Avda. Puerta de Hierro s/n, 28040 Madrid, Spain, [2]Incarlopsa, I+D, Ctra. N-400 km. 95400, 16400 Tarancón, Cuenca, Spain; anarey@ucm.es

The aim of this investigation was to study the effect of vitamin D3 in drinking water given to pigs prior to slaughter on oxidative status and stability of pork (TBARS). Pigs were distributed into different boxes (n=25 per box) at the slaughterhouse (Incarlopsa, Tarancon) and maintained in fasting conditions with free access to drinking water. Water was administered by three tanks with reactor systems: one deposit contained the Vitamin D3 supplement at low dose (VITD1); a second deposit contained the Vitamin D3 supplement at high dose (VITD2) and the third one without supplement acted as control (C). Serum samples (n=8 per experimental group) were obtained at the time of slaughter after blood centrifugation; and longissimus dorsi muscle was taken at the time of cutting after carcass cooling. α-Tocopherol concentration in pigs' serum was measured by an HPLC provided with RP-C18 column (Agilent technologies). Serum and muscle levels of malondialdehyde (MDA) were determined spectrophotometrically after heating sample extracts in presence of thiobarbituric acid. The statistical analysis of data was carried out using the statistical program SAS v. 9.3. Serum α-tocopherol concentration was higher in those groups supplemented with vitamin D3 when compared to control (P<0.05); whereas serum TBARS reached the highest values in control group (P<0.05). There was not found a dose effect on the oxidative status of pigs. In meat, vitamin D3 supplementation prior to slaughter resulted in lower muscle TBARS after 3 days under refrigerated storage (P<0.05). No differences were detected in meat stability between both groups supplemented with vitamin D3; although group that received the highest dose (VITD2) had the lowest TBARS values. In conclusion, vitamin D3 supplementation in drinking water prior to slaughter could be an interesting and easy strategy to enhance vitamin E concentrations and to improve the oxidative status and quality of pork.

The relationship between plasma biomarkers and dark cutting in beef

K.M.W.L. Loudon[1], G.E. Gardner[1], D.W. Pethick[1], P. McGilchrist[2], F.R. Dunshea[3], R. Polkinghorne[4], G. Tarr[5] and I.J. Lean[6]
[1]Murdoch University, Murdoch, WA 6150, Australia, [2]University of New England, Armidale, NSW 2350, Australia, [3]University of Melbourne, Parkville, Vic 3010, Australia, [4]Birkenwood Pty. Ltd, 431 Timor Rd, Murrurundi, NSW 2338, Australia, [5]The University of Sydney, Sydney, NSW 2006, Australia, [6]Scibus, Camden, NSW 2570, Australia; k.loudon@murdoch.edu.au

This study considered the relationship between pre-slaughter stress, plasma biomarkers and dark cutting in 488 pasture fed Australian cattle. Groups consisting of steer only, heifer only, and mixed sex cattle were transported by boat and road, or marketed via saleyard, then slaughtered. The definition of dark cutting was based on ultimate pH of the longissimus thoracis et lumborum, where ≤5.7 were considered acceptable and pH>5.7 classified as a dark cutter (DFD). A K-means cluster analysis identified 3 distinct groups. The DFD incidence for Group 1 (n=129) was 19%, for Group 2 (n=160) 48%, and Group 3 (n=179) 9%. The high DFD group had the smallest eye muscle area and rump fat score, with no difference in carcass weights between groups. The high DFD group had the largest plasma glucose concentration, but the lowest plasma non-esterified fatty acid (NEFA) and beta-hydroxybutyrate (BHB) concentrations. The large glucose response with apparently little response from NEFA may indicate different mechanisms of action. The fat depot may have been proportionately lower, or the stress may have been more chronic with glucose mobilised from glycerol, or the fat stores may have decreased catecholamine sensitivity. The low DFD group had the highest plasma L-lactate and intermediate NEFA and BHB response. The level of creatine kinase (CK) was the greatest in the high DFD group and the smallest in the low DFD group. The results demonstrated no apparent linear continuum relating plasma biomarkers to increasing incidence of DFD apart from CK which overall was the metabolite most consistently associated with DFD, thus most likely to aid pre-slaughter DFD detection. There was variation in DFD between farms, rather than experimental treatments, suggesting further research into on farm factors influencing DFD such as genetics, nutrition and management is required.

DEXA prediction of beef retail cut weights

H.B. Calnan and G.E. Gardner
Murdoch University, 90 South St, Perth 6150, Australia; honor.calnan@murdoch.edu.au

The quantity of retail cuts procured from a beef carcase is an important determinant of value. The ability to predict cut weights from entire beef sides would allow improved valuation of carcases in the beef supply chain. The development of dual energy X-ray absorptiometry (DEXA) to scan beef sides at abattoir line-speed provides an opportunity to improve the prediction of retail cut weights in addition to predicting carcase composition, enabling carcase sorting for optimal fabrication. This project assesses the ability of DEXA to predict beef cut weights. Beef sides (n=264) with a wide range in weight and fatness were DEXA scanned and boned-out into a selection of commercial cuts. DEXA images were analysed to produce a lean composition value for each side via threshold removal of bone-containing pixels and use of previously established relationships between pixel R value, chemical fat % and tissue thickness. The number of pixels and mean tissue thickness of carcase images were used with lean DEXA values to predict the weight of sub-primal cuts including the blade, chuck and cube roll, tenderloin, striploin, rump, knuckle, silverside and topside. Cut weight and hot side weight (HSW) data was converted to natural logarithms for analysis via general linear models in SAS. Prediction models containing HSW and P8 fat depth (the current Australian standard for estimating carcase fatness) were compared to models containing HSW and DEXA image variables as covariates. Data was randomly divided into 5 groups for training and validation of prediction models, with all models trained in 4 groups and validated in the 5th group. Training and validation were repeated 5 times so models were validated in each group, producing mean precision and accuracy values of prediction performance. This analysis showed that DEXA variables predict beef cut weights with better precision and accuracy than P8 fat depth. The precision varied between cuts, for example the striploin was predicted with an R^2 of 0.84 and root mean square error (RMSE) of 0.08, while the topside was predicted with an R^2 of 0.94 and RMSE of 0.05. DEXA cut weight predictions were highly accurate; validated prediction of the striploin and topside having a mean absolute bias of only 0.09 and 0.07. This improved precision and accuracy was also evident through the capacity of predictive models to differentiate cut weights between carcases of the same weight (i.e. 320 kg) yet with 22 mm variation in P8 fatness. In this case DEXA predicted a 2.2 kg difference in striploin weight, contrasting with only a 2 kg difference in striploin weight if P8 was used as the predictor. These results demonstrate that a commercial beef DEXA has the capacity to improve cut weight predictions, with these predictions likely to improve with further calibration of the DEXA system.

Beef eating quality categorisation in Poland based on Meat Standards Australia

G. Pogorzelski[1], J. Wierzbicki[2], A. Onopiuk[1] and A. Wierzbicka[1]
[1]Warsaw University of Life Sciences, Institute of Human Nutrition Sciences, Nowoursynowska 159c, 02-776 Warsaw, Poland, [2]Polish Beef Association, Smulikowskiego 4, 00-389 Warsaw, Poland; grzegorz_pogorzelski@sggw.pl

In Poland and more widely in Europe, meat quality is described using the EUROP system. Like many other systems around the world determining the quality of beef, it has no relationship to sensory quality. Meat Standards Australia (MSA), an eating quality system based on consumer evaluation, uses an interactive prediction model to provide objective descriptions of beef meal outcomes for 33 carcass muscles by multiple cooking methods. The Polish meat market is struggling with the problems of low beef consumption. In order to obtain an answer to the question why there is such a low consumption of beef in Poland, the beef industry decided to ask consumers about it. The MSA methodology was used as a tool to evaluate the sensory categorisation of beef in Poland. In the first steps, it was checked whether the MSA methodology is useful for assessing the eating quality of Polish beef by Polish consumers. Consumers evaluated paired samples of grilled beef cut to either 6 or 25 mm thickness and cooked to medium doneness. A wide quality range was created by utilisation of 3 muscles, from 22 divergent bull carcasses. Consumers rated each sample for four sensory attributes, and selected a grade. An MQ4 score was calculated from a weighted combination of the four sensory scales (tenderness, juiciness, flavour, overall liking) and used to categorise the tested samples. Polish consumers clearly distinguished between grades, defined by MQ4 (Meat Quality, 4 variables) boundaries derived from discriminant analysis. Similar accuracy was achieved for both thicknesses. It was concluded that the MQ4 score could be successfully used to support a meat quality grading system in Poland. Verification that a system developed on similar principles will provide a new, more accurate tool for determining meat quality. This will allow beef to be better aligned with consumer expectations, it will improve the demand for beef in Poland. Encouraged by this information, the Polish beef industry decided to continue its research. As a result, over 9,000 Polish consumers evaluated samples of beef cooked by 4 different methods.

Automated grading of eye muscle area and intramuscular fat in Australian beef

S.M. Stewart[1,2], T. Lauridsen[3], H. Toft[3], D.W. Pethick[1,2], G.E. Gardner[1,2], P. McGilchrist[1,4] and M. Christensen[3]
[1]Advanced Livestock Measurement Technologies (ALMTech), 90 South Street, 6150 Murdoch, Australia, [2]Murdoch University, School of Science, Health and Engineering, 90 South Street, 6150 Murdoch, Australia, [3]Frontmatec A/S, Hassellunden 9, 2765 Smørum, Denmark, [4]University of New England, School of Environmental and Rural Science, Elm Avenue, 2351 Armidale, Australia; s.stewart@murdoch.edu.au

Recent work indicates that marbling and IMF% independently contribute to palatability scores in Australian beef with chemical IMF% reported as a more precise predictor of consumer scores. Therefore, new camera technologies that aim to predict both of these traits may further improve the prediction of beef eating quality. Multispectral imaging is a modality which has been shown to predict a range of meat quality traits in beef and lamb. Fat and lean tissue can be differentiated through the selection of specific wavelengths coupled with modern analytical techniques. Therefore, we hypothesised that a prototype multispectral camera could predict visual eye muscle area (EMA), marbling score, and chemical IMF% in beef. Images and carcass grading data were acquired from 888 beef carcasses from multiple kill group consignments from the Beef Information Nucleus (BIN) herd and were assigned into calibration and independent validation datasets. Carcasses were comprised of Angus, Hereford and *Bos indicus* genotypes and displayed a large phenotypic variation in EMA, MSA marbling and IMF% traits. Preliminary analysis showed that during calibration, the camera described 92% of the variation in EMA (RMSE=3.17 cm^2), 76% of the variation in MSA marbling (RMSE=63.51) and 76% of the variation in chemical IMF% (RMSE=1.98%). During validation, precision was maintained for EMA (R^2=0.73, RMSE=6.90 cm^2), MSA marbling (R^2=0.75, RMSE=65.00) and chemical IMF% (R^2=0.76, RMSE=1.83). Validation accuracy was also demonstrated, with all three traits displaying minimal bias (EMA=-4.57 cm^2, MSA marbling=0.00, IMF%=-0.70) and slopes close to 1 (EMA=0.95, MSA marbling=0.94, IMF%=0.96). Accurate and precise objective measurement of IMF % and marbling will improve the prediction of eating quality, delivering a more consistent product to consumers. Moreover, objective feedback to producers will enhance management decisions on-farm, improve compliance and accelerate genetic gain.

Global collaboration to encourage a consumer focused beef industry: the Meat Research 3G Foundation

R. Polkinghorne[1], L. Farmer[2], D. Troy[3], M. Miller[4], J.F. Hocquette[5], J. Wierzbicki[6] and A. Wierzbicka[7]
[1]Birkenwood Pty Ltd, 431 Timor Rd, Murrurundi, NSW 2338, Australia, [2]AFBI, Newforge Lane, Belfast, United Kingdom, [3]TEAGASC, Food Research Institute, Ashtown, Ireland, [4]Texas Tech University, Lubbock, Texas, USA, [5]INRAE, UMR1213, 63122 Theix, France, [6]Polish Beef Association, Ul. Kruczkowskiego 3, 00-380 Warszawa, Poland, [7]University of Life Sciences, Nowoursynowska 166, 02-787 Warszawa, Poland; rod.polkinghorne@gmail.com

The beef industry is faced with many challenges but also opportunities as a supplier of high quality nutrition. As all industry revenue comes from the consumer, beef must provide an enjoyable consumer experience at prices that reflect attractive value for the industry to be sustainable in addition to meeting societal standards for environmental management and animal welfare. This consumer focus is not delivered through traditional industry trading systems which focus on yield and describe beef in terms that do not equate to a cooked meal experience. Consequently, the consumer is faced with the challenge of predicting a cooked meal result themselves rather than being offered a simple and accurate choice of value points. This problem has been recognised across the globe resulting in research to better understand consumer sensory response to meat and to develop effective systems to predict consumer satisfaction for combinations of muscle and cooking method from each carcase together with production effects from farm to fork. The Meat Standards Australia system is a leading and successful example of this approach. Over 20 years, research and commercial groups in many countries have collaborated in consumer based research using common protocols and have greatly expanded knowledge of global consumers and the eating quality impact of many animal and processing interactions. Multi country collaboration can deliver significant benefit in both cost and enhanced outcomes through shared data and resources. The International Meat Research 3G Foundation (IMR3GF), a not for profit foundation, was formed to facilitate collaboration through the use of common research protocols under the auspices of the UN Economic Commission for Europe, establishing a DATAbank to house and share data and to provide research and commercial services related to eating quality.

BeeQ – building capacity for beef eating quality assessment in Wales

P.K. Nicholas-Davies[1], N.D. Scollan[2], T. Rowe[3], E. Thomas[4], E. Williams[5] and R. Polkinghorne[6]
[1]Aberystwyth University, Institute of Biological, Environmental and Rural Sciences, Gogerddan Campus, Aberystwyth SY233EE, United Kingdom, [2]Queens University Belfast, Institute for Global Food Security, University Road, Belfast BT71NN, United Kingdom, [3]Castell Howell Foods Ltd, Crosshands, Llanelli SA146SX, United Kingdom, [4]Hybu Cig Cymru, Meat Promotions Wales, Tŷ Rheidol, Parc Merlin, Aberystwyth SY23 3FF, United Kingdom, [5]Menter a Busnes, Uned 3, Parc Gwyddoniaeth, Aberystwyth SY233AH, United Kingdom, [6]Birkenwood Pty Ltd, 45 Church Street, Hawthorn, Victoria 3122, Australia; pkn@aber.ac.uk

The BeefQ – Beef Eating Quality Project is a pre-competitive collaboration between international research and industry partners, to develop an eating quality assessment system for Wales. The system being developed is based on eating quality assessment protocols implemented successfully in other countries, such as the Australian Meat Standards Australia (MSA) system. The project consists of four main strands: a survey of PGI Welsh Beef carcasses submitted for slaughter; development of a model for predicting beef eating quality in Wales; training for industry personnel in eating quality assessment and meat science and finally engagement with industry stakeholders to promote the concept of beef eating quality assessment and develop a strategy for taking the BeefQ outputs forward in Wales post project. The survey, to describe and quantify the population of beef carcasses from animals born and reared in Wales, comprised the eating quality grading (using UNECE protocols) of 2,090 carcasses. Four cuts from 90 sides of beef surveyed were selected for testing with 1,200 consumers. Consumers were presented with seven samples of grilled steak and asked to score them according to taste, tenderness and juiciness. This data forms the basis for developing an eating quality prediction system for Welsh beef. The training of processing plant personnel and industry representatives in the various aspects of eating quality grading has not only been valuable for building eating quality assessment expertise in Wales but has enabled practical farmer focussed demonstration and discussion events on eating quality. This type of activity, along with broader industry stakeholder engagement is raising the profile of, and discussion around, the potential benefits of eating quality prediction for the Welsh Beef sector. Establishing and maintaining good relationships with companies processing PGI Welsh Beef has been integral for the successful delivery of BeefQ activities. The consumer events, hosted by Further Education Colleges, provided an unforeseen legacy for BeefQ by allowing the project team to engage directly with the farmers, chefs and consumers on beef eating quality.

Welsh beef carcase survey – exploring proxies for predicting eating quality

S. Siphambili[1], P.K. Nicholas-Davies[1] and R. Polkinghorne[2]
[1]Aberystwyth University, Institute of Biological, Environmental and Rural Sciences, IBERS Gogerddan Campus, Aberystwyth University, SY23 3EE, Aberystwyth, United Kingdom, [2]Birkenwood Pty Ltd, 45 Church Street, Hawthorn, 3122, Victoria, Australia; sis39@aber.ac.uk

The BeefQ project is investigating the potential benefit and challenges associated with using a consumer based beef grading system for PGI Welsh Beef. As part of this, a survey of beef carcasses submitted for slaughter in approved PGI Welsh Beef processors, was conducted. The survey was designed to characterise the population of PGI Welsh Beef animals slaughtered and enable an industry representative sample of the population to be selected for consumer testing in a subsequent piece of work on sensory evaluation. Full UNECE eating quality grading data, EUROP classification, breed and date of birth was collected on a total of 2,090 head of cattle at 8 processing sites – 1,037 in February 2019 and a further 1,053 in August 2019. The two periods were selected to take into account seasonal differences in diet. A total of 69 breeds and crosses were represented in the survey, demonstrating the diversity within the Welsh cattle population and the challenges associated with potentially using breed as an input for predicting eating quality. Whilst beef breeds outnumbered dairy breeds in the sample (78 and 22%, respectively) it is likely, due to peculiarities of the breed identification system, that a large number of the beef animals were in fact beef × dairy. Whilst a relationship between age (available on the cattle passport) and ossification was observed, it was not a strong association with, for example, animals ranging from 400-1,100 days of age showing the same level of ossification. There was some evidence of sex differences in the relationship between age and ossification score, with young bulls reaching a specific ossification level at a much younger age than beef heifers or steers. Some association between EUROP fat class and marbling was observed but the variation makes it unsuitable for use as a proxy. There was no relationship between EUROP carcase conformation score and marbling. A correlation matrix of EUROP and UNECE grading inputs indicated that there were reasonable correlations between age and ossification along with moderate relationships within fat measures and for carcase weight, eye muscle area and hump height. Further analysis of the survey data in conjunction with the BeefQ sensory evaluation data will be required to determine the potential for these relationships and interactions to act as proxies for predicting eating quality.

Contributions of tenderness, juiciness and flavour liking to overall liking of beef

J.J. Liu[1], M.P. Ellies-Oury[2], S. Chriki[3], I. Legrand[4], G. Pogorzelski[5], J. Wierzbicki[6], L. Farmer[7], D. Troy[8], R. Polkinghorne[9] and J.F. Hocquette[1]
[1]INRAE, UMR1213, 63122 Theix, France, [2]Bordeaux Sciences Agro, 1 cours du Général de Gaulle, 33175 Gradignan, France, [3]Isara Agro School for Life, 23 rue Jean Baldassini, 69364 Lyon cedex 07, France, [4]Institut de l'Elevage, MRA-NA, 87060 Limoges Cedex 2, France, [5]Warsaw University of Life Sciences, Nowoursynowska 159C, 02-776 Warsaw, Poland, [6]Polish Beef Association Ul, Kruczkowskiego 3, 00-380 Warszawa, Poland, [7]Agri-Food and Biosciences Institute, Newforge Lane, Belfast BT9 5PX, United Kingdom, [8]Teagasc, Ashtown, D15 KN3K Dublin, Ireland, [9]Birkenwood Pty Ltd, 431 Timor Road, NSW 2338, Australia; jingjing.liu@inrae.fr

Overall liking of beef depends on tenderness, juiciness and flavour liking. This study aimed to evaluate the contributions of these three sensory traits to overall liking by European consumers. Eating quality scores of beef by untrained consumers from France, Poland, Ireland and Northern Ireland were determined using the Meat Standards Australia protocols for 8,216 cuts from 18 muscles of 743 different cattle breeds. Due to strong correlations ($0.79 < R^2 < 0.85$, $P < 0.001$) among the sensory traits, principle components analysis was performed in addition to analysis of variance, multivariate regression model analysis and relative weight analysis. Multiple regression showed that overall liking = 0.32 Tenderness + 0.08 Juiciness + 0.61 Flavour liking ($R^2 = 94\%$). It means that a 1 unit change in flavour liking or a 1.8 unit change in tenderness or a 6.7 unit change in juiciness have the same impact on overall liking. Thus, juiciness has the smallest effect on overall liking. Accordingly, flavour liking was the most important contributor (38.6%) to beef overall liking, followed by tenderness (31.2%) and juiciness (24.3%). Furthermore, with increasing tenderness class, the contribution of flavour liking increased up to 54% and meanwhile the contribution of tenderness decreased up to 7%. We thus hypothesise that the improvement in tenderness over the last decades has increased the relative contribution of flavour liking. Furthermore, one sensory trait with a low score has the potential to be compensated by other traits with higher scores and more emphasis will be placed on the trait with the lowest perception.

The French Meat Dictionary (collective work of the French Meat Academy)

F. Landrieu[1], M. Kombolo-Ngah[2] and J.F. Hocquette[1,2]
[1]*Académie de la Viande, 207 rue de Bercy, 75012 Paris, France,* [2]*INRAE, UMR1213, Recherches sur les Herbivores, Theix, 63122, France; jean-francois.hocquette@inrae.fr*

Many terms are used to describe meat products and meat processing. Many of them are well accepted and routinely used in the scientific and professional jargons. However, many of them are also country-specific or species-specific. Furthermore, technology has improved and the diversity of meat products has become more abundant. In addition, some terms are sometimes misused or voluntary used in a wrong way for commercial purposes (such as the word 'meat' used to name some meat substitutes). In some cases, the original meanings of some meat terms have evolved over time to mean something different. Such inconstancy results sometimes in misinformation regarding meat characteristics, consumption habits, meat product manufacturing, etc. Finally, the increasing trading of meat worldwide has also increased the necessity of providing clear and concise definitions for the terms used to define meat products and meat related terms. The French Meat Academy has published in 2012 a Meat Dictionary to solve this problem of confusion and inconsistency. This dictionary identifies and defines the names of all French cuts, all the professional terms of butchery (including utensils), breeds, as well as technical or peripheral terms, whether it is beef, veal, lamb, pork and even poultry. In addition, it also lists all the vocabulary related to the taste and texture of meat, as well as the inventory of traditionally meat dishes. The French Meat dictionary is available in French and English, and also includes a small glossary in Spanish. About 1000 terms have been defined and are regularly updated. In conclusion, this French Meat Dictionary provides a list of meat terms well defined in a standardised way for use by scientists, nutritionists, food technologists, meat industry staff, journalists, public authorities, and consumers to accurately describe and classify meat and meat products with robust meat related information.

European conformation and fat scores of bovine carcasses are not good indicators of marbling

J.J. Liu[1], S. Chriki[2], M.P. Ellies-Oury[3], I. Legrand[4], G. Pogorzelski[5], J. Wierzbicki[6], L. Farmer[7], D. Troy[8], R. Polkinghorne[9] and J.F. Hocquette[1]
[1]*INRAE, UMR1213, 63122 Theix, France,* [2]*Isara Agro School for Life, 23 rue Jean Baldassini, 69364 Lyon Cedex 07, France,* [3]*Bordeaux Sciences Agro, 1 Cours du Général de Gaulle, 33175 Gradignan, France,* [4]*Institut de l'Elevage, MRA-NA, 87060 Limoges Cedex 2, France,* [5]*Warsaw University of Life Sciences, Nowoursynowska 159C, 02-776 Warsaw, Poland,* [6]*Polish Beef Association Ul, Kruczkowskiego 3, 00-380 Warszawa, Poland,* [7]*Agri-Food and Biosciences Institute, Newforge Lane, Belfast BT9 5PX, United Kingdom,* [8]*Teagasc, Ashtown, Dublin, D15 KN3K, Ireland,* [9]*Birkenwood Pty Ltd, 431 Timor Road, Murrurundi, NSW 2338, Australia; liujingjing1003@126.com*

Marbling is one of the primary carcass attributes utilised in most beef grading systems. In addition, marbling is one of the multiple factors that contribute to determine eating quality of beef. In Europe, the current beef grading system is the EUROP grid, which is used to determine the commercial value of beef carcasses and production yield rather than eating quality evaluation. While the EUROP system description of muscle and fat classes relates to yield, it has no direct relationship to beef eating quality and consumer experience. Therefore, the aim of this study was to investigate the relationship of marbling with fat score and conformation score of the EUROP grid. In case of strong associations between marbling and European classification scores, marbling score may be estimated from the carcass characteristics. However, the results indicate that European classification scores explain only a low proportion of the variability in marbling scores ranging from 0.03 to 0.24. The lack of any strong association between marbling and European classification scores suggests that it is currently not possible for the European beef industry to estimate marbling from carcass characteristics. Therefore, whereas the EUROP grid is well adapted to estimating yield, it does not reflect marbling at all. The European beef industry should include additional indicators regarding beef eating quality to assist the determination of the commercial value of carcasses based on eating quality.

Factors influencing conformation and fatness of bovine carcasses in Italy and France

N. Rampado[1], E. Hocquette[1], D. Roux[2], C. Montanari[3], K. De Roest[3] and J.F. Hocquette[1]
[1]INRAE, VetAgro Sup, UMR1213, Theix, 63122, France, [2]INRAE, Herbipole, Theix, 63122, France, [3]Centro Ricerche Produzioni Animali, S.p.A., Viale Timavo 43/2, 42121 Reggio Emilia, Italy; jean-francois.hocquette@inrae.fr

This study aimed to analyse the differences in conformation and fatness of Charolais and Limousine carcasses using the characteristics of 3,869 carcasses from males (young cattle) or females (heifers) from two databases, from Italy and France. Italian carcasses were obtained from young animals reared for 7 months in France, slaughtered in Italy at a live weight of 453±11 kg for Charolais and 374±34.16 kg for Limousine animals. In France, cattle were slaughtered at a live weight of 402.4±63.5 kg for Charolais and 379.2±63.5 kg for Limousine animals. Analyses confirmed that there were very large differences in carcass conformation and fatness between genders, breeds and countries. In fact, carcasses from females are fatter (mainly 3) than those from males (2 or 3). Carcasses from females have also a lower conformation score (R and U for Charolais and Limousin respectively) than males (U and E for Charolais and Limousin respectively). In general, the conformation is better but fatness is lower for the Limousine breed than for the Charolais breed. The characteristics of carcasses are also strongly dependent on the farming system, which is rather intensive in Italy and extensive in France. Indeed, for a given breed, the carcasses are better shaped and leaner for young cattle slaughtered in Italy compared to those slaughtered in France. These results were confirmed at the national level by analysing all the data on the characteristics of the carcasses of young cattle and heifers slaughtered in France or Italy. The differences in carcass characteristics can therefore be explained by different factors: farming systems, feeding, age or genetics.

Feasibility of visible-near infrared spectroscopy to discriminate between steer and bull beef

L. Moran[1], L.J.R. Barron[1], N. Aldai[1] and A.P. Moloney[2]
[1]University of the Basque Country (UPV/EHU), Lactiker Research Group, Dpt. Pharmacy & Food Sciences, 01006, Vitoria-Gasteiz, Spain, [2]Teagasc, Animal and Grassland Research and Innovation Centre, Dunsany, Co. Meath., C15 PW93, Ireland; lara.moran@ehu.eus

Due to perceived differences in sensory properties, steer beef is considered by many consumers to be a superior product to bull beef and they are willing to pay a premium for this product. There is an urgent need therefore for analytical methods that can reliably ensure that beef labelled as 'steer beef' is genuine. Chemical methods generally are slow, complex, destructive and not suitable for application for on the spot inspection. Rapid and non-destructive spectroscopic methods have been proposed as a useful strategy for the authentication of different foods. In this regard, the objective of the present study was to determine the feasibility of visible-near infrared (Vis-NIR) spectroscopy to authenticate steer beef. Samples of longissimus dorsi, gluteus medius and semitendinosus muscles from bulls (n=30×3 muscles) and steers (n=28×3 muscles) with similar carcass weight were collected, vacuum packed and aged (3±1 °C) for 7 days. Raw steaks were scanned (from 350 to 2,500 nm at 1 nm intervals) in reflectance mode using a portable ASD Labspec 5000 NIR spectrometer. Raw spectral data were recorded in duplicate and exported as absorbance to the Unscrambler 9.7 data treatment software. The calibration model was developed by Partial Least Square Discriminant Analysis (PLS-DA) using 116 samples (2/3 of the whole dataset). The remaining 58 samples were used as an independent validation set. The results indicated that Vis-NIR spectroscopy discriminated between bulls and steers for 100% of the cases, regardless of the muscle. The final PLS-DA model was developed using the NIR spectral data only and the variables identified as important by Martens Uncertainty test. Due to the good performance of the prediction model obtained from raw spectral data (coefficient of determination 0.955, cross-validation standard error 0.106 and standard error of prediction 0.142) no pre-treatment of the spectra was performed. The wavelength values with the highest importance in the prediction model were 1,800 and 2,200 to 2,400 nm that are chemically related with the presence of C-H bonds of the intramuscular fatty acids. In conclusion, the use of NIRS at retailer level seems a feasible tool to authenticate steer meat. The classification results must be confirmed with a larger number of commercial beef samples of different origin.

Using real-time ultrasonography to identify the optimum subcutaneous fat depth in Arouquesa breed

C. Venâncio[1,2], D. Moreira[1], J.C. Almeida[1], R. Vieira[1,2], J.A. Silva[1,3] and S.R. Silva[1,3]
[1]Universidade de Trás-os-Montes e Alto Douro (UTAD), Vila Real, 5001-801, Portugal, [2]Centre for the Research and Technology Agro-Environmental and Biological Sciences (CITAB), VR, 5001, Portugal, [3]Veterinary and Animal Research Centre (CECAV), VR, 5001, Portugal; cvenanci@utad.pt

Traditional cattle production represents an important part of the European culture, identity, and heritage. In the North of Portugal, there are several autochthonous cattle breeds reared in traditional mountain systems. The Arouquesa (ARQ) breed is considered as one of the best native breeds for beef production and the quality of its meat is recognised by the PDO label (Protected Designation of Origin). Recent studies highlighted the need to obtain individual information on animals to identify carcass and meat quality traits in bovines specifically reared in traditional systems. In this sense, this work aimed to apply real-time ultrasound (RTU) technique to identify the optimal level of subcutaneous fat (SF) in ARQ feeder calves. Forty-four feeder calves were evaluated (9.8±1.6 months of age; 146±28 kg of carcass weight). Just before slaughter, the animals were scanned using an ultrasound scanner (A6, SonoScape Co. Ltd.) with a linear probe of 5 MHz. The animals were individually restrained in a crate to minimise movements and ensure they were standing in a similar stance. The probe was placed perpendicular to the backbone over the 13th thoracic vertebra (T13). The ultrasound images were captured and saved on the scanner. Afterwards, the images were analysed for the SF measurements. Image analysis was performed using Fiji software (http://fiji.sc/Fiji, NIH, USA). After slaughtering, a cut which included the T13 was separated from the carcass. This cut was taken to the lab, and the thickness of the SF was measured. Simple descriptive statistics and correlation analysis between RTU and actual SF measurements were performed using the JMP-SAS software. The mean values for SF obtained in the cuts and the RTU images were similar (2.81±0.704 vs 2.80±0.795 cm, respectively). The correlation between RTU and actual SF measurements was significant (r=0.874; P<0.01). This work shows that the determination of SF depth by RTU is accurate, and as such, this technique can be used as a tool to monitor the finishing level of ARQ feeder calves.

Rumen protected fat in early stage altered adipogenic gene expression and carcass quality in cattle

K.H. Kim[1,2], K. Thirugnanasambantham[1], R. Bharanidharan[3], J.Y. Kim[2], G.D. Kim[1,2], X. Panyavong[2,4] and S.H. Beak[3]
[1]Seoul National University, Institute of Green Bio Science and Technology, Pyeongchang-ro 1447, Pyeongchang 25354, Korea, South, [2]Seoul National University, Department of International Agricultural Technology, Pyeongchang-ro 1447, Pyeongchang 25354, Korea, South, [3]Seoul National University, Department of Agricultural Biotechnology, Ganwak-ro 1, Seoul 08826, Korea, South, [4]National Univesity of Laos, Department of Livestock and Fisheries, Vientiane Capital, 856, Lao Peoples Dem. Rep.; khhkim@snu.ac.kr

The objective of this study was to demonstrate that supplementation of Rumen-Protected Fat (RPF) to growing cattle (9 to 13 months of age) will affect gene expression involved in lipid metabolism in early stage and improve meat quality of cattle slaughtered at 30 months of age. Thirty steers (Korean beef cattle) with an average initial BW (Body Weight) of 239±25 kg were allocated to 6 pens, balanced for BW and genetic merit for marbling, and assigned to one of two treatment groups; a control (only basal diet), and a test diet (basal diet with 200 g of RPF per day) for 150 days. After finishing RPF supplementation, samples of intramuscular tissue (*Longissimus lumborum*) were collected by biopsy to study gene expression. Then, both groups were fed the same basal diet without RPF until slaughter (average shrunk BW of 759 kg). Muscle expression of *PPARα*, *ZFP423* and *SREBP1c*, which are genes encoding transcriptional factors regulating pre-adipocyte differentiation, was down-regulated (P<0.05) by RPF supplementation. In addition, RPF induced down-regulation of gene expression involved in lipogenesis (*FASN* and *SCD*), fatty acid uptake (*FABP4*) and fatty acid esterification (*GPAT1* and *DGAT2*) (P<0.05). However, RPF did not affect final BW but improved quality grade (P<0.1) as well as intramuscular fat content (P=0.1). Histology of intramuscular depot demonstrated a larger number and a shorter diameter of adipocytes (P<0.05) with the test diet compared to the control one. It may be attributed to the change in the relative expressions of *SNAP23* and *BSCL* (P<0.05) by the test diet. This study revealed that feeding RPF before the fattening period (which is an early metabolic imprinting) could lead to higher intramuscular fat content with fine marbling when compared to the traditional feeding strategy in Korea. It may be caused by promoting the proliferation of pre-adipocyte rather than their differentiation at growing stage.

Genetic trends in the most important beef cattle breeds in Poland

B. Szymik and G. Skrzyński
National Institute of Animal Production, Department of Cattle Breeding, 1, Krakowska street, 32-083, Poland; bartosz.
szymik@izoo.krakow.pl

The Polish population of purebred beef cattle, which is evaluated based on breeding value, is small and contains approximately 23,000 animals. The most numerous purebred cattle breeds in Poland are Limousin (LM), Charolais (CH) and Hereford (HH). The BLUP Animal Model method is used to estimate breeding value. This is based on the phenotypic value of young bull, and the Modified Total Evaluation Index (ZWOZ). ZWOZ consists of two sub-indices: the Development Index (WR) of bulls in the season of evaluation and the Modified Meatness Index (ZWM). The estimated coefficient of heritability for ZWOZ is equal to 0.58. The evaluation of the breeding value of bulls is based on a one-trait linear model. The model includes the fixed effects of year of birth, season of birth, herd effect and the random genetic animal effect. This paper describes the procedures put into place in Poland for the genetic evaluation of beef cattle and the estimation of genetic trends, and critically analyses the progress made in recording beef cattle data. The analysis took into account 4,284 bulls from the LM, CH, HH breeds of cattle in Poland that were evaluated every year from 2007 to 2019. Genetic trends are estimated as the mean breeding value of bulls per year of birth. All genetic trends were positive, but the dynamics of genetic trends varied between breeds. The linear regression coefficient of mean breeding value per year of evaluation for ZWOZ was 0.98 for LM, 1.57 for HH and 2.19 for CH bulls. Although the dynamics of genetic trends in populations of LM, CH and HH cattle are positive, the rate of genetic trends are moderate. Therefore, there is still a need to improve the methodology currently used in Poland for phenotypic and genetic evaluation of beef cattle.

Essential amino acids in beef from Wagyu, Wangus and European bulls under Spanish production systems

J.C. Gardón[1], J.M. Vázquez-Mosquera[2], C. Sánchez-Romero[1], J. Fayos[1], M. Vázquez-Gomez[3], S.S. Pérez-Garnelo[4],
A. Villagrá[5], D. Martínez[6], F. Sebastian[7] and S. Astiz[4,8]
[1]Universidad Católica Valencia, Veterinaria, Guillem de Castro, 94, 46001 Valencia, Spain, [2]UCM, Veterinaria, Pta Hierro s/n,
28040 Madrid, Spain, [3]UAB, Veterinaria, Edifici V, Travessera dels Turons, 08193 Bellaterra, Barcelona, Spain, [4]INIA, Animal
Reproduction, Avda Pta Hierro, 28040 Madrid, Spain, [5]IVIA, CV-315, km 10,7, Valencia, 46113 Moncada, Spain, [6]EMBRIOVET
SL, Polígono Industrial de Piadela II-8, 15300, Betanzos- A Coruña, Spain, [7]COWVET SL, Avda. País Valenciano 6, 46117
Betera-Valencia, Spain, [8]Granja Mudéjar Wagyu, Calle Mayor, 44150 Aliaga-Teruel, Spain; sgarnelo@inia.es

Wagyu is well-known due to its meat quality. Beef products are from pure Wagyu (W), European Angus crossbreds (EC) and crossbred Wagyu × Angus (WN). Our goal was to compare essential amino acids (EAA) in beef, and its relationship with fat in W, WN and EC-bulls, under Spanish production conditions (farm Mudéjar-Wagyu), fed with a diet enriched in oleins. Rib beef samples from 15 W, 12 WN and 11 EC castrated males were analysed. Age at slaughter was 38.9 mo. in W, 27.9 mo. in WN and 23.3 months in EC-bulls. Percentage content in Fatty Acid (FA) was obtained (Labocor AnalíticaSL, Madrid). Saturated (SFA), monounsaturated (MUFA), polyunsaturated (PUFA) and unsaturated FA (IFA) proportions and Ω6 FA/Ω3 FA ratio were obtained. Amino acids were quantified at the Univ. Católica of Valencia. Contents (mg/100 g of fresh matter). Sum of essential AA in mg/100 g (sEAA) and percentage of EAA (PerEAA) were calculated. Data were processed with IBM SPSS Statistics v. 25.0 with Kruskal Wallis test or ANOVA, after confirmation of normal distribution. PerEAA did not differ among breeds (P=0.54; 37.7±4.03; 39.8±1.68 and 38.8±3.02% for W, WN and EC, respectively). However, sEAA (mg/100 g) was 8,980±2,040.5 in W, 9,951±1,431.2 in WN and 10,843±2,675.9 in EC (P=0.03), due to higher total AA and total proteins (TP) content [23,899±5,061.6; 25,089±4,033 and 28,178±7,504.2 mg total AA/100 g (P=0.17) and 17.4±1.82; 18.2±1.40 and 20.1±1.82 mg TP/100 g for W, WN and EC, respectively (P<0.001)]. The ultimate reason of this was the higher fat content in W and WN bulls vs EC [25.4±7.16; 23.3±4.95 and 13.4±5.49% for W, WN and EC, respectively (P<0.001)]. Correlations among PerEAA and fat content or Ω6 FA/Ω3 FA ratio were not significant. However, in W, PerEAA correlated with IFA (r=0.525; P=0.04) and MUFA (r=0.548; P=0.03) and in EC bulls PerEAA correlated inversely with IFA (r=-0.638; P=0.03). Therefore, the content of essential amino acids of bulls fed with a diet enriched in oleins, did not differ by breed, although the total AEE and protein was greater in European breeds. Funded by CDTI-IDI-20180254.

French management of beef eating quality
I. Legrand and C. Denoyelle
Institut de l'Elevage, MRA-NA, Boulevard des Arcades, 87060 Limoges Cedex 2, France; isabelle.legrand@idele.fr

France has the first cattle herd in Europe with 18.2 million heads, more than 20% of the European herd. Meat production is a major economic sector in France with more than 400,000 direct and indirect jobs. France presents a wide variety of cattle breeds. This is a major asset for French meat production, which provides different kinds of carcasses for different commercial channels. The « label rouge » quality sign is based on technical specifications defining production systems from breeders to retailers. An official technical note defines the criteria to obtain the 'label rouge'. More specifically, each 'label rouge' meat is based on the commitment of a chain including breeders, slaughterhouses, butchers and supermarkets. Each label is led by a defence and management organisation that is responsible for the functioning of the 'label rouge'. Each 'label rouge' is controlled by a certifying body. 'Label rouge' requires a total individual traceability from the animal to the steak. Each animal has a passport with different information and also two tags in each ear with the same national number to ensure the correspondence with the passport. This traceability system provides consumers with assurances on the reliability of the compulsory labelling info given to every single piece of beef on sale: its origin. This system has been implemented since 1978. 'Label rouge' is based on the strict respect and control of the specifications all over the chain. To guarantee a premium quality, the meat is analysed by consumer taste panels according to meat and fat colour, tenderness, odour, flavour and global satisfaction. At the moment, 16 different beef 'label rouge' exist all over France. In 2018, the French meat sector represented by Interbev decided to increase the volume of label rouge production up to 40% of French meat production. New specifications for example on feed and welfare will be added and a new quality measurement for marbling will be introduced. To conclude, the French meat sector has a long experience of meat quality management. Its approach is based on the use of well-known factors that influence eating quality. This approach could be complementary to the Australian one, named MSA, in order to decrease negative eating experiences with 'label rouge' products.

Use of NIR spectroscopy in the characterisation of bovine meat for the preparation of hamburgers
C. Maduro Dias, T. Melo, H. Rosa, C. Silva and A. Borba
University of the Azores, Institute of Agricultural and Environmental Research and Technology (IITAA), Rua Capitão João d'Ávila, 9700-042 Angra do Heroísmo, Açores, Portugal; cristianarodrigues@gmail.com

NIR spectroscopy provides a rapid, easy to use, and non-destructive alternative to the methods traditionally employed in the determination of the chemical composition of food stuff, since it doesn't require reagents, time consuming procedures based on manual labour, and can determine multiple parameters simultaneously. Given this, the use of NIR spectroscopy was evaluated in the quantification of the humidity, fat and crude protein contents of raw bovine meat, in order to support the development for industrial use of Azorean IGP meat hamburgers, with a target fat content of 20 g/100 g. A total of 66 meat samples were studied, among which rib mix, flank, plate ribs, chuck and rib eye, brisket, necks, foreshanks, minced meat with added fat and gourmet hamburgers. The samples were analysed according to traditional methods and their humidity, fat, and crude protein contents were determined. It was observed that the samples had a high fat content heterogeneity. The spectre of all samples was collected (400-2,500 nm) and a calibration was performed using partial least squares regression (PLS) and cross validation. Several spectral pre-treatment techniques were applied, including the standard normal variate (SNV). The best calibration for the chemical properties under evaluation showed coefficients of determination (R^2) and standard errors of cross validation (SECV) values of 0.87 and 2.79%, respectively, for the fat content, 0.80 and 1.15% for the crude protein content, and 0.72 and 2.18% for the humidity content. In conclusion, the use of NIR spectroscopy can provide a good estimate of the humidity, fat and protein contents of raw bovine meat.

Optimisation of solid phase micro-extraction methods (SPME) for beef flavour analysis

F.S. Chong[1,2], J. Kerry[1], M. O'Sullivan[1], J. Legako[3], T.D.J. Hagan[2], A.W. Gordon[2] and L.J. Farmer[2]
[1]University College Cork, Cork, Co. Cork, Ireland, [2]Institute of Agri-Food and Biosciences Institute, Newforge Lane, Belfast BT9 5PX, United Kingdom, [3]Texas Tech University, Box 42141, Lubbock TX 79409, USA; linda.farmer@afbini.gov.uk

Flavour liking of beef is increasingly important for the eating quality of beef. To understand and manage the flavour of cooked beef, it is important to optimise a simple and common measurement method for flavour volatiles. Four SPME methods (M1-4) were compared in Trial 1. Beef samples were collected and separated into four batches. In Trial 2, two striploins were collected and subjected to 5 ageing periods (4, 7, 21, 49, 120 days) and analysed using M2, M3 or M4. Beef samples were grilled to internal temperature of 65 °C. Cores (1.27 cm diameter) were collected from samples for M1 and M2 while samples for M3 and M4 were cut into 0.5 cm^3 cube, immersed in liquid nitrogen and homogenised. Headspace volatile compounds were extracted and desorbed manually in M1 while these processes were automated in M2 to M4 with gas chromatography-mass spectrometry (5977B MSD/7890B GC). CAR/PDMS fibres were used in M1 to M3 while DVB/ CAR/ PDMS fibres were used in M4. Operational and validation criteria were considered for method selection. Trial 1 showed that M1, M2, M3 and M4 detected 13, 24, 38 and 47 of the 57 selected compounds, respectively. Thus, M1 was excluded from Trial 2. Operational criteria included ease of use, amount of sample required and method flexibility. Validation criteria included detection range and quantities of volatile compounds, method reproducibility and ability to differentiate beef subjected to different ageing periods. M4 detected more of the volatile compounds but the quantities detected were lower. M1 had the highest reproducibility followed by M3. M4 differentiated best between beef samples subjected to different ageing periods. The total scores, including all criteria, for M2, M3 and M4 were 52.7, 49.5 and 55.3 out of a maximum of 70. Methods M2 and M4 performed better than methods M3 and, especially, M1, but each had their own strengths and weaknesses. M2 had fewer preparation steps while M4 offered extra flexibility and the ability to conduct multiple analyses per sample.

Classification of beef carcasses according to meat quality using pH/temperature decline descriptors

C. Xavier, U. Gonzales-Barron and V.A.P. Cadavez
Centro de Investigação de Montanha, Instituto Politécnico de Bragança, Animal Science, Campus de Santa Apolónia, 5300-253 Bragança, Portugal; vcadavez@ipb.pt

During beef carcass chilling, the eating quality of meat can be severely affected by either hot- or cold-shortening. With basis on previous knowledge that meat of optimal tenderness can be produced when rigor mortis (pH=6.0) is attained when carcass temperature falls between 12-35 °C, the objective of this study was to predict meat quality from modelled pH/temperature decay descriptors and informative animal/carcass characteristics. Temperature and pH from a total of 126 beef carcasses were logged during 24 h post-mortem, and subsequently modelled by exponential decay equations that estimated temperature (kT) and pH (kpH) decay rates. In addition, a number of pH/temperature decay descriptors were estimated from the fitted models. From linear models adjusted to each of these descriptors, it was found that, generally, hot carcass weight, age, breed, gender, age class (calf, vealer, yearling), fat cover, conformation and transport and lairage time had significant influence on pH/temperature decay. Thus, bringing together the orthogonal variables kT and kpH, and the aforementioned animal/carcass characteristics as linear predictors of discriminant functions, a classification analysis was performed.

Difference in beef features between pure and crossbred Wagyu bulls under Spanish production systems

M. Vázquez-Gomez[1], J.M. Vázquez-Mosquera[2], C. García-Contreras[3], S.S. Pérez-Garnelo[3], A. Villagrá[4], J.C. Gardón[5], D. Martínez[6], F. Sebastian[7] and S. Astiz[3]
[1]UAB, Veterinaria, Barcelona, 08193 Bellaterra, Spain, [2]UCM, Veterinaria, Avda Pta de Hierro s/n, 28040 Madrid, Spain, [3]INIA, Animal Reproduction, Avda Pta Hierro s/n, 28040 Madrid, Spain, [4]IVIA, CV-315, km 10.7, 46113 Moncada (Valencia), Spain, [5]Universidad Católica Valencia, Veterinaria, Guillem de Castro, 94, 46001 Valencia, Spain, [6]EMBRIOVET SL, Polígono Industrial de Piadela II-8, 15300 Betanzos A Coruña, Spain, [7]COWVET SL, Avda. País Valenciano 6, 46117 Betera-Valencia, Spain; astiz.susana@inia.es

The outstanding worldwide-known organoleptic characteristics of Wagyu beef have high production costs and prices. Commercially, beef from pure- and crossbred Wagyu are interesting for consumers. This study aimed to find differences in beef nutritional and quality characteristics between purebred and crossbred Wagyu bulls fed with the same diet enriched in oleins, under Spanish production conditions (farm Mudéjar-Wagyu). Sirloin and rib beef cut samples from 18 pure Wagyu and 18 crossbred Wagyu × Angus (Wangus) castrated males were analysed after slaughtering (age 38.9 mo. in Wagyu; 26.7 m. in Wangus bulls). Both beef cuts were separately analysed (Labocor AnalíticaSL, Madrid) for nutritional and quality parameters, including fatty acids (FA). Intramuscular fat (IMF) in fresh matter (FM) from dry matter (DM) values were also calculated. Saturated (SFA), monounsaturated (MUFA), polyunsaturated (PUFA), unsaturated FA (IFA) proportions and $\Omega6$ FA/$\Omega3$ FA ratio were obtained. There were differences between beef from Wagyu and Wangus bulls in the sirloin, with Wagyu bulls showing greater amount of IMF in MF (24.1±7.8 vs 19.4±6.7%; $P<0.05$) and energy (294.2±71.0 vs 247.3±54.8 Kcal/100 g beef; $P<0.05$), saturated FA (10.8±3.6 vs 8.4±3.0 g/100 g beef; $P<0.05$) and polyunsaturated FA in 100 g of beef than the Wangus group (1.0±0.4 vs 0.8±0.3 g/100 g beef; $P<0.05$). The FA proportions without the effect of IMF amount were similar, with these differences being related to the higher IMF value of Wagyu beef. On the other hand, rib beef from Wagyu also had greater IMF than that from Wangus but in DM (57.0±10.8 vs 48.8±13.0%; $P<0.05$). However, there were no differences in FM data. Hence, both breed groups produce two different commercial beef products, but further evaluations of different beef quality parameters, such as marbling or organoleptic characteristics will be useful to obtain more detailed profiles of beef from purebred or crossbred Wagyu cattle and detect more differences between them. Funded by CDTI-IDI-20180254.

Defining the ability of choline and methionine to modulate methyl donor metabolism in the liver

T.L. Chandler[1,2] and H.M. White[2]
[1]Cornell University, Population Medicine and Diagnostic Sciences, Ithaca, NY, 14853, USA, [2]University of Wisconin-Madison, Dairy Science, Madison, WI, 53706, USA; tlc236@cornell.edu

Potential overlap between choline and methionine (Met) metabolic pathways exists within both methyl donor metabolism and cellular lipid packaging, while additional demands for Met are present due to protein synthesis requirements. Use of *in vitro* models can assist in understanding the cellular priorities for and potential sparing of these nutrients across a range of treatments. Use of primary bovine hepatocytes has allowed examination of these pathways through gene expression and flux experiments, with lipid and LPS challenges. Across a series of experiments, a few patterns seem apparent. choline has a clear role in decreasing triglyceride (TG) accumulation which supports *in vivo* findings of decreased liver TG accumulation with choline supplementation. Increased FA oxidation, observed in parallel with decreased ketone body secretion in flux experiments may also contribute to decreased cellular TG. Supplementation of choline, despite Met concentration, also increases expression of genes associated with transmethylation. Interestingly, cellular glycogen was increased in cells supplemented with choline which has also been observed *in vivo*. Supplementation of Met, despite isomer, decreases expression of genes associated with transmethylation and did not alter TG accumulation or lipid oxidation. Supplementation of Met increased cellular glutathione and ameliorated inflammatory response during LPS challenge which supports improved immune response observed *in vivo*. What has yet to be reconciled is the optimal balance of reactive oxygen species (ROS) and antioxidant concentration. While Met increases glutathione and modulates inflammatory response to an LPS challenge, no change was observed in ROS. Conversely, choline supplementation did not alter glutathione concentration but has decreased ROS secretion. The optimal balance of antioxidants, oxidative stress, and inflammatory response, *in vitro* and *in vivo*, especially during homeorhetic periods, should be further examined. Primary hepatocyte culture research suggests different cellular biological priorities for both choline and Met, regardless of potential biochemical overlap.

Importance of folic acid and vitamin B12 for endogenous production of methyl groups in lactating cow

C.L. Girard and M. Duplessis

Agriculture and Agri-Food Canada, Sherbrooke Research and Development Centre, 2000 College street, J1M 0C8, Canada; christiane.girard@canada.ca

Methyl groups provided by S-adenosylmethionine (SAM) through the transmethylation pathway (TM) are needed in more than 50 metabolic reactions, among them DNA methylation and synthesis of creatine and phosphatidylcholine. Methionine (Met) is the precursor for SAM which after giving its methyl group is converted into S-adenosylhomocysteine (SAH) and then into homocysteine (Hcy). The latter can be catabolised into cysteine (Cys) or remethylated into Met. The role of the remethylation cycle is to ensure a constant supply of SAM, even when Met supply is low. Remethylation of Met can be achieved using preformed labile methyl groups provided by choline and betaine or by methylneogenesis through folate metabolism. For de novo synthesis of methyl groups, tetrahydrofolate (THF) accepts a 1-carbon unit, provided by catabolism of serine, glycine or formate, to form 5,10-methylene-THF which is converted into 5-methyl-THF. The latter transfers its methyl group to Hcy for regeneration of Met and THF under the action of an ubiquitous vitamin B_{12}-dependent enzyme, methionine synthase. In multiparous dairy cows, when Met supply was above 2.2% MP, a supplement of folic acid and vitamin B_{12} had no effect on TM but when Met supply was lower than 1.9% MP, the vitamin supplement increased TM. In the latter, the supplement of vitamins also increased *AHCY* (SAH hydrolase) expression supporting the previous observation that, when Met supply is low, methylneogenesis allows maintaining TM. The fate of Hcy flowing TM, however, varied according to plasma Met concentration. When plasma Met was low, plasma Hcy and Cys were decreased by the vitamin supplement; methylneogenesis promotes Met economy by increasing the number of times that a molecule of Hcy is remethylated within hepatic cells before being catabolised. However, when plasma Met was high, this effect is reduced and plasma Hcy and Cys increased. In lactating mature dairy cows, providing an adequate supply in folic acid and vitamin B_{12} promotes endogenous production of methyl groups when methionine supply is low.

Energy balance, adipokines and milk fatty acids of dairy cows receiving B-vitamin supplements

M. Duplessis[1], R. Gervais[2], H. Sauerwein[3], H. Lapierre[1] and C.L. Girard[1]

[1]Agriculture & Agri-Food Canada, 2000 College St, J1M0C8 Sherbrooke, Canada, [2]Université Laval, 2425 rue de l'Agriculture, G1V0A6, Canada, [3]University of Bonn, Katzenburgweg 7-9, 53115 Bonn, Germany; melissa.duplessis@canada.ca

To evaluate the effect of a supplementation of biotin (B8), folic acid (B9) and vitamin B12 (B12) on postpartum energy balance (EB), adipokines, milk yield and fatty acid (FA) profile, 32 cows were blocked based on their previous 305-d milk yield. Within each block, cows received one of the four treatments from -3 to 3 wk relative to the calving: No vitamin (B8-B9B12-); 20 mg/d of dietary B8 (B8+B9B12-); 2.6 g/d of dietary B9 and weekly intramuscular injections of 10 mg of B12 (B8-B9B12+); B8 with B9B12 (B8+B9B12+). According to a 2×2 factorial arrangement, weekly data on EB, plasma adiponectin and leptin, milk yield and FA profile were analysed as repeated measures with Proc MIXED of SAS. No treatment effect was observed on postpartum EB. Leptin and adiponectin showed a typical postpartum change and concentrations were not affected by treatments (P>0.32). Milk de novo and preformed FA concentrations were respectively lower and greater by 12% for B8-B9B12+ cows compared with B8-B9B12- cows whereas no difference was noted when cows received B8 (B8xB9B12 interaction, P=0.02). Regardless of B9B12 supplementation, feeding B8 decreased milk biohydrogenation intermediates and odd- and branched-chain FA (P<0.05), suggesting that a B8 supplement, not protected from ruminal degradation, altered rumen function. Within the first 3 wk of lactation, milk yield of B8-B9B12+ cows (45.5 SE:1.8 kg/d) was greater by 13% than B8-B9B12- cows (40.1 SE:1.9 kg/d) whereas there was no effect on cows receiving B8 (B8xB9B12 interaction, P=0.06). Although similar postpartum EB was observed, milk of B8-B9B12+ cows had greater concentration of cis-9 18:1 than B8-B9B12- cows whereas there was no effect on cows receiving B8 (B8xB9B12 interaction, P=0.01), suggesting that B8-B9B12+ cows mobilised more body fat reserve to support their greater milk production. However, fat mobilisation was probably not excessive as plasma beta-hydroxybutyrate was not different among, suggesting that B9B12 improved energy metabolism in early lactation.

Methyl donor and fatty acid interactions in the dairy cow: classic viewpoints and new perspectives

J.W. McFadden
Cornell University, Animal Science, 264 Morrison Hall, Ithaca, NY 14853, USA; mcfadden@cornell.edu

Our classic understanding is that methyl donor (e.g. choline and methionine) and fatty acid (FA) nutrition influences milk production and health in dairy cattle. The interactions and potential synergism between methyl donors and FA has received limited attention. This presentation will review new insights to justify such study. First, gastrointestinal degradation of choline by bacteria forms trimethylamine, which is converted to trimethylamine N-oxide (TMAO) in liver. In non-ruminants, TMAO is a risk-factor for metabolic disorders. Evidence suggests that TMAO does not modify metabolism or compromise health in early lactation cows; however, dietary FA may be a means to impact trimethylamine formation and choline bioavailability. Second, hepatic methyl donor metabolism is critical for the synthesis of phosphatidylcholine (PC), which is necessary for very-low-density lipoprotein assembly, triglyceride secretion, and fatty liver prevention in transition cows. It has been postulated that the cytidine diphosphate pathway, which relies on choline, preferentially utilises saturated and monounsaturated FA for PC synthesis. Whereas, the phosphatidylethanolamine (PE) N-methyltransferase pathway, which utilises methyl groups from methionine, prefers PE enriched in polyunsaturated FA. Emerging data suggests that the risk for fatty liver in cows is related to the FA composition of PC and inadequate hepatic polyunsaturated FA may limit the ability of methyl donors to stimulate PC synthesis during the peripartum. Third, an omic-discovery has revealed a potential role for lysophosphatidylcholine (LPC) within the context of immunity in dairy cattle. Circulating LPC predominantly include C16:0-, C18:0- and C18:1-LPC; however, a lipopolysaccharide infection markedly decreases these and other LPC in bovine circulation. Although the importance is still being defined, *in vitro* findings reveal the ability of LPC to influence bovine neutrophil function in a manner dependent upon their acyl-moiety. These contemporary perspectives are evidence that our understanding of bovine methyl donor and FA nutrition remains incomplete but continued investigation will likely result in feeding recommendations that consider methyl donors and FA in unison to optimise cow performance.

The rumen- protected amino acids affect sheep milk chemical composition and antioxidant capacity

E. Tsiplakou, A. Mavrommatis and C. Mitsiopoulou
Agricultural University of Athens, Department of Nutritional Physiology and Feeding, Iera Odos, 11855 Athens, Greece; eltsiplakou@aua.gr

Rumen protected amino acids further to its positive impact on milk yield and chemical composition might involve also in the antioxidant defence system. Moreover, not only the dietary inclusion level of rumen protected amino acids but also the ratio among them might have important role in the aforementioned parameters. In order to test this hypothesis, sixty Chios dairy ewes were assigned to one of the following five dietary treatments: A: basal diet (control group); B: basal diet +3.6 g/head rumen-protected methionine; C: basal diet + 3.4 g/head rumen-protected lysine; D: basal diet +3.6 g/head rumen-protected methionine + 3.4 g/head rumen-protected lysine and E: basal diet +7.2 g/head rumen-protected methionine + 3.4 g/head rumen-protected lysine. The results showed a significant increase in FCM6%, ECM, and both protein and fat milk yield in ewes fed with methionine (B) only compared with the control. The total milk caseins content of ewes fed the C compared with the A diet was significantly higher. The proportion of short chain fatty acids in ewes' milk declined significantly in the case of those fed with the E diet. The dietary treatments revealed minor effects in the individual fatty acids of ewes' milk. The glutathione transferase activity increased significantly in the blood plasma of ewes fed with the B, C and D treatments in comparison with the normally fed. The malondialdehyde content in the blood plasma of ewes fed with B compared with the A diet decreased significantly. A trend for increase was observed in the superoxide dismutase activity in ewes' milk fed with the B, C and D compared with A diet. In conclusion, the dietary supplementation with rumen protected amino acids in ewes, appears to be not only a successful strategy for their milk performance but also an important issue for further research relate with their antioxidant impact in both animals' organism and milk.

The role of rumen-protected amino acids on the expression of genes involved on ewes' immune system

E. Tsiplakou[1], A. Mavrommatis[1], D. Skliros[2], F. Righi[3] and E. Flemetakis[2]
[1]*Agricultural University of Athens, Nutritional Physiology and Feeding, Iera Odos, 11855 Athens, Greece,* [2]*Agricultural University of Athens, Molecular Biology, Iera Odos, 11855 Athens, Greece,* [3]*University of Parma, Veterinary Science, Via del Taglio 10, 43126 Parma, Italy; eltsiplakou@aua.gr*

Rumen protected amino acids inclusion in ewes' diets has been proposed to enhance their innate immunity. To test whether this is accompanied by changes in the expression levels of key-genes involved in these processes (TLR-4, NF-κB, TNFa, IL-1β, IL-2, IL-6, IL-8, CCL-5 and CXCL-16), sixty Chios dairy ewes were assigned to one of the following five dietary treatments: A: basal diet (control group); B: basal diet +3.6 g/head rumen-protected methionine; C: basal diet + 3.4 g/head rumen-protected lysine; D: basal diet +3.6 g/head rumen-protected methionine + 3.4 g/head rumen-protected lysine and E: basal diet +7.2 g/head rumen-protected methionine + 3.4 g/head rumen-protected lysine. The results showed a significant decline in the mRNA expression of the TLR-4 gene in both the macrophages and the neutrophils of ewes fed rumen-protected amino acids. Significantly lower mRNA transcript accumulation was also observed for the TNFa, IL-1β and CXCL-16 genes in the macrophages and for the IL-1β gene in the neutrophils of ewes supplemented with amino acids. Diet D increased the mRNA expression of the IL-2 gene in ewes' neutrophils. In conclusion, our results suggest that the dietary supplementation of ewes with rumen-protected amino acids, down regulate the expression of some genes involved in the pro-inflammatory signalling.

Rumen-protected choline supplementation in dairy ruminants: the current status of knowledge

L. Pinotti, M. Ottoboni, A. Luciano and A. Baldi
Università degli Studi di Milano, Department of Health, Animal Science and Food Safety, Milano, 20134, Italy; luciano.pinotti@unimi.it

The productivity of high-yielding dairy cows has increased over the past decades and the milk yield has doubled. Such increases in milk production have been observed not only in dairy cows but also to some extent in other dairy ruminants such as ewes, goats and buffaloes (Bubalus bubalis). As a consequence, in all specialised dairy ruminants it is essential to optimise the macro and micro-nutrient supply, especially during the most critical period in the animals' production cycle i.e. from parturition until the peak of lactation. In this critical phase, an array of factors can enhance the balance between the intake and demand for nutrients, although the availability and supply of the selected micronutrients is also important. Among these choline (provided in a rumen-protected form) supplementation has been reported to have an effect on milk yield and metabolic health in lactating dairy ruminants. However, while direct and indirect evidence about transition dairy cows has been addressed in a few meta-analyses, in the case of other dairy ruminants, the effects of rumen protected choline have been less investigated. The limited data available indicate that choline seems to be effective in increasing milk production in buffaloes, goats, and ewes especially at the onset of lactation as reported for cows. The effects of rumen-protected choline on metabolic profile in dairy ruminants, other than cows, remain largely inconclusive. This review highlights the key knowledge on choline supplementation in dairy ruminants and its potential role in optimising milk production and metabolic health.

Responses to methionine, lysine plus histidine and energy supply on mammary metabolism in dairy cows

C. Omphalius[1,2], H. Lapierre[3], L. Bernard[4], L. Bahloul[1] and S. Lemosquet[2]
[1]Adisseo France SAS, CERN, 03600 Commentry, France, [2]INRAE, Agrocampus Ouest, PEGASE, 35590 Saint-Gilles, France, [3]Agriculture and Agri-Food Canada, Sherbrooke, QC, Canada, [4]INRAE, UMRH, 63122 Saint-Genès Champanelle, France; c.omphalius@gmail.com

Our objective was to study how balancing Lys, Met and His (LMH) at 2 supplies of net energy of lactation (NE_L) affect milk yield and composition. Four Holstein cows received duodenal infusion of 2 amino acids (AA) mixtures and 2 dietary levels of NE_L (E-: 31.5 vs E+: 35.9 Mcal/d), combined in factorial arrangement, distributed in a 4×4 Latin square, with 3-wk periods. The 2 AA mixtures, both providing 92±2 g/d of PDI (metabolisable protein in INRA, 2018), contained (g/d) Lys (28), Met (9), His (9) and non-essential AA (NEAA) in LMH+ or only NEAA in LMH-. Diets plus infusions provided 5.8, 2.0 and 2.0% of PDI of Lys, Met and His respectively in LMH- compared to 7.1, 2.4 and 2.4% PDI in LMH+; LMH+ met INRA (2018) requirements of 7.0, 2.4 and 2.4% PDI for these 3 AA, respectively. The mammary uptake (MU) of metabolites were measured on day 20 (6 samples at 2-h intervals) of each period, mammary plasma flow being estimated using the Fick principle (Phe+Tyr). Analyses of variance of these data included the effects of LMH, NE_L and interaction using a MIXED procedure (SAS) with cow as random effect. No interaction was observed on milk composition (P>0.22). Balancing LMH tended to increase (P=0.07) and increasing NE_L supply increased (P≤0.02) milk and fat yields averaging respectively 28.2, 29.6, 30.1 and 31.2±1.33 kg/d and 1,289, 1,317, 1,363 and 1,410±55.1 g/d for E-LMH-, E-LMH+, E+LMH- and E+LMH+. Both treatments increased (P<0.05) milk protein yield (828, 916, 927 and 962±39.2 g/d for E-LMH-, E-LMH+, E+LMH- and E+LMH+). The increased milk protein yield were linked to increased MU of LMH in response to LMH supply (+ 8.4 mmol N/h; P=0.01) and to increased MU of essential AA other than LMH in response to NE_L supply (+ 11.8 mmol N/h; P=0.05). Increasing NE_L supplies increased milk fatty acids *de novo* synthesised within the mammary gland (+58 mmol C/h; P=0.02) and MU of β-hydroxybutyrate (+171 mmol C/h; P=0.03). It increased milk yield through an increased lactose yield (+ 97 g/d; P=0.01) while LMH supply had no effect on it (P=0.21). Glucose and acetate MU only tended to be higher in E+LMH+ than in the other treatments (LMH × E: P≤0.06). In conclusion, balancing LMH increased milk, protein and fat yields through different mechanisms than increasing NE_L supply indicating the high mammary gland metabolic flexibility to nutrient availability.

Interobserver- and retest-reliability of different human-animal-relationship tests in dairy cattle

K. Krugmann, C. Möhlmann, J. Krieter and I. Czycholl
Christian-Albrechts-University Kiel, Institute of Animal Breeding and Husbandry, Olshausenstr. 40, 24098 Kiel, Germany; kkrugmann@tierzucht.uni-kiel.de

Animal welfare constitutes an important issue for society and the objective assessment is continuously discussed. An acceptable animal welfare standard includes a good human-animal relationship, which has to be measured reliably. This study examines interobserver and retest-reliability of five human-animal-relationship tests (novel-object-test (NOT), voluntary human-approach test (VHAT), ly pass test (LPT), avoidance distance test at the feeding fence (ADTF) and avoidance distance test in the stable (ADTS)). Therefore, two trained observers tested 924 randomly selected cows of 11 farms three times (each time with one week in between) within six months. To assess the behaviour different variables were used: Scoring systems (score 0 until 4) ranging from disinterest until contact (NOT) respectively until hectically standing up (LPT) or from fleeing to touching head or body (ADTF, ADTS) as well as approach latencies to the human or novel object (NOT, VHAT). The results were examined with regard to interobserver- and retest-reliability using a combination of parameters (Spearman Rank Correlation Coefficients (RS), Intraclass Correlation Coefficients (ICC), Limits of Agreement (LoA). RS respectively ICC equal to or greater than 0.4 was interpreted as an acceptable reliability and greater than or equal to 0.7 as a good reliability. Regarding LoA an interval less than or equal to -0.1 to 0.1 was interpreted as acceptable agreement. All examined tests revealed acceptable to good interobserver reliabilities regarding the individual scores (e.g. NOT: RS=0.71 – 0.98; ICC=0.64 – 0.99; LoA ∈ [-1.28-1.27]), whereas the retest-reliabilities indicated no acceptable results except for score 3 (calmly standing up) of the LPT between times of testing 1 and 2 (RS=0.55; ICC=0.42; LoA = -2.45-2.72). This low retest-reliability could potentially be improved by changes in the test procedure as e.g. a targeted selection of the animals, execution of the tests in a test arena or a clearer gradation of the individual scores.

Ingestive and postural behaviour in frequently regrouped heifers

M. Moratorio[1], F. Silveira[1], A. Ruba[1], M. Carriquiry[2], R. Ungerfeld[3] and C. Fiol[1]
[1]Departamento de Producción Animal, Facultad de Veterinaria, Universidad de la República, Alberto Lasplaces 1550, 11600, Uruguay, [2]Departamento de Producción Animal y Pasturas, Facultad de Agronomía, Universidad de la República, Av. Gral. Eugenio Garzón 780, 12900, Uruguay, [3]Departamento de Biociencias Veterinarias, Facultad de Veterinaria, Universidad de la República, Alberto Lasplaces 1550, 11600, Uruguay; moratoriomariana@gmail.com

The aim was to determine the effects of social regrouping (SR) on ingestive and postural behaviours in dairy heifers. Holstein heifers (153.3±16.1 kg, 11 mo-old) were allocated to 2 homogenous groups according to body weight, age and farmer origin: 1) Control (CON; n=14 + 5 'fixed') and 2) Regrouped (RG; n=14 + 5 'exchanged'), in which 'exchanged' were switched for 5 unknown heifers every 21 d (total=6 SR; 1st regrouping=SR 1). The day after each SR, ingestive (eating, ruminating) and postural (lying, standing) behaviours were registered (n=14/group) through scan sampling every 10 min during 8 h. Data were analysed by Proc Glimmix. Results are presented as proportion of time/8 h. There was no treatment effect in eating behaviour (0.45 vs 0.46±0.008; ns), but CON spent more time eating (P<0.05) in SR 1 (0.37 vs 0.30±0.02) and 4 (0.64 vs 0.55±0.02) than RG heifers, and RG spent more time eating in SR 6 (0.55 vs 0.31±0.02; P<0.05) than CON heifers. Ruminating (0.24 vs 0.21±0.006) and lying (0.18 vs 0.14±0.006) time were greater (P<0.05) in CON than RG heifers, and there was a group×SR effect (P<0.05) for both activities: CON heifers spent more time ruminating than RG in SR 1 (0.26 vs 0.15±0.01), 5 (0.22 vs 0.12±0.01) and 6 (0.28 vs 0.23±0.01), while RG were more time ruminating in SR 3 (0.32 vs 0.22±0.02) than CON heifers. In SR 1, RG were more time lying (0.26 vs 0.17±0.01) than CON, but CON were more time lying in SR 2 (0.19 vs 0.14±0.01), 3 (0.28 vs 0.19±0.02), 5 (0.09±0.01 vs 0.06±0.008) and 6 (0.32 vs 0.15±0.02) than RG. Regrouped heifers tended to spent more time standing than CON heifers (0.07 vs 0.06±0.004; P=0.06), and RG were more time (P<0.05) standing in SR 2, 4 (0.09±0.01 vs 0.06±0.009, in both) and 5 (0.16±0.01 vs 0.07±0.009) than CON heifers. In contrast, CON were more time standing in SR 6 (0.05±0.008 vs 0.02±0.006; P<0.05) than RG heifers. In conclusion, frequently regrouped heifers were less time ruminating and lying, and more time standing than non-regrouped ones, which might be explained by greater social instability in RG heifers.

Evaluation of methods for assessing positive emotions in dairy calves in groups for usage by farmers

M. Guse, J. Nachtsheim, F. Wolbers, J.J. Hayer and J. Steinhoff-Wagner
University of Bonn, Institute of Animal Science, Katzenburgweg 7-9, 53115 Bonn, Germany; jste@itw.uni-bonn.de

Animal welfare has become increasingly important in today's society. However, most attention is focused on the avoidance of negative issues. Methods for assessing positive emotions in young calves are currently in discussion in regard to quality criteria. For this purpose, this study examined and compared four modified methods to be used by farmers with regard to their validity, objectivity and reliability as well as their on-farm applicability. The methods examined, were two Qualitative Behaviour Assessments (QBA, simplified QBA), one Reactions to a Stationary Human Test (RSHT) and one Novel Object Test (NOT). All protocolls were carried out at the experimental farm of the University of Bonn in triplicate (3 observers). The experimental group consisted of 12 Holstein-Friesian calves in group housing with an automatic feeder. For statistical analysis Spearman's rank correlations, generalised linear models and a mixed model were used. The activity behaviour (QBA) and the exploratory behaviour of the calves (approach time, RSHT and NOT) were influenced by the time of performance (P<0.05). Scores were lower between 10-12 am than between 9-10 am and 13-15 pm. This resulted in an overall unreliability. In both QBA methods and the RSHT, the identical trained observers also rated the behaviour different (P<0.05). Validity of RSHT tested trough different test humans and different clothing (P<0.05) was not sufficient. Whereas no influence was found between different objects during the NOT. The simplified QBA were easiest applicable. In summary, evaluated methods showed disadvantages in one or more of the tested criteria. Daily routine of the animals influenced their behaviour and needed to be considered farm specific. The simplified Qualitative Behaviour Assessment and the NOT seemed to be the most promising on-farm method for calf groups.

Relationship between tail biting in pigs and tail posture

T. Wilder[1], J. Krieter[1], N. Kemper[2] and K. Büttner[1]
[1]Christian-Albrechts-University, Institute of Animal Breeding and Husbandry, Olshausenstr. 40, 24098 Kiel, Germany,
[2]University of Veterinary Medicine Hannover, Foundation, Institute for Animal Hygiene, Animal Welfare and Farm Animal
Behaviour, Bischhofsholer Damm 15, 30173 Hannover, Germany; twilder@tierzucht.uni-kiel.de

Tail biting (TB) in pigs affects the animals' health and welfare and causes costs for the farmer. The most common measure against TB is tail docking, which only reduces the frequency of TB. Therefore, early indicators of TB need to be identified to prevent TB before severe tail lesions occur. One of those indicators is the tail posture (TP) and previous studies could show the correlation between a lowered TP and tail lesions. This study investigated the direct impact of TB on TP. Therefore, 6 commercial pens each with 24 individually marked pigs were recorded by video cameras after the weaning for 40 days and the tail lesions were scored twice a week. The observation period for each pen was defined as 4 days before the first large tail lesions (larger than the diameter of the tail) were detected. From 10:00 to 18:00 TB behaviour was analysed using continuous event sampling and a TB event was defined as any manipulating, sucking or chewing on a pen mate's tail. In the same time TP was analysed using scan sampling of all standing pigs every 20 min. A lifted or curled tail was defined as a raised TP and a hanging or tucked tail was defined as a lowered TP. This study analysed, whether TP is affected if the pig was bitten in the 20 min before or not. A generalised linear mixed model was created with the binary response variable TP (0: raised TP; 1: lowered TP) and the fixed effects TB (0: was not bitten; 1: was bitten), pen (1-6), day (1-4), hour (10:00-18:00), the interaction between pen and day and the pig nested in the pen as a random effect. The results show that the probability for a lowered TP was 1.59 times higher for pigs that were tail bitten in the 20 min before (0: 54.8±2.8%, 1: 65.9±2.7%, P<0.05). Besides, the probability for a lowered TP increased from 13:00 to 14:00 (13:00: 54.3±3.7%, 14:00: 65.9±3.1%, P<0.05) and increased from day 1 to day 4 in 4 of 6 pens (exemplary pen 1, day 1: 29.0±6.6%, day 4: 69.1±6.7%, P<0.05). This study shows that TB affects TP and therefore TP can be used as an indicator of TB.

Characterisation of tail biting outbreaks in rearing piglets: identification of early signals

S. Ammer[1], J. Ahlhorn[1], M. Wutke[2], S. Zeidler[2] and I. Traulsen[1]
[1]Livestock Systems, Department of Animal Sciences, Georg-August-University, Albrecht-Thaer Weg 3, 37075 Göttingen,
Germany, [2]Breeding Informatics, Department of Animal Sciences, Georg-August-University, Margarethe von Wrangell-Weg
7, 37075 Göttingen, Germany; stefanie.ammer@uni-goettingen.de

The objective of this study was to characterise various processes of tail biting outbreaks in rearing piglets in order to identify early indicators of a tail biting begin. Therefore, 630 undocked piglets over 9 consecutive batches were examined during their 6-week rearing period. After weaning at day 27 p.p., piglets were randomly grouped in 10 conventional rearing pens with 7 piglets/pen. Twice a week tail lesions (score 0 to 3), losses (score 0 to 4) and posture (score 0 to 5) were assessed animal individually. Additionally, the weight gain during rearing was measured. Furthermore, the water consumption per pen as well as the climatic conditions inside the barn were continuously monitored in 15-min intervals. Continuous video records enabled the analysis of animal behaviour in each pen. Videos were analysed automatically in terms of the number of pixel changes, which is to indicate the activity of the piglets. For 21.3% of the piglets tail lesions were assessed from the first day of rearing, while 3.5% remained uninjured until day 42. Almost 60% of the piglets had the first tail lesion until day 8 of rearing, which were up to 50% assessed as light lesion. However, the later the first tail lesion occurred, the higher was the proportion of severe lesions. The current analysis of specific groups of animals showed that piglets without involvement in tail biting had higher weaning weights (8.2 kg) than piglets which were biter and/or victims in bite events (7.8 kg). The amount female piglets which were assessed as biter was greater (68%) than for male piglets (32%). The automatic activity analysis showed highest pixel changes in pens with severe tail lesions compared to pens assessed with light and medium lesions (P<0.01). However, on the days before tail biting was observed the pixel changes were not different between the pens scored from 0-3. In conclusion, a characterisation of tail biting processes is necessary to develop suitable indicators for a biting begin to ensure early prevention of severe outbreaks.

Fattening pigs display weak behavioural changes due to gastric ulcerations

M. Holinger[1,2], B. Früh[2], P. Stoll[3], M. Kreuzer[1] and E. Hillmann[1,4]
[1]ETH Zürich, Universitätsstrasse 2, 8092 Zürich, Switzerland, [2]Research Institute of Organic Agriculture FiBL, Animal Science, Ackerstrasse 113, 5070 Frick, Switzerland, [3]Agroscope, Rte de la Tioleyre 4, 1725 Posieux, Switzerland, [4]Humboldt-Universität zu Berlin, Philippstrasse 13, 10115 Berlin, Germany; mirjam.holinger@fibl.org

Gastric ulcers and erosions in the stomach are a common health problem in fattening pigs. However, such changes can only be assessed after slaughtering. Potential behavioural indicators that allow for an early detection would be favourable. In a study with 147 male (entire and castrated) fattening pigs kept in groups of three, we conducted behavioural observations for 2 days shortly before slaughtering with an average live weight of 102 kg. We recorded body positions, feeding, drinking and social interactions. After slaughtering, stomachs were inspected for signs of erosions or ulcers. First, behavioural outcomes were modelled using linear mixed effect models with random, but without fixed effects. The residuals obtained were then analysed with linear discriminant analysis (LDA). In a second step, the prevalence of gastric ulceration was analysed using a generalised linear mixed effect model with behavioural variables as fixed effects. From 96 focal pigs, 16 (16.7%) were found to have erosions or ulcers. The LDA did not reveal satisfactory levels of correct classification. While it was possible with this approach to classify healthy pigs as healthy (93.7%), pigs with erosions or ulcers were classified only partially as sick (6.2%). By modelling the gastric ulceration score with behavioural variables, we found that affected pigs tended to lie more in contact with other pigs, to show more manipulations of pen mates and to significantly lie longer in total and feed longer on concentrate. These results indicate that there are small differences in behaviour between healthy and affected pigs. Automated behavioural observations with a higher resolution could provide more insights into how pigs show pain or discomfort due to gastric ulcerations.

Do sows have individual nest building profiles?

A. Wallenbeck[1,2], B. Algers[2], L. Rydhmer[1] and R. Westin[2]
[1]Swedish University of Agricultural Sciences, Department of Animal Breeding and Genetics, Box 7023, 750 07 Uppsala, Sweden, [2]Swedish University of Agricultural Sciences, Department of Animal Environment and Health, Box 234, 532 23 Skara, Sweden; anna.wallenbeck@slu.se

Through evolution, sows have developed the ability to build a nest before farrowing in order to shelter the piglets and increase their chance of survival. Nest building is one of the most important natural behaviours of sows and although domestic sows are kept indoors with resources to keep the piglets warm, they are still performing specific behaviour sequences related to nest building. The overall aim of this study is to explore variations in nest building profiles in sows, including repeatability over parities. Nest building behaviour was analysed from videos during the last 18 hours prior to farrowing. The data include 54 sows farrowing in individual loose housing pens with two farrowings per sow. All nest building events (rooting, pawing, arranging material, etc.) were recorded and summarised per hour in relation to the start of farrowing (defined as first piglet being born). There was a large variation between sows in total amount of nest building and in nest building profiles over time. A preliminary descriptive analysis indicates repeatable individual nest building profiles for the majority (63%) of the sows. Some of them performed more nest building during the first 9 hours and some during the last 9 hours before farrowing, while others performed only a little nest building during all 18 hours. The results thus indicate that there are different types of nest building profiles with regard to timing and amount of nest building behaviour. The nest building profiles will be thoroughly studied with trend analysis.

Relationships between temperament and productive traits in Pyrenean cattle

L.X. Estevez-Moreno[1], G.C. Miranda-De La Lama[1], L. García-Pastor[2], J.A. Abecia[1], P. Santolaria-Blasco[1] and G.A. Maria[1]
[1]*University of Zaragoza, Animal Production and Food Science, Miguel Servet 177, Universidad de Zaragoza, Facultad de Veterinaria, 50013, Spain, [2]Asociación Aragonesa de Ganaderos de Raza Pirenaica, Asociación Pirenaica, Huesca, Spain, 50.000, Spain; levrino@unizar.es*

This study aimed to identify relationships between temperament and performance of Pyrenean breed heifers and cows (n=262). The animals' behaviour in a chute (RC) and their exit mode form the chute (EC) were observed. In RC animals were classed as calm, restless or trying to escape. In EC animals were classed as walking, trotting or running. Reactions to the approach of a handler were classed in two categories: up to 5 m (D5) and 1 m (D1) distances. Behaviour during the test (RH) was rated as 'friendly and exploratory', 'stood still', 'walked away', 'ran away or aggressive attitude'. Bivariate analysis using chi-square, T-Student, U-Mann Whitney, ANOVA or Kruskal Wallis tests, were used to identify associations (P<0.05) between the observed behaviours and productive traits (age, parity, postpartum days, postpartum interval, mean calf live weight, mean carcass weight). Most animals (82.4%) remained calm while restrained in the chute, and 97.3% exited it walking. Additionally, 30.9% did not allow the handler to approach within 5 m and 59.2% allowed the handler to approach within 5 m but walked/ran away before reaching the 1 m distance. When the handler approached the animals 68.7% walked away, 21.4% ran away, and just 9.9% remained still. Significant associations were observed between RC behaviour and parity, where more reactive animals had a higher mean parity, compared to calmer animals (6.28 vs 4.95, P<0.05). Behaviour of running away from the chute were associated with parity ≥5 (P<0.05). D5 animals were more frequent in animals of 0 to 2 years or >8 years (P<0.05). The same relationship was observed in cows of 7 to 182 postpartum days (P<0.01). For RH, a significant association (P<0.05) was found between the behaviour of walking away and a parity of 4 to 6, and of staying still and a parity >6. Cows with less reactive behaviours showed lower mean calf weight at birth and mean carcass weight. Our results suggest that cows' temperament is associated with their performance, which should be taken into account by the system managers.

Tail biting in Switzerland: a retrospective study

C. Ollagnier and G. Bee
Agroscope, Swine production group, Tioleyre, 4, 1725, Switzerland; catherine.ollagnier@agroscope.admin.ch

Tail-biting is an abnormal behaviour, which can be triggered by many different factors like lack of enrichment material, unfavourable environmental conditions, unbalanced diet, and impaired health condition. This study assessed the changes in feeding behaviour during an outbreak of tail biting occurring in a feed efficiency study. Seventy-two castrated male pigs (110.0±10.7 days old; 43.0±9.2 kg) were group housed in a 78 m^2 pen with straw in racks and sawdust on the floor. Pigs had restricted feed access (80% of assumed *ad libitum*) distributed via 7 individual automatic feeders. The dietary crude proteins and essential amino acids were limited to 80% of the Swiss recommendations. A tail-biting outbreak occurred two months after this feeding regimen was set in place. To better characterise this outbreak, the feeding behaviour (e.g. time spent at the feeder, number of feeder visits, daily feed consumption, average daily gain and feed efficiency) was analysed. Data was divided in 3 phases (of one week each): phase A: before any tail lesions; phase B: acute tail lesions; phase C: no visible tail-biting behaviour, after restoring *ad libitum* feeding and removing the tail-biting initiator. The number of feeder visits was different (P<0.01) between each phase, and reached its minimum during phase B with 2.35 and 0.98 less feeder visits per day, compared to phases A and C, respectively. The consumption time (for a fixed quantity of feed eaten) decreased while tail-biting outbreak evolves: Pigs ate faster (P<0.001) in phases B and C, compared to phase A. These findings could be an indicator of a putative increased level of stress. Thus, one can conclude that feeding behaviour traits may be potential precursor indicators of tail-biting outbreaks.

Tail docking or tail biting in pigs?
M. Parreiras, S. Silva, S. Botelho and D. Monteiro
UTAD, Animal Science, Quinta de Prados, 5000-801 Vila Real, Portugal; sbotelho@utad.pt

Tail docking is a usual practice in pig farming in order to reduce the tail biting. However, the EU legislation states that tail-docking should not be done routinely. Although the incidence of tail-biting behaviour is higher in intact-tailed pigs compared to the cut-tailed animals, leading to the conclusion that tail-cutting is a useful measure in reducing bites. On the other hand, it is well documented that tail biting is a multifactorial problem, where factors such as lack of environmental enrichment, animal density, inadequate nutrition, and environmental factors are considered relevant to this problem. In this context, the objective of the present study is to evaluate the effect of uncutting tail has on the productive performance of animals (weight, mortality, the incidence of bite), by following animals since birth until the end of the fattening phase. The animals under study were distributed in two experimental groups: pigs with cut tail and pigs with intact tail. During lactation period: 12 litters, 12 piglets each litter, 6 replicas, 0.23 m^2 piglet-1; rearing: 6 pens with groups of 24 pigs, 3 replicas, 0.35 m^2/pig and; fattening: 6 pens with groups of 24 pigs, 3 replicas, 0.88 m^2/pig. Continuous variables were compared by simple analysis of variance (ANOVA), using the Tukey test to compare means. Significant difference results were considered when P<0.05. Differences in the number of animals bitten or not bitten were analysed using the χ^2 test. The results show the fact that in general there is no difference in the productive parameters in the animals without bite problems. However, it is possible to highlight the high percentages of bite in rearing and fattening, and the incidence of the biting behaviour of animals with the intact tail was 84 times higher than the other animals. Concerning fattening this problem is five times higher when comparing the two groups again. By analysing the rearing and fattening animals as a whole, it was observed that the intact tailed pigs were 27 times more bitten than the animals with the cut tail. The results show the need to improve all the factors related with the tail biting problem in order to avoid the consequences during a suppression of tail cutting.

Welfare assessment and wool cortisol determinations in feedlot Castellana and INRA 401 lambs
R. Bodas[1], T. Peric[2], A. Comin[3], M. Montañés[1] and J.J. García-García[1]
[1]Agrarian Technological Institute (ITACyL), Avda. Burgos, km 119, 47071 Valladolid, Spain, [2]Univerza v Novi Gorici, Vipavska cesta, 5000 Nova Gorica, Slovenia, [3]University of Udine, Via delle Scienze, 206, 33100 Udine, Italy; bodrodra@itacyl.es

The AWIN (welfare assessment protocol for sheep) and wool cortisol (WCC) as a retrospective marker of the hypothalamic-pituitary-adrenal (HPA) axis activity were monthly assessed in 15 Castellana (C) and 15 INRA 401 (I) lambs from 1 to 3 months of age. Lambs were intensively reared in 2 independent farms (one per breed) participating in the EcoLamb project (ERA-Net SusAn funded). Data on qualitative behaviour assessment (QBA, items were being scaled from 0 -absence- to 10 -all the animals fully expressed the evaluated item-), familiar approach and fleece quality tests were subjected to descriptive statistical analyses. At the same time, wool samples from the shoulder of each lamb were taken (the first at 1 month of life and the 2 following as re-growth wool during the 2nd and the 3rd month of life) and assessed for WCC by AlphaLISA® method. Positive (8.3 and 8.9 for C and I) and negative (0.2 and 0.6 for C and I) QBA descriptors averaged similar results between breeds, as well as familiar approach and fleece quality, albeit some scores varied with time: e.g. contrary to expected, the average number of animals ruminating while being observed decreased with time. Conversely, WCC differed between breeds (P=0.01) and time (P<0.01): the mean (±SE) values were particularly high in the 1st month of life (31.1±2.7 and 20.1±2.8 pg/mg for C and I, respectively) while they were decreased for the following 2nd (2.3±0.3 and 1.9±0.3 pg/mg for C and I) and 3rd (2.5±0.5 and 1.5±0.5 pg/mg for C and I) sampling. Beyond the complexity of assessing animal welfare, these first results apparently show that there is no direct relationship between values of QBA descriptors and actual cortisol contents in wool of young lambs. The AWIN is made on the group level while physiological measures as WCC are individually based and give information about the allostatic load of the singular animal.

Behaviour and growth performances in growing rabbits submitted to time restrictions on access to wat

V. Pinheiro, J. Gonçalves and D. Monteiro
UTAD, Animal Science, Quinta de Prados, Vila Real, P.O. Box 1013, 5001-801, Portugal; divanildo@utad.pt

Reducing food intake, directly or indirectly by reducing water intake, can be used in rabbit farming to reduce the incidence of digestive disorders and mortality. The aim of this work was to study the productive performance and behaviour of fattening rabbits, submitted to different times of water availability. 144 New Zealand × Californian rabbits of both sexes were divided into four groups (36 animals in each group, 4 per cage and 9 repetitions). We proceeded to a water restriction comparing a control treatment (C) with permanent access and three different times of access: 3 h/day morning starting 9:00 hours (3AM), 3 h/day evening starting 18:00 hours (3PM) and 6 h/day morning and evening starting 9:00 hours (6AMPM). The rabbits were controlled from 35 days (weaning age) to 57 days, and the weight and food consumption were monitored weekly to determine productive performances. The behaviour was monitored with weekly filming. The animal's behaviour was not significantly affected by the water restriction (P>0.05). However, we observed that the sum of all parameters related to abnormal behaviour tended to be significant (P=0.08) with the control treatment having the lower values. Daily weight gain was significantly affected by water restriction (P<0.05), with a decrease of 23% for 3AM. The 3PM and 6AMPM treatments also decreased 16 and 10% compared with the control group, but without significant effect. Despite this different growth, final body weight was not significantly affected by the treatment. A restricted access to drinking water of 3AM, 3PM and 6AMPM induced feed restriction in growing rabbits to 87, 85 and 86% of the *ad libitum* level, respectively, but without significant effect (P>0.05). Feed conversion ratio was only improved with hydric restriction in 6AMPM, but without significant effect (P>0.05).

The Sustainable Meat project: science based communication to contrasts fake news drift

A. Bertaglio[1], G. Pulina[2], E. Capri[3] and E. Bernardi[4]
[1]Associazione Carni Sostenibili, Online Communication, Piazza di Spagna, 35, 00187, Roma, Italy, [2]Università di Sassari, Agraria, Via De Nicola, 9, 07100, Sassari, Italy, [3]Università Cattolica del Sacro Cuore, Chimica Agraria, Via Emilia Parmense, 84, 29122, Piacenza, Italy, [4]Università degli Studi di Bari Aldo Moro, Bioscienze, Biotecnologie e Biofarmaceutica, Piazza Umberto I, 70121, Bari, Italy; andrea.bertaglio@gmail.com

Meat is a food of primary importance. It is however, at least for the last two decades, subjected to numerous attacks and criticisms. Among the main accusations that stand out are its environmental impact and the supposed health problems connected with it. With the awareness that sustainability in meat is a complex and debated subject, the Sustainable Meat Project (Carni Sostenibili) wants to identify the key issues, the state of knowledge and the latest trends in scientific and technical guidelines. The intent is to show that production and consumption of meat can be sustainable, both for health and for the environment. Various kinds of organisations and stakeholders participate to the debate on the production and consumption of meat, each characterised by different purposes: animal welfare organisations and/or environmental groups, research centres, media. In this context, the point of view of meat producers, at least in Italy, has never been inserted, and who feel instead the necessity to join in the debate by providing information, details and objective data that is useful to correct, where necessary, some opinions, which are sometimes prejudiced if not completely incorrect. From 2012, to achieve this scope, a group of operators in the livestock sector (companies and associations) was organised to support scientific studies that, in a logic of pre-competitive transparency, allowed to achieve, as well as the publication of the report 'The sustainability of meat and cured meats in Italy', the launching of the Sustainable Meat Project and then the Web portals www.carnisostenibili.it and www.thesustainablemeat.com Born from the common purpose of the three main meat industry associations, Assocarni, Assica and Unaitalia, the site aims to cover all topics related to the world of meat: an unprecedented project in Italy, contributing with an instructing and informative approach to a balanced report on health, nutrition and sustainability. A communication project that became a successful case history around Europe and the rest of the world.

Farmpedia: the reference toolbox for teaching animal husbandry in general education

A. Chouteau[1], C. Disenhaus[2], R. Baumont[3] and G. Brunschwig[4]
[1]IDELE, 149 rue de Bercy, 75012 Paris, France, [2]Agrocampus-Ouest, INRAE, UMR PEGASE, 65 Rue de Saint-Brieuc, 35000 Rennes, France, [3]UCA, INRAE, VetAgro Sup, UMR Herbivores, Theix, 63122 Saint-Genès-Champanelle, France, [4]UCA, INRAE, VetAgro Sup, UMR Herbivores, 89 avenue de l'Europe, 63370 Lempdes, France; alizee.chouteau@idele.fr

In the secondary schools of France, mention of agriculture is very poor and even more for livestock production. When these topics are covered in classroom, it's mainly based on the textbooks which often propose biased debates, misinformation and/or unsourced knowledge. Animal husbandry is mainly described as having a high energetic cost and a strong negative impact on climate. As secondary school students will be the citizens of tomorrow and education is still and will continue being an important source of information it seemed relevant to provide their teachers with appropriate teaching resources to deal with livestock production. For this reason, we have created Farmpedia, a reference toolbox to talk about animal husbandry in the classrooms. It contains all the scientific information needed to prepare lessons, popularised for a non-expert public, with the sources of each piece of information clearly referenced. An entire section is dedicated to the study of the links between animal husbandry and environment. Farmpedia was created by a working group composed mainly by major agronomic and veterinary universities professors and researchers, and validated by experts in the topics presented. The idea was that specialised professors should support generalist teachers. In order to make Farmpedia accessible to all, a dedicated website has been created. On www.ressources-elevages.fr, teachers or even curious citizens wishing to learn more will find the Farmpedia tool but also a selection of various teaching resources to be re-used in class. Disseminating the site to reach the target audience is our biggest challenge now: to do this, we are working on strengthening our partnerships with general education institutions.

Serious games to explain animal husbandry

A. Chouteau[1], E. Zanchi[2], R. Baumont[3], C. Disenhaus[4] and G. Brunschwig[5]
[1]IDELE, 149 rue de Bercy, 75012 Paris, France, [2]EPL de Fayl-Billot, 5 ruelle aux Loups, 52500 Fayl-Billot, France, [3]UCA, INRAE, UMR Herbivores, Theix, 63122 Saint-Genes Champanelle, France, [4]Agrocampus-Ouest, INRAE, UMR PEGASE, 65 Rue de Saint-Brieuc, 35000 Rennes, France, [5]VetAgroSup, UMR Herbivores, 89 avenue de l'Europe, 63370 Lempdes, France; alizee.chouteau@idele.fr

Several studies led us to think that it is urgent to better equip teachers of general education to talk about livestock farming in class. One of the teachers' requests was to succeed in making this topic more attractive for secondary school students. Therefore, we decided to develop serious games to talk about animal husbandry in class. We thought interesting to ask students from the agricultural sector to create these games: this way, young future breeders will talk and explain their job to young people from the general society. With the support of the French Ministry of Agriculture, we organised a competition to create serious games on animal husbandry. Students from agricultural secondary and advanced level schools had 5 months to produce a serious game to explain what livestock farming is all about. Twenty-five teams of various ages (from 13 to 21 years old on average) participated to the competition. Among the four themes proposed, the one on 'livestock farming and environment' was one of the most dealt with by the teams: it is a sign that the future breeders have understood the challenge around this subject and wanted to discuss it. The format of the game to be created was free, but most of the 25 games were board games accompanied by quiz cards. The jury was particularly vigilant regarding the technical level of the questions, the vocabulary chosen, and the support provided to the teachers, as the aim was above all to create games intended for the general public. This was the most complicated part for the competitors, who often misunderstood the knowledge gap between them and the general society. The games selected by the jury will then be corrected and validated, in order to be put up on the website www.ressources-elevages.fr, which was built specifically to provide tools for teachers wishing to talk about animal husbandry in their classrooms.

Assessment and valuation of ecosystem services provided by beef farms in Alentejo, Portugal

M.P. Dos Santos, T.G. Morais, T. Domingos and R.F.M. Teixeira
IST, Universidade de Lisboa, MARETEC – Marine, Environment and Technology Centre, LARSyS, Avenida Rovisco Pais n1, 1049-001 Lisboa, Portugal; manueldossantos@tecnico.ulisboa.pt

Pasture-based beef production can have positive effects on the environment as it provides ecosystem services (ES) such as climate change mitigation and adaptation through carbon sequestration in soils. These effects are particularly relevant in Mediterranean areas where livestock are key for managing ecosystems. However, this role is under-studied. Here we perform economic and environmental assessments of the effects of pasture-based beef production as a driver of change in ESS. The assessment is performed using data from 40 case study beef farms in Alentejo, collected by the authors during the Animal Future (SusAn/0001/2016) project. It includes economic data (farm accounting indicators) and agronomic data (materials and energy consumption and management practices). To extrapolate from farm data to a regional accounting of ES, the method used in the Portuguese Mapping and Assessment of ES (PT-MAES) project was applied. The ES considered were those included in PT-MAES, namely: soil protection, crop production, extensive animal production, carbon sequestration, fibre production, and biodiversity. Their assessment required a carbon and nutrient farm budget model. The BalSim mass-balance model, developed by MARETEC/LARSyS (IST), was used for each of the 40 farms to depict how nutrients (C, N and P) flow through ecosystems and are emitted to air, soil and water. C and N are key elements to understand the anthropogenic influence on natural cycles due to land use and its contribution to climate change. P is an crucial nutrient in Alentejo grasslands and in legume-based pastures in general. Economic valuation of ES under this approach was performed using available information and relied mainly on the use of avoided costs (carbon sequestration and soil protection), willingness to pay (biodiversity), and market prices (crop, extensive animal, and fibre production) methods. This study explains how land use for pasture-based meat production systems affects ES at farm-level. It enables a regional, forward-looking assessment of livestock production in Alentejo that could maximise economic and environmental results.

Prediction of nitrogen output in urine and faeces from beef on diets with different protein contents

A. Angelidis[1], L. Crompton[1], T. Misselbrook[2], T. Yan[3], C.K. Reynolds[1] and S. Stergiadis[1]
[1]University of Reading, School of Agriculture, Policy, and Development, Earley gate, Reading, RG6 6AR, United Kingdom, [2]Rothamsted Research, North Wyke, EX20 2SB, Okehampton, United Kingdom, [3]Agri-Food and Biosciences Institute, Large park, BT26 6DR, Hillsborough, United Kingdom; a.angelidis@pgr.reading.ac.uk

Prediction of N outputs in urine (UNO) and faeces (FNO) is crucial for evaluating the environmental impact of beef production. Diet N content influences N use efficiency (NUE), but existing equations may over- or under- predict N outputs from beef that excrete low or high amounts of N, respectively. This study developed prediction models for UNO and FNO, for animals fed contrasting dietary crude protein (CP) contents. A database from animal trials, including bodyweight, feed intake, diet chemical composition, UNO and FNO was used to develop prediction equations: (1) as a whole (n=570); and (2) diet-specific from 3 sub-datasets (n=190) of contrasting diet CP contents (low, 84 -143 g/kg DM; medium, 144-162 g/kg DM; high, 163-217 g/kg DM). Equations were developed using residual maximum likelihood analysis and validated against external databases of corresponding diet CP contents, by mean prediction error (MPE). The most important predictors across all groups were feed intake and diet CP content, which reflect N intake (known to strongly affect NUE). Addition of starch, fibre and/or metabolisable energy (ME) as predictors improved prediction accuracy in most cases; as ME further accounts for energy supply towards rumen microbial protein synthesis and post-absorptive N use. In low CP diets the MPE for UNO and FNO was 0.594 and 0.263, respectively; when diet-specific equations were used, MPE for FNO was reduced to 0.215, but for UNO it was increased to 0.628. In medium-CP diets the MPE for UNO and FNO was 0.391 and 0.394, respectively; diet-specific equations reduced MPE to 0.230 and 0.328, respectively. In high-CP diets MPE for UNO and FNO was 0.154 and 0.144, respectively; diet-specific equations reduced MPE to 0.140 and 0.135, respectively. In conclusion, using diet-specific equations (developed from data with corresponding diet CP contents) can improve prediction accuracy of N outputs in beef cattle.

The emissions balance of the production of Mozzarella di Bufala Campana DOP

R. De Vivo[1], L. Zicarelli[2] and R. Napolano[3]
[1]Istituto Tecnico Industriale E. Fermi, Dipartimento di Chimica, Via Luosi 23, 41124 Modena, Italy, [2]Università di Napoli Federico II, Dipartimento di Medicina Veterinaria e Produzioni Animali, Via Federico Delpino 1, 80137 Napoli, Italy, [3]Fattorie Garofalo, Via Santa Maria Capua Vetere 121, 81043 Capua CE, Italy; robertodevivo@virgilio.it

The environmental impact in terms of emissions has become more and more important with regard to foods, in particular, for those of animal origin. The LCA (Life-Cycle Assessment) method, an internationally standardised method used to calculate the environmental impact of goods or services, does not take into account the carbon fixed and consequently the subtraction of carbon dioxide by plant biomass aimed or not at the production of food of animal origin. Not considering organic carbon in plant biomass alters the carbon balance calculation. This method could overestimate the carbon dioxide that is generated in the production of vegetable and animal products. For the production of Mozzarella di Bufala Campana DOP, in the specific case, the masses of the various fodder and cereal species used were quantified starting from the food rations of the different categories divided by age and production phase (dry, lactation, the young cattle destined to replace the eliminated). The population taken into consideration includes all the animals reared in the areas covered by the DOP specification and with milk production orientation. The fixed carbon and consequently the carbon dioxide removed from the atmosphere was calculated from the mass of food, through the various collection indices and percentages of dry matter. The LCA method has been used for all the other production phases, from the transformation to packaging and transportation of the finished product. The equivalent carbon dioxide from the various production phases has been balanced with the carbon dioxide removed. From this elaboration it emerges that the environmental impact, in terms of greenhouse gas emissions, of the MdBC DOP is negative. The amount of greenhouse gases converted into equivalent carbon dioxide emitted during the production process is less than the carbon dioxide removed from the atmosphere. In total, the production of one kg of MdBC DOP would contribute to the removal of more than 10 kg of carbon dioxide from the atmosphere. It is likely that for agricultural and animal products, if this factor were taken into account, the environmental impact in terms of emissions would be reduced.

Italian livestock emissions balance

R. De Vivo[1] and L. Zicarelli[2]
[1]Pioneer Hi-Bred Italia SRL, Via Pari Opportunità 2, 26030 Gadesco Pieve Delmona, Italy, [2]Università di Naapoli Federico II, Dipartimento di Medicina Veterinaria e Produzioni Animali, Via Federico Delpino 1, 80137 Napoli, Italy; robertodevivo@virgilio.it

Introduction: Among the greenhouse gas emissions due to zootechnical activities there is, in addition to rumen methane, another source of atmospheric pollution represented by what derives from the fermentation of manure of farmed animals and its management. To feed the farmed animals, however, a considerable amount of vegetables is used which subtract carbon dioxide from the atmosphere. Materials and methods: The emissions relating to ruminal fermentations and those relating to the manure of all the heads of the species bred in Italy have been added to the emissions from its management and spreading as well as the manure released by grazing animals. The emissions due to the breathing of the animals as well as the carbon dioxide established by the main crops of zootechnical interest have been calculated using the 'CalvinBenson Cycle' and then drained from the atmosphere. From statistical data, the amount of fodder and cereals produced in Italy and abroad and used for animal husbandry in Italy has been traced. The vegetative biomass has been traced back through the various collection indices, from the quantity of fodder and cereals produced, also calculating the hypogean part left on the ground as a crop residue. The emissions deriving from the cultivation of plant species have also been taken into account; they can be attributed to the tillage, the production of fertilisers and pesticides, electricity, fuels and the operation of the machines. Results and conclusion: From the elaborated data it emerges that the carbon dioxide drained from the atmosphere by the plants grown in Italy and imported to feed the animals raised in Italy is higher than the sum of the equivalent carbon dioxide emitted by agricultural processes, by the physiological rumen fermentations and that due to the management of manure. Ultimately, in Italy, the carbon dioxide absorbed is equal to about 15% more than those released into the atmosphere with a positive balance for GHG purposes. This balance is even more advantageous if the carbon dioxide set by the forage crops in Italy is added to that of crops grown abroad and imported which entails a positive balance of 32%. From the results of this study, it can be said that animal husbandry in Italy, but also in general, probably does not contribute to increasing GHG emissions into the atmosphere.

Could rumen volumes measured by CT scanning help to breed sheep with lower methane emissions?
N.R. Lambe, A. McLaren, K. McLean, J. Gordon and J. Conington
Scotland's Rural College, Hill and Mountain Reserach Centre, Crianlarich, West Perthshire, FK20 8RU, Scotland, United Kingdom; nicola.lambe@sruc.ac.uk

Since the 1990s, whole-body computed tomography (CT) scanning of elite terminal sire ram lambs has been routinely performed in UK breeding programmes, leading to substantial improvements in carcass composition. Combined with research trial outputs, this has resulted in a large archive of CT scans from >10K sheep with associated pedigree and performance records. Previous research studies, in the UK and internationally, have identified links between rumen morphology, measured by CT, and methane emissions from sheep. In particular, larger reticulo-rumen volumes (RRvol) are associated with increased methane emissions. To progress these investigations, CT image archives at SRUC were mined to investigate differences between breeds in RRvol, as a potential predictor of methane emissions. Scottish Blackface (SBF; n=151) and Texel (TEX; n=119) lambs of both sexes, reared together on lowground pastures, were CT scanned in batches pre-slaughter (average age 20 weeks) and RRvol measured. Multiple linear regression, adjusting for live weight, slaughter batch, sex, breed and relevant interactions, found that SBF lambs had reticulo-rumen volumes 26% higher than those of Texels (P<0.001), on average, with males having significantly larger RRvol than females (P<0.001). Significant sire differences were identified, suggesting genetic control of RRvol within breed. Genetic analysis of RRvol measured from 649 commercial Texel ram lambs (~20 weeks of age), within the national breeding programme in 2017-19, was performed using an animal model in ASREML, fitting live weight and relevant fixed effects. RRvol measured from images obtained during routine CT scanning was found to be moderately heritable (h^2=0.5, s.e. 0.13). These results add to our understanding of the relationships between rumen morphology and function and greenhouse gas emissions from livestock. The genetic relationships between rumen measurements and other economically and environmentally important traits should be investigated before recommendations can be made on how to use breeding values for traits such as reticulo-rumen volume to reduce emissions from sheep systems.

Calibrating lamb cut weight data using computed tomography
A. Williams, F. Anderson and G.E. Gardner
Murdoch University, Murdoch, 6150, WA, Australia; andrew.williams@murdoch.edu.au

Within Australia the lamb, beef and pork industries are developing technologies that can measure carcase composition and the weight of commercial cuts, enabling sorting prior to fabrication to optimally meet market end-points. The process of training devices to predict these cut-weights requires the generation of large datasets of commercial cuts from phenotypically diverse animals which are boned-out by trained butchers. Despite close scrutiny, these butchers have systematic variation which is evident in the weights of cuts that they procure. This study details a method for quantifying and correcting for this human variation using computed tomography prior to then training an equation using carcase weight and a standard abattoir measure of GR tissue depth to predict cut weights. Two hundred mixed sex lamb carcasses with weights ranging between 12 and 40 kg were computed tomography scanned, and measurements taken for hot standard carcase weight (HSCW) and GR tissue depth. Computed tomography was used to determine fat% and lean% in the fore, saddle, and hind sections of the carcase. Carcases were also boned-out and commercial cuts weighed. In this paper we use the eye of the shortloin and rump as examples. Firstly, an equation predicting ln(cut weight) was derived using a linear mixed effects model, with ln (carcase weight), CT fat% and CT lean% as continuous variables, and boner ID included as a random term. Cut weight was then corrected for the estimates of the random term, and a general linear model was used to predict both the corrected and non-corrected cut weight data, using carcase weight and GR tissue depth as predictors. For a 26 kg carcase, the root mean square error of the model predicting the non-corrected data was ±79 g for the shortloin and for the ±64 g rump, while for the corrected data it was ±77 g for the shortloin and ±56 g for the rump. This demonstrates that the precision of models predicting cut weight can be substantially improved by adjusting relative to computed tomography to minimise human error. It is likely that this method will become crucial when incorporating future data sets derived from other experiments, where different boners are used and systematic bias due to human error is greater.

Genetic basis of body reserves mobilisation and accretion in Manchega ewes

C. Diaz[1], I. Ureña[1], A. Rubio[1], C. Gonzalez[1], M.D. Perez-Guzman[2], M. Ramon[2], M. Serrano[1] and M.J. Carabaño[1]
[1]INIA, Ctra de la Coruña km 7.5, 28040 Madrid, Spain, [2]IRIAF-CERSYRA, Avda del Vino 10, 13300 Valdepeñas, Spain; cdiaz@inia.es

The effect of climate change on animal production occurs in two ways, effects on the physiology and metabolism of animals and effects on the quality and the availability of raw feedstuff and pasture. This is particularly true in harsh environments where the cost of feeding has a large impact on the sustainability of the production system. In those restricted environments, resilience to the environment reflected in the plasticity of animals to maintain their functions regardless of the environmental constraint, could be a potential breeding objective. In this case, plasticity is understood as the ability of ewes to mobilise and to accrete body reserves (BRMA) in certain periods of the year. This study aims to examine potential indicators of BRMA and their genetic basis. For that, 300 ewes located in the south plateau of Spain were sampled at three physiological stages during their first two lactations. Physiological stages were mating (MAT), 1 to 7 days before AI or natural mating; pregnancy (PRE) about 1 month before lambing; and lactation (LAC), 1 month after lambing. At each stage, body condition score (BCS), body weight (BW) and rectal temperature (RT) were recorded in each ewe (Animal Traits). In addition, blood samples were taken and NEFA (non-esterified fatty acid), Insulin, T3 (triiodothyronine), leptin, glucose and BHB (beta-hydroxybutyrate) were determined in plasma (Blood Traits). All ewes were genotyped with a mixture of 50K and HD Illumina Bead Chips. Genetic parameters of BRMA were estimated and GWA studies were performed. Changes in BCS values along with changes in levels of BHB and NEFA resulted very useful to evaluate individual BRMA throughout the three physiological stages. The most critical period in BRMA seems to be the transition from LAC in primiparous to the subsequent MAT stage. Estimates of heritability ranged from 0.07 (in BW from MAT to PRE) to 0.50 (BW in LAC to MAT) for Animal Traits, and from 0.06 (in insulin from MAT to PRE) to 0.45 (for BHB from LAC to MAT) for Blood Traits. GWA studies provided some chromosome-wise signals pointing out some genes of potential interest to understand the mechanisms involved in BRMA.

Determining factors affecting feed intake in grazing sheep

F. McGovern[1], P. Creighton[1], N. Galvin[2], D. Hennessy[2], M. O'Donovan[2], B. Garry[2], N. McHugh[2] and M. Beecher[2]
[1]Teagasc, Animal and Bioscience, Animal and Grassland Research Centre, Teagasc Athenry, Co. Galway, H65 R718, Ireland, [2]Teagasc, Grassland, Animal and Grassland Research Centre, Teagasc Moorepark, Fermoy Co. Cork, P61 C996, Ireland; fiona.mcgovern@teagasc.ie

Optimising nutrition for animals requires accurate estimates of dry matter intake (DMI) in accordance with the nutritive composition of the diet. Estimating DMI in animal grazing systems can be difficult. The n-alkane technique is routinely used in animal research for estimating DMI. This technique involves dosing the animal with long chain alkanes which are used as internal markers for predicting daily DMI. The aim of this study was to investigate factors, both dietary and animal, affecting DMI and the accuracy of the n-alkane technique for estimating DMI in ovine animals. Detailed total DMI measurements were obtained from over 200 Texel wether animals from 2012 to 2014. *In vivo* DMI was determined using the total collection method where animals were housed in metabolism crates and offered fresh perennial ryegrass *ad libitum*. Simultaneously herbage DMI was estimated using the n-alkane technique whereby all animals were dosed once daily with an n-alkane bolus for 12 days and faeces collected for the final six days of each intake run. The association between *in vivo* DMI and n-alkane DMI was estimated using a fixed effect model in PROC GLM. A multiple regression model for DMI was also created using a stepwise forward-backward regression where P=0.05 was used as the threshold for entry and exit of the independent variables from the model. The relationship between each of the independent variables (live-weight, herbage mass, season, dietary fibre fractions and crude protein) DMI and n-alkane DMI were determined using partial Spearman's rank correlations (PROC CORR) in SAS. The average in-vivo DMI recorded was 1.54 kg DM/day while the estimated n-alkane DMI was 1.45 kg DM/day. The r-squared value between *in vivo* DMI and n-alkane estimated DMI was 0.75. This is dependent on multiple factors, most importantly grass quality. This data highlights the suitability of the n-alkanes technique for measuring grass DMI in sheep.

The estimation of dispersion parameters for growth traits of lambs in Slovenia
M. Bizjak[1], M. Špehar[2], K. Potočnik[1], M. Štepec[1], B. Luštrek[1], G. Gorjanc[1,3] and M. Simčič[1]
[1]University of Ljubljana, Biotechnical Faculty, Department of Animal Science, Jamnikarjeva 101, 1000 Ljubljana, Slovenia, [2]Ministry of Agriculture, Ilica 101, 10000 Zagreb, Croatia, [3]University of Edinburgh, The Roslin Institute and (Dick) School of Veterinary Studies, Easter Bush, Edinburgh, United Kingdom; mojca.simcic@bf.uni-lj.si

The aim of this study was to estimate dispersion parameters for growth traits (birth weight; BW, weaning weight; WW, and daily gain; DG) of lambs in two Slovenian sheep breeds, the Improved Jezersko-Solčava sheep (JSR) and the autochthonous Bela Krajina sheep (BP). These breeds are reared mainly for lamb production. The records of BW, and WW were collected in the period from 2004 to 2019 for 30,373 lambs (26,374; JSR, 3,999; BP) according to breeding programs. Daily gain was calculated from BW and WW considering the age at weighing. A pedigree file for 38,859 animals was prepared from the Central database of small ruminants in Slovenia. The fixed part of the model for all analysed growth traits included breed, gender, litter size, dam parity, and season of weighing. In the model for WW the birth weight and age at weighing as linear regressions were additionally used. The random part of the model included the additive genetic effect and the random effect of the flock. Variance components were estimated using REML method implemented in the VCE-6 program. Analysis of variance showed significant differences in BW and WW as well as in DG between breeds, genders, litter size, dams parity, and season. Lambs of the JSR had higher growth traits than lambs of BP sheep. Male lambs had higher growth traits than female lambs. Single lambs had higher growth traits compared to twins or three lambs in the litter. Heritability estimates for growth traits were 0.28 (WW) and 0.29 (BW, DG), while the flock effect explained 0.30 (WW, DG) and 0.41 (BW) of variability. These dispersion parameters will be used in the genetic evaluation of growth traits from the year 2020 onwards.

Liveweight and body composition responses to differential feeding in a range of Merino genetics
C.J. Byrne, S. Blumer and A.N. Thompson
Murdoch University, College of Science, Health, Engineering and Education, Murdoch University, 90 South Street, Perth, 6150, Australia; colin.byrne@murdoch.edu.au

The aim of this study was to characterise important production variables that could potentially affect the optimum-stocking rate of different genotypes of Merino sheep. Wethers (n=320) were blocked by sire, randomly allocated to individual pens and fed a chaff diet at 100% of maintenance from days 0-35 and *ad libitum* or 60% of maintenance (RES) from days 35-70. Liveweight was recorded thrice weekly and body condition score (BCS) was recorded weekly. On days 0, 35 and 70, muscle and subcutaneous fat depth were measured by ultrasound scanning and body composition assessed by dual energy x-ray absorptiometry scanning. Whole body energy (WBE; MJ) was determined as: (kg fat × 35.9 MJ) + (kg lean × 5.31 MJ). Data were analysed by repeated measures ANOVA, within feeding period, using the proc mixed procedure in SAS 9.4 with sire, day and diet, where appropriate, as fixed effects. There was an effect of sire on liveweight at days 0, 35 and 70 (P<0.001), with a range of 8 kg at days 0 and 35 and 9.5 kg at day 70 between the heaviest and lightest sire group. There was also a range of 0.7 BCS at each time point between highest and lowest sire groups. At day 70, the RES sheep were lighter (57.1 vs 67.8 kg; P<0.001) and leaner (2.7 vs 3.0 BCS; P<0.01) than those offered the *ad lib* diet. There was also an effect of sire on fat and muscle depth at all timepoints (P<0.001) with differences of ~1.5-2.0 mm and 3.5-5.0 mm for fat and muscle, respectively. Diet had no effect on muscle depth (P>0.05) but RES diet had less fat than *ad lib* (2.7 vs 3.1 mm; P<0.01). Sire influenced WBE (P<0.001) and the range in WBE was 90 MJ at days 0 and 35. At day 70, this range was reduced to 75 and 80 MJ for sheep on RES and *ad lib* diets, respectively. There are also differences in fat depletion and hence energy utilisation when wethers were offered a RES diet and not all sires increased their energy reserves in the same manner when feed was plentiful. It is evident that there are sire differences in body composition and WBE. On-going research will elucidate the efficiency of energy utilisation between sire groups and its implications for potential stocking rate and hence profitability per hectare.

Weaning performance of sheep grazing on different swards under cell grazing or continuous stocking

M.J. Rivero[1], H. Fleming[1], O. Lawal-Adebowale[1,2], R. Pywell[3] and J. Storkey[4]
[1]Rothamsted Research, North Wyke, Okehampton, Devon, United Kingdom, [2]Federal University of Agriculture, Abeokuta, Ogun State, Nigeria, [3]Centre for Ecology and Hydrology, Wallingford, Oxfordshire, United Kingdom, [4]Rothamsted Research, Harpenden, Hertfordshire, United Kingdom; jordana.rivero-viera@rothamsted.ac.uk

Given the growing interest in multispecies swards in grazing systems, the joint effect of sward type and grazing strategy should be assessed in terms of animal performance. We studied the performance at weaning of sheep grazing under two grazing strategies (GR; cell grazing, CG; continuous stocking, CO) on three sward types (ST; permanent pasture, PP; grasses mixture, GM; multispecies, MS). The twelve 1-ha plots at Rowden site (Devon, UK) had two levels of drainage (D; enhanced, standard), with six plots each. At lambing, 15 ewes and their 26 lambs were allocated to each of the plots. We recorded the liveweight (LW) of ewes and lambs at lambing and at weaning and monitored the herbage mass (HM) weekly. The ANOVA considered the main effects of all three factors – D, ST, GR – and two 2-factor interactions – GR × ST and D × ST using Genstat. Some ewes and lambs were removed from some GM plots and some were added to some MS plots due to low and high HM, respectively. The average HM was greater on the CG plots (P<0.05) and tended to be greater (P=0.06) on the MS compared with the other two ST. The kg of lamb weaned per ewe and the LW change of the ewes were not affected by the main factors (P>0.05), averaging 56.2 kg and -4.79 kg, respectively. The kg of lamb weaned per kg of ewe (considering only the ewes and lambs that were present from lambing to weaning) was affected by the interaction GR × ST (P<0.05) where MS and PP performed similar between the GR, averaging 0.73 and 0.74, whilst the GM-CO showed a greater value than the GM-CG (0.76 vs 0.69 kg//kg). The total kg LW gained of lambs per ha (considering all the lambs that grazed the plots between lambing and weaning at any timepoint) was greater in the CG (P<0.01) than the CO (691 vs 639 kg), and lower (P<0.05) in the GM (626 kg) than the other two ST (685 kg on average). The CG strategy favoured greater HM with a positive effect on productivity per ha whilst the PP and MS perform similarly under contrasting GR.

Differential mucin gene expression in the ovine cervix contribute to breed differences in fertility

L. Abril-Parreño[1,2], P. Cormican[1], A. Krogenæs[3], X. Druart[4], K.G. Meade[1] and S. Fair[2]
[1]Teagasc, Animal & Bioscience Research Department, Teagasc, Grange, Trim, Co. Meath, C15 PW93, Ireland, [2]University of Limerick, Department of Biological Sciences, University of Limerick, Castletroy, Limerick, Co. Limerick, V94 T9PX, Ireland, [3]Norwegian University of Life Sciences, Department of Production Animal Clinical Sciences, NMBU-School of Veterinary Science P.O. Box 369 sentrum, Oslo, N-0102, Norway, [4]Institut National de la Recherche Agronomique, Physiologie de la Reproduction et des Comportements, UMR PRC, INRA 85, CNRS 7247, Université de Tours, IFCE, 37380 Nouzilly, France; laura.parreno@ul.ie

The outcome of cervical artificial insemination with frozen-thawed semen is limited by the inability of sperm to traverse the cervical mucus of some ewe breeds. The main structural component of cervical mucus are mucins, which are highly glycosylated proteins. The aim of this study was to identify mucin genes differentially expressed between high and low fertility ewe breeds over the oestrous cycle which could potentially explain differences in sperm transit through the cervix. Cervical tissue samples were collected from two Irish ewe breeds (Belclare and Suffolk; high and low fertility, respectively) and from two Norwegian ewe breeds (Norwegian White Sheep and Fur; both with high fertility compared to the Irish ewe breeds) at both the follicular and luteal phases of the oestrus cycle (n=10-15 ewes per breed at each phase). High-quality RNA was analysed by RNA-seq and the profiles of these specific genes were assessed. We identified three gel-forming mucin genes (*MUC5AC*, *MUC5B* and *MUC6*) and seven transmembrane mucin genes (*MUC1*, *MUC3A*, *MUC4*, *MUC13*, *MUC16*, *MUC17* and *MUC20*). Our results showed that *MUC1*, *MUC4*, *MUC16* and *MUC5B* had the highest levels of expression in all ewe breeds (P<0.05). Using the low-fertility Suffolk breed as a baseline; we identified significantly lower expression of *MUC1*, *MUC4* and *MUC20* in the Norwegian ewe breeds compared to the Suffolk breed during the follicular phase (P<0.05). In addition, Suffolk had higher levels of *MUC5B* than Fur breed at the follicular phase (P<0.05), while there were no differences with other ewe breeds (P>0.05). In conclusion, significant differences in mucin gene expression between sheep breeds could explain differences in cervical sperm transit after AI. The work was funded by the ERANET SusAn.

Analysis of the sheep footrot microbiome using whole-metagenome sequencing

A. Usié[1], C. Leão[2], D. Gaspar[1], A. Botelho[2], S. Cavaco[2], M. Monteiro[3], M. Madeira[3], J. Santos[3], L. Tábuas[3], S. Branco[4], E. Bettencourt[4], L. Padre[4], R. Romão[4], P. Caetano[4], P. Damião[4], C. Dias[4], N. Carolino[2], C. Bettencourt[5], C. Matos[3] and A.M. Ramos[1]
[1]CEBAL, Centro de Biotecnologia Agrícola e Agro-Alimentar do Alentejo, Beja, 7800-072 Beja, Portugal, [2]INIAV, Instituto Nacional de Investigação Agrária e Veterinária, Portugal, Oeiras, 2780-157, Portugal, [3]ACOS, Agricultores do Sul, Beja, 7800-072, Portugal, [4]Universidade de Évora, Évora, 7000-645, Portugal, [5]Centro de Experimentação do Baixo Alentejo, Herdade da Abóbada, Vila Nova de S. Bento, 7830, Portugal; marcos.ramos@cebal.pt

In the Alentejo region Merino sheep are the most common breed, reared for the production of meat, dairy and wool. Footrot is responsible for lameness, decreased animal welfare and higher production losses, generating a negative economic impact. The disease is caused by the bacteria *Dichelobacter nodosus*, a process on which it interacts with the sheep foot microbiome, to date largely uncharacterised. To understand and characterise the footrot microbiome dynamics, a whole metagenome sequencing (WMGS) approach was used to study the microbiome of sheep with different footrot affection scores. Foot tissue samples were collected in 214 animals with different footrot degrees, ranging from 0 to 5. DNA was extracted from each sample and used in WMGS. The sequence dataset was analysed in two different ways. First, the reads were mapped directly to the *D. nodosus* and *Fusobacterium necrophorum* reference genomes. Then, a classic metagenomics approach was used, to characterise and quantify the composition of the microbial community present in each sample. The mapping results showed a general positive correlation between the number of mapped reads against the *D. nodosus* and *F. necrophorum* genomes and footrot score. The *D. nodosus* serogroups were also determined for a subset of samples using qPCR. Finally, the microbiome composition showed a higher percentage of *D. nodosus* and *F. necrophorum* in animals affected with footrot. This study showed WMGS to be useful for characterising the sheep footrot microbiome.

A survey of pathogens on lamb carcasses from Portuguese local breeds

S. Coelho-Fernandes, D. Félix-Oliveira, G. Rodrigues, V.A.P. Cadavez and U. Gonzales-Barron
Centro de Investigação de Montanha, Instituto Politécnico de Bragança, Animal Science, Campus de Santa Apolónia, 5300-253 Bragança, Portugal; saraccoelhof@hotmail.com

This study aimed to evaluate the levels of microbial contamination on lamb carcasses from two Portuguese breeds, Bordaleira-de-Entre-Douro-e-Minho (BDM) and Churra-Galega-Bragançana (CGB). Thirty BDM and 30 CGB lambs were reared in a semi-intensive system, and slaughtered at 4 months age. On 11 visits to the abattoir, 400 cm^2 neck/loin/hind pooled areas were swabbed from dressed carcasses. Chilled *L. dorsi* sections were vacuum packed and cold stored. Swabs were analysed for mesophiles, coliforms, *Escherichia coli*, *Salmonella* spp., *Lactobacillus monocytogenes* and *E. coli* O157, while meat samples were analysed for *Salmonella* on the 3rd, 9th, and 15th day post-slaughter. Linear and logistic mixed models were adjusted to assess any effect of breed on microbial occurrence. BDM lamb carcasses presented higher counts ($P<0.05$) of mesophiles (3.52 log cfu/cm^2), coliforms (0.936 log cfu/cm^2) and *E. coli* (0.307 log cfu/cm^2) than CGB carcasses (3.03, 0.633 and 0.079 log cfu/cm^2, respectively). There was no difference between BDM and CGB in the incidences of *Salmonella* spp. (21.4% [95% CI: 10.0-40.2%] versus 16.7% [7.10-34.3%]), *L. monocytogenes* (3.50% [0.50-21.4%] versus 6.70% [1.60-23.1%]) and *E. coli* O157 (32.1% [17.6-51.1%] versus 16.7% [7.10-34.3%]). When *Salmonella* was found in a sampled batch of lamb carcasses, the odds of finding *Salmonella* in meat, at a later processing stage, increased by 8.7 times ($P=0.078$).

Polymorphisms of the melatonin receptor 1A gene determine sexual activity of rams in spring

J.A. Abecia[1], L. Pulinas[2], M.C. Mura[2], M. Carvajal-Serna[1], A. Casao[1], R. Pérez-Pe[1] and V. Carcangiu[2]
[1]IUCA, Univ. Zaragoza, Miguel Servet, 177, 50013 Zaragoza, Spain, [2]Dipart.Medicina Veterinaria, Univ. Sassari, Via Vienna 2, 07100 Sassari, Italy; alf@unizar.es

The melatonin receptor1A (MNTR1A) gene has been proposed as a candidate gene to control of sexual seasonality in sheep. This gene is located on chromosome 26 in sheep, and shows two polymorphic sites that can be evidenced by means of the restriction enzymes MnlI and RsaI, with cleavage sites in position g.17355458C>T and g.17355452A>G (Oar_rambouillet_v1.0), respectively. They have been associated with different reproductive activities in the ewe, although no evidence on ram's reproductive performance has been presented. This work was designed to determine sexual activity of rams in spring -measured by serving capacity tests – related with the above-mentioned polymorphisms. Eighteen Rasa Aragonesa rams were selected for this study according to their RsaI genotype C/C (n=6), C/T (n=6) and T/T (n=6). According to the MnlI allele, rams were divided into G/G (n=9), G/A (n=3) and A/A (n=6). In late March and late May, rams were individually exposed for 2 min to 3 oestrous ewes in a 15 m^2 pen, and the number of flehmen, vulva-sniffing, approaches, attempted mountings, mountings, and total number of activities were recorded. χ^2 tests were used to assess the statistical significance of differences in proportional values of the events performed by each genotype. Both in March and May, T/T rams demonstrated a higher level of sexual activity than the other genotypes, so that 55, 48, 48 and 49% of the vulva-sniffing (P<0.001), approaches (P<0.01), mountings (P<0.05) and total activities (P<0.001) displayed by rams were performed by this group in March, and 50% of every event in May (P<0.001). For the MnlI variant, G/G rams performed a significantly higher proportion (P<0.001) of the vulva-sniffing (41%), approaches (46%) and total activities (40%) observed in March, and 52, 43, 46 and 47% of the vulva-sniffing (P<0.001), approaches (P<0.001), mountings (P<0.05) and total activities (P<0.001) recorded in May. In conclusion, the allelic variants of the MTNR1A gene conditioned the intensity of sexual activity of rams during the seasonal anoestrus, so that selecting rams by their genotype could be a useful tool to improve reproductive results in spring mating.

Breeding value estimates of daily gain in six small Swiss sheep populations

A. Burren[1], C. Hagger[1], C. Aeschlimann[2], P. Beffa[2] and H. Joerg[1]
[1]Bern University of Applied Sciences, School of Agricultural, Forest and Food Sciences, Länggasse 85, 3052 Zollikofen, Switzerland, [2]Swiss Sheep Breeding Association, Industriestrasse 9, 3362 Niederönz, Switzerland; hannes.joerg@bfh.ch

In 2019 breeding value estimates of average daily gain up to 45 days of age were introduced for the Swiss sheep breeds with small population sizes Charollais (CHS, litters=20,968), Rouge de l'Ouest (RDO, litters=3,895), Dorper (DOP, litters=11,006), Shropshire (SHR, litters=7,400), Suffolk (SU, litters=11,823) and Texel (TEX, litters=12,332). Due to the small number, the RDO phenotypes were analysed together with CHS phenotypes. The splitting of direct and maternal components was done in the same way as for the four major Swiss sheep breeds namely White Alpine, Black-Brown Mountain, Valais Blacknose, and Brown Headed Meat sheep. Direct and maternal breeding values for average daily gain up to 45 days of age were estimated using REML and BLUP methodology. The statistical model accounted for fixed effects of sex of lamb, litter size, litter number×age of dam, age at weighing, lambing season and random effects of herd×year, permanent environment, direct component of the animal, maternal component and residual effect. Litter size was taken as a combination of total lambs born and number of lambs alive at 45 days of age. Litter size and age of dam were the two fixed effects with the largest influence on growth of lamb up to 45 days. Estimated heritabilities for the direct component were 0.17, 0.10, 0.04, 0.17 and 0.13 for CHS/RDO, DOP, SHR, SU and TEX breed, respectively. For the maternal component, the estimated heritabilities were 0.02, 0.04, 0.04, 0.02 and 0.05 for the CHS/RDO, DOP, SHR, SU and TEX breed, respectively. The genetic correlations between direct and maternal components were negative and in the same range for all breeds. Estimated values for this parameter were -0.85, -0.80, -0.80, -0.69 and -0.71 for the CHS/RDO, DOP, SHR, SU and TEX breed, respectively. The results seem plausible and correspond to estimates found in the literature for similar traits. The average breeding values per year of birth gave indications on the genetic trends of the two components. The genetic trends differed between the breeds due to different breeding strategies.

Effect of dietary seaweed on growth performance, carcass quality and faecal coccidia oocysts of lamb

D. Martín[1], B. Agudo[2], I. Hurtado[2], C. Bodini[3], J. Viguera[1] and J. Peinado[1]
[1]Imasde Agroalimentaria, S.L., C/ Nápoles, 3, 28224 Pozuelo de Alarcón (Madrid), Spain, [2]EA Group, S.C, Ctra. Ext. 104, km 4.7, 06700 Villanueva de la Serena (Badajoz), Spain, [3]Lombarda Trading S.R.L, Via Progresso 17, 26011 Casalbuttano ed Uniti (Cremona), Italy; dmartin@e-imasde.com

Fattening lambs from Merino sheep extensive farms has important risk factors related to health disorders such as coccidiosis and respiratory diseases. The influence of the inclusion of seaweed extract (Algatan®) in fattening diet on growth performance, carcass quality, and faecal coccidia oocysts of Merino lambs was tested. A total of 120 Merino lambs of 4±1.0 kg of initial BW were randomly allotted according with three diets: (1) Control; (2) Experimental A with seaweed extract at a level of 1% during all the fattening period (15 to 90 d); (3) Experimental B with seaweed extract at a level of 1% from 15 to 50 d and 0.5% from 50 to 90 d. The experimental unit was the animal (20 replicates per treatment). Feeding program was common for all the lambs and consisted of two basal barley-wheat-corn diets offered *ad libitum* (from 15 to 50 d, and from 50 d to slaughter). Individual BW was measured at five moments (0, 15, 50, 75 and 90 d of age) to determine ADG. The oocyst output was recorded at the end of fattening period. Carcass weight and pulmonary lesions were measured in a commercial slaughterhouse. Data were analysed as a completely randomised design by GLM of SPSS, including treatment as main effect. Lambs from Experimental diets grew faster during the first phase of fattening than lambs from Control diet (0.213 vs 0.256 kg/d; P<0.01). However, no differences were found between treatments for ADG during the second phase of fattening or oocyst output recorded at the end of trial. Lambs from Experimental diets showed higher carcass weight than lambs from Control diet (11.5 vs 12.6 kg; P=0.05), however no differences were found between treatments for the carcass yield. Finally, the percentage of carcases with pulmonary inflammation, lymphoid hyperplasia, and bronchial hyperplasia was higher in lambs from Control diet (P<0.05). It is concluded that lambs fed with seaweed extract grew faster and showed carcasses with higher weight and lower pulmonary lesions.

Indirect parameters of resistance and tolerance to worms for sheep breeding in developing countries

L. Sartori Menegatto[1], R. Dutra Do Bem[2], L. Afonso De Freitas[1], L. Vage Coelho Sartori[2], N. Bonvino Stafuzza[2] and C.C. Paro De Paz[2]
[1]University of São Paulo, Department of Genetics, Ribeirão Preto Medical School, Bandeirantes Avenue, 3900, Ribeirão Preto, SP, Brazil, 14049-900, Brazil, Brazil, [2]Animal Science Institute, Beef Cattle Research Center, Carlos Tonani Highway, 94, Sertãozinho, SP, Brazil, 14888-000, Brazil, Brazil; leonardomenegatto@gmail.com

Gastrointestinal nematodes are one of the biggest agents that cause damage to sheep farming, mainly in tropical environments. Sheep breeding programs have been developed to select animals resistant to parasites, but the difficulties related to genetic evaluation are equally relevant in developing countries. Hematophagous endoparasites are those that cause anaemia, mostly the Trichostrongyloidea superfamily, which can be quantified by total faecal egg count (FEG) parameter. We aimed to evince that even indirect parameters can correctly stratify animals, allowing an effective selection for the resistance trait in high and low parasites infections environments. A total of 174 Santa Inês animals from two farms with different levels of infestation were evaluated by eye colour chart (ECC) through Famacha®, FEG and corpuscular volume (CV) by blood samples. The total and Trichostrongyloidea FEG were determined as resistance measures and the populations were sorted from permutations performed in original scripts in R program. With the anaemia attributes of ECC and CV, the non-resistant animals were equally divided by permutation into tolerant (have infestation, but do not exhibit anaemia) and susceptible (exhibit anaemia) and the outputs of the assortments were compared. The results of Pearson's correlations (P<0.05) between the FEG parameters were 0.7496 and 0.5591 to the farms with high and low infestation, respectively, and between anaemia parameters were -0.4882 and -0.3794. With low degree of infestation, the animals' stratification showed differences among the three groups for specific FEG value, while with high degree of infestation there was significant difference only between the susceptible group against the others. The results demonstrated an adequate stratification of the animals, such demonstrate that sheep selection is possible in those herds, even under the lack of resources and at different levels of infestation.

Clustering models and the use of extrinsic and intrinsic parameters as anaemia measurements in sheep

L. Sartori Menegatto[1], L. Afonso De Freitas[1], R. Dutra Do Bem[2], L. Vage Coelho Sartori[2], N. Bonvino Stafuzza[2] and C.C. Paro De Paz[2]
[1]University of São Paulo, Department of Genetics, Ribeirão Preto Medical School, Bandeirantes Avenue, 2900, Ribeirão Preto, SP, Brazil, 14049-900, Brazil, Brazil, [2]Animal Science Institute, Beef Cattle Research Center, Carlos Tonani Highway, 94, Sertãozinho, SP, Brazil, 14888-000, Brazil; leonardomenegatto@gmail.com

The hematophagous nematodes are important damage agents in sheep farming, causing anaemia and production losses. This issue is verified by extrinsic and intrinsic mechanisms that can be used as selection criteria to resistance to worms, like the total plasma protein (TPP) and body condition score (BCS) parameters. Our hypothesis was that significant differences in these parameters could be identified in sheep breeding programs, in order to allow the animal's anaemia evaluation. A total of 628 Santa Inês animals from three Brazilian herds were evaluated through faecal egg count (FEG), corpuscular volume (CV), total plasma protein (TPP), and body condition score (BCS). The FEG and CV parameters were used to stratify the population in resistance and susceptible groups by (1) permutations performed in original scripts in R program; (2) by the 100 animals with the most disparate values (50 at each group), and (3) using only VG threshold to anaemia as 22.0%. The TPP e BCS results between resistance and susceptible groups were compared two by two through Tukey's Test with $\alpha=5\%$. The similarity of the independent classifications between FEG and CV did not exceed 55%, which is probably due to the tolerance capacity of some animals, i.e. those sheep that have infestation, but do not exhibit anaemia. However, the permutation only considering VG and the extremes analysis on this criterion showed significant differences in both parameters, while the use of VG threshold of 22.0% was highly significant. The susceptible TPP average was 6.9 against 6.4 of resistance animals and the BCS was 3.6 against 2.7 in the first grouping, whilst using the threshold the averages obtained were 6.8 to 5.9 and 3.2 to 2.1, respectively. These results demonstrated the correlation between CV and TPP and between CV and BCS, but they showed that ignoring the tolerance phenomenon using FEG to animal stratification can impair the use of these parameters in the selection.

Exploiting most informative markers to predict group membership of North Aegean sheep

A. Kominakis, E. Tarsani, A. Hager and I. Hadjigeorgiou
Agricultural University of Athens, Animal Science, Iera Odos 75, 18855 Athens, Greece; acom@aua.gr

The aim of the present study was to identify a small panel of informative SNPs with high assignment power to: (1) predict group membership of sheep of three North Aegean islands (Lemnos, Lesvos and Agios Efstratios); and (2) assign animals to two agro-ecological zones (rough and flat landscape) on the Lemnos island. A total number of 256 ewes belonging to n=15 herds dispersed on Lemnos (total n=197, rough landscape n=60, flat landscape n=137), Lesvos (n=38) and Agios Efstratios (n=21) genotyped with the 50K SNP array were used. Application of quality criteria at the marker level resulted in 37,201 SNPs retained. Of the 37K SNPs, we searched for minimum number (n=50 or 100) of strongly differentiated SNPs between groups (islands/zones) defined as those with highest values for the FST fixation index. K-th nearest-neighbour Nonparametric Discriminant Analysis (NDA) of Principal Components (PCs) constructed on strongly differentiated SNPs followed. The discriminant criterion derived on the first two PCs of n=50 strongly differentiated SNPs resulted in misclassification error rates as high as 0.167, 0.237 and 0.048 for Lemnos, Lesvos and Agios Efstratios island, respectively (average error rate 0.150). The respective discriminant criterion derived on two PCs of n=100 strongly differentiated SNPs resulted in lowest misclassification error rates of 0.010, 0.001 and 0.048 for Lemnos, Lesvos and Agios Efstratios islands, respectively (average error rate 0.019). Finally, the discriminant criterion derived on one PC of n=50 strongly differentiated SNPs could accurately assign individuals to the two agro-ecological zones with an average misclassification error rate of 0.012 (0.007 and 0.017 for rough and flat landscape, respectively). Current results imply that most informative markers such as those examined here can be implemented to unequivocally determine the origin/habitat of local sheep and possibly their associated specific products.

Intra-day repeatability of methane emissions in sheep

E. O'Connor[1,2], F.M. McGovern[2], T.M. Boland[1], E. Dunne[2] and N. McHugh[2]
[1]University College Dublin, School of Agriculture and Food Science, Belfield, Dublin 4, D04 V1W8, Ireland, [2]Teagasc, Animal and Grassland Research and Innovation Centre, Mellows Campus, Athenry, Co. Galway, H65 R718, Ireland; edel.oconnor@teagasc.ie

Agricultural livestock production accounts for 34% of anthropogenic greenhouse gas emissions in Ireland and therefore the agricultural sector is under significant pressure to reduce methane emissions without compromising animal productivity. However, methane (CH_4) emissions have never been reported for Irish sheep production systems. The study objectives were to quantify CH_4 emissions from Irish sheep and to investigate the within-day variation of CH_4 emissions. A total of 12 dry ewe lambs (10 to 11 months of age) were randomly selected each day from a cohort of 48 animals over a consecutive 9 day period. Methane emissions from the 12 lambs were measured in 12 Portable Accumulation chambers (PAC) at two time points daily, morning (8 to 11 am) and evening (2 to 4 pm). Animals were removed from grass silage for at least one hour prior to measurements in the PAC and animals were assigned randomly to each chamber. Methane (ppm) concentration was measured at two time points (0 minutes and 50 minutes from entry of the animal into the first chamber) using an Eagle 2 monitor. Live weight was also recorded once daily; the average live weight was 46 kg. Factors associated with CH_4 emissions expressed in CH_4 gram/hour were modelled using a mixed model which included the fixed effects of the individual PAC chamber, time of daily measurement, day of measurement, breed, live weight and humidity whilst also accounting for the repeated measurements of animals both within and across days. The average CH_4 emissions recorded across all animals was 0.006 CH_4 g/hr (±0.002 CH_4 g/hr), this would extrapolate to a value 8.6 CH_4 g/day. Factors associated with CH_4 emissions included date of measurement, time of measurement and live weight of the animal (P<0.01); CH_4 emissions did not differ by breed of the animal or chamber of measurement (P>0.01). For each kilogram increase in the live weight of the animal, CH_4 emissions increased by 0.000143 CH_4 g/hr (P<0.01). Greater CH_4 emissions were recorded in the afternoon relative to morning measurements (P<0.001). The estimated correlation between intra-day CH_4 emissions was 0.63 (P<0.001).

Environmental impact of Sardinia dairy goat farms

M.F. Lunesu, A.S. Atzori, P. Sau, A. Ledda, F. Lai, A. Fenu, R. Rubattu, A. Mazza, F. Correddu and A. Cannas
University of Sassari, Dipartimento di Agraria, Viale Italia 39, 07100 Sassari, Italy; mflunesu@uniss.it

The aim of this work was to estimate the carbon footprint (CF) of a sample of 6 Sardinia (Italy) dairy goat farms selected from 2 farming systems identified as: intensive (INT) with cosmopolitan breeds kept indoor and mixed (MX) with mix breeds partially kept indoor. Data were collected by interviewing the farmers for a complete life cycle inventory (LCI) of farm production processes from October 1st 2016 to September 30th 2017. The LCI included information on flock, animal diets, feed purchases, crops, farm stocks, and an energy use audit. Data were analysed with a modified Tier 2 of the IPCC (2006), by using coefficients of IPCC for estimation of animal and manure CH_4 and N_2O emissions and literature coefficients for purchased feeds and for energy. Reported values of CF were allocated 100% to milk yield. Average milk production level (PL) of the flock was equal to 711 and 581 kg of fat and protein corrected milk (FPCM)/head/year for INT and MX farming system, respectively. Total emissions were higher in the INT than in the MX farming systems (342,255 vs 259,104 kg CO_2eq, respectively). In contrast, emission intensity was on average higher in the MX than in the INT farming systems (3.03 vs 1.94 kg CO_2eq kg FPCM^{-1}, respectively). The PL was the best predictor of CF (CF=-1.823ln(PL)+14.001; R^2=0.9188). The contribution to total emissions was about 43% from enteric methane, 27% from purchased feeds, 21% from manure management and 11% from energy use. In conclusion, the results of this work suggested that the impact decreases: (1) increasing milk production level, feed conversion and reproduction efficiency and food self-supply; and (2) going from mixed to intensive dairy goat farming systems but mainly due to production level. Despite, the low number of farms, the short temporal space taken into account and the high variability among farms this dataset allowed to define the drivers explaining the most part of variability of emission intensity in dairy goat farms. Acknowledgements The authors gratefully acknowledge the Europe Union for the financial support (Project Forage4CLimate LIFE 15 CCM/IT/000039).

The effect of rotational grazing on sheep and grassland performance

T. Meeke[1,2] and A. Aubry[2]
[1]Queens University Belfast, School of Biological Sciences, Belfast, United Kingdom, [2]Agri-Food and Biosciences Institute, Hillsborough, Co. Down, United Kingdom; tara.meeke@afbini.gov.uk

The competitive advantage in ruminant livestock production hinges on maximising the contribution of grazed grass in the diet. Grass, our cheapest feed resource, can supply up to 95% of the energy requirements of sheep; thus, the efficient utilisation of herbage in lamb production systems is the key to profitability. Currently, there are inefficiencies in the level of herbage utilised/ha within sheep farms in Northern Ireland. The objective of this study was to examine the effect of 4 vs 8 paddock rotational grazing systems on animal and grassland performance. The study was replicated over two grazing seasons from April to November 2018 and from March to November 2019, at the Agri-Food and Bioscience Institute. The area used for the study consisted of a predominantly perennial ryegrass sward. There were two grazing treatments (4 vs 8 paddock rotational grazing system) which were balanced for ewe live weight, body condition score and lamb sire breed. Each system consisted of 1.6 ha which were rotationally grazed at a stocking rate of 14 ewes ha-1. Lambs were weighed fortnightly from 6 weeks of age using portable electronic scales and were drafted for slaughter on reaching 45 kg of live weight. Lambs were weaned on average at 14 weeks of age. Pre- and post-grazing compressed sward heights were determined on each paddock before and after grazing by taking 30 measurements across the diagonal of the paddock with a rising plate meter. Pre- and post-grazing herbage mass was determined for each paddock by taking four quadrat (0.5×0.5 m) cuts. All harvested herbage was weighed and a sub-sample was retained for DM and quality analysis. Data was analysed using linear mixed models, with ewe as a random effect and lamb sire breed, gender and deviation in lamb age from the treatment mean included as fixed effects. Lambs grazing the 4-paddock system had higher average daily gains from 10 weeks of age to weaning (P<0.001) compared to those grazing the 8-paddock system, which resulted in higher weaning weights (P<0.001) for the 4-paddock lambs. In conclusion, the 8-paddock rotational grazing system produced higher levels of herbage utilisation but resulted in lower lamb performance.

Genomic characterisation of Portuguese native sheep breeds

D. Gaspar[1,2], H. Magalhães[2], A. Usié[2,3], C. Leão[4], C. Ginja[1], C. Matos[5] and A.M. Ramos[2,3]
[1]CIBIO/InBIO, Archaeogenetics, Universidade do Porto, Vairão, 4485-661 Vairão, Portugal, [2]CEBAL, Centro de Biotecnologia Agrícola e Agro-Alimentar do Alentejo, Beja, 7800-295, Portugal, [3]MED-Mediterranean Institute for Agriculture, Environment and Development, Núcleo da Mitra, 7006-554 Évora, Portugal, [4]INIAV, Instituto Nacional de Investigação Agrária e Veterinária, Quinta do Marquês, 2780-157 Oeiras, Portugal, [5]ACOS – Agricultores do Sul, Rua Cidade S. Paulo, 7801-904 Beja, Portugal; danibgaspar@gmail.com

Merino, Campaniça and Serra da Estrela sheep are among the most relevant native breeds reared in Portugal under extensive conditions. Merino and Campaniça are mainly distributed in the south of Portugal, in the Alentejo region, being the basis for the production of certified meat, dairy and wool products. Serra da Estrela is the main Portuguese dairy breed, producing a high-value certified cheese, and is raised in the mountain region that its name originates from. These breeds are classified as threatened genetic resources despite their importance. The purpose of this study was to assess the dynamics of population genomics in these sheep breeds using whole-genome resequencing. Blood samples were collected from 56 sheep across these breeds and a population of crossbred Merino sheep. Following DNA extraction and resequencing, the raw data were filtered by quality and used for variant calling. A total of 31,320,381 high-quality SNPs were kept for downstream analysis. Among these, 11,148,321 SNPs were located in genic regions of which 120,172 were annotated as synonymous and 80,882 as non-synonymous. The remaining 20,172,060 SNPs were identified in intergenic regions. These data were used to determine the genomic diversity and population structure of Portuguese native breeds in the context of worldwide sheep genomic variation. Structural variation was also characterised, yielding a total of 340,188 variants after filtering and 4,197 variants annotated in exonic regions. These variants were used to evaluate the patterns of copy number variation in sheep populations. The results derived from this study will be used to develop a genotyping assay specific for more diverse traditional sheep breeds, including markers useful for genome-wide association studies and traceability analysis.

No such thing as a 'strong' lamb – ewe influence on pre- and post-weaning animal growth

A.G. Jones, T. Takahashi, H. Fleming, B.A. Griffith, P. Harris and M.R.F. Lee
Rothamsted Research, North Wyke, EX20 2SB, United Kingdom; andy.jones@rothamsted.ac.uk

Daily liveweight gain (DLWG) of lambs has a substantial impact on the overall efficiency of pasture-based sheep production systems, as longer grazing times are associated with greater maintenance energy expenditure. Identification of factors affecting DLWG is thus a worthwhile exercise, with studies already showing a positive correlation between lamb's pre-weaning DLWG and the physical condition of their dam during lactation. Little is known, however, about the mechanism linking the growth rates between pre- and post-weaning periods. This study therefore investigated the interrelationship between DLWG from these two distinct stages of physiological development, and how strongly the growth at each stage is influenced by the dam. The study used a dataset of 1,391 lambs and their mothers recorded at the North Wyke Farm Platform, a farm-scale grazing trial in Devon, UK. DLWG was calculated for each lamb, for each period between fortnightly weighing events. Contrary to our expectations, Spearman's rank correlation test indicated no significant association between DLWG of individual lambs before and after weaning (ρ=-0.023, P=0.40). An additional analysis using analysis of variance (ANOVA) supported this result; pre-weaning DLWG, for example, was primarily explained by the fixed effect from the mother (P<0.05 for 50.5% of ewes), while post-weaning DLWG did not exhibit a similar pattern (P<0.05 for 1.8% of ewes). These results suggest that a ewe's impact on her lambs' growth is not primarily caused via a genetic relationship regulating 'strong' lambs but rather via the dam's mothering and lactating abilities that do not persist post-weaning. Finally, when twin-bearing ewes from each year were ranked according to the combined weaning weight of their lambs and grouped into equal thirds, there was a significant difference in next year's combined weaning weight between the top and bottom groups (P=0.03). This finding not only confirms that dams have a substantial impact on lambs' pre-weaning growth but also that this effect is repeatable over multiple years, meaning that selective replacement strategies based on ewes' historic performance data could improve the farm's long-term profitability.

Modelling adaptive capacity of goats to a high concentrate diet using rumen hydrogen ion

M. Taghipoor and S. Giger-Reverdin
INRAE, 16, rue Claude Bernard, 75005, France, Metropolitan; masoomeh.taghipoor@inrae.fr

High producing ruminants need high concentrate diets to satisfy their nutritional requirements and to meet their performance objectives. However, several studies showed that these diets might induce sub-acute ruminal acidosis (SARA), which will affect dry matter intake and decrease production performance. The main objective of this study is to study the individual variability of goats, and their short and long-term adaptive capacities to an abrupt change of diet, from a standard diet (20% concentrate) to a high concentrate diet (50% concentrate). To avoid any calculation bias related of the use of pH, in this study all analyses were performed for hydrogen ion H+(Murphy 1982). Eight dairy goats were fed with a standard diet during the first week of experimentation and then with a high concentrate diet during the 4 following weeks. Rumen fluid was sampled before the morning feed delivery and 1, 2, 4 and 6 hours after feeding. To study the response of goats, firstly the dynamic records of post-prandial H+ was summarised into 3 independent variables, v0 (initial value of H+), Rec (recovery capacity) and Amp (v0-min(H+)). Secondly, a mixed model was developed for these variables to study the short (1st week of experimentation) and long-term adaptive capacity of goats (at the end of experimentation). Results showed that goats were not able to buffer the effect of high concentrate diet at long term. This is illustrated by the significant increase of Amp and v0 at the end of the trial. At short term, no dietary significant effect on either variables was observed. The variable Rec was not affected by the dietary change. Moreover, a high individual variability was observed for variables Amp and v0, while Rec showed small variability between animals. In conclusion, results showed that all animals succeeded to recover after the drop of pH during the first week of experimentation, probably by adapting their feeding behaviour. Nevertheless, at week 5, they were no longer able to resist the negative effect of high concentrate diet. The large variability observed among goats suggests the importance to develop non-invasive methods of detection for SARA.

Effect of essential oils and plant extracts supplementation (Aerorum) on lamb growth performances

P.H. Pomport[1], V. Berthelot[2], J.P. Ricaud[3] and M. Pondet[3]
[1]Ferme expérimentale AgroParis Tech, route de la Ferme, 78850 Thiverval-Grignon, France, [2]Université Paris-Saclay, INRAE, UMR Modélisation Systémique Appliquée aux Ruminants, AgroParisTech, 75005 Paris, France, [3]IDENA, 21 rue du Moulin, 44880 Sautron, France; pierre-henri.pomport@agroparistech.fr

Poor environmental conditions can lead to stunted growth, mortality and economic losses. When ammonia content in the barn is over 35 ppm, lambs average daily gain and dry matter intake is impaired. The objective of the Aerorum product made up of a mixture of essential oils and plant extracts is to possibly enhance respiratory comfort to allow better growth and animal performances. The aim of the study was to evaluate the effect of Aerorum on performances. Lambs were assigned into 2 groups: a control group (C, 134 lambs) and an experimental group (E, 136 lambs) during 2 trials (Trial A: winter 2018 and Trial B: autumn 2018). The lambs were fed *ad libitum* a barley/rapeseed meal based diet supplemented or not with 490 mg/d of active principles (EO: ravintsara, oregano and eucalyptus/ plant extracts: rosemary, fenugreek) from weaning (85.9±5 days- 31.3±6.5 kg of body weight (BW) for E and 32.2±5.5 kg BW for C (P=0.25)) till slaughter at fixed BW (42 to 46 kg). During fattening, lamb ADG was measured individually and lambs concentrate intake was measured by group. To evaluate the barn environmental conditions, its ammonia content was measured 1 m above the litter and at the litter level. ANOVA procedure was used for statistical analyses with SAS. The ammonia contents in the barn ranged from 3.4-5 ppm (1 m above the litter) to 35 ppm (on litter) which can be considered as limiting conditions for respiratory comfort. Concentrate intake was similar in Trial A (E=1.48 vs C=1.47 kg/d/lamb) or decreased in Trial B (E=1.40 vs C=1.55 kg/d/lamb) with Aerorum supplementation. ADG was higher in E group than in C group (386±109 vs 297±89 g/d respectively, P<0.0001) in both trials. The feed conversion ratio with Aerorum supplementation was improved by 28% (E=3.98 vs C=5.10 kg concentrate/kg BW gain) in trial A and 33% (E=4.21 vs C=5.62) in Trial B. The Aerorum supplementation resulted in better growth performance and diet efficiency during the fattening period of indoors lambs.

Screening Mediterranean shrubs selected by browsing goats against gastrointestinal strongyles

L. Padre[1], C. Costa[1] and A.T. Belo[2]
[1]ICAAM-Évora University, Pólo da Mitra, 7006-554 Évora, Portugal, [2]INIAV-IP, Fonta Boa, 2005-048 Vale de Santarém, Portugal; lpadre@uevora.pt

Ethanolic extracts from the main species selected by browsing Charnequeira goats on Mediterranean vegetation were assayed for a rapid screening of their anthelmintic potential against gastrointestinal (GI) strongyles. Shrubs selected included *Olea europeae* var. *sylvestris* (OS), *Quercus coccifera* (QC), *Pistacia lentiscus*, *Rhamnus alaternus* and *Rhamnus lycioides* (RL). Forage species included *Cichorium intybus* (CI) and *Hedysarum coronarium* (HC). Extract concentrations were chosen taking as reference the fecal total phenolics content (TP) from those goats after a five-month browsing period (8.4 mg GAE/g DM). A quantitative coproculture method was adapted using faeces obtained from naturally infected goats, not dewormed or having access to browse (reference group); larvae culture was performed in triplicate for three different TP concentrations (25, 10 and 5 mg GAE/g DM) against a distilled water control, for a 7-day incubation period at 27 °C. The number of larvae developed was counted differentiating total and live larvae per gram of faeces (LDPG). The efficacy of each extract and concentration, as percentage of reduction in LDPG against control, was determined. Data was log transformed and submitted to analysis of variance. Means were compared by Tukey test at 5% significance level. A reduction in LDPG was observed in all extracts being significant (P<0.05) for the highest concentration. The extracts' efficacy over exogenous forms showed not only a reduction of total LDPG but, for some, also a high level of larvae mortality, resulting in reduction of live infective larvae (L3). These results indicate that some extracts, besides their ovicidal activity, have also larvicidal potential. Extracts efficacy for total LDPG was higher for QC (82%), OS (80%), HC and RL (76%) and CI (69%). Reduction of live L3 was higher for OS (90%) and HC (80%). Results obtained on these extracts potential in reducing exogenous forms, namely L3 forms of GI strongyles, is an important aspect in the control of these parasitic populations, as it results in lower pasture contamination and consequently a lower rate of reinfection of the animals. Funding obtained from project VegMedCabras – ALT20-03-0145-FEDER-000009.

Developing cross-boundaries based innovations to design diversified and integrated livestock systems

M.A. Magne[1], M. Duru[2] and B. Dedieu[3]
[1]INRAE / ENSFEA, UMR AGIR, 24 Chemin de Borde Rouge, Auzeville, CS 52627, 31326 Castanet Tolosan Cedex, France, [2]INRAE, UMR AGIR, 24 Chemin de Borde Rouge, Auzeville, CS 52627, 31326 Castanet Tolosan Cedex, France, [3]INRAE, UMR Territoires, Route de Theix, 63122 Saint-Gènes-Champanelle, France; marie-angelina.magne@inrae.fr

There is an urgent need to develop livestock farming systems (LFSs) that ensure human food and nutritional security, reduce negative impacts of agriculture on human health and the environment, and maintain a decent livelihood, while also facing the scarcity of non-renewable and renewable resources. One issue is that innovation has been thought per domain (genetic, technical, etc.), to meet one challenge (e.g. mitigate greenhouse gases, improve animal welfare, etc.), at a given scale (e.g. animal, farm), in one sector (e.g. dairy/beef cattle; livestock/crop system), which does not meet the transformative ambition of the transition to be made. This is partly due to the high specialisation of the LFSs, breeds and territories. Here, we demonstrate that fostering the sustainability transition of LFSs implies developing cross-boundaries innovations to design diversified and integrated LFSs. Based on three case studies, we show that managing agrobiodiversity and biological/ecological interactions in LFSs have benefits for their sustainability and resilience but remains under-used. Through the case of the transition of dairy cattle systems towards rotational crossbreeding management, we highlight that using animal agrobiodiversity is hindered by a sociotechnical lock-in. This one results from organisational, cultural, technological and economic interrelated barriers from upstream to downstream. Based on two cases, the diversification of pig feed through the use of flax and grain legumes and the crop-livestock integration at the territory scale, we show two strategies for unlocking the sociotechnical system, depending on who has been driven it: a downstream stakeholder with the development of a product label or a group of livestock and crop farmers and local stakeholders. We finally suggest some research avenues in animal production research to organise the design of cross-boundaries innovations and foster the transition towards diversified and integrated LFSs.

Preconditioning programs: a solution to reduce bovine respiratory diseases in fattening units?

E. Vanbergue[1], G. Foucras[2], M. Guiadeur[1], N. Cebron[2], G. Meyer[2], R. Maillard[2], B. Mounaix[1] and S. Assié[3]
[1]French Livestock Institute, rue de Bercy, 75012, France, [2]ENVT, INRAE, Chemin des Capelles, 31076 Toulouse, France, [3]Oniris, Rue de la Géraudière, 44300 Nantes, France; elise.vanbergue@idele.fr

Preconditioning programs are sets of management practices to boost young bull's immunity and low weaning and arrival stress, in order to reduce respiratory diseases in fattening units. To evaluate the interest of preconditioning programs in the French context, one hundred and sixty eight young bulls had been monitored from birth to slaughter in a control/case study. Preconditionned young bull (PREP) were weaned indoors 50 d before leaving the farm, received a trivalent vaccine (BRSV, BPI3, *Mannheimia haemolytica*) and were fed with a diet supplemented in vitamins and trace elements. Control bulls (CTRL) remained in pasture with their dams, with no vaccination nor particular diets. Weight, health, behaviour, immune competence, post-mortem lung status has been scored and analysed by variance analysis and ascending hierarchical classification. During the 50d prior leaving, average daily gain, serum zinc and serum glutathione peroxidase concentrations were higher for PREP compared to CTRL (+ 440 g/d; +3 pmol/l; +118 U/gHb; P<0.001), in relation to feeding management. During the preconditioning period, slight and moderate respiratory signs had been reported for certain PREP and CTRL batches, in relation to husbandry and environment factors (pathogens circulation, housing conditions). At the end of the fattening period, carcass weight and quality did not differ between batches. No significant effect has been observed on fattening duration due to beef market situation and experimental constraints. Unexpectedly, clinical scores tended to be higher for PREP (although variable) and lung lesional scores were higher for PREP compared to CTRL (+8 points; P<0.01). Pathogens diversity detected on farms did not match entirely with the vaccine and can partly explain the results. Besides, variability in individual immune competence has been observed in relation to farms, trace elements and vaccine status. In a context of reduction of antibiotics use, preconditioning programs offers promising prospects providing adjustments to take into account local condition and husbandry factors.

Rapeseed-oil as ω3-rich natural sources for livestock feeding to obtain differentiated meat products

N. Mandaluniz[1], I. Olazaran[2], J. Arranz[1], E. Ugarte[1] and R. Ruiz[1]
[1]NEIKER, Animal Production, P.O. Box 46, 01080, Spain, [2]Leartiker, Xemein Etorbidea 12, 48270, Spain; nmandaluniz@neiker.eus

Society is increasingly aware of the food it consumes, the way it is produced and every time demands functional foods. The objective of the operational group KALIKOLZA (https://www.kalikolza.com/) was to implement animal feeding systems, mainly by Omega-3 (Ω3) natural sources, to improve the profile of fatty acids (FA) and obtain differentiated products. The Ω3 PUFA appears in both animal and plant feeds; in the case of vegetables, rapeseed is a Ω3-rich source. Under this context, an on-farm trial was carried out in Kerexara commercial farm. This farm works under circular economy keys. During the milking period sheep were fed with a concentrate that contains on-farm produced rapeseed-cake (25%), which involves rapeseed-oil production. Dairy sheep activity is complemented with pigs to take advantage of whey and beef-calves to take advantage of surplus fodder of sheep. In order to give value to rapeseed-oil (a Ω3-rich source) an on-farm trial was implemented. During 2019 a sample of calves and pigs were fed with a concentrate with 5-6% rapeseed-oil. After slaughtering, calve (n=5, 15-19 months-old) and pig (n=5, 5-6 months-old) chops were sent to Leartiker laboratory in order to determinate nutritional parameters (protein, fat and energy content) and fatty acid profile (saturated fatty acids-SFA, unsaturated-UFA, and Ω3- Ω 6 contents). Commercial chops of similar characteristics were analysed to compare meat quality results. According to the results, average nutritional parameters of calve-chops were: 19.47±1.53% protein, 5.18±2.99% fat and 130.50±20.63 Kcal/100 g meat. Rapeseed-oil fed chops showed higher content of UFA, Ω3 and Ω6 (4.97, 20.17 and 16.29%, respectively) and lower SFA and aterogenic index (4.87 and 19.68%, respectively) compared to control ones. On the other hand, pig-chops average nutritional parameters were: 20.52±2.29% protein, 8.98±1.43% fat and 168.00±14.02 Kcal/100 g meat. Rapeseed-oil fed chops showed higher content of UFA, Ω3 and Ω6 (11.62, 81.97 and 32.46%, respectively) and lower SFA and aterogenic index (12.76 and 20.31% respectively) compared to control ones. The coincidence of these tendencies with those observed in experimental trials shows the possibility to extrapolate accurate information obtained at experimental scale to a commercial on-farm scale. As conclusion, the inclusion of rapeseed-oil in beef-calves and pigs feeding increases UFA, Ω3 and Ω6 content and reduces SFA content and aterogenic index of the meat, being more marked the difference in pig meat. *Acknowledgement:* Authors thank Kerexara farm for their collaboration.

Sustainability of extensive livestock production systems: the challenge of PA technologies

J. Serrano, S. Shahidian, E. Carreira, J. Marques Da Silva, A. Pereira and M. Carvalho
Universidade de Évora, MED- Mediterranean Institute for Agriculture, Environment and Development, UInstituto de Investigação e Formação Avançada, Universidade de Évora, Pólo da Mitra, Ap. 94, Évora, 7006-554 Évora, Portugal; jmrs@uevora.pt

In the extensive livestock production systems of southern Portugal the basis of animal feed is provided by biodiverse pastures, complemented by the availability of acorns in the winter period. Since the turn of the new millennium, greater recognition of the importance of pastures and grasslands in conserving soil, promoting biodiversity, stabilising farming communities, and providing a wealth of natural ecosystem services has led to renewed interest in the dynamics of grasslands function. This important agro-silvo-pastoral ecosystem, known as *Montado*, is characterised by a high complexity resulting from the interactions between climate, soil, pasture, trees, and animals. Its sustainability depends on its capacity to improve soil fertility, to increase pasture productivity and the rational use of production factors (especially soil amendment or fertilisation) and to promote dynamic grazing (livestock numbers and time of grazing in each plot). In order to achieve these objectives, it is essential to calibrate and validate expeditious monitoring tools that provide information to support decision making. This work presents the results of a study carried out between 2015 and 2019 on a 4.0 ha experimental field located at Mitra farm (plot 'ECO-SPAA'). Proximal sensors were used to monitor the electrical conductivity and soil resistance, the productivity and quality of pasture, the leaf area index of trees and the animal grazing preferences. The interest of remote sensing from satellite images was also evaluated and demonstrated to monitor pasture quality and produce a warning system for feed supplementation needs. The results showed the potential of technologies associated with precision agriculture (PA). These tools provide a better understanding of the *Montado* ecosystem what allows the optimisation of the herbaceous stratum and the dynamic management of grazing in a holistic and sustainable form.

A framework to analyse agricultural innovation systems applied to the sheep sector

D. Martin-Collado[1], A. Caudevilla Pérez[2], A. Bernués[1] and I. Casasús[1]
[1]Ctr Invest y Tecnol Agroal Aragon (CITA), Unidad de Producción y Sanidad Animal, Avda. Montañana, 930, 50059, Zaragoza, Spain, [2]Aragonesa de Imagen y Comunicación, Albareda 7, 50004, Spain; dmartin@cita-aragon.es

Agricultural innovation studies have evolved from the Innovations Diffusion model approach to the actual Agriculture Innovation Systems (AIS). AIS perspective considers all stakeholders and factors that affect the development and adoption of innovation. Innovation is not just technology but a comprehensive view of what production systems should look like in the future. Although the AIS concept draws a realistic view of the innovation process, it also highlights its complexity. To tackle this complexity researchers have developed several approaches to analyse and assess AIS inefficiencies, and to suggest improvements. We present a methodology to analyse livestock innovation system by disengaging its components and interactions, and identifying the drivers and constraints for an efficient development and dissemination of innovations. We adapt previous work into a methodology tuned up to the sheep sector. Central to this method is the assessment of the functions that AIS should fulfil: (1) guidance of the search; (2) knowledge development; (3) field experimentation; (4) knowledge diffusion; (5) resources mobilisation; and (6) innovation brokering. The methodology is based on a multi-stakeholder consultative process with the following consecutive steps: (1) 'Sheep Innovation System definition' in which the system boundaries (geographical area, species, innovation types) are defined and stakeholders are identified; (2) 'Stakeholder and social network analysis' which involves an analysis of stakeholder functions and areas of interest, and an indicator-based social network analysis; (3) 'Assessment of AIS function performance' which consists of a stakeholder consultative assessment of functions and identification of key enablers and disablers of the innovation system performance. Finally, we show how the methodology can be implemented in practice using the case of the sheep sector in Spain. The Spanish analysis highlighted the lack of a common vision across the sector stakeholders and the central role of breed associations in improving sector linkage and information flow among AIS stakeholders.

How an on-farm experimentation may improve attitudes and practices of dairy farmers

B. Mounaix[1], M. Guiadeur[1], T. Jozan[2] and S. Assie[3]
[1]Institut de l'Elevage, Health and Welfare, 149 rue de Bercy, 75595 PARIS cedex 12, France, [2]MSD Santé Animale, 12 rue Olivier de Serres, 49070 Beaucouze, France, [3]INRA Oniris, UMR BioEpAR, Site de la Chantrerie – route du Gachet, 44307 Nantes, France; beatrice.mounaix@idele.fr

Vaccination of dairy herds is often too late to be fully efficient. This paper reviews the impact of a full-scale experimentation of dairy cattle vaccination in French farms on farmers' attitude and their practices regarding calves health. Before the experimentation (2017), one-on-one conversations were carried out with 36 farmers involved in the experimental design; 29 of them were interviewed at the end of the experiment (2018). Before the experimentation, farmers had a positive attitude towards vaccination. They associated it to prevention (77% of answers) to limit health issues (64%), to decrease time spent for sick animals (44%) or to save on treatments (11%). Though, vaccination of cows was mainly implemented when several calves already suffered from diarrhoea, and should then be qualified as a curative approach. Hence, cow vaccination was never cited as a mean to improve the quality of calves. The experimentation had a positive effect on the attitude of farmers towards the colostrum. 75% of them now consider that the colostrum is involved into the cow-calf immunity transfer. They now give more importance to the timing between the calving and the distribution of colostrum (72% of answers) and to the quantity taken by the calf (65%). They still consider as less important the quality of colostrum (50%), mainly because of the lack of references and/or practical solutions when quality is low. As a consequence, 41% farmers indicated that, after the experimentation, they changed their way to distribute the colostrum to better control the quantity which is really taken by the calf. A lower impact was observed on famers' attitude towards the vaccination of cows. Although 58% of farmers showed interest in the experimentation and 41% could observe positive impacts on calf health, only 44% of them declared that they would put it in practice, and only during winter as a risky period (curative measure). Growth of calves was not considered by farmers as a motivation to implement cow vaccination although we could measure positive effects, mainly because growth is rarely monitored by dairy farmers.

Planning farm efficiency at territorial scale to improve environmental performance of dairy sheep

A.S. Atzori and A. Cannas

Univeristy of Sassari, Department of Agriculture Section of Animal Science, viale Italia 39, 07100, Italy; asatzori@uniss.it

The Sardinian sheep sector includes about 10.000 dairy farms and 2.7 million ewes, producing about 330 millions l/yr. of sheep milk, mostly from December to July (about 130 l/yr. of milk per present ewe). This work proposes a method to be applied at territorial scale to improve technical and environmental efficiency of the dairy sheep sector and to quantify economic and environmental benefits in Sardinia. Partial correlations and distributions of 58 variables from 490 real sheep farms were calculated, to describe the technical profile of small farms (<300 heads) and medium-large farms (>300 heads). A MonteCarlo simulation (@RiskSoftware-Palisade.com) was run to develop a synthetic database of the Sardinian sheep sector (6,261 farms <300 heads, 41% of heads; 3,744 farms >300 heads, 59% of heads). Several scenarios were tested in a 10 year horizon, assuming stable total Sardinian milk deliveries. Estimated emissions resulted: CO_2eq. from methane (kg) = 87.4 milk (kg/yr per ewe) – 0.765 R^2=0.9 (P<0.01).An improvement of 20 l/yr. per ewe of the 3,744 medium-large farms would reduce the regional flock of 8.5%, saving 15,000 ha of pasture land. Similar improvements would be obtainable increasing by 30 l/yr. the production of the 6,261 small farms. Increasing annual fertility would induce limited advantages. Else, improvement of lambing synchronisation (to reduce delayed lambings and increase milking days in autumn, i.e. from 20 to 60) in the 3,744 medium-large farms would favour: increase the regional milk production level of 17 l/yr. per present ewe, reduce the regional flock of 12%, save resources for ewe maintenance costs equal to 22,000 ha of pasture land (or stored forage for 19 Million of €) and reduce sheep methane emissions of 11%. The same strategy applied to all farms would reduce of 21% the ewes, reducing pasture land by 40.000 ha and methane emissions by 15%. This method might be used to test territorial policies in other production sectors.

Resilience of three farming systems to decline in synthetic nitrogen and feed imports availability

C. Pinsard, F. Accatino, S. Martin and F. Leger

INRAE, TERRA, 16 rue Claude Bernard, 75005, France; corentin.pinsard@inrae.fr

Agriculture in France is predominantly input-intensive and imports feed. As a result, it is dependent on oil, whose production may start to decline from the 2020s onwards and is exposed to potential economic shocks. It seems essential to assess the capacity of such an agriculture to maintain production levels in the context of a decline in oil production, in other words its resilience. To do this, the analysis of nitrogen fluxes whose synthesis requires fossil fuels, is a suitable tool. The objective of this study is to model and compare the variations in animal and plant productions in three French farming systems by 2050 under scenarios of decreasing availability of feed and synthetic fertiliser imports. We developed a dynamical model consisting in a regional-scale soil nitrogen balance with annual time steps, considering nitrogen fluxes between an animal compartment and a soil-plant compartment. Plant production is a function of mineral nitrogen in the soil. Livestock quantities are adjusted as a consequence of feed availability. The three farming systems are characterised by different combinations of cropland, grassland, and livestock: (1) arable crops (Plateau Picard); (2) arable crops and intensive monogastrics (Bretagne centrale); (3) extensive ruminants (Bocage Bourbonnais). The three scenarios are: (a) decrease in synthetic fertiliser, (b) decrease in feed imports, (c) decrease in synthetic fertiliser and feed imports. The results show that production decreases can be as much as 80% for the three regions. Region (ii) reduces its production from 50 to 75% for all three scenarios. Concerning regions (i) and (iii), production falls up to 25% for scenarios (b) and (a) respectively. In scenario (c), the three regions lose 75% of their production, however the trajectories differ: regions (ii) and (iii) have a smaller decrease than the region (i) in the first 10 years. In conclusion, the simulations suggest that livestock plays a buffering role. They also suggest that reducing a region dependence on nitrogen imports by coupling crop and livestock production improves its resilience to future oil decline.

Can collective brands push dairy farms to adopt innovative practices?

T.T.S. Siqueira[1,2], A. Gonçalves[1] and L. Mur[1,2]
[1]French National Research Institute for Agriculture, Food and Environment (INRAE), Agroecologies, Innovations & Ruralities, 24, chemin de Borde-Rouge, CS 52627, 31326 Castanet Tolosan, France, [2]University of Toulouse – INP – Ecole d'Ingénieurs de Purpan, 75, Voie du Toec, 31076 Toulouse, France; tiago.siqueira@purpan.fr

Many farmers and consumers consider that conventional dairy value chains are not able to answer the new societal demands and territorial issues anymore. The development of farmers' collective brands appears as a solution to face these issues and at the same time challenges the innovative capacities of dairy farms. This paper proposes to analyse innovative practices at two levels. First, it aims to study the innovation process at the scale of the collective brand by analysing its trajectory and territorial anchorage though the analysis of the resources mobilised to build the brand. Second, it aims to study if these collective brands push farms to adopt innovative practices to meet territorial and societal issues. In this study we collect data by interviewing farmers and the representatives of this new collective brands. Data collection and analysis will be partly based on a specific approach that comes from the economic sociology: the quantified narratives analysis. We will focus on the initiatives developed in the Occitanie region (France) a territory where we have an important development of farmers collective brands and 32,9% decrease in the number of dairy farms in the last 10 years.

Green Finance: a tool for assessing the environmental impacts of investments in livestock farming

M. Nugues[1], A. Bell[2], C. Kauffmann[1], M. Fert[3], C. Nzally[1], I. Taurou[4], J.-Y. Dourmad[1], H. Guyomard[2] and J.-L. Peyraud[1]
[1]INRAE, AGROCAMPUS OUEST, PEGASE, 16 le clos, 35590 Saint Gilles, France, [2]INRAE, CODIR, 147 rue de l'université, 75007 Paris, France, [3]INRAE, Université de Reims Champagne-Ardennes, FARE, 2 esplanade Roland Garros, 51100 Reims, France, [4]INRAE, Campus de la Grande Ferrade, 71 avenue Edouard Bourlaux, 33883 Villenave d'Ornon, France; melaine.nugues@inrae.fr

As part of the development of green financing for agriculture, we propose a tool for assessing the environmental impacts of farm investments. This tool could allow stakeholders (bankers and investors) to allocate loans according to the 'green' nature of equipment. The work presented here considers specialised dairy and pig farms but also works for crop farming. The impacts of farm equipment were assessed through the impacts of agricultural practices they allow to implement. Firstly, each practice was evaluated for eight environmental performances: soil quality, water quality, air quality, greenhouse gas emissions, energy consumption, water consumption, biodiversity and animal welfare. The notation associated to each performance was based on a qualitative five level scale corresponding to a strongly negative, negative, neutral, positive or strongly positive impact. The practices dealt with feeding and forage production, manure management, housing, herd management, milking, biodiversity management and resource use. Their impacts were evaluated based on scientific literature and expert judgement. Secondly, the pieces of equipment required to implement these practices were listed, resulting in 157 pieces of equipment for dairy farming and 91 for pig farming. When a piece of equipment was used to implement a single practice, it received the grade of the practice; when several pieces of equipment could be used to implement different practices, their impacts were evaluated through the mean value of practice impacts. The final databases gather 162 practices for dairy farming, 72 for pig farming and 248 pieces of equipment evaluated for the eight performances. When a practice or a piece of equipment has both positive and negative environmental impacts, we explain the antagonisms and identify the means of mitigating the most negative impacts.

Short supply chains of Bísaro pig – TREASURE project as a reference of knowledge and networking tool

J. Santos Silva[1], J.L. Cerqueira[2,3], M. Čandek-Potokar[4] and J.P. Araújo[3,5]
[1]Min. Agricultura, Florestas e Desenv. Rural, Guim, 4800, Portugal, [2]CECAV, UTAD, VR, 5001, Portugal, [3]Escola Superior Agrária, Inst. Polit. Viana Castelo, P Lima, 4990-706, Portugal, [4]Agricultural Institute Slovenia, Ljubljana, 1000, Slovenia, [5]Centro Investigação de Montanha (CIMO), ESA-IPVC, VC, 4900, Portugal; pedropi@esa.ipvc.pt

The development of the modern food and catering market, the tourism and gastronomic culture, opens a new and diverse paths for producers of healthy food. Consumers look for food quality, its excellent taste, rarity and unique culinary experience: feel and discover a place through its typical food, as part of the nature, culture and history of the place. Rural world has an excellent opportunity here. Gastronomy and small-scale agriculture (poly-culture, organic farming) have a place and a huge opportunity for growth in our society. Establishment of synergies between rural and urban, brings a better quality of life, greater convergence and territorial cohesion, improved sustainability. The concept 'A farm, a producer, a consumer' offering specific products adapted to needs and expectations of consumers towards intrinsic and extrinsic quality and societal demands for sustainable food system. Traditionally, Bísaro pig breed is adapted to specific local environment and fed with various locally available feedstuffs. So besides their genetic merit for agro-biodiversity, represent the foundations of sustainable local pork chains. Concerning Bísaro Pig, several cases of direct contracts between small producers and customers /users have been observed. Such contracts normally comprise an agreement on the breed, the diet and/or carcass weight. In our work, we anticipate that in the future, supply contracts will be increasingly specific and supported by technical-scientific knowledge and precision livestock farming. TREASURE project has allowed improving the scientific-technological knowledge about local pig breeds throughout the whole chain and attributes of the meat could be included in the contract, like: (1) ante-mortem: genotype, sex, feeding, live weight and production system; (2) post-mortem: carcass weight, meat yield, physical/chemical characteristics of meat and fat, their nutritional value; (3) gastronomic: product history, confectionery, food presentation, taste.

Horizontal arrangements, technologies and productive capacity of Brazilian dairy farmers

R.R. Martinelli[1], J.C. Damasceno[1], V.D.V. Da Costa[1], T.T.S. Siqueira[2,3] and F.I. Bánkuti[1]
[1]State University of Maringá, Department of Animal Sciences, Av. Colombo, 5790, 87020-900, Maringá, PR, Brazil, [2]French National Research Institute for Agriculture, Food and Environment (INRAE), Agroecologies, Innovations & Ruralities, 24, chemin de Borde-Rouge, CS 52627, 31326 Castanet Tolosan, France, [3]University of Toulouse – INP – Ecole d'Ingénieurs de Purpan, 75, Voie du Toec, 31076 Toulouse, France; tiago.siqueira@purpan.fr

In Brazil, dairy farmers are asked to improve scale and the quality of milk. To achieve these demands, technologies are an important issue that can be more easily overcome if the farmer belong to horizontal arrangements, such as cooperatives and associations. Considering this statement, we aim to analyse if the participation in horizontal arrangements can improve the technology input in dairy system. Data were collected in 347 dairy production systems (DPS) located in Paraná and São Paulo, States. Factor Analysis was applied. Two Factors (F1 and F2) were generated. Factor 1 compressed variables related to technology used in dairy systems. Factor 2 composed variables related to production capacity of dairy system. After that, farmers were divided in two groups, G1: farmers who do not participate in horizontal arrangements and G2: farmers that participate in horizontal arrangements. Mean tests showed that G2 were defined by farmers with more technology input and more productivity capacity ($P<0.05$). Pearson's chi-square test showed an association between participation in horizontal arrangements and the use of technologies, such as milking practices, reproduction techniques and racial pattern of the herd ($P<0.05$). This result indicates that participation in cooperatives and farmers associations can have a positive impact on the adoption of technologies by dairy farmers, especially associated with increased performance. Therefore, participation on horizontal arrangements has to be encouraged by public and private agents in order to help small dairy farmers to become more competitive.

Comparison of organic and traditional lamb production systems by meta-analysis

R. Catani[1], J.P. Araújo[2,3] and V. Cadavez[3]
[1]*Universidade Tecnológica Federal do Paraná, Zootecnia, Av. Sete de Setembro, 3165 Rebouças, 80230-901 Curitiba-PR, Brazil,* [2]*Escola Superior Agrária de Ponte de Lima, Rua D. Mendo Afonso, 147 Refóios do Lima, 4990-706 Ponte de Lima, Portugal,* [3]*Centro de Investigação de Montanha, Instituto Politécnico de Bragança, Animal Science, Campus de Santa Apolónia, 5300-253 Bragança, Portugal; vcadavez@ipb.pt*

The objective of this work was to use meta-analysis to study the effects of the production system on lambs growth and carcass quality. Seven independent studies concerning the comparison of organic and traditional lamb production systems were used. In each study the lambs initial age (IA), diet crude protein (CP), metabolisable energy (ME), average daily gain (ADG), subcutaneous fat cover (SF) and carcass yield (CY) were extracted. The data of the mean standardised difference of the two production systems were analysed using a random-effects meta-analysis model. After evaluating the collected data, a large discrepancy in the experimental methodology was observed among studies and between the production systems treatments within the studies. As an example, the duration of the experimental period varied between production systems, and some studies did not present this information, which made it difficult to interpret the results. Lambs reared in the organic system showed a higher (P=0.023) ADG (1.96; 95% CI: 0.2636-3.666) than lambs reared in the traditional system. The variance between studies was very high (I2=95%), and the use of initial age as a moderator explained 9% of the variation between studies. Lambs reared in the traditional system presented higher SF (1.07; 95% CI: -1.156 – 3.296), however the difference was not significant (P=0.346). No differences (P=0.258) were observed between the productions systems in CY (1.27; 95% CI: -0.932 – 3.476). The studies analysed presented high variability in the experimental design, making it difficult to compare the two production systems. This meta-analysis study identified some sources of variation capable of biasing the experimental results. Thus, it is clear that studies of this nature need to be supported by experimental designs able to control the sources of variation, so that the effects of the production systems can be accurately identified and quantified.

Life cycle assessment of the Iberian pig systems in the Spanish dehesa

C. Reyes-Palomo[1], E. Aguilera[1,2], M. Llorente[3], C. Diaz-Gaona[1], G. Moreno[3] and V. Rodriguez-Estévez[1]
[1]*Universidad de Córdoba, Cátedra Ganadería Ecologica, Campus Universitario de Rabanales Ctra. Madrid-Cádiz km 396, 14071 Cordoba, Spain,* [2]*Universidad Politécnica de Madrid, CEIGRAM, Pº Juan XXIII, 11, 28040 Madrid, Spain,* [3]*Universidad de Extremadura, INDEHESA, Calle Virgen Puerto 2, 10600 Plasencia, Spain; ganaderiaecologica@uco.es*

The Iberian pig production has its own quality regulation; which differentiate the handling and feeding system of this breed production. These differences influence production costs and market price. However, although most of the Iberian pigs are intensively raised, consumers associate the dehesa agroforestry system to Iberian pigs. This study has been conducted collecting data from 20 Iberian pig farms with different handling systems and performances: selling piglets with 23 kg or selling fatteners with three different feeding systems (foraging acorns and grass, and eating feed indoors or outdoors). Life Cycle Assessment methodology has been carried out to perform the environmental analysis. The manure handling emissions have been calculated using Tier 2 equations on IPPC guidelines and Spanish National GHG Inventories. The not differentiable emissions have been estimated by economic approach and these environmental impacts have been calculated with Ecoinvent 3, Agri-footprint and ELCD databases. The limit of the system has been 'cradle to gate'. The results obtained for Iberian pigs finished foraging acorns and grass (called 'bellota', meaning fed with acorns) are 5.96 kgCO$_2$eq/kg live weight (LW); while for pigs finished eating feed outdoors (called 'cebo de campo') are 3.97 kg CO$_2$eq/kg of LW and for pigs finished indoors are 13.44 kg CO$_2$eq/kg LW. These results are consistent with previous references, as García-Gudiño *et al.* (2020), with 3.40 and 4.36 kg CO$_2$eq/kg LW respectively for 'bellota' and 'cebo de campo' systems. In conclusion, although there are different production systems, the ones with lower environmental impact are associated with the dehesa grasslands. In fact, the carbon sequestration of this agroforestry system should be considered for a better calculation and to differentiate intensive and extensive carbon footprint.

A study of growth performance of Bordaleira de Entre Douro e Minho lambs breed

R. Catani[1], J. Cerqueira[2], J.P. Araújo[2,3] and V.A.P. Cadavez[3]
[1]Universidade Tecnológica Federal do Paraná, Zootecnia, Av. Sete de Setembro, 3165 Rebouças, 80230-901 Curitiba-PR, Brazil, [2]Escola Superior Agrária de Ponte de Lima, Rua D. Mendo Afonso, 147 Refóios do Lima, 4990-706 Ponte de Lima, Portugal, [3]Centro de Investigação de Montanha, Instituto Politécnico de Bragança, Animal Science, Campus de Santa Apolónia, 5300-253 Bragança, Portugal; vcadavez@ipb.pt

The objective of this work was to investigate the growth performance of Bordaleira-de-Entre-Douro-e-Minho (BDM) breed lambs raised under semi-intensive production system. A total of 15 BDM lambs were used, and remained with their mothers from birth to 4 months of age. During the growth trial, lambs and sheep had *ad libitum* access to water and natural meadow hay. Sheep grazed in the experimental natural pastures of Escola Superior Agrária de Ponte de Lima. The lambs were weighed at birth (BW), and at 15-day intervals until slaughter, which occurred at 4 months of age. The growth data were used to estimate the live weight (LW) and the average daily gain (ADG) of the lambs at 30 (LW30 and ADG30), 60 (LW60 and ADG60), 90 (LW90 and ADG90), and 120 (LW120 and ADG120) days of age. The results showed that the growth of BDM lambs slowed down at 60 days of age. The lambs had an average BW of 2.6 (±0.281) kg and reached 13.5 kg (±1.48) at 4 months of age. The ADG reached a maximum at 30 days of age (0.153±0.021), and decreased about 20 g per day between each standard age. The BW presented a moderate correlation (r=0.564) with LW30, and the correlation decreased as the lambs' age increased, being the lowest (r=0.119) for LW120. The high correlation (r=0.933) between LW60 and LW90 shows that growth after weaning is clearly an indicator of the genotypic value of lambs. The ADG of BDM lambs decreased from 60 days of age, and this information should be used to define the optimal slaughter point of BDM lambs. The performance results of BDM lambs presented in this work is also important to define the supplementation strategies of lambs throughout the growing period.

A study of growth performance of Churra Galega Bragançana lambs breed

R. Catani[1] and V.A.P. Cadavez[2]
[1]Universidade Tecnológica Federal do Paraná, Zootecnia, Av. Sete de Setembro, 3165 Rebouças, 80230-901 Curitiba-PR, Brazil, [2]Centro de Investigação de Montanha, Instituto Politécnico de Bragança, Animal Science, Campus de Santa Apolónia, 5300-253 Bragança, Portugal; vcadavez@ipb.pt

The objective of this work was to investigate the growth performance of Churra-Galega-Bragançana (CGB) breed lambs raised under semi-intensive production system. A total of 15 CGB lambs were used, and remained with their mothers from birth to 4 months of age. During the growth trial, lambs and sheep had *ad libitum* access to water and natural meadow hay. Sheep were fed in a grazing system in the experimental pastures of Escola Superior Agrária de Bragança. The lambs were weighed at birth (BW), and at 15-day intervals until slaughter, which occurred at 4 months of age. The growth data were used to estimate the live weight (LW) and the average daily gain (ADG) of the lambs at 30 (LW30 and ADG30), 60 (LW60 and ADG60), 90 (LW90 and ADG90), and 120 (LW120 and ADG120) days of age. The results showed that the growth of CGB lambs slowed down at 60 days of age. No differences (P>0.05) were found for BW between female and male lambs (3.4±0.191 vs 3.7±0.233 kg). From birth to 120 days of age, the male LW increased linearly about 8.8 times. At 120 days of age, male lambs presented higher (P<0.05) LW than female lambs, being this difference of approximately 16% (29.9 vs 25.8 kg). The males ADG reached the maximum value at 60 days of age, while female lambs reached their maximum at 30 days of age. The performance of pre-weaning, assessed through bi-weekly weighing, provides useful information on the lambs growth performance and can be used as an indication of the herd's health, as well as for the management of the herd. Weaning weight control is useful for evaluating pre- and post-weaning growth and thus monitoring ADG in these two phases. Low ADG values are warning signs for food deficiencies (quality and/or quantity) or for the presence of diseases responsible for reducing lamb growth. This information is essential to define the optimal slaughter point of CGB lambs, as well as to define the strategy of supplementation of male and female lambs, which should be adjusted according to the growth potential.

Eco-efficiency assessment of Iberian pig production in Spanish dehesa: preliminary results

E. Angón[1], J. García-Gudiño[2], I. Blanco-Penedo[3] and J. Perea[1]
[1]Universidad de Córdoba, Animal Production, Campus de Rabanales, 14071, Córdoba, Spain, [2]CICYTEX, Animal Production, Finca La Orden, 06187, Guadajira, Spain, [3]SLU, Department of Clinical Sciences, P.O Box 7054, 75007, Uppsala, Sweden; eangon@uco.es

The interest for improving farm management for a more sustainable and resource use efficient system in Spanish *dehesa* has never been larger. *Dehesa* system is characterised by the maintenance of the extensive livestock farms, which formed this typical ecosystem. The objective of this study was to assess the eco-efficiency of the traditional Iberian pig using the joint approach of Life Cycle Assessment (LCA) and Data Envelopment Analysis (DEA), called LCA +DEA. This research was carried out in 36 traditional farms (decision making unit, DMU) of Iberian pigs located in Extremadura and Andalusian region (southwest of Spain). To obtain more profitability in porcine system with less environmental impact we assess their eco-efficiency. DEA efficiency scores range between 0 and 1 and score close to 1 indicates better levels of eco-efficiency. In addition the best-practice for each farm were identified in order to improve its ranking. The inputs used were land area (hectares of montanera) and sow numbers (either pregnant or lactating), while the output variable was annual porcine sales (€/year). Environmental impact of Iberian pig production on climate change (CC, kg CO_2 eq) was previously calculated by LCA methodology using SimaPro software for inventory and analysis phases. CC was treated as an undesirable output that must be minimised. DEA ranked Iberian pig farms by combining a set of inputs in order to maximise economic results minimising environmental pressure. To obtain the scores, the package 'deaR' in R software was applied. On average the traditional Iberian pig farms showed a high level of eco-efficiency. The estimated mean level of eco-efficiency was 0.92 suggesting that the average farm could decrease its climate change by 8% given the level of inputs. Also, we obtained the best-practice through benchmarking techniques and 15 DMUs were considered efficient. Managers for the inefficient DMUs need to improve their performance by increasing porcine sales (by 18%) and decreasing CC (8%) simultaneously. Finally, further research should be focused towards the identification of the environmental pressure exerted by the different types of management of Iberian pig production in the *dehesa*.

Author index

A

Aamand, G.P.	581, 582, 583	Alfonso, L.	270
Aargaard Poulsen, N.	295	Algers, B.	613
Aass, L.	576	Aliakbari, A.	313, 382
Abdalla, E.A.	103	Al-Izzi, W.	232, 237
Abe, H.	144	Allain, C.	449
Abecia, J.A.	614, 625	Allueva, D.	513, 519
Abid, M.	150	Almeida, A.M.	214, 220, 232, 277
Ablondi, M.	223, 495	Almeida, J.	157, 248, 443
Abreu, H.	326	Almeida, J.C.	602
Abreu, P.	468, 469	Almeida, M.	230, 245, 308, 554
Abril-Parreño, L.	623	Almqvist, V.	514
Abu, O.A.	300	Alòs Saiz, N.	521
Abud, M.J.	538	Altarriba, J.	496, 498
Åby, B.A.	576	Alvarado, A.	160
Accatino, F.	162, 404, 635	Álvarez, I.	374
Acciaro, M.	435, 517, 522	Alvarez, L.A.	456
Achard, C.	546	Alvarez-Oxiley, A.	539
Adams, S.	553	Álvarez-Rodríguez, J.	250, 338, 477
Addis, M.F.	263	Álvaro-Minguito, G.	298
Addo, S.	427	Alves, S.	248, 350
Aditya, S.	121	Alves, S.P.	214, 240, 270, 450, 480
Adriaens, I.	293, 589	Amalfitano, N.	224, 227, 230, 422
Adrian, B.	418	Amaral, A.	499
Aernouts, B.	589	Amat, J.P.	467, 554
Aeschlimann, C.	625	Ambriz Vilchis, V.	220
Afonso De Freitas, L.	626, 627	Amdi, C.	130, 215, 391
Agabriel, J.	162	Amer, P.R.	282, 563
Aganovic, K.	255	Amer, S.A.	122
Agazzi, A.	129, 264	Amiri Roudbar, M.	487
Agea, I.	151, 525	Ammer, S.	433, 549, 612
Agha, S.	456	Amon, T.	294
Agudo, B.	626	Amor, J.	369, 370
Aguerre, S.	555	Ampe, B.	310
Aguilera, E.	490, 638	Ampuero Kragten, S.	299, 517
Ahlhorn, J.	612	Amsler, Z.	593
Ahn, I.	313	Anastasiou, I.	526
Ahvenjärvi, S.	545	Anderson, C.	548
Aibar, J.	132	Anderson, F.	620
Ajmone Marsan, P.	230, 358, 388	Andonovic, I.	364
Åkerfeldt, M.	552	Andonov, S.	488
Alaedin, M.T.	276	Andrade, E.	305
Alameer, A.	586	Andrés, S.	542, 546
Albenzio, M.	124, 277, 420, 447	Andriamandroso, A.L.H.	322
Albernaz-Gonçalves, R.	331	Andueza, D.	508
Albertini, M.	263	Anestis, V.	211
Albertsdóttir, E.	488	Angelidis, A.	618
Alborali, G.L.	330	Angerer, V.	404
Albouy-Kissi, B.	437	Angón, E.	181, 570, 640
Albrecht, E.	540, 541	Annicchiarico, G.	277
Albuquerque, J.	432	Antonakos, G.	556
Albuquerque, L.G.	571	Antunes, R.	157
Aldai, N.	146, 414, 419, 601	Anzolabehere, M.	269, 321
Aldridge, M.N.	288, 292, 575	Aoki, H.	233
Alemu, S.	375	Aoun, M.	308
		Appel, A.K.	286

Apper, E.	546	**B**	
Aquilue, B.	338	Baba, T.	144
Arana, A.	270	Bacardit, J.	586
Aranda, E.	391	Bacchin, E.	311
Arango, J.	483	Bacellar, D.	471, 554
Araújo, J.P.	143, 320, 515, 518, 571, 637, 638, 639	Bach, K.D.	228
Aravindakshan, T.V.	106	Bach Knudsen, K.E.	125, 128
Arbel, I.	294	Bach Larsen, L.B.	295
Arca, P.	212	Baes, C.	101, 387
Arcuri, S.	272	Baes, C.F.	103, 149, 150, 378, 553
Arduin, E.	322	Baessler, S.	224
Arends, D.	316	Baeten, V.	236
Arévalo Sureda, E.	219	Båge, R.	488, 492
Argente, M.J.	151, 283, 525	Bagirov, V.	478
Arias-Álvarez, M.	194	Bagnicka, E.	269, 362, 491
Arias Coronel, M.J.	431	Bahloul, L.	171, 424, 502, 610
Arias, K.D.	194, 344	Baila, C.	173, 175, 338, 422, 477
Arias, R.	181	Bailey, R.	329
Arigoni, M.	271	Bailly, J.	587
Armero, E.	151, 525	Bailoni, L.	311
Arnal, M.	558	Baima, E.	392
Arnin, A.	294	Bain, M.	265
Arranz, J.	482, 633	Baird, H.	478
Arranz, J.J.	179, 421, 556, 561, 564	Bąk, A.	384
Arrebola, F.	345, 477	Baki, C.	593
Arriaga, H.	447	Bakoev, N.	573
Arrigo, Y.	299	Bakoev, S.	573
Arroyo Schnell, A.	183	Bakuła, T.	158
Arsenos, G.	455, 572	Balcells, J.	267
Artuso-Ponte, V.	219	Baldi, A.	231, 352, 609
Ask, B.	284	Balia, R.L.	121
Ask-Gullstrand, P.	492	Ballan, M.	189
Asselstine, V.	356	Bam, J.	334
Assié, S.	632, 634	Banchero, G.	560
Assunção, D.	468	Banchero, M.N.	302
Astessiano, A.L.	269, 538	Bani, P.	388, 435
Astigarraga, L.	211	Bánkuti, F.I.	637
Astiz, S.	498, 499, 569, 603, 606	Banos, G.	428, 565, 572
Astruc, J.M.	197, 374, 458, 556, 557, 561, 564	Bapst, B.	173, 556
Atkinson, R.C.	364	Baptista, F.	507
Atxaerandio, R.	304, 329, 482	Baptista, M.C.	275, 479
Atzori, A.S.	172, 212, 368, 628, 635	Barajas, R.	595
Aubert, H.	587	Barão, L.	206
Aubert, T.	121	Barbano, D.M.	228
Aubry, A.	163, 472, 629	Barbas, J.P.	479
Auclair-Ronzaud, J.	470, 587	Barbat, A.	188
Audano, M.	263	Barbey, S.	363, 532
Auer, F.J.	436, 583	Barbier, N.	512
Auffret, M.	392, 576	Barchilon, N.	201
Aulanniam, A.	121	Barilly, C.	313
Aupiais, A.	510	Barka, E.	312, 526
Aury, J.-M.	463	Barnes, A.	547
Averós, X.	482	Barrajon-Masa, C.	498, 499
Avila, G.	272	Barraso, C.	212
Ávila, G.	265, 273	Barreto, L.	281
Aymerich, P.	127	Barrio, E.	176
Ayrle, H.	593	Barron, L.J.R.	414, 419, 601
Azevedo, J.	230	Barros, M.	518

Blanco-Penedo, I.	320, 520, 570, 640	Branco, S.	624
Blaschikoff, L.	461	Brandt, P.	120
Blavi, L.	592	Brasca, M.	230
Blees, T.	263	Breen, J.	161
Blichfelct, T.	478	Brem, G.	174, 193, 386, 478
Bliebernicht, M.	468, 469, 471	Bremn, C.	401
Bloch, V.	291, 294, 510	Brenig, B.	529
Blumer, S.	475, 622	Brétière, G.	468
Blunk, I.	378, 574	Breves, G.	453
Bock, M.H.	434	Brevini, T.A.L.	272
Bodas, R.	615	Brewer, H.	474
Bodini, C.	626	Briens, M.	235, 497
Bodin, L.	194, 425	Brito, L.F.	192, 382, 486, 556
Bodziony, D.	384	Brito, L.M.	339
Boerner, V.	578	Brito, N.V.	407, 528
Boichard, D.	188, 191, 281, 363, 372, 429, 461, 532	Brocas, C.	204, 205
Boland, T.M.	628	Brochard, M.	191
Bollard, N.	398	Brockmann, G.A.	316
Bompa, J.F.	504, 587	Brodowska, P.	362, 491
Bonagúrio, L.P.	133, 514	Bronzo, V.	263
Bonckaert, C.	329	Brossard, L.	113
Bonekamp, G.	360	Brouard, S.	317
Bonet, J.	127	Brown, A.	318
Bonifazi, R.	579	Brscic, M.	551
Bonilauri, P.	155	Bruckmaier, R.M.	536
Bonilha, S.F.M.	370	Brügemann, K.	123, 222
Bonnet, M.	261, 268, 437	Bruggeman, G.	265
Bonos, E.	211, 312, 526	Brugger, D.	311, 434, 525, 594
Bontempo, V.	129, 264	Brun, A.	520
Bonvino Stafuzza, N.	626, 627	Bruni, G.	556
Borba, A.	237, 245, 604	Bruno De Sousa, C.	461
Borboudaki, K.	246	Brunschwig, G.	403, 617
Bordes, L.	555	Brzakova, M.	178, 380
Borgemeister, C.	256	Buccioni, A.	231, 451
Borges, F.	275	Buckley, F.	496
Bos, J.A.	425	Bueno, D.	243, 244
Bosi, P.	390, 392	Bugno-Poniewierska, M.	346
Bosse, M.	188	Bugueiro, A.	508
Bossis, I.	303, 363	Bui, H.	116, 333
Botelho, A.	624	Buisson, D.	374, 455, 458
Botelho Fontela, S.	217, 298, 519, 523, 615	Buitenhuis, A.J.B.	295
Boudon, A.	506	Burchmore, R.	129, 262, 264, 265
Bouki, C.	246	Bureš, D.	306, 567
Boulot, S.	217	Burfeind, O.	221, 549
Boussaha, M.	188, 190, 191	Burger, P.A.	110, 361
Boutinaud, M.	352, 424	Burgers, E.E.A.	226, 323
Bouwman, A.C.	188, 429, 439, 462	Burgos, A.	588
Bouwmeester-Vosman, J.J.	358	Burgoz Paz, W.O.	427
Bovenhuis, H.	295, 323, 357	Burjel, M.V.	492
Bovmais, C.	494	Burrell, A.	388, 389
Bovolenta, S.	204	Burren, A.	625
Bovo, S.	104, 189	Busch, G.	193
Bowen, E.	474	Butler, G.	169
Boyle, L.A.	138	Buttazzoni, L.	361
Bracarense, A.P.F.R.L.	452	Büttgen, L.	339
Bragolusi, M.	568	Büttner, K.	413, 612
Brajkovic, V.	105, 106, 373, 486	Buys, N.	106, 285, 341, 355
Branco, R.H.	370	Byrne, C.J.	622

Byrne, T.   561

C

Caballero-Villalobos, J.   181
Cabezas, A.   243
Cabiddu, A.   435
Cabrera, C.   544
Cabrita, A.R.J.   153, 220, 297, 326, 351, 432
Caceres, E.   498, 499
Cachucho, L.   240, 242, 243, 244
Cacic, M.   106
Cadavez, V.   638
Cadavez, V.A.P.   473, 476, 504, 605, 624, 639
Caetano, M.C.   494
Caetano, P.   624
Cagide, F.   275
Caimi, C.   155
Caja, G.   170, 171, 172, 174, 198, 199, 420, 424
Calduch, J.   325
Caldwell, A.   452
Calik, J.   109
Calmels, M.   147
Calnan, H.B.   596
Calogero, R.   271
Calus, M.P.L.   192, 313, 425, 426, 485, 579, 580, 581, 582
Calvo, L.   516, 595
Campanile, G.   362
Campbell, S.   296
Campos, R.   241, 242, 394, 454
Canali, E.   294
Canals, R.M.   418, 419
Canario, L.   284, 373, 382, 587
Čandek-Potokar, M.   637
Candrák, J.   345
Cannas, A.   172, 628, 635
Cannas, S.   414
Cañón, J.   456
Cánovas, A.   356
Cantalapiedra-Hijar, G.   114, 280, 383, 508
Cantalapiedra, J.   320
Cantor, M.C.   588
Capdeville, J.   399
Capela, L.   494
Capion, N.   529
Cappelli, J.   306
Cappelloni, M.   104, 189
Cappucci, A.   268
Capri, E.   616
Caraballo, C.   573
Carabaño, M.J.   428, 455, 572, 621
Carballo, D.E.   239
Carcangiu, V.   625
Cardoso, D.F.   382
Cardoso, P.   481
Cardoso, S.F.S.   159
Carè, S.   206
Carolino, I.   481

Carolino, N.   107, 110, 339, 472, 481, 494, 523, 624
Caroprese, M.   124, 277, 420, 447
Carreira, E.   633
Carrière, F.   146
Carrión, D.   127, 132, 298
Carriquiry, M.   116, 117, 269, 273, 611
Carro, M.D.   241, 242, 243, 244
Carta, A.   200
Caruso, D.   263
Carvajal-Serna, M.   625
Carvalheira, J.G.V.   454
Carvalheiro, R.   571
Carvalho, A.   296
Carvalho, G.   388, 389
Carvalho, I.   160
Carvalho, M.   166, 633
Carvalho, M.G.   494
Carvalho, P.   401
Casal, A.   269, 273, 538
Casao, A.   625
Casasús, I.   145, 173, 175, 306, 368, 402, 422, 634
Caselllas, J.   431
Casey, N.H.   542
Cason, E.D.   493
Cassandro, M.   388
Cassar-Malek, I.   281, 371
Cassina, A.   273
Castaneda, C.   470
Castanheira, F.   153
Castellanos-Uribe, M.   463
Castelló, A.   264
Castelo-Branco, A.   256
Castelo, C.   217, 523
Castex, M.   122
Castilha, L.D.   133
Casto-Rebollo, C.   283
Castro-Costa, A.   199
Castro Muñoz, E.   322
Catani, R.   638, 639
Catarino, J.   160
Catillo, G.   361
Catorci, A.   408
Catozzi, C.   273
Caudevilla Pérez, A.   634
Cavaco, S.   624
Cazeau, G.   554
Cazzola, R.   235
Ceacero, F.   306
Cebo, C.   352
Cebron, N.   632
Cecchinato, A.   222, 227, 230, 358, 422
Ceciliani, F.   261, 263, 265, 268, 271, 272, 273, 356
Cedano-Castro, J.I.   103
Cenci, F.   527
Ceniti, C.   229
Cerqueira, J.L.   143, 320, 515, 518, 571, 637, 639
Cervantes, I.   194, 288, 289, 325, 344, 588
Cesarani, A.   180, 181, 522

Chaalia, B.	199, 424	Claeys, J.	260
Chacón, M.J.	194	Clar, M.A.	306
Chadaram, S.	385, 389	Claramunt, M.	269
Chagas, J.C.	449	Clark, K.	355
Chagunda, M.	141	Clasen, J.B.	426, 487, 575
Chamberland, N.	236	Claudio, C.	226
Chanat, E.	424	Clelland, N.	560
Chandler, T.L.	606	Clement, V.	458, 556
Chapoutot, P.	115, 249, 307	Cleveland, M.A.	576
Chardon, C.	470	Cobo, E.	425
Charfeddine, N.	529	Cobos, S.	212
Charlier, C.	188, 355	Coelho-Fernandes, S.C.	473, 624
Charneca, R.	507, 519, 520	Coertze, R.J.	542
Charvolin-Lemaire, E.	459	Coffey, M.	318
Chase-Topping, M.	329	Čolak, S.	326
Chassier, M.	178	Cole, J.	529
Chatziplis, D.	572	Cole, J.B.	100, 101, 355
Chatzizisis, L.	526	Colinet, F.G.	137
Chaves, A.	220	Collins, T.	547
Chavinskaia, L.	187	Coma, J.	127
Cheli, F.	352	Combasteix, A.	374
Chemello, G.	155	Combes, S.	348, 382, 546
Chen, L.L.	338	Comi, M.	129, 233, 247, 264, 352
Chen, L.Y.	305	Comin, A.	615
Chen, S.	124, 295	Conceição, L.A.	494, 569
Chery, I.	202	Conceição, M.A.P.	521
Chesneau, G.	115, 249, 307	Concheri, G.	405
Chesney, L.	321	Conde-Pulgarin, A.	170
Chiarellli, D.D.	247	Condon, T.	359
Chicoteau, P.	333	Conejos, J.R.V.	142
Chillemi, G.	138	Conington, J.	559, 560, 561, 565, 620
Chincarini, M.	200	Conrad, R.	385, 388, 389
Chitneedi, P.K.	179	Constantin, S.	301
Choisis, J.P.	403	Conte, G.	451
Chong, F.S.	605	Contiero, B.	409, 568
Choucair, A.M.	531	Contini, S.	212, 435
Chouteau, A.	617	Contò, M.	138, 268, 527
Chriki, S.	599, 600	Contreras-Jodar, A.	170
Christaki, E.	312, 526	Copland, A.	114
Christen, A.M.	529	Coppieters, W.	188
Christensen, J.M.	488, 492	Corazzin, M.	204
Christensen, M.	597	Corbach, M.	553
Christensen, O.	484	Corbière, F.	550
Christensen, O.F.	285, 379, 578	Corbishley, A.	555
Chrysanthakopoulou, F.	303	Cordeiro, L.	153, 350
Chumkam, S.	327	Cordeiro, P.	479
Ciappesoni, G.	556, 560	Cordero, G.	591
Ciliberti, M.G.	124, 447	Cormican, P.	623
Cimarelli, L.	417, 418	Correa, F.	390, 392
Cipolat-Gotet, C.	222, 223, 227	Correddu, F.	628
Cipriano, J.	205	Correia, A.	153, 351
Cirot, C.	502	Corrent, E.	210
Cisse, S.	116	Cortazzo, N.	321
Citek, J.	348	Cortellari, M.	180, 181
Cittadini, A.	418, 419	Cortés, O.	456
Ciucci, F.	451	Cossu, D.	212
Ciurescu, G.	136, 395	Costa, A.	168, 330, 357, 362
Claerebout, E.	355	Costa, A.L.	468, 469, 471

Delgado, J.V.	456	Djezzar, R.	116
Della Badia, A.	160, 369	Doblas, A.	447
Della Malva, A.	277, 420	Dodgson, J.	509
Della Porta, M.	235	Doekes, H.	427
Delosière, M.	261, 268	Doeschl-Wilson, A.B.	329, 558, 561, 562
Delprat, L.	134, 135	Dohme-Meier, F.	299
De Marchi, M.	168, 330, 362	Dolle, J.B.	204, 205
De Marco, M.	235	Domingos, T.	165, 205, 206, 365, 445, 618
Demars, J.	382	Domingues, J.P.	162
Demey, V.	310	Dominguez, R.	230, 518
De Monte, E.	356	Doornweerd, J.E.	439
De Moraes, A.	401	Doran, M.J.	491, 501
Denis, P.	115	Dore, A.	587
Deniskova, T.	174, 478	Dorenlor, V.	512
Denman, S.E.	261	Dorfner, G.	162, 165
Denoyelle, C.	604	Dorigo, M.	446
Dentinho, M.T.P.	240, 242, 243, 307	Dornier, X.	555
De Olde, E.	162	Dos Santos, A.	454
De Oliveira, D.	548	Dos Santos, M.P.	205, 618
De Oliveira, J.	435	Dotsev, A.V.	174, 193, 386, 478
De Palo, P.	415, 420	Doucet, M.	550
Depuydt, J.	285	Dougherty, H.C.	144
Derno, M.	257	Douguet, M.	317
De Roest, K.	601	Douhard, F.	555, 562
Derrien, C.	410	Douls, S.	146, 557
Déru, V.	382	Dourmad, J.Y.	112, 216, 365, 636
Deruytter, D.	260, 409	Dovc, A.	384
De Souza, F.	401	Dovc, P.	384
Dessauge, F.	354	Doyle, E.	474
Dessie, T.	375	Doyle, R.	474
Dettori, M.L.	422	Dragset, K.I.	148
Devant, M.	508	Draper, J.	560
Devicenzi, T.	117	Drillich, M.	262
De Vivo, R.	619	Drögemüller, C.	287
De Vries, A.	400	Dronne, Y.	162
De Vries, S.	216	Drössler, K.	436, 437, 506, 507, 583
Dewhurst, R.J.	280, 392, 576	Druart, X.	623
Dhakal, R.	148	Druet, T.	188, 489
Dhollande, M.	555	Druyan, S.	201
Dhumez, O.	249, 307	Drzaic, I.	105, 106, 486
Diana, A.	330	Dubarry, E.	522
Dias, C.	624	Duce, P.	212
Díaz-Gaona, C.	428, 455, 490, 521, 566, 572, 621, 638	Ducrocq, V.	187, 279, 281, 282, 372, 489, 558
		Duenk, P.	425, 426
Dibbits, B.W.	288	Dufrasne, I.	209
Di Biagio, R.	467	Dugué, C.	217
Di Corcia, M.	277	Dugué, M.	340
Díez-Ramos, M.	419	Dumitru, M.	136, 395
Di Gianvito, A.	517, 522	Dumont, B.	402, 413
Di Mauro, S.	265, 272	Dumont Saint Priest, B.	340, 343
Dinardo, F.R.	415	Dunisławska, A.	328, 350, 389, 395, 543, 547
Ding, X.	104	Dunne, E.	628
Di Nicola, U.	167	Dunshea, F.R.	274, 596
Dinis, R.	110	Dunston-Clarke, E.	547
Disenhaus, C.	617	Duperray, J.	234
Dittmann, A.	277	Duplessis, M.	500, 607
Dittrich, I.	585	Durand, C.	557
Djemali, M.	459	Duru, M.	632

Garcia, M.L.	151, 525	Ghaffari, M.H.	262, 263, 266
García, A.J.	306	Ghanem, N.	450, 457, 460, 461
García, M.L.	283	Giakoumaki, I.	246
García-Álvarez, A.	288	Giammarco, M.	200
Garcia-Baccino, C.A.	561, 577	Giamouri, E.	231, 247
García-Ballesteros, S.	325	Giannakopoulou, K.	300
García-Casco, J.M.	573	Giannenas, I.	211, 312, 526
García-Contreras, C.	606	Gianni, M.	106
García-García, J.J.	615	Giardino, I.	447
García-Gudiño, J.	520, 570, 640	Gibbons, J.M.	142
Garcia-Herreros, M.	177	Giersberg, M.F.	549
García-Launay, F.	216	Gigante, P.	155
García-Pastor, L.	614	Giger-Reverdin, S.	630
García Pintos, J.	302	Gilbert, H.	115, 125, 313, 382, 448
Garcia-Roche, M.	116, 117, 273	Giller, K.	541
García-Rodríguez, A.	304, 329, 482	Ginja, C.	450, 456, 457, 460, 461, 629
Garcia-Santos, S.	230, 245, 308	Giorello, D.	560
García-Vázquez, L.M.	239	Giorgi, M.	448
Gardner, G.E.	596, 597, 620	Giovambattista, G.	456
Gardón, J.C.	603, 606	Giovanetti, V.	172, 435
Garon, D.	555	Giráldéz, F.J.	239, 394, 542, 546
Garrafa, A.	498, 499	Girard, C.L.	500, 607
Garreau, H.	373, 382	Girardie, O.	373
Garrido, N.	136	Girard, M.	453
Garrido-Varo, A.	118, 367	Giromini, C.	231, 352
Garrod, G.	125	Gispert, M.	520
Garry, B.	621	Givens, D.I.	476
Garzón, A.	181	Givens, I.	137
Gasco, L.	155, 409	Gjerlaug-Enger, E.J.	196, 376
Gaspar, D.	461, 624, 629	Gkagkavouzis, K.	572
Gaspar, P.	207, 335	Glamočić, D.	379
Gauggel, I.	529	Gluchová, M.	238
Gauly, M.	168, 208, 332, 335, 356, 404, 550	Godinho, A.M.	270, 480
Gauthier, R.	112, 365	Goers, S.	540
Gautier, J.M.	198, 200	Gohin, A.	216
Gava, C.	374	Goi, A.	330
Gaynor, C.	483	Goiri, I.	304, 329
Geibel, J.	339, 385	Gold, M.	158
Gelasakis, A.I.	303, 363	Gollé-Leidreiter, F.	436, 437, 506, 507, 583
Gengler, N.	137, 223, 341, 529, 533	Gomaaa, W.M.S.	305
Genovese, P.	544	Gomes, M.J.	298, 308, 443
Gentz, M.	549	Gomez Exposito, R.	450
Georges, M.	188, 355	Gómez, G.	573
Georgiou, M.	246	Gomez, I.	508
Gérard, C.	506	Gomez, M.	498, 499
Gerdts, V.	329	Gómez-Maqueda, I.	367
Gere, J.	211	Gonçalves, A.	636
Germain, A.	249, 307	Gonçalves, J.	259, 616
Germani, S.	356	Gonçalves, V.	468
Germon, P.	363	Gonzales-Barron, U.	473, 476, 605, 624
Gerrard, T.	509	González-Bulnes, A.	136
Gerrits, W.J.J.	216	Gonzalez, C.	621
Gershon, E.	294	González-García, E.	146, 198, 312, 557
Gertz, M.	584, 585	González-González, R.	171
Gervais, R.	176, 607	Gonzalez, J.J.	509
Gesek, M.	158	González-Luna, S.	199, 424
Getmantseva, L.	573	Gonzalez-Martin, J.V.	498, 499
Ghaderi-Zefreh, M.	558	Gonzalez-Prendes, R.	280, 450, 457

González-Recio, O.	329, 564
González-Ronquillo, M.	220, 391, 423
González-Rosales, M.G.	160
Goossens, K.	225, 250, 310
Goossens, S.	254
Gordo, D.G.M.	279
Gordon, A.W.	605
Gordon, D.	475
Gordon, J.	620
Gori, A.S.	489
Gorin, S.	512
Gorjanc, G.	102, 191, 464, 483, 485, 622
Görs, S.	121, 540
Gorssen, W.	106, 285, 355
Goselink, R.M.A.	226
Gotoh, T.	351, 539
Gotoh, Y.	539
Gottardo, F.	409, 530, 568
Götz, K.U.	485, 580
Gouda, A.	122
Gourdine, J.L.	448
Gouva, E.	526
Govignon-Gion, A.	190
Goyache, F.	289, 374
Granado-Tajada, I.	430
Grande, S.	180, 181
Gredler-Grandl, B.	192
Green, M.J.	398
Grelet, C.	147, 219, 223
Griffith, B.A.	630
Griffon, L.	114
Grilli, G.	265
Grindflek, E.	196, 376
Grisendi, A.	155, 235
Grivault, D.	217
Grodkowski, G.	552
Grosperrin, P.	191
Gross, J.J.	536
Große-Brinkhaus, C.	286
Große-Butenuth, K.	584, 585
Grossi, G.	210
Grøva, L.	198
Grubor, M.	309
Gualdron, J.L.	489
Guarino Amato, M.	527
Guarnido, P.	383
Guay, F.	593
Guedes, C.	245, 298, 308, 471
Guedes, C.M.	230
Gueguen, S.	329
Guemez, H.R.	595
Guerreiro, O.	240, 243, 244, 307
Guerrero-Ginel, J.E.	118, 367
Guiadeur, M.	632, 634
Guicho, E.	107
Guidou, C.	155
Guilet, D.	116
Guillard, F.	126

Guimarães, I.C.S.B.	228, 229
Guimarães, J.	153
Guimarães Júnior, R.	309
Guimarães, S.	461
Guinguina, A.	147, 438
Guldbrandtsen, B.	381, 490
Gullstrand, P.	488
Gunnarsson, S.	218
Gunter, C.	385
Guo, K.J.	338
Gurgul, A.	346
Guse, M.	611
Gutiérrez, G.	103, 431
Gutiérrez-Gil, B.	179, 421, 556, 564
Gutiérrez, J.P.	194, 288, 289, 325, 344, 588
Gutiérrez-Rivas, M.	329
Gutman, R.	157
Guyomard, H.	162, 636
Guzzo, N.	311, 324

H

Haak, T.	152
Haberl, H.	165
Habit, S.R.	234
Hachemi, M.A.	235, 497
Hadjigeorgiou, I.	249, 479, 627
Hadjipavlou, G.	459
Hagan, T.D.J.	605
Hager, A.	479, 627
Hagger, C.	625
Hagiya, K.	141, 143, 144
Hahn, A.	340
Halachmi, I.	111, 198, 201, 292, 294
Halas, V.	125
Halbe, J.	165
Halo, M.	345
Hamilton, A.W.	364
Hammadi, M.	150
Hammon, H.M.	261, 268, 354
Hanamure, T.	141
Hannius, L.M.	548
Hanotte, O.	375
Hansen, H.H.	148
Hanušovský, O.	238, 239
Hanzal, V.	528
Harding, J.	329
Hargreaves, P.R.	444, 445
Harlander, A.	150, 553
Haro, A.	131
Haro, M.A.	482
Harrabi, H.	150
Harris, P.	203, 565, 630
Harty, E.	296
Hasan, S.	327, 453
Haskell, M.	529
Hauschild, L.	112
Hawken, R.	429
Hayakawa, H.	141

Hayashi, Y.	233
Hayer, J.J.	221, 316, 337, 399, 611
Hazard, D.	146, 374, 455, 557
Hechavarria, T.	516
Heckenberger, G.	248, 534, 535
Heimo, D.	202
Heinemann, C.	221, 316, 331, 333, 337, 399
Hein, L.	487
Heinzl, E.U.L.	414
Heinz, V.	255
Heirman, T.	504
Heise, J.	387
Heiskanen, A.	396
Hely, F.S.	563
Hennart, S.	161, 163
Henne, H.	286
Hennessy, D.	621
Henrotte, E.	489
Henry, D.	587
Henshall, J.M.	429
Heras, J.	569
Heringstad, B.	148, 360, 529
Herlin, A.	508
Hermans, D.	265
Hernández-García, F.I.	136, 520
Herold, J.	222
Herosimczyk, A.	395
Herrera, P.M.	566
Herskin, M.S.	551
Hertel-Böhnke, P.	436, 506, 507
Hervás, G.	160, 176, 369, 370, 556, 564
Hervás-Rivero, C.	431, 496, 498
Herve, S.	512
Hessle, A.	199, 318
Hess, M.	568
Heuel, M.	158
Heuer, H.	586
Heuß, E.M.	286
Heyrman, E.	355
Hickey, J.	485
Hickey, J.M.	191, 483
Hiemstra, S.J.	184
Hijbeek, R.	167
Hildebrandt, F.	413
Hiller, P.	527
Hillmann, E.	613
Hinrich, D.	490
Hinrichs, D.	427, 580
Hintringer, J.	531
Hirooka, H.	233, 299
Hoang, Q.	385
Hochu, I.	448
Hocquette, E.	601
Hocquette, J.F.	598, 599, 600, 601
Hoelker, M.	262
Hoenen, M.	437
Hofer, A.	107
Hoffman, L.	567

Hoffmann, D.	434
Hofmanova, B.	178, 348
Hofs, I.	276
Högberg, N.	199
Höglund, J.	199
Holinger, M.	173, 334, 613
Hollister, S.M.	497
Holzhauer, M.	529
Hornstra, J.F.	218
Horrillo, A.	207, 335
Horvatić, A.	263
Hosotani, G.	435
Hosseini Ghaffari, M.	276
Hosseini Salekdeh, G.	104
Hötzel, M.J.	331
Houlahan, K.	387
Hozé, C.	188, 190, 191, 429, 460, 557
Hristov, A.N.	502
Huaman, A.	103
Huang, C.H.	195
Huang, Y.	262
Huau, C.	202, 203, 504, 563
Huber, K.	224
Hubin, X.	355, 489
Hudson, C.	398
Huguel, D.	437
Huhtanen, P.	147, 438
Hulsegge, I.	429, 575
Hulst, A.D.	359
Hultgren, J.	514
Hummel, J.	223
Hurdebise, J.	209
Hurtado, I.	626
Hurtaud, C.	352
Hurtaud, J.	373
Hutchison, J.L.	355
Huuki, H.	545
Huws, S.A.	568
Huxley, J.	398
Hu, Y.	589
I	
Iakhno, S.	213
Ibáñez-Escriche, N.	283, 431
Ignatieva, L.P.	386
Ijiri, D.	351
Ingrand, S.	457
Inhuber, V.	311
Insausti, K.	414, 418, 419
Invernizzi, G.	129, 231, 264, 388
Ioannides, I.M.	459
Ishida, S.	143
Ishii, K.	378
Islas-Trejo, A.	356
Iso-Touru, T.	464, 501
Ites, S.	255
Ithurralde, J.	538, 544
Ivanova, S.	517

Ivanov, N.	276
Iversen, M.W.	196
Iwański, H.	526
Iwersen, M.	262
Izquierdo, M.	136, 520
Izumi, K.	143

J

Jacob, P.	481
Jacobs, J.	114
Jacquiet, P.	555
Jagusiak, W.	108, 190, 364, 432, 503
Jahnel, R.E.	574
Jakimowicz, M.	190, 197
Jakobsen, J.H.	478, 556
Jakobsen, N.	215
Jamrozik, J.	387
Jang, S.	578
Janiszewski, P.	528
Janodet, E.	474
Jansman, A.J.M.	125, 156, 216, 313
Jansons, I.	134, 301
Janssens, S.	106, 285, 341, 355
Jansson, A.	257, 258, 259, 412, 469
Janvier, E.	126
Jara, A.	508
Jarousse, A.	161
Jasielczuk, I.	346
Jasinsky, A.	117
Jenko, J.	191
Jensen, A.B.	251
Jensen, J.	195, 493
Jeppsson, K.-H.	119
Jerónimo, E.	240, 242, 243, 244, 248, 307, 450
Jesus, N.	258
Jianlin Han, J.	104
Jibrila, I.	582
Jimenenz, E.	543
Jimenez, E.	301
Jiménez, J.A.	329
Jimenéz, M.J.	452
Jiménez-Moreno, E.	127, 132, 298
Jiménez, R.	103
Jintasataporn, O.	327
Jo, J.H.	142
Jo, Y.H.	142
Joerg, H.	625
Johansson, A.M.	490
Johansson, M.	318
Joly, F.	402, 403, 405
Joly, P.	373
Jonas-Levi, A.	157
Jones, A.G.	565, 630
Jongen, M.	365
Jonkus, D.	134
Jonsson, N.N.	262
Joosten, L.	251
Jousset, T.	416, 587

Joy, M.	173, 175, 306, 338, 422, 477
Jozan, T.	634
Józefiak, A.	154
Judge, M.M.	359
Julé, C.	587
Junge, W.	584, 585
Junnikkala, S.	274
Juráček, M.	238, 239
Jurquet, J.	446
Juska, R.	213
Juskiene, V.	213

K

Kaart, T.	108
Kadarmideen, H.N.	367
Kairenius, P.	209, 532
Kalbe, C.	166
Kaler, J.	398
Källström, H.N.	426
Kalm, E.	342
Kalogianni, A.I.	303, 363
Kamel Oroumieh, S.	271, 544
Kamola, D.	440, 441, 442
Kamphuis, C.	282, 366, 429
Kanitz, E.	286
Kantanen, J.	450, 457, 460, 461
Kapsona, V.	428, 572
Kapusta, A.	269
Kar, S.	156, 349
Karaiskou, C.	175, 180, 570
Karaiskou, N.	572
Karakousis, I.	312
Karaman, E.	279, 379
Karamoutsios, A.	312
Karatzia, M.A.	175, 456
Karger, V.	162, 165
Kargo, M.	317, 426, 428, 487, 500, 575
Karim, L.	188
Karlsson, K.	341
Karrow, N.	356
Karunaratne, S.	114
Kasapidou, E.	175, 180, 570
Kasarda, R.	178, 345, 372, 386, 442
Kaser, G.	99
Kaske, M.	585
Kašná, E.	380, 535
Kasper, C.	125, 129
Kato, K.	299
Katsaganis, G.	303
Katsumata, S.	233
Katzenberger, K.	550
Kauffmann, C.	636
Kaufmann, F.	553
Kautto, A.H.	514
Kavlak, A.T.	584
Kawakami, J.	141, 144
Kawęcka, A.	109
Kawecka-Grochocka, E.	269

Kazama, D.C.S.	309	König Von Borstel, U.	404
Keady, T.	200	Koning, L.	208, 292
Kearney, G.A.	474	Konkol, D.	526
Kebede, F.	375	Koompans, B.	434
Keeling, L.	548	Korczyński, M.	526
Keiller, J.	474	Korzeniowska, M.	526
Keita, A.	512	Kosińska-Selbi, B.	190, 500
Keller, E.	416	Kostyunina, O.	573
Kemp, B.	226, 323, 549, 585	Kotlarz, K.	381
Kemper, N.	527, 553, 612	Kotrba, R.	306, 567
Kemppainen, A.	510	Kouretas, D.	180
Kenéz, Á.	149, 224, 266, 272	Kowalski, M.W.	331
Kenyon, F.	198	Kranjcevicova, A.	380
Kerry, J.	605	Krattenmacher, N.	342
Kesidis, A.	570	Krawczyk, J.	109
Kettunen, H.	274, 327, 396, 453	Kreuzer, M.	158, 349, 541, 613
Khamis, F.M.	256	Kreuzer-Redmer, S.	316
Kharzinova, V.R.	193	Kricka, T.	309
Khayatzadeh, N.	314, 361	Krieter, J.	221, 413, 511, 584, 585, 610, 612
Khelil, H.	301	Kristensen, N.B.	503
Khula, B.	507	Kristensen, T.	290
Kim, E.J.	142, 497	Krizsan, S.J.	449
Kim, G.	497	Krogenæs, A.	623
Kim, G.D.	602	Krogh Jensen, S.	234
Kim, J.E.	142	Krogh, U.	128
Kim, J.Y.	602	Krogmeier, D.	580
Kim, K.H.	602	Krpalkova, L.	296
Kim, W.S.	142	Krsnik, J.	381
Kimura, Y.	299	Krugmann, K.	610
King, E.	475	Krupa, E.	196, 383
Kinoshita, A.	539	Krupinski, J.	109
Kipp, C.	123	Krupová, Z.	196, 383, 535
Kistemaker, G.	387	Kuenz, S.	311, 434, 525
Kistler, T.	466	Kugonza, D.R.	450, 457, 460, 461
Kjærulf, M.	303	Kuhla, B.	121
Kleefisch, M.T.	331	Kühl, S.	208, 356, 404
Klimek, P.	366	Kühn, C.	342
Klop, A.	308	Kühne, P.	248, 534, 535
Klopčič, M.	400	Kuipers, A.	400
Knapp, P.	289, 314, 315	Kumagai, H.	233, 299
Knierim, U.	509	Kunrath, T.	401
Knol, E.F.	113, 314	Kunz, C.	349
Knudsen, K.E.B.	503	Küster, S.	512
Koch, C.	224, 276	Kyriazakis, I.	113, 586, 590
Koch, F.	121		
Köck, A.	436, 583	**L**	
Koeck, A.	293, 529	Labadie, K.	463
Koesling, M.	566	Labrune, Y.	448
Kofler, J.	529	Labussière, E.	122
Koivula, M.	102, 581, 582	Lacetera, N.	119, 210
Kok, A.	323, 585	Lachica, M.	131, 394, 454
Kolenda, M.	440, 441, 442	Lach, Z.	440, 441, 442
Kolkman, I.	287	Lacurto, M.	161
Kolláthová, R.	238, 239	Lærke, H.N.	128
Kombolo-Ngah, M.	600	Laforest, J.P.	452
Komen, H.	375	Lagriffoul, G.	198, 200, 374, 458
Kominakis, A.	479, 556, 627	Lai, F.	628
König, S.	123, 168, 222, 225, 357	Lam, T.	323

Lambe, N.R.	560, 620	Lee, Y.L.	188
Lamberton, P.	202, 203, 502	Leeb, C.	315
Lambert, W.	210	Lefebvre, R.	188, 363, 460, 532
Lambertz, C.	208, 332, 335, 433, 482, 593	Lefebvre, T.	463
Lambrechts, A.	254	Le Floc'h, N.	128, 234, 546
Lamy, A.	417	Lefranc, M.	156
Lamy, P.	396	Lefrançois, M.	134, 135
Landete-Castillejos, T.	306	Legako, J.	605
Landrieu, F.	600	Le Gall, A.	443
Langbein, J.	398	Le Gall, M.	298
Lange, A.	549	Legarra, A.	197, 458, 484, 561, 577, 578
Lantinga, E.	319	Leger, F.	635
Lanza, I.	568	Legrand, I.	599, 600, 604
Laperruque, F.	504	Lehmann, J.O.	290, 319
Lapierre, H.	607, 610	Lehmann, M.	134, 135
Lapointe, J.	452, 593	Lei, T.	385
Lara, L.	131	Leiber, F.	319, 334, 371, 433, 593
Lara, L.A.C.	464	Leino, A.-M.	283
Lara, M.A.	456	Leishman, E.M.	150, 553
Largouët, C.	365	Leite, J.V.	407, 528
Larroque, H.	374, 449, 455, 458, 556, 558	Leiva, B.	471, 554
Larsberg, F.	316	Lejard, A.	205
Larzul, C.	217, 315, 448	Lemaire, G.	401
Lasbleiz, R.	205	Lemonnier, G.	448
Lassalas, J.	506	Lemosquet, S.	424, 502, 610
Lasser, J.	366	Leopold, J.	433
Latorre, M.A.	513, 519	Lepczynski, A.	395
Lauk, C.	165	Le Pichon, D.	317
Launay, C.	126	Leplat, J.J.	448
Lauridsen, T.	597	Lerch, S.	202, 203
Lauvie, A.	406	Leroy, G.	186
Lawal-Adebowale, O.	623	Le Sciellour, M.	448
Lazaridou, E.	408	Leskinen, H.	209
Lazaro, S.F.	382	Lessard, M.	593
Lazzari, B.	465	Lessire, F.	209
Leal, L.N.	538	Letaief, N.	407
Lean, I.J.	596	Létourneau-Montminy, M.P.	591
Leão, C.	624, 629	Leubin, M.	334, 593
Leão, M.T.	469, 471	Leubner, C.D.	331, 333, 337, 399
Lebeau, F.	322	Leung, Y.H.	224
Leblois, J.	137, 147, 533	Levesque, J.	272
Lebret, B.	518	Levit, H.	292, 294
Lecardonnel, J.	448	Lewicz, Ł.	440, 441, 442
Lecchi, C.	263, 265, 272, 273	Li, B.	219
Lechevestrier, Y.	298	Li, Y.	349
Lecler, B.	236	Li, Z.	334, 537, 540, 541
Leclerc, H.	281	Liang, D.	104
Lecocq, A.	251	Lidauer, M.H.	283, 510
Lecomte, C.	147, 533	Lidauer, P.	545
Le Cour Grandmaison, J.	210	Lidfors, L.	199
Le Cozler, Y.	202, 446	Liechti, P.	508
Ledda, A.	628	Liesegang, A.	537
Leduc, A.	424	Ligda, C.	456, 459
Lee, G.I.	128	Ligonesche, B.	315
Lee, H.G.	142, 497	Lihoreau, M.	587
Lee, J.S.	142	Lima, J.	576
Lee, K.	439	Lima, S.B.G.P.N.P.	123
Lee, M.R.F.	164, 203, 207, 476, 565, 630	Lima, T.	351

Marina, H.	421	Mathot, M.	163
Marín, M.F.	117	Mathys, A.	158
Marino, R.	277, 420, 447	Matika, O.	558
Marjanovic, J.	288, 580	Matin, A.	309
Markakis, N.	246	Matos, C.	624, 629
Marko, O.	364	Matos, E.	325
Mármol-Sánchez, E.	264	Matos, G.	573
Marois, D.	195	Matos, R.	160
Maroto-Molina, F.	118, 367	Mattalia, S.	281
Marquardt, S.	541	Matte, J.J.	593
Marques, C.C.	275, 479, 494	Matthews, K.	318
Marques Da Silva, J.	166, 633	Matthiesen, C.F.	303
Marques, G.	305	Matulaitis, R.	213
Marques, J.	519, 520	Matusewičius, P.	158, 528
Marques, M.R.	177, 423, 481	Matzhold, C.	366
Marsot, M.	554	Maugan, L.H.	460
Martel, G.	403	Maupin, M.	373
Martel-Kennes, Y.	452, 593	Maurer, L.	530
Martín, A.	241, 242, 394, 454	Maurer, V.	593
Martín-Alonso, M.J.	338	Mavrommatis, A.	608, 609
Martín-Amores, R.	194	Maxa, J.	567
Martín-Burriel, I.	456	Mayer, A.	165
Martin, C.	280, 508	Mayer, E.	353
Martín-Collado, D.	402, 634	Mayeres, P.	355
Martín, D.	626	Mayer, M.	574
Martín De Hijas-Villalba, M.	431	May, K.	357
Martinelli, R.R.	637	May, S.T.	463
Martínez, A.	456	Mazurkiewicz, J.	328
Martínez-Álvaro, M.	576	Mazza, A.	628
Martinez Boggio, G.	449	Mazzoni, G.	189
Martínez-Castillero, M.	498	McAmmond, B.	451
Martinez, D.	603, 606	McArt, J.A.A.	228
Martinez, J.	157	McAuliffe, G.A.	164, 203, 207
Martínez, R.	427	McCoard, S.	480
Martínez, V.	302	McEwan, J.C.	478
Martín-García, A.I.	261, 301, 536, 543	McFadden, J.W.	608
Martin, L.	164, 521	McGee, M.	145
Martin, N.	210	McGeough, E.	305
Martin, O.	372	McGilchrist, P.	596, 597
Martino, P.	273	McGovern, F.M.	621, 628
Martin, P.	279, 532	McGuire, R.	163
Martin, S.	635	McHugh, N.	559, 563, 621, 628
Martins, A.	401	McKeegan, D.	265
Martins, C.F.	214, 220, 232, 277	McLaren, A.	620
Martins, E.	351	McLaughlin, M.	265
Martinsen, K.H.	376	McLean, K.	620
Martins, P.	504	McNally, J.	474
Martín-Tereso, J.	538	McNeilly, T.N.	555, 565
Martyniuk, E.	109, 187	Mdyogolo, S.	375
Marykutty, T.	106	Meade, K.G.	623
Maselyne, J.	524	Méda, B.	210
Massaro, A.	568	Medrano, J.F.	356
Massender, E.	556	Meehan, D.J.	297
Masuda, Y.	579	Meeke, T.	629
Mateo, J.	239	Mehaba, N.	170
Mateos, I.	239, 241, 242, 394, 454	Mehtiö, T.	283, 510
Mateus, T.L.	159	Meijer, N.P.	252
Mathew, J.	410	Meir, Y.B.	294

O

Obata, A.	233
Obregón, P.	335
O'Brien, A.C.	556
O'Brien, A.C.	559, 564
Obšteter, J.	191
Ocack, S.	200
OcElova, V.	452
O'Connor, E.	628
Octavia, M.	252
Oddy, V.H.	144
Odintsov- Vaintrub, M.	167, 200
O'Doherty, J.V.	390
O'Donovan, M.	621
Oe, M.	539
Oelkers, J.	221
Ogino, A.	299
Ohtsuka, A.	351
Oishi, K.	233, 299
Ojeda, C.	289
Ojima, K.	539
Okamura, T.	378
Okamura, Y.	351
Okereke, C.O.	300
Okeyo, A.M.	110
Olaizola, A.	402
Olaleru, I.F.	300
Olazaran, I.	633
Oliet, A.	498, 499
Oliveira E Sousa, C.	481
Oliveira, G.A.	378
Oliveira, H.M.	153
Oliveira, H.R.	192, 382, 486, 556
Oliveira, J.	110, 515, 571
Oliveira Junior, G.A.	149
Oliveira, M.H.V.	344
Oliveira, P.	275
Oliver, J.	199
Ollagnier, C.	455, 508, 614
Olmos, G.	331
Olonen, A.	396
Olsen, H.F.	213, 320, 376
Olsson, A.-C.	119
Oltjen, J.W.	144
O'Mahony, N.	296
O'Mara, F.	185
Omer, E.A.M.	427
Ominski, K.	305
Omirou, M.	459
Omodei Zorini, F.	231, 388
Omphalius, C.	610
Onken, F.	436, 506, 507
Onopiuk, A.	597
Oppriesnig, T.	329
Opsomer, G.	225
Orcasberro, M.S.	211, 539
Orehovacki, V.	106
O'Riordan, E.G.	145, 161
Ørnsrud, R.	235
Orquera, K.	145, 368
Orsi, M.	406
Ortega, M.S.	355
Ortigues-Marty, I.	145, 280, 281, 371, 383
Ortiz, A.	336
Osawa, T.	143
Oshima, I.	539
Oshima, K.	539
Osorio Avalos, J.	391
Ospina, A.T.	344, 346
Østergaard, S.	426, 575
Ostersen, T.	285
O'Sullivan, M.	491, 605
Otomaru, K.	539
Ottoboni, M.	155, 233, 235, 238, 247, 609
Ouédraogo, D.	110, 361
Ouédraogo-Koné, S.	110, 361
Outor-Monteiro, D.	298
Øverland, M.	213
Ozgo, M.	395

P

Pabiou, T.	559
Pacheco, A.	565
Padre, L.	624, 631
Pagnacco, G.	465
Pahmeyer, C.	161
Pais, J.	494
Paiva Dos Santos, M.	162
Paixão, G.	217, 519, 520, 523
Palhière, I.	458, 556, 563
Pallares, P.	569
Palma-Granados, P.	573
Palma-Hidalgo, J.M.	261, 301, 536, 543
Palma, J.	445
Palucci, V.	490
Panseri, S.	129
Panteli, P.	246
Panyavong, X.	602
Papadaki, A.	246
Papadomichelakis, G.	231, 247, 249, 408
Papadopoulos, V.	570
Papanastasiou, D.K.	211
Pappas, A.C.	231, 247
Paradowska, M.	350
Pardo, Z.	394, 454
Parisot, S.	557
Park, H.Y.	142
Park, J.	497
Park, J.S.	497
Parniakov, O.	255
Paro De Paz, C.C.	626, 627
Parreiras, M.	615
Parroni, A.	310
Parr, T.	463
Pas Schrijver, A.	169
Pastell, M.	291, 294, 510, 568, 584, 589

Pastore, D.	124	Perruchot, M.H.	354
Paternostre, L.	127, 132	Perucho, L.	406
Patil, M.A.	385	Pesaantez-Pacheco, J.L.	569
Paul, A.	411	Pestana, A.	157
Paulos, K.	240, 242, 243	Petani, B.	326
Paul-Victor, C.	561	Peterson, C.	401
Pavlovic, D.	364	Pethick, D.W.	596, 597
Pawlina-Tyszko, K.	269, 346	Petit, L.	325
Payan-Carreira, R.	217, 519, 520, 523	Petrov, S.	478
Payola, F.	126	Pétursdóttir, Á.H.	300
Paz, C.C.P.	123, 440	Peyraud, J.L.	182, 185, 636
Pazzola, M.	422	Pfeiffer, C.	289, 314, 315
Pearston, F.	318	Philibert, A.	446, 510
Pedersen, A.Ø.	303	Philipsson, J.	466
Pedersen, L.V.	284	Phocas, F.	466
Pedersen, M.L.M.	215, 391	Piacere, A.	178
Pedersen, P.	120	Piazza, M.	224
Pedro, A.R.V.	351	Picard, F.	508
Pedro, R.	153	Piccinini, R.	263
Peeters, I.	393	Pieper, H.	553
Pegler, D.	474	Pierce, K.M.	434, 491, 496, 501
Pegolo, S.	230, 324, 358, 422	Pietrzak, E.	328
Peharda Uljevic, M.	327	Pigozzi, G.	343
Peinado Izaguerri, J.	265	Pijl, R.	529
Peinado, J.	336, 626	Pimenta, J.	275, 479, 494
Pelayo, R.	421	Pimentel, E.C.G.	580
Pelegrín-Valls, J.	250, 338, 477	Pinard, T.P.	353
Pellattiero, E.	330	Pineau, C.	404
Pellerin, D.	500	Pineda-Quiroga, C.	197, 455
Peñaranda, A.	508	Pinerio, M.	129, 264
Penasa, M.	168, 330, 357, 362	Pinheiro, V.	298, 305, 308, 616
Penchev, I.	517	Pinho, M.	220, 277
Penedo, M.C.T.	456	Pinloche, E.	235
Penne, G.	433	Piñon, A.	592
Perdomo, C.M.	134	Pinotti, L.	155, 233, 235, 238, 247, 352, 609
Perdomo-González, D.I.	345, 347	Pinsard, C.	404, 635
Perea, J.	181, 570, 640	Pinto, O.	326
Pereira, A.	633	Pinto, S.	294
Pereira, E.	513, 518	Piórkowska, K.	131, 133
Pereira, F.C.	297, 309	Piraux, E.	137
Pereira, L.	494	Pires, A.E.	461
Pereira, R.M.L.N.	275, 479, 494	Pires, B.V.	123, 440
Pérez-Ciria, L.	513, 519	Pires, J.	202, 203, 281, 371, 563
Perez-Clariget, R.	538, 544	Pires, M.A.	217
Perez-Enciso, M.	577	Pirlo, G.	161, 206
Pérez-Garnelo, S.S.	498, 499, 603, 606	Piro, R.	568
Perez-Guzman, M.D.	621	Pitino, R.G.	139, 330, 495
Pérez, H.	243, 244	Pitkänen, T.	583
Pérez, J.F.	592	Pitkänen, T.J.	582
Perez, M.A.	136	Piwczyński, D.	441, 442
Pérez-Marín, D.C.	118, 367	Plante-Dubé, M.	176
Pérez-Pardal, L.	374, 454	Plastow, G.	329
Pérez-Pe, R.	625	Plouzin, D.	446
Pérez-Sánchez, J.	325	Pocrnic, I.	464, 483, 577
Pérez, T.	288	Pogorzelski, G.	597, 599, 600
Peric, T.	615	Pokorny, J.	594
Perri, C.	229	Polak, G.M.	109, 346
Perrot, C.	444	Polák, P.	372

Polasik, D.	524
Pol, F.	522
Polkinghorne, R.	596, 598, 599, 600
Pomar, C.	112, 113, 215, 591
Pomar, J.	125
Pomport, P.H.	251, 631
Pondet, M.	251, 631
Pong-Wong, R.	558, 561
Pook, T.	339, 385, 427
Pooley, C.	329
Popiela, E.	526
Poppe, M.	282, 360
Porcelli, F.	138
Portugal, A.P.V.	240, 242, 248, 450
Posada-Ochoa, S.L.	511
Posta, J.	556
Potočnik, K.	381, 415, 622
Poulopoulou, I.	332, 335
Poulsen, B.	285
Pourazad, P.	121
Pourrat, J.	508
Pozza, P.C.	133, 514
Pradel, J.	557
Pradervand, N.	455
Prates, J.A.M.	277
Pravettoni, D.	272, 273
Prearo, M.	155
Preciosa, P.	518
Press, C.	213
Prevedello, P.	530
Priepke, A.	166
Priester, A.	504
Primi, R.	268
Príncipe-Aguirre, K.	367
Probst, L.	110
Pröll-Cornelissen, M.J.	286
Promp, J.	281
Prunier, A.	217
Pryce, J.	529
Prytkov, Y.	573
Pszczola, M.	377
Ptak, E.	364
Puig, P.	516
Puillet, L.	282
Pulina, G.	616
Pulinas, L.	625
Puntigam, R.	594
Pupillo, G.	155
Puppel, K.	552
Purtscher, S.	322
Purup, S.	476
Pywell, R.	623

Q

Qanbari, S.	104, 109, 377
Qin, N.	274
Quarantelli, A.	139, 495
Quaresma, M.	471, 554

Quénon, J.	457
Quesnel, H.	128, 546
Quigley, O.	496
Quintanilla, R.	521
Quispe-Potosino, S.	588

R

Rabadán, R.	132
Rabaglino, M.B.	367
Radhika, G.	106
Rafiepour, M.	104
Rainard, P.	363
Rainer, V.	353
Räisänen, S.E.	502
Ralão, J.	339
Ramanzin, M.	204, 405, 406
Ramin, M.	449
Ramirez-Agudelo, J.F.	511
Ramirez, I.	336
Ramón, M.	428, 455, 572, 621
Ramos, A.	521
Ramos, A.M.	624, 629
Ramos, I.	351
Rampado, N.	601
Ramsay, S.	470
Ran, T.	433
Ranilla, M.J.	239, 241, 242, 244, 394, 454
Raniolo, S.	405, 406
Rankin, J.	472
Raoul, J.	425, 562
Raspa, F.	415
Rato, A.E.	166
Rau, A.	195
Rauch, E.	550
Ravon, L.	587
Razzaghi, A.	532
Rebollo, C.H.	322
Rebucci, R.	231, 352
Redfearn, A.	474, 480
Rediger, K.	329
Reents, R.	342
Reese, S.	550
Reiber, C.	141
Reilas, T.	450
Reimer, C.	385
Reindl, A.	162, 165
Reinsch, N.	378, 484, 574
Reis, M.	365, 445
Reis, S.	432
Reisinger, N.	353
Remus, A.	112, 113, 215
Renand, G.	114, 383
Renaudeau, D.	122, 234, 448
Renevey, P.	508
Renna, M.	155
Resende, M.	153
Resende, V.	168
Resink, J.W.	589

Reverberi, M.	310	Rodríguez, B.	560
Reverter, A.	421	Rodríguez-Estévez, V.	490, 521, 566, 638
Rey, A.I.	516, 595	Rodríguez, J.	212
Rey, J.	304, 329	Rodríguez, M.	592
Reyer, H.	174, 193, 478	Rodriguez-Ortega, T.	162
Reyes-Palomo, C.	490, 638	Rodríguez, P.L.	212
Reynaud, K.	350	Rodriguez-Ramilo, S.T.	315
Reynolds, C.K.	508, 618	Rodriguez-Zas, S.L.	465
Ribani, A.	104, 189	Roehe, R.	392, 576
Ribeiro, D.M.	214, 220, 232, 277	Rogberg-Muñoz, A.	456
Ribeiro, J.M.B.F.	177, 423	Rogel-Gaillard, C.	448
Ribeiro, J.R.	481	Roh, S.	539
Ribeiro, V.	407, 528	Rolinec, M.	238, 239
Ricard, A.	340, 343, 382	Roman, E.	259
Ricard, E.	504, 587	Romão, R.	494, 624
Ricaud, J.P.	251, 308, 631	Romé, H.	195
Riccioli, C.	118	Romero, P.	301
Richardson, A.	156	Romo, J.A.	595
Rico, D.E.	134, 135, 272	Romo, J.M.	595
Rico, J.	507	Ron, L.	322
Rico, J.E.	134, 135, 272	Roncada, P.	229
Riedel, A.	527	Ronchi, B.	268
Rieke, L.	553	Rønn, M.	436, 503
Rienesl, L.M.	583	Rooni, K.	108
Rieple, A.	138	Roosen, J.	400
Riepl, F.	311	Roozeboom, C.	140
Riggio, V.	558	Roozen, A.	490
Righi, F.	139, 330, 495, 609	Ropka-Molik, K.	133
Ríkharðsson, G.	300	Roque, A.	523
Rikkers, R.S.C.	349	Rosa, E.	447
Ring, S.	359	Rosa, G.J.M.	227
Ringard, P.	435	Rosa, H.	604
Ringmark, S.	469	Rosales Nieto, C.A.	474
Rios, A.	463	Rosas, J.P.	431
Riosa, R.	262	Rosati, A.	198, 561
Ripoll, G.	368, 477, 513, 519	Roseiro, L.C.	240
Rippel, D.	149	Rose, N.	512, 522
Riquet, J.	313, 448	Rosen, B.D.	105
Rito, S.	230, 245	Rosero-Noguera, J.R.	511
Rius-Vilarrasa, E.	488	Rosner, F.	529
Riuzzi, G.	409, 568	Rostellato, R.	281
Riva, F.	265	Rothacher, M.	299
Rivero, M.J.	623	Roux, D.	601
Rizzi, R.	465	Rovai, M.	171
Robert, C.	452	Rovere, N.	233, 238, 247
Robert, R.	373	Rovira, F.	560
Robert-Granié, C.	558	Rowe, T.	598
Robinson, R.	398	Royer, E.	550
Robles-Jimenez, L.E.	220, 391	Różański, H.	526
Rodehutscord, M.	589	Ruba, A.	611
Rodenburg, T.B.	201, 549	Rubattu, R.	628
Rodrigues, G.	624	Rubio, A.	621
Rodrigues, L.Y.	344	Ruesche, J.	373
Rodrigues, M.	305	Ruiz-Ascacibar, I.	125
Rodrigues, N.R.	205, 206, 365, 445	Ruiz, R.	304, 482, 633
Rodrigues, R.	518	Rulli, M.C.	233
Rodrigues, S.S.Q.	513, 518	Rupp, R.	449, 559, 561, 563
Rodriguez, A.I.	595	Russ, I.	342

Schneider, C.	319	Setoguchi, A.	299
Schneider, J.	554	Sevi, A.	124, 277, 420
Schodl, K.	314, 315	Sevillano, C.A.	581
Schokker, D.	156, 366, 393, 429	Sexton, J.R.	497
Scholtz, M.M.	375	Seyedalmossavi, S.M.M.	334
Scholz, H.	248, 534, 535	Shabalina, T.	225
Schönherz, A.	490	Shahidian, S.	166, 633
Schop, M.	216, 574	Sharifi, A.R.	223
Schori, F.	152	Shimamoto, S.	351
Schott, T.	353	Shimshoni, I.	292
Schreefel, L.	169	Shumo, M.	256
Schregel, J.	537, 540, 541	Si, J.	104
Schroyen, M.	219	Sib, E.	333, 337
Schuh, K.	276	Siddavatam, P.	388
Schulte, R.P.O.	169	Signer-Hasler, H.	107
Schulze, H.	545	Sikora, J.	109
Schulze-Schleppinghoff, W.	340	Silacci, P.	397
Schwarz, K.	173	Sild, E.	108
Schwarz, T.	131	Silva, B.N.	476
Sciascia, Q.	537, 540, 541	Silva, C.	604
Scocco, P.	354, 408	Silva, J.	214, 297, 326, 480
Scollan, N.D.	163, 321, 598	Silva, J.A.	602
Scott, M.	475	Silva, J.A.I.I.	344, 346
Sebastian, F.	603, 606	Silva, J.L.	220
Šebek, L.B.	208, 292	Silva, L.H.P.	497
Secchi, G.	422	Silva, M.	153, 235
Seefried, F.R.	287	Silva, S.	230, 245, 308, 443, 615
Segato, S.	409, 568	Silva, S.R.	152, 471, 554, 602
Segura, J.	595	Silva, T.H.	228, 229
Sehested, J.	295	Silva-Hidalgo, G.	595
Seibt, K.D.	267, 275	Silveira, A.	308
Seidel, A.	342	Silveira, F.	611
Seigers, G.	366	Silveira, M.	107, 110
Seiquer, I.	131	Simčič, M.	622
Seleguim Chud, T.	387	Simeone, A.	302, 321, 322, 492
Sele, V.	235	Simianer, H.	186, 339, 385, 427
Selionova, M.	174	Šimko, M.	238, 239
Sell-Kubiak, E.	314	Simm, G.	392
Selmin, F.	235	Simon, G.	512
Semik-Gurgul, E.	346	Simongiovanni, A.	128
Sentandreu, E.	146	Simoni, M.	139, 330, 495
Sentandreu, M.A.	146, 414	Sindic, M.	137
Seoni, E.	299	Sinz, S.	541
Sepchat, B.	145, 281	Siphambili, S.	145, 599
Sequeira, A.	157	Siqueira, T.T.S.	636, 637
Sergheraert, R.	234	Sitanaka, N.Y.	514
Serhan, S.	199, 424	Sitkowska, B.	440, 441, 442
Sermyagin, A.A.	386, 478	Siwek, M.	328, 389, 395, 543, 547
Serpa, R.	339	Skarwecka, M.	190, 380
Serra, A.	451	Skjerve, T.A.	376
Serra, E.	434	Skliros, D.	609
Serra, G.	435	Skoufos, I.	211, 312, 526
Serra, M.G.	212	Skrzyński, G.	603
Serrano, J.	166, 633	Slama, J.	594
Serrano, M.	455, 572, 621	Slawinska, A.	328, 389, 543, 547
Serrano-Pérez, B.	250, 338	Sleator, R.D.	359
Serviento, A.M.	122	Smetana, S.	253, 255
Servin, B.	561	Smidt, H.	393, 450

Tsiotsias, A.	570	Van Den Brand, H.	549
Tsiplakou, E.	231, 247, 495, 608, 609	Vandenbussche, C.	524
Tsousis, G.	585	Van Den Heuvel, J.	485
Tsukahara, T.	233	Vandenplas, J.	579, 581, 582
Tsuruta, S.	578, 579	Van Der Beek, S.	140, 355
Tsutsumi, H.	233	Van Der Fels-Klerx, H.J.	252
Tuchscherer, A.	286, 537, 540	Van Der Heide, E.M.M.	366
Tuchscherer, M.	286	Van Der Linde, J.	553
Turner, S.P.	284	Van Der Linden, A.	162
Tuz, R.	131	Van Der Sluis, M.	201
Tyra, M.	133, 151, 524	Van Der Weide, R.Y.	255
Tzora, A.	211, 312, 526	Vandeweyer, D.	252
		Van Dixhoorn, I.	393
U		Van Dooren, H.J.C.	575
Uerlings, J.	219	Van Dorpe, I.	432
Ugarte, E.	197, 304, 430, 455, 556, 633	Van Duinkerken, G.	296, 308
Uimari, P.	584	Van Gijssel, A.	586
Ungerfeld, R.	611	Van Gorp, A.	254
Urban, P.	594	Van Gosliga, S.P.	294
Ureña, I.	621	Vanhaecke, I.	271, 544
Uriarte, J.M.	595	Van Hamme, J.D.	451
Urrutia, O.	270	Vanhatalo, A.	545
Usié, A.	624, 629	Van Hese, I.	225
Utama, G.L.	121	Van Huis, A.	256
Utsunomiya, Y.T.	105	Van Ittersum, M.K.	167
Utzeri, V.J.	104, 189	Van Kaam, J.B.C.H.M.	223
		Van Knegsel, A.T.M.	226, 323, 585
V		Van Krimpen, M.M.	589
Vaarst, M.	319	Vanlierde, A.	507
Vacca, G.M.	422	Vanloofsvelt, P.	253
Vacherie, B.	463	Van Meulebroek, L.	271, 544
Vaga, M.	257, 258	Van Miert, S.	254
Vage Coelho Sartori, L.	626, 627	Van Milgen, J.	125, 128, 215, 234
Vagnoni, E.	212	Van Niekerk, M.	489
Vahidi, M.F.	104	Van Peer, M.	254
Valada, T.	445	Van Pelt, M.L.	360
Valdés, C.	542	Vanraden, P.M.	100
Valente, I.M.	153	Van Riet, M.M.J.	393
Valente, L.M.P.	325, 326	Van Rozen, K.	255
Valentim, P.	152	Van Selm, B.J.	161, 167
Valentim, R.	504	Van Staaveren, N.	150, 553
Valera, M.	345, 347	Van Weyenberg, S.	524
Valizadeh, R.	271, 544	Van Wyk, J.B.	489, 493
Valkonen, E.	327, 453	Van Zanten, H.H.E.	161, 167, 169
Valle, E.	415	Vanzin, A.	230
Valleix, M.	416, 468	Van Zyl, J.H.C.	179
Valvassori, E.	415	Vargas-Bello-Pérez, E.	148, 391, 423
Van Baal, J.	537, 589	Varona, K.	587
Vanbergue, E.	510, 632	Varona, L.	430, 431, 496, 498, 578
Van Breukelen, A.E.	292	Värv, S.	108
Van Bruggen, S.	586	Vasconcelos, L.	513
Van Campenhout, L.	252	Vasconcelos, M.	232
Vandaele, L.	225, 250, 310	Vašeková, P.	238, 239
Van Dam, E.A.	586	Vasseur, E.	529
Van Damme, M.-A.	341	Vassilev, Z.	437
Van De Gucht, T.	524	Vastardis, I.	303, 363
Vandenberg, G.	409	Vayssier-Taussat, M.	568
Van Den Bossche, T.	250	Vaz, P.S.	515, 528, 571

Wild, C.	433	Yáñez-Ruiz, D.R.	261, 301, 369, 370, 536, 543
Wilder, T.	612	Yang, W.Z.	305, 433
Willam, A.	289, 315	Yates, J.	560
Williams, A.	620	Yin, T.	225
Williams, A.R.	391	Yougbaré, B.	110, 361
Williams, E.	598	Young, J.	475
Williams, S.R.O.	297	Yu, H.	439
Willis, C.	388	Yu, S.J.	338
Willis, R.	547	Yu, X.	478, 580
Willis, R.C.	389		
Wilmot, H.	137	**Z**	
Wilms, J.N.R.	538	Zaalberg, R.M.	295
Wimel, L.	416, 470, 587	Zabalveytia, N.	492
Wimmers, K.	174, 193, 286, 478	Ząbek, T.	269
Winckler, C.	314, 315, 530	Žak, G.	133, 524
Windig, J.	427	Žáková, E.	383
Windisch, W.	311, 434, 525, 594	Zalewska, M.	269
Winkler, A.	540	Založnik, J.	415
Winters, M.	318	Zanchi, E.	617
Witkowska, D.	158, 528	Zanon, T.	168
Wittenburg, D.	109, 377, 574	Zaragoza, P.	456
Woad, K.	398	Żarnecki, A.	190, 290, 380
Wobbe, M.	342	Zavadilová, L.	535
Woelders, H.	349, 393	Zecconi, A.	226
Wolbers, F.	611	Zehetmeier, M.	162, 165
Wolc, A.	483	Zeidler, S.	612
Wolfová, M.	196, 383	Zemb, O.	313, 448
Wollmeister, M.	141	Zervas, G.	231, 247, 408
Wolski, W.	277	Zhang, G.	120
Wood, B.J.	150, 553	Zhang, G.Z.	338
Woodroffe, R.E.	331	Zhang, J.	489
Woodrum Sester, M.M.	588	Zhang, X.	529
Workel, I.	340	Zhang, Y.	104, 219, 486, 539
Wouters, F.	253	Zhao, Y.	540, 541
Wurzinger, M.	103, 110, 361, 431	Ziadi, C.	120, 477
Wuthijaree, K.	332, 335	Zicarelli, L.	619
Wutke, M.	612	Žídek, R.	386
Wuyts, A.	254	Zieger, P.	123
		Zimina, A.A.	386
X		Zinovieva, N.A.	174, 193, 386, 478
Xavier, C.	202, 605	Zjalic, S.	309, 310
Xie, J.	329	Zoma-Traoré, B.	110, 361
Xu, Q.	486	Zonderland, J.	218
Xu, W.	149, 266	Zorc, M.	384
		Zschiesche, M.	223
Y		Žukowski, K.	133, 290, 362, 384, 432, 503
Yamaguchi, S.	143, 144	Zuliani, A.	204
Yamanaka, S.	233	Zupan, I.	327
Yamasaki, T.	144	Zurbrügg, C.	158
Yamazaki, T.	143	Zwaan, B.J.	485
Yan, J.L.	338	Zwane, A.A.	460
Yan, P.	486		
Yan, T.	618